TECHNOLOGY of
MACHINE TOOLS

sixth edition

Steve F. Krar

Arthur R. Gill

Peter Smid

 **Higher Education**

Boston   Burr Ridge, IL   Dubuque, IA   Madison, WI   New York   San Francisco   St. Louis
Bangkok   Bogotá   Caracas   Kuala Lumpur   Lisbon   London   Madrid   Mexico City
Milan   Montreal   New Delhi   Santiago   Seoul   Singapore   Sydney   Taipei   Toronto

Higher Education

TECHNOLOGY OF MACHINE TOOLS, SIXTH EDITION

3 4 5 6 7 8 9 0 QPD/QPD 0 9 8 7 6

ISBN 978–0–07–830722–5
MHID 0–07–830722–8

Publisher, Trades and Engineering Technology: *Thomas E. Casson*
Publisher, Career Education: *David T. Culverwell*
Managing Developmental Editor: *Jonathan Plant*
Editorial Coordinator: *Connie Kuhl*
Outside Developmental Services: *Brian Mackin*
Marketing Manager: *Lynn M. Kalb*
Project Manager: *Jodi Rhomberg*
Senior Production Supervisor: *Sherry L. Kane*
Lead Media Project Manager: *Audrey A. Reiter*
Media Technology Producer: *Janna Martin*
Designer: *Rick D. Noel*
Cover/Interior Designer: *Kaye Farmer*
(USE) Cover Image: *Courtesy of the Author*
Senior Photo Research Coordinator: *John C. Leland*
Photo Research: *David Tietz*
Compositor: *Carlisle Communications, Ltd.*
Typeface: *10/12 Times Roman*
Printer: *Quebecor World Dubuque*

Any photos not credited on page are courtesy of the author.

Library of Congress Cataloging-in-Publication Data

Krar, Stephen F.
 Technology of machine tools / Steve F. Krar, Arthur R. Gill, Peter Smid. — 6th ed.
  p. cm.
 Includes index.
 ISBN 0–07–830722–8 (alk. paper)
 1. Machine-tools. 2. Machine-shop practice. I. Gill, Arthur. II. Smid, Peter, 1930–. III. Title.

TJ1185.K668 2005
621.9'02—dc22   2004053055
                                        CIP

www.mhhe.com

contents

SECTION 6

Layout Tools and Procedures 140

SECTION 7

Hand Tools and Bench Work 160

SECTION 8

Metal-Cutting Technology 194

SECTION 9

Metal-Cutting Saws 272

SECTION 10

Drilling Machines 298

Contents   v

preface

The use of computers continues to change how machine tools are used to manufacture products. Computers have improved until there are now highly sophisticated units capable of controlling the operation of a single machine, a group of machines, or even a complete manufacturing plant. Section 14, "Computer-Age Machining," now includes not only computer numerical control machine tools, such as turning and machining centers, but also newer manufacturing technologies. To increase manufacturing productivity, machine tools have been equipped with modular tooling and work-holding systems, as well as new cutting tools to produce accurate parts faster and at competitive prices.

Today's industries are putting more emphasis on using new manufacturing technologies and manufacturing intelligence systems to improve their productivity and remain competitive in the world. Section 14 gives an overview of 10 technologies or processes that are giving manufacturers an advantage over their competition. A few of these are Artificial Intelligence, Open Architecture CNC, Step NC, Virtual Reality, e-Manufacturing, Nanotechnology, and Cyber Management Systems. Through the use of CDs and videotapes, students can learn how the machine tool trade will be changing in the future.

This book is based on the authors' many years of trade experience and experience as specialists in teaching. To keep up-to-date with technological change, the authors have researched the latest technical information available and have visited industries that are leaders in their field. Key personnel in manufacturing firms and leading educators reviewed many sections of this book, so that accurate and up-to-date information is presented. The authors are grateful to the reviewers for the technical and practical suggestions that were incorporated into the text.

The sixth edition of *Technology of Machine Tools* is presented in unit form; each unit is introduced with a set of objectives, followed by related theory and operational sequence. Dual dimensioning (inch/metric) is used in the book, so that machine tool technicians become familiar with both inch and metric systems of measurement. Each operation is explained in a step-by-step procedure, which students can readily follow. Advanced operations are introduced by problems, followed by step-by-step solutions and matching procedures. Many new illustrations and photographs with color have been included to emphasize important points and clarify the text. End-of-unit questions can be used for review or for homework assignments to prepare students for subsequent operations.

The purposes of this text are to assist instructors in providing the basic training on conventional machine tools; to cover basic programming for CNC machines (such as turning and machining centers); and to introduce new manufacturing technologies and processes. To make this course interesting and challenging for students, videotapes can be used to cover new technologies. They are available on loan or for a small fee from technical societies, manufacturers, and publishers. The instructor's manual includes sources of videotapes, along with answers to the review questions in the text. A student workbook is also available.

Steve F. Krar
Arthur R. Gill
Peter Smid

STEVE F. KRAR

Steve F. Krar spent 15 years in the trade, first as a machinist and finally as a tool and die maker. After this period, he entered Teachers' College and graduated from the University of Toronto with a Specialist's Certificate in Machine Shop Practice. During this 20 years of teaching, Mr. Krar was active in vocational and technical education and served on the executive committee of many educational organizations. For 10 years, he was on the summer staff of the College of Education, University of Toronto, involved in teacher training programs. Active in machine tool associations, Steve Krar is a Life Member of the Society of Manufacturing Engineers and former associate director of the GE Superabrasives Partnership for Manufacturing Productivity.

Mr. Krar's continual research over the past 45 years in manufacturing technology has involved many courses with leading world manufacturers and an opportunity to study under Dr. W. Edwards Deming. Mr. Krar spent a week researching Nanotechnology at leading research centers, universities, and industry in Switzerland. He is coauthor of over 65 technical books, such as *Machine Shop Training, Machine Tool Operations, CNC Simplified, Superabrasives—Grinding and Machining,* and *Exploring Advanced Manufacturing Technologies,* some of which have been translated into 5 languages and used throughout the world.

ARTHUR R. GILL

Arthur R. Gill served an apprenticeship as a tool and die maker. After 10 years in the trade, he entered the Ontario Community College system. Mr. Gill served as a professor and coordinator of precision metal trades and apprenticeship training for 30 years at Niagara College in St. Catharines. He was a member of the Ontario Precision Metal Trades college curriculum committee for apprenticeship training and Heads of Apprenticeship Training. Mr. Gill is a member of the Society of Manufacturing Engineers and worked closely with industry to continually improve manufacturing technology.

Mr. Gill has coauthored a number of textbooks, including *CNC technology and Programming, Computer Numerical Control Simplified,* and *Exploring Advanced Manufacturing Technology* with Steve Krar. In 1991, he was invited by China to assist in developing a Precision Machining and Computer Numerical Control (CNC) training facility at Yueyang University in Hunan Province.

Other Highlights (Krar and Gill)

> Consulting editor, manufacturing technology, for Industrial Press, New York, and contributing editor, *Advanced Manufacturing Magazine,* Burlington, Ontario

> Judge at the annual SkillsUSA competition for CNC programming and machining, held in Kansas City

> Research in new developments in Manufacturing Technology in North America and Nanotechnology at universities, industries, and the IBM Research Lab in Switzerland

PETER SMID

Peter Smid graduated from high school with a specialty in machine shop training. He then entered industry, completed an apprenticeship program, and gained valuable experience as a machinist skilled on all types of machine tools. Mr. Smid emigrated to Canada in 1968 and spent the next 26 years employed in the machine tool industry as a machinist and tool and die maker.

In the early 1970s, he became involved in Computer Numerical Control (CNC) as a programmer/operator and devoted the next 18 years to becoming proficient in all aspects of computerized manufacturing. In 1989, he became an independent consultant, and hundreds of companies have used Mr. Smid's CNC and CAD/CAM skills to improve their manufacturing operations. He also wrote a comprehensive, 500-page CNC programming handbook, which is rapidly becoming the Bible of the trade.

In 1995, he became a consultant/professor of Advanced Manufacturing focusing on industrial and customized training in CNC, CAD/CAM, and Agile Manufacturing. His many years of teaching, training, lecturing, and designing curriculum gives him the opportunity to pass along his vast knowledge of modern manufacturing technology to students of all ages.

ACKNOWLEDGMENTS

The authors wish to express their sincere thanks and appreciation to Alice H. Krar for her untiring devotion in reading, typing, and checking the manuscript for this text. Without her supreme effort, this text could not have been produced.

The authors wish to thank all the teachers and industrial reviewers who took their valuable time to offer concrete suggestions to the manuscript, which we were happy to include. Their suggestions will make this book, which is considered to be a world leader, more valuable as an educational tool.

Our sincere thanks go to the following firms that reviewed sections of the manuscript and offered suggestions that were incorporated to make this text as accurate and up-to-date as possible: ABB Robotics; American Iron & Steel Institute; American Superior Electric Co.; Cincinnati Machine, a UNOVA Co.; Dorian Tool International; GE Superabrasives; Moore Tool Co.; Norton Co.; Nucor Corporation; and Stelco, Inc.

We are grateful to the following firms, which have assisted in the preparation of this text by supplying illustrations and technical information:

3D Systems
300 Below, Inc.
ABB Robotics, Inc.
Able Corporation
Allen Bradley Co.
Allen, Chas. G. & Co.
American Chain & Cable Co. Inc.,
    Wilson Instrument Division
American Iron & Steel Institute
American Machinist Ametek Testing
    Equipment
AMT—The Association for
    Manufacturing Technology

Armstrong Bros. Tool Co.
Ash Precision Equipment Inc.
Automatic Electric Co.
Avco-Bay State Abrasive Company
Bausch & Lomb
Bethlehem Steel Corporation
Blanchette Tool and Gage
Boston Gear Works
Bridgeport Machines, Inc.
Brown & Sharpe Manufacturing Co.
Carboloy, Inc.
Carborundum Co. Div. Saint Gobain
    Abrasives

Carpenter Technology Corp.
Carr Lane Manufacturing Co.
Charmilles Technologies Corp.
Cincinnati Gilbert Co.
Cincinnati Machine, a UNOVA Co.
Clausing Industrial, Inc.
Cleveland Taping Machine Co.
Cleveland Twist Drill Ltd.
CNC Technologies
Cogsdill Tool Products, Inc.
Colchester Lathe Co.
Concentric Tool Corp.
Criterion Machine Works

Deckel Maho, Inc.
Delmia Robotics
Delta File Works
Delta International Machinery Corp.
Denford, Inc.
Deltronic Corp.
DeVlieg Machine Co.
Dillon, WC. & Co., Inc.
DoAll Company
Dorian Tool International
Duo-Fast Corp.
Emco Maier Corp.
Enco Manufacturing Co.
Everett Industries, Inc.
Executive Manufacturing, Inc.
Explosive Fabricators Division
Exxair Corp.
FAG Fisher Bearing Manufacturing Ltd.
Federal Products Corp.
GAR Electroforming Ltd.
GE Superabrasives
General Motors
Giddings & Lewis
Grav-i-Flo Corporation
Greenfield Industries, Inc.
Grinding Wheel Institute
Haas Automation, Inc.
Hanita Cutting Tools, Inc.
Hardinge Brothers, Inc.
Hewlett-Packard Co.
Ingersoll Cutting Tool Co.
Ingersoll Milling Machine Co.
Inland Steel Co.
Jacobs Manufacturing Co.
Jones & Lamson Division of Waterbury
    Farrel

Kaiser Steel Corp.
Kelmar Associates
Keyence Corp.
Kurt Manufacturing
LaserMike, Div Techmet Corp.
Lodge & Shipley
Magna Lock Corp.
Mahr Gage Co., Inc.
Makino Machine Tool Co.
Mazak Corp.
Milacron, Inc.
Modern Machine Shop Magazine
Monarch Machine Tool Co.
Moore Tool Co.
Morse Twist Drill & Machine Co.
MTI Corporation
National Broach & Machine
National Twist Drill & Tool Co.
Neill, James & Co.
Nicholson File Co. Ltd.
Northwestern Tools, Inc.
Norton, Div. Saint Gobain Abrasives
Nucor Corporation
Optical Gaging Products, Inc.
PolyPlan Technologies
POM, Inc.
Powder Metallurgy Parts
    Manufacturers' Association
Pratt & Whitney Co., Inc.
Praxair, Inc.
Precision Diamond Tool Co.
Precision Gage and Tool Co.
QQC, Inc.
Rockford Machine Tool Co.
Rockwell International Machinery
Royal Products

Sheffield Measurement Div. Giddings
    & Lewis
Shore Instrument & Mfg. Co., Inc.
Slocomb, J.T. Co.
Sony Corp.
South Bend Lathe Corp.
Stanley Tools
Starrett, L.S. Co.
STEP Tools, Inc.
Sun Oil Co.
Sunnen Products Co.
Superior Electric Co. Ltd.
Taft-Pierce Manufacturing Co.
Taper Micrometer Corp.
Technomatix Technology, Inc.
TechPlate, Inc.
Thompson Grinder Co.
Thomson Industries, Inc.
Thriller, Inc.
Toolex Systems, Inc.
Union Butterfield Corp.
United States Steel Corporation
Valenite, Inc.
Volstro Manufacturing Co., Inc.
Weldon Tool Co.
Wellsaw, Inc.
Whitman & Barnes
Wilkie Brothers Foundation
Williams, J.H. & Co.
Wilson Instrument Division, American
    Chain & Cable Co.
Woodworth, W. J. & J. D. Woodworth
Worthington Industries

TECHNOLOGY of
MACHINE TOOLS

Section 1

(Emco Maier Corp.

INTRODUCTION TO MACHINE TOOLS

The progress of humanity throughout the ages has been governed by the types of tools available. Ever since primitive people used rocks as hammers or as weapons to kill animals for food, tools have governed our standard of living. The use of fire to extract metals from ore led to the development of newer and better tools. The harnessing of water led to the development of hydropower, which greatly improved humanity's well-being.

With the industrial revolution in the mid-18th century, early machine tools were developed and were continually improved. The development of machine tools and related technologies advanced rapidly during and immediately after World Wars I and II. Since World War II, processes such as computer numerical control, electro-machining, computer-aided design (CAD), computer-aided manufacturing (CAM), and flexible manufacturing systems (FMS) greatly altered manufacturing methods.

Today we are living in a society greatly affected by the development of the computer. Computers affect the growing and sale of food, manufacturing processes, and even entertainment. Although the computer influences our everyday lives, it is still important that you, as a student or an apprentice, be able to perform basic operations on standard machine tools. This knowledge will provide the necessary background for a person seeking a career in the machine tool trade.

UNIT 1

History of Machines

OBJECTIVES

After completing this unit, you will be familiar with:

1 The development of tools throughout history

2 The standard types of machine tools used in shops

3 The newly developed space-age machines and processes

The high standard of living we enjoy today did not just happen. It has been the result of the development of highly efficient machine tools over the past several decades. Processed foods, automobiles, telephones, televisions, refrigerators, clothing, books, and practically everything else we use are produced by machinery.

The history of machine tools began during the stone age (over 50,000 years ago), when the only tools were hand tools made of wood, animal bones, or stone (Fig. 1-1).

Between 4500 and 4000 B.C., stone spears and axes were replaced with copper and bronze implements and power supplied by humans was in a few cases replaced with animal power. It was during this bronze age that human beings first enjoyed "power-operated" tools.

Around 1000 B.C., the iron age dawned, and most bronze tools were replaced with more durable iron implements. After smiths learned to harden and temper iron, its use became widespread. Tools and weapons were greatly improved, and animals were domesticated to provide power for some of these tools, such as the plow. During the iron age, all commodities required by humans, such as housing and shipbuilding materials, wagons, and furniture, were handmade by the skilled craftspeople of that era.

About 300 years ago, the iron age became the machine age. In the 17th century, people began exploring new sources of energy. Water power began to replace human and animal power. With this new power came improved machines and, as production increased, more products became available. Machines continued to be improved, and the boring machine made it possible for James Watt to produce the first steam engine in 1776, beginning the industrial revolution. The steam engine made it possible to provide power to any area where it was needed. With quickening speed, machines were improved and new ones invented. Newly designed pumps reclaimed thousands of acres of the Netherlands from the sea. Mills and plants which had depended on water power were converted to steam power to produce flour, cloth, and lumber more efficiently. Steam engines replaced sails and steel replaced wood in the shipbuilding industry. Railways sprang up, unifying countries, and steamboats connected the continents. Steam-driven tractors and improved farm machinery lightened the farmer's task. As machines improved, further sources of power were developed. Generators were made to produce electricity, and diesel and gasoline engines were developed.

With further sources of energy available, industry grew and new and better machines were built. Progress

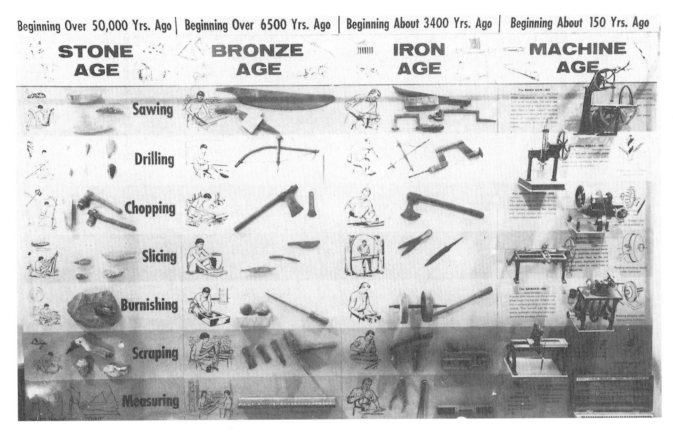

Beginning Over 50,000 Yrs. Ago | Beginning Over 6500 Yrs. Ago | Beginning About 3400 Yrs. Ago | Beginning About 150 Yrs. Ago

STONE AGE   BRONZE AGE   IRON AGE   MACHINE AGE

Sawing

Drilling

Chopping

Slicing

Burnishing

Scraping

Measuring

■ **Figure 1-1** The development of hand tools over the years.   *(DoAll Company)*

continued slowly during the first part of the 20th century except for spurts during the two world wars. World War II sparked an urgent need for new and better machines, which resulted in more efficient production (Fig. 1-2).

Since the 1950s, progress has been rapid and we are now in the space age. Calculators, computers, robots, and automated machines and plants are commonplace. The atom has been harnessed and nuclear power is used to produce electricity and to drive ships. We have traveled to the moon and outer space, all because of fantastic technological developments. Machines can mass produce parts to millionths of an inch accuracy. The fields of measurement, machining, and metallurgy have become sophisticated. All these factors have produced a high standard of living for us. All of us, regardless of our occupation or status, are dependent on machines and/or their products (Fig. 1-3 on p. 6).

Through constant improvement, modern machine tools have become more accurate and efficient. Improved production and accuracy have been made possible through the application of hydraulics, pneumatics, fluidics, and electronic devices such as computer numerical control to basic machine tools.

■ **Figure 1-2** New machine tools were developed during the mid-20th century.   *(DoAll Company)*

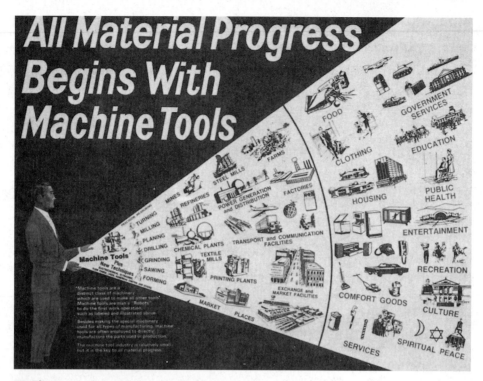

Figure 1-3 Machine tools produce tools and machines for manufacturing all types of products. *(DoAll Company)*

▶▶ Common Machine Tools

Machine tools are generally power-driven metal-cutting or -forming machines used to shape metals by:

> The removal of chips

> Pressing, drawing, or shearing

> Controlled electrical machining processes

Any machine tool generally has the capability of:

> Holding and supporting the workpiece

> Holding and supporting a cutting tool

> Imparting a suitable movement (rotating or reciprocating) to the cutting tool or the work

> Feeding the cutting tool or the work so that the desired cutting action and accuracy will be achieved

The machine tool industry is divided into several different categories, such as the general machine shop, the toolroom, and the production shop. The machine tools found in the metal trade fall into three broad categories:

1. *Chip-producing machines,* which form metal to size and shape by cutting away the unwanted sections. These machine tools generally alter the shape of steel-produced products by casting, forging, or rolling in a steel mill.

2. *Non-chip-producing machines,* which form metal to size and shape by pressing, drawing, punching,

or shearing. These machine tools generally alter the shape of sheet steel products and produce parts which need little or no machining by compressing granular or powdered metallic materials.

3. *New-generation machines,* which were developed to perform operations that would be very difficult, if not impossible, to perform on chip- or non-chip-producing machines. Electro-discharge, electro-chemical, and laser machines, for example, use either electrical or chemical energy to form metal to size and shape.

4. *Multi-tasking machines,* a combined machining and turning center, can produce virtually any shape of part, starting with a rough piece of material to a finished part in a single machine setup. These machines consist of a turning center with two independent spindles and a vertical machining center having a rotary tool spindle. They combine Information Technology (IT) and Manufacturing Technology (MT) for the efficient multiple-face machining of workpieces. Besides the conventional turning and milling operations, it is possible to hob gears, machine molds, and cylindrical grind in the same work setup.

The performance of any machine tool is generally stated in terms of its metal-removal rate, accuracy, and repeatability. *Metal-removal rate* depends upon the cutting speed, feed rate, and depth of cut. *Accuracy* is determined by how precisely the machine can position the cutting tool

Figure 1-4 Common machine tools found in a machine shop. *(DoAll Company)*

to a given location once. *Repeatability* is the ability of the machine to position the cutting tool consistently to any given position.

A general machine shop contains a number of standard machine tools that are basic to the production of a variety of metal components. Operations such as turning, boring, threading, drilling, reaming, sawing, milling, filing, and grinding are most commonly performed in a machine shop. Machines such as the drill press, engine lathe, power saw, milling machine, and grinder are usually considered the *basic machine tools* in a machine shop (Fig. 1-4).

▶▶ Standard Machine Tools

DRILL PRESS

The drill press or drilling machine (Fig. 1-5), probably the first mechanical device developed prehistorically, is used primarily to produce round holes. Drill presses range from the simple hobby type to the more complex automatic and numerical control machines used for production purposes. The function of a drill press is to grip and revolve the cutting tool (generally a twist drill) so that a hole can be produced in a piece of metal or other material. Operations such as drilling, reaming, spot facing, countersinking, counterboring, and tapping are commonly performed on a drill press.

ENGINE LATHE

The engine lathe (Fig. 1-6) is used to produce round work. The workpiece, held by a work-holding device mounted on the lathe spindle, is revolved against a cutting tool, which produces a cylindrical form. Straight turning, tapering, facing, drilling, boring, reaming, and thread cutting are some of the common operations performed on a lathe.

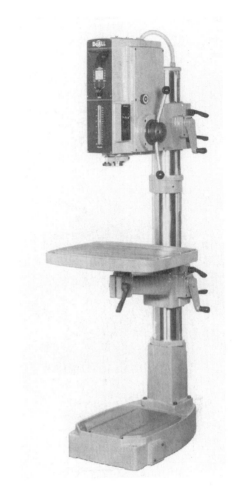

Figure 1-5 A standard upright drill press. *(DoAll Company)*

Figure 1-6 An engine lathe is used to produce round work. *(Clausing Industrial, Inc.)*

METAL SAW

The metal-cutting saws are used to cut metal to the proper length and shape. There are two main types of metal-cutting saws: the bandsaw (horizontal and vertical) and the reciprocating cutoff saw. On the vertical bandsaw (Fig. 1-7 on p. 8) the workpiece is held on the table and brought into

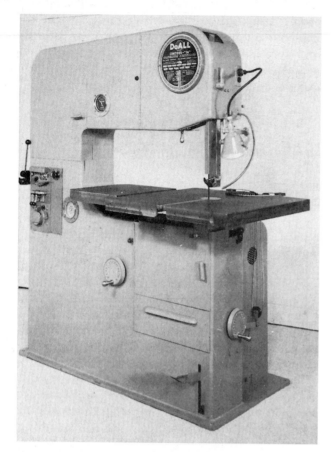

Figure 1-7 A contour-cutting bandsaw. *(DoAll Company)*

Figure 1-8 A vertical milling machine. *(Clausing Industrial, Inc.)*

contact with the continuous-cutting saw blade. It can be used to cut work to length and shape. The horizontal band-saw and the reciprocating saw are used to cut work to length only. The material is held in a vise and the saw blade is brought into contact with the work.

MILLING MACHINE

The horizontal milling machine and the vertical milling machine (Fig. 1-8) are two of the most useful and versatile machine tools. Both machines use one or more rotating milling cutters having single or multiple cutting edges. The workpiece, which may be held in a vise, fixture, accessory, or fastened to the table, is fed into the revolving cutter. Equipped with proper accessories, milling machines are capable of performing a wide variety of operations, such as drilling, reaming, boring, counterboring, and spot facing, and of producing flat and contour surfaces, grooves, gear teeth, and helical forms.

GRINDER

Grinders use an abrasive cutting tool to bring a workpiece to an accurate size and produce a high surface finish. In the grinding process, the surface of the work is brought

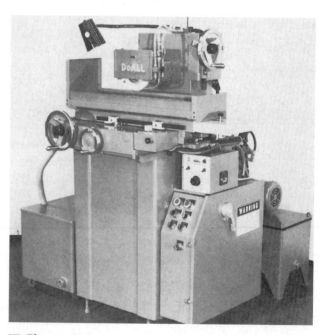

Figure 1-9 A surface grinder is used to grind flat surfaces. *(DoAll Company)*

into contact with the revolving grinding wheel. The most common types of grinders are the surface, cylindrical, cutter and tool, and bench or pedestal.

Surface grinders (Fig. 1-9) are used to produce flat, angular, or contoured surfaces on a workpiece.

Cylindrical grinders are used to produce internal and external diameters, which may be straight, tapered, or contoured.

Cutter and tool grinders are generally used to sharpen milling machine cutters.

Bench or pedestal grinders are used for offhand grinding and the sharpening of cutting tools such as chisels, punches, drills, and lathe and planer tools.

SPECIAL MACHINE TOOLS

Special machine tools are designed to perform all the operations necessary to produce a single component. Special-purpose machine tools include gear-generating machines; centerless, cam, and thread grinders; turret lathes; and automatic screw machines.

▶▶ Computer Numerical Control Machines

Computer numerical control (CNC) has brought tremendous changes to the machine tool industry. New machine tools, controlled by computers, have allowed industry to produce parts quickly and to accuracies undreamed of only a few years ago. The same part can be reproduced, to the exact accuracy, any number of times if the part program has been properly prepared. The operating commands that control the machine tool are executed with amazing speed, accuracy, efficiency, and reliability. In many cases throughout the world, conventional machine tools operated by hand are being replaced by CNC machine tools operated by computers.

Chucking and turning centers (Fig. 1-10a and b), the CNC equivalent of the engine lathe, are capable of machining round parts in a minute or two that would take a skilled machinist an hour to produce. The *chucking center* is designed to machine parts in a chuck or some form of holding and driving device. The *turning center,* similar to a chucking center, is designed mainly for shaft-type workpieces that must be supported by some type of tailstock center.

The machining centers (Fig. 1-11a and b), the CNC equivalent of the milling machine, can perform a variety of operations on a workpiece by changing its own cutting tools. There are two types of machining centers, the vertical and the horizontal. The *vertical machining center* (Fig. 1-11a), whose spindle is in a vertical position, is used primarily for flat parts where three-axis machining is required. The *horizontal machining center,* (Fig. 1-11b) whose spindle is in a horizontal position, allows parts to be machined on any side in one setup if the machine is equipped with an indexing table. Some machining centers have both vertical and horizontal spindles that can change from one to another very quickly.

Electrical discharge machines (EDM) (Fig. 1-12) use a controlled spark erosion process between the cutting tool and the workpiece to remove metal. The two most common EDM machines are the *wire-cut* and the vertical *ram type.* The wire-cut EDM uses a traveling wire to cut the internal and external shapes of a workpiece. The vertical ram-type EDM, commonly called the die sinking machine, generally feeds a form tool down into the workpiece to reproduce its form.

(a)

(b)

■ **Figure 1-10** Chucking (a) and turning centers (b) are capable of producing round parts quickly and accurately. *(Mazak Corp.)*

(a)

Figure 1-11 (a) The vertical machining center is used primarily for flat parts when three-axis machining is required. *(Cincinnati Machine)*

(b)

Figure 1-11 (b) The horizontal machining center can machine any side of a part if it is equipped with an indexing table. *(Cincinnati Machine)*

(a) (b)

■ **Figure 1-12** EDM machines remove metal by an electric spark-erosion process. *(Charmilles Technology Corp.)*

Electro-discharge machining, electrochemical machining, electrolytic grinding, and *laser machining* have made it possible to machine new space-age materials and to produce shapes which were difficult or often impossible to produce by other methods.

The numerical control principle has also been applied to *robots,* which are now capable of handling materials and changing machine tool accessories as easily and probably more efficiently than a person can (Fig. 1-13). *Robotics* has become one of the fastest-growing areas of the manufacturing industry.

Since its development, the *laser* has been applied to several areas of manufacturing. Lasers are now used increasingly for cutting and welding all types of metals—even those that have been impossible to cut or weld by other methods. Laser beams can pierce diamonds and any other known material and are also used in extremely accurate measuring and surveying devices, and as sensing devices.

With the introduction of numerous special machines and special cutting tools, production has increased tremendously over that attained with standard machine methods. Many products are produced automatically by a continuous flow of finished parts from these special machines. Product control and high production rates allow us to enjoy the pleasure and convenience of automobiles, power

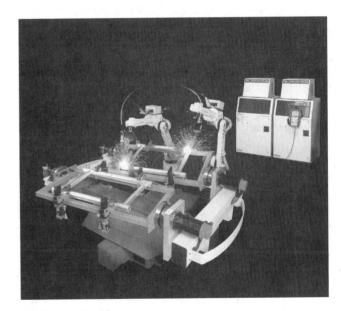

■ **Figure 1-13** Robots are finding ever-increasing applications in industry. *(GE Fanuc)*

lawn mowers, automatic washers, stoves, and scores of other modern products. Without the basic machine tools required for mass production and automation, the costs of many luxuries that we now enjoy would be prohibitive.

History of Machines   **11**

▶▶ Major Developments in Metalworking over the Past Half Century

Prior to the 20th century, manufacturing methods changed very slowly. Mass production had not developed nearly to the stage that we know now. It wasn't until the early 1930s that new and outstanding developments in manufacturing began to affect manufacturing processes. Since then, progress has been so rapid that now some of the newer developments astound most of us. It is because of this progress over the past 60 years or so that we in North America enjoy one of the highest standards of living in the world.

Manufacturing prior to 1932 was done on standard types of machine tools with little or no automation. Engine lathes, turret lathes, drill presses, shapers, planers, and horizontal milling machines were the common machine tools of the day. Most of the cutting tools were made of carbon steel or early grades of high-speed steel, which were not very efficient by today's standards. Production was slow and much of the work was finished by hand. This resulted in the high cost of the items produced in relation to wages paid to the workers.

In the early 1930s, machine tool manufacturers took advantage of the lull in production and sales caused by the Great Depression to upgrade their machines by improving flexibility and controls. Thus began the trend leading to the machines of the present.

According to the Society of Manufacturing Engineers and the AMT—Association for Manufacturing Technology, the following chronology lists the most important developments in metalworking during the past half century.

1942: *Heliarc welding* was a new process developed for welding magnesium on a production basis.

1943: *Air gaging* provided a means of gaging parts more quickly and accurately than was previously possible.

1944: The *60,000 r/min grinder motor* was developed to provide small grinding wheels (1/8 in. diameter or less) with sufficient speed for efficient grinding.

1945: *Man-au-trol control* was the first hydraulic-electric control system introduced for automatic machine control.

1946: The *ENIAC digital computer* was the first all-electronic general-purpose computer introduced and would eventually help with design problems.

1947: *Automatic size control* provided a means of automatically boring, honing, and gauging the size of engine block cylinders.

1948: *Cardamatic milling* used punch cards to automatically control the cycle of a milling machine.

1949: *Ultrasonic inspection* provided a nondestructive method of testing materials by means of extremely high-frequency sound waves.

1950: *Electronic hardness testing* was a quick and accurate type of test based on the magnetism retention of a part against a standard.

1951: *Method X electrical discharge machining* provided a means of removing metal from the workpiece by means of a spark of high density and short duration.

1952: *Numerical control* introduced a system attached to a milling machine whereby the table and cutting-tool movements were controlled by punched tape.

1953: *Project tinkertoy* was a system developed to automatically manufacture and assemble electronic circuit elements.

1954: *Indexible insert tooling* introduced a throwaway type of carbide cutting-tool insert, which could be turned over and used on both sides. This eliminated the need for expensive cutting-tool maintenance.

1955: The *numericord system* was the first completely automatic control for machines, provided by means of electronic control and magnetic tape.

1956: The *gear-honing process* provided a method used after heat treating to remove nicks and burrs from a gear and form it to its correct specifications.

1957: *Manufactured diamond* was developed by the General Electric Co. for grinding and machining hard, abrasive nonferrous and nonmetallic materials. It was produced by subjecting a form of carbon and a metal catalyst to high pressure and high temperature.

1958: The *machining center* introduced a computer-controlled machine with a tape-controlled toolchanger capable of performing milling, drilling, tapping, and boring on a workpiece as large as an 18-in. cube.

1959: The *APT (automatically programmed tool)* programming language was a 107-word computer language used by programmers to write programs using data from engineering drawings.

1960: *Ultra-high-speed machining* was based on the principle that at extremely high cutting speeds (2500 sf/min and higher) the tool temperature and horsepower required to machine a workpiece drop. Speeds of 18,000 sf/min were used and speeds of up to 36,000 sf/min were planned.

1961: The *industrial robot* provided a single-armed device that can manipulate parts or tools through a sequence of operations or motions as controlled by computer programs.

1962: *Computer-controlled steelmaking* introduced a system in which every steelmaking variable, from order and raw-material requirements to the finished product, is computer controlled.

1963: The *ADAPT programming language* provided a program compatible with APT language, used only about half of APT's vocabulary words, and was designed for use with small computers to control machine operations.

1964: *DAC-1, design augmented by computers,* was a computer system that allowed the computer to read drawings from paper or film and to generate new drawings by use of the keyboard and a lightpen.

1965: *System/360* introduced a large mainframe computer capable of responding in billionths of a second and became the standard in industry for the next decade.

1966: The *single-layer metal-bonded diamond grinding wheel* was a diamond-impregnated grinding wheel, contoured to the profile of the workpiece, and reduced the grinding time required for certain parts from 10 hours to 10 minutes.

1967: *Computer numerical control* provided a computer control system that combined the functions of separate tape preparation equipment, numerical control, and program and part verification into one unit.

1968: *Direct numerical control* allowed the operation of machines directly from the mainframe computer without the use of tapes.

1969: The *programmable controller* was a smaller, single-purpose computer that could control as many as 64 machines using APT-created programs.

1969: *CBN (cubic boron nitride),* an extremely hard abrasive, was developed by the General Electric Co. for grinding and machining hard, abrasive ferrous metals. It was produced by subjecting hexagonal boron nitride along with a catalyst to high pressure and high temperature.

1970: *Polycrystalline superabrasive cutting tool blanks* consist of a layer of diamond or CBN (cubic boron nitride) bonded to a cemented carbide substrate. They are used to cut hard, abrasive, nonferrous and nonmetallic materials (diamond) and ferrous materials (CBN).

1970: *System GEMINI* provided a system whereby a supervisory computer and a distribution computer control several machines in the total manufacture of a part. This system was the forerunner of the automatic factory.

1971: *Robotic sensory capabilities* permitted a robot to "feel" for objects by means of a sensor applied to the robot's gripping fingers or vacuum cup.

1972: The *Hummingbird press* was a 60-ton automated press with speeds of up to 1600 strokes per minute and a feed rate of 400 ft/min.

1973: *Robotic vision* was a robot system that utilized a television camera and image processing equipment to permit the robot to "see" and prevent the arm from bumping other parts as it travels to the desired location.

1974: *Remote machine diagnostics* allowed diagnosis of CNC machine problems in a plant by a computer in the manufacturer's head office by tying both computers into the telephone system.

1974: *Group Technology (GT)* is a system of classifying parts on the basis of their similarities and physical characteristics so that they can be grouped together for manufacturing using the same process. This improves manufacturing productivity by the better use of machine tools and the efficient flow of parts through the machines.

1975: *Computer Integrated Manufacturing (CIM)* is an information system where computers integrate all manufacturing functions such as CNC, process planning, resource planning, and CAD/CAM with processes such as finance, inventory, payroll, and marketing. The CIM system controls all data flow within a company.

1975: The *super CIMFORM grinding wheel* was a long-life vitrified aluminum oxide grinding wheel developed for high production. It cut grinding costs by 25%.

1976: *CAM-I automated process planning,* when a part is required, allows the computer to determine the "family" the part belongs to, calls up the drawing, makes any necessary modifications, and then directs the production of the part in the shop.

1977: *Distributed plant management systems* allowed a DNC computer system in a plant to be controlled and programmed by a remote computer that may not even be located in the same plant.

1978: *Automated programmable assembly systems* were designed to increase production by the use of several programmed robots to assemble component parts into a unit.

1979: *YMS-50 flexible manufacturing systems* linked standard NC machine modules with a parts-handling device and provided total computer control of the system.

1980: The *variable mission automatic toolhead changer* stores and installs cutting tools, as programmed, on as many as 18 multiple spindle heads.

1980: *Adaptive control* uses the power of the computer to monitor a machining operation and make adjustments to the speed and feed rates to optimize a machining operation. It can be used to sense tool wear, the geometry of the cut, the hardness and rigidity of the workpiece, and the position of the tool in relation to the workpiece.

1981: The *grinding center* provides computer-controlled grinding that can be programmed for as many as 48 different grinds on a workpiece.

1982: *Just-In-Time manufacturing,* a concept developed to improve productivity, reduce costs, reduce scrap and rework, use machines efficiently, reduce inventory and work-in-process (WIP), and make the best use of manufacturing space. It involves having the right materials, tools, and machines available at the time they are required for production.

1983: *Artificial Intelligence (AI)* is a field of computer science that deals with computers doing humanlike functions such as interpretation and reasoning. It uses robots, vision systems, expert systems, and language and voice recognition to perform operations normally requiring human understanding.

1986: *Manufacturing Automation Protocol (MAP)* is a 7-layer broadband, token-bus communications protocol for the factory floor to achieve real-time cost monitoring, real-time quality monitoring, and real-time production monitoring. It is designed to accommodate a broad range of manufacturing environments and makes communication possible among computer-controlled factory floor equipment.

1988: *Microcrystalline aluminum oxide grinding wheels,* commonly called seeded gel or SG wheels, contain a submicron crystal structure with billions of particles in each grain. This feature allows grains to resharpen themselves, resulting in fewer wheel dressings and increasing productivity while lowering the cost per part.

1989: *Direct ironmaking and direct steelmaking processes* are in the developmental stage to produce iron and steel in one step. The aim is to develop environmentally sound procedures that reduce manufacturing time, require less energy, and lower the manufacturing cost.

1989: *Net shape manufacturing* involves the production of components by billet forming, precision and die casting, sheet forming, injection molding, and die making that are close to the finished size required.

1989: *Rapid prototyping and manufacturing,* also called stereolithography, combines the technologies of CAD, computers, and lasers to produce solid prototype models from a 3-dimensional technical drawing.

1990: *CVD (chemical vapor deposition)* was developed to provide a thin, long-wearing diamond film on cutting tools, wear parts, heat sinks and electronic substrates, optical devices, etc.

1991: *Concurrent engineering* is the integration of product design, manufacturing processes, and related technology to provide early manufacturing input into the design process.

1992: *Agile manufacturing,* the newest form of manufacturing, combines the state-of-the-art fabrication and product-delivery technologies to custom-make products to suit customer's specifications without increasing the price. This process is especially designed to quickly respond to continuously changing market conditions.

1993: The *octahedral hexapod* is a radical machine tool design for machining centers. It consists of a six-legged structure that connects the bed to the head and the spindle virtually floats in space. The hexapod has 6-axis contouring capability, five times the rigidity and is two to 10 times more accurate than conventional machines.

1994: *High-velocity manufacturing* uses a new axis drive system, *high-force linear motors,* to move machine axes. It is 10 times faster than ball screws, increases velocity rates three to four times higher, with greater accuracy and reliability. Spindle speeds range from 0 to 15,000 r/min on machining centers.

1995: *Combination conventional/programmable machines,* such as vertical mills, lathes, and surface grinders, can be used as conventional machine tools and have limited programming features for repetitive steps or operations. These machines can be taught to record a manual path and as a result increase productivity and part accuracy when doing repetitive operations on smaller lot jobs.

1996: *Artificial Intelligence* is a manufacturing tool that is having a great effect in the areas of artificial vision, expert systems, robotics, machine control, natural language understanding, and voice recognition. It is also finding applications in product design, diagnostics, inspection, planning, and scheduling.

1997: *e-Manufacturing* is the technology of networking and controlling machine tools over the Internet through hardware and software technology. Machine tools are turned into Web servers, and they are integrated into the supply chain in ways limited only by the manufacturing imagination. Any process that contributes to a machine tool's effectiveness and productivity can see dramatic efficiency improvements within the InterNetwork.

1998: The *Direct Metal Deposition (DMD)*™ process involves the blending of five common technologies: lasers, computer-aided design (CAD), computer-aided manufacturing (CAM), sensors, and powder metallurgy. It is a form of rapid prototyping that makes parts and molds from metal powder that is melted by moving a laser back and forth, under CNC control, and tracing out a pattern controlled by a computerized CAD design.

1999: *Virtual Reality (VR)* uses computers, hardware, and software to create objects and/or even alternate worlds. VR allows a person to create, manufacture, manipulate, look at, and play with something that is totally real in every respect, except it does not physically exist. Immersion provides the ability to believe that the user is present in the virtual world and can navigate through it and function within the simulated environment as if it were real.

2000: *Step NC* is a worldwide standard, developed by the International Standards Organization (ISO), to extend STEP (**St**andard for the **E**xchange of **P**roduct) model data so that it can be used to define data for NC (numerical control) machine tools. STEP is an extensible, comprehensive, international data standard for product data, created by an international team to give a clear and complete representation of product data throughout its life cycle.

2001: *Nanotechnology* is the technology that applies to the controlling of the structure of materials down to a few atoms or molecules. Researchers have a vision of synthetic molecular nanomachines made of mechanical parts consisting of actual gears and axles on a molecular scale. When these tiny (subminuscule) parts are planned as self-replicating nanorobots, they would push atoms and molecules together to build a wide variety of essential materials.

2002: *Cyber Management Systems* is where all manufacturing information is digitized and accessible to all departments—Engineering, Production, Sales, Distributors, On-line Support, and Administration—through a Central Dataway. Sharing data between all departments and using the Internet realize the absolute minimum time between the start of product design and shipment of finished products to customers.

unit 1 review questions

1. Briefly trace the development of tools from the stone age to the industrial revolution.

2. Why are machine tools so important to our society?

3. How have improved production and accuracy been achieved with basic machine tools?

4. Name four categories of machine tools used in metalworking.

5. List five operations that can be performed on each of the following:
   a. Drill press
   b. Lathe
   c. Milling machine

6. Name four types of grinders found in a machine shop.

7. List four advantages of CNC machine tools.

8. What is the difference between a chucking center and a turning center?

9. Name two types of machining centers.

10. What is the purpose of a:
    a. wire-cut EDM machine?
    b. ram-type EDM machine?

11. What is the importance of the electro-machining processes?

12. What effect has computer numerical and computer control had on manufacturing?

13. State two applications of robots.

14. What is the importance of lasers in modern industry?

Section 2

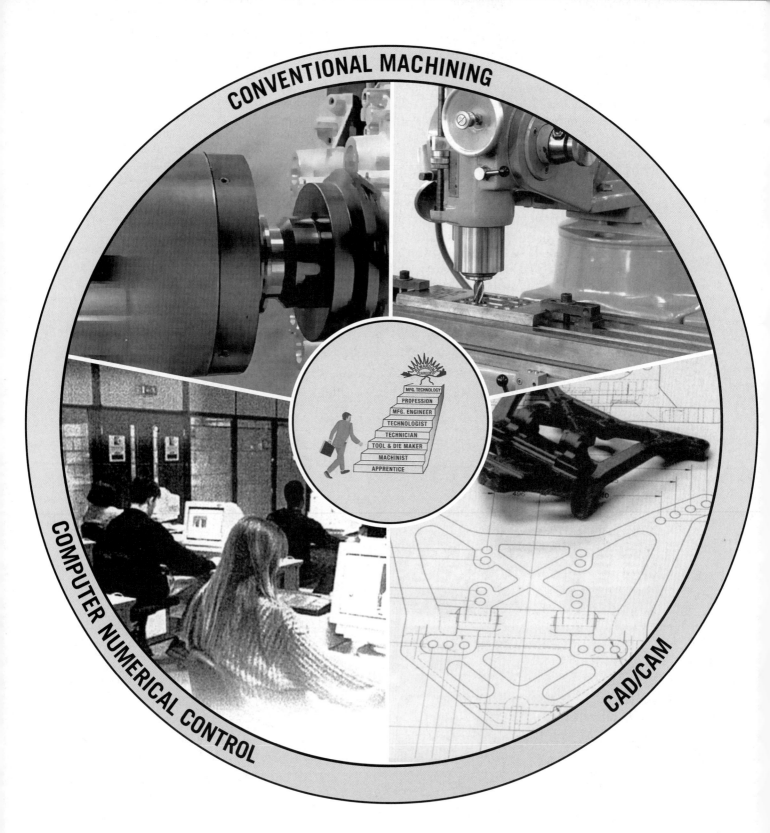

CONVENTIONAL MACHINING

COMPUTER NUMERICAL CONTROL

CAD/CAM

REWARDING CAREERS

MFG. TECHNOLOGY
PROFESSION
MFG. ENGINEER
TECHNOLOGIST
TECHNICIAN
TOOL & DIE MAKER
MACHINIST
APPRENTICE

MACHINE TRADE OPPORTUNITIES

Almost all products used by people, whether in farming, mining, manufacturing, construction, transportation, communication, or the professions, are dependent on machine tools for their manufacture. Constant improvements to and efficient use of machine tools affect the standard of living of any nation. Only through machine tools have we been able to enjoy the automobile, airplane, television, home furnishings, appliances, and many other products on which we rely in our daily lives. Through constant improvement, modern machine tools have become more accurate and efficient. Improved production and accuracy have been made possible through the application of hydraulics and electronic devices such as numerical control, computer numerical control, direct numerical control, and lasers to basic machine tools.

UNIT 2

Careers in the Metalworking Industry

OBJECTIVES

After completing this unit, you will be familiar with:

1 The various types of jobs available in the metal-working industry

2 The type of work each job entails

Advancing technology, new ideas, new products, and special processes and manufacturing techniques are creating new and more specialized jobs. To advance in the machine trade, a person must keep up-to-date with modern technology. A young person leaving school may be employed in an average of five jobs in his or her lifetime, three of which do not even exist today. Industry is always on the lookout for bright young people who are conscientious and do not hesitate to assume responsibility. To be successful, do the job to the best of your ability and never be satisfied with inferior workmanship. Always try to produce quality products, at a reasonable price, in order to compete with foreign products, which are of such serious concern to North American industry.

▶▶ New Technologies

Technology is a tool that makes it possible to produce better quality goods at lower prices. Our standard of living has always been built on the ability to produce products that are in demand throughout the world. Therefore, it can be said that technology can be used to increase the resources of a nation and generate wealth. It seems that the most progressive and wealthiest countries in the world are those who use the latest manufacturing technologies to make them more productive than other countries.

Technology is continually changing and improving, with the amount of technology doubling every three to five years. Machines and manufacturing processes as little as 10 years old can be two generations behind those of the most progressive manufacturing nations of the world. Not only is it important to keep up with the improvements in equipment and processes, it is equally or more important to prepare our students to enter the technological workplace. We cannot expect high technology work from low technology workers; training can make the difference between success and failure. The ever-changing technology means that industrial workers and students in schools must be prepared for continual education (life-long learning) if they expect to survive in the technological world in which we live.

TECHNOLOGY CURRICULUMS

Rapidly changing technology in the metalworking industry makes it imperative for educators to stay abreast of the new improvements and manufacturing processes. To best serve industry and increase the productivity of the nations, educators must continually introduce new material into their curriculum to prepare students to enter the technological world of today. Old "time-proven" methods and processes have been outmoded by new technology. The educational institutions that recognize this and take the appropriate steps will produce graduates who are a credit to their school and make a valuable contribution to the country.

TECHNOLOGY COURSES

Machine shop, which provides the background and groundwork for all manufacturing technologies, is the prerequisite for all students planning to enter the exciting manufacturing world. Along with the knowledge of the basic machining process, a good understanding of computer numerical control (CNC) is essential. In the world today, with approximately 90% of the machine tools manufactured for CNC use, a good knowledge of this area is as important as being able to read and write. A strong background opens up many exciting careers to the progressive student, such as artificial intelligence, computer-aided design, computer-automated manufacturing, flexible manufacturing systems, group technology, just-in-time, lasers, metrology, robotics, statistical process control, superabrasive technology, etc. (Fig. 2-1).

MODULAR TRAINING SYSTEM

A modular training system (Fig. 2-2) offers technical/vocational training covering the skills required for modern manufacturing in the metalworking trade, where computers

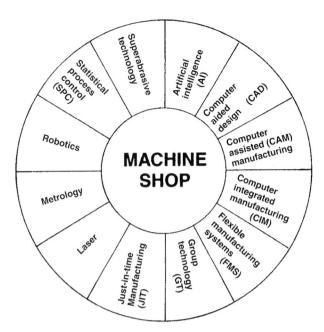

■ **Figure 2-1** A good machine shop background provides the basics for all manufacturing technology careers. *(Kelmar Associates)*

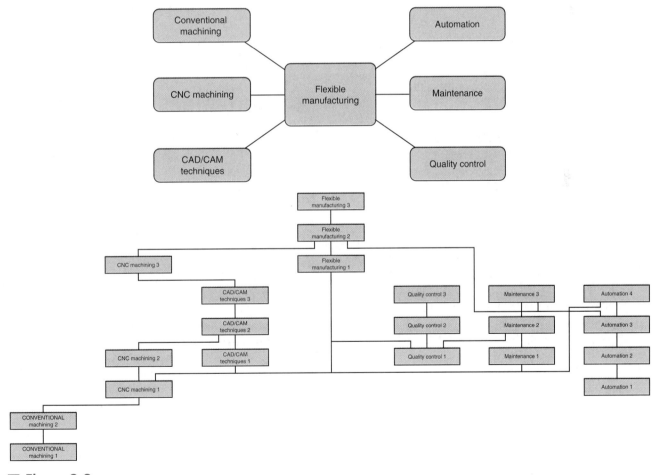

■ **Figure 2-2** Modular training system. *(Emco Maier Corp.)*

are playing an ever-increasing role. The training program, developed by a leading machine tool manufacturer and used by many educators throughout the world, consists of individual modules that can readily be incorporated into a technical education curriculum.

The training modules start with the conventional machine tools and processes; progress through the CNC modules; and incorporate skills relating to quality control, machine maintenance, and low-cost automation. In easy-to-learn stages, these modules prepare a student for the complex task of learning how to operate, program, and build a modern flexible manufacturing system.

A brief summary of the main areas of the modular training system follows.

> *Conventional machining* covers the fundamentals of conventional machine tools and the machining processes associated with each machine tool. These skills are necessary so that students learn the basics of machining metal workpieces and develop a feeling for machining the part most efficiently. No metalworking profession or related technology should be without these basic skills.

> *CNC machining* covers the skills and knowledge necessary to program and operate CNC machine tools. This must include the most cost-effective way of producing error-free CNC part programs from the simple to the complex workpiece. CNC training is essential for anyone looking for a rewarding career in manufacturing technology because of the large number of these machines used in industry.

> *CAD/CAM techniques* provide the skills to design parts on a computer and then use this data for machining the parts on CNC machine tools. Without the flexibility, ease of design, and time saving made possible by CAD/CAM techniques, a manufacturer cannot successfully compete today.

> *Quality control* covers the methods and tools used for measuring dimensions, shapes, and surface textures of finished workpieces. The data collected in this process can be used to improve and eliminate errors from manufacturing processes. Quality control is one of the most important parts in the manufacturing process. It plays a vital role in improving the competitive position of the manufacturer.

> *Maintenance* covers the routine upkeep, aligning and adjusting machines, troubleshooting, and repairs of conventional and CNC machine tools. This maintenance, commonly called *preventive maintenance,* ensures that machine tools operate properly to provide a continuous manufacturing operation, with minimum downtime for maintenance.

> *Automation* provides training in automating the loading and unloading of workpieces for CNC machine tools, material handling, and quality control of the manufacturing process. This low-cost automation is used by many manufacturers who cannot implement the more complex flexible manufacturing technology.

> *Flexible manufacturing*—This module enables students to identify the elements of a flexible manufacturing system, its programming, the planning of manufacturing processes, and design of the system. (All the prior modules of this program provide the fundamental skills for the flexible manufacturing program.) The flexible manufacturing system is able to adapt to changes in the number of parts required, the type of part machined, accept new or different products, make design changes to parts, and allow it to be expanded quickly.

Many different careers are available in the metalworking industry. Choosing the right one depends on the skill, initiative, and qualifications of an individual. Metalworking offers exciting opportunities to any ambitious young person who is willing to accept the challenge of working to close tolerances and producing intricate parts. To be successful in this trade, the individual must also possess characteristics such as care of self, orderliness, accuracy, confidence, and safe work habits.

▶▶ Apprenticeship Training

One of the best ways to learn a skilled trade is through an apprenticeship program. An apprentice is a person who is employed to learn a trade under the guidance of skilled tradespeople (Fig. 2-3). The apprenticeship program is set up as a joint agreement between the sponsoring apprentice company, the Department of Labor Federal Bureau of Apprenticeship, and the trade union. It is usually about two to four years' duration and includes on-the-job training and related theory or classroom work. This period of time may be reduced by the completion of approved courses or because of previous experience in the trade.

To qualify for an apprenticeship, the individual should have completed a high school program or its equivalent. Mechanical ability with a good standing in mathematics, science, English writing skills, and mechanical drawing is desirable. Apprentices earn as they learn; the wage scale increases periodically during the training program.

Upon completion of an apprenticeship program, a certificate is granted, which qualifies a person to apply for journeyman status in the trade. Further opportunities in the trade are limited only by the person's initiative and interest. It is quite possible for an apprentice eventually to become an engineer, tool designer, supervisor, or shop owner.

▶▶ Machine Operator

Machine tool operators are classified as semiskilled tradespeople (Fig. 2-4). They are usually rated and paid according to their job classification, skill, and knowledge. The class A operator possesses more skill and knowledge than class B and C operators. For example, a class A operator should be able to operate the machine and:

> Make necessary machine setups

> Adjust cutting tools

> Calculate cutting speeds and feeds

> Read and understand drawings

> Read and use precision measuring tools

With the continued advancement in computer-controlled machines and programmable robots, gradually operators' jobs will be minimized. However, some machine tool operators may qualify with advanced technology courses and remain employed as operators on computer numerical control (CNC) turning centers, machining centers, and CNC robots.

■ **Figure 2-3** An apprentice learns the trade under the guidance of a skilled tradesperson. *(DoAll Company)*

■ **Figure 2-4** A machine operator generally operates only one type of machine.

▸▸ Maintenance Machinist

The maintenance machinist needs a combination of mechanical, rigging, and carpentry skills. The time for apprenticeship varies but usually ranges from two to four years and includes the necessary related theory for the job. During the apprenticeship, the trainee works with qualified tradespeople. A maintenance machinist may be required to:

> Move and/or install machinery, including production lines

> Read drawings and calculate sizes, fits, and tolerances of machine parts

> Repair machines by replacing and fitting new parts

> Dismantle and install equipment

To become a maintenance machinist apprentice, a student should have good technical training and be a high school graduate. A general knowledge of electricity, carpentry, sheet metal, and the machine tool trade is helpful.

The job outlook is good. Most large industries are expected to employ a workforce of maintenance machinists to maintain and install machinery and production lines for the foreseeable future.

▸▸ Machinist

Machinists (Fig. 2-5) are skilled workers who can efficiently operate all standard machine tools. Machinists must be able to read drawings and use precision measuring instruments and hand tools. They must have acquired enough knowledge and developed sound judgment to perform any bench, layout, or machine tool operation. In addition, they should be capable of making mathematical calculations required for setting up and machining any part. Machinists should have a thorough knowledge of

metallurgy and heat treating. They should also have a basic understanding of welding, hydraulics, electricity, and pneumatics and be familiar with computer technology.

TYPES OF MACHINE SHOPS

A machinist may qualify to work in a variety of shops. The three most common are maintenance, production, and jobbing shops.

A *maintenance shop* is usually connected with a manufacturing plant, lumber mill, or foundry. A maintenance machinist generally makes and replaces parts for all types of setup and cutting tools and production machinery. The machinist must be able to operate all machine tools and be familiar with bench operations such as layout, fitting, and assembly.

A *production shop* may be connected with a large factory or plant that makes many types of identical machined parts, such as pulleys, shafts, bushings, motors, and sheet-metal pieces. A person working in a production shop generally operates one type of machine tool and often produces identical parts.

A *jobbing shop* is generally equipped with a variety of standard machine tools and perhaps a few production machines, such as a turret lathe and punch and shear presses. A jobbing shop may be required to do a variety of tasks, usually under contract to other companies. This work may involve the production of jigs, fixtures, dies, molds, tools, or short runs of special parts. A person working in a jobbing shop generally is a qualified machinist, toolmaker, or mold maker and is required to operate all types of machine tools and measuring equipment.

▸▸ Tool and Diemaker

A *tool and diemaker* is a highly skilled craftsperson who must be able to make different types of dies, molds, cutting tools, jigs, and fixtures (Fig. 2-6). These tools may be used

■ **Figure 2-5** A machinist is skilled in the operation of all machines. *(Cincinnati Lathe & Tool Company)*

■ **Figure 2-6** A tool and diemaker can operate all machine tools and plan procedures for making tools, dies, and fixtures. *(Diamond Innovations, at www.abrasivesnet.com)*

in the mass production of metal, plastic, or other parts. For example, to make a die to produce a 90° bracket in a punch press, the tool and diemaker must be able to select, machine, and heat treat the steel for the die components. He or she should also know what production method will be used to produce the part since this information will help to produce a better die. For a mold used to produce a plastic handle in an injection molding machine, the tool and diemaker must know the type of plastic used, the finish required, and the process used in production.

To qualify as a tool and diemaker, a person should serve an apprenticeship, have above-average mechanical ability, and be able to operate all standard machine tools. This work also requires a broad knowledge of shop mathematics, print reading, machine drafting, principles of design, machining operations, metallurgy, heat treating, computers, and space-age machining processes.

▶▶ CNC Machine Operator/Programmer

The wide use of CNC machine tools in the metalworking industry creates a big demand for a person skilled in computer numerical control (Fig. 2-7). The duties of a CNC machine operator will vary from shop to shop. In some shops, the person will be responsible only for the setup and operation of the machine tool, while in others it may also involve preparing the computer program.

■ **Figure 2-7** CNC machine tool operators and programmers are in high demand. *(Hewlett Packard)*

A *CNC machine operator* must be able to:

> Visualize a CNC program
> Understand machining processes and the sequence of operations
> Make machine setups
> Calculate speeds and feeds
> Select cutting tools

A *CNC programmer* must possess all the skills of a CNC machine operator and must:

> Be skilled in print reading
> Have a good knowledge of computer programming languages and procedures
> Be able to visualize machining processes and operations

▶▶ Technician

A *technician* is a person who works at a level between the professional engineer and the machinist. The technician may assist the engineer in making cost estimates of products, preparing technical reports on plant operation, or programming a numerical control machine.

▶▶ Technologist

A *technologist* works at a level between a graduate engineer and a technician. Most technologists are three- or four-year graduates from a community or technical college. Their studies generally include physics, advanced mathematics, chemistry, engineering graphics, computer programming, business organization, and management.

An *engineering technologist* may do many jobs normally performed by an engineer, such as design studies, production planning, laboratory experiments, and supervision of technicians. This permits the engineer to work in other important areas. Technologists are often employed to serve in a middle management position within a large company. A technologist may be employed in many areas, such as quality and cost control, production control, labor relations, training, and product analysis. A technologist may qualify as an engineer by obtaining further education at the university level and by passing a qualifying examination.

▶▶ Quality Control Inspector

An *inspector* (Fig. 2-8 on p. 24) checks and examines machined parts to determine whether they meet the specifications on the drawing. If the parts are not within the

■ Figure 2-8 Inspectors check the dimensions of finished parts and products. *(The L.S. Starrett Co.)*

limits shown on the drawing, they will not fit together and function properly when assembled. This task is very important since parts made in one country may be assembled in another or be interchanged with worn or broken parts.

An inspector should have a technical or vocational education and be familiar with measuring tools and inspection processes. On-the-job training may take from several weeks to several years, depending on the job, the items to be inspected, and the technical knowledge required. An inspector may need varying degrees of skill depending on the size, cost, type, and tolerances required on the finished workpiece. A good inspector should be able to:

> Understand and read mechanical drawings

> Make basic mathematical calculations

> Use micrometers, gages, comparators, and precision measuring instruments

▶▶ Instrument Makers

Instrument makers are highly skilled tool and diemakers who work directly with scientists and engineers. Instrument makers should be able to operate all precision machine tools in order to make measuring instruments, gages, and special machines for testing purposes. Generally, instrument makers must have more training than a machinist or a tool and diemaker. They must work to closer tolerances and limits than a machinist. Also, they usually must perform all the work on the instrument or gage being produced.

Instrument makers generally serve four or five years of apprenticeship. They may also learn the craft by transferring from a toolmaker or machinist trade and further training on the job.

With new processes in manufacturing, the demand for instrument makers will remain fairly close to the demand for skilled tradespeople. Instrument makers are employed by research centers, scientific laboratories, manufacturers of gages and measuring tools, and government and standards organizations.

▶▶ Professions

Many areas are open to the engineering graduate. Teaching is one of the most satisfying and challenging professions. Graduation from a college, including teacher training course work, is required. On-the-job industrial experience is not always a prerequisite, but it will prove helpful in teaching. However, some states and/or schools require on-the-job experience to qualify for technical and vocational certification. With industrial experience a person may teach industrial arts or in certain subject areas in a technical school, vocational school, or community college.

Engineers in industry are responsible for designing and developing new products and production methods and for redesigning and improving existing products. Most engineers specialize in a specific discipline of engineering, such as:

> Industrial

> Metallurgical

> Aerospace

> Mechanical

> Electrical

> Electronics

A bachelor's degree in engineering is usually required to enter the profession. However, in some branches of engineering, such as tool and manufacturing, the individual progresses through a program of practical experience and obtains a certificate after passing a qualifying examination. Because of the variety of engineering jobs available, many women, as well as men, are entering the profession.

A technician should have completed high school and have at least two additional years of education at a community college, technical institute, or university. A technician must also possess a good knowledge of drafting, mathematics, and technical writing.

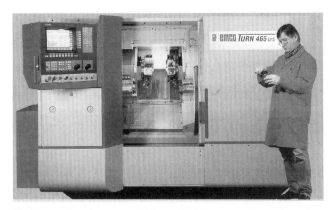

Figure 2-9 A technician is often required to check the setup and operation of a machine program. *(Emco Maier Corp.)*

Figure 2-10 A management conference. *(AMT—The Association for Manufacturing Technology)*

Opportunities for technicians are becoming more plentiful because of the development of machine tools such as computer numerical control, turning centers, and electro-machining processes. Technicians are usually trained in only one area of technology, such as electrical, manufacturing, machine tool, or metallurgy. Some technicians may need knowledge beyond that of their specialty field. For example, a machine tool technician (Fig. 2-9) should have a knowledge of industrial machines and manufacturing processes to know the best method of manufacturing a product. However, it is not necessary for a technician to perform as a skilled machinist. A technician may qualify as a technologist after at least one year of on-the-job training under a technologist or an engineer.

▶▶ Trade Organizations

Society today could not exist without manufacturing, and manufacturing could not exist without machine tools, which are the power base of the modern industrial world. The dramatic changes in technology over the past 25 years have made it imperative to learn about and implement new manufacturing technology in order to stay competitive in the world. Two major machine tool related organizations have been leaders in seeing that their members stay abreast of this rapidly changing world: AMT—The Association for Manufacturing Technology and the Society of Manufacturing Engineers (SME). Both are involved in the continual updating of programs for students, educators, tradespeople, and manufacturing personnel.

AMT—THE ASSOCIATION FOR MANUFACTURING TECHNOLOGY

AMT (formerly the National Machine Tool Builders' Association) is a nonprofit trade association representing U.S. companies in the machine tool building and re-

lated manufacturing industries. This association has been preparing its members for the change created by the ever-changing manufacturing technology. Supported by executives of member companies, AMT develops programs to meet member needs in marketing, technology, production, training, communications, and financial administration (Fig. 2-10).

The association has seen the need to develop skilled personnel for all phases of the machine tool industry. An extensive scholarship program is supported by AMT for technical training.

AMT and its member companies provide the machines, tools, and equipment for young machinists to compete in the VICA National Precision Machining Competition at the Skills U.S.A. Championships. Since 1990, it has hosted the International Machining Trials at its biannual International Manufacturing Technology Show (IMTS), to select contestants who will represent the United States in the International Skills Competition. AMT has also developed effective training procedures and textbooks used in vocational schools and technical colleges.

THE SOCIETY OF MANUFACTURING ENGINEERS

The Society of Manufacturing Engineers (SME) is an international society dedicated to advancing the manufacturing profession through the sharing of technical information. Its goal is to advance the knowledge of manufacturing technology of its over 80,000 members throughout the world.

To encourage engineering as a profession among young people in vocational schools, technical institutions, and community colleges, SME sponsors about 300 student chapters with over 9000 members. The society also assists colleges and technical institutions with curriculum development, career guidance resources, and grants for equipment and software. Over the years SME has become a storehouse for the most up-to-date technological knowledge through its publications of technical papers, textbooks, magazines, films, and videotapes. Membership in an organization such as SME is essential to keep up with the ever-changing technology. It is open to students, teachers, machinists, technicians, and manufacturing engineers who wish to advance their careers through lifelong learning.

unit 2 review questions

1. What four effects does technology have on the country and its ability to manufacture goods?

2. Name four of the most important careers available to those with a good background in CNC.

3. List the seven key elements of a modular training system.

4. Define an apprentice.

5. Name three desirable qualities a person should have for apprenticeship training.

6. Explain the difference between a machinist and a machine operator.

7. Briefly explain the difference between a jobbing and a production shop.

8. Define a tool and diemaker.

9. How can a person become a tool and diemaker?

10. How do the duties of a CNC programmer differ from those of a CNC machine operator?

11. Explain the difference between a technician and a technologist.

12. Briefly define the duties of an inspector and an instrument maker.

13. What qualifications must a person have to become a technical teacher?

14. List four areas in industry that require an engineer's qualifications.

Getting the Job

OBJECTIVES

After completing this unit, you will be able to:

1 Prepare a comprehensive résumé

2 Arrange for a job interview

3 Prepare for and follow up on the job interview

After graduation or leaving school, your most important task probably is that of finding a full-time job. *Choosing the right job is very important.*

After consulting with the school guidance counselor, state employment service, and any other agency that may be helpful, start looking into the job that appeals to you. In most areas, aptitude and interest tests are available to help you decide on the career you wish to explore.

▶▶ Assess Your Abilities

To help determine where your interests lie, ask yourself the following questions:

> What type of work do I like?

> What type of work do I dislike?

> What jobs have I done with some success?

> What skills have I acquired at school?

> What have I done in my part-time work that has been outstanding?

> Do I enjoy taking apart and repairing appliances and items which are not working?

■ **Figure 3-1** Research as much as possible about a job that appeals to you.   *(©Joe Raedle/Newsmakers/Getty Images)*

▶▶ Explore Your Interests

After having narrowed the field, then:

> Research information on the chosen subject(s) (Fig. 3-1).

> Talk to people who do the type of work that you are considering. Ask them about the job and job opportunities.

> Talk further with your school guidance counselor.

> Consult state or federal employment services.

When you have gathered enough information to discuss reasonably the type of work you are interested in, check the classified ads in the newspapers.

▶▶ Writing a Résumé

When you have decided on the job you wish to apply for, prepare a résumé listing the following facts in a logical order:

> Your full name
> Address, including zip code, and telephone number, including area code
> Social security number
> Father's and mother's names and address(es)
> Education—schools attended and grade or level completed
> Other special training that may be helpful, for example, first-aid courses
> Special interests and hobbies
> Sports in which you are active or interested
> Any organizations to which you belong or in which you are active
> Previous employment. List the names of firms and places where you have worked and in reverse chronological order. Include the following information:

Dates of employment

Type of work and equipment operated (if any)

Supervisor's name

Salary

Reason for leaving (be honest because this information is usually available if the prospective employer chooses to phone your former employer)

> A list of at least three persons who can be contacted for character references. Include addresses and telephone numbers. Be sure to ask permission to use their names *before* listing them on an application.

Many state employment services and government agencies have pamphlets that can be used as a guide to prepare résumés. Also available is information on how to get and keep a job, which can be especially valuable to a person looking for that first job.

▶▶ Facts About Interviews

ARRANGING AN INTERVIEW

After completing your résumé, submit it with a cover letter to the personnel manager of the company you are interested in. Be sure to include a request for an interview. In many cases you may phone the company and make an appointment for an interview. In this case, leave the résumé with the person who interviews you.

QUESTIONS COMMONLY ASKED DURING AN INTERVIEW

Prior to the interview, consider the following questions, which are often asked by employers, and the probable reason for asking them. You should be prepared with satisfactory answers.

Question: **Why would you like to work here?**
Reason: To see if you have gathered any information about the company before the interview.

Question: **What were your best subjects at school?**
Reason: This will reveal some of your interests and abilities.

Question: **What sports or activities did you participate in when attending school?**
Reason: To find out your interests and abilities outside of school. Also to see if you can work as part of a team.

Question: **What type of job do you hope to have five years from now?**
Reason: To assess your ambition and initiative.

Question: **At what salary would you expect to start?**
Reason: To see if you are familiar with the going rates and to see how you assess your abilities.

Question: **What do you have to offer for the job?**
Reason: To give you a chance to outline your abilities.

Question: **What type of books or plays are you interested in?**
Reason: To assess your interests and often your environment.

Question: **How did you get along with your previous employer?**
Reason: Your answer may reveal whether you are a complainer and a person who is hard to get along with.

Question: **Why are you applying for this job?**
Reason: To see if you have checked into this particular job and aren't just looking for any job.

THE INTERVIEW

When preparing for the interview, you should consider the following:

> Be sure of the address, the office, and the time.
> Know the name and position of the person who will be interviewing you. This information can be obtained by a phone call to the company prior to the interview.

■ **Figure 3-2** Be at ease while you are interviewed.
(© Bob Daemmrich)

■ **Figure 3-3** Always thank the interviewer—it may
pay off!   (© Bill Aron/PhotoEdit)

> Be neatly dressed and groomed. Remember, a neat applicant usually commands more attention (Fig. 3-2).

> Be punctual.

> Display confidence when you introduce yourself to the interviewer.

> During the interview, be honest. Emphasize your good qualities and abilities, but don't bluff.

> Know enough about the company to enter into a discussion with the interviewer.

AFTER THE INTERVIEW

> Thank the interviewer and ask when you may expect to hear from him or her (Fig. 3-3).

> If you are offered a job, accept it (or reject it) as soon as possible. Never let the prospective employer await your decision indefinitely. If you aren't interested, explain why.

> The next day, send the interviewer a short letter expressing your appreciation for the valuable time he or she took for the interview.

> If you don't hear from the company in a reasonable time (seven to 10 days) call and ask for the person who interviewed you. Ask if he or she has made a decision yet and, if not, when you might expect one.

> If you don't get the first job, apply to other companies. Don't stop looking.

> Try to learn from each interview, which will eventually help you to get the right job.

> After several unsuccessful interviews, you might seek professional counseling from the Department of Labor, a vocational school, or a community college.

▶▶ Points to Remember

> The job will not find you. You must find the job.

> Know the type of work that you want and don't offer to take just any job.

> Look for the type of work that you feel will be interesting. You will be more successful if you like your work.

unit 3 review questions

1. List four things that you should consider when attempting to assess your abilities.

2. Name four ways to obtain further information about a trade or job.

3. Assume that you are applying for a job; prepare a personal résumé that you would submit to an employer.

4. Name three methods of arranging an interview.

5. List four important points that you must consider in preparing for an interview.

6. Name four important actions that you can take to follow up an interview.

HIGH VOLTAGE

DANGER TO HANDS

INFLAMMABLE

POISON

DANGER OF FALLS

SAFETY

There has never been a more meaningful saying than "Safety Is Everyone's Business." To become a skilled craftsperson, it is very important for you to learn to work safely, taking into consideration not only your own safety but the safety of your fellow workers. In general, everyone has a tendency to be careless about safety at times. We take chances every day by not wearing seat belts, walking under ladders, cluttering the work area, and doing many other careless and unsafe things. People tend to feel that accidents always happen to others. However, be sure to remember that a moment of carelessness can result in an accident that can affect you for the rest of your life. A loss of eyesight because of not wearing safety glasses or the loss of a limb because of loose clothing caught in a machine can seriously affect or end your career in the machine tool trade. *Think safe, work safe, and be safe.*

UNIT 4

Safety in the Machine Shop

OBJECTIVES

After completing this unit, you will be able to:

1 Recognize safe and unsafe work practices in a shop

2 Identify and correct hazards in the shop area

3 Perform your job in a manner that is safe for you and other workers

All hand and machine tools can be dangerous if used improperly or carelessly. Working safely is one of the first things a student or an apprentice should learn because the safe way is usually the correct and most efficient way. A person learning to operate machine tools *must first* learn the safety regulations and precautions for each tool or machine. Far too many accidents are caused by careless work habits. It is easier and much more sensible to develop safe work habits than to suffer the consequences of an accident. *Safety is everyone's business and responsibility.*

▸▸ Safety on the Job

The safety programs initiated by accident prevention associations, safety councils, government agencies, and industrial firms are constantly attempting to reduce the number of industrial accidents. Nevertheless, each year accidents that could have been avoided result not only in millions of dollars' worth of lost time and production but also in a great deal of pain, many lasting physical handicaps, or even the death of workers. Modern machine tools are equipped with safety features, but it is still the operator's responsibility to use these machines wisely and safely.

Accidents don't just happen; they are caused. The cause of an accident can usually be traced to carelessness on someone's part. Accidents can be avoided, and a person learning the machine tool trade must first develop safe work habits.

A safe worker should:

> Be neat, tidy, and safely dressed for the job he or she is performing

> Develop a responsibility for personal safety and the safety of fellow workers

> Think safely and work safely at all times

▸▸ Safety in the Shop

Safety in a machine shop may be divided into two broad categories:

> Those practices that will prevent injury to workers

> Those practices that will prevent damage to machines and equipment. Too often damaged equipment results in personal injuries.

When considering these categories, we must consider personal grooming, proper housekeeping (including machine maintenance), safe work practices, and fire prevention.

PERSONAL GROOMING

The following rules should be observed when working in a machine shop.

1. Always wear approved safety glasses in any area of the machine shop. Most plants now insist that all employees and visitors wear safety glasses or some other eye protection device when entering a shop area. Several types of eye protection devices are available for use in the machine shop:

   a. The most common are *plain safety glasses* with side shields (Fig. 4-1a). These glasses offer sufficient eye protection when an operator is operating any machine or performing any bench or assembly operation. The lenses are made of shatterproof glass, and the side shields protect the sides of the eyes from flying particles.

   b. *Plastic safety goggles* (Fig. 4-1b) are generally worn by anyone wearing prescription eyeglasses. These goggles are soft, flexible plastic and fit closely around the upper cheeks and forehead. Unfortunately, they have a tendency to fog up in warm temperatures.

   c. *Face shields* (Fig. 4-1c) may also be used by those wearing prescription glasses. The plastic shield gives full face protection and permits air circulation between the face and the shield, thus preventing fogging up in most situations. These shields, as well as approved protective clothing and gloves, *must* be worn when an operator is heating and quenching materials during heat-treating operations or when there is any danger of hot flying particles. In industry, some companies provide their employees with prescription safety glasses, which eliminate the need for protective goggles or shields.

 SAFETY PRECAUTIONS Never think that because you are wearing glasses your eyes are safe. If the lenses are not made of approved safety shatterproof glass, serious eye injury can still occur.

(a)

(b)

(c)

■ **Figure 4-1** Types of safety glasses: (a) plain; (b) plastic goggles; (c) face shields. *((a) © George Disario/CORBIS; (b) © Dana White/PhotoEdit; (c) Phillip A. Nickerson Jr./U.S. Navy/Getty Images)*

Safety in the Machine Shop   **33**

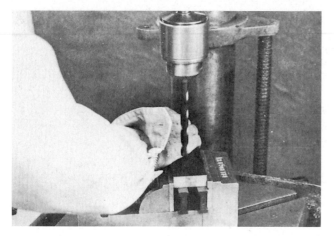

Figure 4-2 Loose clothing can easily be caught in moving parts of machinery. *(Kelmar Associates)*

2. Never wear loose clothing when operating any machine (Fig. 4-2).

   a. Always roll up your sleeves or wear short sleeves.
   b. Clothing should be made of hard, smooth material that will not catch easily in a machine. Loose-fitting sweaters should not be worn for this reason.
   c. Remove or tuck in a necktie before starting a machine. If you want to wear a tie, make it a bow tie.
   d. When wearing a shop apron, *always tie it at the back* and never in front of you so that the apron strings will not get caught in rotating parts (Fig. 4-3).

3. Remove wrist watches, rings, and bracelets; these can get caught in the machine, causing painful and often serious injury (Fig. 4-4).

4. Never wear gloves when operating a machine.

5. Long hair must be protected by a hair net or an approved protective shop cap (Fig. 4-5 on p. 35). One of the most common accidents on a drill press is caused by long, unprotected hair getting caught in a revolving drill.

6. Canvas shoes or open-toed sandals must *never* be worn in a machine shop because they offer no protection to the feet against sharp chips or falling objects. In industry, most companies make it mandatory for employees to wear safety shoes.

HOUSEKEEPING

The operator should remember that *good housekeeping will never interfere with safety or efficiency*; therefore, the following points should be observed.

1. *Always stop the machine before you attempt to clean it.*

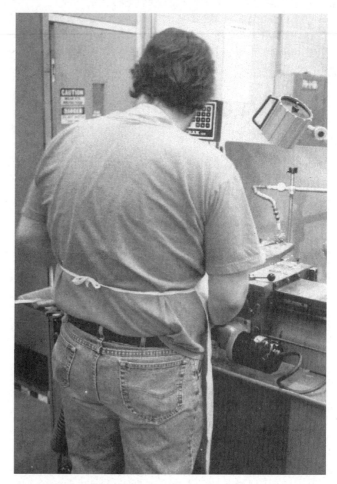

Figure 4-3 Tie aprons behind your back to keep the ties from being caught in machinery.

Figure 4-4 Wearing rings and watches can cause serious injury. *(Kelmar Associates)*

2. Always keep the machine and hand tools clean. Oily surfaces can be dangerous. Metal chips left on the table surface may interfere with the safe clamping of a workpiece.

3. Always use a brush and not a cloth to remove any chips. Chips stick to cloth and can cause cuts when the cloth is used later.

Figure 4-5 Long hair must be protected by a hair net or an approved shop cap. *(Kelmar Associates)*

Figure 4-6 Grease and oil on the floor can cause dangerous falls.

4. Oily surfaces should be cleaned with a cloth.

5. Do not place tools and materials on the machine table—use a bench near the machine.

6. Keep the floor free from oil and grease (Fig. 4-6).

7. Sweep up the metal chips on the floor frequently. They become embedded in the soles of shoes and can cause dangerous slippage if a person walks on a terrazzo or concrete floor. Use a scraper, mounted on the floor near the door, to remove these chips before leaving the shop (Fig. 4-7).

8. Never place tools or materials on the floor close to a machine where they will interfere with the operator's ability to move safely around the machine (Fig. 4-8).

9. Return bar stock to the storage rack after cutting off the required length (Fig. 4-9).

10. Never use compressed air to remove chips from a machine. Not only is it a dangerous practice because of flying metal chips, but small chips and dirt can become wedged between machine parts and cause undue wear.

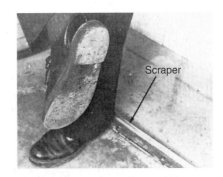

Figure 4-7 Remove chips from the soles of your shoes before leaving the shop. *(Kelmar Associates)*

Figure 4-8 Poor housekeeping can lead to accidents. *(© Prographics)*

Figure 4-9 Store stock safely in a material stock rack. *(Kelmar Associates)*

SAFE WORK PRACTICES

1. Do not operate any machine before understanding its mechanism and knowing how to stop it quickly. Knowing how to stop a machine quickly can prevent a serious injury.

2. Before operating any machine, be sure that the safety devices are in place and in working order. Remember, safety devices are for the operator's protection and should not be removed.

Figure 4-10 Power switches must be locked off before you repair or adjust a machine. *(Allen Bradley Co.)*

Figure 4-11 The machine must be stopped before you measure a workpiece. *(Kelmar Associates)*

Figure 4-12 Follow recommended lifting procedures to avoid back injury. *(Kelmar Associates)*

3. Always disconnect the power and lock it off at the switch box when making repairs to any machine (Fig. 4-10). Place a sign on the machine noting that it is out of order.

4. Always be sure that the cutting tool and the workpiece are properly mounted before starting the machine.

5. *Keep hands away from moving parts.* It is dangerous practice to "feel" the surface of revolving work or to stop a machine by hand.

6. *Always stop a machine before measuring, cleaning, or making any adjustments.* It is dangerous to do any type of work around moving parts of a machine (Fig. 4-11).

7. *Never use a rag near the moving parts of a machine.* The rag may be drawn into the machine, along with the hand that is holding it.

8. *Never have more than one person operate a machine at the same time.* Not knowing what the other person would or would not do has caused many accidents.

9. *Get first aid immediately for any injury, no matter how small.* Report the injury and be sure that the smallest cut is treated to prevent the chance of a serious infection.

10. Before you handle any workpiece, remove all burrs and sharp edges with a file.

11. Do not attempt to lift heavy or odd-shaped objects that are difficult to handle on your own.

12. For heavy objects, follow safe lifting practices:

    a. Assume a squatting position with your knees bent and back straight.
    b. Grasp the workpiece firmly.
    c. Lift the object by straightening your legs and keeping your back straight (Fig. 4-12). This

procedure uses the leg muscles and prevents injury to the back.

13. Be sure the work is clamped securely in the vise or to a machine table.

14. Whenever work is clamped, be sure the bolts are placed closer to the workpiece than to the clamping blocks.

15. Never start a machine until you are sure that the cutting tool and machine parts will clear the workpiece (Fig. 4-13).

16. Use the proper wrench for the job, and replace nuts with worn corners.

17. It is safer to pull on a wrench than to push on it.

FIRE PREVENTION

1. *Always* dispose of oily rags in proper metal containers.

2. Be sure of the proper procedure before lighting a gas furnace.

3. Know the location and the operation of every fire extinguisher in the shop.

4. Know the location of the nearest fire exit from the building.

5. Know the location of the nearest fire-alarm box and its operating procedure.

6. When using a welding or cutting torch, be sure to direct the sparks away from any combustible material.

■ **Figure 4-13** Make sure that the cutting tool and machine parts will clear the workpiece. *(Kelmar Associates)*

unit 4 review questions

1. What must be learned before operating a machine tool for the first time?

2. List three qualities of a safe worker.

Personal Grooming

3. Name three types of eye protection that may be found in a shop.

4. State four precautions that must be observed with regard to clothing worn in a shop.

5. Why should gloves not be worn when operating a machine?

6. How must long hair be protected?

Housekeeping

7. Why must a cloth not be used to remove chips?

8. Why should shoe soles be scraped before leaving the shop?

9. State two reasons why compressed air should not be used for cleaning machines.

Safe Work Practices

10. State three precautions to observe before operating any machine.

11. Describe the procedure to follow for lifting a heavy object.

12. What should you do immediately after receiving any injury?

Fire Prevention

13. What three fire prevention factors should everyone become familiar with before starting to work in a machine shop?

Safety in the Machine Shop   **37**

Section 4

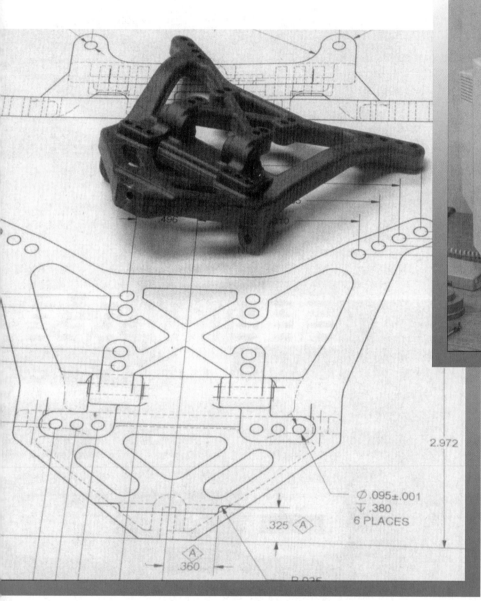

(Hass Automation Inc.)

JOB PLANNING

Machine shop work consists of machining a variety of parts (round, flat, contour) and either assembling them into a unit or using them separately to perform some operation. It is important that the sequence of operations be carefully planned in order to produce a part quickly and accurately. Improper planning or following a wrong sequence of operations often results in spoiled work.

UNIT 5

Engineering Drawings

OBJECTIVES

After completing this unit, you will be able to:

1 Understand the meaning of the various lines used on engineering drawings

2 Recognize the various symbols used to convey information

3 Read and understand engineering drawings or prints

Engineering drawing is the common language by which draftspersons, tool designers, and engineers indicate to the machinist and toolmaker the physical requirements of a part. Drawings are made up of a variety of lines, which represent surfaces, edges, and contours of a workpiece. By adding symbols, dimension lines and sizes, and word notes, the draftsperson can indicate the exact specifications of each individual part. Geometric dimensioning and tolerancing (GD&T) has become a universal language on engineering (technical) drawings for specifying a parts' exact geometry or shape and how the part should be inspected and gaged. The *American ANSI* Y14.5, the American Standard, ASME Y14.5M-1994 (formerly ANSI Y14.5M-1982 R 1988), and the ISO R1 101 standards are very similar with only a few variations.

A complete product is usually shown on an *assembly drawing* by the drafter. Each part or component of the product is then shown on a *detailed drawing,* which is reproduced as copies called *prints.* The prints are used by the machinist or toolmaker to produce the individual parts that eventually will make up the complete product. Some of the more common lines and symbols will be reviewed briefly.

▶▶ Types of Drawings and Lines

To describe the shape of noncylindrical parts accurately on a drawing or print, the draftsperson uses the *orthographic view* or *projection method.* The orthographic view shows the part from three sides: the front, top, and right-hand side (Fig. 5-1). These three views enable the draftsperson to describe a part or an object so completely that the machinist knows exactly what is required.

Cylindrical parts are generally shown on prints in two views: the front and right side (Fig. 5-2). However, if a part contains many details, it may be necessary to use the top, bottom, or left-side views to describe the part accurately to the machinist.

In many cases, complicated interior forms are difficult to describe in the usual manner by a draftsperson. Whenever this occurs, a *sectional view,* which is obtained by making an imaginary cut through an object, is presented. This imaginary cut can be made in a straight line in any direction to best expose the interior contour or form of a part (Fig. 5-3).

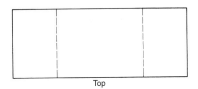

Top

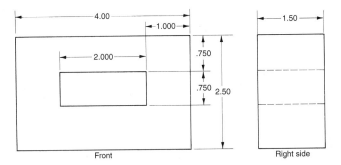

Front

Right side

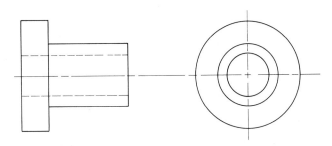

■ Figure 5-1 Three views of orthographic projection make it easier to describe the details of a part.

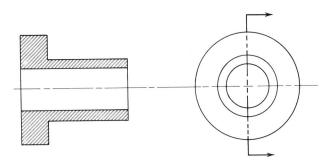

■ Figure 5-2 Cylindrical parts are generally shown in two views.

■ Figure 5-3 Section views are used to show complicated interior forms.

A wide variety of *standard lines* are used in engineering drawings for the designer to indicate to the machinist exactly what is required. Thick, thin, broken, wavy, and section lines are used on shop or engineering drawings. See Table 5.1 on p. 42 for examples, including the description and purpose of some of the more common lines used on shop drawings.

▶▶ Drafting Terms and Symbols

Common drafting terms and symbols are used on shop and engineering drawings for the designer to describe each part accurately. If it were not for the universal use of terms, symbols, and abbreviations, the designer would have to make extensive notes to describe exactly what is required. These notes not only would be cumbersome but could be misunderstood and therefore result in costly errors. Some of the common drafting terms and symbols are explained in the following paragraphs and examples.

Limits (Fig. 5-4) are the largest and the smallest permissible dimensions of a part (the maximum and minimum dimensions). Both sizes would be given on a shop drawing.

EXAMPLE

.751 largest dimension

.749 smallest dimension

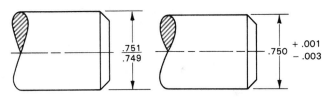

■ Figure 5-4 Limits show the largest and smallest size of a part.

■ Figure 5-5 Tolerance is the permissible variation of a specified size.

Tolerance (Fig. 5-5) is the permissible variation of the size of a part. The basic dimension plus or minus the variation allowed is given on a drawing.

EXAMPLE

$$.750 \begin{array}{c} +.001 \\ -.003 \end{array}$$

The tolerance in this case would be .004 (the difference between +.001 oversize or −.003 undersize).

table 5.1 Common lines used on shop drawings

Example	Name	Description	Use
a	Object lines	Thick black lines appoximately .030 in. wide (the width may vary to suit drawing size).	Indicate the visible form or edge of an object.
b	Hidden lines	Medium-weight black lines of .125 in. long dashes and .060 in. spaces.	Indicate the hidden contours of an object.
c	Center lines	Thin lines with alternating long lines and short dashes. —Long lines from .500 to 3 in. long —Short dashes .060 to .125 in. long, spaces .060 in. long.	Indicate the centers of holes, cylindrical objects, and other sections.
d 1½	Dimension Lines	Thin black lines with an arrowhead at each end and a space in the center for a dimension.	Indicate the dimensions of an object.
e	Cutting-plane lines	Thick black lines make up a series of one long line and two short dashes. Arrowheads show the line of sight from where the section is taken.	Show the imagined section cut.
f	Cross section lines	Fine, evenly spaced parallel lines at 45°. Line spacing is in proportion to the part size.	Show the surfaces exposed when a section is cut.

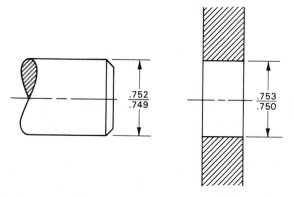

Figure 5-6 Allowance is the intentional difference in the sizes of mating parts.

Allowance (Fig. 5-6) is the intentional difference in the sizes of mating parts, such as the diameter of a shaft and the size of the hole. On a shop drawing, both the shaft and the hole would be indicated with maximum and minimum sizes to produce the best fit.

Fit is the range of tightness between two mating parts. There are two general classes of fits:

1. *Clearance fits,* whereby a part may revolve or move in relation to a mating part
2. *Interference fits,* whereby two parts are forced together to act as a single piece

Scale size is used on most shop or engineering drawings because it would be impossible to draw parts to exact size; some drawings would be too large, and others would be too small. The scale size of a drawing is generally found in the title block and indicates the scale to which the drawing has been made, which is a representative measurement.

Scale	Definition
1:1	Drawing is made to the actual size of the part, or full scale.
1:2	Drawing is made to one-half the actual size of the part.
2:1	Drawing is made to twice the actual size of the part.

⏩ Units of Measurement

Although the metric system of measurement is the international standard of measurement, the inch system is still widely used in the United States and Canada. Therefore, it is important that the machinist understand both dimensioning systems to be able to work in either system.

The dimensions in this book are primarily decimal inch; however, at times dual dimensions are provided, with the inch dimension first and its metric equivalent in parentheses. A note on, or near, the title block of the drawing should identify whether inch, metric, or dual dimension is being used—for example, $\dfrac{\text{MILLIMETER}}{\text{INCH}}$ and/or MILLIMETER (INCH)

⏩ Manufacturing Methods

The drawing should only define a part and not specify how the part is to be made or the operations to produce the part. Generally only the hole diameter is shown without indicating whether it should be drilled, reamed, bored, or produced by any other method. Where a dimension is critical, its tolerance or limits should be provided on the drawing so the craftsperson will use the proper method to produce the part to the required accuracy.

⏩ Basic Dimensioning

Dimensioning is used on working drawings to explain to the machinist the shapes and sizes required to manufacture a part. The type of material for the part, the number of parts required, and special notes are generally found in the title block of the drawing.

⏩ Dimensioning Tolerances

Each dimension on a drawing should have a tolerance to define the accuracy of a specific operation or the part. Common machine trade practice show the tolerance on dimension as a + or − unit of the last digit—for example:

> .12 (two decimal places) indicate a tolerance of ± .010 in.

> .345 (three decimal places) indicate a tolerance of ± .001 in.

> .6789 (four decimal places) indicate a tolerance of ± .0001 in.

Whenever the tolerance varies from these examples, it may be shown as specific limits (high or low) for a dimension, or as plus or minus tolerancing.

⏩ Inch Dimensions

> Fractional sizes—for example, 1/2 in.—are stated to two decimal places (such as .50 in.), indicates it is not a critical size.

> Whole dimensions are shown with a minimum of two zeros to the right of the decimal point—e.g., 5.00 in., not 5 in.

> No zero is used to the left of the decimal for any value of less than 1 inch—for example, .36 in., not 0.36 in., and .625 in., not 0.625 in.

> Sizes that are critical dimensions are shown in three or four decimal places and, where necessary, the tolerance or limit dimensions are included.

⏩ Metric Dimensions

> A zero must be used to the left of the decimal for all sizes less than 1 millimeter—for example, 0.35 mm, not .35 mm.

> Where the dimension is a whole number, no decimal point or zero follows the number—for example, 4 mm, not 4.0 mm.

> Where the dimension is larger than the whole number by a decimal fraction, the last digit to the right of the decimal point is not followed by a zero—for example, 6.5 mm, not 6.50 mm.

SYMBOLS

Some of the symbols and abbreviations used on shop drawings indicate the surface finish, type of material, roughness symbols, and common machine shop terms and operations.

A few of the common symbols used in this book are as follows:

∠ angularity	∨ countersink
60 basic dimension	⟂ depth/deep
∠ between	∅ diameter
→ conical taper	⊥ perpendicularity
⌴ counterbore/spotface	R radius

Countersinks, counterbores, and spotfaces can be shown on drawings by abbreviations or dimension symbols, with the symbols being preferred. Samples of some common symbols in use are shown in Fig. 5-7 on p. 44.

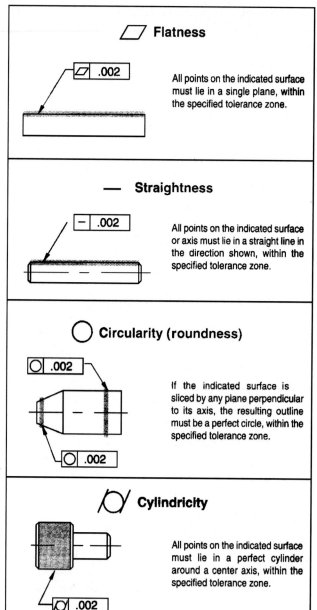

Flatness

.002

All points on the indicated surface must lie in a single plane, within the specified tolerance zone.

Straightness

.002

All points on the indicated surface or axis must lie in a straight line in the direction shown, within the specified tolerance zone.

Circularity (roundness)

.002

.002

If the indicated surface is sliced by any plane perpendicular to its axis, the resulting outline must be a perfect circle, within the specified tolerance zone.

Cylindricity

.002

All points on the indicated surface must lie in a perfect cylinder around a center axis, within the specified tolerance zone.

Linear profile

.005

.005 | A

— A —

All points on any full slice of the indicated surface must be on its theoretical two-dimensional profile, as defined by basic dimensions, within the specified tolerance zone. The profile may or may not be oriented with respect to datums.

Surface profile

.005

.005 | A

— A —

All points on the indicated surface must lie on its theoretical three-dimensional profile, as defined by basic dimensions, within the specified tolerance zone. The profile may or may not be oriented with respect to datums.

Figure 5-7 Common geometric symbols and definitions. *(Carr Lane Mfg. Co.)*

▶▶ Dimensioning Systems

Dimensions are used on prints to give the distance between two points, lines, planes, or some combination of points, lines, and planes.

> The numerical value gives the actual measurement (distance).

> The dimension line indicates the direction in which the value applies.

> The arrowheads indicate the points between which the value applies.

The *decimal system,* used for machine shop and computer numerical control work, uses only decimal fractions for all dimensional values. In computer numerical control work, two types of dimensioning are used:

1. *Incremental system* where all dimensions are given from a previously known point.

2. *Absolute system* where all dimensions or positions are given from a fixed zero or origin point.

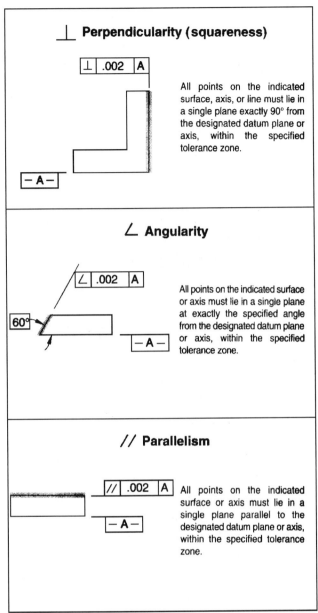

Perpendicularity (squareness)

⊥ | .002 | A

All points on the indicated surface, axis, or line must lie in a single plane exactly 90° from the designated datum plane or axis, within the specified tolerance zone.

– A –

∠ Angularity

∠ | .002 | A

60°

– A –

All points on the indicated surface or axis must lie in a single plane at exactly the specified angle from the designated datum plane or axis, within the specified tolerance zone.

// Parallelism

// | .002 | A

– A –

All points on the indicated surface or axis must lie in a single plane parallel to the designated datum plane or axis, within the specified tolerance zone.

↗ Circular runout

↗ | .002 | A

– A –

↗ | .003 | A

Each circular element of the indicated surface is allowed to deviate only the specified amount from its theoretical form and orientation during 360° rotation about the designated datum axis.

↗↗ Total runout

↗↗ | .002 | A

– A –

↗↗ | .003 | A

The entire indicated surface is allowed to deviate only the specified amount from its theoretical form and orientation during 360° rotation about the designated datum axis.

◎ Concentricity

– A –

◎ | .002 | A

If the indicated surface is sliced by any plane perpendicular to the designated datum axis, every slice's center of area must lie on the datum axis, within the specified cylindrical tolerance zone (controls rotational balance).

⊕ Position (replaces ≡ symmetry)

.005 | – A –

.375

4X ⌀ .125 ± .002

⊕ | .002 Ⓜ | A

The indicated feature's axis must be located within the specified tolerance zone from its true theoretical position, correctly oriented relative to the designated datum plane or axis.

■ **Figure 5-7** (Continued)

▸▸ Workplace Communication

Since the early 1950s, manufacturing has become a part of the global economy. It is not uncommon to have different components of a product made in various countries throughout the world and assembled in another country. In order for these components to be made to specifications, it is important that the graphic represen-

tation (symbols and characteristics) used on engineering drawings be universal and easily understood throughout the world.

The *International Standards Organization (ISO)* was established in 1946, and the goal of one of its technical committees (TC#10) was to develop a set of universally accepted standards for technical drawings. The *American Society of Mechanical Engineers (ASME)* establishes the standards for the United States through its ASME Y14.5 committee, made up of representatives from industry, education, and technical organizations. Members of the ASME Y14.5 committee also serve on the ISO TC10 subcommittee, so that standards acceptable to most countries are developed.

Over the years there has been a constant refinement of drawing standards by modifying the work of the two committees to come up with mutually acceptable standards. The ASME Y14.5—1994 publication on *Dimensioning and Tolerancing* lists the latest standards universally accepted throughout the world. Some of the most common geometric symbols, characteristics, and definitions are shown in the following tables:

Common machine shop abbreviations

CBORE	Counterbore
CSK	Countersink
DIA	Diameter
Ø	Diameter
HDN	Harden
L	Lead
LH	Left hand
mm	Millimeter
NC	National coarse
NF	National fine
P	Pitch
R	Radius
Rc	Rockwell hardness test
RH	Right hand
THD	Thread or threads
TIR	Total indicated runout
TPI	Threads per inch
UNC	Unified national coarse
UNF	Unified national form

SURFACE SYMBOLS

Surface finish is the deviation from the nominal surface caused by the machining operation. Surface finish includes roughness, waviness, lay, and flaws and is measured by a surface finish indicator in microinches (μin.).

The surface finish mark, used in many cases, indicates which surface of the part must be finished. The number in the √ indicates the quality of finish required on the surface (Fig. 5-8). In the example shown in Fig. 5-8, √ the *roughness height* or the measurement of the finely spaced irregularities caused by the cutting tool cannot exceed 40 μin.

If the surface of a part must be finished to exact specifications, each part of the specification is indicated on the symbol (Fig. 5-9) as follows:

40 Surface finish in microinches

.002 Waviness height in thousandths of an inch

.001 Roughness width in thousandths of an inch

⊥ Machining marks run perpendicular to the boundary of the surface indicated

The following symbols indicate the direction of the lay (marks produced by machining operations on work surfaces).

= Parallel to the boundary line of the surface indicated by the symbol

X Angular in both directions on the surface indicated by the symbol

M Multidirectional

C Approximately circular to the center of the surface indicated by the symbol

R Approximately radial to the center of the surface indicated by the symbol

Fig. 5-10 shows the drafting symbols used to indicate some of the most common materials used in a machine shop.

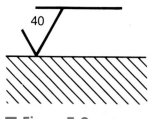

■ **Figure 5-8** Surface finish symbols indicate the type and finish of the surface.

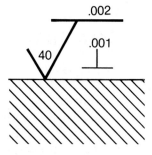

■ **Figure 5-9** Surface finish specifications.

Material Symbols

Represents copper, brass, bronze, etc.

Represents steel and wrought iron

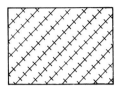

Represents aluminum, magnesium, and their alloys

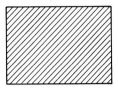

Represents cast iron and malleable iron

■ **Figure 5-10** Symbols used to indicate types of material.

unit 5 review questions

1. How can a drafter indicate the exact specifications required for a part?

2. What is the purpose of:

   a. An assembly drawing?
   b. A detailed drawing?

3. What is the purpose of an orthographic view?

4. Why are section views shown?

5. What lines are used to show:

   a. The form of a part?
   b. The centers of holes, objects, or sections?
   c. The exposed surfaces of where a section is cut?

6. Define:

   a. Limits
   b. Tolerance
   c. Allowance

7. How is half-scale indicated on an engineering drawing?

8. Define each part of the surface-finish symbol

   $$\sqrt[60]{\frac{.003}{.002}}$$

9. What do the following abbreviations mean?

   a. CBORE
   b. HDN
   c. mm
   d. THD
   e. TIR

UNIT 6

Machining Procedures for Various Workpieces

OBJECTIVES

After completing this unit, you should be able to:

1 Plan the sequence of operations and machine round work mounted between lathe centers

2 Plan the sequence of operations and machine round work mounted in a lathe chuck

3 Plan the sequence of operations to machine flat workpieces

Planning the procedures for machining any part so that it can be produced accurately and quickly is very important. Many parts have been spoiled because the incorrect sequence was followed in the machining process. Although it would be impossible to list the exact sequence of operations that would apply to every type and shape of workpiece, some general rules should be followed to machine a part accurately and in the shortest time possible.

▶▶ Machining Procedures for Round Work

Most of the work produced in a machine shop is round and is turned to size on a lathe. In industry, much of the round work is held in a chuck. A larger percentage of work in school shops is machined between centers because of the need to reset work more often. In either case, it is important to follow the correct machining sequence of operations to prevent spoiling work, which often happens when incorrect procedures are followed.

GENERAL RULES FOR ROUND WORK

1. Rough-turn all diameters to within .030 inch (in.) [0.79 millimeter (mm)] of the size required.
   > Machine the largest diameter first and progress to the smallest.
   > If the small diameters are rough-turned first, it is quite possible that the work would bend when the large diameters are machined.

2. Rough-turn all steps and shoulders to within .030 in. (0.79 mm) of the length required (Fig. 6-1).
   > Be sure to measure all lengths from one end of the workpiece.

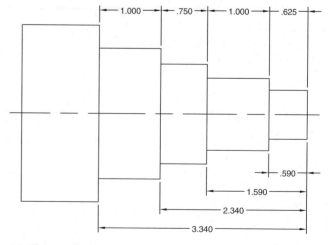

■ **Figure 6-1** A sample part, showing rough- and finish-turned lengths. *(Kelmar Associates)*

> If all measurements are not taken from the end of the workpiece, the length of each step would be .030 in. (0.79 mm) shorter than required. If four steps were required, the length of the fourth step would be .125 in. (3.17 mm) shorter than required (4 × .030 in., or 4 × 0.79 mm) and would leave too much material for the finishing operation.

3. If any special operations such as knurling or grooving are required, they should be done next.

4. Cool the workpiece before starting the finishing operations.

   > Metal expands from the friction caused by the machining process, and any measurements taken while work is hot will be incorrect.
   > When the workpiece is too cold, the diameters of round work will be smaller than required.

5. Finish-turn all diameters and lengths.

   > Finish the largest diameters first and work down to the smallest diameter.
   > Finish the shoulder of one step to the correct length and then cut the diameter to size.

WORKPIECES REQUIRING CENTER HOLES

Sometimes it is necessary to machine the entire length of a round workpiece. When this is required, usually on shorter workpieces, center holes are drilled in each end. The workpiece shown in Fig. 6-2 is a typical part that can be machined between the centers on a lathe.

Machining Sequence

1. Cut off a piece of steel .125 in. (3 mm) longer and .125 in. (3 mm) larger in diameter than required.

   > In this case, the diameter of the steel cut off would be 1.625 in. (41 mm) and its length would be 9.625 in. (409 mm).

2. Hold the workpiece in a three-jaw chuck, face one end square, and then drill the center hole.

3. Face the other end to length and then drill the center hole.

4. Mount the workpiece between the centers on a lathe.

5. Rough-turn the largest diameter to within .030 in. (0.79 mm) finish size or 1.530 in. (39 mm).

Note: The purpose of the rough cut is to remove excess metal as quickly as possible.

6. Finish-turn the diameter to be knurled.

Note: The purpose of the finish cut is to cut work to the required size and produce a good surface finish.

7. Knurl the 1.500-in. (38-mm) diameter.

8. Machine the 45° chamfer on the end.

9. Reverse the work in the lathe, being sure to protect the knurl from the lathe dog with a piece of soft metal.

10. Rough-turn the 1.250-in. (31-mm) diameter (Fig. 6-2) to 1.280 in. (32 mm) (Fig. 6-3 on p. 50).

    > Be sure to leave the length of this section .125 in. (3 mm) short [7.375 in. (327 mm) from the end] to allow for finishing the .125-in. radius.

11. Rough-turn the 1.125-in. (28-mm) diameter (Fig. 6-3) to 1.160 in. (29 mm).

    > Leave the length of this section .030 in. (0.79 mm) short [4.970 in. (177 mm) from the end] to allow for finishing the shoulder.

12. Rough-turn the .875-in. (22-mm) diameter to .905 in. (23 mm).

    > Leave the length of this section .030 in. (0.79 mm) short [1.970 in. (50 mm) from the end] to allow for finishing the shoulder.

13. Rough-turn the .500-in. (13-mm) diameter to .530 in. (13.48 mm).

    > Machine the length of this section to .720 in. (18 mm).

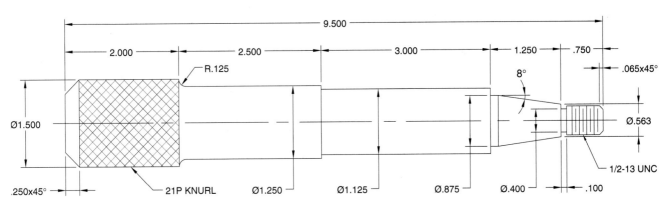

Figure 6-2 A sample of a round shaft that can be machined. *(Kelmar Associates)*

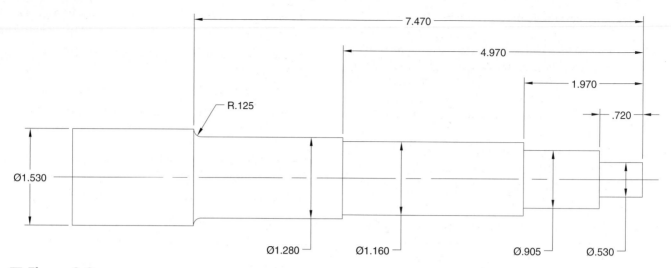

■ Figure 6-3 Rough-turned diameters and lengths of a shaft. *(Kelmar Associates)*

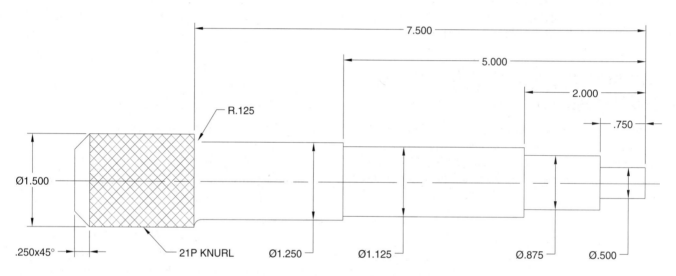

■ Figure 6-4 The shaft turned to diameter and length. *(Kelmar Associates)*

14. Cool the work to room temperature before starting the finishing operations.

15. Finish-turn the 1.250-in. (32-mm) diameter to 7.375 in. (327 mm) from the end.

16. Mount a .125-in. (3-mm) radius tool and finish the corner to the correct length (Fig. 6-4).

17. Finish-turn the 1.125-in. (28-mm) diameter to 5.000 in. (177 mm) from the end.

18. Finish-turn the .875-in. (22-mm) diameter to 2.000 in. (50 mm) from the end.

19. Set the compound rest to 8° and machine the taper to size.

20. Finish-turn the .500-in. (13-mm) diameter to .750 in. (19 mm) from the end.

21. With a cutoff tool, cut the groove at the end of the .500-in. (13-mm) diameter (Fig. 6-5).

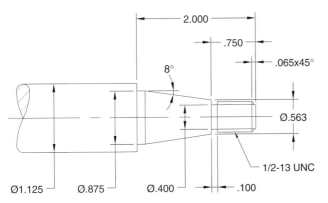

Figure 6-5 Special operations completed on a shaft.
(Kelmar Associates)

22. Chamfer the end of the section to be threaded.
23. Set the lathe for threading and cut the thread to size.

►► Workpieces Held in a Chuck

The procedure for machining the external surfaces of round workpieces held in a chuck (three-jaw, four-jaw, collet, etc.) is basically the same as for machining work held between lathe centers. However, if both external and internal surfaces must be machined on work held in a chuck, the sequence of some operations is changed.

Whenever work is held in a chuck for machining, it is very important that the workpiece be held short for rigidity and to prevent accidents. Never let work extend more than *three times its diameter* beyond the chuck jaws unless it is supported by some means, such as a steady rest or center.

MACHINING EXTERNAL AND INTERNAL DIAMETERS IN A CHUCK

To machine the part shown in Fig. 6-6, the following sequence of operations is suggested.

1. Cut off a piece of steel .125 in. (3 mm) larger in diameter and .500 in. (13 mm) longer than required.

   > In this case, the rough diameter would be 2.125 in. (54 mm).
   > The length would be 3.875 in. (98 mm), to allow the piece to be gripped in the chuck.

2. Mount and center the workpiece in a four-jaw chuck, gripping only .310 to .380 in. (8 to 9.5 mm) of the material in the chuck jaws.

   > A three-jaw chuck would not hold this size workpiece securely enough for the internal and external machining operations.

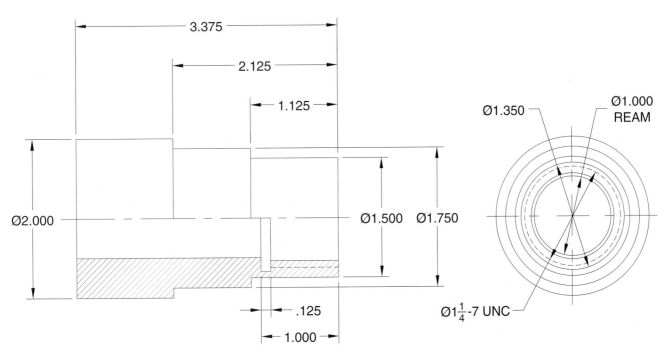

Figure 6-6 A round part requiring internal and external machining. *(Kelmar Associates)*

3. Face the end of the work square.

> Remove only the minimum amount of material required to square the end.

4. Rough-turn the three external diameters, starting with the largest and progressing to the smallest to within .030 in. (0.79 mm) of size and length.

5. Mount a drill chuck in the tailstock spindle and center-drill the work.

6. Drill a ½-in. (13-mm) diameter hole through the work.

7. Mount a ¹⁵⁄₁₆-in. (24 mm) drill in the tailstock and drill through the work.

8. Mount a boring bar in the toolpost and bore the 1-in. (25-mm) ream hole to .968 in. (24.58 mm) in diameter.

9. Bore the 1¼-in.-7 UNC threaded section to the tap-drill size, which is 1.107 in. (28 mm).

$$\text{Tap-drill size} = \text{TDS} = D - P$$

where $\quad D =$ diameter

$\quad\quad\quad\quad P =$ pitch

10. Cut the groove at the end of the section to be threaded to length and a little deeper than the major diameter of the thread.

11. Mount a threading tool in the boring bar and cut the 1¼-in.-7 UNC thread to size.

12. Mount a 1-in. (25-mm) reamer in the tailstock and finish the hole to size.

13. Finish-turn the external diameters to size and length, starting with the largest and working down to the smallest.

14. Reverse the workpiece in the chuck and protect the finish diameter with a piece of soft metal between it and the chuck jaws.

15. Face the end surface to the proper length.

▶▶ Machining Flat Workpieces

Since there are so many variations in size and shape of flat workpieces, it is difficult to give specific machining rules for each. Some general rules are listed, but they may have to be modified to suit particular workpieces.

1. Select and cut off the material a little larger than the size required.

2. Machine all surfaces to size in a milling machine in a proper sequence of surfaces.

3. Lay out the physical contours of the part, such as angles, steps, radii, and so on.

4. Lightly prick-punch the layout lines that indicate the surfaces to be cut.

5. Remove large sections of the workpiece on a contour bandsaw.

6. Machine all forms, such as steps, angles, radii, and grooves.

7. Lay out all hole locations and, with dividers, scribe the reference circle.

8. Drill all holes and tap those which require threads.

9. Ream holes.

10. Surface-grind any surfaces that require it.

OPERATIONS SEQUENCE FOR A SAMPLE FLAT PART

The part shown in Fig. 6-7 is used only as an example to illustrate a sequence of operations that should be followed when machining similar parts. These are not meant to be hard-and-fast rules, only guides.

The sequence of operations suggested for the sample part shown in Fig. 6-7 is different from those suggested for machining a block square and parallel as outlined in Unit 69 because:

1. The part is relatively thin and has a large surface area.

2. Since at least .125 in. (3 mm) of work should be above the vise jaws, it would be difficult to use a round bar between the work and movable jaw for machining the large flat surfaces.

3. A small inaccuracy (out-of-squareness) on the narrow edge would create a greater error when the large surface was machined.

procedure

1. Cut off a piece of steel .625 in. (16 mm) × 3.375 in. (86 mm) × 5.625 in. (143 mm) long.

2. In a milling machine, finish one of the larger surfaces (face) first.

   Note: Leave .010 in. (0.25 mm) on each surface to be ground.

3. Turn the workpiece over and machine the other face to .500 in. (13 mm) thick.

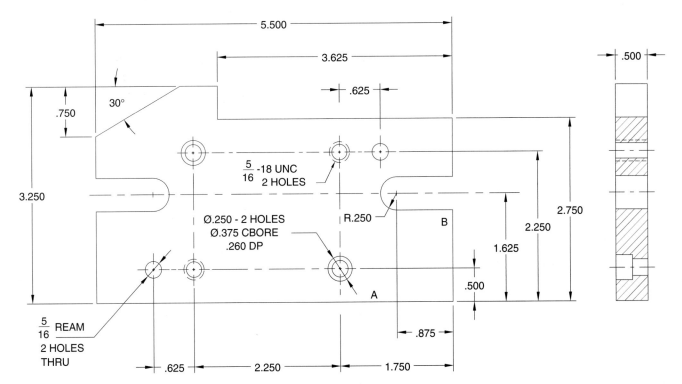

Figure 6-7 A typical flat part that must be laid out and machined. *(Kelmar Associates)*

4. Machine one edge square with the face.

5. Machine an adjacent edge square (at 90°) with the first edge.

6. Place the longest finished edge (A) down in the machine vise and cut the opposite edge to 3.250 in. (83 mm) wide.

7. Place the narrower finished edge (B) down in the machine vise and cut the opposite edge to 5.500 in. (140 mm) long.

8. With edge A as a reference surface, lay out all the horizontal dimensions with an adjustable square, a surface gage, or a height gage (Fig. 6-8).

9. With edge B as a reference surface, lay out all the vertical dimensions with an adjustable square, a surface gage, or a height gage (Fig. 6-9 on p. 54).

10. Use a bevel protractor to lay out the 30° angle on the upper right-hand edge.

11. With a divider set to .250 in. (6 mm), draw the arcs for the two center slots.

12. With a sharp prick punch, lightly mark all the surfaces to be cut and the centers of all hole locations.

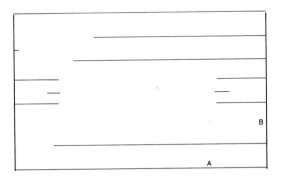

Figure 6-8 Lay out all horizontal lines using edge A as the reference surface. *(Kelmar Associates)*

13. Center-punch and drill 1/2-in. (13-mm) diameter holes for the two center slots.

14. On a vertical bandsaw, cut the 30° angle to within 0.30 in. (0.79 mm) of the layout line.

15. Place the workpiece in a vertical mill and machine the two .500-in. (13-mm) slots.

16. Machine the step on the top edge of the workpiece.

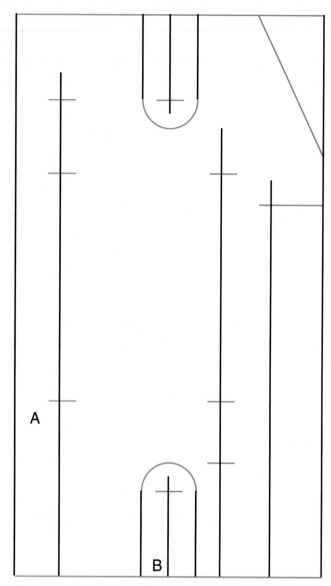

Figure 6-9 Lay out all vertical lines using edge B as the reference surface. *(Kelmar Associates)*

17. Set the work to 30° in the machine vise and finish the 30° angle.

18. Prick-punch the hole locations, scribe reference circles, and then center-punch all hole centers.

19. Center-drill all hole locations.

20. Drill and counterbore the holes for the ¼ in.–20 NC screws.

21. Tap drill the ⁵⁄₁₆ in.-18 thread holes (F drill or 6.5 mm).

22. Drill the ¼-in. (6-mm) ream holes to ¹⁵⁄₆₄ in. (5.5 mm).

23. Countersink all holes to be tapped slightly larger than their finished size.

24. Ream the ¼-in. (6-mm) holes to size.

25. Tap the ⁵⁄₁₆ in.–18 UNC holes.

MACHINE TOOLS

Although machine shops ordinarily have CNC machines, they typically also have manually operated machines. This insert includes the types of machines you will find in machine shops. It will also take a look into the future.

■ Heavy-duty, large-capacity radial drill. *(Clausing Industrial, Inc.)*

■ Contour bandsaw. *(DoALL)*

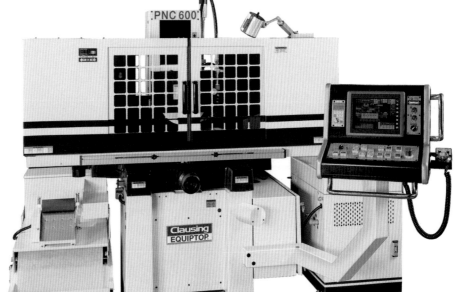

■ PNC precision profile grinder. *(Clausing, Industrial, Inc.)*

LATHES

As mentioned in the text, there are many types of lathes. These include manual and computer numerically controlled (CNC) lathes, as well as CNC turning centers.

■ CNC lathe with Multi-tasking Machining Capabilities—Super Quadrex 250M. *(MAZAK Corp.)*

■ Horizontal turning HLV manual toolroom lathe. *(Hardinge Inc.)*

■ Large CNC lathe. *(Clausing Industrial, Inc.)*

■ Horizontal turning CNC lathe Quest 10/65. *(Hardinge Inc.)*

■ Vertical turning lathe Integrex Series E1850 VIZ. *(MAZAK Corp.)*

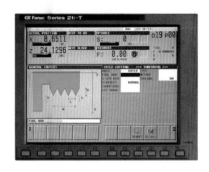

■ The Clausing/METOSA CNC lathe with G.E. Fanuc 21i T Control. *(Clausing Industrial, Inc.)*

MILLING MACHINES

There are many types of milling machines, used for a variety of tasks. Both manual and CNC milling machines are used. Some typical ones are shown here.

■ Manual variable speed mill with extra-heavy knee Model FV60. *(Clausing Industrial, Inc.)*

■ Standard mill Series 1. *(Hardinge, Inc.)*

■ Manual mill. *(Rem Sales, Inc.)*

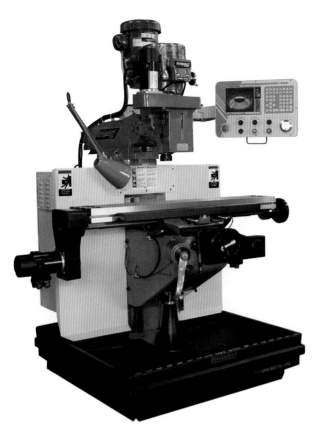

■ EZPLUS automated CNC mill. *(Hardinge, Inc.)*

■ Knee-type CNC milling machine. *(Clausing Industrial Inc.)*

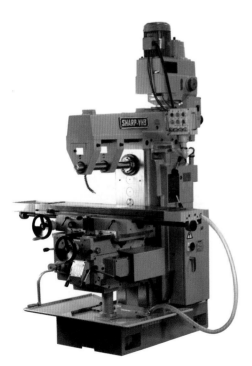

■ Vertical/horizontal combination mill Model VH-3.
(Sharp Industries)

MACHINING CENTERS

Machining centers can perform many operations, including drilling, milling, boring, and tapping. Shown here are horizontal and vertical machining centers.

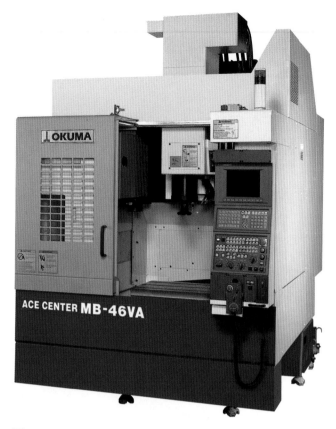

■ Vertical machining center ES-V Series. *(Okuma Corp.)*

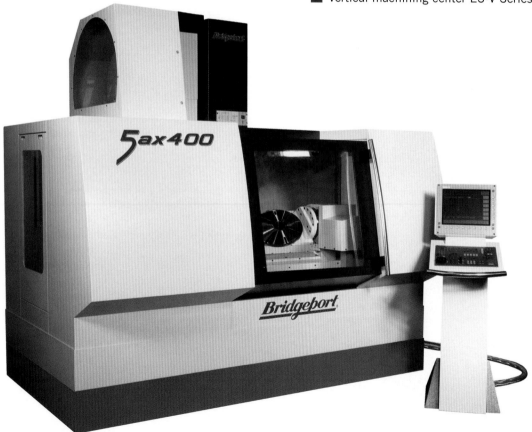

■ 5-axis vertical machining center 5AX 400. *(Hardinge, Inc.)*

■ Horizontal machining center High Velocity F3 660.   *(MAZAK Corp.)*

■ Horizontal machining center MA-H Series.   *(Okuma Corp.)*

WHAT'S NEW

In Section 17, you read about some of these advanced manufacturing technologies.

■ Adaptive machine control. *(GE Fanuc)*

■ Ethernet communications for productivity. *(GE Fanuc)*

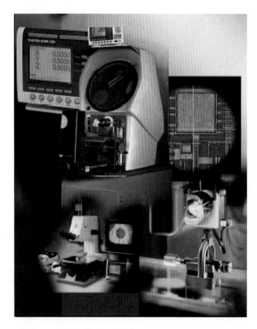

■ Optical measuring instruments.
(Brown & Sharpe)

■ QQC diamond process. *(QQC, Inc.)*

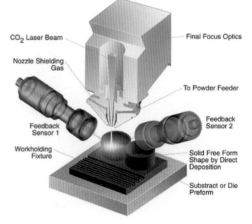

■ DMD rapid tooling process. *(The POM Group, Inc.)*

unit 6 review questions

1. Why is it not advisable to rough-turn small diameters first?

2. To what size should work be rough-turned?

3. Why must all measurements be taken from one end of a workpiece?

4. Why is it important that the workpiece be cooled before finish-turning?

5. Why can a workpiece be bent during the knurling operation?

6. Why is the work cut off longer than required when machining in a chuck?

7. How far should a 6-in. long workpiece, 1 in. in diameter, extend beyond the chuck jaws?

8. How deep should the groove be cut for the internal section to be threaded?

9. How is a finish diameter protected from the chuck jaws?

10. How much material should be left on a flat surface for grinding?

11. When using a bandsaw to remove excess material, how close to the layout line should the cut be made?

12. When machining flat surfaces, what surface should be machined first?

Section 5

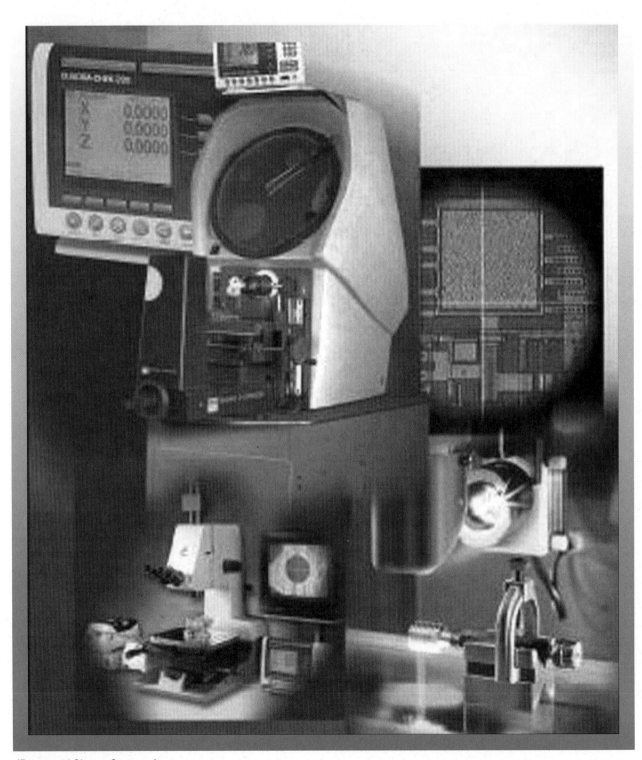

(Brown and Sharpe Company)

MEASUREMENT

The world has depended on some form of measurement system since the beginning of civilization. The Egyptians, for example, used a unit of length called the cubit, a unit equal to the length of the forearm from the middle finger to the elbow. James Watt, on the other hand, improved his steam engine by maintaining its tolerances to the thickness of a thin shilling, an English coin. The days of such crude measurements, however, are gone. Today we live in a demanding world where products must be built to precise tolerances. These same products may start out as components built by several subindustries and then be used by other industries in the manufacture of final consumer products. From start to finish, a product may be utilized by several industries located in separate places, often nations away, and then be sold at still other locations. *Interchangeable manufacture,* world trade, and the need for high precision have all contributed to the need for a *highly accurate international* measurement system. In 1960, the International System of Units (SI) was developed to satisfy this need.

Presently, there are two major systems of measurement used in the world. The inch system, often called the English system of measurement, is still widely used in the United States and Canada. Because over 90% of the world's population uses some form of metric measurement, the need for universal adoption of metric measurement is evident.

▸▸ Metric (Decimal) Systems

On December 8, 1975, the U.S. Senate passed Metric Bill S100 "to facilitate and encourage the substitution of metric measurement units for customary measurement units in education, trade, commerce and all other sections of the economy of the United States. . . ." On January 16, 1970, the Canadian government adopted SI for implementation throughout Canada by 1980.

Although both the United States and Canada are now committed to conversion to the metric system as rapidly as possible, it is likely to be some years before all machine tools and measuring devices are redesigned or converted. The change to the metric system in the machine shop will be gradual because of the long life expectancy of the costly machine tools and measuring equipment involved. It is probable, therefore, that people involved in the machine tool trade will have to be familiar with both the metric and the inch systems during the long changeover period.

▸▸ The Changeover Period

Although precision tool manufacturers produce most measuring tools in metric sizes, the use of these tools is not widespread because of the reluctance of industry to make the costly changeover. Consequently, many present-day machinists will probably have to be familiar with both the inch and the metric systems of measurement.

Since the changeover will be gradual, it is safe to assume that students graduating from schools for the next several years must be taught a dual system of measurement (inch and metric) until the majority of industries have changed to the metric system. With this in mind, we have used dual measurements throughout this book so that the student will be able to work effectively in both systems.

To address these problems, we adopted the following policy for this book to enable you to work effectively in both systems now, while permitting an easy transition to full metric use as new materials and tools become available:

1. Where general measurements or references to quantity are not related specifically to metric standards, tools, or products, inch units are given with the metric equivalent in parentheses.

2. Where the student may be exposed to equipment designed to both metric and inch standards, separate information is given on both types of equipment in *exact* dimensions.

3. Where only inch standards, tools, or products exist, inch measurements are given with a metric conversion to two decimal places provided in parentheses.

▸▸ Inch/Metric Dimensioning Standards

The inch and metric dimensions in this book follow the 1994 American National Standards Institute (ANSI) guidelines. They are as follows.

INCH DIMENSIONS

> A zero *is not used* before the decimal point for values less than 1 inch (.125 in.).

> A dimension is expressed to the same number of decimal places as its tolerance. Zeros are added to the right of the decimal point where necessary (2.350 in., tolerance ±.001).

METRIC DIMENSIONS

> A zero *is used* before the decimal point for values less than 1 millimeter (0.15 mm).

> Where the dimension is a whole number, neither the decimal point nor a zero is shown (12 mm).

> For dimensions larger than a whole number by a decimal fraction of a millimeter, the last digit to the right of the decimal point is not followed by a zero (25.5 mm).

FRACTIONAL/INCH SYSTEM

Due to the ever-increasing use of CNC and the use of digital data, fractional sizes do not lend themselves to most manufacturing operations. ANSI has for many years recommended that fractional dimensions not be used in dimensioning drawings and they be replaced with decimal dimensions.

Fractional dimensions may still be used to identify the sizes of holes produced by drills ordinarily stocked in fraction sizes and for standard taps and screw thread sizes. Therefore, wherever possible, in this book, fractional sizes have been replaced by decimal sizes.

▶▶ Symbols for Use with SI

Following is a list of some common SI quantities, names, and symbols that you are likely to encounter in your work in the machine shop:

Quantity	Name	Symbol
length*	meter	m
volume*	liter	ℓ and l
mass*	gram	g
time	minute	min
	second	s
force	newton	N
pressure, stress*	pascal	Pa
temperature	degree Celsius	°C
area*	square meter	m^2
velocity (speed)	meters per minute and meters per second	m/min and m/s
angles	degrees	°
	minutes	'
	seconds	"
electric potential	volt	V
electric current	ampere	A
frequency	hertz	Hz
electric capacitance	farad	F

Following is a list of prefixes often used with the quantities indicated by an asterisk (*) in the previous list:

Prefix	Meaning	Multiplier	Symbol
micro	one-millionth	.000 001	μ
milli	one-thousandth	.001	m
centi	one-hundredth	.01	c
deci	one-tenth	.1	d
deka	ten	10	da
hecto	one hundred	100	h
kilo	one thousand	1 000	k
mega	one million	1 000 000	M

UNIT 7

Basic Measurement

OBJECTIVES

After completing this unit, you should be able to:

1   Identify several types of steel rules

2   Measure round and flat work to ⅟₆₄-in. accuracy with a rule

3   Measure with spring calipers and a rule

Basic measurement can be termed as those measurements taken by use of a rule or any other nonprecision measuring tool, whether it be on the inch or metric standard.

▶▶ Inch System

The unit of length in the inch system is the inch, which may be divided into fractional or decimal fraction divisions. The fractional system is based on the binary system, or base 2. The binary fractions commonly used in this system are ½, ¼, ⅛, ⅟₁₆, ⅟₃₂, and ⅟₆₄. The decimal-fraction system has base 10, so any number may be written as a product of 10 and/or a fraction of 10.

Value	Fraction	Decimal
one-tenth	⅟₁₀	.1
one-hundredth	⅟₁₀₀	.01
one-thousandth	⅟₁₀₀₀	.001
one ten-thousandth	⅟₁₀,₀₀₀	.0001
one hundred-thousandth	⅟₁₀₀,₀₀₀	.00001
one millionth	⅟₁,₀₀₀,₀₀₀	.000001

▶▶ Metric System

Linear metric dimensions are expressed in multiples and submultiples of the meter. In the machine tool trade, the millimeter is used to express most metric dimensions. Fractions of the millimeter are expressed in decimals.

A brief comparison of common inch and metric equivalents shows:

$$
\begin{aligned}
1 \text{ yd} &= 36 \text{ in.} \\
1 \text{ m} &= 39.37 \text{ in.} \\
1000 \text{ m} &= 1 \text{ km} \\
1 \text{ km} &= 0.621 \text{ mi} \\
1 \text{ mi} &= 1.609 \text{ km}
\end{aligned}
$$

Table 7.1 shows inch-metric comparisons for common inch-metric system measurements.

Note: In machine shop metric measurements, most dimensions will be given in millimeters (mm). Large dimensions will be given in meters (m) and millimeters (mm). For metric-inch conversion tables and decimal equivalents, see the Appendix of Tables at the end of this book.

table 7-1   Inch/Metric conversion

Inch Size	Metric Size			
	Millimeter (mm)	Centimeter (cm)	Decimeter (dm)	Meter (m)
1 in.	25.4	2.54	0.254	0.0254
1 ft	304.8	30.48	3.048	0.3048
1 yd	914.4	91.44	9.144	0.9144

▶▶ Fractional Measurement

Fractional dimensions, often called *scale dimensions,* can be measured with such instruments as rulers or calipers. The steel rules used in machine shop work are graduated either in binary-fractional divisions of 1, ½, ¼, ⅛, ¹⁄₁₆, ¹⁄₃₂, and ¹⁄₆₄ of an inch (Fig. 7-1) or in decimal fractional divisions: decimeters, centimeters, millimeters, and half-millimeters (Fig. 7-2). Divisions of ¹⁄₆₄ of an inch or 0.50 mm are about as fine as can be seen on a rule without the use of a magnifying glass. Precision measuring instruments such as micrometers and verniers are required when metric-drawn prints show any dimensions of less than 0.50 mm or when inch-drawn prints show any dimensions in decimals.

▶▶ Steel Rules

METRIC STEEL RULES

Metric steel rules (Fig. 7-3), usually graduated in millimeters and half-millimeters, are used for making linear metric measurements that do not require great accuracy. A wide variety of metric rules are available in lengths from 15 cm to 1 m. The 15-cm rule shown in Fig. 7-3 is 2.4 mm, or about ³⁄₃₂ in., shorter than a standard 6-in. rule.

FRACTIONAL STEEL RULES

The common binary fractions found on inch steel rules are ¹⁄₆₄, ¹⁄₃₂, ¹⁄₁₆, and ⅛ of an inch. Several varieties of inch steel rules may be used in machine shop work, such as *spring-tempered, flexible, narrow,* and *hook.* Lengths range from 1 to 72 in. Again, these rules are used for measurements that do not require great accuracy. Spring-tempered quick-reading 6-in. rules (Fig. 7-4 on p. 62) with No. 4 graduations are the most frequently used inch rules in machine shop work. These rules have four separate scales, two on each side. The front is graduated in eighths and sixteenths and the back is graduated in thirty-seconds and sixty-fourths of an inch. Every fourth line is numbered to make reading in thirty-seconds and sixty-fourths easier and quicker.

Hook rules (Fig. 7-5 on p. 62) are used to make accurate measurements from a shoulder, step, or edge of a workpiece. They may also be used to measure flanges and circular pieces, and for setting inside calipers to a dimension.

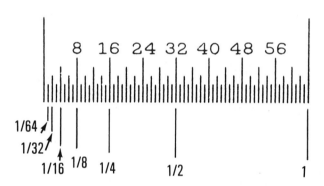

■ **Figure 7-1** Fractional divisions of an inch. *(Kelmar Associates)*

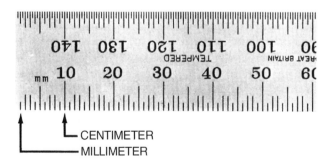

■ **Figure 7-2** Metric rules are usually graduated in millimeters and half-millimeters. *(Kelmar Associates)*

■ **Figure 7-3** A 15-cm metric rule. *(The L. S. Starrett Co.)*

Basic Measurement   **61**

■ Figure 7-4 A spring-tempered (quick-reading) 6-in. rule. *(The L. S. Starrett Co.)*

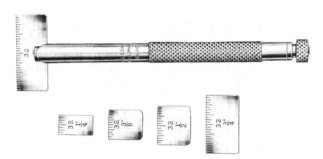

■ Figure 7-5 A hook rule is used to make accurate measurements from an edge or a shoulder. *(The L. S. Starrett Co.)*

■ Figure 7-6 Short-length rules are used for measuring small openings. *(The L. S. Starrett Co.)*

Short-length rules (Fig. 7-6) are useful in measuring small openings and hard-to-reach locations where an ordinary rule cannot be used. Five small rules come to a set; they range between 1/4 and 1 in. in length and can be interchanged in the holder.

Decimal rules (Fig. 7-7) are most often used when it is necessary to make linear measurements smaller than $\frac{1}{64}$ in. Since linear dimensions are sometimes specified on drawings in decimals, these rules are useful to the machinist. The most common graduations found on decimal rules are .100 ($\frac{1}{10}$ of an inch), .050 ($\frac{1}{20}$ of an inch), .020 ($\frac{1}{50}$ of an inch), and .010 ($\frac{1}{100}$ of an inch). A 6-in. decimal rule is shown in Fig. 7-8.

MEASURING LENGTHS

With a reasonable amount of care, fairly accurate measurements can be made using steel rules. Whenever possible, butt the end of a rule against a shoulder or step (Fig. 7-9) to ensure an accurate measurement.

Through constant use, the end of a steel rule becomes worn. Measurements taken from the end, therefore,

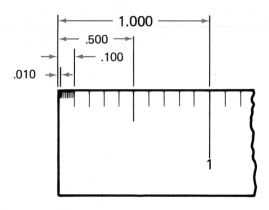

■ Figure 7-7 Decimal graduations on a rule provide an accurate and simple form of measurement. *(Kelmar Associates)*

■ Figure 7-8 Common graduations found on a 6-in. decimal rule. *(The L. S. Starrett Co.)*

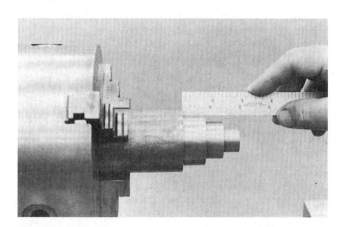

■ Figure 7-9 Butting a rule against a shoulder. *(Kelmar Associates)*

are often inaccurate. Fairly accurate measurements of flat work can be made by placing the 1-in. or 1-cm graduation line on the edge of the work, taking the measurement, and subtracting 1 in. or 1 cm from the reading (Fig. 7-10). When measuring flat work, be sure that the edge of the rule is parallel to the edge of the work. If the rule is placed at an angle to the edge (Fig. 7-11), the measurement will not be accurate. When measuring the diameter of round stock, start from the 1-in. or 1-cm graduation line.

THE RULE AS A STRAIGHTEDGE

The edges of a steel rule are ground flat. The rule may therefore be used as a straightedge to test the flatness of

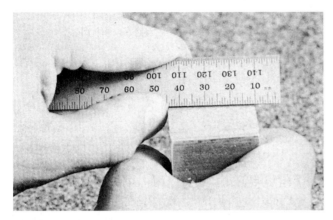

Figure 7-10 Measuring with a rule starting at the 1-cm line. *(Kelmar Associates)*

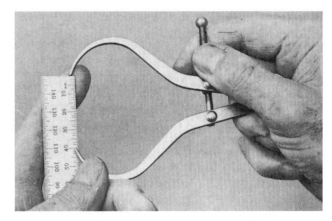

Figure 7-12 Setting an outside caliper to size with a rule. *(Kelmar Associates)*

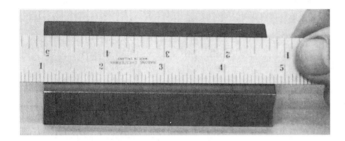

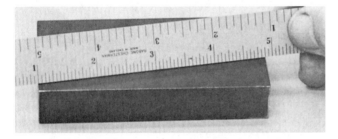

Figure 7-11 The rule must be held parallel to the edge of the workpiece; otherwise, the measurement will not be correct. *(Kelmar Associates)*

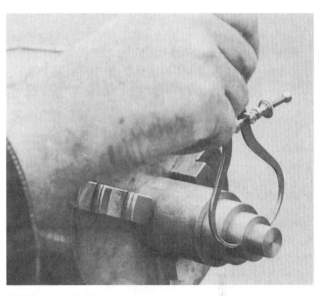

Figure 7-13 Checking a diameter with an outside caliper. *(Kelmar Associates)*

workpieces. The edge of a rule should be placed on the work surface, which is then held up to the light. Inaccuracies as small as a few thousandths of an inch, or 0.05 mm, can easily be seen by this method.

▶▶ Outside Calipers

Outside calipers are not precision tools; however, they can be used to approximately measure the outside surface of either round or flat work. They are made in several styles, such as *spring joint* and *firm joint calipers*. The outside spring joint caliper is most commonly used; however, it cannot be read directly and must be set to a steel rule or a standard-size gage. Calipers should not be used when an accuracy of less than .015 in. (0.39 mm) is required.

USING OUTSIDE CALIPERS

When calipers are set to a rule, it is important that the end of the rule be in good condition, not worn or damaged. Use the following procedure:

1. Hold both legs of the caliper parallel to the edge of the rule. Turn the adjusting nut until the end of the lower leg just splits the desired graduation line on the rule (Fig. 7-12).

2. Place the caliper on the work with both legs of the caliper at right angles to the centerline of the work (Fig. 7-13).

3. The diameter is correct when the caliper just slides over the work by its own weight.

▶▶ Inside Calipers

Inside calipers are used to measure the diameter of holes or the width of keyways and slots. They are made in several styles, such as *spring joint* and *firm joint calipers.*

MEASURING AN INSIDE DIAMETER

Fairly accurate measurements of holes and slots may be made using an inside caliper and a rule. Use the following procedure:

1. Place one leg of the caliper near the bottom edge of the hole (Fig. 7-14).

2. Hold the caliper leg in this position with a finger.

3. Keep the caliper legs vertical or parallel to the hole.

4. Move the top leg in the direction of the arrows and turn the adjusting nut until a slight drag is felt on the caliper leg.

5. Find the size of the setting by placing the end of a rule and one leg of the caliper against a flat surface.

6. Hold the legs of the caliper parallel to the edge of the rule and note the reading on the rule.

TRANSFERRING MEASUREMENTS

When an accurate measurement is required, the caliper setting should be checked with an outside micrometer. Use the following procedure:

1. Check the accuracy of the micrometer (Unit 9).

2. Hold the micrometer in the right hand so that you can easily adjust it with the thumb and forefinger (Fig. 7-15).

3. Place one leg of the caliper on the micrometer anvil and hold it in position with a finger.

4. Rock the top leg of the caliper in the direction of the arrows.

5. Adjust the micrometer thimble until *only a slight drag* is felt as the caliper leg passes over the measuring face.

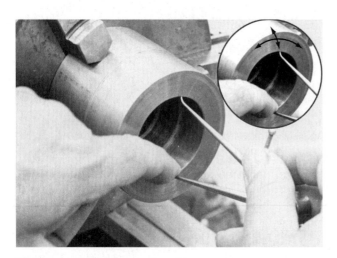

■ **Figure 7-14** Adjusting an inside caliper to the size of a hole. *(Kelmar Associates)*

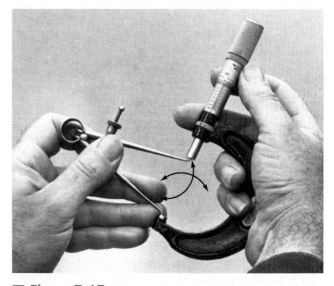

■ **Figure 7-15** An inside caliper setting checked with a micrometer. *(Kelmar Associates)*

1. Name two systems of measurement presently used in North America.

2. What is the common unit of length in the SI system?

3. How are metric rules usually graduated?

4. Name four types of steel rules used in machine shop work.

5. Describe a rule with No. 4 graduations.

6. State the purpose of:

   a. Hook rules   b. Decimal rules

7. Name two types of outside calipers.

8. What is the procedure for setting an outside caliper to a size?

9. Explain how you would know when the work is the same size as the caliper setting.

10. Explain the procedure for setting an inside caliper to the size of a hole.

11. In point form, list the procedure for checking an inside caliper setting with a micrometer.

UNIT 8

Squares and Surface Plates

OBJECTIVES

After completing this unit, you will be able to identify and explain the uses of:

1 The machinist's combination square

2 Three types of solid and adjustable squares

3 Two types of surface plates

The square is a very important tool used by the machinist for layout, inspection, and setup purposes. Squares are manufactured to various degrees of accuracy, ranging from semiprecision to precision squares. Precision squares are hardened and accurately ground.

▶▶ Machinist's Combination Square

The combination square is a basic tool used by the machinist for quickly checking 90° and 45° angles. It is part of a combination set (Fig. 8-1) that includes the square head, the center head, the bevel protractor, and a graduated grooved rule to which the various heads may be attached. In addition to its use for laying out and checking angles, the combination square may also be used as a depth gage (Fig. 8-2) or to measure the length of work to reasonable accuracy. Further uses of the combination set will be discussed in the unit on layout work (Unit 19).

▶▶ Precision Squares

Precision squares are used chiefly for inspection and setup purposes. They are hardened and accurately ground and must be handled carefully to preserve their accuracy. A great variety of squares are manufactured for specific purposes. The squares are all variations of either the *solid square* or *the adjustable square*.

▶▶ Beveled-Edge Squares

The better quality standard squares used in inspection have a beveled-edge blade, which is hardened and ground. The beveled edge allows the blade to make line contact with the work, thereby permitting a more accurate check. Two methods of using a beveled-edge square for checking purposes are illustrated in Figs. 8-3 and 8-4. In Fig. 8-3, if the work is square (90°), both pieces of paper will be tight between the square and the work. In Fig. 8-4 on p. 68, the light is shut out only where the blade makes line contact with the surface of the work. Light shows through where the blade of the square does not make contact with the surface being checked.

▶▶ Toolmaker's Surface Plate Square

The *toolmaker's surface plate square* (Fig. 8-5 on p. 68) provides a convenient method of checking work for squareness on a surface plate. Since it is of one-piece construction, there is little chance of any inaccuracy developing, as is the case with a blade and beam square.

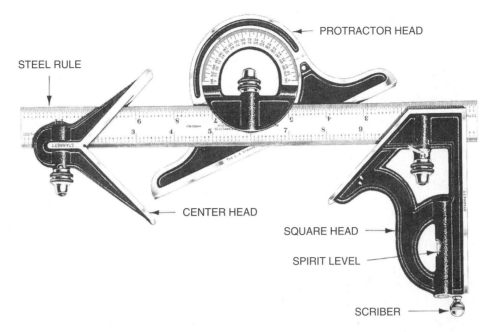

PROTRACTOR HEAD

STEEL RULE

CENTER HEAD

SQUARE HEAD

SPIRIT LEVEL

SCRIBER

■ Figure 8-1 The combination set may be used for laying out and checking work. *(The L.S. Starrett Co.)*

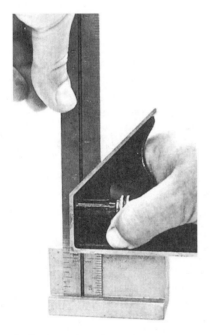

■ Figure 8-2 The combination (adjustable) square used as a depth gage. *(Kelmar Associates)*

PAPER FEELERS

■ Figure 8-3 Using paper between the blade of the square and the workpiece to check for squareness. *(Kelmar Associates)*

▸▸ Cylindrical Squares

Cylindrical squares are commonly used as master squares against which other squares are checked. The square consists of a thick-walled alloy steel cylinder that has been hardened, ground, and lapped. The outside diameter is a nearly true cylinder, and the ends are ground and lapped square with the axis. The ends are recessed and notched to decrease the inaccuracy from dust and to reduce friction. When a cylindrical square is used, it must be set carefully

Squares and Surface Plates **67**

Figure 8-4 Light shines through where the blade of the square does not make line contact with the surface. *(Kelmar Associates)*

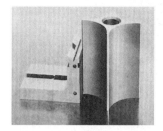

Figure 8-5 A toolmaker's surface plate square.

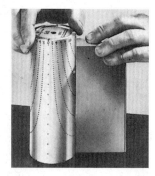

Figure 8-6 A direct-reading cylindrical square being used to check a part for squareness.

Figure 8-7 A diemaker's square is useful for checking die clearance. *(The L. S. Starrett Co.)*

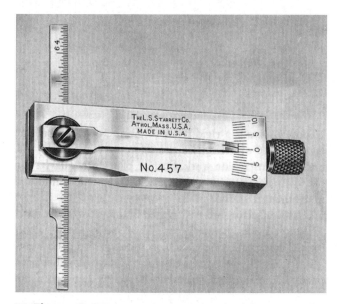

Figure 8-8 The direct-reading diemaker's square indicates the angle at which the blade is set. *(The L. S. Starrett Co.)*

on a clean surface plate and rotated slightly to force particles of dust and dirt into the end notches; the square can then make the proper contact with the surface plate. Cylindrical squares provide perfect line contact with the part being checked.

Another type of cylindrical square is the *direct-reading cylindrical square* (Fig. 8-6), which indicates directly the amount that the part is out of square. One end of the cylinder is lapped square with the axis, while the other end is ground and lapped slightly out of square. The circumference is etched with several series of dots that form elliptical curved lines. Each curve is numbered at the top to indicate the amount, in ten-thousandths of an inch (.0001), that the workpiece is out of square over the length of the square. Absolute squareness, or *zero deviation,* is indicated by an etched, vertical dotted line on the square.

When used, the square is carefully placed in contact with the work and turned until no light is seen between it and the part being inspected. The uppermost curved line in contact with the work is noted and followed to the top, where the number shows the amount the work is out of square. This square may also be used as a conventional cylindrical square if the opposite end, which is ground and lapped square with the axis, is used.

▶▶ Adjustable Squares

The *adjustable square,* while not providing the accuracy of a good solid square, is used by the toolmaker where it would be impossible to use a fixed square.

A *diemaker's square* (Fig. 8-7) is used to check the clearance angle on dies. The blade is adjusted to the angle of the workpiece by means of a blade-adjusting screw. This angular setting must then be checked with a protractor. Another form of diemaker's square is the *direct-reading* type, which indicates the angle at which the blade is set (Fig. 8-8).

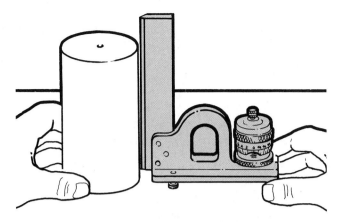

▶▶ Adjustable Micrometer Square

The *adjustable micrometer square* (Fig. 8-9) may be used to check a part for squareness accurately. When a piece of work is being checked and light shows between the blade and the work, turn the micrometer head until the full length of the blade, which may be tilted, touches the work. The amount the part is out of square may be read from the micrometer head. When the micrometer head is set at zero, the blade is perfectly square with the beam.

▶▶ Straightedges

A *straightedge* is used to check surfaces for flatness and to act as a guide for scribing long, straight lines in layout work. Straightedges are generally rectangular bars of hardened and accurately ground steel, having both edges flat and parallel. They are supplied with either plain or beveled edges. Long straightedges are generally made of cast iron with ribbed construction.

▶▶ Surface Plates

A *surface plate* is a rigid block of granite or cast iron, the flat surface of which is used as a reference plane for layout, setup, and inspection work. Surface plates generally have a three-point suspension to prevent rocking when mounted on an uneven surface.

Cast-iron plates are well ribbed and supported to resist deflection under heavy loads. They are made of close-grained cast iron, which has high strength and good wear-resistance qualities. After a cast-iron surface plate

■ **Figure 8-10** Granite surface plates are not affected by changes in humidity and temperature. *(Kelmar Associates)*

has been machined, its surface must be scraped by hand to a flat plane. This operation is long and tedious; therefore, the cost of these plates is high.

Granite surface plates (Fig. 8-10) have many advantages over cast-iron plates and are replacing them in many shops. They may be manufactured from gray, pink, or black granite and are obtainable in several degrees of accuracy. Extremely flat finishes are produced by lapping. The advantages of granite plates are:

1. They are not appreciably affected by temperature change.
2. Granite will not burr, as does cast iron; therefore, the accuracy is not impaired.
3. They are nonmagnetic.
4. They are rustproof.
5. Abrasives will not embed themselves as easily in the surface; thus, they may be used near grinding machines.

CARE OF SURFACE PLATES

1. Keep surface plates clean at all times, and wipe them with a dry cloth before using.
2. Clean them occasionally with solvent or surface-plate cleaner to remove any film.
3. Protect them with a wooden cover when they are not in use.
4. Use parallels whenever possible to prevent damage to plates by rough parts or castings.
5. Remove burrs from the workpiece before placing it on the plate.
6. Slide heavy parts onto the plate rather than place them directly on the plate; a part might fall and damage the plate.
7. Remove all burrs from cast-iron plates by honing.
8. When they are not in regular use, cover cast-iron plates with a thin film of oil to prevent rusting.
9. Center punching or prick punching layout lines should not be done on a surface plate, since these plates will not withstand impact forces.

Precision Squares

1. Name two types of solid squares and state the advantage of each.

2. Why are beveled-edge squares used in inspection work?

3. What procedure should be followed when using a cylindrical square?

4. State the purpose of a diemaker's square.

5. How can the angle of the workpiece be determined using each type of diemaker's square?

Surface Plates

6. What is the purpose of a surface plate?

7. Name three types of granite used in making surface plates.

8. State five advantages of granite over cast-iron surface plates.

9. List eight ways of caring for surface plates.

UNIT 9

Micrometers

OBJECTIVES

After completing this unit, you should be able to:

1 Identify the most common types of outside micrometers and their uses

2 Measure the size of a variety of objects to within .001-in. accuracy

3 Read vernier micrometers to .0001-in. accuracy

4 Measure the size of a variety of objects to within 0.01-mm accuracy

Although fixed gages are convenient for checking the upper and lower limits of external and internal dimensions, they do not measure the actual size of the part. The machinist must use some form of precision measuring instrument to obtain this desired size (Fig. 9-1a). Precision measuring tools——Units

9–18—may be divided into five categories: tools used for outside measurement, inside measurement, depth measurement, thread measurement, and height measurement.

The *micrometer caliper,* usually called the *micrometer,* is the most commonly used measuring instrument when accuracy is required. The *standard inch micrometer,* shown in a cutaway view in Fig. 9-1b (on p. 72), measures accurately to .001 in. Since many phases of modern manufacturing require greater accuracy, the *vernier micrometer,* capable of even finer measurements, is used to an increasing extent.

The standard metric micrometer measures in hundredths of a millimeter, whereas the vernier metric micrometer measures up to 0.002 mm.

The only difference in construction and reading between the standard inch and the vernier micrometer is the addition of the vernier scale on the sleeve above the index or centerline.

▶▶ Principle of the Inch Micrometer

To understand the principle of the inch micrometer, the student should be familiar with two important thread terms:

> *Pitch,* which is the distance from a point on one thread to a corresponding point on the next thread. For inch threads, pitch is expressed as $1/N$ (number of threads). For metric threads, it is expressed in millimeters.

> *Lead,* which is the distance a screw thread advances axially in one complete revolution or turn.

Since there are 40 threads per inch on the micrometer, the pitch is $\frac{1}{40}$ (.025) in. Therefore, one complete revolution of the spindle will either increase or decrease the distance between the measuring faces by $\frac{1}{40}$ (.025) in. The 1-in. distance marked on the micrometer sleeve is divided into 40 equal divisions, each of which equals $\frac{1}{40}$ (.025) in.

If the micrometer is closed until the measuring faces just touch, the zero line on the thimble should line up with the index line on the sleeve (barrel). If the thimble is revolved counterclockwise one complete revolution, one

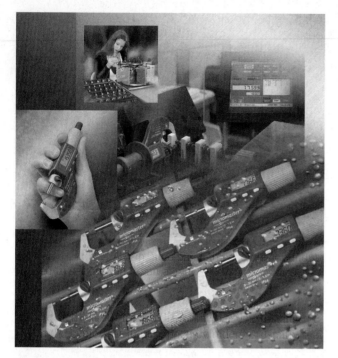

Figure 9-1a Micrometers can be used for a variety of measuring applications. *(Brown & Sharpe)*

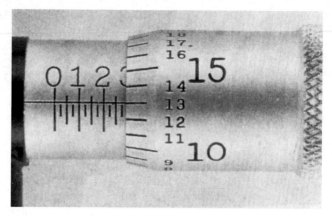

Figure 9-2 An inch micrometer reading of .288 in. *(Kelmar Associates)*

The thimble has 25 equal divisions about its circumference. Since one turn moves the thimble .025 in., one division represents 1/25 of .025, or .001. Therefore, each line on the thimble represents .001 inch.

TO READ A STANDARD INCH MICROMETER

1. Note the last number showing on the sleeve. Multiply that number by .100.

2. Note the number of small lines visible to the right of the last number shown. Multiply that number by .025.

3. Note the number of divisions on the thimble from zero to the line that coincides with the index line on the sleeve. Multiply that number by 001.

4. Add the three products to get the total reading.

In Fig. 9-2:

> #2 is shown on the sleeve $2 \times .100 = .200$

> Three lines are visible past the number $3 \times .025 = .075$

> #13 line on thimble coincides with the index line $13 \times .001 = \underline{.013}$

       Total reading   .288 in.

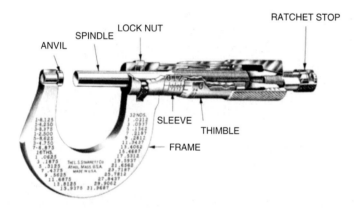

Figure 9-1b Cutaway view of a standard micrometer with a ratchet stop. *(The L.S. Starrett Co.)*

line will appear on the sleeve. Each line on the sleeve indicates .025 in. Thus, if three lines were showing on the sleeve (or barrel), the micrometer would have opened 3 × .025, or .075 in.

Every *fourth* line on the sleeve is longer than the others and is numbered to permit easy reading. Each numbered line indicates a distance of .100 in. For example, #4 showing on the sleeve indicates a distance between the measuring faces of 4 × .100, or .400 in.

▶▶ Vernier Micrometer

The *inch vernier micrometer* (Fig. 9-3) has, in addition to the graduations found on a standard micrometer, a *vernier*

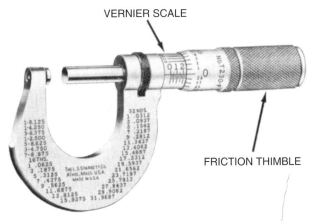

Figure 9-3 An inch vernier micrometer caliper with a friction thimble. *(The L. S. Starrett Co.)*

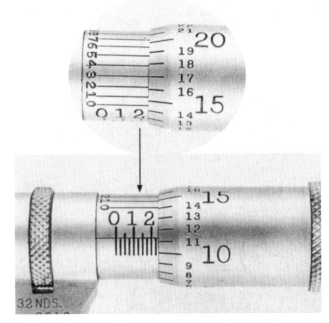

Figure 9-4 An inch vernier micrometer reading of .2363 in. *(Kelmar Associates)*

scale on the sleeve. This vernier scale consists of 10 divisions that run *parallel to and above* the index line. These 10 divisions on the sleeve occupy the same distance as nine divisions (.009) on the thimble. One division on the vernier scale, therefore, represents $\frac{1}{10} \times .009$, or .0009 in. Since one graduation on the thimble represents .001, or .0010 in., the difference between one thimble division and one vernier scale division represents $.0010 - .0009$, or .0001. Therefore, each division on the vernier scale has a value of .0001 in.

TO READ A VERNIER MICROMETER

1. Read the vernier micrometer as you would a standard micrometer.

2. Note the line on the vernier scale that coincides with a line on the thimble. This line will indicate the number of ten-thousandths that must be added to the reading in step 1.

   Refer to Fig. 9-4. The reading of the vernier micrometer is as follows:

   > #2 is shown on the sleeve $2 \times .100 = .200$

   > One line is visible past the number $1 \times .025 = .025$

   > The #11 line on the thimble is just past the index line $11 \times .001 = .011$

   > In Fig. 9-4, the #3 line on the vernier scale coincides with a line on the thimble $3 \times .0001 = \underline{.0003}$

        Total reading   .2363 in.

▸▸ Metric Micrometer

The *metric micrometer* (Fig. 9-5) is similar to the inch micrometer with two exceptions: the pitch of the spindle screw and the graduations on the sleeve and thimble. The pitch of the screw is 0.5 mm; therefore, a complete revolution of the thimble increases or decreases the distance between the measuring faces 0.5 mm. Above the index line on the sleeve, the graduations are in millimeters (from 0 to 25), with every fifth line numbered. Below the index line, each millimeter is subdivided into two equal parts of 0.5 mm, which corresponds to the pitch of the thread. It is apparent, therefore, that two turns of the thimble will be required to move the spindle 1 mm.

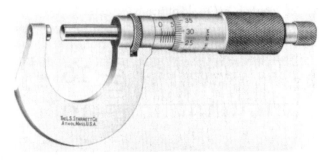

Figure 9-5 A metric micrometer measures in hundredths of a millimeter. *(The L.S. Starrett Co.)*

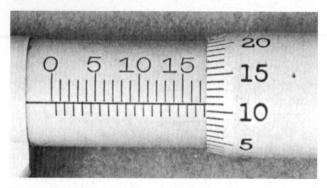

■ Figure 9-6 A metric micrometer reading of 17.61 mm. *(Kelmar Associates)*

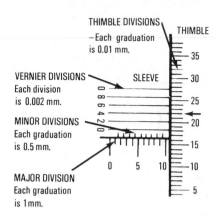

■ Figure 9-7 A metric vernier micrometer reading of 10.164 mm.

The circumference of the thimble is divided into 50 equal divisions, with every fifth line numbered. Since one revolution of the thimble advances the spindle 0.5 mm, each graduation on the thimble equals $\frac{1}{50} \times 0.5$ mm = 0.01 mm.

TO READ A METRIC MICROMETER

1. Note the number of the last main division showing *above the line* to the left of the thimble. Multiply that number by 1 mm.

2. If there is a half-millimeter line showing *below the index line,* between the whole millimeter and the thimble, then add 0.5 mm.

3. Multiply the number of the line on the thimble that coincides with the index line times 0.01.

4. Add these products.

In Fig. 9-6, there are:

> 17 lines above the index
  line   17 × 1   = 17
> 1 line below the index
  line   1 × .5 =   .5
> 11 lines on the thimble   11 × .01 =   .11
       Total reading   17.61 mm

▶▶ Metric Vernier Micrometer

The *metric vernier micrometer,* in addition to the graduations found on the standard micrometer, has five vernier divisions on the barrel, each representing 0.002 mm. In the vernier micrometer reading illustrated in Fig. 9-7, each major division (below the index line) has a value of 1 mm. Each minor division (above the index line) has a

value of 0.5 mm. There are 50 divisions around the thimble, each having a value of 0.01 mm.

TO READ A METRIC VERNIER MICROMETER

1. Read the micrometer as you would a standard metric micrometer.

2. Note the line on the vernier scale that coincides with a line on the thimble. This line will indicate the number of two-thousandths of a millimeter that must be added to the reading from step 1.

Refer to Fig. 9-7. The reading of the metric vernier micrometer would be:

> Major divisions
  (below the index line)   10 × 1   = 10
> Minor divisions
  (above the index line)   0 × 0.5   = 0
> Thimble divisions   16 × 0.01 = 0.16
> The second vernier
  division coincides
  with a thimble line   2 × 0.002 = 0.004
       Total reading   10.164 mm

▶▶ Combination Inch-Metric Micrometer

With the gradual shift to metric measurement, a dual dimension system will be needed for some time. The combination inch-metric micrometer (Fig. 9-8a) will give readings in both inch and metric sizes. It has a digital reading for one system and a standard barrel and thimble reading for the other.

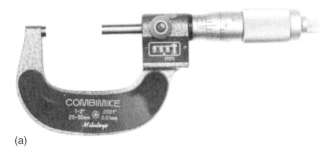

(a)

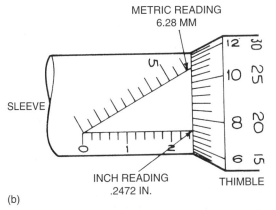

(b)

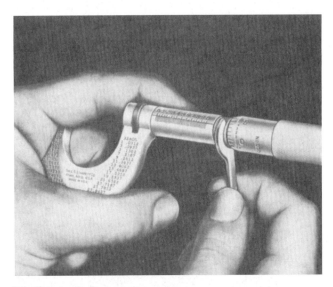

Figure 9-9 Removing the play in micrometer spindle screw threads. *(The L.S. Starrett Co.)*

Figure 9-8 (a) A combination inch–metric micrometer with a digital reading for one system; (b) a combination inch–metric micrometer that has dual scales on the sleeve and the barrel. *(MTI Corp.)*

Another type of inch-metric micrometer (Fig. 9-8b) has dual scales on the sleeve and the thimble. A horizontal scale on the sleeve (usually black) and the left-hand scale on the thimble represent inch readings. The angular scale on the sleeve (usually red) and the right-hand scale on the thimble represent metric readings.

▶▶ Micrometer Adjustments

Proper care and use of a micrometer is necessary to preserve its accuracy and keep adjustments to a minimum. Minor adjustments to micrometers can easily be made; however, it is extremely important that all parts of the micrometer be kept free from dust and foreign matter during any adjustment.

TO REMOVE PLAY IN THE MICROMETER THREADS

To remove play (looseness) in the spindle threads due to wear:

1. Back off the thimble, as shown in Fig. 9-9.
2. Insert the C-spanner into the slot or hole of the adjusting nut.
3. Turn the adjusting nut clockwise until play between the threads has been eliminated.

Note: After the micrometer has been adjusted, the spindle should advance freely while the ratchet stop or friction thimble is being turned.

▶▶ Testing the Accuracy of Micrometers

The accuracy of a micrometer should be tested periodically to ensure that the work produced is the size required. Always make sure that both measuring faces are clean before checking a micrometer for accuracy.

To test a 1-in. or 25-mm micrometer, first clean the measuring faces. Then turn the thimble using the friction thimble or ratchet stop until the measuring faces contact each other. If the zero line on the thimble coincides with the center (index) line on the sleeve, the micrometer is accurate. Micrometers can also be checked for accuracy by measuring a gage block or other known standard.

The reading of the micrometer must be the same as the gage block or standard. Any micrometer that is not accurate should be adjusted by a qualified person.

TO ADJUST THE ACCURACY OF A MICROMETER

Should the accuracy of a micrometer require adjustment, follow this procedure:

1. Clean the measuring faces and inspect them for damage.
2. Close the measuring faces carefully by turning the ratchet stop or friction thimble.

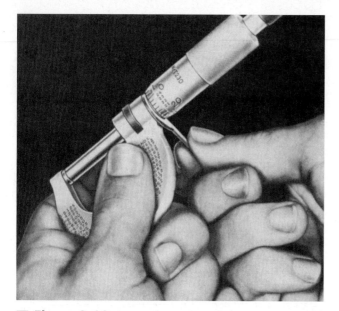

■ **Figure 9-10** Resetting the accuracy of a micrometer. *(The L.S. Starrett Co.)*

3. Insert the C-spanner into the hole or slot provided in the sleeve (Fig. 9-10).

4. Carefully turn the *sleeve* until the index line on the sleeve coincides with the zero line on the thimble.

5. Recheck the accuracy of the micrometer by opening the micrometer and then closing the measuring faces by turning the ratchet stop or friction thimble.

▸▸ Special-Purpose Micrometers

Although the design of most micrometers is fairly standard, certain refinements may be added to the basic design, if desired. Items such as the lock ring, ratchet, friction thimble, carbide measuring faces, and anvil extensions increase the accuracy and range of these instruments. Some of the more common types of micrometers used in the machine tool industry are shown in Figs. 9-11 to 9-17.

The *direct-reading micrometer* (Fig. 9-11) has graduations on the thimble and barrel as in a standard micrometer, in addition to a digital readout built into the frame. The exact micrometer reading at any point within its range is shown in the readout. Some micrometers combine both the standard inch reading on the thimble and barrel with a millimeter reading.

The *large-frame micrometer* (Fig. 9-12) is made for easier, faster precision measuring of large outside diameters (up to 60 in.). The frame is made of special steel to give it extreme rigidity and the lightest possible weight. Interchangeable anvils give each micrometer a range of 6 in.

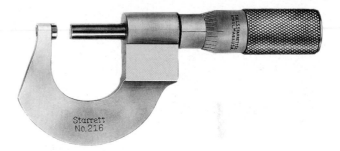

■ **Figure 9-11** A direct-reading micrometer has graduations like those in a standard micrometer and a digital readout built into the frame. *(The L.S. Starrett Co.)*

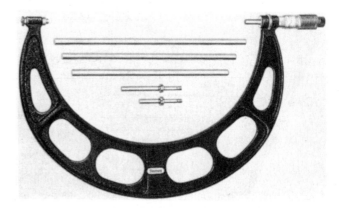

■ **Figure 9-12** A large-frame micrometer has interchangeable anvils that increase the range of the micrometer. *(The L.S. Starrett Co.)*

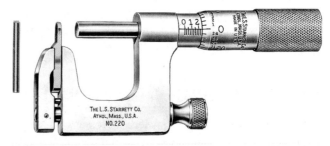

■ **Figure 9-13** The Mul-T-Anvil micrometer is used for measuring tubing and distances from a slot to an edge. *(The L.S. Starrett Co.)*

The *Mul-T-Anvil micrometer* (Fig. 9-13) comes equipped with round and flat anvils, which are interchangeable. The round (rod) anvil is used to measure the wall thickness of tubing and cylinders and for measuring from a hole to an edge. The flat anvil is used to measure the distance from the inside of slots and grooves to an edge.

The *indicating micrometer* (Fig. 9-14) uses an indicating dial and a movable anvil to permit accurate measurements to ten-thousandths of an inch (0.002 mm). This micrometer may be used as a comparator by setting it to a particular size with gage blocks or a standard and

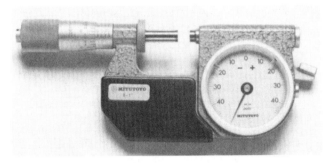

Figure 9-14 An indicating micrometer may be used as a comparator to check parts to ten-thousandths of an inch (0.002 mm). *(MTI Corp.)*

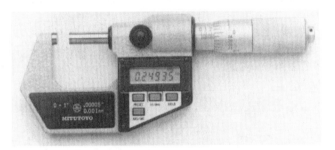

Figure 9-15 The Digi-Matic micrometer provides a digital display of readings accurate to 50 millionths of an inch. *(MTI Corp.)*

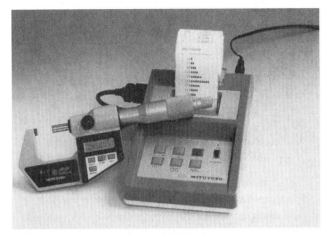

Figure 9-16 The Digi-Matic micrometer with statistical process control is a miniature data processor. *(MTI Corp.)*

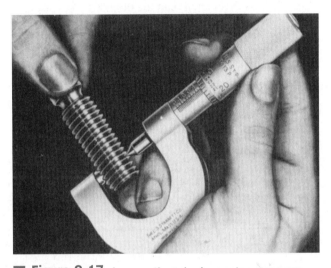

Figure 9-17 A screw thread micrometer measures the pitch diameter of a thread. *(The L.S. Starrett Co.)*

locking the spindle. The tolerance arms are then set to the required limits and each workpiece can be compared with the micrometer setting.

The *Digi-Matic micrometer* (Fig. 9-15) is used as a hand gage for inspecting small parts. It is accurate up to 50 millionths of an inch (0.00127 mm) and displays readings in both inch or metric sizes.

The *Digi-Matic micrometer with statistical process control* (Fig. 9-16) provides a stand-alone inspection system that can be used on the production floor. This unit can be interfaced with a personal or host computer, providing valuable statistics on production quality.

▶▶ Screw Thread Micrometers

Sharp-V, American National, Unified, and International Organization for Standardization (ISO) threads may be measured with reasonable accuracy with a screw thread micrometer. This type of micrometer has a pointed spindle and a double-V swivel anvil, which are shaped to contact the pitch diameter of the thread being measured (Fig. 9-17). The micrometer reading indicates the pitch diameter of the thread, which is equal to the outside diameter less the depth of one thread.

Each thread micrometer is limited to measuring a certain range of threads; this range is stamped on the micrometer frame. One-inch thread micrometers are manufactured in four ranges to cover the following range of threads per inch (TPI):

8 to 13 TPI

14 to 20 TPI

22 to 30 TPI

32 to 40 TPI

Metric-thread micrometers are available in sizes from 0 to 25 mm, 25 to 50 mm, 50 to 75 mm, and 75 to 100 mm. A set of 12 anvil and spindle inserts is available for thread pitches from 0.4 to 6 mm.

To check the accuracy of a thread micrometer, carefully bring the measuring faces into light contact; the micrometer reading for this setting should be zero.

When measuring threads, the micrometer gives a slightly distorted reading because of the helix angle of the thread. To overcome this inaccuracy, set the thread micrometer to a thread plug gage or to a thread that must be duplicated.

TO MEASURE WITH A THREAD MICROMETER

1. Select the correct thread micrometer to suit the number of threads per inch of the workpiece, or pitch in mm for an ISO thread.
2. Thoroughly clean the measuring surfaces.
3. Check the micrometer for accuracy by bringing the measuring faces together; the reading should be zero.
4. Clean the thread to be measured.
5. Set the micrometer to the required thread plug gage and note the reading.
6. Fit the swivel anvil onto the threaded workpiece.

7. Adjust the spindle until the point just bears against the opposite side of the thread.
8. Carefully roll the micrometer over the thread to get the proper "feel."
9. Note the readings and compare them to the micrometer reading of the thread plug gage.

Threads may also be checked by the screw *thread comparator micrometer,* which has two conical measuring surfaces. Since it does not measure the pitch diameter, it is important to set this instrument to a thread plug gage before measuring a threaded workpiece. The screw micrometer is used for a quick comparison of threads, as well as for checking small grooves and recesses where regular micrometers cannot be used.

When thread micrometers or comparators are not available, threads may be accurately checked by the *three-wire method,* which we discuss fully in Unit 55.

unit 9 review questions

Micrometers

1. How many threads per inch are there on a standard inch micrometer?
2. What is the value of:
   a. Each line on the sleeve?
   b. Each numbered line on the sleeve?
   c. Each line on the thimble?
3. Read the following standard micrometer settings.

4. Describe briefly the principle of the vernier micrometer.
5. Describe the procedure for reading a vernier micrometer.
6. Read the following vernier micrometer settings.

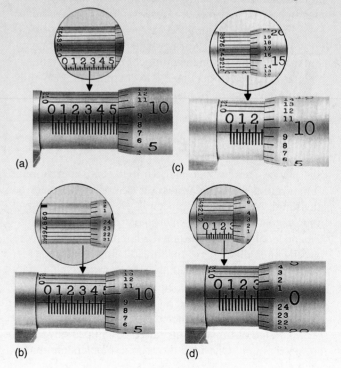

(a)

(c)

(b)

(d)

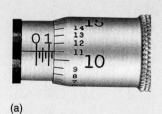

(a)

(c)

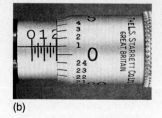

(b)

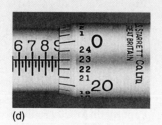

(d)

7. Explain how to adjust a micrometer:
   a. To remove play in the spindle threads.
   b. For accuracy.

Metric Micrometers

8. What are the basic differences between a metric and an inch micrometer?

9. What is the value of one division on:
   a. The sleeve above the index line?
   b. The sleeve below the index line?
   c. The thimble?

10. Read the following metric micrometer settings.

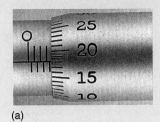

(a)

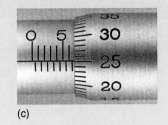

(c)

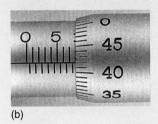

(b)

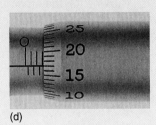

(d)

Indicating Micrometers

11. State two uses for an indicating micrometer.

Screw Thread Micrometers

12. Describe the construction of the contact points of a screw thread micrometer.

13. What dimension of the thread is indicated on a screw thread micrometer reading?

14. List the four ranges covered by screw thread micrometers.

15. How may threads be measured *accurately* with a screw thread micrometer?

UNIT 10

Vernier Calipers

OBJECTIVES

After completing this unit, you will be able to:

1 Measure workpieces to within an accuracy of .001 in. using a 25-division inch vernier caliper

2 Measure workpieces to within an accuracy of .001 in. using a 50-division inch vernier caliper

3 Measure workpieces to within an accuracy of 0.02 mm using a metric vernier caliper

Vernier calipers are precision measuring tools used to make accurate measurements to within .001 in. for inch verniers or to 0.02 mm for metric verniers. The bar and the movable jaw may be graduated on both sides or both edges. One side is used to take outside measurements; the other, to take inside measurements (Fig. 10-1a). Vernier calipers are available in inch and metric graduations; however, some types have both inch and metric graduations on the same caliper.

■ Figure 10-1a Vernier calipers can be used for inside and outside measurements. *(Brown & Sharpe)*

▶▶ Vernier Calipers

Vernier calipers (Fig. 10-1b) are precision tools used to make accurate measurements to within .001 in. or 0.02 mm, depending on whether they are inch or metric vernier calipers.

PARTS OF THE VERNIER CALIPER

The vernier caliper, regardless of the standard of measurement used, consists of an L-shaped frame and a movable jaw.

The L-shaped frame consists of a *bar,* which shows the *main scale graduations,* and the *fixed jaw.* The *movable jaw,* which slides along the bar, contains the *vernier scale.* Adjustments for size are made by means of an *adjusting nut.* Readings may be locked into place by means of the *clamp screws.*

Most bars are graduated on both sides or on both edges, one for outside measurements and the other for inside measurements. The outer tips of the jaws are cut away to form nibs, which permit inside measurements to be

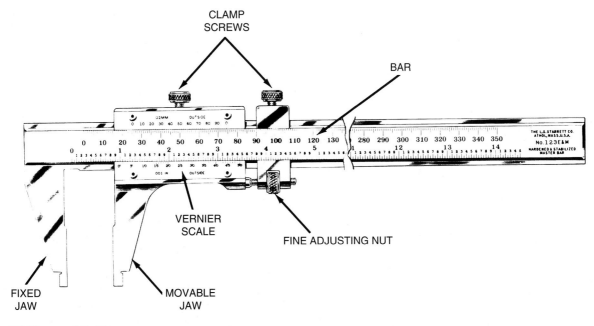

Figure 10-1b Parts of a vernier caliper. *(The L. S. Starrett Company)*

taken. Inch vernier calipers are manufactured with both 25- and 50-division vernier scales. The 50-division scale is much easier to read than the 25-division scale. Metric vernier calipers graduated in millimeters are also available.

Some vernier calipers are provided with two small indentations, or points, on the bar and a movable jaw, which may be used to set dividers accurately to a specific dimension or radius.

The bar of the vernier caliper with the 25-division vernier scale on the movable jaw is graduated exactly the same as a micrometer. Each inch is divided into 40 equal divisions, each having a value of .025 in. Every fourth line, representing ⅒ or .100, is numbered. The vernier scale on the movable jaw has 25 equal divisions, each representing .001. The 25 divisions on the vernier scale, which are .600 in. in length, are equal to 24 divisions on the bar. The difference between *one* division on the bar and one vernier division equals .025 − .024, or .001 in. Therefore, only one line of the vernier scale will line up exactly with a line on the bar at any one setting.

MEASURING A WORKPIECE WITH A 25-DIVISION INCH VERNIER CALIPER

1. Remove all burrs from the workpiece and clean the surface to be measured.

2. Open the jaws enough to pass over the work.

3. Close the jaws against the work and lock the right-hand clamp screw.

4. Turn the adjusting screw until the jaws *just touch* the work surface. Be sure that the jaws are in place

by attempting to move the bar slightly sideways and vertically while turning the adjusting nut.

5. Lock the clamp screw on the movable jaw.

6. Read the measurement shown in Fig. 10-2 as follows:

> The large #1 on the bar   = 1.000
> The small #4 past the #1   4 × .100 =   .400
> One line is visible past
  the #4   1 × .025 =   .025
> The eleventh line of the
  vernier scale coincides
  with a line on the bar   11 × .001 =   .011
       Total reading   1.436 in.

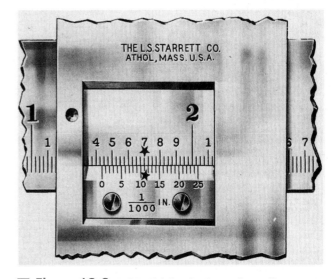

Figure 10-2 A 25-division inch vernier caliper reading of 1.436 in. *(The L. S. Starrett Company)*

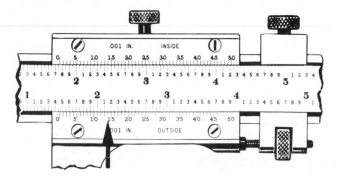

■ Figure 10-3 A 50-division inch vernier caliper reading of 1.464 in.

THE 50-DIVISION INCH VERNIER CALIPER

Because 25-division vernier calipers are often difficult to read, many vernier calipers are now manufactured with 50 divisions (equal to 49 on the main scale) on the vernier scale of the movable jaw. Each of these scales on the bar and movable jaw is equal to 2.450 in. in length. Each division on the bar then equals 2.450 divided by 49 divisions, or .050 in. in length. Each division on the vernier scale would equal 2.450 divided by 50 divisions, or .049 in. in length. The *difference in length* between one main scale division and one vernier division equals .050 − .049, or .001 in.

Each line on the main scale of a 50-division vernier caliper has a value of .050 in. Each line on the vernier scale has a value of .001 in. In Fig. 10-3:

> The large #1 on the bar   = 1.000
> The small #4 past the #1   4 × .100 = .400
> 1 line is visible past #4   1 × .050 = .050
> The fourteenth line on the
  vernier scale coincides
  with a line on the bar   14 × .001 = .014
  Total reading   1.464 in.

▶▶ The Metric Vernier Caliper

Vernier calipers are also made with metric readings, and many have both metric and inch graduations on the same instrument (Fig. 10-4). The parts of metric vernier calipers are the same as those of the inch vernier.

The *main scale* is graduated in millimeters and every main division is numbered. Each numbered division has a value of 10 mm; for example, #1 represents 10 mm, #2 represents 20 mm, etc. There are 50 graduations on the sliding or *vernier scale,* with every fifth one numbered. These 50 graduations occupy the same space as 49 graduations on the main scale (49 mm). Therefore,

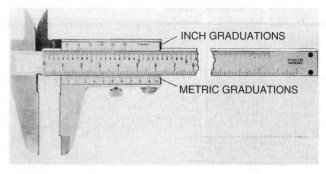

■ Figure 10-4 A vernier caliper with both inch and metric readings.

$$1 \text{ vernier division} = \frac{49}{50}$$
$$= 0.98 \text{ mm}$$

The difference between 1 main scale division and 1 vernier scale division is

$$1 - 0.98 = 0.02 \text{ mm}$$

TO READ A METRIC VERNIER CALIPER

1. The last numbered division on the bar to the left of the vernier scale represents the number of millimeters multiplied by 10.

2. Note how many full graduations are showing between this numbered division and the zero on the vernier scale. Multiply this number by 1 mm.

3. Find the line on the vernier scale that coincides with a line on the bar. Multiply this number by 0.02 mm.

In Fig. 10-5:

> The large #4 graduation
  on the bar   4 × 10 = 40
> Three full lines past the
  #4 graduation   3 × 1 = 3
> The ninth line on the
  vernier scale coincides
  with a line on the bar   9 × 0.02 = 0.18
  Total reading   43.18 mm

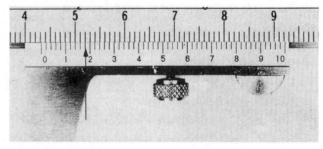

■ Figure 10-5 A metric vernier caliper reading of 43.18 mm.

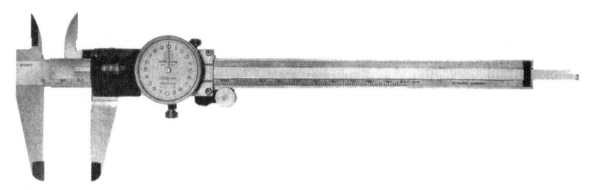

■ Figure 10-6 A dial caliper with digital readout provides a quick and accurate method of measurement. *(MTI Corp.)*

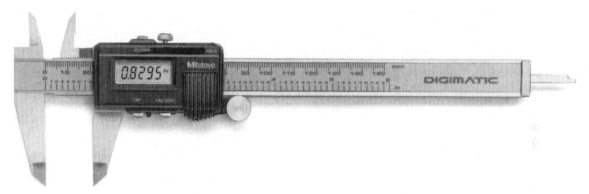

■ Figure 10-7 The digital electronic caliper can make accurate outside diameter, inside diameter, step, and depth measurements. *(MTI Corp.)*

▶▶ Direct-Reading Dial Caliper

Because it is easier to read, the *direct-reading dial caliper* is gradually replacing the standard vernier caliper. Dial calipers are manufactured in inch and/or metric standards and are available with digital readout. A dial indicator, the hand of which is attached to a pinion, is mounted on the sliding jaw. For the metric dial caliper (Fig. 10-6), one revolution of the hand represents 2 mm of travel; one revolution on the inch caliper may represent .100 or .200 in.

of travel, depending on the manufacturer. Most direct-reading calipers have a narrow sliding blade attached to the sliding jaw (and dial). This narrow blade permits the dial caliper to be used as an efficient and accurate depth gage (see Unit 11).

The digital electronic caliper (Fig. 10-7) can provide readings to a resolution of .0005 in., or 0.01 mm, at the touch of a button. It is of rugged construction with no rack, pinion, or glass scale. The digital electronic caliper can make inch or metric outside diameter, inside diameter, step, and depth measurements, and it can be connected to Statistical Process Control (SPC) equipment for inspection purposes.

Vernier Caliper

1. Describe the principle of:

   a. The 25-division vernier
   b. The 50-division vernier

2. Describe the procedure for reading a vernier caliper.

3. What are the following vernier caliper readings?

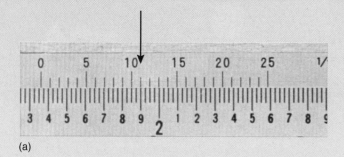

(a)

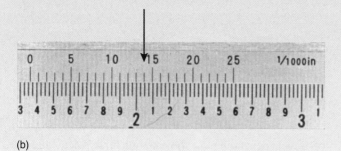

(b)

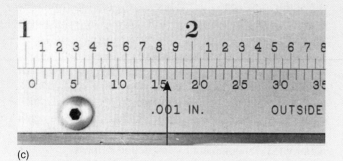

(c)

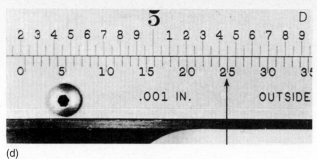

(d)

Metric Vernier Caliper

4. Describe the principle of the metric vernier caliper.

5. Read the following metric caliper settings.

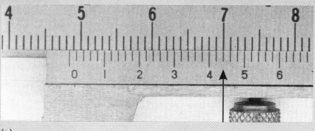

(a)

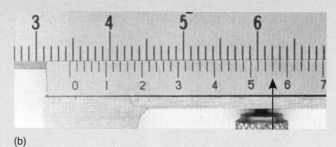

(b)

UNIT 11

Inside-, Depth-, and Height-Measuring Instruments

OBJECTIVES

After completing this unit, you should be able to:

1 Measure hole diameters to within .001-in. (0.02-mm) accuracy using inside micrometers and inside micrometer calipers

2 Measure depths of slots and grooves to an accuracy of .001 in. (0.02 mm)

3 Measure heights to an accuracy of .001 in. (0.02 mm) using a vernier height gage

Because of the great variety of measurements required in machine shop work, many types of measuring tools are available. These permit the machinist to measure not only outside sizes but also inside diameters, depths, and heights (Fig. 11-1a on p. 86). Direct-reading instruments are most commonly used and are generally the most accurate; however, because of the shape or size of the part, transfer-type instruments may be required.

▶▶ Inside-Measuring Instruments

All inside-measuring instruments fall into two categories: direct-reading and transfer-type.

With *direct-reading instruments,* the size of the hole can be read on the instrument being used to measure the hole. The most common direct-reading instruments are the inside micrometer, the Intrimik, and the vernier caliper.

Transfer-type instruments are set to the diameter of the hole and then this size is transferred to an outside micrometer to determine the actual size. The most common transfer-type instruments are inside calipers, small hole gages, and telescope gages.

DIRECT-READING INSTRUMENTS

Inside Micrometer Calipers

The inside micrometer caliper (Fig. 11-1b on p. 86) is designed for measuring holes, slots, and grooves, from

.200 to 2.000 in. in size for the inch-designed instruments, or 5 to 50 mm in size for metric instruments. The nibs or ends of the jaws are hardened and ground to a small radius to permit accurate measurement. A locking nut provided with this micrometer can be used to lock it at any desired size.

The inside micrometer caliper is based on the same principle as a standard micrometer, except that the barrel readings on some calipers are reversed (as shown in Fig. 11-1b). Extreme care must be taken in reading this type of instrument. Other inside micrometer calipers have the readings on the spindle and are read in the same manner as a standard outside micrometer. Inside micrometer calipers are special-purpose tools and are not used in mass production measurement.

To Use an Inside Micrometer Caliper

1. Adjust the jaws to slightly less than the diameter to be measured.

2. Hold the fixed jaw against one side of the hole and adjust the movable jaw until the proper "feel" is obtained.

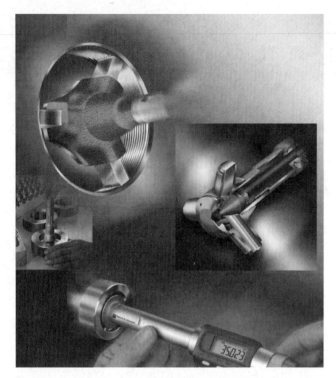

Figure 11-1a Accurate internal measurements can be made with direct-reading instruments. *(Brown & Sharpe)*

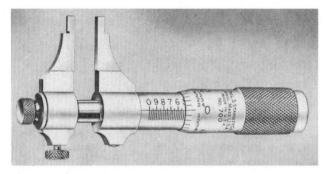

Figure 11-1b An inside micrometer caliper. *(The L. S. Starrett Co.)*

Note: Move the movable jaw back and forth to ensure that the measurement taken is across the true diameter.

3. Set the lock nut, remove the instrument, and check the reading.

Inside Micrometers

For internal measurements larger than 1.500 in. or 40 mm, inside micrometers (Fig. 11-2) are used. The inside micrometer set consists of a micrometer head, having a range

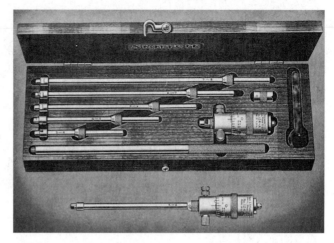

Figure 11-2 An inside micrometer set can measure a large range of sizes. *(The L. S. Starrett Co.)*

.500 or 1 in.; several extension rods of different lengths, which may be inserted into the head; and a .500-in. spacing collar. These sets cover a range from 1.500 to over 100 in., or from 40 to 1000 mm for metric tools. Sets that are used for the larger ranges generally have hollow tubes, rather than rods, for greater rigidity.

The inside micrometer is read in the same manner as the standard micrometer. Since there is no locking nut on the inside micrometer, the thimble nut is adjusted to a tighter fit on the spindle thread to prevent a change in the setting while it is being removed from the hole.

To Measure with an Inside Micrometer

1. Measure the size of the hole with a rule.
2. Insert the correct micrometer extension rod after having carefully cleaned the shoulders of the rod and the micrometer head.
3. Align the zero marks on the rod and micrometer head.
4. Hold the rod firmly against the micrometer head and tighten the knurled set screw.
5. Adjust the micrometer to slightly less than the diameter to be measured.
6. Hold the head in a fixed position and adjust the micrometer to the hole size while moving the rod end in the direction of the arrows (Fig. 11-3).

Note: When a micrometer has been properly adjusted to size, there should be a slight drag when the rod end is moved past the centerline of the hole.

7. Carefully remove the micrometer and note the reading.
8. To this reading, add the length of the extension rod and collar.

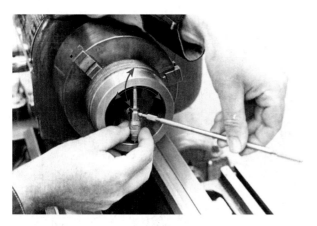

Figure 11-3 Using an inside micrometer to measure the size of a hole. *(Kelmar Associates)*

Intrimik

A difficulty encountered in measuring hole sizes with instruments employing only two measuring faces is that of properly measuring the diameter and not a chord of the circle. An instrument that eliminates this problem is the *Intrimik* (Fig. 11-4).

The Intrimik consists of a head with three contact points spaced 120° apart; this head is attached to a micrometer-type body. The contact points are forced out to contact the inside of the hole by means of a tapered or conical plug attached to the micrometer spindle (Fig. 11-5). The construction of a head with three contact points permits the Intrimik to be self-centering and self-aligning. It is more accurate than other methods because it provides a direct reading, eliminating the necessity of transferring measurements to determine hole size as with telescope or small hole gages.

The range of these instruments is from .275 to 12.000 in., and the accuracy varies between .0001 and .0005 in., depending on the head used. Metric Intrimiks have a range from 6 to 300 mm, with graduations in 0.001 mm. The accuracy of the Intrimik should be checked periodically with a setting ring or master ring gage.

TRANSFER-TYPE INSTRUMENTS

In *transfer measurement* the size of an object is taken with an instrument which is not capable of giving a direct reading. The size is then determined by measuring the setting of the instrument with a direct-reading instrument or gage of a known size.

Small Hole Gages

Small hole gages are available in sets of four, covering a range from .125 to .500 (3 to 13 mm). They are manufactured in two types (Fig. 11-6a and b on p. 88).

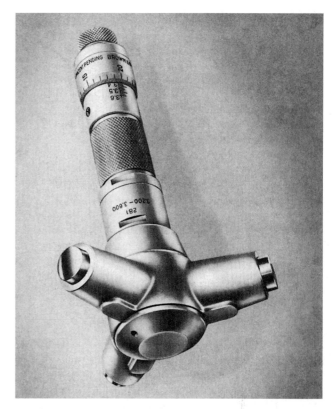

Figure 11-4 The Intrimik, which has three contact points, measures holes accurately. *(Brown & Sharpe)*

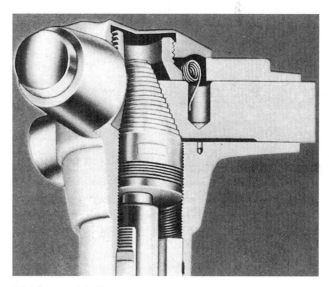

Figure 11-5 Construction of the Intrimik head. *(Brown & Sharpe Company)*

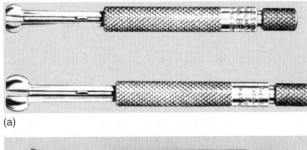

(a)

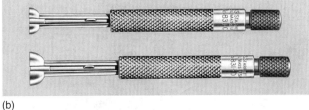

(b)

■ Figure 11-6 (a) Small hole gages with a hardened-ball end; (b) small hole gages with a flat-bottom end. *(The L. S. Starrett Co.)*

The small hole gages shown in Fig. 11-6a have a small, round end, or ball, and are used for measuring holes, slots, grooves, and recesses that are too small for inside calipers or telescope gages. Those shown in Fig. 11-6b have a flat bottom and are used for similar purposes. The flat bottom permits the measurement of shallow slots, recesses, and holes impossible to gage with the rounded type.

Both types are of similar construction and are adjusted to size by turning the knurled knob on the top. This draws up a tapered plunger, causing the two halves of the ball to open up and contact the hole.

To Use a Small Hole Gage

Small hole gages require extreme care in setting, since it is easy to get an incorrect setting when checking the diameter of a hole.

Follow this procedure:

1. Measure the hole to be checked with a rule.
2. Select the proper small hole gage.
3. Clean the hole and the gage.
4. Adjust the gage until it is slightly smaller than the hole and insert it into the hole.
5. Adjust the gage until it can be felt just touching the sides of the hole or slot.
6. Swing the handle back and forth, and adjust the knurled end until the proper "feel" is obtained across the widest dimension of the ball.
7. Remove the gage and check the size with an outside micrometer.

Note: It is important to obtain the same "feel" when transferring the measurement and when adjusting the gage to the hole.

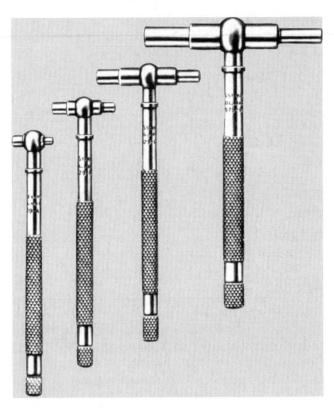

■ Figure 11-7 A set of telescope gages. *(The L. S. Starrett Co.)*

Telescope Gages

Telescope gages (Fig. 11-7) are used to obtain the size of holes, slots, and recesses from .3125 to 6.000 in. (8 to 152 mm). They are T-shaped instruments, each consisting of a pair of telescoping tubes or plungers connected to a handle. The plungers are spring-loaded to force them apart. The knurled knob on the end of the handle locks the plungers into position when it is turned in a clockwise direction.

Note: In some sets, only one plunger moves.

To Measure Using a Telescope Gage

1. Measure the hole size and select the proper gage.
2. Clean the gage and the hole.
3. Depress the plungers until they are slightly smaller than the hole diameter and lightly tighten the knurled knob.
4. Insert it into the hole and, with the handle tilted upward slightly, loosen the knurled knob to release the plungers.
5. *Lightly* snug up the knurled knob.
6. Hold the bottom leg of the telescope gage in position with one hand.
7. Move the handle downward through the center while slightly moving the top leg from side to side (Fig. 11-8).

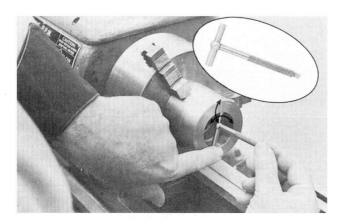

Figure 11-8 Setting a telescope gage to the hole diameter. *(Kelmar Associates)*

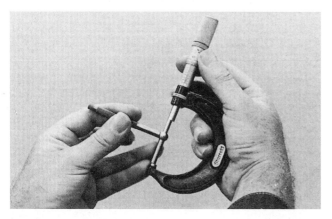

Figure 11-9 Measuring the telescope gage setting with a micrometer. *(Kelmar Associates)*

8. Tighten the knurled knob to lock the plungers in position.

9. Recheck the "feel" on the gage by testing it in the hole again.

10. Check the gage size with outside micrometers, maintaining the same "feel" as in the hole (Fig. 11-9).

Dial Bore Gages

A quick and accurate method of checking hole diameters and bores for size, out-of-round, taper, bell-mouth, hourglass, or barrel shapes is by means of the *dial bore gage* (Fig. 11-10).

Gaging is accomplished by three spring-loaded centralizing plungers in the head, one of which actuates the dial indicator, graduated in ten-thousandths of an inch, or in 0.01-mm graduations for metric tools.

These instruments are available in six sizes to cover a range from 3 to 12 in. or from 75 to 300 mm. Each instrument is supplied with extensions to increase its range. The dial bore gage must be set to size with a master gage; the hole size is then compared to the gage setting. Should the hole size vary, it is not necessary to adjust the gage as long as the size remains within the range of the gage.

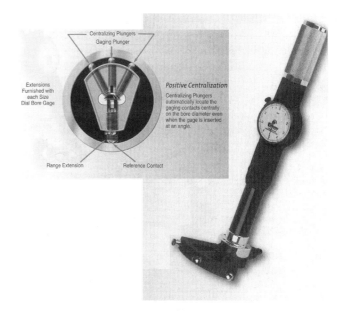

Figure 11-10 Dial bore gages provide a quick and accurate method of measuring hole diameters. *(Brown & Sharpe)*

▸▸ Depth Measurement

Although rules and various attachments can be used for measuring depth, the depth micrometer and the depth vernier are most commonly used where accuracy is required.

MICROMETER DEPTH GAGE

Micrometer depth gages are used for measuring the depth of blind holes, slots, recesses, and projections. Each gage consists of a flat base attached to a micrometer sleeve. An extension rod of the required length fits through the sleeve and protrudes through the base (Fig. 11-11). This rod is held in position by a threaded cap on the top of the thimble.

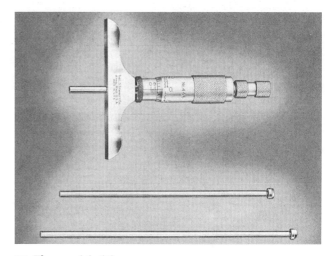

Figure 11-11 A micrometer depth gage and extension rods. *(The L.S. Starrett Co.)*

Inside-, Depth-, and Height-Measuring Instruments

Micrometer extension rods are available in various lengths, providing a range of up to 9.000 in., or 225 mm for metric tools. The micrometer screw has a range of .500 or 1.000 in., or up to 25 mm for metric tools. Depth micrometers are available with either round or flat rods, which are *not interchangeable* with other depth micrometers. The accuracy of these micrometers is controlled by a nut on the end of each extension rod, which can be adjusted if necessary.

To Measure with a Micrometer Depth Gage

1. Remove burrs from the edge of the hole and the face of the workpiece.
2. Clean the work surface and the base of the micrometer.
3. Hold the micrometer base firmly against the surface of the work (Fig. 11-12).
4. Rotate the thimble lightly with the tip of one finger in a clockwise direction until the bottom of the extension rod touches the bottom of the hole or recess.
5. Recheck the micrometer setting a few times to make sure that not too much pressure was applied in the setting.
6. Carefully note the reading.

Note: The numbers on the thimble and the sleeve are the reverse of those on a standard micrometer (Fig. 11-13).

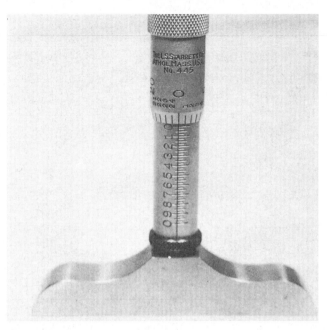

■ **Figure 11-13** Graduations on a depth micrometer are reversed from those on an outside micrometer. *(Kelmar Associates)*

■ **Figure 11-12** Measuring the depth of a shoulder. *(The L. S. Starrett Co.)*

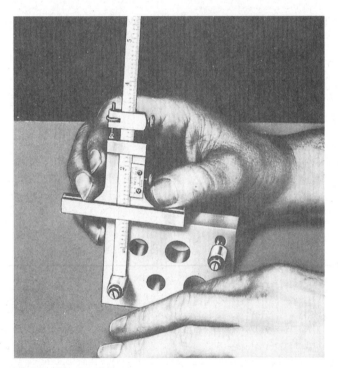

■ **Figure 11-14** Checking the position of toolmaker's buttons using a vernier depth gage. *(The L. S. Starrett Co.)*

VERNIER DEPTH GAGE

The depths of holes, slots, and recesses may also be measured by a vernier depth gage. This instrument is read in the same manner as a standard vernier caliper. Fig. 11-14 illustrates how the toolmaker's button may be set up with this instrument.

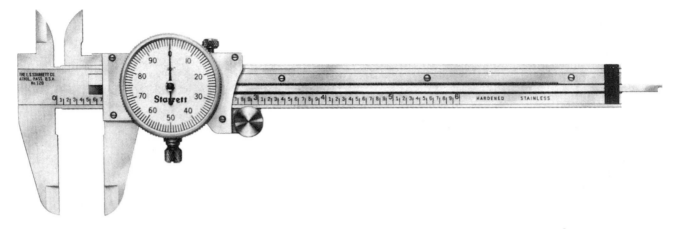

■ Figure 11-15 A dial caliper with a depth gage blade. *(The L. S. Starrett Co.)*

Depth measurements may also be made with certain types of vernier or dial calipers that are provided with a thin sliding blade or depth gage attached to the movable jaw (Fig. 11-15). The blade protrudes from the end of the bar opposite the sliding jaw. The caliper is placed vertically over the depth to be measured, and the end of the bar is held against the shoulder while the blade is inserted into the hole to be measured. Depth readings are identical to standard vernier readings.

▸▸ Height Measurement

Accurate height measurement is very important in layout and inspection work. With the proper attachments, the vernier height gage is a very useful and versatile tool for these purposes. Where extreme accuracy is required, gage blocks or a precision height gage may be used.

VERNIER HEIGHT GAGE

The vernier height gage is a precision instrument used in toolrooms and inspection departments on layout and jig and fixture work to measure and mark off distances accurately. These instruments are available in a variety of sizes—from 12 to 72 in. or from 300 to 1000 mm—and can be accurately set at any height to within .001 in. or 0.02 mm, respectively (Fig. 11-16a). Basically, a vernier height gage is a vernier caliper with a hardened, ground, and lapped base instead of a fixed jaw and is always used with a surface plate or an accurate flat surface. The sliding jaw assembly can be raised or lowered to any position along the beam. Fine adjustments are made by means of an adjusting nut. The vernier height gage is read in the same manner as a vernier caliper.

The vernier height gage is very well suited to accurate layout work and may be used for this purpose if a scriber is mounted on the movable jaw (Fig. 11-16b on p. 92). The

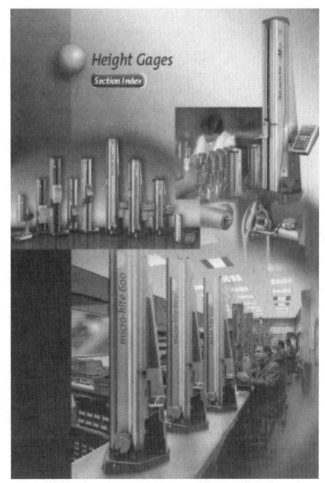

(a)

■ Figure 11-16a Vernier height gages are used for precise layout, inspection, and tool setups. *(Brown & Sharpe)*

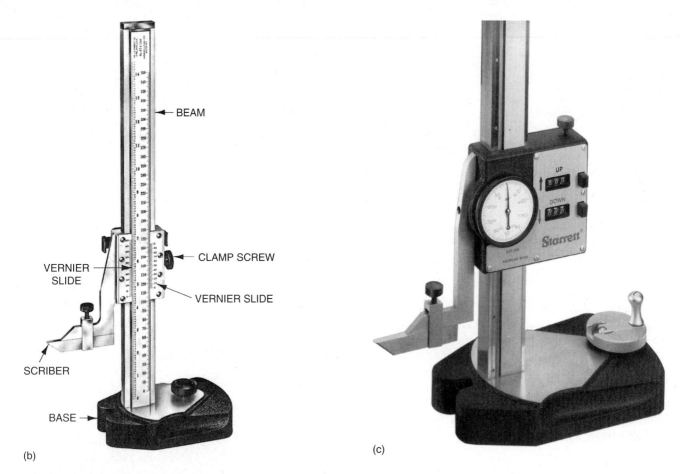

BEAM

CLAMP SCREW

VERNIER
SLIDE

VERNIER SLIDE

SCRIBER

BASE

(b)

(c)

■ Figure 11-16b, c (b) A vernier height gage is used for layout and inspection work; (c) the digital height gage is easy to read and can be set accurately to any dimension. *(The L. S. Starrett Co.)*

scriber height may be set either by means of the vernier scale or by setting the scriber to the top of a gage block buildup of the desired height.

The *digital height gage* (Fig. 11-16c) has an easy-to-read display that can be very quickly set to any dimension. The readout display is in .0001 in. (0.002 mm), and it has a zero function that allows the zero to be set at any position on the job or workpiece. This type of height gage is becoming very popular because it eliminates or reduces the errors common to height gages having vernier scales.

The *offset scriber* (Fig. 11-17) is a vernier height gage attachment that permits setting heights from the face of the surface plate. When using this attachment, it is not necessary to consider the height of the base or the width of the scriber and clamp.

A *depth gage attachment* may be fastened to the movable jaw, permitting the measurement of height differences that may be difficult to measure by other methods.

Another important use for the vernier height gage is in inspection work. A dial indicator may be fastened to the movable jaw of the height gage (Fig. 11-18), and distances between holes or surfaces can be checked to within an accuracy of .001 in. (0.02 mm) on the vernier scale. If

■ Figure 11-17 Offset scribers are used with a vernier height gage for accurate layout work. *(The L. S. Starrett Co.)*

Figure 11-18 Using a height gage and dial indicator to check a height. *(Kelmar Associates)*

greater accuracy (.0001 in. or less) is required, the indicator may be used in conjunction with gage blocks.

In Fig. 11-18, the height gage is being used to check the location of reamed holes in relation to the edges of the plate and to each other.

To Measure with a Vernier Height Gage and Dial Indicator

1. Thoroughly clean the surface plate, height gage base, and work surface.

2. Place a finished edge of the work on the surface plate and clamp it against an angle plate if necessary.

3. Insert a snug-fitting plug into the hole to be checked, with about .500 in. (13 mm) projecting beyond the work.

4. Mount the dial indicator on the movable jaw of the height gage.

5. Adjust the movable jaw until the indicator almost touches the surface plate.

6. Lock the upper slide of the height gage and use the adjusting nut to move the indicator until the dial registers about one-quarter turn.

7. Set the indicator dial to zero.

8. Record the reading of the vernier height gage.

9. Adjust the vernier height gage until the indicator registers zero on the top of the plug. Record this vernier height gage reading.

10. From this reading, subtract the initial reading plus half the diameter of the plug. This will indicate the distance from the surface plate to the center of the hole.

11. Check other hole heights using the same procedure.

To Measure Heights Using Gage Blocks

When hole locations must be accurate to .0005 in. (0.010 mm) or less, a gage block buildup is made for the proper dimension from the surface plate to the top of a plug fitted in the hole.

Follow this procedure:

1. Prepare the required gage block buildup (the center of the hole height plus one-half the hole diameter). (See Unit 12.)

2. Mount a suitable dial indicator on a surface or vernier height gage.

3. Set the dial indicator to register zero on the top of the gage blocks.

4. Move the indicator over the top of the plug. The difference between the gage block buildup and the top of the plug will register on the dial indicator (Fig. 11-18).

PRECISION HEIGHT GAGE

The precision height gage (Fig. 11-19a on p. 94) provides a quick and accurate method of setting any height within the range of the instrument, eliminating the need for calculating and assembling specified gage blocks for comparative measurements. A surface plate is used as the reference surface.

The precision height gage is made from a hardened and ground round steel bar, with ground and lapped measuring steps or disks spaced exactly at 1.000-in. (25-mm) intervals. The measuring bar or column is raised or lowered by turning the large micrometer thimble, which is graduated by steps of .0001 in., or 0.002 mm if the instrument is metric. The column may be raised or lowered a full inch or 25 mm, permitting any reading from zero to the range of the instrument in increments of .0001 in. or 0.002 mm, respectively. Some models have a vernier scale below the micrometer thimble; readings in increments of .000 010 in. or 0.000 25 mm are then possible. It is important that the accuracy of precision height gages be checked periodically with a master set of gage blocks.

The digital height gage (Fig. 11-19b on p. 94) can be equipped with a contact probe for the precision inspection of finished parts. The readout display can provide measurements accurate to .0001 in. or at the touch of a button to 0.002 mm. It can also be connected to statistical process control (SPC) equipment for analysis, data collection, and hard-copy documentation.

Height gages are available in models of 6, 12, 24, and 36 in., and their range may be increased by the use of riser blocks under the base. Metric gages range from 300 to 600 mm.

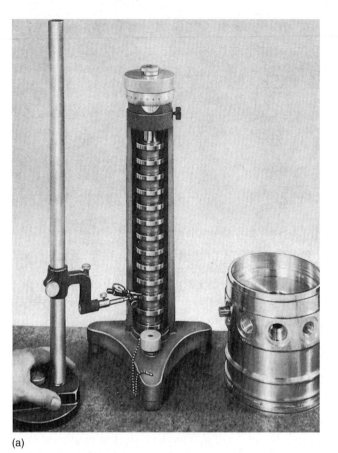

(a)

(b)

■ **Figure 11-19** (a) Checking hole locations with a precision height gage and a height transfer gage *(ExCello Corporation);* (b) the digital height gage with an indicator attachment for inspection purposes. *(The L. S. Starrett Co.)*

To Use a Precision Height Gage

1. Clean the surface plate and the feet of the height gage.

2. Clean the bottom of the work to be checked and place it on the surface plate, using parallels and an angle plate if required.

3. Insert plugs into the holes to be checked.

4. Mount a dial indicator on the movable jaw of a vernier height gage.

5. Adjust the height gage until the dial indicator registers approximately .015 in. (0.4 mm) across the top of the plug.

6. Turn the dial of the indicator to zero.

7. Move the dial indicator over the nearest disk of the precision height gage and raise the column by turning the micrometer until the dial indicator reads zero.

8. Check the micrometer reading. This reading will indicate the distance from the surface plate to the top of the plug.

9. Subtract half the diameter of the plug from this reading.

Note: If the work is set on parallels, this height must be subtracted from the precision height gage reading.

unit 11 review questions

Inside Micrometer Calipers

1. On what type of inside micrometer calipers are the readings reversed to an outside micrometer?

2. What precautions must be observed in taking a measurement with an inside micrometer caliper?

Inside Micrometers

3. What construction feature compensates for a lock nut on inside micrometers?

4. What precautions must be taken when assembling the inside micrometer and extension rod?

5. What is the correct "feel" with an inside micrometer?

Small Hole Gages

6. Name two types of small hole gages and state the purpose of each.

7. What precaution must be observed when using a small hole gage to obtain a dimension?

Telescope Gages

8. List the steps required to measure a hole with a telescope gage.

Dial Bore Gages

9. What hole defects may be conveniently measured with a dial bore gage?

Micrometer Depth Gages

10. How is the accuracy of a micrometer depth gage adjusted?

11. How must the workpiece be prepared prior to measuring the depth of a hole or slot with a micrometer depth gage?

12. Explain the procedure for measuring a depth with a depth micrometer.

13. How does the reading of a depth micrometer differ from that of a standard outside micrometer?

Vernier Height Gages

14. State the two main applications for the vernier height gage.

15. What accessories are required for a vernier height gage to check *accurately* the height of a workpiece?

Precision Height Gages

16. What are the advantages of using a precision height gage instead of a gage block buildup?

17. What dimension(s) must be subtracted from the reading so that the correct reading for the height of a hole being checked will be obtained?

UNIT 12

Gage Blocks

OBJECTIVES

After completing this unit, you will be able to:

1 Explain the use and application of gage blocks

2 Calculate inch and metric gage block buildups

3 Prepare and wring gage blocks together

Interchangeable manufacture requires an accurate standard of measurement to function efficiently. Gage blocks, the acceptable standard of accuracy, have provided industry with a means of maintaining sizes to specific standards or tolerances (Fig. 12-1a). This feature has led to high rates of production and has made interchangeable manufacture possible.

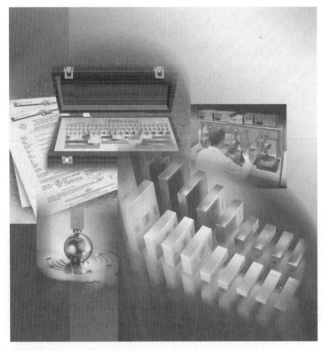

■ **Figure 12-1a** Gage blocks are measurement standards used by industry. *(Brown & Sharpe)*

▸▸ Gage Block Manufacture

Gage blocks are rectangular blocks of hardened and ground alloy steel that have been stabilized through alternate cycles of extreme heat and cold until the crystalline structure of the metal is left without strain. The two mea-suring surfaces are lapped and polished to an optically flat surface and to a specific size accurate within a range of 2 to 8 millionths of an inch (50 to 200 millionths of a millimeter). The size of each block is stamped on one of its surfaces. Chrome-plated gage blocks are also available and, when long wear is desirable, carbide blocks are used. Great care is exercised in their manufacture; the final calibration is made under ideal conditions where the temperature is maintained at 68°F (20°C). Therefore, any gage block is accurate to size *only* when measured at the standard temperature of 68°F (20°C).

Ceramic gage blocks made of zirconia, one of the most durable materials on earth, have recently been introduced. They have 10 times the abrasion resistance of common steel blocks, need no special maintenance, and have a thermal expansion coefficient close to steel. Some of the main features of zirconia ceramic gage blocks are:

> Corrosion resistant

> No detrimental effects as a result of handling

> Superior abrasion resistance

> Thermal expansion coefficient close to steel

> Resistant to impact

> Free from burrs

> Wring together tightly

USES

Industry has found gage blocks to be invaluable tools. Because of their extreme accuracy, they are used for the following purposes:

1. To check the dimensional accuracy of fixed gages to determine the extent of wear, growth, or shrinkage

2. To calibrate adjustable gages, such as micrometers and vernier calipers, imparting accuracy to these instruments

3. To set comparators, dial indicators, and height gages to exact dimensions

4. To set sine bars and sine plates when extreme accuracy is required in angular setups

5. For precision layout with the use of attachments

6. To make machine tool setups

7. To measure and inspect the accuracy of finished parts

▶▶ Gage Block Sets

INCH STANDARD GAGE BLOCKS

Gage blocks are manufactured in sets that vary from a few blocks to as many as 115. The most commonly used is the 83-piece set (Fig. 12-1b). From this set, it is possible to make over 120,000 different measurements, ranging from one hundred-thousandths of an inch to over 25 inches. The blocks that comprise an 83-piece set are listed in Table 12.1 on p. 98.

Two *wear* blocks are supplied with an 83-piece set. Some manufacturers supply .050-in. wear blocks; others provide .100-in. wear blocks. They should be used at each end of a combination, especially if the blocks will be in contact with hard surfaces or abrasives. Thus, the wear which occurs during use will be on the two wear blocks only, rather than on many blocks, prolonging the useful life and accuracy of the set. During use, it is considered good practice always to expose the same face of the wear block to the work surface. A good habit to acquire is always to have the words "Wear Block" appear on the outside of the combination. In this way, all the wear will be on one surface and the wringing quality of the other surface will be preserved.

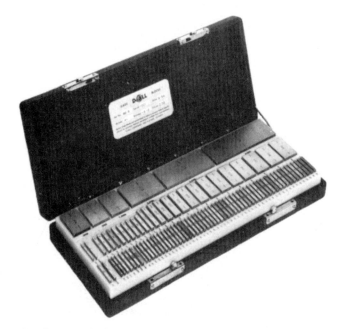

■ **Figure 12-1b** An 83-piece set of gage blocks. *(DoAll Company)*

METRIC GAGE BLOCKS

Metric gage blocks are supplied by most manufacturers in sets of 47, 88, and 113 blocks. The most common is the 88-piece set (Table 12.2 on p. 99). Each of these sets contains a pair of 2-mm wear blocks. These blocks are used at each end of the buildup to prolong the accuracy of the other blocks in the set.

ACCURACY

Gage blocks in inch and metric standards are manufactured to three common degrees of accuracy, depending on the purpose for which they are used.

1. The *Class AA* set, commonly called a *laboratory* or *master* set, is accurate to ±.000002 in. in the inch standard; the metric set, to ±0.00005 mm. These gage blocks are used in temperature-controlled laboratories as references to compare or check the accuracy of working gages.

2. The *Class A* set is used for inspection purposes and is accurate to ±.000004 in. in the inch standard; the metric set, to +0.00015 mm and −0.00005 mm.

3. The *Class B* set, commonly called the *working* set, is accurate to ±.000008 in. in the inch standard; the metric set, to +0.00025 mm and −0.00015 mm. These blocks are used in the shop for machine tool setups, layout work, and measurement.

table 12.1

table 12.1   Sizes in an 83-piece set of inch standard gage blocks

First: .0001-in. Series—9 Blocks								
.1001	.1002	.1003	.1004	.1005	.1006	.1007	.1008	.1009

Second: .001-in. Series—49 Blocks								
.101	.102	.103	.104	.105	.106	.107	.108	.109
.110	.111	.112	.113	.114	.115	.116	.117	.118
.119	.120	.121	.122	.123	.124	.125	.126	.127
.128	.129	.130	.131	.132	.133	.134	.135	.136
.137	.138	.139	.140	.141	.142	.143	.144	.145
.146	.147	.148	.149					

Third: .050-in. Series—19 Blocks									
.050	.100	.150	.200	.250	.300	.350	.400	.450	.500
.550	.600	.650	.700	.750	.800	.850	.900	.950	

Fourth: 1.000-in. Series—4 Blocks			
1.000	2.000	3.000	4.000

Two .050-in. wear blocks

THE EFFECT OF TEMPERATURE

While the effect of temperature on ordinary measuring instruments is negligible, changes in temperature are important when precision gage blocks are handled. Gage blocks have been calibrated at 68°F (20°C), but human body temperature is about 98°F (37°C). A 1°F (0.5°C) rise in temperature will cause a 4-in. (100-mm) stack of gage blocks to expand approximately .000025 in. (0.0006 mm); therefore, these blocks should be handled as little as possible. The following suggestions are offered to eliminate as much temperature-change error as possible.

1. Handle gage blocks only when they must be moved.
2. Hold them by hand for as little time as possible.
3. Hold them between the tips of the fingers so that the area of contact is small, or use insulated tweezers.
4. Have the work and gage blocks at the same temperature. If a temperature-controlled room is not available, both the work and gage blocks may be placed in kerosene until both are at the same temperature.
5. Where extreme accuracy is necessary, use insulating gloves and tweezers to prevent temperature change during handling.

GAGE BLOCK BUILDUPS

Gage blocks are manufactured to great accuracy; they adhere to each other so well when wrung together properly that they can withstand a 200-pound (lb) [890-newton (N)] pull. Many theories have been advanced to explain this adhesion. Scientists have felt that it may be atmospheric pressure, molecular attraction, the extremely flat surfaces of the blocks, or a minute film of oil that gives the blocks this quality. Possibly a combination of any of these factors is responsible.

When the blocks required to make up a dimension are being calculated, the following procedure should be followed to save time, reduce the chance of error, and use as few blocks as possible. For example, if a measurement of 1.6428 in. is required, follow this procedure:

table 12-2   Sizes in an 88-piece set of metric gage blocks

0.001-mm Series—9 Blocks

1.001	1.002	1.003	1.004	1.005	1.006	1.007	1.008	1.009

0.01-mm Series—49 Blocks

1.01	1.02	1.03	1.04	1.05	1.06	1.07	1.08	1.09
1.10	1.11	1.12	1.13	1.14	1.15	1.16	1.17	1.18
1.19	1.20	1.21	1.22	1.23	1.24	1.25	1.26	1.27
1.28	1.29	1.30	1.31	1.32	1.33	1.34	1.35	1.36
1.37	1.38	1.39	1.40	1.41	1.42	1.43	1.44	1.45
1.46	1.47	1.48	1.49					

0.5-mm Series—1 Block

0.5

0.5-mm Series—18 Blocks

1	1.5	2	2.5	3	3.5	4	4.5	5
5.5	6	6.5	7	7.5	8	8.5	9	9.5

10-mm Series—9 Blocks

10	20	30	40	50	60	70	80	90

Two 2-mm wear blocks

step   procedure

step	procedure		Procedure Column	Check Column
1	Write the dimension required on the paper	1.6428		
	1. Dimension required		3.8716 in.	
2	Deduct the size of two wear blocks: 2 × .050 in.	.1000		
	Remainder	1.5428		
	2. Two wear blocks (2 × .050 in.)		.100 3.7716	.100
3	Use a block that will eliminate the right-hand digit	.1008		
	Remainder	1.4420		
	3. Use .1006 in.		.1006 3.6710	.1006
4	Use a block that will eliminate the right-hand digit and at the same time bring the digit to the left of it to a zero (0) or a five (5)	.142		
	Remainder	1.300		
	4. Use .121 in.		.121 3.550	.121
5	Continue to eliminate the digits from the right to the left until the dimension required is attained	.300		
	Remainder	1.000		
	5. Use .550 in.		.550 3.000	.550
6	Use a 1.000-in. block	1.000		
	Remainder	0.000		
	6. Use 3.000 in.		3.000 .000 in.	3.000 3.8716 in.

To eliminate the possibility of subtraction error while making a buildup, use two columns for this calculation. As the following example illustrates, the gage blocks are subtracted from the original dimension in the left-hand column, and the right-hand column is used as a check column. For example, to build up a dimension of 3.8716 in., proceed as follows:

When using metric blocks for buildup of 57.15 mm, proceed as follows:

step	procedure	
1	Write the dimension required on the paper	57.15
2	Deduct the size of 2 wear blocks (2 × 2 mm)	4.00
		Remainder 53.15
3	Use a block that will eliminate the right-hand digit	1.050
		Remainder 52.100
4	Use a block that will eliminate the right-hand digit	1.10
		Remainder 51.00
5	Use a 1-mm block	1.00
		Remainder 50.00
6	Use a 50-mm block	50.00
		Remainder 0.00

For a metric buildup of 27.781 mm, proceed as follows:

	Procedure Column	Check Column
1. Dimension required	27.781 mm	
2. Two wear blocks (2 × 2 mm)	4	4
	23.781	
3. Use 1.001 mm	1.001	1.001
	22.780	
4. Use 1.08 mm	1.08	1.08
	21.7	
5. Use 1.7 mm	1.7	1.7
	20	
6. Use 20 mm	20	20
	0.000 mm	27.781 mm

■ Figure 12-2 Procedure for wringing gage blocks. *(Kelmar Associates)*

To Wring Blocks Together

When wringing blocks together, take care not to damage them. The correct sequence of movement to wring blocks together, illustrated in Fig. 12-2, is as follows:

1. Clean the blocks with a clean, soft cloth.
2. Wipe each of the contacting surfaces on the clean palm of the hand or on the wrist. This procedure removes any dust particles left by the cloth and applies a light film of oil.
3. Place the end of one block over the end of another block, as shown in Fig. 12-2.
4. While applying pressure on the two blocks, slide one block over the other.

Note: If the blocks do not adhere to each other, it is generally because the blocks have not been thoroughly cleaned.

Care of Gage Blocks

1. Gage blocks should always be protected from dust and dirt by being kept in a closed case when not in use.
2. Gages should not be handled unnecessarily, since they absorb heat from the hand. Should this occur, the gage blocks must be permitted to return to room temperature before use.
3. Fingering of lapped surfaces should be avoided to prevent tarnishing and rusting.
4. Care should be taken not to drop gage blocks or scratch their lapped surfaces.
5. Immediately after use, each block should be cleaned, oiled, and replaced in the storage case.

6. Before gage blocks are wrung together, their faces must be free from oil and dust.

7. Gage blocks should never be left wrung together for any length of time. The slight moisture between the blocks can cause rusting, which will permanently damage the blocks.

unit 12 review questions

1. How are gage blocks stabilized, and why is this necessary?

2. State five general uses for gage blocks.

3. For what purpose are wear blocks used?

4. How should wear blocks always be assembled into a buildup?

5. State the difference between a master set and a working set of gage blocks.

6. What precautions are necessary when handling gage blocks to minimize the effect of heat on the blocks?

7. List five precautions necessary for the proper care of gage blocks.

8. Calculate the gage blocks required for the following buildups (use the check column for accuracy):

   a. 2.1743 in.   c. 7.8923 in.   e. 74.213 mm
   b. 6.2937 in.   d. 32.079 mm   f. 89.694 mm

UNIT 13

Angular Measurement

OBJECTIVES

After completing this unit, you will be able to:

1 Make angular measurements to an accuracy of 5′ (minutes) of a degree using a universal bevel protractor

2 Make angular measurements to less than 5′ of a degree using a sine bar, gage blocks, and a dial indicator

Precise angular measurement and setups constitute an important phase of machine shop work. The most commonly used tools for accurately laying out and measuring angles are the universal bevel protractor, sine bar, and sine plate.

▶▶ Universal Bevel Protractor

The *universal bevel protractor* (Fig. 13-1) is a precision instrument capable of measuring angles to within 5′ (0.083°). It consists of a *base* to which a *vernier scale* is attached. A *protractor dial,* graduated in degrees with every tenth degree numbered, is mounted on the circular section of the base. A *sliding blade* is fitted into this dial; it may be extended in either direction and set at any angle to the base. The blade and the dial are rotated as a unit. Fine adjustments are obtained with a small knurled-headed pinion that, when turned, engages with a gear attached to the blade mount. The protractor dial may be locked in any position by means of the *dial clamp nut.*

The vernier protractor shown in Fig. 13-2 is used to measure an obtuse angle, or an angle greater than 90° but less than 180°. An *acute-angle attachment* is fastened to the vernier protractor to measure angles of less than 90° (Fig. 13-3).

The vernier protractor dial, or main scale, is divided into two arcs of 180°. Each arc is divided into two quad-

rants of 90° and has graduations from 0° to 90° to the left and right of the zero (0) line, with every tenth degree numbered.

The vernier scale is divided into 12 spaces on each side of the 0 line, which occupy the same space as 23° on the protractor dial. By simple calculation, it is easy to prove that one vernier space is 5′, or less than two graduations on the main scale. If zero on the vernier scale coincides with a line on the main scale, the reading will be in degrees only. However, if any other line on the vernier scale coincides with a line on the main scale, the number of vernier graduations beyond the zero should be multiplied by 5 and added to the number of full degrees indicated on the protractor dial.

TO READ A VERNIER PROTRACTOR

1. Note the number of whole degrees between the zero on the main scale and the zero on the vernier scale.

2. Proceeding in the *same direction* beyond the zero on the vernier scale, note which vernier line coincides with a main scale line.

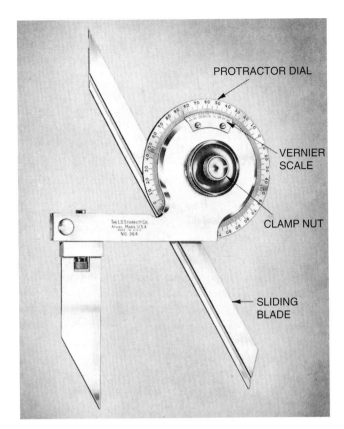

Figure 13-1 The universal bevel protractor can measure angles accurately. *(The L.S. Starrett Co.)*

Labels on figure: PROTRACTOR DIAL, VERNIER SCALE, CLAMP NUT, SLIDING BLADE

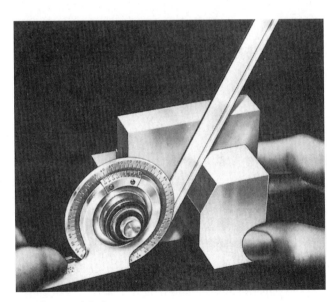

Figure 13-2 Measuring an obtuse angle using a universal bevel protractor. *(The L.S. Starrett Co.)*

3. Multiply this number by 5 and add it to the number of degrees on the protractor dial.

In Fig. 13-4, the angular reading is calculated as follows. The number of degrees indicated on the main scale is 50 plus. The fourth line on the vernier scale *to the left*

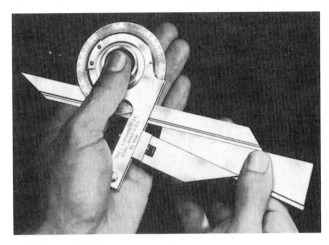

Figure 13-3 Measuring an acute angle. *(The L.S. Starrett Company)*

Figure 13-4 A vernier protractor reading of 50°20′. *(The L.S. Starrett Co.)*

of the zero coincides with a line on the main scale; therefore, the reading is

Number of full degrees	= 50°
Value of vernier scale (4 × 5′)	= 20′
Reading	= 50°20′

Note: A double-check of the reading would locate the vernier scale line on the other side of zero, which coincides with a protractor scale line. This line should always equal the complement of 60′. In Fig. 13-4, the 40′ line to the right of the zero coincides with a line on the protractor scale. This reading, when added to the 20′ on the left of the scale, is equal to 60′, or 1°.

▸▸ Sine Bar

A *sine bar* (Fig. 13-5 on p. 104) is used when the accuracy of an angle must be checked to less than 5′ or work must be located to a given angle within close limits. The sine bar consists of a steel bar with two cylinders of equal diameter fastened near the ends. The centers of these cylinders are on

■ Figure 13-5 A 5-in. sine bar with gage block buildup is used to set up work to an angle. *(Kelmar Associates)*

(a)

(b)

■ Figure 13-6 Setting up for an angle greater than 60°: (a) set the sine bar to the complement of the angle; (b) turn the angle plate 90° on its side. *(Kelmar Associates)*

a line exactly at 90° to the edge of the bar. The distance between the centers of these lapped cylinders is usually 5 or 10 in. on inch sine bars, 125 or 250 mm on metric sine bars. Sine bars are generally made of stabilized tool steel, hardened, ground, and lapped to extreme accuracy. They are used on surface plates, and any desired angle can be set by raising one end of the bar to a predetermined height with gage blocks.

Sine bars are generally made 5 in. or multiples of 5 in. in length; that is, the lapped cylinders are 5 in. ±.0002 or 10 in. ±.00025 between centers. The face of the sine bar is accurate to within .00005 in. in 5 inches. In theory, the sine bar is the hypotenuse of a right-angle triangle. The gage block buildup forms the side opposite, and the face of the surface plate forms the side adjacent in the triangle.

Using trigonometry, it is possible to calculate the side opposite, or gage block buildup, for any angle between 0° and 90° as follows:

$$\text{Sine of the angle} = \frac{\text{side opposite}}{\text{hypotenuse}}$$

$$= \frac{\text{gage block buildup}}{\text{length of sine bar}}$$

When using a 5-in. sine bar, this would become:

$$\text{Sine of the angle} = \frac{\text{buildup}}{5}$$

Therefore, by transposition, the gage block buildup for any required angle with a 5-in. bar is:

$$\text{Buildup} = 5 \times \text{sine of the required angle}$$

EXAMPLE

Calculate the gage block buildup required to set a 5-in. sine bar to an angle of 30°.

$$\begin{aligned} \text{Buildup} &= 5 \sin 30° \\ &= 5(.5000) \\ &= 2.5000 \text{ in.} \end{aligned}$$

Note: This formula is applied only to angles up to 60°.

When an angle greater than 60° is to be checked, it is better to set up the work using the complement of the angle (Fig. 13-6a). The angle plate is then turned 90° to produce the correct angle (Fig. 13-6b). The reason is that when the sine bar is in a near-horizontal position, a small change in the height of the buildup will produce a smaller change in the angle than when the sine bar is in the near-vertical position. This change in gage block height may be shown by calculating the buildups required for both 75° and the complementary angle of 15°.

Buildup required for:

$$75° \, 1' = 5 \sin 75°1' \, (.9660)$$
$$= 4.8300 \text{ in.}$$
$$75° = 5 \sin 75° \, (.96592)$$
$$= 4.82960 \text{ in.}$$

Difference in the buildup for 1′

$$= .00040 \text{ in.}$$

Buildup required for:

$$15° \, 1' = 5 \sin 15°1' \, (.25910)$$
$$= 1.29550 \text{ in.}$$
$$15° = 5 \sin 15° \, (.25882)$$
$$= 1.29410 \text{ in.}$$

Difference in buildup for 1′

$$= .00140 \text{ in.}$$

This example shows that exactly 3.5 times the buildup is required to produce a change of 1′ at 15° than is required for 1′ at 75°. Therefore, a small inaccuracy in setup would result in a smaller error at a smaller angle than it would at a larger one. If the complementary angles of 80° and 10° are used, this ratio increases to over 5:1.

When small angles are to be checked, it is sometimes impossible to get a buildup small enough to place under one end of the sine bar. In such situations, it will be necessary to place gage blocks under both rolls of the sine bar, having a net difference in measurement equal to the required buildup. For example, the buildup required for 2° is .1745 in. Since it is impossible to make this buildup, it is necessary to place the buildup for 1.1745 in. under one roll and a 1.000-in. block under the other roll, giving a net difference of .1745 in.

Before the sine bar is used to check a taper, it is necessary to calculate the angle of the taper so that the proper gage block buildup may be made. Fig. 13-7a and b illustrates how this is done.

In the right-angled triangle *ABD:*

$$\text{Tan } \frac{a}{2} = \frac{\frac{1}{2}}{12}$$
$$= .04166$$
$$= 2°23'10''$$
$$\therefore a = 4°46'20''$$

From this solution, the following formula for solving the included angle, when the taper per foot (*tpf*) is known, is derived.

$$\text{Tan } \frac{a}{2} = \frac{tpf}{24}$$

Note: When calculating the angle of a taper, do *not* use the formula tan *a* = *tpf*/12, since triangle *ABC* is not a right-angle triangle.

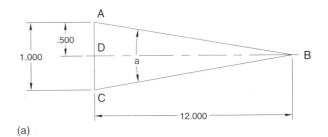

(a)

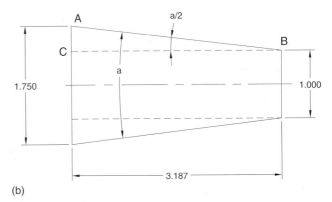

(b)

■ **Figure 13-7** The angle formed by a taper of 1 in./ft.

By transposition, if the included angle is given, the *tpf* may be calculated as follows:

$$tpf = \tan \frac{1}{2} a \times 24$$

If the *tpf* is not known, the angle may be calculated as shown in Fig. 13-7b:

$$AC = \frac{1.750 - 1.000}{2}$$
$$= \frac{.750}{2}$$
$$= .375$$
$$\text{Tan } \frac{a}{2} = \frac{.375}{3.187}$$
$$= .11766$$
$$= 6°42'22''$$
$$\therefore a = 13°24'44''$$

To check the accuracy of this taper using a 5-in. sine bar, we calculate the buildup as follows:

$$\text{Buildup} = 5 \sin 13°24'44'' \, (.23196)$$
$$= 1.1598 \text{ in.}$$

Metric tapers are expressed as a ratio of 1 mm per unit of length; for example, a taper having a ratio of 1:20 would taper 1 mm in diameter in 20 mm of length (see Unit 54).

Tapers can be checked conveniently and accurately with a *taper micrometer.* This measuring instrument measures work to sine bar accuracy while the work is still in the machine. (See "Checking a Taper" in Unit 54.)

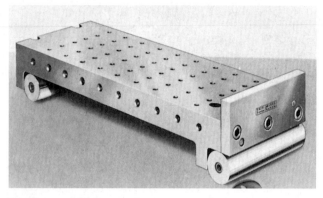

Figure 13-8 A workpiece may be clamped to a sine plate. *(The Taft-Peirce Manufacturing Company)*

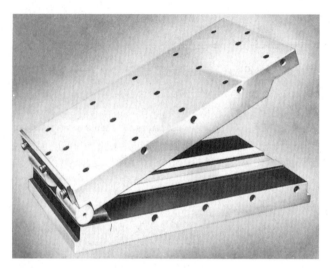

Figure 13-9 A hinged sine plate may be clamped to the machine table.

The *sine plate* (Fig. 13-8) is based on the same principle as the sine bar and is similar in construction except that it is wider. Sine bars are up to 1 in. in width, but sine plates are generally more than 2 in. wide. They have several tapped holes in the surface that permit the work to be clamped to the surface of the sine plate. An end stop on a sine plate prevents the workpiece from moving during machining.

Sine plates may be hinged to a base (Fig. 13-9) and are often called *sine tables.* Both types are supplied in 5- and 10-in. lengths and have a step or groove of .100 or .200 in. deep ground in the base to permit the buildup for small angles to be placed under the free roll.

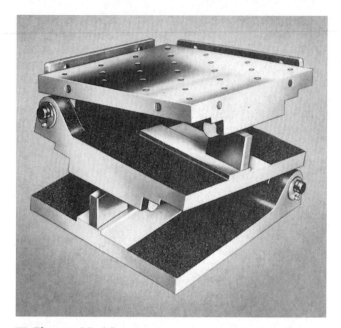

Figure 13-10 A compound sine plate allows the setting of angles in two directions.

▸▸ Compound Sine Plate or Table

The *compound sine plate* (Fig. 13-10) consists of one sine plate superimposed on another sine plate. The lower plate is hinged to a base and may be tilted to any angle from 0° to 60° by placing gage blocks under the free roll or cylinder. The upper base is hinged to the lower base so that its cylinder and hinge are at right angles to those of the lower plate. The upper plate may also be tilted to any angle up to 60°. This feature permits the setting of compound angles (angles in two directions).

Compound sine plates have a side and an end plate on the top table to facilitate setting up work square with the table edge and to prevent movement of the work during machining. A step or groove of .100 or .200 in. deep ground in the base and in the lower table permits the setup for small angles, and the gage block buildup may be placed in the groove.

Since most angles are machined in fixtures, sine plates are not used until the finishing operation, which is generally grinding. To facilitate the holding of parts, both simple and compound sine plates are available with built-in magnetic chucks.

Universal Bevel Protractor

1. Name the parts of a universal bevel protractor and state the purpose of each.

2. Describe the principle of the vernier protractor.

3. Sketch a vernier protractor reading of:

   a. 34°20′ b. 17°45′

Sine Bar

4. Describe the construction and principle of a sine bar.

5. What are the accuracies of the 5-in. and the 10-in. sine bars?

6. Calculate the gage block buildup for the following angles using a 5-in. sine bar:

   a. 7°40′   b. 25°50′   c. 40°10′

7. What procedure should be followed to check an angle of 72° using a sine bar and gage blocks? Why is this procedure recommended?

8. In calculating the angle of a taper, why is the formula

$$\text{Tan}\,\frac{1}{2}\,a = \frac{tpf}{24}$$

used, rather than

$$\text{Tan}\,a = \frac{tpf}{12}\,?$$

Illustrate by means of a suitable sketch.

Sine Plate

9. Describe a sine plate and state its purpose.

10. What is the advantage of a hinged sine plate?

11. What is the purpose of a compound sine plate?

UNIT 14

Gages

OBJECTIVES

After completing this unit, you will be able to:

1 Recognize and describe the uses of three types of plug and ring gages

2 Check the accuracy of a part with a plug or ring gage

3 Check the accuracy of a part using a snap gage

Although modern production processes have reached a high degree of precision, producing a part to an *exact* size would be far too costly. Consequently, industrial production methods, including interchangeable manufacture, permit certain varia-

tions from the exact dimensions specified while ensuring that the component part will fit into the unit at the time of assembly.

To determine the sizes of the various parts, an inspector generally uses some type of gage. Gages used in industry vary from the simpler type of fixed gage to the sophisticated electronic and laser devices used to measure extremely fine variations (Fig. 14-1a).

To ensure that the part will meet the specifications, certain basic terms are used, and these terms are usually added to the drawing of the part. These terms apply to all forms of measurement and inspection and should be understood by both the machinist and the inspector.

▶▶ Basic Terms

The following basic terms are used to express the exact size of a part and allowable variations from that size. An example is provided in Table 14.1.

Basic dimension is the exact size of a part from which all limiting variations are made.

Limits are the maximum and minimum dimensions of a part (the high and low dimensions).

Tolerance is the permissible variation of a part. Tolerance is often shown on the drawing by the basic dimension plus or minus the amount of variation allowed. If a part on a drawing was dimensioned to 3 in. ±.002 (76 mm ±0.05), the tolerance would be .004 in. (0.10 mm).

If the tolerance is in one direction only, that is, plus *or* minus, it is said to be *unilateral tolerance.* However, if

the tolerance is both plus *and* minus, it is referred to as *bilateral tolerance.*

Allowance is the intentional difference in the dimensions of mating parts. For example, it is the difference between the maximum diameter of the shaft and the minimum diameter of the mating bore.

▶▶ Fixed Gages

Fixed gages are used for inspection purposes because they provide a quick means of checking a specific dimension. These gages must be easy to use and accurately finished to the required tolerance. They are generally finished to one-tenth the tolerance they are designed to control. For example, if the tolerance of a piece being checked is to be maintained at .001 in. (0.02 mm), the gage must be finished to within .0001 in. (0.002 mm) of the required size.

▶▶ Cylindrical Plug Gages

Plain *cylindrical plug gages* (Fig. 14-1b) are used for checking the inside diameter of a straight hole and are generally of the "go" and "no-go" variety. This type of gage consists of a handle and plug on each end ground and/or lapped to a specific size. The smaller-diameter plug, or the "go" gage, checks the lower limit of the hole. The larger-diameter plug, or the "no-go" gage, checks the upper limit of the hole (Fig. 14-2). For instance, if a hole size is to be maintained at 1.000 in. ±.0005 (25.4 mm ±0.012), the "go" end of the gage would be designed to fit into a hole .9995 in. (25.38 mm) in diameter. The larger end ("no-go") would not fit into any hole smaller than 1.0005 in. (25.41 mm) in diameter.

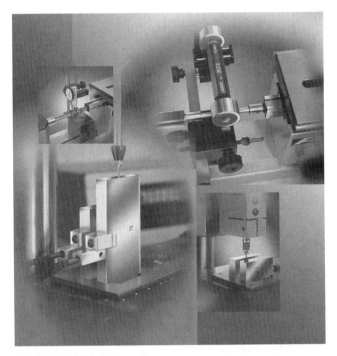

■ **Figure 14-1a** Electronic gages are widely used in industry. *(Brown & Sharpe)*

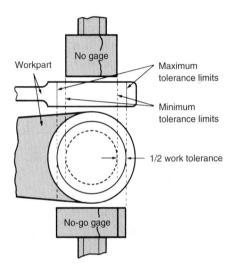

■ **Figure 14-2** The "go" end of the gage checks the minimum tolerance limit, while the "no-go" end checks the maximum tolerance. *(Sheffield Measurement Div.)*

■ **Figure 14-1b** A cylindrical plug gage is used to check hole sizes. *(The Taft-Peirce Manufacturing Company)*

table 14.1	An example of limits and tolerances		
Nominal size	3 in.		
Basic dimension		3.000	(Decimal equivalent of nominal size)
Basic dimension and amount of bilateral tolerance permitted		3.000	±.002
Limits		3.002	(Largest size permitted)
		2.998	(Smallest size permitted)
Tolerance		.004	(Difference between minimum and maximum limits)

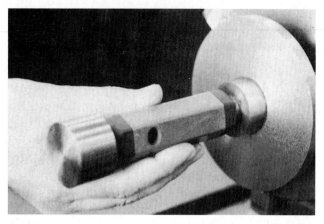

■ Figure 14-3 Checking a hole size with a plug gage. *(Kelmar Associates)*

The dimensions of these gages are usually stamped on the handle at each end adjacent to the plug gage. The "go" end is made longer than the "no-go" end for easy identification. Sometimes a groove is cut on the handle near the "no-go" end to distinguish it from the "go" end.

Due to the wear caused by the constant use of plug gages, many of them are made with carbide tips, which greatly increases gage life.

TO USE A CYLINDRICAL PLUG GAGE

1. Select a plug gage of the correct size and tolerance for the hole being checked.
2. Clean both ends of the gage and the hole in the workpiece with a clean, dry cloth.
3. Check the gage (both ends) and the workpiece for nicks and burrs.
4. Wipe both ends of the gage with an oily cloth to deposit a thin film of oil on the surfaces.
5. Start the "go" end *squarely* into the hole (Fig. 14-3). If the hole is within the limits, the gage will enter easily.

 SAFETY PRECAUTIONS Do not force or turn it.

The plug should enter the hole for its full length, and there should be no excessive play between the plug and the part.

Note: If the gage enters the hole only part way, the hole is tapered. Excessive play or looseness in one direction indicates that the hole is elliptical (out of round).

6. After the hole has been checked with the "go" end, it should be checked with the "no-go" end. This end should not begin to enter the hole. An entry of

■ Figure 14-4 Plain ring gages are used to check the diameter of round workpieces. *(The Taft-Peirce Manufacturing Company)*

more than .060 in. (1.5 mm) indicates an oversize, a bell-mouth, or a tapered hole.

▶▶ Plain Ring Gages

Plain ring gages, used to check the outside diameter of pieces, are ground and lapped internally to the desired size. The size is stamped on the side of the gage. The outside diameter is knurled, and the "no-go" end is identified by an annular groove on the knurled surface (Fig. 14-4). The precautions and procedures for using a ring gage are similar to those outlined for a plug gage and should be followed carefully.

▶▶ Taper Plug Gages

Taper plug gages (Fig. 14-5), made with standard or special tapers, are used to check the size of the hole and taper accuracy. Some of these gages have "go" and "no-go"

■ Figure 14-5 Taper ring and plug gages. *(Kelmar Associates)*

rings scribed on them. If the gage fits into the hole between these two rings, the hole is within the required tolerance. Other taper plug gages have steps ground on the large end to indicate the limits. The rings or steps measure hole-size limits only. A wobble between the plug gage and the hole is evidence of an incorrect taper.

TO CHECK AN INTERNAL TAPER USING A TAPER PLUG GAGE

1. Select the proper taper gage for the hole being checked.

2. Wipe the gage and the hole with a clean, dry cloth.

3. Check both the gage and hole for nicks and burrs.

4. Apply a *thin* coating of Prussian blue to the surface of the plug gage.

5. Insert the plug gage into the hole as far as it will go (Fig. 14-6).

6. Maintaining light end pressure on the plug gage, rotate it *counterclockwise* for approximately one-quarter turn.

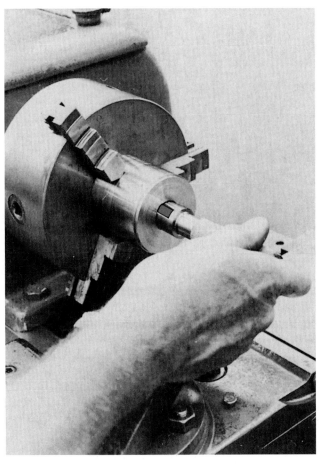

■ **Figure 14-6** Checking a tapered hole with a taper plug gage. *(Kelmar Associates)*

7. Check the diameter of the hole. A proper size is indicated when the edge of the workpiece lies between the limit steps or lines on the gage.

8. Check the taper of the hole by attempting to move the gage radially in the hole. Any error in the taper will be indicated by play at either end between the hole and the gage. Movement or play at the large end indicates excessive taper; movement at the small end indicates insufficient taper.

9. Remove the gage from the hole to see if the bluing has rubbed off evenly along the length of the gage, a result which would indicate a proper fit. A poor fit is evident if the bluing has been rubbed off more at one end than the other.

▸▸ Taper Ring Gages

Taper ring gages (Fig. 14-5) are used to check both the accuracy and the outside diameter of the taper. Ring gages often have scribed lines or a step ground on the small end to indicate the "go" and "no-go" dimensions.

For a taper ring gage, the precautions and procedures are similar to those outlined for a taper plug gage. However, when work that has not been ground or polished is checked, three equally spaced chalk lines around the circumference and extending for the full length of the tapered section may be used to indicate the accuracy of the taper (Fig. 14-7). If the work has been ground or polished, it is advisable to use three thin lines of Prussian blue.

CARE OF PLUG AND RING GAGES

Gage life is dependent on the following factors.

1. Materials from which the gage is made

2. Material of the part being checked

3. Class of fit required

4. Proper care of the gage

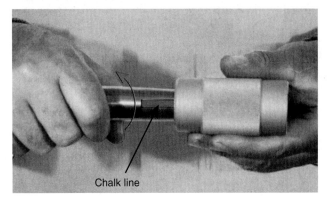

Chalk line

■ **Figure 14-7** Checking the accuracy of a taper using chalk lines. *(Kelmar Associates)*

To preserve the accuracy and life of gages:

1. Store gages in divided wooden trays to protect them from being nicked or burred.

2. Check them frequently for size and accuracy.

3. Correctly align gages with the workpiece to prevent binding.

4. 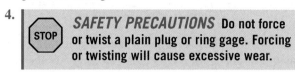 **SAFETY PRECAUTIONS** Do not force or twist a plain plug or ring gage. Forcing or twisting will cause excessive wear.

5. Clean the gage and workpiece thoroughly before checking the part.

6. Use a light film of oil on the gage to help prevent binding.

7. Make provision for air to escape when gaging blind holes with a plug gage.

8. Have gages and work at room temperature to ensure accuracy and prevent damage to the gage.

9. Never use an inspection gage as a working gage.

▶▶ Thread Plug Gages

Internal threads are checked with *thread plug gages* (Fig. 14-8) of the "go" and "no-go" variety and are based on the same principle as cylindrical plug gages.

When a thread plug gage is used, the "go" end, which is the longer end, should be turned in flush to the bottom of the hole. The "no-go" end should just start into the hole and become quite snug before the third thread enters.

Since thread plug gages are quite expensive, certain precautions should be observed in their use.

1. Thread plug gages have a chip groove cut along the thread to clear loose chips. Do not depend on this feature to remove burrs or loose chips. To prolong

■ **Figure 14-8** Thread plug gages are used to check the size and accuracy of an internal thread. *(The Taft-Peirce Manufacturing Company)*

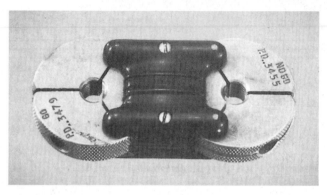

■ **Figure 14-9** "Go" and "no-go" gage thread ring gages in a holder. *(Sheffield Measurement Division)*

the life of the gage, it is advisable to remove burrs and loose chips (wherever possible) by means of an old tap.

2. Before using the thread plug gage, apply a little oil to its surface.

3. Never force the gage.

▶▶ Thread Ring Gages

The most popular gage of this type is the *adjustable thread ring gage.* It is used to check the accuracy of an external thread and has a threaded hole in the center, with three radial slots and a setscrew to permit small adjustments. The outside diameter is knurled, and the "no-go" gage is identified by an annular groove cut into the knurled surface. Both the "go" and "no-go" gages are generally assembled in one holder for ease of checking the part (Fig. 14-9).

When these gages are used, the thread being checked should fully enter the "go" gage but should not enter the "no-go" gage by more than 1½ turns. Before checking a thread, remove any dirt, grit, or burrs. A little oil will help to prolong the life of the gage.

▶▶ Snap Gages

Snap gages, one of the most common types of comparative measuring instruments, are faster to use than micrometers but are limited in their application. They are used to check diameters within certain limits by comparing the part size to the preset dimension of the snap gage. Snap gages generally have a C-shaped frame with adjustable gaging anvils or rolls, which are set to the "go" and "no-go" limits of the part. These gages are supplied in several styles, some of which are shown in Fig. 14-10.

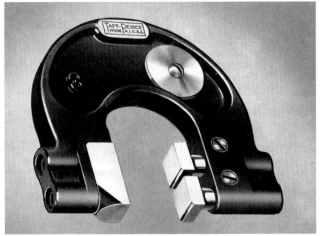

(a)

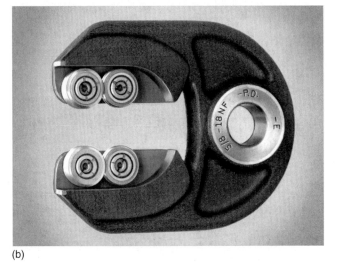

(b)

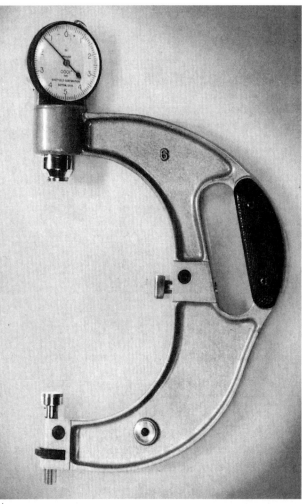

(c)

■ **Figure 14-10** Various types of snap gages: (a) adjustable snap gage *(The Taft-Peirce Manufacturing Company)*; (b) adjustable roll snap gage *(Sheffield Measurement Division)*; (c) dial indicator snap gage. *(The Taft-Peirce Manufacturing Company)*

USING A SNAP GAGE TO CHECK A DIMENSION

Proper use of a snap gage is required to prevent springing the gage and marring the work surface. Follow this procedure:

1. Thoroughly clean the anvils of the gage.

2. Set the "go" and "no-go" anvils to the required limits, using gage blocks or some other standard.

3. Lock the anvils in position and recheck the accuracy of the settings.

Note: If a dial-indicator snap gage is used, set the bezel (outer ring) of the indicator to read 0 and lock it into position.

4. Clean the surface of the work.

5. Hold the gage in the right hand, keeping it square with the work.

6. With the left hand, hold the lower anvil in position on the workpiece.

7. Push the gage over the work surface with a rolling motion. Only light hand pressure should be used to pass the "go" pins.

Note: Do not force the gage; if the work is the correct size, the gage should pass easily over the work.

8. Advance the gage until the "no-go" anvils or rolls contact the work. If the gage stops at this point, the work is within the limits.

1. For each of the following dimensions, indicate the basic diameter, upper limit, lower limit, and tolerance.

   a. 1.750 +.002 in.
            −.000 in.

   b. .625 +.0015 in.
            −.0000 in.

   c. 12.5 ±.02 mm

   d. 20 +0.000 mm
          −0.015 mm

   e. 0.5 ±0.005 mm

2. State whether the tolerances for each size in question 1 are unilateral or bilateral.

Fixed Gages

3. What purpose do fixed gages serve in industry?

4. To what tolerance are fixed gages finished?

5. If a hole size is to be maintained at 1.750 in. ±.002 (44 mm ±0.05), what would be the sizes of the "go" and "no-go" gages?

Cylindrical Plug Gages

6. How are the "go" and "no-go" ends of a cylindrical plug gage identified?

7. What precautions must be observed when using a cylindrical plug gage?

Plain Ring Gages

8. How is a "no-go" ring gage distinguished from a "go" ring gage?

Taper Plug and Ring Gages

9. How may the limits of a taper plug gage be indicated on the gage?

10. List the precautions to observe when checking with a taper plug or ring gage.

11. When may chalk be used to check an external taper and when should bluing be used?

12. Why should a taper plug or ring gage be rotated no more than one-quarter of a turn when checking a taper?

Thread Plug and Ring Gages

13. What types of threads are checked with a plug gage? A ring gage?

14. List three precautions to observe when using a thread plug gage.

15. How should an external thread of the proper size fit into the thread ring gage?

16. Describe a snap gage.

17. What advantage does the dial-indicator snap gage have over an adjustable snap gage?

18. List the precautions necessary when checking a workpiece with a snap gage.

UNIT 15

Comparison Measurement

OBJECTIVES

After completing this unit, you will be able to:

1 Explain the principle of comparison measurement

2 Identify four types of comparators and describe their use

3 Measure to within .0005-in. (0.01-mm) accuracy with a dial indicator, mechanical and optical comparator, or air and electronic gages

Manufacturing processes have now become so precise that component parts are often made in several different places and then shipped to a central location for final assembly. For this process of *interchangeable*

manufacture to be economical, there must be some assurance that these parts will fit on assembly. The components are therefore made to within certain limits, and further inspection or *quality control* ensures that only properly sized parts will be used.

Much of this inspection is done rapidly, accurately, and economically by a process called *comparison measurement.* It consists of comparing the measurement of the part to a known standard or master of the exact dimension required. Basically, *comparators* are gages that incorporate some means of amplification to compare the part size to a set standard, usually gage blocks.

Mechanical, optical, and mechanical-optical comparators and air, electrical, and electronic gages are all utilized for comparison measurement.

▶▶ Comparators

A comparator may be classified as any instrument used to compare the size of a workpiece to a known standard. The simplest form of comparator is a dial indicator mounted on a surface gage. All comparators are provided with some means of amplification by which variations from the basic dimensions can be easily noted (Fig. 15-1a on p. 116).

▶▶ Dial Indicators

Dial indicators are used to compare sizes and measurements to a known standard and to check the alignment of machine tools, fixtures, and workpieces prior to machining. Many types of dial indicators operate on a gear and rack principle (Fig. 15-1b on p. 116). A *rack* cut on the *plunger* or *spindle* is in mesh with a *pinion,* which in turn

is connected to a *gear train.* Any movement of the spindle is then magnified and transmitted to a *hand* or *pointer* over a *graduated dial.* Inch-designed dials may be graduated in thousandths of an inch or less. The dial, attached to a *bezel,* may be adjusted and locked in any position.

During use, the contact point on the end of the spindle bears against the work and is held in constant engagement with the work surface by the rack spring. A hair spring is attached to the gear that meshes with the center pinion. This flat spiral spring takes up the backlash from the gear train and prevents any lost motion from affecting the accuracy of the gage.

Dial indicators are generally of two types: the continuous-reading dial indicator and the dial test indicator.

The *continuous-reading dial indicator* (Fig. 15-2a on p. 116), numbered clockwise for 360°, is available as a regular-range and a long-range indicator. The regular-range dial indicator has only about 2½ revolutions of travel. It is generally used for comparison measurement

■ Figure 15-1a Mechanical and optical instruments are used for comparison measurements. *(Brown & Sharpe)*

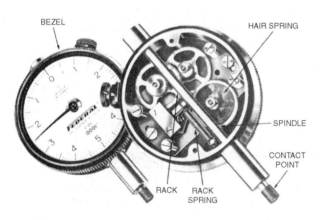

■ Figure 15-1b A balanced-type dial test indicator, showing the internal mechanism. *(Federal Products Corporation)*

■ Figure 15-2a A continuous-reading dial indicator. *(Federal Products Corporation)*

and setup purposes. The digital long-range indicator (Fig. 15-2b) is often used to indicate table travel or cutting tool movement on machine tools.

Dial test indicators (Fig. 15-1b) may have a balanced-type dial, that is, one that reads both to the right and left from 0 and indicates a plus or minus value. Indicators of this type have a total spindle travel of only 2½ revolutions. These instruments may be equipped with tolerance pointers to indicate the permissible variation of the part being measured.

Perpendicular dial test indicators, or back plunger indicators, have the spindle at right angles (90°) to the dial. They are used extensively in setting up lathe work and for machine table alignment.

The *universal* dial test indicator (Fig. 15-3) has a contact point that may be set at several positions through a 180° arc. This type of indicator may be conveniently used to check internal and external surfaces. Fig. 15-4a and b illustrate typical applications of this type of indicator.

Metric dial indicators (Fig. 15-5 on p. 118) are available in both the balanced and continuous-reading types. The type used for inspection purposes is usually graduated in 0.002 mm and has a range of 0.5 mm. The regular indicators are usually graduated in 0.01 mm and have a range up to 25 mm.

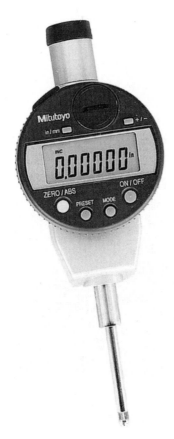

Figure 15-2b A digital direct reading indicator.
(MTI Corporation)

Figure 15-3 The universal dial test indicator.
(Federal Products Corporation)

(a)

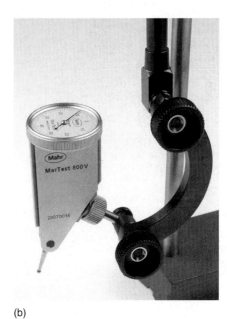

(b)

Figure 15-4 (a) A universal dial test indicator being used to center a workpiece with the machine spindle. *(Federal Products Corporation)*; (b) checking measurements with a dial height gage. *(Federal Products Corporation)*

TO MEASURE WITH A DIAL TEST INDICATOR AND HEIGHT GAGE

1. Clean the face of the surface plate and the vernier height gage.
2. Mount the dial test indicator on the movable jaw of the height gage (Fig. 15-4b).
3. Lower the movable jaw until the indicator point just touches the top of a gage block resting on the surface plate.
4. Tighten the upper locking screw on the vernier and loosen the lower locking screw.
5. Carefully turn the adjusting nut until the indicator needle registers approximately one-quarter turn.

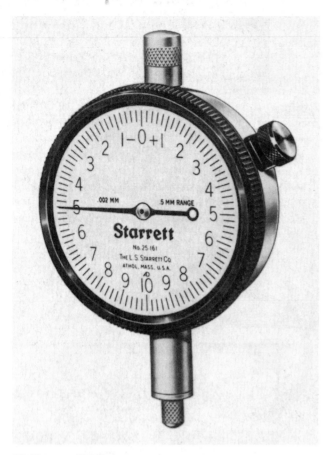

Figure 15-5 A metric dial indicator with a balanced dial. *(The L. S. Starrett Co.)*

Figure 15-6 The Digi-Matic indicator can make readings in increments of .0001 in. or 0.001 mm. *(MTI Corp.)*

6. Turn the bezel to set the indicator to zero.

7. Note the reading on the vernier and record it on a piece of paper.

8. Raise the indicator to the height of the first hole to be measured.

9. Adjust the vernier until the indicator reads zero.

10. Note the vernier reading again and record it.

11. Subtract the first reading from the second and add the height of the gage block.

12. Proceed in this manner to record the location of all other holes.

▸▸ Mechanical and Electronic Comparators

The *mechanical comparator* consists of a base, a column, and a gaging head. Various mechanical comparators operate on different principles. Some are based on the gear and rack principle used in some dial indicators; others use a system of levers similar to the universal dial indicator.

Mechanical comparators are gradually being replaced by electronic indicators and comparators. The *Digi-Matic indicator* (Fig. 15-6) is capable of readings in increments of .0001 in. or 0.001 mm. It can be interfaced with a data recorder, minicomputer, or host computer to provide statistical data based on inspection results.

The *electronic comparator* (Fig. 15-7), a highly accurate form of comparator, uses the Wheatstone bridge circuit to transform minute changes in spindle movement into a relatively large needle movement on the gage. This degree of magnification is controlled by a selector on the front of the *amplifier*. The widely spaced graduations represent zero values from .0001 to .00001 in. (0.002 mm to 0.0002 mm), depending on the scale selected.

When it is not necessary to know the exact dimension of a part, but only whether it falls within the required limits, a signal light attachment may be installed on the gage. When a workpiece is tested, an amber light indicates that it is within the prescribed limits, a red light indicates that the workpiece is too small, and a blue light indicates that it is too large.

Electronic units may also be used as height gages by mounting a rectangular gage head on a height gage stand (Fig. 15-8). This method is particularly suited to the checking of soft, highly polished surfaces because of the light gaging pressure required.

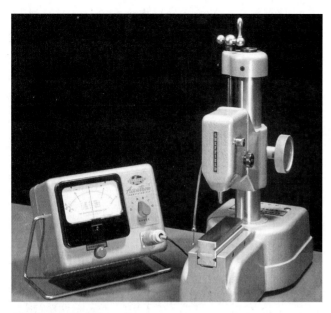

■ **Figure 15-7** The value of the scale divisions on an electronic comparator can be readily changed to suit the accuracy required.

■ **Figure 15-8** The electronic gage with a rectangular head is used to check heights accurately. *(Sheffield Measurement Division)*

All types of comparators are used in inspection to check the size of a part against a master gage. The variation between the part and the master is shown on a scale as a plus or minus quantity.

TO MEASURE WITH A MECHANICAL COMPARATOR

1. Clean the anvil and master gage of the required size.
2. Place the master on the anvil.
3. Carefully lower the gaging head until the stylus touches the master and indicates a movement of the needle.

4. Lock the gaging head to the column.
5. Adjust the needle to zero using the fine adjustment knob and set the limit pointers on the face.
6. Recheck the setting by removing the master and replacing it.
7. Set the tolerance pointers to the upper and lower tolerances of the part being checked.
8. Substitute the work being gaged for the master and note the reading. If the reading is to the right of zero, the work is too large; if to the left, it is too small. If the needle stops between the tolerance pointers, the work is within the permitted limits.

▶▶ Optical Comparators

An *optical comparator* or *shadowgraph* (Fig. 15-9) projects an enlarged shadow onto a screen, where it may be compared to lines or to a master form that indicates the

■ **Figure 15-9** Intricate forms can be checked easily with an optical comparator. *(MTI Corp.)*

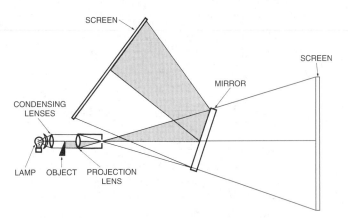

■ Figure 15-10 The principle of the optical comparator.

limits of the dimensions or the contour of the part being checked. The optical comparator is a fast, accurate means of measuring or comparing the workpiece with a master. It is often used when the workpiece is difficult to check by other methods. Optical comparators are particularly suited to checking extremely small or odd-shaped parts, which would be difficult to inspect without the use of expensive gages.

Optical comparators are available in bench and floor models, which are identical in principle and operation (Fig. 15-10). Light from a *lamp* passes through a *condenser lens* and is projected against the workpiece. The shadow caused by the workpiece is transmitted through a *projecting lens* system, which magnifies the image and casts it onto a *mirror*. The image is then reflected to the *viewing screen* and is further magnified in this process.

The extent of the image magnification depends on the lens used. Interchangeable lenses for optical comparators are available in the following magnifications: 5×, 10×, 31.25×, 50×, 62.5×, 90×, 100×, and 125×.

A comparator chart or master form mounted on the viewing screen is used to compare the accuracy of the enlarged image of the workpiece being inspected. Charts are usually made of translucent material, such as cellulose acetate or frosted glass. Many different charts are available for special jobs, but the most commonly used are linear-measuring, radius, and angular charts. A vernier protractor screen is also available for checking angles. *Since charts are available in several magnifications, care must be taken to use a chart of the same magnification as the lens mounted on the comparator.*

Many accessories are available for the comparator, increasing the versatility of the machine. Some of the most common are *tilting work centers,* which permit the workpiece to be tilted to the required helix angle for checking threads; a *micrometer work stage,* which permits quick and accurate measuring of dimensions in both directions; and *gage blocks, measuring rods,* and *dial indicators,* used on comparators for checking measurement.

■ Figure 15-11 Checking a thread form on an optical comparator. *(Kelmar Associates)*

The surface of the workpiece may be checked by a *surface illuminator,* which lights up the face of the workpiece adjacent to the projecting lens system and permits this image to be projected onto the screen.

TO CHECK THE ANGLE OF A 60° THREAD USING AN OPTICAL COMPARATOR

(Refer to Fig. 15-11.)

1. Mount the correct lens in the comparator.
2. Mount the tilting centers on the micrometer cross-slide stage.
3. Set the tilting centers to the helix angle of the thread.
4. Set the workpiece between centers.
5. Mount the vernier protractor chart and align it horizontally on the screen.
6. Turn on the light switch.
7. Focus the lens so that a clear image appears on the screen.
8. Move the micrometer cross-slide stage until the thread image is centralized on the screen.

9. Revolve the vernier protractor chart to show a reading of 30°.

10. Adjust the cross-slides until the image coincides with the protractor line.

11. Check the other side of the thread in the same manner.

Note: If the thread angle is not correct or square with the centerline, adjust the vernier protractor chart to measure the angle of the thread image.

Other dimensions of the thread, such as depth, diameters, and width of flats, may be measured with micrometer-measuring stages or devices such as rods, gage blocks, and indicators.

▶▶ Mechanical-Optical Comparators

The *mechanical-optical comparator* (Fig. 15-12), or the *reed*-type comparator, combines a reed mechanism with a light beam to cast a shadow on a magnified scale to indi-

cate the dimensional variation of the part. It consists of a base, a column, and a gaging head that contains the reed mechanisms and light source.

THE REED MECHANISM

Fig. 15-13 illustrates the principle of the reed mechanism. A fixed steel block A and a movable block B have two pieces of spring steel, or reeds, attached to them (Fig. 15-14a). The upper ends of the reeds are joined and connected to a pointer. Since block A is fixed, any movement of the spindle attached to block B will move this block up or down, causing the pointer to move a much greater distance to the right or left (Fig. 15-14b).

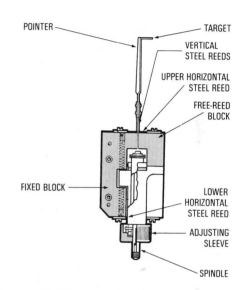

■ **Figure 15-13** Construction and principles of the reed mechanism (mechanical lever).

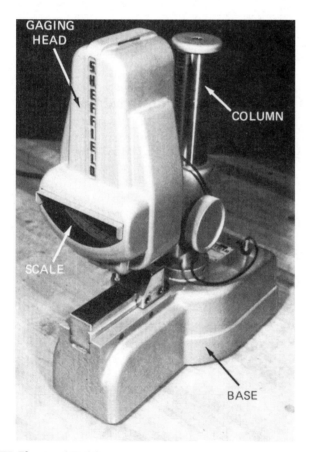

■ **Figure 15-12** A reed-type comparator uses both the mechanical and optical principles of measurement. *(Sheffield Measurement Division)*

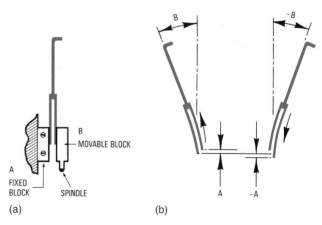

■ **Figure 15-14** (a) Cutaway section showing the blocks and reed mechanism; (b) the upper part of the reed deflects a greater distance than the lower part.

Comparison Measurement   **121**

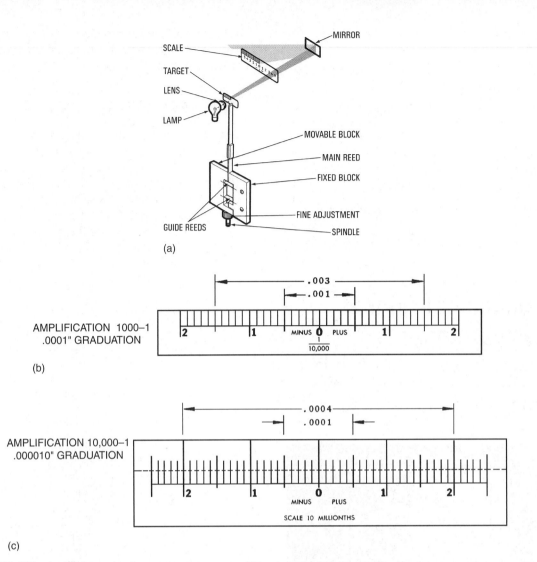

Figure 15-15 (a) The light beam or optical lever amplifies target movement *(Blanchette Tool & Gage Corp.)*; (b, c) different amplifications are available on most visual gage comparators.

A light beam passing through an aperture illuminates the scale (Fig. 15-15a). The pointer, with a *target* attached, is located below the aperture so that a movement of block B will cause the target to interrupt the light beam, casting a highly amplified shadow on the scale. The movement of the pointer and target is obviously greater than the movement of the spindle. Also, the shadow cast on the scale will be larger than the movement of the target. Therefore, the measurement on the scale will be much greater than the movement of the spindle.

Visual gage comparators are available in five amplifications ranging from .0001 on the 500 to 1 scale to .000010 on the 10,000 to 1 scale (Fig. 15-15b and c). The scales for these instruments are graduated in plus and minus, with zero being in the center of the scale. The value of each graduation is marked on the scale of the instrument.

To Measure with a Reed Comparator

1. Raise the gaging head above the required height, and clean the anvil and master thoroughly.

2. Place the master gage or gage block buildup on the anvil.

3. Carefully lower the gaging head until the end of the spindle *just touches the master.*

Note: A shadow will begin to appear on the left side of the scale.

4. Clamp the gaging head to the column.

5. Turn the adjusting sleeve until the shadow coincides with the zero on the scale.

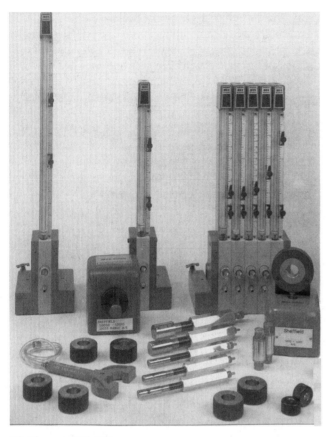

Figure 15-16 A flow- or column-type air gage.

Figure 15-17 Gaging a hole using a pressure-type air gage. *(Federal Products Corporation)*

6. Remove the gage blocks and carefully slide the workpiece between the anvil and the spindle.

7. Note the reading. If the shadow is to the right of zero, the part is oversize; if to the left, it is undersize.

▶▶ Air Gages or Pneumatic Comparators

Air gaging, a form of comparison measurement, is used to compare workpiece dimensions with those of a master gage by means of air pressure or flow. Air gages are of two types: the *flow* or *column type* (Fig. 15-16), which indicates air velocity, and the *pressure type* (Fig. 15-17), which indicates air pressure in the system.

COLUMN-TYPE AIR GAGE

After air has been passed through a filter and a regulator, it is supplied to the gage at about 10 psi (69 kPa) (Fig. 15-18). The air flows through a transparent tapered tube in which a float is suspended by this air flow. The top of the tube is connected to the gaging head by a plastic tube.

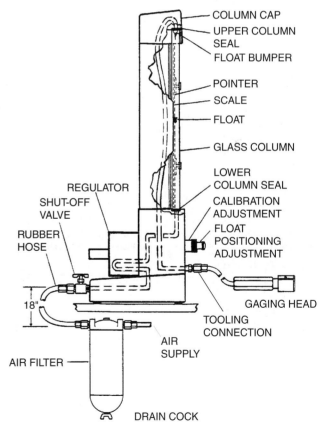

Figure 15-18 Principle of the column-type air gage. *(Precision Gage & Tool Co.)*

The air flowing through the gage exhausts through the passages in the gaging head into the clearance between the head and the workpiece. The rate of flow is proportional to the clearance indicated by the position of the float in the column. The gage is set to a master, and the float is then positioned by means of an adjusting knob. The upper and lower limits for the workpiece are then set. If the hole in the workpiece is larger than the hole size of the master, more air will flow through the gaging head, and the float will rise higher in the tube. Conversely, if the hole is smaller than the master, the float will fall in the tube. Amplification from 1000:1 to 40 000:1 may be obtained with this type of gage. Snap-, ring-, and plug-type gaging heads may be fitted to this type of gaging device.

PRESSURE-TYPE AIR GAGE

In the pressure-type air gage (Fig. 15-17), air passes through a filter and regulator and is then divided into two channels (Fig. 15-19). The air in the *reference channel* escapes to atmosphere through a zero-setting valve. The air in the *measuring channel* escapes to atmosphere through the gage head jets. The two channels are connected by an extremely accurate differential pressure meter.

The master is placed over the gaging spindle and the zero-setting valve is adjusted until the gage needle indicates zero. Any deviation in the workpiece size from the master size changes the reading. If the workpiece is too large, more air will escape through the gaging plug; therefore, pressure in the measuring channel will be less and the dial gage hand will move counterclockwise, indicating how much the part is oversize. A diameter smaller than the master gage indicates a reading on the right side of the dial. Amplification from 2500:1 to 20 000:1 may be obtained with this type of gage. Pressure-type air gages may also be fitted with plug, ring, or snap gaging heads for a wide variety of measuring jobs.

Air gages are widely used, since they have several advantages over other types of comparators:

1. Holes may be checked for taper, out-of-roundness, concentricity, and irregularity more easily than with mechanical gages (Fig. 15-20 on facing page).

2. The gage does not touch the workpiece; therefore, there is little chance of marring the finish.

3. Gaging heads last longer than fixed gages because wear is reduced between the head and the workpiece.

4. Less skill is required to use this type of gaging equipment than other types.

5. Gages may be used at a machine or bench.

6. More than one diameter may be checked at the same time.

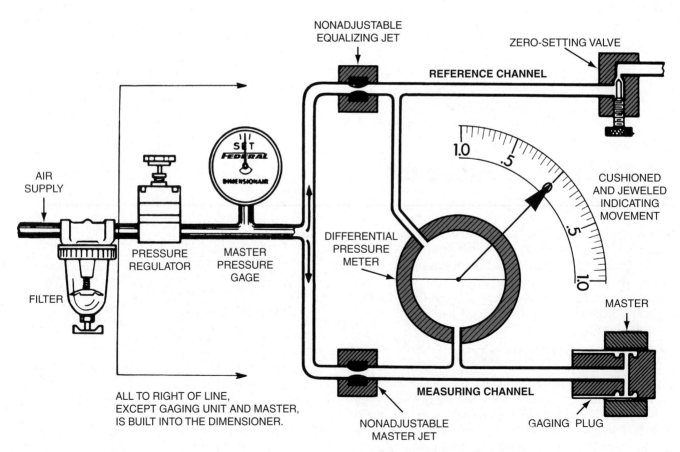

■ **Figure 15-19** Operation of the pressure-type air gaging system. *(Precision Gage & Tool Co.)*

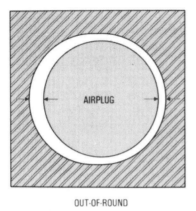

OUT-OF-ROUND

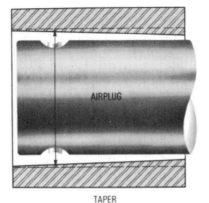

TAPER

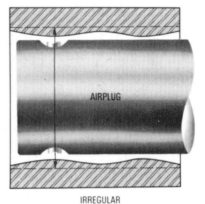

IRREGULAR

CONCENTRICITY

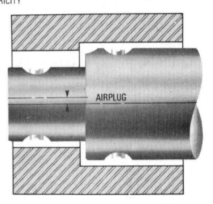

■ **Figure 15-20** Irregularly shaped holes may be easily checked with an air gage. *(Federal Products Corporation)*

unit 15 review questions

Comparison Measurement

1. Describe:
   a. Quality control
   b. Comparison measurement

Dial Indicators

2. What is the difference between a regular-range dial indicator and a long-range indicator?

3. Compare a perpendicular dial indicator with a dial test indicator.

4. How are metric dial indicators usually graduated?

Comparators

5. Define a comparator.

6. List three principles used in mechanical comparators.

7. Why is high amplification necessary in any comparison measurement process?

8. Describe the procedure for measuring a workpiece with a reed-type comparator.

Optical Comparators

9. List the advantages of an optical comparator.

10. Describe the principle of an optical comparator. Illustrate by means of a suitable sketch.

11. What precautions are necessary when charts are used on an optical comparator?

Air Gages

12. Describe the principle of the column-type air gage. Illustrate by means of a suitable sketch.

13. Describe and neatly illustrate the principle of the pressure-type air gage.

14. List six advantages of air gages.

Electronic Comparators

15. What circuit is employed in electronic comparators?

16. Describe the operation of the single light attachment for electrical or electronic gages.

UNIT 16

The Coordinate Measuring System

OBJECTIVES

After completing this unit, you will know:

1 The purpose of and how to apply the coordinate measuring system

2 The main components and operation of the measuring unit

3 The advantages of the coordinate measuring system

Coordinate measurement is a method used to speed up layout, machining, and inspection by having all measurements dimensioned from three rectangular coordinates (surfaces of the workpiece): *X, Y,* and *Z.* Coordinate measuring systems are capable of measuring to .000050 in. or 0.001 mm (Fig. 16-1a). This high degree of accuracy is achieved by the use of a series of dark and light bands called the *moiré fringe pattern.* This system has been applied extensively to machine tools such as lathes, milling machines, and jig borers (Fig. 16-1b and c) to speed up the measurement process, thus reducing overall machining time. Coordinate measuring machines (Fig. 16-1d), widely used in the field of measurement and inspection, provide a quick and accurate means of checking machined parts prior to assembly.

Various units may be mounted on a machine tool to give readings for the *X* (length), *Y* (width), and *Z* (depth) axes. A unit providing direct readout in degrees, minutes, and seconds is available for angular positioning and measurement.

▶▶ Parts of the Measuring Unit

The *measuring unit* (Fig. 16-2) has three basic components: *a machined spar with a calibrated grating* (A), *a reading head* (B), and *a counter with a digital readout display* (C). The main element in this system is an accurately ruled grating (Fig. 16-2a on p. 128) of the desired length of travel. The face of this grating for systems having .0001-in. (0.002-mm) resolution is etched for its entire length with lines spaced .001 in. (0.02 mm) apart. For machines capable of greater accuracy, the grating spar is etched with 2500 lines per inch, so that the digital readout resolution is .000050 in.

(0.001 mm). Metric readings are obtained by pushing a button on the side of the box.

A *transparent index grating,* having the same graduations as the grating spar, is mounted in the reading head (Fig. 16-2b). The index grating is positioned so that its lines are at a slight angle to the lines on the spar. The reading head is mounted so that the index grating is positioned just .002 in. (0.05 mm) above the main grating. Mounted in the reading head are a small light, a collimating lens, and four photoelectric cells (Fig. 16-3).

Note: This electro-optical system "counts" the moiré fringes and produces the high resolution measurement accuracy of the Sheffield Cordax Measuring System.

Figure 16-1a Coordinate measuring machines are widely used for inspection. *(Brown & Sharpe)*

PRINCIPLE OF THE MOIRÉ FRINGE

To illustrate the moiré fringe pattern, we can draw a series of equally spaced lines on two pieces of plastic sheeting (Fig. 16-4 on p. 128). One sheet should then be placed over the other at an angle, as in Fig. 16-4. Where the lines cross, dark bands appear. If the top sheet is moved to the right, the position of these bands will shift down, and the pattern will appear as a series of dark and light bands moving vertically on the sheet. Thus, any longitudinal movement will produce a vertical movement of the light bands. This principle can also be illustrated by placing one comb on another at a small angle and moving one across the face of the other.

▶▶ Operation of the Measuring Unit

Since the etched lines on the reading-head grating are at an angle to the lines on the main spar, a series of bands would appear when viewed from directly above. Let's consider one band only to explain the operation of the system.

(b)

(c)

(d)

Figure 16-1b, c, d (b) A lathe equipped with a coordinate measuring system; (c) the digital readout shows the exact location of the cutting tool or workpiece; (d) the coordinate measuring system being used for inspection purposes. *(Sheffield Measurement Division)*

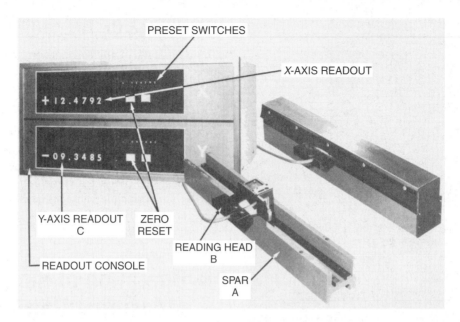

Figure 16-2 Components of a coordinate measuring unit.

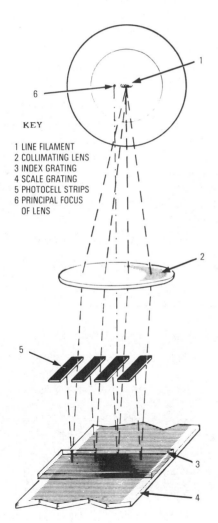

KEY

1 LINE FILAMENT
2 COLLIMATING LENS
3 INDEX GRATING
4 SCALE GRATING
5 PHOTOCELL STRIPS
6 PRINCIPAL FOCUS
  OF LENS

Figure 16-3 Principle of the coordinate measuring unit.

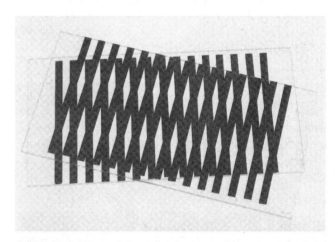

Figure 16-4 Principle of the moiré fringe pattern. *(Sheffield Measurement Div.)*

Light from the lamp (Fig. 16-3) passes through the collimating lens and is converted into a parallel beam of light. This beam strikes the fringe pattern and is reflected back up to one of four photoelectric cells, which converts the fringe pattern into an electrical signal. As the reading head is moved longitudinally, the band moves vertically on the face of the spar and will be picked up by the next photoelectric cell, creating another signal (Fig. 16-5). These signals from the photoelectric cells (output signals) are transmitted to the digital readout display box, where they indicate accurately the head travel at any point. A signal from the next photoelectric cell will increase or decrease the reading on the digital readout box by .0001 in. (0.002 mm), depending on the direction of the movement.

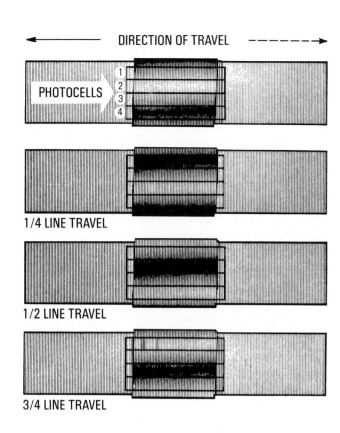

DIRECTION OF TRAVEL

PHOTOCELLS

1
2
3
4

1/4 LINE TRAVEL

1/2 LINE TRAVEL

3/4 LINE TRAVEL

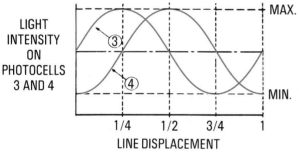

LIGHT
INTENSITY
ON
PHOTOCELLS
3 AND 4

MAX.

③

④

MIN.

1/4   1/2   3/4   1

LINE DISPLACEMENT

■ **Figure 16-5** Lateral movement of the fringe pattern.

Note: The fringe pattern (Fig. 16-4) shifts laterally and continuously across the grating path.

ADVANTAGES OF THE COORDINATE MEASURING SYSTEM

1. Readout boxes provide clear, visible numbers, which eliminate the possibility of misreading a dial gage.

2. It provides a constant readout of the tool (or table) position.

3. The reading indicates the exact position of the tool (or table) and is not affected by machine or lead screw wear.

■ **Figure 16-6** Coordinate measuring machines can be programmed to inspect parts automatically. *(Brown & Sharpe)*

4. This system eliminates the need for gage blocks and measuring rods on jig borers and vertical milling machines.

5. The need for operator calculations and the inherent possibility of errors are eliminated.

6. Machine setup time is greatly reduced.

7. Production is increased, since the workpiece must be checked for one size only. For example, when a workpiece having several diameters is to be machined in a lathe, it is necessary to measure the first diameter only. Once the machine has been set to this size, all other diameters will be accurate.

8. The need for operator skill is reduced.

9. Scrap and rework are almost completely eliminated.

10. Most machine tools can be fitted with this system.

Coordinate measuring machines (Fig. 16-6) were developed to speed up the process of measuring parts produced on numerical control (NC) machine tools. What NC is to manufacturing, the coordinate measuring machine (CMM) is to inspection. It can measure almost any shape with extreme accuracy and without the use of special gages. Coordinate measuring machines, which are computer controlled, have eliminated the problems of operator error; the need for long, complex, and inefficient conventional measuring systems; and the low productivity common to previous inspection methods. These measuring machines can be installed on production lines to automate inspection, minimize operator error, and provide uniform part quality.

1. What is the principle of coordinate measurement?

2. Why have coordinate measuring systems been so widely accepted by industry?

3. State two different applications of coordinate measuring systems.

4. Name the three main parts of the measuring unit.

5. Describe the operation of this measuring system.

6. State seven important advantages of a coordinate measuring system.

UNIT 17

Measuring with Light Waves

OBJECTIVES

After completing this unit, you will be able to:

1 Check pieces for size, flatness, and parallelism using optical flats

2 Describe the operation of a laser interferometer

3 Explain the application of lasers to measurement

Two of the most precise measuring methods are those using optical flats and the laser. Although based on different principles, both use a source of monochromatic light to produce highly accurate measurements.

▶▶ Measuring with Optical Flats

One of the most accurate and reliable means of measuring is with light waves. Optical flats (Fig. 17-1), used with a monochromatic light, are used to check work for flatness (Fig. 17-4 on p. 132), parallelism (Fig. 17-5 on p. 133), and size (Fig. 17-6 on p. 133).

Optical flats are disks of clear fused quartz, lapped to within a few millionths of an inch of flatness. They are generally used with a helium light source, which produces a greenish-yellow light having a wavelength of 23.1323 μin. (0.587 56 μm).

The optical flat, a perfectly flat, transparent disk, is placed on the surface of the work to be checked. The functioning surface of the optical flat is the surface adjacent to the workpiece. It is transparent and capable of reflecting light; therefore, all light waves that strike this surface are split into two parts (Fig. 17-2 on p. 132). One part is reflected back by the lower surface of the flat. The other part passes through this surface and is reflected by the upper surface of the work. Whenever the reflected split portions of two light waves

■ **Figure 17-1** Checking a gage block using an optical flat and a helium light source. *(DoAll Company)*

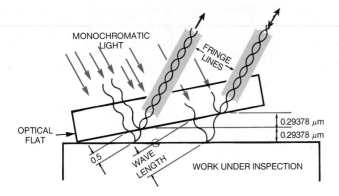

■ Figure 17-2 Principle of the optical flat.

cross each other, or *interfere*, they become visible and produce dark interference bands or fringe lines. This happens whenever the distance between the lower surface of the flat and the upper surface of the workpiece is *only one-half of a wavelength* or multiples thereof (Fig. 17-2).

Since the wavelength of helium light is 23.1323 μin. (0.587 56 μm, or millionths of a meter), each half-wavelength will represent 11.6 μin. (0.293 μm). Each dark band then represents a progression of 11.6 μin. (0.293 μm) above the point of contact between the workpiece and the optical flat. Therefore, when a height is checked, the number of bands between two points on a surface multiplied by 0.000 011 6 (0.293 μm) will indicate the height difference between two surfaces.

For comparison measurements, the difference in height between a master block and the workpiece can be determined as shown in Fig. 17-3. This illustrates the method used for checking accurately the height of an unknown surface by comparing it with a gage block of a known height. It is necessary to first know which block is larger before the unknown block can be measured. To determine which block is larger, apply finger pressure to points *X* and *Y*. If this pressure at *X* does not change the band pattern and the pressure at *Y* causes the bands to separate, the master block (*M*) is larger. If the opposite is true, the unknown block (*U*) is larger. In Fig. 17-3, two bands appear on the low block; therefore, the unknown block is two bands, or 2 × 11.6 = 23.2 μin. (2 × 0.293 = 0.586 μm), less than the master.

HOW TO INTERPRET THE BANDS

Refer to Fig. 17-4. Since the lines show only a slight curve, the block would be convex or slightly higher in the center. The block is 2 × 11.6 μin., or .0000232 in., out of flat.

In Fig. 17-5, since two bands appear on the master block, the workpiece is 2 × 11.6 or 23.2 μin. smaller or larger than the master. If pressure on the master causes the spacing of the bands to widen, the part is smaller. Note the curve in the band, which shows that the workpiece is not exactly parallel. It is about one-half band, or 5.8 μin., out of parallel.

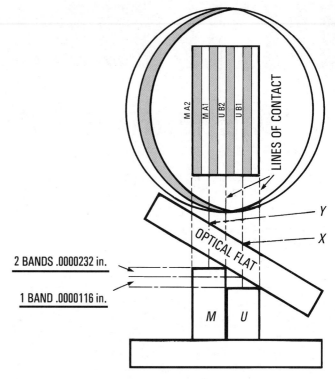

■ Figure 17-3 Checking the height of a block with a master block.

■ Figure 17-4 Checking flatness with an optical flat. *(DoAll Company)*

In Fig. 17-6, the master block is on the left. Note the three bands on the master block and the six bands on the block of unknown size on the right. The unknown block then has three more bands than the master. (The block with more bands is always smaller.) Also note that the lines on the unknown block are straight and evenly spaced, which indicates that it is parallel over its length. Since the lines slope down toward the line of contact, the left side of the unknown block is lower by one-half band, or 5.8 μin. (11.6 ÷ 2).

Figure 17-5 Checking parallelism with an optical flat. *(DoAll Company)*

Figure 17-6 Checking size with an optical flat. *(DoAll Company)*

▶▶ Measurement with Lasers

Besides its many other uses in medicine and industry, the laser provides one of the most accurate means of measurement. A laser-light measuring device, called an *interferometer* (Fig. 17-7), measures changes in position (alignment) by means of light-wave interference.

As Fig. 17-7 shows, the laser beam is split into two parts by the *beam splitter.* One of these beams is transmitted through the beam splitter to a *motion-sensitive mirror* back to the beam splitter at point *X.* From this point, where the two beams rejoin, the recombined beams are transmitted to the *detector.*

If there is no movement, both portions of the beam stay in the same phase and the light reaching the detector will remain constant (Fig. 17-8). If there is any movement at the sensitive mirror, the beam (black) reflected from that mirror will be altered and will fluctuate in and out of phase with the other beam (Fig. 17-9). When this beam

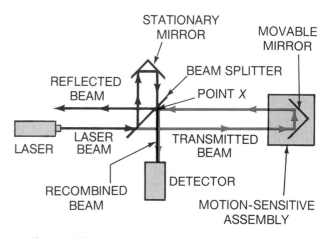

Figure 17-7 Principle of the laser interferometer.

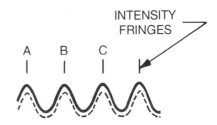

Figure 17-8 The laser beam's wave components are in phase.

Figure 17-9 The laser beam's wave components are out of phase.

(black) is out of phase, it cancels the other beam, causing the light to fluctuate. This fluctuating light pattern registers on the beam detector, where the number of fluctuations are computed relative to the laser wavelength, and the precise movement is displayed on a readout box.

Because of the very thin, straight beam produced by a laser interferometer, these devices are used for precise linear measurement and alignment in the production of large machines. They are also used to calibrate precision machines and measuring devices. Laser devices may also be used to check machine setups (Fig. 17-10 on p. 134). A laser beam is projected against the work and measurements are made by the beam and displayed on a digital readout panel.

Because of their very thin, straight beam characteristics, lasers are used extensively in construction and surveying. They may be used to indicate the exact location for positioning girders on tall buildings or for establishing directional lines for a tunnel being constructed under a river. They are also widely used for establishing grades for sewer and drainage systems.

Lasers have become an established part of the space program. Laser beams are used to indicate how far the

Figure 17-10 A laser interferometer being used to check the alignment of machine parts. *(Cincinnati Machine)*

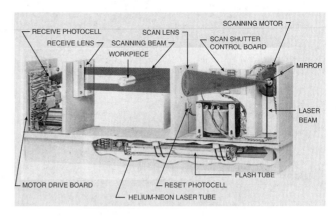

Figure 17-11 The LaserMike, an optical micrometer, makes use of a helium-neon laser beam. *(LaserMike Div. Techmet Co.)*

spaceship is from a given planet. They are also used for military purposes such as missile range finding, guidance, and tracking. Operation of the laser and its other industrial applications will be covered in Unit 94.

▶▶ The Lasermike

The *LaserMike* (Fig. 17-11) is an optical micrometer that is simple in principle yet highly accurate in operation. The heart of the instrument is a helium-neon laser beam that is projected in a straight line with almost no diffusion.

The beam is directed to mirrors mounted on the shaft of a precision electric motor. During rotation, these mirrors "scan" the laser beam through an optical lens, which aligns the beams in parallel and projects them toward a receiving lens. When an object is placed in the center of the laser beam, it creates a shadow segment in the scan path, which is detected by the photocell, enabling the unit to determine the edges of the object. A high-frequency crystal clock times the interval between edges and converts time to linear dimensions.

Without even touching the product, LaserMike provides instant readouts to accuracies never before possible with other conventional measurement techniques. Moreover, measurement speeds permit accurate measurements while the product is in motion as well as at rest.

unit 17 review questions

Optical Flats

1. Describe an optical flat.
2. What light source is used with optical flats and what is its wavelength?
3. Describe and illustrate in detail the principle of an optical flat.
4. When measuring the height of a block using a gage block and an optical flat, how is it possible to determine the higher block?

Lasers

5. Name five measurement applications for an interferometer.
6. Name the four main parts of an interferometer.
7. Briefly describe the operation of an interferometer.

UNIT 18

Surface Finish Measurement

OBJECTIVES

After completing this unit, you will be able to:

1 Interpret the surface finish symbols that appear on a drawing

2 Use a surface finish indicator to measure the surface finish of a part

Modern technology has demanded improved surface finishes to ensure the proper functioning and long life of machine parts. Pistons, bearings, and gears depend to a great extent on a good surface finish for proper functioning and therefore require

little or no break-in period. Finer finishes often require additional operations, such as lapping or honing. These higher finishes are not always required on a part and only result in higher production costs. To prevent overfinishing a part, the desired finish is indicated on the shop drawing. This information specifying the degree of finish is conveyed to the machinist by a system of symbols devised by the American Standards Association (ASA). These symbols provide a standard system of determining and indicating surface finish. The inch unit of surface finish measurement is the *microinch* (μin.). The metric unit for surface finish is the *micrometer* (μm).

The most common instrument used to measure finish is the surface indicator (Fig. 18-1, p. 136). This device consists of a *tracer head* and an *amplifier*. The tracer head houses a diamond stylus, having a point radius of .0005 in. (0.013 mm), that bears against the surface of the work. It may be moved along the work surface by hand or it may be motor-driven. Any movement of the stylus caused by surface irregularities is converted into electrical fluctuations by the tracer head. These signals are magnified by the amplifier and registered on the meter by an indicator hand or needle. The reading shown on the meter indicates the *average* height of surface roughness or the departure of this surface from the reference (center) line.

Readings may be in either *arithmetic average roughness height* (Ra) or *root mean square* (Rq). A highly magnified cross section of a workpiece would appear as shown in Fig. 18-2 on p. 136, with "hills and valleys" above and below the centerline. To calculate the surface finish without a surface indicator, the height of these deviations must be measured and recorded as shown. The Ra or Rq could then be calculated as in Fig. 18-2. The Rq is considered the better method of determining surface roughness since it emphasizes extreme deviations.

For accurate determination of the surface finish, the indicator must first be calibrated by setting it to a precision reference surface on a test block calibrated to ASA standards. The symbols used to identify surface finishes and characteristics are shown in Fig. 18-3 on p. 136.

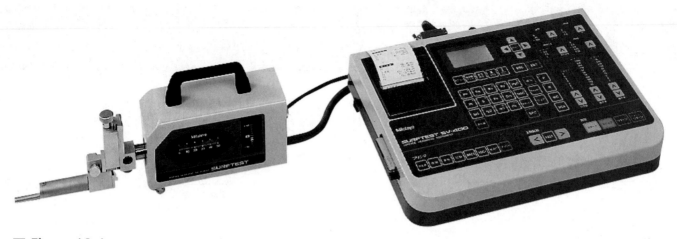

■ **Figure 18-1** A surface indicator can accurately check the roughness of a surface.   *(MTI Corporation)*

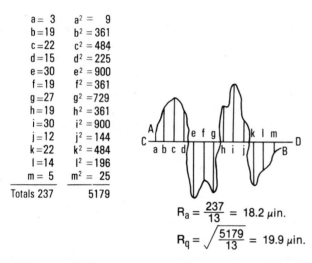

a = 3	a^2 = 9
b = 19	b^2 = 361
c = 22	c^2 = 484
d = 15	d^2 = 225
e = 30	e^2 = 900
f = 19	f^2 = 361
g = 27	g^2 = 729
h = 19	h^2 = 361
i = 30	i^2 = 900
j = 12	j^2 = 144
k = 22	k^2 = 484
l = 14	l^2 = 196
m = 5	m^2 = 25
Totals 237	5179

$$R_a = \frac{237}{13} = 18.2 \ \mu in.$$

$$R_q = \sqrt{\frac{5179}{13}} = 19.9 \ \mu in.$$

■ **Figure 18-2** Calculation of Ra (arithmetic average) and Rq (root mean square) roughness height.

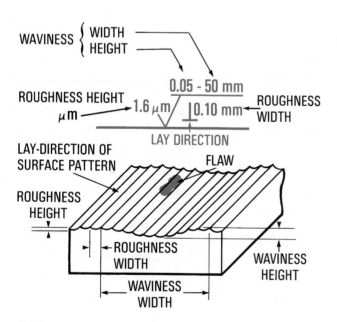

■ **Figure 18-3** Surface characteristics and symbols.

▶▶ Surface Finish Definitions

Surface deviations—any departures from the nominal surface in the form of waviness, roughness, flaws, lay, and profile

Waviness—surface irregularities that deviate from the mean surface in the form of waves; they may be caused by vibrations in the machine or work and are generally widely spaced

Waviness height—the peak-to-valley distance in inches or millimeters

Waviness width—the distance between successive waviness peaks or valleys in inches or millimeters

Roughness—relatively finely spaced irregularities superimposed on the waviness pattern and caused by the cutting tool or the abrasive grain action and the machine feed; these irregularities are much narrower than the waviness pattern

Roughness height—the Ra deviation measured normal to the centerline in microinches or micrometers

Roughness width—the distance between successive roughness peaks parallel to the nominal surface in inches or millimeters

Roughness width cutoff—the greatest spacing of repetitive surface irregularities to be included in the measurement of roughness height; it must always be greater than the roughness width; standard values are .003, .010, .030, .100, .300, and 1 in. (0.075, 0.25, 0.76, 2.54, 7.62, and 25.4 mm)

Flaws—irregularities such as scratches, holes, cracks, ridges, or hollows that do not follow a regular pattern, as in the case of waviness and roughness

Lay—the direction of the predominant surface pattern caused by the machining process

Profile—the contour of a specified section through a surface

Microinch or *micrometer*—the unit of measurement used to measure the surface finish; the microinch is equal to .000001 in.; the micrometer, to 0.000 001 m.

The following symbols indicate the direction of the lay (Fig. 18-4):

∥   parallel to the boundary line of the surface indicated by the symbol

⊥   perpendicular to the boundary line of the surface indicated by the symbol

X   angular in both directions on the surface indicated by the symbol

M   multidirectional

C   approximately circular to the center of the surface indicated by the symbol

R   approximately radial in relation to the center of the surface indicated by the symbol

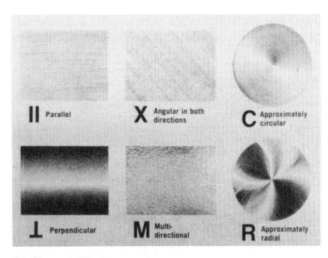

■ **Figure 18-4** Surface symbols used to designate the direction of the lay. *(Reprinted from ASME BY14.36M—1996, by permission of The American Society of Mechanical Engineers)*

3. If necessary, adjust the calibration control so that the instrument registers the same as the test block.

4. Unless otherwise specified, use the .030-in. (0.76-mm) cutoff range for surface roughness of 30 μin. (0.76 μm) or more. For surfaces of less than 30 μin. (0.76 μm), use the .010-in. (0.25-mm) cutoff range.

Note: When measuring a surface with an unknown roughness, set the range switch at a high setting to avoid damaging the instrument. After an initial test, the range switch may be turned to a finer setting for an accurate surface reading.

5. Thoroughly clean the surface to be measured to ensure accurate readings and reduce wear on the rider cap protecting the stylus.

6. With a smooth, steady movement of the stylus, trace the work surface at approximately .125 in./s (3 mm/s).

7. Note the reading from the meter scale.

A more elaborate device for measuring surface finish is the *surface analyzer.* It uses a recording device to reproduce the surface irregularities on a graduated chart, providing an ink-line record.

Although the surface indicator is the most common, other methods may be used to measure surface finish with reasonable accuracy during machining processes, including:

1. *Comparison blocks,* which are used for comparing the finish on the workpiece with the calibrated finish on a test block using the fingernail test

2. *Commercial sets of standard finished specimens,* which have up to 25 different surface finish samples. They consist of blocks or plates having

Average surface roughness produced by standard machining processes		
	Microinches	Micrometers
Turning	100–250	2.5–6.3
Drilling	100–200	2.5–5.1
Reaming	50–150	1.3–3.8
Grinding	20–100	0.5–2.5
Honing	5–20	0.13–0.5
Lapping	1–10	0.025–0.254

▸▸ To Measure Surface Finish with a Surface Indicator

1. Turn the switch on and allow the instrument to warm up for approximately 3 min.

2. Check the machine calibration by moving the stylus over the 125-μin. (3.2-μm) test block at approximately .125 in./s (3 mm/s).

table 18.1 Surface finishes obtained by various machining operations*

Tool	Operation	Material	Speed	Feed	Tool	Analyzer Setting		Surface Finish, Rq
						Cutoff	Range	
Cutoff saw	Sawing	2.50 in. diameter aluminum	320 ft/min (97.5 m/min)	—	10-pitch saw	.030 in. (0.76 mm)	1000	300–400
Vertical milling machine	Fly cutting (flat surface)	Machine steel	820 r/min	.015 in. (0.38 mm)	.060 in. radius Stellite	.030 in. (0.76 mm)	300	125–150
Horizontal milling machine	Slab milling	Cast aluminum	225 r/min	2.50 in./min (63.5 mm/min)	Slab cutter, 4 in. diameter HSS	.030 in. (0.76 mm)	100	40–50
Lathe	Turning	2.50 in. diameter aluminum	500 r/min	.010 in. (0.25 mm)	.046 in. radius HSS	.030 in. (0.76 mm)	300	100–200
	Turning	2.50 in. diameter aluminum	500 r/min	.007 in. (0.18 mm)	.078 in. radius HSS	.030 in. (0.76 mm)	100	50–60
	Facing	2 in. diameter aluminum	600r/min	.010 in. (0.25)mm	.030 in radius HSS	.030 in. (0.76)	300	200-225
	Facing	2 in. diameter aluminum	800 r/min	.005 in. (0.13 mm)	.030 in. radius HSS	.030 in. (0.76 mm)	100	30–40
	Filing	.750 in. diameter machine steel	1200 r/min	—	10-in. lathe file	.010 in. (0.25 mm)	100	50–60
	Polishing	.750 in. diameter machine steel	1200 r/min	—	#120 abrasive cloth	.010 in. (0.25 mm)	30	13–15
	Machine reaming	Aluminum	500 r/min	—	machine reamer HSS, ¾ in. diameter	.030 in. (0.76 mm)	100	25–32
Surface grinder	Grinding a flat surface	Machine steel	—	.030 in. (0.76 mm)	60-grit grinding wheel	.003 in. (0.076 mm)	10	7–9
Cutter and tool grinder	Cylindrical grinding	1 in. machine steel	—	Hand (slow)	46-grit grinding wheel	.010 in. (0.25 mm)	30	12–15
Lapping	Flat lapping	.875 in. × 5.50 in. tool steel (hardened)	—	Hand	600-grit abrasive	.010 in. (0.25 mm)	10	1–2
	Cylindrical lapping	.50 in. diameter tool steel (hardened)	—	Hand	600-grit abrasive	.010 in. (0.25 mm)	10	1–2

Metric figures given represent soft conversions.

surfaces varying from the smoothest to the roughest likely to be required (Fig. 18-5).

These specimens are used to check the finish of the machined part against the sample finish to determine approximately the finish produced on the part. It is often difficult to determine the finish visually. In such cases, the surfaces may be compared by moving the tip of your fingernail over the two surfaces.

Table 18.1 shows the results obtained on pieces of round and flat metal by various machining operations. A model B-1 110 Brush surface analyzer was used to obtain the readings. The speeds, feeds, and tool radii shown are those recommended for a high-speed steel toolbit.

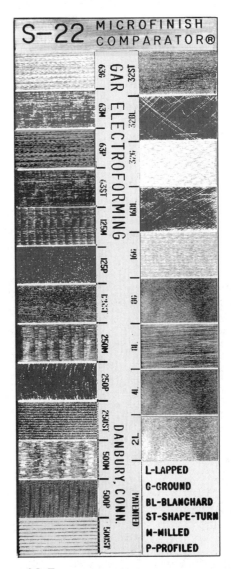

■ **Figure 18-5** A visual surface roughness comparator gage. *(GAR Electroforming Limited)*

unit 18 review questions

1. Explain why present-day standards for surface finish are very important to industry.

2. Define the following surface finish terms: microinch, lay, flaw, roughness, waviness, root mean square (Rq), and arithmetic average roughness height (Ra).

3. Explain each symbol and number (inches) as it applies to surface finish:

$$\sqrt[50]{\frac{.002 - 2}{}} = .020$$

4. Explain what the following lay symbols represent: 11, $\perp$, X, M, C, and R.

5. Describe briefly the principle and operation of a surface indicator and a surface analyzer in measuring surface finish.

Section 6

(The L. S. Starrett Co.)

LAYOUT TOOLS AND PROCEDURES

Laying out is the process of scribing or marking center points, circles, arcs, or straight lines on metal to indicate the shape of the object, the amount of metal to be removed during the machining process, and the position of the holes to be drilled. The layout helps the machinist determine the amount of material to be removed, although the size for rough and finish cuts must be checked by actual measurement.

All layouts should be made from a *baseline* or finished surface to ensure an accurate layout, correct dimensions, and proper location of holes. The importance of proper layout cannot be overemphasized. The accuracy of the finished product depends greatly on the accuracy of the layout. The drawing does not show how much material must be removed from each surface of the casting or workpiece; it merely shows which surfaces are to be finished.

The layout for holes, whether in cored or solid material, is as important as the layout for other dimensions of the workpiece.

Layouts may be of two types—*basic* (or *semiprecision*) and *precision.* A semiprecision layout may involve the use of basic measuring and layout tools, such as a rule and surface gage. It is generally not as accurate as a precision layout, which requires the use of more accurate equipment, such as the vernier height gage. If the part does not have to be precise, time should not be spent making a precision layout. Therefore, keep the layout as simple as workpiece requirements permit.

UNIT 19

Basic Layout Materials, Tools, and Accessories

OBJECTIVES

After completing this unit, you will be able to:

1 Prepare a work surface for layout

2 Use and care for various types of surface plates

3 Identify and use the main basic layout tools and accessories

The accuracy of a layout is very important to the accuracy of the finished product. If the layout is not correct, the workpiece will not be usable. A student should therefore realize that good layout entails the proper and careful use of all layout tools.

▶▶ Layout Solutions

The surface of the metal is usually coated with a layout solution to make layout lines visible. Several types of layout solutions are available. Regardless of the type used, the surface should be clean and free of grease.

The most commonly used layout solution is *layout dye,* or *bluing* (Fig. 19-1). This quick-drying solution, when coated lightly on the surface of any metal, will produce a background for sharp, clearcut lines. Layout dye may be applied with a cloth, brush, or dauber or sprayed on the work surface.

A copper-colored surface can be produced if the clean surface of a steel workpiece is coated with a copper sulfate ($CuSO_4$) solution to which a few drops of sulfuric acid have been added. When this solution is used, the surface of the workpiece must be absolutely clean and free from grease and finger marks.

Note: Copper sulfate should be used on ferrous metal only. It is particularly useful where hot chips are produced, which could affect other layout materials and blur the layout lines.

■ Figure 19-1 The work surface should be coated with layout dye before starting a layout. *(Kelmar Associates)*

A mixture of vermilion powder and shellac is often used for aluminum, since some layout compounds corrode aluminum. Alcohol should be used to thin this solution or to remove it from the workpiece.

The surfaces of castings and hot-rolled steel are often prepared for layout by merely chalking the surface.

Figure 19-2 A granite layout table provides an accurate reference surface for layout work. *(Kelmar Associates)*

Figure 19-3 A cast-iron surface plate. *(Kelmar Associates)*

A mixture of lime and alcohol, which readily clings to the rough surface of the castings, is often used for this purpose.

▶▶ Layout Tables and Surface Plates

Layout work may be performed on a layout table (Fig. 19-2) or on a surface plate (Fig. 19-3) made of granite or cast iron. Granite layout tables and plates are considered better than cast-iron ones because they:

> Do not become burred

> Do not rust

> Are not affected by temperature change

> Do not have internal stresses and therefore will not warp or distort

> Are nonmagnetic

> Can be used for checking near grinding machines, since abrasive particles will not embed in the surface

> Are cheaper than a similar-size cast-iron plate

Granite plates are available in three colors: black, pink, and gray. Black granite plates are considered superior because they are harder, denser, less porous, and therefore less likely to absorb moisture.

CARE OF SURFACE PLATES AND LAYOUT TABLES

Although surface plates are rugged, their accuracy can be easily destroyed. The following precautions should be taken with regard to surface plates:

> Keep the working surface clean.

> Cover the plate or table when it is not in use.

> Carefully place the work on the surface plate—do not drop it onto the plate.

> Use parallels under the workpiece whenever possible.

> Never hammer or punch any layout on a surface plate.

> Remove burrs from cast-iron plates and always protect their surfaces with a thin film of oil and a cover when they are not in use.

▶▶ Scribers

The *scriber* (Fig. 19-4 on p. 144) has a hardened steel point, or points, and may be used in conjunction with a combination square, rule, or straightedge to draw straight lines. On some scribers, one end is bent at an angle to allow marking lines in hard-to-reach places. To be accurate, any layout requires fine lines; therefore, the scriber point must always be sharp. The points of scribers, hermaphrodite calipers, dividers, and

■ Figure 19-4 A pocket and double-end scriber.

■ Figure 19-5 Sharpening a scriber on an oilstone.
(Kelmar Associates)

■ Figure 19-6 A spring divider can be used to scribe arcs and circles.   *(The L.S. Starrett Co.)*

trammels should be honed frequently on a fine oilstone (Fig. 19-5) to maintain their sharpness. When extremely fine lines are required, knife-edge scribers should be used.

▶▶ Dividers and Trammels

Dividers are used for scribing arcs and circles on a layout and for transferring measurements. The spring divider (Fig. 19-6), the most common type, is available in sizes from 3 to 12 in. Larger circles and arcs may be scribed with a trammel (Fig. 19-7).

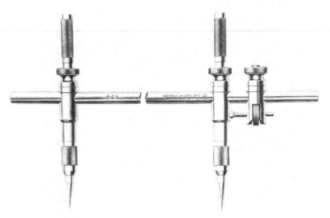

■ Figure 19-7 Trammels are used to scribe large arcs and circles.

A *trammel* consists of a beam on which *two sliding* or *adjustable heads* with *scriber points* are mounted. Some trammels may have an *adjusting screw* for fine adjustment. Rods or beams of different lengths can be used to increase the capacity of the trammel. When a circle is laid out from a hole, a ball attachment may be substituted for one of the scriber points.

▶▶ Hermaphrodite Calipers

The *hermaphrodite caliper* (Fig. 19-8) has one curved leg and one straight leg, which contains a scriber point. It is used for laying out lines parallel to an edge (Fig. 19-8) or for locating the center of round or irregular-shaped stock (Fig. 19-9).

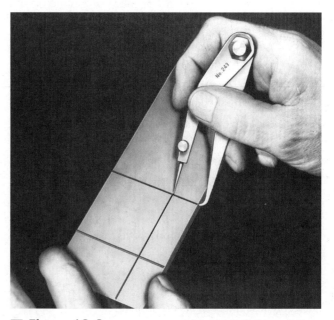

■ Figure 19-8 Scribing a line parallel to an edge with a hermaphrodite caliper.   *(The L.S. Starrett Company)*

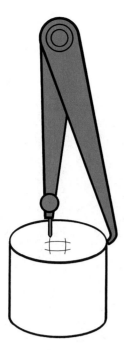

Figure 19-9 Locating the center of a round workpiece with a hermaphrodite caliper. *(Kelmar Associates)*

parts, the beam and the blade, is used where greater accuracy is required. Extremely accurate solid squares called *master squares* are used to check the accuracy of other squares.

▶▶ The Combination Set

The *combination set* (Fig. 19-11), used extensively in layout work, consists of a steel rule, square head, bevel protractor, and center head.

The *steel rule* may be fitted to the other three parts of the combination set for various layout, setup, and checking operations.

The *square head* and the rule or *combination square* may be used to lay out lines parallel to an edge (Fig. 19-12). It is also used to lay out angles at 45° and 90° to an edge (Fig. 19-13 on p. 146). The square head may be moved along the rule to any position. The square head is also used to check 45° and 90° angles and measure depths.

When setting the hermaphrodite caliper to a size, place the bent leg on the end of the rule and adjust the other leg until the scriber point is at the desired graduation. When scribing parallel lines with this tool, be sure always to hold the scriber at 90° to the edge of the workpiece.

▶▶ Squares

Squares are used to lay out lines at right angles (90°) to a machined edge, to test the accuracy of surfaces that must be square (90° to each other), and to set up work for machining.

Adjustable squares are used for general-purpose work. The *solid square* (Fig. 19-10), made up of two

Figure 19-11 The combination set is used for laying out and checking work. *(The L.S. Starrett Co.)*

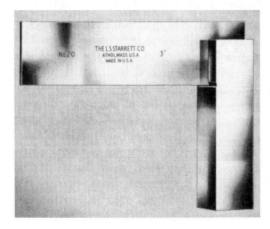

Figure 19-10 The solid square is used for inspection purposes. *(The L.S. Starrett Co.)*

Figure 19-12 Scribing a line parallel to an edge. *(Kelmar Associates)*

Basic Layout Materials, Tools, and Accessories **145**

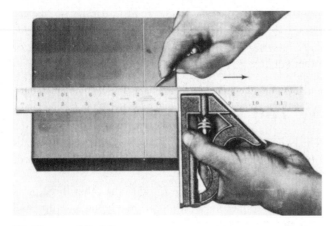

■ **Figure 19-13** Scribing a line at 90° to an edge.
(Kelmar Associates)

When mounted on a rule, the *bevel protractor* is used to lay out and check various angles. The protractor can be adjusted to any angle from 0° to 180°. The accuracy of this protractor is ±0.5° (30′). A universal bevel protractor may be used if an accuracy of 5′ is required.

The *center head* forms a center square when mounted on a rule. It may be used for locating the centers on the ends of round, square, and octagonal stock.

▶▶ Surface Gage

The *surface gage* (Fig. 19-14) is used with a surface plate or any flat surface to scribe layout lines on a workpiece. It consists of a *base, spindle,* and *scriber.*

The surface gage may be set to the required dimension by using a combination square (Fig. 19-14). This

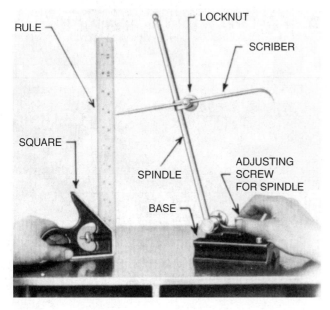

■ **Figure 19-14** Setting a surface gage to a dimension with a combination square. *(Kelmar Associates)*

RULE
LOCKNUT
SCRIBER
SQUARE
ADJUSTING SCREW FOR SPINDLE
SPINDLE
BASE

■ **Figure 19-15** Using a surface gage to lay out lines parallel to the top of a surface plate. *(Kelmar Associates)*

method is usually accurate enough for most layout work. A surface gage may be used on a surface plate to scribe parallel lines on a workpiece (Fig. 19-15). When the workpiece is fastened to an angle plate, horizontal and vertical lines can be laid out in one setup by merely rotating the angle plate 90° on the surface plate and scribing the line(s).

Most surface gages have two pins in the base, which when pushed down are used to guide the surface gage along the edge of the workpiece or surface plate. Some surface gages have a V-groove machined in the base, which allows them to be used on cylindrical work.

▶▶ Layout or Prick Punches and Center Punches

The *layout* or *prick punch* and the *center punch* (Fig. 19-16) differ only in the angle of the point. The prick punch is ground to an angle of 30° to 60° and is used to permanently mark the location of layout lines. The narrower angle of this punch makes a smaller and neater indentation in the metal surface.

The center punch is ground to an angle of 90° and is used to mark the location of the centers of holes. The wider indentation permits easier and more accurate starting of a drill point.

After the layout lines have been scribed on the workpiece, they should be permanently marked by means of layout or prick-punch marks. This step ensures that the layout line location will still be visible, should the line be rubbed off through handling. The intersection of the cen-

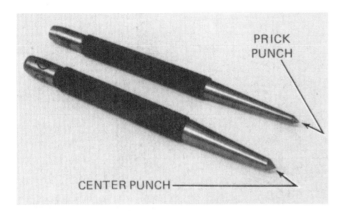

Figure 19-16 A prick punch and a center punch are used in layout work. *(Kelmar Associates)*

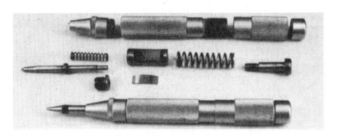

Figure 19-17 An automatic center punch produces uniform indentations on a layout. *(Kelmar Associates)*

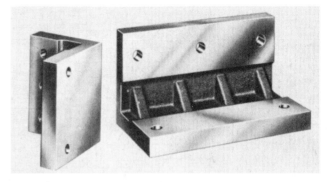

Figure 19-18 Angle plates have sides 90° to each other. *(Kelmar Associates)*

(a)

(b)

Figure 19-19 (a) Scribing horizontal lines with a surface gage; (b) the angle plate turned on its side to scribe vertical lines. *(Kelmar Associates)*

terlines of a circle should be carefully prick-punched and then enlarged with a center punch.

Note: Extreme care should be taken when the intersections of layout lines are being prick-punched. Regardless of how accurate the layout is, it is impossible to mark locations with a prick punch to closer than .003 to .004 in. (0.07 to 0.1 mm).

Uniform layout punch marks may be obtained with an automatic center punch (Fig. 19-17). This punch contains a striking block in the body, which is released by downward pressure; the block then strikes the punch proper, causing it to indent the metal. The impression size can be changed by adjusting the tension on the screw cap on the upper end of the tool. This type of punch produces marks uniform in size, which improves the appearance of the workpiece. Automatic center punches may also have a spacing attachment to provide uniform spacing of layout punch marks.

▶▶ Layout Accessories

In addition to regular layout tools, certain accessories are helpful in layout work. When lines are required on the face of a plate, it is customary to clamp the work to an *angle plate* (Fig. 19-18) with a *toolmaker's clamp*. This will hold the work in a vertical plane so that the layout lines will be accurately positioned. Since an accurate angle plate will have all adjacent surfaces at 90° to each other, it is possible to scribe intersecting 90° lines accurately. This accuracy is achieved by scribing all the horizontal lines on the workpiece, then turning the angle plate on its side and scribing the intersecting lines (Fig. 19-19a and b).

Parallels (Fig. 19-20 on p. 148) may be used when it is necessary to raise the workpiece to a desired height and to maintain the work surface parallel to the top of the surface plate.

Figure 19-20 Parallels keep the bottom surface of the workpiece parallel with the surface plate.

Figure 19-21 Lines may be scribed conveniently at 90° with a special V-block. *(Kelmar Associates)*

V-blocks are used to hold round work for layout and for inspection. They may be used singly or in pairs. Some blocks are so constructed that they may be rotated 90° on their sides without the work having to be removed. This feature permits laying out lines at 90° on a shaft without changing the position of the work in the V-block (Fig. 19-21).

Keyseat rules are used to lay out keyseats on shafts or to draw lines parallel to the axis of the shaft. A solid keyseat rule resembles two straightedges machined at 90° to each other (Fig. 19-22a). *Keyseat clamps,* when attached to a rule or straightedge, will convert it to a keyseat rule (Fig. 19-22b).

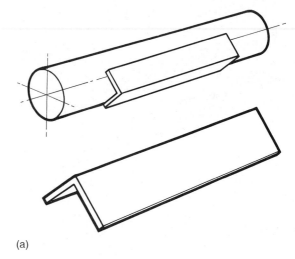

(a)

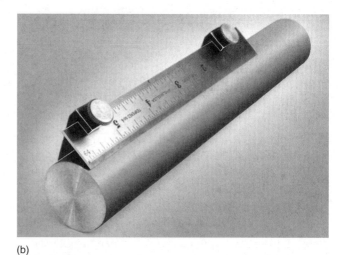

(b)

Figure 19-22 Keyseat rules are used to scribe lines parallel to the axis of a cylinder: (a) solid keyseat rules; (b) keyseat rule and a clamp.

unit 19 review questions

1. State two reasons a layout is necessary.

2. Why should a layout be as simple as possible?

Layout Materials

3. What is the purpose of a layout solution?

4. Name four layout solutions and state one application of each.

5. State two methods of preparing the surface of a casting prior to laying out the workpiece.

Layout Tables and Surface Plates

6. State five reasons granite surface plates are considered better than cast-iron ones.

7. List five precautions to be observed in the care of surface plates.

Scribers

8. Why must the point of a scriber always be sharp?

Dividers and Trammels

9. For what purpose are dividers used when laying out a workpiece?

10. What is the purpose of a trammel?

11. How can a circle be laid out concentric with a hole?

Hermaphrodite Calipers

12. List two uses for hermaphrodite calipers.

Squares

13. Name two types of squares used in layout work.

14. List four main parts of a combination set.

15. State three uses for the combination square.

16. What is the accuracy of the bevel protractor?

17. For what purpose is the center head used?

Surface Gage

18. Name the three main parts of the surface gage.

19. What is the purpose of the two pins in the base of the surface gage?

Layout or Prick Punches and Center Punches

20. How may the layout be made "permanent"?

21. What is the purpose of:
    a. A prick punch?
    b. A center punch?

22. To what angle is each punch ground?

Layout Accessories

23. List four layout accessories and state the purpose of each.

UNIT 20

Basic or Semiprecision Layout

OBJECTIVES

After completing this unit, you will be able to:

1 Lay out a workpiece to an accuracy of ±.007 in.

2 Lay out straight lines using the combination square and surface gage

3 Lay out hole centers, arcs, and circles

Basic or semiprecision layout requires the use of the basic tools described in Unit 19. Remember that the layout should be kept as simple as possible to save time and reduce the chance of error. Since the accuracy of the layout will affect the accuracy of the finished workpiece, care should be exercised in the layout procedure.

Although the layout required will naturally not be the same for each workpiece, certain procedures should be followed in any layout. The jobs described in this unit are intended to acquaint the reader with basic layout procedures.

▶▶ To Lay Out Hole Locations, Slots, and Radii

1. Study Fig. 20-1 and select the proper stock.
2. Cut off the stock, allowing enough material to square the ends if required.
3. Remove all burrs.
4. Clean the surface thoroughly and apply layout dye.
5. Place a suitable angle plate on a surface plate.

Note: Clean both plates (Fig. 20-2).

6. Clamp the work to the angle plate with a finished edge of the part against the surface plate or on a parallel. Leave one end of the angle protruding beyond the workpiece.
7. With the surface gage set to the proper height, scribe a centerline for the full length of the workpiece (Fig. 20-3a).

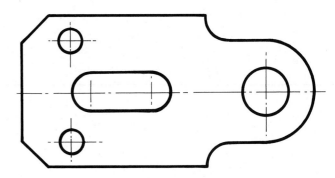

■ **Figure 20-1** A layout exercise.

8. Using the centerline as a reference, set the surface gage for each horizontal line and scribe the center lines for all hole and radii locations (Fig. 20-3b).
9. With the work still clamped to the angle plate, turn the angle plate 90° with edge B down and scribe the baseline at the bottom of the workpiece.

■ Figure 20-2 Place a suitable angle plate on the clean surface.

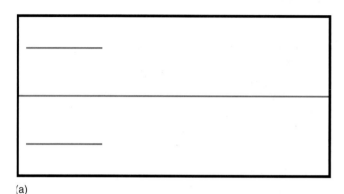

(a)

(b)

■ Figure 20-3 (a) Centerlines are scribed parallel to the base; (b) using a surface gage to scribe parallel lines.

10. Using the baseline as a reference line, locate and scribe the other centerlines for each hole or arc (Fig. 20-4b).

Note: All measurements for any location must be taken from the baseline or finished edge.

11. Locate the starting points for the angular layout (Fig. 20-4a).

12. Remove the workpiece from the angle plate.

13. Carefully prick-punch the center of all hole or radii locations.

BASELINE

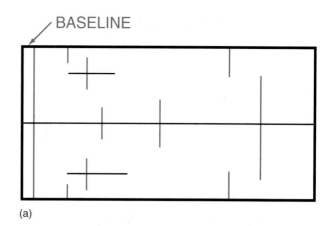

(a)

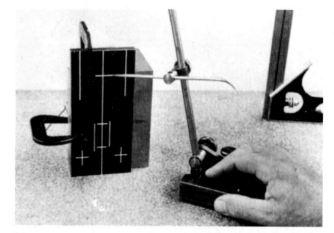

(b)

■ Figure 20-4 (a) Centerlines scribed; (b) the angle plate turned to 90° to scribe vertical lines.

14. Using a divider set as required, scribe all circles and arcs (Fig. 20-5a, p. 152).

15. Scribe any lines required to connect the arcs or circles (Fig. 20-6, p. 152).

16. Draw in the angular lines.

▶▶ To Lay Out a Casting Having a Cored Hole

When a casting that requires a hole in it is molded in a foundry, a core is used to produce the rough hole, which may have to be machined later. Often the core shifts out of place and the hole is cast off center, as shown in Fig. 20-7 on p. 152. If the hole must be machined concentric with the outside of the casting, it may be necessary to lay out the location of the hole. Follow this procedure:

1. Grind the scale off the surface to be laid out.

2. Tap a tightly fitting wooden piece into the cast hole (Fig. 20-7).

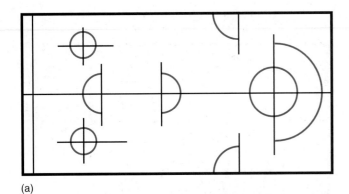

(a)

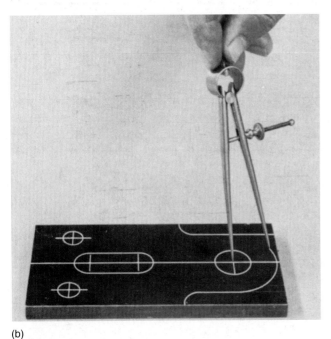

(b)

■ Figure 20-5 (a) Arcs and circles are scribed; (b) arcs and circles are scribed with a divider. *(Kelmar Associates)*

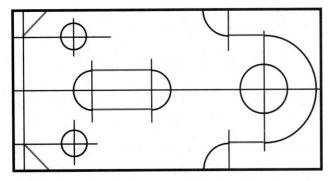

■ Figure 20-6 Arcs and circles are connected. *(Kelmar Associates)*

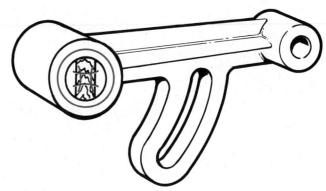

■ Figure 20-7 Method of centering a hole location on a casting. *(Kelmar Associates)*

3. Coat the surface to be laid out and the wooden piece with a solution of slaked lime and alcohol or layout dye.

4. With the hermaphrodite calipers, scribe four arcs as shown, using the outside diameter of the shoulder as the reference surface.

5. Using the intersection of these arcs as a center, scribe a circle of the required diameter on the casting. The hole should be concentric with the outside of the casting (Fig. 20-7).

6. Prick-punch the layout line at about eight equidistant points around the layout circle.

▶▶ To Lay Out a Keyseat in a Shaft

A keyseat is a recessed groove cut in a shaft and into which a key is fitted to prevent a mating part, such as a pulley or gear, from turning on the shaft. Laying out a keyseat requires much care, particularly if the mating parts must maintain a certain relative position. Follow this procedure:

1. Apply layout dye to the end of the shaft and to the area where the keyseat is to be laid out.

2. Mount the workpiece in a V-block.

3. Set the surface gage scriber to the center of the shaft.

4. Scribe a line across the end and continue it along the shaft to the keyseat location (Fig. 20-8).

5. Rotate the workpiece in the V-block and mark the length and the position of the keyseat on the shaft.

6. Set the dividers to half the width of the keyseat, and scribe a circle at each end of the layout (Fig. 20-8).

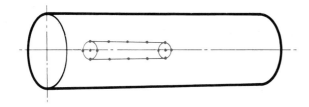

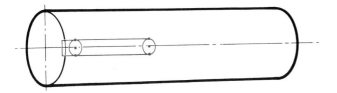

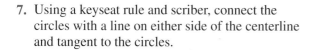

■ Figure 20-8 Keyseat layouts. *(Kelmar Associates)*

■ Figure 20-9 Aligning the keyseat layout in a vise. *(Kelmar Associates)*

7. Using a keyseat rule and scriber, connect the circles with a line on either side of the centerline and tangent to the circles.

Note: If a keyseat rule is not available, the circles may be connected using the surface gage.

8. Prick-punch the layout of the keyseat, and center-punch the centers of the circles.

9. If it is necessary to drill holes at the end of the keyseat, set up the shaft by aligning the end layout line in a vertical position with a square (Fig. 20-9).

unit 20 review questions

1. Outline the main steps that should be followed in making a layout consisting of straight lines, slots, and radii.

2. Name three common tools used to make basic or semiprecision layouts.

3. Describe how to lay out a cored hole concentric with the outside shoulder of a casting.

4. List the main steps for laying out a keyseat on a shaft.

UNIT 21

Precision Layout

OBJECTIVES

After completing this unit, you will be able to:

1 Make a precision layout using the vernier height gage

2 Use the Woodworth Coordinate Factors and Angles tables to calculate equidistant hole locations

3 Make precision layouts using the sine bar and gage blocks

The accuracy of the finished workpiece is generally determined by the accuracy of the layout; therefore, great care must be used when laying out. To make a precision layout, a person must be able to read and understand drawings, select and use the proper layout tools for the job, and accurately transfer measurements from the drawing to the workpiece. After completing a layout, check all layout work against the sizes on the working drawings to make sure that the layout is accurate. When the layout lines must be accurate to within .001 in. (0.02 mm), a *vernier height gage* may be used (Fig. 21-1).

When hole locations and dimension lines are made on a layout, they are generally made from two machined edges called *reference surfaces* using X and Y coordinates. Any layout composed of holes, angles, and lines may be calculated using trigonometry to determine the coordinate measurements. Once the coordinates have been determined, they can be used for setting up the workpiece and accurately positioning the holes for machining. Another method of calculating hole locations is by means of the Woodworth Coordinate Factors and Angles tables. See the Appendix of Tables in this book.

▶▶ The Vernier Height Gage

The *vernier height gage* may be used to measure or mark off vertical distances to ±.001-in. (0.02-mm) accuracy. The main parts of the vernier height gage (Fig. 21-2) are the *base, beam, vernier slide,* and *scriber,* which is attached to the vernier slide when making layouts. Other accessories, such as a dial indicator or a depth gage attachment, may be added to the slide for measurement and inspection work. The graduations on the beam and the vernier slide are the same as those on a vernier caliper and the readings are made in the same manner as with a vernier caliper.

TO MAKE A PRECISION LAYOUT USING A VERNIER HEIGHT GAGE

It is required to lay out the position of five equally spaced holes on a 5-in. diameter circle located in the center of a 7-in.-*square* steel plate (Fig. 21-3). Follow this procedure:

1. Refer to the drawing of the required workpiece (Fig. 21-3).
2. Remove all burrs from the workpiece.
3. Apply layout dye to the surface and mount it on an angle plate.

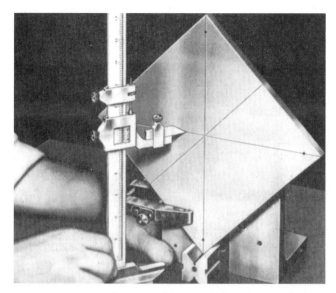

Figure 21-1 A vernier height gage is used when an accurate layout is required. *(The L.S. Starrett Co.)*

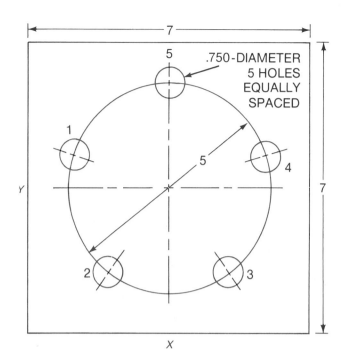

Figure 21-3 Layout of five equally spaced holes on a circle. *(Kelmar Associates)*

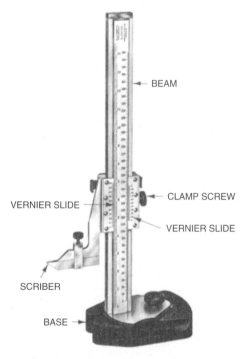

BEAM

CLAMP SCREW

VERNIER SLIDE

VERNIER SLIDE

SCRIBER

BASE

Figure 21-2 The main parts of a vernier height gage. *(The L.S. Starrett Co.)*

Figure 21-4 Offset scribers. *(The L.S. Starrett Co.)*

4. Clean the surface of the layout table, the angle plate, and the base of the height gage.

5. Mount an offset scriber (Fig. 21-4) on the vernier slide and clamp it in position.

6. Move the vernier slide and scriber down until the scriber touches the top of the surface plate.

7. Check the reading on the vernier scale. The zero mark on the vernier should align exactly with the zero graduation on the beam. If zero on the vernier does not coincide with zero on the beam, recheck the assembly of the scriber and the vernier slide.

8. Refer to the Coordinate Factors and Angles tables in the appendix for the coordinates for five equidistant holes.

9. Calculate the location of all five holes as follows:

Hole 1

Horizontal distance from left-hand edge Y

= (diameter of circle × factor for A) + 1.000

= (5 × .024472) + 1.000

= .12236 + 1.000

= 1.122

Vertical distance from upper edge X

= (diameter × factor for B) + 1.000

= (5 × .345492) + 1.000

= 1.72746 + 1.000

= 2.727

Hole 2

Distance from edge Y

= (5 × .206107) + 1.000

= 1.030535 + 1.000

= 2.031

Distance from edge X

= (5 × .904508) + 1.000

= 4.52254 + 1.000

= 5.523

Hole 3

Distance from edge Y

= (5 × .793893) + 1.000

= 3.969465 + 1.000

= 4.969

Distance from edge X

= (5 × .904508) + 1.000

= 4.52254 + 1.000

= 5.523   (same as for hole 2)

Hole 4

Distance from edge Y

= (5 × .975528) + 1.000

= 4.87764 + 1.000

= 5.878

Distance from edge X

= (5 × .345492) + 1.000

= 1.72746 + 1.000

= 2.727   (same as for hole 1)

Hole 5

Distance from edge Y

= (5 × .5000) + 1.000

= 2.5000 + 1.000

= 3.500

Distance from edge X

= (5 × .000) + 1.000

= .000 + 1.000

= 1.000

10. Place edge Y on the layout table surface.

11. Set the height gage to 1.122 and scribe a line by *drawing* the scriber across the face of the workpiece at the location for hole 1.

12. Set the height gage to each of the following settings. After each setting is made, scribe the line for the appropriate hole location (Fig. 21-5).

Hole 2—2.031

Hole 3—4.969

Hole 4—5.878

Hole 5—3.500

13. Rotate the angle plate and the work 90° and place edge X on the layout table.

14. Set the vernier height gage to 2.727.

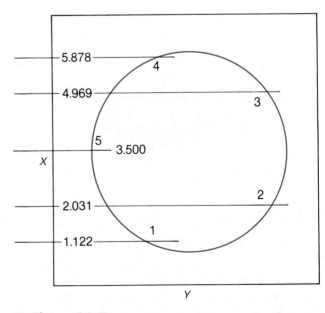

■ **Figure 21-5** The horizonal positions of five holes.

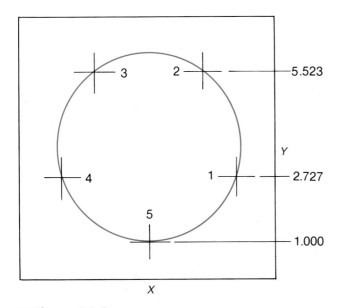

■ **Figure 21-6** The vertical distance of each hole.

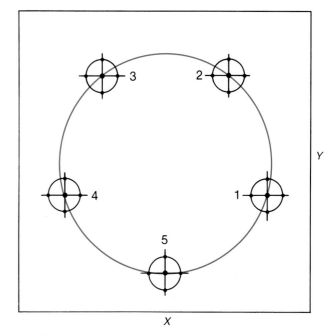

■ **Figure 21-7** Prick-punch marks on the circles ensure the permanency of the layout.

15. Scribe the intersecting lines at the centers of holes 1 and 4.

16. Set the height gage to 5.523.

17. Scribe the intersecting lines for the centers of holes 2 and 3.

18. Set the height gage to 1.000 and scribe the intersecting line for hole 5 (Fig. 21-6).

19. Remove the workpiece from the angle plate and place it on the bench with the layout surface up.

20. Using a sharp prick punch and a magnifying glass, carefully mark the centers of the holes at the intersecting lines.

21. Set the dividers to .375 in. and scribe the five .750 in. circles.

22. Carefully prick-punch the circumference of each circle at four equidistant points to ensure the permanency of the layout (Fig. 21-7).

▶▶ To Make a Precision Layout Using a Sine Bar, Gage Blocks, and a Vernier Height Gage

If a more accurate layout is required for hole positions and angles (Fig. 21-8), it may be done by using a sine bar, gage

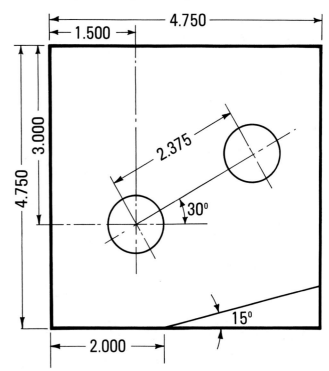

■ **Figure 21-8** An accurate layout for precision work.

blocks, and height gage to accurately establish the hole locations and their X and Y axes. The use of coordinates is preferred to locate hole centers, since the work can be set up on a jig borer or vertical mill using these same coordinate measurements to position the holes for machining.

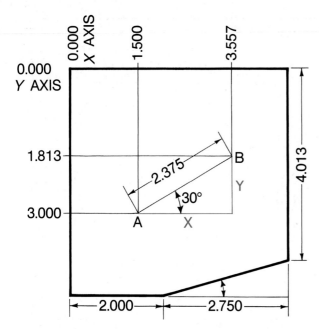

Figure 21-9 The position of hole B is calculated using trigonometry.

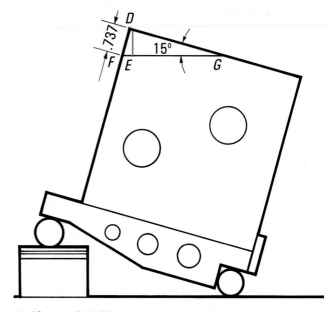

Figure 21-10 To locate line *EG* accurately, distance *DE* must be known. *(Kelmar Associates)*

This type of layout example is shown in Fig. 21-9. Follow this procedure:

1. Check the drawing for the required dimensions (Fig. 21-8).
2. Machine and grind a plate square to size 4.750 in. × 4.750 in.
3. Clean the surface of the workpiece and coat it with layout dye.
4. Set the work on edge on a surface plate and clamp it to an angle plate.
5. Using a vernier height gage, scribe the centerlines for hole A (Fig. 21-9).
6. Calculate the position of hole B as follows:

Length of side X
$$\frac{X}{2.375} = \cos 30°$$
$$X = 2.375 \cos 30°$$
$$= 2.375 \times .86603$$
$$= 2.0568 \text{ in.}$$

Length of side Y
$$\frac{Y}{2.375} = \sin 30°$$
$$Y = 2.375 \sin 30°$$
$$= 2.375 \times .5000$$
$$= 1.1875 \text{ in.}$$

The position of the center of hole B is then 2.0568 in. to the right of the centerline for hole A and

1.1875 in. above the centerline for hole A. It is now possible to position the hole using the coordinate method:

The vertical centerline for hole B would then be located 1.500 + 2.0568 = 3.5568 in. along the *X* axis.

The horizontal centerline for hole B would be located 3.000 − 1.1875 = 1.8125 in. along the *Y* axis.

7. Using a vernier height gage, mark off the centerline for hole B.
8. Remove the workpiece from the angle plate.
9. Carefully prick-punch the intersection of the centerlines of these holes using a sharp punch and a magnifying glass.
10. To lay out the 15° angle at the corner of the plate, calculate the buildup for 15° using a 5-in. sine bar.

$$\text{Buildup} = 5 \sin 15°$$
$$= 5 \times .25882$$
$$= 1.2941 \text{ in.}$$

11. Place the buildup under one end of a sine bar (on a surface plate).
12. Place the workpiece, angled edge up, on a sine bar and clamp to an angle plate.
13. Calculate length *DF* (Fig. 21-10) as follows:

$$\frac{DF}{2.750} = \tan 15°$$
$$= .26795$$
$$DF = .26795 \times 2.750$$
$$= .7368 \text{ in.}$$

14. Calculate length *DE* (Fig. 21-10). To scribe the line accurately at 15° as required in Fig. 21-8, it is necessary first to calculate the length of line *DE*, since this is the vertical distance below *D* that the line must be scribed. By previous calculations, it was determined that *DF* = .7368 in. In the triangle *DEF*, the angle *FDE* is 15°.

$$\therefore \frac{DE}{DF} = \cos 15°$$

$$DE = \cos 15° \times DF$$

$$= .96592 \times .737$$

$$= .7118$$

$$= .712 \text{ in.}$$

15. Set the scriber on the vernier height gage to the uppermost corner of the plate (point *D*).

16. Lower the scriber .712 in. and scribe line *GF*. This will locate point *G* at a position 2.000 in. from the side of the plate, as required in Fig. 21-8.

17. Remove the workpiece from the angle plate.

18. Lightly prick-punch the layout line using a magnifying glass and a sharp prick punch.

unit 21 review questions

1. Name three requirements that a machinist must meet to make a precision layout.

2. List four main parts of a vernier height gage.

3. How can the vertical and horizontal distances of equally spaced holes on a circle be calculated from the edges of a workpiece?

4. Calculate the vertical and horizontal distances for three equally spaced holes on a 4-in. diameter circle located in the center of a 6-in.-square plate.

5. How can the hole centers at intersecting lines be accurately marked?

6. How can accurate angular lines be laid out?

7. Calculate the gage block buildup for setting a sine bar to an 18° angle.

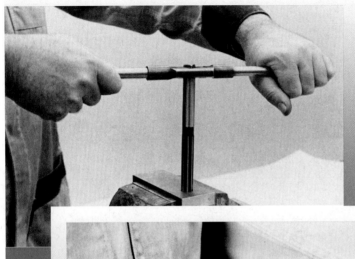

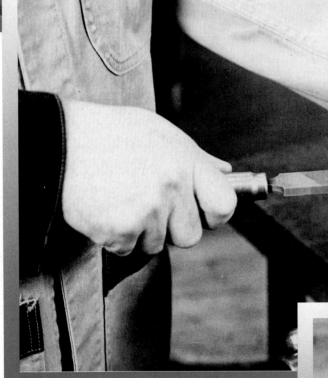

(Kelmar Associates)

HAND TOOLS AND BENCH WORK

The machine tool trade may be divided into two categories: hand tool and machine tool operations.

Although this era is looked on as the machine age, the importance of hand tool operations or bench work should not be overlooked. Bench work includes the operations of laying out, fitting, and assembling. These operations involve sawing, chipping, filing, polishing, scraping, reaming, and threading. A good machinist should be capable of using all hand tools skillfully. Effective selection and use of these tools is possible only with continued practice.

UNIT 22

Holding, Striking, and Assembling Tools

OBJECTIVES

After completing this unit, you will be able to:

1 Select various tools used for holding, assembling, or dismantling workpieces

2 Properly use these tools for holding, assembling, and dismantling workpieces

Hand tools may be divided into two classes: *noncutting* and *cutting.* Noncutting tools include vises, hammers, screwdrivers, wrenches, and pliers, which are used basically for holding, assembling, or dismantling parts.

▶▶ The Bench Vise

The *machinist's,* or *bench, vise* (Fig. 22-1) is used to hold small work securely for sawing, chipping, filing, polishing, drilling, reaming, and tapping operations. Vises are mounted close to the edge of the bench; they permit long work to be held in a vertical position. Vises may be made of cast iron or cast steel. Vise size is determined by the width of the jaws.

A machinist's vise may be of the solid-base or swivel-base type. The swivel-base vise (Fig. 22-1) differs from the solid-base vise because it has a swivel plate attached to the bottom of the vise. This plate allows the vise to be swung into any circular position. To grip finished work or soft material, use jaw caps made of brass, aluminum, or copper to protect the work surface from being marred or damaged.

▶▶ Hammers

Many different types of hammers are used by the machinist, the most common being the *ball-peen hammer* (Fig. 22-2). The larger striking surface is called the *face,*

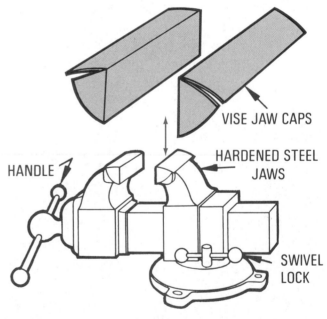

■ **Figure 22-1** A swivel-base bench vise can be rotated to any position. *(Kelmar Associates)*

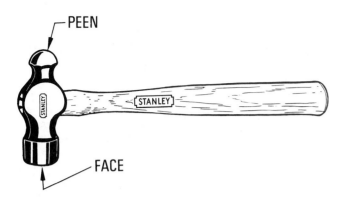

PEEN

FACE

■ **Figure 22-2** A ball-peen hammer. *(Stanley Tools Division of Stanley Works)*

(a)

(b)

■ **Figure 22-3** (a) A soft-faced hammer; (b) cracked hammer handles are dangerous to use. *(Kelmar Associates)*

■ **Figure 22-4** A standard screwdriver.

■ **Figure 22-5** A Phillips screwdriver.

and the smaller, rounded end is the *peen*. Ball-peen hammers are made in a variety of sizes, with head masses ranging from approximately 2 ounces (oz) to 3 pounds (lb) (55 to 1400 g). The smaller sizes are used for layout work and the larger ones for general work. The peen is generally used in riveting or peening operations.

Soft-faced hammers (Fig. 22-3a) have heads made of plastic, rawhide, copper, or lead. These heads are fastened to a steel body and can be replaced when worn. Soft-faced hammers are used in assembling or dismantling parts so the finished surface of the work will not be marred. Lead hammers are often used to seat the workpiece properly on parallels when setting up work in a vise for machining operations. Plastic hammer heads, which have been filled with lead or steel shot, are gradually replacing the lead hammer, since they do not lose their shape and last much longer than the lead hammer heads.

When using a hammer, always grasp it at the end of the handle to provide better balance and greater striking force. This grip also tends to keep the hammer face flat on the work and reduces the chance of damage to the workface.

The following safety precautions should always be observed when using a hammer:

1. Be sure that the handle is solid and not cracked (Fig. 22-3b).

2. See that the head is tight on the handle and secured with a proper wedge to keep the handle expanded in the head.

3. Never use a hammer with a greasy handle or when your hands are greasy.

4. Never strike two hammer faces together. The faces have been hardened and a metal chip may fly off, causing an injury.

▶▶ Screwdrivers

Screwdrivers are manufactured in a variety of shapes, types, and sizes. The two most common types used in a machine shop are the standard or flat blade (Fig. 22-4) and the Phillips screwdriver (Fig. 22-5). Both types are manufactured in various sizes and styles, such as standard shank, stubby shank, and offset (Fig. 22-6 on p. 164).

Holding, Striking, and Assembling Tools **163**

■ **Figure 22-6** An offset screwdriver is often used in confined spaces.

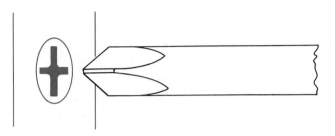

■ **Figure 22-7** The tip of a Phillips screwdriver fits into the socket in the screw. *(Kelmar Associates)*

Phillips screwdrivers have a +-shaped tip for use with Phillips-type recessed screw heads (Fig. 22-7). These screwdrivers are manufactured in four sizes: #1, #2, #3, and #4, to suit the various-sized recesses in the heads of fasteners. Care must be taken to use the proper size screwdriver. Too small a screwdriver will damage both the tip and the recess in the screw head. The screwdriver should be held firmly in the recess and square with the screw.

Blades for smaller standard screwdrivers are generally made of round stock, and blades for larger ones are often square, so that a wrench may be applied for leverage.

CARE OF A SCREWDRIVER

1. Choose the correct size of screwdriver for the job. If too small a screwdriver is used, both the screw slot and the tip of the screwdriver may become damaged.
2. Do not use the screwdriver as a pry, chisel, or wedge.
3. When the tip of a standard screwdriver becomes worn or broken, it should be redressed to shape (Fig. 22-8).

REGRINDING A STANDARD SCREWDRIVER BLADE

When regrinding a screwdriver tip, make the sides of the blade slightly concave by holding the side of the blade tangential to the periphery of the grinding wheel (Fig. 22-8).

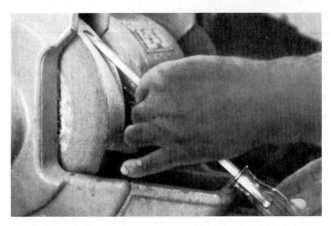

■ **Figure 22-8** Regrinding a standard screwdriver blade. *(Kelmar Associates)*

Grind an equal amount off each side of the blade. This shape will enable the blade to maintain a better grip in the slot. Be sure to retain the original taper, width, and thickness of the tip and grind the end square with the centerline of the blade.

Note: When grinding, remove a minimum amount of metal so you do not grind past the hardened zone in the tip. Quench the tip frequently in cold water so as not to draw the temper from the blade.

▶▶ Wrenches

Many types of wrenches are used in machine shop work, each suited for a specific purpose. The name of the wrench is derived from its use, shape, or construction. The following types of wrenches are commonly used in a machine shop.

Open-end wrenches may be single-ended (Fig. 22-9) or double-ended (Fig. 22-10). The openings on these wrenches are usually offset at a 15° angle to permit turning the nut or bolt head in limited spaces by "flopping" the wrench.

Double-ended wrenches usually have a different-size opening at each end to accommodate two different sizes of bolt heads or nuts. These wrenches are available in both inch and metric sizes.

■ **Figure 22-9** A single-ended open-end wrench.

■ Figure 22-10 A double-ended open-end wrench.

■ Figure 22-11 A box-end or 12-point wrench.

■ Figure 22-12 A set of socket wrenches. *(Kelmar Associates)*

■ Figure 22-13 The correct method of using an adjustable wrench. *(Kelmar Associates)*

■ Figure 22-14 An Allen wrench or hex key. *(Kelmar Associates)*

Box-end 12-point wrenches (Fig. 22-11) completely surround the nut and are useful in close quarters where only a small rotation of the nut can be obtained at one time. The box end has 12 precisely cut notches around the inside face. These notches fit closely over the points on the outside of the nut. Because this wrench cannot slip when the proper size is used, it is preferred over most other styles of wrenches. These wrenches usually have a different size at each end and are available in inch and metric sizes.

Socket wrenches (Fig. 22-12) are similar to box wrenches because they are usually made with 12 points and surround the nut. These sockets are also available in inch and metric sizes. Several types of drives, including ratchet and torque-wrench handles, are available for the various sockets. When nuts or bolts must be tightened to within certain limits to prevent warping, socket wrenches are used in conjunction with a torque-wrench handle.

Adjustable wrenches (Fig. 22-13) may be adjusted to within a certain range to fit several sizes of nuts or bolt heads. This wrench is particularly useful for odd-size nuts or when another wrench of the proper size is not available. Unfortunately, this type of wrench can slip when not properly adjusted to the flats of the nut. This may result in injury to the operator and damage to the corners of the nut.

When using an adjustable wrench, it should be tightened securely to the faces of the nut and the turning force applied in the direction indicated in Fig. 22-13.

Allen setscrew wrenches (Fig. 22-14), commonly called *hex keys,* are hexagonal and fit into the recesses of socket head setscrews. They are made of tool steel and are available in sets to fit a wide variety of screw sizes. The indicated size of the wrench is the distance across the flats of the wrench. Usually this distance is one-half the outside diameter of the Allen setscrew in which it is used. These wrenches are available in both inch and metric sizes.

Pin spanner wrenches are specialized wrenches generally supplied by the machine tool manufacturer for use on specific machines. They are supplied in various types. Fixed-face spanners and adjustable-face spanners (Fig. 22-15a on p. 166) are positioned in two holes on the face of a special nut or threaded fitting on a machine.

A *hook-pin spanner* (Fig. 22-15b) is used on the circumference of a round nut. The pin of the spanner fits into a hole in the periphery of the nut.

Holding, Striking, and Assembling Tools

(a)

(b)

■ **Figure 22-15** (a) A pin spanner wrench; (b) a hook-pin spanner wrench. *(Bahco North America)*

HINTS ON USING WRENCHES

1. Always select a wrench that fits the nut or bolt properly. A wrench that is too large may slip off the nut and cause an accident.
2. Whenever possible, *pull* rather than push on a wrench to avoid injury if the wrench should slip.
3. Always be sure that the nut is fully seated in the wrench jaw.
4. Use a wrench in the same plane as the nut or bolt head.
5. When tightening or loosening a nut, give it a sharp, quick jerk, which is more effective than a steady pull.
6. Put a drop of oil on the threads when assembling a bolt and nut to ensure easier removal later.

▸▸ Pliers

Pliers are useful for gripping and holding small parts for certain machining operations (such as drilling small holes) or when assembling parts. Pliers are made in many types and sizes and are named by their shape, their function, or their construction. The following types of pliers are commonly used in a machine shop.

Combination or *slip-joint pliers* (Fig. 22-16) are adjustable, to grip both large and small workpieces. They may be used to grip certain work when small holes must be drilled or for bending or twisting light, thin materials.

Side-cutting pliers (Fig. 22-17) are used mainly for cutting, gripping, and bending of small diameter (⅛ in. or less) rods or wire.

■ **Figure 22-16** Slip-joint or combination pliers.

■ **Figure 22-17** Side-cutting pliers.

■ **Figure 22-18** Needle-nose pliers.

■ **Figure 22-19** Diagonal cutters.

Needle-nose pliers (Fig. 22-18) are available in both straight- and bent-nose types. They are useful for holding very small parts, positioning them in hard-to-get-at places, and bending or forming wire.

Diagonal cutters (Fig. 22-19) are used solely for cutting wire and small pieces of soft metal.

Vise-grip pliers (Fig. 22-20) provide extremely high gripping power because of the adjustable lever action. The screw in the handle allows adjustment to various sizes. This type of plier is available in several different styles, such as standard jaws, needle jaws, and C-clamp jaws.

Figure 22-20 Vise-grip pliers.

HINTS ON USING PLIERS

The following points should be observed if pliers are to give the proper service.

1. Never use a plier instead of a wrench.

2. Never attempt to cut large-diameter or heat-treated material with pliers. This may cause the jaws to distort or the handle to break.

3. Always keep pliers clean and lubricated.

unit 22 review questions

Bench Vise

1. What is the advantage of the swivel-base vise over the solid-base vise?

2. How may finished work be held in a vise without the surface being marred or damaged?

Hammers

3. Describe the most common hammer used by a machinist.

4. For what purpose are soft-faced hammers used?

5. State three safety rules that should be observed when using a hammer.

Screwdrivers

6. List three important ways to take care of a screwdriver.

7. Explain the procedure for regrinding the tip of a screwdriver blade.

8. List two precautions that should be observed in using a Phillips screwdriver.

Wrenches

9. Why are open-end wrenches offset about 15° to the handle?

10. Why is a properly sized box wrench preferred to other types of wrenches?

11. What advantage does a socket wrench have over a box wrench?

12. What precaution should be observed when using an adjustable wrench?

13. What will happen if excess pressure is applied to an adjustable wrench or pressure applied on the wrong jaw?

14. What is the cross-sectional shape of an Allen wrench and for what purpose is an Allen wrench used?

15. Where is a hook-pin spanner used?

16. State four useful hints for using any wrench.

Pliers

17. Name four types of pliers and state one use for each.

18. What advantage do vise grips have over other types of pliers?

UNIT 23

Hand-Type Cutting Tools

OBJECTIVES

After completing this unit, you will be able to:

1 Select and use the proper hacksaw blade for sawing a variety of materials

2 Select and use a variety of files to perform various filing operations

3 Identify and know the purpose of rotary files, ground burrs, and scrapers

Although most metal cutting can be done more easily, quickly, and accurately on a machine, it is often necessary to perform certain metal-cutting operations at a bench or on a job. Such operations include sawing, filing, scraping, reaming, and tapping. It is therefore important that the prospective machinist knows how to use hand-type cutting tools properly.

▶▶ Sawing, Filing, and Scraping

Hacksaws, files, and scrapers are very common tools in the machine shop and often the most incorrectly used and abused. The proper use of these tools will not come immediately. It is only through practice that the student or apprentice will become proficient in their use.

▶▶ Hand Hacksaw

The *pistol-grip hand hacksaw* (Fig. 23-1) is composed of three main parts: the *frame,* the *handle,* and the *blade.* The frame can be either solid or adjustable. The solid frame is more rigid and will accommodate blades of only one specific length. The adjustable frame is more commonly used and will take blades which range from 10 to 12 in. (250 to 300 mm) long. A wing nut at the back of the frame provides adjustment for blade tensioning.

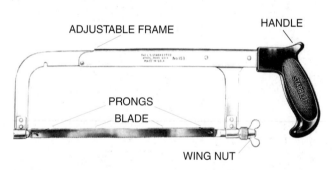

■ **Figure 23-1** Parts of a hand hacksaw. *(The L.S. Starrett Co.)*

Hacksaw blades are made of high-speed molybdenum or tungsten-alloy steel that has been hardened and tempered. There are two types: the solid, or all-hard, blade and the flexible blade. Solid blades are hardened throughout and are very brittle. They break easily if not used properly. Only the teeth of the flexible blade are hardened, while the back of the blade is soft and flexible. Although

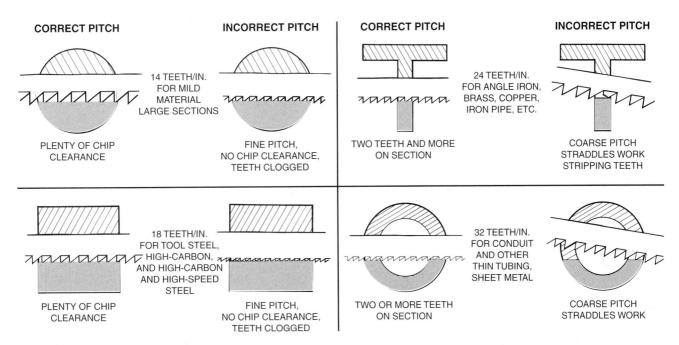

Figure 23-2 Selection of the proper blade pitch is very important.

this type of blade will stand more abuse than the all-hard blade, it will not last as long in general use.

Solid blades are used on brass, tool steel, cast iron, and larger sections of mild steel, since they do not run out of line when pressure is applied. Flexible blades may be used on channel iron, tubing, copper, and aluminum, since they do not break as easily on material with thin cross sections.

Blades are manufactured in various pitches (number of teeth per inch), such as 14, 18, 24, and 32. The pitch is the most important factor to consider when selecting the proper blade for a job. An 18-tooth blade (18 teeth per inch) is recommended for general use. When selecting a blade, choose as coarse a blade as possible to provide plenty of chip clearance and to cut through the work as quickly as possible. The blade selected should have at least *two teeth in contact with the work* at all times. This will prevent the work from jamming between the teeth and stripping the teeth from the blade. Fig. 23-2 provides a guide for proper blade selection.

TO USE THE HAND HACKSAW

1. Check to make sure that the blade is of the proper pitch for the job and that the teeth point *away* from the handle.

2. Adjust the blade tension so that the blade cannot flex or bend.

3. Mount the stock in the vise so that the cut will be about .250 in. (6 mm) from the vise jaws.

4. Grasp the hacksaw as shown in Fig. 23-3 on p. 170. Assume a comfortable stance, standing erect with the left foot slightly ahead of the right foot.

5. Start the saw cut just outside and parallel to a previously scribed line.

Note: File a V-shaped nick at the starting point to help start the saw blade at the right spot.

6. After the cut has started, apply pressure only on the forward stroke. Use about 50 strokes per minute.

7. When cutting thin material, hold the saw at an angle to have at least two teeth in contact with the work at all times. Thin work is often clamped between two pieces of wood, and the cut is made through all pieces (Fig. 23-3).

8. When nearing the end of the cut, slow down to control the saw as it breaks through the material.

Note: If a saw blade breaks or becomes dull in a partly finished cut, replace the blade and rotate the work one-half turn so that the old cut is at the bottom. A new blade will bind in an old cut and the "set" of the new teeth will be ruined quickly.

■ **Figure 23-3** The correct method of holding a hacksaw. *(Kelmar Associates)*

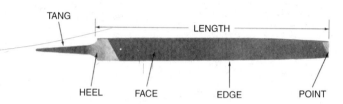

■ **Figure 23-4** The main parts of a file. *(Kelmar Associates)*

▶▶ Files

A file is a hand cutting tool made of high-carbon steel, having a series of teeth cut on its body by parallel chisel cuts. The parts of a file are shown in Fig. 23-4. Files are used to remove surplus metal and to produce finished surfaces. Files are manufactured in a variety of types and shapes, each for a specific purpose. They may be divided into two classes: single-cut and double-cut (Fig. 23-5).

Single-cut files have a single row of parallel teeth running diagonally across the face. They include *mill, long-angle lathe,* and *saw files.* Single-cut files are used when a smooth finish is desired or when hard materials are to be finished.

Double-cut files have two intersecting rows of teeth. The first row is usually coarser and is called the *overcut.* The other row is called the *upcut.* These intersecting rows produce hundreds of cutting teeth, which provide for fast removal of metal and easy clearing of chips.

DEGREES OF COARSENESS

Both single- and double-cut files are manufactured in various degrees of coarseness, such as *rough, coarse, bastard, second-cut, smooth,* and *dead smooth.* Those most commonly used by the machinist are the bastard, second cut, and smooth (Fig. 23-5).

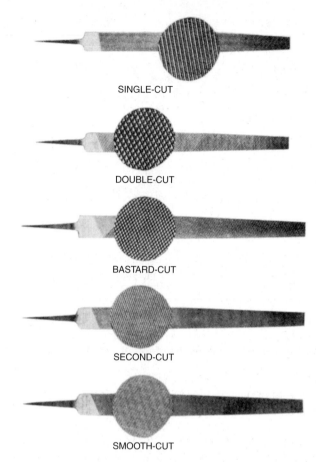

■ **Figure 23-5** File classification. *(Delta File Works)*

MACHINIST FILES

The types of files most commonly used by machinists are the *flat, hand, round, half-round, square, pillar, three-quarter* (triangular), *warding,* and *knife* (Fig. 23-6).

Care of Files

Because files are relatively inexpensive hand tools, they are often abused. Proper care, selection, and use are most important if good results are to be obtained with files. The following points should be observed in the care of files.

1. Do not store files where they will rub together. Hang or store them separately.

2. Never use a file as a pry or a hammer. Since the file is hard, it snaps easily, often causing small pieces to fly that may result in a serious eye injury.

3. Do not knock a file on a vise or other metallic object to clean it. Always use a file card or brush for this purpose (Fig. 23-7).

4. Apply pressure only on the forward stroke when filing. Pressure on the return stroke will dull the file.

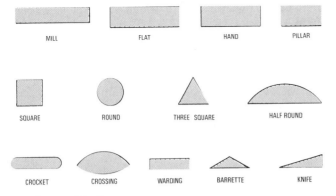

Figure 23-6 Cross-sectional views of machinists' files. *(Kelmar Associates)*

Figure 23-8 Always hold a file level when filing. *(Kelmar Associates)*

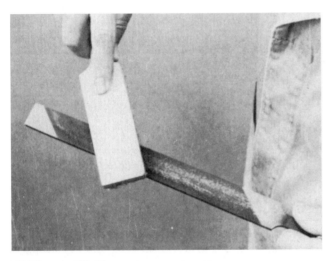

Figure 23-7 Cleaning a file with a file brush. *(Kelmar Associates)*

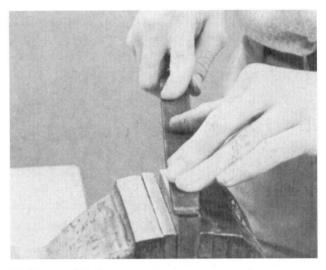

Figure 23-9 Cross-filing will show any high spots on the workpiece surface. *(Kelmar Associates)*

5. Do not press too hard on a new file. Too much pressure tends to break off the cutting edges and shorten the life of the file.

6. Too much pressure also results in "pinning" (*small particles being wedged between the teeth*), which causes scratches on the surface of the work. Keep the file clean. A piece of brass, copper, or wood pushed through the teeth will remove the "pins." Applying chalk to the face of the file will lessen the tendency for the file to become clogged ("pinned").

FILING PRACTICE

Filing is an important hand operation and one that can be mastered only through patience and practice. The following points should be observed when cross-filing:

1. **STOP** ***Never*** **use a file without a handle. Ignoring this rule is a dangerous practice. Serious hand injury may result, should the file slip.**

2. Fasten the work to be filed, at about elbow height, in a vise.

3. To produce a flat surface, hold the right hand, right forearm, and left hand in a horizontal plane (Fig. 23-8). Push the file across the work face in a straight line and do not rock the file.

4. Apply pressure only on the forward stroke.

5. Never rub the fingers or hand across a surface being filed. Grease or oil from the hand causes the file to slide over instead of cutting the work. Oil will also clog the file.

6. Keep the file clean by using a file card frequently.

For rough filing, use a double-cut file and cross the stroke at regular intervals to help keep the surface flat and straight (Fig. 23-9). When finishing, use a single-cut file and take shorter strokes to keep the file flat.

Test the work for flatness occasionally by laying the edge of a steel rule across its surface. Use a steel square to test the squareness of one surface to another.

DRAW FILING

Draw filing is used to produce a smooth, flat surface on the workpiece. This method of filing removes file marks and scratches left by cross-filing.

Note: When draw filing, hold the file as shown in Fig. 23-10 and move the file back and forth along the length of the work.

POLISHING

After a surface has been filed, it may be finished with abrasive cloth to remove small scratches left by the file. This may be done with a piece of abrasive cloth

■ **Figure 23-10** Draw filing is used to produce a flat, smooth surface. *(Kelmar Associates)*

held under the file, which is moved back and forth along the work.

SPECIAL FILES

Long-angle lathe files are used for filing on a lathe because they provide a better shearing action than mill files. The long angle of the teeth tends to clean the file, helps eliminate chatter, and reduces the possibility of tearing the metal.

Aluminum files are designed for soft, ductile metals, such as aluminum and white metal, because regular files tend to clog quickly when used on this type of material. The modified tooth construction on aluminum files tends to reduce clogging. The upcut tooth is deep, and the overcut is fine. This produces small scallops on the upcut, which breaks up the chips, and permits them to clear more easily.

Brass files have a small upcut angle and a fine, long-angle overcut, which produces small, easily cleared chips. The almost straight upcut prevents grooving the surface of the work.

Shear tooth files combine a long angle and a single-cut coarse tooth for filing materials such as brass, aluminum, copper, plastics, and hard rubber.

PRECISION FILES

Precision files include swiss pattern, needle, and riffler files. *Swiss pattern* and *needle files* (Fig. 23-11) are small files with fine tooth cuts and round integral handles. They are made in several shapes and are generally used in tool

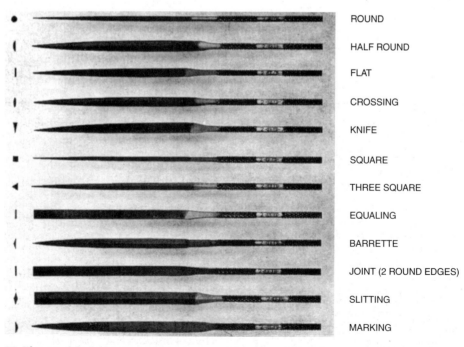

ROUND
HALF ROUND
FLAT
CROSSING
KNIFE
SQUARE
THREE SQUARE
EQUALING
BARRETTE
JOINT (2 ROUND EDGES)
SLITTING
MARKING

■ **Figure 23-11** Needle files are used for intricate work. *(Nicholson File Co. Ltd.)*

■ Figure 23-12 The broken-line teeth of rotary files tend to dissipate heat quickly. *(Nicholson File Co. Ltd.)*

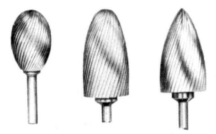

■ Figure 23-13 Ground burrs are used on nonferrous metals. *(Nicholson File Co. Ltd.)*

and die shops for finishing delicate and intricate pieces. *Die sinker rifflers* are curved up at the ends to permit filing the bottom surface of a die cavity.

ROTARY FILES AND BURRS

The increased use of portable electric and pneumatic power tools has developed the widespread application of rotary files and burrs. The wide range of shapes and sizes available makes these tools particularly suitable for metal pattern making and die sinking.

Rotary file teeth (Fig. 23-12) are cut and form broken lines in contrast to the unbroken flutes of the ground burr (Fig. 23-13). The teeth of the rotary file tend to dissipate the heat of friction, making this tool particularly useful for work on tough die steels, forgings, and scaly surfaces.

Ground burrs (Fig. 23-13) may be made of high-speed steel or carbide. The flutes of a burr are generally machine ground to a master burr to ensure uniformity of tooth shape and size. Ground high-speed burrs are used more efficiently on nonferrous metals, such as aluminum, brass, bronze, and magnesium, since they have better chip clearance than rotary files.

Carbide burrs may be used on hard or soft materials with good results and will last up to 100 times longer than a high-speed steel burr.

USING HIGH-SPEED STEEL ROTARY FILES AND BURRS

For best results, rotary files or burrs should be used in the following manner:

1. Move the file or burr at an even rate to produce a smooth surface. An uneven rate of pressure produces surfaces with ridges and hollows.

2. Use the proper speed for the burr diameter or file, as recommended by the manufacturer.

3. Use only sharp burrs or files.

4. For more accurate control of the burr or file, grip the grinder as close as possible to its end.

5. Medium-cut burrs and files generally provide satisfactory metal removal and finish for most jobs. If greater stock removal is required, use a coarse burr or file. For an extra-smooth finish, use a fine burr or file.

▶▶ Scrapers

When a truer surface is required than can be produced by machining, the surface may be finished by scraping. However, this is a long and tedious process. Most bearing surfaces (flat and curved) are now finished by grinding, honing, or broaching.

Scraping is a process of removing small amounts of metal from specific areas to produce an accurate bearing surface. It is used to produce flat surfaces or in fitting brass and babbitt bearings to shafts.

Scrapers are made in various shapes, depending on the surface to be scraped (Fig. 23-14). They are generally made of high-grade tool steel, hardened and tempered. Carbide-tipped scrapers are very popular because they maintain the cutting edge longer than other types.

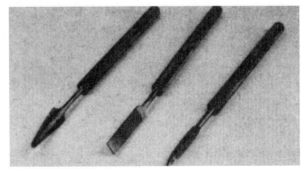

■ Figure 23-14 A set of hand scrapers. *(Kelmar Associates)*

unit 23 review questions

Hand Hacksaw

1 Compare the flexible blade and the solid, or all-hard, hacksaw blade.

2. What pitch hacksaw blade should be selected to cut:
   a. Tool steel?
   b. Thin-wall tubing?
   c. Angle iron and copper?

3. What procedure is recommended if a saw blade breaks or becomes dull in a partially finished cut?

Files

4. Describe and state the purpose of:
   a. Single-cut files   b. Double-cut files

5. Name the most commonly used degrees of coarseness in which files are manufactured.

6. List four important aspects of file care.

7. How can pinning of a file be kept to a minimum?

8. Describe and state the purpose of:
   a. Long-angle lathe files
   b. Aluminum files
   c. Shear tooth files

9. Describe and state the purpose of:
   a. Swiss pattern files
   b. Die sinker rifflers

10. Compare rotary files and ground burrs.

11. List three important considerations in the use of rotary files or ground burrs.

UNIT 24

Thread-Cutting Tools and Procedures

OBJECTIVES

After completing this unit, you will be able to:

1 Calculate the tap drill size for inch and metric taps

2 Cut internal threads using a variety of taps

3 Know the methods used to remove broken taps from a hole

4 Cut external threads using a variety of dies

Threads may be cut internally using a tap and externally using a die. The proper selection and use of these threading tools is an important part of machine shop work.

▸▸ Hand Taps

Taps are cutting tools used to cut internal threads. They are made from high quality tool steel, hardened and ground. Two, three, or four flutes are cut lengthwise across the threads to form cutting edges, provide room for the chips, and admit cutting fluid to lubricate the tap. The end of the shank is square so that a tap wrench (Fig. 24-1a and b) can be used to turn the tap into a hole. For inch taps, the major diameter, number of threads per inch, and type of thread are usually found stamped on the shank of a tap. For example, ½ in.—13 UNC represents:

½ in. = major diameter of the tap

13 = number of threads per inch

UNC = Unified National Coarse (a type of thread)

Hand taps are usually made in sets of three: *taper, plug,* and *bottoming* taps (Fig. 24-2a on p. 176).

A *taper tap* is tapered from the end approximately six threads and is used to start a thread easily. It can be used for tapping a hole that goes *through* the work, as well as for starting a *blind* hole (one that does not go all the way through).

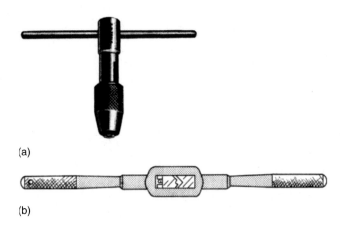

(a)

(b)

■ **Figure 24-1** (a) A T-handle tap wrench; (b) a double-ended adjustable tap wrench. *(Kelmar Associates)*

A *plug tap* is tapered for approximately three threads. Sometimes the plug tap is the only tap used to thread a hole going through a workpiece.

A *bottoming tap* is not tapered but chamfered at the end for one thread. It is used for threading to the bottom of a blind hole. When tapping a blind hole, first use the taper tap, then the plug tap, and complete the hole with a bottoming tap.

TAPER PLUG BOTTOMING

(a)

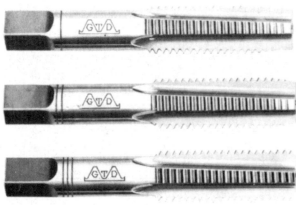

(b)

■ **Figure 24-2** (a) A set of hand taps *(F. B. Tools, Inc.)*; (b) taps may be identified by an annular ring on the shank. *(Kennametal Industrial Products Group)*

Taps may also be identified by an annular ring or rings cut around the shank of the tap. One ring around the shank indicates that it is a taper tap, two rings indicate a plug tap, and three rings indicate a bottoming tap (Fig. 24-2b).

TAP DRILL SIZE

Before a tap is used, the hole must be drilled to the correct tap drill size (Fig. 24-3). This is the drill size that would leave the proper amount of material in the hole for a tap to cut a thread. The tap drill is always smaller than the tap and leaves enough material in the hole for the tap to produce 75% of a full thread.

When a chart is not available, the tap drill size for any American, National, or Unified thread can be found easily by applying this simple formula:

$$TDS = D - \frac{1}{N}$$

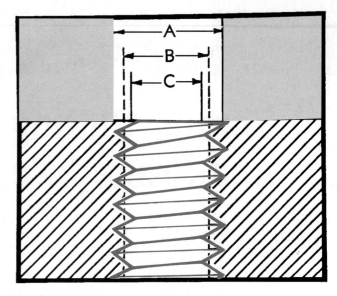

■ **Figure 24-3** Cross section of a tapped hole: A = body size; B = tap drill size; C = minor diameter. *(Kelmar Associates)*

where
$$TDS = \text{tap drill size}$$
$$D = \text{major diameter of tap}$$
$$N = \text{number of threads per inch}$$

EXAMPLE

Find the tap drill size for a ⅞-in.—9 NC tap.

$$TDS = ⅞ − ⅑$$
$$= .875 − .111$$
$$= .764 \text{ in.}$$

The nearest drill size to .764 in. is .765 in. (⁴⁹⁄₆₄). Therefore, ⁴⁹⁄₆₄ in. is the tap drill size for a ⅞-in.—9 NC tap.

METRIC TAPS

Although there are several thread forms and standards in the metric thread system, the International Standards Organization (ISO) has adopted a standard metric thread, which will be used in the United States, Canada, and many other countries throughout the world. This new series will have only 25 thread sizes, ranging from 1.6 to 100 mm diameter. See Table 6 in the appendix for the size and pitch of the threads in this series. Also see Unit 55 for the thread form and dimensions of the ISO metric thread.

Like inch taps, metric taps are available in sets of three: *taper, plug,* and *bottoming* taps. They are identified by the letter *M,* followed by the nominal diameter of the thread in millimeters times the pitch in millimeters. Thus, a tap with the markings M 4—0.7 indicates:

M—a metric thread

4—the nominal diameter of the thread in millimeters

0.7—the pitch of the thread in millimeters

TAP DRILL SIZES FOR METRIC TAPS

The tap drill size for metric taps is calculated in the same manner as for U.S. Standard threads.

$$TDS = \text{major diameter (mm)} - \text{pitch (mm)}$$

EXAMPLE

Find the tap drill size for a 22 − 2.5 mm thread.

$$TDS = 22 - 2.5$$
$$= 19.5 \text{ mm}$$

TAPPING A HOLE

Tapping is the operation of cutting an internal thread using a tap and tap wrench. Because taps are hard and brittle, they are easily broken. *Extreme* care must be used when tapping a hole to prevent breakage. A broken tap in a hole is difficult to remove and often results in scrapping the work.

To Tap a Hole by Hand

1. Select the correct taps and tap wrench for the job.
2. Apply a suitable cutting fluid to the tap.

Note: No cutting fluid is required for tapping brass or cast iron.

3. Place the tap in the hole as vertically as possible; press downward on the wrench, applying equal pressure on both handles; and turn clockwise (for right-hand thread) for about two turns.
4. Remove the tap wrench and check the tap for squareness.

Note: Check at two positions at 90° to each other (Fig. 24-4).

5. If the tap has not entered squarely, remove it from the hole and restart it by applying pressure in the direction from which the tap leans. *Be careful* not

■ **Figure 24-4** Checking a tap for squareness while holding the workpiece in a vise. *(Kelmar Associates)*

to exert too much pressure in the straightening process.

6. When a tap has been properly started, feed it into the hole by turning the tap wrench.
7. Turn the tap clockwise one-quarter turn, and then turn it backward about one-half turn to break the chip. Turning must be done with a steady motion to prevent the tap from breaking.

Note: When tapping blind holes, use all three taps in order: taper, plug, and then the bottoming tap. Before using the bottoming tap, remove all the chips from the hole and be careful not to hit the bottom of the hole with the tap.

REMOVING BROKEN TAPS

If extreme care is not used when cutting a thread, particularly in a blind hole, the tap may break in the hole and considerable work will be required to remove it. In some cases, it may not be possible to remove it, and another piece of work must be started.

Several methods may be used to remove a broken tap; some may be successful, others may not.

Tap Extractor

The tap extractor (Fig. 24-5 on p. 178) is a tool that has four fingers that slip into the flutes of a broken tap. It is adjustable in order to support the fingers close to the broken tap, even when the broken end is below the surface of the work. A wrench is fitted to the extractor and turned counterclockwise to remove a right-hand tap. Tap extractors are made to fit all sizes of taps.

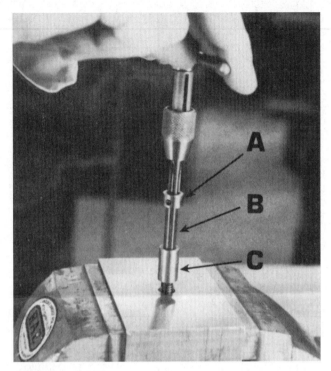

Figure 24-5 A tap extractor being used to remove a broken tap. *(Kelmar Associates)*

To Remove a Broken Tap Using a Tap Extractor

1. Select the proper extractor for the tap to be removed.
2. Slide *collar A,* to which the fingers are attached, down *body B* so that the fingers project well below the end of the body (Fig. 24-5).
3. Slide the fingers into the flutes of the broken tap, making sure they go down into the hole as far as possible.
4. Slide the body down until it rests on top of the broken tap. This will give the maximum support to the fingers.
5. Slide *collar C* down until it rests on top of the work. This also provides support for the fingers.
6. Apply a wrench to the square section on the top of the body.
7. Turn the wrench *gently* in a counterclockwise direction.

Note: Do not force the extractor because this will damage the fingers. It may be necessary to turn the wrench back and forth carefully to free the tap sufficiently to back it out.

Drilling

If the broken tap is made of carbon steel, it may be possible to drill it out. Follow this procedure:

1. Heat the broken tap to a bright red color and allow it to cool *slowly.*

2. Center-punch the tap as close to the center as possible.
3. Using a drill considerably smaller than the distance between opposite flutes, proceed *carefully* to drill a hole through the broken tap.
4. Enlarge this hole to remove as much of the metal between the flutes as possible.
5. Collapse the remaining part with a punch and remove the pieces.

Acid Method

If the broken tap is made of high-speed steel and cannot be removed with a tap extractor, it is sometimes possible to remove it by the acid method. Follow this procedure:

1. Dilute one part nitric acid with five parts water.
2. Inject this mixture into the hole. The acid will act on the steel and loosen the tap.
3. Remove the tap with an extractor or a pair of pliers.
4. Wash the remaining acid from the thread with water so that the acid will not continue to act on the threads.

Tap Disintegrators

Taps may sometimes be removed successfully by a tap disintegrator, which may be held in the spindle of a drill press. The disintegrator uses the *electrical discharge principle* to cut its way through the tap, using a hollow brass tube as an electrode. Taps may also be removed using the same method on any electrical discharge machine (see Unit 92).

▶▶ Threading Dies

Threading dies are used to cut external threads on round work. The most common threading dies are the solid, the adjustable split, and the adjustable screw plate die.

The *solid die* (Fig. 24-6) is used for chasing or recutting damaged threads and may be driven by a suitable wrench. It is not adjustable.

The *adjustable split die* (Fig. 24-7) has an adjusting screw that permits an adjustment over or under the standard depth of thread. This type of die fits into a die stock (Fig. 24-8).

The *adjustable screw plate die* (Fig. 24-9) is probably a more efficient die, since it provides for greater adjustment than the split die. Two die halves are held securely in a collet by means of a threaded plate, which also acts as a guide when threading. The plate, when tightened into the collet, forces the die halves with tapered sides into the tapered slot of the collet. Adjustment is provided by means of two adjusting screws that bear against each die half. The threaded

Figure 24-6 A solid die nut.

Figure 24-7 An adjustable, round split die.

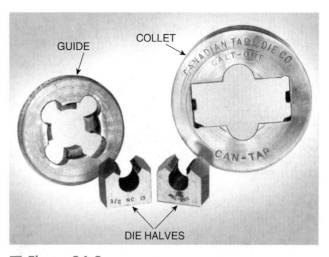

Figure 24-9 An adjustable screw plate die.

Figure 24-8 A die stock is used to turn the die onto the workpiece.

section at the bottom of each die half is tapered to provide for easy starting of the die. Note that the upper side of each die half is stamped with the manufacturer's name, and the lower side of each is stamped with the same serial number. Care should be taken in assembling the die to make sure that both serial numbers are facing down. *Never* use two die halves with different serial numbers.

TO THREAD WITH A HAND DIE

1. Chamfer the end of the workpiece with a file or on a grinder.
2. Fasten the work securely in a vise.
3. Select the proper die and die stock.
4. Lubricate the tapered end of the die with a suitable cutting lubricant.
5. Place the tapered end of the die squarely on the workpiece (Fig. 24-10).
6. Press down on the die stock handles and turn clockwise several turns.
7. Check the die to see that it has started squarely with the work.
8. If it is not square, remove the die from the work and restart it squarely.
9. Turn the die forward one turn and then reverse it approximately one-half turn to break the chip.

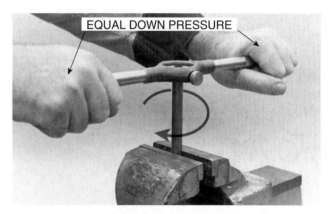

Figure 24-10 Start the tapered end of the die on the work. *(Kelmar Associates)*

10. During the threading process, apply cutting fluid frequently.

> ⚠ **CAUTION** When cutting a long thread, keep the arms and hands clear of the sharp threads coming through the die.

If the thread must be cut to a shoulder, remove the die and restart it with the tapered side of the die facing up. Complete the thread, being careful not to hit the shoulder; otherwise, the work may be bent and the die broken.

Thread-Cutting Tools and Procedures　　**179**

Hand Taps

1. Name, describe, and state the purpose of the three taps in a set.

2. Define tap drill size.

3. Use the formula and calculate the tap drill size for:

   a. ½ in.—13 UNC tap   b. M 42—4.5 mm tap

4. Why should care be used when a hole is being tapped?

5. Explain the procedure for correcting a tap that has not started squarely.

6. Briefly outline the procedure for tapping a blind hole.

7. Briefly explain the method of removing a broken tap using a tap extractor.

Threading Dies

8. For what purpose are threading dies used?

9. State the purpose of the adjustable split die and the solid die.

10. Explain the procedure for starting a die on the work.

11. What procedure should be followed when it is necessary to cut a thread to a shoulder?

UNIT 25

Finishing Processes—Reaming, Broaching, and Lapping

OBJECTIVES

After completing this unit, you will be able to:

1 Identify and explain the purpose of several types of hand reamers

2 Ream a hole accurately with a hand reamer

3 Cut a keyway in a workpiece using a broach and arbor press

4 Lap a hole or an external diameter of a workpiece to size and finish

Hand cutting tools are generally used to remove only small amounts of metal and are designed to do specific operations.

Reamers, available in a wide variety of types and sizes, are used to bring a hole to size and produce a good finish.

Broaches, when used in a machine shop, are generally used with an arbor press to produce special shapes in the workpiece. The broach, which is a multi-tooth cutting tool of the exact shape and size desired, is forced through a hole in the workpiece to reproduce its shape in the metal.

Lapping is a process whereby very fine abrasive powder, embedded in a proper tool, is used to remove minute amounts of material from a surface.

▶▶ Hand Reamers

A hand reamer is a tool used to finish drilled holes accurately and provide a good finish. Reaming is generally performed by machine, but there are times when a hand reamer must be used to finish a hole. Hand reamers, when used properly, will produce holes accurate to size, shape, and finish.

TYPES OF HAND REAMERS

The *solid hand reamer* (Fig. 25-1) may be made of carbon steel or high-speed steel. These straight reamers are available in inch sizes from .125 to 1.500 in. in diameter and in metric sizes from 1 to 26 mm in diameter. For easy start-

■ **Figure 25-1** A solid hand reamer. *(Kennametal Inc.)*

ing, the cutting end of the reamer is ground to a slight taper for a distance equal to the diameter of the reamer. Solid reamers are not adjustable and may have straight or helical flutes. Straight-fluted reamers should not be used on work with a keyway or any other interruption, since chatter and poor finish will result. Since hand reamers are designed to remove only small amounts of metal, no more than .005 in., or 0.12 mm, should be left for reaming, depending on the diameter of the hole. A square on the end

■ Figure 25-2 An expansion hand reamer.
(Kennametal Inc.)

■ Figure 25-3 An adjustable hand reamer.
(Kennametal Inc.)

■ Figure 25-4 A roughing taper reamer. *(Whitman and Barnes)*

■ Figure 25-5 A finishing taper reamer. *(Whitman and Barnes)*

of the shank provides the means of driving the reamer with a tap wrench.

The *expansion hand reamer* (Fig. 25-2) is designed to permit an adjustment of approximately .006 in. (0.15 mm) above the nominal diameter. The reamer is made hollow and has slots along the length of the cutting section. A tapered threaded plug fitted into the end of the reamer provides for limited expansion. If the reamer is expanded too much, it will break easily. For *inch expansion hand reamers,* the limit of adjustment is .006 in. over the nominal size on reamers up to .500 in. and about .015 in. on reamers over .500 in. *Metric expansion hand reamers* are available in sizes from 4 to 25 mm. The maximum amount of expansion on these reamers is 1% over the nominal size. For example, a 10-mm diameter reamer can be expanded to 10.01 mm (10 + 1%). The cutting end of the reamer is ground to a slight taper for easy starting.

The *adjustable hand reamer* (Fig. 25-3) has tapered slots along the entire length of the body. The inner edges of the cutting blades have a corresponding taper so that the blades remain parallel for any setting. The blades are adjusted to size by upper and lower adjusting nuts.

The blades on inch adjustable hand reamers have an adjustment range of 1/32 in. on the smaller reamers to almost 5/16 in. on the larger ones. They are manufactured in sizes 1/4 to 3 in. in diameter. *Metric adjustable hand reamers* are available in sizes from #000 (adjustable from 6.4 to 7.2 mm) to #16 (adjustable from 80 to 95 mm).

Taper reamers are made to standard tapers and are used to finish tapered holes accurately and smoothly. They may be made with either spiral or straight teeth. Because of its shearing action and its tendency to reduce chatter, the spiral-fluted reamer is superior to the straight one. A *roughing reamer* (Fig. 25-4), with nicks ground at intervals along the teeth, is used for more rapid removal of surplus metal. These nicks or grooves break up the chips into smaller sections; they prevent the tooth from cutting and overloading along its entire length. When a roughing reamer is not available, an old taper reamer is often used before finishing the hole with a finishing reamer.

The *finishing taper reamer* (Fig. 25-5) is used after the roughing reamer to finish the hole smoothly and to size. This reamer, which has either straight or left-hand spiral flutes, is designed to remove only a small amount of metal [about .010 in. (0.25 mm)] from the hole. Since taper reamers do not clear themselves readily, they should be removed frequently from the hole and the chips cleared from their flutes.

REAMING PRECAUTIONS

1. Never turn a reamer backward (counterclockwise) because it will dull the cutting teeth.

2. Use a cutting lubricant where required.

3. Always use a helical-fluted reamer in a hole that has a keyway or an oil groove cut in it.

4. Never attempt to remove too much material with a hand reamer; about .010 in. (0.25 mm) is the maximum.

5. Frequently clear a taper reamer (and the hole) of chips.

TO REAM A HOLE WITH A STRAIGHT HAND REAMER

1. Check the size of the drilled hole. It should be between .004 and .005 in. (0.10 and 0.12 mm) smaller than the finished hole size.

2. Place the end of the reamer in the hole and place the tap wrench on the square end of the reamer.

3. Rotate the reamer clockwise to allow it to align itself with the hole (Fig. 25-6).

4. Check the reamer for squareness with the work by testing it with a square at several points on the circumference.

5. Brush cutting fluid over the end of the reamer if required.

6. Rotate the reamer slowly in a clockwise direction and apply downward pressure. Feed should be fairly rapid and steady to prevent the reamer from chattering.

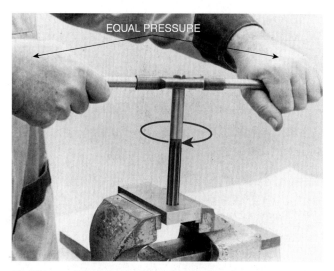

EQUAL PRESSURE

Figure 25-6 Turn the reamer clockwise when starting it in a hole. *(Kelmar Associates)*

Note: The rate of feed should be about one-quarter the diameter of the reamer for each turn.

▶▶ Broaching

Broaching is a process in which a special tapered multi-toothed cutter is forced through an opening or along the outside of a piece of work to enlarge or change the shape of the hole or to form the outside to a desired shape.

Broaching was first used for producing internal shapes, such as keyways, splines, and other odd internal shapes (Fig. 25-7). Its application has been extended to exterior surfaces, such as the flat face on automotive engine blocks and cylinder heads. Most broaching is now performed on special machines that either pull or push the broach through or along the material. Hand broaches are used in the machine shop for operations such as keyway cutting.

The cutting action of a broach is performed by a series of successive teeth, each protruding about .003 in. (0.07 mm) farther than the preceding tooth (Fig. 25-8).

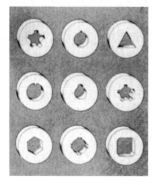

Figure 25-7 Examples of internal broaching.

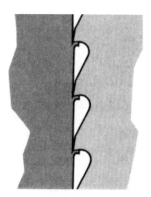

Figure 25-8 The cutting action of a broach.

The last three teeth are generally of the same depth and provide the finish cut.

Broaching has many advantages and an extremely wide range of applications:

1. Machining almost any irregular shape is possible, providing it is parallel to the broach axis.

2. It is rapid; the entire machining process is usually completed in one pass.

3. Roughing and finishing cuts are generally combined in the same operation.

4. A variety of forms, either internal or external, may be cut simultaneously and the entire width of a surface may be machined in one pass, thus eliminating the need for a machining operation.

CUTTING A KEYWAY WITH A BROACH

Keyways may be cut by hand in the machine shop quickly and accurately by means of a broach set and an arbor press (Fig. 25-9 on p. 184). A broach set (Fig. 25-10 on p. 184) covers a wide range of keyways and is a particularly useful piece of equipment when many keyways must be cut. The equipment necessary to cut a keyway is a bushing (Fig. 25-10a) to suit the hole size in the workpiece, a broach (Fig. 25-10b) the size of the keyway to be cut, and shims (Fig. 25-10c) to increase the depth of the cut of the broach.

Follow this procedure:

1. Determine the keyway size required for the size of the workpiece.

2. Select the proper broach, bushing, and shims.

3. Place the workpiece on the arbor press. Use an opening on the base smaller than the opening in the workpiece so that the bushing will be properly supported.

4. Insert the bushing and the broach into the opening. Apply cutting fluid if the workpiece is made of steel.

■ **Figure 25-9** Using an arbor press to cut a keyway with a broach. *(Kelmar Associates)*

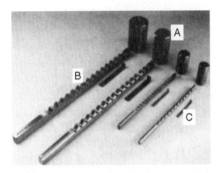

■ **Figure 25-10** A broach set for cutting internal keyways. *(Kelmar Associates)*

5. Check the broach to be sure that it has started squarely in the hole.

6. Press the broach through the workpiece, maintaining constant pressure on the arbor-press handle.

7. Remove the broach, insert one shim, and press the broach through the hole.

8. Insert the second shim, if required, and press the broach through again. This will cut the keyway to the proper depth (Fig. 25-11).

9. Remove the bushing, broach, and shims.

▶▶ Lapping

Lapping is an abrading process used to remove minute amounts of metal from a surface that must be flat, accu-

■ **Figure 25-11** Two shims are used for making the final pass with a broach. *(Kelmar Associates)*

rate to size, and smooth. Lapping may be performed for any of the following reasons:

1. To increase the wear life of a part
2. To improve accuracy and surface finish
3. To improve surface flatness
4. To provide better seals and eliminate the need for gaskets

Lapping may be performed by hand or machine, depending on the nature of the job. Lapping is intended to remove only about .0005 in. (0.01 mm) of material. Lapping by hand is a long, tedious process and should be avoided unless absolutely necessary.

LAPPING ABRASIVES

Both natural and artificial abrasives are used for lapping. Flour of emery and fine powders made of silicon carbide or aluminum oxide are used extensively. Abrasives used for rough lapping should be no coarser than 150 grit; fine powders used for finishing run up to about 600 grit. For fine work, diamond dust, generally in paste form, is used.

TYPES OF LAPS

Laps may be used to finish flat surfaces, holes, or the outside of cylinders. In each case, the lap material must be *softer* than the workpiece.

(a)   (b)

■ **Figure 25-12** (a) A roughing lapping plate; (b) a finishing lapping plate. *(Kelmar Associates)*

Flat Laps

Laps for producing flat surfaces are made from close-grained cast iron. For the roughing operation or "blocking down," the lapping plate should be scored with narrow grooves about .500 in. (13 mm) apart, both lengthwise and crosswise or diagonally to form a square or diamond pattern (Fig. 25-12a). Finish lapping is done on a smooth cast-iron plate (Fig. 25-12b).

Charging the Flat Lapping Plate

Spread a thin coating of abrasive powder over the surface of the plate and press the particles into the surface of the lap with a hardened steel block or roll. Rub as little as possible. When the entire surface appears to be charged, clean the surface with varsol and examine it for bright spots. If any bright spots appear, recharge the lap and continue until the entire surface assumes a gray appearance after it has been cleaned.

Lapping a Flat Surface

If work is to be roughed down, oil should be used on the roughing plate as a lubricant. As the work is rubbed over the lap, the abrasive powder will be washed from the grooves and act between the surface of the work and the lap. If the work has been surface-ground, rough lapping or "blocking down" is not required.

Follow this procedure:

1. Place a little varsol on a finish-lapping plate that has been properly charged.

2. Place the work on top of the plate and gently push it back and forth over the full surface of the lap using an irregular movement. *Do not stay in one spot.*

3. Continue this movement with a light pressure until the desired surface finish is obtained.

Precautions to Be Observed

1. Do not stay in one area; cover the full surface of the lap.

2. Never add a fresh supply of loose abrasive. If required, recharge the lap.

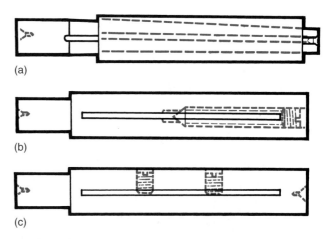

(a)

(b)

(c)

■ **Figure 25-13** Various types of internal laps: (a) lead lap; (b) copper lap; (c) adjustable lap. *(Kelmar Associates)*

3. Never press too hard on the work because the lap will become stripped in places.

4. Always keep the lap moist.

Internal Laps

Holes may be accurately finished to size and smoothness by lapping. Internal laps may be made of brass, copper, or lead and may be of three types.

The *lead lap* (Fig. 25-13a) is made by pouring lead around a tapered mandrel that has a groove along its length. The lap is turned to a running fit into the hole and is then sometimes slit on the outside to trap the loose abrasive during the lapping operation. Adjust by lightly tapping the large end of the mandrel on a soft block. This will cause the lead sleeve to move along the mandrel and expand.

The *internal lap* (Fig. 25-13b) may be made of copper, brass, or cast iron. A threaded-taper plug fits into the end of the lap, which is slit for almost its entire length. The lap diameter may be adjusted by the threaded-taper plug.

The *adjustable lap* (Fig. 25-13c) may be made from copper or brass. The lap is split for almost its full length, but both ends remain solid. Slight adjustment is provided by means of two setscrews in the center section of the lap.

Charging and Using an Internal Lap

Before charging, the lap should be a running fit in the hole. Follow this procedure:

1. Sprinkle some lapping powder evenly on a flat plate.

2. Roll the lap over the powder, applying sufficient pressure to embed the abrasive into the surface of the lap.

3. Remove any excess powder.

4. Mount a lathe dog on the end of the lap.

5. Fit the workpiece over the end of the lap.

Note: The lap should now be a wringing fit in the hole of the work and about 2.5 times the length of the workpiece.

6. Place some oil or varsol on the lap.

7. Mount the lap and the work between lathe centers.

8. Set the machine to run at a slow speed, 150 to 200 r/min for a 1-in. (25-mm) diameter.

9. Hold the work securely and start the machine.

10. Run the work back and forth along the entire length of the lap.

11. Remove the work and rinse it in varsol to remove the abrasive and to bring it to room temperature.

12. Gage the hole for size.

Note: Always keep the lap moist and never add loose abrasive to the lap. Loose abrasive will cause the work to become bell-mouthed at the ends. If more abrasive is necessary, recharge the lap and adjust as required.

External Laps

External laps are used to finish the outside of cylindrical workpieces. They may be of several forms (Fig. 25-14); however, the basic design is the same. External laps may be made of cast iron or they may have a split brass bushing mounted inside by means of a setscrew. There must be some provision for adjusting the lap.

Charging and Using an External Lap

1. Mount the workpiece in a three-jaw chuck on the lathe or drill press.

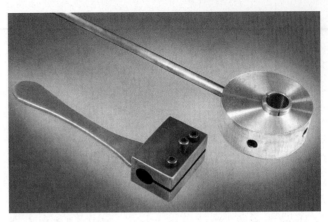

■ **Figure 25-14** External laps. *(Kelmar Associates)*

2. Adjust the lap until it is a running fit on the workpiece.

3. Grip the end of the lap in a vise.

4. Sprinkle abrasive powder in the hole.

5. With a hardened steel pin, roll the abrasive evenly around the inside surface of the lap.

6. Remove any excess lapping powder.

7. Place the lap on the workpiece. It should now be a wringing fit.

8. Set the machine to run at a slow speed [150 to 200 r/min for a 1-in.-diameter (25-mm) workpiece].

9. Add some varsol to the workpiece and the lap.

10. Hold the lap securely and start the machine.

11. Move the lap back and forth along the work.

Note: Always keep the lap moist.

12. To gage the work, remove the lap and clean the workpiece with varsol.

Hand Reamers

1. What is the purpose of a hand reamer?

2. Describe and state the purpose of:

   a. The solid hand reamer
   b. The expansion hand reamer
   c. The taper finishing reamer

3. How much metal should be removed with a hand reamer?

4. List four important precautions to be observed while reaming.

Broaching

5. Define broaching.

6. Describe the cutting action of a broach.

7. State three advantages of broaching.

8. Briefly describe the procedure for broaching a keyway on an arbor press.

Lapping

9. State three reasons for lapping.

10. What abrasives are generally used for lapping?

11. Why must the lap be softer than the workpiece?

12. Explain the procedure for charging a flat lapping plate.

13. Briefly describe the process of lapping a flat surface.

14. How are internal laps charged?

UNIT 26

Surface Finishing Processes

OBJECTIVES

After completing this unit, you should be able to:

1 Identify and explain the purposes of surface finishing processes

2 Explain the benefits of honing

3 Be familiar with oxide coatings and their purposes

All metal products must receive some type of finishing process to improve the appearance and sales value of the manufactured product. Surface treatment used on most metals is used to resist wear, electrolytic decomposition, and corrosive wear due to weather. The treatment process consists of the transformation of the metal surface, by chemical or electrical processes, to produce an oxide of the original metal on its surface. A few of the most common methods used are burnishing, electropolishing, honing, and tumbling.

▶▶ Burnishing

Burnishing is a cold-working process that sizes, finishes, and work hardens internal and external metal surfaces by pressure contact of hardened rolls. This process displaces rather than removes the small peaks and valleys of irregular height and spacing. They are primarily used to polish (by cold working) internal surfaces such as holes.

The burnishing tool (Fig. 26-1a), incorporates a planetary system of tapered rolls that are evenly spaced by a retaining cage. When a tool contacts the workpiece, a hardened mandrel, which is tapered inversely to the taper of the rolls, forces them against the surface of the part.

As the tool, adjusted slightly larger than the part, passes through the part, slight pressure is created that exceeds the yield point of the softer part surface. This results in a small plastic deformation of the surface structure of the part surface. The result is an accurately sized part with a mirrorlike finish and a tough, work-

hardened, and wear- and corrosion-resistant surface (Fig. 26-1b).

▶▶ Electropolishing

Electropolishing, often referred to as a *reverse plating* process, uses a combination of rectified current and a blended chemical electrolyte bath to remove flaws from the surface of a metal part.

THE PROCESS

A power source converts AC (Alternating Current) to DC (Direct Current) at low voltages. A tank typically fabricated from steel and rubber-lined is used to hold the chemical bath. A series of lead, copper, or stainless steel cathode plates are lowered into the bath and installed to the negative ($-$) side of the power source (Fig. 26-2a). A part or group of parts is fixtured to a rack made of tita-

(a)

(b)

■ Figure 26-1 (a) An internal roller burnishing tool that sizes, finishes, and work hardens the material; (b) an external roller burnishing tool. *(Cogsdill Tool Products, Inc.)*

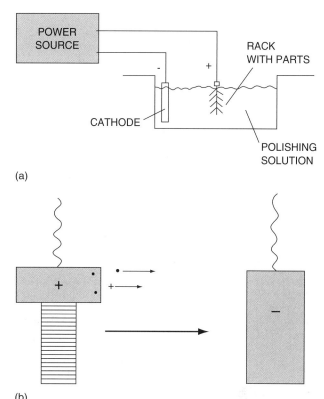

(a)

(b)

■ Figure 26-2 (a) A schematic diagram of the electropolishing process; (b) in electropolishing, the metal part is charged and immersed into the chemical bath. *(Able Electropolishing Co.)*

nium, copper, or bronze. That rack, in turn, is fixtured to the positive (+) side of the power source.

As shown in Fig. 26-2b, the metal part is charged positive and immersed into the chemical bath. When current is applied, the electrolyte acts as a conductor to allow metal ions to be removed from the part. While the ions are drawn toward the cathode, the electrolyte maintains the dissolved metals in solution. Gassing in the form of oxygen occurs at the metal surface, furthering the cleaning process.

Once the process is completed, the part is run through a series of cleaning and drying steps to remove clinging electrolyte. While the process is best known for the bright polish left on a surface, there are some important, often overlooked benefits of this metal removal method. These benefits include deburring, size control, and microfinish improvement.

▶▶ Honing

Honing is an abrasive finishing operation where a small amount of stock can be removed from the internal or external surface of a part to improve its flatness and surface finish.

There are basic differences between honing and grinding (Fig. 26-3 on p. 190); honing is a low-speed operation (85 to 300 sf/min, or 25 to 95 m/min), while grinding is a high-speed operation (5000 to 6500 sf/min, or 25 to 33 m/s).

> Chips produced by grinding are short, hot sparks due to the intermittent contact of each abrasive particle on the workpiece surface. Grinding tends to cause thermal damage to the surface of the workpiece, often to a depth of up to .002 in. (0.05 mm) or more (Fig. 26-3a).

> A continuous cooler chip is produced by honing due to the continuous contact of the abrasive hone with the work (Fig. 26-3b). It is a gentle, cooler finishing operation, with little or no damage or distortion of the workpiece surface.

Surface Finishing Processes   **189**

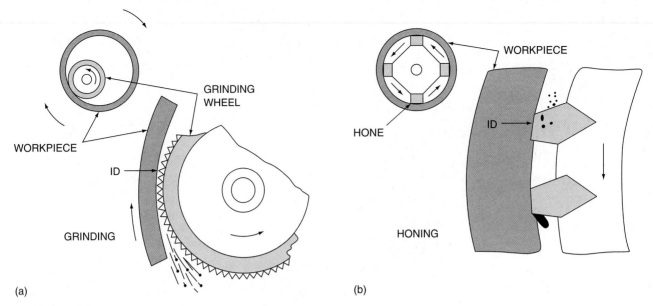

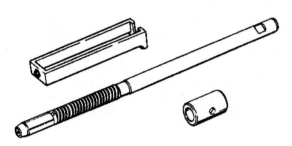

Figure 26-3 Internal finishing by (right) honing and grinding (left). *(GE Superabrasives)*

Figure 26-4 Honing with a single-stroke tool is faster and more accurate than conventional honing. *(Sunnen Products Co.)*

TYPES OF HONING PROCESSES

The two general types of honing processes in industrial use are conventional and single-stroke honing.

During *conventional honing,* cutting pressure (expansion) is applied when the mandrel forces the stone and the guide-shoe surfaces into contact with the bore. The rotational and reciprocal action of the hone in the bore causes thousands of small cutting edges on the stone to shear minute chips from the workpiece.

The *single-stroke* honing tool, (Fig. 26-4), consists of an expandable, diamond-plated abrasive sleeve on a tapered arbor that is expanded to size by a calibrated adjusting screw. It is faster and maintains more consistent size accuracy than conventional honing.

HONING STONES

There are various types of honing stones and mandrels available to suit a variety of honing applications. *Single-stone hones* (Fig. 26-5a), generally used on small bores [less than

3.00 in. (75 mm)]; *four-point-contact hones* (Fig. 26-5b), capable of fast metal removal; and *multi-point-contact hones* that are generally used on production for honing large parts.

MACHINE REQUIREMENTS

Most existing manual-stroke and power-stroke honing machines work well with CBN (Cubic Boron Nitride) hones, provided they are in good condition. To obtain the fullest potential that superabrasives such as CBN offer, the machine should have the following qualities:

> Be rigid, no spindle vibration, and sufficient power for high material-removal rates

> Have high-spindle speeds—about three times the speed for conventional hones and closely controlled increments of stone feed and pressure

> Have a coolant system capable of supplying adequate volumes of filtered honing fluids

HONING GUIDELINES

To select the proper CBN hone for the job, consider the following factors:

> Type and hardness of the work material

> Amount of material to be removed

The following general rules apply to any CBN hone, regardless of the bond:

1. For maximum removal rates, use coarse grit sizes, lower concentration, and narrow stones at high speed.

2. For fine finishes, use fine grit sizes and higher concentrations.

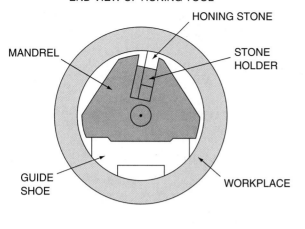

END VIEW OF HONING TOOL

HONING STONE

MANDREL

STONE HOLDER

GUIDE SHOE

WORKPLACE

(a)

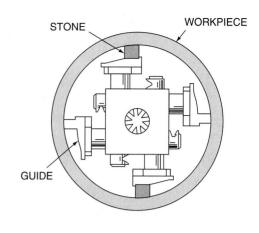

STONE

WORKPIECE

GUIDE

(b)

■ **Figure 26-5** (a) Honing an internal diameter with a single-stone hone; (b) honing an internal diameter with a four-point-contact hone. *(Sunnen Products Co.)*

3. For hard materials, use a soft bond; for soft materials, use a hard bond.

4. Always use an adequate supply of coolant to clean the stone, clear the chip, and cool the workpiece.

▶▶ Tumbling

Tumbling is an important production process that is used for cleaning, polishing, and removing sharp corners and burrs from metal parts. In many cases, this may be the only economical method of surface conditioning, since the material is handled in bulk. Many types of tumbling barrels in various sizes are available, depending on the material and shape of the part to be cleaned, the depth and hardness of the scale or rust, or the previous finish to be removed (Fig. 26-6). In addition to cleaning rust and scales from work, tumbling operations are often used to homogenize the surface mechanically and to remove burrs and sharp edges.

The continuous improvement in tumbling equipment and the abrasives used is steadily extending the range of applications and the precision obtainable. It has been reported that in a typical case an improvement in reading from 10 down to 6 microinches was obtained on the outside diameter of a small steel bushing after tumbling 3 hr in a suitable abrasive, without affecting the 1- to 1.5-microinches honed inside-diameter surface.

SELECTING THE EQUIPMENT AND TUMBLING MEDIA

Choosing the proper type of barrel, abrasive, lubricant, and ratio of volume of work to total volume of batch and barrel, various surfacing operations can be efficiently carried out. These range from deburring and polishing to honing and mirror finishing of metallic parts before and after electroplating.

The most commonly used tumbling abrasives are aluminum oxide, ceramic, plastic, stone chips, crushed corncob, and steel balls (Fig. 26-7 on p. 192). Ground ivory nut or wooden balls have application for fine burnishing and polishing.

Production parts that can withstand large-barrel loading can generally be more quickly processed with aluminum oxide; delicate parts may require the use of a finer abrasive in a smaller barrel.

The size of the media to be used must be determined to avoid wedging of the abrasive material in the work. Smaller chips make more numerous contact points with the workpiece than do the larger chips, with lower resultant pressure exerted at each point. This produces a smoother surface finish with shallower surface scratches.

■ **Figure 26-6** A tumbling barrel used to remove burrs and finish parts. *(Grav-I-Flo Corp.)*

Abrasive

Corncob

Steel

Ceramic

■ **Figure 26-7** Types of abrasive media used for the tumbling operation. *(Grav-I-Flo Corp.)*

THE TUMBLING PROCESS

The parts are loaded into a suitable tumbling barrel and the appropriate tumbling media added. The water level in the loaded barrel bears a definite relation to the type of finishing desired. The higher the water level, the less likelihood of parts being nicked when falling into the barrel; abrasive action is increased as less water is used. Therefore, the water level should be just above the top line of loading for burnishing, about one-third to halfway below the load line for honing and light deburring, and two-thirds or even farther below the load line for fast deburring.

▶▶ Black-Oxide Coatings

Black-oxide coatings are a chemical conversion process produced by the reaction of iron in ferrous metal with the oxidizing salts present in the black oxide. These oxidizing salts include penetrates, catalysts, activators, and proprietary additives that all take part in the chemical reaction. The result of this chemical reaction is the formation of black iron oxide, magnetite on the surface of the metal being coated. This coating is most commonly used for decorative and corrosion prevention purposes on bearings, gears, small components, and firearms (Fig. 26-8).

Black-Oxide Process

Black oxide can be produced on ferrous metal using a molten salt bath operating at 600°F (315°C) and above,

■ **Figure 26-8** Black-oxide coatings are used to protect metal surfaces. *(Techplate, Inc.)*

a cold black solution operating at room temperature, or a hot alkaline aqueous solution operating between 285°F and 300°F (140°C and 148°C).

The hot alkaline aqueous black-oxide solution is most commonly used and is the processing bath referred to when black-oxide coating is specified. This process produces a deep black finish that is consistent and uniform in appearance. Because the process is strictly a chemical reaction, there are no high or low current density areas to cause uneven coating thickness. The oxide formation penetrates into the metal surface up to 5 to 10 millionths of an inch.

1. Why are surface treatments used on metals?

2. Define burnishing.

3. To what is electropolishing often referred?

4. Name four benefits of electropolishing.

5. What is the purpose of honing?

6. List three factors that should be considered when honing.

7. Why is tumbling an important production process?

8. Name five media used in the tumbling process.

9. What is formed on the surface of a part during the black-oxide process?

Section 8

(Cincinnati Machine)

METAL-CUTTING TECHNOLOGY

Industry recognizes that, to operate economically, metals used in the manufacture of its products must be machined efficiently. To cut metals efficiently requires not only a knowledge of the metal to be cut, but also how the cutting-tool material and its shape will perform under various machining conditions. Cutting-tool angles, rakes, and clearances have assumed increasing importance in metal cutting. Many new cutting-tool materials have been introduced in the last few decades. These new materials have led to improved machine construction, higher cutting speeds, and increased productivity.

Many materials are machined most efficiently with cutting fluids; others are not. With the appearance of new and varying alloys, new cutting fluids are constantly being developed. All these factors make metal-cutting theory a challenging and constantly researched field in the machine tool industry.

Physics of Metal Cutting

OBJECTIVES

After completing this unit, you will be able to:

1 Define the various terms that apply to metal cutting

2 Explain the flow patterns of metal as it is cut

3 Recognize the three types of chips produced from various metals

Human beings have been using tools to cut metal for hundreds of years without really understanding how the metal was cut or what was occurring where the cutting tool met the metal. For many years, it was thought that the metal ahead of the cutting tool splits in a manner similar to the way that wood splits in front of an ax (Fig. 27-1). According to the original theory, this accounted for the wear that occurred on the cutting-tool face some distance away from the cutting edge. An early cutting-fluid advertisement illustrated this same theory by showing the metal splitting in front of a cutting tool during a lathe-turning operation (Fig. 27-2).

▸▸ Need for Metal-Cutting Research

The manufacture of dimensionally accurate, closely fitting parts is essential to interchangeable manufacture. The accuracy and wearability of mating surfaces are directly proportional to the surface finish produced on the part. Every year in the United States alone, over 15 million tons (13.6 million tonnes) of metal are cut into chips at a cost of over $10 billion. To reduce the cost of machining, prolong the life of cutting tools, and maintain high surface finishes, it was essential for research to be done in the area of metal cutting. Since World War II, a great deal of research has been conducted in areas such as the theory of

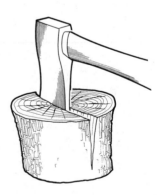

■ **Figure 27-1** The simile of an ax splitting wood was often used incorrectly to illustrate the action of a cutting tool.

■ Figure 27-2 A false conception of metal cutting.

■ Figure 27-3 Chip-tool interface. *(Cincinnati Machine)*

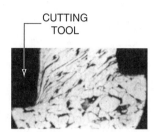

■ Figure 27-4 Photomicrograph of a chip, showing crystal elongation and plastic deformation. *(Cincinnati Machine)*

metal cutting, measurement of cutting forces and temperatures, machinability of metals, machining economics, and theory of cutting-fluid action. This research has found that the metal of a workpiece, instead of rupturing or breaking a little before the cutting tool, is compressed and then flows up the face of the cutting tool. New cutting tools, speeds and feeds, cutting-tool angles and clearances, and cutting fluids have been developed as a result of this research. These developments have greatly assisted in the economical machining of metals; however, much work remains to be done before all the factors affecting surface finish, tool life, and machine output can be controlled.

▸▸ Metal-Cutting Terminology

Many new terms resulted from the research conducted on metal cutting:

A *built-up edge* is a layer of compressed metal from the material being cut, which adheres to and piles up on the face of the cutting tool edge during a machining operation (Fig. 27-3).

The *chip-tool interface* is that portion of the face of the cutting tool on which the chip slides as it is cut from the metal (Fig. 27-3).

Crystal elongation is the distortion of the crystal structure of the work material that occurs during a machining operation (Fig. 27-4).

The *deformed zone* is the area in which the work material is deformed during cutting.

Plastic deformation is the deformation of the work material that occurs in the shear zone during a cutting action (Fig. 27-4).

Plastic flow is the flow of metal that occurs on the shear plane, which extends from the cutting-tool edge to the corner between the chip and the work surface in Fig. 27-4.

A *rupture* is the tear that occurs when brittle materials, such as cast iron, are cut and the chip breaks away from the work surface. Ruptures generally occur when discontinuous or segmented chips are produced.

The *shear angle* or *plane* is the angle of the area of material where plastic deformation occurs (Fig. 27-3).

The *shear zone* is the area where plastic deformation of the metal occurs. It is along a plane from the cutting edge of the tool to the original work surface (Fig. 27-4).

▸▸ Plastic Flow of Metal

To understand more fully what occurs in the metal while it is being deformed in the shear zone, researchers made many tests on various types of materials. They used flat punches on ductile material to study the stress pattern, direction of material flow, and distortion created in the metal. To observe what occurs when pressure is applied to

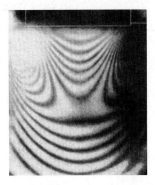

■ Figure 27-5 The stress distribution created by a flat punch in photoelastic material. *(Cincinnati Machine)*

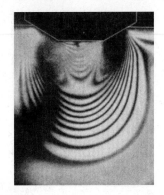

■ Figure 27-6 The stress distribution created by a narrow-faced punch. *(Cincinnati Machine)*

one spot, they used blocks of photoelastic materials, such as celluloid and Bakelite. Researchers also used polarized light to observe stress lines created when pressure is exerted on the punch. Using a suitable analyzer, they saw a series of colored bands known as *isochromatics*. They used three different types of punches (flat, narrow-faced, and knife-edge) on photoelastic material to create the various stresses observed.

FLAT PUNCH

When a flat punch is forced into a block of photoelastic material, the lines of constant maximum shear stress appear, indicating the distribution of the stress. In Fig. 27-5, the shape of these stress lines, or isochromatics, appears as a family of curves almost passing through the corners of the flat punch. The greatest concentration of stress lines occurs at each corner of the punch, and larger circular stress lines appear farther away from the punch. The spacing of the isochromatics is relatively wide.

NARROW-FACED PUNCH

When a narrow-faced punch is forced into a block of photoelastic material, the stress lines are still concentrated at the punch corners and where the punch meets the top surface of the work. As you can see in Fig. 27-6, the isochromatics are spaced closer than with the flat punch.

KNIFE-EDGE PUNCH

When a knife-edge punch is forced into the block of photoelastic material (Fig. 27-7), the isochromatics become a series of circles tangent to the two faces of the punch. In this case, the flow of material occurs upward from the point toward the free area along the faces of the punch.

When a cutting tool engages a workpiece, this flow of material takes place, and the compressed material escapes up the tool face. As the tool advances, opposition to

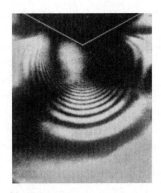

■ Figure 27-7 The stress lines created by a knife-edge punch. *(Cincinnati Machine)*

the upward flow of material creates stresses in the material ahead of the tool because of the friction of the chip flow up the tool face. These stresses are somewhat relieved by the plastic flow or rupture of the material along a plane leading from the cutting edge of the tool to the surface of the unmachined metal. From Figs. 27-5 to 27-7, it can be concluded that internal stresses are created during metal-cutting operations and:

1. Because of the forces exerted by the cutting tool, compression occurs in the work material.

2. As the cutting tool or work moves forward during a cut, the stress lines concentrate at the cutting-tool edge and radiate from there in the material (Fig. 27-7).

3. This concentration of stresses causes the chip to shear from the material and flow along the chip-tool interface.

4. By either plastic flow or rupture, the metal tries to flow along the chip-tool interface. Since most metals are ductile to some degree, a plastic flow generally occurs.

The plastic flow or rupture that occurs as the metal flows along the chip-tool interface determines the type of

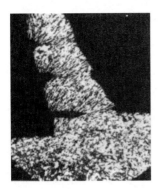

Figure 27-8 A discontinuous chip. *(Cincinnati Machine)*

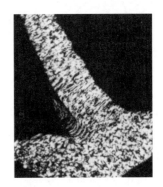

Figure 27-10 A continuous chip with a built-up edge. *(Cincinnati Machine)*

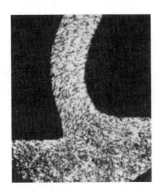

Figure 27-9 A continuous chip. *(Cincinnati Machine)*

chip produced. When brittle materials such as cast iron are being cut, the metal has a tendency to rupture and produce discontinuous or segmented chips. When relatively ductile metals are being cut, a plastic flow occurs, and continuous or flow-type chips are produced.

▶▶ Chip Types

Machining operations performed on lathes, milling machines, or similar machine tools produce chips of three basic types: discontinuous (Fig. 27-8), continuous (Fig. 27-9), and continuous with a built-up edge (Fig. 27-10).

TYPE 1—DISCONTINUOUS (SEGMENTED) CHIP

Discontinuous or segmented chips (Fig. 27-8) are produced when brittle metals such as cast iron and hard bronze are cut, and even when some ductile metals are cut under poor cutting conditions. As the point of the cutting tool contacts the metal (Fig. 27-11a), some compression occurs, as can be noted in Fig. 27-11b and c, and the chip begins flowing along the chip-tool interface. As more

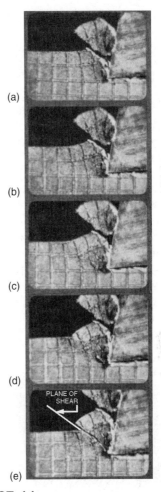

Figure 27-11 Formation of a discontinuous chip.

stress is applied to brittle metal by the cutting action, the metal compresses until it reaches a point where rupture occurs (Fig. 27-11d), and the chip separates from the unmachined portion (Fig. 27-11e). This cycle is repeated indefinitely during the cutting operation, with the rupture of each segment occurring on the shear angle or plane. Generally, as a result of these successive ruptures, a poor surface is produced on the workpiece.

Machine vibration or tool chatter sometimes causes discontinuous chips to be produced when ductile metal is cut.

The following conditions favor the production of a Type 1 discontinuous chip:

1. Brittle work material

2. Small rake angle on the cutting tool

3. Large chip thickness (coarse feed)

4. Low cutting speed

5. Excessive machine chatter

TYPE 2—CONTINUOUS CHIP

The Type 2 chip is a continuous ribbon produced when the flow of metal next to the tool face is not greatly retarded by a built-up edge or friction at the chip-tool interface. The continuous ribbon chip is considered ideal for efficient cutting action because it results in better surface finishes.

When ductile materials are cut, plastic flow in the metal takes place by the deformed metal sliding on a great number of crystallographic slip planes. As is the case with the Type 1 chip, fractures or ruptures do not occur because of the ductile nature of the metal.

In Fig. 27-9, the crystal structure of the ductile metal is elongated when it is compressed by the action of the cutting tool and as the chip separates from the metal. The process of chip formation occurs in a single plane extending from the cutting tool to the unmachined work surface. The area where plastic deformation of the crystal structure and shear occurs is called the *shear zone* (Fig. 27-4). The angle on which the chip separates from the metal is called the *shear plane* or *shear angle*.

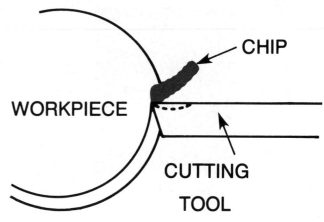

■ **Figure 27-13** A continuous chip is generally formed when a carbide or high-speed toolbit is used to machine steel.

The mechanics of chip formation can best be understood with the aid of the schematic diagram in Fig. 27-12. As the cutting action progresses, the metal immediately ahead of the cutting tool is compressed, with a resultant deformation (elongation) of the crystal structure. This elongation does not take place in the direction of shear. As this process of compression and elongation continues, the material above the cutting edge is forced along the chip-tool interface and away from the work.

Machine steel generally forms a continuous (unbroken) chip with little or no built-up edge when machined with a cemented-carbide cutting tool or a high-speed steel toolbit and cutting fluid (Fig. 27-13). To reduce the amount of resistance occurring as the compressed chip slides along the chip-tool interface, a suitable rake angle is ground on the tool, and cutting fluid is used during the cutting operation. These features allow the compressed chip to flow relatively freely along the chip-tool interface. A shiny layer on the back of a continuous-type chip indicates ideal cutting conditions with little resistance to chip flow.

The conditions favorable to producing a Type 2 chip are:

1. Ductile work material

2. Small chip thickness (relatively fine feeds)

3. Sharp cutting-tool edge

4. A large rake angle on the cutting tool

5. High cutting speeds

6. Cutting tool and work kept cool by use of cutting fluids

7. A minimum of resistance to chip flow by:
   a. A high polish on the cutting-tool face
   b. Use of cutting fluids to prevent the formation of a built-up edge

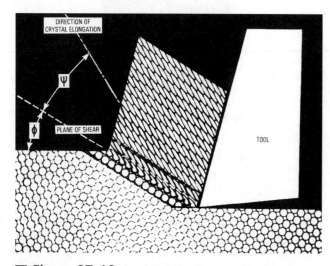

■ **Figure 27-12** A schematic diagram showing formation of a continuous chip and deformation of the crystal structure. *(Cincinnati Machine)*

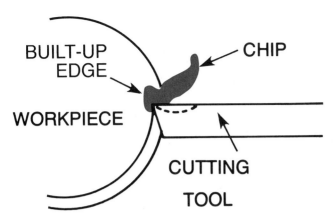

Figure 27-14 A built-up edge is formed when workpiece fragments become welded to the tool face.

c. Use of cutting-tool materials, such as cemented carbides, which have a low coefficient of friction
d. Free-machining materials (those alloyed with elements such as lead, phosphor, and sulphur)

TYPE 3—CONTINUOUS CHIP WITH A BUILT-UP EDGE

Low-carbon machine steel and many high-carbon alloyed steels, when cut at a low cutting speed with a high-speed steel cutting tool and without the use of cutting fluids, generally produce a continuous-type chip with a built-up edge (Fig. 27-10).

The metal ahead of the cutting tool is compressed and forms a chip, which begins to flow along the chip-tool interface (Fig. 27-14). As a result of the high temperature, high pressure, and high frictional resistance against the flow of the chip along the chip-tool interface, small particles of metal begin adhering to the edge of the cutting tool while the chip shears away. As the cutting process continues, more particles adhere to the cutting tool; a larger buildup results, which affects the cutting action. The built-up edge increases in size and becomes more unstable; eventually a point is reached where fragments are torn off. Portions of the fragments that break off stick to both the

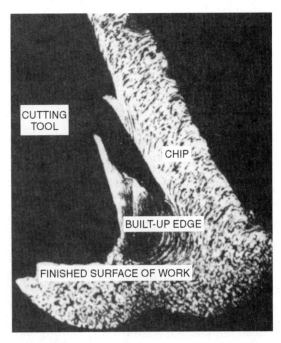

Figure 27-15 A Type 3 continuous chip with a built-up edge formed. *(Cincinnati Machine)*

chip and the workpiece (Fig. 27-15). The buildup and breakdown of the built-up edge occur rapidly during cutting action and cover the machined surface with a multitude of built-up fragments usually identified by a rough, grainy surface. These fragments adhere to and score the machined surface, resulting in a poor surface finish.

The continuous chip with the built-up edge, as well as being the main cause of surface roughness, also shortens cutting-tool life. When a cutting tool starts to dull, it creates a rubbing or compressing action on the workpiece, which generally produces work-hardened surfaces. This type of chip affects cutting-tool life in two ways:

1. The fragments of the built-up edge abrade the tool flank as they escape with the workpiece and chip.

2. A cratering effect is caused a short distance back from the cutting edge where the chip contacts the tool face. As this cratering continues, it eventually extends closer to the cutting edge until fracture or breakdown occurs.

unit 27 review questions

1. Explain the original theory of what occurrs during a metal-cutting operation.

2. Why was extensive research into metal cutting carried out?

3. Explain the new theory of the metal-cutting operation.

Metal-Cutting Terminology

4. Define the following metal-cutting terms:

   a. Built-up edge   c. Plastic deformation
   b. Chip-tool interface   d. Shear angle or plane

Plastic Flow of Metal

5. Why was research conducted to determine the plastic flow in metal?

6. Briefly describe what occurs when the following are forced into a block of photoelastic material:

   a. A flat punch
   b. A narrow-faced punch
   c. A knife-edge punch

7. Describe the two fundamental processes involved in metal cutting.

Chip Formation

8. Describe briefly how each of the following chip types is produced:

   a. Discontinuous
   b. Continuous
   c. Continuous with a built-up edge

9. Which is the most desirable type of chip? Give reasons for your answer.

10. What conditions must be present to produce the Type 2 chip?

11. Explain how a built-up edge is formed and state its effect on cutting-tool life.

UNIT 28

Machinability of Metals

OBJECTIVES

After completing this unit, you will be able to:

1 Explain the factors that affect the machinability of metals

2 Describe the difference between high-carbon steel and alloy steel.

3 Assess the effects of temperature and cutting fluids on the surface finish produced

Machinability describes the ease or difficulty with which a metal can be machined. Such factors as cutting-tool life, surface finish produced, and power required must be considered. Machinability has been measured by the length of cutting-tool life in minutes, or by the rate of stock removal in relation to the cutting speed employed, that is, depth of cut. For finish cuts, *machinability* refers to the life of the cutting tool and the ease with which a good surface finish is produced.

▶▶ Grain Structure

The machinability of a metal is affected by its microstructure and will vary if the metal has been annealed. The ductility and shear strength of a metal can be modified greatly by operations such as annealing, normalizing, and stress relieving. Certain chemical and physical modifications of steel will improve its machinability. Free-machining steels have generally been modified in the following manner by:

1. The addition of sulfur
2. The addition of lead
3. The addition of sodium sulfite
4. Cold working, which modifies the ductility

By making these (free-machining) modifications to the steel, three main machining characteristics become evident:

1. Tool life is increased.
2. A better surface finish is produced.
3. Lower power consumption is required for machining.

LOW-CARBON (MACHINE) STEEL

The microstructure of *low-carbon steel* may have large areas of ferrite (iron) interspersed with small areas of pearlite (Fig. 28-1a and b on p. 204). Ferrite is soft, with high ductility and low strength, whereas pearlite, a combination of ferrite (iron) and iron carbide, has low ductility and high strength. When the amount of ferrite in steel is greater than pearlite—or the ferrite is arranged in alternate layers with pearlite (Fig. 28-1c and d)—the amount of power required to remove material increases and the surface finish produced is poor. Fig. 28-2 on p. 204 illustrates a more desirable microstructure in steel because the pearlite is well distributed, and the material is therefore better for machining purposes.

HIGH-CARBON (TOOL) STEEL

A greater amount of pearlite is present in *high-carbon (tool) steel* because of the higher carbon content. The greater the amount of pearlite (low ductility and high strength) present in the steel, the more difficult it becomes

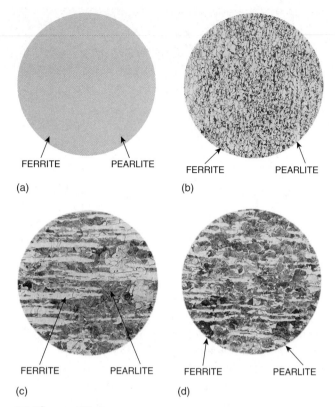

FERRITE — PEARLITE
(a)

FERRITE — PEARLITE
(b)

FERRITE — PEARLITE
(c)

FERRITE — PEARLITE
(d)

■ **Figure 28-1** Photomicrographs indicating undesirable steel microstructures. *(Cincinnati Machine)*

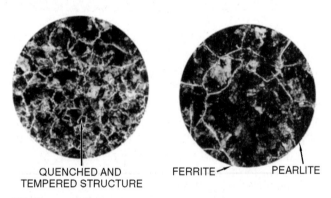

QUENCHED AND TEMPERED STRUCTURE

FERRITE — PEARLITE

■ **Figure 28-2** Photomicrographs showing desirable microstructures in steel. *(Cincinnati Machine)*

to machine the steel efficiently. It is therefore desirable to anneal these steels to alter their microstructures and, as a result, improve their machining qualities.

ALLOY STEEL

Alloy steels are combinations of two or more metals. These steels generally are slightly more difficult to machine than low- or high-carbon steels. To improve their machining qualities, combinations of sulfur and lead or

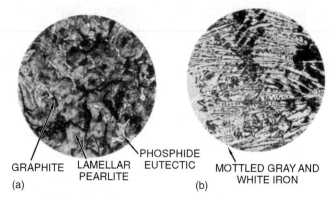

GRAPHITE   LAMELLAR   PHOSPHIDE
PEARLITE   EUTECTIC
(a)

MOTTLED GRAY AND WHITE IRON
(b)

■ **Figure 28-3** (a) The microstructure of white cast iron; (b) the microstructure of gray cast iron. *(Cincinnati Machine)*

sulfur and manganese in proper proportions are sometimes added to alloy steels. A combination of normalizing and annealing is also used with some types of alloy steels to create desirable machining characteristics. The machining of *stainless steel*, generally difficult because of its work-hardening qualities, can be greatly eased by the addition of selenium.

CAST IRON

Cast iron, consisting generally of ferrite, iron carbide, and free carbon, forms an important group of materials used by industry. The microstructure of cast iron can be controlled by the addition of alloys, the method of casting, the rate of cooling, and heat treating. *White cast iron* (Fig. 28-3a), cooled rapidly after casting, is usually hard and brittle because of the formation of hard iron carbide. *Gray cast iron* (Fig. 28-3b) is cooled gradually; its structure is composed of compound pearlite, a mixture of fine ferrite and iron carbide, and flakes of graphite. Because of the gradual cooling, it is softer and therefore easier to machine.

Iron carbide and the presence of sand on the outer surface of the casting generally make cast iron a little difficult to machine. Through annealing, the microstructure is altered. The iron carbide is broken down into graphitic carbon and ferrite; thus, the cast iron is easier to machine. The addition of silicon, sulfur, and manganese gives cast iron different qualities and improves its machinability.

ALUMINUM

Pure *aluminum* is generally more difficult to machine than most aluminum alloys. It produces long, stringy chips and is much harder on the cutting tool because of its abrasive nature.

Most aluminum alloys can be cut at high speeds, yielding a good surface finish and long tool life. Hardened and tempered alloys are generally easier to machine than annealed alloys and produce a better surface finish. Alloys containing silicon are more difficult to machine, since the chips tear, rather than shear, from the work, thus producing a poorer surface finish. Cutting fluid is generally used when heavy cuts and feeds are used for machining aluminum or its alloys.

COPPER

Copper is a heavy, soft, reddish-colored metal refined from copper ore (copper sulfide). It has high electrical and thermal conductivity, good corrosion resistance, and strength and is easily welded, brazed, or soldered. It is very ductile and easily drawn into wire and tubing. Since copper work hardens readily, it must be heated at about 1200°F (648.8°C) and quenched in water to anneal.

Because of its softness, copper does not machine well. The long chips produced in drilling and tapping tend to clog the flutes of the cutting tool and they must be cleared frequently. Sawing and milling operations require cutters with good chip clearance. Coolant should be used to minimize heat and aid the cutting action.

Copper-Base Alloys

Brass, an alloy of copper and zinc, has good corrosion resistance and is easily formed, machined, and cast. There are several forms of brass. *Alpha brasses* containing up to 36% zinc are suitable for cold working. *Alpha + beta brasses* containing 54% to 62% copper are used in hot working of this alloy. Small amounts of tin or antimony are added to alpha brasses to minimize the pitting effect of salt water on this alloy. Brass alloys are used for water and gasoline line fittings, tubing, tanks, radiator cores, and rivets.

Bronze, the term that originally referred to an alloy of copper and tin, has now been extended to include all alloys except copper–zinc alloys, which contain up to 12% of the principal alloying element.

Phosphor-bronze contains about 90% copper, 10% tin, and a very small amount of phosphorus, which acts as a hardener. This metal has high strength, toughness, and corrosion resistance and is used for lock washers, cotter pins, springs, and clutch discs.

Silicon-bronze (a copper–silicon alloy) contains less than 5% silicon and is the strongest of the work-hardenable copper alloys. It has the mechanical properties of machine steel and the corrosion resistance of copper. It is used for tanks, pressure vessels, and hydraulic pressure lines.

Aluminum-bronze (a copper–aluminum alloy) contains between 4% and 11% aluminum. Other elements such as iron, nickel, manganese, and silicon are added to aluminum bronzes. Iron (up to

5%) increases the strength and refines the grain. The addition of nickel (up to 5%) has effects similar to those of iron. Silicon (up to 2%) improves machinability. Manganese promotes soundness in casting. Aluminum-bronzes have good corrosion resistance and strength and are used for condenser tubes, pressure vessels, nuts, and bolts.

Beryllium-bronze (copper and beryllium), containing up to about 2% beryllium, is easily formed in the annealed condition. It has a high tensile strength and fatigue strength in the hardened condition. Beryllium-bronze is used for surgical instruments, bolts, nuts, and screws.

▶▶ The Effects of Temperature and Friction

In the process of cutting metals, heat is created by:

1. The plastic deformation occurring in the metal during the process of forming a chip

2. The friction created by the chips sliding along the cutting-tool face

The cutting temperature varies with each type of metal and increases with the cutting speed employed and the rate of metal removal. The *greatest heat* is generated when ductile material of high tensile strength, such as steel, is cut. The *lowest heat* is generated when soft material of low tensile strength, such as aluminum, is cut. The maximum temperature attained during the cutting action will affect cutting-tool life, quality of the surface finish, rate of production, and accuracy of the workpiece.

At times, the temperature of the metal immediately ahead of the cutting tool comes close to the melting temperature of the metal being cut. This great heat affects the life of the cutting tool. High-speed steel cutting tools cannot withstand the same high temperatures that cemented-carbide tools can without the cutting edges breaking down.

High-speed steel cutting tools can cut metal even when the cutting tools turn red from the cutting action. This property is known as *red hardness* and occurs at temperatures above 900°F (482°C). However, when the temperature exceeds 1000°F (538°C), the edge of the cutting tool will begin to break down.

Cemented-carbide cutting tools can be used efficiently at temperatures up to 1600°F (871°C). The carbide tools are harder than high-speed steel tools, have greater wear resistance, and perform extremely well under red-hardness operating conditions. Therefore, much higher cutting speeds can be used with carbide tools than with high-speed tools.

FRICTION

For efficient cutting action, it is important that the friction between the chip and tool face be kept as low as possible. As the coefficient of friction increases, there is a greater possibility of a built-up edge forming on the cutting edge. The larger the built-up edge, the more friction is created, which results in the breakdown of the cutting edge and poor surface finish. Every time the machine must be stopped to regrind or replace a cutting tool, production rates decrease.

The temperature created by friction also affects the accuracy of the machined part. Even though the workpiece does not reach the same temperature as the cutting-tool point, it is still high enough to cause the metal to expand. If a part heated by the cutting action is machined to size, the part will be smaller than required when it cools to room temperature. A good supply of cutting fluid will help reduce friction at the chip–tool interface and help maintain efficient cutting temperatures.

▶▶ Surface Finish

Many factors affect the surface finish produced by a machining operation, the most common being the feed rate, the nose radius of the tool, the cutting speed, the rigidity of the machining operation, and the temperature generated during the machining process.

If a high temperature is created during the cutting action, there is a marked tendency for a rough surface finish to result. The reason for this is that at high temperatures metal particles tend to adhere to the cutting tool and form a built-up edge. A direct relationship between the temperature of the workpiece and the quality of the surface finish is illustrated in Fig. 28-4.

Fig. 28-4a shows the results of machining a piece of aluminum without cutting fluid at 200°F (93°C). The rough surface finish indicates the presence of a built-up edge on the cutting tool. The same piece of aluminum was machined under the same conditions, but at a room temperature of 75°F (24°C) (Fig. 28-4b). A considerable improvement can be noted between the surface finishes of the samples in Fig. 28-4a and b. When the piece of aluminum was cooled to −60°F (−50°C) and machined, the surface finish improved further. By cooling the work material to −60°F (−50°C), the temperature of the cutting-tool edge was reduced considerably and resulted in a much better surface finish (Fig. 28-4c) than that produced at 200°F (93°C).

▶▶ Effects of Cutting Fluids

Cutting fluids are important to most machining operations because they make it possible to cut metals at higher rates of speed. They perform three important functions:

1. They reduce the temperature of the cutting action.
2. They reduce the friction of the chips sliding along the tool face.
3. They decrease tool wear and increase tool life.

There are three types of cutting fluids: cutting oils, emulsifiable (soluble) oils, and chemical (synthetic) cutting fluids. Some cutting fluids form a nonmetallic film on the metal surface, which prevents the chip from sticking to the cutting edge. This prevents a built-up edge from forming and results in a better surface finish being produced. The surface finish of most metals can be improved considerably by the use of the proper cutting fluids.

Cutting fluids are generally used for machining steel, alloy steel, brass, and bronze with high-speed steel cutting tools. As a rule, cutting fluids are not generally used with cemented-carbide tools unless a great quantity of cutting fluid can be applied to ensure uniform temperatures to prevent the carbide inserts from cracking. Cast iron, aluminum, and magnesium alloys are generally machined dry; however, cutting fluids have been used with good results in some cases.

(a) (b) (c)

Figure 28-4 (a) Aluminum machined at 200°F (93°C); (b) aluminum machined at 75°F (24°C); (c) aluminum machined after cooling to −60°F (−50°C).

unit 28 review questions

Machinability of Metals

1. Define machinability.
2. What factors affect the machinability of a metal?
3. Compare the microstructure of low-carbon and high-carbon steels with respect to their machinability.
4. How can the machining qualities of alloy steels be improved?
5. Why is pure aluminum more difficult to machine than most aluminum alloys?
6. What can be done to improve the machining of aluminum and its alloys?

Effects of Temperature and Friction

7. Name two methods by which heat is created during machining.

8. How does high temperature affect a machining operation?
9. Why is it important that friction between the chip and tool be kept to a minimum?

Surface Finish

10. What common factors determine surface finish?
11. Why does high temperature affect the surface finish produced?

Effects of Cutting Fluids

12. List four ways in which cutting fluids assist the machining of metals.
13. What precaution should be taken when cutting fluids are used with carbide tools?

UNIT 29

Cutting Tools

OBJECTIVES

After completing this unit, you will be able to:

1 Use the nomenclature of a cutting-tool point

2 Explain the purpose of each type of rake and clearance angle

3 Identify the applications of various types of cutting-tool materials

4 Describe the cutting action of different types of machines

One of the most important components in the machining process is the cutting tool, the perfor-mance of which will determine the efficiency of the operation. Consequently, much thought should be given not only to the selection of the cutting-tool material but also to the cutting-tool angles required to machine a workpiece material properly.

There are basically two types of cutting tools (excluding abrasives): single- and multi-point tools. Since both types must have rake and clearance angles ground or formed on them to cut, the nomenclature of cutting-tool points will apply to both types. The lathe tool, which is the most common single-point tool, will be discussed in greater detail. The principles involved in this type of cutting tool will then be related to multi-point cutting tools for ease of understanding.

▶▶ Cutting-Tool Materials

Lathe toolbits are generally made of five materials: high-speed steel, cast alloys (such as stellite), cemented carbides, ceramics, and cermets. More exotic cutting-tool materials—such as polycrystalline cubic boron nitride (PCBN), commonly called Borazon, and polycrystalline diamond (PCD)—are finding wide use in the metalworking industry because of the increased productivity they offer. Borazon is used to machine hardened alloy steels and tough superalloys. Polycrystalline diamond cutting tools are used to machine nonferrous and nonmetallic materials requiring close tolerances and a high surface finish. The properties possessed by each of these materials are different and the application of each depends on the material being machined and the condition of the machine.

Lathe toolbits should possess the following properties:

1. They should be hard.

2. They should be wear-resistant.

3. They should be capable of maintaining a *red hardness* during the machining operation. (*Red hardness* is the ability of a cutting tool material to maintain a sharp cutting edge even when it turns red because of the high heat produced at the work-tool interface during the cutting operation.)

4. They should be able to withstand shock during the cutting operation.

5. They should be shaped so that the edge can penetrate the work. (The shape will be determined by the cutting-tool material, the material being cut, and the angle of keenness.)

HIGH-SPEED STEEL TOOLBITS

Probably the toolbit most commonly used in schools for lathe operations is the high-speed steel toolbit. High-speed steels may contain combinations of tungsten, chromium, vanadium, molybdenum, and cobalt. They are capable of taking heavy cuts, withstand shock, and maintain a sharp cutting edge under red heat.

High-speed steel toolbits are generally of two types: molybdenum-base (Group M) and tungsten-base (Group T). The most widely used tungsten-base toolbit is known as T1, which is sometimes called 18-4-1, since it contains about 18% tungsten, 4% chromium, and 1% vanadium.

A general-purpose molybdenum-base high-speed steel toolbit is known as M_1 or 8-4-1. This alloy contains about 8% molybdenum, 4% chromium, and 1% vanadium.

These two types are general-purpose tools; if more red hardness is desired, a tool containing more cobalt should be selected. Since there are many different grades of high-speed steel toolbits, one should refer to the manufacturer's recommendations for a toolbit for a specific job. Table 29.1 indicates the properties imparted to a high-speed steel toolbit by the various alloying elements.

CAST ALLOY TOOLBITS

Cast alloy (stellite) toolbits usually contain 25% to 35% chromium, 4% to 25% tungsten, and 1% to 3% carbon; the remainder is cobalt. These toolbits have high hardness, high resistance to wear, and excellent red-hardness qualities. Since they are cast, they are weaker and more brittle than high-speed steel toolbits. Stellite toolbits are capable of high speeds and feeds on deep uninterrupted cuts. They may be operated at about two to two-and-a-half times the speed of a high-speed steel toolbit.

Note: When grinding stellite toolbits, apply only light pressure and do not quench the toolbit in water.

CEMENTED-CARBIDE TOOLBITS

Cemented-carbide toolbits (Fig. 29-1) are capable of cutting speeds three to four times those of high-speed steel toolbits. They have low toughness but high hardness and excellent red-hardness qualities.

Figure 29-1 A variety of cemented-carbide tool inserts.

Cemented carbide consists of tungsten carbide sintered in a cobalt matrix. Sometimes other materials such as titanium or tantalum may be added before sintering to give the material the desired properties.

Straight tungsten carbide toolbits are used to machine cast iron and nonferrous materials. Since they crater easily and wear rapidly, they are not suitable for machining steel. Crater-resistant carbides, which are used for machining steel, are made by adding titanium and/or tantalum carbides to the tungsten carbide and cobalt.

Different grades of carbides are manufactured for different work requirements. Those used for heavy roughing cuts contain more cobalt than those used for finishing cuts, which are more brittle and have greater wear resistance at higher finishing speeds.

COATED CARBIDE TOOLBITS

Coated carbide cutting tools are made by depositing a thin layer of wear-resistant titanium nitride, titanium carbide, or aluminum oxide (ceramic) on the cutting edge of the tool. This fused layer increases lubricity, improves the cutting edge wear resistance by 200% to 500%, and lowers the breakage resistance by up to 20% while providing longer life and increased cutting speeds.

Titanium-coated inserts offer greater wear resistance at speeds below 500 sf/min; ceramic coated tips are best suited for higher cutting speeds. Both types of insert are used for cutting steels, cast irons, and nonferrous materials.

CERAMIC TOOLBITS

A *ceramic* is a heat-resistant material produced without a metallic bonding agent such as cobalt. Aluminum oxide is the most popular material used to make ceramic cutting tools. Titanium oxide or titanium carbide may be used as an additive, depending on the cutting tool application.

Ceramic tools (Fig. 29-2 on p. 211) permit higher cutting speeds, increased tool life, and better surface finish than do carbide tools. However, they are much weaker than carbide or coated carbide tools and must be used in shock-free or low-shock situations.

table 29.1   The effect of alloying elements on steel

Effect	Carbon	Chromium	Cobalt	Lead	Manganese	Molybdenum	Nickel	Phosphorus	Silicon	Sulfur	Tungsten	Vanadium
Increases tensile strength	X	X			X	X	X					
Increases hardness	X	X										
Increases wear resistance	X	X			X		X				X	
Increases hardenability	X	X			X	X	X					X
Increases ductility					X							
Increases elastic limit		X				X						
Increases rust resistance		X					X					
Increases abrasion resistance		X			X							
Increases toughness		X				X	X					X
Increases shock resistance		X					X					X
Increases fatigue resistance												X
Decreases ductility	X	X										
Decreases toughness			X									
Raises critical temperature		X	X								X	
Lowers critical temperature					X		X					
Causes hot shortness										X		
Causes cold shortness								X				
Imparts red hardness			X			X					X	
Imparts fine grain structure					X							X
Reduces deformation					X		X					
Acts as deoxidizer					X				X			
Acts as desulphurizer					X							
Imparts oil-hardening properties		X			X	X	X					
Imparts air-hardening properties					X	X						
Eliminates blow holes									X			
Creates soundness in casting										X		
Facilitates rolling and forging					X					X		
Improves machinability				X						X		

Figure 29-2 A variety of ceramic cutting-tool inserts.

CERMET TOOLBITS

A *cermet* is a cutting-tool insert composed of ceramics and metal. Most cermets are made from aluminum oxide, titanium carbide, and zirconium oxide compacted and compressed under intense heat. The advantages of cermet toolbits are:

> They exceed the equivalent tool life of coated and uncoated carbides.

> They can be used for machining at high temperatures.

> They produce an improved surface finish, which eliminates the need for grinding and provides greater dimensional control.

> They may be used to machine steels up to 45 Rc hardness.

DIAMOND TOOLBITS

Diamond tools are used mainly to machine nonferrous metals and abrasive nonmetallics. Single-crystal natural diamonds have high-wear but low shock-resistant factors. The new type of diamond tooling (polycrystalline diamonds) consists of tiny manufactured diamonds fused together and bonded to a suitable carbide substrate. Polycrystalline cutting tools offer greater wear and shock resistance and greatly increased cutting speeds. Polycrystalline diamond tools offer improved surface finish, better part-size control, up to 100 times greater tool life than carbide tools, and increased productivity.

CUBIC BORON NITRIDE TOOLBITS

Cubic boron nitride (Borazon) is next to diamond on the hardness scale. These cutting tools are made by bonding a layer of polycrystalline cubic boron nitride to a cemented-carbide substrate, which provides good shock resistance. They offer exceptionally high wear resistance and edge life and may be used to machine high-temperature alloys and hardened ferrous alloys.

▶▶ Cutting-Tool Nomenclature

Cutting tools used on a lathe are generally single-pointed, and although the shape of the tool is changed for various applications, the same nomenclature applies to all cutting tools (Fig. 29-3).

The *base* is the bottom surface of the tool shank.

The *cutting edge* is the leading edge of the toolbit that does the cutting.

The *face* is the surface against which the chip bears as it is separated from the work.

The *flank* is the surface of the tool adjacent to and below the cutting edge.

The *nose* is the tip of the cutting tool formed by the junction of the cutting edge and the front face.

The *nose radius* is the radius to which the nose is ground. The size of the radius will affect the finish. For rough turning, a small nose radius [about .015 in. (0.38 mm)] is used. A larger radius [about .060 to .125 in. (1.5 to 3 mm)] is used for finish cuts.

The *point* is the end of the tool that has been ground for cutting purposes.

The *shank* is the body of the toolbit or the part held in the toolholder.

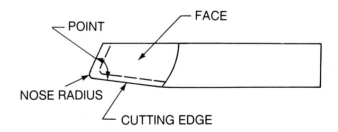

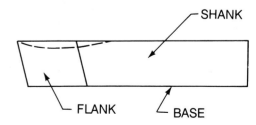

Figure 29-3 Nomenclature of a general-purpose lathe toolbit.

table 29.2 Recommended angles for high-speed steel tools

Material	Side Relief	End Relief	Side Rake	Back Rake	Angle of Keenness
Aluminum	12	8	15	35	63
Brass	10	8	5 to −4	0	75 to 84
Bronze	10	8	5 to −4	0	75 to 84
Cast iron	10	8	12	5	68
Copper	12	10	20	16	58
Machine steel	10 to 12	8	12 to 18	8 to 15	60 to 68
Tool steel	10	8	12	8	68
Stainless steel	10	8	15 to 20	8	72

▸▸ Lathe Toolbit Angles and Clearances

Proper toolbit performance depends on the clearance and rake angles, which must be ground on the toolbit. Although these angles vary for different materials, the nomenclature is the same for all toolbits (Table 29.2).

The *side cutting edge angle* is the angle the cutting edge forms with the side of the tool shank (Fig. 29-4). Side cutting angles for a general-purpose lathe cutting tool may vary from 10° to 20°, depending on the material cut. If this angle is too large (over 30°), the tool will tend to chatter.

The *end cutting edge angle* is the angle formed by the end cutting edge and a line at right angles to the centerline of the toolbit (Fig. 29-4). This angle may vary from 5° to 30°, depending on the type of cut and finish desired. An angle of 5° to 15° is satisfactory for roughing cuts; angles between 15° and 30° are used for general-purpose turning tools. The larger angle permits the cutting tool to be swiveled to the left for taking light cuts close to the dog or chuck, or when turning to a shoulder.

The *side relief (clearance) angle* is the angle ground on the flank of the tool below the cutting edge (Figs. 29-4 and 29-5). This angle is generally 6° to 10°. The side clearance on a toolbit permits the cutting tool to advance lengthwise into the rotating work and prevents the flank from rubbing against the workpiece.

The *end relief (clearance) angle* is the angle ground below the nose of the toolbit, which permits the cutting tool to be fed into the work. It is generally 10° to 15° for general-purpose tools (Figs. 29-4 and 29-5). This angle

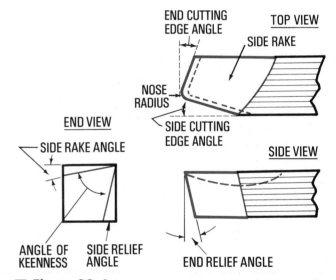

■ **Figure 29-4** Angles and clearances for lathe cutting tools.

must be measured when the toolbit is held in the toolholder. The end relief angle varies with the hardness and type of material and the type of cut. The end relief angle is smaller for harder materials, providing support under the cutting edge.

The *side rake angle* is the angle at which the face is ground away from the cutting edge. For general-purpose toolbits, the side rake is generally 14° (Figs. 29-4 and 29-5). Side rake creates a keener cutting edge and allows the chips to flow away quickly. For softer materials, the side rake angle is generally increased. Side rake may be either positive or negative, depending on the material being cut.

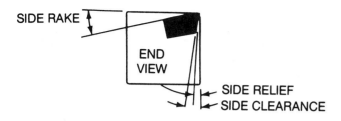

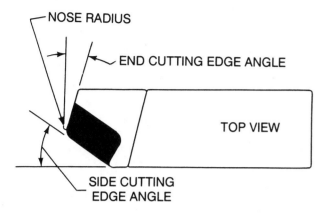

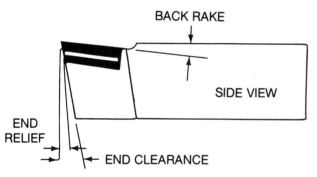

Figure 29-5 Lathe cutting-tool angles and clearances.

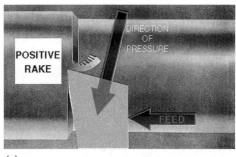

(a)

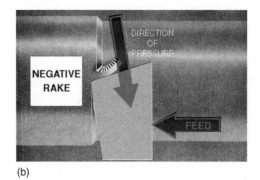

(b)

■ **Figure 29-6** (a) Positive rake angle ground on a cutting tool; (b) negative rake angle ground on a cutting tool.

Each type of rake angle serves a specific purpose. The type used depends on the machining operation performed and the characteristics of the work material. Rake angles can be ground on cutting tools or, in the case of cutting-tool inserts, they can be held in suitable holders, which provide the rake angle desired.

POSITIVE RAKE ANGLE

A positive rake angle (Fig. 29-6a) is considered best for the efficient removal of metal. It creates a large shear angle at the shear zone, reduces friction and heat, and allows the chip to flow freely along the chip-tool interface. Positive-rake-angle cutting tools are generally used for continuous cuts on ductile materials that are not too hard or abrasive. Even though positive-rake-angle tools remove metal efficiently, they are not recommended for all work materials or cutting applications. The following factors must be considered when the type and the amount of rake angle for a cutting tool are being determined:

1. The hardness of the metal to be cut

2. The type of cutting operation (continuous or interrupted)

3. The material and shape of the cutting tool

4. The strength of the cutting edge

The *angle of keenness* is the included angle produced by grinding side rake and side clearance on a toolbit (Fig. 29-4). This angle may be altered, depending on the type of material machined, and will be greater (closer to 90°) for harder materials.

The *back (top) rake angle* is the backward slope of the tool face away from the nose. The back rake angle is generally about 20° and is provided for in the toolholder (Fig. 29-5). Back rake permits the chips to flow away from the point of the cutting tool. Two types of back or top rake angles are provided on cutting tools and are always found on the top of the toolbit:

> *Positive rake* (Fig. 29-6a), where the point of the cutting tool and the cutting edge contact metal first and the chip moves *down the face* of the toolbit

> *Negative rake* (Fig. 29-6b), where the face of the cutting tool contacts the metal first and the chip is forced *up the face* of the toolbit

NEGATIVE RAKE ANGLE

A negative rake angle (Fig. 29-6b) is used for interrupted cuts and when the metal is tough or abrasive. A negative rake angle on the tool creates a small shear angle and a long shear zone; therefore, more friction and heat are created. Although the increase in heat may seem to be a disadvantage, it is desirable when tough metals are machined with carbide cutting tools. Face-milling cutters with carbide tool inserts are a good example of the use of negative rake for interrupted and high-speed cutting.

The advantages of negative rake on cutting tools are:

> The shock from the work meeting the cutting tool is on the tool's face, not its point or edge, which prolongs the life of the tool.

> The hard outer scale on the metal does not come into contact with the cutting edge.

> Surfaces with interrupted cuts can be readily machined.

> Higher cutting speeds can be used.

The shape of a chip can be altered in a number of ways to improve the cutting action and reduce the amount of power required. A continuous *straight* ribbon chip on a lathe can be changed to a continuous curled ribbon by:

1. Changing the *angle of the keenness* (the included angle produced by grinding the side rake and side clearance) on a toolbit (Fig. 29-4).

2. Grinding a chip breaker behind the cutting edge of the toolbit.

A helix angle on a milling cutter affects the cutting performance by providing a shearing action when the chip is removed.

▶▶ Cutting-Tool Shape

The shape of the cutting tool is very important for the efficient removal of metal. Every time a machine must be stopped to recondition or replace a worn cutting tool, production rates decrease. The life of a cutting tool is generally reported as:

1. The number of minutes that the tool has been cutting

2. The length of material cut

3. The number of cubic inches or cubic centimeters (cm^3) of material removed

4. In the case of drills, the number of inches or millimeters of hole depth drilled

To prolong cutting-tool life, reduce the friction between the chip and the tool as much as possible. This re-

duction can be accomplished by providing the cutting tool with a suitable rake angle and by highly polishing the cutting-tool face with a honing stone. The polished cutting face reduces the friction on the chip-tool interface, reduces the size of the built-up edge, and generally results in better surface finish. The rake angle on cutting tools allows chips to flow away freely and reduces friction and the amount of power required for the machining operation.

The rake angle of the cutting tool also affects the shear angle or plane of the metal, which in turn determines the area of plastic deformation. If a *large rake angle* is ground on the cutting tool, a large shear angle is created in the metal during the cutting action (Fig. 29-6a). The results of a large shear angle are:

1. A thin chip is produced.

2. The shear zone is relatively short.

3. Less heat is created in the shear zone.

4. Good surface finish is produced.

5. Less power is required for the machining operation.

A *small* or a *negative rake angle* on the cutting tool (Fig. 29-6b) creates a small shear angle in the metal during the cutting process with the following results:

1. A thick chip is produced.

2. The shear zone is long.

3. More heat is produced.

4. The surface finish is not quite as good as with large-rake-angle cutting tools.

5. More horsepower is required for the machining operation.

▶▶ Tool Life

Tool life, or the number of parts produced by a cutting-tool edge before regrinding is required, is an important cost factor in manufacturing a part or product. Consequently, cutting tools must be reground at the first sign of dullness. If a tool is used beyond this point, it will break down rapidly and much more tool material will have to be removed when regrinding, thus shortening the tool's life.

To determine the time when a cutting tool should be changed, most modern machines are equipped with indicators that show the horsepower used during the machining operation. When a tool becomes dull, more horsepower is required for the operation. This will show on the indicator, and the tool should be reconditioned immediately.

The wear or abrasion of the cutting tool will determine its life. Three types of wear are generally associated with cutting tools: *flank wear, nose wear,* and *crater wear* (Fig. 29-7).

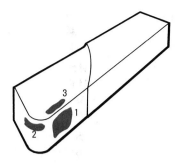

Figure 29-7 The tool wear areas of a cutting tool: (1) flank wear; (2) nose wear; (3) crater wear. *(Kelmar Associates)*

Flank wear occurs on the side of the cutting edge as a result of friction between the side of the cutting-tool edge and the metal being machined. Too much flank wear increases friction and makes more power necessary for machining. When the flank wear is .015 to .030 in. (0.38 to 0.76 mm) long, the tool should be reground.

Nose wear occurs on the nose or point of the cutting tool as a result of friction between the nose and the metal being machined. Wear on the nose of the cutting tool affects the quality of the surface finish on the workpiece.

Crater wear occurs a slight distance away from the cutting edge as a result of the chips sliding along the chip-tool interface, which is a result of a built-up edge on the cutting tool. Too much crater wear eventually breaks down the cutting edge.

The following factors affect the life of a cutting tool:

1. The type of material being cut
2. The microstructure of the material
3. The hardness of the material
4. The type of surface on the metal (smooth or scaly)
5. The material of the cutting tool
6. The profile of the cutting tool
7. The type of machining operation being performed
8. The speed, feed, and depth of cut

▶▶ Principles of Machining

TURNING

A high proportion of the work machined in a shop is turned on a lathe. The workpiece is held securely in a chuck or between lathe centers. A turning tool, mounted in a holder and set to a given depth of cut, is fed parallel to the axis of the work to reduce the diameter of the workpiece (Fig. 29-8).

As the workpiece revolves and the cutting tool is fed along the axis, material is separated by the edge of the cut-

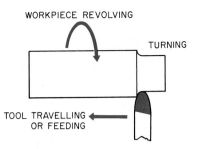

Figure 29-8 The cutting action of a lathe or turning center. *(Kelmar Associates)*

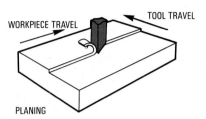

Figure 29-9 The cutting action of a planer. *(Kelmar Associates)*

ting tool. A chip forms and slides along the cutting tool's upper surface created by the side rake. The *angle of keenness* (Fig. 29-4), which permits the tool to remove the metal as the tool is fed along the workpiece, is formed by the side relief angle clearance and the side rake ground on the toolbit. Let's assume that we are cutting a piece of machine steel; if the rake and relief clearance angles are correct and the proper speed and feed are used, a continuous chip should be formed. If the angle of keenness is too small, the edge of the tool will break down too soon. If the angle is too great (that is, little or no side rake), the metal will not be removed as effectively and greater torque will be required to remove it. In either case, a built-up edge and rough surface finish will result.

PLANING

The cutting tool used in the planer is basically the same shape as the lathe tool for machining similar materials. It should have the proper rake and clearance angles ground on it to machine the workpiece efficiently. The cutting action of a planer is illustrated in Fig. 29-9. The workpiece is moved back and forth under a cutting tool, which is fed sideways a set amount at the end of each table reversal.

PLAIN MILLING

A milling cutter is a multi-tooth tool having several equally spaced cutting edges (teeth) around its periphery.

■ **Figure 29-10** The type of chip produced by a helical milling cutter. *(Cincinnati Machine)*

■ **Figure 29-11** As the workpiece is fed into the revolving cutter, each tooth removes material from the work. *(Cincinnati Machine)*

Each tooth may be considered to be a single-point cutting tool and must have proper rake and clearance angles to cut effectively. Fig. 29-10 shows the chip formation produced by a helical milling cutter.

The workpiece, held in a vise or fastened to the table, is fed into the horizontal revolving cutter by hand or by automatic table feed. As the work is fed into the cutter, each tooth makes successive cuts, which produce a smooth, flat, or profiled surface, depending on the shape of the cutter used (Fig. 29-11). The nomenclature of a plain milling cutter is shown in Fig. 29-12.

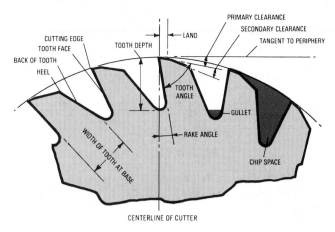

■ **Figure 29-12** The nomenclature of a plain milling cutter.

■ **Figure 29-13** End mills generally cut on the periphery.

END AND FACE MILLING

End mills (Fig. 29-13) are multi-tooth cutters held vertically in a vertical milling machine spindle or attachment. They are used primarily for cutting slots or grooves, whereas shell end mills and face mills are used primarily for producing flat surfaces. The workpiece is held in a vise or clamped to the table and is fed into the revolving cutter

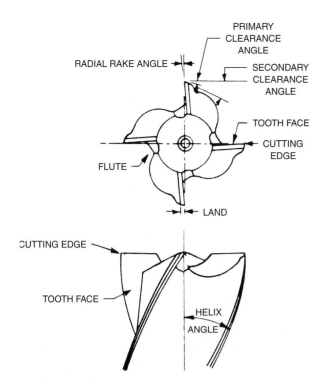

Figure 29-14 The nomenclature of an end mill.

Figure 29-15 The corner of each tooth on face milling cutters is generally chamfered to provide strength and prevent the teeth from chipping.

by hand or automatically. When end milling, the cutting is done by the periphery of the teeth. The nomenclature of an end mill is shown in Fig. 29-14.

The inserted blade face mill consists of a body that holds several equally spaced inserts at the required rake angle. The lower edge of each insert has a relief or clearance angle ground on it. Since the cutting action occurs at the lower corner of the insert, generally each corner is chamfered to give it strength and to prevent the corners from breaking off (Fig. 29-15).

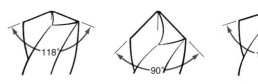

Figure 29-16 The characteristics of a drill point.
(Kelmar Associates)

Figure 29-17 The chip formation of a drill.

DRILLING

The drill is a multi-edge cutting tool that cuts on the point. The drill's cutting edges, or lips, are provided with lip clearance (Fig. 29-16) to permit the point to penetrate the workpiece as the drill revolves. It may be fed into the work manually or automatically. The rake angle is provided by helical-shaped flutes that slope away from the cutting edge. As in other cutting tools, the included angle between the rake angle and the clearance angle is known as the *angle of keenness*. Cutting-point angles for a standard drill are shown in Fig. 29-16; chip formation is shown in Fig. 29-17.

unit 29 review questions

Cutting-Tool Materials

1. What properties should a cutting tool possess?

2. What elements are found in high-speed steel toolbits?

3. State the precaution that should be taken when grinding stellite toolbits.

4. List three qualities of a carbide toolbit.

5. Why are straight tungsten carbide toolbits used?

6. What two substances may be added to tungsten carbide to make it crater-resistant?

7. List four advantages of coated carbide toolbits over conventional carbide toolbits.

8. State the uses of:

   a. Titanium-coated inserts
   b. Ceramic-coated inserts

9. List three advantages of ceramic toolbits.

10. For what application should ceramic toolbits not be used?

11. Describe a polycrystalline diamond toolbit.

12. What are the main applications of polycrystalline diamond tools?

13. List six advantages of polycrystalline diamond toolbits.

14. How are polycrystalline cubic boron nitride toolbits made?

15. State two advantages of polycrystalline cubic boron nitride toolbits.

16. List two applications of polycrystalline cubic boron nitride toolbits.

Cutting-Tool Nomenclature

17. Make neat sketches of a single-point cutting tool and label the following parts:

    a. Face   e. Flank
    b. Cutting edge f. Shank
    c. Nose   g. Radius
    d. Point

Lathe Toolbit Angles and Clearances

18. State the purpose of the following:

    a. Side cutting edge angle   d. Side rake angle
    b. Side relief angle   e. Back rake angle
    c. End relief angle   f. Angle of keenness

19. Draw sketches to illustrate the angles in question 18.

Cutting-Tool Shape

20. Name two methods that can be used to improve the life of a cutting tool.

21. What results can be expected from grinding a large rake angle on a cutting tool?

22. How does negative rake on a cutting tool affect the cutting process?

Tool Life

23. Define flank wear, nose wear, and crater wear.

24. List six important factors that affect the life of a cutting tool.

25. Name the two types of cutting tools.

Cutting in a Lathe

26. What will the effect be if the angle of keenness is:

    a. Too small?   b. Too large?

27. List two results of question 26a and b.

Plain Milling

28. Describe a plain milling cutter.

29. Describe the cutting action of a milling machine.

30. What is another name for the tooth angle in Fig. 29-12?

End and Face Milling

31. What is the purpose of:

    a. End milling?   b. Face milling?

32. How is the rake angle achieved on the blades of an inserted-tooth face mill?

33. Why are inserted-tooth cutters chamfered on the corners?

Drilling

34. Why is lip clearance ground on a drill?

35. What two surfaces provide the angle of keenness for the drill?

UNIT 30

Operating Conditions and Tool Life

OBJECTIVES

After completing this unit, you will be able to:

1 Describe the effect of cutting conditions on cutting-tool life

2 Explain the effect of cutting conditions on metal-removal rates

3 State the advantages of new cutting-tool materials

4 Calculate the economic performance and cost analysis for a machining operation

Intense global competition and the productivity gap between North American and overseas workers are forcing many companies to renew their commitment to product quality, while at the same time reducing manufacturing costs. During the past decade or so, marked gains in productivity have resulted from high technology automation, numerical control machine tools, flexible manufacturing systems, and other innovations. These new, more rigid machine tools and systems with higher horsepower are much more productive than the machines they have replaced. However, realizing their full potential depends on the use of reliable *high-efficiency cutting tools* to repeatedly produce accurate parts at prices that make them competitive with overseas production.

▶▶ Operating Conditions

There has been a continual search for new cutting-tool materials to increase productivity. High-speed steels were gradually replaced by cast alloys, carbides, ceramics, and cermets. Superabrasive high-efficiency cutting tools such as polycrystalline diamonds (PCD) and polycrystalline cubic boron nitride (PCBN) are now finding wide use in the metalworking industry. The productivity of these new high-efficiency cutting tools far surpasses that of other cutting tools for certain applications.

To produce parts efficiently, optimum operating conditions must be achieved during production. Three operating variables—cutting speed, feed rate, and depth of cut—influence metal-removal rate and tool life (Fig. 30-1). This unit will explain how to achieve these optimum

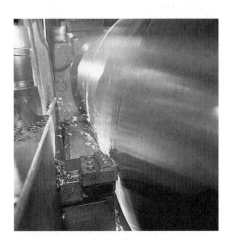

■ **Figure 30-1** The best operating conditions result in the best metal-removal rates and an increase in productivity. *(Diamond Innovations Inc., at www.abrasivesnet.com)*

conditions, which will result in maximum productivity and minimum cost per piece produced.

DEPTH OF CUT, FEED RATE, AND CUTTING SPEED

The *metal-removal rate* (MRR) is the rate at which metal is removed from an unfinished part and is measured in cubic inches or cubic centimeters per minute. Whenever any one of the three variables (cutting speed, feed, and depth of cut) is changed, MRR will change accordingly. For example, if the cutting speed or the depth of cut is increased by 25%, MRR will increase by 25%, but the life of the cutting tool will be reduced. However, a change in each of the variables will affect cutting-tool life differently. This difference can be proven by setting up a test piece on a lathe.

EFFECTS OF CHANGING OPERATING CONDITIONS

Let's assume that a test piece is being machined. The lathe is set to the proper r/min for the material being cut. The feed rate has been selected and the depth of cut has been set at 10 times the rate of feed, which is generally accepted as the minimum depth. After the test cut has been made and the tool life established, increase each variable by 50% and note the effects on tool life (Fig. 30-2). The results are approximately as follows:

> Increasing the depth of cut by 50% reduces tool life by 15%.

> Increasing the feed rate by 50% reduces tool life by 60%.

> Increasing the cutting speed by 50% reduces tool life by 90%.

Based on Fig. 30-2, it can be assumed that:

> Changes in the depth of cut have the least effect on tool life.

> Changes in the feed rate have a greater effect than depth-of-cut changes on tool life.

> Changes in the cutting speed of any material have a greater effect than either depth-of-cut or feed-rate changes on tool life.

GENERAL OPERATING CONDITION RULES

From the previous test, it is evident that the selection of the proper cutting speed is the most critical factor to consider when establishing the optimum or ideal operating conditions. If the cutting speed is too low, fewer parts will be produced. At very low speeds, a built-up edge may occur on the toolbit, requiring tool changes. If the cutting

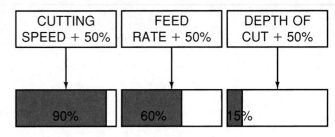

Figure 30-2 Conditions which affect single-point machining operations. *(Kelmar Associates)*

speed is too high, the tool will break down quickly, requiring frequent tool changes. Thus,

The optimum cutting speed for any job should balance the metal-removal rate and cutting-tool life.

When considering the best feed rate and depth of cut, always choose the heaviest depth of cut and feed rate possible, because they will reduce tool life much less than too high a cutting speed. Thus,

The optimum feed rate should balance the metal-removal rate and cutting-tool life.

In summary, the maximum production rate is achieved by a combination of cutting speed, feed, and depth of cut that attains the highest output for which the sum of machining time and tool-change time is a minimum. Factors that affect production rate include:

1. Inadequate horsepower, which limits the metal-removal rate

2. Surface finish requirements, which may limit the feed rate

3. Machine rigidity, which may not be sufficient to withstand cutting forces, feed rate, and depth of cut

4. Rigidity of the part being machined, which may limit the depth of cut

▶▶ Economic Performance

When the cost of any machining operation is analyzed, many factors must be considered to arrive at a true cost. The most important factor affecting the metal-removal rate is the type of cutting tool used. The *cost of a conventional cutting tool* may be very low, but the *cost of using the tool* may be very high. On the other hand, the *cost of a superabrasive cutting tool* such as polycrystalline diamond or cubic boron nitride may be relatively high, but the *cost of using the tool* may be low enough to justify its use.

To arrive at a true picture of whether a cutting tool will be economical or not, two factors must be considered for the total machining cost equation: the cost of using the

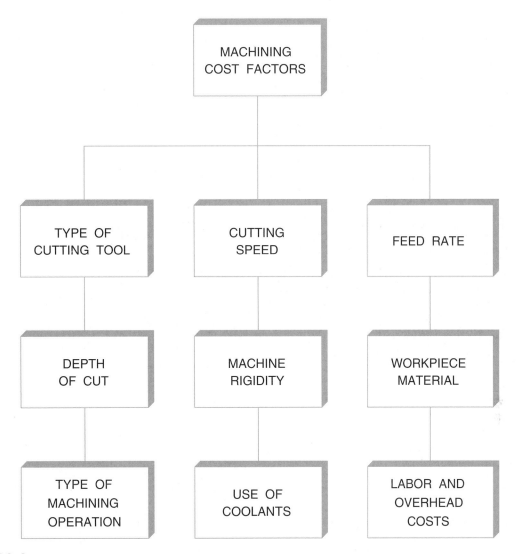

■ Figure 30-3 Factors which affect the cost of machining a part. *(Kelmar Associates)*

cutting tool and the price of the cutting tool. Let us examine what factors must be considered to arrive at the total machining costs per part produced. (See Fig. 30-3).

Cost of Using the Tool

1. The ability of a cutting tool to remove stock will determine the production rate and the amount of labor and investment required to produce parts.

2. Any tool's ability to remove stock is governed by the number of times that a tool must be reconditioned or replaced to produce accurate work and a good surface finish.

3. The rate that a cutting tool wears will influence how often a worn tool must be removed from a machine and replaced.

4. The cutting tool must be reconditioned and stored in inventory, which affects the total machining cost.

▶▶ Machining Cost Analysis

Many factors must be considered to analyze the true cost of any machining operation. One of the most important elements which govern the rate at which metal is removed is the choice of the cutting tool used. Even though one cutting tool may be *more expensive* than another tool, it can be *more economical* in the long run if it removes material faster and at less tool cost per part.

To obtain the lowest possible costs in any machining operation, tool engineers and methods personnel must look at the total operation and carefully examine each element that can affect the cost picture. The many variables which affect a machining operation are listed in Fig. 30-3. All of these factors play a part in determining how a cutting tool performs during a machining operation.

Operating Conditions

1. What are the three factors that influence cutting-tool life?

Depth of Cut, Feed Rate, and Cutting Speed

2. How is metal-removal rate measured?

Effects of Changing Operating Conditions

3. What is the generally accepted minimum depth of cut?

4. Which factor increase causes the greatest reduction in cutting-tool life?

General Operating Condition Rules

5. What may occur if too slow a cutting speed is used?

6. What is the optimum or ideal cutting speed for any job?

7. List four factors that affect production rates.

Economic Performance

8. What is the most important factor affecting metal-removal rate?

9. What two factors must be considered when arriving at the total cost of a part?

Cost of Using the Tool

10. What determines the production rate of a toolbit?

Machining Cost Analysis

11. List six of the most important cost factors in machining a part.

UNIT 31

Carbide Cutting Tools

OBJECTIVES

After completing this unit, you will be able to:

1 Identify and state the purpose of the two main types of carbide grades

2 Select the proper grade of carbide for various workpiece materials

3 Select the proper speeds and feeds for carbide tools

Carbide was first used for cutting tools in Germany during World War I as a substitute for diamonds. During the 1930s, various additives were discovered that generally improved the quality and performance of carbide tools. Since that time, various types of cemented (sintered) carbides have been developed to suit different materials and machining operations. Cemented carbides are similar to steel in appearance, but they are so hard that the diamond is almost the only material that will scratch them. They have been accepted by industry because they have good wear resistance and can operate efficiently at cutting speeds ranging from 150 to 1200 sf/min (46 to 366 m/min). Carbide tools can machine metals at speeds that cause the cutting edge to become red hot without losing its hardness or sharpness.

▶▶ Manufacture of Cemented Carbides

Cemented carbides are products of the powder metallurgy process; they consist primarily of minute particles of tungsten and carbon powders cemented together under heat by a metal of lower melting point, usually cobalt. Powdered metals such as tantalum, titanium, and niobium are also used in the manufacture of cemented carbides to provide cutting tools with various characteristics. The entire operation of producing cemented-carbide products is illustrated in Fig. 31-1 on p. 224.

BLENDING

Five types of powders are used in the manufacture of cemented-carbide tools: tungsten carbide, titanium carbide, tantalum carbide, niobium carbide, and cobalt. One or any combination of these carbide powders and cobalt (the binder) is blended in different proportions, depending on the grade of carbide desired. This powder is mixed in alcohol; the mixing process takes anywhere from 24 to 190 h. After the powder and alcohol have been thoroughly mixed, the alcohol is drained off, and paraffin is added to simplify the pressing operation.

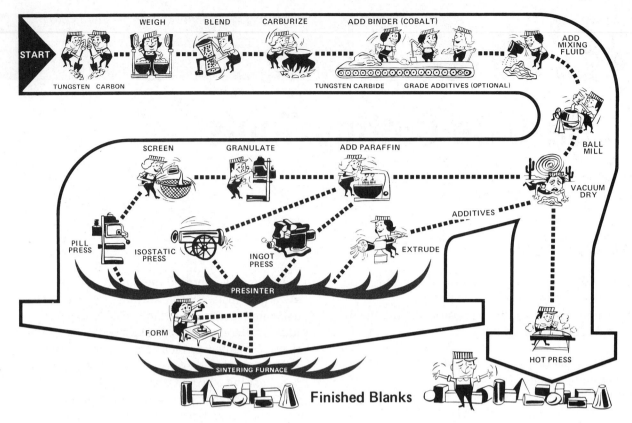

■ Figure 31-1 The process of manufacturing cemented-carbide products. *(Carboloy, Inc.)*

COMPACTION

After the powders have been thoroughly mixed, they must be molded to shape and size. Five different methods may be used to compact the powder to shape (Fig. 31-2): the extrusion process, the hot press, the isostatic press, the ingot press, or the pill press. The green (pressed) compacts are soft and must be presintered to dissolve the paraffin and lightly bond the particles so that they may be handled easily.

PRESINTERING

The green compacts are heated to about 1500°F (815°C) in a furnace under a protective atmosphere of hydrogen. After this operation, carbide blanks have the consistency of chalk and may be machined to the required shape and approximately 40% oversize to allow for the shrinkage that occurs during final sintering.

SINTERING

Sintering, the last step in the process, converts the presintered machined blanks into cemented carbide. Sintering is carried out in either a hydrogen atmosphere or a vacuum,

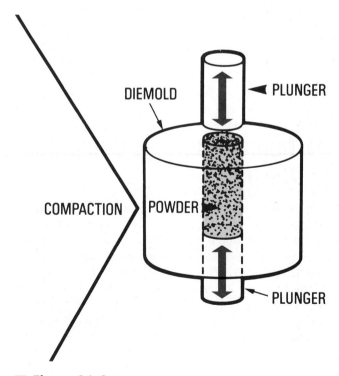

■ Figure 31-2 Compacting carbide powders to shape and size. *(Carboloy, Inc.)*

■ Figure 31-3 A variety of brazed carbide tools. *(Carboloy, Inc.)*

depending on the grade of carbide manufactured, at temperatures between 2550° and 2730°F (1400° and 1500°C). During the sintering operation, the binder (cobalt) unites and cements the carbide powders into a dense structure of extremely hard carbide crystals.

▶▶ Cemented-Carbide Applications

Because of the extreme hardness and good wear-resistance properties of cemented carbide, it has been used extensively in the manufacture of metal-cutting tools. Drills, reamers, milling cutters, and lathe cutting tools are only a few examples of uses for cemented carbides. These tools may have cemented-carbide tips, either brazed or held mechanically, on the cutting edge.

Cemented carbides were first used successfully in machining operations as lathe cutting tools. The majority of cemented-carbide tools in use are single-point cutting tools used on machines such as lathes and milling machines.

TYPES OF CARBIDE LATHE CUTTING TOOLS

Cemented-carbide lathe cutting tools are available in two types: the brazed-tip type and the indexable insert type.

Brazed-Tip Carbide Tools

Cemented-carbide tips can be brazed to steel shanks and are available in a wide variety of styles and sizes (Fig. 31-3).

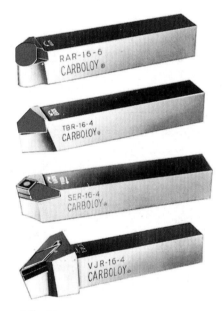

■ Figure 31-4 A variety of indexable cemented-carbide inserts. *(Carboloy, Inc.)*

Brazed-tip carbide tools are rigid and are generally used for production turning purposes.

Indexable Inserts

Cemented-carbide indexable throwaway inserts (Fig. 31-4) are made in a wide variety of shapes, such as triangular, square, diamond, and round. These inserts are held mechanically in a special holder (Fig. 31-5 on p. 226). When one cutting edge becomes dull, it may be quickly indexed or turned in the holder and a new cutting edge will be presented.

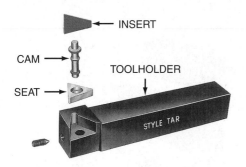

Figure 31-5 A toolholder for indexable cemented-carbide inserts. *(Carboloy, Inc.)*

A triangular insert has three cutting edges on the top surface and three on the bottom, for a total of six cutting edges. When all six cutting edges are dull, the carbide insert is generally discarded and replaced with a new insert.

Cemented-carbide indexable inserts are more popular than the brazed-tip carbide tools because:

1. Less time is required to change to a new cutting edge.
2. The amount of machine downtime is reduced considerably and production is thus increased.
3. The time normally spent in regrinding a tool is eliminated.
4. Faster speeds and feeds can be used.
5. The cost of diamond wheels, required for grinding carbide tools, is eliminated.
6. Indexable inserts are cheaper than brazed-tip tools.

CEMENTED-CARBIDE INSERT IDENTIFICATION

The American Standards Association has developed a system by which indexable inserts can be identified quickly and accurately (Table 31.1). This system has been generally adopted by manufacturers of cemented-carbide inserts.

▶▶ Grades of Cemented Carbides

There are two main groups of carbides from which various grades can be selected: the straight tungsten carbide grades and the crater-resistant grades containing titanium and/or tantalum carbide.

The *straight tungsten carbide grades,* containing only tungsten carbide and cobalt, are the strongest and most wear-resistant. Generally they are used for machining cast iron and nonmetals. Tungsten carbide grades are

not very satisfactory for steel because of their tendency to crater and rapid tool failure.

The size of the tungsten carbide particles and the percentage of cobalt used determine the qualities of tungsten carbide tools:

1. The finer the grain particles, the lower the tool toughness.
2. The finer the grain particles, the higher the tool hardness.
3. The higher the hardness, the greater the wear resistance.
4. The lower the cobalt content, the lower the tool toughness.
5. The lower the cobalt content, the higher the hardness.

For maximum tool life, always select a tungsten carbide grade with the lowest cobalt content and the finest grain size possible that gives satisfactory performance without breakage.

Crater-resistant grades contain titanium carbide and tantalum carbide in addition to the basic components of tungsten carbide and cobalt. These grades are used for machining most steels.

The addition of tantalum carbide and/or titanium carbide provide tools with various characteristics:

1. The addition of titanium carbide provides resistance to tool cratering. The higher the titanium content, the greater is the resistance to cratering.
2. As the titanium carbide content is increased, the toughness of the tool is decreased.
3. As the titanium carbide content is increased, the abrasive wear resistance at the cutting edge is lowered.
4. Tantalum carbide additions have effects similar to tungsten carbide on the resistance to cratering and strength.
5. Tantalum carbide gives good crater resistance without affecting the abrasive wear resistance.
6. The addition of tantalum carbide increases the tool's resistance to deformation.

Indexable inserts are classified by the ANSI standards for ease in selecting carbide inserts from various manufacturers. Each manufacturer may have slight variations from the ANSI coding system. Follow the manufacturer's recommendations as to the type and grade of carbide for each specific application. A few general rules will assist in the selection of the proper cemented-carbide grade:

1. Always use a grade with the lowest cobalt content and the finest grain size (strong enough to eliminate breakage).

0—Sharp corner   3—³⁄₆₄ Radius
1—¹⁄₆₄ Radius   4—¹⁄₁₆ Radius
2—¹⁄₃₂ Radius   6—³⁄₃₂ Radius

8—⅛ Radius
A—Square insert with 45° chamfer
B—Square insert with 45° chamfer and
    4° sweep angle, R.H. or Neg.
C—Square insert with 45° chamfer and
    4° sweep angle, L.H.
D—Square insert with 30° chamfer,
    R.H. or Neg.
E—Square insert with 15° chamfer,
    R.H. or Neg.
F—Square insert with 5° chamfer,
    F.H. or Neg.
G—Square insert with 30° chamfer, L.H.
H—Square insert with 15° chamfer, L.H.
J—Square insert with 5° chamfer, L.H.
K—Square insert with 30° double chamfer
L—Square insert with 15° double chamfer
M—Square insert with 5° double chamfer
N—Truncated triangle insert
P—Flatted corner triangle, R.H. or Neg.
R—Flatted corner triangle, L.H.

R—Round
S—Square
T—Triangle
L—Rectangle   CUTTING POINT
D—Diamond 55°
C—Diamond 80°
M—Diamond 86°
P—Pentagon
B—Parallelogram 82°
A—Parallelogram 85°
H—Hexagon
O—Octagon

CUTTING POINT
A = ±.0002 . . . . ±.001
B = ±.0002 . . . . ±.005
C = ±.0005 . . . . ±.001
D = ±.0005 . . . . ±.005
E = ±.001 . . . . ±.001
G = ±.001 . . . . ±.005
*M = ±.002 ±.004 . . . . ±.005
**U = ±.005 ±.012 . . . . . ±.005
R Blank with grind stock . .

THICKNESS
Number of ¹⁄₃₂nds on
inserts less than ¼ I.C.
Number of ⅛ths on in-
serts
¼ I.C. and over.
Rectangle and
Parallelogram Inserts
require two digits:
1st Digit—Number of
⅛ths in width.
2nd Digit—Number of
¼ths in length.

Shape		Tolerances		Size			Point Radius, Flats	
T	N	G	A	–	4	3	3	A

Relief Angle	Type		Thickness	Finish
N—0°	A—With hole	F—Clamp-on type with chipbreaker	Number of ¹⁄₃₂nds on inserts less than ¼ I.C.	A—Ground all over—light honed
A—3°	B—With hole and one countersink	G—With hole and chipbreaker		B—Ground all over—heavy honed
B—5°	C—With hole and two countersinks	H—With hole, one countersink and chipbreaker	Number of ¹⁄₁₆ths on inserts ¼ I.C. and over.	C—Ground top and bottom only—light honed
C—7°	D—Smaller than ¼ I.C. with hole	J—With hole, two countersinks and chipbreaker	Use width dimension in place of I.C. on Rectangle and Parallelogram inserts.	D—Ground top and bottom only—heavy honed
P—10°	E—Smaller than ¼ I.C. without hole			E—Unground insert—honed
D—15°				F—Unground insert—not honed
E—20°				
F—25°				
G—30°				

*Shall be used only when required. **Exact tolerance is determined by size of insert. (Carboloy, Inc.)*

2. To combat abrasive wear only, use straight tungsten carbide grades.

3. To combat cratering, seizing, welding, and galling, use titanium carbide grades.

4. For crater and abrasive wear resistance, use tantalum carbide grades.

5. For heavy cuts in steel, when heat and pressure might deform the cutting edge, use tantalum carbide grades.

COATED CARBIDE INSERTS

A recent development in the search for better tooling has been the coating of carbide cutting tools with titanium nitride. Coated carbide inserts give longer tool life, greater productivity, and freer-flowing chips. The coating acts as a permanent lubricant, greatly reducing cutting forces, heat generation, and wear. This permits higher speeds to be used during the machining process, particularly when a good surface finish is required. The lubricity and antiweld characteristics of the coating greatly reduce the amount of heat and stress generated when making a cut.

The use of hard, wear-resistant coatings of carbides, nitrides, and oxides to carbide inserts have greatly improved the performance of carbide-cutting tools. Inserts are available with a combination of two or three materials in the coating to give the tool special qualities. Strong wear-resistant titanium carbide forms the innermost layer. This layer is followed by a thick layer of aluminum oxide, which provides toughness, shock resistance, and chemical stability at high temperatures. A third, very thin layer composed of titanium nitride is applied over the aluminum oxide. This provides a lower coefficient of friction and reduces the tendency to form a built-up edge.

Coatings increase tool life and manufacturing productivity while reducing machining costs. Some of the coatings used for cemented-carbide tools that have been successful are:

> *Titanium carbide*—high wear and abrasion resistance at moderate speeds; used for roughing and finishing applications

> *Titanium nitride*—an extremely hard (Rockwell 80) coating with excellent lubricating properties. It has good crater resistance and minimizes edge buildup. Titanium carbide inserts are generally used for heavy roughing cuts at higher speeds

> *Aluminum oxide*—provides chemical stability and maintains hardness at high temperatures. It is generally used for roughing and finishing operations at high speeds

See Table 31.2 for a list of coated and uncoated carbide inserts and their applications.

▶▶ Tool Geometry

The geometry of cutting tools refers to the various angles and clearances machined or ground on the tool faces. Although the terms and definitions relating to single-point cutting tools vary greatly, those adopted by the American Society of Mechanical Engineers (ASME) and currently in general use are illustrated in Fig. 31-6.

CUTTING-TOOL TERMS

Front, or End, Relief (Clearance)

The front (end) relief, or clearance (Fig. 31-6) allows the end of the cutting tool to enter the work. This clearance should be just enough to prevent the tool from rubbing. Too much front clearance will reduce the support under the point and cause rapid tool failure.

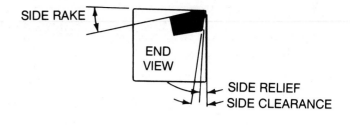

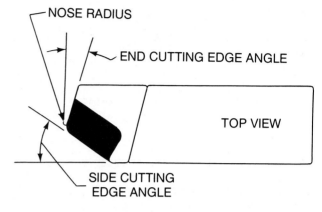

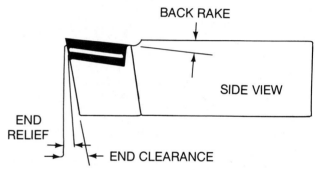

■ **Figure 31-6** Single-point tool nomenclature.

Side Relief (Side)

The side relief, or side (Fig. 31-6), permits the side of the tool to advance into the work. Too little side clearance will prevent the tool from cutting, and excessive heat will be generated by the rubbing action. Too much side clearance will weaken the cutting edge and cause it to chip.

Side Cutting Edge Angle

The angle of the side of the cutting edge that meets the work may be either positive or negative. A negative side cutting angle (Fig. 31-6) is preferred because it protects the point of the tool at both the start and the end of a cut; this is especially useful on work that has a hard abrasive scale.

table 31.2 Carbide grades for metal cutting

Uncoated Carbides

Valenite Grade	ISO Class	Industry Class	Application	Materials	Working Methods and Conditions
VC2	M10-20 K10-20	C2	Turning, boring, and milling	Cast iron, copper, brass, non-ferrous alloys, high temperature exotics, stone, and plastics	General-purpose grade of high toughness and resistance against flank wear at low to medium cutting speeds
VC3	K10-05	C3,C4	Precision turning, boring, and milling	Cast iron, aluminum, high temperature exotics, and nonferrous materials	Wear-resistant grade for finishing cuts, low to medium feed rates under rigid conditions
VC5	P20-30 M20-40	C5	Turning, boring, and milling	Steel, cast steel, malleable cast iron, and 400/500 series stainless steels	General-purpose grade covering a wide range of applications, low to medium cutting speed, high feeds and depths of cut. Has good deformation resistance.
VC7	P05-15	C7	Turning, boring, threading, and grooving	Steel, cast steel, malleable cast iron, and 400/500 series stainless steels	Light roughing to finishing at low to moderate feeds. Good crater and deformation resistance.
VC27	P15-30 M15-30 K20-30	C2	Turning and milling	Steel, cast steel, alloyed cast irons, cast alloys, and exotics	General-purpose fine-grain grade with improved toughness and wear resistance for turning and milling
VC28	M20-30 K15-30	C2	Milling	Cast and alloy irons	General-purpose grade for roughing to finishing in cast irons
VC29	M10-20 K10-20	C2, C3	Turning, boring, and milling	Stainless steels, irons, exotics, and nonferrous materials	Fine-grain grade for finishing of exotic irons and nonferrous metals
VC35	P20-35	C5	Milling	Carbon, alloy steel, and stainless steel	General-purpose steel milling grade for moderate roughing to finishing
VC101	M30-40 K30-40	C1	Turning, boring, and milling	Iron, 200/300 stainless steel, and exotics	Fine-grain, heavy-duty grade for roughing at low to moderate speeds

Coated Carbides

Valenite Grade	ISO Class	Industry Class	Application	Materials	Working Methods and Conditions
VN5	P10-25 M15-20	C5-C7	Turning and boring	Steel, cast steel, malleable cast iron, and stainless steel	A TiN-coated grade for roughing and finishing. Has excellent crater and deformation resistance.
VN8	P10-30 M20-30 K10-30	C5-C7 C2-C3	Turning, boring, milling, threading, and grooving	Cast iron, steels, 300 and 400 series stainless steels, and PH stainless steel	A very-well-balanced TiN-coated grade suitable for a broad range of applications. Has outstanding crater and impact resistance at low to high speeds.
VO1	P01-30 M10-30 K01-30	C8-C5 C4-C2	Turning, boring, and milling	Cast iron, stainless steels, alloyed steels, and carbon steels	A composite ceramic-coated grade providing maximum resistance to built-up edge. Suitable for operations ranging from roughing to finishing at medium to high speeds.
V1N	P30-45 M30-40 K25-45	C5 C1, C2	Milling, threading, and grooving	Cast iron, steels, high temperature exotics, 300 and 400 series stainless steels, and PH stainless steels	A TiN-coated heavy-duty grade used in severe roughing and interrupted cuts at slow speeds.
V88	P05-10 M10-30 K05-30	C6-C7 C3-C1	Turning, boring, and milling	Cast iron, steel, and alloyed steel	A TiC-coated grade with excellent flank wear resistance for use in applications where abrasive wear is the primary failure mode
VX8	P15-30 M15-30 K15-30	C6-C7 C2	Turning, boring, threading, and grooving	Cast iron, steel, high temperature exotics, and stainless steels	A TiC- and TiN-coated grade for moderate to heavy cuts with medium to heavy feeds. Optimized for flank wear resistance.

(Valenite, Inc.)

table 31.3   Nose radius nomograph*

Feed per Rev.		Nose Radius		Depth of Cut		
in.	mm	mm	mm		in.	mm
.050	1.3	4.8 3/16			1–1/4	32
.045	1.1		3.2 1/8		1–1/8	29
.040	1.0	4.0 5/32			1	25
.035	0.9	3.2 1/8	2.4 3/32		7/8	22
.030	0.8				3/4	19
.025	0.6	2.4 3/32	1.6 1/16		5/8	16
.020	0.5				1/2	12
.015	0.4	1.6 1/16	1.2 3/64	7/16		11
.010	0.3			3/8		10
.0075	0.2	1.2 3/64	0.8 1/32	5/16		8
					1/4	6
.005	0.1	0.8 1/32	0.40 1/64	3/16		5
.0025	0.05				1/8	3
		0.40 1/64	0.13 to 0.40 .005 to .015	1/16		1.6
.000	0.00				0	0.00

Column labels between the two Nose Radius columns, reading vertically: ECCENTRIC FORGINGS, INTERRUPTED OR SCALE CUTS   REGULAR CUTS IN CLEAN METAL

*To obtain a maximum efficiency with carbide tools, the nose radius should be kept small. Use chart as a guide. Use straightedge to join feed with depth of cut. Use nose radius where line crosses. (Carboloy, Inc.)

Nose Radius

The nose radius strengthens the finishing point of the tool and improves the surface finish on the work. The nose radius of most cutting tools should be approximately twice the amount of feed per revolution. Too large a nose radius may cause chatter; too small a radius weakens the point of the cutting tool.

To obtain maximum efficiency with carbide tools, keep the nose radius as small as possible. Use Table 31.3 to find the proper nose radius for the depth of cut and feed used.

Side Rake

The side rake angle should be as large as possible, without weakening the cutting edge, to allow the chips to escape readily (Fig. 31-7a). The amount of side rake will be determined by the type and grade of the cutting tool, the type of material being cut, and the feed per revolution. The included angle formed by the side rake and side clearance is called the angle of keenness. This angle will

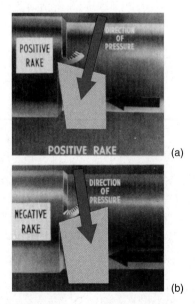

■ **Figure 31-7** (a) Positive rake side angle on a cutting tool; (b) negative rake side angle on a cutting tool. *(Kelmar Associates)*

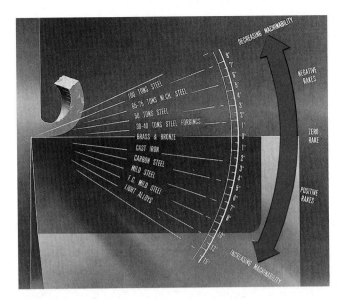

Figure 31-8 Recommended side rake angles for various workpiece materials. *(Kelmar Associates)*

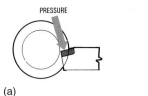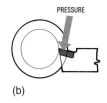

Figure 31-9 (a) Negative back rake; (b) positive back rake. *(Kelmar Associates)*

from the point (Fig. 31-9a). Negative back rake protects the tool point from cutting pressure and from the abrasive action of hard materials and scale. When a tool has *positive back rake* (Fig. 31-9b), the top face of the tool slopes downward away from the point. This allows the chips to flow away freely from the cutting edge.

Note: Because throwaway inserts are generally flat, the required side and back rake angles are built into the tool-holder by the manufacturer.

vary depending on the material being cut. For difficult-to-machine metals, it may be advisable to use a small side rake angle or at times even negative side rake (Fig. 31-7b). Fig. 31-8 shows suggested side rakes for a variety of materials.

Back Rake

The back rake angle is the angle formed between the top face of the tool and the top of the tool shank. It may be positive, negative, or neutral. When a tool has *negative back rake,* the top face of the tool slopes upward away

CEMENTED-CARBIDE CUTTING-TOOL ANGLES AND CLEARANCES

The angles and clearances of single-point carbide tools vary greatly and generally depend on three factors:

1. The hardness of the cutting tool
2. The workpiece material
3. The type of cutting operation

Table 31.4 lists recommended cutting-tool angles and clearances for a variety of materials. These may have to be

table 31.4 Recommended angles for single-point carbide tools*

Material	End Relief (Front Clearance)	Side Relief (Side Clearance)	Side Rake	Back Rake	Angle of Keenness
Aluminum	6° to 10°	6° to 10°	10° to 20°	0° to 10°	60° to 74°
Brass, bronze	6° to 8°	6° to 8°	+8° to −5°	0° to −5°	76° to 87°
Cast iron	5° to 8°	5° to 8°	+6° to −7°	0° to −7°	79° to 89°
Machine steel	5° to 10°	5° to 10°	+6° to −7°	0° to −7°	79° to 87°
Tool steel	5° to 8°	5° to 8°	+6° to −7°	0° to −7°	79° to 89°
Stainless steel	5° to 8°	5° to 8°	+6° to −7°	0° to −7°	79° to 89°
Titanium alloys	5° to 8°	5° to 8°	+6° to −5°	0° to −5°	79° to 87°

*Use the lower range of these figures for hard-to-machine metals and interrupted cuts.

table 31.5 — Recommended cutting speeds and feeds for single-point carbide tools*

Material	Depth of Cut		Feed per Revolution		Cutting Speed	
	in.	mm	in.	mm	ft/min	m/min
Aluminum	.005–.015	0.15–0.4	.002–.005	0.05–0.15	700–1000	215–305
	.020–.090	0.5–2.3	.005–.015	0.15–0.4	450–700	135–215
	.100–.200	2.55–5.1	.015–.030	0.4–0.75	300–450	90–135
	.300–.700	7.6–17.8	.03–.090	0.75–2.3	100–200	30–60
Brass, bronze	.005–.015	0.15–0.4	.002–.005	0.05–0.15	700–800	215–245
	.020–.090	0.5–2.3	.005–.015	0.15–0.4	600–700	185–215
	.100–.200	2.55–5.1	.015–.030	0.4–0.75	500–600	150–185
	.300–.700	7.6–17.8	.03–.090	0.75–2.3	200–400	60–120
Cast iron (medium)	.005–.015	0.15–0.4	.002–.005	0.05–0.15	350–450	105–135
	.020–.090	0.5–2.3	.005–.015	0.15–0.4	250–350	75–105
	.100–.200	2.55–5.1	.015–.030	0.4–0.75	200–250	60–75
	.300–.700	7.6–17.8	.03–.090	0.75–2.3	75–150	25–45
Machine steel	.005–.015	0.15–0.4	.002–.005	0.05–0.15	700–1000	215–305
	.020–.090	0.5–2.3	.005–.015	0.15–0.4	550–700	170–215
	.100–.200	2.55–5.1	.015–.030	0.4–0.75	400–550	120–170
	.300–.700	7.6–17.8	.03–.090	0.75–2.3	150–300	45–90
Tool steel	.005–.015	0.15–0.4	.002–.005	0.05–0.15	500–750	150–230
	.020–.090	0.5–2.3	.005–.015	0.15–0.4	400–500	120–150
	.100–.200	2.55–5.1	.015–.030	0.4–0.75	300–400	90–120
	.300–.700	7.6–17.8	.03–.090	0.75–2.3	100–300	30–90
Stainless steel	.005–.015	0.15–0.4	.002–.005	0.05–0.15	375–500	115–150
	.020–.090	0.5–2.3	.005–.015	0.15–0.4	300–375	90–115
	.100–.200	2.55–5.1	.015–.030	0.4–0.75	250–300	75–90
	.300–.700	7.6–17.8	.03–.090	0.75–2.3	75–175	25–55
Titanium alloys	.005–.015	0.15–0.4	.002–.005	0.05–0.15	300–400	90–120
	.020–.090	0.5–2.3	.005–.015	0.15–0.4	200–300	60–90
	.100–.200	2.55–5.1	.015–.030	0.4–0.75	175–200	55–60
	.300–.700	7.6–17.8	.03–.090	0.75–2.3	50–125	15–40

Inch and millimeter speeds are approximate equivalents. (Carboloy, Inc.)

altered slightly to suit various conditions encountered while machining.

▸▸ Cutting Speeds and Feeds

Many variables influence the speeds, feeds, and depth of cut that should be used with cemented-carbide tools. Some of the most important factors are:

1. The type and hardness of the work material
2. The grade and shape of the cutting tool
3. The rigidity of the cutting tool
4. The rigidity of the work and machine
5. The power rating of the machine

Table 31.5 gives the recommended cutting speeds and feeds for single-point carbide tools. These should be used as a guide and may have to be altered slightly to suit the machining operation.

Table 31.6 illustrates the nomograph method of determining the cutting speed in feet or meters per minute when the Brinell hardness of the steel is known.

table 31.6 Carbide cutting speed nomograph*

DEPTH OF CUT		FEED		SPEED	
Inches	mm	in./rev.	mm/rev.	ft/min	m/min

REFERENCE LINE

BRINELL HARDNESS

DEPTH OF CUT

Inches / mm:
1.000 / 25
.900 / 23
.800 / 20
.700 / 18
.600 / 15
.500 / 12
.450 / 11
.400 / 10
.350 / 9
.300 / 8
.250 / 6
.200 / 5
.180 / 4.5 G
.160 / 4
.140 / 3.5
.120 / 3
E
.100 / 2.5 X
.090 / 2.3
.080 / 2
.070 / 1.8
.060 / 1.5
.050 / 1.2
.045 / 1.1
.040 / 1
.035 / 0.9
.030 / 0.8
.025 / 0.6
.020 / 0.5
.018 / 0.5
.016 / 0.4
.014 / 0.35
.012 / 0.3
.010 / 0.25
.009 / 0.23
.008 / 0.2
.007 / 0.18
.006 / 0.15
.005 / 0.13

FEED

in./rev. / mm/rev.:
.070 / 1.8
.060 / 1.5
.050 / 1.2
.045 / 1.1
.040 / 1
.035 / 0.9
.030 / 0.8 F
.025 / 0.6
.020 / 0.5
.018 / 0.4
.016 / 0.4
.014 / 0.35
.012 / 0.3
.010 / 0.25
.008 / 0.2
.006 / 0.15
.005 / 0.13
.004 / 0.1
.003 / 0.08
.002 / 0.05

SPEED

ft/min / m/min:
30 / 9
35 / 11
40 / 12
45 / 14
50 / 15
60 / 18
70 / 21
80 / 24
90 / 27
100 / 30
120 / 35
140 / 45
160 / 50
180 / 55
200 / 60
250 / 75
300 / 90
350 / 110
400 / 120
500 / 150
600 / 185
700 / 215
800 / 245
900 / 275
1000 / 305
1200 / 365
1400 / 430
1600 / 490
1800 / 550
2000 / 610
2500 / 760
3000 / 915

BRINELL HARDNESS

600
550
500
480
460
440
420
400
380
360
340
320
300
280
260
240
220
200 H
180
160
140

PROBLEM: WANTED: SPEED IN m/min TO TURN STEEL, WHICH HAS A HARDNESS OF 200 BRINELL, AT 3.2 mm DEPTH OF CUT AND A 0.6 mm FEED.

METHOD OF SOLUTION:

1. CONNECT 3.2 mm DEPTH OF CUT AND 0.6 mm FEED WITH LINE E-F, WHICH WILL INTERSECT THE REFERENCE LINE AT POINT X.

2. CONNECT THE POINT X AND 200 BRINELL HARDNESS WITH LINE G-H.

ANSWER:

WHERE LINE G-H CROSSES THE SPEED LINE, READ THE DESIRED SPEED OF 90 m/min.

Inch and millimeter speeds are approximate equivalents. (Carboloy, Inc.)

▶▶ Machining with Carbide Tools

To obtain maximum efficiency with cemented-carbide cutting tools, certain precautions in machine setup and the cutting operation should be observed. The machine used should be rigid and free from vibrations, equipped with heat-treated gears, and have sufficient power to maintain a constant cutting speed. The work and cutting tool should be held as rigidly as possible to avoid chatter or keep it to a minimum.

Single-point carbide cutting tools are more commonly used on a lathe, and therefore the setups and precautions for this machine will be outlined. The same basic precautions and setups should be applied when carbide tools are used on other machines.

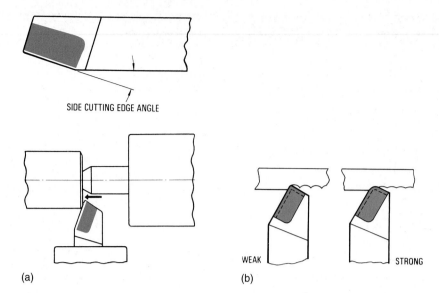

SIDE CUTTING EDGE ANGLE

(a)

WEAK   STRONG

(b)

■ **Figure 31-10** (a) A side cutting edge angle protects the point of the cutting tool; (b) a large nose radius strengthens the tool point and produces a better surface finish. *(Carboloy, Inc.)*

▶▶ Suggestions for Using Cemented-Carbide Cutting Tools

WORK SETUP

1. Work mounted in a chuck or other work-holding device must be held firmly enough to prevent it from slipping or chattering.

2. A revolving center should be used in the tailstock for turning work between centers.

3. The tailstock spindle should be extended only a minimum distance and locked securely to ensure rigidity.

4. The tailstock should be clamped firmly to the lathe bed to prevent loosening.

TOOL SELECTION

1. Use a cutting tool with the proper rake and clearances for the material being cut.

2. Hone the cutting edge for good performance and longer tool life.

3. If the work permits, use a side cutting edge angle (Fig. 31-10a) large enough that the tool can be eased into the work. This helps to protect the nose of the cutting tool (weakest part) from shock and wear as it enters or leaves the work.

4. Use the largest nose radius (Fig. 31-10b) that operating conditions will permit. Too large a nose radius causes chattering; too small a nose radius causes the point to break down quickly. Use Table 31.3 to determine the correct nose radius to use for the amount of feed and depth of cut.

TOOL SETUP

1. Carbide tools preferably should be held in a sturdy, turret-type holder (Fig. 31-11). The amount of tool overhang should be just enough for chip clearance.

2. The cutting tool should be set exactly on center; when it is above or below center, the tool angles and clearances change in relation to the job and result in poor cutting action.

3. Carbide tools are designed to operate while the bottom of the tool shank is in a *horizontal position* (Fig. 31-12).

4. If a rocker-type toolpost is used (Fig. 31-13):

   a. Remove the rocker.
   b. Invert the rocker base.
   c. Shim the tool to the correct height.
   d. Use a special carbide toolholder (having no rake) when machining with carbide toolbits (Fig. 31-14).

5. When setting up a carbide tool, always keep it from touching the work and machine parts to avoid damaging the tool point.

Figure 31-11 Turret-type toolholders hold carbide tools securely.

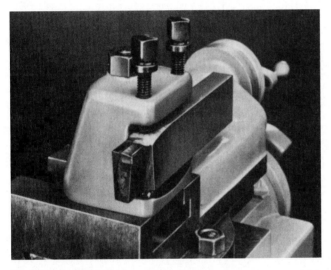

Figure 31-12 The heavy-duty toolpost permits rigid setting of the cutting tool for heavy cuts.

MACHINE SETUP

1. Always make sure that the machine has an adequate power rating for the machining operation and that there is no slippage in the clutch and belts.

2. Set the correct speed for the material cut and the operation performed.

   a. Too high a speed will cause rapid tool failure; too low a speed will result in inefficient cutting action and poor production rates.

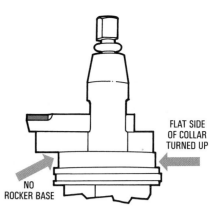

Figure 31-13 To hold carbide cutting tools in a rocker-type toolpost, remove the rocker, invert the rocker base, and shim the tool to the correct height. *(Carboloy, Inc.)*

Figure 31-14 The hole in a carbide toolholder is straight and not slanted as in standard toolholders. *(Bahco North America)*

   b. Use Table 31.5 or 31.6 to calculate the correct speed for the type of material machined.

3. Set the machine at a feed that will give good metal-removal rates and still provide the surface finish desired.

   a. Too light a feed causes rubbing, which may result in hardening of the material being cut.
   b. Too coarse a feed slows down the machine, creates excessive heat, and results in premature tool failure.

CUTTING OPERATION

1. Never bring the tool point against work that is stationary; this will damage the cutting edge.

2. Always use the heaviest depth of cut possible for the machine and size of cutting tool.

3. Never stop a machine while the feed is engaged; this will break the cutting edge. Always stop the feed and allow the tool to clear itself before stopping the machine.

4. Never continue to use a dull cutting tool.

5. A dull cutting tool may be recognized by:

   a. Work being produced oversize and with a glazed finish

table 31.7 Tool selection and application guide

Follow these fundamental techniques for more efficient metal removal.

1. Always select standard products whenever the operating conditions allow.
   Benefits:
   - Less expensive
   - Proven design
   - Availability
   - Interchangeable

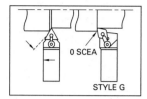

2. Always select the largest side cutting (or end cutting) edge angle the workpiece will allow.
   Benefits:
   - Thin chips
   - Dissipates heat
   - Protects nose radius
   - Reduces insert notching

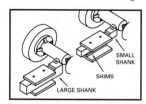

3. Always select the largest toolholder shank the machine tool will allow.
   Benefits:
   - Minimizes deflection
   - Reduces overhang ratio

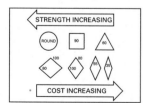

4. Always select the strongest insert shape the workpiece will allow.
   Benefits:
   - Productivity
   - Optimum grade selection
   - Lower cost/edge

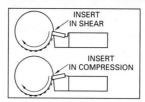

5. Always select negative rake geometry whenever the workpiece or the machine tool will allow.
   Benefits:
   - Double the cutting edges
   - Greater strength
   - Dissipates heat

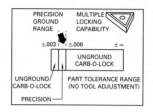

6. Always select a Carb-O-Lock® insert whenever possible.
   Benefits:
   - Less expensive
   - May be multiple locked
   - Strong unground cutting edge
   - Chip control groove
   - Covers most workpiece tolerances

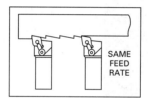

7. Always select the largest insert nose radius that either the workpiece or the operating conditions will allow.
   Benefits:
   - Improves finish
   - Thins chips
   - Dissipates heat
   - Greater strength

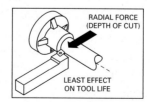

8. Always select the largest depth of cut that either the workpiece or the machine tool will allow.
   Benefits:
   - Greater productivity
   - Negligible effect on tool life

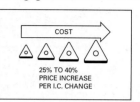

9. Always select the smallest insert size the operating conditions will allow.
   Benefit:
   - Less expensive

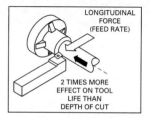

10. Always select the highest feed rate that either the workpiece or the machine tool will allow.
    Benefits:
    - Greater productivity
    - Minimal effect on tool life

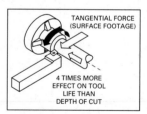

11. Always select speed within the HI-E surface footage range.
    Benefits:
    - Minimum cost
    - Maximum production

The correct selection of tooling and operating conditions is the first step toward lowering tool costs and increasing productivity.

(Carboloy, Inc.)

b. A rough and ragged finish

c. A change in the shape or color of the chips

6. Apply cutting fluid only if:

a. It can be applied under pressure

b. It can be directed at the point of cutting and kept there at all times

Tool Selection and Application Guide

To obtain the best results from cemented-carbide tools, the proper type and style of insert must be used, and the cutting tool must be set up and used properly. The points illustrated in Table 31.7 should be followed as closely as possible to obtain the most efficient metal-removal rates and to produce the most cost-effective machining operation.

Other factors that can affect the optimum life of carbide tools are:

1. The horsepower available on the machine tool

2. The rigidity of the machine tool and toolholders

3. The shape of the workpiece and the setup

4. The speed and feed rates used for the machining operation

Grinding Cemented-Carbide Tools

The efficiency of a cutting tool determines to a large extent the efficiency of the machine tool on which it is used. A tool that has been improperly ground cannot perform well, and the cutting edge will soon break down. To grind cemented-carbide cutting tools successfully, the type of grinding wheel used and the grinding procedure followed are important.

GRINDING WHEELS

1. An 80-grit silicon carbide wheel should be used for rough grinding carbides.

2. A 100-grit silicon carbide wheel should be used for finish grinding carbides.

Note: The silicon carbide wheels should be dressed with a 1/16-in. (1.5-mm) crown (Fig. 31-15) to minimize the amount of heat generated during grinding. These wheels should be used to grind only the carbide, not the steel tool shank of a brazed carbide toolbit.

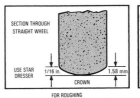
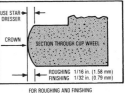

■ Figure 31-15 A crown dressed on a silicon carbide wheel reduces heat when carbide tools are being ground.

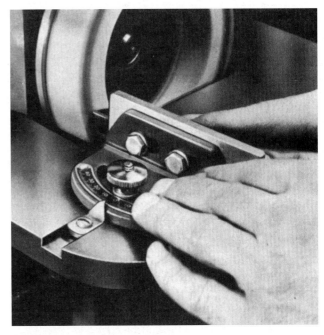

■ Figure 31-16 A carbide tool grinder equipped with an adjustable table and a protractor. *(Carboloy, Inc.)*

3. Aluminum oxide grinding wheels should be used if it is necessary to grind the steel shank of a brazed carbide tool.

4. Diamond grinding wheels (100-grit) are excellent for finish grinding of carbides for general work. Where high finishes on the tool and work are desired, a 220-grit diamond wheel is recommended.

TYPE OF GRINDER

1. A heavy-duty grinder should be used for grinding carbides because the cutting pressures required to remove carbide are from 5 to 10 times as great as those for grinding high-speed steel tools.

2. The grinder should be equipped with an adjustable table and a protractor (Fig. 31-16) so that the necessary tool angles and clearances may be ground accurately.

Carbide Cutting Tools   **237**

TOOL GRINDING

1. Regrind the cutting tool to the angles and clearances recommended by the manufacturer.

2. Use silicon carbide wheels for rough grinding. Diamond wheels should be used when high surface finishes are required.

3. When grinding, move the carbide tool back and forth over the grinding wheel face to keep the amount of heat generated to a minimum.

4. *Never quench carbide tools* that become hot during grinding; allow them to cool gradually. The shock of quenching carbide tools creates heat checks and results in rapid tool failure.

HONING

After carbide tools have been ground, it is important that the cutting edge be honed. The purpose of honing is to remove the fine, ragged edge left by the grinding wheel. This fine, nicked edge is fragile and will break down quickly under machining conditions. Honing often means the difference between tool success or failure. Follow these suggestions for successful honing:

1. A 320-grit silicon carbide or diamond hone is recommended for carbide tools.

2. On carbides used for cutting steel, a 45° chamfer (Fig. 31-17) .002 to .004 in. (0.05 to 0.1 mm) wide should be honed on the cutting edge.

3. Carbide tools used for aluminum, magnesium, and plastics should not be chamfered with a hone. The fine, nicked edge should be honed to produce a sharp, keen cutting edge.

▶▶ Cemented-Carbide Tool Problems

When problems occur during machining with carbide cutting tools, consult Table 31.8 for their possible causes and remedies. It is wise to change only one thing at a time until the problem is corrected. In this way, the real cause of the problem can be determined and steps taken to guard against its recurrence.

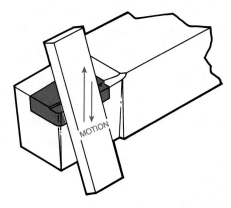

■ **Figure 31-17** Honing a slight chamfer on the edge of a carbide tool will produce a stronger cutting edge.

table 31.8 Carbide tool problems, causes, and remedies

Problem	Cause	Remedy
Brazing failure and cracks	Improper braze material Too much heat Improper cooling Tip not wiped on Contact surfaces dirty	Use the sandwich braze. Silver solder at 1400°F (760°C). Cool slowly. Slide tip back and forth and tap gently. Thoroughly clean both surfaces.
Built-up edge (on the tool)	Cutting speed too slow Insufficient rake angle Wrong grade of carbide	Increase the speed. Increase rake angle. Change to titanium and/or tantalum grade.
Chipped and broken cutting edge	Incorrect tool geometry Improper carbide grade Cutting edge not honed Improper speed, feed, or depth of cut Too much tool overhang Lack of rigidity in setup or machine Machine stopped during cut	Increase side cutting edge angle. Decrease end cutting edge angle. Use a negative rake. Increase the nose radius. Decrease the relief angles. Use a tougher grade. Hone cutting edge at 45°. Change one or all as required. Reduce overhang or use a larger shank. Locate and correct. Disengage feed before stopping the machine.
Cratering	Improper grade Speed too high Wrong tool geometry	Select harder or crater-resistant grade. Reduce the speed. Increase side rake.
Grinding cracks	Improper wheel Wheel glazed Tool quenched while hot Grinding carbide and steel shank with the same wheel	Select proper wheel. Dress often to a 1/16-in. crown. *Do not quench;* allow tool to cool slowly. Relieve the steel shank with an aluminum oxide wheel first.
Tool wear (excessive)	Improper carbide grade Cutting speed too high Feed too light Too little relief angle	Select a wear-resistant grade. Reduce the speed. Increase the feed rate. Increase the relief angle.

unit 31 review questions

Carbide Cutting Tools

1. State four reasons cemented-carbide tools have been widely accepted by industry.

Manufacture of Cemented Carbides

2. Name five powders used in the manufacture of cemented carbides.

3. What purpose does cobalt serve?

4. Describe the process of compaction.

5. What is the purpose of presintering?

6. Describe the sintering process.

Cemented-Carbide Applications

7. Why is cemented carbide used extensively in the manufacture of cutting tools?

8. Name two types of cemented-carbide lathe cutting tools, and state the advantages of each.

9. Identify the following indexable insert:

SNG—321—A

Grades of Cemented Carbides

10. Describe straight tungsten carbide tools, and state why they are used.

11. Explain how the size of the tungsten carbide particles and the percentage of cobalt affect the grade of carbide.

12. How does the addition of titanium carbide affect the cutting tool?

13. What properties does the addition of tantalum carbide provide?

14. List four important rules to observe for selecting a cemented-carbide grade.

Coated Carbide Inserts

15. List three advantages that coated carbide inserts have over standard carbide inserts.

16. What material is deposited on the carbide substrate when a single coating is applied?

17. Describe the process of triple coating.

Tool Geometry

18. Define and state the purpose of:

a. Front or end relief
b. Side relief
c. Side rake
d. Angle of keenness

19. a. What is the purpose of the side cutting edge angle?

b. How is the angle of keenness formed?

20. State the purpose of the nose radius on a cutting tool. How large should it be?

21. Name two types of back rake and state the purpose of each.

22. What three factors determine the angles and clearances of single-point carbide tools?

Cutting Speeds and Feeds

23. Name four important factors that influence the speed, feed, and depth of cut for carbide tools.

24. Use Table 31-6 to determine what cutting speed should be used for:

a. Taking a .090-in. depth of cut on 260 Brinell steel with a .015-in. feed
b. A cut 1.5 mm deep on 300 Brinell steel using a 0.2-mm feed

Machining with Carbide Tools

25. List two important precautions that should be observed when work is set up on a lathe.

26. Why should the cutting edge of a carbide tool be honed?

27. What occurs if the nose radius on the cutting tool is

a. Too large?   b. Too small?

28. Using Table 31.3, determine the nose radius for a carbide tool taking:

a. A .250-in. deep cut at .020-in. feed
b. A 3.2-mm depth of cut using a 0.4-mm feed

29. How should carbide tools be set up for machining?

30. Explain how carbide tools should be set up in a rocker-type toolpost.

31. What precautions should be taken when setting up a machine for cutting with carbide tools?

32. Discuss the effects of:

a. Too light a feed   b. Too coarse a feed

33. Why should a machine never be stopped while the tool is engaged in a cut?

34. Explain how a dull cutting tool may be recognized.

Grinding Cemented-Carbide Tools

35. What type of grinding wheels are used to grind carbide tools?

36. How should silicon carbide wheels be dressed? Explain why.

37. List three important precautions that should be observed in the grinding of carbide tools.

38. Why is it important not to quench carbide tools?

39. What is the purpose of honing carbide tools?

40. How should carbide tools used for steel be honed?

Cemented-Carbide Tool Problems

41. List the factors that would cause the following problems:

a. Cratering
b. Grinding cracks
c. Chipped or broken cutting edge

UNIT 32

Diamond, Ceramic, and Cermet Cutting Tools

OBJECTIVES

After completing this unit, you will be able to:

1 Explain the purpose and application of diamond cutting tools

2 State the uses of two types of ceramic tools

3 Describe the types and application of cermet tools

Since the development of carbide cutting tools, industry has continued to research and develop better cutting tools capable of operating at greater speeds, feeds, and depths of cut. Although no tool has been found that will perform all jobs perfectly, great strides have been made in the development of cutting tools.

▶▶ Diamond Cutting Tools

Since diamond is the hardest known material, it would be natural to assume that it could be used to cut other materials effectively. Two types of diamonds are used in industry: *natural* (or mined) and *manufactured*. Natural, or mined, diamonds were once widely used for machining nonmetallic and nonferrous materials, but they are being replaced by manufactured diamonds, which are superior in performance in most cases. These diamonds are used to machine hard-to-finish materials and produce excellent surface finishes.

MANUFACTURED DIAMONDS

Diamond was used primarily in machine shop work for truing and dressing grinding wheels. Because of the high cost of natural diamonds, industry began to look for cheaper, more reliable sources. In 1954, the General Electric (GE) Company, after four years of research, produced manufactured diamonds in its laboratory. In 1957, GE, after more research and testing, began the commercial production of these diamonds.

Many forms of carbon were used in experiments to manufacture diamonds. After much experimentation with various materials, the first success came when carbon and iron sulfide in a granite tube closed with tantalum disks were subjected to a pressure of approximately 1.5 million psi (10,342,500 KPa) and temperatures of between 2550°F and 4260°F (1400°C and 2350°C) in the "Belt" furnace (Fig. 32-1 on p. 242). Various diamond configurations are produced by using other metal catalysts, such as chromium, manganese, tantalum, cobalt, nickel, or platinum, in place of iron. The temperatures used must be high enough to melt the metal saturated with carbon and start the diamond growth.

Types of Manufactured Diamonds

Because the temperature, pressure, and catalyst-solvent can be varied, it is possible to produce diamonds of the size, shape, and crystal structure best suited to a particular need.

Type RVG Diamond

This manufactured diamond is an elongated, friable crystal with rough edges (Fig. 32-2a on p. 242). The letters RVG

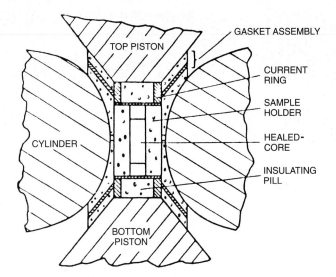

Figure 32-1 Diamonds are manufactured in a high-temperature and high-pressure belt-type apparatus. *(GE Superabrasives)*

It is used in metal-bonded saws (MBS) to cut concrete, marble, tile, granite, stone, and masonry. The diamonds may be coated with nickel or copper to provide a better holding surface in the bond and to prolong the life of the wheel.

ADVANTAGES OF DIAMOND CUTTING TOOLS

Diamond-tipped cutting tools (Fig. 32-3) are used to machine nonferrous and nonmetallic materials that require a high surface finish and extremely close tolerances. They are used primarily as finishing tools because they are brittle and do not resist shock or cutting pressure as well as carbide cutting tools. Single-point diamond-tipped cutting tools are available in various shapes for turning, boring, grooving, and special forming.

The main advantages of diamond-tipped cutting tools are:

1. They can be operated at high cutting speeds, and production can be increased to 10 to 15 times that of other cutting tools.

2. Surface finishes of 5 min. (0.127 μm) or less can be obtained easily. In many cases, the necessity of finishing operations on the workpiece is eliminated.

3. They are very hard and resist abrasion. Much longer runs are therefore possible on abrasive materials.

4. Closer tolerance work can be produced with diamond-tipped cutting tools.

5. Minute cuts, as low as .0005 in. (0.012 mm) deep, can be taken from the inside or outside diameter.

6. Metallic particles do not build up (weld) on the cutting edge.

indicate that this type may be used with a resinoid or vitrified bond for grinding ultrahard materials such as tungsten carbide, silicon carbide, and space-age alloys. The RVG diamond may be used for wet and dry grinding.

Type MBG-II Diamond

This tough, blocky-shaped crystal is not as friable as the RVG type and is used in metal-bonded grinding (MBG) wheels (Fig. 32-2b). It is used for grinding cemented carbides, sapphires, and ceramics, as well as in electrolytic grinding.

Type MBS Diamond

This diamond is a blocky, extremely tough crystal with a smooth, regular surface that is not very friable (Fig. 32-2c).

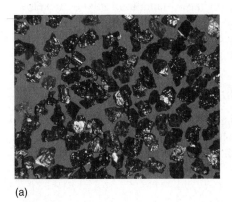

(a)

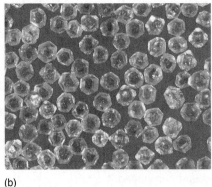

(b)

(c)

Figure 32-2 (a) Type RVG diamond is used to grind ultrahard materials; (b) Type MBG-II is a tough diamond crystal used in metal-bonded grinding wheels; (c) Type MBS diamond is a very tough, large crystal used in metal-bonded saws. *(Diamond Innovations Inc., at www.abrasivesnet.com)*

■ Figure 32-3 Diamond-tipped cutting tools are used to make shallow cuts at very high speeds.

USE OF DIAMOND CUTTING TOOLS

Some of the most successful applications of diamond cutting tools have been in the turning of metallic (nonferrous) and nonmetallic materials. The most common of these materials are listed.

Metallic Materials

1. *Light metals,* such as aluminum, duraluminum, and magnesium alloys
2. *Soft metals,* such as copper, brass, and zinc alloys
3. *Bearing metals,* such as bronze and babbitt
4. *Precious metals,* such as silver, gold, and platinum

For finishing cuts in these metallic materials, diamond tools can be expected to increase production from 10 to 15 times that possible with any other cutting tool.

Nonmetallic Materials

Some of the most common materials machined with diamond cutting tools are hard and soft rubber, all types of cemented carbides, plastics, carbon, graphite, and ceramics. In some cases, diamond tools will increase production 20 to 50 times that of carbide tools.

CUTTING SPEEDS AND FEEDS

In general, diamond-tipped cutting tools operate most efficiently with shallow cuts at high cutting speeds and fine feeds. They are not recommended for materials in which the temperature of the chip or the heat generated at the chip-tool interface exceeds 1400°F (760°C). High cutting speeds, fine feeds, and shallow cuts are used. The heat generated is entirely dissipated in the chip as it leaves the cutting edge. Low cutting speeds and heavy cuts create more heat and damage the diamond tip.

There is an ideal cutting speed for each type of material-machine combination. The minimum cutting speed for diamond tools should be 250 to 300 sf/min (76 to 91 m/min). The conditions of the machine will determine the maximum cutting speed for each job. Cutting speeds as high as 10,000 sf/min (3048 m/min) have been used for some applications. Table 32.1 on p. 244 lists cutting speed, feed, and depth of cut ranges for various material groups.

table 32.1   Diamond cutting-tool data

Material	Cutting Speed		Feed (per Rev.)		Depth of Cut	
	ft/min	m/min	in.	mm	in.	mm
Metallic (nonferrous)	250–10,000	75–3050	.0008–.004	0.002–0.1	.0005–.024	0.001–0.6
Nonmetallic	250–3300	75–1005	.0008–.024	0.002–0.6	.0008–.060	0.002–1.5

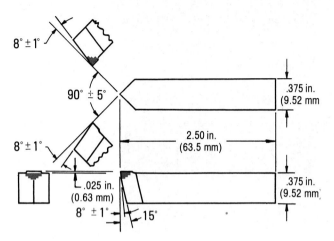

■ **Figure 32-4** Diamond-tipped cutting tool angles and clearances.

HINTS ON THE USE OF DIAMOND TOOLS

Diamond cutting tools will perform more efficiently and have longer life if the following hints and precautions are observed:

1. Diamond-tipped points should be designed with maximum included point angle and radius for added strength (Fig. 32-4).

2. Diamond tools should always be handled with care, especially when setups are being made. The cutting edges should *never* be checked with a micrometer or bumped with a height gage.

3. Diamond tools should always be stored in separate containers, with rubber protectors over the tips, so that they will not be fractured by coming into contact with other tools.

4. The machine tool should be as free of vibration as possible. Any vibration can result in tool failure.

5. A very rigid setup, with the diamond tip set exactly on center, should be used.

6. The work should be roughed out with a carbide tool if possible. This step will establish an even work surface and bring the work close to size.

7. Diamond tools should always be fed into the work while the work is revolving. Never stop a machine during a cut.

8. Interrupted cuts, especially on hard metals, will shorten tool life.

▶▶ Ceramic Cutting Tools

The first ceramic (cemented-oxide) cutting-tool inserts were put on the market in 1956; they were the result of many years of research. At first these ceramic inserts were inconsistent. Weak and unsatisfactory results were obtained because of lack of knowledge and improper use. Since then, the strength of ceramic cutting tools has nearly doubled, their uniformity and quality have been greatly improved, and they are now widely accepted by industry. Ceramic cutting tools are used successfully in the machining of hard ferrous materials and cast iron. As a result, lower costs, increased productivity, and better results are being gained. In some operations, ceramic toolbits can be operated at three to four times the speed of carbide toolbits.

MANUFACTURE OF CERAMIC TOOLS

Most ceramic or cemented-oxide cutting tools are manufactured primarily from aluminum oxide. Bauxite (a hydrated alumina form of aluminum oxide) is chemically processed and converted into a denser, crystalline form called *alpha alumina*. Fine grains (micron size) are obtained from the precipitation of the alumina or from the precipitation of the decomposed alumina compound.

Ceramic tool inserts are produced by either *cold* or *hot pressing*. In cold pressing, the fine alumina powder is compressed into the required form and then sintered in a furnace at 2912°F to 3092°F (1600°C to 1700°C). Hot pressing combines forming and sintering, with pressure and heat being applied simultaneously. Certain amounts of titanium oxide or magnesium oxide are added for certain types of ceramics to aid in the sintering process and to retard growth. After the inserts have been formed, they are finished with diamond-impregnated grinding wheels.

Further research has led to the development of stronger ceramic cutting tools. Aluminum oxide (Al_2O_3)

■ Figure 32-5 Indexable ceramic inserts are available in a variety of sizes.

■ Figure 32-6 Ceramic tips may be cemented to the toolshank.

and zirconium oxide (ZrO_2) are mixed in powder form, cold-pressed into the required shape, and sintered. This white ceramic insert has proved successful in machining high-tensile-strength materials (up to 42 Rockwell c hardness) at speeds of up to 2000 sf/min (609 m/min). However, the most common applications of ceramic inserts are in the general machining of steel, where there are no heavy, interrupted cuts and where negative rakes can be used. This type of toolbit has the highest hot-hardness strength of any cutting-tool material and gives excellent surface finish. No coolant is required with ceramic tools, since most of the heat goes into the chip and not into the workpiece. Ceramic inserts should be used with toolholders that have a fixed or adjustable chipbreaker.

The most common ceramic cutting tool is the *indexable insert* (Fig. 32-5), which is fastened in a mechanical holder. Indexable inserts are available in many styles, such as triangular, square, rectangular, and round. These inserts are indexable; when a cutting edge becomes dull, a sharp edge can be obtained by indexing (turning) the insert in the holder.

Cemented ceramic tools (Fig. 32-6) are the most economical, especially if the tool shape must be altered from the standard shape. The ceramic insert is bonded to a steel shank with an epoxy glue. This method of holding the ceramic inserts almost eliminates the strains caused by clamping inserts in mechanical holders.

CERAMIC TOOL APPLICATIONS

Ceramic tools were intended to supplement rather than replace carbide tools. They are extremely valuable for specific applications and must be carefully selected and used. Ceramic tools can be used to replace carbide tools that wear rapidly in use, but they should never replace carbide tools that are breaking.

Ceramics are used successfully for:

1. High-speed, single-point turning, boring, and facing operations, with continuous cutting action

2. Finishing operations on ferrous and nonferrous materials

3. Light, interrupted finishing cuts on steel or cast iron; heavy, interrupted cuts on cast iron only if there is adequate rigidity in the machine and tool

4. Machining castings when other tools break down because of the abrasive action of sand, inclusions, or hard scale

5. Cutting hard steels up to a hardness of Rockwell c 66, which previously could be machined only by grinding

6. Any operation in which size and finish of the part must be controlled and in which previous tools have not proved satisfactory

FACTORS AFFECTING CERAMIC TOOL PERFORMANCE

As a result of an extensive research and testing program, several factors have been found to significantly affect the performance of ceramic tools. The following factors must be considered for optimum results to be derived from ceramic cutting tools:

1. Accurate and rigid machine tools are essential when ceramic tools are used. Machines with loose bearings, inaccurate spindles, slipping clutches, or any imbalance will result in the ceramic tools chipping and in premature failure.

Diamond, Ceramic, and Cermet Cutting Tools   **245**

2. The machine tool must be equipped with ample power and be capable of maintaining the high speeds necessary for ceramics.

3. Tool mounting and toolholder rigidity are as important as machine rigidity. An intermediate plate between the toolholder clamp and ceramic insert should be used to distribute clamping pressure.

4. The overhang of the toolholder should be kept to a minimum: no more than 1½ times the shank thickness.

5. Negative rake inserts give the best results because less force is applied directly to the ceramic tip.

6. A large nose radius and a large side cutting edge angle on the ceramic insert reduces its tendency to chip.

7. Cutting fluids are generally not required because ceramics remain cool during the machining operation. If cutting fluids are required, a continuous and copious flow should be used to prevent thermal shock to the tool.

8. As the cutting speed or the hardness of the workpiece increases, the ratio of feed to the depth of cut must be checked. Always make a deeper cut with a light feed rather than a light cut with a heavy feed. Most ceramic tools are capable of cuts as deep as one-half the width of the cutting surface on the insert.

9. Toolholders with fixed or adjustable chipbreakers are best for use with ceramic inserts. Adjust the chipbreakers to produce a figure 6 or 9 curled chip.

ADVANTAGES AND DISADVANTAGES OF CERAMIC TOOLS

Advantages

Ceramic cutting tools, when properly mounted in suitable holders and used on accurate, rigid machines, offer the following advantages:

1. Machining time is reduced because of the higher cutting speeds possible. Speeds ranging from 50% to 200% higher than those used for carbide tools are common.

2. High stock removal rates and increased productivity result because heavy depths of cut can be made at high surface speeds.

3. A ceramic tool used under proper conditions lasts from 3 to 10 times longer than a plain carbide tool and will exceed the life of coated carbide tools.

4. Ceramic tools retain their strength and hardness at high machining temperatures [in excess of 2000°F (1093°C)].

5. More accurate size control of the workpiece is possible because of the greater wear resistance of ceramic tools.

6. Ceramic cutting tools withstand the abrasion of sand and of inclusions found in castings.

7. A better surface finish than is possible with other types of cutting tools is produced.

8. Heat-treated materials as hard as Rockwell c 66 can be readily machined.

Disadvantages

Although ceramic cutting tools have many advantages, they have the following limitations:

1. Ceramic tools are brittle and therefore tend to chip easily.

2. They are satisfactory for interrupted cuts only under ideal conditions.

3. The initial cost of ceramics is higher than that of carbides.

4. They require a more rigid machine than is necessary for other cutting tools.

5. Considerably more power and higher cutting speeds are required for ceramics to cut efficiently.

CERAMIC TOOL GEOMETRY

The geometry of ceramic tools depends on five main factors:

1. The material to be machined
2. The operation performed
3. The condition of the machine
4. The rigidity of the work setup
5. The rigidity of the toolholding device

Although the final geometry of a ceramic cutting tool depends on these specific factors, some general considerations should also be mentioned.

Rake Angles

Because ceramic tools are brittle, negative rake angles are generally preferred. A negative rake angle allows the shock of the cutting force to be absorbed behind the tip, thus protecting the cutting edge. Negative rake angles from 2° to 30° are used for machining ferrous and nonferrous metals.

table 32.2 Suggested rake and relief angles for ceramic tools*

Workpiece Material	Rake Angles (Degrees)	Relief Angles (Degrees)
Carbon and alloy steels: Annealed and heat-treated	Neg. 2 to 7	2 to 7
Cast iron: Hard or chilled Gray or ductile	0 to Neg. 7	2 to 7
Nonferrous: Hard or soft	0 to Neg. 7	2 to 7
Nonmetallics: Wood, paper, green ceramics, fiber, asbestos, rubber, carbon, graphite	0 to 10	6 to 18

*For uninterrupted turning with tool point on the centerline of the workpiece. (The Carborundum Company)

Positive rake angles are used for machining nonmetals such as rubber, graphite, and carbon (Table 32.2).

Side Clearance

A side clearance angle is desirable for ceramic cutting tools. The side clearance angle must not be too great; otherwise, the cutting edge will be weakened and tend to chip.

Front Clearance

The front clearance angle should be only large enough to prevent the tool from rubbing on the workpiece. If this angle is too great, the ceramic tool becomes susceptible to chipping.

End Cutting Edge Angle

The end cutting edge angle governs the strength of the tool and the area of contact between the work and the end of the cutting tool. If properly designed, it will remove the crests resulting from feed lines (Fig. 32-7) and will improve the surface finish.

Nose Radius

The nose radius (Fig. 32-7) has two important functions: to strengthen the weakest part of the tool and to improve the surface finish of the workpiece. It should be as large as possible without producing chatter or vibration.

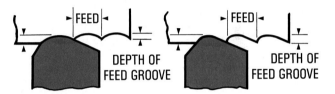

■ **Figure 32-7** The relationship of the nose radius and end cutting edge angle to the surface finish produced.

Cutting Edge Chamfer

A small chamfer, or radius, on the cutting edge is recommended for ceramic tools, especially on heavy cuts and hard materials. This strengthens and protects the cutting edge. A .002- to .008-in. (0.05- to 0.2-mm) radius, or chamfer, is recommended for machining steel. For heavy roughing cuts and hard materials, a .030- to .060-in. (0.76- to 1.52-mm) chamfer gives satisfactory results.

CUTTING SPEEDS

When ceramics are used in machining, the highest cutting speed possible, considering the machine tool's limitations, that gives reasonable tool life should be used. Less heat is generated when machining is done with ceramics than with any other type of cutting tool because there is a lower coefficient of friction between the chip, work, and tool surface. Since most of the heat generated escapes with the chip, the cutting speed can be from 2 to 10 times higher than with other cutting tools. Table 32.3 on p. 248

table 32.3 Recommended cutting speeds for ceramic cutting tools

Workpiece Material	Material Condition or Type	Roughing Cut		Finishing Cut		Recommended Tool Geometry (Type of Rake Angle)	Recommended Coolant
		Depth > .062 in. Feed .015–.030 in.	Depth > 1.6 mm Feed .4–0.75 mm	Depth < .062 in. Feed .010 in.	Depth < 1.6 mm Feed 0.25 mm		
Carbon and tool steels	Annealed	300–1500	90–455	600–2000	185–610	Neg.	None
	Heat-treated	300–1000	90–305	500–1200	150–365	Neg.	
	Scale	300–800	90–245			Neg. honed edge	
Alloy steels	Annealed	300–800	90–245	400–1400	120–425	Neg.	None
	Heat-treated	300–800	90–245	300–1000	90–305	Neg. honed edge	
	Scale	300–600	90–185			Neg. honed edge	
High-speed steel	Annealed	100–800	30–245	100–1000	30–305	Neg.	None
	Heat-treated	100–600	30–185	100–600	30–385	Neg. honed edge	
	Scale	100–600	30–185			Neg. honed edge	
Stainless steel	300 series	300–1000	90–305	400–1200	120–365	Pos. and neg.	Sulfur base oil
	400 series	300–1000	90–305	400–1200	120–365	Neg.	
Cast iron	Gray iron	200–800	60–245	200–2000	60–610	Pos. and neg.	None
	Pearlitic	200–800	60–245	200–2000	60–610	Neg.	
	Ductile	200–600	60–185	200–1400	60–427	Neg.	
	Chilled	100–600	30–185	200–1400	60–427	Neg. honed edge	
Copper and alloys	Pure	400–800	120–245	600–1400	185–425	Pos. and neg.	Mist coolant
	Brass	400–800	120–245	600–1200	185–365	Pos. and neg.	Mist coolant
	Bronze	150–800	45–245	150–1000	45–305	Pos. and neg.	Mist coolant
Aluminum alloys*		400–2000	120–610	600–300	185–915	Pos.	None
Magnesium alloys		800–10,000	245–3050	800–10,000	245–3050	Pos.	None
Nonmetallics	Green ceramics	300–600	90–185	500–1000	150–305	Pos.	None
	Rubber	300–1000	90–305	400–1200	120–365	Pos.	None
	Carbon	400–1000	120–305	600–2000	185–610	Pos.	None
Plastics		300–1000	90–305	400–1200	120–365	Pos.	None

*Alumina-based cutting rods have a tendency to develop a built-up cutting edge on certain aluminum alloys.

lists the recommended speeds for various materials under ideal conditions. Should lower speeds be necessary to suit various machine or setup conditions, ceramic cutting tools will still perform well.

CERAMIC TOOL PROBLEMS

Ceramic tools remove metal faster than most types of cutting tools and therefore should be selected carefully for the type of material and the operation. Some points to consider when selecting ceramic tools are:

1. The tool should be large enough for the job. It cannot be too large but can easily be too small.

2. The style (tool geometry) should be right for the type of operation and material. A tool designed for one job is not necessarily right for another.

Some of the more common problems with ceramic cutting tools and their possible causes are listed in Table 32.4.

table 32.4 Ceramic tool problems and possible causes

Problem	Possible Causes
Chatter	1. Tool not on center 2. Insufficient end relief and/or clearance 3. Too much rake angle 4. Too much overhang or tool too small 5. Nose radius too large 6. Feed too heavy 7. Lack of rigidity in the tool or machine 8. Insufficient power or slipping clutch
Chipping	1. Lack of rigidity 2. Saw-toothed or too keen a cutting edge 3. Chipbreaker too narrow or too deep 4. Chatter 5. Scale or inclusions in the workpiece 6. Improper grinding 7. Too much end relief 8. Defective toolholder
Cratering	1. Chipbreaker set too close to the edge 2. Nose radius too large 3. Side cutting edge angle too great
Torn finish	1. Lack of rigidity 2. Dull tool 3. Speed too slow 4. Chipbreaker too narrow or too deep 5. Improper grinding
Wear	1. Speed too high or feed too light 2. Nose radius too large 3. Improper grinding
Cracking or breaking	1. Insert surfaces not flat 2. Insert not seated solidly 3. Stopping workpiece while tool is engaged 4. Worn or chipped cutting edges 5. Feed too heavy 6. Improperly applied coolant 7. Too much rake angle or end relief 8. Too much overhang or too small 9. Lack of rigidity in setup 10. Speed too slow 11. Too much variation in cut for size of tool 12. Chatter 13. Grinding cracks

GRINDING CERAMIC TOOLS

It is not recommended that ceramic tools be ground; however, with the proper care, these tools may be resharpened successfully. Resinoid-bonded, diamond-impregnated wheels are recommended for grinding ceramic tools. A coarse-grit wheel should be used for rough grinding, and at least a 220-grit wheel should be used for finish grinding. Because ceramic tools are extremely notch-sensitive, all surfaces should be ground as smoothly as possible to avoid notches or grinding lines at the cutting edge. The cutting edge should be honed or lapped after grinding to remove any notches left by grinding and to avoid the wedging action that would occur if material were to enter these notches.

▶▶ Cermet Cutting Tools

As a result of continued research aimed at improving the strength of ceramic cutting tools, cermet tools were developed in about 1960. Cermets are toolbits made of various ceramic and metallic combinations.

TYPES OF CERMET TOOLS

There are two types of cermet tools: those composed of titanium carbide (TiC)–based materials and those containing titanium nitride (TiN)–based materials.

Titanium carbide (TiC) cermets have a nickel and molybdenum binder and are produced by cold-pressing and sintering in a vacuum. They are used extensively for finishing cast irons and steels that require high speeds and light to moderate feeds.

Recently titanium nitride has been added to titanium carbide to produce titanium carbide–titanium nitride (TiC-TiN) cermets. Other materials such as molybdenum carbide, vanadium carbide, zirconium carbide, and others may be added, depending on the application.

Because of their high productivity, cermets are considered a cost-effective replacement for coated and uncoated carbide and ceramic toolbits (Fig. 32-8). However, cermets are not recommended for use with hardened ferrous metals (over 45 Rc) or nonferrous metals.

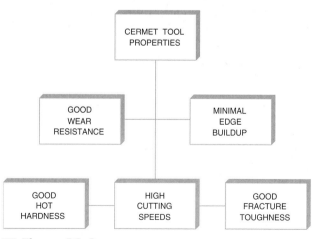

■ **Figure 32-8** The characteristics of cermet tools make them cost-effective. *(Kelmar Associates)*

CHARACTERISTICS OF CERMET TOOLS

The main characteristics of cermet tools are:

> They have great wear resistance and permit higher cutting speeds than do carbide tools.

> Edge buildup and cratering are minimal, which increases tool life.

> They possess high hot-hardness qualities greater than carbide tools but less than ceramic tools.

> They have a lower thermal conductivity than carbide tools because most of the heat goes into the chip, and they can therefore operate at higher cutting speeds.

> Fracture toughness is greater than for ceramic tools but less than for carbide tools.

ADVANTAGES OF CERMET TOOLS

Cermet tools have the following advantages:

1. The surface finish is better than that produced with carbides under the same conditions, which often eliminates the need for finish grinding.

2. High wear resistance permits close tolerances for extended periods, ensuring accuracy of size for larger batches of parts.

3. Cutting speeds can be higher than with carbides for the same tool life.

4. When operated at the same cutting speed as carbide tools, cermet tool life is longer.

5. The cost per insert is less than that of coated carbide inserts and equal to that of plain carbide inserts.

USE OF CERMET TOOLS

Titanium carbide cermets are the hardest cermets and are used to fill the gap between tough tungsten carbide inserts and the hard, brittle ceramic tools. They are used mainly for machining steels and cast irons where high speeds and moderate feeds may be used (Table 32.5).

Titanium carbide–titanium nitride inserts are used for semifinish and finish machining of harder cast irons and steels (less than 45 Rc) such as alloy steel, stainless steel, armor plate, and powder metallurgy parts.

table 32.5	Suggested cutting conditions for cermet cutting tools			
Material	Hardness (Brinell)	Cutting Speed	Feed	Depth of Cut
Cast irons	100–250	200–1200 sf/min	.002–.016 in.	.187–.250 in.
		(60–366 m/min)	(0.05–0.4 mm)	(4.74–6.35 mm)
Steel, carbon	100–250	160–1200 sf/min	.002–.016 in.	.200–.300 in.
		(48–366 m/min)	(0.05–0.4 mm)	(5.08–7.62 mm)
Steels, alloys	250–400	150–1000 sf/min	.002–0.016 in.	.187–.300 in.
and stainless		(46–305 m/min)	(0.05–0.4 mm)	(4.74–7.62 mm)

unit 32 review questions

Diamond Cutting Tools

1. Name two types of diamonds used in industry.
2. List four main advantages of diamond cutting tools.
3. Explain where diamond cutting tools may be used successfully.

Cutting Speeds and Feeds

4. What speed, feed, and depth of cut should be used for diamond tools? Explain why.
5. Why should the diamond tool be as rigid as possible and the machine free from vibration?
6. List five important precautions to be observed when using diamond cutting tools.

Manufacture of Ceramic Tools

7. Briefly explain how cemented-oxide cutting tools are manufactured.
8. What is the difference between ceramic and cermet cutting tools?

Ceramic Tool Applications

9. Name four important applications for ceramic tools.
10. List four main factors which affect the performance of ceramic tools.

Advantages and Disadvantages of Ceramic Tools

11. Name five advantages of ceramic tools.
12. What are the disadvantages of using ceramic cutting tools?

Ceramic Tool Geometry

13. List five factors that determine the geometry of ceramic tools.
14. Fully explain the rake angle of ceramic tools.
15. Define cutting edge chamfer and state its purpose.

Cutting Speeds

16. Explain why high cutting speeds can be used with ceramic tools.
17. What two points should be kept in mind when ceramic tools are selected?
18. List the factors that may cause the following problems:
    a. Chipping
    b. Chatter
    c. Wear

Grinding Ceramic Tools

19. Explain how ceramic tools should be ground.
20. What is the purpose of honing?

Cermet Cutting Tools

21. What is a cermet toolbit?
22. Name the two basic types of cermets.
23. What is a TiC-TiN toolbit?
24. State four characteristics of a cermet cutting tool.
25. List four advantages of cermet cutting tools.
26. State the conditions under which titanium carbide toolbits may be used.
27. What are the applications of TiC-TiN toolbits?

UNIT 33

Polycrystalline Cutting Tools

OBJECTIVES

After completing this unit, you will be able to:

1 Explain the manufacture and properties of polycrystalline tools

2 Select the proper type and size of polycrystalline cutting tools

3 Set up a tool and machine for cutting with polycrystalline tools

The metalworking industry has a relatively short history of developing large-scale cutting tools, starting in 1860, when Henry Bessemer invented the commercial method of making steel. Before then, steel blades were produced and forged by hand processes that were kept secret and passed on through a father-and-son tradition. Carbon-steel cutting tools were replaced gradually by high-speed steel cutting tools in the late 1800s. While successful and still used today, high-speed steel tools, in turn, gave way to the more productive cemented-carbide and coated carbide cutting tools for most production and many toolroom cutting operations.

A new cutting-tool technology was born in about 1954, when the General Electric Company produced manufactured diamond. Later, a polycrystalline layer composed of thousands of small diamond or cubic boron nitride abrasive particles was fused to a cemented-carbide substrate (base) to produce a cutting tool having a long-wearing, superior cutting edge. Because of its excellent abrasion resistance, it soon became a highly efficient cutting tool, which rewrote all production records for machining abrasive nonmetallic, nonferrous materials.

▶▶ Manufacture of Polycrystalline Cutting Tools

There are two distinct types of polycrystalline cutting tools: polycrystalline cubic boron nitride and polycrystalline diamond. Each type is used to machine a certain class of materials. The manufacture of polycrystalline cutting-tool blanks involves basically the same process for both types of tool.

The manufacture of polycrystalline cutting-tool blanks (Fig. 33-1) was a major step in the production of new and more efficient superabrasive cutting tools. A layer of polycrystalline diamond or cubic boron nitride— approximately .020 in. (0.5 mm) thick—is fused on a cemented-carbide substrate or base by a high temperature [3090°F to 3275°F (1700°C to 1800°C)], high-pressure process [about 1 million psi (6,895,000 kPa)]. The substrate is composed of tiny grains of tungsten carbide cemented tightly together by a cobalt metal binder. Under the high-heat, high-pressure conditions, the cobalt liquifies, flows up and sweeps around the diamond or cubic boron

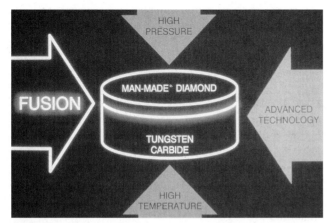

Figure 33-1 Polycrystalline tool blanks consist of a layer of polycrystalline diamond or cubic boron nitride abrasive fused to a cemented-carbide base. *(GE Superabrasives)*

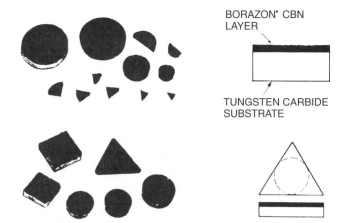

Figure 33-2 The types and sizes of PCBN tool blank insert shapes available. *(GE Superabrasives)*

nitride abrasive particles, and serves as a catalyst that promotes intergrowth (fusing the abrasive particles). This process forms what is known as a *polycrystalline mass.*

▶▶ Polycrystalline Cubic Boron Nitride (PCBN) Tools

The polycrystalline structure of cubic boron nitride (CBN) features nondirectional, consistent properties that resist chipping and cracking and provide uniform hardness and abrasion resistance in all directions. Therefore, it was inevitable that experiments would be conducted to see if these qualities could be built into turning and milling cutting-tool blanks and inserts. These experiments were successful, making possible the application of superabrasive cutting tools to machining operations that previously were difficult, if not impossible, to perform by conventional methods. Polycrystalline cubic boron nitride (PCBN) blanks and inserts (Fig. 33-2) can be operated at higher cutting speeds, take deeper cuts, and machine hardened steels and high temperature alloys (Rc 35 and harder) such as Iconel, Rene, Waspaloy, Stellite, and Colmonoy.

PROPERTIES OF POLYCRYSTALLINE CUBIC BORON NITRIDE

Cubic boron nitride is a synthetic material not found in nature. PCBN cutting tools contain the four main properties that cutting tools must have to cut extremely hard or abrasive materials at high metal-removal rates: hardness, abrasion resistance, compressive strength, and thermal conductivity (Fig. 33-3).

Hardness

Fig. 33-4 (on p. 254) indicates the Knoop hardness of the common abrasives. Cubic boron nitride, next to diamond in hardness, is about twice as hard as silicon carbide and aluminum oxide. PCBN cutting tools have good impact resistance, high strength, and high hardness in all directions because of the random orientation of the tiny CBN crystals.

Abrasion Resistance

Fig. 33-5 (on p. 254) shows the abrasion resistance of CBN in relation to conventional abrasives and diamond. Compared to conventional cutting tools, PCBN tools maintain their sharp cutting edges much longer, thereby increasing productivity while at the same time producing dimensionally accurate parts.

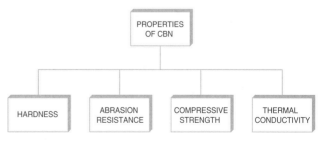

Figure 33-3 The main properties of polycrystalline cubic boron nitride cutting tools. *(Kelmar Associates)*

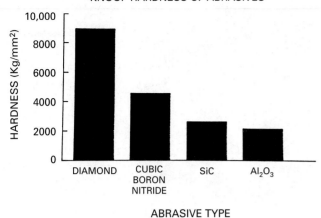

KNOOP HARDNESS OF ABRASIVES

Figure 33-4 The hardness comparisons of various abrasive materials. *(GE Superabrasives)*

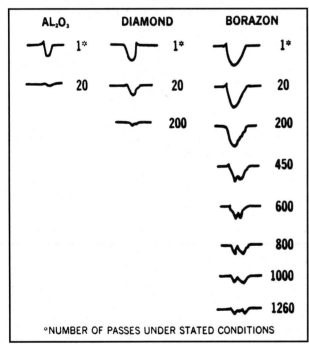

COMPARATIVE WEAR OF ABRASIVE GRAIN ON HARDENED M-2 STEEL

SURFACE SPEED	4000 SFPM	(20 M/SEC)
WORK SPEED	40 FT/MIN	(12.5 M/MIN)
DEPTH OF CUT	.0008 IN	(.02 MM)

*NUMBER OF PASSES UNDER STATED CONDITIONS

Figure 33-5 The comparative abrasion or wear resistance of CBN (Borazon) in relation to other abrasives. *(GE Superabrasives)*

Compressive Strength

Compressive strength is defined as the maximum stress in compression that a material will take before it ruptures or breaks. The high compressive strength of CBN crystals give PCBN tools excellent qualities for withstanding the forces created during high metal-removal rates and the shock of severe interrupted cuts.

Thermal Conductivity

Because they have excellent thermal conductivity, PCBN cutting tools allow greater heat dissipation or transfer, especially when they are used to cut hard, abrasive, tough materials at high material-removal rates. The high cutting temperatures created at the cutting-tool–workpiece interface would weaken or soften conventional cutting-tool materials.

▶▶ Types of PCBN Tools

Finished, ready-to-use tools are available from cutting-tool suppliers and fall into three categories: tipped inserts, full-faced inserts, and brazed-shank tools (Fig. 33-6):

> *Tipped inserts* (Fig. 33-6a) are available in most carbide insert shapes and are usually the most economical. These inserts are manufactured by machining a pocket in the carbide and brazing the PCBN insert in place. They have the same wear life per cutting edge as a PCBN insert but have only one cutting edge. When an edge becomes dull, it can be reground and have the same life as a new cutting tool.

> *Full-faced inserts* (Fig. 33-6b) consist of a layer of PCBN bonded to a cemented-carbide substrate. PCBN inserts are available as triangles, squares, and rounds. They are generally very cost-effective because the manufacturer can downsize them repeatedly, providing new cutting edges.

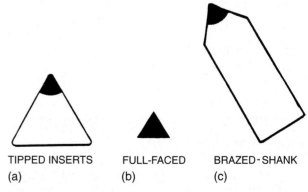

TIPPED INSERTS (a)   FULL-FACED (b)   BRAZED-SHANK (c)

Figure 33-6 The common types of polycrystalline cubic boron nitride (PCBN) cutting tools: (a) tipped inserts; (b) full-faced inserts; (c) brazed-shank tools. *(GE Superabrasives)*

> *Brazed-shank tools* (Fig. 33-6c) are available in most common tool forms. They are made by machining a pocket in the proper-style of tool shank and brazing a PCBN blank in place. They can be specially ordered from the manufacturer to suit most machining applications.

TYPES OF METAL CUT

Polycrystalline cubic boron nitride cutting tools are used on lathes and turning centers to machine round surfaces and on milling machines and machining centers to machine flat surfaces. These cutting tools have been used successfully for straight turning, facing, boring, grooving, profiling, and milling operations.

The four general types of metals that best lend themselves to the use of PCBN cutting tools and have proven to be cost-effective are:

1. *Hardened ferrous metals* of greater than 45 Rc hardness, including:

   > Hardened steels, such as 4340, 8620, M-2, and T-15

   > Cast irons, such as chilled iron and Ni-Hard.

2. *Abrasive ferrous metals,* such as cast irons ranging from 180 to 240 Brinell hardness, including:

   > Pearlistic gray cast iron and Ni-Resist.

3. *Heat-resistant alloys,* which most frequently are high-cobalt ferrous alloys used as flame-applied hard surfacing

4. *Superalloys,* such as high-nickel alloys used in the aerospace industry for jet engine parts

The best applications for PCBN cutting tools are on materials that cause conventional cutting-tool edges of cemented carbides and ceramics to break down too quickly.

ADVANTAGES OF PCBN CUTTING TOOLS

The advantages that PCBN cutting tools offer the metal-working industry more than offset their higher initial costs (Fig. 33-7). These tools greatly improve efficiency, reduce scrap, and increase product quality. Because of the tough, hard microstructure of PCBN tools, their cutting edges last longer. They also efficiently remove material at high rates that would cause conventional cutting tools to break down rapidly.

High Material-Removal Rates

Because PCBN cutting tools are so hard and resist abrasion so well, cutting speeds in the range of 250 to 900 ft/min (274 m/min) and feed rates of .010 to .020 in. (0.25 to 0.5 mm) are possible. These rates result in higher material-removal rates (as much as three times that of carbide tools) with less tool wear.

Cutting Hard, Tough Materials

Polycrystalline cubic boron nitride tools are capable of efficiently machining all ferrous materials with a Rockwell C hardness of 45 and above. They are also used to machine cobalt-base and nickel-base high temperature alloys with a Rockwell c hardness of 35 and above.

High Quality Products

Because the cutting edges of PCBN cutting tools wear very slowly, they produce high quality parts faster and at a lower cost per piece than do conventional cutting tools. The need to inspect the parts produced is greatly reduced, as is the adjustment of the machine tool to compensate for cutting-tool wear or maintenance.

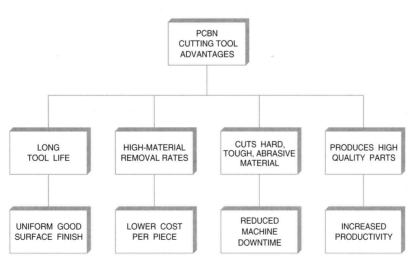

■ Figure 33-7 The main advantages of polycrystalline cubic boron nitride cutting tools. *(Kelmar Associates)*

Uniform Surface Finish

Surface finishes in the range of 20 to 30 μin. are possible during roughing operations with PCBN cutting tools. On finishing operations, surface finishes in single-digit micro-inches are possible.

Lower Cost per Piece

Polycrystalline cubic boron nitride cutting tools stay sharp and cut efficiently throughout long production runs. This results in consistently smoother surface finishes, better control over workpiece shape and size, and fewer cutting-tool changes.

Reduced Machine Downtime

Since PCBN cutting tools stay sharp much longer than carbide or ceramic cutting tools, less time is required to index, change, or recondition cutting tools.

Increased Productivity

All the advantages that PCBN cutting tools offer, such as increased speeds and feeds, long tool life, longer production runs, consistent part quality, and savings in labor costs, combine to increase overall production rates and decrease the manufacturing cost per piece.

▶▶ Polycrystalline Diamond (PCD) Tools

A new cutting tool technology was introduced about 1954 when General Electric produced synthetic manufactured diamond. Later, a polycrystalline diamond (PCD) layer, approximately .020 in. (0.5 mm) thick, was fused to a cemented-carbide substrate (base) to produce a cutting tool with a long-wearing, superior cutting edge. Because of its excellent abrasion resistance, it soon became a highly efficient cutting tool that increased production when machining abrasive nonmetallic, nonferrous materials.

Although PCD and PCBN tools have many similar properties, each has specific applications. The main difference is that diamond tools are not suitable for machining steels and other ferrous materials. Since diamonds are composed purely of carbon, and steel is a "carbon-seeking" metal, it has been found that at high cutting speeds, where much heat is produced, the steel develops an affinity for the carbon. In effect, the carbon molecules are attracted to the steel, which causes the edge of the diamond tool to break down rapidly.

TYPES AND SIZES OF PCD TOOLS

Catalyst-bonded PCD, which is the variety most widely used, is available in three microstructure series for various machining applications. The basic difference between the three series is the size of the diamond particle used to manufacture the polycrystalline blank. Manufacturers use different trade names to identify each series; we will discuss them as coarse, medium-fine, and fine.

> *Coarse PCD blanks* (Fig. 33-8a) are made up of coarse diamond crystals and are designed to cut a wide variety of abrasive nonferrous, nonmetallic materials. They are highly recommended for machining cast aluminum alloys, especially those containing more than 16% silicon. Coarser-grade PCD tools are generally more durable than the other types.

> *Medium-fine PCD blanks* (Fig. 33-8b) are composed of fine to medium-fine crystals with a greater size distribution than coarse PCD blanks. These tools are used for machining highly abrasive nonferrous and nonmetallic materials where short cutting edges are acceptable.

> *Fine PCD blanks* (Fig. 33-8c) are manufactured from fine diamond crystals that are fairly uniform in size. The fine grain structure allows production of tools having extremely sharp cutting edges and high finishes on the rake and flank sides. These tools are recommended for applications requiring very fine surface finishes and long cutting edges.

The diamond microstructure plays a major part in determining the characteristics of a PCD tool blank, its applications, its wear resistance, and its cutting-tool life. Remember, a coarse crystal PCD tool should be used when durability is a factor, while a fine crystal PCD tool should be used when high surface finishes are required. PCD and PCBN cutting tools are generally available as tipped inserts, full-faced inserts, and brazed-shank tools (Fig. 33-6).

PROPERTIES OF PCD TOOLS

Polycrystalline diamond cutting tools, which consist of a layer of diamond fused to a cemented tungsten carbide substrate, have properties that make them superabrasive cutting tools. The composite materials found in the *cemented tungsten carbide substrate* (base) provide mechanical properties. This is due to its relatively high thermal conductivity and a relatively low coefficient of thermal expansion. The substrate also provides excellent mechanical support for the polycrystalline diamond layer and imparts a toughness to the finished PCD tool.

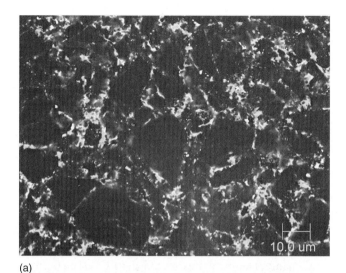

(a)

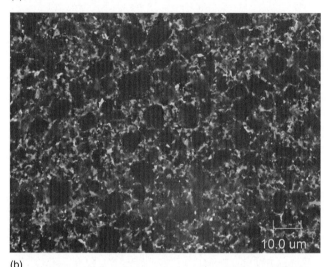

(b)

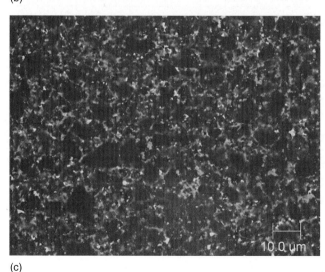

(c)

■ **Figure 33-8** The difference in diamond particle size of the three series of PCD tool blanks: (a) coarse crystals; (b) medium-fine crystals; (c) fine crystals.   *(Diamond Innovations Inc., at www.abrasivesnet.com)*

The main properties of the *diamond layer* of PCD tools are hardness, abrasion resistance, compressive strength, and thermal conductivity. Diamond, the hardest material known, is about 7000 on the Knoop hardness scale. The hardness and abrasion resistance of the polycrystalline diamond layer are due to its inner structure. Unlike that of a single-crystal natural diamond, the PCD crystal arrangement is uniform in all directions. There are no hard or soft planes or weak-bound planes that can lead to severe cracking. Also, no special orientation is required to optimize cutting during machining operations.

The *compressive strength* of the diamond layer is the highest of any cutting tool. This is due to its dense structure that enables PCD cutting tools to withstand the forces created during high metal-removal rates and the shock of interrupted cuts. The *thermal conductivity* of the PCD layer, the highest of any cutting tool, is almost 60% higher than that of polycrystalline cubic boron nitride. This high thermal conductivity allows greater dissipation or transfer of the heat created at the chip-tool interface, especially when tough, abrasive materials are cut at high metal-removal rates.

ADVANTAGES OF PCD TOOLS

The advantages that PCD cutting tools offer industry more than offset their higher initial cost. Primarily used to machine nonferrous and nonmetallic materials, PCD tools are capable of greatly improving efficiency, reducing scrap, and increasing product quality. Some of the main advantages of polycrystalline diamond cutting tools are:

1. Long tool life
2. Cuts tough, abrasive material
3. High quality parts
4. Fine surface finishes
5. Reduced machine downtime
6. Increased productivity

TYPES OF MATERIAL CUT

Polycrystalline diamond cutting tools are used to machine nonferrous or nonmetallic materials, primarily where the workpiece is abrasive. These materials are generally considered difficult to machine because of their abrasive character.

The largest group of nonferrous metals consists of materials that typically are soft but have hard particles dispersed in them, such as silicon suspended in silicon aluminum or glass fibers in plastic. It is the hard abrasive particle that destroys the cutting edge of conventional tools. Diamond is harder than the abrasive particle and tends to shear the hard particle rather than pushing it out of the way or dulling the cutting edge. PCD blank tools often reach a wear life of 100 times that of cemented-carbide tools in such an abrasive machining application.

A growing new category of nonmetallic materials is the ceramics and composites. These materials are hard and abrasive, and they rely on the hardness of the diamond to overcome their own abrasive character. PCD cutting tools can cut the hard, abrasive inclusions in these materials cleanly without the rapid dulling of the cutting edge.

The materials most successfully machined with PCD tools fall into five general categories: silicon-aluminum alloys, copper alloys, tungsten carbide, advanced composites, ceramics, and wood composites. Table 33.1 lists the more common materials that can be cost-effectively machined with PCD tools.

▶▶ Diamond-Coated Tools

Industry has used diamond cutting tools for many years for machining nonferrous materials and grinding ultra-hard materials. Diamond is well suited to these applications because of its properties of unequalled hardness, extreme wear resistance, low coefficient of friction, and high thermal conductivity.

In the early 1980s, a new process of chemical vapor deposition (CVD) was developed to produce a diamond coating of a few microns thick on a variety of products, including cutting tools, to prolong their useful life. The CVD process consists of breaking down hydrogen gas by a heating element or plasma stream into elemental hydrogen. The elemental hydrogen is dissolved in hydrocarbon gas, such as methane, at temperatures around 1330°C. When the mix-

table 33.1   PCD cutting-tool applications

Nonferrous Metals	Nonmetallic Materials	Composites
Aluminum alloys	Alumina, fired	Asbestos
Babbitt alloys	Bakelite	Fiberglass epoxy
Brass alloys	Beryllia	Filled carbons
Bronze alloys	Ceramics	Filled nylon
Copper alloys	Epoxy	Filled phenolic
Lead alloys	Glass	Filled PVC
Manganese alloys	Graphite	Filled silica
Silver, platinum	Macor	Filled teflon
Tungsten carbides	Rubber, hard	Wood, manufactured
Zinc alloys	Various plastics	

ture contacts the cooler (770°C to 900°C) metal to be coated, the carbon precipitates (condenses) in pure crystalline form and coats the metal with a diamond film. In the early experimentation on coating carbide and high-speed cutting tools, there were problems with the diamond coating sticking to the cutting-tool material. The process had a slow diamond deposition rate, from 1 to 5 microns per hour.

THE QQC PROCESS

The QQC process developed by Pravin Mistry in the mid-1990s was a major breakthrough in the application of diamond coatings to a variety of materials by eliminating the problems of adhesion, adjusting to various substrates, coating thickness, and cost. It can quickly and relatively inexpensively apply a superior coating on a wide variety of materials and forms that were difficult (or impossible) by the CVD process.

The QQC process creates diamond film through the use of laser energy and carbon dioxide as the source of carbon (Fig. 33-9).

1. Laser energy is directed at the substrate (base material to the coated) to mobilize, vaporize, and react with the primary element (such as carbon) to change the crystalline structure of the substrate.

2. A conversion zone is created beneath the substrate surface with changes metallurgically from the composition of the underlying substrate to a composition of the diamond coating being formed on its surface.

3. This results in diffusion bonding of the diamong coating to the substrate.

MAJOR ADVANTAGES OF THE QQC PROCESS

> Superior bonding and reduced stress form a metallurgical bond between the diamond and the substrate.

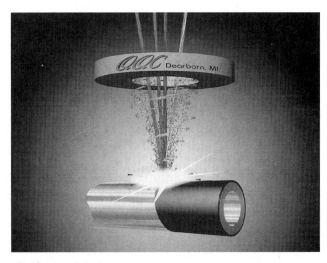

■ Figure 33-9 The QQC process uses the combination of four lasers to produce the diamond reaction. *(QQC, Inc.)*

> The diamond-coating process can be carried out in the atmosphere and doe snot require a vacuum.

> The parts to be coated to not require pretreatment or preheating.

> Only carbon dioxide is the primary or secondary source for carbon; nitrogen acts as a shield.

> Diamond deposition rates exceed 1 micron per second as compared to 1 to 5 microns per hour with the chemical vapor deposition (CVD) process.

> The process can be used for a wide variety of materials such as stainless steel, high-speed steel, iron, plastic, glass, copper, aluminum, titanium, and silicon.

Diamond-coated cutting tool life can be up to 60 times better than tungsten carbide and 240 times better than high-speed steel for machining hard, abrasive nonferrous and nonmetallic materials.

Manufacture of Polycrystalline Tools

1. Describe a polycrystalline cutting tool.

2. How are polycrystalline cutting tools manufactured?

PCBN Tools

3. What advantages do PCBN cutting tools offer industry?

Properties of PCBN

4. Name the four properties that PCBN cutting tools possess.

5. Why is abrasion resistance important to PCBN cutting tools?

Types of Metal Cut

6. List the four general types of metals for which PCBN cutting tools have proven to be cost-effective.

Advantages of PCBN Tools

7. Why do the cutting edges of PCBN tools last longer and remove material at higher rates than do conventional cutting tools?

8. Name four important advantages of PCBN cutting tools.

9. Explain why increased productivity is possible with PCBN cutting tools.

Polycrystalline Diamond Tools

10. Explain the term *carbon-seeking* and how it affects PCD cutting tools.

Types and Sizes of PCD Tools

11. What type of PCD blank is recommended for machining highly abrasive nonferrous and nonmetallic materials?

12. List the three types of PCD cutting tools.

Properties of PCD Tools

13. Name the four main properties of the diamond layer of PCD tools.

14. What is the importance of the high compressive strength of the diamond layer of PCD tools?

Advantages of PCD Tools

15. List four important advantages of PCD cutting tools.

Types of Materials Cut

16. What three general categories of materials can be successfully machined with PCD tools?

UNIT 34

Cutting Fluids—Types and Applications

OBJECTIVES

After completing this unit, you will be able to:

1 State the importance and function of cutting fluids

2 Identify three types of cutting fluids and state the purpose of each

3 Apply cutting fluids efficiently for a variety of machining operations

Cutting fluids are essential in metal-cutting operations to reduce the heat and friction created by the plastic deformation of metal and the chip sliding along the chip-tool interface. This heat and friction cause metal to adhere to the tool's cutting edge, causing the tool to break down; the result is a poor finish and inaccurate work.

The use of cutting fluids is not new; some types have been used for hundreds of years. Centuries ago, it was discovered that water kept a grindstone from glazing and produced a better finish on the part, although it caused the ground part to rust. About 100 years ago, machinists found that wiping tallow on parts before machining helped to produce smoother and more accurate parts. Tallow, the forerunner of cutting fluids, lubricated but did not cool. Lard oils, developed later, lubricated well and had some cooling properties but became rancid quickly. In the early 20th century, soaps were added to water to improve cutting, prevent rust, and provide some lubrication.

The development of soluble oils in 1936 was a great improvement over the previously used cutting oils. These milky-white emulsions combined the high cooling ability of water with the lubricity of petroleum oil. Although economical, they failed to control rust and tended to become rancid.

Chemical cutting fluids were introduced in about 1944. Containing relatively little oil, they depended on chemical agents for lubrication and friction reduction. Chemical emulsions mix easily with water and reduce as well as remove the heat created during machining. Chemical cutting fluids are rapidly increasing in popularity because they provide good rust resistance, do not become rancid quickly, and have good cooling and lubricating qualities.

Cutting fluids cool and lubricate the tool and workpiece. Their use can result in the following economic advantages.

1. *Reduction of tool costs.* Cutting fluids reduce tool wear. Tools last longer, and less time is spent resharpening and resetting them.

2. *Increased speed of production.* Because cutting fluids help reduce heat and friction, higher cutting speeds can be used for machining operations.

3. *Reduction of labor costs.* Because cutting tools last longer and require less regrinding when cutting fluids are used, there is less downtime, reducing the cost per part.

4. *Reduction of power costs.* Since friction is reduced by a cutting fluid, less power is required for machining operations, and a corresponding saving in power costs is possible.

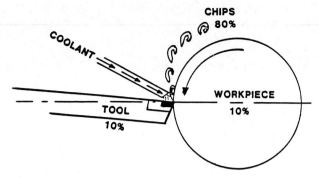

Figure 34-1 Most of the heat generated during an ideal machining operation is carried away by the chips. *(GE Superabrasives)*

Heat Generated During Machining

The heat generated at the chip-tool interface must find its way into one of three places: the workpiece, the cutting tool, or the chips (Fig. 34-1). If the *workpiece* receives too much heat, its size will change and a taper will automatically occur as the workpiece expands with the heat. Too much heat will also cause thermal damage to the surface of the workpiece. If the *cutting tool* receives too much heat, the cutting edge will break down rapidly, reducing tool life. The ideal cutting tool is one that can transfer the heat quickly from the cutting zone to some form of cooling system.

Ideally, most of the heat is taken off in the *chips,* which act as a disposable heat sink or reservoir. This heat transfer is indicated by the change in chip color as the heat causes the chips to oxidize. If too little material is removed to form a chip of sufficient mass, as happens when light feeds and depth of cuts are used, the heat cannot be absorbed by the small chip. It is then forced into the workpiece and the cutting tool.

Cutting fluids assist machining operations by taking away, or dissipating, the heat created at the chip-tool interface. Fig. 34-1 shows how the heat is dissipated during a typical machining operation. The proper use of some form of cutting fluid or coolant system can dissipate at least 50% of the heat created during machining.

Characteristics of a Good Cutting Fluid

For a cutting fluid to function effectively, it should possess the following desirable characteristics:

1. *Good cooling capacity*—to reduce the cutting temperature, increase tool life and production, and improve dimensional accuracy

2. *Good lubricating qualities*—to prevent metal from adhering to the cutting edge and forming a built-up edge, resulting in a poor surface finish and shorter tool life

3. *Rust resistance*—to prevent stain, rust, or corrosion to the workpiece or machine

4. *Stability (long life)*—to hold up both in storage and in use

5. *Resistance to rancidity*—to not become rancid easily

6. *Nontoxic*—to not cause skin irritation to the operator

7. *Transparent*—to allow the operator to see the work clearly during machining

8. *Relatively low viscosity*—to permit the chips and dirt to settle quickly

9. *Nonflammable*—to avoid burning easily and should preferably be noncombustible; in addition, it should not smoke excessively, form gummy deposits that may cause machine slides to become sticky, or clog the circulating system

Types of Cutting Fluids

The need for a cutting fluid that possesses as many desirable characteristics as possible has resulted in the development of many different types. The most commonly used cutting fluids are either aqueous- (water-) based solutions or cutting oils. Cutting fluids fall into three categories: cutting oils, emulsifiable oils, and chemical (synthetic) cutting fluids.

CUTTING OILS

Cutting oils are classified as active or inactive. These terms relate to the oil's chemical activity or ability to react with the metal surface at elevated temperatures to protect it and improve the cutting action.

Active Cutting Oils

Active cutting oils may be defined as those that will darken a copper strip immersed in them for 3 h at a temperature of 212°F (100°C). These oils are generally used when steel is machined and may be either dark or transparent. Dark oils usually contain more sulfur than the transparent types and are considered better for heavy-duty jobs.

Active cutting oils fall into three general categories:

1. *Sulfurized mineral oils* contain from 0.5% to 0.8% sulfur. They are generally light-colored and transparent and have good cooling, lubricating, and antiweld properties. They are useful for the cutting

of low-carbon steels and tough, ductile metals. Sulfurized mineral oils stain copper and its alloys and, therefore, are not recommended for use with these metals.

2. *Sulfochlorinated mineral oils* contain up to 3% sulfur and 1% chlorine. These oils prevent excessive built-up edges from forming and prolong the life of the cutting tool. Sulfochlorinated mineral oils are more effective than sulfurized mineral oils in cutting tough low-carbon and chrome–nickel alloy steels. They are extremely valuable in cutting threads in soft, draggy steel.

3. *Sulfochlorinated fatty oil blends* contain more sulfur than the other types and are effective cutting fluids for heavy-duty machining.

Inactive Cutting Oils

Inactive cutting oils may be defined as those that will not darken a copper strip immersed in them for 3 h at 212°F (100°C). The sulfur contained in an active oil is the natural sulfur of the oil and has no chemical value in the cutting fluid's function during machining. These fluids are termed *inactive* because the sulfur is so firmly attached to the oil that very little is released to react with the work surface during the cutting action.

Inactive cutting oils fall into four general categories:

1. *Straight mineral oils,* because of their low viscosity, have faster wetting and penetrating factors. They are used for the machining of nonferrous metals, such as aluminum, brass, and magnesium, where lubricating and cooling properties are not essential. Straight mineral oils are also recommended for use in cutting leaded (free-machining) metals and tapping and threading white metal.

2. *Fatty oils,* such as lard and sperm oil, once widely used, find limited applications as cutting fluids today. They are generally used for severe cutting operations on tough nonferrous metals where a sulfurized oil might cause discoloration.

3. *Fatty and mineral oil blends* are combinations of fatty and mineral oils, resulting in better wetting and penetrating qualities than straight mineral oils. These qualities result in better surface finishes on both ferrous and nonferrous metals.

4. *Sulfurized fatty–mineral oil blends* are made by sulfur being combined with fatty oils and then mixed with certain mineral oils. Oils of this type provide excellent antiweld properties and lubricity when cutting pressures are high and tool vibration excessive. Most sulfurized fatty–mineral oil blends can be used when nonferrous metals are cut to

produce high surface finishes. They may also be used on machines when ferrous and nonferrous metals are machined at the same time.

EMULSIFIABLE (SOLUBLE) OILS

An effective cutting fluid should possess high heat conductivity; neither mineral nor fatty oils are effective as coolants. Water is the best cooling medium known; however, used as a cutting fluid, water alone would cause rust and have little lubricating value. By adding a certain percentage of soluble oil to water, it is possible to add rust resistance and lubricating qualities to water's excellent cooling capabilities.

Emulsifiable, or soluble, oils are mineral oils containing a soaplike material (emulsifier) that makes them soluble in water and causes them to adhere to the workpiece during machining. These emulsifiers break the oil into minute particles and keep them separated in the water for a long time. Emulsifiable, or soluble, oils are supplied in concentrated form. From *1 to 5 parts* of concentrate are added to *100 parts* of water. Lean mixtures are used for light machining operations and when cooling is essential. Denser mixtures are used when lubrication and rust prevention are essential.

Soluble oils, because of their good cooling and lubricating qualities, are used when machining is done at high cutting speeds, at low cutting pressures, and when considerable heat is generated.

Three types of emulsifiable, or soluble, oils are manufactured for use under various machining conditions:

1. *Emulsifiable mineral oils* are mineral oils to which various compounds have been added to make the oil soluble in water. These oils are low in cost, provide good cooling and lubrication qualities, and are widely used for general cutting applications.

2. *Superfatted emulsifiable oils* are emulsifiable mineral oils to which some fatty oil has been added. These mixtures provide better lubrication qualities and, therefore, are used for tougher machining operations. Often these soluble oils are used when aluminum is machined.

3. *Extreme-pressure emulsifiable oils* contain sulfur, chlorine, and phosphorus, as well as fatty oils, to provide the added lubrication qualities required for tough machining operations. Extreme-pressure oils are generally mixed with water at the rate of *1 part* oil to *20 parts* water.

CHEMICAL CUTTING FLUIDS

Chemical cutting fluids, sometimes called *synthetic fluids,* have been widely accepted since they were first introduced in about 1945. They are stable, preformed emulsions that contain very little oil and mix easily with water. Chemical cutting fluids depend on chemical agents for lubrication and friction reduction. Some types of chemical cutting fluids contain *extreme-pressure* (EP) lubricants, which react with freshly machined metal under the heat and pressure of a cut to form a solid lubricant. Fluids containing EP lubricants reduce both the *heat of friction* between the chip and tool face and the *heat caused by plastic deformation* of the metal.

The chemical agents found in most synthetic fluids include:

1. *Amines* and *nitrites* for rust prevention
2. *Nitrates* for nitrite stabilization
3. *Phosphates* and *borates* for water softening
4. *Soaps* and *wetting agents* for lubrication
5. *Phosphorus, chlorine,* and *sulfur compounds* for chemical lubrication
6. *Glycols* to act as blending agents
7. *Germicides* to control bacteria growth

As a result of the chemical agents that are added to the cooling qualities of water, synthetic fluids provide the following advantages:

1. Good rust control
2. Resistance to rancidity for long periods of time
3. Reduction of the amount of heat generated during cutting
4. Excellent cooling qualities
5. Longer durability than cutting or soluble oils
6. Nonflammable, nonsmoking
7. Nontoxic
8. Easy separation from the work and chips, which makes them clean to work with
9. Quick setting of grit and fine chips so they are not recirculated in the cooling system
10. No clogging of the machine cooling system due to detergent action of the fluid

Three types of chemical cutting fluids are manufactured:

1. *True solution fluids* contain mostly rust inhibitors and are used primarily to prevent rust and provide rapid heat removal in grinding operations. These are generally clear solutions (sometimes a dye is added to color the water) and are mixed *1 part* solution to *50 to 250 parts* water, depending on the application. Some true solution fluids have a tendency to form hard, crystalline deposits when the water evaporates. These deposits may interfere with the operation of chucks, slides, and moving parts.

2. *Wetting-agent types* contain agents to improve the wetting action of water, providing more uniform heat dissipation and antirust action. They also contain mild lubricants, water softeners, and antifoaming agents. Wetting-type chemical cutting fluids are versatile; they have excellent lubricating qualities and provide rapid heat dissipation. They can be used when machining is done with either high-speed steel or carbide cutting tools.

3. *Wetting-agent types with EP lubricants* are similar to wetting-agent types but have chlorine, sulfur, or phosphorus additives to provide EP or boundary lubrication effects. They are used for tough machining jobs with either high-speed steel or carbide cutting tools.

 CAUTION Although chemical cutting fluids have been widely accepted and used for many types of metal-cutting operations, certain precautions should be observed regarding their use. Chemical cutting fluids are generally used on ferrous metals; however, many aluminum alloys can be machined successfully with them. Most chemical cutting fluids are not recommended for use on alloys of magnesium, zinc, cadmium, or lead. Certain types of paint (generally poor quality) may be affected by some chemical cutting fluids, which may mar the machine's appearance and allow paint to get into the coolant and clog the system. Before changing to any type of cutting fluid, it is wise to contact suppliers for the right cutting fluid for the machining operation and the metal being cut.

▸▸ Functions of a Cutting Fluid

The prime functions of a cutting fluid are to provide both cooling and lubrication. In addition, good cutting fluids prolong cutting-tool life, provide rust control, and resist rancidity.

COOLING

Laboratory tests have proved that the heat produced during machining has a definite bearing on cutting-tool wear. Reducing cutting-tool temperature is important to tool life. Even a small reduction in temperature will greatly extend the life of a cutting tool. For example, if tool temperature were reduced only 50°F (28°C) from 950°F to 900°F (510°C to 482°C), cutting-tool life would be increased by five times, from 19.5 to 99 min.

There are two sources of heat generated during a cutting action:

1. Plastic deformation of the metal, which occurs immediately ahead of the cutting tool, accounts for approximately two-thirds to three-quarters of the heat generated.

2. Friction resulting from the chip sliding along the cutting-tool face also produces heat.

Water is the most effective agent for reducing the heat generated during machining. Since water alone causes rusting, soluble oils or chemicals that prevent rust and provide other essential qualities are added to make it a good cutting fluid.

An abundant supply of cutting fluid should be applied to the machining area, at a very low pressure. This will ensure that the machining area will be well covered and that little splashing will occur. The flow of the cutting fluid will help to wash the chips away from the machining area.

LUBRICATION

The lubrication function of a cutting fluid is as important as its cooling function. Heat is generated by the plastic deformation of the metal and by the friction between the chip and tool face. The plastic deformation of metal occurs along the shear plane (Fig. 34-2). Any way of short-ening the length of the shear plane would result in a reduction in the amount of heat generated.

The only known method of shortening the length of the shear plane for any given shape of a cutting tool and work material is to reduce the friction between the chip and tool face. Fig. 34-3 shows a chip sliding along the face of the cutting tool. The enlarged illustration shows the irregularities on the tool face that create areas of friction and tend to cause a built-up edge to form. Note also that because of this friction there is a long shear plane and a small shear angle. The most heat is created at the cutting edge when there is a small shear angle and a long shear plane.

Fig. 34-4 illustrates the same depth of cut as in Fig. 34-3 but shows cutting fluid being used to reduce the friction at the chip-tool interface. As soon as the friction is reduced, the shear plane becomes shorter, and the area where plastic deformation occurs is correspondingly smaller. Therefore, by reduction of the friction at the chip-tool interface, both sources of heat (plastic deformation and friction at the chip-tool interface) can be reduced.

The effective life of a cutting tool can be greatly lengthened if the friction and the resultant heat generated are reduced. When steel is machined, the temperature and pressure at the chip-tool interface may reach 1000°F

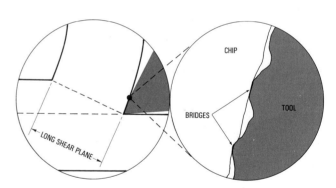

■ Figure 34-3 A long shear plane results in great heat at the shear zone. *(Milacron, Inc.)*

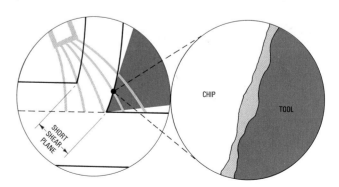

■ Figure 34-4 Cutting fluid reduces friction and produces a shorter shear plane. *(Milacron, Inc.)*

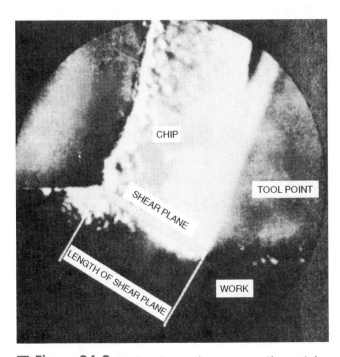

■ Figure 34-2 During the cutting process, the metal is deformed along the shear plane, producing heat. *(Cincinnati Machine)*

Cutting Fluids—Types and Applications 265

(538°C) and 200,000 psi (1,379,000 kPa), respectively. Under such conditions, some oils and other liquids tend to vaporize or be squeezed from between the chip and tool. Extreme-pressure lubricants reduce the amount of heat-producing friction. The EP chemicals of synthetic fluids combine chemically with the sheared metal of the chip to form solid compounds. These solid compounds or lubricants can withstand high pressure and temperature, and they allow the chip to slide up the tool face easily even under these conditions.

CUTTING-TOOL LIFE

Heat and friction are the prime causes of cutting-tool breakdown. Decreasing the amount of heat and friction created during a machining operation can greatly increase the life of a cutting tool. Laboratory tests have proved that if the temperature at the chip-tool interface is reduced by as little as 50°F (28°C), the life of the cutting tool increases fivefold. As a result, when cutting fluids are used, faster speeds and feeds can be used; increased production and a reduction in the cost per piece result.

During a cut, pieces of metal tend to weld themselves to the tool face, causing a built-up edge to form (Fig. 34-5). If the built-up edge becomes large and flat along the tool face, the effective rake angle of the cutting tool is decreased and more power is required to cut the metal. The built-up edge keeps breaking off and re-forming; the result is a poor surface finish, excessive flank wear, and cratering of the tool face. Almost all the roughness of a machined surface is caused by tiny fragments of metal that have been left behind by the built-up edge.

The use of an effective cutting fluid affects the action of a cutting tool in the following ways:

1. It lowers the heat created by the plastic deformation of the metal, thereby increasing cutting-tool life.
2. Friction at the chip-tool interface is decreased, reducing the resultant heat.
3. Less power is required for machining because of the reduced friction.
4. It prevents a built-up edge from forming, resulting in longer tool life.
5. The surface finish of the work is greatly improved.

RUST CONTROL

Cutting fluids used on machine tools should prevent rust from forming; otherwise, machine parts and workpieces will be damaged. Cutting oil prevents rust from forming but does not cool as effectively as water. Water is the best and most economical coolant but causes parts to rust unless rust inhibitors are added.

Rust is oxidized iron, or iron that has reacted chemically with oxygen, water, and minerals in the water. Water alone on a piece of steel or iron acts as a medium for the electro-chemical process to start causing corrosion or rust.

Chemical cutting fluids contain rust inhibitors, which prevent the electro-chemical process of rusting. Some types of cutting fluids form a *polar film* on metals, which prevents rusting. This polar film (Fig. 34-6) consists of negatively charged, long, thin molecules that are attracted to and firmly bond themselves to the metal. This invisible film, only molecules thick, is sufficient to prevent the electro-chemical action of rusting. Other types of cutting fluids contain rust inhibitors that form an insulating blanket known as a *passivating film* (Fig. 34-6) on the metal surface. These inhibitors combine chemically with the metal and form a nonporous, protective coating that prevents rust.

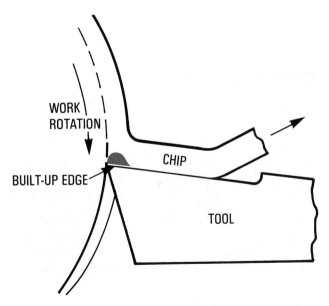

■ **Figure 34-5** A built-up edge is caused by chip fragments being pressure-welded to the cutting-tool face.

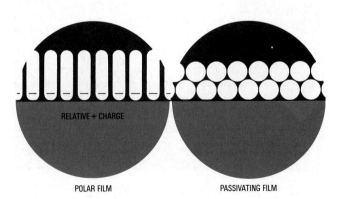

■ **Figure 34-6** Polar and passivating films on metal surfaces prevent rust. *(Milacron, Inc.)*

RANCIDITY CONTROL

When lard oil was the only cutting fluid used, after a few days, it would start to spoil and give off an offensive odor. This rancidity is caused by bacteria and other microscopic organisms, growing and multiplying almost anywhere and eventually causing bad odors to form. Today, any cutting fluid that has an offensive smell is termed *rancid*.

Most cutting fluids contain bactericides that control the growth of bacteria and make the fluids more resistant to rancidity. The bactericide, which is added to the fluid by the manufacturer, must be strong enough to control the growth of bacteria but weak enough not to be harmful to the skin of the operator.

▶▶ Application of Cutting Fluids

Cutting-tool life and machining operations are greatly influenced by the way that the cutting fluid is applied. It should be supplied in a copious stream under low pressure so that the work and cutting tool are well covered. The rule of thumb is that the inside diameter of the supply nozzle should be about three-quarters the width of the cutting tool. The fluid should be directed to the area where the chip is being formed to reduce and control the heat created during the cutting action and to prolong tool life.

Another method used to cool the chip-tool interface in various machining operations is refrigerated air. The *refrigerated air system* (Fig. 34-7a) is an effective, inexpensive, and readily available cooling system where dry machining is necessary or preferred. It uses compressed air that enters a vortex generation chamber (Fig. 34-7b), where it is cooled by as much as 100°F (38°C) below the incoming compressed air temperature. Cool air as low as −40°F (−40°C) can be directed to cool the chip-tool interface and blow the chips away.

LATHE-TYPE OPERATIONS

On horizontal-type turning and boring machines, cutting fluid should be applied to that portion of the cutting tool that is producing the chip. For general turning and facing operations, cutting fluid should be supplied directly over the cutting tool, close to the zone of chip formation (Fig. 34-8 on p. 268). In heavy-duty turning and facing operations, cutting fluid should be supplied by two nozzles, one directly above and the other directly below the cutting tool (Fig. 34-9 on p. 268). See Table 34.1 (on p. 269) for recommended cutting fluids for various materials.

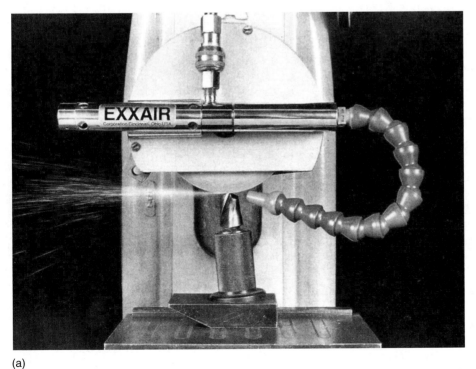

(a)

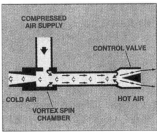

(b)

■ **Figure 34-7** (a) A refrigerated air system used for cooling purposes during surface grinding; (b) the vortex tube converts compressed air to refrigerated air. *(Exxair Corp.)*

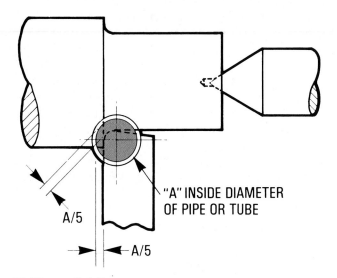

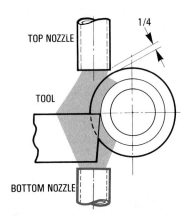

"A" INSIDE DIAMETER OF PIPE OR TUBE

A/5

A/5

■ **Figure 34-8** Cutting fluid supplied by one nozzle for general facing and turning operations. *(Milacron, Inc.)*

TOP NOZZLE

1/4

TOOL

BOTTOM NOZZLE

■ **Figure 34-9** Top and bottom nozzles are used to supply cutting fluid for heavy-duty turning operations. *(Milacron, Inc.)*

DRILLING AND REAMING

The most effective method of applying cutting fluids for these operations is to use "oil-feed" drills and hollow-shank reamers. Tools of this type transmit the cutting fluid directly to the cutting edges and at the same time flush the chips out of the hole (Fig. 34-10). When conventional drills and reamers are used, an abundant supply of fluid should be applied to the cutting edges.

MILLING

In *slab milling*, cutting fluid should be directed to both sides of the cutter by fan-shaped nozzles approximately three-quarters the width of the cutter (Fig. 34-11).

For *face milling*, a ring-type distributor (Fig. 34-12) is recommended to flood the cutter completely. Keeping each tooth of the cutter immersed in cutting fluid at all times can increase cutter life almost 100%.

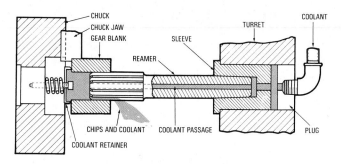

■ **Figure 34-10** Cutting fluid is often supplied through a hole in the center of a reamer. *(Milacron, Inc.)*

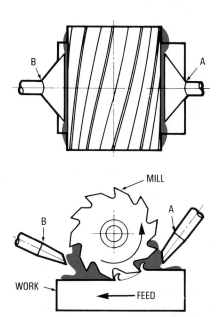

■ **Figure 34-11** Cutting fluid being supplied to both sides of the cutter during slab milling. *(Milacron, Inc.)*

■ **Figure 34-12** Cutting fluid being applied by a ring-type distributor during a face-milling operation. *(Cincinnati Machine)*

table 34.1 Recommended cutting fluids for various materials*

Material	Drilling	Reaming	Threading	Turning	Milling
Aluminum	Soluble oil Kerosene Kerosene and lard oil	Soluble oil Kerosene Mineral oil	Soluble oil Kerosene and lard oil	Soluble oil	Soluble oil Lard oil Mineral oil Dry
Brass	Dry Soluble oil Kerosene and lard oil	Dry Soluble oil	Soluble oil Lard oil	Soluble oil	Dry Soluble oil
Bronze	Dry Soluble oil Lard oil	Dry Soluble oil Lard oil	Soluble oil Lard oil	Soluble oil	Dry Soluble oil Lard oil
Cast iron	Dry Air jet Soluble oil	Dry Soluble oil Mineral lard oil	Dry Sulfurized oil Mineral lard oil	Dry Soluble oil	Dry Soluble oil
Copper	Dry Soluble oil Mineral lard oil Kerosene	Soluble oil Lard oil	Soluble oil Lard oil	Soluble oil	Dry Soluble oil
Mallealble iron	Dry Soda water	Dry Soda water	Lard oil Soda water	Soluble oil	Dry Soda water
Monel metal	Soluble oil Lard oil	Soluble oil Lard oil	Lard oil	Soluble oil	Soluble oil
Steel alloys	Soluble oil Sulfurized oil Mineral lard oil	Soluble oil Sulfurized oil Mineral lard oil	Sulfurized oil Lard oil	Soluble oil	Soluble oil Mineral lard oil
Steel, machine	Soluble oil Sulfurized oil Mineral lard oil	Soluble oil Mineral lard oil	Soluble oil Mineral lard oil	Soluble oil	Soluble oil Mineral lard oil
Steel, tool	Soluble oil Sulfurized oil Mineral lard oil	Soluble oil Sulfurized oil Lard oil	Sulfurized oil Lard oil	Soluble oil	Soluble oil Lard oil

*Chemical cutting fluids can be used successfully for most of the above cutting operations. These concentrates are diluted with water in proportions ranging from 1 part cutting fluid to 15 and as high as 100 parts of water, depending on the metal being cut and the type of machining operation. When using chemical cutting fluids, follow the manufacturer's recommendations for use and mixture. (Milacron, Incorporated.)

GRINDING

Cutting fluid is very important in a grinding operation; it cools the work and keeps the grinding wheel clean. Cutting fluid should be applied in large quantities and under very little pressure.

1. *Surface grinding.* Three methods may be used to apply cutting fluid for surface-grinding operations:

   a. The *flood method* is the one most commonly used. A steady flow of cutting fluid is applied through a nozzle. Because of the reciprocating table action, most surface-grinding operations can be greatly improved if the fluid is supplied through two nozzles, as in Fig. 34-11.

b. In the *through-the-wheel* method, the coolant is fed to a special wheel flange and forced to the periphery of the wheel and to the area of contact by centrifugal force.

c. The *spray-mist system* is one of the most effective cooling systems. It uses the atomizer principle, where compressed air passing through a T connection syphons a small amount of coolant from a reservoir and discharges it at the chip-tool interface. Cooling results from the action of the compressed air and evaporation of the mist vapor. The compressed air also blows steel chips away from the cutting tool area, permitting the machine operator to clearly see the operation.

2. *Cylindrical grinding.* For cylindrical-grinding operations, it is important that the entire contact area between the wheel and work be flooded with a steady stream of clean, cool cutting fluid. A fan-shaped nozzle (Fig. 34-13) somewhat wider than the wheel should be used to direct the cutting fluid.

3. *Internal grinding.* During internal grinding, the cutting fluid must flush the chips and the abrasive wheel particles out of the hole being ground. Because internal-grinding practices call for using as large a wheel as possible, it is sometimes difficult to get enough cutting fluid into the hole. A compromise must be made between grinding wheel size and the amount of fluid entering the hole. As much fluid as possible should be applied during internal-grinding operations.

■ **Figure 34-13** Cutting fluid being applied to the contact zone during cylindrical grinding. *(Cincinnati Machine)*

unit 34 review questions

1. What is the function of a modern cutting fluid?
2. Briefly trace the development of cutting fluids.
3. Explain the causes of heat and friction during a machining process.
4. Name four economical advantages of applying correct cutting fluids.

Characteristics of a Good Cutting Fluid

5. List six *important* characteristics that a good cutting fluid should possess.

Types of Cutting Fluids

6. Name three categories into which cutting fluids fall.
7. Describe active and inactive cutting oils.
8. What type of cutting oil should be used for:
   a. Tough, ductile metals?
   b. Heavy-duty machining?
   c. Nonferrous metals?
   d. Threading white metal?
9. Describe the composition of an emulsifiable oil and state its advantages.

10. State the purpose of:

    a. Emulsifiable mineral oil
    b. EP emulsifiable oil

11. Discuss and state six important advantages of chemical cutting fluids.

12. State the purpose of:

    a. True solution fluids
    b. Wetting-agent types with EP lubricants

Functions of a Cutting Fluid

13. Name five functions of a cutting fluid.

14. Discuss the importance of cooling and lubrication as applied to cutting fluids.

15. Explain how a cutting fluid can change the length of the shear plane.

16. For what purpose are EP lubricants used?

17. What occurs at the cutting-tool face during a cut?

18. What is the main cause of surface roughness?

19. How does the application of cutting fluid affect the cutting tool?

20. Why is the control of rust important?

21. Describe a polar film, a passivating film.

22. Define rancidity and state the purpose of bactericides.

Application of Cutting Fluids

23. What is the general recommendation for applying cutting fluids?

24. How should cutting fluid be applied for lathe operations?

25. State two methods of applying cutting fluid for drilling or reaming.

26. Explain how cutting fluid should be applied for:

    a. Slab milling
    b. Face milling
    c. Cylindrical grinding

27. Describe three methods of applying cutting fluid for surface-grinding operations.

28. Why is it sometimes difficult to apply cutting fluid during internal-grinding operations?

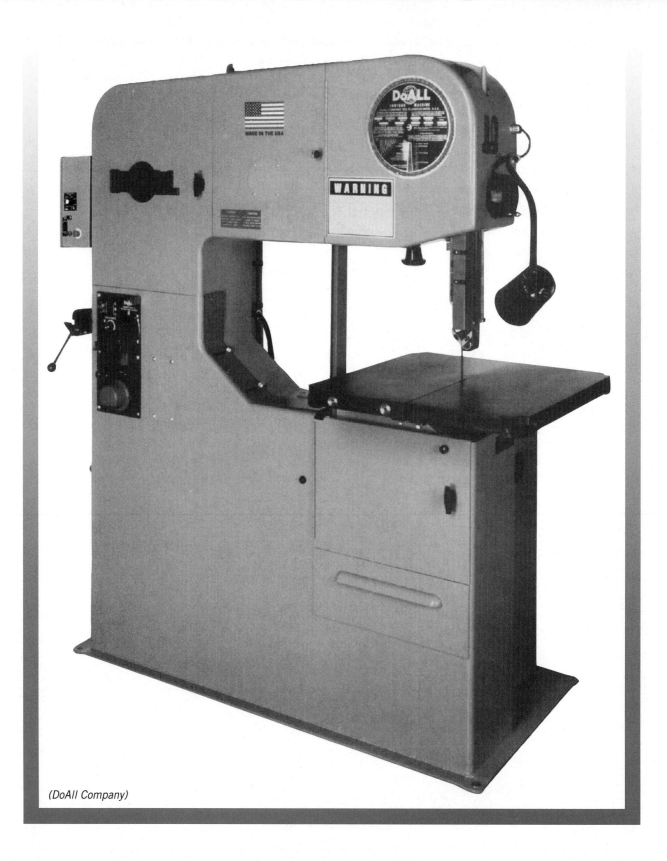

(DoAll Company)

METAL-CUTTING SAWS

Archaeological discoveries show that development of the first crude saw closely followed the origin of the stone ax and knife. The sharp edges of stones were serrated or toothed. This instrument cut by scraping away particles of the object being cut. A great improvement in the quality of saws followed the appearance of copper, bronze, and ferrous metals. With today's steels and hardening methods, many different types of saw blades are available for hand hacksaws and machine power saws.

UNIT 35

Types of Metal Saws

OBJECTIVES

After completing this unit, you will be able to:

1 Name five types of cutting-off machines and state the advantage of each

2 Select the proper blade to use for cutting various cross sections

3 Install a saw band on a horizontal bandsaw

4 Use a bandsaw to cut off work to an accurate length

Metal-cutting saws are available in a wide variety of models to suit various cutting-off operations and materials.

▶▶ Methods of Cutting Off Material

Five of the most common methods of cutting off material are *hacksawing, bandsawing, abrasive cutting, cold sawing,* and *friction sawing.* A brief description of each method and its advantages follows.

The *power hacksaw,* which is a reciprocating type of saw, is usually permanently mounted to the floor. The saw frame and blade travel back and forth, with pressure being applied automatically only on the forward stroke. The power hacksaw finds limited use in machine shop work, since the saw cuts only on the forward stroke, resulting in considerable wasted motion.

The *horizontal bandsaw* (Fig. 35-1) has a flexible, beltlike "endless," or "one-way," blade that cuts continuously in one direction. The thin, continuous blade travels over the rims of two pulley wheels and passes through roller guide brackets that support the blade and keep it running true. Horizontal bandsaws are available in a wide variety of types and sizes and are becoming increasingly popular because of their high production and versatility.

■ **Figure 35-1** A horizontal bandsaw cuts continuously in one direction. *(Clausing Industrial, Inc.)*

■ Figure 35-2 An abrasive cutoff saw will cut hardened metals, glass, and ceramics. *(Everett Industries, Inc.)*

■ Figure 35-3 A cold circular saw is used for cutting soft or unhardened metals. *(Everett Industries, Inc.)*

■ Figure 35-4 The main parts of a horizontal bandsaw. *(DoAll Company)*

The *abrasive cutoff saw* (Fig. 35-2) cuts material by means of a thin, abrasive wheel revolving at high speed. This type of saw is especially well suited for cutting most metals and materials such as glass and ceramics. It can cut material to close tolerances, and hardened metal does not have to be annealed to be cut. Abrasive cutoff can be performed under dry conditions or with a suitable cutting fluid. The use of cutting fluid keeps the work and saw cooler and produces a better surface finish.

The *cold circular cutoff saw* (Fig. 35-3) uses a circular blade similar to the one used on a wood-cutting table saw. The saw blade is generally made of chrome–vanadium steel, but carbide-tipped blades are used for some applications. Cold circular saws produce very accurate cuts and are especially suited for cutting aluminum, brass, copper, machine steel, and stainless steel.

Friction sawing is a burning process by which a saw band, with or without saw teeth, is run at high speeds [10 000 to 25 000 sf/min (3048 to 7620 m/min)] to burn or melt its way through the metal. Friction sawing cannot be used on solid metals because of the amount of heat generated; however, it is excellent for cutting structural and honeycombed parts of machine or stainless steel.

▶▶ Horizontal Bandsaw Parts

The horizontal bandsaw (Fig. 35-4) is the most popular machine used to cut off work. The main operative parts of this saw are as follows:

> The *frame,* hinged at the motor end, has two pulley wheels mounted on it, over which the continuous blade passes.

> The *step pulleys* at the motor end are used to vary the speed of the continuous blade to suit the type of material cut.

> The *roller guide brackets* provide rigidity for a section of the blade and can be adjusted to accommodate various widths of material. These brackets

should be adjusted to just clear the width of the work being cut.

> The *blade tension handle* is used to adjust the tension on the saw blade. The blade should be adjusted to prevent it from wandering or twisting.

> The *vise,* mounted on the table, can be adjusted to hold various sizes of workpieces. It can also be swiveled for making angular cuts on the end of a piece of material.

▶▶ Saw Blades

High-speed tungsten and high-speed molybdenum steel are commonly used in the manufacture of saw blades, and for the power hacksaw, they are usually hardened completely. Flexible blades used on bandsaws have only the saw teeth hardened.

Saw blades are manufactured in various degrees of coarseness, ranging from 4 to 14 pitch. When cutting large sections, use a coarse, or 4-pitch, blade, which provides the greatest chip clearance and helps to increase tooth penetration. For cutting tool steel and thin material, a 14-pitch blade is recommended. A 10-pitch blade is recommended for general-purpose sawing. *Metric saw blades* are available in similar sizes, but in teeth per 25 mm of length rather than teeth per inch. Therefore, the pitch of a blade having 10 teeth per 25 mm would be 10 ÷ 25 mm, or 0.4 mm. Always select a saw blade as coarse as possible, but make sure that *two teeth* of the blade will be in contact with the work at all times. If fewer than two teeth are in contact with the work, the work can be caught in the tooth space (gullet), which will cause the teeth of the blade to strip or break.

INSTALLING A BLADE

When replacing a blade, always make sure that the teeth are pointing in the direction of saw travel or toward the motor end of the machine. The blade tension should be adjusted to prevent the blade from twisting or wandering during a cut. If it is necessary to replace a blade before a cut is finished, rotate the work one-half turn in the vise. This will prevent the new blade from jamming or breaking in the cut made by the worn saw.

To install a saw blade, use the following procedure:

1. Loosen the blade tension handle (Fig. 35-5).
2. Move the adjustable pulley wheel forward slightly.
3. Mount the new saw band over the two pulleys.

Note: *Be sure that the saw teeth are pointing toward the motor end of the machine.*

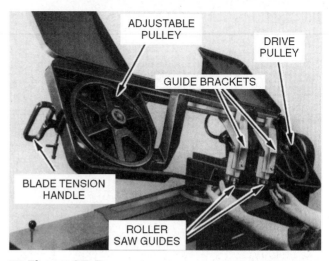

■ **Figure 35-5** Installing a new blade on a horizontal bandsaw. *(Kelmar Associates)*

4. Place the saw blade between the rollers of the guide brackets (Fig. 35-5).
5. Tighten the blade tension handle only enough to hold the blade on the pulleys.
6. Start and quickly stop the machine to make the saw blade revolve a turn or two. This will seat (track) the blade on the pulley.
7. Tighten the blade tension handle as tightly as possible with *one hand.*

▶▶ Sawing

For the most efficient sawing, it is important that the correct type and pitch of saw blade be selected and run at the proper speed for the material being cut. Use finer tooth blades when cutting thin cross sections and extra-hard materials. Coarser tooth blades should be used for thick cross sections and material that is soft and stringy. The blade speed should suit the type and thickness of the material cut. Too fast a blade speed or excessive feeding pressure will dull the saw teeth quickly and cause an inaccurate cut.

To saw work to length, use the following procedure:

1. Check the solid vise jaw with a square to make sure it is at right angles to the saw blade.
2. Place the material in the vise, supporting long pieces with a floor stand (Fig. 35-6).
3. Lower the saw blade until it just clears the work. Keep it in this position by engaging the ratchet lever or by closing the hydraulic valve.
4. Adjust the roller guide brackets until they *just clear* both sides of the material to be cut (Fig. 35-5).

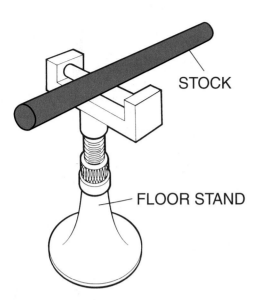

■ Figure 35-6 A floor stand is used to support long pieces while they are cut. *(Kelmar Associates)*

5. Hold a steel rule against the edge of the saw blade and move the material until the correct length is obtained.

6. Always allow .060 in. (1.5 mm) for each 1 in. (25 mm) of thickness longer than required to compensate for any *saw run-out* (slightly angular cut caused by hard spots in steel or a dull saw blade).

7. Tighten the vise and recheck the length from the blade to the end of the material to make sure that the work has not moved.

8. Raise the saw frame slightly, release the ratchet lever or open the hydraulic valve, and then start the machine.

9. Lower the blade slowly until it just touches the work.

10. When the cut has been completed, the machine will shut off automatically.

SAWING HINTS

1. Never attempt to mount, measure, or remove work unless the saw is stopped.

2. Guard long material at both ends to prevent anyone from coming in contact with it.

3. Use cutting fluid whenever possible to help prolong the life of the saw blade.

4. When sawing thin pieces, hold the material flat in the vise to prevent the saw teeth from stripping. Thin material can also be placed

■ Figure 35-7 A stop gage is used when many pieces of the same length must be cut.

between two pieces of wood or soft materials when sawing.

5. Use caution when applying extra force to the saw frame because this generally causes work to be cut out of square.

6. When several pieces of the same length are required, set the stop gage supplied with most cutoff saws (Fig. 35-7).

7. When holding short work in a vise, be sure to place a short piece of the same thickness in the opposite end of the vise. This short piece will prevent the vise from twisting when it is tightened (Fig. 35-8).

■ Figure 35-8 A spacer block should be used to clamp a short workpiece in a vise. *(Kelmar Associates)*

unit 35 review questions

Methods of Cutting Off Material

1. Name and describe the cutting action of five methods of cutting off material.

Horizontal Bandsaw Parts

2. What is the purpose of the roller guide brackets?
3. How tight should the blade tension handle be adjusted?

Saw Blades

4. Of what material are saw blades made?
5. What pitch blade is recommended for:
   a. Large sections?
   b. Thin sections?
   c. General-purpose sawing?

6. Why should two teeth of a saw blade be in contact with the work at all times?
7. In what direction should the teeth of a saw blade point?
8. How should the workpiece be set for sawing partially cut work with a new saw blade?

Sawing

9. What may happen if too fast a blade speed is used?
10. How can a vise be checked for squareness?
11. How should the roller guide brackets be set for cutting work?
12. Explain how thin work should be held in a vise.
13. What precaution should be used for holding short work in a vise?

UNIT 36

Contour Bandsaw Parts and Accessories

OBJECTIVES

After completing this unit, you will be able to:

1 Name the main operative parts of a contour-cutting bandsaw and state the purpose of each

2 Select the proper tooth form and set for any cutting application

3 Calculate the length of a saw band for a two-pulley machine

The vertical bandsaw is the latest machine tool to be developed. Since its development in the early 1930s, it has been widely accepted by industry as a fast and economical method of cutting metal and other materials.

Band cutoff saws differ from power hacksaws because they have a continuous cutting action on the workpiece, whereas the latter cuts only on the forward stroke and on a limited section of the blade.

The contour-cutting bandsaw (vertical bandsaw) offers several features not found on other metal-cutting machines. These advantages are illustrated in Fig. 36-1 (on p. 280).

▶▶ Contour Bandsaw Parts

The construction of contour-band machines differs from most other machines because the band machine is generally fabricated from steel rather than being cast. The contour bandsaw consists of three basic parts: *base, column,* and *head* (Fig. 36-2 on p. 280).

BASE

The base of the contour bandsaw supports the column and houses the drive assembly, which provides a drive for the saw blade.

> The *lower pulley,* which supports and drives the saw band, is driven by a variable-speed pulley that

can be adjusted to various speeds by the *variable speed handwheel.*

> The *table* is attached to the base by means of a *trunnion* (Fig. 36-3 on p. 281). It can be tilted 10° to the left and 45° to the right for making angular cuts by turning the *table tilt handwheel.*

> The *lower saw guide* is attached to the trunnion, and it supports the blade to keep it from twisting.

> A removable *filler plate slide* and *center plate* are mounted in the table.

COLUMN

The column supports the *head, left-hand blade guard, welding unit,* and *variable speed handwheel.*

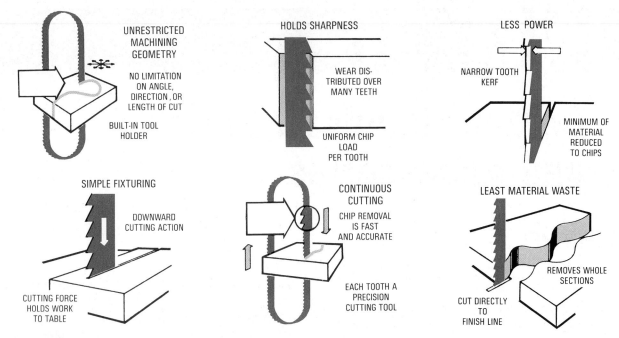

UNRESTRICTED MACHINING GEOMETRY

NO LIMITATION ON ANGLE, DIRECTION, OR LENGTH OF CUT

BUILT-IN TOOL HOLDER

HOLDS SHARPNESS

WEAR DISTRIBUTED OVER MANY TEETH

UNIFORM CHIP LOAD PER TOOTH

LESS POWER

NARROW TOOTH KERF

MINIMUM OF MATERIAL REDUCED TO CHIPS

SIMPLE FIXTURING

DOWNWARD CUTTING ACTION

CUTTING FORCE HOLDS WORK TO TABLE

CONTINUOUS CUTTING

CHIP REMOVAL IS FAST AND ACCURATE

EACH TOOTH A PRECISION CUTTING TOOL

LEAST MATERIAL WASTE

REMOVES WHOLE SECTIONS

CUT DIRECTLY TO FINISH LINE

■ **Figure 36-1** Advantages of a contour bandsaw. *(DoAll Company)*

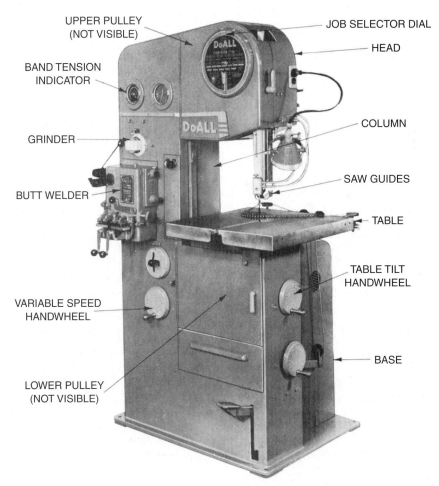

UPPER PULLEY (NOT VISIBLE)

JOB SELECTOR DIAL

HEAD

BAND TENSION INDICATOR

GRINDER

BUTT WELDER

COLUMN

SAW GUIDES

TABLE

TABLE TILT HANDWHEEL

VARIABLE SPEED HANDWHEEL

BASE

LOWER PULLEY (NOT VISIBLE)

■ **Figure 36-2** A contour bandsaw provides an economical means of cutting metals to shape. *(DoAll Company)*

TRUNNION

■ **Figure 36-3** A three-carrier-wheel contour bandsaw with table removed. *(DoAll Company)*

> The *variable speed handwheel* is used to regulate the speed of the bandsaw blade.

> The *blade tension indicator* and the *speed indicator* are located in the column.

> The *welding unit* is used to weld, anneal, and grind the saw blade.

HEAD

The parts found in the head of a saw are generally used to guide or support the saw band.

> The *upper pulley* supports the saw band, which is adjusted by the tension and tracking controls.

> The *upper saw guide,* attached to the saw guidepost, supports and guides the saw blade to keep it from twisting. It can be adjusted vertically to accommodate various sizes of work.

> The *saw guard* and the *air nozzle* for keeping the area being cut free from chips are also found in the head.

Band machines may be of two types. The machine used in many toolrooms has two band carrier wheels; the larger capacity saws have three carrier wheels (Fig. 36-3). On the larger capacity saws, both upper wheels may be tilted so that the band will track properly. When the blade in this type of machine becomes too short, it need not be discarded. It may be shortened to fit over the upper and lower band carrier wheels (Fig. 36-3). The capacity of the machine is reduced, but this adjustment permits the economical use of the saw blade.

▸▸ Bandsaw Applications

Many operations can be performed faster and easier on the contour bandsaw than on any other machine. In addition to saving time, material is also saved because large sections of a workpiece can be removed as a solid instead of being reduced to chips, as on conventional machines. Some of the more common operations performed on a bandsaw are shown in Fig. 36-4a to f (on p. 282)

> *Notching* (Fig. 36-4a)—sections of metal can be removed in one piece rather than in chips.

> *Slotting* (Fig. 36-4b)—this operation can be done quickly and accurately without expensive fixtures.

> *Three-dimensional shaping* (Fig. 36-4c)—complicated shapes may be cut; simply follow the layout lines.

> *Radius cutting* (Fig. 36-4d)—internal or external contours may be cut easily. Internal sections are generally removed in one piece, as shown.

> *Splitting* (Fig. 36-4e)—this operation can be accomplished quickly with a minimum waste of material.

> *Angular cutting* (Fig. 36-4f)—the work may be clamped at any angle and fed through the saw. The table may be tilted for compound angles.

▸▸ Coolants

Some machines, particularly the power-feed models, have a cooling system that circulates and discharges coolant against the faces of the blade and work.

In fixed-table machines where coolant is required, a mist coolant system is generally employed. The mist system uses air to atomize the coolant and direct it onto the faces of the blade and the work. This method is efficient and is recommended for the high-speed machining of nonferrous metals such as aluminum and magnesium alloys. Tough, hard-to-machine alloys can also be cut successfully using the mist coolant system.

Grease-type lubricants and coolants may be applied directly to the blade to assist in cutting on machines having no coolant system.

▸▸ Power Feed

Some of the heavier band cutting machines are equipped with power-feed tables. The work and the table are fed toward the blade by means of a hydraulic system.

On fixed-table machines, power feeding is accomplished by means of a device that uses gravity to provide

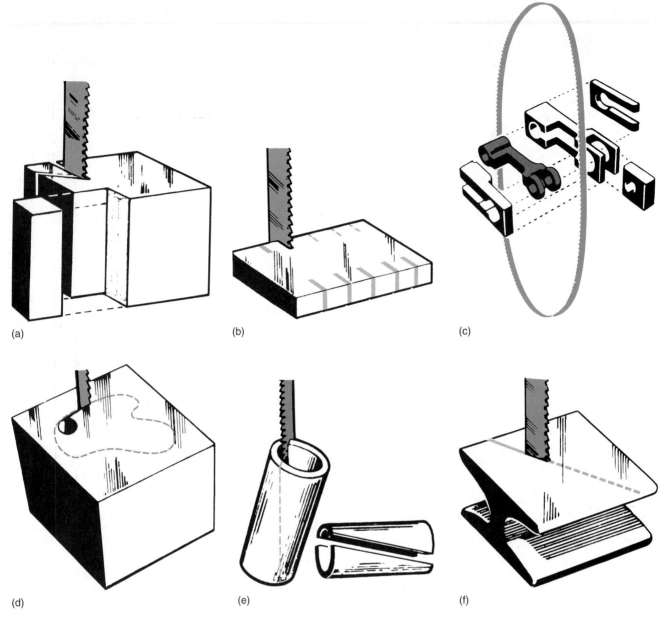

(a)

(b)

(c)

(d)

(e)

(f)

■ **Figure 36-4** (a) Notching; (b) slotting; (c) three-dimensional shaping; (d) radius cutting; (e) splitting; (f) angular cutting. *(DoAll Company)*

a steady mechanical feeding pressure. This allows the operator to use both hands to guide the work into the saw. The workpiece is held against a work jaw and is forced into the blade by means of cables, pulleys, and weights.

The force applied to the blade can be varied up to about 80 lb (356 N). For regular sawing, a feeding force of about 30 to 40 lb (133 to 178 N) should be used.

Greater feed force may be used for sawing straight lines than for cutting contours. To determine the feed to use for any particular job or operation, consult the job selector dial on the bandsaw.

▶▶ Bandsaw Blade Types and Applications

Three kinds of blades are commonly used in bandsawing: carbon-alloy, high-speed steel, and tungsten-carbide–tipped blades. To obtain the best results from any bandsaw, it is necessary to select the proper blade for the job. Consideration must be given to the kind of saw-blade material and tooth form, pitch set, width, and gage needed for the material being cut.

■ Figure 36-5 Bandsaw blade tooth forms. *(DoAll Company)*

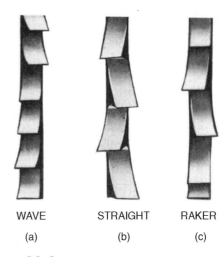

WAVE   STRAIGHT   RAKER
(a)   (b)   (c)

■ Figure 36-6 Common set patterns: (a) wave; (b) straight; (c) raker. *(DoAll Company)*

TOOTH FORMS

Carbon and high-speed steel blades are available in three types of tooth forms (Fig. 36-5). The *precision,* or *regular, tooth* is the most generally used type of tooth. It has a 0° rake angle and about a 30° back clearance angle. It is used when a fine finish and an accurate cut are required.

The *claw,* or *hook, tooth* has a positive rake on the cutting face and slightly less back clearance than the precision or buttress blade. It has the same general application as the buttress-tooth form. It is faster cutting and longer lasting than the buttress tooth but will not produce as smooth a finish.

The *buttress,* or *skip, tooth* is similar to the precision tooth; however, the teeth are spaced farther apart to provide more chip clearance. The tooth angles are the same as those of precision teeth. Buttress, or skip-tooth, blades are used to advantage on thick work sections and for making deep cuts in soft material.

PITCH

Each of the tooth forms is available in various pitches, or numbers of teeth per standard reference length. Inch sawblade pitch is determined by the number of teeth per inch; metric blade pitch, by the number of teeth per 25 mm.

The thickness of the material to be cut determines the pitch of the blade to be used. When cutting thick materials, use a coarse-pitch blade; thin materials require a fine-pitch blade. When selecting the proper pitch, remember that at least two teeth of the saw blade must be in contact with the material being cut at all times.

SET

The set of a blade is the amount that the teeth are offset on either side of the center to produce clearance for the back of the band or blade.

The three common set patterns are (Fig. 36-6):

> *Wave set,* which has a group of teeth offset to the right and the next group to the left, a pattern that produces a wavelike appearance. Wave-set blades are generally used when the cross section of the workpiece changes, such as on structural steel sections or on pipe.

> *Straight set,* which has one tooth offset to the right and the next to the left. It is used for cutting light nonferrous castings, thin sheet metal, tubing, and Bakelite.

> *Raker set,* which has one tooth offset to the right, one offset to the left, and the third tooth straight. This pattern is the most common and is used for most sawing applications.

WIDTH

When making straight, accurate cuts, select a wide blade. Narrow blades are used to cut small radii. Radius charts, showing the proper width of blade to use for contour sawing, are generally found on all bandsaws. When selecting a blade for contour cutting, choose the widest blade that can cut the smallest radius on the workpiece.

GAGE

The gage is the thickness of the saw blade and has been standardized according to blade width. Blades up to ½ (.500) in. (13 mm) wide are .025 in. (0.64 mm) thick, ⅝- (.625) in. (16-mm) and ¾- (.750)-in. (19-mm) blades are .032 in. (0.81 mm) thick, and 1-in.-wide (25-mm) blades are .035 in. (0.89 mm) thick. Since thick blades are stronger than thin blades, the thickest blade possible should be used for sawing tough material.

table 36.1 Job requirement chart

To Increase	Faster Tool Velocity (More Teeth per min)	Slower Tool Velocity (Fewer Teeth per min)	Finer Pitch Band Tool (More Teeth and Smaller Gullets)	Coarser Pitch Band Tool (Fewer Teeth and Larger Gullets)	Slower Feeding Rate (Decrease Chip Load)	Faster Feeding Rate (Increase Chip Load)	Medium Feeding Rate	Claw Tooth (Positive Rake Angle)	Precision and Buttress (0° Rake Angle)
Try One or More of the Following									
Cutting rate	✓			✓		✓		✓	
Tool life		✓	✓				✓	✓	
Finish	✓		✓		✓				✓
Accuracy	✓				✓				

▶▶ Job Requirements

The bandsaw operator should be familiar with the various types of blades and able to select the one that will do the job to the specified finish and accuracy at the lowest cost. Table 36.1 provides a guide for efficient cutting.

▶▶ Blade Length

Metal-cutting saw bands are usually packaged in coils about 100 to 150 ft (30 to 45 m) in length. The length required to fit each machine is cut from the coil and its two ends are then welded together to form a continuous band.

To calculate the length required for a two-pulley bandsaw, take twice the center distance (CD) between each pulley and add to it the circumference of one pulley (PC). The result is the total length of the saw band (Fig. 36-7).

EXAMPLE

Calculate the length of a saw blade for a bandsaw which has:

a. Two 24-in. diameter pulleys and a center-to-center distance of 48 in.

b. Two 600-mm diameter pulleys and a center-to-center distance of 1200 mm.

Solution

a. Blade length = 2(CD) + PC

$$= 2(48) + (24 \times 3.1416)$$
$$= 96 + 75.4$$
$$= 171.4 \text{ in.}$$

b. Blade length = 2(CD) + PC

$$= 2(1200) + (600 \times 3.1416)$$
$$= 2400 + 1885$$
$$= 4285 \text{ mm}$$

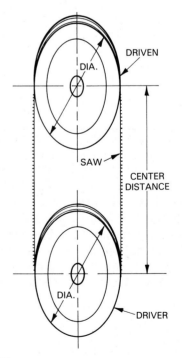

Figure 36-7 Dimensions required to calculate the length of a saw blade.

1. List six advantages of the contour bandsaw.

Bandsaw Parts

2. How is the blade speed adjusted on a contour bandsaw?

3. What supports and guides the blade to keep it from twisting?

Saw Types

4. Why are the following types of saw blades used?
   a. Precision
   b. Claw
   c. Buttress

5. What general rule applies when selecting the pitch of a saw blade?

6. Describe the following blade sets:
   a. Raker   b. Wave   c. Straight

Job Requirements

7. How can the following sawing factors be increased?
   a. Tool life
   b. Accuracy

Blade Length

8. Calculate the length of saw band required for the following contour bandsaws:
   a. Two 30-in. diameter pulleys with a center-to-center distance of 50 in.
   b. Two 750-mm diameter pulleys with a center-to-center distance of 1250 mm.

UNIT 37

Contour Bandsaw Operations

OBJECTIVES

After completing this unit, you will be able to:

1 Set up the machine and saw external sections to within .030 in. (0.8 mm) of layout lines

2 Saw internal sections to within .030 in. (0.8 mm) of layout lines

3 Set up a contour bandsaw to file to a layout

4 Know the purpose of special cutting tools used on a contour bandsaw

The contour bandsaw provides a machinist with the ability to cut material close to the form required quickly, while at the same time removing large sections that can be used for other jobs. The versatility of a bandsaw can be increased by using various attachments and cutting tools. Operations such as sawing, filing, polishing, grinding, and friction and high-speed sawing are all possible on a bandsaw with the proper attachments and cutting tools.

▶▶ Sawing External Sections

With the proper machine setups and attachments, a wide variety of operations can be performed on a contour bandsaw. The most common operation is that of sawing external sections. To perform an operation quickly and accurately, a person must be able to select, weld, and mount the correct saw to suit the size and type of work material.

TO MOUNT A SAW BAND

1. Select and mount the proper saw inserts in the upper and lower saw guides using the proper gage for the thickness of the blade used (Fig. 37-1). Allow .001- to .002-in. (0.02- to 0.5-mm) clearance to ensure that the blade will not bind.

2. Lower the upper saw band carrier wheel to ensure that the blade will slide over the wheels when installed.

3. Mount the blade over the upper and lower wheels with the teeth pointing down toward the table.

4. Adjust the upper wheel until some tension is registered on the blade-tension gage.

5. Set the gearshift lever in the neutral position and turn the upper saw band wheel by hand to make sure that the saw band rides on the center of the crown. If the saw is not tracking properly, the upper wheel must be tilted until the saw band rides on the center of the crown. When the blade is tracking properly, it should be very close to but not touching the backup bearings when the saw is not cutting.

6. Re-engage the gearshift lever and close the doors on the upper and lower carrier housings.

7. Replace the filler plate in the table.

8. Lower the upper saw guide as close as possible to the work to ensure a straight and accurate cut.

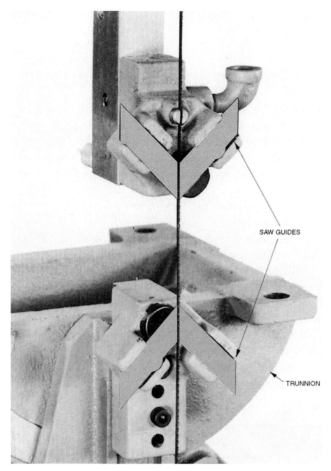

Figure 37-1 Upper and lower saw guides of bandsaw with table removed. Note table support, or trunnion. *(DoAll Company)*

SAW GUIDES

TRUNNION

Figure 37-2 The job selector dial indicates the correct saw speed for the material being cut. *(DoAll Company)*

9. Start the machine and adjust the saw band to the proper tension. This is indicated on the band-tension gage and chart.

TO CUT AN EXTERNAL SECTION

1. Study the print and check the layout on the workpiece for accuracy.

2. Check the job selector (Fig. 37-2) and determine the proper blade to use for the job. Consider the material, its thickness, the type of cut (straight or curved) required, and the finish desired.

Note: When making curved cuts, use as wide a blade as possible.

3. Mount the proper saw blade and guides on the machine.

4. Place the work on the table and lower the upper saw guide until it just clears the work by .250 in. (6 mm). Clamp the guide in place.

Note: It is good practice to drill a hole at every point where a sharp turn must be made to allow the workpiece to be turned easily.

5. Start the machine and make sure that the band is tracking properly.

6. Consult the job selector dial and set the proper speed.

7. Place the material against the work-holding jaw or against a block of wood (Fig. 37-3 on p. 288).

8. Carefully bring the workpiece up to the blade and start the cut.

Note: When the workpiece is to be finished by some other machining operation, the cut should be made about .030 in. (0.8 mm) outside the layout line.

9. Carefully feed the workpiece into the saw. *Do not use too much force. Keep your fingers clear of the moving blade.*

10. Saw to the layout lines.

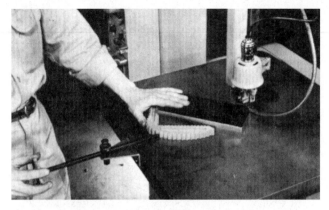

Figure 37-3 The work-holding jaw is used to feed the workpiece into the saw. *(DoAll Company)*

▶▶ Accuracy and Finish

As with any other machine, the accuracy and finish produced on a bandsaw depend on the operator's ability to set up and operate the machine properly. Table 37.1 indicates some of the most common sawing problems encountered and their possible causes.

▶▶ Sawing Internal Sections

The bandsaw is well suited for removing internal sections of a workpiece. A starting hole must be drilled through the section to be removed to allow the saw blade to be inserted and welded. It is also good practice to drill a hole at every point where a sharp turn must be made to allow the workpiece to be turned easily. The saw cut should be made close to the layout line, leaving enough material for the finishing operation.

The butt welder (Fig. 37-4) adds greatly to the versatility of the bandsaw. It permits convenient welding of the blade for the removal of internal sections. Blades can be cut from coil stock and welded into a continuous band; broken blades may be welded and used again. Welders on vertical bandsaws are resistance-type welders that fuse the ends of the blade when the correct current is applied and the blade is held properly.

In addition to knowing how to select the proper blade for the job, the operator must also be able to weld

table 37.1   Sawing problems and possible causes

Problem	Too Heavy Feed	Too Light Feed	Improper Blade Tracking	Improper Blade Tension	Saw Guides Too Far Apart	Incorrect Saw Speed	Pitch Too Coarse	Pitch Too Fine	Wrong Type of Blade	Blade Dull on One Side	Blade Dull on Both Sides	Machine Too Light
Blade wanders	X		X	X	X					X		
Blade not cutting		X		X	X	X		X	X		X	
Blade dulls quickly	X					X			X			
Poor finish	X			X	X	X	X		X	X	X	X
Severe diagonal waviness	X		X	X	X	X						X
Saw teeth chipping	X						X		X			
Saw teeth clogging						X		X	X			

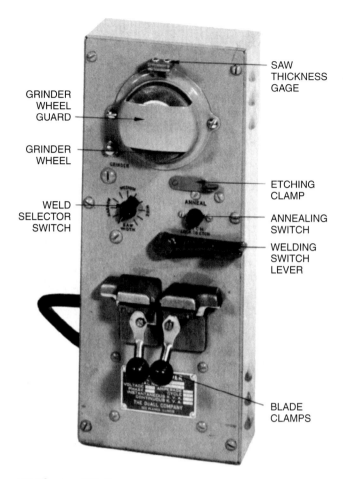

GRINDER WHEEL GUARD

GRINDER WHEEL

WELD SELECTOR SWITCH

SAW THICKNESS GAGE

ETCHING CLAMP

ANNEALING SWITCH

WELDING SWITCH LEVER

BLADE CLAMPS

■ **Figure 37-4** Parts of a butt welder. *(DoAll Company)*

the blade. Improper welds will cause the blade to break and require rewelding, which is costly and time-consuming. Unless the weld is as strong as the band itself, it is not a good weld. The butted ends of the welded blade must not overlap in width, set, or pitch of the teeth.

TO WELD A BANDSAW BLADE

1. Select the proper blade for the job by checking the chart on the machine.

2. Determine the length of the blade required.

   The blade length for a two-wheel machine is determined by adding twice the center-to-center distance of the wheels plus the circumference of one wheel. If the upper wheel is extended the full distance, deduct 1 in. (25 mm) to allow for the stretch of the blade.

3. Place the blade in the cutoff shear, and cut it to the required length.

   Make sure that the blade is held straight and against the blade-squaring bar on the shear.

4. If the ends of the blade are straight but not square, hold them firmly with the teeth reversed, as in Fig. 37-5, and grind both ends in one operation.

   If the ends are not square after grinding, it will not matter. When the blade ends are placed together for welding, they will match perfectly.

5. To get the proper tooth spacing after the blade has been welded, grind off some teeth at the ends of the blade to the depth of the gullet prior to welding. This step is necessary, since the welding operations on a DoAll welder consume about .250 in. (6 mm) of the blade length. Other types of welders may consume varying blade lengths during welding. The number of teeth that are ground off will depend on the pitch of the blade. Fig. 37-6 (on p. 290) illustrates the amount to grind off each type of blade from 4 to 10 pitch.

6. Clean the welder jaws and position the inserts for the blade size and pitch.

7. Adjust the jaw pressure by turning the selector handle for the width of the blade being welded.

8. Clamp the blade as shown in Fig. 37-7 (on p. 290). Make sure that the blade is against the aligning surface on the back of the jaws and centered between the two jaws. If the ends of the blades do not butt against each other for the full width, they should be removed and reground.

9. Depress the welding switch or lever and hold it until the weld has cooled.

STOP **SAFETY PRECAUTIONS** Stand to one side and wear safety glasses to avoid injury from the welding flash.

10. Release the movable jaw clamp and then release the welding lever.

GRIND HERE

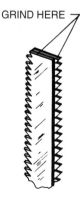

■ **Figure 37-5** Method of grinding blade ends prior to welding. *(DoAll Company)*

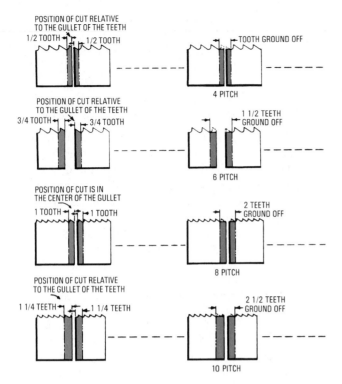

Figure 37-6 Amounts to grind off various blades to achieve proper tooth spacing. *(DoAll Company)*

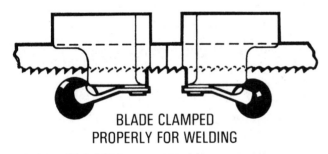

BLADE CLAMPED PROPERLY FOR WELDING

■ **Figure 37-7** Blade clamped properly for welding.

11. Remove the blade and check the weld for the following points.

    a. The flash material should be uniform on both sides.

    b. The spacing of the teeth should be uniform.

    c. The weld should be in the center of the gullet.

    d. The back of the blade should be straight.

Note: If the blade does not meet all of these requirements, it should be broken, prepared again, and rewelded.

12. Move the reset lever to the anneal position and clamp the blade, with the weld in the center of the jaws and the teeth to the rear and clearing the jaws.

13. Set the selector switch at the proper setting for annealing the blade. Push in the anneal switch button and jog it intermittently until the band reaches a dull red color.

> ⚠ **CAUTION:** Do not permit the blade to become too hot at this time because it will air-harden on cooling. As the weld starts to cool, jog the anneal switch occasionally to permit the blade to cool slowly.

14. Remove the blade and grind off the welding flash. Grind the weld to the same thickness as the blade, being careful not to grind the teeth. Continually check the thickness of the blade in the thickness gage located on the welder. When properly ground, it should just slide through the thickness gage (Fig. 37-8).

15. After the blade has been ground to the proper thickness, it is advisable to anneal it to a blue color.

TO SAW INTERNAL SECTIONS

1. Drill a hole slightly larger than the width of the saw blade near the edge of the section to be cut.

■ **Figure 37-8** A thickness gage is used to check the thickness of the blade at the weld. *(DoAll Company)*

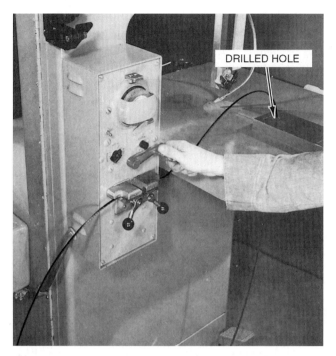

DRILLED HOLE

■ Figure 37-9 Welding a blade for sawing internal sections. *(DoAll Company)*

■ Figure 37-10 The internal section can be used to make a die punch. *(DoAll Company)*

Note: It is good practice to drill a hole at every point where a sharp turn must be made to allow the workpiece to be turned easily.

2. Cut the saw blade and thread it through one of the drilled holes in the workpiece (Fig. 37-9).

3. Weld the blade and then grind off the weld bead to fit the saw thickness gage.

4. Anneal the weld section to remove the brittleness and prevent the blade from breaking.

5. Mount the saw band on the upper and lower pulleys and apply the proper tension for the size of the blade.

6. Insert the table filler plate.

7. Set the machine to the proper speed for the type and thickness of material being cut.

8. Cut out the internal section, staying within 1/32 in. (0.8 mm) of the layout line.

9. Remove the saw band from the pulleys.

10. Cut the blade at the weld point on the cutoff shear.

11. Remove the workpiece and the blade.

TRIMMING AND BLANKING DIES

It is possible to use this internal cutting technique to make short-run trimming and blanking dies. By this process, the internal section, or slug, becomes the punch, while the external material forms the die (Fig. 37-10). When doing work of this type, the table must be tilted to provide the proper clearance for the die.

▶▶ Friction Sawing

Friction sawing (Fig. 37-11 on p. 292) is the fastest means of sawing ferrous metals up to 1 in. (25 mm) in thickness. In this process, the metal is fed into a bandsaw traveling at a high velocity [up to 15,000 sf/min (4572 m/min)]. The tremendous heat generated by friction brings the metal immediately ahead of the saw teeth to a plastic state, and the teeth easily remove the softened metal. Since the thermal conductivity of steel is very low, the depth to which the metal is softened is only about .002 in. (0.5 mm).

The temperature of the saw blade remains quite low, since each tooth is only momentarily in contact with the metal and has time to cool as it travels around the carrier

■ Figure 37-11 Friction sawing uses a high saw-band speed. *(DoAll Company)*

greater strength. They may be obtained in widths of 1/4 (.250), 3/4 (.750), and 1 in. (13, 19, and 25 mm) and in 10 and 14 pitch with raker set only.

The teeth on a standard saw blade traveling at the high velocity required for friction sawing would dull very quickly; therefore, the teeth on friction saws are not sharpened. Since dull teeth create more friction than sharp teeth, they are more efficient for friction sawing.

The procedure for setting up a machine for friction sawing is basically the same as for conventional sawing.

▶ High-Speed Sawing

High-speed bandsawing is performed at speeds ranging from 2000 to 6000 sf/min (609 to 1827 m/min). It is merely a standard sawing procedure performed at higher than standard speeds on nonferrous metals, such as aluminum, brass, bronze, magnesium, and zinc, and other materials such as wood, plastic, and rubber.

The same machine setups and procedures apply as for conventional sawing. In high-speed sawing, the chips must be removed rapidly. Consequently, buttress or claw-tooth blades are the most efficient for this operation.

wheels before contacting the metal again. This method of sawing leaves a small burr that is easily removed from the workpiece.

Friction sawing is used on hardened ferrous alloys, armor plate, and parts having a thin wall or section that would be damaged by other sawing methods. Friction sawing is particularly suited to the cutting of stainless-steel alloys, which are difficult to cut because they work-harden so quickly. It is not suitable for sawing most cast irons because grains break off before the metals soften. Aluminum and brass cannot be cut by friction sawing because the high thermal conductivity of the metal does not let the heat concentrate just ahead of the blade. These metals also melt almost immediately after softening and will weld to the blade and clog the teeth. Most thermoplastics react in the same manner as the nonferrous metals.

Friction-sawing machines resemble the standard vertical band machines but are of heavier construction in the frame, bearings, spindles, and guides to withstand the vibrations created by the high speeds required. The saw band is almost completely covered to protect the operator from the shower of sparks produced by friction sawing.

Friction-sawing bands are made of standard carbon-alloy steel but are thicker than standard blades to provide

▶ Band Filing

When a better finish than that produced by conventional sawing is required on the edge of the workpiece, it may be produced by means of a *band file*.

The band file consists of a steel band onto which are riveted a number of short, interlocking file segments (Fig. 37-12 on p. 293). The ends of the band are locked together to form a continuous loop. Band files may be obtained in flat, oval, and half-round cross sections; in bastard and medium cuts; and in widths of (.250, .375, and .500) in. (6, 9.5, and 13 mm). Special file guides and a file-adapter table-filler plate are used for band filing.

TO SET UP FOR BAND FILING

1. Select the proper band file for the job by consulting the job selector dial. Consideration must be given to the material being filed and the shape, size, and cut of the file.
2. Set the gearshift lever into neutral position.
3. Remove the saw guides and filler plate.
4. Mount the proper file guide and backup support.
5. Lock the ends of the file blade together.

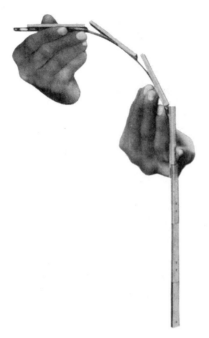

Figure 37-12 Segments of a band file.

Figure 37-13 Mounting a file band on a contour bandsaw. *(DoAll Company)*

6. Mount the file band, with the teeth pointing in the proper direction, on the bandsaw carrier wheels (Fig. 37-13).

Note: On some makes of machines, or when filing internal sections, it may be necessary to mount the file band and then join the ends.

7. Lightly add tension to the band.

8. Check the alignment and tracking of the file band.

9. Lower the upper guidepost to the proper work thickness. The distance should not exceed 2 in. (50 mm) for a .250-in. (6-mm) file band and 4 in. (100 mm) for a .375- or .500-in. (9.5- or 13-mm) band.

10. Mount the proper table-filler plate.

11. Set the gearshift lever into low gear and start the machine.

12. Adjust the band file to the proper tension.

13. Adjust the machine to the proper speed for the material being filed.

TO FILE ON A CONTOUR BANDSAW

1. Consult the job selector (Fig. 37-2) and set the machine to the proper speed. The best filing speeds are between 50 and 100 ft/min (15 and 30 m/min).

2. Apply light work pressure to the file band. It not only gives a better finish but prevents the teeth from becoming clogged.

3. Keep moving the work back and forth against the file to prevent filing grooves in the work.

4. Use a file card to keep the file clean. Loaded files cause bumpy filing and scratches in the work.

SAFETY PRECAUTIONS Be sure to stop the machine before attempting to clean the file.

▶▶ Additional Band Tools

Although the saw blade and band file are the most commonly used, several other band tools make this machine particularly versatile.

> *Knife-edge blades* (Fig. 37-14 on p. 294) are available with knife, wavy, and scalloped edges

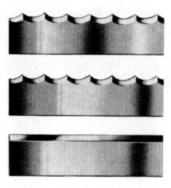

Knife-edge

■ **Figure 37-14** Knife-edge blades. *(DoAll Company)*

■ **Figure 37-15** Spiral-edge blades. *(DoAll Company)*

■ **Figure 37-16** Intricate forms may be cut with spiral-edge blades. Note the special guides. *(DoAll Company)*

and are used for cutting soft, fibrous materials such as cloth, cardboard, cork, and rubber. Scalloped-edge blades are particularly suited for cutting thin corrugated aluminum. Special roller guides must be used with knife-edge blades.

> *Spiral-edge blades* (Fig. 37-15) are round and have a continuous helical cutting edge around the circumference. This construction provides a cutting edge of 360° and permits the machining of intricate contours and patterns (Fig. 37-16) without the workpiece having to be turned. Spiral-edge blades require special guides.

Spiral-edge blades are made in two types: the spring-tempered blade, which is used for plastics and wood, and the all-hard blade, which is used for light metals. These blades are manufactured in diameters of .020, .040, .050, and .074 in. (0.5, 1, 1.3, and 1.9 mm). Special guides (Fig. 37-16) are used on the machine with this blade. When spiral-edge blades are welded, sheet copper is used to protect the cutting edges from the welder jaws.

Line-grinding bands (Fig. 37-17) have an abrasive (either aluminum oxide or silicon carbide) bonded to the thin edge of the steel band. These bands are used to cut

■ **Figure 37-17** Line-grinding bands are used to cut hardened metals. *(DoAll Company)*

hardened steel alloys and other materials, such as brick, marble, and glass, which could not be cut by bandsawing.

This type of machining requires a high speed—3000 to 5000 sf/min (914 to 1524 m/min)—and the use of coolant because of the heat generated. A diamond dressing stick is used to dress line-grinding bands.

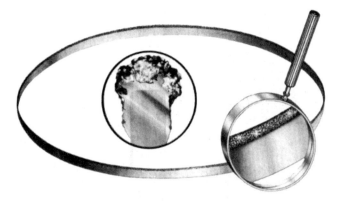

Figure 37-18 Diamond-edge blades are used to cut superhard materials. *(DoAll Company)*

Figure 37-19 Cutting a heat exchanger by electro-band machining. *(DoAll Company)*

Diamond-edge blades (Fig. 37-18) are used to cut superhard space-age materials, as well as ceramics, glass, silicon, and granite. This type of blade has diamond particles fused to the edges of the saw teeth. These blades operate at about 3000 sf/min (914 m/min) and generally require coolant for most efficient operation. Although diamond-edge bands are very expensive, they will outlast 200 steel blades when cutting asbestos-cement pipe.

Polishing bands are used to remove burrs and provide a good finish to surfaces which have been sawed or filed. They may also be used for sharpening carbide toolbits. A polishing band is a continuous loop of 1-in.-wide (25-mm) abrasive cloth manufactured to a specific length to fit the machine. They are available in several grain sizes in both aluminum-oxide and silicon-carbide abrasive.

The polishing band is mounted in the same manner as a saw band. The special polishing guide uses the same backup support as the file bands. A special polishing-band center plate is used during band polishing. Most polishing bands are marked with an arrow on the back to indicate the direction of travel.

Electro-band machining (Fig. 37-19) is the latest development in band machining and is used to machine materials such as thin-wall tubing, stainless steel, aluminum, and titanium honeycombing.

By this process, a low-voltage, high-amperage current is fed into the saw blade. The workpiece is connected to the opposite pole of the circuit. When the work comes close to the fast-moving band [6000 sf/min (1827 m/min)], a continuous electric spark passes from the knife edge of the saw to the work. This arc acts on the material and disintegrates it. The blade does not touch the work. Coolant is flooded onto the cutting area to prevent damage to the material by the heat created. Power feed must be used in this type of machining operations.

Figure 37-20 A work-holding jaw is used to guide a workpiece into a saw blade. *(DoAll Company)*

▶▶ Bandsaw Attachments

Several standard attachments can be obtained to increase the scope of the band machine. Some of the most common are:

> The *work-holding jaw* (Fig. 37-20) is a device used by the operator to hold and guide the work into the saw. Because it is usually connected to the weight-type power feed, the operator merely steers the

work with the work-holding jaw and does not have to apply any feed force.

> The *disk-cutting attachment* (Fig. 37-21) permits the cutting of accurate circles from approximately 2.5 to 30 in. (64 to 760 mm) in diameter.

> The *cutoff and mitering attachment* is used to support the work when square or angular cuts are made.

> The *ripping fence* provides a means for cutting long sections of flat bar stock or plate into narrow parallel sections.

When special work-holding devices are required, they are generally made in the shop to suit the specific job. These devices, called *fixtures,* are usually attached to the machine table.

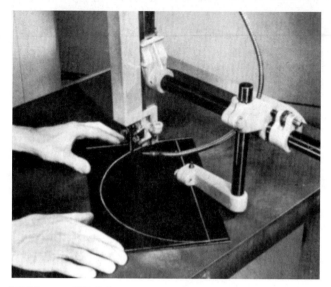

■ **Figure 37-21** The disk-cutting attachment is used to cut circular shapes. *(DoAll Company)*

unit 37 review questions

1. Name five operations that can be performed on a contour bandsaw.

Sawing External Sections

2. Describe how to make the blade track properly.

3. Why must the upper saw guide be close to the top of the work?

4. What factors should be considered when selecting a blade for a contour cut?

5. How should the work be fed into a revolving saw?

Accuracy and Finish

6. How can the following bandsawing problems be corrected?

   a. Blade wander   b. Poor finish

Sawing Internal Sections

7. How is the length of a blade calculated when the upper carrier wheel is extended the full distance?

8. Make a sketch to show how the ends of the bandsaw are positioned for grinding prior to welding.

9. To obtain the correct tooth spacing, how many teeth should be ground off each end of a 10-pitch blade? A 14-pitch blade?

10. What are the characteristics of a good weld on a bandsaw blade?

11. List the main steps required to remove an internal section from a piece of work.

Friction and High-Speed Sawing

12. Describe the principle of friction sawing.

13. Why is friction sawing particularly suited to cutting stainless steel?

14. How does high-speed sawing differ from friction sawing?

Band Filing

15. List the main steps required to set up the machine for band filing.

16. What is the recommended filing speed?

17. How can the band file be kept clean?

Additional Band Tools

18. Name five materials that can be cut satisfactorily with knife-edge bands.

19. Name two types of spiral-edge bands and state two uses for each.

20. Describe a line-grinding band and state its purpose.

21. Name five materials that may be cut with a diamond-edge band.

22. Describe the principle of electro-band machining.

Bandsaw Attachments

23. Describe and state the purpose of:
    a. A work-holding jaw
    b. A disk-cutting attachment
    c. A cutoff and mitering attachment

Section 10

(© Content Mine/Robertstock.com)

DRILLING MACHINES

Probably one of the first mechanical devices developed prehistorically was a drill to bore holes in various materials. A bow string was wrapped around an arrow and then rapidly sawed back and forth. This process not only produced fire but also wore a hole in the wood. The principle of a rotating tool making a hole in various materials is the one on which all drill presses operate. From this basic principle evolved the drill press—one of the most common and useful machines in industry for producing, forming, and finishing holes. Modern drilling machines are made in several types and sizes, ranging from sensitive hand-fed drill presses to the sophisticated computer-controlled production machine of the computer age.

UNIT 38

Drill Presses

OBJECTIVES

After completing this unit, you will be able to:

1 Identify six standard operations that may be performed on a drill press

2 Identify four types of drill presses and their purposes

3 Name and state the purpose of the main parts of an upright and a radial drill

The *drilling machine,* or *drill press,* is essential in any metalworking shop. Basically, a drilling machine consists of a spindle (which turns the drill and which can be advanced into the work, either automatically or by hand) and a work table (which holds the workpiece rigidly in position as the hole is drilled). A drilling machine is used primarily to produce holes in metal; however, operations such as tapping, reaming, counterboring, countersinking, boring, and spot-facing can also be performed.

▶▶ Standard Operations

Drilling machines may be used for performing a variety of operations besides drilling a round hole. A few of the more standard operations, cutting tools, and work setups will be briefly discussed.

> *Drilling* (Fig. 38-1a) may be defined as the operation of producing a hole by removing metal from a solid mass using a cutting tool called a *twist drill.*

> *Countersinking* (Fig. 38-1b) is the operation of producing a tapered or cone-shaped enlargement to the end of a hole.

> *Reaming* (Fig. 38-1c) is the operation of sizing and producing a smooth, round hole from a previously drilled or bored hole with the use of a cutting tool having several cutting edges.

> *Boring* (Fig. 38-1d) is the operation of truing and enlarging a hole by means of a single-point cutting tool, which is usually held in a boring bar.

> *Spot-facing* (Fig. 38-1e) is the operation of smoothing and squaring the surface around a hole to provide a seat for the head of a cap screw or a nut. A boring bar, with a pilot section on the end to fit into the existing hole, is generally fitted with a double-edged cutting tool. The pilot on the bar provides rigidity for the cutting tool and keeps it concentric with the hole. For the spot-facing operation, the work being machined should be securely clamped and the machine set to approximately one-quarter of the drilling speed.

> *Tapping* (Fig. 38-1f) is the operation of cutting internal threads in a hole with a cutting tool called a tap. Special machine or gun taps are used with a tapping attachment when this operation is performed by power in a machine.

> *Counterboring* (Fig. 38-1g) is the operation of enlarging the top of a previously drilled hole to a given depth to provide a square shoulder for the head of a bolt or capscrew.

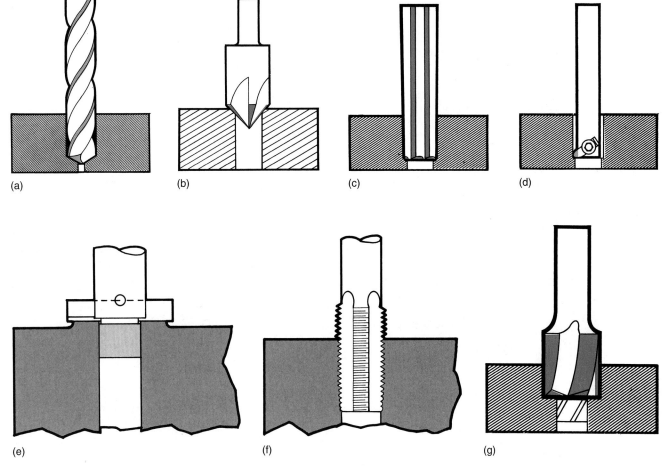

■ Figure 38-1 (a) Drilling produces a straight hole; (b) countersinking produces a cone-shaped hole; (c) reaming is used to finish a hole; (d) boring is used to true and enlarge a hole; (e) spot-facing produces a square surface; (f) tapping produces internal threads; (g) counterboring produces square shoulders in a hole. *(Kelmar Associates)*

▶▶ Principal Types of Drilling Machines

A wide variety of drill presses are available, ranging from the simple sensitive drill to highly complex automatic and numerically controlled machines. The size of a drill press may be designated in different ways by different companies. Some companies state the size as the distance from the center of the spindle to the column of the machine. Others specify the size by the diameter of the largest circular piece that can be drilled in the center.

SENSITIVE DRILL PRESSES

The simplest type of drilling machine is the sensitive drill press (Fig. 38-2 on p. 302). This type of machine has only a hand feed mechanism, which enables the operator to "feel" how the drill is cutting and to control the downfeed pressure accordingly. Sensitive drill presses are generally light, high-speed machines and are manufactured in bench and floor models.

Sensitive Drill Press Parts

Although drill presses are manufactured in a wide variety of types and sizes, all drilling machines contain certain basic parts. The main parts on the bench and floor models are the *base, column, table,* and *drilling head* (Fig. 38-2). The floor model is larger and has a longer column than the bench type.

Base

The base, usually made of cast iron, provides stability for the machine and rigid mounting for the column. The base is usually provided with holes so that it may be bolted to a table or bench. The slots or ribs in the base allow the workholding device or the workpiece to be fastened to the base.

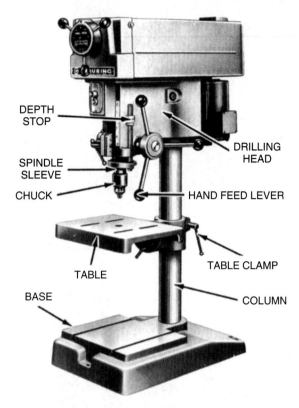

Figure 38-2 A bench-type sensitive drill press. *(Clausing Industrial, Inc.)*

Column

The column is an accurate cylindrical post that fits into the base. The table, which is fitted to the column, may be adjusted to any point between the base and head. The drill press head is mounted near the top of the column.

Table

The table, either round or rectangular in shape, is used to support the workpiece to be machined. The table, whose surface is at 90° to the column, may be raised, lowered, and swiveled around the column. On some models, it is possible to tilt the table in either direction for drilling holes on an angle. Slots are provided in most tables to allow jigs, fixtures, or large workpieces to be clamped directly to the table.

Drilling Head

The head, mounted close to the top of the column, contains the mechanism used to revolve the cutting tool and advance it into the workpiece. The *spindle,* which is a round shaft that holds and drives the cutting tool, is housed in the *spindle sleeve,* or *quill.* The spindle sleeve does not revolve but slides up and down inside the head to provide a downfeed for the cutting tool. The end of the spindle may have a tapered hole to hold taper shank tools or may be threaded or tapered for attaching a *drill chuck* (Fig. 38-2).

The *hand feed lever* is used to control the vertical movement of the spindle sleeve and the cutting tool. A *depth stop* attached to the spindle sleeve can be set to control the depth that a cutting tool enters the workpiece.

UPRIGHT DRILLING MACHINE

The standard *upright drilling machine* (Fig. 38-3) is similar to the sensitive-type drill except that it is larger and heavier. The basic differences are as follows:

1. It is equipped with a gearbox to provide a greater variety of speeds.

Figure 38-3 A standard upright drilling machine with a square, or production-type, table. *(Do All Company)*

2. The spindle may be advanced by three methods:

   a. Manually with a hand lever
   b. Manually with a handwheel on most models
   c. Automatically by the feed mechanism

3. The table may be raised or lowered by means of a table-raising mechanism.

4. Some models are equipped with a reservoir in the base for the storage of coolant.

For high-speed production work, a number of spindles may be mounted on a single head. This *multispindle head* (Fig. 38-4) may incorporate 20 or more spindles on a single head driven by a drilling machine spindle. Several heads equipped with multispindle attachments may be combined and controlled automatically to drill as many as 100 holes in a single operation. This type of automated drilling is used, for example, by the automotive industry for the drilling of engine blocks.

RADIAL DRILLING MACHINE

The *radial drilling machine* (Fig. 38-5 on p. 304), sometimes called a radial-arm drill, has been developed primarily for the handling of larger workpieces than is possible

■ **Figure 38-4** A multispindle drilling head.

on upright machines. The advantages of this machine over the upright drill are:

1. Larger and heavier work may be machined.

2. The drilling head may be easily raised or lowered to accommodate various heights of work.

3. The drilling head may be moved rapidly to any desired location while the workpiece remains clamped in one position; this feature permits greater production.

4. The machine has more power; thus, larger cutting tools can be used.

5. On universal models, the head may be swiveled so that holes can be drilled on an angle.

Radial Drilling Machine Parts

Base

The base is made of heavy, box-type, ribbed cast iron or of welded steel. The base, or pedestal, is used to bolt the machine to the floor and to provide a coolant reservoir. Large work may be clamped directly to the base for drill press operations. For convenience in drilling smaller work, a *table* may be bolted to the base.

Column

The column is an upright cylindrical member fitted to the base, which supports the radial arm at right angles.

Radial Arm

The arm is attached to the column and may be raised and lowered by means of a *power-driven elevating screw.* The arm may also be swung about the column and may be clamped in any desired position. It also supports the drive motor and drilling head.

Drilling Head

The drilling head is mounted on the arm and may be moved along the length of the arm by means of a *traverse handwheel.* The head may be clamped at any position along the arm. The head houses the change gears and controls for the spindle speeds and feeds. The drill spindle may be raised or lowered manually by means of the spindle-feed handles. When the spindle-feed handles are brought together, automatic feed is provided to the drill spindle.

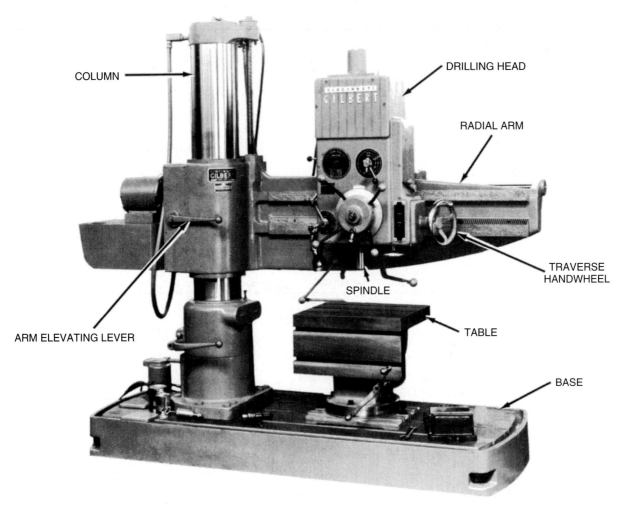

COLUMN

DRILLING HEAD

RADIAL ARM

TRAVERSE
HANDWHEEL

SPINDLE

ARM ELEVATING LEVER

TABLE

BASE

■ **Figure 38-5** A radial drilling machine permits large parts to be drilled. *(Cincinnati Gilbert)*

unit 38 review questions

Standard Operations

1. Define drilling, boring, and reaming.

2. What is the difference between spot-facing and counterboring?

Types of Drilling Machines

3. State two methods by which the size of a drill press may be determined.

4. Compare a sensitive drill press with an upright drilling machine.

5. Describe and state the purpose of the following parts of a sensitive drill press:

   a. Base   d. Table
   b. Column   e. Depth stop
   c. Drilling head

6. What are the advantages of a radial drilling machine?

7. Compare the construction of the radial drilling machine to that of a standard upright drill press.

UNIT 39

Drilling Machine Accessories

OBJECTIVES

After completing this unit, you will be able to:

1 Identify and use three types of drill-holding devices

2 Identify and use work-holding devices for drilling

3 Set up and clamp work properly for drilling

The versatility of the drill press is greatly increased by the various accessories available. Drill press accessories fall into two categories:

1. *Tool-holding devices,* which are used to hold or drive the cutting tool

2. *Work-holding devices,* which are used to clamp or hold the workpiece

▶▶ Tool-Holding Devices

The drill press spindle provides a means of holding and driving the cutting tool. It may have a tapered hole to accommodate taper shank tools, or its end may be tapered or threaded for mounting a drill chuck. Although there are a variety of tool-holding devices and accessories, the most common found in a machine shop are drill chucks, drill sleeves, and drill sockets.

DRILL CHUCKS

Drill chucks are the most common devices used on a drill press for holding straight-shank cutting tools. Most drill chucks contain three jaws that move simultaneously when the outer sleeve is turned, or on some types of chucks when the outer collar is raised. The three jaws hold the straight shank of a cutting tool securely and cause it to run accurately. There are two common types of drill chucks: the key type and the keyless type.

CHUCKS

Straight-shank drills are held in a drill chuck. This chuck may be mounted on the drill press spindle by means of a taper (Fig. 39-1a) or a thread (Fig. 39-1b). Chucks used in larger drill presses are usually held in the spindle by means of a self-holding taper.

There are four types of drill chucks, all of which will retain their accuracy for years when properly used:

> *Key-type drill chucks* (Fig. 39-2 on p. 306) are the most common. They have three jaws that move in or out simultaneously when the outer sleeve is turned. The drill is placed in the chuck and the outer sleeve turned by hand until the jaws are snug on the drill shank. The sleeve is then tightened with the key, causing the drill to be held securely and accurately.

> *Keyless drill chucks* are used more in production work, since the chuck may be loosened or tightened by hand without a key.

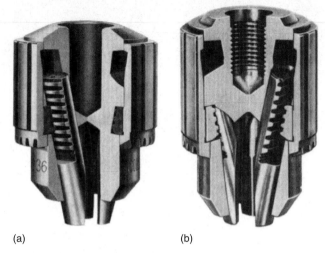

(a)   (b)

■ **Figure 39-1** Methods of mounting drill chucks:
(a) taper-mounted; (b) thread-mounted. *((a) Kennametal Inc.; (b) Jacobs® Chuck Manufacturing Company)*

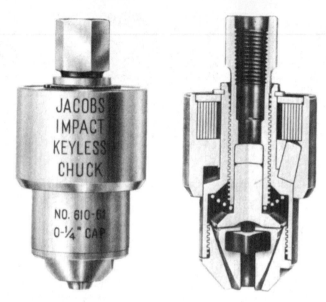

■ **Figure 39-4** A Jacobs impact keyless chuck.
(Jacobs® Chuck Manufacturing Company)

■ **Figure 39-2** A key-type drill chuck.   ■ **Figure 39-3** A keyless drill chuck.

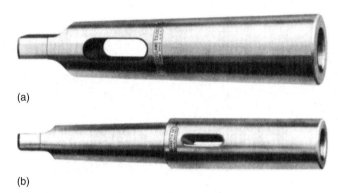

(a)

(b)

■ **Figure 39-5** (a) A drill sleeve; (b) a drill socket.
(Kennametal Inc.)

> The *precision keyless chuck* (Fig. 39-3) is designed to hold smaller drills accurately. The drill is changed by turning the outer knurled sleeve.

> The *Jacobs impact keyless chuck* (Fig. 39-4) will hold small or large drills (within the range of the chuck) securely and accurately by means of Rubber-Flex collets. The drill is gripped or released quickly and easily by means of a built-in impact device in the chuck.

DRILL SLEEVES AND SOCKETS

The size of the tapered hole in the drill press spindle is generally in proportion to the size of the machine: the larger the machine, the larger the spindle hole. The size of the tapered shank on cutting tools is also manufactured in proportion to the size of the tool. *Drill sleeves* (Fig. 39-5a) are used to adapt the cutting-tool shank to the machine spindle if the taper on the cutting tool is smaller than the tapered hole in the spindle.

A *drill socket* (Fig. 39-5b) is used when the hole in the spindle of the drill press is too small for the taper shank of the drill. The drill is first mounted in the socket and then the socket is inserted into the drill press spindle. Drill sockets may also be used as extension sockets to provide extra length.

A flat, wedge-shaped tool called a *drill drift* is used to remove tapered-shank drills or accessories from the drill press spindle. When using a drill drift (Fig. 39-6), always place the rounded edge up so that this edge will bear against the round slot in the spindle. A hammer is used to tap the drill drift and loosen the tapered drill shank in the spindle. A board or piece of masonite should be used to protect the table in case the drill drops when it is being removed.

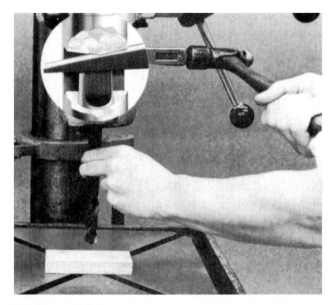

Figure 39-6 Removing a tapered-shank drill with a drill drift. Note the board that prevents damage to the table if the drill drops. *(Kelmar Associates)*

Figure 39-7 A drill vise should be clamped to the table when larger holes are drilled. *(Rockwell Tool Division)*

Figure 39-8 An angle vise permits holes to be drilled at an angle in workpieces.

Figure 39-9 A contour vise adjusts automatically to the shape of the workpiece. *(Volstro Manufacturing Company)*

▸▸ Work-Holding Devices

All workpieces must be fastened securely before cutting operations are performed on a drilling machine. If the work moves or springs during drilling, the drill usually breaks. Serious accidents can be caused by work becoming loose and spinning around during a drilling operation. Some of the common work-holding devices used on drill presses are as follows:

> A *drill vise* (Fig. 39-7) may be used to hold round, rectangular, square, or odd-shaped pieces for any operation that can be performed on a drill press. It is a good practice to clamp or bolt the vise to the drill table when drilling holes over 3⁄8 in. (9.5 mm) in diameter or provide a table stop to prevent the vise from swinging during the drilling operation.

> An *angle vise* (Fig. 39-8) has an angular adjustment on its base to allow the operator to drill holes at an angle without tilting the drill press table.

> A *contour vise* (Fig. 39-9) has special movable jaws consisting of several free-moving, interlocking segments that automatically adjust to the shape of odd-shaped workpieces when the vise is tightened. These vises are valuable when one operation must be performed on many similar odd-shaped workpieces.

> *V-blocks* (Fig. 39-10a), made out of cast iron or hardened steel, are generally used in pairs to support round work for drilling. A U-shaped strap may be used to fasten the work in a V-block, or work may be held with a T-bolt and a strap clamp (Fig. 39-10b).

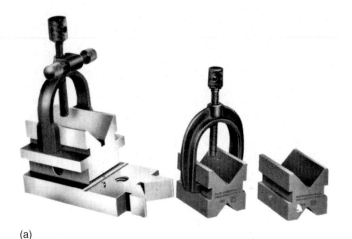

(a)

■ Figure 39-11 Step blocks support the end of the clamp used to hold the workpiece. *(Northwestern Tools, Inc.)*

(b)

■ Figure 39-10 (a) V-blocks are used to hold round workpieces for drilling; (b) a workpiece clamped in V-blocks. *(The L. S. Starrett Company)*

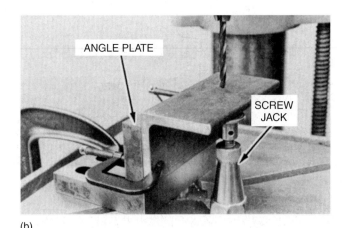

(a)

> *Step blocks* (Fig. 39-11) are used to provide support for the outer end of the strap clamps when work is fastened for drilling machine operations. They are made in various sizes and steps to accommodate different work heights.

> The *angle plate* (Fig. 39-12a) is an L-shaped piece of cast iron or hardened steel machined to an accurate 90° angle. It is made in a variety of sizes and has slots or holes (clearance and tapped) that provide a means for fastening work for drilling. The angle plate may be bolted or clamped to the table (Fig. 39-12b).

> *Drill jigs* (Fig. 39-13) are used in production for drilling holes in a large number of identical parts. They eliminate the need for laying out a hole location, avoid incorrectly located holes, and allow holes to be drilled quickly and accurately.

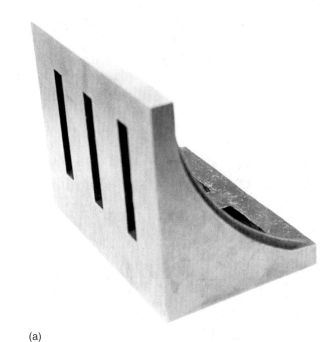

ANGLE PLATE

SCREW JACK

(b)

■ Figure 39-12 (a) Workpieces may be clamped to an angle plate for drilling; (b) a workpiece clamped to an angle plate and supported by a screw jack. *(Kelmar Associates)*

Figure 39-13 Identical parts are drilled quickly and accurately in a drill jig.

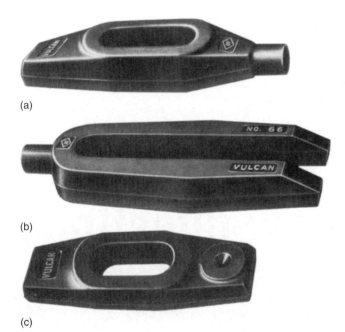

(a)

(b)

(c)

Figure 39-14 Types of clamps used to fasten work on a drill press: (a) finger clamp; (b) U-clamp; (c) straight clamp. *(Bahco North America)*

> *Clamps* or *straps* (Fig. 39-14), used to fasten work to the drill table or an angle plate for drilling, are made in various sizes. They are usually supported at the end by a step block and bolted to the table by a T-bolt that fits into the table T-slot. It is a good practice to place the T-bolt in the clamp or strap as close to

the work as possible so that pressure will be exerted on the workpiece. Modifications of these clamps are the *double-finger* and *gooseneck* clamps.

▶▶ Clamping Stresses

Whenever work is clamped for any machining operation, stresses are created. It is important that these clamping stresses should not be great enough to cause springing or distortion of the workpiece.

When work is to be held for drilling, reaming, or any machining operation, it is important that the workpiece be held securely. The clamps, bolts, and step blocks should be properly located and the work clamped firmly enough to prevent movement, yet not enough to cause the workpiece to spring or distort. It is important that the clamping pressures be applied to the work, not to the packing or step block.

Figure 39-15a (on p. 310) illustrates the correct clamping procedure, the main pressure being applied to the workpiece.

Note: The step block is slightly higher than the workpiece and the bolt is close to the work.

Figure 39-15b shows a piece of work incorrectly clamped. The bolt is located close to the step block, which is slightly lower than the workpiece. The main clamping pressure with this type of setup is applied to the step block, not the workpiece.

CLAMPING HINTS

The following suggestions are made for clamping work so that good clamping pressure will be obtained and work distortion avoided:

1. Always place the bolt as close as possible to the workpiece (Fig. 39-16 on p. 310).

2. Have the packing or step block slightly higher than the work surface being clamped.

3. Insert a piece of paper between the machine table and the workpiece to prevent the work from shifting during the machining process.

4. Place a metal shim between the clamp and workpiece to spread the clamping force over a wider area.

5. Use a sub-base or liner under a rough casting to prevent damage to the machine table.

6. Parts that do not lie flat on a machine table should be shimmed to prevent the work from rocking, thus presenting distortion when the work is clamped.

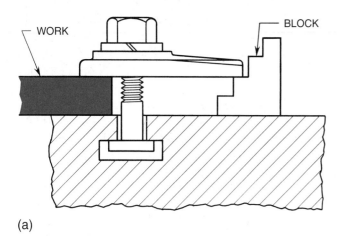

(a)

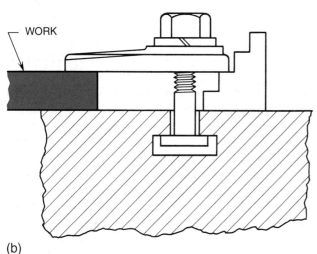

(b)

■ **Figure 39-15** (a) Work is clamped correctly when the bolt is close to the work; (b) work is clamped incorrectly because the clamping pressure is on the step block. *(Kelmar Associates)*

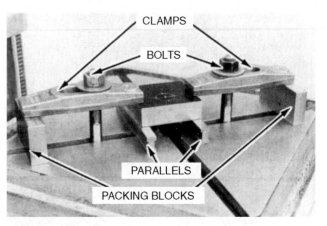

■ **Figure 39-16** The bolts and the packing blocks placed correctly for maximum clamping pressure. *(Kelmar Associates)*

unit 39 review questions

Tool-Holding Devices

1. What is the purpose of a drill chuck?
2. How may chucks be secured to the spindle in:
   a. Small drill presses?   b. Large drill presses?
3. Name three types of drill chucks.
4. What is the purpose of a:
   a. Drill sleeve?   b. Drill socket?
5. How is a tapered-shank drill removed from a drill press spindle?

Work-Holding Devices

6. Name three types of drill vises and state the purpose of each.

7. What is the purpose of:
   a. V-blocks?   b. Step blocks?
8. Describe an angle plate and state its purpose.
9. What is the advantage of a drill jig?

Clamping Stresses

10. Why is it important that work be clamped properly for any machining operation?
11. Explain the procedure for clamping a workpiece properly.
12. List four important clamping hints.

UNIT 40

Twist Drills

OBJECTIVES

After completing this unit, you will be able to:

1 Identify the parts of a twist drill

2 Identify four systems of drill sizes and know where each is used

3 Grind the proper angles and clearances on a twist drill

*T*wist drills are end-cutting tools used to produce holes in most types of material. On standard drills, two helical grooves, or flutes, are cut lengthwise around the body of the drill. They provide cutting edges and space for the cuttings to escape during the drilling process. Since drills are among the most efficient cutting tools, it is necessary to know the main parts, how to sharpen the cutting edges, and how to calculate the correct speeds and feeds for drilling various metals to use them most efficiently and prolong their life.

▶▶ Twist Drill Parts

Most twist drills used in machine shop work today are made of high-speed steel. High-speed steel drills have replaced carbon-steel drills, since they can be operated at double the cutting speed and the cutting edge lasts longer. High-speed steel drills are always stamped with the letters "H.S." or "H.S.S." Since the introduction of *carbide-tipped drills,* speeds for production drilling have increased up to 300 % over high-speed steel drills. Carbide drills have made it possible to drill certain materials that would not be possible with high-speed steels.

A drill (Fig. 40-1 on p. 312) may be divided into three main parts: *shank, body,* and *point.*

SHANK

Generally drills up to ½ in. or 13 mm in diameter have straight shanks, while those over this diameter usually have tapered shanks. *Straight-shank drills* (Fig. 40-2a on

p. 312) are held in a drill chuck; *tapered-shank drills* (Fig. 40-2b) fit into the internal taper of the drill press spindle. A tang is provided on the end of tapered-shank drills to prevent the drill from slipping while it is cutting and to allow the drill to be removed from the spindle or socket without the shank being damaged.

BODY

The *body* is the portion of the drill between the shank and the point. It consists of a number of parts important to the efficiency of the cutting action.

1. The *flutes* are two or more helical grooves cut around the body of the drill. They form the cutting edges, admit cutting fluid, and allow the chips to escape from the hole.

2. The *margin* is the narrow, raised section on the body of the drill. It is immediately next to the flutes and extends along the entire length of

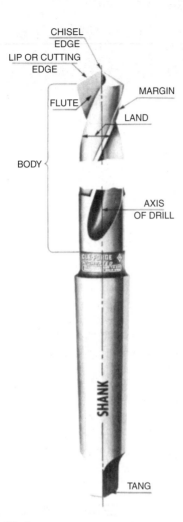

CHISEL EDGE

LIP OR CUTTING EDGE

MARGIN

FLUTE

LAND

BODY

AXIS OF DRILL

SHANK

TANG

■ **Figure 40-1** The main parts of a twist drill. *(Kennametal Inc.)*

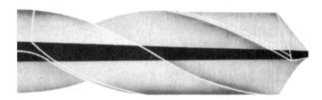

(a)

(b)

■ **Figure 40-2** Types of drill shanks: (a) straight; (b) tapered. *(Regal Deloit)*

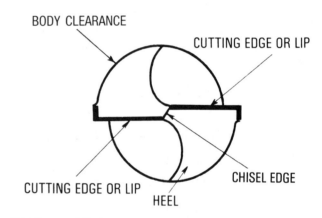

■ **Figure 40-3** The web is the tapered metal column that separates the flutes. *(Kennametal Inc.)*

BODY CLEARANCE

CUTTING EDGE OR LIP

CUTTING EDGE OR LIP

HEEL

CHISEL EDGE

■ **Figure 40-4** The point of a twist drill.

the flutes. Its purpose is to provide a full size to the drill body and cutting edges.

3. The *body clearance* is the undercut portion of the body between the margin and the flutes. It is made smaller to reduce friction between the drill and the hole during the drilling operation.

4. The *web* (Fig. 40-3) is the thin partition in the center of the drill that extends the full length of the flutes. This part forms the chisel edge at the cutting end of the drill (Fig. 40-4). The web gradually increases in thickness toward the shank to give the drill strength.

POINT

The *point* of a twist drill (Fig. 40-4) consists of the chisel edge, lips, lip clearance, and heel. The *chisel edge* is the chisel-shaped portion of the drill point. The *lips* (cutting edges) are formed by the intersection of the flutes. The lips must be of equal length and have the same angle so

that the drill will run true and will not cut a hole larger than the size of the drill.

The *lip clearance* is the relief ground on the point of the drill extending from the cutting lips back to the *heel* (Fig. 40-5). The average lip clearance is from 8° to 12°, depending on the hardness or softness of the material to be drilled.

▶▶ Drill Point Characteristics

Efficient drilling of the wide variety of materials used by industry requires a great variety of drill points. The most important factors determining the size of the drilled hole are the characteristics of the drill point.

A drill is generally considered a roughing tool capable of removing metal quickly. It is not expected to finish

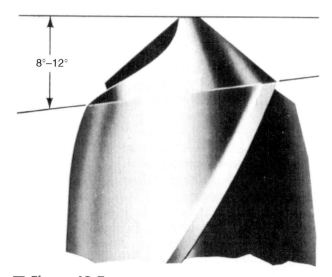

■ Figure 40-5 The lip clearance angle of the cutting edge should be 8° to 12°. *(Kennametal Inc.)*

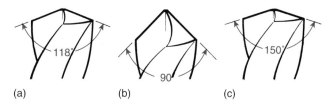

■ Figure 40-6 (a) The drill point angle of 118° is suitable for most general work; (b) a drill point angle of 60° to 90° is used for soft material; (c) a drill point angle of 135° to 150° is best for hard materials. *(Kelmar Associates)*

a hole to the accuracy possible with a reamer. However, a drill can often be made to cut more accurately and efficiently by proper drill point grinding. The use of various point angles and lip clearances, in conjunction with the thinning of the drill web, will:

1. Control the size, quality, and straightness of the drilled hole
2. Control the size, shape, and formation of the chip
3. Control the chip flow up the flutes
4. Increase the strength of the drill's cutting edges
5. Reduce the rate of wear at the cutting edges
6. Reduce the amount of drilling pressure required
7. Control the amount of burr produced during drilling
8. Reduce the amount of heat generated
9. Permit the use of various speeds and feeds for more efficient drilling

DRILL POINT ANGLES AND CLEARANCES

Drill point angles and clearances are varied to suit the wide variety of material that must be drilled. Three general drill points are commonly used to drill various materials; however, there may be variations of these to suit various drilling conditions.

The *conventional point* (118°), shown in Fig. 40-6a, is the most commonly used drill point and gives satisfactory results for most general-purpose drilling. The 118° point angle should be ground with 8° to 12° lip clearance for best results. Too much lip clearance weakens the cutting edge and causes the drill to chip and break easily. Too

little lip clearance results in the use of heavy drilling pressure; this pressure causes the cutting edges to wear quickly because of the excessive heat generated and places undue strain on the drill and equipment.

The *long angle point (60° to 90°)*, shown in Fig. 40-6b, is commonly used on low helix drills for the drilling of nonferrous metals, soft cast irons, plastics, fibers, and wood. The lip clearance on long angle point drills is generally from 12° to 15°. On standard drills, a flat may be ground on the face of the lips to prevent the drill from drawing itself into the soft material.

The *flat angle point (135° to 150°)*, shown in Fig. 40-6c, is generally used to drill hard and tough materials. The lip clearance on flat angle point drills is generally only 6° to 8° to provide as much support as possible for the cutting edges. The shorter cutting edge tends to reduce the friction and heat generated during drilling.

▶▶ Systems of Drill Sizes

Drill sizes are designated under four systems: fractional, number, letter, and millimeter (metric) sizes.

> The *fractional* size drills range from 1/64 to 4 in., varying in steps of 1/64 in. from one size to the next.

> The *number* size drills range from #1, measuring .228 in., to #97, which measures .0059 in.

> The *letter* size drills range from A to Z. Letter-A drill is the smallest in the set (.234 in.) and Z is the largest (.413 in.).

> *Millimeter (metric)* drills are produced in a wide variety of sizes. Miniature metric drills range from 0.04 to 0.09 mm, in steps of 0.01 mm. Straight-shank standard metric drills are available in sizes from 0.5 to 20 mm. Taper-shank metric drills are manufactured in sizes from 8 up to 80 mm.

Drill sizes may be checked by using a drill gage (Fig. 40-7 on p. 314). These gages are available in fractional, letter, number, and millimeter sizes. The size of a drill may also be checked by measuring the drill, *over the margins,* with a micrometer (Fig. 40-8 on p. 314).

Twist Drills   **313**

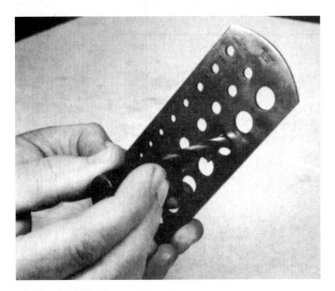

Figure 40-7 Checking a drill for size using a drill gage. *(Kelmar Associates)*

Figure 40-8 Checking the size of a twist drill using a micrometer. *(Kelmar Associates)*

▶▶ Types of Drills

A variety of twist-drill styles are manufactured to suit specific drilling operations, types and sizes of material, high production rates, and special applications. The design of drills may vary in the number and width of the flutes, the amount of helix or rake angle of the flutes, or the shape of the land or margin. In addition, the flutes may be straight or helical, and the helix may be a right-

hand or left-hand helix. Only commonly used drills are covered in this text. For special-purpose drills, consult a manufacturer's catalog.

Twist drills are manufactured from carbon tool steel, high-speed steel, and cemented carbides. *Carbon-steel drills* are generally used in hobby shops and are not recommended for machine shop work, since the cutting edges tend to wear down quickly. *High-speed steel drills* are commonly used in machine shop work because they can be operated at twice the speed of carbon-steel drills and the cutting edges can withstand more heat and wear. *Cemented-carbide drills,* which can be operated much faster (up to three times faster) than high-speed steel drills, are used to drill hard materials. Cemented-carbide drills have found wide use in industry because they can be operated at high speeds, the cutting edges do not wear rapidly, and they are capable of withstanding higher heat.

The most commonly used drill is the *general-purpose drill,* which has two helical flutes (Fig. 40-2). This drill is designed to perform well on a wide variety of materials, equipment, and job conditions. The general-purpose drill can be made to suit different conditions and materials by varying the point angle and the speeds and feeds used. Straight-shank drills are commonly known as *general-purpose jobbers length drills.*

The *low-helix drill* was developed primarily to drill brass and thin materials. This type of drill is used to drill shallow holes in some aluminum and magnesium alloys. Because of its design, the low-helix drill can remove the large volume of chips formed by high rates of penetration when it is used on machines such as turret lathes and screw machines.

High-helix drills (Fig. 40-9) are designed for drilling deep holes in aluminum, copper, die-cast material, and other metals where the chips have a tendency to jam in a hole. The high helix angle (35° to 45°) and the wider flutes of these drills assist in clearing chips from the hole.

A *core drill* (Fig. 40-10), designed with three or four flutes, is used primarily to enlarge cored, drilled, or punched holes. This drill has advantages over the two-fluted drills in productivity and finish. In some cases, a core drill may be

Figure 40-9 A high-helix drill. *(Kennametal Inc.)*

Figure 40-10 A core drill. *(Kennametal Inc.)*

Figure 40-11 An oil hole drill.

Figure 40-13 Deep hole, or gun, drill.
(Kennametal Inc.)

Figure 40-12 A straight-fluted drill.

Figure 40-14 A spade drill. (DoAll Company)

Figure 40-15 A hard-steel drill. (DoAll Company)

Figure 40-16 A step drill. (Regal Deloit)

used in place of a reamer for finishing a hole. Core drills are produced in sizes from ¼ to 3 in. (6 to 76 mm) in diameter.

Oil hole drills (Fig. 40-11) have one or two oil holes running from the shank to the cutting point through which compressed air, oil, or cutting fluid can be forced when deep holes are being drilled. These drills are generally used on turret lathes and screw machines. The cutting fluid flowing through the oil holes cools the drill's cutting edges and flushes the chips out of the hole.

Straight-fluted drills (Fig. 40-12) are recommended for drilling operations on soft materials such as brass, bronze, copper, and various types of plastic. The straight flute prevents the drill from drawing itself into the material (digging in) while cutting.

If a straight-fluted drill is not available, a conventional drill can be modified by grinding a small flat approximately .060 in. (1.5 mm) wide on the face of both cutting edges of the drill.

Note: The flat must be ground parallel to the axis of the drill.

Deep hole, or *gun, drills* (Fig. 40-13) are used for producing holes from approximately ⅜ to 3 in. (9.5 to 76 mm) in diameter and as deep as 20 ft (6 m). The most common gun drill consists of a round, tubular stem, on the end of which is fastened a flat, two-fluted drilling insert. Cutting fluid is forced through the center of the stem to flush the chips from the hole. When the drilling insert becomes dull, it can be replaced quickly by loosening one screw, which holds it to the tubular stem.

Spade drills are similar to gun drills because the cutting end is a flat blade with two cutting lips. Spade drills are usually clamped in a holder (Fig. 40-14) and are easily replaced or sharpened. Spade drills are available in a wide range of sizes from very small microdrills to drills up to 12 in. in diameter. Some of the smaller spade drills have replaceable carbide inserts.

A unique drill is the *hard-steel drill* (Fig. 40-15) used for drilling hardened steel. These drills are made from a heat-resistant alloy. As the drill is brought into contact with the workpiece, the fluted, triangular point softens the metal by friction and then removes the softened metal ahead of it, in chip form.

Step drills (Fig. 40-16) are used to drill and countersink or drill and counterbore different sizes of holes in one operation. The drill may have two or more diameters ground on it. Each size or step may be separated by a square or angular shoulder, depending on the purpose of the hole. For example, holes to be tapped should have a slight chamfer at the top of the hole to ease the starting of the tap, protect the thread, and leave the tapped hole free of burrs caused by the tap.

A *saw-type hole cutter* (Fig. 40-17 on p. 316) is a cylindrical-diameter cutter with a twist drill in the center to provide a guide for the cutting teeth on the hole cutter. This type of cutter is made in various diameters and is generally used for drilling holes in thin materials. It is especially valuable for drilling holes in pipe and sheet metal because little burr is produced and the cutter does not have a tendency to jam as it is breaking through.

■ Figure 40-17 A saw-type hole cutter. *(The L. S. Starrett Co.)*

▸▸ Drilling Facts and Problems

The most common drill problems encountered are illustrated in Fig. 40-18. Study these various problems to ensure that the amount of drill breakage, regrinding, and downtime is kept to a minimum.

DRILL GRINDING

The cutting efficiency of a drill is determined by the characteristics and condition of the point of the drill. Most new drills are provided with a general-purpose point (118° point angle and an 8° to 12° lip clearance). As a drill is used, the cutting edges may wear and become chipped, or the drill may break. Drills are generally resharpened by hand. However, small drill point grinders or drill-sharpening attachments are inexpensive, readily available, and provide more consistent quality than hand grinding.

To ensure that a drill will perform properly, examine the drill point carefully before mounting the drill in the drill press. A properly ground drill should have the following characteristics:

> The length of both cutting lips should be the same. Lips of unequal length will force the drill point off center, causing one lip to do more cutting than the other and producing an oversize hole (Fig. 40-19a).

> The angle of both lips should be the same. If the angles are unequal, the drill will cut an oversize hole because one lip will do more cutting than the other (Fig. 40-19b).

> The lips should be free from nicks or wear.

> There should be no sign of wear on the margin.

If the drill does not meet all of these requirements, it should be resharpened. If the drill is not resharpened, it will give poor service, will produce inaccurate holes, and may break because of excessive drilling strain.

EXCESSIVE SPEED WILL CAUSE WEAR AT OUTER CORNERS OF DRILL; THIS PERMITS FEWER REGRINDS OF DRILL DUE TO AMOUNT OF STOCK TO BE REMOVED IN RECONDITIONING. DISCOLORATION IS WARNING SIGN OF EXCESS SPEED.

AMOUNT OF GRINDING NECESSARY TO REPOINT. THE USE OF MACHINE-POINT GRINDING IS RECOMMENDED.

EXCESSIVE FEED SETS UP ABNORMAL END THRUST, WHICH CAUSES BREAKDOWN OF CHISEL POINT AND CUTTING LIPS. FAILURE INDUCED BY THIS CAUSE WILL BE BROKEN OR SPLIT DRILL.

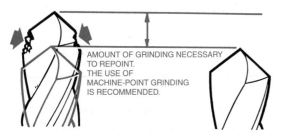

EXCESSIVE CLEARANCE RESULTS IN LACK OF SUPPORT BEHIND CUTTING EDGE WITH QUICK DULLING AND POOR TOOL LIFE, DESPITE INITIAL FREE CUTTING ACTION. CLEARANCE ANGLE BEHIND CUTTING LIP FOR GENERAL PURPOSES IS 8 TO 12°.

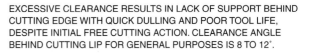

LIP CLEARANCE ANGLE

A
B

A — CUTTING LIP LINE
B — HEEL LINE

INSUFFICIENT CLEARANCE CAUSES THE DRILL TO RUB BEHIND THE CUTTING EDGE. IT WILL MAKE THE DRILL WORK HARD, GENERATE HEAT, AND INCREASE END THRUST. THIS RESULTS IN POOR HOLES AND DRILL BREAKAGE.

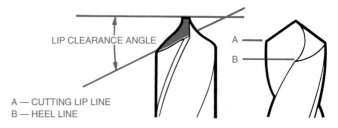

LIP CLEARANCE ANGLE

A
B

A — CUTTING LIP LINE
B — HEEL LINE

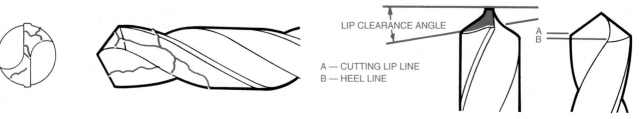

■ Figure 40-18 Facts about drills and drilling. *(Greenfield Industries, Inc.)*

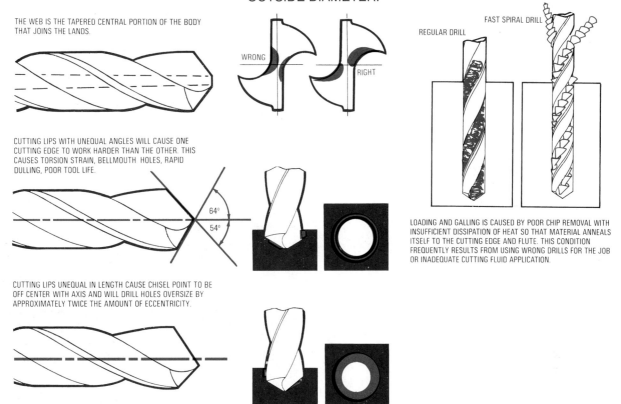

IMPROPER WEB THINNING IS THE RESULT OF TAKING MORE STOCK FROM ONE CUTTING EDGE THAN FROM THE OTHER, THEREBY DESTROYING THE CONCENTRICITY OF THE WEB AND OUTSIDE DIAMETER.

THE WEB IS THE TAPERED CENTRAL PORTION OF THE BODY THAT JOINS THE LANDS.

WRONG   RIGHT

FAST SPIRAL DRILL
REGULAR DRILL

CUTTING LIPS WITH UNEQUAL ANGLES WILL CAUSE ONE CUTTING EDGE TO WORK HARDER THAN THE OTHER. THIS CAUSES TORSION STRAIN, BELLMOUTH HOLES, RAPID DULLING, POOR TOOL LIFE.

64°
54°

LOADING AND GALLING IS CAUSED BY POOR CHIP REMOVAL WITH INSUFFICIENT DISSIPATION OF HEAT SO THAT MATERIAL ANNEALS ITSELF TO THE CUTTING EDGE AND FLUTE. THIS CONDITION FREQUENTLY RESULTS FROM USING WRONG DRILLS FOR THE JOB OR INADEQUATE CUTTING FLUID APPLICATION.

CUTTING LIPS UNEQUAL IN LENGTH CAUSE CHISEL POINT TO BE OFF CENTER WITH AXIS AND WILL DRILL HOLES OVERSIZE BY APPROXIMATELY TWICE THE AMOUNT OF ECCENTRICITY.

■ **Figure 40-18** Facts about drills and drilling *(continued)*. *(Greenfield Industries, Inc.)*

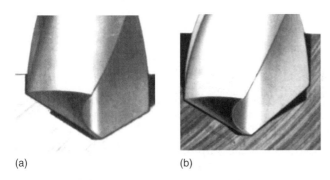

(a)   (b)

■ **Figure 40-19** (a) An incorrect point with lips of unequal lengths; (b) lips with unequal angles produce oversize holes. *(Kennametal Inc.)*

While a drill is being used, there will be signs to indicate that the drill is not cutting properly and should be resharpened. If the drill is not sharpened at the first sign of dullness, it will require extra power to force the slightly dulled drill into the work. This causes more heat to be generated at the cutting lips and results in a faster rate of wear.

When any of the following conditions arise while a drill is in use, it should be examined and reground:

> The color and shape of the chips change.
> More drilling pressure is required to force the drill into the work.
> The drill turns blue because of the excessive heat generated while drilling.
> The top of the hole is out of round.
> A poor finish is produced in the hole.
> The drill chatters when it contacts the metal.
> The drill squeals and may jam in the hole.
> An excessive burr is left around the drilled hole.

CAUSES OF DRILL FAILURE

Drills should not be allowed to become so dull that they cannot cut. Overdulling of any metal-cutting tool generally results in poor production rates, inaccurate work, and

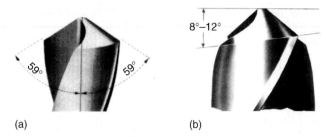

(a)   (b)

Figure 40-20 (a) The point angle for a general-purpose drill is 118°; (b) the lip clearance is 8° to 12°. *(Kennametal Inc.)*

the shortening of the tool life. Premature dulling of a drill may be caused by any one of a number of factors:

> The drill speed may be too high for the hardness of the material being cut.

> The feed may be too heavy and overload the cutting lips.

> The feed may be too light and cause the lips to scrape rather than cut.

> There may be hard spots or scale on the work surface.

> The work or drill may not be supported properly, resulting in springing and chatter.

> The drill point may be incorrect for the material being drilled.

> The finish on the lips may be poor.

To Grind a Drill

A general-purpose drill has an included point angle of 118° and a lip clearance of from 8° to 12° (Fig. 40-20a and b). Follow these steps to grind a drill:

1. Be sure to wear approved safety glasses.

2. Check the grinding wheel and dress it, if necessary, to sharpen and/or straighten the wheel face.

3. Adjust the grinder tool rest so that it is within .060 in. (1.5 mm) of the wheel face.

4. Examine the drill point and the margins for wear. If there is any wear on the margins, it will be necessary to grind the point of the drill back until all margin wear has been removed.

5. Hold the drill near the point with one hand, and with the other hand hold the shank of the drill slightly lower than the point (Fig. 40-21).

6. Move the drill so that it is approximately 59° to the face of the grinding wheel (Fig. 40-22).

Note: A line scribed on the toolrest at 59° to the wheelface will assist in holding the drill at the proper angle.

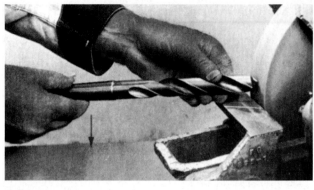

Figure 40-21 To provide lip clearance, lower the shank of the drill before grinding. *(Kelmar Associates)*

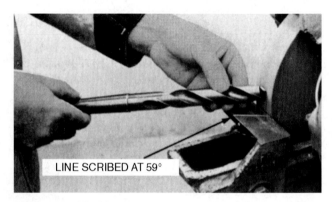

LINE SCRIBED AT 59°

Figure 40-22 Hold the drill at 59° to the face of the grinding wheel. *(Kelmar Associates)*

7. Hold the lip or cutting edge of the drill parallel to the grinder toolrest.

8. Bring the lip of the drill against the grinding wheel and slowly lower the drill shank.

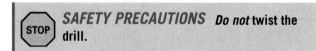

SAFETY PRECAUTIONS *Do not* twist the drill.

9. Remove the drill from the wheel without moving the position of the body or hands, rotate the drill one-half turn, and grind the other cutting edge.

10. Check the angle of the drill point and length of the lips with a drill point gage (Fig. 40-23).

11. Repeat operations 6 to 10 until the cutting edges are sharp and the lands are free from wear nicks.

WEB THINNING

Most drills are manufactured with webs that gradually increase in thickness toward the shank to give the drill

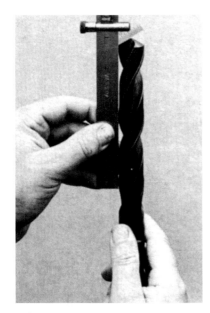

Figure 40-23 Check the drill point angle with a drill point gage. *(Kelmar Associates)*

Figure 40-24 Note the difference in web thickness between the drill cross section near the point (left) and the shank (right). *(Kennametal Inc.)*

strength. As the drill becomes shorter, the web becomes thicker (Fig. 40-24) and more pressure is required to cut. This increase in pressure results in more heat, which shortens the drill life. To reduce the amount of drilling pressure and resultant heat, the web of a drill is generally thinned. Webs can be thinned on a special web-thinning grinder, on a tool and cutter grinder, or freehand on a conventional grinder. It is important when thinning a web to grind equal amounts off each edge; otherwise, the drill point will be off center (Fig. 40-19a).

unit 40 review questions

Twist Drills

1. Name three materials used to manufacture drills and state the advantage of any two.

2. Define the *body, web,* and *point* of a twist drill.

3. State the purpose or purposes of each of the following parts:

   a. Tang   c. Margin
   b. Flutes   d. Body clearance

Drill Point Characteristics

4. Why are various drill points and clearances used for drilling operations?

5. Describe and state the purpose of the following drill points:

   a. Conventional   b. Long angle   c. Flat angle

6. Why is it necessary to thin the web of a drill?

Drill Sizes

7. List four systems of drill sizes and give the range of each.

8. State the purpose of the:

   a. High-helix drill   d. Straight-fluted drill
   b. Core drill   e. Gun drill
   c. Oil hole drill   f. Hard-steel drill

Drilling Facts and Problems

9. What problems generally result from the use of excessive speed and excessive feed?

10. Discuss excessive lip clearance and insufficient lip clearance.

11. What is the effect of:

    Drills with unequal angles on the cutting lips?
    Drills with cutting lips of unequal length?

Drill Grinding

12. What are the characteristics of a properly ground drill?

13. List the main steps in grinding a general-purpose drill.

UNIT 41

Cutting Speeds and Feeds

OBJECTIVES

After completing this unit, you will be able to:

1 Calculate the revolutions per minute (r/min) for inch and metric size drills

2 Select the feed to be used for various operations

3 Calculate the revolutions per minute for the reaming operation

The most important factors that the operator must consider when selecting the proper speeds and feeds are the diameter and material of the cutting tool and the type of material being cut. These factors will determine what speeds and feeds should be used and therefore will affect the amount of time it takes to perform the operation. Production time will be wasted if the speed and/or feed is set too low; the cutting tool will show premature wear if the speed and/or feed is too high. The ideal speed and feed for any job is when a combination of the best production rate and the best tool life is attained.

▸▸ Cutting Speed

The speed of a twist drill is generally referred to as *cutting speed, surface speed,* or *peripheral speed.* It is the distance that a point on the circumference of a drill will travel in 1 min.

A wide range of drills and drill sizes is used to cut various metals; an equally wide range of speeds is required for the drill to cut efficiently. For every job, there is the problem of choosing the drill speed that will result in the best production rates and the least amount of downtime for regrinding the drill. Recommended cutting speeds for drilling various types of materials are shown in Table 41.1. The most economical drilling speed depends on many variables, such as:

1. The type and hardness of the material
2. The diameter and material of the drill

3. The depth of the hole
4. The type and condition of the drill press
5. The efficiency of the cutting fluid employed
6. The accuracy and quality of the hole required
7. The rigidity of the work setup

Although all these factors are important in the selection of economical drilling speeds, the type of work material and the diameter of the drill are the most important.

When reference is made to the speed at which a drill should revolve, the cutting speed of the material in *surface feet per minute (sf/min)* or *meters per minute (m/min)* is implied unless otherwise stated. The number of revolutions of the drill necessary to attain the proper cutting speed for the metal being machined is called the *revolutions per minute (r/min)*. A small drill operating at the same r/min as a larger drill will travel fewer feet per minute; it naturally would cut more efficiently at a higher number of r/min.

REVOLUTIONS PER MINUTE

To determine the correct number of r/min of a drill press spindle for a given size drill, the following should be known:

1. The type of material to be drilled
2. The recommended cutting speed of the material
3. The type of material from which the drill is made

Formula (Inch)

$$r/min = \frac{CS \text{ (feet per minute)} \times 12}{\pi D \text{ (drill circumference in inches)}}$$

where CS = the recommended cutting speed in *feet per minute* for the material being drilled

D = the diameter of the drill being used

Note: The CS will vary depending on the material from which the drill is made.

Since not all machines can be set to the exact calculated speed, π (3.1416) is divided into 12 to arrive at a simplified formula, which is accurate enough for most drilling operations.

$$r/min = \frac{CS \times 4}{D}$$

EXAMPLE

Calculate the r/min required to drill a ½-in. hole in cast iron (CS 80) with a high-speed steel drill.

$$
\begin{aligned}
r/min &= \frac{CS \times 4}{D} \\
&= \frac{80 \times 4}{1/2} \\
&= \frac{320}{1/2} \\
&= 640
\end{aligned}
$$

table 41.1 Speeds for high-speed steel drills*

Drill Size		Steel Casting		Tool Steel		Cast Iron		Machine Steel		Brass and Aluminum	
		\multicolumn{10}{c}{Cutting Speeds in Feet per Minute or Meters per Minute}
in.	mm	40 ft/min	12 m/min	60 ft/min	18 m/min	80 ft/min	24 m/min	100 ft/min	30 m/min	200 ft/min	60 m/min
⅟₁₆	2	2445	1910	3665	2865	4890	3820	6110	4775	12 225	9550
⅛	3	1220	1275	1835	1910	2445	2545	3055	3185	6110	6365
³⁄₁₆	4	815	955	1220	1430	1630	1910	2035	2385	4075	4775
¼	5	610	765	915	1145	1220	1530	1530	1910	3055	3820
⁵⁄₁₆	6	490	635	735	955	980	1275	1220	1590	2445	3180
⅜	7	405	545	610	820	815	1090	1020	1365	2035	2730
⁷⁄₁₆	8	350	475	525	715	700	955	875	1195	1745	2390
½	9	305	425	460	635	610	850	765	1060	1530	2120
⅝	10	245	350	365	520	490	695	610	870	1220	1735
¾	15	205	255	305	380	405	510	510	635	1020	1275
⅞	20	175	190	260	285	350	380	435	475	875	955
1	25	155	150	230	230	305	305	380	380	765	765

*There is no direct relationship between the inch and metric drill sizes.

Formula (Metric)

$$r/min = \frac{CS \ (m)}{\pi D \ (mm)}$$

It is necessary to convert the meters in the numerator to millimeters so that both parts of the equation are in the same unit. To accomplish this, multiply the CS in meters per minute by 1000 to bring it to millimeters per minute.

$$r/min = \frac{CS \times 1000}{\pi D}$$

Not all machines have a variable speed drive and therefore cannot be set to the exact calculated speed. By dividing π (3.1416) into 1000, a simplified formula is derived that is accurate enough for most drilling operations.

$$r/min = \frac{CS \times 320}{D}$$

EXAMPLE

Calculate the r/min required to drill a 15-mm hole in tool steel (CS 18) using a high-speed steel drill.

$$r/min = \frac{CS \times 320}{D}$$
$$= \frac{18 \times 320}{15}$$
$$= \frac{5760}{15}$$
$$= 384$$

▶▶ Feed

Feed is the distance that a drill advances into the work for each revolution. Drill feeds may be expressed in decimals, fractions of an inch, or millimeters. Since the feed rate is a determining factor in the rate of production and the life of the drill, it should be carefully chosen for each job. The rate of feed is generally governed by:

1. The diameter of the drill
2. The material of the workpiece
3. The condition of the drilling machine

A general rule of thumb is that the feed rate increases as the drill size increases. For example, a ¼-in. (6-mm) drill will have a feed of only .002 to .004 in. (0.05 to 0.1 mm), while a 1-in. (25-mm) drill will have a feed of .010 to .025 in. (0.25 to 0.63 mm) per revolution. Too coarse a feed may chip the cutting edges or break the drill. Too light a feed causes a chattering or scraping noise, which quickly dulls the cutting edges of the drill.

The drill feeds listed in Table 41.2 are recommended for general-purpose work. When drilling alloy or hard steels, use a somewhat slower feed. Softer metals such as aluminum, brass, or cast iron can usually be drilled with a faster feed. Whenever steel chips coming from the hole turn blue, stop the machine and examine the drill. Blue chips indicate too much heat at the cutting edge. This heat is caused by either a dull cutting edge or too high a speed.

▶▶ Cutting Fluids

Drilling work at the recommended cutting speeds and feeds causes considerable heat to be generated at the drill point. This heat must be dissipated as quickly as possible; otherwise, it will cause the drill to dull rapidly.

The purpose of a *cutting fluid* is to provide both cooling and lubrication. For a liquid to be most effective in dissipating heat, it must be able to absorb heat rapidly, have a good resistance to evaporation, and have a high thermal conductivity. Unfortunately, oil has poor cooling qualities. Water is the best coolant; however, it is rarely

table 41.2 Drill feeds

Drill Size		Feed per Revolution	
in.	mm	in.	mm
⅛ and smaller	3 and smaller	.001 to .002	0.02 to 0.05
⅛ to ¼	3 to 6	.002 to .004	0.05 to 0.1
¼ to ½	6 to 13	.004 to .007	0.1 to 0.18
½ to 1	13 to 25	.007 to .015	0.18 to 0.38
1 to 1½	25 to 38	.015 to .025	0.38 to 0.63

used by itself because it promotes rust and has no lubricating value. Basically, a good cutting fluid should:

1. Cool the workpiece and tool
2. Reduce friction
3. Improve the cutting action
4. Protect the work against rusting
5. Provide antiweld properties
6. Wash away the chips

See Unit 34, Table 34.1, for the recommended cutting fluids for various metals.

unit 41 review questions

Drilling Speeds

1. Why is it important that a drill be operated at the correct speed?
2. Explain the difference between cutting speed and r/min.
3. What factors determine the most economical drilling speed?
4. Calculate the r/min required to drill the following holes using a high-speed drill:

   a. A ⅜-in. diameter hole in tool steel
   b. A 1-in. diameter hole in aluminum
   c. a 9-mm hole in a steel casting
   d. a 20-mm hole in cast iron

Cutting Fluids

5. What is the purpose of a cutting fluid?
6. Name four important qualities that a cutting fluid should have.

UNIT 42

Drilling Holes

OBJECTIVES

After completing this unit, you will be able to:

1 Measure the size of inch and metric drills

2 Drill the correct size center holes in workpieces

3 Drill small and large holes to an accurate location

Early humans used an arrow wrapped with a bow string to drill holes in bone and wood (Fig. 42-1). Although the technique for drilling holes is much different today, the principle remains the same— pressing a rotating cutting tool into the workpiece. Some of the important factors in drilling are securely fastening the workpiece, observing proper safety precautions, and setting the proper speeds and feeds. The pressure applied to the feed handle and the appearance of the chips are good indicators as to how effective the drilling operation is being performed.

▶▶ Drill Press Safety

Before starting any drill press operation, observe some basic safety precautions. Not only will these precautions ensure your safety, but they also will prevent damage to the machine, cutting tool, and workpiece:

1. *Do not operate any machine before understanding its mechanism and knowing how to stop it quickly. This can prevent serious injury.*

2. Always wear approved safety glasses to protect your eyes.

3. Never attempt to hold the work by hand; a table stop or clamp should be used to prevent the work from spinning (Fig. 42-2).

4. Never set speeds or adjust the work unless the machine is stopped.

5. Keep your head well back from revolving parts of a drill press to prevent your hair from being caught.

■ **Figure 42-1** Primitive peoples used a bow drill to produce holes.

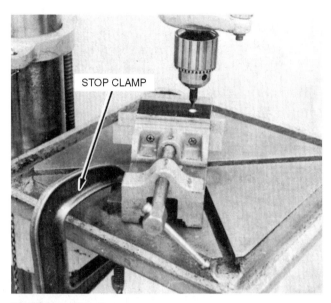

Figure 42-2 A clamp or table stop should be fastened to the left-hand side of the table. *(Kelmar Associates)*

Figure 42-3 Poor housekeeping can lead to accidents. *(Kelmar Associates)*

6. As the drill begins to break through the work, ease up on the drill pressure and allow the drill to break through gradually.

7. Always remove burrs from a drilled hole with a file or deburring tool.

8. *Never* leave a chuck key in the drill chuck.

9. *Never* attempt to grab work that may have caught in the drill. *Stop the machine first.*

10. Always keep the floor around a drill press clean and free of tools, chips, and oil (Fig. 42-3). They can cause serious accidents.

▸▸ Drilling Hints

The following hints should help prevent many problems that could affect the accuracy of the hole and the efficiency of the drilling operation:

1. Treat cutting tools with care; they can be damaged through careless use, handling, and storage.

2. Always examine the condition of the drill point before use and, if necessary, resharpen it. *Do not use dull tools.*

3. Make sure that the drill point angle is correct for the type of material to be drilled.

4. Set the correct revolutions per minute (r/min) for the size of the drill and the workpiece material. Too high a speed quickly dulls a drill, and too low a speed causes a small drill to break.

5. Set up the work so that the drill will not cut into the table, parallels, or drill vise as it breaks through the workpiece.

6. The work should always be clamped securely for the drilling operation. For small-diameter holes, a clamp or stop fastened to the left-hand side of the table will prevent the work from spinning (Fig. 42-2).

7. The end of the workpiece farthest from the hole should be placed on the left-hand side of the table so that, if the work catches, it will not swing toward the operator.

8. Always clean a tapered drill shank, the sleeve, and the machine spindle before inserting a drill.

9. Use the shortest drill length possible and/or hold it short in the chuck to prevent breakage.

10. It is a good practice to start each hole with a center drill. The small point of the center drill will pick up a center-punch mark accurately; the center-drilled hole will provide a guide for the drill to follow.

11. Thin workpieces, such as sheet metal, should be clamped to a hardwood block for drilling. This prevents the work from catching and steadies the drill point as it breaks through the workpiece.

12. The chips from each flute should be the same shape; if the chip turns blue during drilling, check the drill point condition and the speed of the drill press.

13. A drill squeak usually indicates a dull drill. Stop the machine and examine the condition of the drill point; regrind the drill if necessary.

14. When increased pressure must be applied during a drilling operation, the reason is generally a dull drill or a chip caught in the hole between the drill and the work; correct these conditions before proceeding.

MEASURING THE SIZE OF A DRILL

To produce a hole to size, the correct size of drill must be used to drill the hole. It is good practice to always check a drill for size before using it to drill a hole. Drills may be checked for size with a drill gage (Fig. 42-4a) or with a micrometer (Fig. 42-4b). Although the size of most drills is stamped on the drill shank, the micrometer is still the most accurate way of measuring the exact size of a drill. When checking a drill for size with a micrometer, always be sure to take the measurement across the margin of the drill.

▶▶ Lathe Center Holes

Work to be turned between the centers on a lathe must have a hole drilled in each end so that the work may be supported by the lathe centers. A *combination drill and countersink* (Fig. 42-5), more commonly called a *center drill*, is used for this operation.

To ensure an adequate bearing surface for the work on the lathe center, center holes must be drilled to the correct size and depth (Fig. 42-6a).

(a)

(b)

■ Figure 42-4 (a) Checking a drill for size using a drill gage; (b) checking the size of a twist drill using a micrometer. *(Kelmar Associates)*

(a)

(b)

■ Figure 42-5 Two types of center drills: (a) regular-type; (b) bell-type. *(Kennametal Inc.)*

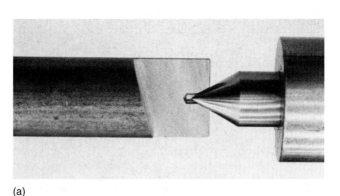

(a)

(b)

(c)

■ Figure 42-6 (a) A center hole drilled to the proper depth; (b) a center hole drilled too shallow; (c) a center hole drilled too deep. *(Kelmar Associates)*

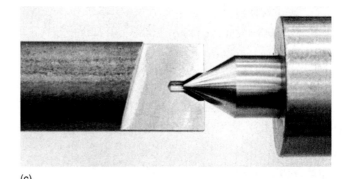

table 42.1 **Center drill sizes**

Size		Work Diameter		Diameter of Countersink	Drill Point Diameter	Body Size
Regular Type	Bell Type	in.	mm	in.	in.	in.
1	11	3/16–5/16	3–8	3/32	3/64	1/8
2	12	3/8–1/2	9.5–12.5	9/64	5/64	3/16
3	13	5/8–3/4	15–20	3/16	7/64	1/4
4	14	1–1½	25–40	15/64	1/8	5/16
5	15	2–3	50–75	21/64	3/16	7/16
6	16	3–4	75–100	3/8	7/32	1/2
7	17	4–5	100–125	15/32	1/4	5/8
8	18	6 and over	150 and over	9/16	5/16	3/4

A center hole that is too shallow is illustrated in Fig. 42-6b. This results in poor support for the work and possible damage to both the lathe center and the work.

Figure 42-6c shows a center hole that has been drilled too deep. The taper on the lathe center cannot contact the taper of the center hole; the result is poor support for the work.

DRILLING LATHE CENTER HOLES

1. Select the proper size of center drill to suit the diameter of the work (see Table 42.1).

2. Fasten the center drill in the drill chuck, having it extend beyond the chuck only about .500 in. (13 mm).

3. Place the work to be center drilled in the drill vise, as shown in Fig. 42-7.

4. Set the drill press at the proper speed and start the machine.

5. Locate the center-punch mark in the work directly below the center drill point.

6. Carefully feed the center drill into the center-punch mark in the work for about .060 in. (1.5 mm).

7. Raise the center drill, apply a few drops of cutting fluid, and continue drilling.

8. Frequently remove the drill from the hole to apply cutting fluid, remove the chips, and measure the diameter of the top of the center hole.

9. Continue drilling until the top of the hole is the proper size.

DRILL CHUCK

VISE ON ITS SIDE

WORK

DRILL PRESS TABLE

■ **Figure 42-7** Work set up for center-hole drilling. (*Kelmar Associates*)

▶▶ Spotting a Hole Location with a Center Drill

The chisel edge at the end of the web on most drills is generally wider than the center-punch mark on the work, and

Drilling Holes **327**

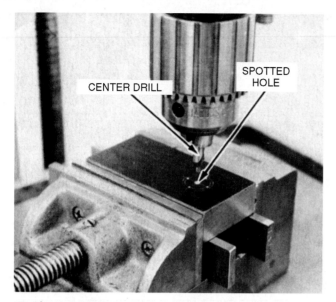

■ Figure 42-8 Spotting a hole location with a center drill. *(Kelmar Associates)*

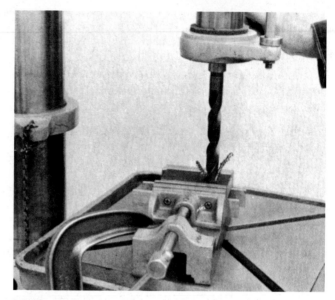

■ Figure 42-9 Tighten the table clamp while the drill is revolving in the hole. *(Kelmar Associates)*

therefore it is difficult to start a drill at the exact location. To prevent a drill from wandering off center, it is considered good practice to first spot every center-punch mark with a center drill. The small point on the center drill will accurately follow the center-punch mark and provide a guide for the larger drill which will be used.

1. Mount a small-size center drill in the drill chuck.

2. Mount the work in a vise or set it on the drill press table. *Do not clamp the work or the vise.*

3. Set the drill press speed to about 1500 r/min.

4. Bring the point of the center drill into the center-punch mark and allow the work to center itself with the drill point.

5. Continue drilling until about one-third of the tapered section of the center drill has entered the work (Fig. 42-8).

6. Spot all the holes to be drilled.

▶▶ Drilling Work Held in a Vise

The most common method of holding small workpieces is by means of a vise, which may be held by hand against a table stop or clamped to the table. When drilling holes larger than ½ in. (13 mm) in diameter, the vise should be clamped to the table.

1. Spot the hole location with a center drill.

2. Mount the correct-size drill in the drill chuck.

3. Set the drill press to the proper speed for the size of drill and the type of material to be drilled.

4. Fasten a clamp or stop on the left side of the table (Fig. 42-2).

5. Mount the work on parallels in a drill vise and tighten it securely.

6. With the vise against the table stop, locate the spotted hole under the center of the drill.

7. Start the drill press spindle and begin to drill the hole.
   a. For holes up to ½ in. (13 mm) in diameter, hold the vise against the table *or* stop by hand (Fig. 42-2).
   b. For holes over ½ in. (13 mm) in diameter:
      • Lightly clamp the vise to the table with a clamp.
      • Drill until the full drill point is into the work.
      • With the drill revolving, keep the drill point in the work and tighten the clamp holding the vise securely (Fig. 42-9).

8. Raise the drill occasionally and apply cutting fluid during the drilling operation.

9. Ease up on the drilling pressure as the drill starts to break through the workpiece.

▶▶ Drilling to an Accurate Layout

If a hole must be drilled to an exact location, the position of the hole must be accurately laid out as shown in Fig. 42-10a. During the drilling operation, it may be necessary to draw the drill point over so that it is concentric with the layout (Fig. 42-10b).

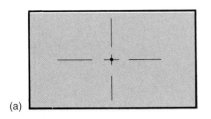 (a)

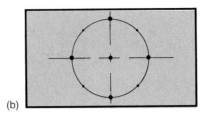 (b)

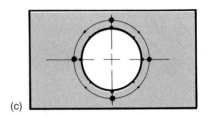 (c)

Figure 42-10 Layout for a hole to be drilled.

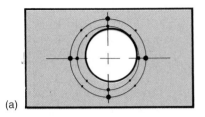 (a)

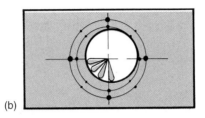 (b)

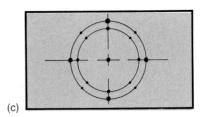 (c)

Figure 42-11 Drawing the drill point to the layout. *(Kelmar Associates)*

1. Clean and coat the surface of the work with layout dye.

2. Locate the position of the hole from *two machined edges* of the workpiece and scribe the lines as shown in Fig. 42-10a.

3. Lightly prick-punch where the two lines intersect.

4. Check the accuracy of the punch mark with a magnifying glass and correct if necessary.

5. With a pair of dividers, scribe a circle to indicate the diameter of the hole required (Fig. 42-10b).

6. Scribe a test circle .060 in. (1.5 mm) smaller than the hole size.

7. Punch four witness marks on circles up to .750 in. (19 mm) in diameter and eight witness marks on larger circles (Fig. 42-10c).

8. Deepen the center of the hole location with a center punch to provide a larger indentation for the drill to follow.

9. Center drill the work to *just beyond the depth of the drill point.*

10. Mount the proper size drill in the machine and drill a hole to a depth equal to one-half to two-thirds of the drill diameter.

11. Examine the drill indentation; it should be concentric with the inner proof circle (Fig. 42-11a).

12. If the spotting is off center, cut shallow V-grooves with a cape or diamond-point chisel on the side toward which the drill must be moved (Fig. 42-11b).

13. Start the drill in the spotted and grooved hole. The drill will be drawn toward the direction of the grooves.

14. Continue cutting grooves into the spotted hole until the drill point is drawn to the center of the scribed circles as shown in Fig. 42-11c.

Note: *The drill point must be drawn to the center of the scribed circle before the drill has cut or spotted to the drill's diameter.*

15. Continue to drill the hole to the desired depth.

▶▶ Drilling Large Holes

As drills increase in size, the thickness of the web also increases to give the drill added strength. The thicker the web, the thicker will be the point or chisel edge of the drill. As the chisel edge becomes larger, poorer cutting action results and more pressure must be applied for the drilling operation. A thick web will not follow the center-punch mark accurately and the hole may not be drilled in the proper location. Two methods are generally employed to overcome the poor cutting action of a thick web on large drills.

1. The web is thinned.

2. A lead, or pilot, hole is drilled.

The usual procedure for drilling large holes is that first a lead, or pilot, hole (Fig. 42-12), the diameter of which is slightly larger than the thickness of the web, is drilled. *Care must be taken to drill the pilot hole on center.* The pilot hole is then followed with a larger drill. This method may also be used to drill average-size holes when the drill press is small and does not have sufficient power to drive the drill through the solid metal.

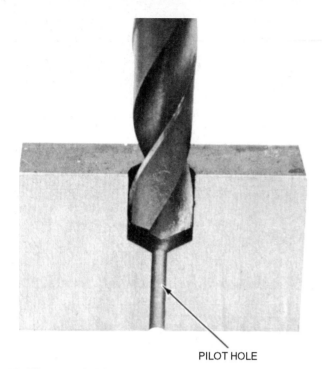

Figure 42-12 Drilling a lead, or pilot, hole helps the larger drill cut easily and accurately. *(Kelmar Associates)*

Never drill a pilot hole any bigger than necessary; otherwise, the larger drill may:

1. Cause chattering
2. Drill the hole out-of-round
3. Spoil the top (mouth) of the hole

The following drilling procedure is recommended:

1. Check the print and select the proper drill for the hole required.
2. Measure the thickness of the web at the point. Select a pilot drill with a diameter slightly larger than the web thickness (Fig. 42-13).
3. Mount the workpiece on the table.
4. Adjust the height and position of the table so that the drill chuck can be removed and the larger drill placed in the spindle without having to lower the table after the pilot hole is drilled. Lock the table securely in this position.
5. Place a center drill in the drill chuck, set the proper spindle speed, and accurately drill a center hole.

Note: The center drill should be used first, since it is short, rigid, and more likely to follow the center-punch mark.

6. Using the proper-size pilot drill and correct spindle speed, drill the pilot hole to the required depth (Fig. 42-12). The work may be lightly clamped at this time.

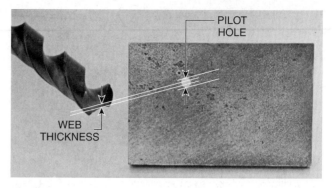

Figure 42-13 The size of the pilot drill should be slightly larger than the thickness of the drill web. *(Kelmar Associates)*

7. Shut off the machine, leaving the pilot drill in the hole.
8. Clamp the work securely to the table.
9. Raise the drill spindle and remove the drill and drill chuck.
10. Clean the taper shank of the drill and the drill press spindle hole. Remove any burrs on the drill shank with an oilstone.
11. Mount the large drill in the spindle.
12. Set the proper spindle speed, and feed and drill the hole to the required depth. If hand feed is used, the feed pressure should be eased as the drill breaks through the work.

▶▶ Drilling Round Work in a V-Block

V-blocks may be used to hold round work for drilling. Round work is seated in the accurately machined V-groove. Small diameters may be held in place with a U-shaped clamp, and larger diameters are fastened with strap clamps.

1. Select a V-block to suit the diameter of round work to be drilled. If the work is long, use a pair of V-blocks.
2. Mount the work in the V-block and then rotate it until the center-punch mark is in the center of the workpiece. With a rule and square, check that the distance from both sides is equal (Fig. 42-14).
3. Tighten the U-clamp securely on the work in the V-block *or* hold the work and V-block in a vise, as shown in Fig. 42-15.
4. Spot the hole location with a center drill.
5. Mount the proper drill size and set the machine to the correct speed.
6. Drill the hole, being sure that the drill does not hit the V-block or vise when it breaks through the work.

■ Figure 42-14 Using a square and a rule to align the center punch mark on a round workpiece. *(Kelmar Associates)*

■ Figure 42-15 Turn the workpiece until the line on the end of the work is in line with the blade of the square. *(Kelmar Associates)*

unit 42 review questions

Safety

1. Select three of the most important safety suggestions and explain why they should be observed.

Drilling Hints

2. What result might be expected from:

   a. Too high a drill speed?
   b. Too low a drill speed?

3. Why is it good practice to start each hole with a center drill?

Lathe Center Holes

4. State three reasons a center drill should be removed from the work frequently.

5. Why should center holes not be drilled:

   a. Too shallow?   b. Too deep?

Spotting a Hole

6. What is the purpose of spotting a hole before drilling?

7. How deep should each hole be spotted?

Drilling Work in a Vise

8. What is the purpose of fastening a clamp or table stop to the left side of the drill press table?

9. When should drilling pressure be eased?

Drilling to an Accurate Layout

10. List the procedure for laying out a hole before drilling to a layout.

11. How deep should the hole be drilled before the drill point indentation is examined?

12. Explain how a drill may be drawn over to a layout.

Drilling Large Holes

13. List three disadvantages of a thick web found on large drills.

14. What are pilot holes and why are they necessary?

15. Why should pilot holes not be drilled any larger than necessary?

16. How high should the drill press table be adjusted when drilling large holes?

UNIT 43

Reaming

OBJECTIVES

After completing this unit, you will be able to:

1 Identify and state the purpose of hand reamers and machine reamers

2 Explain the advantages of carbide-tipped reamers

3 Calculate the reaming allowance required for each reamer

4 Ream a hole by hand in a drill press

5 Machine ream a hole

Each component in a product must be made to exact standards for that product to function properly. Since it is impossible to produce holes that are round, smooth, and accurate to size by drilling, the reaming operation is very important. Reamers are used to enlarge and finish a hole previously formed by drilling or boring. Speed, feed, and reaming allowances are the three main factors which will affect the accuracy and finish of the hole and the life of the reamer.

▶▶ Reamers

A reamer is a rotary cutting tool with several straight or helical cutting edges along its body. It is used to accurately size and smooth a hole that has been previously drilled or bored. Some reamers are operated by hand (hand reamers), while others may be used under power in any type of machine tool (machine reamers).

REAMER PARTS

Reamers generally consist of three main parts: *shank, body,* and *angle of chamfer* (Fig. 43-1).

The *shank,* which may be straight or tapered, is used to drive the reamer. The shank of machine reamers may be round or tapered, while hand reamers have a square on the end to accommodate a tap wrench.

The *body* of a reamer contains several straight or helical grooves, or flutes, and lands (the portion between the flutes). A margin (the top of each tooth) runs from the *angle of chamfer* to the end of the flute. The *body clearance angle* is a relief or clearance behind the margin that reduces friction while the reamer is cutting. The *rake angle* is the angle formed by the face of the tooth when a line is drawn from a point on the front marginal edge through the center of the reamer (Fig. 43-1). If there is no angle on the face of the tooth, the reamer is said to have *radial land*.

The *angle of chamfer* is the part of the reamer that actually does the cutting. It is ground on the end of each tooth and there is clearance behind each chamfered cutting edge. On rose reamers, the angle of chamfer is ground on the end only and the cutting action occurs at this point. On fluted reamers, each tooth is relieved and most of the cutting is done by the reamer teeth.

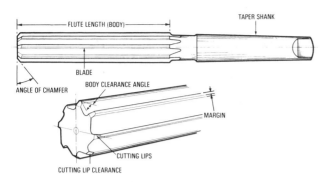

Figure 43-1 The main parts of a reamer.

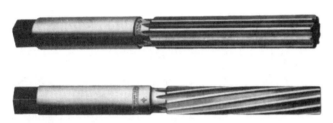

Figure 43-2 Straight and helical-fluted hand reamers. *(Kennametal Inc.)*

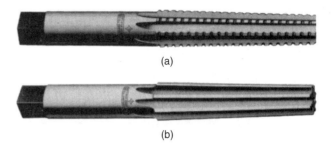

Figure 43-3 Taper hand reamers: (a) roughing; (b) finishing. *(Kennametal Inc.)*

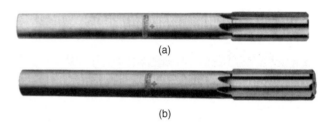

Figure 43-4 (a) A rose reamer cuts on the end angle only; (b) a fluted reamer has more teeth than a rose reamer and cuts on the sides and end. *(Kennametal Inc.)*

TYPES OF REAMERS

Reamers are available in a variety of designs and sizes; however, they all fall into two general classifications: *hand* and *machine* reamers.

Hand Reamers

Hand reamers (Fig. 43-2) are finishing tools used when a hole must be finished to a high degree of accuracy and finish. Holes to be hand reamed should be bored to within .003 to .005 in. (0.07 to 0.12 mm) of the finish size. *Never* attempt to remove more than .005 in. (0.12 mm) with a hand reamer.

A square on the shank end allows a wrench to be used for turning the reamer into the hole. The teeth on the end of the reamer are tapered slightly for a distance equal to the reamer diameter so that it can enter the hole to be reamed.

A hand reamer should never be used under mechanical power and should never be turned backward. When using a hand reamer, keep it true and straight with the hole. The dead center in a lathe or a stub center in a drill press will help keep the reamer aligned during the hand-reamer operation.

Taper hand reamers (Fig. 43-3), both roughing and finishing, are available for all standard-size tapers. Because chips do not fall out readily, a taper reamer should be removed from the hole and the flutes cleaned frequently.

Machine Reamers

Machine reamers may be used in any machine tool for both roughing and finishing a hole. They are also called *chucking reamers* because of the method used to hold them for the reaming operation. Machine reamers are available in a wide variety of types and styles. Only some of the more common types will be discussed.

Rose reamers (Fig. 43-4a) can be purchased with straight or tapered shanks and with straight or helical flutes. The teeth on the end have a 45° chamfer that is backed off to produce the cutting edge. The lands are nearly as wide as the flutes and are not backed off. Rose reamers cut on the end angle only and can be used to remove material quickly and bring the hole fairly close to the size required. Rose reamers are usually made .003 to .005 in. (0.07 to 0.12 mm) under the normal size.

Fluted reamers (Fig. 43-4b) have more teeth than rose reamers for a comparable diameter. The lands are relieved for the entire length, and fluted reamers therefore cut along the side as well as at the chamfer on the end. These reamers are considered finishing tools and are used to bring a hole to size.

Carbide-tipped reamers (Fig. 43-5 on p. 334) were developed to meet the ever-increasing demand for high production rates. They are similar to rose or fluted reamers, except that carbide tips have been brazed to their cutting edges. Because of the hardness of the carbide tips,

■ Figure 43-5 A carbide-tipped reamer.
(Kennametal Inc.)

■ Figure 43-6 Shell reamers are economical for reaming large holes. *(Kennametal Inc.)*

■ Figure 43-7 An adjustable reamer with inserted blades. *(Kennametal Inc.)*

■ Figure 43-8 An expansion reamer can be expanded slightly. *(Kennametal Inc.)*

these reamers resist abrasion and maintain sharp cutting edges even at high temperatures. Carbide-tipped reamers outlast high-speed steel reamers, especially on castings where hard scale or sand is a problem. Because carbide-tipped reamers can be run at higher speeds and still maintain their size, they are used extensively for long production runs.

Shell reamers (Fig. 43-6) are reamer heads mounted on a driving arbor. The shank of the driving arbor may be straight or tapered, depending on the size and type of shell reamer used. Two slots in the end of this reamer fit into lugs on the driving arbor. Sometimes a locking screw in the end of the arbor holds the shell reamer in place. The advantages of shell reamers are:

1. They are economical for larger holes.
2. Various head sizes can be easily interchanged on one arbor.
3. When a reamer becomes worn, it may be thrown away, and the driving arbor can be used with other reamers.

Adjustable reamers (Fig. 43-7) have inserted blades that can be adjusted approximately at .015 in. (0.38 mm) over or under the nominal reamer size. The threaded body has a series of tapered grooves cut lengthwise into which blades are fitted. Adjusting nuts on either end can be used to increase or decrease the diameter of the reamer. Hand- or machine-adjustable reamers can be readily sharpened and are available with either high-speed steel or carbide inserts.

Expansion reamers (Fig. 43-8) are similar to adjustable reamers; however, the amount they can be expanded is limited. The body of this reamer is slotted, and a tapered, threaded plug is fitted into the end. Turning this plug will allow a 1-in. (25-mm) reamer to expand up to

.005 in. (0.12 mm). Expansion reamers are not meant to be oversize reamers but to give longer life to finishing reamers.

Emergency reamers, drills whose corners (at the lip and land) have been slightly rounded and honed may be used with fairly good results if a reamer of a particular size is not available. First, drill the hole as close as possible to the required size. Then run the reaming drill at a fairly high speed and feed it into the hole slowly.

REAMER CARE

The accuracy and surface finish of a hole, as well as the life of a reamer, depend greatly on the care a reamer receives. Remember that a reamer is a finishing tool and should be handled carefully.

1. Never turn a reamer backward; this will ruin the cutting edges.
2. Always store reamers in separate containers to prevent the cutting edges from being nicked or burred. Plastic or cardboard tubes make excellent reamer containers.
3. Never roll or drop reamers on metal surfaces, such as bench tops, machines, and plates.
4. When not in use, a reamer should be oiled, especially on the cutting edges, to prevent rusting.
5. A fine, free-cutting grinding wheel should be used for resharpening reamers. Burring of the cutting edges destroys the life of the reamer; a rough cutting edge produces a rough hole and the reamer dulls quickly.

▸▸ Reaming Allowances

The amount of material left in a hole for the reaming operation depends on a number of factors. If a hole has been punched, rough-drilled, or bored, it requires more metal for reaming than a hole that has already been reamed with

table 43.1 — Recommended stock allowances for reaming

Hole Size		Allowance	
in.	mm	in.	mm
¼	6.35	.010	0.25
½	12.7	.015	0.38
¾	19.05	.018	0.45
1	25.4	.020	0.5
1¼	31.75	.022	0.55
1½	38.1	.025	0.63
2	50.8	.030	0.76
3	76.2	.045	1.14

table 43.2 — Recommended reaming speeds for high-speed steel reamers

Material	Speed	
	ft/min	m/min
Aluminum	130–200	39–60
Brass	130–180	39–55
Bronze	50–100	15–30
Cast iron	50–80	15–24
Machine steel	50–70	15–21
Steel alloys	30–40	9–12
Stainless steel	40–50	12–15
Magnesium	170–270	52–82

a roughing reamer. The type of machining operation prior to reaming must be considered as well as the material being reamed.

General rules for the amount of material that should be left in a hole for machine reaming are:

1. For holes up to .500-in. diameter, allow .015 in. for reaming.
2. For holes over .500-in. diameter, allow .030 in. for reaming.

Note: Never leave more than .005 in. in a hole for hand reamers up to .500 in. in diameter. On larger holes, a proportional allowance should be left to make a good finish possible.

For metric-size reamers, allow 0.1 mm for holes up to 12 mm in diameter. For holes over 12 mm, allow 0.2 to 0.78 mm for reaming. See Table 43.1 for the recommended allowances for various-size holes.

▶▶ Reaming Speeds and Feeds

SPEEDS

The selection of the most efficient speed for machine reaming depends on the following factors:

1. The type of material being reamed
2. The rigidity of the setup
3. The tolerance and finish required in the hole

Generally, reaming speeds should be from one-half to two-thirds the speed used for drilling the same material.

Higher reaming speeds can be used when the setup is rigid; slower speeds should be used when the setup is less rigid. A hole requiring close tolerances and a fine finish should be reamed at slower speeds. The use of coolants improves the surface finish and allows higher speeds to be used.

Reamers do not work well when they chatter; the speed selected should always be low enough to eliminate chatter.

Table 43.2 gives the recommended reaming speeds for high-speed steel reamers. Carbide reamers may be operated at higher speeds.

FEEDS

The feed used for reaming is usually two to three times greater than that used for drilling. The feed rate will vary with the material reamed; however, it should be approximately .001 to .004 in. (0.02 to 0.1 mm) per flute per revolution. Feeds that are too low generally result in glazing, excessive reamer wear, and sometimes chatter. Too high a feed tends to reduce the hole accuracy and sometimes results in poor surface finish. Generally, feeds should be the highest possible that will still produce the hole accuracy and finish required.

An exception to these feed rates occurs when tapered holes are being reamed. Because tapered reamers cut along their entire length, a light feed is necessary. The reamer should be removed occasionally and the flutes cleaned.

▶▶ Reaming Hints

1. Examine a reamer and remove all burrs from the cutting edges with a hone so that good surface finishes will be produced.

2. Cutting fluid should be used in the reaming operation to improve the hole finish and prolong the life of the reamer.

3. Helical-fluted reamers should always be used when long holes and those with keyways or oil grooves are reamed.

4. Straight-fluted reamers are generally used when extreme accuracy is required.

5. To obtain hole accuracy and good surface finish, use a roughing reamer first and then a finishing reamer. An old reamer that is slightly undersize may be used as a roughing reamer.

6. *Never, under any circumstances,* turn a reamer backward.

7. *Never* attempt to start a reamer on an uneven surface; the reamer will go toward the point of least resistance and will not produce a straight, round hole.

8. If chatter occurs, stop the machine, reduce the speed, and increase the feed. To overcome the chatter marks, it may be necessary to restart the reamer slowly by pulling the drill press belt by hand.

9. To avoid chatter, select a reamer with an incremental cut (unequally spaced teeth).

10. When hand reaming in a drill press, always use a stub center in the drill press spindle to keep the reamer aligned.

▶▶ Reaming a Straight Hole

Two types of reamers are used in machine shop work—hand reamers and machine reamers. Hand reamers have a square on one end and are used to remove no more than .005 in. (0.12 mm) from a hole. Machine reamers may have straight shanks that are held and driven by a drill chuck or tapered shanks that fit directly into the drill press

■ **Figure 43-9** A workpiece clamped to the table correctly. *(Kelmar Associates)*

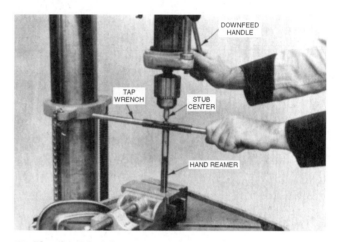

■ **Figure 43-10** Using a stub center to keep the reamer aligned. *(Kelmar Associates)*

spindle. They are generally used to remove from .015 in. (0.4 mm) to .030 in. (0.8 mm) of metal from a hole, depending on the hole diameter.

HAND REAMING A STRAIGHT HOLE

1. Mount the work on the parallels in a vise and clamp it securely to the table (Fig. 43-9).

2. Drill the hole to the proper size, leaving an allowance for the hand reamer to be used.

Note: Reaming allowance should be no more than .005 in. (0.12 mm) for a 1-in. (25-mm) diameter reamer.

3. Do not move the location of the work or the table; remove the drill and mount a stub center in the drill chuck (Fig. 43-10).

4. Start the end of the reamer in the drilled hole.

5. Fasten a tap wrench on the reamer.

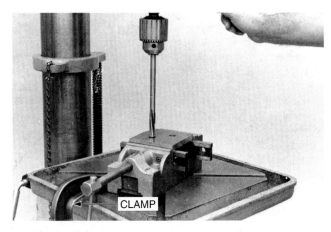

■ Figure 43-11 Reaming a hole in the drill press. *(Kelmar Associates)*

6. Engage the stub center in the center hole on the end of the reamer.

7. With the downfeed lever, apply slight pressure while turning the reamer clockwise by hand.

8. Apply cutting fluid and ream the hole.

9. When removing the reamer, turn it clockwise, never counterclockwise.

MACHINE REAMING A STRAIGHT HOLE

A hole that must be finished to size should be reamed immediately after it has been drilled and while the hole is still aligned with the drill press spindle. This will ensure that the reamer follows the same location as the drill.

1. Mount the work on parallels in a vise and fasten it securely to the table.

2. Select the proper-size drill for the reaming allowance required and drill the hole.

Note: *Do not move the work or drill press table at this time.*

3. Mount the proper reamer in the drill press.

4. Adjust the spindle speed to suit the reamer and the work material.

5. Start the drill press and carefully lower the spindle until the chamfer on the reamer starts to cut (Fig. 43-11).

6. Apply cutting fluid and feed the reamer by applying enough pressure to keep the reamer cutting.

7. Remove the reamer from the hole by raising the downfeed handle.

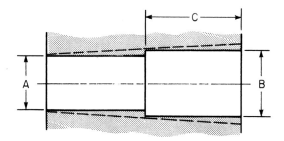

A. (SMALL DRILL SIZE) SHOULD BE THE SMALLEST DIAMETER OF THE TAPERED HOLE MINUS 1/64 in.

B. (LARGE DRILL SIZE) SHOULD BE THE DIAMETER AT MIDDLE OF TAPERED HOLE MINUS 1/64 in.

C. (DEPTH OF LARGE HOLE) ONE-HALF THE LENGTH OF THE TAPER MINUS 1/16 in.

■ Figure 43-12 Drill sizes for tapered holes. *(Kelmar Associates)*

8. Shut off the machine and remove the burr from the edge of the hole.

▶▶ Reaming a Tapered Hole

When reaming a tapered hole in a workpiece, step-drill the hole prior to reaming. The recommended drills to be used can be determined by referring to Fig. 43-12.

1. Mount the work on parallels in a vise or on the drill table and fasten it securely.

2. Align the hole center under the drill point.

3. Drill a hole .015 in. (0.4 mm) smaller than the smallest diameter (A) of the tapered hole (Fig. 43-12).

4. Obtain a drill ⅟₆₄ in. (0.4 mm) smaller than the size of the finished hole measured at a diameter B (one-half the length of the tapered section).

5. Drill the hole .060 in. (1.5 mm) less than the C dimension (Fig. 43-12).

6. Mount a roughing tapered reamer in the drill spindle.

7. Adjust the spindle speed to about one-half the speed used to ream a straight hole.

8. Rough-ream the hole to about .005 in. (0.12 mm) undersize while applying cutting fluid.

9. Mount the finish reamer, apply cutting fluid, and finish-ream the hole to size.

Reamers

1. What is the purpose of a reamer?

2. Define the following reamer parts:

   a. Body   c. Body clearance angle
   b. Angle of chamfer   d. Radial land

3. How may a hand reamer be recognized, and for what purpose is it used?

4. Compare a rose reamer with a fluted reamer.

5. State the advantages of carbide-tipped reamers.

6. Describe briefly the following types of reamers and state their purpose:

   a. Shell   b. Adjustable

Reamer Care

7. List three important points which should be observed in the care of reamers.

Reaming Allowances

8. What is the general rule for the amount of material left in a hole for machine reaming?

9. How much should be left in the following holes for machine reaming?

   a. 1¼ in.   b. 19.05 mm

Reaming Hints

10. List seven of the most important reaming hints.

Reaming a Straight Hole and a Tapered Hole

11. How much material should be left in a hole for hand reaming?

12. How can the reamer be kept in alignment when hand reaming on a drill press?

13. Why should the work or drill press table not be moved before a hole is reamed?

14. Make a neat sketch to show the procedure for drilling in preparation for reaming a tapered hole.

UNIT 44
Drill Press Operations

OBJECTIVES

After completing this unit, you should be able to:

1 Counterbore and countersink holes

2 Select and use the proper tap to thread a hole in a drill press

3 Use three methods to transfer hole locations

The drill press is a versatile machine tool and can be used to perform a variety of operations other than the drilling of holes. The variety of cutting and finishing tools available allow operations such as counterboring, countersinking, tapping, and spot-facing to be performed on a drill press.

▶▶ Counterboring

Counterboring is the operation of enlarging the end of a hole that has been drilled previously. A hole is generally counterbored to a depth slightly greater than the head of the bolt, cap screw, or pin it is to accommodate.

Counterbores (Fig. 44-1) are supplied in a variety of styles, each having a pilot in the end to keep the tool in line with the hole being counterbored. Some counterbores are available with interchangeable pilots to suit a variety of hole sizes.

TO COUNTERBORE A HOLE

1. Set up and fasten the work securely.
2. Drill the proper size of hole in the workpiece to suit the body of the pin or screw.
3. Mount the correct size of counterbore in the drill press (Fig. 44-2 on p. 340).

■ **Figure 44-1** A set of counterbores. *(Kelmar Associates)*

■ Figure 44-2 A counterbore is used to enlarge the end of a hole. *(Kelmar Associates)*

■ Figure 44-3 Countersinking produces a tapered hole to fit a flat-head machine screw. *(Kelmar Associates)*

4. Set the drill press speed to approximately one-quarter that used for drilling.

5. Bring the counterbore close to the work and see that the pilot turns freely in the drilled hole.

6. Start the machine, apply cutting fluid, and counterbore to the required depth.

▶▶ Countersinking

Countersinking is the process of enlarging the top end of a hole to the shape of a cone to accommodate the conical-shaped heads of fasteners so that the head will be flush with or below the surface of the part. These cutting tools, called *countersinks* (Fig. 44-3), are available with various included angles, such as 60°, 82°, 90°, 100°, 110°, and 120°.

An 82° countersink is used to enlarge the top of a hole so that it will accommodate a flat-head machine screw (Fig. 44-4). The hole is countersunk until the head of the machine screw is flush with or slightly below the top of the work surface (Fig. 44-4). All holes to be

■ Figure 44-4 Countersink until the top of the hole is slightly larger than the diameter of the screw head. *(Kelmar Associates)*

threaded should be countersunk slightly larger than the tap diameter to protect the start of the thread.

The speed recommended for countersinking is approximately one-quarter of the drilling speed.

TO COUNTERSINK A HOLE FOR A MACHINE SCREW

1. Mount an 82° countersink in the drill chuck.

2. Adjust the spindle speed to about one-half that used for drilling.

3. Place the workpiece on the drill table.

4. With the spindle stopped, lower the countersink into the hole. Clamp the work if necessary. If a pilot-type countersink is used, the pilot should be a slip fit in the drilled hole. The pilot will center the cutting tool and the work.

5. Raise the countersink slightly, start the machine, and feed the countersink by hand until the proper depth is reached. The diameter may be checked by placing an inverted screw in the countersunk hole (Fig. 44-4).

6. If several holes are to be countersunk, set the depth stop so that all the holes will be the same depth. The gage should be set when the spindle is stationary and the countersink is in the hole.

7. Countersink all the holes to the depth set on the gage.

▶▶ Tapping

Tapping in a drill press may be performed either by hand or under power with the use of a tapping attachment. The advantage of using a drill press for tapping a hole is that the tap can be started squarely and maintained that way through the entire length of the hole.

Machine tapping involves the use of a tapping attachment mounted in the drill press spindle. The tapping operation should be done immediately after the drilling operation to obtain the best accuracy and to avoid duplication of the setup. This sequence is especially important when tapping by hand.

The most common taps used for tapping holes in a drill press are hand taps and machine taps. Hand taps (Fig. 44-5) are available in sets containing the taper, plug, and bottoming taps. When using hand taps in a drill press, it is important that the tap be guided by holding it in a drill chuck or supporting it with a stub center and turning it by hand.

Machine taps are designed to withstand the torque required to thread the hole and to clear the chips out of the hole quickly. The most common machine taps are the gun, stub-flute, and spiral-flute taps (Fig. 44-6).

The *fluteless tap* (Fig 44-7) is actually a forming tool used to produce internal threads in ductile materials such as copper, brass, aluminum, and leaded steels.

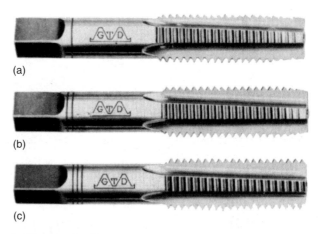

(a)

(b)

(c)

Figure 44-5 A set of hand taps: (a) taper; (b) plug; (c) bottoming. *(Kennametal Industrial Products Group)*

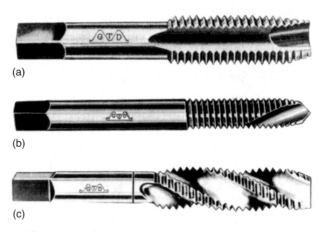

(a)

(b)

(c)

Figure 44-6 Types of machine taps: (a) gun; (b) stub-flute; (c) spiral-flute. *(Kennametal Industrial Products Group)*

(a)

(b)

Figure 44-7 (a) A fluteless tap; (b) lobes of the tap. *(Kennametal Industrial Products Group)*

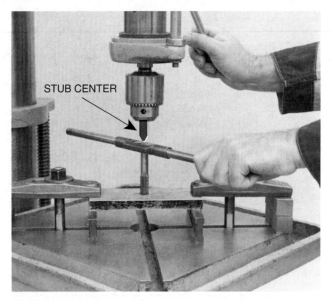

STUB CENTER

■ Figure 44-8 Guiding a tap into the workpiece using a stub center held in a drill chuck. *(Kelmar Associates)*

TO TAP A HOLE BY HAND IN A DRILL PRESS

1. Mount the work on parallels with the center-punch mark on the work in line with the spindle, and clamp the work securely to the drill press table.

2. Adjust the drill press table height so that the drill may be removed after the hole has been drilled without moving the table or work.

3. Center drill the hole location.

4. Drill the hole to the correct *tap drill size* for the tap to be used.

Note: The work or table *must not* be moved after the drilling.

5. Mount a stub center in the drill chuck (Fig. 44-8),

OR

Remove the drill chuck and mount a special center in the drill press spindle.

6. Fasten a suitable tap wrench on the end of the tap.

7. Place the tap in the drilled hole, and lower the drill press spindle until the center fits into the center hole in the tap shank.

8. Turn the tap wrench clockwise to start the tap into the hole, and at the same time keep the center in light contact with the tap.

■ Figure 44-9 A tapping attachment held in a drill press. *(Kelmar Associates)*

■ Figure 44-10 Transferring hole locations by spotting with a drill. *(Kelmar Associates)*

9. Continue to tap the hole in the usual manner; keep the tap aligned by applying light pressure on the drill press downfeed lever.

A *tapping attachment* (Fig. 44-9) may be mounted in a drill press spindle to rotate the tap by power. It has a built-in friction clutch that drives the tap clockwise when the drill press spindle is fed downward. If there is excessive pressure against the tap because it is stuck or jammed in a hole, the clutch will slip before the tap breaks. The tapping attachment has a reversing mechanism, engaged by the drill press spindle being raised, to back a tap out of the hole.

■ Figure 44-11 Transferring hole locations using a transfer punch. *(Kelmar Associates)*

Two- or three-fluted machine or gun taps are used for tapping under power because of their ability to clear the chips. Tapping speed for most materials ranges from 60 to 100 r/min.

▶▶ Transferring Hole Locations

During the construction of dies, jigs, fixtures, and machine parts, it is often necessary to transfer the location of holes accurately from one part to another. Three common methods of transferring hole locations are:

1. Spotting with a twist drill
2. Using transfer punches
3. Using transfer screws

Regardless of which method is used, the holes in the existing part are used as a master, or guide, to transfer the hole locations to another part.

TO SPOT WITH A TWIST DRILL

1. Remove the burrs from the mating surfaces on both parts.
2. Align both parts accurately and clamp them together.
3. Mount a drill, the same diameter as the hole to be transferred, in the drill press spindle.
4. Start the drill into the hole of the guide part and spot-drill the second part (Fig. 44-10).

Note: Never spot-drill deeper than the diameter of the drill.

5. Spot-drill all the holes to be transferred.
6. Remove the original part.
7. Drill the spotted holes to the required diameter.

TO USE TRANSFER PUNCHES

1. Remove the burrs from the mating surfaces on both parts.
2. Align both parts accurately and clamp them together.
3. Secure a transfer punch (Fig. 44-11) of the same diameter as the hole to be transferred.
4. Place the punch in the hole and *lightly* strike it with a hammer to mark the hole location.
5. Use the correct size of transfer punch on all the holes to be transferred.
6. Remove the original part.
7. Use a divider to lay out proof circles for the holes to be drilled.
8. With a center punch, deepen the existing transfer-punch marks.
9. Use the method outlined in Drilling to a Layout (Unit 42) to drill the holes to location accurately.

TO USE TRANSFER SCREWS

It is often necessary to transfer the location of threaded holes. This may be easily accomplished by the use of transfer screws (Fig. 44-12) that have been hardened and sharpened to a point. Two flats are ground on the point to

■ Figure 44-12 Transfer screws used to transfer threaded hole locations. *(Kelmar Associates)*

allow the screws to be threaded into a hole with a small wrench or a pair of needle-nose pliers.

1. Remove all burrs from the mating surfaces.

2. Thread transfer screws into the holes to be transferred, allowing the points to extend beyond the work surface approximately .030 in. (0.8 mm).

3. Align both parts accurately and then sharply strike one part with a hammer.

4. Remove the original part, and deepen the marks left by the transfer screws with a center punch.

5. Drill all holes to the required size.

▸▸ Drill Jigs

A drill jig is used whenever it is necessary to drill holes to an exact location in many identical parts. Drill jigs are used to save layout time, avoid incorrectly located holes, and produce holes accurately and economically. The advantages of using a drill jig are as follows:

1. Since it is not necessary to lay out the hole locations, layout time is eliminated.

2. Each part is quickly and accurately aligned.

3. The part is held in position by a clamping mechanism.

4. The drill jig bushings provide a guide for the drill.

5. The hole locations in each part will be exactly the same; therefore, the parts produced are interchangeable.

6. Unskilled labor can be used.

■ **Figure 44-13** A drill jig positions any number of identical parts so that a hole or holes can be drilled to an accurate location. *(Kennametal Inc.)*

A drill jig (Fig. 44-13) is designed so that the part to be drilled may be fastened into it and drilled immediately. Hardened drill jig bushings, used to guide and keep the drill positioned, are located in the drill jig wherever holes must be drilled. When two or more different sizes of holes are to be drilled in the same part, it is preferable to have a gang or multispindle drill press set up. A different size of drill is mounted in each spindle, and the drill jig is passed from one spindle to the next for each hole.

Counterboring and Countersinking

1. List the procedures for counterboring a hole.

2. Why should holes that are to be tapped be countersunk?

Tapping

3. What is the advantage of tapping a hole by hand in a drill press?

4. Describe the procedure for tapping a hole by hand in a drill press.

5. Explain how a tapping attachment operates.

Transferring Hole Locations

6. Name three methods of transferring the location of holes from one part to another.

7. Explain the procedure for spotting with a twist drill.

8. What are transfer punches and how are they used?

9. Describe transfer screws and explain how they are used.

Section 11

(Colchester Lathe Co.; Cincinnati Machine; South Bend Lathe)

THE LATHE

Historically, the lathe is the forerunner of all machine tools. The first application of the lathe principle was probably the potter's wheel. This machine rotated a mass of clay and enabled the clay to be formed into a cylindrical shape.

The modern lathe operates on the same basic principle. The work is held and rotated on its axis while the cutting tool is advanced along the lines of a desired cut (Fig. 1 on p. 348). The lathe is one of the most versatile machine tools used in industry. With suitable attachments, the lathe may be used for turning, tapering, form turning, screw cutting, facing, drilling, boring, spinning, grinding, and polishing operations. Cutting operations are performed with a cutting tool fed either parallel or at right angles to the axis of the work. The cutting tool may also be fed at an angle relative to the axis of the work for machining tapers and angles.

Modern production has led to the development of many special types of lathes, such as the engine, turret, single- and multiple-spindle automatic, tracer, and numerically controlled lathes, and now computer-controlled turning centers.

The *engine lathe* (Fig. 2 on p. 348), basically not a production lathe, is found in jobbing shops, school shops, and toolrooms.

When many duplicate parts are required, the *turret lathe* (Fig. 3 on p. 348) may be used. This lathe is equipped with a multisided toolpost, called a turret, to which several different cutting tools may be mounted. Different cutting tools are employed in a given sequence to perform a series of operations on each part. This same sequence may be repeated on many parts without having to change or reset the cutting tools.

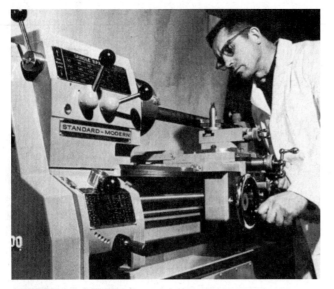

■ Figure 1 The main purpose of a lathe is to machine round work. *(Standard-Modern Tool Company)*

■ Figure 2 The engine lathe is the most common lathe found in a machine shop. *(South Bend Lathe Corp.)*

■ Figure 3 A turret lathe is used to mass-produce parts. *(Sheldon Machine Tool Co.)*

■ Figure 4 A single-spindle automatic lathe.

When hundreds or thousands of identical small parts are required, they may be produced on *single-* and *multiple-spindle automatic lathes* (Fig. 4). On these machines, six or eight different operations may be performed on as many parts at the same time. Once set up, the machine will produce the parts for as long as required.

Tracer lathes (Fig. 5) are used where a few duplicate parts are required. A hydraulically operated cross-slide (and cutting tool) is controlled by a stylus bearing against a round or flat template. Tracer attachments are available for converting most engine lathes into tracer lathes.

■ Figure 5 A hydraulic tracer attachment mounted on a lathe. *(Cincinnati Machine, a UNOVA Co.)*

(a)

INTELLIGENT "DRO"
(b)

DUAL ELECTRONIC HANDWHEELS
(c)

MANUAL 8-STATION TURRET
(d)

■ Figure 6 The EZ Path® lathe with a two-axis control can be operated manually or automatically.

An innovative feature in lathes is the development of the conventional/programmable lathe (Fig. 6a). This lathe can be operated as a standard lathe or a programmable lathe to automatically repeat machining operations or machine entire parts. It is equipped with a 2-axis, 32-bit control that allows quick switching from the manual to the programmable mode. The intelligent "DRO" (digital readout), Fig. 6b shows the exact location of the cutting tool and the workpiece dimensions in the X and Z axes in inches or millimeters.

■ Figure 7 A numerically control lathe. *(Mazak Corporation)*

■ Figure 8 A lathe equipped with a digital readout system. *(Sheffield Measurement Div.)*

The programmable control can remember the steps in an operation performed manually in its memory bank. If necessary, this information can be edited, should there be a need to change any steps or information in the program.

Computerized numerically controlled lathes and *turning centers* (Fig. 7) have come into widespread use. With these machines, the cutting-tool movements are controlled by a computer-controlled program to perform a sequence of operations automatically on the workpiece once the machine has been set up.

A useful and popular device that may be added to an engine lathe (or any other machine) is the digital readout system. Engine lathes equipped with these devices (Fig. 8) can produce duplicate parts to within .001 in. (0.02 mm) or less on the diameters and lengths of a workpiece.

Only the engine lathe, which is basic to all lathes, will be discussed in detail in Units 45 through 58.

UNIT 45

Engine Lathe Parts

OBJECTIVES

After completing this unit, you will be able to:

1 Identify and state the purposes of the main operative parts of the lathe

2 Set the lathe to run at any required speed

3 Set the proper feed for the cut required

Most work turned on a lathe, other than in production shops, will be turned on an engine lathe. This lathe is an accurate and versatile machine on which many operations such as turning, tapering, form turning, threading, facing, drilling, boring, grinding, and polishing may be performed. Three common engine lathes are the toolroom, heavy-duty, and gap-bed lathes.

▶▶ Lathe Size and Capacity

Lathe size is designated by the *largest work diameter* that can be swung over the lathe ways and generally the *maximum distance between centers* (Fig. 45-1). Some manufacturers designate the lathe size by the largest work diameter that can be swung over the ways and the overall bed length.

Lathes are manufactured in a wide range of sizes, the most common being from 9- to 30-in. swing, with a capacity of 16 in. to 12 ft between centers. A typical lathe may have a 13-in. swing, a 6-ft-long bed, and a capacity to turn work 36 in. long between centers (Fig. 45-1).

The average metric lathe used in school shops may have a 230- to 330-mm swing and have a bed length of from 500 to 3000 mm.

▶▶ Parts of the Lathe

The main parts of a lathe are the bed (and ways), headstock, quick-change gearbox, carriage, and tailstock (Fig. 45-2).

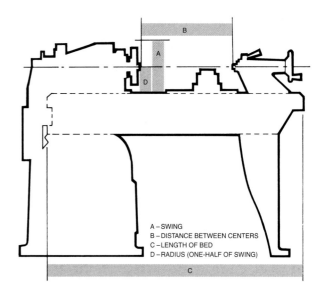

A – SWING
B – DISTANCE BETWEEN CENTERS
C – LENGTH OF BED
D – RADIUS (ONE-HALF OF SWING)

■ **Figure 45-1** Lathe size is indicated by the swing and the length of the bed. *(South Bend Lathe Corp.)*

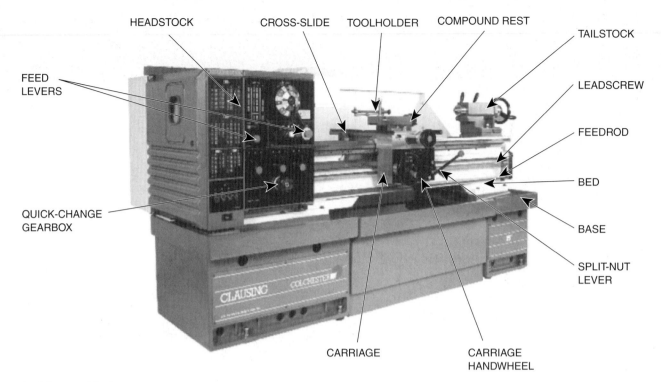

Labels, clockwise from top:
HEADSTOCK   CROSS-SLIDE   TOOLHOLDER   COMPOUND REST   TAILSTOCK   LEADSCREW   FEEDROD   BED   BASE   SPLIT-NUT LEVER   CARRIAGE HANDWHEEL   CARRIAGE   QUICK-CHANGE GEARBOX   FEED LEVERS

■ **Figure 45-2** The parts of an engine lathe.   *(South Bend Lathe)*

BED

The *bed* is a heavy, rugged casting made to support the working parts of the lathe. On its top section are machined ways that guide and align the major parts of the lathe.

HEADSTOCK

The *headstock* (Fig. 45-3) is clamped on the left-hand side of the bed. The *headstock spindle,* a hollow, cylindrical shaft supported by bearings, provides a drive through gears from the motor to work-holding devices. A live center, a faceplate, or a chuck can be fitted to the spindle nose to hold and drive the work. The live center has a 60° point that provides a bearing for the work to turn between centers.

Headstock spindles can be driven either by a stepped pulley and a belt, or by transmission gears in the headstock (Fig. 45-3). The lathe with a stepped pulley drive is generally called a belt-driven lathe; the gear-driven lathe is referred to as a geared-head lathe. Some light- and medium-duty lathes are equipped with a variable speed drive. On this type of drive, the speed is changed while the headstock spindle is revolving. However, when the lathe is shifted from a high- to low-speed range or vice versa, the lathe spindle must be stopped before the speed change is made. The *feed reverse lever,* mounted on the headstock, reverses the rotation of the feed rod and lead screw.

■ **Figure 45-3** A gear-drive headstock.

QUICK-CHANGE GEARBOX

The *quick-change gearbox* (Fig. 45-4), containing a number of different-size gears, provides the *feed rod* and *lead-screw* with various speeds for turning and thread-cutting operations. The feed rod advances the carriage for turning operations when the *automatic feed lever* is engaged. The lead screw advances the carriage for thread-cutting operations when the *split-nut lever* is engaged.

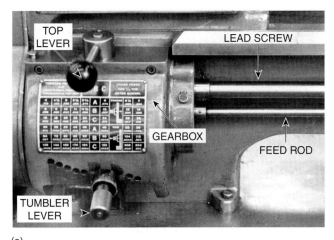

(a)

(b)

■ **Figure 45-4** (a) A quick-change gearbox; (b) a cutaway section of a quick-change gearbox. *(Kelmar Associates)*

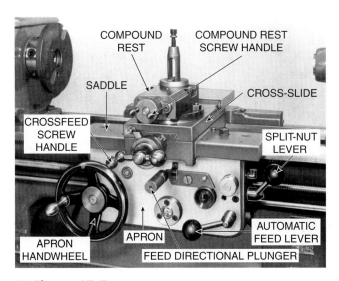

■ **Figure 45-5** The main parts of an engine lathe carriage. *(Kelmar Associates)*

CARRIAGE

The *carriage* (Fig. 45-5)—consisting of three main parts: the saddle, cross-slide, and apron—is used to move the cutting tool along the lathe bed. The *saddle,* an H-shaped casting mounted on the top of the lathe ways, provides a means of mounting the cross-slide and the apron.

The *cross-slide,* mounted on top of the saddle, provides a manual or an automatic cross movement for the cutting tool. The *compound rest,* fitted on top of the cross-slide, is used to support the cutting tool. It can be swiveled to any angle for taper-turning operations and is moved manually. The cross-slide and compound rest both have graduated collars that ensure accurate cutting-tool settings in thousandths of an inch or hundredths of a mil-

limeter. The apron, fastened to the saddle, houses the gears and mechanism required to move the carriage or cross-slide automatically. A locking-off lever inside the apron prevents engaging the split-nut lever and the automatic feed lever at the same time.

The *apron handwheel* can be turned manually to move the carriage along the lathe bed. This handwheel is connected to a gear that meshes in a rack fastened to the lathe bed.

The *automatic feed lever* engages a clutch that provides automatic feed to the carriage. The *feed-change lever* can be set for longitudinal feed or for crossfeed. When in the neutral position, the feed-change lever permits the split-nut lever to be engaged for thread cutting. For thread-cutting operations, the carriage is moved automatically when the split-nut lever is engaged. This causes the threads of the split-nut to engage into the threads of the revolving lead screw and move the carriage at a predetermined rate.

TAILSTOCK

The *tailstock* (Fig. 45-6 on p. 354)—consisting of the upper and lower tailstock castings—can be adjusted for taper or parallel turning by two screws set in the base. The tailstock can be locked in any position along the bed of the lathe by the *tailstock clamp.* The *tailstock spindle* has an internal taper to receive the *dead center,* which provides support for the right-hand end of the work. Other standard tapered-shank tools, such as reamers and drills, can be held in the tailstock spindle. A *spindle clamp* is used to hold the tailstock spindle in a fixed position. The *tailstock handwheel* moves the tailstock spindle in or out of the tailstock casting. It can also be used to provide a hand feed for drilling and reaming operations.

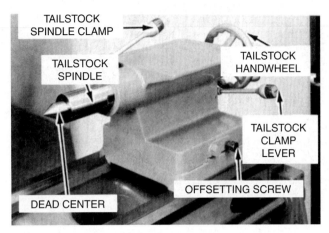

Figure 45-6 The tailstock assembly. *(Cincinnati Machine)*

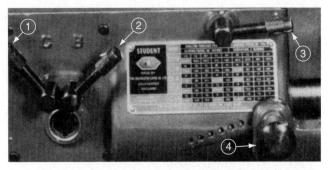

(a)

(b)

Figure 45-7 (a) A quick-change gearbox permits fast setting of feeds; (b) a quick-change gearbox chart—note the setting for metric threads. *(Colchester Lathe Company)*

Setting Speeds on a Lathe

Engine lathes are designed to operate at various spindle speeds for machining of different materials. These speeds are measured in revolutions per minute (r/min) and are changed by the stepped pulleys or gear levers.

On a *belt-driven lathe,* various speeds are obtained by changing the flat belt and the back gear drive.

On the *geared-head lathe* (Fig. 45-3), speeds are changed by moving the speed levers into proper positions according to the revolutions per minute chart fastened to the headstock. While shifting the lever positions, place one hand on the faceplate or chuck, and turn the lathe spindle slowly by hand. This will enable the levers to engage the gear teeth without clashing.

 SAFETY PRECAUTIONS Never change speeds when the lathe is running. On lathes equipped with *variable speed drives,* the speed is changed by turning a dial or handle *while the machine is running.*

Setting Feeds

The feed of an engine lathe, or the distance the carriage will travel in one revolution of the spindle, depends on the speed of the feed rod or lead screw. This is controlled by the change gears in the quick-change gearbox (Fig. 45-7a). This quick-change gearbox obtains its drive from the headstock spindle through the end gear train (Fig. 45-4b). A chart mounted on the front of the quick-change gearbox indicates the various feeds and metric pitches or threads per inch that may be obtained by setting levers to the positions indicated (Fig. 45-7b).

TO SET THE FEED FOR THE APRON (CARRIAGE DRIVE)

1. Select the desired feed on the chart.
2. Move tumbler lever #4 (Fig. 45-7a) into the hole directly below the selected feed.
3. Follow the row in which the selected feed is found to the left, and set the *feed-change levers* (#1 and #2) to the letters indicated.
4. Set lever #3 to disengage the lead screw.

Note: Before turning on the lathe, be sure all levers are fully engaged by turning the headstock spindle by hand, and see that the feed rod turns.

Shear Pins and Slip Clutches

To prevent damage to the feed mechanism from overload or sudden torque, some lathes are equipped with either *shear pins* or *slip clutches* (Fig. 45-8a, b). Shear pins, usually made of brass, may be found on the feed rod, lead screw, and end gear train. Spring-loaded slip clutches are found only on feed rods. When the feed mechanism is overloaded, either the shear pin will break or the slip clutch will slip, causing the automatic feed to stop. This prevents damage to the gears or shafts of the feed mechanism.

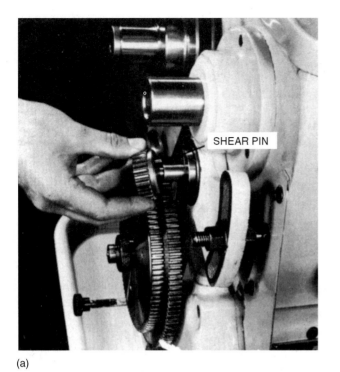

SHEAR PIN

(a)

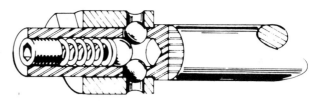

(b)

Figure 45-8 (a) A shear pin in the end gear train prevents damage to the gears in case of an overload; (b) a spring-ball clutch will slip when too much strain is applied to the feed rod. *(Colchester Lathe Company)*

unit 45 review questions

1. List the operations that can be performed on a lathe.

2. How is the size of a lathe designated?

Parts of the Engine Lathe

3. Name four *main* units of a lathe.

4. State the purpose of the following:

   a. Headstock spindle
   b. Lead screw and feed rod
   c. Quick-change gearbox
   d. Split-nut lever
   e. Feed-change lever
   f. Cross-slide
   g. Compound rest

Setting Speeds and Feeds

5. List three types of lathe drives.

6. Explain how speeds on a geared-head lathe are changed.

7. List the steps to set a feed of .010 in. (0.25 mm).

8. What is the purpose of:

   a. A shear pin?
   b. A slip clutch?

UNIT 46

Lathe Accessories

OBJECTIVES

After completing this unit, you will be able to:

1 Identify and state the purpose of the common work-holding and -driving accessories

2 Identify and state the purpose of the common cutting-tool-holding accessories

3 Identify and state the purpose of modular and quick-change tooling

Many lathe accessories are available to increase the versatility of the lathe and the variety of work that

can be machined. Lathe accessories may be divided into two categories:

1. Work-holding, -supporting, and -driving devices
2. Cutting-tool-holding devices

Work-holding, -supporting, and -driving devices include lathe centers, chucks, faceplates, mandrels, steady and follower rests, lathe dogs, and drive plates. Cutting-tool-holding devices include various types of straight and offset toolholders, threading toolholders, boring bars, turret-type toolposts, and quick-change toolpost assemblies.

▸▸ Work-Holding Devices

LATHE CENTERS

Most turning operations can be performed between centers on a lathe. Work to be turned between centers must have a center hole drilled in each end (usually 60°) to provide a bearing surface, which allows the work to turn on the centers. The centers merely support the work while the cutting operations are performed. A lathe dog, fitted into a driving plate, provides a drive for the work (Fig. 46-1).

A variety of lathe centers are used to suit various operations or workpieces. Probably the most commonly used centers in school shops were the solid 60° centers with a Morse taper shank (Fig. 46-2). These are generally made from high-speed steel or a good grade of machine steel with carbide inserts or tips. Care must be taken when using these centers to adjust and lubricate

LATHE DOG

■ **Figure 46-1** A workpiece mounted between centers is usually driven by a lathe dog. *(Kelmar Associates)*

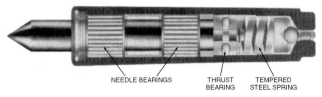

NEEDLE BEARINGS   THRUST   TEMPERED
                          BEARING   STEEL SPRING

(a)

(b)

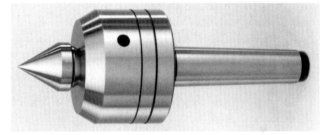

(c)

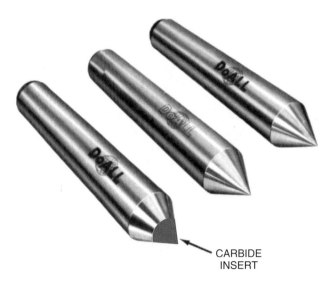

Figure 46-2 A variety of 60° lathe centers. *(DoAll Company)*

CARBIDE
INSERT

them occasionally as the work heats up and expands. If this step is not done, both the center and the workpiece may be damaged. The damage to the workpiece will be the loss of concentricity, which will prevent future operations from being performed using the center holes. The lathe center must also be reground to remove the damaged section before it can be used.

Revolving tailstock centers (Fig. 46-3), sometimes called *live dead centers,* have generally replaced solid dead centers for most machining operations. They are commonly used to support work held in a chuck or when work is being machined between centers. This type of center usually contains antifriction bearings, which allow the center to revolve with the workpiece. No lubrication is required between the center and the workpiece, and the center tension is not affected by workpiece expansion during the cutting action. A variety of centers are available to accommodate different types and sizes of workpieces, provide clearance for cutting tools, and special purposes (Fig. 46-3).

A *microset adjustable center* (Fig. 46-4) fits into the tailstock spindle and provides a means of aligning lathe centers or producing slight tapers on work machined between centers. An eccentric, or sometimes a dovetail, slide allows this type of center to be adjusted a limited amount to each side of center. Lathe centers are quickly and easily aligned with this type of center.

The *self-driving live center* (Fig. 46-5 on p. 358), mounted in the headstock spindle, is used when the entire length of a workpiece is being machined in one operation and when a chuck or lathe dog could not be used to drive the work. Grooves ground around the circumference of the lathe center point provide the drive for the workpiece. The work (usually a soft material, such as aluminum) is forced onto the driving center; a revolving dead center is

(d)

Figure 46-3 Types of revolving tailstock centers used to support work between lathe centers: (a) revolving dead center; (b) long point center; (c) changeable point center; (d) types of changeable points. *(Royal Products)*

.006 in.

.006 in.

Figure 46-4 A microset adjustable dead center.

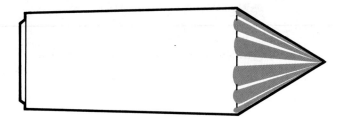

Figure 46-5 A self-driving live center.

used to support the work and hold it against the grooves of the driving center.

CHUCKS

Because of their size and shape, some workpieces cannot be held and machined between lathe centers. Lathe chucks are used extensively for holding work for machining operations. The most commonly used lathe chucks are the three-jaw universal, four-jaw independent, and collet chuck.

The *three-jaw universal chuck* (Fig. 46-6) holds round and hexagonal work. It grasps the work quickly and within a few thousandths of an inch or hundredths of a millimeter of accuracy because the three jaws move simultaneously when adjusted by the chuck wrench. This simultaneous motion is caused by a scroll plate into which all three jaws fit. Three-jaw chucks are made in various sizes, from 4 to 16 in. (100 to 400 mm) in diameter. They are usually provided with two sets of jaws, one for outside chucking and the other for inside chucking.

The *four-jaw independent chuck* (Fig. 46-7) has four jaws, each of which can be adjusted independently by a chuck wrench. They are used to hold round, square,

Figure 46-6 A three-jaw universal geared scroll chuck.

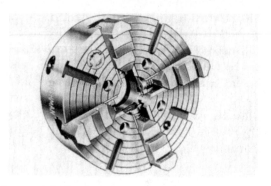

Figure 46-7 A four-jaw independent chuck.

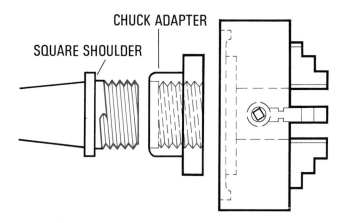

Figure 46-8 A threaded spindle nose. *(Kelmar Associates)*

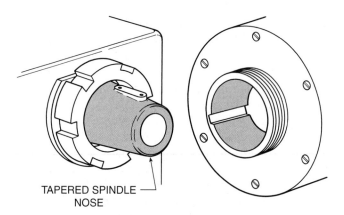

Figure 46-9 An American standard lathe spindle nose. *(Kelmar Associates)*

hexagonal, and irregularly shaped workpieces. The jaws can be reversed to hold work by the inside diameter.

Universal and independent chucks can be fitted to the three types of headstock spindles. Fig. 46-8 shows a threaded spindle nose; Fig. 46-9, a tapered spindle nose; and Fig. 46-10, a cam-lock spindle nose. The threaded type screws on in a clockwise direction; the tapered type is held by a lock nut that tightens on the chuck. The cam-lock is held by tightening the cam-locks using a T-wrench.

On the taper and cam-lock types, the chuck is aligned by the taper on the spindle nose.

The *collet chuck* (Fig. 46-11) is the most accurate chuck and is used for high-precision work. Spring collets are available to hold round, square, or hexagon-shaped workpieces. Each collet has a range of only a few thousandths of an inch or hundredths of a millimeter over or under the size stamped on the collet.

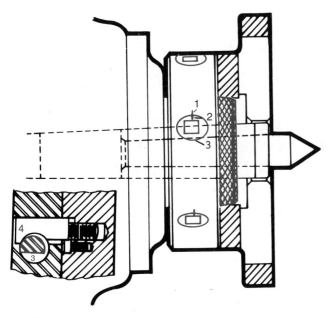

■ **Figure 46-10** The parts of a cam-lock spindle nose: (1) registration lines on spindle nose; (2) registration lines on cam-lock; (3) cam-locks; (4) cam-lock mating stud on chuck or faceplate.

A special adapter is fitted into the taper of the headstock spindle, and a hollow draw bar having an internal thread is inserted in the opposite end of the headstock spindle. As the handwheel (and draw bar) is rotated, it draws the collet into the tapered adapter, causing the collet to tighten on the workpiece. This type of chuck is also referred to as a *spring-collet chuck.* Another form of spring-collet chuck uses a chuck wrench to tighten the collet on the workpiece. This type is mounted on the spindle nose in the same manner as standard chucks and can hold larger work than the draw-in type.

The *Jacobs collet chuck* (Fig. 46-12 on p. 360) has a wider range than the spring-collet chuck. Instead of a draw bar, it utilizes an impact-tightening handwheel to close the collets on the workpiece. A set of 11 Rubber-Flex collets, each having an adjustment range of almost .125 in. (3 mm), makes it possible to hold a wide range of work diameters. When the handwheel is turned clockwise, the Rubber-Flex collet is forced into a taper, causing it to tighten on the workpiece. When the handwheel is turned counterclockwise, the collet opens and releases the workpiece.

Magnetic chucks (Fig. 46-13 on p. 360) are used to hold iron or steel parts that are too thin or that may be damaged if held in a conventional chuck. These chucks are fitted to an adapter mounted on the headstock spindle. Work is held lightly for aligning purposes by turning the chuck wrench approximately one-quarter turn. After the work has been trued, the chuck is turned to the full-on position to hold the work securely. This type of chuck is used only for light cuts and for special grinding applications.

Faceplates are used to hold work that is too large or of such a shape that it cannot be held in a chuck or between centers. Faceplates are usually equipped with several slots

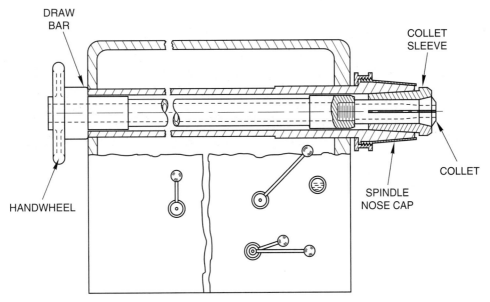

■ **Figure 46-11** A cross-sectional view of a headstock showing the construction of a draw-in collet assembly. *(Kelmar Associates)*

■ **Figure 46-12** The Jacobs collet chuck has a wider range than other collet chucks. *(Jacobs® Chuck Manufacturing Company)*

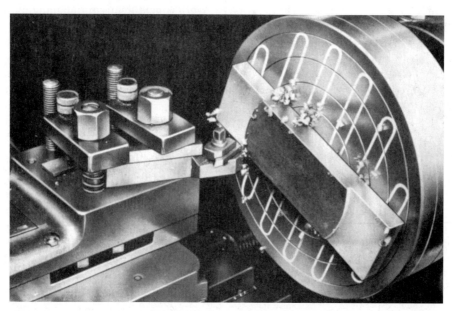

■ **Figure 46-13** Workpieces may be held on a magnetic chuck for turning operations.

to permit the use of bolts to secure the work or angle plate (Fig. 46-14) so that the axis of the workpiece may be aligned with the lathe centers. When work is mounted off center, a counterbalance (Fig. 46-14) should be fastened to the faceplate to prevent imbalance and the resultant vibrations when the lathe is in operation.

A *steadyrest* (Fig. 46-15) is used to support long work held in a chuck or between lathe centers. It is located on, and aligned by, the ways of the lathe and may be positioned at any point along the lathe bed, provided it clears the carriage travel. The three jaws, tipped with plastic, bronze, or rollers, may be adjusted to support any work di-

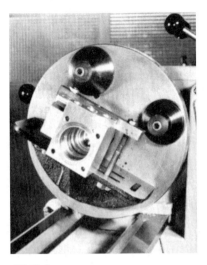

Figure 46-14 An angle plate fastened to a faceplate is used to hold a workpiece for machining. *(Colchester Lathe Company)*

Figure 46-15 A steadyrest is often used to support a long or slender workpiece during machining.

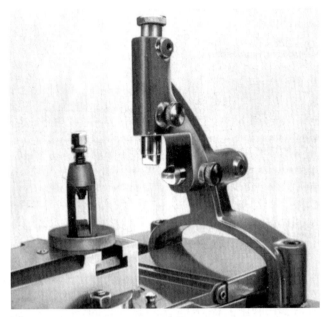

Figure 46-16 A follower rest mounted on the saddle may be used to support a long, slender workpiece during machining.

Figure 46-17 A plain mandrel. *(Ash Precision Equipment Inc.)*

ameter within the steadyrest capacity. During machining operations performed on or near the end of a workpiece, a steadyrest supports the end of work held in a chuck, when the work cannot be supported by the tailstock center. A steadyrest also supports the center of long work to prevent springing when the work is machined between centers.

A *follower rest* (Fig. 46-16), mounted on the saddle, travels with the carriage to prevent work from springing up and away from the cutting tool. The cutting tool is generally positioned just ahead of the follower rest to provide a smooth bearing surface for the two jaws of the follower rest.

A *mandrel* holds an internally machined workpiece between centers so that further machining operations are concentric with the bore. There are several types of mandrels, the most common being the *plain mandrel* (Fig. 46-17), *expanding mandrel, gang mandrel,* and *stub mandrel.*

LATHE DOGS

When work is machined between centers, it is generally driven by a lathe dog. The lathe dog has an opening to receive the work and a setscrew to fasten the dog to the work. The tail of the dog fits into a slot on the driveplate and provides the drive to the workpiece. Lathe dogs are made in a variety of sizes and types to suit various workpieces.

The *standard bent-tail lathe dog* (Fig. 46-18a on p. 362) is the most commonly used dog for round workpieces. These dogs are available with square-head setscrews or headless setscrews, which are safer, since there is no protruding head.

The *straight-tail dog* (Fig. 46-18b) is driven by a stud in the driveplate. Since this is a more balanced type of dog

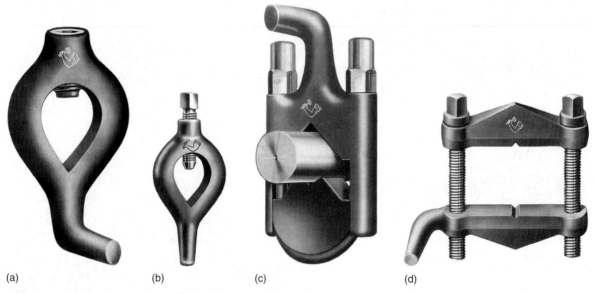

■ Figure 46-18 Common types of lathe dogs: (a) standard bent-tail; (b) straight tail; (c) safety clamp; (d) clamp type. *(Armstrong Tools)*

than the bent-tail dog, it is used in precision turning where the centrifugal force of a bent-tail dog may cause inaccuracies in the work.

The *safety clamp lathe dog* (Fig. 46-18c) may be used to hold a variety of work, since it has a wide range of adjustment. It is particularly useful on finished work where the setscrew of a standard lathe dog may damage the finish.

The *clamp lathe dog* (Fig. 46-18d) has a wider range than the other types and may be used on round, square, rectangular, and odd-shaped workpieces.

▶▶ Cutting-Tool-Holding Devices

Most toolbits used in lathe-turning operations in school shops are square and are generally held in a standard toolholder (Fig. 46-19). These toolholders are made in various styles and sizes to suit different machining operations. Toolholders for turning operations are available in three styles: left-hand offset, right-hand offset, and straight.

Each of these toolholders has a square hole to accommodate the square toolbit, which is held in place by a setscrew. The hole in the toolholder is at an angle of approximately 15° to 20° to the base of the toolholder (Fig. 46-19c). When the cutting tool is set on center, this angle provides the proper amount of back rake in relation to the workpiece.

The *left-hand offset toolholder* offset to the right (Fig. 46-19a) is designed for machining work close to the chuck or faceplate and for cutting from right to left. This type of toolholder is designated by the letter L to indicate the direction of cut.

The *right-hand offset toolholder* offset to the left (Fig. 46-19b) is designed for machining work close to the tailstock, for cutting from left to right, and for facing operations. This type of toolholder is designated by the letter R.

The *straight toolholder* (Fig. 46-19c) is a general-purpose type. It can be used for taking cuts in either direction and for general machining operations. This type of toolholder is designated by the letter S.

The *carbide toolholder* (Fig. 46-19d) has a square hole parallel to the base of the toolholder to accommodate carbide-tipped toolbits. When using carbide-tipped toolbits, hold the toolbit so that there is little or no back rake (Fig. 46-19d). Toolholders of this type are designated by the letter C. The correct methods of using this and other types of carbide toolholders are fully explained in Unit 31.

Toolholders for indexable carbide inserts are shown in Fig. 46-20. The insert in Fig. 46-20a is held in the holder by a cam action. The insert in Fig. 46-20b is secured by means of a clamp. These types of toolholders are available for use with conventional, turret-type, and heavy-duty toolposts.

Cutting-off, or *parting, tools* are used when the work must be grooved or parted off. The long, thin cutting-off blade is locked securely in the toolholder by means of either a cam lock or a locking nut. Three types of parting toolholders are shown in Fig. 46-21.

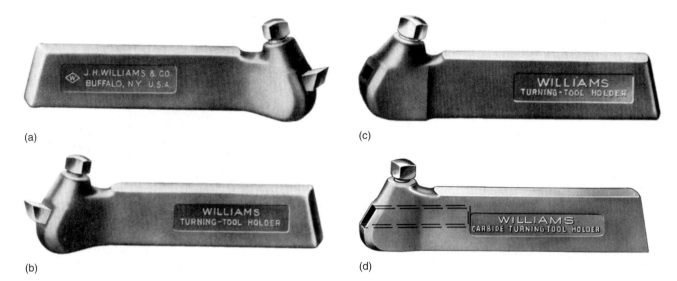

■ **Figure 46-19** Common lathe toolholders: (a) left-hand offset; (b) right-hand offset; (c) straight; (d) carbide. *(Bahco North America)*

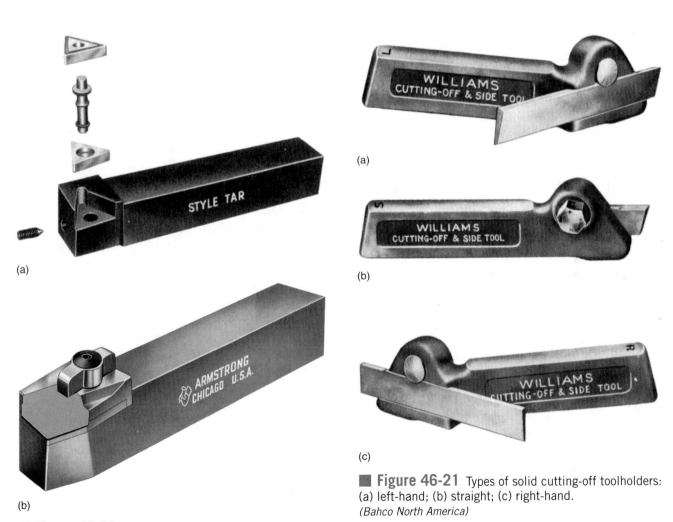

■ **Figure 46-20** (a) Carbide inserts are held securely in a holder by means of cam action; (b) carbide inserts are held securely by a clamp. *(Armstrong Tools)*

■ **Figure 46-21** Types of solid cutting-off toolholders: (a) left-hand; (b) straight; (c) right-hand. *(Bahco North America)*

Figure 46-22 A preformed thread-cutting tool and toolholder. *(Bahco North America)*

A *threading toolholder* (Fig. 46-22) is designed to hold a special form-relieved thread-cutting tool. Although most machinists grind their own thread-cutting tools, this tool is convenient because it has an accurately ground 60° angle. This angle is maintained throughout the life of the tool because only the top of the cutting surface is sharpened when it becomes dull.

Boring toolholders are made in several styles. The *light boring toolholder* (Fig. 46-23a) is held in a standard toolpost and is used for small holes and light cuts. The *medium boring toolholder* (Fig. 46-23b) is suitable for heavier cuts and is held in a standard toolpost. The cutting tool may be held at 45° or 90° to the axis of the bar.

The *heavy-duty boring bar holder* (Fig. 46-23c) is mounted on the compound rest of the lathe. It has three bars of different diameters to suit the diameter of the hole being bored. With this type of boring bar, use the largest bar possible to get the maximum rigidity and to avoid chatter. The toolbit may be held at 45° or 90° to the axis.

▸▸ Compound Rest Tooling Systems

The *standard,* or *round, toolpost* (Fig. 46-24) is generally supplied with the conventional engine lathe. This toolpost fits into the T-slot of the compound rest and provides a means of holding and adjusting the type of toolholder or cutting tool required for an operation. The concave ring and the wedge or rocker provide for the adjustment of the cutting-tool height.

MODULAR, OR QUICK-CHANGE, TOOLING

Modular, or *quick-change, tooling systems* were initially developed for CNC machine tools to improve accuracy, reduce tool-change time, and increase productivity. These same benefits can also be realized on conventional lathes through the use of compound rest tooling systems designed specifically for these machines.

Modular tooling, sometimes called the *complete tooling system,* can provide the flexibility and versatility to build a series of cutting tools necessary to manufacture any part. A modular tooling system must be rigid, accu-

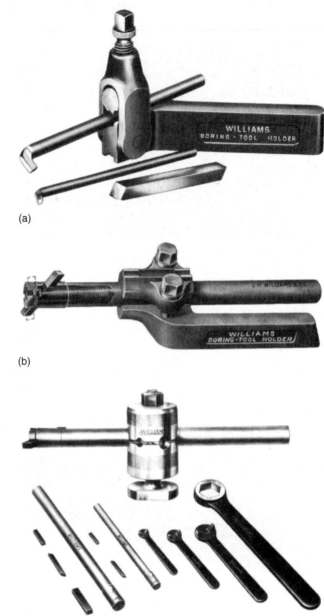

(a)

(b)

(c)

Figure 46-23 (a) A light boring toolholder; (b) a medium boring toolholder; (c) a heavy-duty boring tool. *(Bahco North America)*

rate, and have quick-change capabilities to provide productivity increases.

The principal function of a modular tooling system is to reduce the cost of keeping a large tool inventory. The basic clamping unit or turret for conventional lathes, which fits into the T-slot of the compound rest, can hold a variety of cutting tool modules. Any combination of cutting tools (turning, grooving, threading, knurling, cut-off, drilling, boring, etc.) can be mounted on the dovetailed turret quickly and accurately. Tools can be specifically mounted to suit the characteristics of the workpiece to be

■ Figure 46-24 A standard toolpost.
(Kelmar Associates)

Labels in figure: TOOLPOST SCREW, CONCAVE RING, TOOLPOST, ROCKER

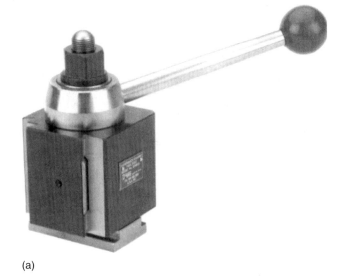

(a)

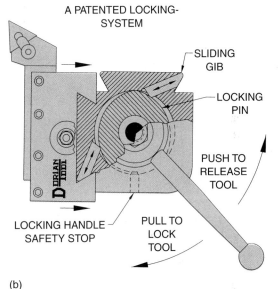

(b)

■ Figure 46-25 (a) The quick-change toolpost is a fast, accurate, and reliable method of changing tools; (b) the lock mechanism activates two sliding gibs that maximize the locking forces to ensure rigidity.
(Dorian Tool International)

Labels in figure (b): A PATENTED LOCKING-SYSTEM, SLIDING GIB, LOCKING PIN, PUSH TO RELEASE TOOL, PULL TO LOCK TOOL, LOCKING HANDLE SAFETY STOP

machined. This enables the operator to change tools quickly and accurately simply by releasing the quick-change locking mechanism, replacing the cutting unit, and locking the new tool in place.

Some of the more common quick-change tooling systems available for conventional lathes are:

1. *The Super Quick-Change Toolpost* (Fig. 46-25a) provides a fast, accurate, and reliable method of quickly changing and setting various toolholders for different operations. Its locking system has two sliding gibs that are forced out against the tool-holder when the handle is pulled into the lock position (Fig. 46-25b). This construction provides a rigid, positive lock with zero backlash.

   Various types of cutting tools can be mounted and preset in the dovetailed holders to machine a specific part. The cutting tool, which is held in the holder by means of setscrews, is generally sharpened and preset (in the holder) in the toolroom. After a cutting tool becomes dull, the unit can be quickly replaced with another sharpened and preset unit. This procedure ensures accuracy of size, since each preset toolbit will be in exactly the same position as the previous cutting tool. Each unit may be adjusted vertically on the toolpost by means of the knurled adjusting nut, then locked in position by means of a clamp.

2. *The Quadra* Index Toolpost* (Fig. 46-26a on p. 366) allows four tools to be mounted on the turret at the same time. Each tool is locked independently,

which provides the flexibility to use from one to four tools simultaneously. The unique indexing system of the turret allows it to be set in 24 positions, at every 15°, for the widest range of machining operations (Fig. 46-26b). It is possible to index from one cutting tool to another in less than one second with a repeatability of millionths of an inch.

3. *The Super-Six Index Turret* (Fig. 46-27 on p. 366) is designed to simplify and increase the machining productivity on engine lathes when multi-operation jobs require the use of more than one tool. The rotary index turret can be fitted with as many as six tools for external and internal

*Trademark Dorian Tool International

(a)

■ **Figure 46-27** The super-six index turret can increase the machining productivity of a lathe when machining multi-operation jobs.

OPERATING

Index in Position in less than 1 Second Tool to Tool

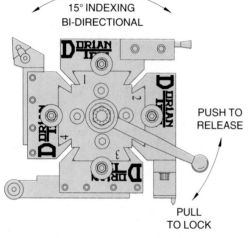

15° INDEXING
BI-DIRECTIONAL

PUSH TO
RELEASE

PULL
TO LOCK

Push the handle to release the QITP,
index into the desired position, then pull
the handle to lock the toolpost.

(b)

■ **Figure 46-26** (a) The Quadra index toolpost can hold up to four tools at once; (b) the Quadra index toolpost can be set at 24 positions, at every 15°.
(Dorian Tool International)

■ **Figure 46-28** The vertical index turret can give close to CNC performance for an engine lathe.
(Dorian Tool International)

machining operations. This unit allows height adjustments for each tool, and tool changes can be made in less than one second.

4. *The Vertical Index Turret (VIT)* (Fig. 46-28) is designed to give the highest accuracy, fastest tool change, and greatest rigidity of any tool system available for engine lathes. It operates on the same concept as the indexing turrets on CNC lathes and can hold up to six or eight tools for various machining operations. Its performance comes closest to the performance of CNC lathes in speed, precision, and rigidity.

■ **Figure 46-29** The universal knurling tool can use interchangeable heads to produce a variety of knurling patterns. *(Dorian Tool International)*

Other tooling systems developed for engine lathes are the *universal knurling tool* (Fig. 46-29) and the *spherical tool* (Fig. 46-30), designed for spherical machining operations. Both of these tools will be described in detail in Unit 53.

Note: The turret-type and quick-change toolposts have replaced the standard or round toolposts in most industrial shops and in some school shops, where carbide tools are used. These toolholders are more suitable for use with carbides. However, many training programs and jobbing shops still use high-speed steel toolbits and standard toolholders. The authors decided to continue with the use of standard toolposts, since more attention must be paid to the setup using this type of toolpost.

■ **Figure 46-30** The spherical tool is used to produce spherical concave and convex forms.

unit 46 review questions

Work-Holding Devices

1. Name the three types of lathe centers and state the purpose of each.

2. What precautions must be taken when turning work supported by solid centers?

3. Describe and state the purpose of the following:
   a. Three-jaw universal chuck
   b. Four-jaw independent chuck
   c. Collet chuck
   d. Magnetic chuck

4. What advantage does a Jacobs collet chuck have over a spring-collet chuck?

5. What is the purpose of:
   a. A steadyrest?   b. A follower rest?

6. Name three types of lathe dogs.

7. What is the disadvantage of the square-head setscrew in a lathe dog?

8. What is the advantage of a straight-tail dog?

Cutting-Tool-Holding Devices

9. Describe three types of standard toolholders and state the purpose of each.

10. How does a carbide toolholder differ from a standard toolholder?

11. State two methods by which indexable carbide inserts are held in a toolholder.

12. What procedure should be followed when using a heavy-duty boring bar set?

13. Name and state the purpose of four types of toolposts.

UNIT 47

Cutting Speed, Feed, and Depth of Cut

OBJECTIVES

After completing this unit, you will be able to:

1 Calculate the speed at which a workpiece should revolve

2 Determine the proper feed for roughing and finishing cuts

3 Estimate the time required to machine a part in a lathe

To operate a lathe efficiently, the machinist must consider the importance of cutting speeds and feeds. Much time may be lost if the lathe is not set at the proper spindle speed and if the proper feed rate is not selected.

▶▶ Cutting Speed

Lathe work *cutting speed (CS)* may be defined as the rate at which a point on the work circumference travels past the cutting tool. For instance, if a metal has a CS of 90 ft/min, the spindle speed must be set so that 90 ft of the work circumference will pass the cutting tool in 1 min. Cutting speed is always expressed in feet per minute (ft/min) or in meters per minute (m/min). Do not confuse the CS of a metal with the number of turns the workpiece will make in 1 min (r/min).

Industry demands that machining operations be performed as quickly as possible; therefore, correct CS must be used for the type of material cut. If the CS is too high, the cutting-tool edge breaks down rapidly, resulting in time lost to recondition the tool. With too slow a CS, time will be lost for the machining operation, resulting in low production rates. Based on research and testing by steel and cutting-tool manufacturers, the CS for high-speed steel tools listed in Table 47.1 are recommended for efficient metal-removal rates. These speeds may be varied slightly to suit factors such as the condition of the machine, the

type of work material, and sand or hard spots in the metal. The CS for cemented-carbide and ceramic cutting tools may be found in Units 31 and 32, Tables 31.5 and 32.3.

To calculate the lathe spindle speed in revolutions per minute (r/min), the CS of the metal and the diameter of the work must be known. The proper spindle speed can then be set by dividing the CS (in inches per minute) by the circumference of the work (in inches). The calculation for determining the spindle speed (r/min) is as follows.

INCH CALCULATIONS

$$\text{Formula:} \qquad \text{r/min} = \frac{\text{CS} \times 12}{\pi D}$$

where
$$\text{CS} = \text{cutting speed}$$
$$D = \text{diameter of work to be turned}$$

However, because most lathes provide only a limited number of speed settings, a simpler formula is usually used:

$$\text{r/min} = \frac{\text{CS} \times 4}{D}$$

table 47.1 — Lathe cutting speeds in feet and meters per minute using a high-speed steel toolbit

| Material | Turning and Boring | | | | Threading | |
| | Rough Cut | | Finish Cut | | | |
	ft/min	m/min	ft/min	m/min	ft/min	m/min
Machine steel	90	27	100	30	35	11
Tool steel	70	21	90	27	30	9
Cast iron	60	18	80	24	25	8
Bronze	90	27	100	30	25	8
Aluminum	200	61	300	93	60	18

Thus, to calculate the r/min required to rough-turn a 2-in. diameter piece of machine steel (CS 90):

$$r/min = \frac{CS \times 4}{D}$$

$$= \frac{90 \times 4}{2}$$

$$= 180$$

Note: The recommended speeds and feeds for carbide cutting tools are found in Unit 31. The same formula for calculating spindle speed will be used for these tools.

METRIC CALCULATIONS

$$\text{Formula:} \quad r/min = \frac{CS \times 320}{D}$$

EXAMPLE

Calculate the r/min required to turn a 45-mm diameter piece of machine steel (CS 40 m/min).

$$r/min = \frac{CS \times 320}{D}$$

$$= \frac{40 \times 320}{45}$$

$$= \frac{12,800}{45}$$

$$= 284$$

▸▸ Lathe Feed

The *feed* of a lathe may be defined as the distance the cutting tool advances along the length of the work for every revolution of the spindle. For example, if the lathe is set for a .015-in. (0.4-mm) feed, the cutting tool will travel along the length of the work .015 in. (0.4 mm) for every complete turn that the work makes. The feed of an engine lathe is dependent on the speed of the lead screw or feed rod. The speed is controlled by the change gears in the *quick-change gearbox* (Fig. 45-4a).

Whenever possible, only two cuts should be taken to bring a diameter to size: a roughing cut and a finishing cut. Since the purpose of a roughing cut is to remove excess material quickly and surface finish is not too important, a coarse feed should be used. The finishing cut is used to bring the diameter to size and produce a good surface finish, and therefore a fine feed should be used.

For general-purpose machining, a .010- to .015-in. (0.25- to 0.4-mm) feed for roughing and a .003- to .005-in. (0.07- to 0.012-mm) feed for finishing is recommended. Table 47.2 lists the recommended feeds for cutting various materials when a high-speed steel cutting tool is used.

▸▸ Depth of Cut

The *depth of cut* may be defined as the depth of the chip taken by the cutting tool and is one-half the total amount removed from the workpiece in one cut. Fig. 47-1 on p. 370 shows a .125-in. depth of cut being taken on a 2-in. diameter workpiece. Note that the diameter has been reduced by .250 in. to 1.750 in. When machining a workpiece, take only one roughing and one finishing cut if

table 47.2 Feeds for various materials (using a high-speed steel cutting tool)

Material	Rough Cuts		Finish Cuts	
	in.	mm	in.	mm
Machine steel	.010–.020	0.25–0.5	.003–.010	0.07–0.25
Tool steel	.010–.020	0.25–0.5	.003–.010	0.07–0.25
Cast iron	.015–.025	0.4–0.65	.005–.012	0.13–0.3
Bronze	.015–.025	0.4–0.65	.003–.010	0.07–0.25
Aluminum	.015–.030	0.4–0.75	.005–.010	0.13–0.25

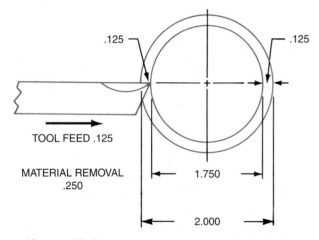

.125   .125

TOOL FEED .125

MATERIAL REMOVAL
.250

1.750

2.000

■ **Figure 47-1** The depth of cut on a lathe.

■ **Figure 47-2** Micrometer collars on the compound rest and crossfeed handles permit machining the workpiece to an accurate size. The thumbscrew, A, is used to lock the collar in place. *(South Bend Lathe Corp.)*

possible. If much material must be removed, the roughing cut should be as deep as possible to reduce the diameter to within .030 to .040 in. (0.76 to 1 mm) of the size required. The depth of a rough-turning cut will depend on the following factors:

> The condition of the machine
> The type and shape of the cutting tool used
> The rigidity of the workpiece, the machine, and the cutting tool
> The rate of feed

The depth of a finish-turning cut will depend on the type of work and the finish required. In any case, it should not be less than .005 in. (0.13 mm).

▶▶ Graduated Micrometer Collars

When the diameter of a workpiece must be turned to an accurate size, *graduated micrometer collars* should be used. Graduated micrometer collars are sleeves or bushings that are mounted on the compound rest and crossfeed screws (Fig. 47-2). They assist the lathe operator to set the

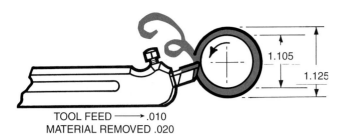

TOOL FEED ──→ .010
MATERIAL REMOVED .020

■ **Figure 47-3** On machines where the workpiece revolves, the cutting tool should be set in for only half the amount to be removed from the diameter. *(Kelmar Associates)*

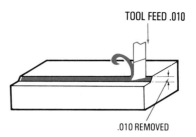

TOOL FEED .010

.010 REMOVED

■ **Figure 47-4** On machines where the workpiece does not revolve, the cutting tool should be set in for the amount of material to be removed. *(Kelmar Associates)*

cutting tool accurately to remove the required amount of material from the workpiece. The collars on lathes using the inch system of measurement are usually graduated in thousandths of an inch (.001). The micrometer collars on lathes using the metric system of measurement are usually graduated in steps of two hundredths of a millimeter (0.02 mm). The circumference of the crossfeed and compound rest screw collars on lathes using the inch system of measurement are usually divided into 100 or 125 equal divisions, each having a value of .001 in. Therefore, if the crossfeed screw is turned *clockwise* 10 graduations, the cutting tool will be moved .010 in. toward the work. Because the work in a lathe revolves, a .010-in. depth of cut will be taken from the entire work circumference, thereby reducing the diameter .020 in. (2 × .010 in.; see Fig. 47-3). On some lathes, however, the graduations are such that, when the graduated collar is moved 10 divisions, the cutting tool will move only .005 in. and the diameter will be reduced by .010 in., or by an amount equal to the reading on the collar.

Note: Some industrial machines may have collars with 250 or even 500 graduations, which make a much finer infeed possible.

Machine tools equipped with graduated collars generally fall into two classes:

1. *Machines in which the work revolves.* These machines include lathes, vertical boring mills, and cylindrical grinders.

2. *Machines in which the work does not revolve.* These machines include milling machines and surface grinders.

When the *circumference of work* is cut on machines in which the work revolves, remember that, since material is removed from the entire circumference, the cutting tool should be moved in only half the amount of material to be removed.

On machines in which the work does not revolve, the material removed from a workpiece is equal to the amount set on the graduated collar because the machining is taking place only on one surface. Therefore, if a .010-in. depth of cut is set, .010 in. will be removed from the work (Fig. 47-4).

HINTS ON GRADUATED COLLAR USE

1. If the graduated collar has a locking screw, make sure the collar is secure before setting a depth of cut (Fig. 47-2).

2. All depths of cut must be made by feeding the cutting tool *toward the workpiece.*

3. If the graduated collar is turned past the desired setting, it must be turned backward a half-turn and then fed into the proper setting to remove the backlash (the play between the feed screw and the nut).

4. Never hold a graduated collar when setting a depth of cut. Graduated collars with friction devices can be moved easily if held when a depth of cut is being set.

5. The graduated collar on the *compound rest* can be used for accurately setting the depth of cut for the following operations:

> *Shoulder turning.* When a series of shoulders must be spaced accurately along a piece of work, the compound rest should be set at 90° to the cross-slide. With the carriage locked in position, the graduated collar of the compound rest can be used for the spacing of shoulders to within .001-in. (0.02-mm on metric collars) accuracy.

> *Facing.* The graduated collars may also be used to face a workpiece to length. When the compound rest is swung to 30°, the amount

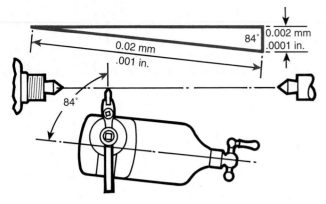

■ Figure 47-5 The compound rest is set at 84°16′ for making fine settings.

removed from the length of the work during facing will be one-half the amount of feed on the graduated collar.

> *Machining accurate diameters.* When accurate diameters must be machined or ground in a lathe, the compound rest should be set to 84°16′ to the cross-slide. On inch-calibrated compound rests, a .001-in. movement would result in a .0001-in. infeed movement of the tool (Fig. 47-5). Similarly, on metric lathes, a 0.02-mm movement of the compound rest results in a 0.002-mm infeed movement of the cutting tool.

▸▸ Calculating Machining Time

Any machinist should be able to estimate the time required to machine a workpiece. Factors such as spindle speed, feed, and depth of cut must be considered.

The following formula may be applied to calculate the time required to machine a workpiece:

$$\text{Time} = \frac{\text{distance}}{\text{rate}}$$

where   distance = length of cut
                 rate = feed × r/min

EXAMPLE

Calculate the time required to machine a 2-in. diameter machine-steel shaft 16 in. long to 1.850-in. diameter finish size.

Solution

Roughing cut—see Table 47.1:

$$\text{r/min} = \frac{CS \times 4}{D}$$

$$= \frac{90 \times 4}{2}$$

$$= 180$$

Roughing feed—see Table 47.2:

$$\text{Feed} = .020$$

$$\text{Roughing cut time} = \frac{\text{length of cut}}{\text{feed} \times \text{r/min}}$$

$$= \frac{16}{.020 \times 180}$$

$$= 4.4 \text{ min}$$

Finishing cut:

$$\text{r/min} = \frac{100 \times 4}{1.850}$$

$$= 216$$

$$\text{Finishing feed} = .003$$

$$\text{Finishing cut time} = \frac{16}{.003 \times 216}$$

$$= 24.7 \text{ min}$$

Total machine time: Roughing cut time + finishing cut time:

$$\text{Total time} = 4.4 + 24.7$$

$$= 29.1 \text{ min}$$

Cutting Speeds and Feeds

1. Define CS and state how it is expressed.

2. Why is proper CS important?

3. At what r/min should the lathe revolve to rough-turn a piece of cast iron 101 mm in diameter when using a high-speed steel toolbit?

4. Calculate the r/min to turn a piece of 3 3/4-in. diameter machine steel using a high-speed steel toolbit.

5. Define lathe feed.

Depth of Cut

6. Define depth of cut.

7. How deep should a roughing cut be?

8. What factors determine the depth of a roughing cut?

9. A 2.500 in. diameter workpiece must be machined to a 2.375-in. finished diameter. What should be the depth of:

   a. The roughing cut?   b. The finishing cut?

Graduated Micrometer Collars

10. Name and state the differences between two classes of machines equipped with graduated collars.

11. What precautions should be taken when setting a depth of cut?

12. What is the value of one graduation on a metric graduated micrometer collar?

13. What size will a 75-mm diameter workpiece be after a 6.25-mm deep cut is taken from the work?

Machining Time

14. Calculate the time required to machine a 3.125 in. diameter piece of tool steel 14 in. long to a diameter of 3.000 in.

UNIT 48

Lathe Safety

OBJECTIVES

After completing this unit, you will be able to:

1 State the importance of safety standards in a shop

2 List the necessary safety precautions required to run a lathe

3 Point out any safety infractions by other workers

A good worker will be aware of safety requirements in any area of a shop and will always attempt to observe safety rules. Failure to do so may result in serious injury, with the resultant loss of time and pay and in the loss of production to the company.

▶▶ Safety Precautions

The lathe, like most other machine tools, can be hazardous if not operated properly. A good lathe operator is a safe operator who realizes the importance of keeping the machine and the surrounding area clean and tidy. Accidents on any machine do not just happen; they are usually caused by carelessness and can generally be avoided. To minimize the chance of accidents when operating a lathe, the following precautions should be observed:

1. Always wear approved safety glasses. When operating a lathe, chips fly and it is important to protect your eyes.

2. Roll up your sleeves, remove your tie, and tuck in loose clothing. Short sleeves are preferable because loose clothing can get caught by revolving lathe dogs, chucks, and rotating parts of the lathe. You can be drawn into the machine and be seriously injured.

3. Never wear a ring or watch (Fig. 48-1).

   a. Rings or watches can be caught in revolving work or lathe parts and cause serious injuries.

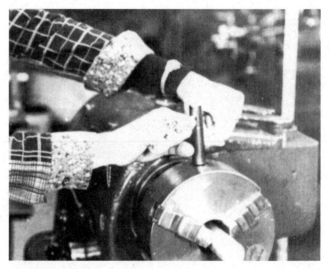

■ Figure 48-1 Wearing rings and watches in a machine shop can be dangerous. *(Kelmar Associates)*

   b. A metal object falling on the hand would bend or break a ring, causing much pain and suffering until the ring could be removed.

4. Do not operate a lathe until you fully understand its controls.

    a. Not knowing what might happen when levers or switches are turned on can be very dangerous.

    b. Be sure that you can stop the machine quickly in case something unexpected happens.

5. Never operate a machine if safety guards are removed or not closed properly.

    a. Safety guards are installed by the manufacturer to cover revolving gears, belts, and shafts.

    b. Loose clothing or a hand can be drawn into the revolving parts if guards are not replaced.

6. Stop the lathe before you measure the work or clean, oil, or adjust the machine. Measuring revolving parts can result in broken tools or personal injury.

7. Do not use a rag to clean the work or the machine when the lathe is in operation. The rag may get caught and be drawn in along with your hand.

8. Never attempt to stop a lathe chuck or driveplate by hand. Your hand can be injured or fingers broken if they get caught in the slots and projections of the driveplate or chuck.

9. Be sure that the chuck or faceplate is mounted securely before starting the lathe.

    a. If the lathe is started with a loose spindle accessory, the rotation would loosen the accessory and cause it to be thrown from the lathe.

    b. A heavy accessory, with the velocity created by the revolving spindle, can turn into a dangerous missile.

10. Always remove the chuck wrench after use (Fig. 48-2). *Never leave it in the chuck at any time!* If the lathe is started with a chuck wrench in the chuck, the following could result:

    a. The wrench could fly out and injure someone.

    b. The wrench could be jammed against the lathe bed, damaging the wrench, lathe bed, chuck, and lathe spindle.

11. Move the carriage to the farthest position of the cut and revolve the lathe spindle one complete turn by hand before you start the lathe.

    a. This will ensure that all parts will clear without jamming.

    b. It will also prevent an accident and damage to the lathe.

12. Keep the floor around the machine free from grease, oil, metal cuttings, tools, and workpieces (Fig. 48-3).

    a. Oil and grease can cause falls, which can result in painful injuries.

    b. Objects on the floor are hazards that can cause tripping accidents.

13. Avoid horseplay at all times, especially when operating any machine tool. Horseplay can result in tripping accidents or being pushed into the revolving spindle or workpiece.

■ **Figure 48-3** Good housekeeping can prevent tripping and slipping accidents. *(Kelmar Associates)*

■ **Figure 48-2** Never leave a chuck wrench in a chuck. *(Kelmar Associates)*

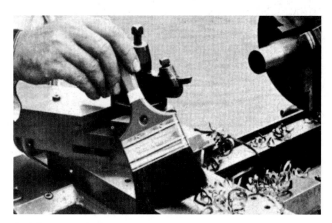

■ **Figure 48-4** Always remove chips with a brush or hook, never by hand. *(Kelmar Associates)*

14. Always remove the chips with a brush and never with your hand or a cloth (Fig. 48-4). Steel chips are sharp and can cause cuts if handled manually or with a cloth that has chips embedded in it.

15. Whenever polishing, filing, cleaning, or making adjustments to the workpiece or machine, always remove the sharp toolbit from the toolholder to prevent serious cuts to your arms or hands.

unit 48 review questions

1. State three possible results of not observing the safety rules in a shop.

2. What is generally the most important cause of an accident?

3. Why are the following precautions important when operating a lathe?

   a. Wearing safety glasses
   b. Not wearing loose clothing
   c. Not wearing watches and rings
   d. Removing a chuck wrench
   e. Keeping the machine and surrounding floor area clean
   f. Stopping the lathe to measure work or clean the machine

4. Why should chips not be cleaned from a lathe with a cloth?

UNIT 49

Mounting, Removing, and Aligning Lathe Centers

OBJECTIVES

After completing this unit, you will be able to:

1 Mount and/or remove the lathe centers properly

2 Align the lathe centers by visual, trial-cut, and dial-indicator methods

Any work machined between lathe centers is generally turned for some portion of its length, then reversed, and the other end finished. It is important, when machining work between centers, that the live center runs absolutely true. If the live center does not run true, the two turned diameters will not be concentric when the work is reversed to machine the opposite end and the part may have to be scrapped.

▶▶ To Mount Lathe Centers

1. Remove any burrs from the lathe spindle, centers, or spindle sleeves (Fig. 49-1).

2. Thoroughly clean the tapers on the lathe centers and in the headstock and tailstock spindles (Fig. 49-2).

Note: Never attempt to clean the taper in the headstock spindle while the lathe is running.

3. Partially insert the cleaned center in the lathe spindle.

4. With a quick snap, force the center into the spindle. When mounting a tailstock center, follow the same procedure.

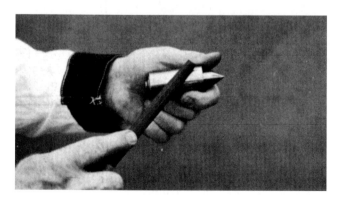

■ **Figure 49-1** Remove burrs from the spindle sleeve before inserting it in a lathe. *(Kelmar Associates)*

■ **Figure 49-2** Clean the headstock spindle before inserting a center. *(Kelmar Associates)*

Figure 49-3 Using a dial indicator to check the trueness of the live center. *(Kelmar Associates)*

After a center has been mounted in the headstock spindle, it should be checked for trueness. Start the lathe and observe if the center runs true. Whenever accuracy is required, check the trueness of the center with a dial indicator (Fig. 49-3). If the center is not running true and has been mounted properly, it should be ground while mounted in the lathe spindle.

▶▶ To Remove Lathe Centers

The *live center* may be removed by using a *knockout bar* that is pushed through the headstock spindle (Fig. 49-4). A slight tap is required to remove the center. When removing the live center with a knockout bar, place a cloth over the center and hold it with one hand to prevent an accident or damage to the center (Fig. 49-5).

The *dead center* can be removed by turning the tailstock handwheel to draw the spindle back into the tailstock. The end of the screw contacts the end of the dead center, forcing it out of the spindle.

Figure 49-4 Removing the live center with a knockout bar. *(Kelmar Associates)*

Figure 49-5 When removing the lathe center, hold a cloth over its point to prevent hand injury. *(Kelmar Associates)*

▶▶ Alignment of Lathe Centers

To produce a parallel diameter when machining work between centers, the lathe center must be aligned; that is, the two lathe centers must be in line with each other and true with the centerline of the lathe. If the centers are not aligned, the work being machined will be tapered.

Three common methods are used to align lathe centers:

1. By aligning the centerlines on the back of the tailstock with each other (Fig. 49-6); this is only a visual check and therefore not too accurate

2. By using the trial-cut method (Fig. 49-7), where a small cut is taken from each end of the work and the diameters are measured with a micrometer

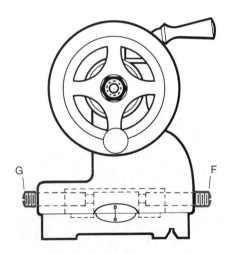

Figure 49-6 The lines on the tailstock must be in line to produce a parallel diameter. *(South Bend Lathe Corp.)*

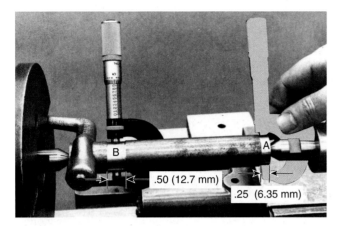

■ Figure 49-7 A trial cut at each end of the workpiece is used to check the lathe's center alignment. *(Kelmar Associates)*

3. By using a parallel test bar and dial indicator (Fig. 49-8); this is the fastest and most accurate method of aligning lathe centers

TO ALIGN CENTERS BY ADJUSTING THE TAILSTOCK

1. Loosen the tailstock clamp nut or lever.

2. Loosen one of the adjusting screws, G or F (Fig. 49-6), depending on the direction the tailstock must be moved. Tighten the other adjusting screw until the line on the top half of the tailstock aligns exactly with the line on the bottom half.

3. Tighten the loosened adjusting screw to lock both halves of the tailstock in place.

4. Make sure that the tailstock lines are still aligned; adjust them if necessary.

5. Lock the tailstock clamp nut or lever.

TO ALIGN CENTERS BY THE TRIAL-CUT METHOD

1. Take a light cut [approximately .005 in. (0.12 mm)] to true the diameter from section A at the tailstock end for .250 in. (6 mm) long, Fig. 49-7.

2. Stop the feed and note the reading on the graduated collar of the crossfeed handle.

3. Move the cutting tool away from the work with the crossfeed handle.

4. Bring the cutting tool close to the headstock end.

5. Return the cutting tool to the same graduated collar setting as at section A.

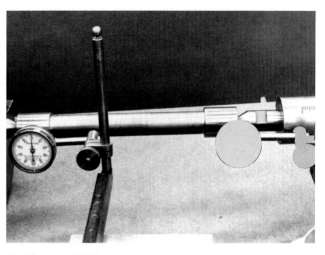

■ Figure 49-8 Tailstock alignment may be accurately checked with a parallel test bar and dial indicator. *(Kelmar Associates)*

6. Cut a .500-in. (13-mm) length at section B and then stop the lathe.

7. Measure both diameters with a micrometer (Fig. 49-7).

8. If both diameters are not the same size, adjust the tailstock either toward or away from the cutting tool one-half the difference of the two readings.

9. Take another light cut at A and B at the same crossfeed graduated collar setting. Measure these diameters and adjust the tailstock, if required.

TO ALIGN CENTERS USING A DIAL INDICATOR AND TEST BAR

1. Clean the lathe and work centers and mount the test bar.

2. Adjust the test bar snugly between centers and tighten the tailstock spindle clamp.

3. Mount a dial indicator on the toolpost or lathe carriage. Make sure that the indicator plunger is parallel to the lathe bed and that the contact point is set on center.

4. Adjust the cross-slide so that the indicator registers approximately .025 in. (0.65 mm) at the tailstock end, and set the indicator bezel to 0.

5. Move the carriage by hand so that the indicator registers on the diameter at the headstock end (Fig. 49-8) and note the indicator reading.

6. If both indicator readings are not the same, adjust the tailstock with the adjusting screws until the indicator registers the same reading at both ends.

7. Tighten the adjusting screw that was loosened.

8. Tighten the tailstock clamp nut.

9. Adjust the tailstock spindle until the test bar is snug between the lathe centers.

10. Recheck the indicator readings at both ends and adjust the tailstock, if necessary.

unit 49 review questions

1. Briefly describe the proper procedure for mounting a lathe center.

2. How may the live center be checked for trueness after it has been mounted in the spindle?

3. What precautions should be observed when removing a live center?

4. Briefly describe how to align the lathe centers by the trial-cut method.

5. Why must the test bar be snug between centers when aligning centers with a dial indicator?

Grinding Lathe Cutting Tools

OBJECTIVES

After completing this unit, you will be able to:

1 Explain the importance of various rakes and clearance cutting-tool angles

2 Grind a general-purpose cutting tool from a blank

3 Resharpen a dull cutting tool

Because of the number of turning operations that can be performed on a lathe, a wide variety of cutting tools are used. For these cutting tools to perform effectively, they must possess certain angles and clearances for the material being cut (see Table 29.2). All lathe tools cut if they have front and side clearance. The addition of side and top rake enables the chips to escape quickly from the cutting edge, making the tool cut better. All lathe cutting tools must have certain angles and clearances regardless of shape; therefore, only the grinding of a general-purpose cutting tool (Fig. 50-1) will be explained in detail.

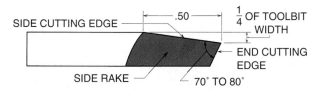

■ **Figure 50-1** The shape and dimensions of a general-purpose lathe toolbit. *(Kelmar Associates)*

▶▶ To Grind a General-Purpose Toolbit

1. Dress the face of the grinding wheel.
2. Grip the toolbit firmly, supporting the hands on the grinder toolrest (Fig. 50-2 on p. 382).
3. Hold the toolbit at the proper angle to grind the cutting edge angle. At the same time, tilt the bottom of the toolbit in toward the wheel and grind the 10° side relief or clearance angle on the cutting edge.

Note: The cutting edge should be approximately ½ in. (13 mm) long and should extend over about one-quarter the width of the toolbit (Fig. 50-2).

4. While grinding, move the toolbit back and forth across the face of the wheel. This speeds up grinding and prevents grooving the wheel.
5. The toolbit must be cooled frequently during the grinding operation.

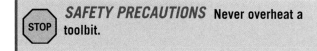

SAFETY PRECAUTIONS Never overheat a toolbit.

Note: *Never* quench stellite or cemented-carbide tools and *never* grind carbides with an aluminum oxide wheel.

6. Grind the end cutting edge so that it forms an angle of a little less than 90° with the side cutting edge

(a)

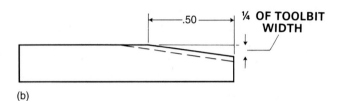

(b)

■ **Figure 50-2** Grinding the side cutting edge and side relief angles on a toolbit. *(Kelmar Associates)*

(Fig. 50-3). Hold the tool so that the end cutting edge angle and end relief angle of 15° are ground at the same time.

7. Using a toolbit grinding gage, check the amount of end relief when the toolbit is in the toolholder (Fig. 50-4).

8. Hold the top of the toolbit at approximately 45° to the axis of the wheel and grind the side rake to approximately 14° (Fig. 50-5).

Note: When grinding the side rake, *be sure that the top of the cutting edge is not ground below the top of the toolbit.* If a step is ground in the top of the toolbit, a chip trap is formed, which greatly reduces the efficiency of the cutting tool.

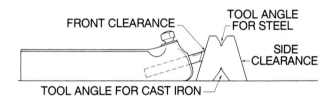

■ **Figure 50-4** Checking the end relief angle of a toolbit while it is in a toolholder. *(South Bend Lathe Corp.)*

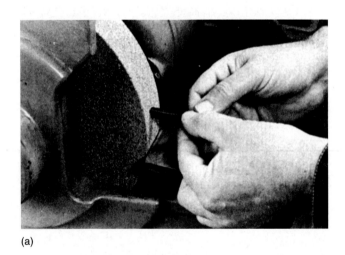

(a)

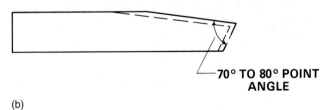

(b)

■ **Figure 50-3** Grinding the end relief angle on a lathe toolbit. *(Kelmar Associates)*

(a)

(b)

■ **Figure 50-5** Grinding the side rake on a lathe toolbit. *(Kelmar Associates)*

9. Grind a slight radius on the point of the cutting tool, *being sure to maintain the same front and side clearance angle.*

10. With an oilstone, hone the cutting edge of the toolbit slightly. This will lengthen the life of the toolbit and enable it to produce a better surface finish.

unit 50 review questions

1. What are two requirements that must be met to enable a lathe toolbit to cut?

2. Why should the top of the cutting edge not be ground below the top of the toolbit when the side rake is ground?

3. How should the point of a toolbit be conditioned?

4. Briefly state the procedure for grinding a general-purpose toolbit.

UNIT 51

Facing Between Centers

OBJECTIVES

After completing this unit, you will be able to:

1 Set up a workpiece for machining between centers

2 Set up a workpiece to face the ends

3 Face a workpiece to an exact length

Work mounted between centers can be machined, removed, or set up for additional machining and still maintain the same degree of accuracy. Facing on a lathe is one of the most important machining operations in a machine shop. It is very important that the cutting tool and work be properly set up, or damage to the machine, work, and lathe centers will result.

▶▶ To Set Up a Cutting Tool for Machining

1. Move the toolpost to the *left-hand side* of the T-slot in the compound rest.

2. Mount a toolholder in the toolpost so that the setscrew in the toolholder is approximately 1 in. (25 mm) beyond the toolpost (Fig. 51-1).

3. Insert the proper cutting tool into the toolholder, having the toolbit extend .500 in. (13 mm) beyond the toolholder, but never more than twice its thickness.

4. Tighten the toolholder setscrew with only two-finger pressure on the wrench to hold the toolbit in the toolholder.

Note: If the toolholder setscrew is tightened too tightly, it will break the toolbit, which is very hard and brittle.

5. Set the cutting-tool point to center height. Check it against the lathe center point (Fig. 51-1).

6. Tighten the toolpost securely to prevent it from moving during a cut.

■ **Figure 51-1** A toolholder and toolbit being set up for a machining operation. *(Kelmar Associates)*

▶▶ To Mount Work Between Centers

USING A REVOLVING TAILSTOCK CENTER

1. Check the live center by holding a piece of chalk close to it while it is revolving. If the live center is not running true, the chalk will mark only the high spot.

2. If this chalkmark shows, remove the live center from the headstock and clean the tapers on the center and the headstock spindle.

3. Replace the center and check for trueness.

4. Adjust the tailstock spindle until it extends about 2.50 to 3 in. (63 to 75 mm) beyond the tailstock.

5. Mount a revolving center into the tailstock spindle (Fig. 51-2).

6. Loosen the tailstock clamp nut or lever.

7. Place the lathe dog on the end of the work with the tail pointing to the left.

8. Place the end of the work with the lathe dog on the live center and slide the tailstock toward the headstock until the dead center supports the other end of the work.

9. Tighten the tailstock clamp nut or lever (Fig. 51-3).

10. Adjust the tail of the dog in the slot of the driveplate and tighten the lathe dog screw.

11. Tighten the tailstock handwheel using *thumb and finger pressure* only to snug up the work between centers.

■ Figure 51-3 When the tailstock is in the correct position, tighten the clamp nut. *(Kelmar Associates)*

12. Tighten the tailstock spindle clamp.

13. Move the carriage to the farthest position (left-hand end) of the cut and revolve the lathe spindle by hand to make sure that the dog does not hit the compound rest.

USING A DEAD CENTER

1. Follow steps 1 to 4 of the revolving center method.

2. Mount the dead center in the tailstock spindle and then check the center alignment.

3. Place the dog on the end of the work and lubricate the dead center hole (Fig. 51-4).

4. Mount the work between centers and tighten the tailstock clamp nut or lever (Fig. 51-3).

5. Adjust the tail of the dog in the driveplate and tighten the setscrew.

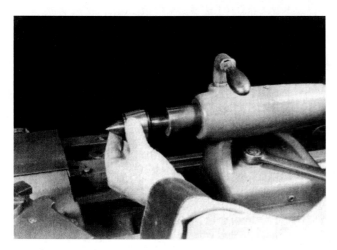

■ Figure 51-2 Clean both the internal and external tapers before mounting a center in the tailstock spindle. *(Kelmar Associates)*

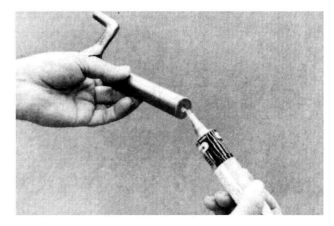

■ Figure 51-4 Applying lubricant to a center hole that will be supported by a dead tailstock center. *(Kelmar Associates)*

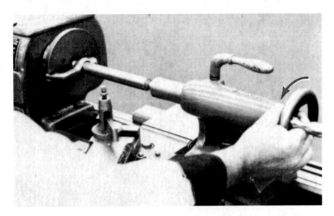

■ Figure 51-5 Reverse the tailstock handwheel until the tail of the dog drops into the slot. *(Kelmar Associates)*

■ Figure 51-6 The center tension is correct when the tail of the dog drops of its own weight and there is no end play between centers. *(Kelmar Associates)*

6. Turn the driveplate until the slot is in a horizontal position.

7. Hold the tail of the dog *up* in the slot and tighten the tailstock handwheel.

8. Turn the handwheel backward until the tail of the dog *just drops* and tighten the tailstock spindle clamp (Fig. 51-5).

9. Check the center tension of the work. The tail of the dog should drop of its own weight, and there should be no end play (Fig. 51-6).

▶▶ Facing Between Centers

Workpieces to be machined are generally cut a little longer than required and then end-faced to the proper length. Facing is an operation of machining the ends of a workpiece square with its axis. To produce a flat, square surface when material is being faced between centers, the lathe centers must be in line. Work is often held in a chuck,

faced to length, and center drilled in one setup. This operation is covered in Unit 57.

The purposes of facing are:

1. To provide a true, flat surface, square with the axis of the work

2. To provide an accurate surface from which to take measurements

3. To cut the work to the required length

TO FACE WORK BETWEEN CENTERS

1. Move the toolpost to the *left-hand side* of the compound rest, and set the right-hand facing toolbit to the height of the lathe center point (Fig. 51-7).

2. Clean the lathe and work centers and mount the work between centers.

Note: Use a half-center in the tailstock if one is available (Fig. 51-8).

3. Set the facing toolbit pointing left, as shown in Fig. 51-8.

Note: The point of the toolbit must be closest to the work and a space must be left along the side.

4. Set the lathe to the correct speed for the diameter and type of material being cut.

5. Start the lathe and bring the toolbit as close to the center of the work as possible.

■ Figure 51-7 Grip the toolholder short and set the toolbit to center. *(Kelmar Associates)*

■ Figure 51-8 A half-center allows the entire surface to be faced. *(Kelmar Associates)*

SPACE

■ Figure 51-9 Lock the carriage to produce a flat surface. *(Kelmar Associates)*

6. Move the carriage to the left, using the apron handwheel, until a small cut is started.

7. Feed the cutting tool out by turning the crossfeed handle and cut from the center outward. If the automatic crossfeed is used for feeding the cutting tool, the carriage should be locked in position (Fig. 51-9).

8. Repeat operations 5, 6, and 7 until the work is cut to the correct length. (Before facing, mark the correct length with center-punch marks and then face until the punch marks are cut in half.)

Note: When facing, finishing cuts should begin at the center of the workpiece and feed toward the outside.

FACING THE WORKPIECE TO AN ACCURATE LENGTH

When work is to be machined accurately to length, the compound rest graduated micrometer collars may be used. When facing, the compound rest may be set to 30°

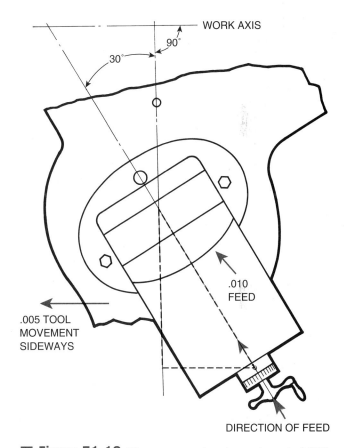

■ Figure 51-10 The compound rest may be set at 30° for accurate end facing. *(Kelmar Associates)*

(to the cross-slide). The side movement of the cutting tool is always half the amount of the compound feed. For example, if the compound rest is fed in .010 in. (0.25 mm), the side movement of the tool, or the amount of material removed from the end of the workpiece, will be .005 in. (0.12 mm) (Fig. 51-10).

Mounting Work Between Centers

1. Explain the procedure for setting up a cutting tool for turning.

2. List the main steps for mounting work between centers.

Facing Between Centers

3. State three purposes of facing a workpiece.

4. Explain how a facing toolbit is set up.

UNIT 52

Machining Between Centers

OBJECTIVES

After completing this unit, you should be able to:

1 Set up the cutting tool for turning operations

2 Turn parallel diameters to within ±.002-in. (0.05-mm) accuracy for size

3 Produce a good surface finish by filing and polishing

4 Machine square, filleted, and beveled shoulders to within .015-in. (0.3-mm) accuracy

In school shops or training programs, where the length of each session is fixed, much of the work machined on a lathe is mounted between centers. When both the headstock and tailstock are aligned, work can be machined, removed from the lathe at the end of the work period, and replaced for additional machining with the assurance that any machined diameter will run true (concentric) with other diameters. In training programs, it is often necessary to remove and replace work in a lathe many times before it is completed; therefore, machining between centers saves much valuable time in setting up work accurately in comparison to other work-holding methods. The most common operations performed on work mounted between centers are facing, rough and finish-turning, shoulder turning, filing, and polishing.

▶▶ Setting Up a Cutting Tool

When machining between centers, it is very important that the workpiece and cutting tool be set up properly; otherwise, the workpiece could be ruined or the lathe centers damaged. Carelessness in setting up work properly could also result in work being thrown out of a lathe and causing injury to the operator.

1. Move the toolpost to the *left-hand side* of the T-slot in the compound rest.

2. Mount a toolholder in the toolpost so that the setscrew in the toolholder is approximately 1 in. (25 mm) beyond the toolpost (Fig. 52-1 on p. 390).

When taking heavy cuts, set the toolholder at right angles to the work (Fig. 52-2a on p. 390). If the toolholder should move under pressure of the cut, the cutting tool would swing away from the work and make the diameter larger.

If the toolholder were set as in Fig. 52-2b and it moved under pressure of the cut, the toolbit would swing into the work, causing the diameter to be cut undersize. A toolholder can be set as in Fig. 52-2b for light finishing cuts.

3. Insert the proper cutting tool into the toolholder, having the tool extend .500 in. (13 mm) beyond the toolholder and never more than twice its thickness.

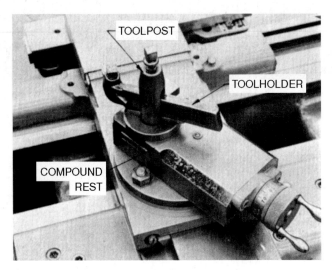

Figure 52-1 Grip the toolholder short in the toolpost. *(Kelmar Associates)*

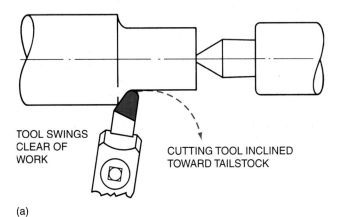

(a)

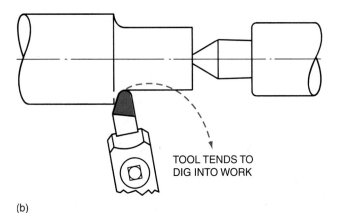

(b)

Figure 52-2 (a) The toolholder set to prevent the toolbit from digging into the work; (b) the toolholder improperly set for making a heavy cut. *(Kelmar Associates)*

4. Set the cutting-tool point to center height. Check it against the lathe center point (Fig. 52-3).

5. Tighten the toolpost *securely* to prevent it from moving during a cut.

Figure 52-3 Set the point of the toolbit even with the center. *(Kelmar Associates)*

▸▸ Mounting Work Between Centers

Since the procedure for mounting the work between lathe centers for machining is the same as for facing, the explanation of this operation will not be repeated. See Unit 51 for the proper procedure for mounting work between centers.

▸▸ Parallel Turning

Work is generally machined on a lathe for two reasons: to cut it to size and to produce a true diameter. Work that must be cut to size and have the same diameter along the entire length of the workpiece involves the operation of parallel turning. Many factors determine the amount of material that can be removed on a lathe at one time. Whenever possible, a diameter should be cut to size in two cuts: a roughing cut and a finishing cut (Fig. 52-4).

Note: To remove metal from a cylindrical piece of work and have the same diameter at each end, *the lathe centers must be in line.* (See Unit 49 for the methods of aligning centers.) Before either the rough or finish cut is taken, the cutting tool must be set accurately for the depth of cut desired.

SETTING AN ACCURATE DEPTH OF CUT

In order to machine any diameter on a lathe accurately to a size, it is important that a trial cut be taken off the diameter to be turned before setting *any depth of cut* on the crossfeed micrometer graduated collar. The purposes of a trial cut are to:

> Produce an accurate turned diameter, which can be measured with a micrometer

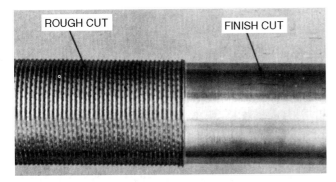

Figure 52-4 A rough cut and a finish cut. *(Kelmar Associates)*

> Set the cutting-tool point to the diameter

> Set the crossfeed micrometer collar to the diameter

Only after the trial cut has been taken is it possible to set an accurate depth of cut.

To Take a Trial Cut

1. Set up the workpiece and cutting tool as for turning.

2. Set the proper speeds and feeds to suit the material being cut.

3. Start the lathe and position the toolbit over the work approximately .125 in. (3 mm) from the end.

4. Turn the compound rest handle *clockwise* one-quarter of a turn to remove any backlash.

5. Feed the toolbit into the work by turning the crossfeed handle clockwise until a light ring appears around the entire circumference of the work (Fig. 52-5).

6. **DO NOT MOVE THE CROSSFEED HANDLE SETTING.** If the crossfeed is moved, two out of the three purposes of a trial cut are destroyed; the cutting tool and the crossfeed micrometer collar are no longer set to the diameter. Therefore, the starting point of setting an accurate cut is lost.

7. Turn the carriage handwheel until the toolbit clears the end of the workpiece by about .060 in. (1.5 mm).

8. Turn the crossfeed handle clockwise about .010 in. (0.25 mm) and take a trial cut .250 in. (6 mm) along the length of the work (Fig. 52-6).

9. Disengage the automatic feed and clear the toolbit past the end of the work with the carriage handwheel.

10. Stop the lathe.

Figure 52-5 The first step in setting an accurate depth of cut is to machine a light ring around the circumference of the workpiece. *(Kelmar Associates)*

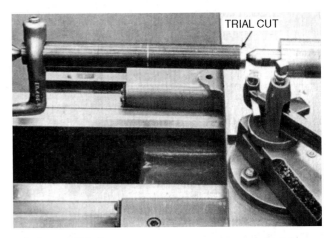

Figure 52-6 Make a light trial cut about .250 in. long to clean up the diameter. *(Kelmar Associates)*

11. Test the accuracy of the micrometer by cleaning and closing the measuring faces and then measure the trial-cut diameter (Fig. 52-7 on p. 392).

12. Calculate how much material must still be removed from the diameter of the work.

13. Turn the crossfeed handle clockwise one-half the amount of material to be removed (the full amount if the micrometer collar indicates material being removed from the work diameter).

Note: In some cases, especially on older machines, setting a long-range dial indicator to bear against the cross-slide movement is a more accurate method of setting the depth of a cut.

14. Take another trial cut .250 in. (6 mm) long and stop the lathe.

Note: If the diameter is too small, note the graduated collar setting, and turn the crossfeed handle *counterclockwise* one-half turn and then clockwise to the required setting.

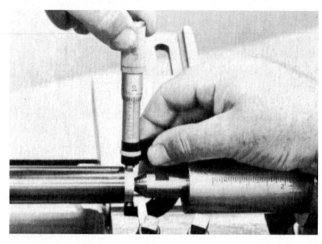

Figure 52-7 With the cutting tool clear of the workpiece, measure the diameter. *(Kelmar Associates)*

15. Clear the toolbit over the end of the work with the carriage handwheel.

16. Measure the diameter and, if necessary, readjust the crossfeed handle until the diameter is correct.

17. Machine the diameter to length.

▶▶ Rough Turning

Rough turning removes as much metal as possible in the shortest length of time. Accuracy and surface finish are not important in this operation; therefore, a .020- to .030-in. (0.5- to 0.76-mm) feed is recommended. Work is generally rough-turned to within .030 in. (0.8 mm) of the finished size when removing up to .500 in. (13 mm) from the diameter; within .060 in. (1.6 mm) when removing more than .500 in.

Follow this procedure:

1. Set the lathe to the correct speed for the type and size of material being cut (Table 47.1 on p. 369).

2. Adjust the quick-change gearbox for a .010- to .030-in. (0.25- to 0.76-mm) feed, depending on the depth of cut and condition of the machine.

3. Move the toolholder to the left-hand side of the compound rest and set the toolbit height to center.

4. Tighten the toolpost *securely* to prevent the toolholder from moving during the machining operation.

5. Take a light trial cut at the right-hand end of the work for a .250-in. (6-mm) length (Fig. 52-6).

6. Measure the work and adjust the toolbit for the proper depth of cut.

7. Cut along for .250 in. (6 mm), stop the lathe, and check the diameter for size. The diameter should

be approximately .030 in. (0.8 mm) over the finish size (Fig. 52-7).

8. Readjust the depth of cut, if necessary.

▶▶ Finish Turning

Finish turning, which follows rough turning, produces a smooth surface finish and cuts the work to an accurate size. Factors such as the condition of the cutting tool, the rigidity of the machine and work, and the lathe speeds and feeds affect the type of surface finish produced.

Follow this procedure:

1. Make sure that the cutting edge of the toolbit is free from nicks, burrs, etc. It is good practice to hone the cutting edge before taking a finish cut.

2. Set the toolbit on center; check it against the lathe center point.

3. Set the lathe to the recommended speed and feed. The feed used depends on the surface finish required.

4. Take a light trial cut .250 in. (6 mm) long at the right-hand end of the work to:

    a. Produce a true diameter
    b. Set the cutting tool to the diameter
    c. Set the graduated collar to the diameter

5. Stop the lathe and measure the diameter.

6. Set the depth of cut for half the amount of material to be removed.

7. Cut along for .250 in. (6 mm), stop the lathe, and check the diameter.

8. Readjust the depth of cut, if necessary, and finish-turn the diameter.

Note: To produce the truest diameter possible, finish-turn work to the required size. Should it be necessary to finish a diameter by filing or polishing, *never* leave more than .002 to .003 in. (0.05 to 0.07 mm) for this operation.

▶▶ Filing in a Lathe

Work should be filed in a lathe only to remove a small amount of stock, remove burrs, or round off sharp corners. Work should always be turned to within .002 to .003 in. (0.05 to 0.07 mm) of size, if the surface is to be filed. When larger amounts must be removed, the work should be machined because excessive filing will produce work that is out of round and not parallel. The National Safety Council recommends filing with the left hand, so that the arms and hands can be kept clear of the revolving chuck or driveplate. Always remove the toolbit from the toolholder before filing, unless the machining operation does

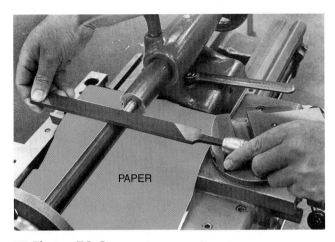

Figure 52-8 Always hold the file handle in your left hand to avoid injury. *(Kelmar Associates)*

not permit it. In this case, move the carriage so that the toolbit is as far as possible from the area being filed.

Note: Before attempting to file or polish in a lathe, cover the lathe bed with a piece of paper to prevent filings from getting into the slides and causing excessive wear and damage to the lathe (Fig. 52-8). Cloth is not suitable for this purpose because it tends to get caught in the revolving work or the lathe.

TO FILE IN A LATHE

1. Set the spindle speed to approximately twice that used for turning.

2. Mount the work between centers, lubricate, and carefully adjust the dead center in the workpiece. Use a revolving dead center if one is available.

3. Move the carriage as far to the right as possible and remove the toolpost.

4. Disengage the lead screw and feed rod.

5. Select a 10- or 12-in. (250- or 300-mm) *mill file* or a *long-angle lathe file*.

Note: Be sure that the file handle is properly secured on the tang of the file.

6. Start the lathe.

7. Grasp the file handle in the left hand and support the file point with the fingers of the right hand (Fig. 52-8).

8. Apply light pressure and push the file forward to its full length. Release the pressure on the return stroke.

9. Move the file about half the width of the file for each stroke and continue filing, using 30 to 40 strokes per minute, until the surface is finished.

10. When filing in a lathe, take the following precautions:
    a. Roll sleeves up above the elbows—short sleeves are preferable.
    b. Remove watches and rings.
    c. Never use a file without a properly fitted handle.
    d. Never apply too much pressure to the file. Excessive pressure produces out-of-roundness and causes the file teeth to clog and damage the work surface.
    e. Clean the file frequently with a file brush. Rub a little chalk into the file teeth to prevent clogging and facilitate cleaning.

▶▶ Polishing in a Lathe

After the work surface has been filed, the finish may be improved by polishing with abrasive cloth. Proceed as follows:

1. Select the correct type and grade of abrasive cloth for the finish desired. Use a piece about 6 to 8 in. (150 to 200 mm) long and 1 in. (25 mm) wide. For ferrous metals, use aluminum oxide abrasive cloth. Silicon carbide abrasive cloth should be used for nonferrous metals.

2. Set the lathe to run at high speed.

3. Disengage the feed rod and lead screw.

4. Remove the toolpost and toolholder.

5. Lubricate and adjust the dead center. Use a revolving dead center if one is available.

6. Roll sleeves up above the elbows and tuck in any loose clothing.

7. Start the lathe.

8. Hold the abrasive cloth on the work (Fig. 52-9).

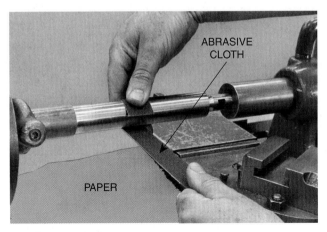

Figure 52-9 A high surface finish can be produced with abrasive cloth. *(Kelmar Associates)*

9. With the right hand, press the cloth firmly on the work while *tightly* holding the other end of the abrasive cloth with the left hand. (*Caution:* Do not let the short end of the abrasive cloth wrap around the work.)

10. Move the cloth slowly back and forth along the work.

Note: For normal finishes, 80- to 100-grit abrasive cloth should be used. For better finishes, use a finer-grit abrasive cloth.

▶▶ Turning to a Shoulder

When turning more than one diameter on a piece of work, the change in diameters, or step, is known as a *shoulder.* Three common types of shoulders are illustrated in Fig. 52-10.

TO TURN A SQUARE SHOULDER

1. With the work mounted in a lathe, lay out the shoulder position from the finished end of the work. In case of filleted shoulders, allow sufficient length to permit the proper radius to be formed on the finished shoulder.

2. Place the point of the toolbit at this mark and cut a *small* groove around the circumference to mark off the length.

3. With a turning tool, rough- and finish-turn the work to within .060 in. (1.5 mm) of the required length (Fig. 52-11).

4. Set up an end-facing tool, chalk the small diameter of the work, and bring the cutting tool up until it just removes the chalk mark (Fig. 52-12).

5. Note the reading on the graduated crossfeed handle.

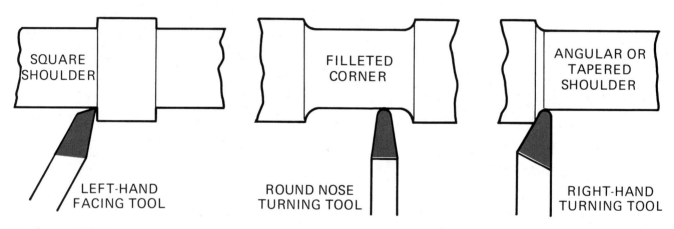

■ Figure 52-10 Types of shoulders. *(Kelmar Associates)*

■ Figure 52-11 Turn the small diameter to within .060 in. of the finished length. *(Kelmar Associates)*

■ Figure 52-12 Note the reading on the graduated collar when the toolbit just touches the small diameter. *(Kelmar Associates)*

Figure 52-13 The length of the shoulder is indicated by a center-punch mark. *(Kelmar Associates)*

Figure 52-14 Setting the radius toolbit to the small diameter. *(Kelmar Associates)*

6. Face (square) the shoulder, cutting to the line *using hand feed.*

7. For successive cuts, return the crossfeed handle to the same graduated collar setting.

TO MACHINE A FILLETED SHOULDER

Fillets are used at a shoulder to overcome the sharpness of a corner and to strengthen the part at this point. If a filleted corner is required, a toolbit having the same radius is used to finish the shoulder. Follow this procedure:

1. Lay out the length of the shoulder with a center-punch mark or by cutting a light groove (Fig. 52-13)

2. Rough- and finish-turn the small diameter to the correct length *minus the radius to be cut.* For example, a 3-in. (75-mm) length with a .125-in. (3-mm) radius should be turned 2.875 in. (73 mm) long.

3. Mount the correct radius toolbit and set it to center.

4. Set the lathe for one-half the turning speed.

5. Coat the small diameter near the shoulder with chalk or layout dye.

6. Start the lathe and feed the cutting tool in until it lightly marks the small diameter near the shoulder (Fig. 52-14).

7. Slowly feed the cutting tool sideways with the carriage handwheel until the shoulder is cut to the correct length.

TO MACHINE A BEVELED (ANGULAR) SHOULDER

Beveled, or angular, shoulders are used to eliminate sharp corners and edges, to make parts easier to handle, and to

Figure 52-15 Using a protractor to set the toolbit side cutting edge to an angle. *(Kelmar Associates)*

improve the appearance of the part. They are sometimes used to strengthen a part by eliminating the sharp corner of a square shoulder. Shoulders are beveled at angles ranging from 30° to 60°; however, the most common is the 45° bevel. Follow this procedure:

1. Turn the large diameter to size.

2. Lay out the position of the shoulder along the length of the workpiece.

3. Rough- and finish-turn the small diameter to size.

4. Mount a side cutting tool in the toolholder and set it to center.

5. Use a protractor and set the side cutting edge of the toolbit to the desired angle (Fig. 52-15).

6. Apply chalk or layout dye to the small diameter as close as possible to the shoulder location.

7. Set the lathe spindle to approximately one-half the turning speed.

8. Bring the point of the toolbit in until it just removes the chalk or layout dye.

Figure 52-16 Machining a beveled shoulder with the side of a cutting tool. *(Kelmar Associates)*

9. Turn the carriage handwheel by hand to feed the cutting tool into the shoulder (Fig. 52-16).

10. Apply cutting fluid to assist the cutting action and to produce a good surface finish.

11. Machine the beveled shoulder until it is the required size.

If the size of the shoulder is large and chatter occurs during cutting with the side of the toolbit, it may be necessary to cut the beveled shoulder using the compound rest (Fig. 52-17). If so, follow this procedure:

Figure 52-17 The compound rest swiveled to cut a large beveled shoulder. *(Kelmar Associates)*

1. Set the compound rest to the desired angle.

2. Adjust the toolbit so that only the point does the cutting.

3. Machine the bevel by feeding the compound rest by hand.

unit 52 review questions

Setting Up a Cutting Tool

1. How should the toolholder and cutting tool be set for machining between centers?

2. What might happen if the toolholder were set to the left and moved under the pressure of the cut?

Parallel Turning

3. What precaution must be taken before starting a parallel turning operation?

4. Explain the procedure for setting an accurate depth of cut.

5. State the purpose of rough and finish turning.

6. How many cuts should be taken to turn a diameter to size?

7. What is the purpose of a light trial cut at the right-hand end of the work?

Filing and Polishing

8. How much material should be left on a diameter for filing to size?

9. How should the file be held for filing in a lathe?

10. List two of the most important things to remember about filing in a lathe.

11. List the main steps for polishing a diameter in a lathe.

Shoulder Turning

12. What type of cutting tool should be used to machine a square shoulder?

13. How close to finish length should a diameter that requires a filleted shoulder be cut?

14. Name two methods of cutting angular shoulders on a lathe.

UNIT 53

Knurling, Grooving, and Form Turning

OBJECTIVES

After completing this unit, you should be able to:

1 Set up and use knurling tools to produce diamond-shaped or straight patterns on diameters

2 Cut square, round, and V-shaped grooves on work between centers or in a chuck

3 Machine convex or concave forms on diameters freehand

Operations such as knurling, grooving, and form turning alter either the shape or the finish of a round workpiece. These operations are normally performed on work mounted in a chuck; however, they can also be performed on work mounted between lathe centers if certain precautions are observed. Knurling is used to improve the surface finish on the work and provide a hand grip on the diameter. Grooving is used to provide a relief at the end of a thread or a seat for snap or O-rings. Form turning produces a concave or convex form on internal or external surfaces of a workpiece.

▶▶ Knurling

Knurling is a process of impressing a diamond-shaped or straight-line pattern into the surface of the workpiece to improve its appearance or to provide a better gripping surface. Straight knurling is often used to increase the workpiece diameter when a press fit is required.

Diamond- and straight-pattern rolls are available in three styles: fine, medium, and coarse (Fig. 53-1 on p. 398).

The knurling tool (Fig. 53-2a on p. 398) is a toolpost-type toolholder on which a pair of hardened-steel rolls are mounted. These rolls may be obtained in diamond and straight-line patterns and in coarse, medium, and fine pitches. Some knurling tools are made with the three various pitched rollers on one holder (Fig. 53-2b).

The *universal knurling tool system* (Fig. 53-3 on p. 398) consists of a dovetailed shank and as many as seven interchangeable knurling heads that can produce a wide range of knurling patterns. This tooling system combines versatility, rigidity, ease of handling, and simplicity in one tool.

TO KNURL IN A LATHE

1. Mount the work between centers and mark the required length to be knurled. If the work is held in a chuck for knurling, the right end of the work should be supported with a revolving tailstock center.

2. Set the lathe to run at *one-quarter* the speed required for turning.

3. Set the carriage feed to *.015 to .030 in.* (0.38 mm to 0.76 mm).

4. Set the center of the floating head of the knurling tool even with the dead-center point (Fig. 53-4 on p. 398).

Figure 53-1 Fine, medium, and coarse diamond- and straight-pattern knurling rolls. *(Bahco North America)*

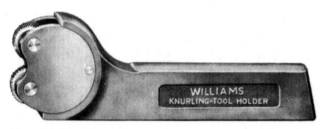

(a)

(b)

Figure 53-2 (a) A knurling tool with one set of rolls in a self-centering head; (b) a knurling tool with three sets of rolls in a revolving head. *(Bahco North America)*

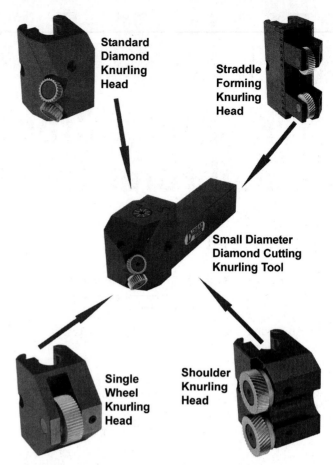

Standard Diamond Knurling Head

Straddle Forming Knurling Head

Small Diameter Diamond Cutting Knurling Tool

Single Wheel Knurling Head

Shoulder Knurling Head

Figure 53-3 The universal knurling tool system allows interchangable heads to be mounted on the shank for different knurling patterns. *(Dorian Tool International)*

Figure 53-4 Setting a knurling tool to center. *(Kelmar Associates)*

5. Set the knurling tool at right angles to the workpiece and tighten it securely in this position (Fig. 53-5).

6. Start the machine and lightly touch the rolls against the work to make sure that they are tracking properly (Fig. 53-6). Adjust if necessary.

7. Move the knurling tool to the end of the work so that only half the roll face bears against the work. If the knurl does not extend to the end of the workpiece, set the knurling tool at the correct limit of the section to be knurled.

Figure 53-5 The knurling tool set at 90° and moved near the end of the workpiece. *(Kelmar Associates)*

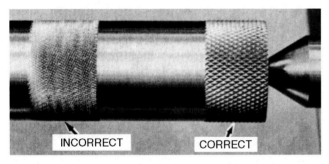

Figure 53-6 Correct and incorrect knurling patterns. *(Kelmar Associates)*

8. Force the knurling tool into the work approximately .025 in. (0.63 mm) and start the lathe.

OR

Start the lathe and then force the knurling tool into the work until the diamond pattern comes to a point.

9. Stop the lathe and examine the pattern. If necessary, reset the knurling tool.

    a. If the pattern is incorrect (Fig. 53-6), it is usually because the knurling tool is not set on center.

    b. If the knurling tool is on center and the pattern is not correct, it is generally because of worn knurling rolls. In this case, it will be necessary to set the knurling tool off square slightly so that the corner of the knurling rolls can start the pattern.

Figure 53-7 Disengaging the automatic feed will damage the knurling pattern. *(Kelmar Associates)*

10. Once the pattern is correct, engage the automatic carriage feed and apply cutting fluid to the knurling rolls.

11. Knurl to the proper length and depth.

Note: Do not disengage the feed until the full length has been knurled; otherwise, rings will be formed on the knurled pattern (Fig. 53-7).

12. If the knurling pattern is not to a point after the length has been knurled, reverse the lathe feed and take another pass across the work.

▶▶ Grooving

Grooving, commonly called *recessing, undercutting,* or *necking,* is often done at the end of a thread to permit full travel of the nut up to a shoulder or at the edge of a shoulder to ensure a proper fit of mating parts. Grooves are generally square, round, or V-shaped (Fig. 53-8 on p. 400).

Rounded grooves are usually used where there is a strain on the part and where a square corner would lead to fracturing of the metal at this point.

TO CUT A GROOVE

1. Grind a toolbit to the desired size and shape of the groove required. If a parting tool is used to cut a groove, never grind the width of the tool.

2. Lay out the location of the groove.

3. Set the lathe to half the speed for turning.

4. Mount the workpiece in the lathe.

5. Set the toolbit to center height (Fig. 53-9 on p. 400).

6. Locate the toolbit on the work at the position where the groove is to be cut.

7. Start the lathe and feed the cutting tool toward the work using the crossfeed handle until the toolbit marks the work lightly.

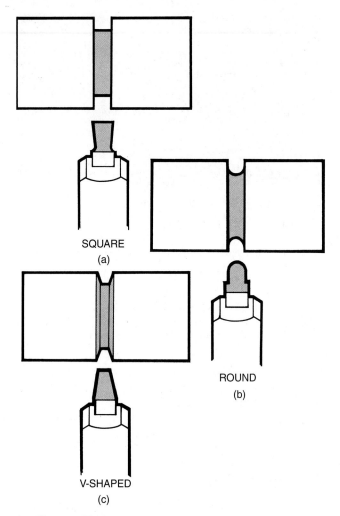

SQUARE
(a)

ROUND
(b)

V-SHAPED
(c)

■ Figure 53-8 Three types of common grooves.
(Kelmar Associates)

■ Figure 53-9 A grooving tool set to center. *(Kelmar Associates)*

■ Figure 53-10 The graduated collar should be set to zero when the tool just touches the diameter of the workpiece. *(Kelmar Associates)*

8. Hold the crossfeed handle in position and then set the graduated collar to zero (Fig. 53-10).

9. Calculate how far the crossfeed screw must be turned to cut the groove to the proper depth.

10. Feed the toolbit into the work *slowly* using the crossfeed handle.

11. Apply cutting fluid to the point of the cutting tool. To ensure that the cutting tool will not bind in the groove, move the carriage slightly to the left and to the right while grooving. Should chatter develop, reduce the spindle speed of the lathe.

12. Stop the lathe and check the depth of groove with outside calipers or knife-edge verniers.

> **STOP** **SAFETY PRECAUTIONS** Always wear safety goggles when grooving on a lathe.

▸▸ Form Turning on a Lathe

It is often necessary to form irregular shapes or contours on a workpiece. Form turning may be done on a lathe by four methods:

1. Freehand
2. Form-turning tool
3. Spherical tool
4. Hydraulic tracer attachment

TURNING A FREEHAND FORM OR RADIUS

Freehand form turning probably presents the greatest problem to the beginning lathe operator. Coordination of both hands is required and practice is important in mastering this skill.

To Turn a .500-in. or 13-mm Radius on the End of a Workpiece

1. Mount the workpiece in a chuck and face the end.

2. With the work revolving, mark a line .500 in. (or 13 mm) from the end using a pencil (Fig. 53-11).

3. Mount a round-nose turning tool on center.

4. Start the lathe and adjust the toolbit until it touches the diameter about .250 in. (6 mm) from the end.

5. Place one hand on the crossfeed handle and the other on the carriage handwheel.

6. Turn the carriage *handwheel* (*not* the handle) to feed the toolbit slowly toward the end of the work; at the same time, turn the crossfeed handle to move the tool into the work.

Note: It will take practice to coordinate the movement of the carriage in relation to the crossfeed. For the first .250 in. (6 mm) of the radius, the carriage must be moved faster than the crossfeed handle. However, for the second .250 in. (6 mm), the crossfeed handle must be moved faster than the carriage.

7. Back the toolbit out and move the carriage to the left.

8. Take successive cuts as in step 6 until the toolbit starts to cut close to the .500-in. (13-mm) line.

9. Test the radius with a .500-in. (13-mm) gage.

10. If the radius is not correct, it may have to be recut. It is often possible to finish the radius to the required shape by filing.

■ **Figure 53-11** Turning a .500-in. (13-mm) radius on the end of the workpiece. *(Kelmar Associates)*

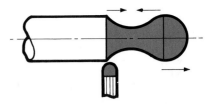

■ **Figure 53-12** Turning concave and convex radii on a workpiece. *(Kelmar Associates)*

Note: Follow the same procedure as in step 6 when cutting internal radii (Fig. 53-12). Always start at the large diameter, feeding along and in until the proper radius and diameter are obtained.

FORM-TURNING TOOLS

Smaller radii and contours are conveniently formed on a workpiece by a form-turning tool. The lathe toolbit is ground to the desired radius and used to form the contour on the workpiece. Toolbits may also be ground to produce a concave radius (Fig. 53-13a).

This method of forming radii and contours eliminates the need for checking with a gage or template once the toolbit is ground to the desired shape. Duplicate contours may also be formed on several workpieces when the same toolbit is used.

When producing a convex radius, it is necessary to leave a collar of the desired size on the workpiece (Fig. 53-13b).

To produce a good finish by this method, the work should be revolved slowly. The tool should be fed into the work slowly while cutting oil is applied. To eliminate chatter during the cutting operation, the cutting tool should be moved slightly back and forth (longitudinally).

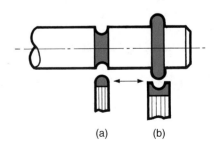

(a)   (b)

■ **Figure 53-13** Toolbits ground to cut (a) a concave radius; (b) a convex radius. *(Kelmar Associates)*

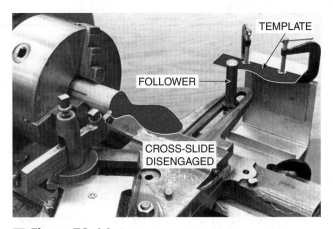

Figure 53-14 A follower on the cross-slide follows the template to produce a special form. *(Kelmar Associates)*

Figure 53-15 The cross-slide must be disconnected to follow the template. *(Kelmar Associates)*

FORM TURNING USING A TEMPLATE AND FOLLOWER

When only a few pieces of a special form are required, they can be accurately produced by using a template and a follower fastened to the cross-slide of the lathe. The accuracy of the template, which must be made, determines the accuracy of the form produced. Follow this procedure:

1. Make an accurate template to the form desired.

2. Mount the template on a bracket fastened to the back of the lathe (Fig. 53-14).

3. Position the template lengthwise in relation to the workpiece.

4. Mount a round-nose cutting tool in the toolpost.

5. Fasten a follower, the face of which must have the same form as the point of the cutting tool, on the cross-slide of the lathe.

6. Rough-out the form on the workpiece freehand by keeping the follower close to the template by manually operating the carriage and cross-slide controls. While taking roughing cuts, keep the distance between the follower and the template fairly constant. For the final roughing cut, the follower should be kept within approximately .030 in. (0.8 mm) of the template at all times.

7. Disconnect the cross-slide from the crossfeed screw (Fig. 53-15).

8. Apply light hand pressure on the cross-slide to keep the follower in contact with the template.

9. Engage the automatic carriage feed and take a finish cut from the workpiece, while keeping the follower in contact with the template.

SPHERICAL TOOL

The *spherical tool* (Fig. 53-16) can make a perfect spherical ball or spherical cavity to within .0001-in. (0.002-mm) accuracy. The tool consists of a dovetailed turret that holds the cutting tool and a drive mechanism that can be hand fed or power driven. The dial microscrew indicates the depth of cut and the diameter of the sphere. The spherical tool is easy to set up and provides a fast, accurate method of producing spherical forms (Fig. 53-16a). When the internal toolholder is installed on the dovetail turret, concave forms can be accurately cut (Fig. 53-16b).

HYDRAULIC TRACER ATTACHMENT

When many duplicate parts having several radii or contours that may be difficult to produce are required, they may be easily made on a hydraulic tracer lathe or on a lathe equipped with a hydraulic tracer attachment (Fig. 53-17).

Hydraulic tracer lathes incorporate a means of moving the cross-slide by controlled oil pressure supplied by a hydraulic pump. A flat template of the desired contour of the finished piece, or a circular template identical to the finished piece, is mounted in an attachment on the lathe. Automatic control of the tool slide and duplication of the part is achieved by a stylus, which bears against the template surface. As the carriage is fed along automatically, the stylus follows the contour of the template. The stylus arm actuates a control valve regulating the flow of oil into a cylinder incorporated in the tool slide base. A piston connected to the tool slide is moved in or out by the flow of oil to the cylinder. This movement causes the tool slide (and toolbit) to move in or out as the carriage moves along, duplicating the profile of a template on the workpiece.

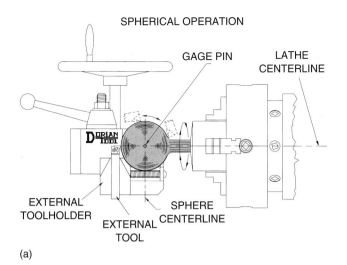

SPHERICAL OPERATION

GAGE PIN

LATHE CENTERLINE

EXTERNAL TOOLHOLDER

EXTERNAL TOOL

SPHERE CENTERLINE

(a)

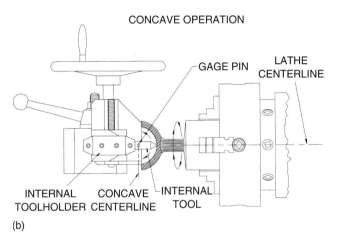

CONCAVE OPERATION

GAGE PIN

LATHE CENTERLINE

INTERNAL TOOLHOLDER

CONCAVE CENTERLINE

INTERNAL TOOL

(b)

■ **Figure 53-16** The spherical tool can be used to machine (a) a precise spherical ball or (b) a concave form. *(Dorian Tool International)*

Advantages of a Tracer Attachment

> Intricate forms, difficult to produce by other means, can be readily produced.

> Various forms, tapers, and shoulders can be produced in one cut.

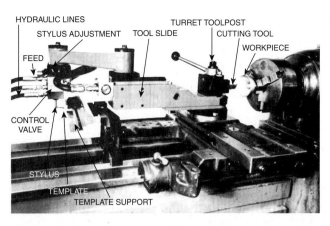

HYDRAULIC LINES
STYLUS ADJUSTMENT
FEED
CONTROL VALVE
TOOL SLIDE
TURRET TOOLPOST
CUTTING TOOL
WORKPIECE
STYLUS
TEMPLATE
TEMPLATE SUPPORT

■ **Figure 53-17** A hydraulic tracer attachment used to machine an intricate form. *(Retor Developments Ltd.)*

> Duplicate parts can be produced rapidly and accurately.

> Accuracy and finish of the part do not depend on the skill of the operator.

Hints on the Use of a Tracer Attachment

1. The toolbit point and stylus should have the same form and radius.

2. The radius on the toolbit should be smaller than the smallest radius on the template.

3. The stylus must be set to the point on the template giving the smallest diameter of the work.

4. The centerline of the template must be parallel to the ways of the lathe.

5. The form of the template must be smooth.

6. No angle larger than 30°, or the equivalent radius, should be incorporated in the form of the template.

7. Duplicate parts produced between centers must be the same length and have the center holes drilled to the same depth.

8. Duplicate parts held in a chuck should project the same distance from the chuck jaws.

9. The included angle of the tool point should be less than the smallest angle on the template.

Knurling

1. Define the process of knurling.
2. Explain how to set up the knurling tool.
3. Why is it important not to disengage the feed during the knurling operation?

Grooving

4. For what purposes are grooves used?
5. How can the depth of the cut be gaged during grooving?
6. What should be done to prevent the cutting tool from binding in a deep groove?

Form Turning

7. Name four methods by which form turning may be done on a lathe.
8. Briefly describe the procedure for turning a .500-in. (13-mm) radius on the end of a workpiece.
9. What is a template?
10. What types of templates may be used with a tracer lathe?
11. List three advantages of a tracer lathe or tracer attachment.
12. List six points to observe when using a tracer lathe.

UNIT 54

Tapers and Taper Turning

OBJECTIVES

After completing this unit, you will be able to:

1 Identify and state the purpose of self-holding and self-releasing tapers

2 Cut short, steep tapers using the compound rest

3 Calculate and cut tapers on work between centers by offsetting the tailstock

4 Calculate and machine tapers with a taper attachment

A *taper* may be defined as a uniform change in the diameter of a workpiece measured along its axis. Tapers in the inch system are expressed in taper per foot, taper per inch, or degrees. Metric tapers are expressed as a ratio of 1 mm per unit of length; for example, 1:20 taper has a 1-mm change in diameter in 20 mm of length. A taper provides a rapid and accurate method of aligning machine parts and an easy method of holding tools such as twist drills, lathe centers, and reamers.

Machine tapers (those used on machines and tools) are now classified by the American Standards Association as *self-holding tapers* and *steep, or self-releasing, tapers.*

Self-Holding Tapers

Self-holding tapers, when seated properly, remain in position because of the wedging action of the small taper angle. The most common forms of self-holding tapers are the Morse, the Brown & Sharpe, and the .750-in.-per-foot machine taper. See Table 54.1 on p. 406.

The smaller sizes of self-holding tapered shanks are provided with a tang to help drive the cutting tool. Larger sizes employ a tang drive with the shank held in by a key, or a key drive with the shank held in with a draw bolt.

Steep Tapers

Steep (self-releasing) tapers have a 3.500 in. taper per foot (*tpf*). This was formerly called the standard milling machine taper. It is used mainly for alignment of milling machine arbors and accessories. A steep taper has a key drive and uses a draw-in bolt to hold it securely in the milling machine spindle.

Standard Tapers

Although many of the tapers referred to in Table 54.1 are taken from the Morse and Brown & Sharpe taper series, those not listed in this table are classified as nonstandard machine tapers.

The *Morse taper,* which has approximately .625-in. *tpf,* is used for most drills, reamers, and lathe center shanks. Morse tapers are available in eight sizes ranging from #0 to #7.

The *Brown & Sharpe taper,* available in sizes from #4 to #12, has approximately .502-in. *tpf,* except #10,

table 54.1   Basic dimensions of self-holding tapers

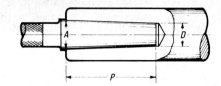

Number of Taper	Taper per Foot	Diameter at Gage Line (A)	Diameter at Small End (D)	Length (P)	Series Origin
1	.502	.2392	.200	$^{15}/_{16}$	
2	.502	.2997	.250	$1^3/_{16}$	Brown & Sharpe taper series
3	.502	.3752	.3125	$1^1/_2$	
0*	.624	.3561	.252	2	
1	.5986	.475	.369	$2^1/_8$	
2	.5994	.700	.572	$2^9/_{16}$	
3	.6023	.938	.778	$3^3/_{16}$	
4	.6233	1.231	1.020	$4^1/_{16}$	Morse taper series
4½	.624	1.500	1.266	$4^1/_2$	
5	.6315	1.748	1.475	$5^3/_{16}$	
6	.6256	2.494	2.116	$7^1/_4$	
7	.624	3.270	2.750	10	
200	.750	2.000	1.703	$4^3/_4$	
250	.750	2.500	2.156	$5^1/_2$	
300	.750	3.000	2.609	$6^1/_4$	
350	.750	3.500	3.063	7	
400	.750	4.000	3.516	$7^3/_4$	¾-in.-*tpf* series
450	.750	4.500	3.969	$8^1/_2$	
500	.750	5.000	4.422	$9^1/_4$	
600	.750	6.000	5.328	$10^3/_4$	
800	.750	8.000	7.141	$13^3/_4$	
1000	.750	10.000	8.953	$16^3/_4$	
1200	.750	12.000	10.766	$19^3/_4$	

*Taper #0 is not part of the self-holding taper series. It has been added to this table to complete the Morse taper series.

which has a taper of .516-in./ft. This self-holding taper is used on Brown and Sharpe machines and drive shanks.

The *Jarno taper*, .600-in. *tpf*, was used on some lathe and drill spindles in sizes from #2 to #20. The taper number indicates the large diameter in eighths of an inch and the small diameter in tenths of an inch. The taper length is indicated by the taper number divided by 2.

The *standard taper pins* used for positioning and holding parts together have ¼-in. *tpf*. Standard sizes in these pins range from #6/0 to #10.

▸▸ Lathe Spindle Nose Tapers

Two types of tapers are used on lathe spindle noses. The *Type D-1* has a short tapered section (3.000-in. *tpf*) and is used on cam-lock spindles (Fig. 54-1). The *Type L* lathe spindle nose has a taper of 3.500-in./ft and has a considerably longer taper than the Type D-1. The chuck or driveplate is held on by a threaded lock ring fitted on the spindle behind the taper nose. A key drive is employed in this type of taper (Fig. 54-2).

■ **Figure 54-1** Tapered lathe spindle nose, Type D-1. *(Kelmar Associates)*

■ **Figure 54-2** Tapered lathe spindle nose, Type L. *(Kelmar Associates)*

▸▸ Taper Calculations

To machine a taper, particularly by the tailstock offset method, it is often necessary to make calculations to ensure accurate results. Since tapers are often expressed in *taper per foot, taper per inch,* or *degrees,* it may be necessary to calculate any of these dimensions.

TO CALCULATE THE *TPF*

To calculate the *tpf,* it is necessary to know the large diameter, the small diameter, and the length of taper (*l*). The *tpf* can be calculated by applying the following formula:

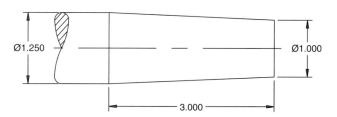

■ **Figure 54-3** The main part of an inch taper. *(Kelmar Associates)*

$$tpf = \frac{D - d}{\text{length of taper}} \times 12$$

To calculate the *tpf* for the workpiece in Fig. 54-3:

$$tpf = \frac{(1.250 - 1)}{3} \times 12$$

$$= .250 \times \frac{1}{3} \times 12$$

$$= 1$$

TO CALCULATE THE TAILSTOCK OFFSET

When calculating the tailstock offset, the *tpf* and the total length of the work (*L*) must be known (Fig. 54-4):

$$\text{Tailstock offset} = \frac{tpf \times \text{length of work}}{24}$$

$$tpf = \frac{1.125 - 1}{3} \times 12$$

$$= .125 \times \frac{1}{3} \times 12$$

$$= .500 \text{ in.}$$

$$\text{Tailstock offset} = \frac{.500 \times 6}{24}$$

$$= \frac{.500}{1} \times \frac{1}{24} \times 6$$

$$= .125 \text{ in.}$$

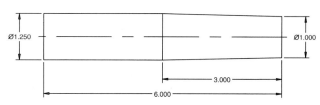

■ **Figure 54-4** Dimensions of a workpiece having a taper. *(Kelmar Associates)*

A simplified formula can be used to calculate the tailstock offset if the taper per inch t/in. is known:

$$\text{Taper per inch} = \frac{\text{taper per foot}}{12}$$

$$\text{Tailstock offset} = \frac{\text{taper per inch} \times OL}{2}$$

where OL = overall length of work.

In cases where it is not necessary to find the *tpf*, the following simplified formula can be used to calculate the amount of tailstock offset.

$$\text{Tailstock offset} = \frac{OL}{TL} \times \frac{(D - d)}{2}$$

where OL = overall length of work
TL = length of the tapered section
D = diamter of the large end
d = diameter at the small end

For example, to find the tailstock offset required to cut the taper for the work in Fig. 54-4:

$$\text{Tailstock offset} = \frac{6}{3} \times \frac{1}{8} \times \frac{1}{2}$$

$$= .125 \text{ in.}$$

INCH TAPER ATTACHMENT OFFSET CALCULATIONS

Most tapers cut on a lathe with the taper attachment are expressed in *tpf*. If the *tpf* of the taper on the workpiece is not given, it may be calculated by using the following formula:

$$tpf = \frac{(D - 2) \times 12}{TL}$$

EXAMPLE

Calculate the *tpf* for a taper with the following dimensions: large diameter (D), 1.375; small diameter (d), .9375; length of tapered section (TL), 7 in.

$$tpf = \frac{1.375 - .9375 \times 12}{7}$$

$$= \frac{.4375 \times 12}{7}$$

$$= .750 \text{ in.}$$

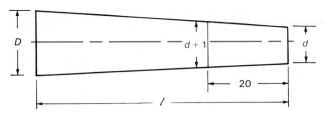

■ Figure 54-5 Characteristics of a metric taper. *(Kelmar Associates)*

METRIC TAPERS

Metric tapers are expressed as a ratio of 1 mm per unit of length. In Fig. 54-5, the work would taper 1 mm in a distance of 20 mm. This taper would then be expressed as a ratio of 1:20 and would be indicated on a drawing as "taper = 1:20."

Since the work tapers 1 mm in 20 mm of length, the diameter at a point 20 mm from the small diameter (d) will be 1 mm larger ($d + 1$).

Some common metric tapers are:

Milling machine spindle	1:3.429
Morse taper shank	1:20 (approx.)
Tapered pins and pipe threads	1:50

Metric Taper Calculations

If the small diameter (d), the unit length of taper (k), and the total length of taper (l) are known, the large diameter (D) may be calculated.

In Fig. 54-6, the large diameter (D) will be equal to the small diameter plus the amount of taper. The amount of taper for the unit length (k) is ($d + 1$) − (d) or 1 mm. Therefore, the amount of taper per millimeter of unit length = $1/k$.

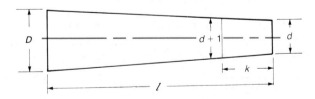

■ Figure 54-6 Dimensions of a metric taper. *(Kelmar Associates)*

The *total amount of taper* is the taper per millimeter (1/k) multiplied by the total length of taper (l):

$$\text{Total taper} = \frac{1}{k} \times l \text{ or } \frac{l}{k}$$

$$D = d + \text{total amount of taper}$$

$$= d + \frac{l}{k}$$

EXAMPLE

Calculate the large diameter D for a 1:30 taper having a small diameter of 10 mm and a length of 60 mm.

Solution

Since the taper is 1:30, k = 30.

$$D = d + \frac{l}{k}$$

$$= 10 + \frac{60}{30}$$

$$= 10 + 2$$

$$= 12 \text{ mm}$$

METRIC TAILSTOCK OFFSET CALCULATIONS

If the taper is to be turned by offsetting the tailstock, the amount of offset O is calculated as follows (Fig. 54-7):

$$\text{Offset} = \frac{D - d}{2 \times l} \times L$$

where   D = large diameter
         d = small diameter
         l = length of taper
         L = length of work

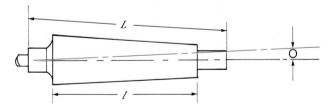

■ **Figure 54-7** Metric taper turning by the tailstock offset method. *(Kelmar Associates)*

EXAMPLE

Calculate the tailstock offset required to turn a 1:30 taper × 60 mm long on a workpiece 300 mm long. The small diameter of the tapered section is 20 mm.

Solution

$$D = d + \frac{l}{k}$$

$$= 20 + \frac{60}{30}$$

$$= 20 + 2$$

$$= 22 \text{ mm}$$

$$\text{Tailstock offset} = \frac{D - d}{2 \times l} \times L$$

$$= \frac{22 - 20}{2 \times 60} \times 300$$

$$= \frac{2}{120} \times 300$$

$$= 5 \text{ mm}$$

METRIC TAPER ATTACHMENT OFFSET CALCULATIONS

When the taper attachment is used to turn a taper, the amount the guide bar is set over may be determined as follows:

1. If the angle of taper is given on the drawing, set the guide bar to one-half the angle (Fig. 54-8).

2. If the angle of taper is not given on the drawing, use the following formula to find the amount of guide bar setover.

$$\text{Guide bar setover} = \frac{D - d}{2} \times \frac{GL}{l}$$

where D = large diameter of taper
       d = small diameter of taper
       l = length of taper
     GL = length of taper attachment guide bar

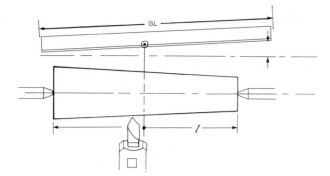

Figure 54-8 Metric taper turning by the taper attachment method. *(Kelmar Associates)*

EXAMPLE

Calculate the amount of setover for a 500-mm long guide bar to turn a 1:50 × 250-mm long taper on a workpiece. The small diameter of the taper is 25 mm.

Solution

$$D = d + \frac{l}{k}$$

$$= 25 + \frac{250}{50}$$

$$= 30 \text{ mm}$$

$$\text{Guide bar setover} = \frac{D - d}{2} \times \frac{GL}{l}$$

$$= \frac{30 - 25}{2} \times \frac{500}{250}$$

$$= \frac{5}{2} \times 2$$

$$= 5 \text{ mm}$$

▸▸ Taper Turning

Taper turning on a lathe can be performed on work held between centers or in a lathe chuck. There are three methods of producing a taper:

1. By offsetting the tailstock
2. By means of a taper attachment set to the proper *tpf* or the proper taper angle of the workpiece; on inch tapers by means of the taper attachment set to the *tpf* or taper angle of the workpiece; or on metric tapers by calculating the guide bar offset

3. By adjusting the compound rest to the angle of the taper

The method used to machine any taper depends on the work length, taper length, taper angle, and number of pieces to be machined.

TAILSTOCK OFFSET METHOD

The tailstock offset method is generally used to cut a taper when no taper attachment is available. This involves moving the tailstock center out of line with the headstock center. However, the amount that the tailstock may be offset is limited. This method will not permit steep tapers to be turned or standard tapers to be turned on the end of a long piece of work.

Methods of Offsetting the Tailstock

The tailstock may be offset by three methods:

1. By using the graduations on the end of the tailstock (visual method)
2. By means of the graduated collar and feeler gage
3. By means of a dial indicator

To Offset the Tailstock by the Visual Method

1. Loosen the tailstock clamp nut.
2. Offset the upper part of the tailstock by loosening one setscrew and tightening the other until the required amount is indicated on the graduated scale at the end of the tailstock (Fig. 54-9).

Note: Before machining the work, make sure that both setscrews are snugged up to prevent any lateral movement of the tailstock.

Figure 54-9 Measuring the amount of offset with a rule. *(Kelmar Associates)*

Figure 54-10 Offsetting a tailstock using a dial indicator. *(Kelmar Associates)*

Figure 54-11 Offsetting a tailstock using the crossfeed graduated collar and a feeler. *(Kelmar Associates)*

To Offset the Tailstock Accurately

The tailstock may be accurately offset by using a dial indicator (Fig. 54-10):

1. Adjust the tailstock spindle to the distance it will be used in the machining setup and lock the tailstock spindle clamp.
2. Mount a dial indicator in the toolpost with the plunger in a horizontal position and on center.
3. Using the crossfeed handle, move the indicator so that it registers approximately .020 in. (0.5 mm) on the work, and set the indicator and crossfeed graduated collars to 0.
4. Loosen the tailstock clamp nut.
5. With the tailstock adjusting setscrews, move the tailstock until the required offset is shown on the dial indicator.
6. Tighten the tailstock setscrew that was loosened, making sure that the indicator reading does not change.
7. Tighten the tailstock clamp nut.

The tailstock may also be offset fairly accurately by using a feeler gage between the toolpost and the tailstock spindle in conjunction with the crossfeed graduated collar (Fig. 54-11).

To Turn a Taper by the Tailstock Offset Method

1. Loosen the tailstock clamp nut.
2. Offset the tailstock the required amount.
3. Set up the cutting tool as for parallel turning.

Note: The cutting tool *must* be on center.

4. Starting at the small diameter, take successive cuts until the taper is .050- to .060-in. (1.27- to 1.52-mm) oversize.
5. Check the taper for accuracy using a taper ring gage, if required (see Taper Ring Gages, Unit 14).
6. Finish-turn the taper to the size and fit required.

Taper Turning Using the Taper Attachment

The use of a taper attachment for taper turning provides several advantages:

1. The lathe centers remain in alignment, preventing the distortion of centers on the workpiece.
2. The setup is simple and permits changing from taper to parallel turning with no time lost to align the centers.
3. The length of the workpiece does not matter, since duplicate tapers may be turned on any length of work.
4. Tapers may be produced on work held between centers, in a chuck, or in a collet.
5. Internal tapers can be produced by this method.
6. Metric taper attachments are graduated in millimeters and degrees, while inch attachments are graduated in both degrees and inches of *tpf.* This eliminates the need for lengthy calculations and setup.
7. A wider range of tapers may be produced.

There are two types of taper attachments:

1. The plain taper attachment (Fig. 54-12)
2. The telescopic taper attachment (Fig. 54-13)

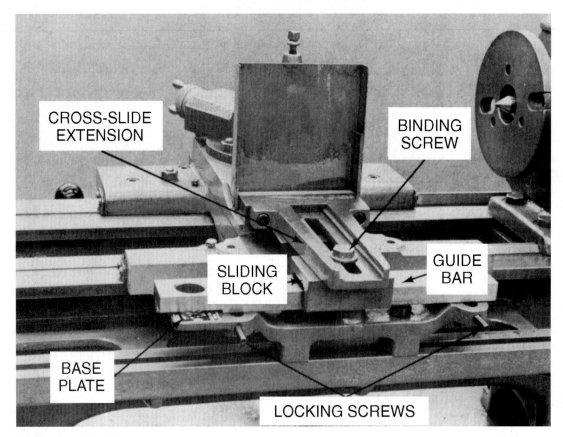

Figure 54-12 The parts of a plain taper attachment. *(Kelmar Associates)*

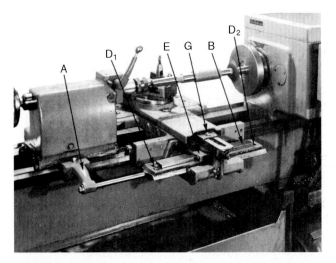

Figure 54-13 The telescopic taper attachment.

When using the *plain taper attachment,* remove the binding screw that holds the cross-slide to the crossfeed screw nut. The binding screw is then used to connect the sliding block to the slide of the taper attachment. With the plain taper attachment, the depth of cut is made by using the compound rest feed handle.

When a *telescopic taper attachment* is used, the crossfeed screw is not disengaged and the depth of cut can be set by the crossfeed handle.

To Cut a Taper Using a Telescopic Taper Attachment

1. Clean and oil the guide bar B (Fig. 54-13).

2. Loosen the lock screws D_1 and D_2 and offset the end of the guide bar the required amount or, for inch attachments, set the bar to the required taper in degrees or *tpf.*

3. Tighten the lock screws.

4. With the compound rest set at 90°, set up the cutting tool on center.

5. Set the workpiece in the lathe and mark the length of taper.

6. Tighten the connecting screw G on the sliding block E.

Note: If a plain taper attachment is used, remove the binding screw in the cross-slide and use it to connect the sliding block and the connecting slide. The compound rest must also be set at right angles to the lathe bed.

7. Move the carriage until the center of the attachment is opposite the length to be tapered.

8. Lock the anchor bracket A to the lathe bed.

9. Take a cut .060 in. (1.5 mm) long, stop the lathe, and check the end of the taper for size.

10. Set the depth of the roughing cut to .050 to .060 in. (1.27 to 1.52 mm) oversize, and machine the taper.

Note: Start the feed about .500 in. (13 mm) before the start of the cut to remove any play in the taper attachment.

11. Readjust the taper attachment if necessary, take a light cut, and recheck the taper fit.

12. Finish-turn and fit the taper to a gage.

When standard tapers must be produced on a piece of work, a taper plug gage may be mounted between centers and the taper attachment adjusted to this angle by using a dial indicator mounted on center in the toolpost.

When an *internal taper* is cut, the same procedure is followed, except that the guide bar is set to the side of the centerline opposite to that used when turning an external taper.

When mating external and internal tapers must be cut, it is advisable first to machine the internal taper to a plug gage. The external taper is then fitted to the internal taper.

TAPER TURNING USING THE COMPOUND REST

To produce short or steep tapers stated in degrees, the compound rest method is used. The tool must be fed in by hand, using the compound rest feed handle. Follow this procedure:

1. Refer to the drawing for the amount of taper required in degrees. However, if the angle on the drawing is not given in degrees, calculate the compound rest setting as follows:

$$\text{Tan } \frac{1}{2} \text{ angle} = \frac{tpf}{24} \text{ or } \frac{tpi}{2}$$

For example, for the workpiece illustrated in Fig. 54-3, the calculations would be:

$$tpf = .250 \times \frac{12}{3}$$
$$= 1 \text{ in.}$$
$$\text{Tan angle} = \frac{1}{24}$$
$$= 0.04166$$

By referring to the trigonometric tables in any handbook, you will find that one-half the angle of this taper (and the compound rest setting) is 2°23′. If a set of trigonometric tables is not available, use the simplified formula to calculate the angle of the taper and the compound rest setting. Tan angle = $tpf \times 2°33′$.

2. Loosen the compound rest lock screws.

3. Swivel the compound rest as follows:
   a. Where included angles are given on the drawing, swivel the compound rest to one-half the angle (Fig. 54-14, top).
   b. Where angles are given on one side only (Fig. 54-14, bottom), swivel the compound rest to that angle.

4. Tighten the compound rest lock screws using only two-finger pressure on the wrench to avoid stripping the lock screw threads.

5. Set the cutting tool to center with the toolholder at right angles to the taper to be cut.

6. Tighten the toolpost securely.

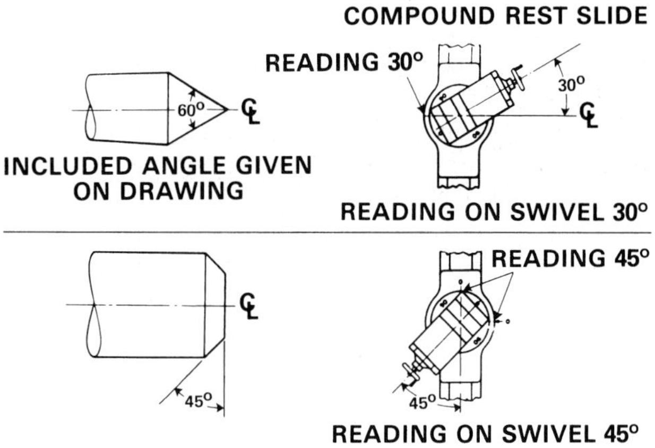

Figure 54-14 Direction to swing the compound rest for cutting various angles. *(Kelmar Associates)*

Figure 54-15 Cutting a short taper using a compound rest. *(Kelmar Associates)*

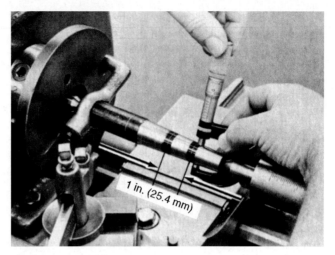

Figure 54-16 Checking the accuracy of a taper with a micrometer. *(Kelmar Associates)*

7. Back off the top slide of the compound rest so that there will be enough travel to machine the length of the taper.

8. Move the carriage to position the cutting tool near the start of the taper and then lock the carriage.

9. Rough turn the taper by feeding the cutting tool using the *compound rest feed handle* (Fig. 54-15).

10. Check the taper for accuracy and readjust the compound rest setting if necessary.

11. Finish-turn and check the taper for size and fit.

▸▸ Checking a Taper

Inch tapers can be checked by scribing two lines exactly 1 in. apart on the taper and carefully measuring the taper at these points with a micrometer (Fig. 54-16). The difference in readings will indicate the *tpi* of the workpiece. Tapers may be more accurately checked by using a sine bar (see Unit 13).

To obtain a more accurate taper, a taper ring gage is used to check external tapers. A taper plug gage is used to check internal tapers (see Unit 14).

The *taper micrometer* (Fig. 54-17) measures tapers quickly and accurately while the workpiece is still in the machine. This instrument includes an adjustable anvil and a 1-in. sine bar attached to the frame, which is adjusted by the micrometer thimble. The micrometer reading indicates the *tpi*, which can be readily converted to *tpf* or angles. The anvil can be adjusted to accommodate a wide range of work sizes.

Figure 54-17 Using a taper micrometer to check an external taper. *(Tape Micrometer Corporation)*

Taper micrometers are available in various models for measuring internal tapers and dovetails (Fig. 54-18) and in bench models incorporating two indicators for quickly checking the accuracy of tapered parts.

Figure 54-18 A taper micrometer for measuring internal tapers. *(Tape Micrometer Corporation)*

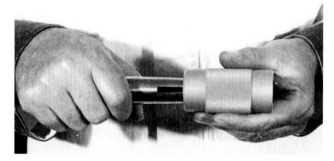

Figure 54-19 Checking the accuracy of a taper using chalk lines. *(Kelmar Associates)*

3. Remove the workpiece and examine the chalk marks. If the chalk has spread along the whole length of the taper, the taper is correct. If the chalk lines are rubbed from only one end, the taper setup must be adjusted.

4. Make a slight adjustment to the taper attachment and, taking trial cuts, machine the taper until the fit is correct.

TO CHECK A METRIC TAPER

1. Check the drawing for the taper required.

2. Clean the tapered section of the work and apply layout dye.

3. Lay out two lines on the taper that are the same distance apart as the second number in the taper ratio. For example, if the taper were 1:20, the lines would be 20 mm apart.

Note: If the work is long enough, lay out the lines at double or triple the length of the tapered section and increase the difference in diameters by the appropriate amount. For instance, on a 1:20 taper, the lines may be laid out 60 mm apart, or three times the unit length of the taper. Therefore, the difference in diameters would then be 3 × 1, or 3 mm, which gives a more accurate check of the taper.

4. Measure the diameters carefully with a metric micrometer at the two lines. The difference between these two diameters should be 1 mm for each unit of length.

5. If necessary, adjust the taper attachment setting to correct the taper.

The advantages of taper micrometers are:

> The taper accuracy can be checked while the workpiece is still in the machine.

> They provide a quick and accurate means of checking tapers.

> They are simple to operate.

> The need for costly gaging equipment is eliminated.

> They can be used for measuring external tapers, internal tapers, and dovetails.

TO FIT AN EXTERNAL TAPER

1. Make three equally spaced lines with chalk or mechanics blue along the taper (see Unit 14, Taper Ring Gages).

2. Insert the taper into the ring gage and turn *counterclockwise* for one-half turn (Fig. 54-19).

 CAUTION Do *not* force the taper into the ring gage.

Tapers

1. Define *taper*.

2. Explain the difference between self-holding and steep tapers.

3. State the *tpf* for the following tapers:

   a. Morse
   b. Brown & Sharpe
   c. Jarno
   d. Standard taper pin

4. Describe the Type D-1 and Type L spindle nose and state where each is used.

Taper Calculations

5. Calculate the *tpf* and tailstock offset for the following work:

   a. D = 1.625 in., d = 1.425 in., TL = 3 in., OL = 10 in.
   b. D = .875 in., d .4375 in., TL = 6 in., OL = 9 in.

6. Calculate the tailstock offset for the following using the simplified tailstock offset formula:

   a. D = .750 in., d = .531 in., TL = 6 in., OL = 18 in.
   b. D = .875 in., d = .781 in., TL = 3.50 in., OL = 10.50 in.

7. Explain what is meant by a metric taper of 1:50.

8. Calculate the large diameter of a 1:50 taper having a small diameter of 15 mm and a length of 75 mm.

9. Calculate the tailstock offset required to turn a 1:40 taper × 100 mm long on a workpiece 450 mm long. The small diameter is 25 mm.

Taper Turning

10. Name three methods of offsetting the tailstock for taper turning.

11. List the advantages of a taper attachment.

12. List the main steps required to cut an external taper using the taper attachment.

13. Describe a taper micrometer and state its advantages.

14. Explain in point form how to fit an external taper.

15. Calculate the amount of setover for a 480-mm long guide bar to turn a 1:40 taper × 320 mm long on a workpiece. The small diameter of the taper is 37.5 mm.

16. At what angle should the compound rest be set to machine a workpiece with a large diameter of 1.250 in., small diameter of .750 in., and length of taper of 1 in.?

UNIT 55

Threads and Thread Cutting

OBJECTIVES

After completing this unit, you should be able to:

1 Recognize and state the purposes of six common thread forms

2 Set up a lathe to cut inch external Unified threads

3 Set up an inch lathe to cut metric threads

4 Set up a lathe and cut internal threads

5 Set up a lathe and cut external Acme threads

Threads have been used for hundreds of years for holding parts together, making adjustments to tools and instruments, and transmitting power and motion. A thread is basically an inclined plane or wedge that spirals around a bolt or nut. Threads have progressed from the early screws, which were filed by hand, to the highly accurate ball screws used on the precision machine tools of today. Although the purpose of a thread is basically the same as when the early Romans developed it, the art of producing threads has continually improved. Today, threads are mass-produced by taps, dies, thread rolling, thread milling, and grinding to exacting standards of accuracy and quality control. Thread cutting is a skill that every machinist should possess because it is still necessary to cut threads on an engine lathe, especially if a special size or form of thread is required.

▶▶ Threads

A *thread* may be defined as a helical ridge of uniform section formed on the inside or outside of a cylinder or cone. Threads are used for several purposes:

1. To fasten devices such as screws, bolts, studs, and nuts
2. To provide accurate measurement, as in a micrometer
3. To transmit motion; the threaded lead screw on the lathe causes the carriage to move along when threading
4. To increase force; heavy work can be raised with a screw jack

THREAD TERMINOLOGY

To understand and calculate thread parts and sizes, the following definitions relating to screw threads should be known (Fig. 55-1 on p. 418):

> A *screw thread* is a helical ridge of uniform section formed on the inside or outside of a cylinder or cone.

> An *external thread* is cut on an external surface or cone, such as on a cap screw or a wood screw.

> An *internal thread* is produced on the inside of a cylinder or cone, such as the thread on the inside of a nut.

> The *major diameter* is the largest diameter of an external or internal thread.

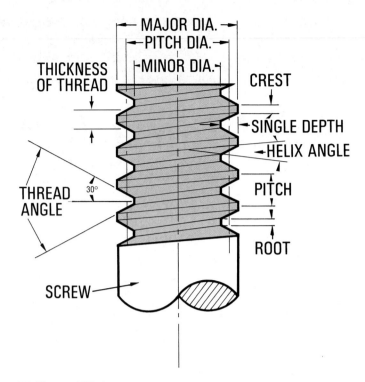

THICKNESS OF THREAD

MAJOR DIA.

PITCH DIA.

MINOR DIA.

CREST

SINGLE DEPTH

HELIX ANGLE

THREAD ANGLE

30°

PITCH

ROOT

SCREW

Figure 55-1 The parts of a screw thread. *(Kelmar Associates)*

> The *minor diameter* is the smallest diameter of an external or internal thread. This was formerly known as the root diameter.

> The *pitch diameter* is the diameter of an imaginary cylinder that passes through the thread at a point where the groove and thread widths are equal. The pitch diameter is equal to the major diameter minus a single depth of thread. The tolerance and allowances on threads are given at the pitch diameter line. The pitch diameter is also used to determine the outside diameter for rolled threads. The diameter of the blank is always equal to the pitch diameter of the thread to be rolled. Thread rolling is a displacement operation and the amount of metal displaced is forced up to form the thread above the pitch line.

Note: Pitch diameter is not used as a basis for determining International Organization for Standardization (ISO) metric thread dimensions.

> The *number of threads per inch* is the number of crests or roots per inch of threaded section. This term does not apply to metric threads.

> The *pitch* is the distance from a point on one thread to a corresponding point on the next thread, measured parallel to the axis. Pitch is expressed in millimeters for metric threads.

> The *lead* is the distance a screw thread advances axially in one revolution. On a single-start thread, the lead and the pitch are equal.

> The *root* is the bottom surface joining the sides of two adjacent threads. The root of an external thread is on its minor diameter. The root of an internal thread is on its major diameter.

> The *crest* is the top surface joining two sides of a thread. The crest of an external thread is on the major diameter, while the crest of an internal thread is on the minor diameter.

> A *flank* (side) is a thread surface that connects the crest with the root.

> The *depth of thread* is the distance between the crest and root measured perpendicular to the axis.

> The *angle of thread* is the included angle between the sides of a thread measured in an axial plane.

> The *helix angle* (lead angle) is the angle that the thread makes with a plane perpendicular to the thread axis.

> A *right-hand thread* is a helical ridge of uniform cross section onto which a nut is threaded in a clockwise direction. When the thread is held in a horizontal position with its axis pointing from right to left, a right-hand thread will slope *down* and to the right (Fig. 55-2a). When a right-hand thread is cut on a lathe, the toolbit advances from right to left.

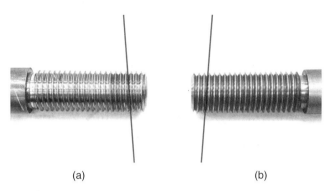

(a)   (b)

■ **Figure 55-2** (a) Right-hand screw threads; (b) left-hand screw threads.

> A *left-hand thread* is a helical ridge of uniform crosssection onto which a nut is threaded in a counterclockwise direction. When the thread is held in a horizontal position with its axis pointing from right to left, the thread will slope *down* and to the left (Fig. 55-2b). When a left-hand thread is cut on a lathe, the toolbit advances from left to right.

THREAD FORMS

Over the past several decades, one of the world's major industrial problems has been the lack of an international thread standard whereby the thread standard used in any country could be interchanged with that of another country. In April 1975, the ISO drew up an agreement covering a standard metric thread profile, specifying the sizes and pitches for the various threads in the new ISO Metric Thread Standard. The new series has only 25 thread sizes, which range in diameter from 1.6 to 100 mm. Countries throughout the world have been encouraged to adopt the ISO series (Table 55.1).

These metric threads are identified by the letter M, the nominal diameter, and the pitch. For example, a metric thread with an outside diameter of 5 mm and a pitch of 0.8 mm would be identified as follows: M 5 × 0.8.

The ISO series will not only simplify thread design but will generally produce stronger threads for a given diameter and pitch and will reduce the large inventory of fasteners now required by industry.

The *ISO metric thread* (Fig. 55-3) has a 60° included angle and a crest equal to 0.125 times the pitch, similar to the National Form thread. The main difference, however, is the depth of thread (*D*), which is *0.6134* times the pitch. Because of these dimensions, the flat on the root of the thread (*FR*) is wider than the crest (*FC*). The root of the ISO metric thread is one-fourth of the pitch (0.250*P*).

$$D \text{ (external)} = 0.54127 \times P$$

$$FC = 0.125 \times P$$

$$FR = 0.250 \times P$$

table 55.1	ISO metric pitch and diameter combinations		
Nominal Diameter (mm)	Thread Pitch (mm)	Nominal Diameter (mm)	Thread Pitch (mm)
1.6	0.35	20	2.5
2	0.4	24	3
2.5	0.45	30	3.5
3	0.5	36	4
3.5	0.6	42	4.5
4	0.7	48	5
5	0.8	56	5.5
6.3	1	64	6
8	1.25	72	6
10	1.5	80	6
12	1.75	90	6
14	2	100	6
16	2		

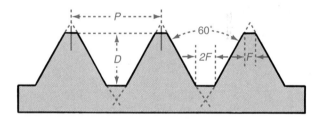

■ **Figure 55-3** ISO metric thread.

The most commonly used thread forms in North America at the present time are the following.

The *American National Standard thread* (Fig. 55-4 on p. 420) is divided into four main series, all having the same shape and proportions: National Coarse (NC), National Fine (NF), National Special (NS), and National Pipe (NPT). This thread has a 60° angle with a root and crest truncated to one-eighth the pitch. This thread is used in fabrication, machine construction and assembly and for components where easy assembly is desired.

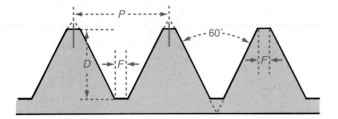

■ Figure 55-4 American National Standard thread.

The formula for calculating the depth of a 100% thread is 0.866/*N*. However, since this thread would be very difficult to cut (especially internally), the following formula, which gives approximately 75% of a thread, is generally a standard in the industry:

$$D = .61343 \times P \text{ or } \frac{.61343}{N}$$

$$F = .125 \times P \text{ or } \frac{.125}{N}$$

The *British Standard Whitworth (BSW) thread* (Fig. 55-5) has a 55° V-form with rounded crests and roots. This thread application is the same as for the American National form thread:

$$D = .6403 \times P \text{ or } \frac{.6403}{N}$$

$$R = .1373 \times P \text{ or } \frac{.1373}{N}$$

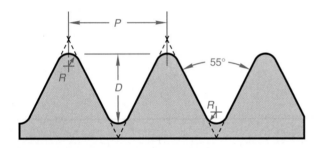

■ Figure 55-5 British Standard Whitworth thread.

The *Unified thread* (Fig. 55-6) was developed by the United States, Britain, and Canada so that equipment produced by these countries would have a standardized thread system. Until this thread was developed, many problems were created by the noninterchangeability of threaded parts used in these countries. The Unified thread is a combination of the British Standard Whitworth and the American National Standard thread. This thread has a 60° angle with a rounded root, and the crest may be rounded or flat.

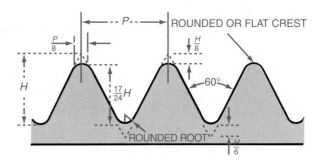

■ Figure 55-6 Unified thread.

$$D \text{ (external thread)} = .6134 \times P \text{ or } \frac{.6134}{N}$$

$$D \text{ (internal thread)} = .5413 \times P \text{ or } \frac{.5413}{N}$$

$$F \text{ (external thread)} = .125 \times P \text{ or } \frac{.125}{N}$$

$$F \text{ (internal thread)} = .250 \times P \text{ or } \frac{.250}{N}$$

The *American National Acme thread* (Fig. 55-7) is replacing the square thread in many cases. It has a 29° angle and is used for feed screws, jacks, and vises.

$$D = \text{minimum } .500P$$

$$= \text{maximum } .500P + 0.010$$

$$F = .3707P$$

$$C = .3707P - .0052$$
$$\text{(for maximum depth)}$$

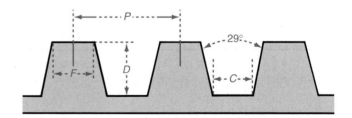

■ Figure 55-7 American National Acme thread.

The *Brown & Sharpe worm thread* (Fig. 55-8) has a 29° included angle, as does the Acme thread; however, the depth is greater and the widths of the crest and root are different. This thread is used to mesh with worm gears and transmit motion between two shafts at right angles to each other but not in the same plane. The self-locking feature makes it adaptable to winches and steering mechanisms.

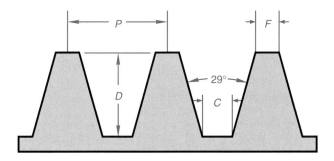

Figure 55-8 Brown & Sharpe worm thread.

$$D = .6866P$$

$$F = .335P$$

$$C = .310P$$

The *square thread* (Fig. 55-9) is being replaced by the Acme thread because of the difficulty in cutting it, particularly with taps and dies. Square threads were often found on vises and jack screws.

$$D = .500P$$

$$F = .500P$$

$$C = .500P + .002$$

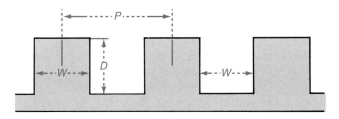

Figure 55-9 Square thread.

The *international metric thread* (Fig. 55-10) is a standardized thread used in Europe. This thread has a 60° included angle with a crest and root truncated to one-eighth the depth. Although this thread is used extensively

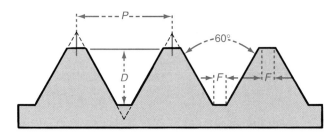

Figure 55-10 International metric thread.

throughout Europe, its use in North America has been confined mainly to spark plugs and the manufacture of instruments.

$$D = 0.7035P \text{ (maximum)}$$

$$= 0.6855P \text{ (minimum)}$$

$$F = 0.125P$$

$$R = 0.0633P \text{ (maximum)}$$
$$0.054P \text{ (minimum)}$$

THREAD FITS AND CLASSIFICATIONS

Certain terminology is used when referring to thread classifications and fits. To understand any thread system properly, the terminology relating to thread fits should be understood.

Fit is the relationship between two mating parts. It is determined by the amount of clearance or interference when they are assembled.

Allowance is the intentional difference in size of the mating parts or the minimum clearance between mating parts (Fig. 55-11). With threads, the allowance is the permissible difference between the largest external thread and the smallest internal thread. This difference produces the tightest fit acceptable for any given classification.

The allowance for a 1 in.—8 UNC (Unified National Coarse) Class 2A and 2B fit is:

Minimum pitch diameter of the
    internal thread (2B)   = .9188 in.

Maximum pitch diameter of the
    external thread (2A)   = .9168 in.

Allowance or intentional difference   = .002 in.

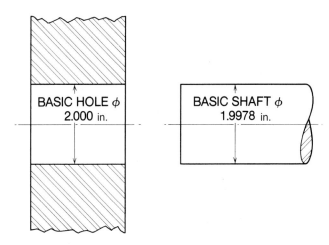

Figure 55-11 The allowance (intentional difference) between the shaft and hole is .0022 in.

Tolerance is the variation permitted in part size. The tolerance may be expressed as plus, minus, or both. The total tolerance is the sum of the plus and minus tolerances. For example, if a size is 1.000 ± .002 (a bilateral tolerance), the total tolerance is .004. In the Unified and National systems, the tolerance is plus on external threads and minus on internal threads. Thus, when a thread varies from the basic or nominal size, it will ensure a freer rather than a tighter fit.

The tolerance for a 1 in.—8 UNC Class 2A thread is:

Maximum pitch diameter of the
    external thread (2A)   = .9168 in.

Minimum pitch diameter of the
    external thread (2A)   = .9100 in.

Tolerance or permitted variation   = .0068 in.

Limits are the maximum and minimum dimensions of a part. The limits for a 1 in.—8 UNC Class 2A thread are:

Maximum pitch diameter of the
    external thread (2A)   = .9168 in.

Minimum pitch diameter of the
    external thread (2A)   = .9100 in.

The pitch diameter of this thread must be between .9168 in. (upper limit) and .9100 in. (lower limit).

Nominal size is the designation used to identify the size of the part. For example, in the designation of 1 in.—8 UNC, the number 1 indicates a 1-in. diameter thread.

Actual size is the measured size of a thread or part. The basic size is the size from which tolerances are set. Although the basic major diameter of a 1 in.—8 UNC Class 2A thread is 1.000 in., the actual size may vary from .998 to .983 in.

Classification of Thread Fits

With wide use of threads, it became necessary to establish certain limits and tolerances to identify classes of fit properly.

ISO Metric Tolerances and Allowances

The ISO metric screw thread tolerance system provides for allowances and tolerances defined by tolerance grades, tolerance positions, and tolerance classes.

Tolerance grades are specified numerically. For example, a medium tolerance, used on a general-purpose thread, is indicated by the number 6. Any number below 6 indicates a finer tolerance and any number above 6 indicates a greater tolerance. The tolerance for the thread at the pitch line and for the major diameter may be shown on the drawing.

Symbols are used to indicate the *allowance*. For external threads:

    e indicates a large allowance

    g indicates a small allowance

    h indicates no allowance

For internal threads:

    G indicates a small allowance

    H indicates no allowance

EXAMPLE

An external metric thread may be designated as follows:

Metric ↓	Nominal Size ↓		Pitch ↓	Pitch Dia. Tolerance ↓	Outside Dia. Tolerance ↓
M	6	×	0.75 −	5g	6g

The thread fit between mating parts is indicated by the internal thread designation followed by the external thread tolerance:

$$M\ 20 \times 2 - 6H/5g\ 6g$$

Unified thread fits have been divided into three categories and the applications of each have been defined by the Screw Thread Committee. External threads are classified as 1A, 2A, and 3A and internal threads as 1B, 2B, and 3B.

Classes 1A and 1B include those threads for work that must be readily assembled. They have the loosest fit, with no possibility of interference between the mating external and internal threads when the threads are dirty or bruised.

Classes 2A and 2B are used for most commercial fasteners. These threads provide a medium or free fit and permit power wrenching with minimum galling and seizure.

Classes 3A and 3B are used where a more accurate fit and lead are required. No allowance is provided, and the tolerances are 75% of those used for 2A and 2B fits.

By classifying the tolerances of threads, the cost of threaded parts is reduced, since manufacturers may use any combination of mating threads that suits their needs. With the former system of identifying classes of tolerances (Classes 1, 2, 3, 4), it was felt, for example, that a Class 3 internal thread should be used with a Class 3 external thread.

With reference to the Unified system, it should be noted that "Class" refers to tolerance or tolerance and allowance; it does not refer to fit. The fit between the mating parts is determined by the selected combination used for a specific application. For example, if a closer-than-normal fit is required, a Class 3B nut may be used on a Class 2A bolt. The basic dimensions, tolerances, and allowances for these threads may be found in any machine handbook.

THREAD CALCULATIONS

To cut a correct thread on a lathe, it is necessary first to make calculations so that the thread will have the proper dimensions. The following formulas will be helpful when calculating thread dimensions. The symbols used in these formulas are:

$$D = \text{single depth of thread}$$

$$P = \text{pitch}$$

EXAMPLE

Calculate the pitch, depth, minor diameter, and width of flat for a ¾—10 UNC thread.

$$
\begin{aligned}
P &= \frac{1}{tpi} \\
&= \frac{1}{10} \\
&= .100 \text{ in.} \\
D &= .61343 \times P \\
&= .61343 \times \frac{1}{10} \\
&= .061 \text{ in.} \\
\text{Minor diameter} &= \text{major diameter} - (D + D) \\
&= .750 - (.061 + .061) \\
&= .628 \text{ in.} \\
\text{Width of flat} &= \frac{P}{8} \\
&= \frac{1}{8} \times \frac{1}{10} \\
&= .0125 \text{ in.}
\end{aligned}
$$

EXAMPLE

What are the pitch, depth, minor diameter, width of crest, and width of root for an M 6.3 × 1 thread?

$$
\begin{aligned}
P &= 1 \text{ mm} \\
D &= 0.54127 \times 1 \\
&= 0.54 \text{ mm}
\end{aligned}
$$

$$
\begin{aligned}
\text{Minor diameter} &= \text{major diameter} - (D + D) \\
&= 6.3 - (0.54 + 0.54) \\
&= 5.22 \text{ mm} \\
\text{Width of crest} &= 0.125 \times P \\
&= 0.125 \times 1 \\
&= 0.125 \text{ mm} \\
\text{Width of root} &= 0.25 \times P \\
&= 0.25 \times 1 \\
&= 0.25 \text{ mm}
\end{aligned}
$$

To Set the Quick-Change Gearbox for Threading

The quick-change gearbox provides a means of quickly setting the lathe for the desired pitch of the thread in number of threads per inch on inch-system lathes or millimeters on metric lathes. This unit contains a number of different sizes of gears, which vary the ratio between headstock spindle revolutions and the rate of carriage travel when thread cutting.

Follow this procedure:

1. Check the drawing for the thread pitch required.

2. From the chart on the quick-change gearbox, find the *whole number* that represents the pitch in threads per inch or in millimeters.

3. With the lathe stopped, engage the *tumbler lever* in the hole, which is in line with the pitch (*tpi* or in millimeters) (Fig. 55-12).

4. Set the *top lever* in the proper position as indicated on the chart.

5. Engage the *sliding gear* in or out as required.

Note: Some lathes have two levers on the gearbox that take the place of the top lever and sliding gear, and these should be set as indicated on the chart.

SLIDING GEAR

TOP LEVER

GEAR-BOX

TUMBLER LEVER

■ **Figure 55-12** Quick-change gear mechanism is used for setting the number of threads per inch to be cut.

6. Turn the lathe spindle by hand to ensure that the lead screw revolves.

7. Recheck the lever settings to avoid errors.

▶▶ Thread-Chasing Dial

To cut a thread on a lathe, the lathe spindle and the lead screw must be in the same relative position for each successive cut. Most lathes have a thread-chasing dial either built into or attached to the carriage for this purpose. The chasing dial indicates when the split nut should be engaged with the lead screw for the cutting tool to follow the previously cut groove.

The thread-chasing dial is connected to a worm gear, which meshes with the threads of the lead screw (Fig. 55-13). The dial is graduated into eight divisions, four numbered and four unnumbered, and it revolves as the lead screw turns. Fig. 55-14 indicates when the split-nut lever should be engaged for cutting various numbers of threads per inch (*tpi*).

> Even threads use any division.

> Odd threads must stay on either numbered or unnumbered lines: cannot use both.

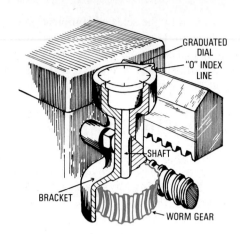

■ **Figure 55-13** Thread-chasing dial mechanism.

▶▶ Thread Cutting

Thread cutting on a lathe is a process that produces a helical ridge of uniform section on a workpiece. This is performed by taking successive cuts with a threading toolbit of the same shape as the thread form required. Work to be threaded may be held between centers or in a chuck.

THREADS PER INCH TO BE CUT	WHEN TO ENGAGE SPLIT NUT	READING ON DIAL
EVEN NUMBER OF THREADS	ENGAGE AT ANY GRADUATION ON THE DIAL 1 1½ 2 2½ 3 3½ 4 4½	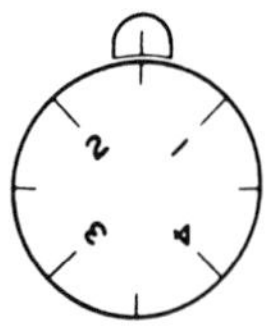
ODD NUMBER OF THREADS	ENGAGE AT ANY MAIN DIVISION 1 2 3 4	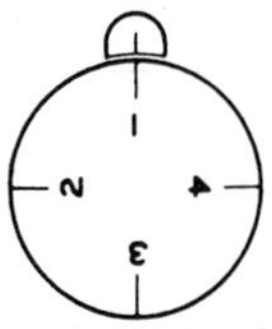
FRACTIONAL NUMBER OF THREADS	1/2 THREADS, E.G.,1-1½ ENGAGE AT EVERY OTHER MAIN DIVISION 1 & 3, OR 2 & 4 OTHER FRACTIONAL THREADS ENGAGE AT SAME DIVISION EVERY TIME	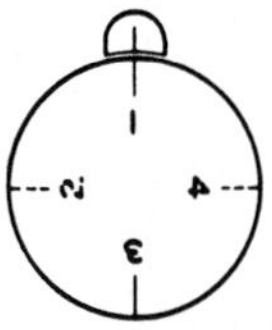
THREADS WHICH ARE A MULTIPLE OF THE NUMBER OF THREADS PER INCH IN THE LEAD SCREW	ENGAGE AT ANY TIME THAT SPLIT NUT MESHES	USE OF DIAL UNNECESSARY

■ **Figure 55-14** Split-nut engagement rules for thread cutting.

If work is held in a chuck, it should be turned to size and threaded before the work is removed.

TO SET UP A LATHE FOR THREADING (60° THREAD)

1. Set the lathe speed to about one-quarter the speed used for turning.
2. Set the quick-change gearbox for the required pitch in threads per inch or in millimeters.
3. Engage the lead screw.
4. Secure a 60° threading toolbit and check the angle using a thread center gage.
5. Set the compound rest at 29° to the right (Fig. 55-15); set it to the left for a left-hand thread.

Note: With the compound rest set at 29°, a slight shaving action occurs on the following edge of the thread (right side) each time that the threading tool is fed in with the compound rest handle.

6. Set the cutting tool to the height of the lathe center point.
7. Mount the work between centers. Make sure the lathe dog is tight on the work. If the work is mounted in a chuck, it must be held tightly.
8. Set the toolbit at right angles to the work, using a thread center gage (Fig. 55-16).

Note: Never jam a toolbit into a thread center gage. This can be avoided by aligning only the cutting edge (leading side) of the toolbit with the gage. A piece of paper on the

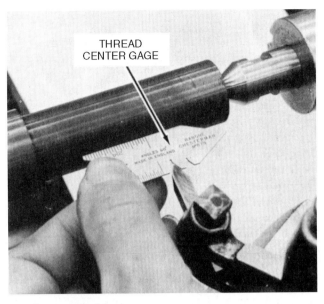

■ **Figure 55-16** Setting a threading tool square with a thread center gage. *(Kelmar Associates)*

cross-slide under the gage and toolbit makes it easier to check tool alignment.

9. Arrange the apron controls to allow the split-nut lever to be engaged.

THREAD-CUTTING OPERATION

Thread cutting is one of the more interesting operations performed on a lathe. It involves manipulation of the lathe parts, coordination of the hands, and strict attention to the operation. Before proceeding to cut a thread for the first time on any lathe, take several trial passes, without cutting, to get the feel of the machine.

To Cut a 60° Thread

1. Check the major diameter of the work for size. It is good practice to have the diameter .002 in. (0.05 mm) undersize.
2. Start the lathe and chamfer the end of the workpiece with the side of the threading tool to just below the minor diameter of the thread.
3. Mark the length to be threaded by cutting a light groove at this point with the threading tool while the lathe is revolving.
4. Move the carriage until the point of the threading tool is near the right-hand end of the work.
5. Turn the *crossfeed handle* until the threading tool is close to the diameter, but stop when the handle is at the 3 o'clock position (Fig. 55-17).

■ **Figure 55-15** Compound rest set at 29° for thread cutting. *(Kelmar Associates)*

■ Figure 55-17 The handle of the crossfeed screw set at the 3 o'clock position for thread cutting. *(Kelmar Associates)*

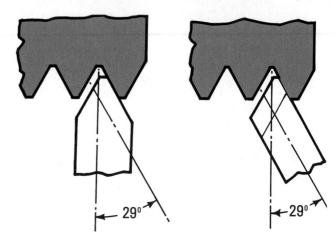

■ Figure 55-19 When the tool is fed in at 29°, most of the cutting is done by the leading edge of the toolbit. *(Kelmar Associates)*

6. Hold the crossfeed handle in this position and set the graduated collar to zero (0).

7. Turn the compound rest handle until the threading tool *lightly marks the work.*

8. Move the carriage to the right until the toolbit clears the end of the work.

9. Feed the compound rest *clockwise* about .003 in. (0.08 mm).

10. Engage the split-nut lever on the correct line of the thread-chasing dial (Fig. 55-14) and take a trial cut along the length to be threaded.

11. At the end of the cut, turn the crossfeed handle *counterclockwise* to move the toolbit away from the work and then disengage the split-nut lever.

12. Stop the lathe and check the number of *tpi* with a thread pitch gage, rule, or center gage (Fig. 55-18).

If the pitch (*tpi* or millimeters) produced by the trial cut is not correct, recheck the quick-change gearbox setting.

13. After each cut, turn the carriage handwheel to bring the toolbit to the start of the thread and return the crossfeed handle to zero (0).

14. *Set the depth of all threading cuts with the compound rest handle* (Fig. 55-19). For National Form threads, use Table 55.2; for ISO metric threads, see Table 55.3.

15. Apply cutting fluid and take successive cuts until the top (crest) and the bottom (root) of the thread are the same width.

16. Remove the burrs from the top of the thread with a file.

17. Check the thread with a master nut and take further cuts, if necessary, until the nut fits the thread freely with no end play (Fig. 55-20).

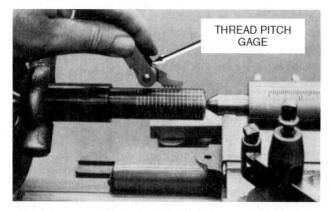

■ Figure 55-18 Checking the number of threads per inch using a thread pitch gage. *(Kelmar Associates)*

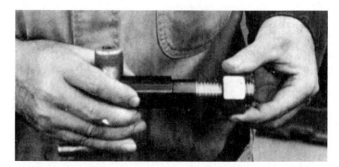

■ Figure 55-20 Checking a thread with a master nut. *(Kelmar Associates)*

table 55.2 — Depth settings for cutting 60° national form threads*

tpi	Compound Rest Setting		
---	0°	30°	29°
24	.027	.031	.0308
20	.0325	.0375	.037
18	.036	.0417	.041
16	.0405	.0468	.046
14	.0465	.0537	.0525
13	.050	.0577	.057
11	.059	.068	.0674
10	.065	.075	.074
9	.072	.083	.082
8	.081	.0935	.092
7	.093	.1074	.106
6	.108	.1247	.1235
4	.1625	.1876	.1858

When using this table for cutting National Form threads, the correct width of flat (.125P) must be ground on the toolbit; otherwise, the thread will not be the correct width.

table 55.3 — Depth setting for cutting 60° ISO metric threads

Pitch (mm)	Compound Rest Setting (mm)		
---	0°	30°	29°
0.35	0.19	0.21	0.21
0.4	0.21	0.25	0.24
0.45	0.24	0.28	0.27
0.5	0.27	0.31	0.31
0.6	0.32	0.37	0.37
0.7	0.37	0.43	0.43
0.8	0.43	0.5	0.49
1	0.54	0.62	0.62
1.25	0.67	0.78	0.77
1.5	0.81	0.93	0.93
1.75	0.94	1.09	1.08
2	1.08	1.25	1.24
2.5	1.35	1.56	1.55
3	1.62	1.87	1.85
3.5	1.89	2.19	2.16
4	2.16	2.5	2.47
4.5	2.44	2.81	2.78
5	2.71	3.13	3.09
5.5	2.98	3.44	3.4
6	3.25	3.75	3.71

There are six ways to check threads, depending on the accuracy required:

1. Master nut or screw
2. Thread micrometer
3. Three wires
4. Thread roll or snap gage
5. Thread ring or plug gage
6. Optical comparator

To Reset a Threading Tool

A threading tool must be reset whenever it is necessary to remove partly threaded work from the lathe and finish it later, if the threading tool is removed for regrinding, or if the work slips under the lathe dog. Follow this procedure:

1. Set up the lathe and work for thread cutting.
2. Start the lathe and, with the toolbit clear of the work, engage the split-nut lever on the correct line.
3. Allow the carriage to travel until the toolbit is opposite any portion of the unfinished thread (Fig. 55-21 on p.428).

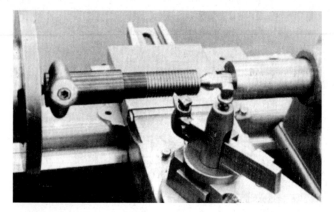

Figure 55-21 With the split-nut lever engaged, stop the machine when the threading tool is over the thread. *(Kelmar Associates)*

4. Stop the lathe, *leaving the split-nut lever engaged.*

5. Feed the toolbit into the thread groove using *only* the compound rest and crossfeed handles until the right-hand edge of the toolbit touches the rear side of the thread (Fig. 55-22).

Note: Do not let the cutting edge of the toolbit contact the thread at this time.

6. Set the crossfeed graduated collar to zero (0).

7. Back out the threading tool using the crossfeed handle, disengage the split-nut lever, and move the carriage until the toolbit clears the start of the thread.

8. Set the crossfeed handle back to zero (0) and take a trial cut without setting the compound rest.

9. Set the depth of cut using the compound rest handle and finish the thread to the required depth.

Figure 55-22 Resetting the threading tool in a partially cut groove using only the crossfeed and compound rest handles. *(Kelmar Associates)*

To Convert an Inch-Designed Lathe to Metric Threading

Metric threads may be cut on a standard quick-change gear lathe by using a pair of change gears having 50 and 127 teeth, respectively. Since the lead screw has inch dimensions and is designed to cut threads per inch, it is necessary to convert the pitch in millimeters to centimeters and then into threads per inch. To do this, it is first necessary to understand the relationship between inches and centimeters.

$$1 \text{ in.} = 2.54 \text{ cm}$$

Therefore, the ratio of inches to centimeters is 1:2.54, or 1/2.54.

To cut a metric thread on an inch lathe, it is necessary to install certain gears in the gear train that will produce a ratio of 1/2.54. These gears are:

$$\frac{1}{2.54} \times \frac{50}{50} = \frac{50 \text{ teeth}}{127 \text{ teeth}}$$

To cut metric threads, two gears having 50 and 127 teeth must be placed in the gear train of the lathe. The 50-tooth gear is used as the spindle or drive gear, and the 127-tooth gear is placed on the lead screw.

To Cut a 2-mm Metric Thread on a Standard Quick-Change Gear Lathe

1. Mount the 127-tooth gear on the lead screw.

2. Mount the 50-tooth gear on the spindle.

3. Convert the 2-mm pitch to threads per centimeter:

$$10 \text{ mm} = 1 \text{ cm}$$

$$P = \frac{10}{2} = 5 \text{ threads/cm}$$

4. Set the quick-change gearbox to 5 *tpi*. By means of the 50- and 127-tooth gears, the lathe will now cut 5 threads/cm or 2-mm pitch.

5. Set up the lathe for thread cutting. See To Set Up a Lathe for Threading (60° Thread) on p. 425.

6. Take a light trial cut. At the end of the cut, back out the cutting tool and stop the machine but *do not disengage the split nut.*

7. Reverse the spindle rotation until the cutting tool has just cleared the start of the threaded section.

8. Check the thread with a metric screw pitch gage.

9. Cut the thread to the required depth (Table 55-3).

Note: Never disengage the split nut until the thread has been cut to depth.

To Cut a Left-Hand Thread (60°)

A left-hand thread is used to replace a right-hand thread on certain applications where the nut may loosen because of the rotation of a spindle. The procedure for cutting left-hand threads is basically the same as for right-hand threads, with a few exceptions:

1. Set the lathe speed and the quick-change gearbox for the pitch of the thread to be cut.

2. Engage the feed-direction lever so that the lead screw will revolve in the *opposite* direction to that for a right-hand thread.

3. Set the compound rest to 29° to the *LEFT* (Fig. 55-23).

4. Set up the left-hand threading tool and square it with the work.

5. Cut a groove at the left end of the section to be threaded. This gives the cutting tool a starting point.

6. Proceed to cut the thread to the same dimensions as for a right-hand thread.

Cutting a Thread on a Tapered Section

When a tapered thread, such as a pipe thread, is required on the end of a workpiece, either the taper attachment or the offset tailstock may be used for cutting the taper. The same setup is then used as for regular thread cutting. When setting up the threading tool, it is most important that it be set at 90° to the axis of the work and not square with the tapered surface (Fig. 55-24).

■ **Figure 55-23** Set the compound rest at 29° to the left for left-hand threads. *(Kelmar Associates)*

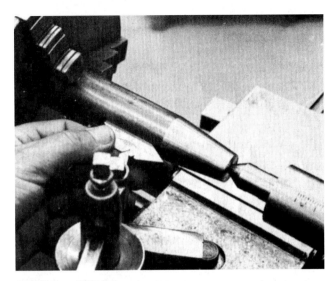

■ **Figure 55-24** The toolbit must be set square with the axis of the workpiece for cutting a thread on a tapered section. *(Kelmar Associates)*

THREAD MEASUREMENT

Interchangeable manufacture demands that all parts be made to certain standards so that, on assembly, they will fit the intended component properly. This fit is especially important for threaded components, and therefore the measurement and inspection of threads is important.

Threads may be measured by a variety of methods; the most common are:

1. A thread ring gage
2. A thread plug gage
3. A thread snap gage
4. A screw thread micrometer
5. A thread comparator micrometer
6. An optical comparator
7. The three-wire method

The description and use of the ring gage, plug gage, snap gage, screw thread micrometer, thread comparator micrometer, and optical comparator for checking threads are described fully in Section 5.

Three-Wire Method of Measuring Threads

The three-wire method of measuring threads is recommended by the Bureau of Standards and the National Screw Thread Commission. It is recognized as one of the best methods of checking the pitch diameter because the results are least affected by an error that may be present in the included thread angle. For threads that require an accuracy of .001 in. or 0.02 mm, a micrometer can be used

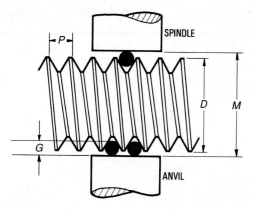

Figure 55-25 The three-wire method of measuring 60° threads. *(Kelmar Associates)*

to measure the distance over the wires. An electronic comparator should be used to measure the distance over the wires for threads requiring greater accuracy.

Three wires of equal diameter are placed in the thread, two on one side and one on the other side (Fig. 55-25). The wires used should be hardened and lapped to three times the accuracy of the thread to be inspected. A standard micrometer may then be used to measure the distance over the wires (*M*). Different sizes and pitches of threads require different sizes of wires. For the greatest accuracy, the *best-size wire* should be used, one that will contact the thread at the pitch diameter (middle of the sloping sides). If the best-size wire is used, the pitch diameter of the thread can be calculated by subtracting the wire constant (found in any handbook) from the measurement over the wires.

To Calculate the Measurement over the Wires

The measurement over the wires for American National Threads (60°) can be calculated by applying the following formula:

$$M = D + 3G - \frac{1.5155}{N}$$

where M = measurement over the wires
D = major diameter of the thread
G = diameter of the wire size used
N = number of *tpi*

Any of the following formulas can be used to calculate *G*:

$$\text{Largest wire} = \frac{1.010}{N} \text{ or } 1.010P$$

$$\text{Best-size wire} = \frac{.57735}{N} \text{ or } .57735P$$

$$\text{Smallest wire} = \frac{.505}{N} \text{ or } .505P$$

For the most accurate thread measurement, the best-size wire (.57735P) should be used because this wire will contact the thread at the pitch diameter.

EXAMPLE

Find *M* (measurement over the wires) for a 3/4—10 NC thread.

1. Calculate *G* (wire size):

$$G = \frac{.57735}{10}$$
$$= .0577$$

2. Calculate *M* (measurement over the wires):

$$M = D + 3G - \frac{1.5155}{N}$$
$$= .750 + (3 \times .0577) - \frac{1.5155}{10}$$
$$= .750 + .1731 - .1516$$
$$= .9231 - .1516$$
$$= .7715$$

MULTIPLE THREADS

Multiple threads are used when it is necessary to obtain an increase in lead and a deep, coarse thread cannot be cut. Multiple threads may be double, triple, or quadruple, depending on the number of starts around the periphery of the workpiece (Fig. 55-26).

The *pitch* of a thread is always the distance from a point on one thread to the corresponding point on the next thread. The *lead* is the distance a nut advances lengthwise in one complete revolution. On a single-start thread, the pitch and lead are equal. On a double-start thread, the lead will be twice the pitch. On triple-start threads, the lead will be three times the pitch.

Multiple-start threads are not as deep as single-start threads and therefore have a more pleasing appearance. For example, a double-start thread having the same lead as a single-start thread would be cut only half as deep.

Multiple threads may be cut on a lathe by:

1. Using an accurately slotted driveplate or faceplate

2. Disengaging the intermediate gear of the end gear train and rotating the spindle the desired amount

3. Using the thread-chasing dial (only for double-start threads with an odd-number lead)

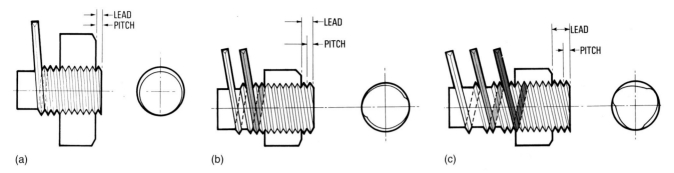

(a)   (b)   (c)

■ **Figure 55-26** The relationship between the pitch and the lead for single-start and multiple-start threads: (a) single thread; (b) double thread; (c) triple thread. *(Kelmar Associates)*

To Cut an 8-tpi Double Thread

1. Set up the lathe and cutting tool as for cutting a single-start thread.

2. Set the quick-change gearbox to 4 *tpi*. (The lead of this thread is .250 in.).

3. Cut the first thread to half the depth required for 4 *tpi*.

4. Leave the crossfeed handle set to the depth of the thread and *note the reading on the compound rest graduated collar.*

5. Withdraw the threading tool from the work using the *compound rest handle.*

6. Revolve the work exactly one-half turn by either of the following methods.

   a. 1. Remove the work from the lathe with the lathe dog attached.
      2. Replace the work in the lathe with the tail of the dog in the slot exactly opposite the one used for the first thread.

Note: An accurately slotted driveplate or faceplate must be used for this method of indexing. A special indexing plate may also be used for this purpose.

OR

   b. 1. Turn the lathe by hand until a tooth of the spindle gear is exactly between two teeth of the intermediate gear.
      2. With chalk, mark both the spindle tooth and the space in the intermediate gear (Fig. 55-27).
      3. Disengage the intermediate gear from the spindle gear.
      4. Starting with the tooth *next* to the marked tooth on the spindle gear, count the number required for a half-revolution of the spindle. For example, if the spindle gear has 24 teeth, count 12 teeth and mark this one with chalk.

■ **Figure 55-27** Marking the spindle and intermediate gear before indexing the workpiece exactly one-half turn for cutting a double thread. *(Kelmar Associates)*

      5. Revolve the lathe spindle by hand one-half turn to bring the marked tooth in line with the chalk mark on the intermediate gear.
      6. Reengage the intermediate gear.

7. Reset the crossfeed handle to the same position as when cutting the first thread.

8. Cut the second thread, feeding the compound rest handle until the graduated collar is at the same setting as for the first thread.

The Thread-Chasing Dial Method of Cutting Multiple Threads

Double-start threads with an odd-numbered lead (for example, ⅕, ⅐, etc.) may be cut using the thread-chasing dial.

1. Take one cut on the thread by engaging the split nut at a *numbered* line on the chasing dial.

2. Without changing the depth of cut, take another cut at an *unnumbered* line on the chasing dial. The second thread will be exactly in the middle of the first thread.

3. Continue cutting the thread to depth, taking two passes (one on a numbered line, the other on an unnumbered line) for every depth-of-cut setting.

SQUARE THREADS

Square threads were often found in vise screws, jacks, and other devices where maximum power transmission was required. Because of the difficulty of cutting this thread with taps and dies, it is being replaced by Acme thread. With care, square threads can be readily cut on a lathe.

The Shape of a Square Threading Tool

The square threading tool looks like a short cutting-off tool. It differs from it because both sides of the square threading tool must be ground at an angle to conform to the helix angle of the thread (Fig. 55-28).

The helix angle of a thread, and therefore the angle of the square threading tool, depends on two factors:

1. The helix angle changes for each *different lead* on a given diameter. The greater the lead of the thread, the greater the helix angle.

2. The helix angle changes for each *different diameter* of thread for a given lead. The larger the diameter, the smaller the helix angle.

The helix angle of either the leading or following side of a square thread can be represented by a right-angle triangle (Fig. 55-28). The side opposite equals the *lead* of the thread, and the side adjacent equals the circumference of either the major or minor diameter of the thread. The angle between the hypotenuse and the side adjacent represents the helix angle of the thread.

To Calculate the Helix Angles of the Leading and Following Sides of a Square Thread

$$\text{Tan leading angle} = \frac{\text{lead of thread}}{\text{circumference of minor diameter}}$$

$$\text{Tan following angle} = \frac{\text{lead of thread}}{\text{circumference of major diameter}}$$

Clearance

If a square toolbit were ground to the same helix angles as the leading and following sides of the thread, it would have no clearance and the sides would rub. To prevent the tool from rubbing, it must be provided with approximately 1° clearance on each side, making it thinner at the bottom (Fig. 55-29). For the leading side of the tool, *add* 1° to the calculated helix angle. On the following side, *subtract* 1° from the calculated angle.

EXAMPLE

Find the leading and following angles of a threading tool to cut a 1 1/4 in.—4 square thread.

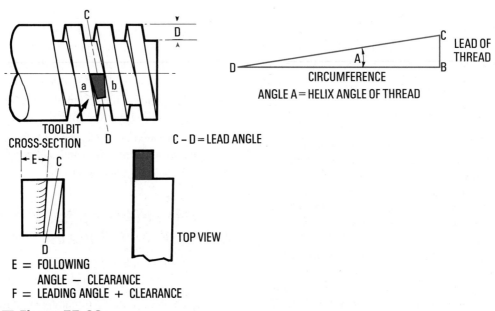

TOOLBIT CROSS-SECTION

C – D = LEAD ANGLE

E = FOLLOWING ANGLE − CLEARANCE
F = LEADING ANGLE + CLEARANCE

ANGLE A = HELIX ANGLE OF THREAD

TOP VIEW

■ Figure 55-28 The shape of a square threading tool.

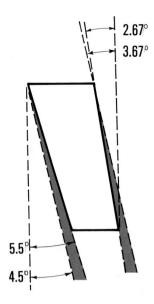

Figure 55-29 Helix angles of the thread and clearance angles necessary for a square threading tool.

Solution

$$\text{Lead} = .250 \text{ in.}$$

$$\text{Single depth} = \frac{.500}{4}$$

$$= .125 \text{ in.}$$

$$\text{Double depth} = 2 \times .125$$

$$= .250 \text{ in.}$$

$$\text{Minor diameter} = 1.250 - .250$$

$$= 1.000 \text{ in.}$$

$$\text{Tan leading angle} = \frac{\text{lead}}{\text{minor dia. circumference}}$$

$$= \frac{.250}{1.000 \times \pi}$$

$$= \frac{.250}{3.1416}$$

$$= .0795 \text{ in.}$$

$$\therefore \text{ the angle of the thread} = 4°33'$$

$$\text{The toolbit angle} = 4°33' \text{ plus } 1° \text{ clearance}$$

$$= 5°33'$$

$$\text{Tan following angle} = \frac{\text{lead}}{\text{major dia. circumference}}$$

$$= \frac{.250}{1.250\pi}$$

$$= \frac{.250}{3.927}$$

$$= .0636 \text{ in.}$$

$$\therefore \text{ the angle of the thread} = 3°38'$$

$$\text{The toolbit angle} = 3°38' \text{ minus } 1° \text{ clearance}$$

$$= 2°38'$$

To Cut a Square Thread

1. Grind a threading tool to the proper leading and following angles. The width of the tool should be approximately .002 in. (0.05 mm) wider than the thread groove. This will allow the completed screw to fit the nut readily. Depending on the size of the thread, it may be wise to grind two tools: a roughing tool .015 in. (0.38 mm) undersize and a finishing tool .002 in. (0.05 mm) oversize.

2. Align the lathe centers and mount the work.

3. Set the quick-change gearbox for the required number of *tpi*.

4. Set the compound rest at 30° to the right, which will provide side movement if it becomes necessary to reset the cutting tool.

5. Set the threading tool square with the work and on center.

6. Cut the right-hand end of the work to the minor diameter for approximately .060 in. (1.58 mm) long. This will indicate when the thread is cut to the full depth.

7. If the work permits, cut a recess at the end of the thread to the minor diameter. This will provide room for the cutting tool to "run out" at the end of the thread.

8. Calculate the single depth of the thread as

$$\frac{.500}{N}$$

9. Start the lathe and just touch the tool to the work diameter.

10. Set the *crossfeed graduated collar* to zero (0).

11. Set a .003-in. (0.08-mm) depth of cut with the *crossfeed screw* and take a trial cut.

12. Check the thread with a thread pitch gage.

13. Apply cutting fluid and cut the thread to depth, moving the *crossfeed* in from .002 to .010 in. (0.05 to 0.25 mm) for each cut. The depth of the cut will depend on the thread size and the nature of the workpiece.

Note: Since the thread sides are square, *all cuts* must be set using the *crossfeed screw*.

ACME THREAD

The *Acme thread* is gradually replacing the square thread because it is stronger and easier to cut with taps and dies. It is used extensively for lead screws because the 29° angle formed by its sides allows the split nut to be engaged readily during thread cutting.

The Acme thread is provided with .010-in. clearance for both the crest and root on all sizes of threads. The hole for an internal Acme thread is cut .020 in. larger than the minor diameter of the screw, and the major diameter of a tap or internal thread is .020 in. larger than the major diameter of the screw. This provides .010-in. clearance between the screw and nut on both the top and bottom.

To Cut an Acme Thread

1. Grind a toolbit to fit the end of the Acme thread gage (Fig. 55-30). Be sure to provide sufficient side clearance so that the tool will not rub while cutting the thread.
2. Grind the point of the tool flat until it fits into the slots of the gage indicating the number of threads per inch to be cut.

Note: If a gage is not available, the width of the toolbit point may be calculated as follows:

$$\text{Width of point} = \frac{.3707}{N} - .0052 \text{ in.}$$

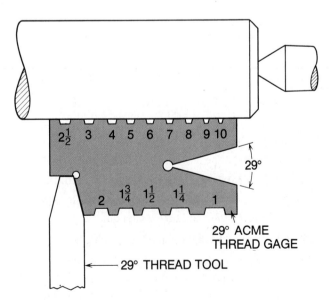

■ **Figure 55-30** The toolbit is squared with the workpiece by an Acme thread gage. *(Kelmar Associates)*

3. Set the quick-change gearbox to the required number of threads per inch.
4. Set the compound rest 14½° to the right (half the included thread angle).
5. Set the Acme threading tool on center and square it with the work using the gage shown in Fig. 55-30.
6. At the right-hand end of the work, cut a section .060 in. long to the minor diameter. This will indicate when the thread is to the full depth.
7. Cut the thread to the proper depth by feeding the cutting tool, using the *compound rest*.

Measuring Acme Threads

For most purposes, the *one-wire method* of measuring Acme threads is accurate enough. A single wire or pin of the correct diameter is placed in the thread groove (Fig. 55-31) and measured with a micrometer. The thread is the correct size when the micrometer reading over the wire is the same as the major diameter of the thread and the *wire is tight in the thread*.

Note: It is important that the burrs be removed from the diameter before using the one-wire method.

The diameter of the wire to be used can be calculated as follows:

$$\text{Wire diameter} = .4872 \times \text{pitch}$$

For example, if 6 threads per inch are being cut, the wire diameter should be:

$$\text{Wire size} = .4872 \times \frac{1}{6}$$

$$= .081 \text{ in.}$$

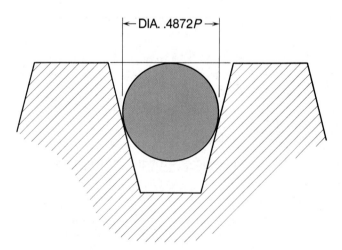

■ **Figure 55-31** Using one wire to measure the accuracy of an Acme thread. *(Kelmar Associates)*

INTERNAL THREADS

Most internal threads are cut with taps; however, sometimes a tap of a specific size is not available and the thread must be cut on a lathe. Internal threading, or cutting threads in a hole, is an operation performed on work held in a chuck or collet or mounted on a faceplate. The threading tool is similar to a boring toolbit, except the shape is ground to the form of the thread to be cut.

To Cut a 1⅜ in.—6 NC Internal Thread

1. Calculate the tap drill size of the thread.

$$\text{Tap drill size} = \text{major diameter} - \frac{1}{N}$$

$$= 1.375 - \frac{1}{6}$$

$$= 1.375 - .166$$

$$= 1.209 \text{ in.}$$

2. Mount the work to be threaded in a chuck or collet or on a faceplate.

3. Drill a hole approximately ¹⁄₁₆ in. smaller than the tap drill size in the workpiece. For this thread, it would be 1.209 − .062 = 1.147, or a 1⁵⁄₃₂-in. hole.

4. Mount a boring tool in the lathe and bore the hole to the tap drill size (1.209 in.). The boring bar should be as large as possible and held short. The boring operation cuts the hole to size and makes it true.

5. Recess the start of the hole to the major diameter of the thread (1.375 in.) for .060-in. length. During the thread-cutting operation, this will indicate when the thread is cut to depth.

6. If the thread does not go through the workpiece, a recess should be cut at the end of the thread to the major diameter (Fig. 55-32). This recess should be wide enough to allow the threading tool to "run out" and permit time to disengage the split-nut lever.

7. Set the compound rest at 29° to the left (Fig. 55-32), to the right for left-hand threads.

8. Mount a threading toolbit into the boring bar and set it to center.

9. Square the threading tool with a thread center gage (Fig. 55-33).

10. Place a mark on the boring bar, measuring from the threading tool, to indicate the length of hole to be threaded. This will show when the split-nut lever should be disengaged.

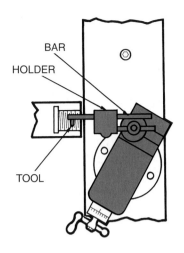

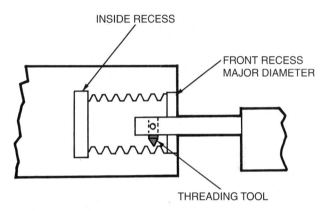

■ Figure 55-32 The compound rest is set at 29° to the left for cutting right-hand internal threads. *(Kelmar Associates)*

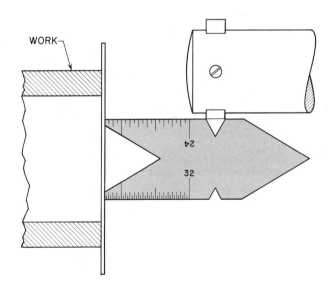

■ Figure 55-33 Using a thread center gage to square the threading tool with the workpiece. *(Kelmar Associates)*

11. Start the lathe and turn the crossfeed handle *out* until the threading tool just scratches the internal diameter.

12. Set the crossfeed graduated collar to zero (0).

13. Set a .003-in. depth of cut by feeding the compound rest *out* and take a trial cut.

14. At the end of *each* cut on an internal thread, disengage the split-nut lever and feed the crossfeed handle *in* to clear the thread.

15. Clear the threading tool from the hole and check the pitch of the thread.

16. Return the crossfeed handle back to zero (0) and set the depth of cut by turning the compound rest *out* the desired amount.

17. Cut the thread to depth; check the fit with a screw or threaded plug gage.

Note: The last few cuts should not be deeper than .001 in. each to eliminate the spring of the boring bar.

unit 55 review questions

Threads

1. Define thread.
2. List four purposes of threads.

Thread Terminology

3. Define pitch diameter, pitch, lead, root, and crest.
4. How is the pitch designated for:
   a. UNC threads?   b. Metric threads?
5. Why is the diameter of the blank for a rolled thread equal to the pitch diameter?
6. How may a right-hand thread be distinguished from a left-hand thread?

Thread Fits and Classifications

7. Define fit, allowance, tolerance, and limits.
8. How are external UNC threads classified?
9. Name and describe three classifications of UNC fits.
10. How are thread fits designated for the ISO threads?
11. Why was the ISO metric system of threads adopted?
12. Describe a thread designated as M 56 × 5.5.
13. Name five thread forms used in North America and state the included angle of each.
14. What are the principal differences between the American and Unified threads?

Thread Calculations

15. For an M 20 × 2.5 thread, calculate:
    a. Pitch   b. Depth   c. Minor diameter
    d. Width of crest   e. Width of root
16. Sketch a UNC thread and show the dimensions of the parts.
17. How does a Brown & Sharpe worm thread differ from an Acme thread?
18. For a 1 in.—8 NC thread, calculate:
    a. Minor diameter
    b. Width of toolbit point
    c. Amount of compound rest feed
19. What is the purpose of the quick-change gearbox?
20. Describe a thread-chasing dial and state its purpose.
21. The lead screw on a lathe has 6 *tpi;* at what point or points on the thread-chasing dial may the split-nut lever be engaged to cut the following threads: 8, 9, 11½, 12?

Thread Cutting

22. List the main steps required to set up the lathe for cutting a 60° thread.
23. List the main steps required to cut a 60° thread.
24. It has been necessary to remove a partially threaded piece of work from the lathe and to finish it later. Describe how to reset the threading tool to "pick up" the thread.

Metric Threads

25. a. What two change gears are required to cut a metric thread on a standard lathe?

    b. Where are these gears mounted?

26. Describe how a lathe (having a quick-change gearbox) is set up to cut a 2.5-mm thread.

27. What precaution should be taken when cutting a metric thread?

Thread Measurement

28. Make a sketch to illustrate how a thread is measured using the three-wire method.

29. Calculate the best wire size and the measurement over the wires for the following threads:

    a. 1/4 in.—20 NC
    b. 5/8 in.—11 NC
    c. 1 1/4 in.—7 NC

Multiple Threads

30. What is the purpose of a multiple thread?

31. If the pitch of a multiple thread is ⅛ in., what will be the lead for a double-start thread? A triple-start thread?

32. List three methods by which multiple threads can be cut.

Square Threads

33. Name two factors that affect the helix angle of a thread.

34. Calculate the leading and following angles of a square threading toolbit required to cut a 1 1/2 in.—6 square thread.

35. List the main steps required to cut a square thread.

Acme Threads

36. If the width of the root of an Acme thread is .3707P at minimum depth, why is the Acme threading tool ground to .3707P − .0052?

37. Describe how an Acme thread can be measured.

Internal Threads

38. List the steps required to cut a 1 1/4 in.—7 UNC thread 2 in. deep in a block of steel 3 in. × 3 in. × 3 in. Show all necessary calculations.

UNIT 56

Steady Rests, Follower Rests, and Mandrels

OBJECTIVES

After completing this unit, you should be able to:

1 Set up and use a steady rest for machining a long shaft

2 Set up and use a follower rest when machining a long shaft

3 Mount and machine work on a mandrel

4 Lay out and machine an eccentric on work held between centers

Various accessories make it possible to machine different lengths and shapes of workpieces in a lathe. Steady and follower rests are used to support long or slender workpieces and prevent them from springing during a machining operation either between lathe centers or in a chuck. Steady rests are attached to the lathe bed and are generally set in about the middle of a workpiece. Follower rests are attached to the lathe carriage and are set up immediately to the right of the cutting tool. Because the relationship of the attachment and the cutting tool remains the same, the workpiece is supported throughout the cut.

Mandrels generally fit the hole of a thin part, such as gears, flanges, and pulleys, and they allow the outside diameter of the work to be machined between centers or in a chuck.

▶▶ Steady Rest

A *steady rest* (Fig. 56-1) is used to support long, slender work and prevent it from springing while being machined between centers. A steady rest may also be used when it is necessary to perform a machining operation on the end of a workpiece held in a chuck. The steady rest is fastened to the lathe bed and its three jaws are adjusted to the surface of the work to provide a supporting bearing. The jaws on a steady rest are generally made of soft material, such as fiber or brass, to prevent damaging the work surface. Other steady rests have rollers attached to the jaws to provide good support for the work.

■ **Figure 56-1** Using a steady rest to support the end of a long workpiece held in a chuck. *(South Bend Lathe Corp.)*

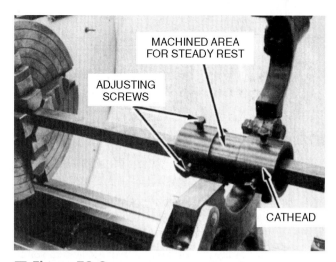

Figure 56-2 The cathead can be adjusted to provide a true bearing surface for the steady rest, even if a square workpiece is machined. *(Kelmar Associates)*

TO SET UP A STEADY REST

1. Mount the work between centers.

   OR

   Set up and true the work in a chuck.

2. **a.** If the work diameter is not round, turn a true spot on the diameter (slightly wider than the steady rest jaws) at the point where the steady rest will be supporting the work. Long work in a chuck should be supported first by the tailstock center. If the diameter is rough, turn a section for the steady rest and one near the chuck to the same diameter.

   b. If it is impossible to turn a true diameter (because of the shape of the workpiece), mount and adjust a *cathead* (Fig. 56-2) on the work.

3. Move the carriage to the tailstock end of the lathe.

4. Place the steady rest on the lathe bed at the desired position. If the work diameter is turned and held in a chuck, slide the steady rest up close to the chuck.

5. Adjust the lower two jaws to the work diameter, using a paper feeler to provide clearance between the jaws and the work.

6. Slide the steady rest to the desired position and fasten it in place.

7. Close the top section of the steady rest and adjust the top jaw, using a paper feeler.

8. Apply a suitable lubricate to the diameter at the steady rest jaws.

9. Start the lathe and carefully adjust each jaw until it just touches the diameter.

Figure 56-3 A 60° spotting tool set up to recut a damaged center. *(Kelmar Associates)*

Note: The lubricant will smear just as the jaw contacts the work.

10. Tighten the lock screw on each jaw and then apply a suitable lubricant.

11. Before machining, indicate the top and front of the turned diameter at the chuck and at the steady rest to check for alignment. If the indicator reading varies, adjust the steady rest until it is correct.

TO TRUE A DAMAGED CENTER HOLE

1. Mount and true the work in a chuck and steady rest, if necessary.

2. Grind a 60° spotting tool (Fig. 56-3) and mount it on center in the toolholder.

3. Start the lathe and gradually bring the spotting tool into the damaged center hole.

4. Recut the center hole until the damaged section is removed.

5. Remove the workpiece, mount it between centers, and turn the diameter as required.

▶▶ Follower Rest

A *follower rest,* mounted on the saddle, moves along with the carriage to prevent work from springing up and away from the cutting tool. The follower rest, positioned immediately behind the cutting tool, can be used to support long work for successive operations, such as thread cutting (Fig. 56-4 on p. 440).

TO SET UP A FOLLOWER REST

1. Mount the work between centers.

2. Fasten the follower rest to the saddle of the lathe.

Figure 56-4 A follower rest used to support a long, slender workpiece between centers during thread cutting.

3. Position the cutting tool in the toolpost so that it is just to the left of the follower rest jaws.

4. Turn the work diameter, for approximately 1.50 in. (38 mm) long, to the desired size.

5. Adjust both jaws of the follower rest until they lightly contact the turned diameter.

6. Tighten the lock screw on each jaw.

7. Lubricate the work and the follower rest jaws to prevent marring the finished diameter.

8. If successive cuts are required to reduce the diameter of a workpiece, readjust the follower rest jaws as in steps 4 to 7.

▶▶ Mandrel

A *mandrel* (Fig. 56-5) is a precision tool that, when pressed into the hole of a workpiece, provides centers for a machining operation. They are especially valuable for thin work, such as flanges, pulleys, and gears, where the outside diameter must run true with the inside diameter and it would be difficult to hold the work in a chuck.

CHARACTERISTICS OF A STANDARD MANDREL

1. Mandrels are usually hardened and ground and are tapered .006 to .008 in./ft (0.5 to 0.66 mm/m).

2. The nominal size is near the middle, and the small end is usually .001 in. (0.02 mm) under; the large end is usually .004 in. (0.1 mm) over the nominal size.

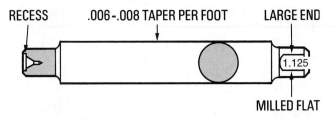

Figure 56-5 A standard solid mandrel. *(Kelmar Associates)*

3. Both ends are turned smaller than the body and provided with a flat so that the lathe dog does not damage the accuracy of the mandrel.

4. The size of the mandrel is stamped on the large end.

5. The center holes, which are recessed slightly, are large enough to provide a good bearing surface and to withstand the strain caused by machining a workpiece.

TYPES OF MANDRELS

Many types of mandrels are used with various types of workpieces and machining operations. Some of the more common types of mandrels are:

> The *solid mandrel* (Fig. 56-5) is available for most of the standard hole sizes. It is a general-purpose mandrel that can be used for a variety of workpieces.

> The *expansion mandrel* (Fig. 56-6) consists of a sleeve, with four or more slots cut lengthwise, fitted over a solid mandrel. A taper pin fits into the sleeve to expand it to hold work that does not have a standard-size hole. Another form of expansion mandrel has a slotted bushing fitting over a tapered mandrel. Bushings of various sizes can be used with this mandrel, increasing its range.

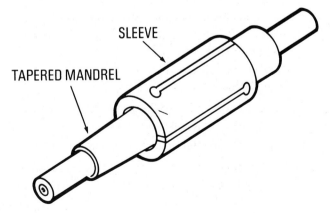

Figure 56-6 An expansion mandrel. *(Kelmar Associates)*

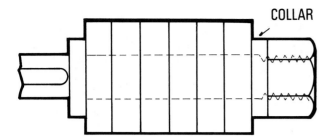

Figure 56-7 A gang mandrel. *(Kelmar Associates)*

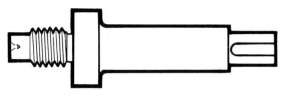

Figure 56-8 A threaded mandrel. *(Kelmar Associates)*

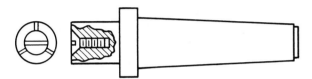

Figure 56-9 A taper-shank mandrel. *(Kelmar Associates)*

> The *gang mandrel* (Fig. 56-7) is used to hold a number of identical parts for a machining operation. The body of the mandrel is parallel (no taper) and has a shoulder or flange on one end. The other end is threaded for a locking nut.

> The *threaded mandrel* (Fig. 56-8) is used for holding workpieces having a threaded hole. An undercut at the shoulder ensures that the workpiece will seat squarely and is not canted on the threads.

> The *taper-shank mandrel* (Fig. 56-9) may be fitted to the tapered hold in the headstock spindle. The projecting portion may be machined to any desired form to suit the workpiece. This type of mandrel is often used for small workpieces or those that have blind holes.

TO MOUNT WORK ON A PLAIN MANDREL

1. Secure a mandrel to fit the hole in the workpiece.
2. Thoroughly clean the mandrel and apply a thin film of oil on the diameter.
3. Clean and remove any burrs from the hole in the workpiece (Fig. 56-10).

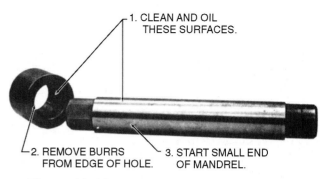

1. CLEAN AND OIL THESE SURFACES.

2. REMOVE BURRS FROM EDGE OF HOLE.   3. START SMALL END OF MANDREL.

Figure 56-10 Preparing the mandrel and workpiece before mounting. *(Kelmar Associates)*

Figure 56-11 Using an arbor press to force a mandrel into a workpiece. *(Kelmar Associates)*

4. Start the small end of the mandrel (the large end has the size stamped on it) into the hole by hand.
5. Place the work on an arbor press with a machined surface down so that the hole is at right angles to the table surface (Fig. 56-11).
6. Press the mandrel firmly into the workpiece.

TO TURN WORK ON A MANDREL

Work pressed on a mandrel is held in position by friction; therefore, cutting operations should be toward the large end of the mandrel. This will tend to keep the work tight on the mandrel.

Follow this procedure:

1. Fasten the lathe dog on the *large end* of the mandrel (where the size is stamped).
2. Clean the lathe and mandrel centers and then mount the work.
3. If the entire side of the work must be faced, it is good practice to use a paper feeler between the toolbit point and mandrel for setting the toolbit. This will prevent marring or scoring the surface of the mandrel.

4. When turning the outside diameter of work, always cut toward the large end of the mandrel.

5. On large-diameter work, it is advisable to take light cuts to prevent the work from slipping on the mandrel or chattering.

▶▶ Eccentrics

An *eccentric* (Fig. 56-12) is a shaft that may have two or more turned diameters parallel to each other but not concentric with the normal axis of the work. Eccentrics are used in locking devices, in the feed mechanism on some machines and in the crank shaft of automobiles, where it is necessary to *convert rotary motion into reciprocating motion,* or vice versa.

■ **Figure 56-12** The axes of an eccentric are parallel, not aligned. *(Kelmar Associates)*

The amount of eccentricity, or *throw,* of an eccentric is the distance that a set of center holes has been offset from the normal work axis. If the center holes were offset .250 in. from the work axis, the amount of throw would be .250 in. (6 mm), the total travel of the eccentric would be .500 in. (12 mm).

There are three types of eccentrics, which are generally cut on a lathe in the following ways:

1. When the throw enables all centers to be located on the ends of the workpiece

2. When the throw is too small to allow all centers to be located on the workpiece at the same time

3. When the throw is so great that all centers cannot be located on the workpiece

TO TURN AN ECCENTRIC WITH A .375-IN. OR 10-MM THROW

1. Place the work in a chuck and face it to length. If the center holes are to be removed later, leave the work .750 in. (19 mm) longer.

2. Place the work in a V-block on a surface plate and apply layout dye to both ends of the work.

3. Set a vernier height gage to the top of the work and note the vernier reading.

■ **Figure 56-13** Laying out the centers of an eccentric. *(Kelmar Associates)*

4. Subtract half the work diameter from the reading and set the gage to this dimension.

5. Scribe a centerline on both ends of the work.

6. Rotate the work 90° and scribe another centerline on both ends at the same height-gage setting (Fig. 56-13).

7. Lower or raise the height-gage setting .375 in. (10 mm) and scribe the lines for the offset centers on both ends.

8. Carefully center-punch the four scribed centers and drill the center holes in each end.

9. Mount the work in a lathe and turn the diameter with the true centers.

10. Set the work on the offset centers and turn the eccentric (center section) to the required diameter.

TO CUT AN ECCENTRIC WITH A SMALL THROW

This procedure should be followed when the centers are too close to be located on the workpiece at the same time:

1. Cut the work .750 in. (19 mm) longer than required.

2. Face the ends and drill one set of center holes in the lathe.

3. Mount the work between centers and turn the large diameter to size.

4. Cut off the ends to remove the center holes.

5. Lay out and drill a new set of center holes, offsetting them from the center position to the required throw.

6. Turn the eccentric diameter to size.

TO TURN AN ECCENTRIC WITH A LARGE THROW

1. Set the work on the normal centers and turn both ends to size.

2. Secure or make a set of support blocks as shown in Fig. 56-14. The hole in the support block should fit the turned ends of the work snugly. A setscrew in each block is used to securely fasten the support blocks to the work. The number of centers required should be laid out and drilled in the support blocks (Fig. 56-14).

3. Align both support blocks parallel on the work and lock them in position.

4. Counterbalance the lathe to prevent undue vibration.

5. Turn the various diameters as required.

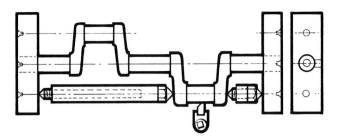

■ **Figure 56-14** The setup required for turning an eccentric with a large throw. *(Kelmar Associates)*

unit 56 review questions

Follower Rests and Steady Rests

1. State the purpose of a follower rest.

2. Explain how to set up a steady rest for turning a long shaft held between centers.

3. Describe a cathead and state when it is used.

4. Explain how a damaged center hole may be trued.

Mandrels

5. State the purpose of a mandrel.

6. Draw a 1-in. (25-mm) standard mandrel and include all specifications.

7. Name and describe four types of mandrels and state their purpose.

8. List the precautions that must be taken when a mandrel is used (include mounting and turning).

Eccentrics

9. Define an *eccentric* and state its purpose.

10. Explain the difference between the throw and total travel of an eccentric.

11. A crankshaft 6 in. (150 mm) long having equal-length journals is required to produce a travel of 1.500 in. (38 mm) on a piston. The journal (finished shaft) size is 1 in. (25 mm).

    a. Describe how to lay out this eccentric.
    b. What size material would be required if .125 in. (3 mm) is allowed for "cleaning up"? This calculation is easy with the aid of a sketch.

12. What precautions must be taken when turning an eccentric having a large throw?

UNIT 57

Machining in a Chuck

OBJECTIVES

After completing this unit, you should be able to:

1 Mount and machine work in a three-jaw chuck

2 Mount and machine work in a four-jaw chuck

3 Face, groove, and cut off work held in a chuck

Spindle-mounted accessories, such as chucks, centers, driveplates, and faceplates, are fitted to and driven by the headstock spindle. The most versatile and commonly used spindle accessories are the lathe chucks. The jaws of a chuck are adjustable and therefore workpieces which might be difficult to hold by another method can be held securely. Three-jaw chucks, whose jaws move simultaneously, are generally used to hold round finished diameters. Four-jaw chucks, whose jaws move independently, are generally used to hold odd-shaped pieces and when greater holding power and accuracy are required.

All the machining operations that can be performed on work between centers, such as turning, knurling, and threading, can be performed on work held in a chuck. Most chuck work is generally fairly short; however, if workpieces longer than three times their diameter must be machined in a chuck, the end of the work should be supported to prevent it from springing and being thrown out of the lathe.

▶▶ Mounting and Removing Lathe Chucks

The proper procedure for mounting and removing chucks must be carefully followed to prevent damage to the lathe spindle and/or the chuck and to preserve the accuracy of the lathe. Three types of spindle noses are found on engine lathes on which lathe chucks may be mounted. They are the threaded spindle nose, the tapered spindle nose, and the cam-lock spindle nose. The procedures for mounting a chuck on each type of spindle nose follow.

TO MOUNT A CHUCK

1. Set the lathe to the slowest speed. *SHUT OFF THE ELECTRICAL SWITCH.*

2. Remove the driveplate and live center.

3. Clean all surfaces of the spindle nose and the mating parts of the chuck.

Note: Steel chips or dirt will destroy the accuracy of the spindle nose and the mating taper in the chuck.

4. Place a cradle block on the lathe bed in front of the spindle and place the chuck on the cradle (Fig. 57-1).

Figure 57-2 Align the keyway in the chuck with the key on the spindle. *(Kelmar Associates)*

Figure 57-1 A properly fitted cradle block makes mounting and removal of chucks easy and safe. *(Kelmar Associates)*

5. Slide the cradle close to the lathe spindle nose and mount the chuck.

   a. *Threaded Spindle Nose Chucks*
     (1) Revolve the lathe spindle *by hand* in a counterclockwise direction and bring the chuck up to the spindle. *NEVER START THE MACHINE.*
     (2) If the chuck and spindle are correctly aligned, the chuck should easily thread onto the lathe spindle.
     (3) When the chuck adapter plate is within .060 in. (1.5 mm) of the spindle shoulder, give the chuck a quick turn to seat it against the spindle shoulder.
     (4) Do not jam a chuck against the shoulder too tightly because it may damage the threads and make the chuck difficult to remove.

   b. *Taper Spindle Nose Chucks*
     (1) Revolve the lathe spindle by hand until the key on the spindle nose aligns with the keyway in the tapered hole of the chuck (Fig. 57-2).
     (2) Slide the chuck onto the lathe spindle.
     (3) Turn the lock ring in a counterclockwise direction until it is hand-tight.
     (4) Tighten the lock ring securely with a spanner wrench by striking it sharply downward (Fig. 57-3).

   c. *Cam-Lock Spindle Nose Chucks*
     (1) Align the registration line of each cam lock with the registration line on the lathe spindle nose.

Figure 57-3 Strike the spanner wrench sharply to tighten the lock ring. *(Kelmar Associates)*

     (2) Revolve the lathe spindle by hand until the holes in the spindle align with the cam-lock studs of the chuck (Fig. 57-4 on p. 446).
     (3) Slide the chuck onto the spindle.
     (4) Tighten each cam lock in a *clockwise* direction (Fig. 57-5 on p. 446).

TO REMOVE A CHUCK

The following procedures for removing a lathe chuck also apply to other spindle-mounted accessories such as drive plates and faceplates.

1. Set the lathe in the slowest speed. *STOP THE MOTOR.*

2. Place a cradle block under the chuck (Fig. 57-1).

3. Remove the chuck by the following methods:

   a. *Threaded Spindle Nose Chucks*
     (1) Turn the chuck until a wrench hole is in the top position.

Machining in a Chuck   **445**

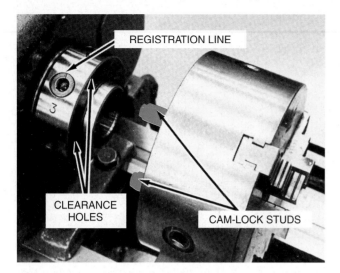

■ Figure 57-4 Mounting a chuck on a cam-lock spindle nose. *(Kelmar Associates)*

■ Figure 57-6 A hardwood block can be used to remove a chuck from a threaded spindle nose. *(South Bend Lathe Corp.)*

■ Figure 57-5 Tighten the cam locks in a clockwise direction. *(Kelmar Associates)*

(2) Insert the chuck wrench into the hole and pull it *sharply* toward the front of the lathe.

OR

(1) Place a block or short stick under the chuck jaw as shown in Fig. 57-6.
(2) Revolve the lathe spindle by hand in a *clockwise* direction until the chuck is loosened on the spindle.
(3) Remove the chuck from the spindle and store it where it will not be damaged.

b. *Taper Spindle Nose Chucks*
(1) Secure the proper C-spanner wrench.
(2) Place it around the lock ring of the spindle with the handle in an upright position.
(3) Place one hand on the curve of the spanner wrench to prevent it from slipping off the lock ring (Fig. 57-7).

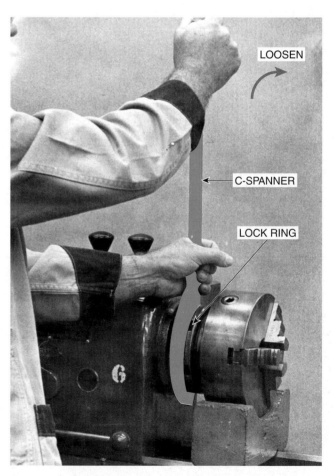

■ Figure 57-7 Using a C-spanner wrench to loosen the lock ring that holds the chuck on the taper nose spindle. *(Kelmar Associates)*

(4) With the palm of the other hand, *sharply* strike the handle of the wrench in a clockwise direction.

(5) Hold the chuck with one hand, and with the other hand, remove the lock ring from the chuck.

Note: The lock ring may turn a few turns and then become tight. It may be necessary to use the spanner wrench again to loosen the taper contact between the chuck and the spindle nose.

(6) Remove the chuck from the spindle and store it with the jaws in the up position.

c. *Cam-Lock Spindle Nose Chucks*

(1) With the chuck wrench, turn each cam lock in a counterclockwise direction until its registration line coincides with the registration line on the lathe spindle nose (Fig. 57-8).

(2) Place one hand on the chuck face, and with the palm of the other hand, *sharply* strike the top of the chuck. This action is necessary to break the taper contact between the chuck and the lathe spindle (Fig. 57-9).

Note: Sometimes it may be necessary or desirable to use a soft-faced hammer for this operation.

4. Slide the chuck clear of the spindle and place it carefully in a storage compartment.

▶▶ Mounting Work in a Chuck

For workpieces to be held securely for machining operations in either a three- or four-jaw chuck, certain procedures should be followed. Since the construction and

■ **Figure 57-8** Loosen each cam lock until the registration lines on the lock and spindle match. *(Kelmar Associates)*

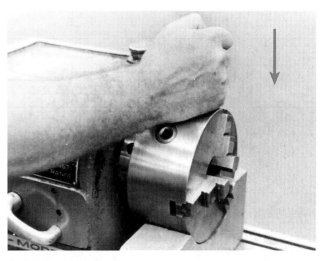

■ **Figure 57-9** Strike the top of the chuck sharply with your hand to break the taper contact. *(Kelmar Associates)*

purpose of three- and four-jaw chucks differ, the method of mounting work in each chuck also differs in various ways.

TO MOUNT WORK IN A THREE-JAW CHUCK

1. Clean the chuck jaws and the surfaces of the workpiece.
2. Use the proper size chuck wrench and open the chuck jaws slightly more than the diameter of the work.
3. Place the work in the chuck, leaving no more than three times the diameter extending beyond the chuck jaws.
4. Tighten the chuck jaws with the wrench in the left hand while slowly rotating the workpiece with the right hand.
5. Tighten the chuck jaws securely by *using the chuck wrench only.*

ASSEMBLING JAWS IN A THREE-JAW UNIVERSAL CHUCK

Three-jaw chucks are supplied with two sets of chuck jaws: one set for outside gripping and one set for inside gripping (Fig. 57-10 on p. 448). The jaws supplied with each chuck are marked with the same serial number and should *never be used with another chuck.* When it is necessary to change the chuck jaws, the proper sequence must be followed; otherwise, the work held in the jaws will not run true.

1. Thoroughly clean the jaws and the jaw slides in the chuck.
2. Turn the chuck wrench clockwise until the start of the scroll thread is almost showing at the back edge of slide 1.

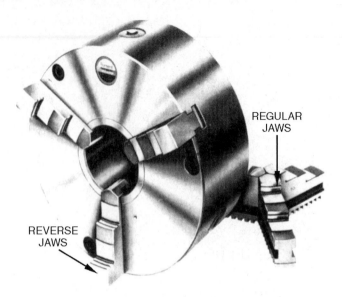

REGULAR JAWS

REVERSE JAWS

■ Figure 57-10 A three-jaw universal chuck with reversible jaws.

■ Figure 57-12 The scroll plate turned to start the second jaw. *(Kelmar Associates)*

■ Figure 57-11 Inserting the first jaw into the start of the scroll plate. *(Kelmar Associates)*

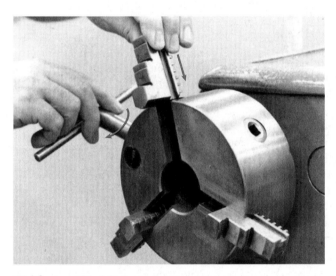

■ Figure 57-13 Inserting the third jaw into a three-jaw chuck. *(Kelmar Associates)*

3. Insert jaw 1 (in slot 1) and press down with one hand while turning the chuck wrench clockwise with the other (Fig. 57-11).

4. After the scroll thread has engaged in the jaw, continue turning the chuck wrench clockwise until the start of the scroll is near the back edge of groove 2.

5. Insert the second jaw and repeat steps 3 and 4 (Fig. 57-12).

6. Insert the third jaw in the same manner (Fig. 57-13).

Some chucks are equipped with one set of top jaws, which are fastened on by allen screws. To reverse these jaws, it is necessary to remove the screws, clean the mat-ing parts, and replace the jaws in the reverse direction. The screws should be tightened uniformly and securely so that the jaws are not distorted.

TO MOUNT WORK IN A FOUR-JAW CHUCK

Work that must run absolutely true must be mounted in a four-jaw chuck because each jaw can be adjusted independently. The chuck jaws, which are reversible, can hold round, square, or irregularly shaped workpieces securely. The work can be adjusted to be either concentric or off center.

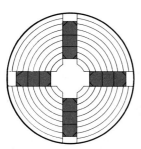

Figure 57-14 The lines on a four-jaw chuck can be used as a guide for truing a workpiece. *(Kelmar Associates)*

1. Measure the diameter of the work to be chucked.

2. With a chuck wrench, adjust the jaws to the approximate size according to the ring marks on the face of the chuck (Fig. 57-14).

3. Set the work in the chuck and tighten the jaws snugly against the work surface.

4. True the workpiece by any of the following methods:

   a. *Chalk Method*
   (1) Start the lathe and, with a piece of chalk, lightly mark the high spot on the diameter (Fig. 57-15).
   (2) Stop the lathe and check the chalk mark. If it is an even, lightly marked line around the work, the work is true.
   (3) If there is only one mark, loosen the jaw *opposite* the chalk mark and tighten the jaw *next to* the chalk mark.

Figure 57-15 Using chalk to indicate the high spot on a workpiece. *(Kelmar Associates)*

Figure 57-16 Using a surface gage to true a workpiece in a four-jaw chuck. *(Kelmar Associates)*

   (4) Continue this operation until the chalk marks lightly around the work or leaves two marks opposite each other.

   b. *Surface Gage Method*
   (1) Place a *surface gage* on the lathe bed and adjust the point of the scriber so that it is close to the work surface (Fig. 57-16).
   (2) Revolve the lathe by hand to find the low spot on the work.
   (3) Loosen the jaw *nearest* the low spot and tighten the jaw *opposite* the low spot to adjust the work closer to center.
   (4) Repeat steps (2) and (3) until the work is running true.

TO TRUE WORK IN A FOUR-JAW CHUCK USING A DIAL INDICATOR

A dial indicator should be used whenever a machined diameter must be aligned to within a few thousandths of an inch or hundredths of a millimeter.

1. Mount the work and true it approximately, using either the chalk or surface gage method.

2. Mount an indicator, with a range of at least .100 in. (2.5 mm), in the toolpost of the lathe (Fig. 57-17 on p. 450).

3. Set the indicator spindle in a *horizontal position* with the contact point set to center height.

4. Bring the indicator point against the work diameter so that it registers approximately .020 in. (0.5 mm) and rotate the lathe *by hand*.

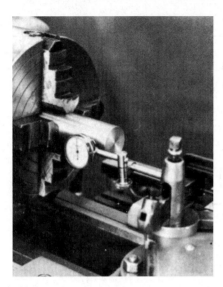

Figure 57-17 Truing a workpiece in a chuck using a dial indicator. *(Kelmar Associates)*

5. Note the highest and the lowest reading on the dial indicator.

6. Slightly loosen the chuck jaw at the lowest reading and tighten the jaw at the high reading until the work is moved half the difference between the two indicator readings.

7. Continue to adjust *only these two opposite jaws* until the indicator registers the same at both jaws.

Note: Disregard the indicator readings on the work between these two jaws.

8. Adjust the other set of opposite jaws in the same manner until the indicator registers the same at any point on the work circumference.

9. Tighten all jaws evenly to secure the workpiece firmly.

10. Rotate the lathe spindle by hand and recheck the indicator reading.

TO FACE WORK HELD IN A CHUCK

The purpose of facing work in a chuck is the same as that of facing between centers: to obtain a true, flat surface and to cut the work to length.

1. True up the work in a chuck using the chalk or dial indicator method. (This is not necessary if a three-jaw universal chuck is used.)

2. Have at least an amount equal to the diameter of the work projecting from the chuck jaws.

3. Swivel the compound rest at 90° (right angles) to the cross-slide when facing a series of shoulders (Fig. 57-18).

Figure 57-18 The compound rest set at 90° for facing accurate lengths. *(Kelmar Associates)*

Figure 57-19 Feed the compound rest twice the amount to be faced off a surface. *(Kelmar Associates)*

OR

Swivel the compound rest 30° to the right if only one surface on the work must be faced (Fig. 57-19).

4. Set up the facing toolbit to the height of the dead center and pointing slightly to the left.

5. Lock the carriage in position (Fig. 57-18).

6. Set the depth of cut by using the graduated collar on the compound rest screw:

   a. Twice the amount to be removed if the compound rest is set at 30°

   b. The same as the amount to be removed if the compound rest is set at 90° to the cross-slide

7. Face the work to length.

▶▶ To Rough- and Finish-Turn Work in a Chuck

Although much work is turned between centers in a lathe in school shops and training programs because of the ease of resetting the work accurately, most work machined in industry is held in a chuck. Since most work held in a chuck is relatively short, in many cases there is no need to support the end of the work; it is held securely enough by the lathe chuck. Workpieces that extend more than three times their diameter must be supported to prevent the work from springing during the machining operation. The most common methods of supporting the end of long workpieces is by the use of a revolving tailstock center or with a steady rest.

The operations of rough and finish turning in a chuck are the same as turning between centers (see Unit 52) and will not be detailed here. Whenever possible, a diameter should be machined to size in two cuts: one roughing and one finishing cut.

ROUGH AND FINISH TURNING

1. Mount the work securely in a chuck, with no more than three times the diameter extending beyond the chuck jaws.
2. Move the toolpost to the left of the compound rest and grip the toolholder short (Fig. 57-20).
3. Fasten a general-purpose toolbit in the toolholder and set the point to center.
4. Tighten the toolpost screw securely.
5. Set the lathe to the correct speed and feed for rough turning.

■ **Figure 57-20** The toolpost set on the left-hand side of the compound rest and the toolholder held short. *(Kelmar Associates)*

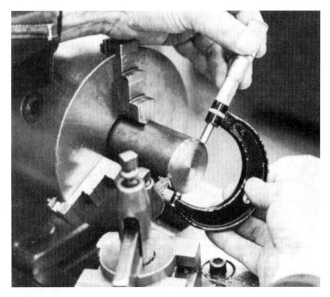

■ **Figure 57-21** Measuring the diameter of a trial cut. *(Kelmar Associates)*

6. Take a light trial cut at the end of the work and measure the diameter (Fig. 57-21).
7. Adjust the crossfeed graduated collar to one-half the amount of metal to be removed.
8. Rough-turn the diameter to the correct length.
9. Set the lathe speed and feed for finish turning.
10. Take a light trial cut at the end of the workpiece.
11. Set the graduated collar in one-half the amount of metal to be removed.
12. Take another trial cut, measure the diameter, and if it is correct, finish-turn the diameter.

▶▶ Turning Hardened Steel with PCBN Tools

Polycrystalline cubic boron nitride (PCBN) cutting tools have been widely used in the automotive and aerospace industries for machining hard, abrasive workpieces. They are also finding use in general machine shops and tool and die shops for the same purpose, and for machining metals that are difficult to machine. Their characteristics of superior hardness, wear resistance, compressive strength, and thermal conductivity make them ideal choices where conventional tools such as carbides fail or are not cost-effective. PCBN tools are cost-effective for machining materials from 45 Rc and higher, such as hardened tool and alloy steels, abrasive cast irons, and tough high-temperature alloys. They can remove material at high metal-removal rates, with long tool life even under interrupted cuts.

GUIDELINES FOR PCBN TURNING

In a turning operation, any cutting tool will work better on a machine in good condition, providing the tool is used properly. Since PCBN tools are used to cut hard, abrasive materials where the cutting pressures are higher, certain conditions relating to the tool, cutting-tool setup, and operating procedures should be followed to ensure ideal cutting conditions and the most cost-effective operation:

1. Machines should have good bearings, tight slides, and enough power to provide a constant surface speed.

2. The proper cutting tool should be selected to suit the type of material being cut and the machining operation.

   > Use negative-rake tools with a large lead angle wherever possible.
   > Use as large a tool nose radius as the machining operation will allow.

Note: Follow the manufacturer's recommendation for each type and grade of PCBN tool.

3. Follow the manufacturer's recommendation for the proper speed, feed, and depth of cut.

4. Set cutting tools on center and keep the overhang as short as possible to prevent chatter and vibration.

5. Use cutting fluid wherever possible.

The *Interactive Superabrasive Machining Advisor* software program, supplied by GE Superabrasives, was used on the following machining operation to select the PCBN tool and the machining conditions. To use this software program effectively, a person must know the material hardness, machine horsepower, maximum spindle speed, maximum carriage feed, depth of cut, and surface finish. The computer program makes recommendations for the tool grade, tool edge preparation, tool geometry, speed, feed, and type of coolant. If this software program is not available, be sure to follow the tool manufacturer's recommendations.

TURNING THE PART

The specifications for the part, cutting tool, and machining conditions are as follows:

1. *Workpiece material*—1-in. diameter, hardened tool steel 55-60 Rc

2. *Machine*—a conventional lathe with a 5 HP (horsepower) drive to the spindle

3. *Tool*—BZN* 8100 tipped insert, CNMA 433

*Trademark GE Superabrasives.

4. *Machining conditions*—
   > .010 in. (0.25 mm) depth of cut
   > .004 in. (0.1 mm) feed rate
   > 400 sf/min (120 m/min) cutting speed

procedure

1. Set the machine spindle to the correct speed. Check it for accuracy with a portable tachometer, if one is available.

2. Set the lathe for the recommended feed per revolution.

3. Lock the PCBN insert in the proper toolholder.

4. Fasten the toolholder in the toolpost and adjust the tool to centerline height.

5. Use a .010-in. (0.25-mm) plastic shim and set the tool tip to the outside diameter of the part. The plastic shim will prevent chipping or cracking the tool point.

6. Turn the carriage handwheel *clockwise* so that the tool clears the end of the part.

7. Turn the crossfeed handle clockwise the thickness of the shim plus the desired depth of cut.

8. Engage the automatic feed to machine the diameter toward the lathe headstock (Fig. 57-22).

9. At the end of the cut, turn the crossfeed handle *counterclockwise* to withdraw the tool from the diameter, to prevent chipping the cutting edge.

10. Return to the starting position of the cut, reset the tool for depth, and take additional passes until the part is machined to size.

■ Figure 57-22 Turning a piece of hardened tool steel with a PCBN cutting tool on a conventional lathe. *(G. E. Superabrasives)*

▶▶ Cutting Off Work in a Chuck

Cutoff tools, often called parting tools, are used for cutting off work projecting from a chuck, for grooving, and for undercutting. The inserted blade-type parting tool is the one most commonly used, and it is provided in three holders (Fig. 57-23).

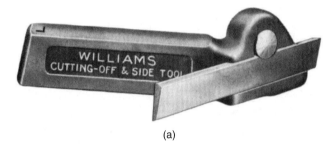
(a)

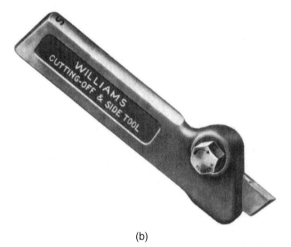
(b)

(c)

■ Figure 57-23 Inserted blade cutoff, or parting, tools: (A) left-hand offset; (B) straight; (C) right-hand offset. *(J.H. Williams & Co.)*

1. Mount the work in the chuck with the part to be cut off as close to the jaws as possible.
2. Mount the cutoff tool on the left-hand side of the compound rest, with the cutting edge set on center (Fig. 57-24).
3. Place the holder as close to the toolpost as possible to prevent vibration and chatter.
4. Extend the cutting blade beyond the holder half the diameter of the work to be cut, plus .125 in. (3 mm) for clearance.
5. Set the lathe to approximately one-half the turning speed.
6. Move the cutting tool into position (Fig. 57-25).

■ Figure 57-24 The tip of the cutoff blade set to center height. *(Kelmar Associates)*

■ Figure 57-25 The cutoff blade being moved to the correct position. *(Kelmar Associates)*

7. Start the lathe and feed the cutoff tool into the work by hand, keeping a steady feed during the operation. Cut brass and cast iron dry, but use cutting fluid for steel.

8. When grooving or cutting off deeper than .250 in. (6 mm), it is good practice to move the parting tool sideways slightly. This may be accomplished by moving the carriage handwheel back and forth a few thousandths of an inch, or hundredths of a millimeter, during the cutting operation. This side motion cuts a little wider groove and prevents the tool from jamming.

9. Before the cut is completed, remove the burrs from each side of the groove with a file.

Note: To avoid chatter, keep the tool cutting and apply cutting fluid constantly during the operation. Feed slowly when the part is almost cut off.

unit 57 review questions

Mounting and Removing Chucks

1. What safety precautions should be observed before attempting to mount or remove chucks?

2. Why is it necessary that the taper on the lathe spindle nose and the mating taper in the chuck be thoroughly cleaned before mounting?

3. Explain the procedure for mounting a chuck on a taper spindle nose.

4. Explain the procedure for removing a chuck from a cam-lock spindle nose.

Mounting Work in a Chuck

5. How far may work extend beyond the chuck jaws before it must be supported?

6. Briefly list the procedure for assembling jaws in a three-jaw universal chuck.

7. Why are four-jaw chucks used? Explain.

8. Briefly explain how to true a piece of work to within .001-in. (0.02-mm) accuracy in a four-jaw chuck.

Facing in a Chuck

9. How should the cutting tool be set for facing?

10. How much should the compound rest graduated collar be turned in for facing:

    a. .050 in. with the compound rest set at 90° to the cross-slide?
    b. .025 in. when the compound rest is set at 30°?

Rough and Finish Turning

11. How should the toolholder and the cutting tool be set for machining work in a chuck?

12. What is the purpose of a light trial cut at the end of a workpiece?

Cutting Off

13. How far should the blade extend beyond the holder for cutting off?

14. What procedure should be used for cutting grooves deeper than .250 in. (6 mm)?

UNIT 58

Drilling, Boring, Reaming, and Tapping

OBJECTIVES

After completing this unit, you should be able to:

1 Drill small- and large-diameter holes in a lathe

2 Ream holes to an accurate size and good surface finish

3 Bore a hole to within .001-in. (0.02-mm) accuracy

4 Use a tap to produce an internal thread that is concentric with the outside diameter

Internal operations such as drilling, boring, reaming, and tapping can be performed on work held in a chuck. Boring tools are mounted in the toolpost, while drills, reamers, and taps may be held either in a drill chuck mounted in the tailstock spindle or directly in the tailstock spindle. Since the work held in a chuck is generally machined true, these operations are usually machined concentric with the outside diameter of the workpiece.

▶▶ To Spot and Drill Work in a Chuck

Spotting ensures that a drill will start in the center of the work. A spotting tool is used to make a shallow, V-shaped hole in the center of the work, which provides a guide for the drill to follow. In most cases, a hole can be spotted quickly and fairly accurately by using a center drill (Fig. 58-1). Where extreme accuracy in spotting is necessary, a spotting toolbit should be used.

■ **Figure 58-1** Using a center drill to spot a hole. *(Kelmar Associates)*

Figure 58-2 A tapered shank-drill mounted in the spindle is prevented from turning by a lathe dog. *(Kelmar Associates)*

Figure 58-3 Supporting the end of a drill will prevent it from wobbling when starting a hole. *(Kelmar Associates)*

procedure

1. Mount the work true in a chuck.
2. Set the lathe to the proper speed for the type of material to be drilled.
3. Check the tailstock and make sure it is in line.
4. Spot the hole with a center drill or spotting tool.
5. Mount the twist drill in the tailstock spindle, in a drill chuck, or in a drill holder (Figs. 58-2 to 58-4).

Note:

a. When a tapered-shank drill is mounted directly in the tailstock spindle, a dog should be used to stop the drill from turning and scoring the tailstock spindle taper (Fig. 58-2).

b. The end of the drill may be supported with the end of a toolholder so that the drill will start on center (Fig. 58-3).

c. Tapered-shank drills are often mounted in a drill holder. The point of the drill is positioned in the hole, while the end of the holder is supported by the dead center. The handle of the holder rests on the toolholder and against the toolpost to prevent the drill from turning and from pulling into the work (Fig. 58-4).

6. Start the lathe and drill to the desired depth, applying cutting fluid frequently.

7. To gage the depth of the hole, use the graduations on the tailstock spindle, or measure the depth with a steel rule (Fig. 58-5).

8. Withdraw the drill frequently to remove the chips and measure the depth of the hole.

Figure 58-4 Drilling a hole with a drill mounted in a drill holder and supported on a dead center. *(Kelmar Associates)*

 CAUTION Always ease the force on the feed as the drill breaks through the work.

▶▶ Boring

Boring is the operation of enlarging and truing a drilled or cored hole with a single-point cutting tool. Special-diameter holes, for which no drills are available, can be produced by boring.

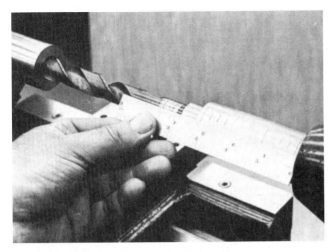

Figure 58-5 Measuring the depth of a drilled hole using a rule. *(Kelmar Associates)*

Figure 58-6 The boring bar should be held short and the toolbit set on center to machine inside diameters. *(R. K. LeBlond-Makino Machine Tool Company)*

Holes may be drilled in a lathe; however, such holes are generally not considered accurate, even though the drill may have started straight. During the drilling process, the drill may become dull or hit a hard spot or blowhole in the metal, which will cause the drill to wander or run off center. If such a hole is reamed, the reamer will follow the drilled hole, and, as a result, the hole will not be straight. Therefore, if it is important that a reamed hole be straight and true, the hole should first be drilled, then bored and reamed.

TO BORE WORK IN A CHUCK

1. Mount the work in a chuck; face, spot, and drill the hole approximately .060 in. (1.5 mm) undersize.
2. Select a boring bar as large as possible and have it extend beyond the holder only enough to clear the depth of the hole to be bored.
3. Mount the boring bar holder in the toolpost on the left-hand side of the compound rest.
4. Set the boring toolbit to center (Fig. 58-6).
5. Set the lathe to the proper speed and select a medium feed.
6. Start the lathe and bring the boring tool into contact with the inside diameter of the hole.
7. Take a light trial cut [approximately .005 in. (0.12 mm) or until a true diameter is produced] .250 in. (6 mm) long at the right-hand end of the work.

8. Stop the lathe and measure the hole diameter with a telescopic gage or inside micrometer.
9. Determine the amount of material to be removed from the hole.

Note: Leave approximately .010 to .020 in. (0.25 to 0.5 mm) for a finish cut.

10. Set the depth of cut for half the amount of metal to be removed.
11. Start the lathe and take the roughing cut.

Note: If chatter or vibration occurs during machining, slow the lathe speed and increase the feed until it is eliminated.

12. Stop the lathe and bring the boring tool out of the hole without moving the crossfeed handle.
13. Set the depth of the finish cut and bore the hole to size. For a good surface finish, a fine feed is recommended.

▶▶ Reaming

Reaming may be performed in a lathe to quickly obtain an accurately sized hole and to produce a good surface finish. Reaming may be performed after a hole has been drilled or bored. If a true, accurate hole is required, it should be bored before the reaming operation.

Drilling, Boring, Reaming, and Tapping   **457**

TO REAM WORK IN A LATHE

1. Mount the work in a chuck; face, spot, and drill the hole to size. For holes under .500 in. (13 mm) in diameter, drill .015 in. (0.4 mm) undersize; for holes over .500 in. (13 mm) in diameter, drill .030 in. (0.8 mm) undersize. If the holes must be true, they should be bored .010 in. (0.25 mm) undersize.

2. Mount the reamer in a drill chuck or drill holder (Fig. 58-7). When reaming holes .625 in. (16 mm) in diameter and larger, fasten a lathe dog near the reamer shank and support the tail on the compound rest to prevent the reamer from turning.

3. Set the lathe to approximately half the drilling speed.

4. Bring the reamer close to the hole and lock the tailstock in position.

5. Start the lathe, apply cutting fluid to the reamer, and slowly feed it into the drilled or bored hole with the tailstock handwheel.

6. Occasionally remove the reamer from the hole to clear chips from the flutes and apply cutting fluid.

7. Once the hole has been reamed, stop the lathe and remove the reamer from the hole.

 CAUTION Never turn the lathe spindle or reamer backward for any reason. This will damage a reamer.

8. Clean the reamer and store it carefully to prevent it from being nicked or damaged.

▶▶ Tapping

Tapping is one method of producing an internal thread on a lathe. The tap is aligned by placing the point of the lathe dead center in the shank end of the tap to guide it while the tap is turned by a tap wrench. A standard tap may be used for this operation; however, a gun tap is preferred because the chips are cleared ahead of the tap. When tapping a hole in a lathe, lock the spindle and turn the tap by hand (Fig. 58-8).

TO TAP A HOLE IN A LATHE

1. Mount the work in the chuck; face and center drill.

2. Select the proper tap drill for the tap to be used.

3. Set the lathe to the proper speed.

4. Drill with the tap drill to the required depth. Use cutting fluid if required.

5. Chamfer the edge of the hole slightly larger than the tap diameter.

6. Stop the lathe and lock the spindle, or put the lathe in its lowest speed.

7. Place a taper tap in the hole and support the shank with the tailstock center.

8. With a suitable wrench, turn the tap, keeping the dead center snug into the shank of the tap by turning the tailstock handwheel.

9. Apply cutting fluid while tapping the hole (Fig. 58-8).

10. Back off the tap frequently to break the chip.

11. Remove the taper tap and finish tapping the hole with a plug or bottoming tap.

■ **Figure 58-7** Setup for reaming in a lathe. *(Kelmar Associates)*

■ **Figure 58-8** Using a tap to cut internal threads. *(Kelmar Associates)*

▸▸ Grinding on a Lathe

Cylindrical and internal grinding may be done on a lathe if a proper grinding machine is not available. A toolpost grinder, mounted on a lathe, may be used for cylindrical and taper grinding as well as an angular grinding of lathe centers. An internal grinding attachment for the toolpost grinder permits the grinding of straight and tapered holes. Grinding should be done on a lathe only when no other machine is available, or when the cost of performing a small grinding operation on a part would not warrant setting up a regular grinding machine. Since the work should rotate in an opposite direction to the grinding wheel, the lathe must be equipped with a reversing switch.

TO GRIND A LATHE CENTER

1. Remove the chuck or driveplate from the lathe spindle.
2. Mount the lathe center to be ground in the headstock spindle.
3. Set a slow spindle speed.
4. Swing the compound rest to 30° (Fig. 58-9) with the centerline of the lathe.
5. Protect the ways of the lathe with cloth or canvas and place a pan of water below the lathe center.
6. Mount the toolpost grinder and adjust the center of the grinding spindle to center height.
7. Mount the proper grinding wheel; true and dress.
8. Start the lathe, with the spindle revolving in reverse.

■ **Figure 58-9** The compound rest set at 30° to grind a lathe center. *(Kelmar Associates)*

9. Start the grinder and adjust the grinding wheel until it sparks lightly against the revolving center.
10. Lock the carriage in this position.
11. Feed the grinding wheel in .001 in. (0.02 mm), using the crossfeed handle.
12. Move the grinder along the face of the center using the compound rest feed.
13. Check the angle of the center using a center gage, and adjust the compound rest if necessary.
14. Finish-grind the center.

Note: If a high finish is desired, polish the center with abrasive cloth at a high spindle speed.

A SUMMARY OF LATHE OPERATIONS

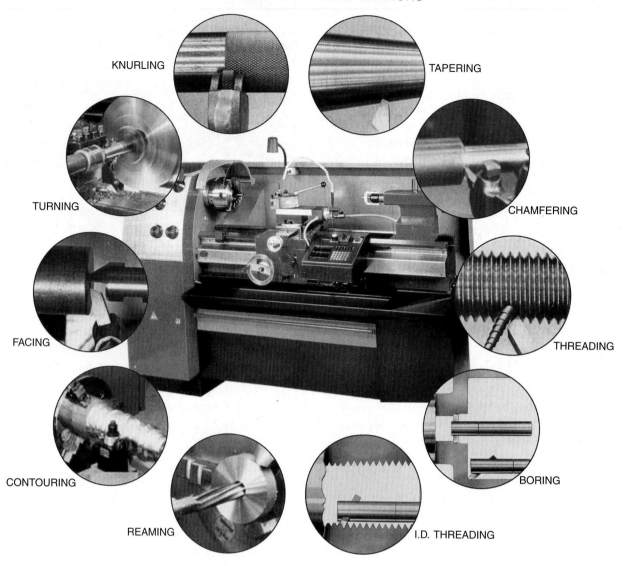

KNURLING

TAPERING

TURNING

CHAMFERING

FACING

THREADING

CONTOURING

BORING

REAMING

I.D. THREADING

unit 58 review questions

Drilling

1. Why is spotting important before a hole is drilled?

2. Name three methods of holding various sizes of drills in a tailstock.

3. How can the depth of a drilled hole be gaged?

Boring

4. Define the boring process.

5. Why should a hole be bored before it is reamed?

6. How should the boring bar and cutting tool be set up for boring?

7. Briefly explain how to bore a hole to 1.750 in. (44 mm).

Reaming

8. What is the purpose of reaming in a lathe?

9. How much material should be left in a hole for reaming a:
   a. $\frac{7}{16}$-in. hole?   b. 1-in. hole?

10. Why should a reamer never be turned backward?

Tapping

11. How is the tap started and guided so that the thread will be true to the bored hole?

Section 12

MILLING MACHINES

Milling machines are machine tools used to produce one or more machined surfaces accurately on a piece of material, *the workpiece;* this is done by one or more rotary milling cutters having single or multiple cutting edges. The workpiece is held securely on the *work table* of the machine or in a holding device clamped to the table. It is then brought into contact with a revolving cutter.

The vertical milling machine, one of the most common and versatile machine tools, uses end-milling-type cutters to machine horizontal, vertical, and angular surfaces, grooves, slots, and keyways plus a wide variety of other machining operations. It is easy to set up and operate; equipped with the proper accessories, it can machine circular forms and perform jig-boring operations. Many operations usually performed on a horizontal milling machine can be produced faster and easier on a vertical milling machine.

The horizontal milling machine is a versatile machine tool that can handle a variety of operations normally performed by other machine tools. It is used not only for the milling of flat and irregularly shaped surfaces but also for gear and thread cutting, drilling, boring, reaming, and slotting operations.

UNIT 59

The Vertical Milling Machine

OBJECTIVES

After completing this unit, you should be able to:

1 List four main uses of a vertical milling machine

2 Describe how angular surfaces can be machined

3 List three types of vertical milling machines

4 State the purposes of the main parts of a knee and column machine

The vertical milling machine, developed in the 1860s, combines the vertical spindle of a drill press with the longitudinal and traverse (crossfeed) movements of a milling machine. It is used to accurately machine flat, angular, rounded, and multi-shaped surfaces which can be in one, two, or three planes (*X, Y, Z* axes). The milling process may be vertical, horizontal, angular, or helical. Since the spindle is usually in a vertical position, the vertical milling machine can be used for milling, drilling, boring, and reaming operations.

Modern vertical milling machines are usually equipped with variable speed spindle drives that permit a wide range of speeds to accommodate various cutter sizes and machining operations.

The spindle head housing can be swiveled as much as 90° to the right and left, as well as a limited amount backward and forward for angular operations.

The longitudinal and crossfeed movements of the table may be operated by hand or by automatic feeds. Some table-feed mechanisms are equipped with a rapid drive that eliminates wasted time during setup purposes. The versatility of the vertical milling machine is increased by the addition of various drive heads, horizontal milling attachments, cutter-holding devices, and digital readout systems.

The vertical milling machine has become popular in industry, offering versatility not found in any other machine. Some of the operations that can conveniently be carried out on these machines are face milling, end milling, keyway cutting, dovetail cutting, T-slot and circular slot cutting, gear cutting, drilling, boring, and jig boring. Because of the machine construction (vertical spindle), many of the facing operations can be done with a fly cutter, which reduces the cost of cutters considerably. Also, since most cutters are much smaller than for the horizontal mill, the cost of the cutters for the same job is usually much less for a vertical milling machine.

▸▸ Types of Vertical Milling Machines

The *ram-type vertical milling machine* (Fig. 59-1) is most commonly found in industry because of its simplicity and ease of setup. It has all the construction features of a plain horizontal milling machine, except that the cutter spindle is mounted in a vertical position. The spindle head may be swiveled, which allows the machining of angular surfaces.

The *2-axis control vertical milling machine* (Fig. 59-2) a step between the standard milling machine and the CNC machining center, can be operated manually or automatically to perform operations, and even to machine an entire part. The PC-based control allows an operator to store thousands of individual operations in its memory without any knowledge of CNC programming. The control can be taught to produce any part by manually going through each move, such as moving the table to a position and pushing the **ENTER** button at the end of each move to store the machine position in memory. This 2-axis control machine is capable of machining angles and radii without a rotary table; routine milling of arcs, slots, pockets; and drilling bolt-hole circles, at a fraction of the time it takes to do them manually.

■ **Figure 59-2** A standard vertical milling machine. *(Bridgeport Machines, Inc.)*

▸▸ Heavy-Duty Vertical Milling Machine

The heavy-duty industrial vertical milling machine (Fig. 59-3 on p. 466) is used for machining operations that require larger cutters and good rigidity. This machine is especially adapted to operations with end mills and face mills; for profiling interior and exterior surfaces; for milling dies and metal molds; and for locating and boring holes in jigs and fixtures. On most machines, automatic feeds are available for the vertical movements of the spindle and the movements of the machine table.

▸▸ Parts of the Ram-Type Vertical Mill

The *base* is made of ribbed cast iron. It may contain a coolant reservoir.

The *column* is often cast integrally with the base. The machined face of the column provides the ways for vertical movement of the knee. The upper part of the column is machined to receive a *turret,* on which the overarm is mounted.

The *overarm* is round, or of the ram type, as illustrated in Fig. 59-1. It may be adjusted toward or away from the column to increase the capacity of the machine.

The *head* is attached to the end of the ram. Provision is made to swivel the head in one plane. On universal-type machines, the head may be swiveled in two planes. Mounted on top of the head is the *motor,* which provides drive to the *spindle,* usually through V-belts. Spindle speed changes are effected by means of a variable-speed

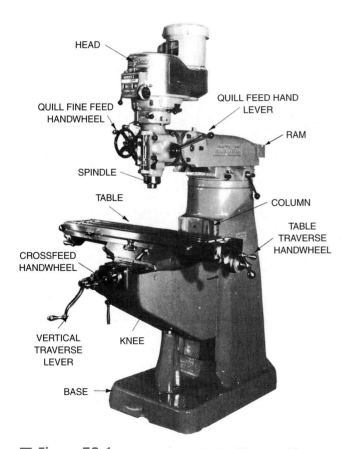

■ **Figure 59-1** A ram-type vertical milling machine. *(Bridgeport Machines, Inc.)*

pulley and crank or by belt changes and a reduction gear. The spindle may be fed by means of a hand lever or a handwheel, or by automatic power feed. Most machines are equipped with a micrometer quill stop for precision drilling and boring to depth.

The *knee* moves up and down on the face of the column and supports the saddle and the table. The knee in this type of machine does *not* contain the gears for the automatic feed, as in the horizontal milling machine and the standard vertical milling machine. The automatic feed on most of these types of machines is not a standard feature. It is an external device and controls the longitudinal and crossfeeds of the table. Most cutting on the vertical milling machine is done by end mill cutters; therefore, it is not necessary to swing the table even when cutting a helix. As a result, vertical milling machines are equipped with plain tables only.

■ **Figure 59-3** An industrial-size vertical milling machine. *(Cincinnati Machine, a UNOVA Co.)*

unit 59 review questions

1. State three reasons a vertical milling machine is such a versatile machine tool.

2. Name seven operations that can be performed on a vertical milling machine.

3. What are the three axes in which work may be machined?

4. How far may the vertical head be swiveled for machining angular surfaces?

Ram-Type Vertical Mill Parts

5. State the purpose of five main parts of a vertical mill.

Heavy-Duty Vertical Mill

6. For what purpose are heavy-duty mills used?

UNIT 60

Cutting Speed, Feed, and Depth of Cut

OBJECTIVES

After completing this unit, you will be able to:

1 Select cutting speeds and calculate the r/min for various cutters and materials

2 Select and calculate the proper feeds for various cutters and materials

3 Follow the correct procedure for taking roughing and finishing cuts

The most important factors affecting the efficiency of a milling operation are cutting speed, feed, and depth of cut. If the cutter is run too slowly, valuable time will be wasted, while excessive speed results in loss of time in replacing and regrinding cutters. Somewhere between these two extremes is the efficient *cutting speed* for the material being machined.

The rate at which the work is fed into the revolving cutter is important. If the work is fed too slowly, time will be wasted, and cutter chatter, which shortens the life of the cutter, may occur. If the work is fed too fast, the cutter teeth can be broken. Much time will be wasted if several shallow cuts are taken instead of one deep or roughing cut. Therefore, *speed, feed,* and *depth of cut* are three important factors in any milling operation.

▶▶ Cutting Speed

One of the most important factors affecting the efficiency of a milling operation is cutter speed. The cutting speed of a metal may be defined as *the speed, in surface feet per minute (sf/min) or meters per minute (m/min) at which the metal may be machined efficiently.* When the work is machined in a lathe, it must be turned at a specific number of revolutions per minute (r/min), *depending on its diameter,* to achieve the proper cutting speed. When work is machined in a milling machine, the *cutter* must be revolved at a specific number of r/min, *depending on its diameter,* to achieve the proper cutting speed.

Since different types of metals vary in hardness, structure, and machinability, different cutting speeds must be used for each type of metal and for various cutter materials. Several factors must be considered when determining the proper r/min at which to machine a metal. The most important are:

> The type of work material
> The cutter material
> The diameter of the cutter
> The surface finish required
> The depth of cut taken
> The rigidity of the machine and work setup

The cutting speeds for the more common metals are shown in Table 60.1 on p. 468.

table 60.1 Milling machine cutting speeds

Material	High-Speed Steel Cutter		Carbide Cutter	
	ft/min	m/min	ft/min	m/min
Alloy steel	40–70	12–20	150–250	45–75
Aluminum	500–1000	150–300	1000–2000	300–600
Bronze	65–120	20–35	200–400	60–120
Cast iron	50–80	15–25	125–200	40–60
Free machining steel	100–150	30–45	400–600	120–180
Machine steel	70–100	21–30	150–250	45–75
Stainless steel	30–80	10–25	100–300	30–90
Tool steel	60–70	18–20	125–200	40–60

INCH CALCULATIONS

To get optimum use from a cutter, the proper speed at which the cutter should be revolved must be determined. When cutting machine steel, a high-speed steel cutter would have to achieve a surface speed of about 90 ft/min (27 m/min). Since the diameter of the cutter affects this speed, it is necessary to consider its diameter in the calculation. The following example illustrates how the formula is developed.

EXAMPLE

Calculate the speed required to revolve a 3-in. diameter high-speed steel milling cutter for cutting machine steel (90 sf/min).

Solution

1. Determine the circumference of the cutter or the distance a point on the cutter would travel in one revolution. The circumference of the cutter = 3 × 3.1416.

2. To determine the proper cutter speed, or *r/min*, it is necessary only to divide the cutting speed (CS) by the circumference of the cutter.

$$r/min = \frac{CS \text{ (ft)}}{\text{circumference (in.)}}$$
$$= \frac{90}{3 \times 3.1416}$$

Since the numerator is in feet and the denominator in inches, the numerator must be changed to inches. Therefore,

$$r/min = \frac{12 \times CS}{3 \times 3.1416}$$

Because it is usually impossible to set a machine to the exact r/min, it is permissible to consider that 3.1416 will divide into 12 approximately 4 times. The formula now becomes:

$$r/min = \frac{4 \times CS}{D}$$

Using this formula and Table 60.1, you can calculate the proper cutter or spindle speed for any material and cutter diameter.

EXAMPLE

At what speed should a 2-in. diameter carbide cutter revolve to mill a piece of cast iron (CS 150)?

Solution

$$r/min = \frac{4 \times 150}{2}$$
$$= 300$$

METRIC CALCULATIONS

The r/min at which the milling machine should be set when using metric measurements is as follows:

$$r/min = \frac{CS(m) \times 1000}{\pi \times D\,(mm)}$$

$$= \frac{CS \times 1000}{3.1416 \times D}$$

Since only a few machines are equipped with variable speed drives, which allow them to be set to the exact calculated speed, a simplified formula can be used to calculate r/min. The π (3.1416) on the bottom line of the formula will divide into the 1000 of the top line approximately 320 times. This results in a simplified formula, which is close enough for most milling operations:

$$r/min = \frac{CS\,(m) \times 320}{D\,(mm)}$$

EXAMPLE

Calculate the r/min required for a 75-mm diameter high-speed steel milling cutter when cutting machine steel (CS 30 m/min).

Solution

$$r/min = \frac{30 \times 320}{75}$$

$$= \frac{9600}{75}$$

$$= 128$$

Although these formulas are helpful in calculating the cutter (spindle) speed, it should be remembered that they are approximate only and the speed may have to be altered because of the hardness of the metal and/or the machine condition. Best results may be obtained if the following rules are observed:

1. For longer cutter life, use the lower CS in the recommended range.

2. Know the hardness of the material to be machined.

3. When starting a new job, use the lower range of the CS and gradually increase to the higher range if conditions permit.

4. If a fine finish is required, reduce the feed rather than increase the cutter speed.

5. The use of coolant, properly applied, will generally produce a better finish and lengthen the life of the cutter. The coolant absorbs heat, acts as a lubricant, and washes chips away.

▸▸ Milling Feed and Depth of Cut

The two other factors which affect the efficiency of a milling operation are the *milling feed,* or the rate at which the work is fed into the milling cutter, and the *depth of cut* taken on each pass.

FEED

Milling machine feed may be defined as the distance in inches (or millimeters) per minute that the work moves into the cutter. On most milling machines, the feed is regulated in inches (or millimeters) per minute and is independent of the spindle speed. This arrangement permits faster feeds for larger, slowly rotating cutters.

Feed is the rate at which the work moves into the revolving cutter, and it is measured either in inches per minute or in millimeters per minute. The *milling feed* is determined by multiplying the chip size (chip per tooth) desired, the number of teeth in the cutter, and the r/min of the cutter.

Chip, or *feed, per tooth* (*CPT,* or *FPT*) is the amount of material that should be removed by each tooth of the cutter as it revolves and advances into the work. See Tables 60.2 and 60.3 on p. 470 or the recommended CPT for some of the more common metals.

The feed rate used on a milling machine depends on a variety of factors, such as:

1. The depth and width of cut

2. The design or type of cutter

3. The sharpness of the cutter

4. The workpiece material

5. The strength and uniformity of the workpiece

6. The type of finish and accuracy required

7. The power and rigidity of the machine, the holding device, and the tooling setup

As the work advances into the cutter, each successive tooth advances into the work an equal amount, producing chips of equal thickness. This thickness of the chips or the *feed per tooth,* along with the number of teeth in the cutter, forms the basis for determining the rate of feed. The ideal rate of feed may be determined as follows:

Feed = no. of cutter teeth × feed/tooth × cutter r/min

table 60.2 — Recommended feed per tooth (high-speed cutters)

Material	Face Mills in.	Face Mills mm	Helical Mills in.	Helical Mills mm	Slotting and Side Mills in.	Slotting and Side Mills mm	End Mills in.	End Mills mm	Form-Relieved Cutters in.	Form-Relieved Cutters mm	Circular Saws in.	Circular Saws mm
Alloy steel	.006	0.15	.005	0.12	.004	0.1	.003	0.07	.002	0.05	.002	0.05
Aluminum	.022	0.55	.018	0.45	.013	0.33	.011	0.28	.007	0.18	.005	0.13
Brass and bronze (medium)	.014	0.35	.011	0.28	.008	0.2	.007	0.18	.004	0.1	.003	0.08
Cast iron (medium)	.013	0.33	.010	0.25	.007	0.18	.007	0.18	.004	0.1	.003	0.08
Free machining steel	.012	0.3	.010	0.25	.007	0.17	.006	0.15	.004	0.1	.003	0.07
Machine steel	.012	0.3	.010	0.25	.007	0.18	.006	0.15	.004	0.1	.003	0.08
Stainless steel	.006	0.15	.005	0.13	.004	0.1	.003	0.08	.002	0.05	.002	0.05
Tool steel (medium)	.010	0.25	.008	0.2	.006	0.15	.005	0.13	.003	0.08	.003	0.08

table 60.3 — Recommended feed per tooth (cemented-carbide-tipped cutters)

Material	Face Mills in.	Face Mills mm	Helical Mills in.	Helical Mills mm	Slotting and Side Mills in.	Slotting and Side Mills mm	End Mills in.	End Mills mm	Form-Relieved Cutters in.	Form-Relieved Cutters mm	Circular Saws in.	Circular Saws mm
Aluminum	.020	0.5	.016	0.40	.012	0.3	.010	0.25	.006	0.15	.005	0.13
Brass and bronze (medium)	.012	0.3	.010	0.25	.007	0.18	.006	0.15	.004	0.1	.003	0.08
Cast iron (medium)	.016	0.4	.013	0.33	.010	0.25	.008	0.2	.005	0.13	.004	0.1
Machine steel	.016	0.4	.013	0.33	.009	0.23	.008	0.2	.005	0.13	.004	0.1
Tool steel (medium)	.014	0.35	.011	0.28	.008	0.2	.007	0.18	.004	0.1	.004	0.1
Stainless steel	.010	0.25	.008	0.2	.006	0.15	.005	0.13	.003	0.08	.003	0.08

Inch Calculations

The formula used to find the work feed in inches per minute is:

$$\text{Feed (in./min)} = N \times \text{CPT} \times \text{r/min}$$

N = number of teeth in the milling cutter

CPT = chip per tooth for a particular cutter and metal, as given in Tables 60.2 and 60.3

r/min = revolutions per minute of the milling cutter

Note: The calculated feeds would be possible only under ideal conditions. Under average operating conditions, and *especially in school shops,* the milling machine feed should be set to approximately one-third or one-half the amount calculated. The feed can then be gradually increased to the capacity of the machine and the finish desired.

EXAMPLE

Find the feed in inches per minute using a 3.5-in. diameter, 12-tooth helical cutter to cut machine steel (CS 80).

Solution

First, calculate the proper r/min for the cutter:

$$\text{r/min} = \frac{4 \times \text{CS}}{D}$$

$$= 4 \times \frac{80}{3.5}$$

$$= 91$$

$$\text{Feed (in./min)} = N \times \text{CPT} \times \text{r/min}$$

$$= 12 \times .010 \times 91$$

$$= 10.9 \quad \text{or} \quad 11 = 10.9 \text{ in/min}$$

Metric Calculations

The formula used to find work feed in millimeters per minute is the same as the formula used to find the feed in inches per minute, except that mm/min is substituted for in./min.

EXAMPLE

Calculate the feed in millimeters per minute for a 75-mm diameter, six-tooth helical carbide milling cutter when machining a cast-iron workpiece (CS 60).

Solution

First, calculate the r/min of the cutter:

$$\text{r/min} = \frac{\text{CS} \times 320}{D}$$

$$= \frac{60 \times 320}{75}$$

$$= \frac{19,200}{75}$$

$$= 256$$

$$\text{Feed (mm/min)} = N \times \text{CPT} \times \text{r/min}$$

$$= 6 \times 0.25 \times 256$$

$$= 384$$

$$= 384 \text{ mm/min}$$

Tables 60.2 and 60.3 give suggested feed per tooth for various types of milling cutters for roughing cuts under average conditions. For finishing cuts, the feed per tooth would be reduced to one-half or even one-third of the value shown.

Direction of Feed

One final consideration concerning feed is the direction in which the work is fed into the cutter. The most commonly used method is to feed the work against the rotation direction of the cutter (*conventional,* or *up milling*) (Fig. 60-1). However, if the machine is equipped with a backlash eliminator, certain types of work can best be milled by *climb milling* (Fig. 60-2 on p. 472).

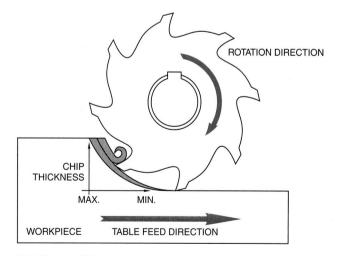

■ **Figure 60-1** Conventional milling. *(Hanita Cutting Tools)*

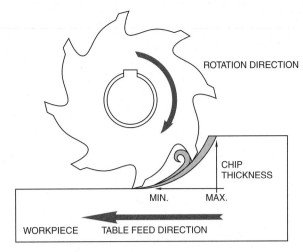

Figure 60-2 Climb milling. *(Hanita Cutting Tools)*

ROTATION DIRECTION

CHIP
THICKNESS

MIN.   MAX.

WORKPIECE   TABLE FEED DIRECTION

Climb milling, which can increase cutter life up to 50%, is effective for most milling applications. To know whether climb or conventional milling is being used, look at the relationship between the cutter rotation and the direction of the machine table/work feed. Climb milling is being used when the cutter and the workpiece are going in the same direction (Fig. 60-2). Conventional milling is when the cutter and the workpiece are going in opposite directions.

Advantages of Climb Milling

> *Increased tool life.* Since the chips pile up behind or to the left of the cutter, tool life can be increased by as much as 50%.

> Less costly fixtures required. Climb milling forces the workpiece down instead of trying to lift as with conventional milling; therefore, simpler holding devices are required.

> *Improved surface finishes.* Chips are less likely to be carried into the workpiece by the cutter teeth, which prevents damage to the work surface.

> *Less edge breakout.* Because the thickness of the chip tends to get smaller as it nears the edge of a workpiece, there is less chance of breaking, especially with brittle materials.

> *Easier chip removal.* The cutter teeth force the chips to fall behind the cutter and make it easier to remove the chips.

> *Lower power requirements.* Since a cutter with a higher rake angle can be used, approximately 20% less power is required to remove the same amount of metal.

Disadvantages of Climb Milling

> *This method cannot be used unless the machine has a backlash eliminator* and the table gibs have been tightened.

> It cannot be used for machining castings or hot-rolled steel, since the hard outer scale will damage the cutter.

Conventional milling is recommended when machining castings or forgings, where there is a hard or abrasive surface due to scale or sand. It should always be used on machines which do not have a backlash eliminator.

DEPTH OF CUT

Where a smooth, accurate finish is desired, it is good milling practice to take roughing and finishing cuts. Roughing cuts should be deep, with a feed as heavy as the work and the machine will permit. Heavier cuts may be taken with helical cutters having fewer teeth, since they are stronger and have a greater chip clearance than cutters with more teeth.

Finishing cuts should be light, with a finer feed than is used for roughing cuts. The depth of the cut should be at least .015 in. (0.4 mm). Lighter cuts and extremely fine feeds are not advisable, since the chip taken by each tooth will be thin, and the cutter will often rub on the surface of the work, rather than bite into it, thus dulling the cutter. When a fine finish is required, the feed should be reduced rather than the cutter speeded up; more cutters are dulled by high speeds than by high feeds.

Note: To prevent damage to the finished surface (dwell marks), *never* stop the feed when the cutter is revolving over the workpiece. For the same reason, stop the cutter before returning the work to the starting position on completion of the cut.

1. Name two common materials from which end mills are made.

2. State the purpose of 2-, 3-, and multiple-fluted end mills.

3. Describe climb milling, conventional milling.

UNIT 61

End Mills

OBJECTIVES

After completing this unit, you should be able to:

1. Name two types of material of which end mills are made and state their application

2. Describe the purpose of two-flute and multiple-flute end mills

3. Know the purpose of climb and conventional milling

End mills have greatly improved since the days of carbon-steel cutting tools. High-speed steel (HSS) cutting tools are considered the old-timers today, yet they still maintain a very important place in the metal-cutting industry.

The machinist or CNC programmer must select the cutting-tool material with the properties that will provide efficient metal-removal rates for the machining application. All the variables involved such as part shape, machine condition, workpiece material, relative wear resistance of the cutting tool, red hardness, and toughness influence the decision on the type of cutter that should be selected.

▶▶ High-Speed End Mills

Although there is an ever-increasing use of carbides for conventional and CNC metal removal, the traditional high-speed steel (HSS) end mills are still used for many milling operations (Fig. 61-1). They are relatively inexpensive, are easy to get, and do many jobs quite well. The term *high-speed steel* does not suggest much productivity nowadays, particularly when compared with carbide cutters. The relatively low cost of HSS tools and their capability to machine to close tolerances make them a primary choice for many applications. They are probably the single most versatile rotary tools used on conventional and CNC machines.

Although solid-carbide end mills and end mills with replaceable carbon spiral flutes or inserts are used for many different jobs, particularly those requiring high metal-removal rates, the HSS end mill is still a common choice.

Many machining applications call for a harder tool than high-speed steel, but not as hard as carbide. The frequent solution is to use an end mill with additional hardeners—for example, a cobalt end mill. A little more expensive than a high-speed steel tool, but far less expensive than carbide, cobalt-based end mills have longer tool life and can be used the same way as a standard end mill.

▶▶ Carbide End Mills

Competition to produce more quality parts per shift is the driving force behind the ever-increasing use of carbide tools (Fig. 61-2). Carbide has higher hardness, has greater

Figure 61-1 A variety of high-speed steel end mills. *(Niagara Cutter)*

Figure 61-2 A variety of solid carbide end mills. *(Niagara Cutter)*

rigidity, and can withstand higher cutting temperatures, compared with HSS tool materials. These physical properties allow carbide tools to run at higher speeds and feeds, increasing production rates while providing long tool life. Carbide is a high-performance tool material, that can be used for a wide range of machining applications to reduce metal-removal costs.

▶▶ Machining Operations

Some of the most common machining operations (Fig. 61-3) that can be performed with HSS, cobalt, solid carbide, or an indexable insert type end mill are:

> Peripheral end milling

> Milling of slots and keyways

> Channel groves, face grooves, and recesses

> Open and closed pockets

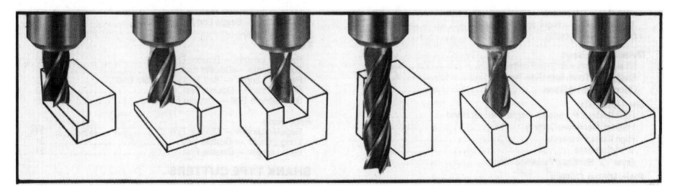

Figure 61-3 Common operators performed on a vertical milling machine. *(Niagara Cutter)*

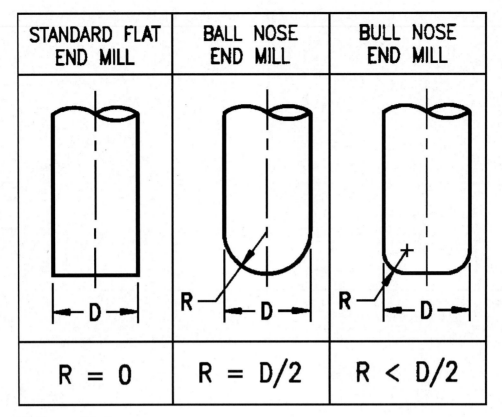

STANDARD FLAT END MILL	BALL NOSE END MILL	BULL NOSE END MILL
R = 0	R = D/2	R < D/2

■ **Figure 61-4** Three common shapes ground on end mills. *(Niagara Cutter)*

> Facing operations for small areas
> Counterboring and spotfacing
> Chamfering

END MILL FORMS

End mills can be ground into required shapes, mainly a flat bottom end mill (the most common type), an end mill with a full radius (often called a ball nose end mill), and an end mill with a corner radius (often called a bull nose end mill).

Each type of end mill is used for a specific type of machining operation. Standard flat end mills are used for all operations requiring a flat bottom and a sharp corner between the wall and the bottom. A ball nose end mill is used for 3D machining on various surfaces, and a bull nose end mill is used for either 3D work, or for flat surfaces that require a corner radius between the wall and the bottom. Other shapes are also popular in some industries—for example, a taper ball nose end mill. Fig. 61-4 shows the three common types and the relationship of the radius to the tool diameter.

TYPES OF END MILLS

There are many types, sizes, and shapes of end milling cutters being used for a wide variety of applications. Fig. 61-5 shows a few of the more common types.

■ **Figure 61-5** End mills are available with two or more flutes and in different lengths and diameters. *(Niagara Cutter)*

Two-Flute End Mills

The two spiral flute design end mills have large, open flutes that provide excellent chip flow. They are recommended for general-purpose milling because they provide good chip clearance and fast metal-removal rates, where chip removal could be a problem. Always select the shortest end mill possible for the job to obtain maximum tool rigidity. This will help to reduce chatter, tool deflection, and tool breakage. Two-flute end mills can have different length lips on the end that allows them to mill slots, keyways, plunge cut, and drill shallow holes.

Three-Flute End Mills

Three-flute end mills with end teeth are used to plunge into the workpiece. These cutters can be used to mill slots, pockets, and keyways and to drill into a workpiece as a start for a milling operation. The three-flute end mill design provides the advantages of both the two-flute and the multiple-flute end mills. They have more teeth than a two-flute end mill to minimize chatter and more space between the flutes than a multiple-flute end mill, providing better chip removal.

Multiple-Flute End Mills

Multiple-flute end mills are those with four or more flutes. The multiple-flute design produces a fine finish after a roughing cut has been taken. Center-cutting end teeth allow these end mills to drill into the workpiece to start a machining operation. These center-cutting end mills are recommended for pocketing, tracer milling, cam milling, die sinking, and slotting. Since most of these end mills are not end cutting, they require a starting hole before a slot can be milled in the center of a workpiece.

Roughing End Mills

Roughing end mills (Fig. 61-6) are designed to provide the best performance while machining a broad range of materials. Their design breaks chips into small pieces for fast metal removal, which reduces heat, friction, and the horsepower required to cut the metal. The chip-breaking action of the roughing end mill allows deeper cuts at faster feed rates.

DIRECTION OF CUT

The two methods of end milling are climb, or down, milling and conventional, or up, milling. When a milling operation is performed or a CNC program is being written and executed for a right-hand cutter, the spindle rotates clockwise (M03). In climb milling, cutter-radius compensation is to the left of the workpiece (G41). The opposite, G42 compensation, to the right of the workpiece, will result in conventional milling. Fig. 61-7 on p. 478 illustrates the two milling modes.

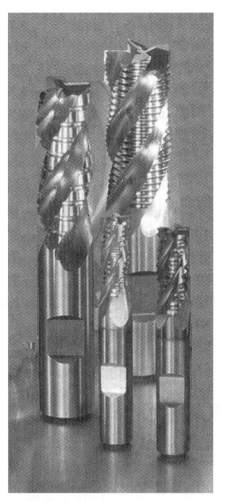

■ **Figure 61-6** Roughing end mills are used for rapid metal removal. *(Niagara Cutter)*

Climb milling is when the cutter rotation and the table (work) feed are going in the same direction (Fig. 61-7). When vertical milling, the cutter has a tendency to pull the workpiece into the cutting flutes. During horizontal milling, the rotation of the cutter has the tendency to push the workpiece against the table or the fixture. Maximum thickness of the chip occurs at the beginning of the cut and as it exits the part the chip is very thin. The practical result is that the chip absorbs most of the heat generated, and hardening of the workpiece is largely prevented. When climb milling, the workpiece should be held securely and the machining table should be equipped with a backlash eliminator.

Conventional milling is when the cutter rotation and the table (work) feed are moving in opposite directions (Fig. 61-7). When horizontal milling, the rotation of the cutter is against the feed direction; this has the tendency to pull or lift the workpiece up from the table or the fixture; therefore, it is important that the work be held securely.

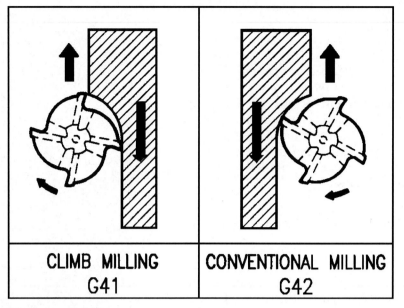

CLIMB MILLING G41	CONVENTIONAL MILLING G42

■ **Figure 61-7** The difference between climb and conventional milling. *(Niagara Cutter)*

▶▶ Milling Cutter Failure

The problem of obtaining maximum efficiency in the use of milling cutters is one that every user faces. The solution to this problem lies in constant awareness of factors that contribute to tool failure or poor performance, other than the style or type of end mill or cutter used.

EXCESSIVE HEAT

Excessive heat (Fig. 61-8) is one of main causes of total cutting edge failure or shortened tool life. Heat, which is present in all milling operations, is caused by cutting edges rubbing on the workpiece and by chips sliding along the faces of teeth when they are separated from the part being milled. When this heat becomes excessive, the hardness of the cutting edges is affected, resulting in less resistance to wear and subsequent rapid dulling.

The heat problem is one of an ever-expanding cycle. As tool-wear lands increase, cutting temperatures rise because of increased friction. As cutting edges dull, more force is needed to part chips and the sliding pressure of chips on the faces of teeth increases, resulting in still more heat. As more heat is generated, the coolant, which is used to reduce temperature, becomes less efficient, temperatures rise higher, the hardness and strength of the cutting edges are further affected, and the cycle is repeated. Heat cannot be totally eliminated but it can be minimized by using properly designed and sharpened tools, by operating at the speeds and feeds recommended for the workpiece material, and by efficiently applying a suitable coolant.

■ **Figure 61-8** Excessive heat shortens cutter life. *(The Weldon Tool Company)*

ABRASION

Abrasion (Fig. 61-9) is a wearing-away action caused by the metallurgy of the workpiece. It dulls cutting edges and causes "wear lands" to develop around the periphery of a cutter. As dulling increases and wear lands grow wider, friction increases and greater force is required to keep the cutter cutting. The rapid rise in friction, heat, and rota-

Figure 61-9 Abrasion dulls cutting edges. *(The Weldon Tool Company)*

Figure 61-10 Too heavy a load on a cutting edge will cause it to chip. *(The Weldon Tool Company)*

tional forces resulting from the abrasive nature of workpiece material can reach a point where the cutter ceases to function effectively or is totally destroyed. Since heat and abrasion are related, the suggestions for minimizing heat also apply to abrasion. Because some materials are much more abrasive than others, for good tool life it is extremely important to follow recommendations specifically for correct CS and feeds.

CHIPPING OR CRUMBLING OF CUTTING EDGES

When cutting forces impose a greater load on cutting edges than their strength can withstand, small fractures occur and small areas of the cutting edges chip out (Fig. 61-10). The material that is left uncut by these chipped-out portions imposes a still greater cutting load on the following cutter tooth, aggravating the problem. This condition is progressive and, once started, will lead to total cutter failure, since chipped edges are dull edges that increase friction, heat, and horsepower requirements.

The major causes of chipping and fracturing of cutting edges are:

> Excessive feed per tooth (FPT)

> Poor cutter design

> Brittleness of the end mill due to improper heat treatment

> Running cutters backward

> Chattering due to a nonrigid condition in the fixture, workpiece, or machine

Figure 61-11 Clogging reduces chip space and can cause a cutter to break. *(The Weldon Tool Company)*

> Inefficient chip washout, which permits chips to be recut or compressed between workpiece surfaces and cutting edges

> Built-up edge break-away

CLOGGING

Some workpiece materials have a "gummy" composition that causes chips to be long, stringy, and compressible (Fig. 61-11). Chips from still other materials may have a tendency to cold-weld or gall to cutting edges and/or the faces of teeth. During the milling of these materials, the chips often clog or jam into the flute area, resulting in

End Mills **479**

■ Figure 61-12 Built-up edges result in poor cutting action. *(The Weldon Tool Company)*

■ Figure 61-13 Work hardening of the workpiece can cause cutter failure. *(The Weldon Tool Company)*

cutter breakage. This condition can be minimized by reducing the depth or width of cut; reducing the FPT; using tools with fewer teeth, creating more chip space; and more effectively applying a coolant under pressure that lubricates the tooth face and is directed so that it flushes out the flute area, keeping it free from chips. It might be necessary at times to use two coolant nozzles to accomplish this.

BUILT-UP EDGES

Built-up edges (Fig. 61-12) occur when particles of the material being machined cold-weld, gall, or otherwise adhere to the faces of teeth adjacent to the cutting edges. This process will continue until the "build-up" itself functions as the cutting edge. When this happens, more power is required and a poor surface finish is usually produced on the workpiece. Periodically, the built-up material will break away from the tooth face and go with either the chip or the workpiece. This intermittent break-away often takes with it a portion of the cutting edge, causing a complete tool breakdown with all of its attendant problems. This condition is quite prevalent on a tool such as a milling cutter because, as each tooth engages the cut in an intermittent manner, the possibility of built-up material break-away and edge chipping is increased.

The built-up edge problem can often be moderated by reducing feed and/or depth of cut. However, the most effective solution to the problem is usually found in a forceful application of a good cutting fluid that gets into the area where the chip is being formed. Optimum results are gained when the coolant coats the cutting edges with a thin fluid or oxide layer that forms a cushion between the chip and the tool to prevent a built-up edge from forming.

WORK HARDENING OF THE WORKPIECE

Work hardening of the workpiece (Fig. 61-13) can cause milling cutter failure. This condition, sometimes called strain hardening, cold working, or glazing, is the result of the action of cutting edges deforming or compressing the surface of the workpiece, causing a change in the work material structure that increases its hardness. Usually, this increase in hardness is evidenced by a smooth, highly glazed surface that resists the wetting action of coolants and offers extreme resistance to the penetration of cutting edges. Fortunately, not all materials are subject to work hardening, but this condition can occur during machining most of the high-temperature and high-strength superalloys, all of the austenitic steels, and many of the highly alloyed carbon tool steels.

Since it is basically the rubbing action of cutting edges that causes work hardening on the surface of these materials, it is extremely important to use sharp tools operating at generous power feeds to keep rubbing contact between tool and work at a minimum. The use of proper CS is important, and climb cutting, with a generous application of activated cutting oil, is highly recommended. Avoid the use of dull milling cutters and light finishing cuts, and never permit a tool to dwell or rotate in contact with the workpiece without feeding.

Machines and work-holding fixtures should be massive and rigid and tool overhang as short as possible to keep deflection during cut at a minimum. Should a surface to be milled be already glazed or work hardened from some previous cutting operation, it is very beneficial to break up the glaze by vapor honing or abrading the surface with coarse emery cloth. This will reduce the slipperiness of the surface and make it easier for tool cutting edges to bite in. Also, breaking the glaze on the surface will permit the coolant to give better wetting action, which is very helpful in prolonging tool life.

CRATERING

Cratering (Fig. 61-14) is caused by chips sliding on the tooth face adjacent to the cutting edge. This is the area of high heat and extreme abrasion due to high chip pressures. The sliding and curling of the chips erodes a narrow hollow or a groove into the tooth face. Once this cratering starts, it can get progressively worse until it results in total tool failure. This problem can be minimized by efficiently applying a coolant that provides a high-pressure fluid film or a chemical oxide film on the tool to prevent metal-to-metal contact between chips and tooth face. Also beneficial are properly applied tool surface treatments that impart a high-abrasion-resisting superficial hardness to faces of the teeth.

■ **Figure 61-14** The use of cutting fluid can reduce cratering. *(The Weldon Tool Company)*

unit 61 review questions

1. Name three factors that affect the efficiency of a milling operation.

Cutting Speed

2. List six factors to be considered when selecting the proper r/min for milling.

3. At what maximum speed should a 3.50-in. cemented-carbide milling cutter revolve to machine cast iron?

4. At what r/min should a 115-mm, high-speed steel cutter revolve to machine a piece of tool steel?

5. What rules should be observed to obtain best results when using milling cutters?

Feed

6. Name three factors that determine milling feed.

7. Define chip per tooth.

8. Calculate the feed (in./min) for an eight-tooth, 3-in. diameter cemented-carbide face milling cutter to cut cast iron.

9. Calculate the feed (m/min) for a 12-tooth 90-mm diameter, high-speed steel helical milling cutter to machine aluminum.

10. List six advantages of climb milling.

Depth of Cut

11. What is considered to be a proper roughing cut?

12. Why is a light cut not desirable as a finish cut?

Milling Cutter Failure

13. List seven main causes of milling cutter failure and state how each can be minimized.

UNIT 62

Vertical Mill Operations

OBJECTIVES

After completing this unit, you should be able to:

1 Align the vertical head and vise to within ±.001 in. (0.02 mm)

2 Insert and remove end mills from spring collets

3 Accurately machine a block square and parallel

4 Drill holes to an accurate location

The vertical milling machine is suitable for performing a wide variety of machining operations because of its versatility and ease of setup. Some common operations performed on this machine include end milling, face milling, keyway and dovetail cutting, T-slot and circular slot cutting, gear cutting, drilling, boring, and reaming. Since most of the cutting tools used on this machine are relatively small, their cost is usually not too high.

▶▶ Aligning the Vertical Head

Proper alignment of the head is of the utmost importance when machining holes or when face milling. If the head is not at an angle of 90° to the table, the holes will not be square with the work surface when the cutting tool is fed by hand or automatically. When face milling, the machined surface will be stepped if the head is not square with the table. Although all heads are graduated in degrees and some have vernier devices used for setting the head, it is wise to check the spindle alignment as follows.

■ **Figure 62-1** An indicator assembly fastened to a rod in a chuck. *(Kelmar Associates)*

procedure

1. Mount a dial indicator, with a large contact button, on a suitable rod, bent at 90° and held in the spindle (Fig. 62-1).

2. Position the indicator over the front *Y* axis of the table.

   Note: If a flat ground plate is available, place it on the clean table and position the indicator near the outer edge.

3. Carefully lower the spindle until the indicator button touches the table and the dial indicator registers no more than one-quarter revolution; then set the bezel to zero (0) (Fig. 62-2). Lock the spindle in place.

Figure 62-2 Setting the indicator to 0 on the right-hand side of the table. *(Kelmar Associates)*

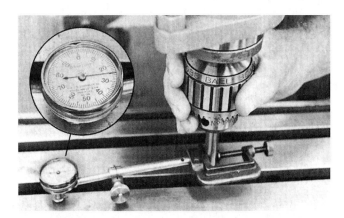

Figure 62-3 The spindle rotated 180°. *(Kelmar Associates)*

4. Carefully rotate the vertical mill spindle 180° by hand until the button bears on the opposite side of the table (Fig. 62-3). Compare the readings.

5. If there are any differences, loosen the locking nuts on the swivel mounting and adjust the head until the indicator registers approximately one-half the difference between the two readings. Tighten the locking nuts.

6. Recheck the accuracy of the alignment; it will probably be necessary to readjust the head because its pivot point is not in the center of the machine spindle in the *Y* axis.

7. Rotate the vertical mill spindle 90° (*X* axis) and set the dial indicator as in step 3.

8. Rotate the machine spindle 180°, and check the reading at the other end of the table.

9. If the two readings do not coincide, repeat step 5 until the readings are the same.

10. Tighten the locking nuts on the swivel mount.

11. Recheck the readings and adjust if necessary.

Note: When readings are taken, it is important that the indicator button does not catch in the T-slots of the table. To prevent this, it is advisable to work from the high reading first and then rotate to the low reading. It should be apparent that, the longer the rod used, the more accurate the setting will be.

▸▸ Aligning the Vise

When the vise is aligned on a vertical milling machine, the dial indicator may be attached to the quill or the head by any convenient means, such as clamps (Fig. 62-4) or a magnetic base. The same method of alignment should be followed as is outlined in Unit 63 for aligning the vise on a horizontal milling machine.

Figure 62-4 Indicator assembly for checking vise alignment.

▸▸ Mounting and Removing Cutters

A variety of cutting tools and accessories (such as end mills, shell mills, T-slot cutters, fly cutters, drill chucks, boring heads and tools, etc.) can be inserted into a vertical mill spindle to allow this machine to perform a wide range of operations. End mills, cutting tools, and accessories are generally held in the machine spindle by either a *spring collet* or a *solid collet or adapter.*

The *spring collet* (Fig. 62-5a on p. 484) is pulled into the spindle by a draw-bar that closes on the cutter shank and drives it by means of friction between the collet and cutter. On large-diameter cutters, it is important to tighten the draw-bar securely to prevent the cutter from moving up or down during cutting.

The *solid collets* (Fig. 62-5b), pulled up into the machine spindle by a draw-bar, are more rigid than spring

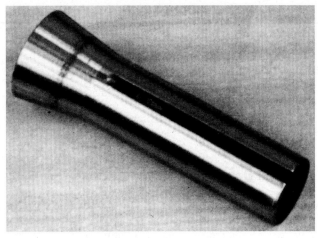

(a)

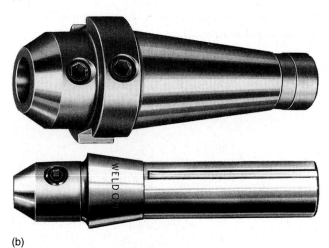

(b)

■ **Figure 62-5** Types of vertical milling machine collets: (a) spring *(Kelmar Associates)*; (b) solid *(Weldon Tool Co.)*

■ **Figure 62-6** The spring collet is held in the vertical mill spindle by a draw-bar. *(Kelmar Associates)*

■ **Figure 62-7** Tighten the draw-bar *clockwise* to secure the collet and cutter in the spindle. *(Kelmar Associates)*

collets and hold the cutting tools securely. They may be driven by a key in the spindle and keyway in the collet or by two drive keys on the spindle. The cutter is driven and prevented from turning by one or two setscrews in the collet that bear against the flats of the cutter shank and prevent any end movement.

TO MOUNT A CUTTER IN A SPRING COLLET

1. Shut off the electric power to the machine.
2. Place the proper cutter, collet, and wrench on a piece of masonite, wood, or soft plastic on the machine table (Fig. 62-6).
3. Clean the taper in the machine spindle.
4. Place the draw-bar into the hole in the top of the spindle.

5. Clean the taper and keyway on the collet.
6. Insert the collet into the bottom of the spindle, press up, and turn it until the keyway aligns with the key in the spindle.
7. Hold the collet up with one hand and, with the other, thread the draw-bar clockwise into the collet for about four turns.
8. Hold the cutting tool with a cloth and insert it into the collet for the full length of the shank.
9. Tighten the draw-bar into the collet (clockwise) by hand.
10. Hold the spindle brake lever and tighten the draw-bar as tightly as possible with a wrench, using hand pressure only (Fig. 62-7).

TO REMOVE A CUTTER FROM A COLLET

The operation for removing cutting tools is similar to mounting, but in the reverse order:

1. Shut off the electric power to the machine.
2. Place a piece of masonite, wood, or soft plastic on the machine table to hold the necessary tools.
3. Pull on the spindle brake lever to lock the spindle, and loosen the draw-bar with a wrench (counter-clockwise).
4. Loosen the draw-bar, by hand, only about three full turns.

Note: Do not unscrew the draw-bar from the collet.

5. Hold the cutter with a cloth.
6. With a soft-faced hammer, strike down sharply on the head of the draw-bar to break the taper contact between the collet and spindle.
7. Remove the cutter from the collet.
8. Clean the cutter and replace it in its proper storage place where it will not be damaged by other tools.

TO MOUNT A CUTTER IN A SOLID COLLET

1. Shut off the electric power to the machine.
2. Place the cutter, collet, and necessary tools on a piece of masonite, wood, or soft plastic on the machine table.
3. Slide the draw-bar through the top hole in the spindle.
4. Clean the spindle taper and the taper on the collet.
5. Align the keyway or slots of the collet with the keyway or drive keys in the spindle, and insert the collet into the spindle.
6. Hold the collet up in the spindle and thread the draw-bar clockwise with the other hand.
7. Pull on the brake lever and tighten the draw-bar as tightly as possible with a wrench, using hand pressure only.
8. Insert the end mill into the collet until the flat(s) align with the setscrew(s) of the collet.
9. Tighten the setscrews securely using hand pressure only.

■ **Figure 62-8** Tap the workpiece down with a soft-faced hammer until all paper feelers are tight. *(Kelmar Associates)*

TO MACHINE A FLAT SURFACE

1. Clean the vise and mount the work securely in the vise, on parallels if necessary (Fig. 62-8).
2. Check that the vertical head is square with the table.
3. If possible, select a cutter that will just overlap the edges of the work. This will then require only one cut to be taken to machine the surface. If the surface to be machined is fairly narrow, an end mill slightly larger in diameter than the width of the work should be used. If the surface is large and requires several passes, a shell end mill or a suitable fly cutter should be used.

Note: It is not advisable to use a facing cutter that is too wide, since the head may be thrown out of alignment if the cutter should jam.

4. Set the proper spindle speed for the size and type of cutter and the material being machined; check cutter rotation.
5. Tighten the quill clamps.
6. Start the machine, and adjust the table until the end of the work is under the edge of the cutter.
7. Raise the table until the work surface just touches the cutter. Move the work clear of the cutter.
8. Raise the table about .030 in. (0.8 mm) and take a trial cut for approximately .250 in. (6 mm).
9. Move the work clear of the cutter, *stop the cutter,* and measure the work (Fig. 62-9 on p. 486).
10. Raise the table the desired amount, and lock the knee clamp.
11. Mill the surface to size using the automatic feed, if the machine is so equipped.

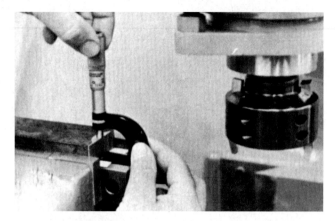

Figure 62-9 Measuring a workpiece after a light trial cut. *(Kelmar Associates)*

▶▶ Machining a Block Square and Parallel

In order to mill the four sides of a piece of work so that its sides are square and parallel, it is important that each side be machined in a definite order. It is important that dirt and burrs be removed from the work, vise, and parallels, since they can cause inaccurate work.

MACHINING SIDE 1

1. Clean the vise thoroughly and remove all burrs from the workpiece.
2. Set the work on parallels in the center of the vise with the largest surface (side 1) facing up (Fig. 62-10).
3. Place short paper feelers under each corner between the parallels and the work.

Figure 62-10 Side 1, or the surface with the largest area, should be facing up. *(Kelmar Associates)*

4. Tighten the vise securely.
5. With a soft-faced hammer, tap the workpiece down until all paper feelers are tight.
6. Mount a fly cutter in the milling machine spindle.
7. Set the machine for the proper speed for the size of the cutter and the material to be machined.
8. Start the machine and raise the table until the cutter just touches near the right-hand end of side 1.
9. Move work clear of the cutter.
10. Raise the table about .030 in. (0.76 mm) and machine side 1 using a steady feed rate.
11. Take the work out of the vise and remove all burrs from the edges with a file.

MACHINING SIDE 2

12. Clean the vise, work, and parallels thoroughly.
13. Place the work on parallels, if necessary, with side 1 against the solid jaw and side 2 up (Fig. 62-11).
14. Place short paper feelers under each corner between the parallels and the work.
15. Place a round bar between side 4 and the movable jaw.

Note: The round bar must in the *center* of the amount of work held inside the vise jaws.

16. Tighten the vise securely and tap the work down until the paper feelers are tight.
17. Follow steps 8 to 11 and machine side 2.

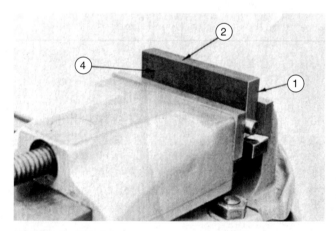

Figure 62-11 The finished side (side 1) should be placed against the cleaned solid jaw for machining side 2. *(Kelmar Associates)*

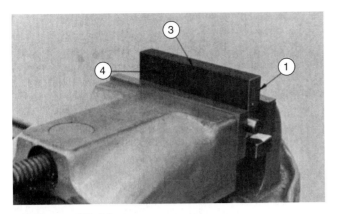

■ Figure 62-12 Place side 1 against the solid jaw and side 2 down for machining side 3. *(Kelmar Associates)*

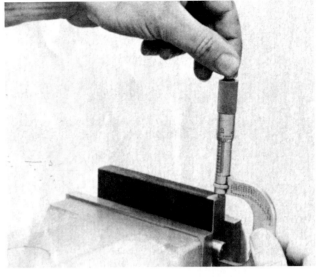

■ Figure 62-13 Measuring the trial cut on side 3. *(Kelmar Associates)*

MACHINING SIDE 3

18. Clean the vise, work, and parallels thoroughly.

19. Place side 1 against the solid vise jaw, with side 2 resting on parallels if necessary (Fig. 62-12).

20. Push the parallel to the left so that the right edge of the work extends about .250 in. (6 mm) beyond the parallel.

21. Place short paper feelers under each end or corner between the parallels and the work.

22. Place a round bar between side 4 and the movable jaw, making sure that the round bar is in the center of the amount of work held *inside the vise*.

23. Tighten the vise securely and tap the work down until the paper feelers are tight.

24. Start the machine and raise the table until the cutter just touches near the right-hand end of side 3.

25. Move the work clear of the cutter and raise the table about .010 in. (0.25 mm).

26. Take a trial cut about .250 in. (6 mm) long, stop the machine, and measure the width of the work (Fig. 62-13).

27. Raise the table the required amount and machine side 3 to correct width.

28. Remove the work and file off all burrs.

MACHINING SIDE 4

29. Clean the vise, work, and parallels thoroughly.

30. Place side 1 down on the parallels with side 4 up and tighten the vise securely (Fig. 62-14).

■ Figure 62-14 The workpiece is set up properly for machining side 4. Note the paper feelers between the parallels and the workpiece. *(Kelmar Associates)*

Note: With three finished surfaces, the round bar is not required when machining side 4.

31. Place short paper feelers under each corner between the parallels and work (Fig. 62-14).

32. Tighten the vise securely.

33. Tap the work down until the paper feelers are tight.

34. Follow steps 24 to 27 and machine side 4 to the correct thickness.

■ **Figure 62-15** Setting the workpiece edge square for machining its ends. *(Kelmar Associates)*

■ **Figure 62-16** One end machined square with the side of the workpiece. *(Kelmar Associates)*

▸▸ Machining the Ends Square

Two common methods are used to square the ends of workpieces in a vertical mill. Short pieces are generally held vertically in the vise and are machined with an end mill or a fly cutter (Fig. 62-15). Long pieces are generally held flat in the vise, with one end extending past the end of the vise. The end surface is cut square with the body of an end mill.

PROCEDURE FOR SHORT WORK

1. Set the work in the center of the vise with one of the ends up and tighten the vise lightly.

2. Hold a square down firmly on top of the vise jaws and bring the blade into light contact with the side of the work.

3. Tap the work until its edge is aligned with the blade of the square (Fig. 62-15).

4. Tighten the vise securely and recheck the squareness of the side.

5. Take about a .030-in. deep (0.76-mm) cut and machine the end square (Fig. 62-16).

6. Remove the burrs from the end of the machined surface.

7. Clean the vise and set the machined end on paper feelers in the bottom of the vise.

8. Tighten the vise securely and tap the work down until the paper feelers are tight.

9. Take a trial cut from the end until the surface cleans up.

■ **Figure 62-17** Measuring the length of the workpiece with a depth micrometer. *(Kelmar Associates)*

10. Measure the length of the workpiece with a depth micrometer (Fig. 62-17).

11. Raise the table the required amount and machine the work to length.

TO MACHINE AN ANGULAR SURFACE

1. Lay out the angular surface.

2. Clean the vise.

3. Align the vise with the direction of the feed. *This is of the utmost importance.*

4. Mount the work on parallels in the vise.

Figure 62-18 Head swiveled to machine an angle. *(Kelmar Associates)*

Figure 62-19 Machining an angle by adjusting the workpiece. *(Kelmar Associates)*

5. Swivel the vertical head to the required angle (Fig. 62-18).

6. Tighten the quill clamp.

7. Start the machine and raise the table until the cutter touches the work. Carefully raise the table until the cut is of the desired depth.

8. Take a trial cut for about .50 in. (13 mm).

9. Check the angle with a protractor.

10. If the angle is correct, continue the cut.

Note: It is always advisable to feed the work into the rotation of the cutter, rather than with the rotation of the cutter, which may draw the work into the cutter and cause damage to the work, the cutter, or both.

11. Machine to the required depth, taking several cuts if necessary.

Alternate Method

Angles may sometimes be milled by leaving the head in a vertical position and setting the work on an angle in the vise (Fig. 62-19). This will depend on the shape and size of the workpiece.

procedure

1. Check that the vertical head is square with the table.

2. Clean the vise.

3. Lock the quill clamp.

4. Set the workpiece in the vise with the layout line parallel to the top of the vise jaws and about .250 in. (6 mm) above them.

5. Adjust the work under the cutter so that the cut will start at the narrow side of the taper and progress into the thicker metal.

6. Take successive cuts of about .125 to .150 in. (3 to 4 mm), or until the cut is about .030 in. (0.8 mm) above the layout line.

7. Check to see that the cut and the layout line are parallel.

8. Raise the table until the cutter just touches the layout line.

9. Clamp the knee at this setting.

10. Take the finishing cut.

Milling Hardened Steel with PCBN Tools

Polycrystalline cubic boron nitride (PCBN) blank tools and inserts are widely used in general machining and tool and die work, and in the automotive and aerospace industries because of their consistent part accuracy, cool-cutting features, and high productivity. PCBN tools can be used on abrasive and difficult-to-cut (DTC) metals such as alloy and tool steel hardened above 45 Rc. In many cases, steels for die punches can be hardened and turned to size, eliminating the warpage of heat treating and the need for grinding operations.

Before using a PCBN or any other type of superabrasive cutting tool, it is important to consider the following factors to ensure successful superabrasive machining:

> Select a milling machine that is in good working order.
> Know the machine's capabilities.
> Select the proper toolholder and tool grade, and follow tool preparation procedures.
> Calculate proper speed, feed, and depth of cut.
> Follow the recommended setup and machining sequence.

PREPARATION FOR PCBN MILLING

The *Interactive Superabrasive Machining Advisor* software program supplied by GE Superabrasives was used for the selection of the PCBN tool and the machining conditions. In order to use this software effectively, a person must know the material hardness, machine horsepower, maximum spindle speed, maximum table feed, depth of cut, and surface finish required. The computer software gives recommendations for the application, such as tool grade, tool edge preparation, tool geometry, surface speed and r/min, feed per tooth and in./min, and type of coolant.

MACHINING THE PART

A piece of alloy steel hardened to 53-55 Rc was used for this exercise. The machine was a standard vertical milling machine with a 2 HP spindle. A 2.0-in. (50-mm) diameter facemill with an R-8 shank, containing a single PCBN insert with a geometry of CNMA 433 was used. The insert had an edge chamfer of 15° × .008 in. (0.2 mm), with a .001 in. (0.025 mm) hone.

Note: *Edge preparations should be made only by fully qualified, skilled toolmakers or by the tool manufacturer.*

The insert grade was a BZN®* 6000 used without coolant. A depth of cut of .020 in. (0.05 mm) at 500 sf/min (150 m/min) and a feed rate of .006 in. (0.15 mm) was used.

MACHINE SETUP

1. Set the machine spindle to the recommended r/min. Check the speed with a portable tachometer to see that it is correct.

2. Set the table feed to the recommended feed in in./min.

3. Fasten a single PCBN insert into the proper toolholder. If the holder has additional pockets, use dull carbide inserts to fill the pockets to provide balance for the cutter.

Note: The tips must be ground an amount greater than the cutting depth so that they do not cut.

4. Fasten the cutter into the machine spindle, and lock the quill in the UP position to ensure rigidity.

5. Tighten all table locks except the table machining axis.

6. Use a .010-in. (0.25-mm) plastic shim to set the tool to the top of the part surface. The plastic shim will prevent chipping the cutting edge of the insert.

7. Move the table so that the cutter clears the part, and raise the knee the thickness of the shim plus the desired depth of cut.

8. Feed the table in the direction shown in Fig. 62-20 and machine the part surface.

9. *Do not* traverse back across the part until the knee is lowered or the table moved off the part to prevent tool chipping.

10. Reposition and take additional passes as required until the part is machined to size.

Producing and Finishing Holes

Drilling, reaming, boring, and tapping are some of the more common operations that can be performed accurately on a vertical mill. The wide variety of tools that can be inserted and driven by the spindle allow these and many other types of operations to be performed. Not only can the workpiece be held securely, but the location of holes, slots, etc. can be located accurately by using the mi-

*®Trademark Diamond Innovations Inc.

■ Figure 62-20 Milling hardened steel with a PCBN fly cutter. *(Diamond Innovations Inc.)*

■ Figure 62-21 Center drilling the hole location. *(Kelmar Associates)*

crometer collars on the table feed screws. Holes can be drilled at 90° to the work surface, or at any angle to which the head can be swiveled.

TO DRILL HOLES AT 90° TO THE WORK SURFACE

Note: Before proceeding with this operation, be sure that the vertical head is aligned at 90° to the table.

1. Mount the work in a vise, or clamp it to the table. Work must be supported by parallels that are positioned so that they will not interfere with the drill.
2. Mount a drill chuck in the spindle.
3. Mount a center finder in the drill chuck.
4. Adjust the table until the center-punch mark of the hole to be drilled is in line with the tip of the rotating center finder.

OR

Locate the center of the machine spindle on two adjacent edges of the workpiece and move the table to the hole location using the table feed screw micrometer collars.

5. Tighten the table and saddle clamps.
6. Stop the machine and remove the center finder.
7. Mount a center drill in the drill chuck, having it extend only .50 in. to prevent breakage.
8. Set the spindle speed to approximately 1000 r/min, and feed the center drill with the quill handfeed lever until about half of the angular portion enters the workpiece (Fig. 62-21).
9. Mount the drill size required in the drill chuck.
10. Set the depth stop so that the drill just clears the bottom of the workpiece to prevent drilling into the machine table.
11. Set the machine to the proper speed and feed if the automatic feed is to be used for drilling the hole.
12. Apply cutting fluid and use the quill handfeed lever, or the automatic feed, to drill the hole (Fig. 62-22 on p. 492).

TO DRILL ANGULAR HOLES

Angular holes may be drilled on a vertical mill by swiveling the head to the required angle and feeding the drill into the workpiece by using the quill handfeed lever.

1. Mount the work in a vise, or clamp it to the machine table.
2. Loosen the lock nuts and swivel the milling head to the required angle of the hole to be drilled.
3. Check the angle of the head by the graduations on the housing, or with a bevel or vernier protractor (Fig. 62-23 on p. 492).

Figure 62-22 Using the drill handfeed lever to drill a hole in a vertical mill. *(Kelmar Associates)*

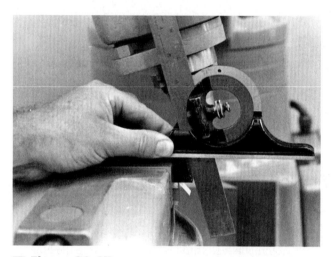

Figure 62-23 Checking the angle of the head for drilling holes on an angle. *(Kelmar Associates)*

4. Tighten the lock nuts and recheck the accuracy of the setting.

5. Mount a drill chuck in the spindle.

6. Mount a center finder in the drill chuck.

7. Locate the center of the spindle as close as possible to where the hole is to be drilled.

8. Lower the table so that there is enough room between the spindle in the UP position and the top of the work to mount the longest drill required for the drilling operation.

9. Lock the knee of the machine in this position and *do not move it;* otherwise, the hole location will be lost.

10. Adjust the table until the center-punch mark of the hole to be drilled is in line with the tip of the rotating center finder.

OR

Locate the center of the machine spindle on two adjacent edges of the workpiece and move the table to the hole location using the table feed screw micrometer collars.

11. Tighten the table and saddle clamps.

12. Stop the machine, raise the spindle to the top position with the quill handfeed lever, and remove the center finder.

13. Insert a large center drill or spotting tool into the drill chuck.

14. Set the spindle speed for the size of spotting tool to be used.

15. Spot each hole location so that the top of the spotted hole is slightly larger than the size of hole to be drilled.

> This is necessary so that the edge of the drill is not deflected when drilling on an angle.
> To prevent any possible deflection, use an end mill (size of drill diameter) to counterbore the top of each hole until it reaches its full diameter.

16. When more than one angular hole must be drilled, it is wise to record the crossfeed and table micrometer collar locations of each hole.

> This will make it easy to return to each hole location later.

17. Stop the machine and remove the spotting tool.

18. Insert the correct-size drill into the drill chuck.

19. Set the speed and feed for each hole to be drilled.

20. Drill each hole to the required depth.

21. Remove all burrs from the hole edges with a file and scraper.

TO REAM ON A VERTICAL MILL

The purpose of reaming is to bring a drilled or bored hole to size and shape and to produce a good surface finish in the hole. Speed, feed, and reaming allowance are three factors that can affect the accuracy of a reamed hole. Approximately .250 in. (0.4 mm) is left for reaming holes up to .500 in. (12.7 mm) in diameter; .030 in. (0.8 mm) is recommended for holes over .500 in. (12.7 mm) diameter. The speed for reaming is generally about one-half of the drilling speed.

Reamers are usually held in a drill chuck or adapter. Two types of reamers—the rose, or fluted, and the precision end-cutting—are used for bringing a hole to size quickly. The location of a workpiece should not be moved

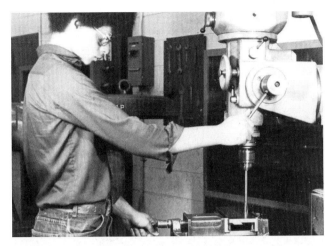

Figure 62-24 Holes should be reamed after drilling but *before* the table location has been changed. *(Kelmar Associates)*

after drilling in order to keep the spindle in line for the reaming operation.

1. Mount the reamer in the spindle or drill chuck.

2. Set the speed and feed for reaming (approximately one-quarter the drilling speed). Too high a speed will quickly dull the reamer.

3. Apply cutting fluid as the reamer is fed steadily into the hole with the downfeed lever (Fig. 62-24).

4. Stop the machine spindle.

5. Remove the reamer from the hole. (Do not turn the reamer backward; otherwise, the cutting edges will be ruined.)

TO BORE ON A VERTICAL MILL

Boring is the operation of enlarging and truing a drilled or cored hole with a single-point cutting tool. Many holes are bored on a milling machine to bring them to accurate size and location. The offset boring chuck is especially useful because it allows accurate settings to be made for removing material from a hole (Fig. 62-25).

1. Align the vertical head square (at 90°) to the table.

2. Set up and align the work parallel to the table travel.

3. Align the center of the milling machine spindle with the reference point or edges of the work.

4. Set the graduated micrometer dials on the crossfeed and table screws to zero.

5. Calculate the coordinate location for the hole to be bored.

6. Move the table so that the center of the spindle aligns with the hole location required.

7. Lock all table clamps to keep the table in this position.

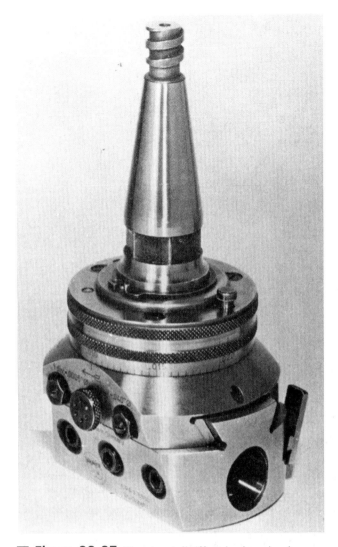

Figure 62-25 The dovetail offset boring chuck allows the cutting tool to be set accurately when boring a hole.

Note: It is important that the table location not be changed until the drilling and boring operations are completed.

8. Spot the hole location with a center drill or spotting tool.

9. Drill holes under .500 in. (12.7 mm) diameter to within .015 in. (0.39 mm) of finish size. Drill holes over .500 in. (12.7 mm) diameter to within .030 in. (0.8 mm) of size.

10. Mount the boring chuck using the largest boring bar or tool possible.

11. Rough bore the hole to within .005 to .007 in. (0.12 to 0.17 mm) of finish size.

12. Finish boring the hole to the required size (Fig. 62-26 on p. 494).

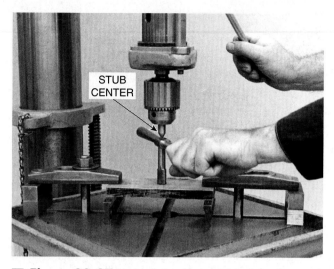

■ **Figure 62-27** A stub center is used to keep the tap square during the tapping operation.
(Cincinnati Milacron, Inc.)

Note: The work or table must not be moved after drilling the hole; otherwise, the alignment will be disturbed and the tap will not enter squarely.

5. Mount a stub center in the drill chuck (Fig. 62-27).

6. Fasten a tap wrench on the correct-size tap and place it into the hole.

7. Lower the spindle using the quill handfeed lever until the stub center point fits into the center hole in the end of the tap shank.

8. Turn the tap wrench clockwise to start the tap into the hole. At the same time, keep the stub center in light contact with the tap by applying pressure on the handfeed lever.

9. Continue to tap the hole while keeping the tap aligned by applying light pressure on the handfeed lever.

A tapping attachment may be mounted in the vertical mill spindle to rotate the tap by power. Special two- or three-fluted gun taps are used for tapping under power because of their ability to clear chips. The speed for tapping under power generally ranges from 60 to 100 r/min.

▶▶ Cutting Slots and Keyseats

Slots and keyseats with one or two blind ends may be cut in shafts more easily on a vertical milling machine, using a two- or three-fluted end mill, than with a horizontal mill and a side facing cutter.

■ **Figure 62-26** Boring a hole with an offset boring chuck to produce a true round hole.   *(Moore Tool Co.)*

TO TAP A HOLE ON THE VERTICAL MILL

Tapping a hole on the vertical mill can be performed either by hand or with the use of a tapping attachment. The advantage of tapping a hole on a vertical mill is that the tap can be started squarely and kept that way throughout the entire length of the hole being threaded.

1. Mount the work in a vise, or clamp it to the machine table. If parallels are used in the work setup, be sure that they clear the hole to be tapped.

2. Mount a center drill in the drill chuck, and adjust the machine table until the center-punch mark on the work aligns with the point of the center drill.

3. With a center drill spot each hole to be tapped to slightly larger than the tap diameter.

4. Drill the hole to the correct tap drill size for the size of tap to be used.

1. Lay out the position of the keyseat on the shaft, and scribe reference lines on the end of the shaft (Fig. 62-28).

2. Secure the workpiece in a vise on a parallel. If the shaft is long, it may be clamped directly to the table by placing it in one of the table slots or in V-blocks.

3. Using the layout lines on the end of the shaft, set up the shaft so that the keyseat layout is in the proper position on top of the shaft.

4. Mount a two- or three-fluted end mill, of a diameter equal to the width of the keyseat, in the milling machine spindle.

   Note: If the keyseat has two blind ends, a two- or three-lip end mill must be used, since they may be used as a drill to start the slot. If the slot is at one end of the shaft (one blind end only), a four-fluted end mill may be used, although a two- or three-lip end mill will give better chip clearance.

5. Center the workpiece by carefully touching the cutter to one side of the shaft (Fig. 62-29). This may also be done by placing a piece of thin paper between the shaft and the cutter.

   Note: Paper may be made to adhere to shafts or work surfaces by wetting it with coolant or oil before applying it to the surface. This eliminates the necessity of holding paper between the cutter and the work, thus making it a safer operation.

6. Lower the table until the cutter clears the workpiece.

7. Move the table over an amount equal to half the diameter of the shaft plus half the diameter of the cutter plus the thickness of the paper (Fig. 62-30). For example, if a .250-in. (6-mm) slot is required in a 2-in. (50-mm) shaft and the thickness of the paper used is .002 in. (0.05 mm), the table would be moved over 1.000 + .125 + .002 = 1.127 in. (25 + 3 + 0.05 = 28.05 mm).

8. If the keyseat being cut has two blind ends, adjust the work until the end of the keyseat is aligned with the edge of the cutter.

9. Feed the cutter down (or the table up) until the cutter *just* cuts to its full diameter. If the keyseat has one blind end only, the work is adjusted so that this cut is taken at the end of the work. The work would now be moved clear of the cutter.

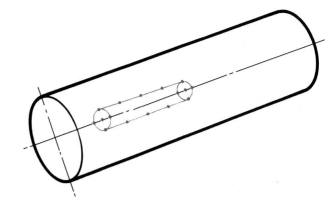

■ **Figure 62-28** Layout of a keyseat on a shaft. *(Kelmar Associates)*

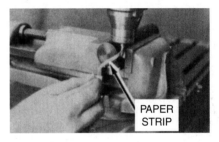

■ **Figure 62-29** Setting a cutter to the side of the workpiece. *(Kelmar Associates)*

PAPER STRIP

■ **Figure 62-30** Cutter centered on the layout. *(Kelmar Associates)*

10. Adjust the depth of cut one-half the thickness of the key, and machine the keyseat to the proper length (Fig. 62-31 on p. 496).

■ Figure 62-31 Keyseat milled to depth and length. *(Kelmar Associates)*

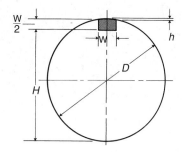

■ Figure 62-32 Keyseat calculations.

CALCULATING THE DEPTH OF KEYSEATS

If the keyseat is at the end of a shaft, the proper depth of the keyseat is checked by measuring diametrically from the bottom of the keyseat to the opposite side of the shaft. This distance may be calculated as follows (Fig. 62-32):

$$H = D - \left(\frac{W}{2} + h\right)$$

$$h = .5(D - \sqrt{D^2 - W^2})$$

H = distance from the bottom of the keyseat to the opposite side of the shaft

D = diameter of the shaft

W = width of the keyseat

h = height of the segment above the width of the keyseat

Calculate the measurement H if the shaft is 2 in. in diameter and a .500-in. is to be used.

$$h = 500(D - \sqrt{D^2 - W^2}$$
$$= .500(2 - \sqrt{4 - .250}$$
$$= .500(50 - \sqrt{3.750}$$
$$= .500(2 - 1.9365)$$
$$= .500(.0635)$$
$$= .0318 \text{in.}$$

$$H = D - \left(\frac{W}{2} + h\right)$$

$$= 2 - \left(\frac{.500}{2} + .0318\right)$$
$$= 2 - (.250 + .0318)$$
$$= 2 - .2818$$
$$= 1.7182$$

WOODRUFF KEYS

Woodruff keys are used when keying shafts and mating parts (Fig. 62-33a). Woodruff keyseats can be cut more quickly than square keyseats, and the key should not require any fitting after the keyseat has been cut. These keys are semicircular in shape and can be purchased in standard sizes, which are designated by E numbers. They can be conveniently made from round bar stock of the required diameter.

Woodruff keyseat cutters (Fig. 62-33b) have shank diameters of ½ in. for cutters up to 1½ in. in diameter. The shank is undercut adjacent to the cutter to permit the cutter to go into the proper depth. The sides of the cutter are slightly tapered toward the center to permit clearance during cutting. Cutters over 2 in. in diameter are mounted on an arbor.

The size of the cutter is stamped on the shank. The last two digits of the number indicate the nominal diameter in eighths of an inch. The digit or digits preceding the last two

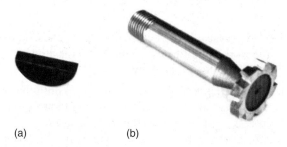

(a)   (b)

■ Figure 62-33 (a) Woodruff key; (b) keyseat cutter. *(Kelmar Associates)* numbers indicate the nominal width of the cutter in thirty-seconds of an inch. Thus, a cutter marked 608 is 8 × ⅛, or 1 in., in diameter and 6 × 1/32, or 3/16 in., wide. The key would be a semicircle cross section to fit the groove exactly.

To Cut a Woodruff Keyseat

1. Align the spindle of the vertical milling machine 90° to the table.

2. Lay out the position of the keyseat.

3. Set the shaft in the vise of the milling machine or on V-blocks. Be sure that the shaft is level (parallel to the table).

4. Mount the cutter of the proper size in the spindle.

5. Start the cutter, and touch the bottom of the cutter to the top of the workpiece. Set the vertical graduated feed collar to zero (0) and check cutter rotation.

6. Move the work clear of the cutter. Raise the table half the diameter of the work plus half the thickness of the cutter. Lock the knee at this setting.

7. Position the center of the slot with the center of the cutter. Lock the table in this position.

8. Touch the revolving cutter to the work. Use a strip of paper between the cutter and the work if desired. Set the crossfeed screw collar to zero (0).

9. Cut the keyseat to the proper depth.

Note: Keyseat proportions may be found in any handbook.

unit 62 review questions

Aligning the Vertical Head and Vise

1. Why is it necessary to align the vertical head square with the table?

2. Describe briefly how the vertical head may be aligned with the table surface.

3. Why is it important that the vise be aligned with the table travel?

4. Describe one method of aligning the vise parallel to the table travel.

Machining a Flat Surface

5. What precautions should be observed before machining a flat surface?

6. What type of cutter should be used to machine a large surface?

7. How can the edge of a flat piece be machined square with a flat finished surface?

Machining a Block Square and Parallel

8. How should the work be set in a vise to machine side 1?

9. How should the work be set in a vise to machine side 2?

10. Draw a sketch of the work setup required for machining side 3 and side 4.

Machining Angular Surfaces

11. Describe briefly two methods of machining angular surfaces.

Producing and Finishing Holes

12. What precaution should be taken before drilling holes on a vertical mill?

13. How deep should a spotting hole be drilled?

14. How can the setting of the vertical head be checked when drilling angular holes?

15. How much material should be left for reaming:
    a. Holes up to .500 in. (12.7 mm) in diameter?
    b. Holes over .500 in. (12.7 mm) in diameter?

16. List three important points to observe when preparing a machine for boring.

17. How can the tap be kept in alignment when tapping on a vertical mill?

Cutting Slots and Keyways

18. What type of cutter should be used to cut slots or keyseats having one or two blind ends?

19. How can round work be held for machining slots and keyseats?

20. Explain one method of aligning the end mill with the center of a shaft.

21. How may the depth of a keyseat be measured?

22. Calculate the measurement for a ½-in. wide keyseat from the bottom of a 2-in. diameter shaft to the bottom of the keyseat (½ × ½ key).

UNIT 63

Special Milling Operations

OBJECTIVES

After completing this unit, you will be able to:

1 Set up and use the rotary table to mill a circular slot

2 Set up and mill internal and external dovetails

3 Jig bore holes on a vertical mill

The versatility of the vertical milling machine is further increased by the use of specially shaped cutters and machine accessories. Operations such as milling of T-slots and dovetails are possible because special milling cutters are manufactured for each purpose. The rotary table accessory permits the milling of radii and circular slots. A boring head mounted in the machine spindle and measuring rods on the table allow the vertical milling machine to be used for the accurate jig boring of holes.

▸▸ The Rotary Table

The *rotary table* or the *circular milling attachment* (Fig. 63-1) can be used on plain universal vertical milling machines and slotters. It may provide rotary motion to the workpiece in addition to the longitudinal and vertical motion provided by the machine. With this attachment, it is possible to cut radii, circular grooves, and circular sections not possible by other means. The drilling and boring of holes that have been designated by angular measurements, as well as other indexing operations, are easily accomplished with this accessory. This attachment is also suitable for use with the slotting attachment on a milling machine.

Rotary tables may be of two types: those having hand feed and those having power feed. The construction of these is basically the same, the only exception being the automatic feed mechanism.

ROTARY TABLE CONSTRUCTION

The rotary table unit has a *base,* which is bolted to the milling machine table. Fitting into the base is the *rotary*

■ **Figure 63-1** A hand-feed rotary table.

table, on the bottom of which is mounted a *worm gear* (Fig. 63-1). A *worm shaft* mounted in the base meshes with and drives the worm gear. The worm shaft may be quickly disengaged when rapid rotation of the table is

Figure 63-2 A rotary table with an indexing attachment. *(Cincinnati Milacron, Inc.)*

Figure 63-3 A rotary table with power feed. *(Cincinnati Milacron, Inc.)*

required, as when setting work concentric with the table. A *handwheel* is mounted on the outer end of the worm shaft. The bottom edge of the table is graduated in half degrees. On most rotary table units, there is a *vernier scale* on the handwheel collar, which permits setting to within two minutes of a degree. The table has T-slots cut into the top surface to permit the clamping of work.

A hole in the center of the table accommodates test plugs to permit easy centering of the table with the machine spindle. Work may be centered with the table by means of test plugs or arbors.

Some rotary tables have an *indexing attachment* instead of a handwheel (Fig. 63-2). This attachment is often supplied as an accessory to the standard rotary table. It not only serves the same purpose as a handwheel but also permits the indexing of work with dividing head accuracy. The worm and wormwheel ratio of rotary tables is not necessarily 40:1, as on most dividing heads. The ratio may be 72:1, 80:1, 90:1, 120:1, or any other ratio. Larger ratios are usually found on larger tables. The method of calculating the indexing is the same as for the dividing head, except that the number of teeth in the wormwheel is used rather than 40, as in the dividing head calculations.

EXAMPLE

Calculate the indexing for five equally spaced holes on a circular plate using a rotary table with an 80:1 ratio.

$$\text{Indexing} = \frac{\text{number of teeth in wormwheel}}{\text{number of divisions required}}$$
$$= \frac{80}{5}$$
$$= 16 \text{ turns}$$

Accurately spaced holes on circles and segments (such as clutch teeth and teeth in gears too large to be held between index centers) are some of the applications of this type of rotary table. It may be supplied with a power feed mechanism.

The third type of rotary table has provision for power rotation of the table (Fig. 63-3). Here, the worm shaft is connected to the milling machine lead screw drive gear by a special shaft and an end gear train. The rate of rotation is controlled by the feed mechanism of the milling machine. When operated by power, the rotation of the table may be controlled by the operator using a feed lever attached to the unit or by trip dogs located on the periphery of the table. The operation of the table may be controlled by the handwheel or the automatic feed lever as required.

This type of attachment is particularly suited for production work and continuous milling operations when large numbers of small, identical parts are required. In this operation, the parts are mounted in suitable fixtures on the table, and the rotary feed moves the parts under the cutter. After the piece has passed beneath the cutter, the finished piece is removed and replaced with an unfinished workpiece.

To Center the Rotary Table with the Vertical Mill Spindle

1. Square the vertical head with the machine table.

2. Mount the rotary table on the milling machine.

3. Place a test plug in the center hole of the rotary table.

4. Mount an indicator with a grasshopper leg in the machine spindle.

5. With the indicator just clearing the top of the test plug, rotate the machine spindle by hand and approximately align the plug with the spindle.

6. Bring the indicator into contact with the diameter of the plug, and rotate the spindle by hand.

7. Adjust the machine table by the longitudinal and crossfeed handles until the dial indicator registers no movement.

8. Lock the machine table and saddle, and recheck the alignment.

9. Readjust if necessary.

To Center a Workpiece with the Rotary Table

Often it is necessary to perform a rotary table operation on several identical workpieces, each having a machined hole in the center. To align each workpiece quickly, a special plug can be made to fit the center hole of the workpiece and the hole in the rotary table. Once the machine spindle has been aligned with the rotary table, each succeeding piece can be aligned quickly and accurately by placing it over the plug.

If there are only a few pieces, which would not justify the manufacture of a special plug, or if the workpiece does not have a hole through its center, the following method can be used to center the workpiece on the rotary table:

1. Align the rotary table with the vertical head spindle.

2. Lightly clamp the workpiece to the rotary table in the approximate center.

Note: Do *not* move the crossfeed or longitudinal feed handles.

3. Disengage the rotary table worm mechanism.

4. Mount an indicator in the machine spindle or on the milling machine table, depending on the workpiece.

5. Bring the indicator into contact with the surface to be indicated, and revolve the rotary table by hand.

6. With a soft metal bar, tap the work (away from the indicator movement) until no movement is registered on the indicator in a complete revolution of the rotary table.

7. Clamp the workpiece tightly, and recheck the accuracy of the setup.

Note: If a center-punch mark must be aligned, a wiggler instead of an indicator is mounted in the milling machine spindle and the punch mark is positioned under the point of the wiggler.

▶▶ Radius Milling

When it is required to mill the ends on a workpiece to a certain radius or to machine circular slots having a definite radius, a certain sequence should be followed. Fig. 63-4 illustrates a typical setup.

■ **Figure 63-4** Milling a circular slot in a workpiece. *(Kelmar Associates)*

procedure

1. Align the vertical milling machine spindle at 90° to the table.

2. Mount a circular milling attachment (rotary table) on the milling machine table.

3. Center the rotary table with the machine spindle, using a test plug in the table and a dial indicator in the spindle.

4. Set the longitudinal feed dials and the crossfeed dial to zero (0).

5. Mount the work on the rotary table, aligning the center of the radial cuts with the center of the table. A special arbor may be used for this purpose. Another method is to align the center of the radial cut with a wiggler mounted in the machine spindle.

6. Move either the crossfeed or the longitudinal feed (whichever is more convenient) an amount equal to the radius required.

7. *Lock both the table and the saddle,* and remove the handles if convenient.

8. Mount the proper end mill.

9. Rotate the work, using the rotary table feed handwheel, to the starting point of the cut.

10. Set the depth of cut and machine the slot to the size indicated on the drawing, using hand or power feed.

MILLING A T-SLOT

T-slots are machined in the tops of machine tables and accessories to receive bolts for clamping workpieces. They are machined in two operations.

procedure

1. Consult a handbook for the T-slot dimensions.

2. Lay out the position of the T-slot.

3. Square the vertical milling machine spindle with the machine table.

4. Mount the work on the milling machine. If the work is to be held in a vise, the vise jaw must be aligned with the table travel. If the work is clamped to the table, the position of the slot must be aligned with the table travel.

5. Mount an end mill having a diameter slightly larger than the diameter of the bolt body. The size of the end mill to be used is shown in the T-slot tables.

6. Machine the center slot to the proper depth of the T-slot, using the end mill.

7. Remove the end mill, and mount the proper T-slot cutter.

8. Set the T-slot cutter depth to the bottom of the slot.

9. Machine the lower part of the slot.

MILLING DOVETAILS

Dovetails are used to permit reciprocating motion between two elements of a machine. They are composed of an external or an internal part and are adjusted by means of a gib. Dovetails may be machined on a vertical milling machine or on a horizontal mill equipped with a vertical milling attachment. A dovetail cutter is a special single-angle end mill type of cutter ground to the angle of the dovetail required.

Milling an Internal Dovetail

1. Refer to a handbook for the method of measuring a dovetail.

2. Check the measurements of the workpiece in which the dovetail is to be cut. Remove all burrs.

(a)   (b)

■ **Figure 63-5** Milling a dovetail: (a) center section roughed out; (b) both sides of the dovetail machined. *(Kelmar Associates)*

3. Lay out the position of the dovetail.

4. Mount the workpiece in a vise and clamp it on a rotary table, or if the work is long, bolt it directly to the machine table.

5. Indicate the side of the workpiece or the slot layout to see that it is parallel to the line of table travel.

6. Mount an end mill of a diameter narrower than the center section of the dovetail (Fig. 63-5a).

7. Start the end mill and touch up to the side of the work, after checking cutter rotation.

8. Set the crossfeed dial to zero (0).

9. Move the work over until the end mill is in the center of the dovetail. In this case, it will be the distance from the side of the workpiece to the center of the dovetail plus half the diameter of the cutter (plus the thickness of paper, if used).

10. Lock the saddle in this position, and set the crossfeed dial to zero (0).

11. Touch the edge of the cutter to the top of the work.

12. Move the work clear of the cutter, and set the depth of cut. Lock the knee in this position. The depth of this slot should be .030 to .050 in. (0.76 to 1.27 mm) deeper than the bottom of the dovetail to prevent drag and to provide clearance for dirt and chips between the mating dovetail parts.

13. Mill the channel to the width of the cutter (Fig. 63-5a).

14. Move the work over an amount equal to half the difference between the machined slot size and the size of the dovetail at the top. *Check for backlash.*

15. Take this finish cut along the one side of the work.

16. Check the width of the slot.

17. Move the work over to the finished width of the top of the dovetail. *Check for backlash.*

18. Cut the second side and check the width of the slot.

19. Return the saddle to zero (0).

20. Mount a dovetail cutter.

21. Set the depth for a roughing cut. This should be about .005 to .010 in. (0.12 to 0.25 mm) less than the finish depth.

22. Calculate the width of the dovetail at the bottom.

23. Move the work over .010 in. (0.25 mm) less than the finished size of this side. This will leave enough for the finish cut. Note the readings on the crossfeed dial.

24. Rough out the angle on the first side.

25. Move the work over to the other side the same amount from the centerline, and rough-cut the other side.

26. Set the cutter to the proper depth.

27. Machine the bottom surface (both sides) of the dovetail to the finished depth.

28. Using two rods, measure the dovetail for size.

29. Move the table over half the difference between the rough dovetail and the finished size. *Check for backlash.*

30. Take the finish cut on one side.

31. Move the table over the required amount, and finish the other side (Fig. 63-5b).

32. Check the finished size of the dovetail.

Note: If the work is mounted centrally on a rotary table, it is possible to rotate the work a half-turn (180°) after step 18 and take the same cuts on each side of the block for each successive step.

To Mill an External Dovetail

1. Center the cutter over the dovetail position.

2. Remove as much material as possible from each side of the external dovetail; that is, cut it to the largest size of the dovetail. In this operation, it will be necessary to note the readings from the centerline and remove the backlash for each side.

3. Mount a dovetail cutter, and center it with the workpiece.

4. Move the work over one-half the width of the dovetail plus half the diameter of the cutter. Allow .010 in. (0.25 mm) for a finish cut.

5. Take this roughing cut.

6. Move the work over an equal amount to the other side of the centerline, taking up the backlash.

7. Rough-cut the second side.

8. Adjust the work over, and take the finish cut on the one side.

9. Measure the dovetail using two rods.

10. Adjust for the finish cut on the second side, and take this cut.

11. Measure the width of the finished dovetail.

Note: If the work is mounted centrally on a rotary table, the work may be rotated a half-turn and the same cuts taken on each side.

▶▶ Jig Boring on a Vertical Milling Machine

If a jig borer is not available, the vertical milling machine may be used for jig-boring purposes. When the vertical milling machine is used for accurate hole location, the same setup, locating, and machining methods used in a jig borer apply. The coordinate system of hole location is used in each case.

Since the vertical milling machine does not have the same lead screw accuracy as a jig borer, it must have some external measuring system to ensure the accuracy of the table setting. Measuring rods and dial indicators, a vernier scale, or optical measuring devices, such as digital read-out boxes, may be used for this purpose.

Inch precision end-measuring rods (Fig. 63-6a) are generally supplied in sets of 11 rods, including two micrometer heads capable of measuring from 4 to 5 in. to an accuracy of .0001 in. These micrometer heads (Fig. 63-6b) are usually furnished with a red identifying ring on one and a black identifying ring on the other. One is used

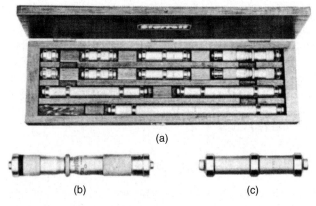

(a)

(b)   (c)

■ **Figure 63-6** Precision end-measuring rods: (a) a set of measuring rods; (b) a micrometer head; (c) a 3-in. measuring rod. *(The L.S. Starrett Co.)*

■ Figure 63-7 Using measuring rods to position the table accurately. *(The L.S. Starrett Co.)*

for longitudinal settings and the other for transverse settings. The solid rods are made of hardened and ground tool steel and have several concentric collars (Fig. 63-6c). The nine rods in a standard set include two each in lengths of 1, 2, and 6 in. and one 12-in. rod. Other rods are available in lengths of 4, 5, 7, 8, 10, and 15 in. The rod ends (and the micrometer head ends) are hardened, ground, and precision lapped, providing extreme accuracy.

The rods are held in V-shaped troughs, one mounted on the milling machine table and the other on the saddle (Fig. 63-7). A stop rod is mounted at one end of the trough, while at the other end a long-range dial indicator with .0001 in. graduations is mounted.

A *metric precision end-measuring rod* set consists of two micrometer heads, capable of measuring to an accuracy of 0.002 mm, and a number of solid measuring rods.

A standard measuring rod set consists of two rods in the following lengths: 25, 50, 75, 150, and 300 mm. Other rods are available in lengths of 100, 125, 175, 200, 250, and 375 mm.

The table is positioned accurately from the X and Y coordinates by adding (or subtracting) specified lengths or buildups of various rod combinations and micrometer settings.

TO POSITION THE MILLING MACHINE TABLE USING MEASURING RODS

1. Set the spindle to the edge of the work.
2. Clean the trough and ends of the stop rods and indicator rods.
3. Check the indicator for free operation.
4. Place the required number of rods, including the micrometer head, in the trough to take up the space between the stop rod and the indicator rod.
5. Adjust the micrometer until it has extended enough to cause the indicator needle to move a half-turn.
6. Lock the table.
7. Set the indicator bezel to zero (0).
8. Increase the rod and micrometer buildup by the length of the measurement between the side of the workpiece and the hole location.
9. Move the table along more than this required distance.
10. Insert the rods and the micrometer head.
11. Move the table back until the needle moves the half-turn and registers zero (0).
12. Lock the table.
13. Recheck the setting and adjust if necessary.

DIGITAL READOUT BOXES

Electronically controlled measuring devices, suitably mounted on the table and the saddle, indicate the table travel to an accuracy of .0005 or .0001 in. (0.012 or 0.002 mm). A digital readout display (Fig. 63-8) resembles the odometer on an automobile dashboard and indicates the distance traveled through a series of numbers visible through a glass front on the display. This arrangement permits quick and accurate setting of the machine table in two directions: horizontal and transverse, or on the X and Y axes.

VERNIER SCALE

The milling machine may be equipped with scales on the table and saddle, with pointers suitably mounted on the

■ Figure 63-8 A digital readout display. *(Sheffield Measurement Div.)*

machine. The scales are generally graduated in increments of tenths of an inch or 2 mm. A vernier arrangement is mounted adjacent to the feed screw collars, which permits the reading in .0001 in. or 0.002 mm.

Note: Settings on this machine must be made in one direction only in order to remove the backlash.

Vertical milling machines may be used to position holes without the use of any of the aforementioned equipment; however, this method is not too accurate.

The table is positioned by means of the graduated feed collars. This method may be used if the accuracy of the location must not be less than .002 in. or 0.05 mm. Again, in order to eliminate backlash, all settings must be made in one direction only.

▸▸ Vertical Milling Machine Attachments

The versatility of the vertical milling machine may be further increased by the use of certain attachments.

The *rack milling attachment* permits the machining of racks and broaches on the vertical mill. The spindle of this attachment is fitted into the spindle of the machine while the housing is clamped to the machine quill. It operates on the same principle as a rack milling attachment on the horizontal milling machine.

The *slotting attachment* (Fig. 63-9) is generally fitted to the back end of the overarm, which may be rotated

180° to permit the use of this device. The slotting attachment operates independently of the machine drive on most vertical milling machines. It transmits a reciprocating motion to a single-pointed tool by means of a motor-driven eccentric. This attachment may be used for cutting keyways and slotting out small blanking dies.

DIE SINKING

One important application of the vertical milling machine is that of *die sinking*. Dies used in drop forging and die casting have impressions or cavities cut in them by means of die sinking. The operation of machining a die cavity on a vertical mill is generally done by hand control of the machine, using various end-mill-type cutters. This is usually followed by considerable filing, scraping, and polishing to produce the highly finished, properly curved and contoured surfaces required on the die.

Complicated shapes and patterns on molds and dies are made more conveniently and accurately on a vertical milling machine equipped with tracer or computer control. In addition to the regular cutting head, the machine is equipped with a tracer head. Electrical discharge machining is finding ever-increasing use in the manufacture of molds and dies. Intricate shapes and forms that would be difficult or impossible to produce on conventional machines can be quickly and easily reproduced in even supertough, space-age metals.

On a tracer-controlled machine (Fig. 63-10), the form of a master or pattern is transferred to the workpiece by means of a hydraulic tracer unit actuated by a stylus or

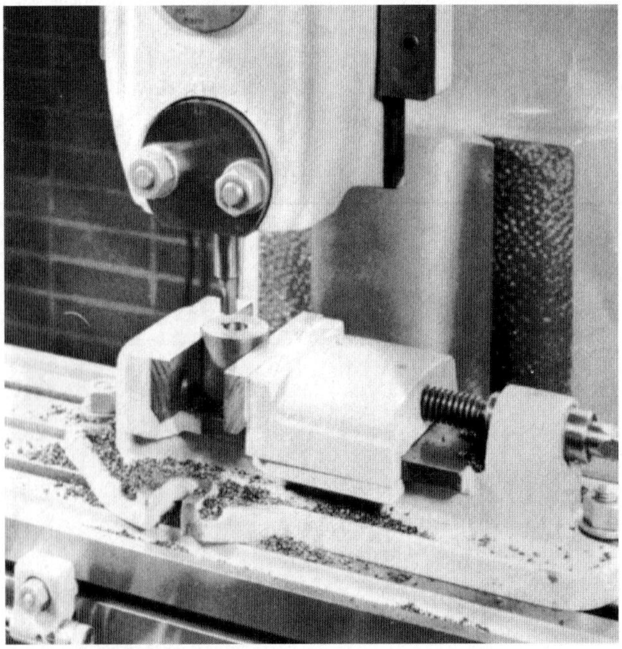

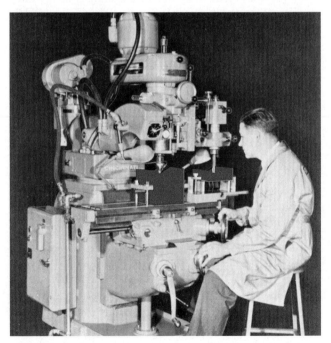

■ **Figure 63-9** Cutting an internal keyway with a slotting attachment. *(Cincinnati Milacron, Inc.)*

■ **Figure 63-10** A tracer-controlled vertical milling machine. *(Cincinnati Milacron, Inc.)*

tracer finger. This contacts the master and moves the cutter up or down with the vertical or side travel of the stylus.

Since both the work and the master are fastened to the table, they both travel at the same rate. As the pattern moves under the stylus, the workpiece assumes the identical form of the master. Some machines may be equipped with ratio arms or devices, and the master can be made considerably larger than the finished workpiece. If a master is made 10 times larger, the arms on the machine are set at a 10:1 ratio and then the part is cut. An error of .010 in. (0.25 mm) on the master will only result in a .001-in. (0.025-mm) error on the workpiece.

Only a light pressure on the stylus, in contact with the master or template, is necessary to deflect the tracer arm and actuate the control valve, which controls the movement of the cutting head.

unit 63 review questions

Rotary Table

1. Why are rotary tables used?

2. Describe briefly the construction of a rotary table.

3. What is the purpose of the hole in the center of a rotary table?

4. What common ratios are found on rotary tables?

5. Describe briefly how a rotary table may be centered with the vertical mill spindle.

6. Explain how a number of identical parts, having a hole in the center, can be quickly aligned on a rotary table.

Radius and T-Slot Milling

7. Explain how a large radius may be cut using a rotary table.

8. What purpose do T-slots serve?

9. List the two operations necessary in order to cut a T-slot.

Milling Dovetails

10. What is the purpose of a dovetail?

11. What is the procedure for machining the center section of an internal dovetail?

12. Explain how the first angular side of a dovetail is cut.

13. How can an internal dovetail be measured accurately for size?

Jig Boring on a Vertical Mill

14. Name and describe three types of measuring systems used on vertical mills for jig boring.

15. Name three methods of locating an edge.

16. How may a table be moved exactly 2.6836 in. (68.163 mm) from one location to another by using measuring rods?

Vertical Milling Machine Attachments

17. What are the purposes of the rack milling and slotting attachments?

18. How can a vertical mill be used for various die-sinking operations:

    a. Manually?   b. Automatically?

19. Explain the purpose of ratio arms or devices in die sinking.

UNIT 64

Horizontal Milling Machines and Accessories

OBJECTIVES

After completing this unit, you should be able to:

1 Recognize and explain the purposes of four milling machines

2 Know the purposes of the main operational parts of a horizontal and a vertical milling machine

3 Recognize and state the purposes of four milling machine accessories and attachments

A wide variety of milling machines are required by industry to meet the job requirements of the many parts that must be machined. To make milling machines more versatile, a large variety of accessories and attachments are available so that each machine can perform more operations on each workpiece.

▶▶ Horizontal Milling Machines

To meet many different industrial requirements, milling machines are made in a wide variety of types and sizes. They are classified under the following headings:

1. *Manufacturing-type,* in which the cutter height is controlled by vertical movement of the headstock

2. *Special-type,* designed for specific milling operations

3. *Knee-and-column-type,* in which the relationship between the cutter height and the work is controlled by vertical movement of the table

▶▶ Manufacturing-Type Milling Machines

Manufacturing-type milling machines are used primarily for quantity production of identical parts. This type of machine may be either semiautomatic or fully automatic and is of simple but sturdy construction. Fixtures clamped to the table hold the workpiece for a variety of milling operations, depending on the type of cutters or special spindle attachments used. Some of the distinctive features of manufacturing-type machines are the *automatic cycle* of cutter and work approach, the *rapid movement* during the noncutting part of the cycle, and the *automatic spindle stop.* After this machine has been set up, the operator is required only to load and unload the machine and start the automatic cycle controlled by cams and preset *trip dogs.* Two

Figure 64-1 A plain manufacturing-type milling machine. *(Cincinnati Milacron, Inc.)*

Figure 64-2 A small plain automatic knee-and-column-type milling machine. *(Cincinnati Milacron, Inc.)*

of the common manufacturing-type milling machines are the *plain manufacturing type* (Fig. 64-1) and the *small plain automatic knee-and-column type* (Fig. 64-2).

▶▶ Special-Type Milling Machines

Special-type milling machines are designed for individual milling operations and are used for only one particular type of job. They may be completely automatic and are used for production purposes when hundreds or thousands of similar pieces are to be machined.

MACHINING CENTERS

A special type of machine used widely in the industrial world is the computer numerical control horizontal machining center (Fig. 64-3 on p. 508). This machine is capable of handling a wide variety of work such as straight and contour milling, drilling, reaming, tapping, and boring—all in one setup. Because of the rugged construction, reliable controls, and antibacklash ball screws, it is capable of high production rates while still maintaining a high degree of accuracy. This machine and its contribution to manufacturing will be discussed in greater detail in Unit 78.

KNEE-AND-COLUMN-TYPE MILLING MACHINES

Machines in this class fall into three categories:

1. Plain horizontal milling machines
2. Universal horizontal milling machines
3. Vertical milling machines

Universal Horizontal Milling Machines

The universal horizontal milling machine is essential for advanced machine shop work. The difference between this machine and the plain horizontal mill will be dealt with in this unit.

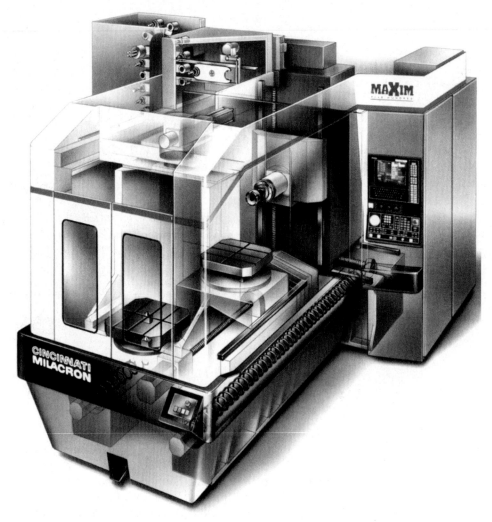

■ **Figure 64-3** A computer numerical control horizontal machining center.
(Cincinnati Milacron, Inc.)

Fig. 64-4 shows the parts of a universal horizontal mill. The only difference between this mill and the plain horizontal machine is the addition of a *table swivel housing* between the table and the saddle. This housing permits the table to be swiveled 45° in either direction in a horizontal plane for such operations as the milling of helical grooves in twist drills, milling cutters, and gears.

▶▶ Parts of the Milling Machine

> The *base* gives support and rigidity to the machine and acts as a reservoir for the cutting fluids.

> The *column face* is a precision-machined and scraped section used to support and guide the knee when it is moved vertically.

> The *knee* is attached to the column face and may be moved vertically on the column face either manually or automatically. It houses the feed mechanism (Fig. 64-4).

> The *saddle* is fitted on top of the knee and may be moved in or out either manually by means of the crossfeed handwheel or automatically by the crossfeed engaging lever (Fig. 64-5 on p. 510).

> The *swivel table housing,* fastened to the saddle on a universal milling machine, enables the table to be swiveled 45° to either side of the centerline.

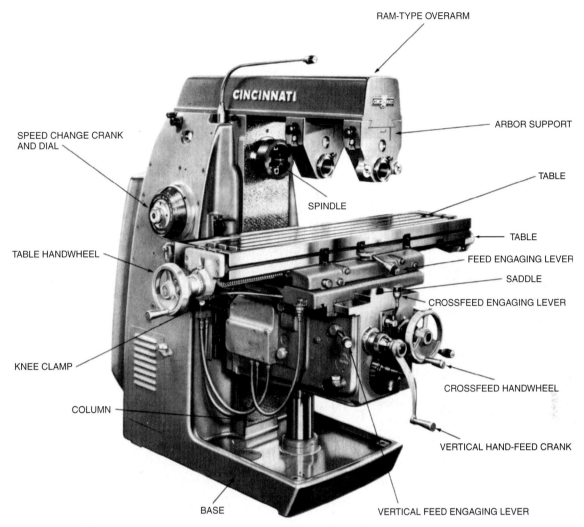

RAM-TYPE OVERARM

ARBOR SUPPORT

SPEED CHANGE CRANK
AND DIAL

TABLE

SPINDLE

TABLE

TABLE HANDWHEEL

FEED ENGAGING LEVER

SADDLE

CROSSFEED ENGAGING LEVER

KNEE CLAMP

CROSSFEED HANDWHEEL

COLUMN

VERTICAL HAND-FEED CRANK

BASE

VERTICAL FEED ENGAGING LEVER

■ **Figure 64-4** Universal horizontal milling machine. *(Cincinnati Milacron, Inc.)*

> The *table* rests on guideways in the saddle and travels longitudinally in a horizontal plane. It supports the vise and the work (Fig. 64-5).

> The *crossfeed handwheel* is used to move the table toward or away from the column.

> The *table handwheel* is used to move the table horizontally left and right in front of the column.

> The *feed dial* is used to regulate the table feeds.

> The *spindle* provides the drive for arbors, cutters, and attachments used on a milling machine.

> The *overarm* provides for correct alignment and support of the arbor and various attachments. It can be adjusted and locked in various positions, depending on the length of the arbor and the position of the cutter.

> The *arbor support* is fitted to the overarm and can be clamped at any location on the overarm. Its purpose is to align and support various arbors and attachments.

> The *elevating screw* is controlled by hand or an automatic feed. It gives an upward or downward movement to the knee and the table.

> The *spindle speed dial* is set by a crank that is turned to regulate the spindle speed. On some milling machines, the spindle speed changes are made by means of two levers. When making speed changes, always check to see whether the change can be made when the machine is running or if it must be stopped.

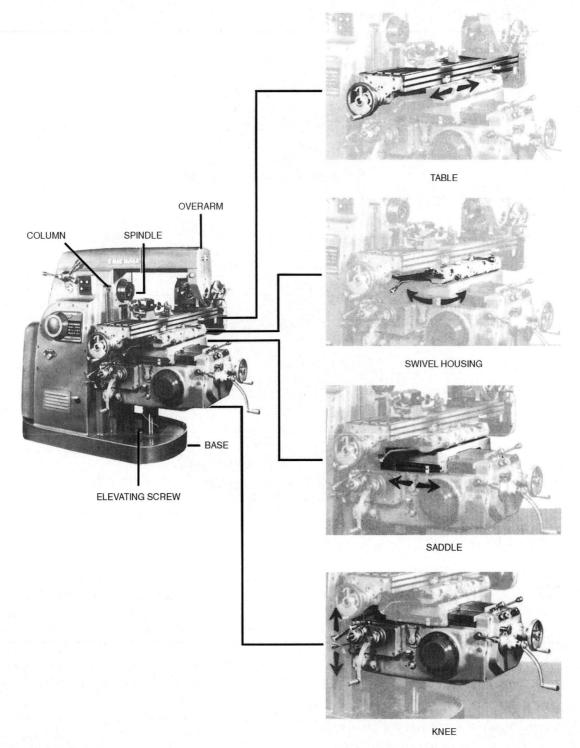

■ Figure 64-5 The main parts of a milling machine. *(Cincinnati Milacron, Inc.)*

The labeled parts in the figure are: COLUMN, SPINDLE, OVERARM, BASE, ELEVATING SCREW, TABLE, SWIVEL HOUSING, SADDLE, KNEE.

BACKLASH ELIMINATOR

A feature on most milling machines is the addition of a *backlash eliminator.* This device, when engaged, eliminates the backlash (play) between the nut and the table lead screw, permitting the operation of *climb* (down)

milling. Fig. 64-6 shows a diagrammatic sketch of the Cincinnati Machine backlash eliminator.

The backlash eliminator works as follows. Two independent nuts are mounted on the lead screw. These nuts engage a common crown gear, which in turn meshes with a rack. Axial movement of the rack is controlled by the

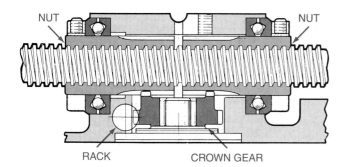

Figure 64-6 Cross section of a Cincinnati Machine backlash eliminator. *(Cincinnati Milacron, Inc.)*

backlash eliminator engaging knob on the front of the saddle. By turning the knob in, the nuts are forced to move along the lead screw in opposite directions, removing all backlash. The nuts-gear-rack arrangement is shown in Fig. 64-6.

▶▶ Milling Machine Accessories

A wide variety of accessories, which greatly increase the milling machine's versatility and productivity, are available. These accessories may be classed as *fixtures* or as *attachments*.

FIXTURES

A fixture (Fig. 64-7) is a work-holding device fastened to the table of a machine or to a machine accessory, such as a rotary table. It is designed to hold workpieces that cannot be readily held in a vise or to be used in production work when large quantities are to be machined. The fixture must be designed so that identical parts, when held in the fixture, will be positioned exactly and held securely.

Figure 64-7 Milling a workpiece held in a fixture. *(Cincinnati Milacron, Inc.)*

Fixtures may be constructed to hold one or several parts at one time and should permit the quick changing of workpieces. The work may be positioned by stops, such as pins, strips, or setscrews, and held in place by clamps, cam-lock levers, or setscrews. To produce uniform workpieces, clean the chips and cuttings from a fixture before mounting a new workpiece.

ATTACHMENTS

Milling machine attachments may be divided into three classes:

1. Those designed to hold special attachments; they are attached to the spindle and column of the machine. They are the *vertical, high-speed, universal, rack milling,* and *slotting* attachments. These attachments are designed to increase the versatility of the machine.

2. *Arbors, collets,* and *adapters,* which are designed to hold standard cutters.

3. Those designed to hold the workpiece, such as a *vise, a rotary table,* and an *indexing,* or *dividing, head.*

Vertical Milling Attachment

The *vertical milling attachment* (Fig. 64-8), which may be mounted on the face of the column or the overarm, enables a plain or universal milling machine to be used as a vertical milling machine. Angular surfaces may be machined by swiveling the head, parallel to the face of the

Figure 64-8 A vertical milling attachment. *(Cincinnati Milacron, Inc.)*

(a)   (b)

■ **Figure 64-9** (a) A rack milling attachment; (b) a rack indexing attachment. *(Cincinnati Milacron, Inc.)*

column, to any angle up to 45° on either side of the vertical position. On some models, the head may be swiveled as much as 90° to either side. Vertical attachments enable the horizontal milling machine to be used for such operations as face milling, end milling, drilling, boring, and T-slot milling.

A modification of the vertical milling attachment is the *universal milling attachment,* which may be swiveled in two planes, parallel to the column and at right angles to it, for the cutting of compound angles. The vertical and universal attachments are also manufactured in *high-speed* models, which permit the efficient use of small- and medium-size end mills and cutters for such operations as die sinking and keyseating.

Rack Milling Attachment

The *rack milling attachment* (Fig. 64-9a) and the *rack indexing attachment* (Fig. 64-9b) are used to mill longer gear racks (flat gears) than could be cut with the standard horizontal milling machine. These attachments will be discussed in Unit 69 with gear-cutting accessories.

Slotting Attachment

The *slotting attachment* (Fig. 64-10) converts the rotary motion of the spindle into reciprocating motion for cutting keyways, splines, templates, and irregularly shaped surfaces. The length of the stroke is controlled by an adjustable crank. The tool slide may be swiveled to any angle in a plane parallel to the face of the column, making the slotting attachment especially valuable in die work.

Arbors, Collets, and Adapters

Arbors, used for mounting the milling cutter, are inserted and held in the main spindle by a draw bolt or a special quick-change adapter (Fig. 64-11).

Shell-end mill arbors may fit into the main spindle or the spindle of the vertical attachment. These devices permit face milling to be done either horizontally or vertically.

Collet adapters are used for mounting drills or other tapered-shank tools in the main spindle of the machine or the vertical milling attachment.

■ Figure 64-10 A slotting attachment. *(Cincinnati Milacron, Inc.)*

ADAPTERS   ARBOR

SLEEVES   COLLET   ADAPTER

■ Figure 64-11 Arbors, collets, and adapters.

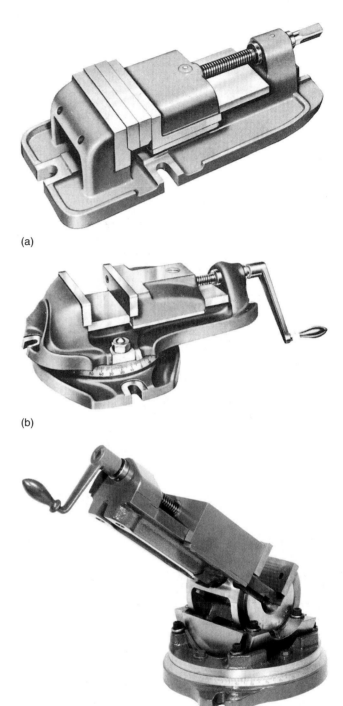

(a)

(b)

(c)

■ Figure 64-12 (a) A plain vise. (b) a swivel base vise; (c) a universal vise. *(Cincinnati Milacron, Inc.)*

A *quick-change adapter,* mounted in the spindle, permits such operations as drilling, boring, and milling without a change in the setup of the workpiece.

Vises

Milling machine vises are the most widely used work-holding devices for milling. They are manufactured in three styles.

The *plain vise* (Fig. 64-12a) may be bolted to the table so that its jaws are parallel or at right angles to the axis of the spindle. The vise is positioned quickly and accurately by keys on the bottom of the base that fit into the T-slots on the table.

The *swivel base vise* (Fig. 64-12b) is similar to the plain vise, except that it has a swivel base that enables the vise to be swiveled through 360° in a horizontal plane.

The *universal vise* (Fig. 64-12c) may be swiveled through 360° in a horizontal plane and may be tilted from 0° to 90° in a vertical plane. It is used chiefly by toolmakers,

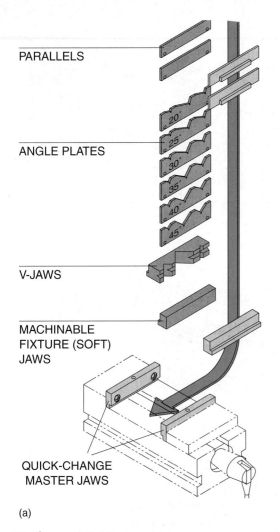

PARALLELS

ANGLE PLATES

20°
25°
30°
35°
40°
45°

V-JAWS

MACHINABLE
FIXTURE (SOFT)
JAWS

QUICK-CHANGE
MASTER JAWS

(a)

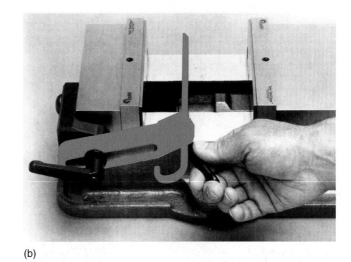

(b)

■ **Figure 64-13** Plain vise fixturing systems: (a) vise jaws; (b) vise stop. *(Toolex Systems, Inc.)*

moldmakers, and diemakers, since it permits the setting of compound angles for milling.

Fixturing Systems

Quick-change, self-locking systems (Fig. 64-13a) can be incorporated into a plain vise for use in the machine shop and toolroom, for manual machining. Accuracy can be maintained, and valuable reductions in setup time can be realized. For basic setups and for repeat positioning of parts, an easy-to-adjust stop (Fig. 64-13b) can be attached to either side of the vise.

Indexing, or Dividing, Head

The *indexing,* or *dividing, head* is a very useful accessory that permits the cutting of bolt heads, gear teeth, ratchets, and so on. When connected to the lead screw of the milling machine, it will revolve the work as required to cut helical gears and flutes in drills, reamers, and other tools. The dividing head will be fully discussed in Unit 68.

Horizontal Milling Machine

1. Name six operations that can be performed on a milling machine.

Manufacturing-Type Milling Machine

2. List four features of a manufacturing-type milling machine.

Knee-and-Column-Type Milling Machine

3. What is the difference between a plain horizontal and a universal horizontal milling machine?

4. What is the purpose of the backlash eliminator?

Milling Machine Accessories and Attachments

5. What is the purpose of a fixture?

6. List three milling machine attachments and give examples of each.

7. Describe the purpose of the following devices:
   a. Vertical milling attachment
   b. Rack milling attachment
   c. Slotting attachment

8. Name three methods of holding cutters on a milling machine.

9. What are the features of the:
   a. Plain vise?
   b. Swivel base vise?
   c. Universal vise?

UNIT 65

Milling Cutters

OBJECTIVES

After completing this unit, you should be able to:

1 Identify and state the purposes of six standard milling cutters

2 Identify and state the purposes of four special-purpose cutters

3 Use high-speed steel and carbide cutters for proper applications

The proper selection, use, and care of milling cutters must be practiced if the best results are to be achieved with the milling machine. The operator or apprentice not only must be able to determine the proper spindle speed for any cutter but should also constantly observe how the milling machine performs with different cutters and clearances.

Milling cutters are manufactured in many types and sizes. Only the most commonly used cutters will be discussed.

▶▶ Milling Cutter Materials

In the milling process, as with most metal-cutting operations, the cutting tool must possess certain qualities to function satisfactorily. Cutters must be harder than the metal being machined and strong enough to withstand pressures developed during the cutting operation. They must also be tough to resist the shock resulting from the contact of the tooth with the work. To maintain keen cutting edges, they must be able to resist the heat and abrasion of the cutting process.

Most milling cutters today are made of high-speed steel or tungsten carbide. Special-purpose cutters, made in the plant for a special job, may be made from plain carbon steel.

High-speed steel, consisting of iron with various amounts of carbon, tungsten, chromium, molybdenum, and vanadium, is used for most solid milling cutters, since it possesses all the qualities required for a milling cutter. In this steel, carbon is the hardening agent, while tungsten and molybdenum enable the steel to retain its hardness up

to red heat. Vanadium increases the tensile strength, and chromium increases the toughness and wear resistance.

When a higher rate of production is desired and when harder metals are machined, cemented carbides replace high-speed steel cutters. Cemented-carbide cutters (Fig. 65-1a), although more expensive, may be operated from 3 to 10 times faster than high-speed steel cutters. Either cemented-carbide tips may be brazed to a steel body or inserts may be held in place by means of locking or clamping devices (Fig. 65-1b).

When cemented-carbide cutters are to be used, care must be taken to select the proper type of carbide for the job. Straight tungsten carbide is used for machining cast iron, most nonferrous alloys, and plastics. Tantalum carbide is used to machine low- and medium-carbon steels, and tungsten-titanium carbide is used for high-carbon steels.

Although cemented carbides have many advantages as cutting tools, several disadvantages limit extensive use:

1. Cemented-carbide cutters are more costly to buy, maintain, and sharpen.

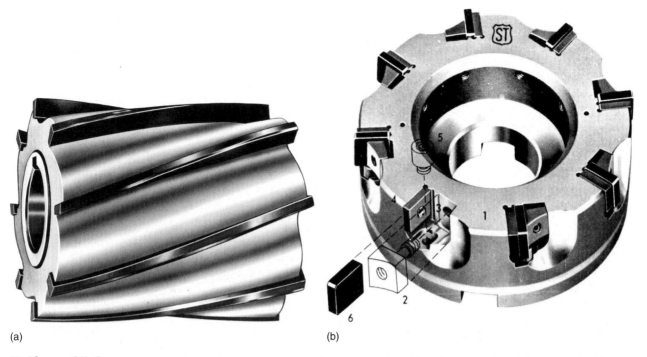

(a)

(b)

■ **Figure 65-1** (a) A cemented-carbide milling cutter; (b) cemented-carbide tips are held in place by a locking device.

2. Efficient use of these cutters requires that machines must be rigid and have greater horse-power and speed than are required for high-speed cutters.

3. Carbide cutters are brittle; the edges break easily if misused.

4. Special grinders with silicon carbide and diamond wheels are required to sharpen carbide cutters properly.

▸▸ Plain Milling Cutters

Probably the most widely used milling cutter is the plain milling cutter, which is a cylinder made of high-speed steel with teeth cut on the periphery; it is used to produce a *flat surface*. These cutters may be of several types, as shown in Fig. 65-2.

Light-duty plain milling cutters (Fig. 65-2a), which are less than ¾ in. (19 mm) wide, will usually have straight

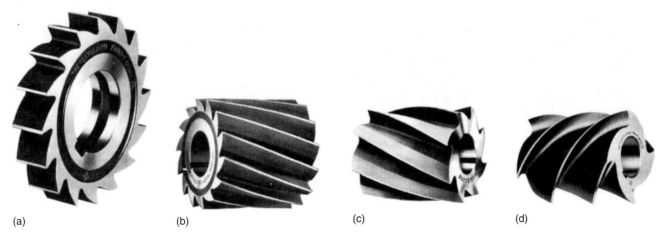

(a)   (b)   (c)   (d)

■ **Figure 65-2** Plain milling cutters: (a) light-duty; (b) light-duty helical; (c) heavy-duty; (d) high-helix.

teeth; those over ¾ in. (19 mm) wide have a helix angle of about 25° (Fig. 65-2b). This type of cutter is used only for light milling operations, since it has too many teeth to permit the chip clearance required for heavier cuts.

Heavy-duty plain milling cutters (Fig. 65-2c) have fewer teeth than the light-duty type, which provide for better chip clearance. The helix angle varies up to 45°. This helix angle on the teeth produces a smoother surface because of the shearing action and reduced chatter. Less power is required with this type of cutter than with straight-tooth and small-helix-angle cutters.

High-helix plain milling cutters (Fig. 65-2d) have helix angles from 45° to over 60°. They are particularly suited to the milling of wide and intermittent surfaces on contour and profile milling. Although this type of cutter is usually mounted on the milling machine arbor, it is sometimes shank-mounted with a pilot on the end and used for milling elongated slots.

▶▶ Standard Shank-Type Helical Milling Cutters

Standard shank-type helical milling cutters (Fig. 65-3), also called *arbor-type cutters,* are used for milling forms from solid metal; for example, they are used when making yokes or forks. Shank-type milling cutters are also used for removing inner sections from solids. These cutters are inserted through a previously drilled hole and supported at the outer end with type A arbor supports. Special spindle adapters are used to hold these cutters.

■ **Figure 65-3** Standard shank-type helical milling cutter. *(The Union Butterfield Corp.)*

▶▶ Side Milling Cutters

Side milling cutters (Fig. 65-4) are comparatively narrow cylindrical milling cutters with teeth on each side as well as on the periphery. They are used for cutting slots and for face and straddle milling operations. Side milling cutters may have straight teeth (Fig. 65-4a) or staggered teeth (Fig. 65-4b). Staggered-tooth cutters have each tooth set alternately to the right and left with an alternately opposite helix angle on the periphery. These cutters have free cutting action at high speeds and feeds. They are particularly suited for milling deep, narrow slots.

Half-side milling cutters (Fig. 65-4c) are used when only one side of the cutter is required, as in end facing. These cutters are also made with interlocking faces so that two cutters may be placed side by side for slot milling. The interlocking type is more suited for slot cutting than the solid-type, staggered-tooth cutter, since the amount ground from the side of the cutter during regrinding may be compensated by a washer between the cutters. Half-side milling cutters have considerable rake and, therefore, are able to take heavy cuts.

▶▶ Face Milling Cutters

Face milling cutters (Fig. 65-5) are generally over 6 in. (150 mm) in diameter and have *inserted teeth* held in place by a wedging device. The teeth may be of high-speed steel or cast tool steel, or they may be tipped with sintered-carbide cutting edges. The corners of this type of cutter are beveled; most of the cutting action occurs at these points and the periphery of the cutter. The face of the tooth removes a small amount of stock left by the spring of the work or cutter. To prevent chatter, only a small portion of the tooth face near the periphery is in contact with the work; the remainder is ground with a suitable clearance (8° to 10°).

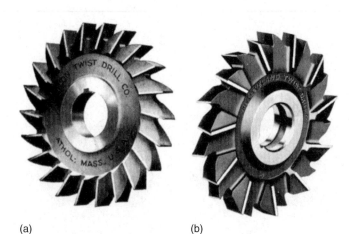

(a)　　　　(b)　　　　　　　(c)

■ **Figure 65-4** Side milling cutters: (a) plain *(Cleveland Twist Drill Co.)*; (b) staggered-tooth *(Cleveland Twist Drill Co.)*; (c) half-side *(The Union Butterfield Corp.)*

Figure 65-5 A face milling cutter. *(The Union Butterfield Corp.)*

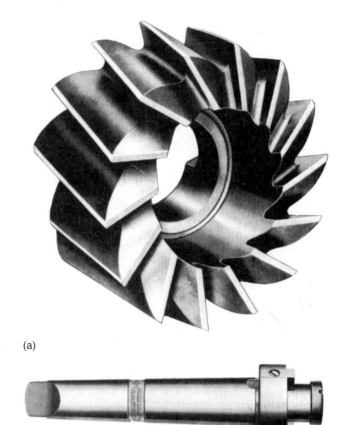

(a)

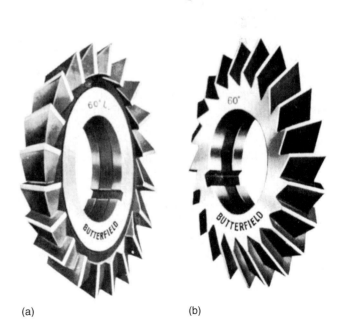

(b)

Figure 65-6 Shell end mill and adapter. *(Cleveland Twist Drill Co.)*

This type of cutter is often used as a combination cutter, making the roughing and finishing cuts in one pass. The roughing and finishing blades are mounted on the same body, with a limited number of finishing blades set to a smaller diameter and extending slightly farther from the face than the roughing blades. The finishing blades have a slightly wider cutting face surface, which creates a better surface finish.

Face milling cutters under 6 in. (150 mm) are called *shell end mills* (Fig. 65-6). They are solid, multiple-tooth cutters with teeth on the face and the periphery. They are usually held on a stub arbor, which may be threaded or may use a key in the shank to drive the cutter. Shell end mills are more economical than large solid end mills because they are cheaper to replace when broken or worn out.

▶▶ Angular Cutters

Angular cutters have teeth that are neither parallel nor perpendicular to the cutting axis. They are used for milling angular surfaces, such as grooves, serrations, chamfers, and reamer teeth. They may be divided into two groups:

1. *Single-angle milling cutters* (Fig. 65-7a) have teeth on the angular surface and may or may not have teeth on the flat side. The included angle between the flat face and the conical face designates the cutters, such as 45° or 60° angular cutter.

(a)

(b)

Figure 65-7 (a) Single-angle milling cutter; (b) double-angle milling cutter. *(The Union Butterfield Corp.)*

Milling Cutters **519**

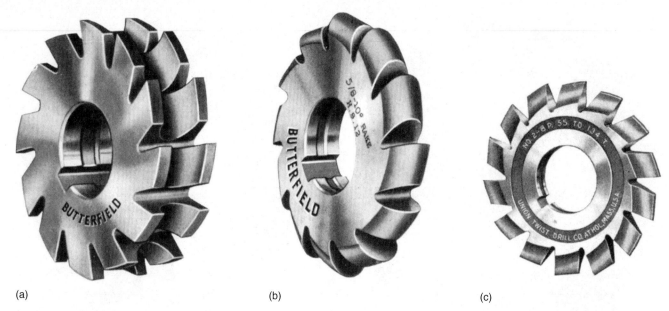

(a)   (b)   (c)

■ **Figure 65-8** Types of formed cutters; (a) concave; (b) convex; (c) gear tooth.   *(The Union Butterfield Corp.)*

2. *Double-angle milling cutters* (Fig. 65-7b) have two intersecting angular surfaces with cutting teeth on both. When these cutters have equal angles on both sides of the line at a right angle to the axis (symmetrical), they are designated by the size of the included angle. When the angles formed with this line are not the same (asymmetrical), the cutters are designated by specifying the angle on either side of the plane or line, such as 12° to 48° double-angle milling cutter.

▶▶ Formed Cutters

Formed cutters (Fig. 65-8) incorporate the exact shape of the part to be produced, permitting exact and more economical duplication of irregularly shaped parts than most other means. Formed cutters are particularly useful for the production of small parts. Each tooth of a formed cutter is identical in shape, and the clearance is machined on the full thickness of each tooth by the form or master tool in a cam-controlled relieving machine. Examples of *formed-relieved cutters* are concave, convex, and gear cutters.

Formed cutters are sharpened by grinding the tooth face. Tooth faces are radial and may have positive, zero, or negative rake, depending on the cutter application. *It is important* that the original rake on the tooth be maintained so that the profiles of the tooth and of the work are not changed. If the tooth face rake is maintained exactly, the cutter may be resharpened until the teeth are too thin for use; thus, the exact, original shape of the tooth can be maintained. These cutters are sometimes produced with an angular face, which

causes a shearing action and reduces chatter during the cutting process. They are, however, more difficult to sharpen.

▶▶ Metal-Slitting Saws

Metal-slitting saws (Fig. 65-9) are basically thin plain milling cutters with sides relieved or "dished" to prevent rubbing or binding when in use. Slitting saws are made in widths from $\frac{1}{32}$ to $\frac{3}{16}$ in. (0.8 to 5 mm). Because of their thin cross section, they should be operated at approximately one-quarter to one-eighth of the feed per tooth used for other cutters. For nonferrous metals, their speed can be increased. Unless a special driving flange is used for slitting saws, it is *not* advisable to key the saw to the milling arbor. When the arbor nut is tightened, it should be *hand tight* with a wrench. Since slitting saws are so easily broken, some operators find it desirable to "climb" or

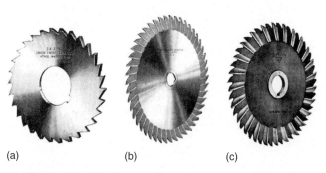

(a)   (b)   (c)

■ **Figure 65-9** Metal-slitting saws.   *(The Union Butterfield Corp.)*

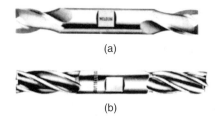

(a)

(b)

■ **Figure 65-10** (a) A two-flute end mill; (b) a four-flute end mill.

"down" mill when sawing (see Unit 60). However, to overcome the play between the lead screw and nut, the backlash eliminator should be engaged.

▶▶ End Mills

End mills have cutting teeth on the end as well as on the periphery and are fitted to the spindle by a suitable adapter. They are of two types: the *solid end mill,* in which the shank and the cutter are integral, and the *shell end mill,* which, as previously stated, has a separate shank.

Solid end mills, generally smaller than shell end mills, may have either straight or helical flutes. They are manufactured with straight and tapered shanks and with two or more flutes. The two-flute type (Fig. 65-10a) has flutes that meet at the cutting end, forming two cutting lips across the bottom. These lips are of different lengths, one extending beyond the center axis of the cutter, which eliminates the center and permits the two-flute end mill to be used in a milling machine for drilling (plunge milling) a hole to start a slot that does not extend to the edge of the metal. When a slot is being cut with a two-flute end mill, the depth of cut should not exceed one-half the diameter of the cutter. When the four-flute end mill (Fig. 65-10b) is used for slot cutting, it is usually started at the edge of the metal.

▶▶ T-Slot Cutter

The *T-slot cutter* (Fig. 65-11a) is used to cut the wide horizontal groove at the bottom of a T-slot after the narrow vertical groove has been machined with an end mill or a side milling cutter. It consists of a small side milling cutter with teeth on both sides and an integral shank for mounting, similar to an end mill.

▶▶ Dovetail Cutter

The *dovetail cutter* (Fig. 65-11b) is similar to a single-angle milling cutter with an integral shank. Some dovetail

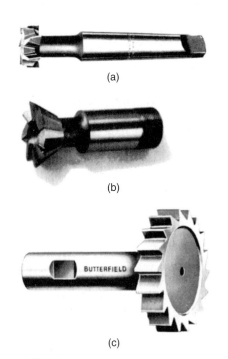

(a)

(b)

(c)

■ **Figure 65-11** (a) A T-slot cutter; (b) a dovetail cutter; (c) a Woodruff keyseat cutter. *(The Union Butterfield Corp.)*

cutters are manufactured with an internal thread and are mounted on a special threaded shank. They are used to form the sides of a dovetail after the tongue or the groove has been machined with another suitable cutter, usually a side milling cutter. Dovetail cutters may be obtained with 45°, 50°, 55°, or 60° angles.

▶▶ Woodruff Keyseat Cutter

The *Woodruff keyseat cutter* (Fig. 65-11c) is similar in design to plain and side milling cutters. Smaller sizes up to approximately 2 in. (50 mm) in diameter are made with a solid shank and straight teeth; larger sizes are mounted on an arbor and have staggered teeth on both the sides and the periphery. They are used for milling semicylindrical keyseats in shafts.

Woodruff cutters are designated by a number system. The right-hand two digits give the nominal diameter in eighths of an inch, while the preceding digits give the width of the cutter in thirty-seconds of an inch. For example, a #406 cutter would have dimensions as follows:

Diameter 06 × 1/8 = 3/4 in.
Width   4 × 1/32 = 1/8 in.

▸▸ Flycutters

The *flycutter* (Fig. 65-12) is a single-pointed cutting tool with the cutting end ground to the desired shape. It is mounted in a special adapter or arbor. Since all the cutting is done with one tool, a fine feed must be used. Flycutters are used in experimental work and when the high cost of a specially shaped cutter would not be warranted.

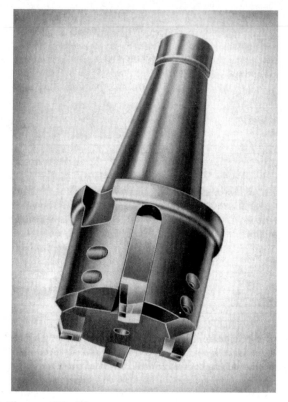

■ **Figure 65-12** A flycutter.

unit 65 review questions

1. Name two materials used to make milling cutters.
2. Name five elements that may be added to iron or steel to produce high-speed steel, and state the purpose of each additive.
3. Discuss the advantages and disadvantages of cemented-carbide tools.
4. Describe three types of plain milling cutters and explain their uses.
5. Describe three types of side milling cutters, and state when each is used.
6. List two ways in which formed cutters differ from plain milling cutters.
7. What advantage does a two-flute end mill have?

UNIT 66

Milling Machine Setups

OBJECTIVES

After completing this unit, you should be able to:

1 Mount and remove a milling machine arbor

2 Mount and remove a milling cutter

3 Align the milling machine table and vise

Before any operation is performed on a milling machine, it is important that the machine be properly set up. Proper setup of the machine will prolong the life of the machine and its accessories and will produce accurate work. The operator should always follow safety procedures to prevent injury and to avoid damaging the machines and spoiling the workpiece.

▶▶ Milling Machine Safety

The milling machine, like any other machine, demands the total attention of the operator and a thorough understanding of the hazards associated with its operation. The following precautions should be taken during operation of a milling machine:

1. Be sure that the work and the cutter are mounted securely before taking a cut.

2. Always wear safety glasses.

3. When mounting or removing milling cutters, always hold them with a cloth to avoid being cut.

4. When setting up work, move the table as far as possible from the cutter to avoid cutting your hands.

5. Be sure that the cutter and machine parts clear the work (Fig. 66-1).

6. *Never* attempt to mount, measure, or adjust work until the cutter has *completely* stopped.

7. Keep your hands, brushes, and rags away from a revolving milling cutter *at all times*.

■ Figure 66-1 Before making a cut, be sure that the arbor and arbor support will clear the workpiece. *(Kelmar Associates)*

8. When using milling cutters, do not use an excessively heavy cut or feed. This can cause a cutter to break, and the flying pieces could cause serious injury.

Figure 66-2 Poor housekeeping can cause tripping and slipping accidents. *(Kelmar Associates)*

9. Always use a *brush*, not a rag, to remove the cuttings after the cutter has stopped revolving.

10. Never reach over or near a revolving cutter; keep hands at least 12 in. (300 mm) from a revolving cutter.

11. Keep the floor around the machine free from chips, oil, and cutting fluid (Fig. 66-2).

▶▶ Milling Machine Setups

To prolong the life of a milling machine and its accessories and to produce accurate work, the following actions should be taken when milling machine setups are made:

1. Prior to mounting any accessory or attachment, check to see that both the machine surface and the accessory are free from dirt and chips.

2. Do not place the tools, cutters, or parts on the milling machine table. Place them on a piece of masonite, a board, or a bench kept for this purpose to prevent damaging the table or machined surfaces (Fig. 66-3).

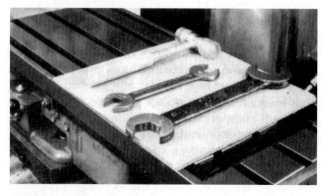

Figure 66-3 Place tools on a piece of masonite or plywood to protect the machine table. *(Kelmar Associates)*

3. When mounting cutters, be sure to use keys on all but slitting saws.

4. Check that the arbor spacers and bushings are clean and free from burrs.

5. When tightening the arbor nut, take care to make it only hand tight with a wrench.

 NEVER use a hammer on the nut or holding device on any machine.

The use of a hammer and wrench to tighten nuts will strip the threads and bend or damage the accessory or part.

6. When work is mounted in a vise, tighten the vise securely by hand and tap it into place with a LEAD OR SOFT-FACED HAMMER.

▶▶ Milling Hints

The following hints should help prevent many problems that could affect the accuracy and the efficiency of a milling operation:

1. Treat milling cutters with care; they can be damaged through careless use, handling, and storage.

2. Always examine the condition of the milling cutter cutting edges before use; if necessary, replace the cutters.

 NEVER use dull milling cutters.

3. Use the correct speeds and feeds for the size of the cutter and the type of work material to be cut. Too high a speed quickly dulls a cutter; too low a speed reduces productivity and may break the cutter teeth.

4. For maximum cutter life, use lower cutting speeds and increase the feed rate as much as possible.

5. Mount end mills as short as possible to prevent breakage.

6. Position arbor supports as close as possible to milling cutters for maximum rigidity.

7. Climb mill wherever possible, since it permits the use of higher speeds and generally improves the surface finish.

Note: The machine must be equipped with a backlash eliminator for climb milling.

8. Always direct the cutting force toward the solid part of the machine, vise, or fixture.

9. Make sure that the milling cutter is rotating in the proper direction for the cutting action of the teeth.

10. Use coarse tooth cutters for roughing cuts; they provide better chip clearance.

11. Stop the machine whenever there is any discoloration of the chips; it generally indicates poor cutting action. Examine the cutter.

12. Use a good supply of cutting fluid for all milling operations. Cutting fluid prolongs the life of the cutter and generally produces more accurate work.

▶▶ Mounting and Removing a Milling Machine Arbor

The milling arbor is used to hold the cutter during the machine operation. When mounting or removing an arbor, follow the proper procedure to preserve the accuracy of the machine. An improperly mounted arbor may damage the taper surfaces of the arbor or machine spindle, cause the arbor to bend, or make the cutter run out of true.

THE ARBOR ASSEMBLY

The milling cutter is driven by a key that fits into the keyways on the arbor and cutter (Fig. 66-4). This prevents the cutter from turning on the arbor. Spacer and bearing bushings hold the cutter in position on the arbor after the nut has been tightened. The tapered end of the arbor is held securely in the machine spindle by a draw-in bar (Fig. 66-5). The outer end of the arbor assembly is supported by the bearing bushing and the arbor support.

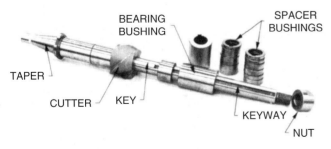

■ **Figure 66-4** The milling machine arbor holds and drives the milling cutter. *(Kelmar Associates)*

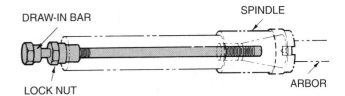

■ **Figure 66-5** The draw-in bar holds the arbor firmly in place in the spindle. *(Kelmar Associates)*

■ **Figure 66-6** Removing a milling machine arbor. *(Kelmar Associates)*

TO MOUNT AN ARBOR

1. Clean the tapered hole in the spindle and the taper on the arbor using a clean cloth.

2. Check to make sure that there are no cuttings or burrs in the taper that would prevent the arbor from running true.

3. Check the bearing bushing and remove any burrs using a honing stone.

4. Place the tapered end of the arbor in the spindle and align the spindle driving lugs with the arbor slots.

5. Place the right hand on the draw-in bar (Fig. 66-6) and turn the thread into the arbor approximately 1 in. (25 mm).

6. Tighten the draw-in bar lock nut securely against the back of the spindle (Fig. 66-7 on p. 526).

TO REMOVE AN ARBOR

1. Remove the milling machine cutter.

2. Loosen the lock nut on the draw-in bar approximately two turns.

3. With a soft-faced hammer, sharply strike the end of the draw-in bar until the arbor taper is free (Fig. 66-8 on p. 526).

Figure 66-7 Locking the arbor into the spindle with a draw-in bar. *(Kelmar Associates)*

Figure 66-8 Removing the arbor from the spindle. *(Kelmar Associates)*

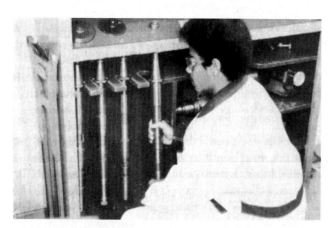

Figure 66-9 Storing the arbor in a suitable rack. *(Kelmar Associates)*

4. With one hand, hold the arbor; unscrew the draw-in bar from the arbor with the other hand (Fig. 66-6).

5. Carefully remove the arbor from the tapered spindle to avoid damage to the spindle or arbor tapers.

6. Leave the draw-in bar in the spindle for further use.

7. Store the arbor in a suitable rack to prevent it from being damaged or bent (Fig. 66-9).

▶▶ Mounting and Removing Milling Cutters

Milling cutters must be changed frequently to perform various operations, so it is important that certain sequences be followed to prevent damaging the cutter, the machine, or the arbor.

TO MOUNT A MILLING CUTTER

1. Remove the arbor nut and collar, and place them on a piece of masonite (Fig. 66-3).

2. Clean all spacing collar surfaces of cuttings and burrs.

3. Set the machine to the slowest spindle speed.

4. Check the direction of the arbor rotation by starting and stopping the machine spindle.

5. Slide the spacing collars on the arbor to the position desired for the cutter (Fig. 66-10).

6. Fit a key into the arbor keyway at the position where the cutter is to be located.

7. Hold the cutter with a cloth and mount it on the arbor. Make sure that the cutter teeth point in the direction of the arbor rotation (Fig. 66-10).

8. Slide the arbor support in place and be sure that it is on a bearing bushing on the arbor.

9. Put on additional spacers, leaving room for the arbor nut. *TIGHTEN THE NUT BY HAND.*

10. Lock the arbor support in position (Fig. 66-11).

11. Tighten the arbor nut firmly with a wrench, using only hand pressure.

12. Lubricate the bearing collar in the arbor support.

13. Make sure that the arbor and arbor support will clear the work (Fig. 66-1).

Figure 66-10 Use spacing collars to set the cutter in position. *(Kelmar Associates)*

Figure 66-11 Tightening the arbor support.
(Kelmar Associates)

TO REMOVE A MILLING CUTTER

1. Be sure that the arbor support is in place and supporting the arbor on a bearing bushing before using a wrench on the arbor nut. This precaution will prevent bending the arbor.
2. Clean all cuttings from the arbor and cutter.
3. Set the machine to the lowest spindle speed.
4. Loosen the arbor nut with a properly fitting wrench.

Note: Most threads on an arbor are left-hand; therefore, loosen in a clockwise direction.

5. Loosen the arbor support and remove it from the overarm.
6. Remove the nut, spacers, and cutter. Place them on a board (Fig. 66-12) and not on the table surface.
7. Clean the spacer and nut surfaces, and replace them on the arbor. *Do not use a wrench to tighten the arbor nut at this time.*
8. Store the cutter in a proper place to prevent the cutting edges from being damaged.

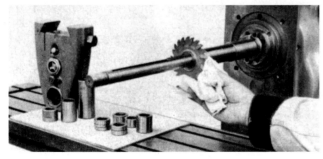

Figure 66-12 When changing a cutter, protect the milling machine table with a piece of masonite. *(Kelmar Associates)*

▶▶ Aligning the Table on a Universal Milling Machine

If the workpiece is to be machined accurately to a layout or have cuts made square or parallel to a surface, it is always good practice to align the table of a universal milling machine prior to aligning the vise or fixture.

Note: If a long keyway is to be milled in a shaft, it is of utmost importance to align the table. If it is not aligned properly, the keyway will not be milled parallel to the axis of the shaft.

step	procedure
1.	Clean the table and the face of the column thoroughly.
2.	Mount a dial indicator *on the table* by means of a magnetic base or any suitable mounting device (Fig. 66-13).
3.	Move the table toward the column until the dial indicator registers approximately one-quarter of a revolution.
4.	Set the indicator bezel to zero (0) by turning the dial.
5.	Using the table feed handwheel, move the table along the width of the column.
6.	Note the reading on the dial indicator and compare it to the zero (0) reading at the other end of the column.

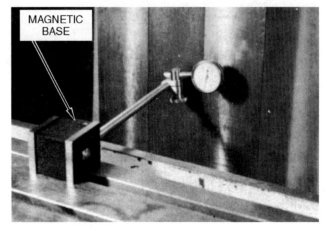

MAGNETIC BASE

Figure 66-13 Aligning a universal milling machine table. *(Kelmar Associates)*

7. If there is any movement of the indicator hand, loosen the locking nuts on the *swivel table housing.*

8. Adjust the table for half the difference of the needle movement and lock the table housing.

9. *Recheck* the table for alignment and adjust, if necessary, until there is no movement in the indicator as it is moved along the column.

Note: *Always indicate from the table to the face of the column,* never from the column to the table.

▸▸ Aligning the Milling Machine Vise

Whenever accuracy is required on the workpiece, it is necessary to align the device which holds the workpiece. This may be a vise, an angle plate, or a special fixture. Since most work is held in a vise, the method of aligning this accessory will be outlined.

TO ALIGN THE VISE PARALLEL TO THE TABLE TRAVEL

1. Clean the surface of the table and the base of the vise.

2. Mount and fasten the vise on the table.

3. Swivel the vise until the solid jaw is approximately parallel with the table slots.

4. Mount an indicator on the arbor or cutter (Fig. 66-14).

5. Make sure the solid jaw is clean and free from burrs.

6. Adjust the table until the indicator registers about one-quarter of a revolution against a parallel held between the jaws of the vise.

7. Set the bezel to zero (0). See Fig. 66-15a.

8. Move the table along for the length of the parallel and note the reading of the indicator (Fig. 66-15b). Compare it to the zero (0) reading at the other end of the parallel.

9. Loosen the nuts on the upper or swivel part of the vise.

10. Adjust the vise to half the difference of the indicator readings by tapping it with your hand or with a *soft-faced* hammer in the appropriate direction.

Note: *Never tap the vise so that the parallel moves against the indicator plunger.* This will damage the indicator.

11. *Recheck* the vise for alignment and adjust, if necessary, until there is no movement in the indicator as it is moved along the parallel.

Methods of aligning an angle plate are shown in Fig. 66-16.

■ **Figure 66-14** A dial indicator mounted on the arbor for alignment purposes. *(Kelmar Associates)*

(a)

(b)

■ **Figure 66-15** Accurately aligning a vise parallel to the table travel. *(Kelmar Associates)*

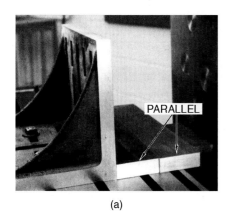
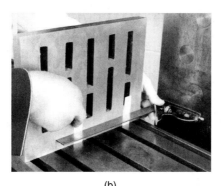
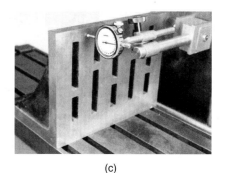

| (a) | (b) | (c) |

Figure 66-16 (a) Aligning an angle plate parallel to the column; (b) and (c) aligning an angle plate at right angles to the column. *(Kelmar Associates)*

unit 66 review questions

1. Why is the proper setup of the machine important?

Safety

2. What can happen if an excessively heavy cut is taken or too fast a feed used?

3. How far should the hands be kept from a revolving cutter?

Mounting and Removing Arbors

4. What is the purpose of an arbor?

5. How is the arbor held securely in the machine spindle?

6. How is the outer end of the arbor supported?

7. How should the arbor be stored when it is not in the machine?

8. In which direction should the cutter teeth point when it is mounted on an arbor?

9. Why should the arbor support be in place before attempting to remove a cutter from the arbor?

Aligning the Table and Vise

10. Why is the table alignment so important to most milling operations?

11. Why must the indicator be mounted on the table and not the column or the arbor when aligning a universal table?

12. By means of suitable sketches, show how the vise may be aligned:

    a. Parallel to the line of longitudinal table travel
    b. At right angles to the milling machine column

UNIT 67

Horizontal Milling Operations

OBJECTIVES

After completing this unit, you will be able to:

1 Set a cutter to the proper depth

2 Mill a flat surface on a workpiece

3 Perform operations such as face, side, straddle, and gang milling

The milling machine is one of the most versatile machine tools found in industry. The wide variety of operations that can be performed depends on the type of machine used, the cutter used, and the attachments and accessories available for the milling machine. The two basic types of milling are:

1. *Plain milling,* where the surface cut is parallel to the periphery of the cutter. These surfaces may either be flat or formed (Fig. 67-1).

2. *Face milling,* where the surface cut is at right angles to the axis of the cutter (Fig. 67-2).

■ **Figure 67-1** Milling the surface of a workpiece held in a fixture. *(Cincinnati Milacron, Inc.)*

■ **Figure 67-2** Face milling on a manufacturing-type milling machine. *(Cincinnati Milacron, Inc.)*

▸▸ Setting the Cutter to the Work Surface

Before setting a depth of cut, the operator should check that the work and the cutter are properly mounted and that the cutter is revolving in the proper direction.

TO SET THE CUTTER TO THE WORK SURFACE

1. Raise the work to within .250 in. (6 mm) of the cutter and directly under it.
2. Hold a long piece of thin paper on the surface of the work (Fig. 67-3).

Note: Have the paper long enough to prevent the fingers from coming in contact with the revolving cutter.

3. Start the cutter rotating.
4. With the left hand on the elevating screw handle, move the work *slowly* until the cutter grips the paper.
5. Stop the spindle.
6. Move the machine table so that the cutter just clears the end of the workpiece (Fig. 67-4).
7. Raise the knee .002 in. (0.05 mm) for paper thickness.
8. Set the graduated collar on the elevating screw handle to zero (0). Do not move the elevating screw handle.
9. Move the work clear of the cutter and raise the table to the desired depth of cut.

■ Figure 67-4 Make sure that the cutter clears the end of the workpiece before setting the depth of cut. *(Kelmar Associates)*

Note: If the knee is moved up beyond the desired amount, turn the handle backward one-half turn and then come up to the required line. This will take up the backlash in the thread movement.

This method can also be used when the edge of a cutter is set to the side of a piece of work. In this case, the paper will be placed between the side of the cutter and the side of the workpiece.

▸▸ Milling a Flat Surface

One of the most common operations performed on a milling machine is that of machining a flat surface. Flat surfaces are generally machined on a workpiece with a helical milling cutter. The work may be held in a vise or clamped to a table. Follow this procedure:

1. Remove all burrs from all edges of the work with a mill file (Fig. 67-5).
2. Clean the vise and workpiece.

■ Figure 67-3 Setting the cutter to the workpiece. *(Kelmar Associates)*

■ Figure 67-5 Remove all burrs before setting a workpiece in a vise. *(Kelmar Associates)*

3. Align the vise to the column face of the milling machine using a dial indicator.

4. Set the work in the vise using parallels and paper feelers under each corner to make sure that the work is seated on the parallels.

5. Tighten the vise securely by hand.

> 🛑 **STOP** **DO NOT** use a hammer on the vise wrench.

6. Tap the work *lightly* at the four corners with a soft-faced hammer until the paper feelers are tight between the work and the parallels (Fig. 67-6).

7. Select a plain helical cutter wider than the work to be machined.

8. Mount the cutter on the arbor for conventional milling (Fig. 67-7).

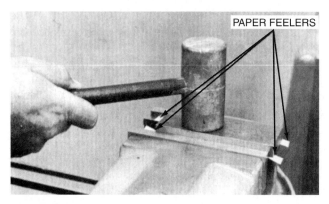

PAPER FEELERS

■ **Figure 67-6** Tap the workpiece down with a soft-faced hammer until all paper feelers are tight. *(Kelmar Associates)*

■ **Figure 67-7** Most stock is machined by conventional milling. *(Kelmar Associates)*

■ **Figure 67-8** After setting the cutter to the workpiece, set the graduated collar to zero.

9. Set the proper speed for the size of cutter and the type of work material.

10. Set the feed to approximately .003 to .005 in. (0.08 to 0.13 mm) chip per tooth (CPT).

11. Start the cutter and raise the work, using a paper feeler between the cutter and the work (Fig. 67-3).

12. Stop the spindle when the cutter just cuts the paper.

13. Raise the knee .002 in. (0.05 mm) for paper thickness.

14. Set the graduated collar on the elevating screw handwheel to zero (0) (Fig. 67-8).

15. Move the work clear of the cutter and set the depth of cut using the graduated collar.

16. For roughing cuts, use a depth of not less than .125 in. (3 mm) and .010 to .025 in. (0.25 to 0.63 mm) for finish cuts.

17. Set the table dogs for the length of cut.

18. Engage the feed and cut side 1.

19. Set up and cut the remaining sides, as required.

Note: See Unit 62 for the setup procedures if the four sides of a block must be machined.

▸▸ Face Milling

Face milling is the process of producing a flat vertical surface at right angles to the cutter axis (Fig. 67-9). The cutters used for face milling are generally inserted-tooth cutters or shell end mills. *Face milling cutters* are made in sizes of 6-in. (150-mm) diameter and over; cutters less than 6-in. (150-mm) diameter are usually called *shell end mills*.

1. When face milling a large surface, use an inserted-tooth cutter and mount it into the spindle of the machine.

■ Figure 67-9 Face milling produces flat, vertical surfaces.

■ Figure 67-10 Side milling a vertical surface. *(Kelmar Associates)*

2. When milling smaller surfaces, the face milling cutter should be about 1 in. (25 mm) larger than the width of the workpiece.

3. Set the speeds and feeds for the type of cutter and the material being cut.

4. Set up the work on the milling machine, making sure that the work-holding clamps do not interfere with the cutting action.

5. Use cutting fluid if the cutter material or work will allow.

6. Take as large a roughing cut as possible to bring the surface close to within .030 to .060 in. (0.8 to 1.5 mm) of the finish size.

7. Set the depth of the finish cut and machine the surface to size.

8. After completing the operation, clean and store the cutter and clamping equipment in their proper places.

▶▶ Side Milling

Side milling is often used to machine a vertical surface on the sides or the ends of a workpiece (Fig. 67-10).

Follow this procedure:

1. Set up the work in a vise and on parallels.

Note: Be sure that the layout line on the surface to be cut extends about .50 in. (13 mm) beyond the edge of the vise and the parallels to prevent the cutter, vise, or parallels from being damaged.

2. Tighten the vise securely by hand.

 ***DO NOT* use a hammer on the vise wrench.**

3. Tap the work *lightly* at the four corners with a soft-faced hammer until the paper feelers are tight between the work and the parallels.

4. Mount a side milling cutter as close to the spindle bearing as possible to provide maximum rigidity when milling.

5. Set the proper speed and feed for the cutter being used.

6. Start the machine and move the table until the top corner of the work just touches the revolving cutter. Make sure that the cutter is rotating in the proper direction.

7. Set the crossfeed graduated collar to zero (0).

8. With the table handwheel, move the work clear of the cutter.

9. Set the required depth of cut with the crossfeed handle.

10. Lock the saddle to prevent movement during the cut.

11. Take the cut across the surface.

CENTERING A CUTTER TO MILL A SLOT

1. Locate the cutter as close to the center of the work as possible.

2. Using a steel square and rule (Fig. 67-11 on p. 534) or a gage block, adjust the work to the center by using the crossfeed screw dial.

3. Lock the saddle to prevent movement during the cut.

4. Move the work clear of the cutter and set the depth of cut.

5. Proceed to cut the slot using the same methods as for milling a flat surface.

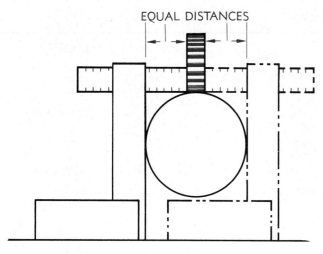

Figure 67-11 Centering the cutter over a round shaft. *(Kelmar Associates)*

Figure 67-12 Straddle milling. *(Cincinnati Milacron, Inc.)*

▶▶ Straddle Milling

Straddle milling (Fig. 67-12) involves the use of two side milling cutters to machine the opposite sides of a workpiece parallel in one cut. The cutters are separated on the arbor by a spacer or spacers of the required length so that the distance between the inside faces of the cutters is equal to the desired size. An adjustable micrometer ar-

bor spacer may be used to vary the distance between the two cutters and to compensate for the wear or the regrinding of the side milling cutters. Applications of straddle milling are the milling of square and hexagonal heads on bolts.

The following points should be observed when straddle milling:

1. Select two sharp side milling cutters, preferably staggered-tooth, of suitable size.
2. Mount the cutters, with suitable arbor spacers, as close to the column as the work will permit.
3. Mount the arbor support as close to the cutters as possible to provide rigidity for the cutters and arbor.
4. Center the cutters on the workpiece in the proper location.
5. Tighten the saddle lock to prevent any movement during a cut.
6. Set the cutter to the work surface.
7. Move the table so that the cutter clears the end of the workpiece.
8. Set the depth of cut required and tighten the knee clamp.
9. Set the proper speeds and feeds for the cutter size and the type of work material; check cutter rotation.
10. Use a good supply of cutting fluid and complete the straddle milling operation in one cut.

▶▶ Gang Milling

Gang milling (Fig. 67-13) is a fast method of milling used a great deal in production work. It is performed by using two or more cutters on the arbor to produce the desired shape. The cutters may be a combination of plain and side milling cutters. If more than one helical milling cutter is used, a right- and left-hand helix cutter should be used to offset the thrust created by this type of cutter and to minimize the possibility of chatter. If several cutters are used and the slots are to be the same size, it is important that the diameter and width of each cutter be the same.

The following points should be observed to avoid problems during a gang milling operation:

1. Select cutters that are as close as possible to the same size and the same number of teeth. This will allow maximum speeds and feeds to be used.

■ Figure 67-13 Gang milling. *(Cincinnati Milacron, Inc.)*

■ Figure 67-14 Sawing and slitting. *(Kelmar Associates)*

2. Mount the correct cutters on the arbor as close to the machine column as possible.

3. Fasten the arbor support as close to the cutters as the work will permit.

4. Set the spindle speed to suit the largest-diameter cutter.

5. Be sure that the work is fastened securely and that the work-holding devices will not come in contact with the cutters.

6. Use a good flow of cutting fluid to assist the cutting action and produce a good surface finish.

▶▶ Sawing and Slitting

Metal-slitting saws may be used for milling narrow slots and for cutting off work (Fig. 67-14). Plain slitting saws, because of their thin cross section, are rather fragile cutting tools. If they are not used carefully, they will break easily.

To get the maximum life from a slitting saw, take the following precautions:

1. Never key a slitting saw on the arbor unless it is mounted in a special flanged mounting collar. If the slitting saw is keyed and jams in the work, it will rotate and cut through the key, making it difficult to remove the broken saw.

Note: The arbor nut must be drawn up as tightly as possible by *hand* only.

2. When selecting a slitting saw, choose one with the smallest diameter that will permit adequate clearance between the arbor collars or supports and the clamping bolts, holding device, or workpiece.

3. Mount the saw close to the column face and have the outer arbor support as close as possible to the saw.

4. Always use a sharp cutter.

5. Be sure that the table gibs are drawn up to eliminate any play between the table and the saddle.

6. Operate saws at approximately one-quarter to one-eighth the feed per tooth used for side milling cutters.

7. When sawing or slitting fairly long or deep slots, it is advisable to climb-mill to prevent the cutter from crowding sideways and breaking. The table should be fed carefully by hand and the backlash eliminator engaged.

Setting Cutter to Work Surface

1. Why should a long, thin piece of paper be used for setting a cutter to the work surface?

2. What should be done when the knee is moved beyond the desired amount?

Milling a Flat Surface

3. List the steps required to set work in a vise for milling.

4. Explain how the cutter should be set to the surface of the work.

Face Milling

5. Describe the process of face milling.

6. How large should a face milling cutter be in relation to the width of the work?

Side Milling

7. Why is the side milling operation often used?

8. How should the work be set up in a vise for side milling?

9. Why should the table saddle be locked before machining?

10. Explain how a cutter can be aligned with the center of a round workpiece.

Straddle and Gang Milling

11. What is the difference between straddle and gang milling?

12. Why should the arbor support be mounted as close to the cutters as possible?

Sawing and Slitting

13. List five of the most important precautions for sawing or slitting.

UNIT 68

The Indexing, or Dividing, Head

OBJECTIVES

After completing this unit, you will be able to:

1 Calculate and mill flats by simple and direct indexing

2 Calculate the indexing necessary with a wide-range divider

3 Calculate the indexing necessary for angular and differential indexing

The *indexing,* or *dividing, head* is one of the most important attachments for the milling machine. It is used to divide the circumference of a workpiece into equally spaced divisions when milling gears, splines, squares, and hexagons. It may also be used to rotate the workpiece at a predetermined ratio to the table feed rate to produce cams and helical grooves on gears, drills, reamers, and other parts.

▶▶ Index Head Parts

The universal dividing head set consists of the *headstock* with *index plates,* headstock *change gears* and *quadrant, universal chuck, footstock,* and *center rest* (Fig. 68-1 on p. 538). A *swiveling block* mounted in the *base* enables the headstock to be tilted from 5° below the horizontal position to 10° beyond the vertical position. The side of the base and the block are graduated to indicate the angle of the setting. Mounted in the swiveling block is a *spindle,* with a *40-tooth worm wheel* attached, that meshes with a worm (Fig. 68-2 on p. 538). The worm, at right angles to the spindle, is connected to the *index crank,* the pin of which engages in the *index plate* (Fig. 68-3 on p. 538). A *direct indexing plate* is attached to the front of the spindle.

A 60° center may be inserted into the front of the spindle, and a *universal chuck* may be threaded onto the end of the spindle.

The *footstock* is used in conjunction with the headstock to support work held between centers or the end of work held in a chuck. The footstock center may be adjusted longitudinally to accommodate various lengths of work and may be raised or lowered off center. It may also be tilted out of parallel with the base when cuts are being made on tapered work.

Long, slender work held between centers is prevented from bending by the *adjustable center rest.*

▶▶ Methods of Indexing

The main purpose of the indexing, or dividing, head is to divide the workpiece circumference accurately into any number of divisions. This may be accomplished by the following indexing methods: direct, simple, angular, and differential.

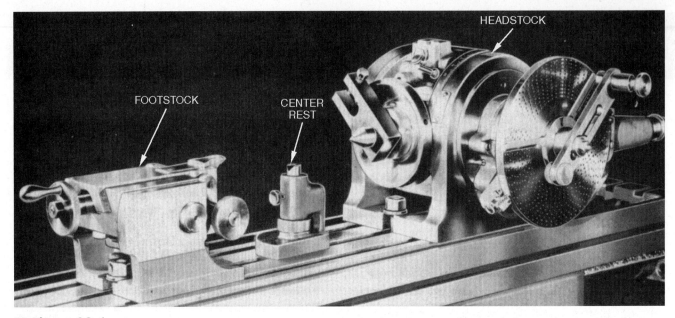

Figure 68-1 A universal dividing head set. *(Cincinnati Milacron, Inc.)*

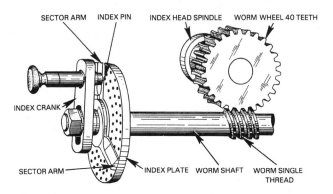

Figure 68-2 Section through a dividing head, showing the worm wheel and worm shaft.

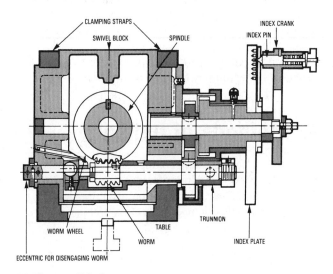

Figure 68-3 Section through a dividing head, showing the spindle and index plate.

DIRECT INDEXING

Direct indexing is the simplest form of indexing. It is performed by disengaging the worm shaft from the worm wheel by means of an eccentric device in the dividing head. Some direct dividing heads do not have a worm and worm wheel but rotate on bearings. The index plates contain slots, which are numbered, and a spring-loaded tongue lock is used to engage in the proper slot. Direct indexing is used for quick indexing of the workpiece when cutting flutes, hexagons, squares, and other shapes.

The work is rotated the required amount and held in place by a pin that engages into a hole or slot in the *direct indexing plate* mounted on the end of the dividing head spindle. The direct indexing plate usually contains three sets of hole circles, or slots: 24, 30, and 36. The number of divisions it is possible to index is limited to numbers that are factors of 24, 30, or 36. The common divisions that can be obtained by direct indexing are listed in Table 68.1.

table 68.1   Direct indexing divisions

Plate Hole Circles	
24	2 3 4 _ 6 8 _ _ 12 _ _ 24 _ _
30	2 3 _ 5 6 _ _ 10 _ 15 _ _ 30 _
36	2 3 4 _ 6 _ 9 _ 12 _ 18 _ _ 36

What direct indexing is necessary to mill eight flutes on a reamer blank?

Solution

Since the 24-hole circle is the only one divisible by 8 (the required number of divisions), it is the only circle that can be used in this case.

$$\text{Indexing} = \frac{24}{8} = 3 \text{ holes on a 24-hole circle}$$

 NEVER count the hole or slot in which the index pin is engaged.

To Mill a Square by Direct Indexing

1. Disengage the worm and worm shaft by turning the worm disengaging shaft lever if the dividing head is so equipped.
2. Adjust the plunger behind the index plate into the 24-hole circle, or slot (Fig. 68-4).
3. Mount the workpiece in the dividing head chuck or between centers.
4. Adjust the cutter height and cut the first side.
5. Remove the plunger pin using the plunger pin lever (Fig. 68-5).

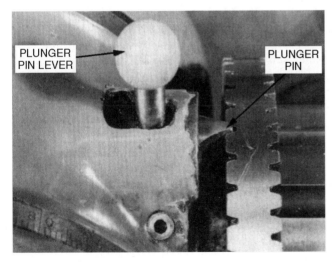

Figure 68-4 Adjusting the plunger pin to fit into the proper hole circle, or slot. *(Kelmar Associates)*

Figure 68-5 The plunger pin and the direct indexing plate are used for indexing a limited number of divisions. *(Kelmar Associates)*

6. Turn the plate attached to the dividing head spindle one-half turn (12 holes or slots) and engage the plunger pin.
7. Take the second cut.
8. Measure the work across the flats and adjust the work height if required.
9. Cut the remaining sides by indexing every six holes until all surfaces are cut.

SIMPLE INDEXING

In *simple indexing,* the work is positioned by means of the crank, index plate, and sector arms. The worm attached to the crank must be engaged with the worm wheel on the dividing head spindle. Since there are 40 teeth on the worm wheel, one complete turn of the index crank will cause the spindle and the work to rotate one-fortieth of a turn. Similarly, 40 turns of the crank will rotate the spindle and work 1 turn. Thus, there is a ratio of 40:1 between the turns of the index crank and the dividing head spindle.

To calculate the indexing or the number of turns of the crank for most divisions, it is necessary only to divide 40 by the number of divisions (*N*) to be cut, or

$$\text{Indexing} = \frac{40}{N}$$

EXAMPLE

The indexing required to cut eight flutes would be:

$$\frac{40}{8} = 5 \text{ full turns of the index crank}$$

If, however, it were necessary to cut seven flutes, the indexing would be

$$\frac{40}{7} = 5\tfrac{5}{7} \text{ turns}$$

Five complete turns are easily made; however, the five-sevenths of a turn involves the use of the index plate and sector arms.

Index Plate and Sector Arms

The *index plate* is a circular plate provided with a series of equally spaced holes into which the index crank pin engages. The *sector arms* fit on the front of this plate and may be set to any portion of a complete turn.

To get five-sevenths of a turn, choose any hole circle (Table 68.2) that is divisible by the denominator 7, such as 21, then take five-sevenths of 21 = 15 holes on a 21-hole circle. Therefore, the indexing for seven flutes would be $\tfrac{40}{7} = 5\tfrac{5}{7}$ turns, or 5 complete turns plus 15 holes on the 21-hole circle. When extreme accuracy is required for indexing, choose the circle with the most holes.

The procedure for cutting seven flutes would be as follows:

1. Mount the proper index plate on the dividing head.

2. Loosen the index crank nut and set the index pin into a hole on the 21-hole circle.

3. Tighten the index crank nut and check to see that the pin enters the hole easily.

4. Loosen the setscrew on the sector arm.

5. Place the narrow edge of the left arm against the index pin.

6. Count 15 holes on the 21-hole circle. *Do not include the hole in which the index crank pin is engaged.*

table 68.2	Index-plate hole circles
Brown & Sharpe	
Plate 1	15-16-17-18-19-20
Plate 2	21-23-27-29-31-33
Plate 3	37-39-41-43-47-49
Cincinnati Standard Plate	
One side	24-25-28-30-34-37-38-39-41-42-43
Other side	46-47-49-51-53-54-57-58-59-62-66

■ **Figure 68-6** Setting the cutter to the top of a workpiece using a paper feeler. *(Kelmar Associates)*

7. Move the right sector arm slightly beyond the fifteenth hole and tighten the sector arm setscrew.

8. Align the cutter with the workpiece.

9. Start the machine and set the cutter to the top of the work by using a paper feeler (Fig. 68-6).

10. Move the table so that the cutter clears the end of the work.

11. Tighten the friction lock on the dividing head before making each cut and loosen the lock when indexing for spaces.

12. Set the depth of cut and take the first cut.

13. After the first flute has been cut, return the table to the original starting position.

14. Withdraw the index pin and turn the crank clockwise five full turns plus the 15 holes indicated by the right sector arm. Release the index pin between the fourteenth and fifteenth holes, and gently tap it until it drops into the fifteenth hole.

15. Turn the sector arm *farthest from the pin* clockwise until it is against the index pin.

Note: It is important that the arm *farthest from the pin* be held and turned. If the arm *next* to the pin were held and turned, the spacing between both sector arms could be increased when the other arm hits the pin. This could result in an indexing error that would not be noticeable until the work was completed.

16. Lock the dividing head; then continue machining and indexing for the remaining flutes. Whenever the crank pin is moved past the required hole, *remove the backlash* between the worm and worm wheel by turning the crank *counterclockwise* approximately one-half turn and then carefully *clockwise* until the pin engages the proper hole.

ANGULAR INDEXING

When the angular distance between divisions is given instead of the number of divisions, the setup for simple indexing may be used; however, the method of calculating the indexing is changed.

One complete turn of the index crank turns the work one-fortieth of a turn, or one-fortieth of 360°, which equals 9°.

When the angular dimension is given in *degrees,* the indexing is then calculated as follows:

$$\text{Indexing in degrees} = \frac{\text{no. of degrees required}}{9}$$

EXAMPLE

Calculate the indexing for 45°.

$$\text{Indexing} = \frac{45}{9}$$

$$= 5 \text{ complete turns}$$

EXAMPLE

Calculate the indexing for 60°.

$$\text{Indexing} = \frac{60}{9}$$

$$= 6\,\tfrac{2}{3}$$

$$= 6 \text{ complete turns plus 12 holes on an 18-hole circle}$$

If the dimensions are given in *degrees* and *minutes,* it will be necessary to convert the degrees into minutes (number of degrees × 60′) and add this sum to the minutes required. The indexing in minutes is calculated as follows:

$$\text{Indexing in minutes} = \frac{\text{no. of minutes required}}{540}$$

EXAMPLE

Calculate the indexing necessary for 24′.

$$\text{Indexing} = \frac{24}{540}$$

$$= \frac{4}{90}$$

$$= \frac{1}{22.5}$$

The indexing for 24′ would be 1 hole on the 22.5-hole circle. As the 23-hole circle is the nearest hole circle, the indexing would be one hole on the 23-hole circle. Since in this case there is a slight error (approximately one-half minute) in indexing, it is advisable to use this method only for a few divisions if extreme accuracy is required.

EXAMPLE

Calculate the indexing for 24°30′. First, convert 24°30′ into minutes:

$$(24 \times 60') = 1440'$$
$$\text{Add } 30' = 30'$$
$$\text{Total} = 1470'$$
$$\text{Indexing} = \frac{1470}{540}$$
$$= 2\,\tfrac{13}{18} \text{ turns}$$
$$= 2 \text{ complete turns plus 13 holes on an 18-hole circle}$$

When indexing for degrees and half degrees (30′), use the 18-hole circle (Brown & Sharpe).

$$\tfrac{1}{2}° \,(30') = 1 \text{ hole on the 18-hole circle}$$

$$1° = 2 \text{ holes on the 18-hole circle}$$

When indexing for ⅓° (20′) and ⅔° (40′), the 27-hole circle should be used (Brown & Sharpe).

$$\tfrac{1}{3}° \,(20') = 1 \text{ hole on the 27-hole circle}$$

$$\tfrac{2}{3}° \,(40') = 2 \text{ holes on the 27-hole circle}$$

When indexing for minutes using a Cincinnati dividing head, note that one space on the 54-hole circle will rotate the work 10′ ($\tfrac{1}{54} \times 540$).

DIFFERENTIAL INDEXING

When it is impossible to calculate the required indexing by the simple indexing method, that is, when the fraction 40/N cannot be reduced to a factor of one of the available hole circles, it is necessary to use *differential indexing.*

With this method of indexing, the index plate must be revolved either forward or backward a part of a turn while the index crank is turned to attain the proper spacing or indexing.

In differential indexing, as in simple indexing, the index crank rotates the dividing head spindle. The spindle rotates the index plate, after the locking pin has been disengaged, by means of change gears connecting the dividing head spindle and the worm shaft (Fig. 68-7 on p. 542). The rotation of the plate may be either in the same direction (positive) or in the opposite direction (negative) of the index crank. This change of rotation is effected by an idler gear or gears in the gear train.

Figure 68-7 Headstock geared for differential indexing. *(Kelmar Associates)*

by using an idler gear. However, if the approximate number is smaller than the required number, the resulting fraction is minus and the index plate must move in a counterclockwise direction. This *negative rotation* requires the use of two idler gears. The numerator of the fraction represents the driving (spindle) gear or gears, and the denominator represents the driven (worm) gear or gears. The gearing may be either simple or compound and the rotation is as follows:

> *Simple gearing*—one idler for a positive rotation of the index plate and two idlers for a negative rotation of the index plate

> *Compound gearing*—one idler for a negative rotation of the index plate and two idlers for a positive rotation of the index plate

EXAMPLE

Calculate the indexing and change gears required for 57 divisions.

The change gears supplied with the dividing head are as follows: 24, 24, 28, 32, 40, 44, 48, 56, 64, 72, 86, 100.

The available index plate hole circles are as follows:

Plate 1:   15, 16, 17, 18, 19, 20
Plate 2:   21, 23, 27, 29, 31, 33
Plate 3:   37, 39, 41, 43, 47, 49

Solution

1. Indexing $= \dfrac{40}{N} = \dfrac{40}{57}$

    Since there is no 57-hole circle and it is impossible to reduce this fraction to suit any hole circle, it is necessary to select an approximate number close to 57, for which simple indexing may be calculated.

2. Let the approximate number of divisions equal 56.

3. Indexing for 56 divisions $= \dfrac{40}{56} = \dfrac{5}{7}$

    or 15 holes on the 21-hole circle.

4. Gear ratio $= (A - N) \times \dfrac{40}{A}$

    $= (56 - 57) \times \dfrac{40}{56}$

    $= -1 \times \dfrac{40}{56}$

    $= -\dfrac{5}{7}$

When it is necessary to calculate the indexing for a required number of divisions by the differential method, a number is chosen close to the required divisions that can be indexed by simple indexing.

To illustrate the principle of differential indexing, assume that the index crank has to be rotated one-ninth of a turn and there is only an 8-hole circle available.

If the crank is moved one-ninth of a turn, the index pin will contact the plate at a spot before the first hole on the 8-hole circle. The exact position of this spot would be the difference between one-eighth and one-ninth of a revolution of the crank. This would be

$$\tfrac{1}{8} - \tfrac{1}{9} = \frac{9 - 8}{72} = \tfrac{1}{72}$$

of a turn *less* than one-eighth of a turn, or one-seventy-second of a turn short of the first hole. Since there is no hole at this point into which the pin could engage, it is necessary to cause the plate to rotate backward by means of change gears one-seventy-second of a turn so that the pin will engage in a hole. At this point, the index crank will be locked at exactly one-ninth of a turn.

The method of calculating the change gears (Fig. 68-7) required to rotate the plate the proper amount is as follows:

$$\text{Change gear ratio} = (A - N) \times \frac{40}{A}$$

$$= \frac{\text{driver (spindle) gear}}{\text{driven (worm) gear}}$$

A = approximate number of divisions
N = required number of divisions

When the approximate number of divisions is larger than the required number, the resulting fraction is plus and the index plate must move in the same direction as the crank (clockwise). This *positive rotation* is accomplished

$$\text{Change gears} = -\frac{5}{7} \times \frac{8}{8}$$

$$= -\frac{40 \text{ (spindle gear)}}{56 \text{ (worn gear)}}$$

Therefore, for indexing 57 divisions, a 40-tooth gear is mounted on the dividing head spindle, and a 56-tooth gear is mounted on the worm shaft. Since the fraction is a negative quantity and simple gearing is to be used, the index plate rotation is negative, or counterclockwise, and two idlers must be used. After the proper gears are installed, the simple indexing procedure for 56 divisions should be followed.

▶▶ The Wide-Range Dividing Head

Although simple or differential indexing is satisfactory for most indexing problems, there may be certain divisions that cannot be indexed by either of these methods. Cincinnati Machine, Inc., manufactures a *wide-range divider* that may be applied to a Cincinnati universal dividing head. With this attachment, it is possible to obtain from 2 up to 400,000 divisions.

The wide-range divider consists of a large index plate *A* (Fig. 68-8); sector arms; and crank *B*, which engages in the plate *A*. This large plate contains 11-hole circles on each side. Mounted in front of the large index plate is a small index plate *C* containing a 54-hole and a 100-hole circle. The crank *D* operates through a reduction of gears having a ratio of 100:1. These gears are mounted in the housing *G*. The ratio between the worm (and the crank *B*) and the dividing head spindle is 40:1.

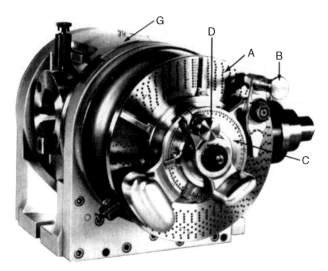

■ Figure 68-8 A wide-range dividing head.
(Cincinnati Milacron, Inc.)

INDEXING FOR DIVISIONS

The ratio of the large index crank to the dividing head is 40:1, as in simple indexing. The ratio of the small index crank, which drives the large crank by planetary gearing, is 100:1. Therefore, *one turn of the small crank* drives the index head spindle $\frac{1}{100}$ of $\frac{1}{40}$, or $\frac{1}{4000}$ of a turn. One hole on the 100-hole circle of the small index plate $C = \frac{1}{100} \times \frac{1}{4000}$, or $\frac{1}{400,000}$ of a turn. Therefore, the formula for indexing *divisions* with a wide-range divider is 400,000/N and is applied as follows.

No. of turns of large index crank	No. of holes on 100-hole circle of large plate	No. of holes on 100-hole circle of small plate
	$\dfrac{4\ 0\ |\ 0\ 0\ |\ 0\ 0}{N}$	

$$\text{For 1250 divisions: } \frac{400,000}{1250}$$

Holes required on the 100-hole circle of the large plate	Holes required on the 100-hole circle of the small plate

$$\frac{4\ 0\ \big|\ \overset{3}{0\ 0}\ \big|\ \overset{2\ 0}{0\ 0}}{1250}$$

Since the ratio of the large index crank is 40:1, any number that divides into 40 (the two numbers to the left of the short vertical line) represents full turns of the large index crank. If a 100-hole circle is used with the large crank, *one hole* on this circle will produce $\frac{1}{100}$ of $\frac{1}{40}$, or $\frac{1}{4000}$ of a turn. Therefore, any numbers that divide into 4000 (the two numbers to the left of the long vertical line) are indexed on the 100-hole circle of the large plate. The numbers to the right of the long vertical line are indexed on the 100-hole circle of the small plate. Thus, for the 1250 divisions, the indexing would be: 3 holes on the 100-hole circle of the large plate plus 20 holes on the 100-hole circle of the small plate.

ANGULAR INDEXING WITH THE WIDE-RANGE DIVIDER

The wide-range divider is especially suited for accurate angular indexing. Indexing in degrees, minutes, and seconds is easily accomplished without the complicated calculations necessary with standard dividing heads.

For angular indexing, both the large and small index cranks are set on the 54-hole circle of each plate. Each space on the 54-hole circle of the large plate will cause the dividing head spindle to rotate 10′. Each space on the 54-hole circle of the small plate will cause the work to rotate 6″. Therefore, for indexing angles with a wide-range divider, the following formulas are used.

$$\text{Degrees} = \frac{N}{9} \text{ (indexed on the large plate)}$$

$$\text{Minutes} = \frac{N}{10} \text{ (indexed on the large plate)}$$

$$\text{Seconds} = \frac{N}{6} \text{ (indexed on the small plate)}$$

EXAMPLE

Index for an angle of 17°36′18″.

Solution

1. $\text{Degrees} = \dfrac{17}{9}$

    $= 1\,\%$ turns

    OR

    one turn plus 48 holes on the 54-hole circle of the large index plate.

2. $\text{Minutes} = \dfrac{36}{10}$

    $= 3$ holes on a 54-hole circle of the large index plate, leaving a remainder of 6′.

3. Convert the 6′ into seconds (6 × 60 = 360″) and add it to the 18″ still required.

4. $\text{Seconds} = \dfrac{378}{6}$

    $= 63$ holes on a 54-hole circle of the small plate

    OR

    one turn and 9 holes on the 54-hole circle.

Therefore, to index for 17°36′18″ would require one turn and 51 holes (48 + 3) on the 54-hole circle of the large plate, plus one turn and 9 holes on the 54-hole circle of the small plate.

▶▶ Linear Graduating

The operation of producing accurate spaces on a piece of flat stock, or that of *linear graduating,* is easily accomplished on the horizontal milling machine (Fig. 68-9).

In this process, the work may be clamped to the table or held in a vise, depending on the shape and size of the part. Care must be taken to align the workpiece parallel with the table travel.

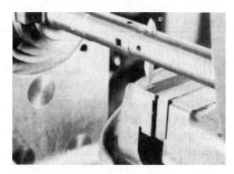

■ **Figure 68-9** Linear graduating. *(Kelmar Associates)*

■ **Figure 68-10** Gearing required for linear graduating. *(Kelmar Associates)*

To produce an *accurate* longitudinal movement of the table, the dividing head spindle is geared to the lead screw of the milling machine (Fig. 68-10).

If the dividing head spindle and the lead screw were connected with gears with equal number of teeth and the index crank turned one revolution, the spindle and lead screw on an inch milling machine would revolve one-fortieth of a revolution. This rotation of the lead screw (having 4 threads per inch [*tpi*]) would cause the table to move $\frac{1}{40} \times \frac{1}{4}$ (one turn of the lead screw) $= \frac{1}{160} = .00625$ in. (0.15 mm). Thus, five turns of the index crank would move the table 5 × .00625, or $\frac{1}{32}$ in. (0.78 mm).

The formula for calculating the indexing for linear graduations in thousandths of an inch is

$$\frac{N}{.00625}$$

Very small movements of the table, such as .001 in., may be obtained by applying the formula:

$$\frac{.001}{.00625} = \frac{1}{6\frac{1}{4}} \text{ turns}$$

($\frac{4}{25}$ turn), or 4 holes on the 25-hole circle.

If the lead screw of a metric milling machine had a pitch of 5 mm, one turn of the index crank would move the table one-fortieth of 5 mm, or 0.125 mm. Therefore, it would require four complete turns of the index crank to move the table 0.5 mm.

The formula for calculating the indexing for linear graduations in millimeters is

$$\frac{N}{0.125}$$

For a small movement of the table, such as 0.025 mm, apply the formula:

$$\frac{0.025}{0.125} = \text{⅕ turn, or 5 holes on a 25-hole circle}$$

Other suitable table movements may be obtained by using the appropriate hole circle and/or different change gears.

The point of the toolbit used for graduating is generally ground to a V-shape, although other special forms may be desired. The tool is mounted vertically in a suitable arbor that is of sufficient length to extend the toolbit over the workpiece (Fig. 68-9).

The uniformity of the length of the lines is controlled by the *accurate* movement of the crossfeed handwheel or by stops suitably mounted on the ways of the knee.

When graduating, position the starting point on the workpiece under the point of the *stationary* vertical toolbit. The work is moved clear of the tool by the crossfed handwheel and the proper depth is set by means of the vertical feed crank. The table is then locked in place. For a uniform width of the lines to be maintained, the work must be held absolutely flat and the table height must never be adjusted.

unit 68 review questions

Indexing, or Dividing, Head

1. Name four parts of a dividing head set.
2. Name four methods of indexing that may be performed using the dividing head.
3. For what purpose is direct indexing used?

Simple Indexing

4. Explain how the ratio of 40:1 is determined on a standard dividing head.
5. Calculate the simple indexing, using a Brown & Sharpe dividing head, for the following divisions: 37, 41, 22, 34, and 120.
6. What procedure should be followed in order to set the sector arms for 12 holes on an 18-hole circle?

Angular Indexing

7. Explain the principle of angular indexing.
8. Calculate the indexing, using a Cincinnati dividing head, for the following angles: 21°, 37°, 21°30′, and 37°40′.

Differential Indexing

9. For what purpose is differential indexing used?
10. What is meant by positive rotation and negative rotation of the index plate?

11. Using a Brown & Sharpe dividing head, calculate the indexing and change gears for the following divisions: 53, 59, 101, and 175. A standard set of change gears, having the following numbers of teeth, is supplied: 24, 24, 28, 32, 40, 44, 48, 56, 64, 72, 86, 100.

Wide-Range Dividing Head

12. How does the wide-range dividing head differ from a standard dividing head?
13. What two ratios are found in a wide-range dividing head?
14. Calculate the indexing for (a) 1000 and (b) 1200 divisions, using the wide-range dividing head.
15. Calculate the angular indexing for the following, using a wide-range dividing head: 20°45′, 25°15′32″.

Linear Graduating

16. How are the dividing head and the milling machine geared for linear graduating?
17. What indexing would be required to move the table .003 in. when using equal gearing on the dividing head and the lead screw?

UNIT 69

Gears

OBJECTIVES

After completing this unit, you should be able to:

1 Identify and state the purposes of six types of gears used in industry

2 Apply various formulas for calculating gear-tooth dimensions

When it is required to transmit rotary motion from one shaft to another, several methods, such as belts, pulleys, and gears, may be used. If the shafts are parallel to each other and quite a distance apart, a flat belt and large pulleys may be used to drive the second shaft, the speed of which may be controlled by the size of the pulleys.

When the shafts are closer together, as in the case of the sensitive drill press, a V-belt, which tends to reduce the excessive slippage of a flat belt, may be used. Here, the speed of the driven shaft may be controlled by means of stepped or variable speed pulleys. When the shafts are close together and parallel, some power may be transmitted by two rollers in contact, with one roller mounted on each shaft. Slippage is the main problem here, and the desired speed of the driven shaft cannot be maintained.

The methods outlined in this unit are means by which power may be transmitted from one shaft to another, but the speed of the driven shaft may not be accurate in all cases due to slippage between the driving and driven members (belts, pulleys, or rollers). In order to eliminate slippage and produce a positive drive, gears are used.

▶▶ Gears and Gearing

Gears are used to transmit power positively from one shaft to another by means of successively engaging teeth (in two gears). They are used in place of belt drives and other forms of friction drives when exact speed ratios and power transmission must be maintained. Gears may also be used to increase or decrease the speed of the driven shaft, thus decreasing or increasing the *torque* of the driven member.

Shafts in a gear drive or train are generally parallel. They may, however, be driven at any angle by means of suitably designed gears.

▶▶ Types of Gears

Spur gears (Fig. 69-1) are generally used to transmit power between two parallel shafts. The teeth on these gears are straight and parallel to the shafts to which they are attached. When two gears of different sizes are in mesh, the larger is called the *gear,* while the smaller is called the *pinion.* Spur gears are used where slow- to moderate-speed drives are required.

Internal gears (Fig. 69-2) are used where the shafts are parallel and the centers must be closer together than could be achieved with spur or helical gearing. This arrangement provides for a stronger drive, since there is a

■ **Figure 69-1** A spur gear and pinion are used for slower speeds.

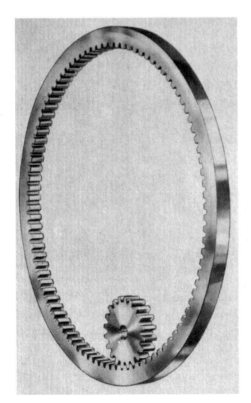

■ **Figure 69-2** Internal gears provide speed reductions with a minimum space requirement.

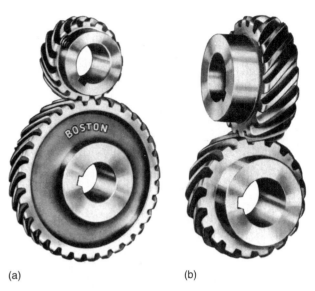

(a)   (b)

■ **Figure 69-3** Helical gears: (a) for drives that are parallel to each other; (b) for drives that are at right angles to each other.

greater area of contact than with the conventional gear drive. It also provides speed reductions with a minimum space requirement. Internal gears are used on heavy-duty tractors where much torque is required.

Helical gears (Fig. 69-3) may be used to connect parallel shafts or shafts that are at an angle. Because of the progressive rather than intermittent action of the teeth, helical gears run more smoothly and quietly than spur gears. Since there is more than one tooth in engagement at any one time, helical gears are stronger than spur gears of the same size and pitch. However, special bearings (thrust bearings) are often required on shafts to overcome the end thrust produced by these gears as they turn.

On most installations where it is necessary to overcome end thrust, *herringbone gears* (Fig. 69-4 on p. 548) are used. This type of gear resembles two helical gears placed side by side, with one-half having a left-hand helix and the other half a right-hand helix. These gears have a smooth, continuous action and eliminate the need for thrust bearings.

When two shafts are located at an angle, with their axial lines intersecting at 90°, power is generally transmitted by means of *bevel gears* (Fig. 69-5a on p. 548). When the shafts are at right angles and the gears are of the same size, they are called *miter gears* (Fig. 69-5b). However, it is not necessary that the shafts be only at right angles in order to transmit power. If the axes of the shafts intersect at any angle other than 90°, the gears are known as *angular bevel gears* (Fig. 69-5c). Bevel gears have straight teeth very similar to spur gears. Modified bevel gears having helical teeth are known as *hypoid gears* (Fig. 69-5d). The shafts of these gears, although at right angles,

Gears   **547**

■ Figure 69-4 Herringbone gears eliminate end thrust on shafts.

are not in the same plane and, therefore, do not intersect. Hypoid gears are used in automobile drives.

When shafts are at right angles and considerable reduction in speed is required, a *worm* and *worm gear* (Fig. 69-6) may be used. The worm, which meshes with the worm gear, may be single- or multiple-start thread. A worm with a double-start thread will revolve the worm gear twice as fast as a worm with a single-start thread and the same pitch.

When it is necessary to convert rotary motion to linear motion, a *rack* and *pinion* (Fig. 69-7) may be used. The rack, which is actually a straight or flat gear, may have straight teeth to mesh with a spur gear, or angular teeth to mesh with a helical gear.

▶▶ **Gear Terminology**

A knowledge of the more common gear terms is desirable to understand gearing and to make the calculations necessary to cut a gear (Fig. 69-8). Most of these terms are applicable to either inch or metric gearing, although the method of calculating dimensions may differ (Table 69.1 on p. 550). These methods, as applicable to inch and metric gear cutting, are explained in Unit 70.

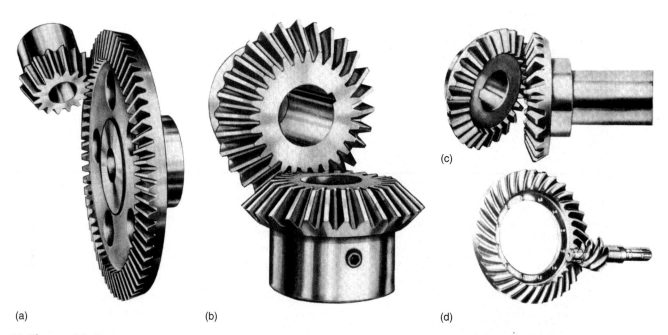

(a)   (b)   (d)

■ Figure 69-5 (a) Bevel gears transmit power at 90°; (b) the driver and driven miter gears are the same size; (c) angular bevel gears are used for shafts that are not at right angles; (d) hypoid gears are used in automotive drives.

Figure 69-6 A worm and worm gear are used for speed reduction.

Figure 69-7 A rack and pinion convert rotary motion to linear motion.

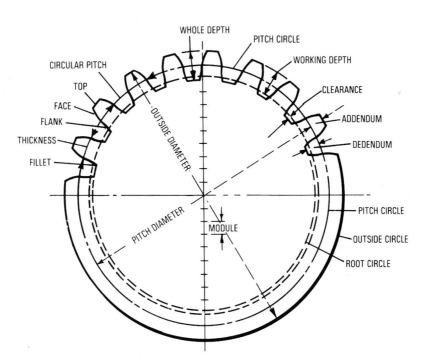

Figure 69-8 The parts of a gear.

table 69.1   Rules and formulas for spur gears*

To Obtain	Knowing	Rule	Formula
Addendum (A)	Circular pitch	Multiply the circular pitch by 0.3183.	$A = \text{CP} \times .3183$
Center distance (CD)	Diametral pitch	Divide 1 by diametral pitch.	$A = \dfrac{1}{\text{DP}}$
Chordal (corrected) addendum (CA)	Circular pitch	Multiply the number of teeth in both gears by the circular pitch, and divide the product by 6.2832.	$\text{CD} = \dfrac{(N + n) \times \text{CP}}{6.2832}$
	Diametral pitch	Divide the total number of teeth in both gears by twice the diametral pitch.	$\text{CD} = \dfrac{N + n}{2 \times \text{DP}}$
Chordal thickness (CT)	Pitch diameter addendum and number of teeth	Subtract from 1 the cosine of the result of 90° divided by the number of teeth. Multiply this result by half the pitch diameter. To this product, add the addendum.	$\text{CA} = \left[\left(1 - \cos\dfrac{90}{N}\right)\dfrac{\text{PD}}{2}\right] + A$
Circular pitch (CP)	Pitch diameter and number of teeth	Divide 90 by the number of teeth; find the sine of this result and multiply by the pitch diameter.	$\text{CT} = \sin\dfrac{90}{N} \times \text{PD}$
	Center-to-center distance	Multiply the center-to-center distance by 6.2832, and divide the product by the total number of teeth in both gears.	$\text{CP} = \dfrac{\text{CD} \times 6.2832}{N + n}$
	Diametral pitch	Divide 3.1416 by the diametral pitch.	$\text{CP} = \dfrac{3.1416}{\text{DP}}$
	Pitch diameter and number of teeth	Multiply pitch diameter by 3.1416 and divide by the number of teeth.	$\text{CP} = \dfrac{\text{PD} \times 3.1416}{N}$
Clearance (CL)	Circular pitch	Divide circular pitch by 20.	$\text{CL} = \dfrac{\text{CP}}{20}$
	Diametral pitch	Divide .157 by the diametral pitch.	$\text{CL} = \dfrac{.157}{\text{DP}}$
Dedendum (D)	Circular pitch	Multiply the circular pitch by .3683.	$D = \text{CP} \times .3683$
	Diametral pitch	Divide 1.157 by the diametral pitch.	$D = \dfrac{1.157}{\text{DP}}$
Diametral pitch (DP)	Circular pitch	Divide 3.1416 by the circular pitch.	$\text{DP} = \dfrac{3.1416}{\text{CP}}$
	Number of teeth and outside diameter	Add 2 to the number of teeth and divide the sum by the outside diameter.	$\text{DP} = \dfrac{N + 2}{\text{OD}}$
	Number of teeth and pitch diameter	Divide the number of teeth by the pitch diameter.	$\text{DP} = \dfrac{N}{\text{DP}}$

There are three commonly used gear-tooth forms that have pressure angles of 14½°, 20°, and 25°. The 20° and 25° tooth forms are now replacing the 14½° tooth form because the base of the tooth is wider and stronger. For the formula regarding the 20° and 25° tooth forms, see Machinery's or the American Machinist's handbooks.

table 69.1 (continued)

To Obtain	Knowing	Rule	Formula
Number of teeth (N)	Outside diameter and diametral pitch	Multiply the outside diameter by the diametral pitch and subtract 2.	$N = OD \times DP - 2$
	Pitch diameter and circular pitch	Multiply the pitch diameter by 3.1416 and divide by the circular pitch.	$N = \dfrac{PD \times 3.1416}{CP}$
	Pitch diameter and diametral pitch	Multiply the pitch diameter by the diametral pitch.	$N = PD \times DP$
Outside diameter (OD)	Number of teeth and circular pitch	Add 2 to the number of teeth and multiply the sum by the circular pitch. Divide this product by 3.1416.	$OD = \dfrac{(N + 2) \times CP}{3.1416}$
	Number of teeth and diametral pitch	Add 2 to the number of teeth and divide the sum by the diametral pitch.	$OD = \dfrac{N + 2}{DP}$
	Pitch diameter and diametral pitch	To the pitch diameter add 2 and divide by the diametral pitch.	$OD = PD + \dfrac{2}{DP}$
Pitch diameter (PD)	Number of teeth and circular pitch	Multiply the number of teeth by the circular pitch and divide by 3.1416.	$PD = \dfrac{N \times CP}{3.1416}$
	Number of teeth and diametral pitch	Divide the number of teeth by the diametral pitch.	$PD = \dfrac{N}{DP}$
	Outside diameter and number of teeth	Multiply the number of teeth by the outside diameter, and divide the product by the number of teeth plus 2.	$PD = \dfrac{N \times OD}{N + 2}$
Tooth thickness (T)	Circular pitch	Divide the circular pitch by 2.	$T = \dfrac{CP}{2}$
	Circular pitch	Multiply the circular pitch by .5.	$T = CP \times .5$
	Diametral pitch	Divide 1.5708 by the diametral pitch.	$T = \dfrac{1.5708}{DP}$
Whole depth (WD)	Circular pitch	Multiply the circular pitch by .6866.	$WD \times CP \times .6866$
	Diametral pitch	Divide 2.157 by the diametral pitch.	$WD = \dfrac{2.157}{DP}$

> *Addendum* is the radial distance between the pitch circle and the outside diameter or the height of the tooth above the pitch circle.

> *Center distance* is the shortest distance between the axes of two mating gears or the distance equal to one-half the sum of the pitch diameters.

> *Chordal addendum* is the radial distance measured from the top of the tooth to a point where the chordal thickness and the pitch circle intersect the edge of the tooth.

> *Chordal thickness* is the thickness of the tooth measured at the pitch circle or the length of the chord that subtends the arc of the pitch circle.

> *Circular pitch* is the distance from a point on one tooth to a corresponding point on the next tooth measured on the pitch circle.

> *Circular thickness* is the tooth thickness measured on the pitch circle; it is also known as the *arc thickness*.

> *Clearance* is the radial distance between the top of one tooth and the bottom of the mating tooth space.

> *Dedendum* is the radial distance from the pitch circle to the bottom of the tooth space. The dedendum is equal to the addendum plus the clearance.

> *Diametral pitch* (inch gears) is the ratio of the number of teeth for each inch of pitch diameter of the gear. For example, a gear of 10 diametral pitch and a 3-in. *pitch diameter* would have 10 × 3, or 30, teeth.

> *Involute* is the curved line produced by a point of a stretched string when it is unwrapped from a given cylinder (Fig. 69-9).

> *Linear pitch* is the distance from a point on one tooth to a similar point on the next tooth of a gear rack.

> *Module* (metric gears) is the pitch diameter of a gear divided by the number of teeth. It is an actual dimension, unlike diametral pitch, which is a ratio of the number of teeth to the pitch diameter.

> *Outside diameter* is the overall diameter of the gear, which is the pitch circle plus two addendums.

> *Pitch circle* is a circle that has the radius of half the pitch diameter with its center at the axis of the gear.

> *Pitch circumference* is the circumference of the pitch circle.

> *Pitch diameter* is the diameter of the pitch circle that is equal to the outside diameter minus two addendums.

> *Pressure angle* is the angle formed by a line through the point of contact of two mating teeth and tangent to the two *base circles* and a line at right angles to the centerline of the gears.

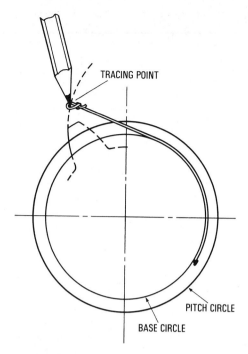

■ **Figure 69-9** Generating an involute.

> *Root circle* is the circle formed by the bottoms of the tooth spaces.

> *Root diameter* is the diameter of the root circle.

> *Tooth thickness* is the thickness of the tooth measured on the pitch circle.

> *Whole depth* is the full depth of the tooth or the distance equal to the addendum plus the dedendum.

> *Working depth* is the distance that a gear tooth extends into the tooth space of a mating gear, which is equal to two addendums.

unit 69 review questions

1. List six types of gears, and state where each may be used.

2. Define the following gear terms, and state the formula used to determine each. Use the formulas involving diametral pitch, *not* circular pitch, where applicable.

   a. Pitch diameter   e. Clearance
   b. Diametral pitch   f. Outside diameter
   c. Addendum   g. Number of teeth
   d. Dedendum

3. Calculate the pitch diameter, outside diameter, and whole depth of tooth for the following gears:

   a. 8 DP with 36 teeth
   b. 12 DP with 81 teeth
   c. 16 DP with 100 teeth
   d. 6 DP with 23 teeth
   e. 4 DP with 54 teeth

UNIT 70

Gear Cutting

OBJECTIVES

After completing this unit, you should be able to:

1 Select the proper cutter for any gear to be cut

2 Calculate gear-tooth dimensions for inch gears

3 Calculate gear-tooth dimensions for metric gears

4 Set up and cut a spur gear

Most gears cut on a milling machine are generally used to repair or replace a gear that has been broken or lost or is no longer carried in inventory. Industry generally mass-produces gears on special machines designed for this purpose. The most common types of *gear-generating machines* are the *gear-shaping machines* and the *gear-hobbing machines.* It is generally more economical to buy gears from a firm which specializes in gear manufacture unless it is important that the machine be back in operation soon. It is quite possible that a machinist would be called on to cut a gear in order to repair the machine quickly and get it back into production.

▸▸ Involute Gear Cutters

Gear cutters are an example of formed cutters. This type of cutter is sharpened on the face and ensures exact duplication of the shape of the teeth, regardless of how far back the face of the tooth has been ground.

Gear cutters are available in many sizes, ranging from 1 to 48 diametral pitch (DP). Cutters for teeth smaller than 48 DP are available as special cutters. Comparative sizes of teeth ranging from 4 to 16 DP are shown in Fig. 70-1.

When gear teeth are cut on any gear, a cutter must be chosen to suit both the DP and the number of teeth (N). The tooth space for a small pinion cannot be of the same shape as the tooth space for a large mating gear. The teeth on smaller gears must be more "curved" to prevent binding of meshing gear teeth. Therefore, sets of gear cutters are made in a series of slightly different shapes to permit the cutting of any desired number of teeth in a gear with

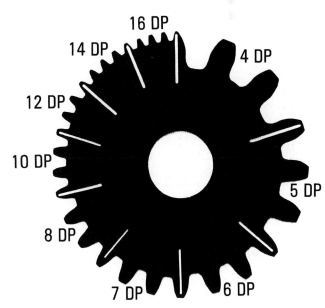

■ **Figure 70-1** Comparative gear-tooth sizes: 4 to 16 DP.

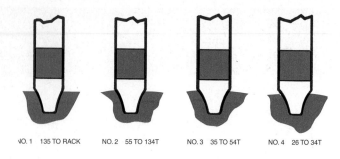

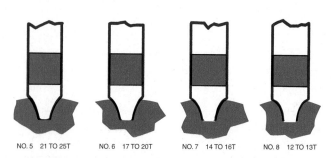

Figure 70-2 Involute profiles for a set of gear cutters.

the assurance that the teeth will mesh properly with those of another gear of the *same* DP.

These cutters are generally made in sets of eight and are numbered from 1 to 8 (Fig. 70-2). Notice the gradual change in shape from the #1 cutter, which has almost straight sides, to the much more curved sides of the #8 cutter. As shown in Table 70.1, the #1 cutter is used for cutting any number of teeth in a gear from 135 teeth to a rack, while the #8 cutter will cut only 12 and 13 teeth. It should be noted that, for gears to mesh, they must be of the same DP; the cutter number permits *only* a more accurate meshing of the teeth.

Some gear-cutter manufacturers have augmented the set of eight cutters with seven additional cutters in half sizes, making a total of 15 cutters in the set, numbered 1, 1½, 2, 2½, etc. In the half series, a #1½ cutter would be used to cut from 80 to 134 teeth, but a 7½ cutter would cut only 13 teeth (Table 70.1).

EXAMPLE

A 10-DP gear and a pinion in mesh have 100 teeth and 24 teeth, respectively. What cutters should be used to cut these gears?

Cutter Selection

Since the gears are in mesh, both must be cut with a 10-DP cutter.

A #2 cutter should be used to cut the teeth on the gear, since it will cut from 55 to 134 teeth.

table 70.1	Involute gear cutters
Cutter Number	**Range**
1	135 teeth to a rack
1½	80 to 134 teeth
2	55 to 134 teeth
2½	42 to 54 teeth
3	35 to 54 teeth
3½	30 to 34 teeth
4	26 to 34 teeth
4½	23 to 25 teeth
5	21 to 25 teeth
5½	19 and 20 teeth
6	17 to 20 teeth
6½	15 and 16 teeth
7	14 to 16 teeth
7½	13 teeth
8	12 and 13 teeth

A #5 cutter should be used to cut the pinion, since it will cut from 21 to 25 teeth.

TO CUT A SPUR GEAR

The procedure for machining a spur gear is outlined in the following example.

EXAMPLE

A 52-tooth gear with an 8 DP is required.

Procedure

1. Calculate all the necessary gear data (Table 70.1).

   a. Outside diameter $= \dfrac{N + 2}{DP}$

   $= \dfrac{54}{8}$

   $= 6.750$ in.

b. Whole depth of tooth $= \dfrac{2.157}{DP}$

$\qquad\qquad\qquad = \dfrac{2.157}{8}$

$\qquad\qquad\qquad = .2697$ in.

c. Cutter number $= 3$ (35 to 54 teeth)

d. Indexing (using Cincinnati standard plate)

$\qquad = \dfrac{40}{N}$

$\qquad = \dfrac{40}{52}$

$\qquad = \dfrac{10}{13} \times \dfrac{3}{3} = 30$ holes on the 39-hole circle

2. Turn the gear blank to the proper outside diameter (6.750 in.).

3. Press the gear blank firmly onto the mandrel.

Note: If the blank was turned on a mandrel, be sure that it is tight because the heat caused by turning might have expanded the blank slightly.

4. Mount the index head and footstock, and check the alignment of the index centers (Fig. 70-3).

5. Set the dividing head so that the index pin fits into a hole on the 39-hole circle and the sector arms are set for 30 holes.

Note: Do not count the hole in which the pin is engaged.

6. Mount the mandrel (and workpiece), with the large end toward the indexing head, between the index centers.

Note:

a. The footstock center should be adjusted up tightly into the mandrel and locked in position.

b. The dog should be tightened properly on the mandrel and the tail of the dog should not bind in the slot.

c. The tail of the dog should then be locked in the driving fork of the dividing head by means of the setscrews. This will ensure that there will be no play between the dividing head and the mandrel.

d. The dog should be far enough from the gear blank to ensure that the cutter will not hit the dog when the gear is being cut.

7. Move the table close to the column to keep the setup as rigid as possible.

8. Mount an 8 DP#3 cutter on the milling machine arbor over the approximate center of the gear. Be sure to have the cutter rotating in the direction of the indexing head.

9. Center the gear blank with the cutter by either of the following methods:

a. Place a square against the outside diameter of the gear (Fig. 70-4). With a pair of inside calipers or a rule, check the distance between the square and the side of the cutter. Adjust the table until the distances from both sides of the gear blank to the sides of the cutter are the same.

b. A more accurate method of centralizing the cutter is to use gage blocks instead of the inside calipers or rule.

■ **Figure 70-3** Checking the alignment of index centers with a dial indicator. *(Kelmar Associates)*

■ **Figure 70-4** Centering a gear cutter and the workpiece. *(Kelmar Associates)*

Figure 70-5 Setting a gear cutter to the diameter of the workpiece. *(Kelmar Associates)*

10. *LOCK THE CROSS-SLIDE.*

11. Start the milling cutter and run the work under the cutter.

12. Raise the table until the cutter *just* touches the work. This can be done by using a chalk mark on the gear blank or a piece of paper between the gear blank and the cutter to indicate when the cutter is just touching the work (Fig. 70-5).

13. Set the graduated feed collar on the vertical feed to zero (0).

14. Move the work clear of the cutter by means of the longitudinal feed handle and raise the table to about two-thirds the depth of the tooth (.180 in.); then *lock the knee clamp.*

Note: A special stocking cutter is sometimes used to rough out the teeth.

15. Slightly notch all gear teeth on the end of the work to check for correct indexing (Fig. 70-6).

16. Rough-out the first tooth and set the automatic feed trip dog after the cutter is clear of the work.

17. Return the table to the starting position.

Figure 70-6 Notching all gear teeth eliminates errors. *(Kelmar Associates)*

Note: Clear the end of the work with the cutter.

18. Cut the remaining teeth and return the table to the starting position.

19. Loosen the knee clamp, raise the table to the full depth of .270 in., and *lock the knee clamp.*

Note: It is advisable to remove the crank from the knee-elevating shaft so that it will not be moved accidentally and change the setting.

20. Finish-cut all teeth.

Note: After each tooth has been cut, the cutter should be stopped before the table is returned to prevent marring the finish on the gear teeth.

▶▶ Metric Gears and Gear Cutting

Countries that have been using the metric system of measurement usually use the *module* system of gearing. The *module* (M) of a gear equals the pitch diameter (PD) divided by the number of teeth (N), or $M = PD/N$, whereas the DP of a gear is the ratio of N to the PD, or $DP = N/PD$. The DP of a gear is the *ratio* of the number of teeth per inch of diameter, whereas *M is an actual dimension*. Most of the terms used in DP gears remain the same for module gears; however, the method of calculating the dimensions has changed in some instances. Table 70.2 gives the necessary rules and formulas for metric spur gears.

METRIC MODULE GEAR CUTTERS

The most common metric gear cutters are available in modules ranging from 0.5 to 10 mm (Table 70.3). However, metric module gear cutters are available in sizes up to 75 mm. Any metric module size is available in a set of eight cutters, numbered from #1 to #8. The range of each cutter is the reverse of that of a DP cutter. For instance, a #1 metric module cutter will cut from 12 to 13 teeth; a #8 DP cutter will cut from 135 teeth to a rack. Table 70.3 shows the cutters available and the range of each cutter in the set.

EXAMPLE

A spur gear has a PD of 60 mm and 20 teeth. Calculate:

1. Module
2. Circular pitch
3. Addendum
4. Outside diameter

table 70.2 Rules and formulas for metric module spur gears

To Obtain	Knowing	Rule	Formula
Addendum (A)	Normal module	Addendum equals module.	$A = M$
Center-to-center distance (CD)	Pitch diameters	Divide the sum of the pitch diameters by 2.	$CD = \dfrac{PD_1 + PD_2}{2}$
Circular pitch (CP)	Module	Multiply module by π.	$CP = M \times 3.1416$
	Pitch diameter and number of teeth	Multiply pitch diameter by π and divide by number of teeth.	$CP = \dfrac{PD \times 3.1416}{N}$
Chordal thickness (CT)	Outside diameter and number of teeth	Multiply outside diameter by π and divide by number of teeth minus 2.	$CP = \dfrac{OD \times 3.1416}{N - 2}$
	Module and outside diameter	Divide 90° by the number of teeth. Find the sine of this angle and multiply by the pitch diameter.	$CT = PD \times \sin \dfrac{90\mathbf{i}}{N}$
	Module	Multiply module by π and divide by 2.	$CT = \dfrac{M \times 3.1416}{2}$
	Circular pitch	Divide circular pitch by 2.	$CT = \dfrac{CP}{2}$
Clearance (CL)	Module	Multiply module by 0.166 mm.	$CL = M \times 0.166$
Dedendum (D)	Module	Multiply module by 1.166 mm.	$D = M \times 1.166$
Module (M)	Pitch diameter and number of teeth	Divide pitch diameter by the number of teeth.	$M = \dfrac{PD}{N}$
	Circular pitch	Divide circular pitch by π.	$M = \dfrac{CP}{3.1416}$
Number of teeth (N)	Outside diameter and number of teeth	Divide outside diameter by number of teeth plus 2.	$M = \dfrac{OD}{N + 2}$
	Pitch diameter and module	Divide pitch diameter by the module.	$N = \dfrac{PD}{M}$
Outside diameter (OD)	Pitch diameter and circular pitch	Multiply pitch diameter by π and divide product by circular pitch.	$N = \dfrac{PD \times 3.1416}{CP}$
	Number of teeth and module	Add 2 to the number of teeth and multiply sum by module.	$OD = (N + 2) \times M$
Pitch diameter (PD)	Pitch diameter and module	Add 2 modules to pitch diameter.	$OD = PD + 2M$
	Module and number of teeth	Multiply module by number of teeth.	$PD = M \times N$
	Outside diameter and module	Subtract 2 modules from outside diameter.	$PD = OD - 2M$
	Number of teeth and outside diameter	Multiply number of teeth by outside diameter and divide product by number of teeth plus 2.	$PD = \dfrac{N \times OD}{N + 2}$
Whole depth (WD)	Module	Multiply module by 2.166 mm.	$WD = M \times 2.166$

table 70.3 Metric module gear cutters

Module Size (mm)		Milling Cutter Numbers	
		Cutter No.	For Cutting
0.5	3.5		
0.75	3.75	1	12 to 13 teeth
1	4	2	14 to 16 teeth
1.25	4.5	3	17 to 20 teeth
1.5	5	4	21 to 25 teeth
1.75	5.5	5	26 to 34 teeth
2	6	6	35 to 54 teeth
2.25	6.5	7	55 to 134 teeth
2.5	7	8	135 teeth to rack
2.75	8		
3	9		
3.25	10		

5. Dedendum
6. Whole depth
7. Cutter number

Solution

1. $M = \dfrac{PD}{N}$

   $= \dfrac{60}{20}$

   $= 3$ mm

2. $CP = M \times \pi$

   $= 3 \times 3.1416$

   $= 9.425$ mm

3. $A = M$

   $= 3$ mm

4. $OD = (N + 2) \times M$

   $= 22 \times 3$

   $= 66$ mm

5. $D = M \times 1.166$

   $= 3 \times 1.166$

   $= 3.498$ mm

6. $WD = M \times 2.166$

   $= 3 \times 2.166$

   $= 6.498$ mm

7. Cutter number (see Table 70.3) $= 3$

EXAMPLE

Two identical gears in mesh have a CD of 120 mm. Each gear has 24 teeth. Calculate:

1. Pitch diameter
2. Module
3. Outside diameter
4. Whole depth
5. Circular pitch
6. Chordal thickness

Solution

1. $PD = \dfrac{2 \times CD}{2}$ (equal gears)

   $= \dfrac{2 \times 120}{2}$

   $= \dfrac{240}{2}$

   $= 120$ mm

2. $M = \dfrac{PD}{N}$

   $= \dfrac{120}{24}$

   $= 5$

3. $OD = (N + 2) \times M$

   $= 26 \times 5$

   $= 130$ mm

4. $WD = M \times 2.166$

   $= 5 \times 2.166$

   $= 10.83$ mm

5. $CP = M \times \pi$

   $= 5 \times 3.1416$

   $= 15.708$ mm

6. $CT = \dfrac{M \times \pi}{2}$

$= \dfrac{5 \times 3.1416}{2}$

$= 7.85$ mm

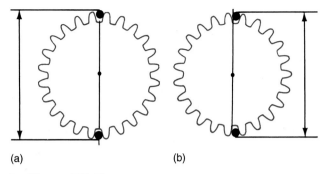

(a)   (b)

■ Figure 70-7 Wires or pins are used to check gear sizes accurately.

▶▶ **Gear-Tooth Measurement**

To ensure that the gear teeth are of the proper dimensions, they should be measured with a gear tooth vernier caliper. The caliper should be set to the corrected addendum, a dimension that may be found in most handbooks.

Gear sizes may also be accurately checked by measuring over wires or pins of a specific diameter that have been placed in two diametrically opposite tooth spaces of the gear (Fig. 70-7a). For gears having an odd number of teeth, the wires are placed as nearly opposite as possible (Fig. 70-7b). A measurement taken over these wires is checked against tables found in most handbooks. These tables indicate the measurement over the wires for any gear having a given number of teeth and a specific pressure angle. Since these tables are far too extensive to be printed in this book, you may refer to any handbook for them.

In order to measure inch gears accurately, the diametral pitch and number of teeth of the gear must be known. In order to measure metric gears, the module must be known.

The wire or pin size to be used is determined as follows:

1. For external inch spur gears, the wire or pin size is equal to 1.728 divided by the diametral pitch of the gear.

2. For internal inch spur gears, the wire size is equal to 1.44 divided by the diametral pitch of the gear.

3. Metric module gears are measured using a wire size equal to 1.728 multiplied by the module of the gear. The measurement over the wires should equal the value shown in the handbook tables multiplied by the module.

EXAMPLE (INCH)

Determine the wire size and the measurement over the wires for a 10-DP external gear having 28 teeth and a 14.50° pressure angle.

$$\text{Wire size} = \dfrac{1.728}{10}$$
$$= .1728 \text{ in.}$$

In handbook tables, the size over the wires for a gear having 28 teeth and a 14.50° pressure angle should be 30.4374 in. divided by the DP. Therefore, the measurement over the wires should be:

$$\dfrac{30.4374}{10} = 3.0437 \text{ in.}$$

If the measurement is larger than this size, the PD is too large and the depth of cut will have to be increased. If it is less than the determined size, the gear is undersize. Gears having an odd number of teeth are calculated in a similar manner but using the proper tables for these gears.

1. What cutter numbers should be used to cut the following gears?

   a. 8 DP—36 teeth   c. 16 DP—100 teeth
   b. 12 DP—81 teeth   d. 6 DP—23 teeth

2. Describe two methods of centering the gear blank with the cutter for machining a spur gear.

3. What precautions should be observed when mounting the gear blank between the index head and centers?

4. What is the purpose of notching before gear teeth are cut?

5. Compare the terms *module* and *diametral pitch*.

6. How does the metric module numbering system for gear cutters differ from that for diametral pitch systems?

7. For a 40-tooth spur gear, 240-mm pitch diameter, calculate:

   a. Module   e. Dedendum
   b. Circular pitch   f. Whole depth
   c. Outside diameter   g. Cutter number
   d. Addendum

Gear-Tooth Measurement

8. State two methods of measuring gear teeth.

9. How is the wire size determined for:

   a. External gears?
   b. Internal gears?

UNIT 71

Helical Milling

OBJECTIVES

After completing this unit, you will be able to:

1 Calculate the lead and helix angle of a helical gear

2 Set up a milling machine to machine a helix

3 Make the calculations and setup for milling a helical gear

The process of milling helical grooves, such as flutes in a drill, teeth in helical gears, or the worm thread on a shaft, is known as *helical milling.* It is performed on the universal milling machine by gearing the dividing head through the worm shaft to the lead screw of the milling machine.

▶▶ Helical Terms

The term *spiral* is often used, incorrectly, in place of *helix.*

> A *helix* is a theoretical line or path generated on a *cylindrical* surface by a cutting tool that is fed lengthwise at a uniform rate, while the cylinder is also rotated at a uniform rate (Fig. 71-1). The flutes on a drill or the threads on a bolt are examples of helices.

> A *spiral* is the path generated by a point moving at a fixed rate of advance along the surface of a *rotating cone* or *plane* (Fig. 71-2 on p. 562). Threads on a wood screw and pipe threads are examples of conical spirals. Watch springs and scroll threads on a universal lathe chuck are examples of plane or flat spirals.

In order to cut either an inch or a metric helix, any two of the following must be known:

1. *The lead of the helix,* which is the longitudinal distance the helix advances axially in one complete revolution of the work

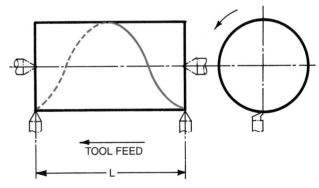

■ **Figure 71-1** A helix will be generated if the work is turned as the tool is moved along uniformly.

2. *The angle of the helix,* which is formed by the intersection of the helix with the axis of the workpiece

3. *The diameter (and circumference) of the workpiece*

In comparing two different helices, it will be noticed that the greater the angle with the centerline, the shorter will be the lead. However, if the diameter is increased but

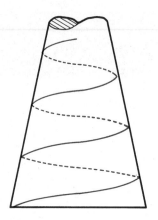

Figure 71-2 A spiral is produced on a conical surface.

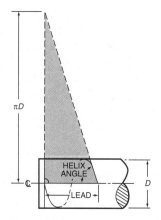

Figure 71-3 Relationship of lead circumference to helix angle.

the helix angle remains the same, the greater the lead will be. Thus, it is evident that the lead of a helix varies with:

1. The diameter of the work
2. The angle of the helix

The relationship between the diameter (and circumference), the helix angle, and the lead is shown in Fig. 71-3. Note that if the surface of the cylinder could be unwound to produce a flat surface, the helix would form the hypotenuse of a right-angle triangle, with the circumference forming the side opposite and the lead the side adjacent.

▶▶ Cutting a Helix

To cut a helix on a cylinder, the following steps are necessary:

1. Swing the table in the proper direction to the angle of the helix to ensure that a groove of the same contour as the cutter is produced.

2. The work must rotate one turn while the table travels lengthwise the distance equal to the lead. This is achieved by installing the proper change gears between the worm shaft on the dividing head and the milling machine lead screw.

▶▶ Determining the Helix Angle

To ensure that a groove of the same contour as the cutter is produced, the table must be swung to the angle of the helix (Fig. 71-4a). The importance of this is shown in Fig. 71-4b.

Note that when the table is not swung (Fig. 71-4b), a helix having the proper lead but an improper contour will be generated. By referring to Fig. 71-3, it can easily be seen that the angle may be calculated as follows:

$$\text{Tangent of helix angle} = \frac{\text{circumference of work}}{\text{lead of helix}}$$

$$= \frac{3.1416 \times \text{diameter } (D)}{\text{lead of helix}}$$

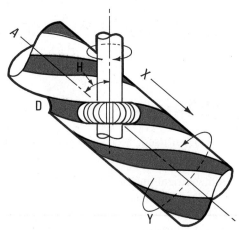

(a)

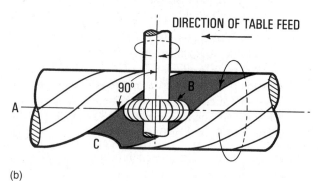

(b)

Figure 71-4 (a) When the table is swiveled to the helix angle, the exact profile of the cutter will be generated; (b) an incorrect angle produces an incorrect profile.

EXAMPLE (INCH)

To what angle must the milling machine table be swiveled to cut a helix having a lead of 10.882 in. on a piece of work 2 in. in diameter?

$$\text{Tangent of helix angle} = \frac{3.1416 \times D}{\text{lead of helix}}$$
$$= \frac{3.1416 \times 2}{10.882}$$
$$= \frac{6.2832}{10.882}$$
$$= .57739$$
$$\text{Helix angle} = 30°$$

After the helix angle has been calculated, it is necessary to determine the *direction* in which to swivel the table to produce the proper hand of helix (that is, right- or left-hand).

EXAMPLE (METRIC)

To what angle must a milling machine table be swiveled to cut a helix having a lead of 450 mm on a workpiece 40 mm in diameter?

$$\text{Tangent of helix angle} = \frac{3.1416D \,(\text{mm})}{\text{Lead of helix (mm)}}$$
$$= \frac{3.1416 \times 40}{450}$$
$$= 0.2792$$
$$\text{Helix angle} = 15°36'$$

▶▶ Determining the Direction to Swing the Table

In order to determine the hand of a helix, hold the cylinder on which the helix is cut in a horizontal plane with its axis running in a right-left direction.

If the helix slopes *down* and to the right, it is a right-hand helix (Fig. 71-5). A left-hand helix slopes *down* and to the left. When a *left-hand helix* is to be cut, the table of the milling machine must be swiveled in a clockwise direction (operator standing in front of the machine). A right-hand helix may be produced similarly by moving the right end of the table in toward the column or by moving it in a counterclockwise direction.

■ **Figure 71-5** The grooves of a right-hand helical cutter slope down to the right. *(Kelmar Associates)*

■ **Figure 71-6** The worm shaft and the lead screw are connected for helical milling. *(Kelmar Associates)*

▶▶ Calculating the Change Gears to Produce the Required Lead

To cut a helix, it is necessary to have the work move lengthwise and rotate at the same time. The amount the work (and table) travels lengthwise as the work revolves one complete revolution is the *lead*. The rotation of the work is caused by gearing the worm shaft of the dividing head to the lead screw of the machine (Fig. 71-6).

Helical Milling   **563**

INCH CALCULATIONS

To cut a helix on an inch milling machine, it is necessary first to understand how to calculate the required change gears for any desired lead. Assume that the dividing head worm shaft is geared to the table lead screw with equal gears (for example, both having 24-tooth gears). The dividing head ratio is 40:1, and a standard milling machine lead screw has 4 threads per inch (*tpi*). The lead screw, as it revolves one turn, would revolve the dividing head spindle one-fortieth of a revolution. In order for the dividing head spindle to revolve one turn, it would be necessary for the lead screw to revolve 40 times. Thus, the table would travel 40 × ¼ in., or 10 in., while the work revolves one turn. Therefore, the lead of a milling machine is said to be 10 in. when the lead screw (4 *tpi*) is connected to the dividing head (40:1 ratio) with equal gears.

In calculating the change gears required to cut any lead, the following formula may be used:

$$\frac{\text{Lead of helix}}{\text{Lead of machine (10)}} = \frac{\text{product of driven gears}}{\text{product of driver gears}}$$

The ratio of gears required to produce any lead on a milling machine having a lead screw with 4 *tpi* is always equal to a fraction having the lead of the helix for the numerator and 10 for the denominator.

Note: The preceding formula may be inverted if preferred.

$$\frac{\text{Lead of the machine}}{\text{Lead of the helix}} = \frac{\text{product of driver gears}}{\text{product of driven gears}}$$

EXAMPLE

Calculate the change gears required to produce a helix having a lead of 25 in. on a piece of work. The available change gears have the following number of teeth: 24, 24, 28, 32, 36, 40, 44, 48, 56, 64, 72, 86, 100.

Solution

$$\text{Gear ratio} = \frac{\text{lead of helix (driven gears)}}{\text{lead of machine (driver gears)}}$$

$$= \frac{25}{10}$$

Since 10- and 25-tooth gears are not supplied with standard dividing heads, it is necessary to multiply the 25:10 ratio by any number that will suit the change gears available.

$$\text{Gear ratio} = \frac{25}{10} \times \frac{4}{4}$$

$$= \frac{100 \text{ (driven gear)}}{40 \text{ (driver gear)}}$$

Since both 100- and 40-tooth gears are available, simple gearing may be used.

EXAMPLE

Calculate the change gears required to produce a helix having a lead of 27 in. The available change gears are the same as those in the preceding example.

Solution

$$\text{Gear ratio} = \frac{\text{lead of helix (driven gears)}}{\text{lead of machine (driver gears)}}$$

$$= \frac{27}{10}$$

Since there are no gears in the set that are multiples of both 27 and 10, it is impossible to use simple gearing. Compound gearing must therefore be used, and it becomes necessary to factor the fraction ²⁷⁄₁₀ as follows:

$$\text{Gear ratio} = \frac{27}{10}$$

$$= \frac{3}{2} \times \frac{9 \text{ (driven)}}{5 \text{ (driver)}}$$

It is now necessary to multiply both the numerator and denominator of each fraction by the same number to bring the ratio into the range of the gears available.

Note: This does not change the value of the fraction.

$$\frac{3 \times 16}{2 \times 16} = \frac{48}{32}$$

$$\frac{9 \times 8}{5 \times 8} = \frac{72}{40}$$

$$\text{The gear ratio} = \frac{48 \times 72 \text{ (driven gears)}}{32 \times 40 \text{ (driver gears)}}$$

∴ the driven gears are 48 and 72, and the driver gears are 32 and 40. The gears would be placed in the train as follows (Fig. 71-6):

Gear on worm	72	(driven)
First gear on stud	32	(driver)
Second gear on stud	48	(driven)
Gear on lead screw	40	(driver)

The preceding order is not absolutely necessary; the two driven gears may be interchanged and/or the two driver gears may be interchanged, *provided a driver gear is not interchanged with a driven gear.*

METRIC CALCULATIONS

The pitch of the lead screw on a metric milling machine is stated in millimeters. Most milling machine lead screws have a 5-mm pitch and the dividing head has a ratio of 40:1. As the lead screw revolves one turn, it would revolve the dividing head spindle one-fortieth of a revolution. In order for the dividing head spindle (and work) to revolve one full turn, the lead screw must make 40 complete revolutions. Therefore, the lead of the machine would be 40 times the pitch of the lead screw.

For metric calculations, the change gears required are calculated as follows:

$$\frac{\text{Lead of helix (mm)}}{\text{(Lead of machine (mm))}} = \frac{\text{product of driven gears}}{\text{product of driver gears}}$$

The normal change gears in a set are 24, 24, 28, 32, 36, 40, 44, 48, 56, 64, 72, 86, 100.

EXAMPLE

Calculate the change gears required to cut a helix having a lead of 500 mm on a workpiece using a standard set of gears. The milling machine lead screw has a pitch of 5 mm.

Solution

$$\frac{\text{Driven gears}}{\text{driver gears}} = \frac{\text{lead of helix}}{\text{pitch of lead screw} \times 40}$$

$$= \frac{500}{5 \times 40}$$

$$= \frac{500}{200}$$

$$= \frac{5}{2} \times \frac{20}{20}$$

$$= \frac{100}{40}$$

Driven gear = 100

Driver gear = 40

▸▸ Direction of Spindle Rotation

Fig. 71-6 illustrates the setup required to cut a left-hand helix on most machines. Note that the gear on the lead screw and the worm gear revolve in the same direction. To cut a right-hand helix, the spindle must revolve in the opposite direction, and therefore another idler must be inserted, as in Fig. 71-7. In this case, the idler acts as neither a driven nor a driver gear and is not considered in the calculation of the gear train. It acts merely as a means of changing the direction of rotation of the dividing head spindle. It should also be noted that the direction of spindle rotation for simple gearing will be opposite to that for compound gearing.

■ **Figure 71-7** A second idler reverses the direction of rotation. *(Kelmar Associates)*

▶▶ Cutting Short Lead Helices

When it is necessary to cut leads smaller than those shown in most machinery handbooks, it is advisable to disengage the dividing head worm and wormwheel and connect the change gears directly from the table lead screw to the dividing head spindle, rather than to the worm shaft. This method permits machining leads to one-fortieth of the leads shown in the handbook tables. Thus, if the machine is geared to cut a lead of 4.000 in., by connecting the worm shaft and the lead screw, the same gearing would produce a lead of ¼₀ × 4.000 in., or .100 in., when geared directly to the dividing head spindle.

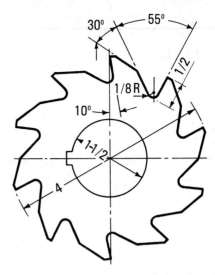

■ **Figure 71-8** Dimensions of a helical milling cutter.

EXAMPLE

A plain helical milling cutter to the following specifications is required:

> Diameter: 4 in.
>
> Number of teeth: 9
>
> Helix: right-hand
>
> Helix angle: 25°
>
> Rake angle: 10° positive radial rake
>
> Angle of flute: 55°
>
> Depth of flute: ½ in.
>
> Length: 4 in.
>
> Material: tool steel

Procedure

1. Turn blank to sizes indicated (Fig. 71-8).
2. Apply layout die to the end of the blank and lay out as in Fig. 71-9.
3. Lay out a line on the periphery to indicate the direction of the *right-hand helix* (Fig. 71-10).
4. Press the cutter blank firmly on the mandrel. If a threaded mandrel is used, be sure to tighten the nut securely.
5. Mount the dividing head and footstock.
6. Calculate the indexing for nine divisions.

> Indexing $= \dfrac{40}{9}$
>
> $= 4\frac{4}{9}$
>
> $= 4$ turns $+ 8$ holes on an 18-hole circle

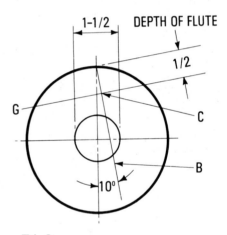

■ **Figure 71-9** Locating the first tooth on the cutter.

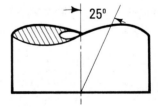

■ **Figure 71-10** Laying out the direction of the flute.

7. Set the sector arms to 8 holes on the 18-hole circle.

Note: Do not count the hole in which the pin is engaged.

8. Disengage the index plate locking device.

9. Calculate the lead of the helix.

$$\text{Lead} = \frac{3.1416 \times D}{\tan \text{ helix angle}}$$
$$= 3.1416 \times D \cot \text{ helix angle}$$
$$\left(\text{since } \frac{1}{\tan} = cot\right)$$
$$= 3.1416 \times 4 \times 2.1445$$
$$= 26.949 \text{ in.}$$

10. Consult any handbook for the change gears to cut the lead closest to 26.949 in. Obviously, this is 27.

11. If a handbook is not available, change gears can be calculated for the closest lead, which is 27 in.

12. Change gears required for a 27-in. lead are calculated as follows:

$$\frac{\text{Required lead}}{\text{Lead of machine}} = \frac{27}{10}$$
$$= \frac{9}{5} \times \frac{3}{2}$$
$$\frac{9 \times 8}{5 \times 8} = \frac{72}{40} \quad \frac{3 \times 16}{2 \times 16} = \frac{48}{32}$$
$$\text{Change gears} = \frac{72 \times 48 \text{ (driven gears)}}{40 \times 32 \text{ (driver gears)}}$$

13. Mount the change gears, allowing a slight clearance between mating teeth.

14. Mount the work between the centers, with the large end of the mandrel against the dividing head.

15. Swivel the table 25° counterclockwise.

16. Adjust the crossfeed handwheel until the table is about 1 in. from the face of the column. This is to ensure that the table clears the column when the cutter is being machined.

17. Swing the table back to zero (0).

18. Mount a 55° double-angle cutter so that it revolves toward the dividing head, and center it approximately over the flute layout.

19. Rotate the blank until the flute layout is aligned with the cutter edge. This may be checked with a rule or straightedge (Fig. 71-11).

20. Move the blank over, using the crossfeed for the distance of *m* (Fig. 71-11) or until the point C (Fig. 71-9) is in line with the centerline of the cutter.

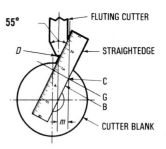

Figure 71-11 Aligning the cutter blank with the cutter.

21. With the work clear of the cutter, set the depth to .500 in.

22. Rotate the table 25° (counterclockwise), and lock securely (right end in toward the column).

23. Carefully cut the first tooth space, checking the accuracy of the location and the depth.

24. Index for and cut the remaining flutes.

25. Remove the fluting cutter and mount a plain helical milling cutter.

26. Rotate the work (using the index crank) until a line at 30° to the side of the flute is parallel to the table (Fig. 71-9). This may be checked by means of a surface gage. The blank, however, may be rotated by indexing an amount equal to

$$90 - \left(30 + \frac{55}{2}\right) = 90 - 57.5$$
$$= 32.5° \ (32°30')$$

Indexing for 32°30':
$$32° \times 60' = 1920'$$
$$30' = 30'$$
$$32°30' = 1950'$$
$$= \frac{1950}{540}$$
$$= \frac{3330}{540}$$
$$= 3^{11}\!/_{18}$$
$$= 3 \text{ turns} + 11 \text{ holes on the 18-hole circle}$$

27. Adjust the workpiece under the cutter.

28. With the cutter rotating, raise the table until the width of the land on the workpiece is about .030 in. wide.

29. Cut the secondary clearance (30° angle) on all teeth of the workpiece.

1. Define:
   a. Helix   c. Lead
   b. Spiral   d. Angle of helix

2. Make a sketch to illustrate the relationship among the lead, circumference, and helix angles.

3. List two factors that affect the lead of a helix.

4. To what angle must the table be swung to cut the following helices?

   a. Lead 10.290 in., work diameter 3.250 in.
   b. Lead 12.000 in., work diameter 2.750 in.
   c. Lead = 600 mm, work diameter = 100 mm
   d. Lead = 232 mm, work diameter = 25 mm

5. How may a right- and left-hand helix be recognized, and in which direction should the table be swiveled for each?

6. a. Calculate the change gears to cut the following leads:

   1. 6.000 in.
   2. 7.500 in.
   3. 9.600 in.

   b. The lead screw of a milling machine has a pitch of 5 mm. The available change gears are 24, 24, 28, 32, 36, 40, 44, 48, 56, 64, 72, 86, 100. Calculate the change gears required for leads of 800 mm and 560 mm.

7. It is required to make a helical milling cutter having the following specifications:

   Diameter: 3.475 in.
   Helix: left-hand
   Rake angle: 5° positive
   Depth of flute: 0.5 in.
   Material: tool steel
   Number of teeth: 7
   Helix angle: 20°
   Angle of flute: 55°
   Length: 3 in.

   Calculate:
   a. Indexing
   b. Lead
   c. Change gears required to cut this lead

UNIT 72

Cam, Rack, Worm, and Clutch Milling

OBJECTIVES

After completing this unit, you will be able to:

1 Calculate and cut a uniform-motion cam

2 Set up the machine and cut a rack

3 Understand how a worm is cut

4 Set up the machine and cut a clutch

The variety of attachments available makes the milling machine a versatile machine tool. Besides the standard operations generally performed on a milling machine, with the proper setups and attachments, it is possible to cut cams, racks, worms, and clutches. Although a machinist may not be called on frequently to perform them, it is wise to be familiar with these operations so that you will be able to cut the form required.

▶▶ Cams and Cam Milling

A *cam* is a device generally applied to a machine to change rotary motion into straight-line or reciprocating motion and to transmit this motion to other parts of the machine through a follower. The cam shaft on an automobile engine incorporates several cams, which control the opening and closing of the intake and exhaust valves. Many machine operations, especially on automatic machines, are controlled by cams, which transmit the desired motion to the cutting tool through a follower and some type of push rod.

Cams are also used to transform linear motion into a reciprocating motion of the follower. Cams of this type are called *plate,* or *bar* cams or are often referred to as *templates.* Templates are often used on tracer-type milling machines and lathes where parts must be produced to the profile of the template.

Cams may also be used as locking devices. Extensive applications are found in jig and fixture design and in quick-locking clamps.

CAMS USED TO IMPART MOTION

Cams used to impart motion are generally found on machines and may be of two types: the *positive* and the *nonpositive.*

Positive-type cams, such as the cylindrical and grooved plate (Fig. 72-1 on p. 570), control the follower at all times; that is, the follower remains engaged in the groove on the face or the periphery of the cam and uses no other means to maintain this engagement.

Examples of the *nonpositive-type cams* are the plate, toe and wiper, and crown (Fig. 72-2). In the nonpositive types, the cam pushes the follower in a given direction and depends on some external force, such as gravity or springs, to keep the follower bearing against the cam surface.

Followers may be of several types:

> The *roller type* (Fig. 72-3a on p. 570) has the least frictional drag and requires little or no lubrication.

> The *tapered roller type* (Fig. 72-3b) is used with grooved plate or cylindrical cams.

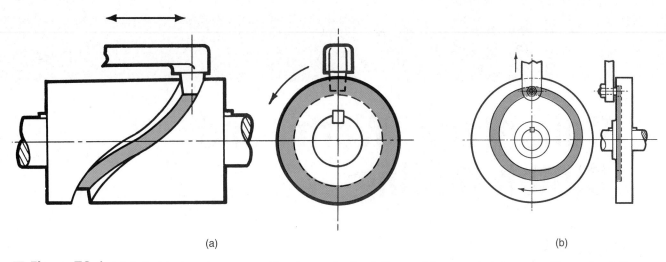

Figure 72-1 (a) Cylindrical or drum cam with a tapered roller follower; (b) grooved-plate cam with a roller follower.

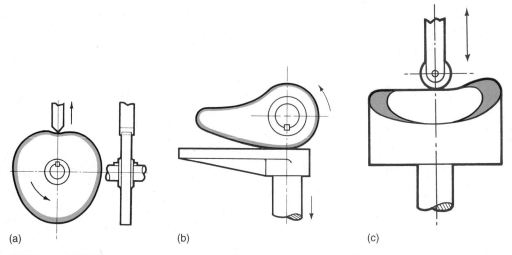

Figure 72-2 (a) Plate cam with a knife-edge follower; (b) toe and wiper cam; (c) crown cam.

> The *flat,* or *plunger, type* (Fig. 72-3c) is used to transmit large forces and requires lubrication.

> The *knife-edge,* or *pointed, type* (Fig. 72-3d) is used on more intricate cams because it permits sharp contours to be followed more readily than with a roller cam.

CAM MOTIONS

There are three standard types of motions imparted by cams to followers and machine parts:

1. Uniform motion
2. Harmonic motion
3. Uniformly accelerated and decelerated motion

The *uniform-motion cam* moves the follower at the same rate from the beginning to the end of the stroke.

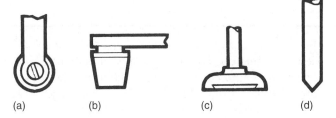

Figure 72-3 Types of cam followers: (a) roller; (b) tapered roller; (c) flat, or plunger; (d) knife-edge, or pointed.

Since the movement goes from zero to full speed instantaneously and ends in the same abrupt way, there is a distinct shock at the beginning and the end of the stroke. Machines using this type of cam must be rigid and sturdy enough to withstand this constant shock.

The *harmonic-motion cam* provides a smooth start and stop to the cycle. It is used when uniformity of motion is not essential and where high speeds are required.

The *uniformly accelerated and decelerated cam* moves the follower slowly at first, then accelerates or decelerates at a uniform rate. It then gradually decreases in speed, permitting the follower to come to a slow stop before reversal takes place. This type is considered the smoothest of the three motions and is used on high-speed machines.

RADIAL CAM TERMS

A *lobe* is a projecting part of the cam that imparts a reciprocal motion to the follower. Cams may have one or several lobes (Figs. 72-4 and 72-5), depending on the application to the machine.

The *rise* is the distance one lobe will raise or lower the follower as the cam revolves.

The *lead* is the total travel that would be imparted to the follower in one revolution of a uniform-rise cam, having only one lobe in 360°. In Fig. 72-5, the lead for a double-lobe cam is twice the lead of a single-lobe cam having the same rise. It is the lead of the cam and not the rise that controls the gear selection in cam milling.

Uniform rise is the rise generated on a cam that moves inward at an even rate around the cam, assuming the shape of an Archimedes spiral. This rise is caused by uniform feed and rotation of the work when a cam is being machined.

CAM MILLING

In the majority of plate cams that do not have a uniform rise, the cam must be laid out and machined by incremental cuts. By this method, the blank is rotated through an angular increment and the cut is taken to the layout line or a predetermined point. This process is repeated until the outline of the cam is produced as closely as possible. The ridges left between each successive cut are then removed by filing and polishing (Fig. 72-6).

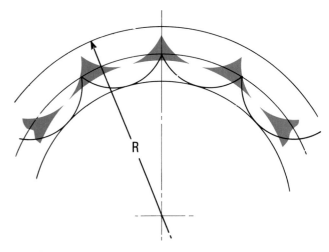

■ **Figure 72-6** Ridges left by incremental cuts.

■ **Figure 72-7** Machine set up for cam milling using a short-lead milling attachment. *(Cincinnati Milacron, Inc.)*

■ **Figure 72-4** Single-lobe uniform-rise cam.

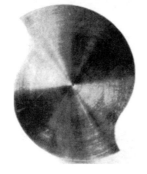

■ **Figure 72-5** Double-lobe uniform-rise cam.

Uniform-rise cams may be produced in the milling machine with a vertical head, by the combined *uniform rotation* of the cam blank, held in the spindle of a dividing head, and the *uniform feed* of the table.

When a cam is machined by this method, the work and the vertical head are usually swung at an angle so that the axis of the work and the axis of the mill are parallel (Fig. 72-7).

If the work and the vertical milling attachment are maintained in a vertical position, only a cam having the same lead for which the machine is geared can be cut. When the work and the vertical milling attachment are inclined, any desired lead may be produced, providing that the desired lead is *less* than the lead for which the machine is geared. In other words, the required lead to be cut on the cam must always be less than the forward feed of the table during one revolution of the work.

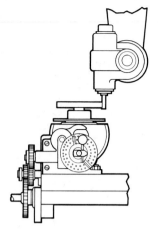
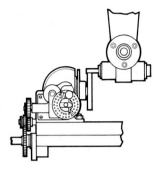

■ **Figure 72-8** Vertical head and workpiece set at 90°.

■ **Figure 72-9** Vertical attachment and workpiece at zero.

The principle involved in swinging the head is as follows: if a milling machine were set up to cut a cam and had equal gears on the dividing head and the lead screw, with the work and the vertical attachment in a vertical position (Fig. 72-8), the table would advance 10 in. while the work revolved one turn. An Archimedes spiral would be generated, and the cam would have a lead of 10 in., which is also a rise of 10 in. in 360°.

Rotate both the work and cutter so that they are parallel to the table and reading zero (Fig. 72-9). If a cut is taken, the work will move along the length of the end mill (which in this case would have to be 10 in.), and a circle having no lead will be generated.

From these two examples, it can be seen that it is possible to mill any lead between zero (0) and that for which the machine is geared, if the cutter and the work are inclined to any given angle between 0° and 90°. If this method were not used, it would be necessary to have a different change gear combination for every lead to be cut. This would be impossible because of the large number of change gears and the time required to change the gears for each different lead.

Calculations Required

From the drawing, determine the *lead* of the lobe or lobes of the cam; that is, determine the amount of rise of each lobe if it were continued for the full circumference of the cam.

If the space occupied by the lobe is indicated in degrees on the drawing, the lead is calculated as follows:

$$\text{Lead} = \frac{\text{rise of lobe in inches} \times 360}{\substack{\text{number of degrees of circumference} \\ \text{occupied by the lobe}}}$$

However, if the circumference is divided into 100 equal parts, it will be necessary to calculate the lead as follows:

$$\text{Lead} = \frac{\text{rise of lobe in inches} \times 100}{\text{percentage of circumference occupied by the lobe}}$$

EXAMPLE

A uniform-rise cam having a rise of .375 in. in 360° is to be cut. Calculate the required lead, the inclination of the work, and the vertical head.

Solution

$$\text{Lead of cam} = .375 \text{ in.}$$

The machine and the dividing head should then be geared to .375 in. This is impossible, since the shortest lead that can be cut with regular change gears on a milling machine is generally .670 in.

Note: Any handbook will contain these milling tables.

To cut a lead of .375 in., it will be necessary to gear the machine to something more than .375 in.; in this case, it will be .670 in. By consulting a handbook, it will be noted that the gears required for a .670-in. lead are 24, 86, 24, 100. It will also be necessary to swing the work and the vertical head to a definite angle. Figure 72-10 illustrates how this is calculated.

In the diagram,

L = lead to which the machine is geared

H = lead of the cam

i = angle of inclination of the dividing head spindle in degrees

$$\sin i = \frac{H}{L}$$

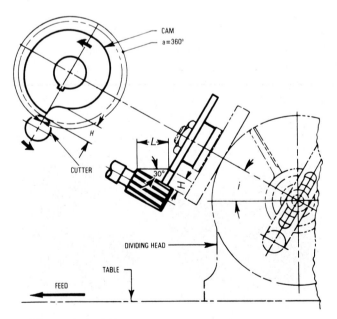

■ **Figure 72-10** Relationship between cam lead and table lead. *(Cincinnati Milacron, Inc.)*

Therefore,

$$\frac{\text{Sine angle}}{\text{of inclination}} = \frac{\text{lead of cam}}{\text{lead to which machine is geared}}$$

$$\text{Sine angle} = \frac{.375}{.670}$$

$$= .5597$$

$$\text{Angle} = 34°2'$$

Fig. 72-10 illustrates that, when the work travels along the distance L (.670), it will rotate one turn and reach a point on the cutter that is .375 higher than the starting point. This will produce a rise and a lead of .375 in. on the cam.

EXAMPLE

Specify the steps required to cut a uniform-rise cam having three lobes, each lobe occupying 120° and each having a rise of .150 in.

Procedure

1. Lead of cam $= \dfrac{.150 \times 360}{120}$

$$= .450 \text{ in.}$$

2. The smallest lead over .450 in. to which a machine can be geared is .670 in. (handbook).

3. Change gears required to produce a lead of .670 in.

$$\text{Gear ratio} = \frac{24}{86} \times \frac{24 \text{ (driven)}}{100 \text{ (drivers)}}$$

4. Mount the dividing head, and connect the worm shaft and the lead screw with the above gears. *Disengage the index plate locking device.*

5. Scribe three marks 120° apart on the periphery of the cam.

6. Mount the work in the dividing head chuck.

7. Mount an end mill of sufficient length in the vertical attachment.

8. Calculate the offset of the work and the vertical head:

$$\text{Sine of angle} = \frac{.450}{.670}$$

$$= .67164$$

$$\therefore \text{ angle} = 42° 12'$$

9. Swivel the dividing head to 42°12′.

10. Swivel the vertical milling attachment to 90° − 42°12′ = 47°48′.

11. Centralize the work and the cutter.

12. Rotate the work with the index crank until one scribed mark on the cam blank is exactly at the bottom dead center.

13. Adjust the table until the cutter is touching the lower side of the work and the center of the cutter is in line with the mark.

Note: When the cutter is below the work, the setup will be more rigid and there will be no chance of chips obscuring any layout lines.

14. Set the vertical feed collar to zero (0).

15. Start the machine.

16. Using the index head crank, rotate the work one-third turn or until the second scribed line is in line with the forward edge of the cutter.

Note: If the machine is geared to a lead over 2.500 in., the automatic feed may be used. If it is geared to a lead of less than 2.500 in., a short lead attachment should be used or the table should be fed along manually by means of the index head crank.

17. Lower the table slightly and disengage the gear train, or disengage the dividing head worm.

18. Return the table to the starting position.

19. Rotate the work until the next line on the circumference is in line with the center of the cutter.

20. Re-engage the gear train or the dividing head worm.

21. Cut the second lobe.

22. Repeat steps 17, 18, 19, and 20, and cut the third lobe.

Note: When calculating the offset as in step 8, the lead of the machine is often set to the whole number nearest to *twice* the cam lead. When the lead of the machine is exactly twice the lead of the cam, the offset will always equal 30° because the lead of the machine, which is the hypotenuse (Fig. 72-10), is twice the lead of the cam. Thus, the sine of the angle of inclination is equal to ½, or .500, which then makes the angle equal to 30°.

▶▶ Rack Milling

A *rack*, in conjunction with a gear (pinion), is used to convert rotary motion into longitudinal motion. Racks are found on lathes, drill presses, and many other machines in a shop. A rack may be considered as a spur gear that has been straightened out so that the teeth are all in one plane. The circumference of the pitch circle of this gear would now become a straight line, which would just touch the pitch circle of a gear meshing with the rack. Thus, the pitch line of a rack is the distance of one addendum (1/DP) below the top of the tooth.

The pitch of a rack is measured in linear (circular) pitch, which is obtained by dividing 3.1416 by the diametral pitch; that is,

$$\text{Rack pitch} = \frac{3.1416}{\text{DP}}$$

The method used to cut a rack will depend generally on the length of the rack. If the rack is reasonably short [10 in. (250 mm) or less], it may be held in the milling machine vise in a position parallel to the cutter arbor. On short racks, the teeth may be cut by accurately moving the cross-slide of the machine an amount equal to the circular pitch of the gear and then moving the table longitudinally to cut each tooth. If the rack is longer than the cross travel of the milling machine table, it must be held longitudinally on the table and is generally held in a special fixture.

The milling cutter is held in a *rack milling attachment* (Fig. 72-11). When cutting a straight tooth, the cutter is held at 90° to the position used when cutting a spur gear.

It is possible to mount slotting or narrow side milling cutters on the rack milling attachment for milling operations that can be handled more easily by using the machine crossfeed.

RACK INDEXING ATTACHMENT

When cutting a rack using the rack milling attachment, the table is often moved (indexed) for each tooth by means of the rack indexing attachment (Fig. 72-11). This consists of an indexing plate with two diametrically opposed notches and a locking pin. Two change gears selected from a set of 14 are mounted as shown in Fig. 72–11. Different combinations of change gears permit the machine table to be moved accurately in increments, corresponding to the linear (circular) pitch of the rack, by making either a half-turn or one complete turn of the plate. For indexing requiring one complete turn only, provision is made to close off one of the slots, thus preventing any error in indexing.

This attachment permits the indexing of all diametral pitches from 4 to 32, as well as all circular pitches from 1/8 in. to 3/4 in., varying by sixteenths. The following table movements can also be produced: 1/7, 1/6, 1/5, 2/7, 1/3, and 2/5 in.

Figure 72-11 Cutting the teeth on a helical rack using the rack milling and indexing attachments. *(Cincinnati Milacron, Inc.)*

Worms and Worm Gears

Worms and *worm gears* are used when a great ratio reduction is required between the driving and driven shafts. A worm is a cylinder on which is cut a single- or multiple-start Acme-type thread. The angle of this thread ranges from a 14.5° to 30° pressure angle. As the lead angle of the worm increases, the greater the pressure angle should be on the side of the thread. The teeth on a worm gear are machined on a peripheral groove, which has a radius equal to half the root diameter of the worm. The drive ratio between a worm and worm gear assembly depends on the number of teeth in the worm gear and whether the worm has a single- or double-start thread. Thus, if a worm gear had 50 teeth, the ratio would be 50:1, providing the worm had a single-start thread. If it had a double-start thread, the ratio would be 50:2, or 25:1.

TO MILL A WORM

Worms are often cut on a milling machine with a rack milling attachment and a thread milling cutter (Fig. 72-12). The setup of the cutter is similar to that for rack milling. The work is held between index centers and is rotated by suitable gears between the worm shaft and the lead screw of the milling machine. This is similar to the setup for helical milling. A short lead (less than 1 in.) attachment is used when a worm is being milled because the thread usually has a short lead. If a short lead attachment is not available, the work may be rotated and the table moved lengthwise by means of the index crank on the dividing head.

procedure

1. Calculate all dimensions of the thread, that is, lead, pitch, depth, and angle of thread.

   Note: The angle of the thread is calculated using the pitch diameter.

Figure 72-12 Milling machine set up for machining a worm. *(Cincinnati Milacron, Inc.)*

2. Mount the worm blank between the dividing head centers located at the end of the milling machine table.

3. Determine the proper gears for the lead, and mount them so that they connect the worm shaft and the lead screw.

4. Disengage the index plate locking device.

5. Mount the proper thread milling cutter on the rack milling attachment.

6. Swing the rack milling attachment to the required helix angle of the worm thread and in the proper direction for the lead of the worm.

7. Center the work under the cutter.

8. Raise the work up to the cutter.

9. Move the work clear of the cutter, and raise the table to the required depth of thread.

10. Cut the thread using the automatic feed or by turning the index crank handle to feed the table manually.

▶▶ Clutches

Positive-drive clutches are used extensively to drive or disconnect gears and shafts in machine gearboxes. The headstocks on most lathes use clutches, machined on the hubs of gears, to engage or disengage gears to provide different spindle speeds. The positive drive on this type of clutch is produced by means of interlocking teeth or projections on the driving and driven parts and does not rely on friction drive, as in the case of friction-type clutches.

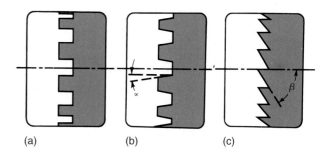

Figure 72-13 Types of clutch teeth: (a) straight-tooth; (b) inclined-tooth; (c) sawtooth.

Three forms of positive-drive clutches are shown in Fig. 72-13. The *straight-tooth clutch* (Fig. 72-13a) permits rotation in either direction. This type is more difficult to engage, since the mating teeth and grooves must be in perfect alignment before engagement is possible.

The *inclined-tooth clutch* (Fig. 72-13b) provides an easier means of engaging or disengaging the driving and driven members because of an 8° or 9° angle machined on the faces of the teeth. Since this type of clutch tends to disengage more rapidly, it must be provided with a positive means of locking it in engagement. Clutches of this type permit the shafts to run in either direction without backlash.

The *sawtooth clutch* (Fig. 72-13c) permits drive in only one direction but is more easily engaged than the other two types of clutches. The angle of the teeth in this type is generally 60°.

TO MACHINE A STRAIGHT-TOOTH CLUTCH HAVING THREE TEETH

Note: This method applies to all clutches having an odd number of teeth.

1. Mount the dividing head on the milling machine table.

2. Mount a three-jaw chuck in the dividing head.

3. Mount the workpiece in the chuck.

Note: The spindle and the chuck of the dividing head may be positioned either horizontally or vertically. For this operation, they are assumed to be in a vertical position.

4. Set the sector arms to the proper indexing, that is, $^{40}/_3 = 13^1/_3$ turns or 13 turns and 13 holes on a 39-hole circle.

5. Mount a side milling cutter on the milling machine arbor. The cutter should be no wider than the narrowest or innermost part of the groove.

Figure 72-14 Adjusting a workpiece to the cutter. *(Kelmar Associates)*

6. Set the proper spindle speed and table feed.

7. Start the cutter and adjust the work until the edge nearest the front of the machine *just* touches the inner side of the cutter (Fig. 72-14). Set the crossfeed graduated feed collar to zero (0).

8. Move the table longitudinally until the work is clear of the cutter.

9. Move the table laterally half the diameter of the work plus about .001 in. (0.02 mm) for clearance. Lock the saddle in this position.

10. Set the depth of cut, and lock the knee clamps.

11. Take a cut across the full width of the workpiece, as in Fig. 72-15.

12. Return the table to the starting position.

13. Index for the next tooth and take the second cut as shown in Fig. 72-16.

Figure 72-15 Cutting the first tooth. *(Kelmar Associates)*

Figure 72-16 Cutting the second tooth. *(Kelmar Associates)*

Figure 72-17 Cutting the third tooth. *(Kelmar Associates)*

14. Return the table to the starting position.

15. Index the next tooth.

16. Take the third cut as shown in Fig. 72-17.

TO MACHINE A STRAIGHT-TOOTH CLUTCH HAVING FOUR TEETH

When machining a straight- or inclined-tooth clutch having an even number of teeth, it is necessary to machine one side of each tooth first and then machine the second side of each. This procedure obviously requires more time; therefore, clutches should be designed with an odd number of teeth to reduce machining time and the chance of error.

procedure

1. Mount the work as in the previous example and set the proper indexing.

2. Start the cutter and adjust the workpiece until the edge nearest the front of the machine just touches the inner edge of the cutter (Fig. 72-14).

3. With the work clear of the cutter, move the saddle over half the diameter of the work plus the thickness of the cutter minus .001 in. (0.02 mm) for clearance.

4. Adjust the work until the cutter is over the center hole of the clutch. It is advisable to cut from the center to the outside of the clutch in order to minimize the vibration.

5. Set the depth and lock the knee clamp.

6. Take the first cut.

Note: When cutting an even number of clutch teeth, it is absolutely necessary that the cut be made through one wall only.

7. Index for and cut the remaining teeth on the one side. (Indexing = 10 turns for each tooth.)

8. Revolve the work one-eighth of a turn (five turns of the index crank).

9. Touch the opposite side of the work to the other side of the cutter, that is, the outside of the cutter to the inner edge of the work.

10. Repeat operations 3, 4, 5, 6, and 7 until all the teeth have been cut. If any pie-shaped pieces of metal remain in the tooth spaces, they must be removed with an additional cut through the center of the space.

unit 72 review questions

Cams and Cam Milling

1. Define a cam.

2. Name four types of cams.

3. Name four types of followers and state how each is used.

4. List three cam motions, and describe the type of motion imparted to the follower in one revolution of the cam.

5. In a single-lobe cam and a double-lobe cam, what is the relationship of the lead to the rise?

6. Calculate:
   a. Lead of cam
   b. The change gears required to produce the lead
   c. The angle at which to swing the dividing head
   d. The angle at which to swing the vertical head for each of the following examples of uniform rise:
      1. A single-lobe cam having a rise of .125 in. in 360°
      2. A two-lobe cam, each lobe having a rise of .187 in. in 180°
      3. A three-lobe cam, each lobe having a rise of .200 in. in 120°

Rack Milling

7. Define rack and state its purpose.

8. A 10-DP gear is in mesh with a rack. The gear has 42 teeth. The rack is 1 in. thick from the top of the tooth to the bottom of the rack. Calculate the distance from the center of the gear to the bottom of the rack.

9. Calculate the linear pitch of a 5-pitch, an 8-pitch, and a 14-pitch rack.

Worms and Worm Gears

10. Define *worm* and *worm gear*, and describe their use.

11. Describe briefly how a worm is cut on a milling machine.

Clutches

12. List three types of clutches, and state the application of each.

13. Why are clutches with an odd number of teeth preferred to those with an even number of teeth?

14. After the outside surface of a 3-in. diameter (or 76-mm diameter) clutch blank has been touched up to a .50-in. wide (or 13-mm wide) side cutter, how far must the table be moved over when cutting:
   a. A clutch having five teeth?
   b. A clutch having six teeth?

Section 13

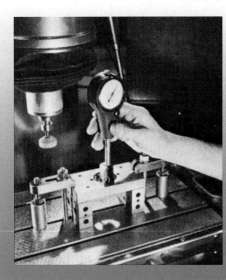

(Moore Tool Company)

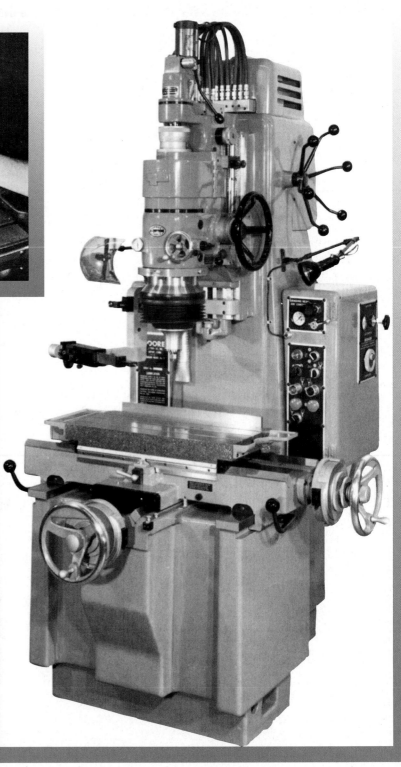

THE JIG BORER
AND JIG GRINDER

The jig borer was developed primarily to overcome the toolmaker's problem of accurately locating and producing holes at precise locations. This machine was a very important machine tool before the age of computer-controlled machining centers that can do similar operations faster and more accurately. The latest machining centers can repeatedly position holes to within 20 millionths of an inch. This accuracy, built into CNC machine tools, has sharply reduced the use of jig borers to the point at which today it is no longer manufactured. It is mentioned only briefly in this section because many jig borers may still be used in industry.

UNIT 73

The Jig Borer

OBJECTIVES

After completing this unit, you will be able to:

1 Identify and state the purposes of the main operative parts of a jig borer

2 Use various accessories and work-holding devices for setting up and boring holes

A jig borer (Fig. 73-1), although similar to a vertical milling machine, is much more accurate and built closer to the floor so that the worker can operate it while seated. The precision-ground lead screws controlling the table movements are capable of infinitely fine incremental divisions and permit simultaneous measurement and positioning to within an accuracy of .0001 in. (0.002 mm) over the table length. This machine must be not only rugged for the heavy cuts necessary for roughing purposes but also sensitive for the more accurate finishing cuts.

▶▶ Jig Borer Parts

> The *variable pitch pulley drive* is operated by pushing a button on the *electric control panel* to provide the spindle with a variable speed range from 60 to 2250 revolutions per minute (r/min).

> The *quill housing* can be raised or lowered to accommodate various sizes of work if first the *quill housing clamp* is loosened and then the *quill housing vertical positioning handle* is turned.

> The *brake lever* is manually operated to stop the rotation of the spindle. It is especially useful while various tools are removed or replaced in the spindle.

> The *rapid feed handwheel* allows the spindle to be raised or lowered rapidly by hand.

> The *friction clutch* may be used to engage or disengage the handfeed of the quill.

> The *graduated downfeed dial* reads the distance of vertical spindle travel, by means of a vernier, in thousandths of an inch or hundredths of a millimeter for metric dials.

> The *adjustable stop for hole depths* can be adjusted to allow the spindle to move to a predetermined depth for drilling or boring a hole.

> The *spindle* revolves inside the quill and supplies drive for the cutting tools. An internal taper in the spindle allows a variety of tools to be rapidly and accurately interchanged.

> The *reference scales* (longitudinal and crossfeed) serve as reference points in moving the table into position. They determine the position of the starting or reference point of the job.

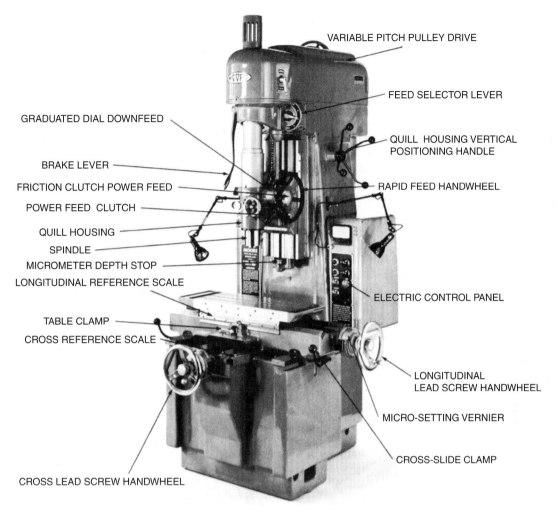

■ Figure 73-1 A #3 Moore precision jig borer. *(Moore Tool Company, Inc.)*

Labels in figure:
- VARIABLE PITCH PULLEY DRIVE
- FEED SELECTOR LEVER
- GRADUATED DIAL DOWNFEED
- QUILL HOUSING VERTICAL POSITIONING HANDLE
- BRAKE LEVER
- FRICTION CLUTCH POWER FEED
- RAPID FEED HANDWHEEL
- POWER FEED CLUTCH
- QUILL HOUSING
- SPINDLE
- MICROMETER DEPTH STOP
- LONGITUDINAL REFERENCE SCALE
- ELECTRIC CONTROL PANEL
- TABLE CLAMP
- CROSS REFERENCE SCALE
- LONGITUDINAL LEAD SCREW HANDWHEEL
- MICRO-SETTING VERNIER
- CROSS-SLIDE CLAMP
- CROSS LEAD SCREW HANDWHEEL

> The *graduated dials,* with micro-setting verniers on the *longitudinal* and *crossfeed screw handwheels,* allow the table to be positioned quickly and accurately to within .0001 in. or to 0.002 mm for metric dials.

TO INSERT SHANKS IN THE SPINDLE

Toolroom work, with a variety of hole sizes, requires frequent changing of tools (Fig. 73-2 on p. 582). The tapered hole in the spindle allows this to be done quickly and accurately if certain precautions are taken.

1. The taper shank on the tool being inserted and the hole in the spindle must be *perfectly clean;* otherwise, the tapers will be damaged and inaccuracy will occur.

2. Protect the taper shanks from finger perspiration, especially if the shank will be in the machine for some time. This may cause both the shank and the spindle to rust.

3. Avoid inserting shanks too tightly, especially when the spindle is warmer than the inserted shank. When the shank warms up, it expands and may jam in the spindle.

4. When removing or replacing tools in the spindle, apply the brake firmly with the left hand. The wrench should be held carefully to prevent it from slipping out of the hand and damaging the machine table.

▶▶ Accessories and Small Tools

A wide variety of accessories enable a jig borer to meet three basic requirements: *accuracy, versatility,* and *productivity.*

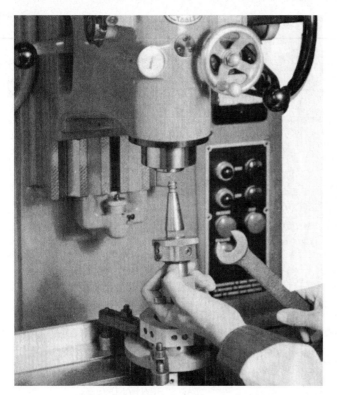

Figure 73-2 Replacing a tool in the jig borer spindle. *(Moore Tool Company, Inc.)*

Figure 73-3 A set of solid boring tools.

Figure 73-4 A swivel block boring chuck. *(Moore Tool Company, Inc.)*

Only accessories concerned with drilling, boring, and reaming will be dealt with in this unit.

DRILLING

Key-type and keyless chucks are used to hold the smaller straight-shank spotting tools, drills, and reamers. Special collets are used to hold the larger straight-shank spotting tools, drills, and reamers. A setscrew in the collet is tightened against a flat on the tool shank, providing a positive grip, which eliminates twisting and scuffing of the tool shank.

BORING

Single-point boring, the most accurate method of attaining locational accuracy in jig boring, makes it necessary to have a wide variety of boring tools. The most commonly used boring tools are a solid boring bar, a swivel block boring chuck, an offset boring chuck, and a DeVlieg microbore boring bar.

The *solid boring bar* (Fig. 73-3) is fitted with an adjusting screw that, when adjusted, advances the toolbit over a relatively short range. Solid boring bars are rigid, making them especially useful in boring deep holes; a number of solid boring bars may be left set at a specific size for repetitive boring.

The *swivel block boring chuck* (Fig. 73-4) provides a greater range of adjustment than do other types, in proportion to its diameter, and better visibility to the operator while boring. One disadvantage of this type of chuck is that, since the tool swings in an arc, the graduations for adjusting the tool travel vary depending on the length of cutting tool used.

The *offset boring chuck* (Fig. 73-5) is a versatile tool that permits the cutting tool to be moved outward at 90° to the spindle axis of the machine. This allows the use of a wide variety of cutting tools without altering the value of the adjusting graduations. This chuck makes it possible to perform operations such as boring, counterboring, facing, undercutting, and machining outside diameters.

The *DeVlieg microbore boring bar* (Fig. 73-6) is equipped with a micrometer vernier adjustment, which permits the cutting tool to be accurately adjusted to within a tenth of a thousandth of an inch.

Single-Point Boring Tools

Since single-point boring is the most accurate method of generating accurate hole location, a wide variety of cutting tools are available for this operation. The toolbits

■ **Figure 73-5** A rugged, versatile boring chuck designed for making heavy cuts.

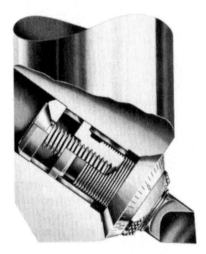

■ **Figure 73-6** A DeVlieg microbore boring bar. *(DeVlieg Machine Company)*

■ **Figure 73-7** Single-point boring tools. *(Moore Tool Company, Inc.)*

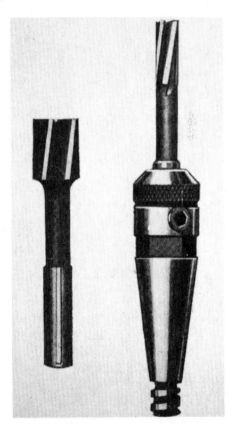

■ **Figure 73-8** A hardened and ground collet holding a precision end-cutting reamer. *(Moore Tool Company, Inc.)*

shown in Fig. 73-7 are generally used for small holes; however, with the necessary chuck attachments, larger holes may be bored. These toolbits are available in high-speed steel and with brazed, cemented-carbide tips.

Collets and Chucks

An assortment of collets and chucks (Fig. 73-8) is available for a jig borer spindle to hold straight-shank spotting tools, drills, and precision end-cutting reamers.

REAMERS

Two types of reamers, the rose, or fluted, and the precision end-cutting, are used in jig boring for bringing a hole to size quickly. The *rose, or fluted, reamer* is used after a hole has been bored and provides an accurate method of sizing a hole. If handled carefully and used to remove only about .001 to .003 in. (0.02 to 0.07 mm), these reamers maintain fairly accurate hole sizes.

Precision end-cutting reamers, with short, sturdy shanks (Fig. 73-8), provide the fastest method of locating and sizing holes to within an accuracy of ±.0005 in. (0.01 mm). The end-cutting reamer, held rigidly and running true with the spindle, acts as a boring tool and reamer, locating and sizing the hole at the same time. End-cutting reamers eliminate the use of boring tools if the accuracy of the hole diameter and location does not require closer tolerance than ±.0005 in. (0.01 mm).

Jig Borer

1. For what purpose were jig borers developed?

2. For what type of work are they especially valuable?

3. Name the operations that can be performed on a jig borer.

4. Explain the difference between a jig borer and a vertical milling machine.

Accessories and Small Tools

5. Name four common boring tools and explain the advantages of each.

6. Name two types of reamers used in jig boring and explain the advantages of each.

UNIT 74

The Jig Grinder

OBJECTIVES

After completing this unit, you will be able to:

1 Select the wheels and grinding methods required for jig-grinding holes

2 Set up work and jig-grind a straight hole to within a tolerance of ±.0002 in. (0.005 mm)

The need for accurate hole locations in hardened material led to the development of the jig grinder in 1940. While it was originally developed to position and grind accurately straight or tapered holes, many other uses have been found for the jig grinder over the years. The most important of these has been the grinding of contour forms, which may include a combination of radii, tangents, angles, and flats (Fig. 74-1).

The advantages of jig grinding are:

1. Holes distorted during the hardening process can be accurately brought to correct size and position.

2. Holes and contours requiring taper or draft may be ground. Mating parts, such as punches and dies, can be finished to size, eliminating the tedious job of hand fitting.

3. Because more accurate fits and better surface finishes are possible, the service life of the part is greatly prolonged.

4. Many parts requiring contours can be made in a solid form, rather than in sections, as was formerly necessary.

■ **Figure 74-1** A flanged punch represents an ideal example of jig grinding. *(Moore Tool Company, Inc.)*

▶▶ Jig Grinder Parts

The *jig grinder* (Fig. 74-2 on p. 586) is similar to a jig borer. Both have precision-ground lead screws capable of positioning the table within .0001-in. (0.002-mm) accuracy over its entire length. Both are vertical spindle machines and are based on the same basic cutting principle encountered in single-point boring. The main difference between these two machines is in the spindles.

The jig grinder is equipped with a high-speed pneumatic turbine grinding spindle for holding and driving the grinding wheel. The spindle construction permits outfeed grinding (Fig. 74-3 on p. 586), as well as the grinding of tapered holes (Fig. 74-4 on p. 586).

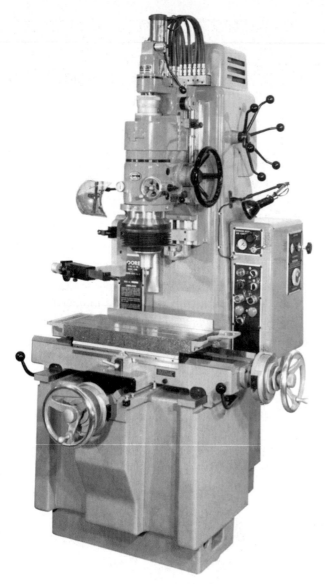

■ Figure 74-2 A #3 Moore jig grinder. *(Moore Tool Company, Inc.)*

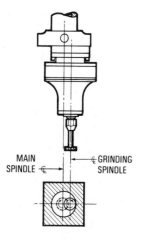

MAIN SPINDLE ℄   ℄ GRINDING SPINDLE

■ Figure 74-3 The grinding spindle may be offset from the main spindle. Lower view shows the planetary path of rotation.

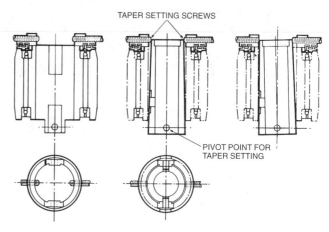

TAPER SETTING SCREWS

PIVOT POINT FOR TAPER SETTING

■ Figure 74-4 Main spindle assembly. *(Moore Tool Company, Inc.)*

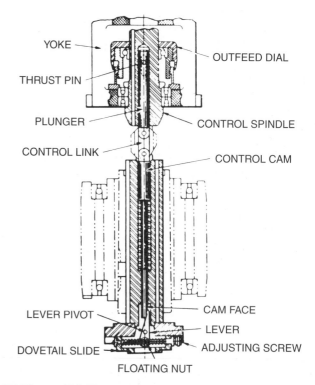

YOKE

OUTFEED DIAL

THRUST PIN

PLUNGER

CONTROL SPINDLE

CONTROL LINK

CONTROL CAM

LEVER PIVOT

CAM FACE

DOVETAIL SLIDE

LEVER

ADJUSTING SCREW

FLOATING NUT

■ Figure 74-5 Assembly for controlling size by outfeed. Dial setting outfeed is graduated in tenths. *(Moore Tool Company, Inc.)*

▸▸ Grinding Head Outfeed

A horizontal dovetail slide connects the grinding head to the main spindle of the jig grinder. The grinding head may be offset from the center of the main spindle to grind various-size holes. The amount of eccentricity (offset) of the grinding head can be *accurately* controlled by the internally threaded *outfeed dial*, which is mounted on the non-rotating yoke at the top of the jig-grinding spindle (Fig. 74-5). The dial is graduated in steps of .0001 in.

DOWNFEED GRADUATED DIAL AND VERNIER

(a)

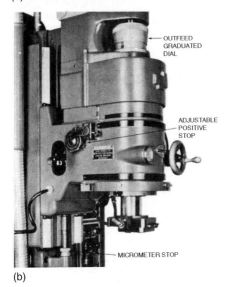

OUTFEED GRADUATED DIAL

ADJUSTABLE POSITIVE STOP

MICROMETER STOP

(b)

■ **Figure 74-6** Depth-measuring devices. *(Moore Tool Company, Inc.)*

(0.002 mm), permitting accurate control of the hole size during grinding.

Coarse adjustment of the grinding wheel position is attained by a fine pitch-adjusting screw within the dovetail slide (Fig. 74-5). This coarse-adjusting screw is accessible only when the machine spindle is stopped.

DEPTH-MEASURING DEVICES

The Moore jig grinder has three distinct features for controlling and measuring the depth of holes (Fig. 74-6a and b):

1. The *adjustable positive stop* is on the left-hand end of the pinion shaft. Microadjustments can be made by a limiting screw.

■ **Figure 74-7** Dressing a wheel with a diamond dressing arm.

2. The *graduated dial* on the downfeed handwheel indicates the travel of the quill. It can be set to zero (0) at any position and reads the travel depth in steps of .001 in. (0.02 mm).

3. The *micrometer stop* (Fig. 74-6), fastened to the column of the grinder, controls hole depth.

DIAMOND DRESSING ARM

Jig grinders must rapidly dress grinding wheels without disturbing the setup and location of the grinding spindle. The diamond dressing arm (Fig. 74-7) may be quickly swung into the approximate grinding wheel location and then locked into position. The final approach to the grinding wheel is done by a fine-adjusting knurled screw, which advances the diamond through the dressing arm.

▶▶ Grinding Methods

The removal of material from a hole with a conventional grinding wheel is carried out by two methods: outfeed and plunge grinding. Each method has its advantages, and at times both can be used effectively to grind the same hole. Small holes, less than ¼ in. (6 mm) in diameter, can be effectively ground by using diamond-charged mandrels. Holes larger in diameter than the normal machine range can be ground effectively if an extension plate (Fig. 74-8) is used between the grinding spindle and the main spindle. With the use of an extension plate, holes up to 9 in. (225 mm) in diameter may be ground.

OUTFEED GRINDING

Outfeed grinding is similar to internal grinding where the wheel is fed radially into the work with passes as fine as .0001 in. (0.002 mm) at a time. The cutting action takes place with the periphery of the grinding wheel. Outfeed

Figure 74-8 Grinding a large hole using an extension plate. *(Moore Tool Company, Inc.)*

Figure 74-9 Hole size may be increased by .060 in. (1.6 mm) in one cut by plunge grinding.

grinding is generally used to remove small amounts of stock when high finish and accurate hole size are required.

PLUNGE GRINDING

Plunge grinding with a grinding wheel can be compared to the cutting action of a boring tool. The grinding wheel is fed radially to the desired diameter and then into the work (Fig. 74-9). Cutting is done with the bottom corner of the wheel only. It is a rapid method of removing excess stock, and if the wheel is properly dressed, it produces satisfactory finishes for some jobs. The sharp cutting action

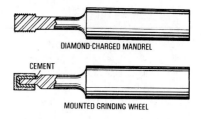

Figure 74-10 The strength and rigidity of a diamond-charged mandrel exceed those of a mounted grinding wheel. *(Moore Tool Company, Inc.)*

that results from the small contact area of the wheel keeps the work cooler than outfeed grinding.

Diamond-charged mandrels (Fig. 74-10) are used instead of conventional grinding wheels for grinding holes of less than .250 in. (6 mm) in diameter. These mandrels should be made of cold-rolled steel that has been turned to the correct size and shape in relation to the hole to be ground. The grinding end of the mandrel is placed in diamond dust and is tapped sharply with a small hardened hammer to embed the diamond dust in the mandrel surface.

The advantages of this tool over a conventional grinding wheel are:

1. Mandrels have maximum strength and rigidity.
2. Mandrels can be made the ideal diameter and length for each hole.
3. The velocity required for efficient grinding is approximately one-quarter of that for a wheel.
4. The cost per hole is less due to the greater efficiency.

▶▶ Grinding Wheels

SELECTION

Selection of the proper wheel is necessary for satisfactory grinding performance. Since many specific factors influence the selection of the grinding wheel to be used, a few general principles will be helpful:

1. The shank or mandrel of mounted wheels should be as short as possible to ensure rigidity.
2. Wherever possible, the grinding wheel diameter should be approximately three-quarters of the diameter of the hole to be ground.
3. Widely spaced abrasive grains in the bond increase the penetrating power of the wheel.
4. When soft, low-tensile-strength materials are ground, a hard abrasive grain with a fairly strong or hard bond should be used.
5. For grinding high-alloy hardened steels, a hard abrasive grain in a soft or weak bond is recommended.

WHEEL SPEED

The majority of grinding wheels are operated most efficiently at about 6000 surface feet per minute (sf/min) (1828 m/min). Diamond-charged mandrels used for small hole grinding should be operated at approximately 1500 sf/min (457 m/min). The spindle speed can be varied for the different types and diameters of wheels used; three grinding heads are available for the Moore jig grinder. With these heads, a range from 12,000 to 60,000 revolutions per minute (r/min) is possible. The speed of each head may be varied within its range by adjustment of the pressure regulator, which controls its air supply.

WHEEL DRESSING

For a grinding wheel to perform efficiently, it must be dressed or trued properly. An improperly dressed wheel will tend to produce the following conditions:

1. Poor surface finish on the hole
2. Surface burns
3. Out-of-round holes
4. Taper or bell-mouth holes
5. Locational error

Care in dressing a grinding wheel can prevent many of these undesirable conditions. To develop the best cutting characteristics of the wheel, use the following recommended techniques for dressing a grinding wheel on a jig grinder:

1. While the wheel is running at a reduced rate, dress the top and bottom face with an abrasive stick held in the hand.
2. Dress the diameter of the wheel with a *sharp* diamond (Fig. 74-11).

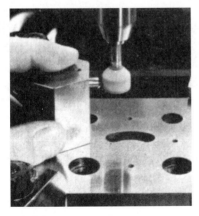

■ Figure 74-11 The hand-held diamond dresser is convenient and effective.

■ Figure 74-12 The width of the wheel face is reduced to avoid excessive side pressure during grinding. *(Moore Tool Company, Inc.)*

3. Repeat steps 1 and 2 with the wheel at the proper operating speed.
4. Relieve the upper portion of the diameter (Fig. 74-12) so that approximately .250 in. (6 mm) of the cutting face remains.
5. The bottom face of the wheel should be concaved slightly with an abrasive stick for grinding to a shoulder on the bottom of a hole.

In *outfeed grinding*, only the diameter of the wheel should be dressed when required. When *plunge grinding*, dress the bottom face of the wheel with an abrasive stick.

▶▶ Grinding Allowances

Many factors determine the amount of material that should be left in a hole for the jig-grinding operation. Some of the more common factors are:

1. Type of surface finish in the bored hole
2. Size of the hole
3. Material of the workpiece
4. Distortions that occur during the hardening process

It is difficult to set specific rules on the amount of material that should be left for grinding because of the many variable factors involved. However, general rules which would apply in most cases are:

1. Holes of up to .500 in. (13 mm) in diameter should be .005 to .008 in. (0.12 to 0.2 mm) undersize for the grinding operation.
2. Holes of over .500 in. (13 mm) in diameter should be .010 to .015 in. (0.25 to 0.4 mm) undersize for the grinding operation.

▶▶ Setting Up Work

When setting up work for jig grinding, take care to avoid distortion of the workpiece or machine table due to clamping pressures. Keep the following points in mind while mounting the work:

1. When bolts or strap clamps are used, keep the bolts as close as possible to the work.

2. Strap clamps should be placed exactly over the parallels supporting the work. Distortion of the work can occur if a strap clamp is tightened over a part of the work not supported by parallels.

3. Bolts should *not* be tightened any more than is required to hold the workpiece. There is less pressure exerted during jig grinding than during jig boring.

4. Do not clamp work too tightly in the precision vise clamp, since this may spring the stationary jaw, dislocating the aligned edge of the work.

5. Set up work on parallels high enough to allow the bottom of the hole being ground to be measured.

TO LOCATE THE WORKPIECE

The same basic methods as in jig boring are used to locate accurately a workpiece on the jig grinder; the straight-edge, edgefinder, and indicator are used, for example. Distortion of the workpiece during the heat-treating process may necessitate "juggling" during the setting-up process to ensure that all holes will "clean up." The workpiece may be set up parallel to the table travel by three methods:

1. Indicate an edge of the workpiece.

2. Set the work against the table straightedge; then check the alignment with an indicator.

3. On a heat-treated piece, indicate two or more holes and set up the work to suit the average location of a group of holes.

▶▶ Grinding Sequence

When a series of holes in a workpiece must be accurately related to each other, consideration must be given to the sequence of grinding operations. The following sequence is suggested when a number of different holes are required—for example, straight, tapered, or blind—or holes with shoulders:

1. Rough-grind all holes first. When a high degree of accuracy is required, allow the work to cool to room temperature before proceeding to finish-grind.

2. Finish-grind all holes that can be ground with the same grinding head. This avoids continually changing grinding heads.

3. Holes whose relationship to others is most important should be ground in one continuous period of time.

4. Grind holes with shoulders or steps only once to avoid having to make accurate depth settings twice.

TO GRIND A TAPERED HOLE

The grinder spindle can be set for grinding tapers in either direction by loosening one adjusting screw and tightening the other. For most taper hole grinding, it is accurate enough to set the grinder spindle to the degrees indicated on the taper-setting plate (Fig. 74-13).

When an *extremely accurate* angular setting is required, the following steps are suggested:

1. Convert the angle into thousandths taper per inch (or into hundredths of a millimeter per 25 mm of length) by mathematical calculations.

2. Mount an indicator in the machine spindle.

3. Set an angle plate or a master square on the machine table.

4. Move the indicator through 1 in. (25 mm) of vertical movement as read on the downfeed dial.

5. Set the adjusting screws until the desired taper is attained.

Note: If the adjusting screws are too tight, they will bind the vertical movement of the spindle; if too loose, the machine may grind out of round.

By reversing this procedure, one can accurately reset the machine spindle for straight-hole grinding.

■ **Figure 74-13** The taper angle is indicated on the taper-setting plate. *(Moore Tool Company, Inc.)*

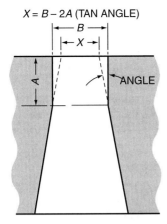

$X = B - 2A \text{ (TAN ANGLE)}$

Figure 74-14 Formula for calculating *X* at the top of a tapered hole to produce the desired length of section *A*. *(Moore Tool Company, Inc.)*

Most tapered holes also have a straight section, and the taper is ground to a certain distance from the top. Sometimes it is difficult to see exactly where the taper begins because of the high finish in the hole and the gradual advance of most tapers. Two methods are generally used to show where the tapered section begins:

1. Apply layout dye to the top portion of the hole with a pipe cleaner. The dye will be removed from the taper portion during the grinding operation, allowing the length of the straight hole to be measured.

2. On holes too small or difficult to see and measure, it is recommended that the taper be ground first to dimension *X* in Fig. 74-14. This involves the use of the formula in Fig. 74-14 in order to calculate what size the hole would be at dimension *X*. Once the tapered hole has been correctly ground to size, the straight hole is ground to the proper diameter. This will automatically produce the proper length of the straight hole *A*.

TO GRIND SHOULDERED HOLES

Many times it is necessary to grind not only the diameter of a hole but also the bottom of blind or shouldered holes. Grinding shouldered holes presents a few problems not encountered in straight or taper grinding, and the following suggestions are offered:

1. Select the proper-size grinding wheel for the hole size (Fig. 74-15).

2. Make the bottom of the wheel slightly concave with an abrasive stick.

3. Set the depth stop so that the wheel just touches the bottom or shoulder of the hole.

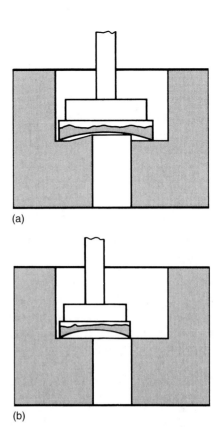

(a)

(b)

Figure 74-15 (a) Grinding wheel is too large and cannot grind a flat surface; (b) wheel is small enough to clear the opposite side of the hole.

4. Rough-grind the sides and shoulder of the hole at the same time. This eliminates leaving a slight step near the bottom of the hole.

5. Dress the wheel and proceed to finish-grind the hole.

▶▶ Jig Grinding with Cubic Boron Nitride Wheels

Cubic boron nitride (CBN) wheels and pins are widely used for jig grinding because they are twice as hard as aluminum oxide wheels, resist breakdown, last much longer, and require fewer wheel changes. CBN wheels perform best when jig grinding tool and die, carbon, and alloy steels hardened to Rc 50 and above; hard abrasive cast iron; and superalloys with a hardness of Rc 35 and above. The advantages of CBN wheels and pins over aluminum oxide wheels in jig grinding are shown in Fig. 74-16. They give the best performance on machines with good spindle bearings and constant speeds and feeds.

WHEEL SELECTION GUIDELINES

CBN wheels and pins are available in a variety of types and sizes for jig-grinding operations. The proper wheel

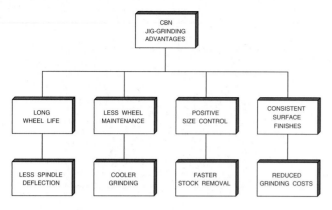

Figure 74-16 The advantages of jig grinding with CBN wheels and pins.

selection to suit the workpiece material and type of jig-grinding operation is very important to successfully jig grind with CBN wheels. The most important factors which should be considered when selecting CBN wheels are the abrasive type, wheel bond, and grit size.

1. *Abrasive type.* Since there is a wide variety of CBN abrasives available for jig grinding purposes, select the one that best suits the material to be ground.

2. *Bond types.* CBN jig-grinding wheels are available in resin, vitrified, metal, and electro-plated bonds.

   > *Resin- and vitrified-bond wheels* have crystals distributed evenly throughout the wheel matrix. These wheels offer high material-removal rates, long wheel life, and high surface finishes.
   > *Metal-bond wheels* are widely used because they offer long wheel life and excellent form retention.
   > *Electroplated wheels* have a single layer of CBN abrasive plated to a steel core or shank. These wheels offer several advantages such as low cost, high stock removal, free cutting, and cool grinding.

3. *Grit size.* The size of the abrasive grit determines the material-removal rate and the type of surface finish produced.

   > *Choose the largest abrasive size* that will produce the surface finish required because it generally results in the best material-removal rates and longest wheel life.
   > *Choose finer abrasive size* to produce good surface finishes and to reduce grinding heat.

WHEEL MOUNTING AND CONDITIONING

Correct mounting and conditioning (truing and dressing) procedures are important for the best performance of CBN jig-grinding wheels and pins. Both factors affect the life of the jig-grinding wheels, the rate at which material is removed, and the quality of the workpiece.

Mounting

> Have the shank extending as little as possible to avoid overhang, which can result in chatter, vibration, and spindle deflection.

> Use an indicator on the wheel shank and rotate the spindle slowly by hand to check the runout of the spindle; it should not be more than .001 in. (0.02 mm).

Truing

Resin-, vitrified-, and metal-bond jig-grinding wheels must be trued. *Never true or condition electroplated wheels.* Always use 150-grit-size diamond-impregnated nibs for truing jig-grinding wheels. *Never use a single-point diamond.*

> Mount a diamond-impregnated truing nib in a sturdy holder.
> Position the diamond nib close to the CBN wheel (Fig. 74-17).

Figure 74-17 When truing a CBN wheel, the wheel should be fed at a steady rate, past a rigidly supported impregnated diamond. *(Diamond Innovations, at www.abrasivesnet.com)*

> Take light infeed truing increments of .0005 in. (0.01 mm) or less.

> Vertically (up and down) feed the full wheel length past the diamond truing nib at a feed rate of 30 to 40 in./min (0.75 to 1 m/min).

> Continue truing until the wheel is true.

Dressing

The truing operation glazes CBN wheels and leaves the abrasive crystals and the bond on the same plane. The wheel cannot grind in this condition because there is no chip space (gullet) beside the abrasive grain for the chip to enter. The dressing operation removes some of the wheel bond and exposes the sharp edges of the CBN crystal.

1. Use a 220-grit, medium hardness aluminum oxide dressing stick.

2. Soak the dressing stick in water-soluble oil so that a slurry is created while dressing.

3. Start the wheel and force the dressing stick in horizontally for the depth of the CBN abrasive section (Fig. 74-18). *NEVER use an up-and-down motion for dressing.*

4. Withdraw the dressing stick and move it up a distance equal to the wheel length.

5. Again force the dressing stick in to the depth of the wheel's abrasive section.

6. Continue this procedure until the wheel seems to draw the dressing stick in effortlessly.

■ **Figure 74-18** Resin-, vitrified-, and metal-bond CBN wheels should be dressed after truing with an aluminum oxide dressing stick. *(Diamond Innovations, at www.abrasivesnet.com)*

As the bond material is exposed, the sharp crystals will consume the dressing stick rapidly. Do not overdress because this may remove too much bond material and result in the loss of valuable abrasive crystals.

JIG-GRINDING GUIDELINES

The accuracy and efficiency of any jig-grinding operation and the quality of the surface finish depends on variables such as wheel speed, reciprocal and planetary speed, grinding method, and grinding mode.

1. *For best results, grind wet.*

2. *Use the proper wheel speeds, outfeeds, and reciprocal and planetary speeds.*

 STOP | ***NEVER* use a wheel at higher speeds than those recommended by the manufacturer.**

> *Outfeeds* depend on the spindle r/min, wheel diameter, wheel bond, and workpiece material.
> *Reciprocal and planetary speeds* should be fairly fast at a continuous light outfeed rate.

3. *Use the proper grinding mode to suit the operation required.*

> *Hole,* or *outfeed, grinding* (Fig. 74-19a), the most common method of grinding holes, removes stock by a continuous outfeed of the wheel while grinding.
> *Wipe grinding* (Fig. 74-19b) removes stock by moving the workpiece past a revolving wheel that is in a stationary position.
> *Chop grinding* (Fig. 74-19c) removes stock by reciprocating the revolving wheel up and down as the work is fed past it in controlled or preset increments.
> *Shoulder,* or *bottom, grinding* (Fig. 74-19d) is done with a wheel whose bottom surface has been relieved so that it can grind a sharp corner.

SUMMARY

CBN jig-grinding wheels and pins are twice as hard and more abrasion-resistant than aluminum oxide wheels. CBN wheels cost more; however, they are cost-effective due to their longer wheel life, shorter grinding time, less wheel maintenance, cool cutting action, and improved part quality. They remove material 30% to

50% faster than aluminum oxide wheels, resulting in higher productivity.

Jig-Grinding Hints

1. Calculate all coordinate hole locations first.

2. Clamp work just enough to hold it in place. Clamping too tightly may cause distortion.

3. Select a grinding wheel that is three-quarters the diameter of the hole to be ground.

4. A wheel with widely spaced grains should be selected for rough grinding.

5. Relieve the wheel diameters so that only .250 in. (6 mm) of the cutting face remains.

6. Never use a glazed wheel for grinding.

7. Rough-grind all holes by plunge grinding.

8. Allow the work to cool before finish grinding.

9. Finish-grind holes with a freshly dressed wheel by outfeed grinding.

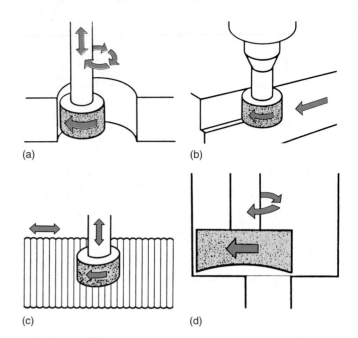

(a) (b) (c) (d)

■ **Figure 74-19** (a) Hole, or outfeed, grinding is the most common method of grinding holes; (b) the wipe grinding mode is generally used for form grinding; the workpiece is fed past the revolving wheel, which is in a stationary position; (c) in chop grinding, the grinding reciprocates up and down and the work is indexed slightly at the end of each cycle; (d) the bottom of the wheel must be relieved when bottom, or shoulder, grinding.

1. Why was the jig grinder developed?
2. State the advantages of jig grinding.

Jig-Grinding Parts

3. How is the jig grinder spindle constructed to allow for the grinding of taper holes?
4. Explain how the grinding wheel may be positioned to the hole diameter.
5. What three methods may be used for controlling and measuring the depth of a hole?
6. Name two methods of dressing a wheel on a jig grinder.

Grinding Methods

7. Compare outfeed and plunge grinding.
8. How may large holes be ground on a jig grinder?
9. State the advantages of diamond-charged mandrels over grinding wheels for small hole grinding.

Grinding Wheels

10. List four general principles that should be followed in selecting a grinding wheel.
11. At what speed should the following be operated:
    a. Grinding wheels?
    b. Diamond-charged mandrels?
12. What undesirable conditions will an improperly dressed wheel cause?
13. Explain the procedure for dressing a grinding wheel.

Grinding Allowances

14. Name the factors that determine the amount of material that should be left in a hole for jig grinding.
15. State the grinding allowance that would apply in most cases for holes:
    a. Under .500 in. (13 mm)
    b. Over .500 in. (13 mm)

Setting Up Work

16. Explain how bolts and clamps should be placed when work is being set up.
17. Why is it important that bolts not be tightened too tightly?
18. Name three methods used to set up work parallel to the table travel.

Grinding Sequence

19. List the sequence suggested when grinding a variety of holes.
20. Explain how the grinding head may be set to an accurate angle.
21. Calculate the X dimension for a .188-in. diameter hole with a .200-in. straight section. The included angle of the tapered hole is $2°$ (see Fig. 74-14).
22. Explain the procedure for grinding shouldered holes.

(Mazak Corp.)

COMPUTER-AGE MACHINING

No single invention since the industrial revolution has made such an impact on society as the computer. Today computers can guide and direct spaceships to the moon and outer space and bring them back safely to earth. They route long-distance telephone calls, schedule and control train and plane operations, predict the weather, and produce an instantaneous report of your bank balance. In chain stores, the cash register (connected with a central computer) totals bills, posts sales, and updates the inventory with every entry. These are but a few of the applications of the computer in our society.

During the past three to four decades, basic computers have been applied to machine tools to program and control the machine operations. These devices have been steadily improved until they are now highly sophisticated units capable of totally controlling the programming, maintenance, troubleshooting, and operation of a single machine, a group of machines, or soon even a complete manufacturing plant.

UNIT 75

The Computer

OBJECTIVES

After completing this unit, you will be able to:

1 Describe generally the development of computers over the ages

2 Explain briefly the effect of computers on everyday life

Ever since primitive people became aware of the concept of quantity, people have used some device to count and perform calculations. Primitive people used their fingers, toes, and stones to count (Fig. 75-1). In about 4000 B.C., the abacus, really the first computer, was developed in the Orient. It uses the principle of moving beads on several wires to make calculations (Fig. 75-2). The abacus is very accurate when properly used. It may still be found in some of the older and smaller Asian businesses today.

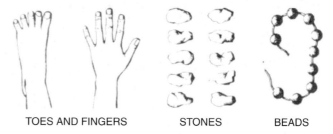

TOES AND FINGERS   STONES   BEADS

■ **Figure 75-1** Primitive means of counting.

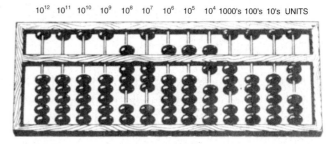

10^{12} 10^{11} 10^{10} 10^{9} 10^{8} 10^{7} 10^{6} 10^{5} 10^{4} 1000's 100's 10's UNITS

■ **Figure 75-2** The abacus was the first real computer.

▶▶ History of the Computer

In 1642, the first mechanical calculator was constructed by a Frenchman named Blaise Pascal. It consisted of eight wheels, or dials, each with the numbers 0 to 9, and each wheel representing units, tens, hundreds, thousands, etc. It could, however, only add or subtract. Multiplication or division was done by repeated additions or subtractions.

In 1671, a German mathematician added the capability of multiplication and division. However, this advanced machine could only do arithmetical problems.

Charles Babbage, a 19th-century English mathematician, produced a machine called the difference engine that could rapidly and accurately calculate long lists of various functions, including logarithms.

In 1804, a French mechanician, J. M. Jacquard, introduced a punch card system to direct the operations of a weaving loom. In the United States, Herman Hollerith introduced the use of punched cards to record personal information, such as age, sex, race, and marital status, for the 1890 U.S. census (Fig. 75-3). The information was encoded on cards and read and tabulated by electric sensors.

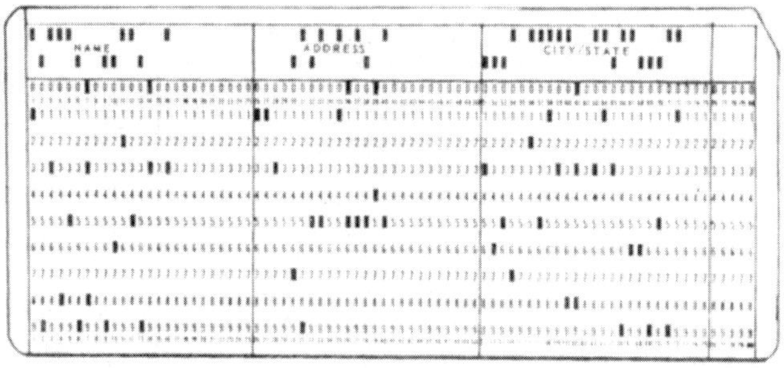

Figure 75-3 Punched cards were our first method of data processing.

This use of punched cards led to the development of the early office machines for the tabulation of data.

In the 1930s, a German named Konrad Zuse built a simple computer that, among other things, was used to calculate wing designs for the German aircraft industry.

A mathematician named George Stibitz produced a similar device in 1939 for the Bell Telephone Laboratories in the United States. This machine was capable of doing calculations over telephone wires; thus was born the first remote data processing machine.

During World War II, the British built a computer called the Colossus I, which helped break the German military codes.

The earliest digital computers used electromechanical on-off switches or relays. The first large computer, the Mark I, assembled at Harvard University by IBM, could multiply two 23-digit numbers in about 5 seconds—a very slow feat compared to today's machines.

In 1946, the world's first electronic digital computer, the *ENIAC* (*e*lectronic *n*umerical *i*ntegrating *a*utomatic *c*omputer) was produced. It contained more than 19,000 vacuum tubes, weighed almost 30 tons, and occupied more than 15,000 ft^2 of floor space. It was a much faster computer—able to add two numbers in $\frac{1}{5000}$ of a second. A machine of this size had many operational problems, particularly with tubes burning out and with circuit wiring.

In 1947, the first transistor was produced by the Bell Laboratories. These were used as switches to control the flow of electrons. They were much smaller than vacuum tubes, had fewer failures, gave off less heat, and were much cheaper to make. Computers were then assembled using transistors, but still the problem of extensive hand wiring existed. This problem led to the development of the printed circuit.

In the late 1950s, Kilby of Texas Instruments and Noyce of Fairchild discovered that any number of transistors, along with the connections between them, could be etched on a small piece of silicon (about .250 × .250 × .030 in. thick). These chips, called *integrated circuits* (*ICs*), contained entire sections of the computer, such as a logic circuit or a memory register. These chips have been

Figure 75-4 Thousands of bits of information can be stored on a tiny silicon chip. *(Rockwell International)*

further improved, and today thousands of transistors and circuits are crammed into this tiny silicon chip (Fig. 75-4). The only problem with this advanced chip was that the circuits were rigidly fixed and the chips could do only the duties for which they were designed.

In 1971, the Intel Corporation produced the microprocessor—a chip that contained the entire *central processing unit (CPU)* for a single computer. This single chip could be programmed to do any number of tasks, from steering a spacecraft to operating a watch or controlling the new personal computers.

▶▶ The Role of the Computer

Although today's computers amaze us (the older generation, particularly), they have become part of everyday life. They will become an even greater influence in the years to come.

We are amazed that some computers today can perform 1 million calculations per second because of the thousands of transistors and circuits jammed into the tiny chips (ICs). Computer scientists can foresee the day when 1 billion transistors or electronic switches (with the necessary connections) will be crowded into a single chip. A single chip will have a memory large enough to store the text of 200 long novels. Advances of this type will decrease the size of computers considerably.

The prototype of a thinking computer incorporating *Artificial Intelligence (AI)* was introduced in the 1990s. The commercial product followed about five years later. This machine is able to recognize natural speech and written language and to translate and type documents automatically. Once a verbal command is given, the computer acts, unless it does not understand the command. At this point the machine asks questions until it is able to form its own judgments and act. It also learns by recalling and studying its errors.

Computers today are used in larger medical centers to catalog all known diseases with their symptoms and known cures. This knowledge is more than any doctor could remember. Doctors now are able to patch their own computer into the central computer and get an immediate and accurate diagnosis of the patient's problem, thus saving many hours and even days of awaiting the results of routine tests.

Because of the computer, children will learn more at a younger age and, as may be expected, future generations will have a much broader and deeper knowledge than those of past generations. It was said that in the past we doubled our knowledge every 25 years. Now with the computer, the amount of knowledge is said to double every 3 years. With this greater knowledge, the human race will explore and develop new sciences and areas that are unknown to us today (much the same as the computer has affected the older people of this era).

In other areas, the computer has been and will continue to be used to predict weather; guide and direct planes, spaceships, missiles, and military artillery; and monitor industrial environments.

In everyday life, everyone is and will continue to be affected by the computer. Department store computers list and total your purchases, at the same time keeping the inventory up to date and advising the company of people's buying habits. Thus, the computer permits the company to buy more wisely. Credit bureau computers know how much every adult owes, to whom, and how the debt is being repaid. School computers record students' courses, grades, and other information. Hospital and medical records are kept on anyone who has been admitted to a hospital.

Police agencies have access to a national computer that can produce the police records of any known offender. The census bureau and tax department of any country have information on all its citizens on computer.

On the office front, computers have relieved the accountants of the drudgery of repetitive jobs such as payroll processing. Many office workers work at home on a company computer. This eliminates the necessity of traveling

■ Figure 75-5 CAD systems are invaluable to engineers who research and design products.

long distances to work and the need for baby-sitting services required by so many young working families today.

In air defense control systems, the position and course of all planes from the network of radar stations are fed into the computer along with the speed and direction of each. The information is stored and the future positions of the planes are calculated.

In the manufacturing industry, the computer has contributed to the efficient manufacture of all goods. It appears that the impact of the computer will be even greater in the years to come. Computers will continue to improve productivity through *computer-aided design (CAD),* by which the design of a product can be researched, fully developed, and tested before production begins (Fig. 75-5). *Computer-assisted manufacturing (CAM)* results in less scrap and more reliability through the computer control of the machining sequence and the cutting speeds and feeds.

Robots, which are computer-controlled, are used by industry to an increasing extent. Robots can be programmed to paint cars, weld, feed forges, load and unload machinery, assemble electric motors, and perform dangerous and boring tasks formerly done by humans.

unit 75 review questions

1. Name three methods of counting used by primitive people.

2. What was the first computer ever developed?

3. For what purpose were the first punched cards used in the United States?

4. How many transistors and circuits can be found on a silicon chip?

5. How are computers used in medical centers?

6. How are computers affecting the manufacturing industry?

UNIT 76

Computer Numerical Control

OBJECTIVES

After completing this unit, you should be able to:

1 Identify the types of systems and controls used in computer numerical control

2 List the steps required to prduce a part by computer numerical control

3 Discuss the advantages and disadvantages of computer numerical control

Numerical control (NC) may be defined as a method of accurately controlling the operation of a machine tool by a series of coded instructions, consisting of numbers, letters of the alphabet, and symbols that the *machine control unit (MCU)* can understand. These instructions are converted into electrical pulses of current, which the machine motors and controls follow to carry out machining operations on a workpiece. The numbers, letters, and symbols are coded instructions that refer to specific distances, positions, functions, or motions that the machine tool can understand as it machines the workpiece. The measuring and recording devices incorporated into computer numerical control machine tools ensure that the part being manufactured will be accurate. *Computer numerical control (CNC)* machines minimize human error.

▶▶ Theory of CNC

Computer numerical control (CNC) and the computer have brought tremendous changes to the metalworking industry. New machine tools, in combination with CNC, enable industry to consistently produce parts to accuracies undreamed of only a few years ago. The same part can be reproduced to the same degree of accuracy any number of times if the CNC program has been properly prepared, the computer properly programmed, and the machine properly set up. The operating commands that control the machine tool are executed automatically with amazing speed, accuracy, efficiency, and repeatability.

▶▶ Role of a Computer in CNC

The computer has also found many uses in the overall manufacturing process. It is used for part design using computer-aided design (CAD), testing, inspection, quality control, planning, inventory control, gathering of data, work scheduling, warehousing, and many other functions in manufacturing. The computer is having profound effects on manufacturing techniques and will continue to have in the future. Computers fill three major roles in CNC:

1. Almost all machine control units (MCUs) include or incorporate a computer in their operation. These

units are generally called computer numerical controls (CNC).

2. Most of the part programming for CNC machine tools is done with off-line computer assistance.

3. An increasing number of machine tools are controlled or supervised by computers that may be in a separate control room or even in another plant. This is more commonly known as *direct numerical control (DNC)*.

▶▶ Types of Computers

Most computers fall into two basic types, either analog or digital. The *analog computer* is used primarily in scientific research and problem solving. Analog computers have been replaced in most cases by the digital computer. Most computers used in industry, business, and at home are of the electronic digital type. The *digital computer* accepts an input of digital information in numerical form, processes that information according to prestored or new instructions, and develops output data (Fig. 76-1).

There are generally three categories of computers and computer systems: the mainframe, the minicomputer, and the microcomputer.

> The *mainframe computer* (Fig. 76-2), which can be used to do more than one job at the same time, is large and has a huge capacity. It is generally a company's main computer, the one that performs general-purpose data processing such as CNC part programming, payroll, cost accounting, inventory, and many other applications.

> The *minicomputer* (Fig. 76-3) is generally smaller in size and capacity than the mainframe computer. This type of computer is generally a "dedicated" type, which means that it performs specific tasks.

> The *microcomputer* (Fig. 76-4) generally has one chip (a microprocessor) that contains at least the arithmetic-logic and control-logic functions of the central processing unit (CPU). The microprocessor is usually designed for simple applications and must be accompanied by other electronic devices, on a printed circuit board, for more complex applications.

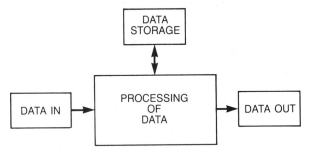

■ Figure 76-1 The main function of the computer is to accept, process, and output data.

■ Figure 76-2 The mainframe computer, which can do many jobs simultaneously, is larger and has more capacity than other computers.   *(Hewlett-Packard Company)*

■ Figure 76-3 The minicomputer is generally a dedicated type and performs specific tasks.   *(Hewlett-Packard Company)*

■ Figure 76-4 The microcomputer generally contains only one chip and is designed for simple applications. *(Hewlett-Packard Company)*

▸▸ Computer Functions

The function of a computer is to receive coded instructions (input data) in numerical form, process this information, and produce output data that causes a machine tool to function. Many methods are used to put information on a computer, such as magnetic tape, floppy disks, CD/DVD, and specially designed sensors (Fig. 76-5). The most commonly used method to input data is directly through the computer.

▸▸ CNC Performance

CNC has made great advances since NC was first introduced in the mid-1950s as a means of guiding machine tool motions automatically, without human assistance. The early machines were capable only of point-to-point positioning (straight-line motions), were very costly, and required highly skilled technicians and mathematicians to produce the tape programs. Not only have the machine

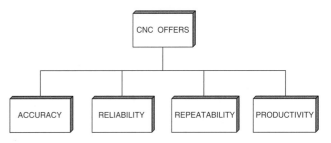

■ Figure 76-6 CNC offers industry many advantages. *(Kelmar Associates)*

tools and controls been dramatically improved, but the cost has been continually lowered. CNC machines are now within the financial reach of small manufacturing shops and educational institutions. Their worldwide acceptance has been a result of their accuracy, reliability, repeatability, and productivity (Fig. 76-6).

ACCURACY

CNC machine tools would not have been accepted by industry if they were not capable of machining to very close tolerances. At the time CNC was being developed, industry was looking for a way to improve production rates and achieve greater accuracy on its products. A skilled machinist is capable of working to close tolerances, such as ±.001 in. (0.025 mm), or even less, on most machine tools. It has taken the machinist many years of experience to develop this skill, but this person may not be capable of working to this accuracy every time. Some human error means that some of the product may have to be scrapped.

Modern CNC machine tools are capable of consistently producing workpieces that are accurate to within a tolerance of .0001 to .0002 in. (0.0025 to 0.005 mm). The machine tools have been built better, and the electronic control systems ensure that parts within the tolerance allowed by the engineering drawing will be produced.

RELIABILITY

The performance of CNC machine tools and their control systems had to be at least as reliable as the machinists, toolmakers, and diemakers for industry to accept this machining concept. Since consumers throughout the world were demanding better and more reliable products, there was a great need for equipment that could machine to closer tolerances and be counted on to repeat this time and time again. Improvements in machine slides, bearings, ball screws, and machine tables all helped to make machines sturdier and more accurate. New cutting tools and toolholders were developed that matched the accuracy of the machine tool and made it possible to consistently produce accurate parts.

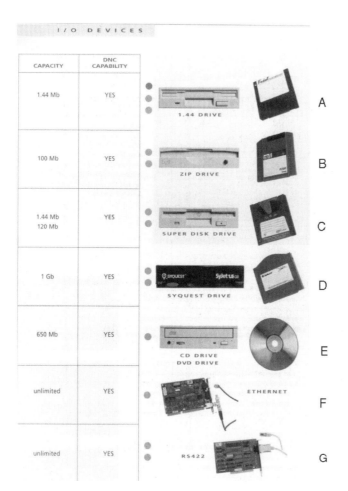

■ Figure 76-5 Common methods used to provide input data into and out of the computer. *(Modern Machine Shop)*

REPEATABILITY

Repeatability and reliability are very difficult to separate because many of the same variables affect each. Repeatability of a machine tool involves comparing each part produced on that machine to see how it compares to other parts for size and accuracy. The repeatability of a CNC machine should be at least one-half the smallest tolerance on the part. Machine tools capable of greater accuracy and repeatability naturally cost more because of the accuracy built into the machine tool and/or the control system.

PRODUCTIVITY

The goal of industry has been to produce better products at competitive or lower prices to gain a bigger share of the market. To meet foreign competition, manufacturers must produce higher quality products, while also improving return on capital invested and lowering manufacturing and labor costs. These factors alone are justification for using CNC and automating factories. They provide the opportunity to produce goods of better quality faster and at a lower cost.

The modern CNC machine control unit has several features that were not found on the pre-1970 hard-wired control units (Fig. 76-7).

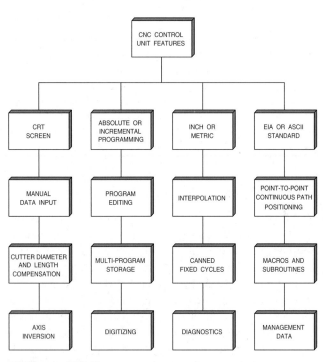

■ **Figure 76-7** CNC machine control units contain many features not found on earlier controls. *(Kelmar Associates)*

▶▶ Advantages of CNC

CNC has grown at an ever-increasing rate, and its use will continue to grow because of the many advantages that it has to offer industry. Some of the most important advantages of CNC are illustrated in Fig. 76-8.

1. *Greater operator safety.* CNC systems are generally operated from a console away from the machining area, which is enclosed on most machines. Therefore, the operator is exposed

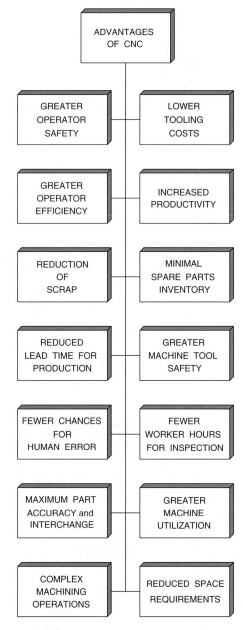

■ **Figure 76-8** CNC offers industry many advantages that increase productivity and produce quality products.

2. *Greater operator efficiency.* A CNC machine does not require as much attention as a conventional machine, allowing the operator to perform other jobs while it is running.

3. *Reduction of scrap.* Because of the high degree of accuracy of CNC systems, scrap has been drastically reduced.

4. *Reduced lead time for production.* The program preparation and setup time for computer numerically controlled machines is usually short. Many jigs and fixtures formerly required are not necessary.

5. *Fewer chances for human error.* The CNC program reduces or eliminates the need for an operator to take trial cuts, make trial measurements, make positioning movements, or change tools.

6. *Maximum part accuracy and interchange.* CNC ensures that all parts produced will be accurate and of uniform quality.

7. *Complex machining operations.* Complex operations can be performed quickly and accurately with CNC and electronic measuring equipment.

8. *Lower tooling costs.* CNC machines generally use simple holding fixtures, which reduce the cost of tooling by as much as 70%. Standard turning and milling tools eliminate the need for special form tools.

9. *Increased productivity.* Because the CNC system controls all the machine functions, parts are produced faster and with less setup and lead time.

10. *Minimal spare parts inventory.* A large inventory of spare parts is no longer necessary, since additional parts can be made to the same accuracy when the same program is used again.

11. *Greater machine tool safety.* The damage to machine tools as a result of operator error is virtually eliminated, since there is less operator intervention.

12. *Fewer worker hours for inspection.* Because CNC machines produce parts of uniform quality, less inspection time is required.

13. *Greater machine utilization.* Production rates could increase as much as 80% because less time is required for setup and operator adjustments.

14. *Reduced space requirements.* A CNC system requires fewer jigs and fixtures and therefore less storage space.

▶▶ Cartesian Coordinates

Almost everything that can be produced on a conventional machine tool can be produced on a numerical control machine tool, with its many advantages. The machine tool movements used in producing a product are of two basic types: *point-to-point* (straight-line movements) and *continuous-path* (contouring movements).

The Cartesian, or rectangular, coordinate system allows any specific point on a job to be described in mathematical terms in relation to any other point along three perpendicular axes. This fits machine tools perfectly, since their construction is generally based on three axes of motion (X, Y, Z) plus an axis of rotation. On a vertical milling machine, the X axis is the horizontal movement (right or left) of the table, the Y axis is the table cross movement (toward or away from the column), and the Z axis is the vertical movement of the knee or the spindle. CNC systems rely on the use of rectangular coordinates because the programmer can precisely locate every point on a job.

When points are located on a workpiece, two straight intersecting lines, one vertical and one horizontal, are used. These lines must be at right angles to each other, and the point where they cross is called the *origin,* or *zero point* (Fig. 76-9).

The three-dimensional coordinate planes are shown in Fig. 76-10.

> The X and Y planes (axes) are horizontal and represent horizontal machine table motions.

> The Z plane, or axis, represents the vertical tool motion.

> The plus ($+$) and minus ($-$) signs indicate the direction of movement from the zero point (origin) along the axis.

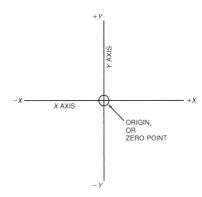

■ **Figure 76-9** The zero point is established where intersecting lines form right angles. *(Allen Bradley)*

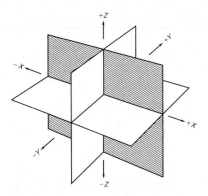

Figure 76-10 The three-dimensional coordinate planes (axes) used in CNC. *(Allen Bradley)*

> The four quadrants formed when the *X*-*Y* axes cross are numbered in a counterclockwise direction (Fig. 76-11).

1. All positions located in quadrant 1 are positive *X* (+*X*) and positive *Y* (+*Y*).
2. In the second quadrant, all positions are negative *X* (−*X*) and positive *Y* (+*Y*).
3. In the third quadrant, all locations are negative *X* (−*X*) and negative *Y* (−*Y*).
4. In the fourth quadrant, all locations are positive *X* (+*X*) and negative *Y* (−*Y*).

In Fig. 76-11, point *A* is 2 units to the right of the *Y* axis and 2 units above the *X* axis. Assume that each unit equals 1 in. The location of point *A* is *X* + 2.000 and *Y* + 2.000. For point *B*, the location is *X* + 1.000 and *Y* − 2.000. In CNC programming, it is not necessary to indicate plus (+) values, since these are assumed. However, the minus (−) values must be indicated. For example, the locations of both *A* and *B* are indicated as follows:

A	*X*2.000	*Y*2.000
B	*X*1.000	*Y*−2.000

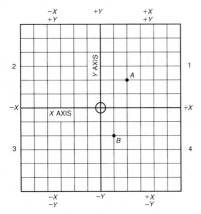

Figure 76-11 The quadrants formed when the *X* and *Y* axes cross allow any point to be accurately located from the *X*-*Y* zero point, or origin.

GUIDELINES

Since CNC is so dependent on the system of rectangular coordinates, it is important to follow some guidelines. In this way, everyone involved in the manufacture of a part—the engineer, draftsperson, programmer, and machine operator—will understand exactly what is required.

1. Use reference points on the part itself, if possible. This makes it easier for quality control personnel to check the accuracy of the part later.
2. Use Cartesian coordinates—specifying *X*, *Y*, and *Z* planes—to define all part surfaces.
3. Establish reference planes along part surfaces that are parallel to the machine axes.
4. Establish the allowable tolerances at the design stage.
5. Describe the part so that the cutter path may be easily determined and programmed.
6. Dimension the part so that it is easy to determine its shape without calculations or guessing.

MACHINE AXES

Every CNC machine tool has sliding and rotary controllable axes. In order to control these axes, letters (called *addresses*) are used to identify each direction of table or spindle movement. Combined with a number to form a word, it establishes the distance the axis moves. These words are necessary for the programmer to pass along information about the job to persons responsible for the setup and operation of the CNC machine. Machine tool builders usually follow standards set by the Electronics Industries Association (EIA), which assigns the coding system for CNC machine axes. The primary axes are *X*, *Y*, and *Z*, and these apply to most machine tools, with some exceptions. The EIA standard states that the longest horizontal axis movement, which is parallel to the work table, is the *X* axis. Movement along the machine spindle is the *Z* axis, and the *Y* axis is assigned to the motion perpendicular (at right angles) to both the *X* and *Z* axes.

In addition to the primary axes, there are secondary axes that are parallel to the *X*, *Y*, and *Z* axes. The addresses (letters) *A*, *B*, and *C* refer to rotary motion axes around the primary axes. *I*, *J*, and *K* words are also used for rotary axes on some machines when circular interpolation is used for programming circles or partial arcs, while on other machines, an *R* word represents the radius of a circle. Some chucking and turning centers also use *U* and *W* words for incremental movement parallel to the *X* and *Z* primary axes.

MACHINES USING CNC

CNC is used on all types of machine tools, from the simplest to the most complex. The most common machine tools, the chucking center (lathe) and the machining center (milling machine), are covered in this book.

1. *Chucking centers* (Fig. 76-12) were developed in the mid-1960s after studies showed that about 40% of metal-cutting operations were performed on lathes. These computer numerically controlled machines are capable of greater accuracy and higher production rates than were possible on the engine lathe. The basic chucking center operates on two axes:

   > The X axis controls the cross motion of the turret (tool) head.
   > The Z axis controls the lengthwise travel (toward or away from the headstock) of the turret head.

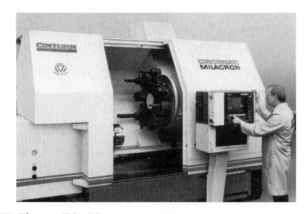

■ **Figure 76-12** The chucking center can produce round parts quickly and accurately. *(Cincinnati Milacron)*

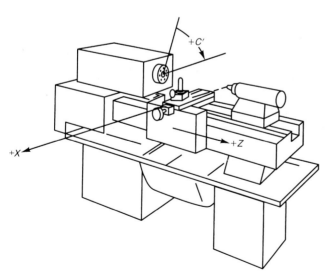

■ **Figure 76-13** The lathe cutting tool moves only along the X and Z axes.

■ **Figure 76-14** Machining centers can perform a wide variety of machining operations. *(Cincinnati Milacron)*

2. The *engine lathe* (Fig. 76-13) has always been a very efficient way of producing round parts. Most lathes operate on two axes:

   > The X axis controls the cross motion (in or out) of the cutting tool.
   > The Z axis controls the carriage travel toward or away from the headstock.

3. *Machining centers* (Fig. 76-14), developed in the 1960s, allow more operations to be done on a part in one setup instead of moving from machine to machine for various operations. These machines greatly increase productivity because the time formerly used to move a part from machine to machine is eliminated. Two main types of machining centers are the *horizontal* and the *vertical spindle* models. The vertical machining center operates on three axes:

   > The X axis controls the table movement left or right.
   > The Y axis controls the table movement toward or away from the column.
   > The Z axis controls the vertical movement (up or down) of the spindle or knee.

4. The *milling machine* (Fig. 76-15) can perform operations such as milling, contouring, gear cutting, drilling, boring, and reaming. The milling machine operates on three axes:

   > The X axis controls the table movement left or right.
   > The Y axis controls the table movement toward or away from the column.
   > The Z axis controls the vertical (up or down) movement of the knee or spindle.

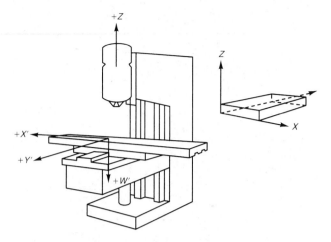

■ Figure 76-15 The main movements on a milling machine are along the *X, Y,* and *Z* axes.

▸▸ Programming Systems

Two types of programming modes, the incremental system and the absolute system, are used for CNC (Fig. 76-16). Both systems have applications in CNC programming, and no system is right or wrong all the time. Most controls on machine tools are capable of handling both incremental and absolute programming by altering the code between the G90 (absolute) and G91 (incremental) commands.

INCREMENTAL SYSTEM

In the incremental system, program dimensions or positions are given from the current point. Incremental dimensioning on a job print is shown in Fig. 76-17. As will be noted, the dimensions for each hole are given from the previous hole. A disadvantage of incremental positioning or programming is that, if an error has been made in any location, this error is automatically carried over to all following locations. The G91 command tells the computer and the MCU that the program is to be in the incremental mode.

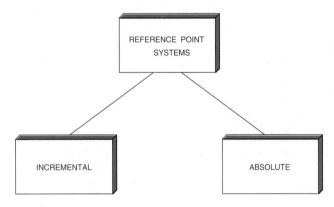

■ Figure 76-16 CNC programming uses two systems, the incremental and the absolute.

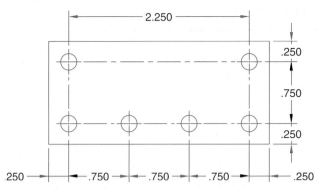

■ Figure 76-17 Incremental system mode showing workpiece dimensions. *(Kelmar Associates)*

Command codes that tell the machine to move the table, spindle, and knee are explained here using a vertical milling machine as an example:

> A "plus *X*" (+*X*) command causes the cutting tool to be located to the right of the last point.

> A "minus *X*" (−*X*) command causes the cutting tool to be located to the left of the last point.

> A "plus *Y*" (+*Y*) command causes the cutting tool to be located toward the column.

> A "minus *Y*" (−*Y*) command causes the cutting tool to be located away from the column.

> A "plus *Z*" (+*Z*) command causes the cutting tool or spindle to move up or away from the workpiece.

> A "minus *Z*" (−*Z*) command moves the cutting tool down or into the workpiece.

ABSOLUTE SYSTEM

In the absolute system, all dimensions or positions are given from one reference point on the job or machine. In Fig. 76-18, the same workpiece is used as in Fig. 76-17, but all dimensions are given from the zero or reference point, which in this case is the upper left-hand corner on the workpiece. Therefore, in the absolute system of dimensioning or programming, an error in any dimension is still an error, but the error is not carried on to any other location.

In absolute programming, the G90 command indicates to the computer and MCU that the program is to be in the absolute mode.

> A "plus *X*" (+*X*) command causes the cutting tool to be located to the right of the zero point, or origin.

> A "minus *X*" (−*X*) command causes the cutting tool to be located to the left of the zero point, or origin.

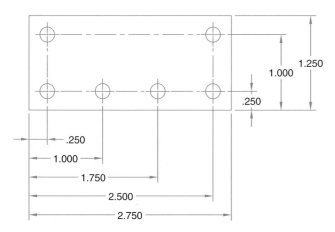

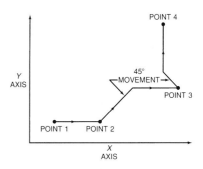

Figure 76-20 The path followed by point-to-point positioning between various programmed points.

Figure 76-18 In the absolute system mode, all dimensions are given from the same point of reference. *(Kelmar Associates)*

> A "plus Y" (+Y) command causes the cutting tool to be located toward the column (above the zero point, or origin).

> A "minus Y" (−Y) command causes the cutting tool to be located away from the column (below the zero point, or origin).

> A "plus Z" (+Z) command causes the cutting tool to move above the program Z0 (usually the top surface of the part).

> A "minus Z" (−Z) command causes the cutting tool to move below the program Z0.

▶▶ CNC Positioning Systems

CNC programming falls into two distinct categories, point-to-point and continuous-path (Fig. 76-19), which can be handled by most control units. A knowledge of both programming methods is necessary to understand what application each has in CNC.

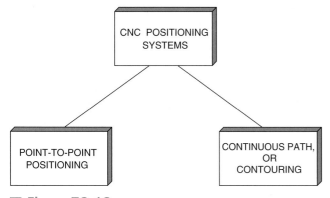

Figure 76-19 Types of CNC positioning systems.

POINT-TO-POINT POSITIONING

Point-to-point positioning consists of any number of programmed points joined together by straight lines. This method is used to accurately locate the spindle, or the workpiece mounted on the machine table, at one or more specific locations to perform operations such as drilling, reaming, boring, tapping, and punching (Fig. 76-20). Point-to-point positioning (G00, rapid positioning) is the process of positioning from one coordinate (X-Y) position or location to another, performing the machining operation, clearing the tool from the work, and moving to the next location until all the operations have been completed at all programmed locations.

Drilling machines, or point-to-point machines, are ideally suited for positioning the machine tool to an exact location or point, performing the machining operation (drilling a hole), and then moving to the next location (where another hole could be drilled). As long as each point or hole location in the program is identified, this operation can be repeated as many times as required.

Point-to-point machining moves from one point to another as fast as possible (*rapids*) while the *cutting tool is above the work surface. Rapid travel* is used to quickly position the cutting tool or workpiece between location points before a cutting action is started. The rate of rapid travel is usually between 200 and 800 in./min (5 and 20 m/min). Both axes (X and Y) move simultaneously and at the same rate during rapid traverse. This results in a movement along a 45° angle line until one axis is reached, and then there is a straight line movement to the other axis.

CONTINUOUS-PATH (CONTOURING)

Continuous-path, or *contouring machining,* involves work produced on a lathe or milling machine where the cutting tool is usually in contact with the workpiece as it travels from one programmed point to the next. Continuous-path positioning is the ability to control motions on two or more machine axes simultaneously, to keep a constant cutter-workpiece relationship. The information in the CNC program must accurately position the cutting

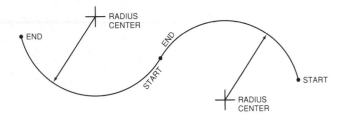

Figure 76-21 Complex forms on a number of axes can be generated by circular interpolation. *(Kelmar Associates)*

tool from one point to the next and follow a predefined accurate path at a programmed feed rate in order to produce the form or contour required (Fig. 76-21).

▸▸ Control Systems

Open and closed loop are the two main types of control systems used for computer numerical control machines. Most machine tools manufactured contain a closed loop system because it is very accurate and as a result better quality work can be produced on the machine. However, open loop systems can still be found on some older NC machines; therefore, an explanation of this system is necessary.

OPEN LOOP SYSTEM

In the open loop system (Fig. 76-22), input data is fed into the machine control unit (MCU). This decoded information, in the form of a program, is sorted until the CNC machining cycle is started by the operator. Program commands are automatically converted into electric pulses, or signals, which are sent to the MCU to energize the servo control units. The servo control units direct the servomotors to perform certain functions according to the information supplied by the program input data. The amount each servomotor will move the lead screw of the machine de-

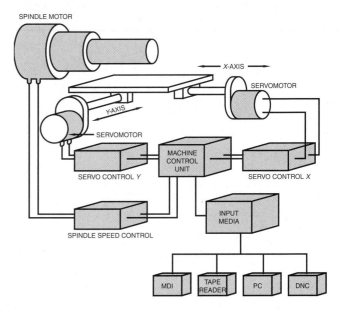

Figure 76-22 An open loop computer numerical control system does not have a means of checking the accuracy of a move. *(Kelmar Associates)*

pends on the number of electric pulses it receives from the servo control unit.

This type of system is fairly simple; however, since there is no means of checking to determine whether the servomotor has performed its function correctly, it is not generally used where an accuracy greater than .001 in. (0.02 mm) is required. The open loop system may be compared with a gun crew that has made all the calculations necessary to hit a distant target but does not have an observer to confirm the accuracy of the shot.

CLOSED LOOP SYSTEM

The closed loop system (Fig. 76-23) can be compared to the same gun crew that now has an observer to confirm the accuracy of the shot. The observer relays the information

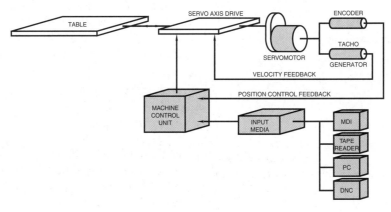

Figure 76-23 A closed loop system verifies the accuracy of every move. *(Kelmar Associates)*

regarding the accuracy of the shot to the gun crew, which then makes the necessary adjustments to hit the target.

The closed loop system is similar to the open loop system, with the exception that a feedback unit is added to the electric circuit (Fig. 76-23). This feedback unit, in the form of a rotary resolver or encoder, can be mounted to the back of the servomotor shaft or to the opposite end of the lead screw. It is used for absolute position control and/or for velocity feedback. Another type of feedback system, a *linear encoder,* consists of a scale mounted to a stationary part of a machine, such as the knee of a milling machine. The linear encoder uses a reader head or slider that is mounted to the moving part of the machine, such as the machine table. Both parts are connected magnetically or optically and operate the same as the rotary resolver. Regardless of the system used, the feedback unit compares the amount the machine table has been moved by the servomotor with the signal sent by the control unit. The control unit instructs the servomotor to make whatever adjustments are necessary until both the signal from the control unit and the signal from the servo unit are equal. In the closed loop system, 10,000 electric pulses are required to move the machine slide 1.000 in. (25.4 mm). Therefore, in this type of system, one pulse will cause .0001 in. (0.002 mm) of movement of the machine slide. Closed loop computer numerical control systems are very accurate because the accuracy of the command signal is recorded and there is an automatic compensation for error.

▶▶ Input Media

As computer numerical control evolved, the input media used to load data into the machine computer also evolved. The main media for many years were the 1-in. wide, 8-track punched tape. Other types of input media, such as magnetic tape, punched cards, magnetic disks, and manual data input (MDI), were also used to a lesser degree. Fig. 76-24 illustrates the different input media. Punched tape use is rapidly being replaced by other methods.

Modern CNC machines use a computer keyboard formatted to the *American Standard Code for Information Interchange* (ASCII) standard to input program information directly into the machine control unit. For proper operation, keyboard use required some type of communications software and compatible connection between the computer keyboard and the machine control unit. Direct numerical control (DNC), which uses a microcomputer along with communications software, is becoming the preferred input method. With DNC, the program data can be sent to the CNC for machining parts. An *alphanumeric* keyboard on the operator's control panel is needed for manual data input. If editing is performed to the program, this new information can also be sent back through the DNC link to be stored for future use.

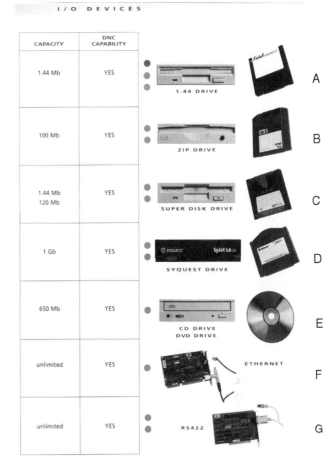

I/O DEVICES

CAPACITY	DNC CAPABILITY	
1.44 Mb	YES	1.44 DRIVE — A
100 Mb	YES	ZIP DRIVE — B
1.44 Mb 120 Mb	YES	SUPER DISK DRIVE — C
1 Gb	YES	SYQUEST DRIVE — D
650 Mb	YES	CD DRIVE DVD DRIVE — E
unlimited	YES	ETHERNET — F
unlimited	YES	RS422 — G

■ **Figure 76-24** Various types of media that can be used to input data into the MCU. *(Modern Machine Shop)*

▶▶ Types of Computer Control

There are two types of control units used in industry for numerical control work. The *CNC control,* which evolved from DNC applications in the early 1970s, is generally used to control individual machines. The *DNC control* is generally used where six or more CNC machines are involved in a complete manufacturing program, such as a flexible manufacturing system.

COMPUTER NUMERICAL CONTROL

There are four main parts, or elements, of a computer numerical control system:

1. A general-purpose computer, which gathers and stores the programmed information

2. A control unit, which communicates and directs the flow of information between the computer and the machine control unit

3. The machine logic, which receives information and passes it on to the machine control unit

4. The machine control unit, which contains the servo units, speed and feed controls, and machine operations such as spindle and table movements and automatic tool changer (ATC)

The CNC system (Fig. 76-25), built around a powerful minicomputer, contains a large memory capacity and has many features to assist in programming. These could include operations such as program editing at the machine, machine setup, operation, and maintenance. Many of these features are sets of machine and control instructions, stored in memory, that can be called into use by the part program or by the machine operator.

Some older CNC systems still use tape readers to read the part program, which is prepared in an office on an off-line unit and delivered to the machine in the form of punched tape. In this system, the tape is read once and the part program is stored in memory for repetitive machining. CNC does not require rereading the tape for each part, as was the case with NC. As CNC machines evolved, minicomputers and then microcomputers were incorporated into the controls. This allows the machine operator to manually input the program required to produce the part at the CNC machine. The program is stored in the computer memory for the production of parts. The main advantage of this system is its ability to operate in a *live,* or *conversational, mode,* with direct communication between the machine and the computer. This feature enables a programmer to make program changes at the machine, or even develop a program at the machine, and the input to the computer is transformed immediately into machine motions. Therefore, the changes to the program can be seen immediately and re-visions made, if necessary. This concept of machine control allows programs to be tried, corrected, and revised quickly.

Advantages of CNC Programming

> More flexible because changes can be made to the program rather than making a new tape, as was required with conventional controls

> Can diagnose programs on a graphic display screen, which shows the machine and control functions before the part is produced; other machines use the *dry run mode,* which usually bypasses the Z axis movement and spindle rotation

> Can be integrated with DNC systems in complex manufacturing systems by using a communications loop

> Increases productivity because of easy programming

> Makes corrections on the first part possible, which cuts costs on the entire batch by using offsets and cutter radius compensation

> Practical, and even profitable, to produce short-run lots

DIRECT NUMERICAL CONTROL

In a DNC system (Fig. 76-26), a number of CNC equipped machines are controlled from a mainframe computer. This can handle the scheduling of work and can download a complete program into the machine's memory when new parts are required. Since most CNC machines are equipped with their own minicomputer or microcomputer, it is possible to operate each machine individually

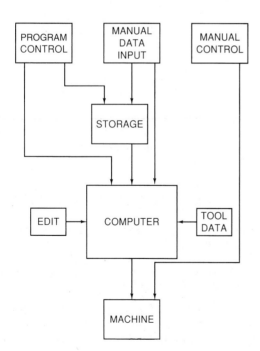

■ **Figure 76-25** The components of a CNC system of machine control.

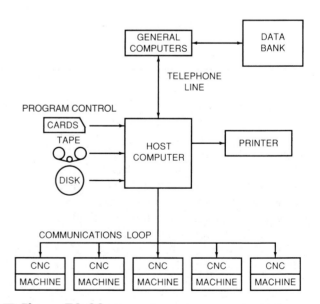

■ **Figure 76-26** The main components of a DNC system.

by CNC in case the mainframe computer should break down. In a smaller manufacturing facility, a microcomputer can be used for DNC purposes.

Advantages of DNC

> A single computer can control many machine tools at the same time.

> Time is saved in eliminating program errors or revising the program. The programmer can make revisions or corrections on a keyboard right at the machine tool.

> Programming is faster, simpler, and more flexible.

> The computer can record any production, machining, or time data required.

> The main control unit can be kept in a clean processing room, away from dirty shop conditions.

> When three or more machines are DNC-controlled, the initial cost is lower than for conventional NC.

> Operating costs are lower than with NC.

▸▸ Programming Format

The most common type of programming format used for CNC programming systems is the *word address format.* This format contains a large number of different *codes* to transfer program information to machine servos, relays, micro-switches, etc. to carry out the machine movements required to manufacture a part. These codes, which conform to established standards, are then put together in a logical sequence called a *block of information.* Each block should contain only enough information to carry out one step of a machining operation.

WORD ADDRESS FORMAT

Part programs must be put in a format that the machine control unit can understand. The format used on a CNC system is determined by the machine tool builder and is based on the control unit of the machine. A variable-block format that uses words (letters) is most commonly used. Each instruction word consists of an address character, such as *S, X, Y, T, F,* or *M.* This alphabetical character is followed by numerical data used to identify a specific function from a word group or to give the distance, feed rate, or speed value.

CODES

The most common codes used for CNC programming are the *G*-codes (preparatory commands) and *M*-codes (miscellaneous functions) (Fig. 76-27). Codes *F, S, D, H, P,*

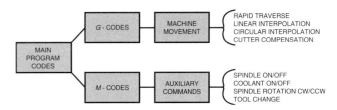

■ Figure 76-27 Types of codes used in CNC programming. *(Kelmar Associates)*

and *T* are used to represent functions such as feed, speed, cutter diameter offset, tool length compensation, subroutine call, tool number, etc. Codes A (angle) and R (radius) are used to locate points on arcs and circles that involve angles and radii.

G-codes, sometimes called *cycle codes,* refer to some action occurring on the *X, Y,* and/or *Z* axis of a machine tool. These codes are grouped into categories, such as group number 01, that contain codes G00, G01, G02, and G03. These codes cause some movement of the machine table or the head.

> A G00 code is used to rapidly position the cutting tool or workpiece from one point on a job to another point. During rapid traverse, either the *X* or *Y* axis can be moved, or both axes can be moved at the same time. The rate of rapid travel may vary from machine to machine and may range between 200 and 800 in./min (5 and 20 m/min).

> The G01, G02, and G03 codes move the axes at a controlled feed rate.

1. G01 is used for linear interpolation (straight-line movement).
2. G02 (clockwise) and G03 (counterclockwise) are used for circular interpolation (arcs and circles).

Some *G*-codes are classified as either modal or nonmodal. *Modal codes* stay in effect in the program until they are changed by another code from the same group. *Nonmodal codes* stay in effect for one operation only and must be programmed again whenever required. In group 01, for example, only one of the four codes in this group can be used at any one time. If a program begins with a G00 and a G01 is entered next, the G00 is cancelled from the program until it is entered again. If a G02 or a G03 code is entered into the program, the G01 will be cancelled, and so on. Fig. 76-28 (on p. 614) shows many of the common *G*-codes that conform to the EIA standards.

M-codes are used to turn either ON or OFF different functions that control certain machine tool operations. *M*-codes are usually not grouped by categories, although several codes may control the same type of operations for components of the machine. For example, three codes, M03, M04, and M05, all control some function of the machine tool spindle:

Group	G-code	Function
01	G00	Rapid positioning
01	G01	Linear interpolation
01	G02	Circular interpolation clockwise (CW)
01	G03	Circular interpolation counterclockwise (CCW)
00	G04	Dwell
00	G10	Offset value setting
02	G17	*XY* plane selection
02	G18	*ZX* plane selection
02	G19	*YZ* plane selection
06	G20	Inch input (in.)
06	G21	Metric input (mm)
00	G27	Reference point return check
00	G28	Return to reference point
00	G29	Return from reference point
07	G40	Cutter compensation cancel
07	G41	Cutter compensation left
07	G42	Cutter compensation right
08	G43	Tool length compensation in positive (+) direction
08	G44	Tool length compensation in minus (−) direction
08	G49	Tool length compensation cancel
09	G80	Canned cycle cancel
09	G81	Drill cycle, spot boring
09	G82	Drilling cycle, counterboring
09	G83	Peck drilling cycle
09	G84	Tapping cycle
09	G85	Boring cycle #1
09	G86	Boring cycle #2
09	G87	Boring cycle #3
09	G88	Boring cycle #4
09	G89	Boring cycle #5
03	G90	Absolute programming
03	G91	Incremental programming
00	G92	Setting of program zero point
05	G94	Feed per minute

■ **Figure 76-28** Commonly used EIA preparatory codes, according to standard EIA–274–D.

> M03 turns on the machine spindle in a clockwise direction.

> M04 turns on the machine spindle in a counterclockwise direction.

> M05 turns off the spindle.

All three of these codes are considered modal because they stay operational until another code is entered in its place. Fig. 76-29 shows some of the most common M-codes used on CNC machine tools.

On a turning center, the function of some G-codes and *M*-codes may differ from the functions of those on a machining center. Other codes not listed for machining centers are used exclusively for turning centers. See Unit 77 for the appropriate turning center *G*-code and *M*-code charts. Several *G*-codes and *M*-codes are unassigned and

M-Code	Function
M00	Program stop
M01	Optional stop
M02	End of program
M03	Spindle start (forward CW)
M04	Spindle start (reverse CCW)
M05	Spindle stop
M06	Tool change
M07	Mist coolant on
M08	Flood coolant on
M09	Coolant off
M19	Spindle orientation
M30	End of program (return to top of memory)
M48	Override cancel release
M49	Override cancel
M98	Transfer to subprogram
M99	Transfer to main program (subprogram end)

■ **Figure 76-29** The most common EIA *M*-codes that control miscellaneous machine functions.

may be used by CNC manufacturers for special functions on their machines.

BLOCK OF INFORMATION

Each block of information should contain only enough information to carry out one step of a machining operation. In the milling example (Fig. 76-30), the tool first moves from point A to point B. This block should be written as G01 F8.0 X3.0; *absolute position* (G90) movement will be from X.500 to X3.000, at a feed rate of 8.0 in./min. The next move is from point B to point C, which should be written as Y−1.250, to move from Y0 to Y−1.250. These two blocks cannot be combined as G01 F8.0 X2.5 Y−1.25; the machine control unit must be told to make each move separately by creating a block for each move.

▶▶ Interpolation

Interpolation, or the generation of data points between given coordinate positions of the axes, is necessary for any type of programming. Within the MCU, a device called an *interpolator* causes the drives to move simultaneously from the start of the command to its completion. Linear and circular interpolation are most commonly used in CNC programming applications (Fig. 76-31).

> Linear interpolation is used for straight-line machining between two points.

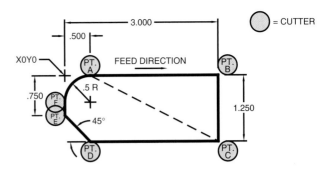

EXAMPLE 1   STARTING FROM ABSOLUTE X.5 Y0
               FIRST PATH FROM POINT A TO B   G01 F8.0 X3.0
                                 B TO C   Y −1.25

EXAMPLE 2   SECOND PATH FROM POINT A TO C   G01 F8.0 X3.0 Y −1.25

■ **Figure 76-30** A sample part to illustrate programming procedures. *(Kelmar Associates)*

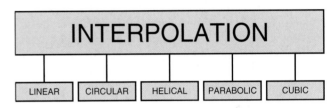

■ **Figure 76-31** The most common types of interpolation used on CNC machines. *(Kelmar Associates)*

> Circular interpolation is used for circles and arcs.

> Helical interpolation, used for threads and helical forms, is available on many CNC machines.

> Parabolic and cubic interpolation are used by industries that manufacture parts having complex shapes, such as aerospace parts, and dies for car bodies.

Interpolation is always performed under programmed feed rates.

LINEAR INTERPOLATION

Linear interpolation consists of any programmed points joined together by *straight lines*. These include horizontal, vertical, or angular lines, where the points may be close together or far apart. In Fig. 76-30, they include the lines from points A to B, B to C, C to D, and D to E (or from A to C, if it was necessary to make an angular cut).

CIRCULAR INTERPOLATION

Circular interpolation makes the process of programming arcs and circles easy. On some CNC systems, only one-quarter of a circle or quadrant (90°) can be programmed at a time. However, recent MCUs have full circle capability within the same command, which helps to shorten the program. It also improves the quality of the part because

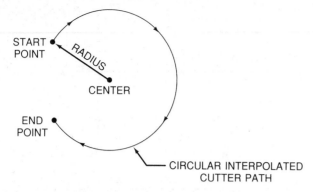

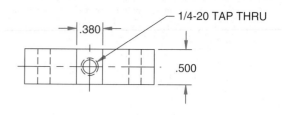

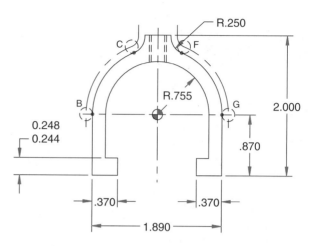

Figure 76-32 The circle center position, radius, start point, end point, and direction of cut are required for circular interpolation. *(Kelmar Associates)*

there is a smooth transition throughout the entire circle, with no hesitation or dwell between quadrants.

Fig. 76-32 shows the basic information required to program a circle. This must include the position of the circle center, the start and end points of the arc being cut, the direction of the cut, and the feed rate for the tool. An example of an arc and the block of information required to program it is shown in Fig. 76-33. Several methods can be used to write the block for the arc:

> One method uses the I and J command to identify the coordinates of the center of the arc.

> A simpler method uses the R (radius of the arc) command, which the MCU uses to calculate the arc center.

▶▶ Program Planning

Program planning is a very important part of CNC machining. Important information must be gathered, analyzed, and calculated before writing the program. The programmer must also consider the capabilities of the machine by referring to the programming and operation manual, which lists the capacity, tooling requirements, programming format, etc.

MACHINING PROCEDURES

To become a good CNC programmer, a person should have a good fundamental background in conventional machining processes and procedures. The CNC programmer must consider all the variables that are required for the conventional manufacture of parts. Referring to the part print and finding the answers to the following questions will be helpful in successfully programming a part:

1. What are the proper cutting speeds and feeds for the type of material being machined?

2. How will the part be held, in a simple vise or in a special fixture? Will clamps interfere with movement of the axes?

NOTE: X0Y0 AT BORE CENTER - TOOLPATH IS THE PART EDGE WHEN CUTTER RADIUS COMPENSATION IS USED

EXAMPLE 1 — POINT B TO POINT C:

G02 X-.3479 Y.8786 I.945 J0 F5.0
OR
G02 X-.3479 Y.8786 R.945 F5.0

EXAMPLE 2 — POINT F TO POINT G:

G02 X.945 Y0 I-.3479 J-.8786 F5.0
OR
G02 X.945 Y0 R.945 F5.0

Figure 76-33 Any form on two axes can be generated by circular interpolation. *(Kelmar Associates)*

3. Are the required tools and holders available?

4. Will special coolant be required, or are the present type and concentration correct?

5. What is the table feed direction? Keep in mind that climb milling is preferable and not a problem, since most CNC machines have ballscrews (Fig. 76-34).

6. How fast can the tool be moved to location: rapid traverse or at a feed rate?

7. What will the tool do when it reaches its location— for example, drill a hole or mill a pocket?

8. Where will the part zero point, or origin, be located, on the part or the machine?

It is wise to remember that the procedure for machining a part, whether by conventional or CNC machining, is basically the same. In conventional machining, a skilled operator moves the machine slides manually, while in CNC

Figure 76-34 Ballscrews and backlash eliminators are used on CNC machines to eliminate end play between the screw and nut. *(Cincinnati Milacron)*

machining, the machine slides are moved automatically from the information supplied by the CNC program.

TOOL LIST

The programmer should make a list of all the tools required for the machining process. Correct speeds and feeds must be calculated for each tool based on tool material type, the type of material being cut, depth of cut, etc. Some CNC machine systems require presetting the tool length for the purpose of offsets. If this is necessary, a special gage may be needed, and all lengths must be recorded for entry into the proper offset registers during machine setup. When using an automatic tool changer (ATC), tools must be assigned to a pocket for both sequencing and balance of the ATC. Fig. 76-35 shows an example of a developed tool list.

Part No. CNB-140-1		Material Stainless Steel			Operation		N/C	
Part Name Upper Nose (Mach)		**Program No.** 01160	**Programmed By:** M.C.			**Date** 10/19/03		**Sheet** 1 of 3
Tool	**Tool**	**Tool Diameter**		**Tool Length**		**Operation**		
Seq. #	**Description**	**Value**	**"D"**	**Value**	**"H"**	**Description**		
01	⁵⁄₁₆ Dia. (.312) Long Flute H.S.S. End Mill			4.250	H01	Set to End		
02	²¹⁄₆₄ Dia. (.328) H.S.S. Reamer			5.000	H02	Set to End		
03	²⁹⁄₆₄ Dia. (.453) H.S.S. C'Bore			5.000	H03	Set to End		
04	⅞ Dia. (.875) H.S.S. End Mill			4.500	H04	Set to End		
05	½ Dia. (.500) H.S.S. End Mill			4.250	H05	Set to End		
06	¼ Dia (.250) H.S.S. Spot Drill			5.000	H06	Set to Point		
07	³⁄₁₆ Dia. (.187) 6 in. Long H.S.S. Drill			6.500	H07	Set to End		
08	#7 Dia. (.201) H.S.S. Drill			4.250	H08	Set to Point		

Figure 76-35 The tool list must include all the information about each tool, such as tool number, speeds, feeds, and offsets. *(Duo-Fast Corporation)*

MANUSCRIPT

Before a program for a workpiece to be cut on a CNC machine tool is written, the job print must be studied carefully. In order to determine the sequence of operations, the programmer must decide which surfaces of the workpiece must be machined, the special operations required, and the dimensional tolerances for the part. It is also the programmer's responsibility to see that the machine tool receives the proper information to cut the part to the proper shape and size. Using alphanumeric language (letters and numbers), the programmer must record on a prepared form (*manuscript*) all the instructions that the machine tool must have to complete the job. The manuscript must contain all the machine tool movements, cutting tools, speeds, feeds, and any other information that is required to machine the part (Fig. 76-36). This information should be in a uniform format and as clear as possible to give the CNC machine operator a good understanding of what is required. Fig. 76-37 shows the type of information that the programmer must include in, or supply with, the manuscript.

1. *Part sketch*
   > A rough sketch of the part should be made. While absolute positioning is most often used, the programmer should give the location for each axis from the zero, or reference, point, either in incremental or absolute dimensions, depending on the positioning system to be used for the job.

2. *Zero (or reference) point*
   > A zero (or reference) point should be set either on the workpiece or on the machine tool.
   > Machines not equipped with an automatic tool changer require the selection of a tool change position, which provides enough room to change cutting tools and to load and unload parts.

3. *Work-holding device*
   > The device or fixture best suited to hold the part securely and not interfere with the machining operations must be selected.

Part no. CNB-140-1	Rev. P	Note												Date 10-20-04			Page 115
Program number O(;) 1160				(Note) End of block code (;) is CR in EIA or LF in ISO							Programmer M.C.						

/	N	G	X	Y	Z	A/B/C	R/I	J	K	B	P	Q	L	H/D	F/S	T	M	;
	%																	
	Ō 11	60	(CNB-140-1)															
	N10		G0 G17	G20 G40	G49 G64		G80	G90	G54									
		G43	X.407	Y-7.094	Z 5.									H1	S1000		M3	
	G81	G98			Z -.1		R.3								F3.		M8	
	N20															T2	M6	
		G43	X.407	Y-7.094	Z 5.									H2	S1000		M3	
	G81	G98			Z-.1		R.3								F3.		M8	
	N30															T3	M6	
		G43	X1.5	Y -6.	Z 5.									H3	S450			
	G82	G98			Z 3.26		R3.9				P0700				F2.			
	G0	G80		Y -5.														
					Z 3.25													
		G1		Y -7.														
		G0			Z 5.													
			X 6.141	Y -5.2														
					Z 2.072													
	G1			Y -2.8														
		G0			Z 5.													
				Y -5.2														
					Z 2.062													
		G1		Y -2.8														
		G0	X 7.16	Y -6.														
					Z .473													

Figure 76-36 The manuscript must contain all the information required to machine a part to size and shape.

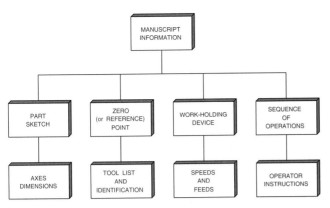

Figure 76-37 The main points that must be considered when preparing a manuscript.

> The fixture must not have any of its components much higher than the part to be machined.
> The setup instructions for the fixture must be included in the manuscript.

4. *Sequence of operations*

> A knowledge of basic machining procedures is essential for listing the operation sequence properly.
> Select the operations that should be done first and the sequence of machine movements (events) that will machine the part in the shortest time possible.

5. *Axes dimensions*

> The data for every movement of the table or the cutting tool must be listed.
> The data must include axis locations for every surface to be machined or every hole to be drilled, tapped, reamed, etc.

6. *Tool list and identification*

> The required tools should be indicated in the "Remarks" or "Comments" column of the manuscript.
> The tool identification number should indicate the order in which each tool will be used, the tool diameter and length, and the tool and offset number.

7. *Speeds and feeds*

> The speed in revolutions per minute (r/min) for each cutting tool must be included in the manuscript.
> The feed rate, in inches per minute or meters per minute (in./min or m/min), that will give the best cutting-tool life while maintaining efficient machining rates should be listed for each tool.

8. *Operator instructions*

> Special instructions for the operator should be included in the "Remarks" or "Comments" column of the manuscript.
> These should include information such as specifications about selecting cutting tools, loading or unloading the workpiece, and changing cutting tools if it is not a CNC function.

unit 76 review questions

1. What type of coded instructions is used in numerical control work?

Types of Computers

2. Name two basic types of computers and state where each is generally used.

3. What are three important functions of a computer?

4. For what purpose are the following computers used?
   a. Mainframe
   b. Minicomputer
   c. Microcomputer

CNC Performance

5. Name four reasons CNC machines have been so widely accepted throughout the world.

6. Compare the accuracy of a machinist with that of a CNC machine tool.

Advantages of CNC

7. List seven of the most important advantages of CNC as they relate to part accuracy and productivity.

Cartesian Coordinate System

8. Why is the coordinate system so important to the machine tool trade?

9. Give the *X* and *Y* coordinate values for the four quadrants, starting at the upper left and proceeding clockwise.

10. Give the coordinate locations of each point in the diagram.

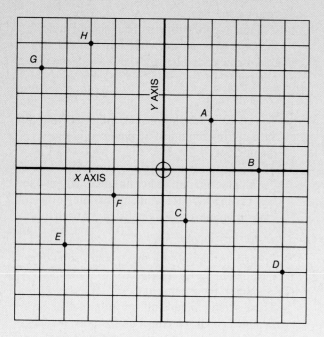

Machines Using CNC

11. Name the two axes on a lathe (chucking center) and state what each controls.

12. Name the three axes on a vertical milling machine (machining center) and state what each controls.

Programming Modes

13. How are incremental dimensions or positions given?

14. What code would the MCU understand for incremental positioning?

15. How are absolute dimensions or positions given?

16. What code would the MCU understand for absolute positioning?

CNC Positioning Systems

17. When is point-to-point positioning used?

18. Name three common operations that may use point-to-point positioning.

19. For what type of work is continuous-path positioning used?

Control Systems

20. Compare the open loop system with the closed loop system.

21. How many electrical pulses are required to move a machine slide 1.000 in. (25.4 mm)?

Types of Computer Control

22. Name four main components of a CNC system.

23. Explain live, or conversational, mode.

24. Name three main advantages of a DNC system.

Programming Systems

25. Name the two most common codes used for CNC programming.

26. What code is used for:

    a. Linear interpolation?
    b. Circular interpolation?

27. What miscellaneous code should be used to:

    a. Turn the spindle on clockwise?
    b. Turn the spindle off?

28. How much information should each program block contain?

Interpolation

29. For what purpose are the following interpolations used? Give examples of each.

    a. Linear
    b. Circular
    c. Helical

30. What information is required to program a circle?

31. Name two methods of programming arcs and circles.

Program Planning

32. List and briefly explain the eight types of information that should be included in a manuscript.

UNIT 77

CNC Turning Center

OBJECTIVES

After completing this unit, you will be able to:

1 State the purpose and functions of chucking, turning, and turning/milling centers

2 Identify the applications of computer numerical control (CNC) for turning centers

3 Name the machining operations that may be performed simultaneously

Extensive studies during the mid-1960s showed that about 40% of all metal-cutting operations were performed on lathes. Up until that time, most work was done on engine or turret lathes, which were not very efficient by present-day standards. Intensive research led to the development of numerically controlled turning centers and chucking lathes that could produce round work of almost any contour (shape) automatically and efficiently (Fig. 77-1). In recent years, these have been updated to more powerful computer-controlled units capable of greater precision and higher production rates than their predecessors.

■ **Figure 77-1** Turning and chucking centers are versatile and very productive.
(Cincinnati Milacron)

▶▶ Types of Turning Centers

The three main types of turning centers are:

1. The CNC chucking center (Fig. 77-2), which holds an individual part in some form of a jaw chuck. Some machines also have dual spindles (Fig. 77-3), which allow single cycle machining on both ends of a workpiece.

2. The CNC universal turning center can use a continuous bar feed system to machine and cut off the parts from the bar or use a tailstock to support long parts. Some machines are also equipped with dual tool turrets (Fig. 77-4).

3. A more current addition to the industry is the combination turning/milling center, which uses a combination of turning tools (Fig. 77-5). Typical machining- center tools include drills, end mills, and taps to perform operations that previously required another setup on a different machining center.

■ **Figure 77-2** A chucking center holds work in some form of chuck or fixture. *(Cincinnati Milacron)*

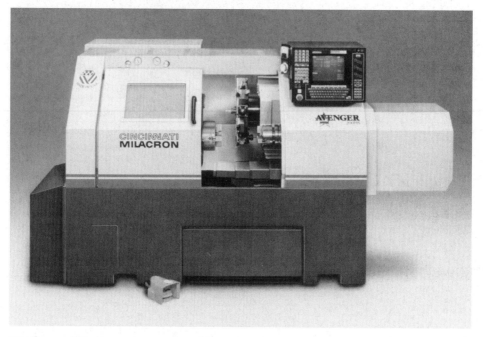

■ **Figure 77-3** A dual spindle chucking center allows both ends of a part to be machined. *(Cincinnati Milacron)*

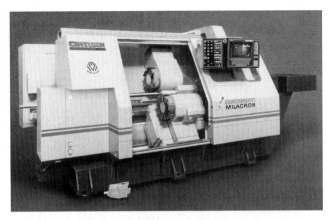

■ **Figure 77-4** Dual tool turrets increase the number of tools available and can cut with two tools at the same time. *(Cincinnati Milacron)*

■ **Figure 77-6** Turning and boring operations may be carried on simultaneously. *(Cincinnati Milacron)*

■ **Figure 77-5** Combination turning/milling centers allow drilling and milling operations to be performed on a turning center. *(Cincinnati Milacron)*

■ **Figure 77-7** Both turrets being used for internal operations. *(Cincinnati Milacron)*

CNC CHUCKING CENTER

Chucking centers (Fig. 77-2 on p. 622), designed to machine most work held in a chuck, are available in a variety of chuck sizes from 8 to 36 in. in diameter. Many types of chucking centers are manufactured and all perform similar functions; therefore, only a four-axis chucking center will be covered in this unit.

The four-axis chucking center has two turrets, each on separate slides, that can machine the workpiece at the same time. While the seven-tool upper turret is machining the inside diameter (ID), the lower turret (also containing seven different cutting tools) may be machining the outside diameter (OD) (Fig. 77-6). If the workpiece requires mainly internal operations, both turrets can be used to machine the inside of the workpiece at the same time (Fig. 77-7). This type of operation is suitable for large-diameter parts that require boring, chamfering, threading, internal radii, or retaining grooves.

For parts requiring mainly OD operations, the upper turret can be equipped with turning tools so that both turrets can be used to machine the OD. Fig. 77-8 shows a diameter and a chamfer being machined at the same time.

When longer parts must be machined, the right-hand end of the shaft should be supported with a tailstock center mounted in the upper turret while the lower turret performs the external machining operations (Fig. 77-9). Other operations that may be performed simultaneously are turning and facing (Fig. 77-10) and internal threading (Fig. 77-11).

■ Figure 77-8 Both turrets being used for external turning. *(Cincinnati Milacron)*

■ Figure 77-10 Simultaneous turning and facing operations using both turrets. *(Cincinnati Milacron)*

■ Figure 77-9 A long shaft being supported by a center mounted in the upper turret. *(Cincinnati Milacron)*

■ Figure 77-11 Cutting internal and external threads simultaneously. *(Cincinnati Milacron)*

Other chucking centers are mainly two-axis models. They may have a single turret on which both ID and OD tools are mounted, or they may have two turrets (usually on the same slide). In the latter case, one turret is normally designed for OD tools and the other for ID tools. Whatever the arrangement, the two-axis control will drive only one turret at a time.

Construction

The main operative parts of all turning centers are basically the same. They consist of the framework components and those that give the framework its CNC capabilities (Fig. 77-12). Because of the high spindle speeds (up to 5000

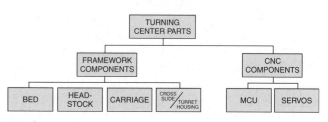

■ Figure 77-12 Turning center parts are divided into two categories. *(Kelmar Associates)*

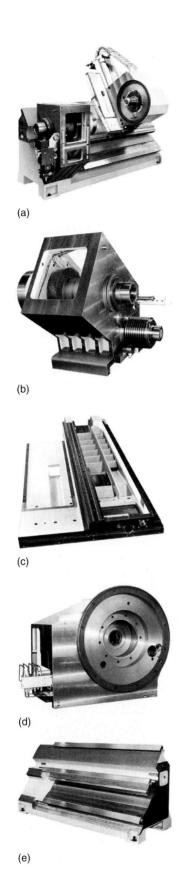

(a)

(b)

(c)

(d)

(e)

Figure 77-13 The main components of a turning center: (a) framework components assembled; (b) headstock; c) carriage; (d) cross-slide/turret housing; (e) bed. *(Cincinnati Milacron)*

r/min) and the high horsepower requirements (up to 75 HP), machines of this type must be rugged to absorb the high cutting forces. The bed and machine frame can be a heavy, one-piece cast-iron casting. The four framework components (Fig. 77-13) are shown separately and as a machine assembly. Another type of construction is the polymer composite, which is cast around premachined inserts (Fig. 77-14a), containing tubes, fittings, or fasteners and threads for use during machine assembly. Fig. 76-14b shows a polymer cast base. To permit easy chip and coolant removal, and easy loading and unloading of the workpieces, the bed is slanted 40° from the vertical plane.

Direct-current servo drives provide accurate positioning and travel of the cutting tools. The servomotors position the turret cross-slides by means of high-precision preloaded ball screws (Fig. 77-15 on p. 626) that ensure slide positioning repeatability to ± .000 060 in. on the X axis and ± .0001 on the Z axis. These turret slides

(a)

(b)

Figure 77-14 (a) Premachined insert cast in a polymer base; (b) polymer composite machine tool base. *(Hardinge Brothers, Inc.)*

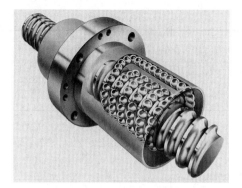

Figure 77-15 Precision ballscrew ensures accurate slide movements. *(Cincinnati Milacron)*

Figure 77-16 A gage for presetting cutting tools for turning centers. *(Cincinnati Milacron)*

operate at speeds from .100 to 400 in./min (2.54 to 10,160 mm/min) using programmed feed rates, and up to 944 in./min (24,000 m/min) on rapid traverse.

Tooling

Both the upper and lower turrets can accommodate seven different types of tools. The toolholders for machining outside diameters are located in the lower turret and are qualified (preset). They require only the changing of the carbide insert when the tool must be replaced. Tools for machining the inside diameter are mounted in a dovetailed block and preset off the machine by means of a tool-setting gage (Fig. 77-16). The dovetailed block tool assembly is then mounted in the upper turret, ensuring proper tool positioning every time. Some machines are equipped with an automatic tool-setting probe (Fig. 77-17) for presetting tools. Each tool is rotated into machining position, the tool point is touched off, and each tool setting is recorded in the control unit for use during the machining cycle. This makes sure that the length of each tool is recorded so that parts are machined to the size required.

Figure 77-17 A tool setter calculates the tool offset value, which is automatically recorded into the MCU. *(Cincinnati Milacron)*

Computer Numerical Control

The control shown in Fig. 77-18 is the "brains" of the chucking center and is usually mounted on the machine (Fig. 77-19). It has a 32-bit microprocessor, cathode ray tube (CRT) video display, program input unit such as a diskette, part program storage, part program edit, tool data management, and maintenance diagnostic display.

The microprocessor controls the logic calculations, mechanism control, and input-output control. The CRT provides a visual display of slide positions, spindle operating condition, sequence numbers, preparatory functions, system fault conditions (diagnostics), operator instructions, and keyboard data.

The keyboard input is used to communicate with the CNC system, to enter setup and tooling data for new programs, and to correct tooling data. The keyboard can also be used to make diagnostic checks on the systems. Information such as oil pressure and spindle conditions is shown on the CRT.

Part program storage with a 20-megabyte (MB) hard disk can store data equivalent to 75,000 ft (22,865 m) of

Figure 77-18 A chucking center control panel contains the computer, video display, and machine control unit. *(Cincinnati Milacron)*

Figure 77-19 The control panel is attached directly to the machine tool. *(Cincinnati Milacron)*

part program tape. A part program edit feature allows the program to be changed at any time. Other features include a constant surface speed monitor, which checks the part diameter and controls the spindle speed accordingly.

CNC TURNING CENTER

Computer numerical control turning centers (Fig. 77-20), while similar to chucking centers, are designed mainly for machining shaft-type workpieces supported by a chuck and a heavy-duty tailstock center.

On four-axis machines, two opposed turrets, each capable of holding seven different tools, are mounted on separate cross-slides, one above and one below the centerline of the work. Because the turrets balance the cutting forces applied to the work, extremely heavy cuts can be taken on a workpiece when it is supported by the tailstock. The dual turrets can also perform other operations such as:

> Roughing and finishing cuts in one pass
> Machining different diameters on a shaft simultaneously (Fig. 77-21)

> Finish-turning and threading simultaneously
> Cutting two different sections of a shaft at the same time

Because the lower turret is designed for ID tools, parts may be gripped in the chuck and machined inside and outside at the same time (Fig. 77-22). When the turning center is equipped with a *steadyrest* (Fig. 77-23), operations such as facing and threading may be performed

Figure 77-21 Both turrets being used to machine a shaft having different diameters. *(Cincinnati Milacron)*

Figure 77-20 A CNC turning center is designed for maximum productivity of shaft-type workpieces. *(Cincinnati Milacron)*

Figure 77-22 The inside and outside diameters of a part machined at the same time. *(Cincinnati Milacron)*

CNC Turning Center   627

on the end of a shaft. When turning long, thin shafts, a *follower rest* (Fig. 77-24) can be used for additional support.

A *bar-feeding mechanism* (Fig. 77-25) permits the machining of shafts and parts from bar stock smaller than the diameter of the spindle through-hole. Some machines are equipped with a bar puller (Fig. 77-26) mounted in the turret. The machine slides move into position and fingers grip the bar, then the Z axis slide moves to position for the correct length. The chuck closes, the puller releases the part, and the turret moves to a safe location for tool indexing. When individual precut shafts are machined, a *production part loader* (Fig. 77-27) can complete a part changeover in as few as seven seconds.

■ **Figure 77-25** A bar-feeding attachment allows numerous parts to be manufactured from bar stock. *(Cincinnati Milacron)*

■ **Figure 77-23** A steadyrest provides support when operations do not allow the use of a tailstock center for end support. *(Cincinnati Milacron)*

■ **Figure 77-26** Bar pullers advance the bar stock from the tool turret. *(Cincinnati Milacron)*

■ **Figure 77-24** Two follower rests being used to support a long, thin shaft. *(Cincinnati Milacron)*

■ **Figure 77-27** Part loaders can improve productivity by making part changes in seven seconds. *(Cincinnati Milacron)*

COMBINATION TURNING/MILLING CENTER

Because of the constant improvement of machine tools and the need to improve manufacturing productivity, the turning/milling center has been developed. In the past, after the operations in a turning center were completed, it was required to take the part out of the machine and wait until a machining center was available to complete the part.

The turning/milling center allows operations such as drilling, milling, and tapping to be performed on the part while it is still in the machine. This is possible because of a special tool turret which contains pockets that have their own drive for *live tools*. Fig. 77-28 shows a carbide insert end mill held in this special tool pocket, which can rotate like the spindle of a machining center. If a shaft requires parallel flats to be cut, the tool turret can be programmed to move to the correct location. The machine spindle, which drives the part for turning operations, is locked in the proper position where the flat is required. With the end mill rotating, the *X* axis slide moves to the required depth, and then the *Z* axis slide feeds the tool to the correct length. Once the first flat is machined, the *X* axis backs the tool out of the part. The spindle rotates 180° and the process is repeated to cut the flat on the opposite side. The tool then returns to the tool indexing position.

Operations such as drilling and tapping (Fig. 77-29) can be performed on the part if the machine has a *contouring* spindle, which can be indexed to exact locations around the circumference of the workpiece.

■ **Figure 77-28** Live tool pockets provide rotation for drilling and milling tools in the turning center. *(Hardinge Brothers, Inc.)*

■ **Figure 77-29** A hole can be drilled in a turning center on the circumference of a workpiece, eliminating the need for a secondary setup. *(Hardinge Brothers, Inc.)*

▶▶ Programming Considerations

One of the main requirements for a good programmer must be the ability to analyze the part print and decide on the sequence of machining operations. It is a good practice to develop the habit of labeling the start and end points for both roughing and finishing operations, if required.

> ⚠ **CAUTION** *Be sure* the programming format suits your equipment before machining parts.

TOOLING SYSTEMS AND CUTTING TOOLS

Tooling systems for CNC turning centers may vary with an individual manufacturer's specifications. It is important to remember that the success of any turning operation depends on the accuracy of the tooling system and the cutting tools being used. A typical tooling system (Fig. 77-30 on page 630) consists of toolholders, boring bar holders, facing and turning holders, and drill sockets.

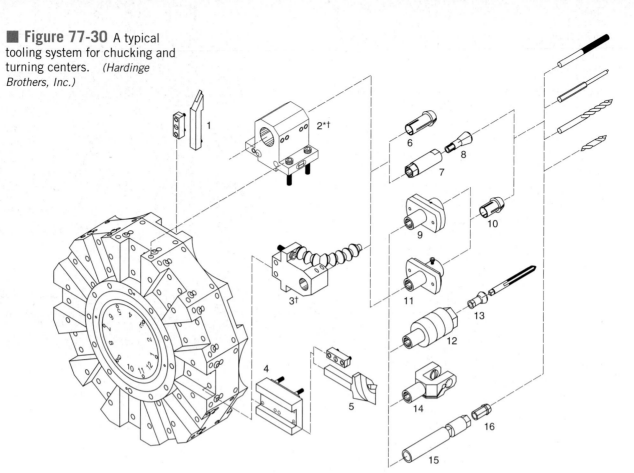

■ Figure 77-30 A typical tooling system for chucking and turning centers. *(Hardinge Brothers, Inc.)*

Item Number	Tool Number	Description
1	—	³/₄″ square shank insert cutting tools
	—	³/₄″ square shank 35° and 55° insert profiling tools
2	SG-46	Double round shank turning toolholder (³/₄″ ID)
	SG-47M	Double round shank turning toolholder (25 mm ID)
3	SG-38	Round shank toolholder (³/₄″ ID)
	SG-39M	Round shank toolholder (25 mm ID)
4	SG-CE	Square shank cut-off toolholder (³/₄″)
	SG-CM	Square shank cut-off toolholder (20 mm)
5	G23	Hardinge-Belcar insert cut-off tool (³/₄)″
6	HDB-6	Precision bushings (³/₄″ OD)
7	T-17 ³/₄	1C collet-type holder (³/₄″ shank)
8	1C	1C collets
9	T-19 ³/₄	Floating reamer holder (³/₄″ shank)
10	HDB-2	Precision bushings (¹/₂″ OD)
11	00D ³/₄	Adjustable holder (³/₄″ shank)
12	TT-³/₄	Releasing TT tap holder (3/4″ shank)
13	—	TT tap collets
14	T-8 ³/₄	"Crush-type" knurling tool
15	DAH-235	Double-angle collet extension toolholder (³/₄″ shank)
16	—	200-series double-angle collets

Inserts

Cutting tool inserts are made from many types of material. Modern turning centers are rigid in construction and have the speeds, feeds, and horsepower available to use all types of cutting tools, even superabrasives.

There is a great variety of cutting-tool materials to suit any workpiece material or machining operation. They include carbide, coated carbide, ceramic, cermet, cubic boron nitride, and diamond tooling. Because toolholders and insert shapes are standardized, most inserts generally fit in the same holders. Refer to Units 31 through 33 for detailed information regarding cutting tools.

Tool Nose Radius Compensation

Cutting-tool inserts are available with a wide variety of tool nose radii, starting with a sharp point and increasing in $\frac{1}{64}$-in. increments from $\frac{1}{64}$ to $\frac{1}{8}$ in. When programming, the *theoretical sharp point* of the tool is programmed; however, this will not position the tool at the correct location. For finish cuts, it is common practice to turn ON the tool nose radius compensation (G41 or G42). The radius of each insert must be stored in the numbered tool list of the control tool management system.

Tool Offsets

The programmer must also provide a tool setup sheet (Fig. 77-31) for the setup operator. For example, a command of T01 will index the tool turret to tool number 1. Other information can be added to the T01 code to indicate offset and compensation information about the tool. The MCU will calculate the correct position at which the tool should be located to accurately machine the part.

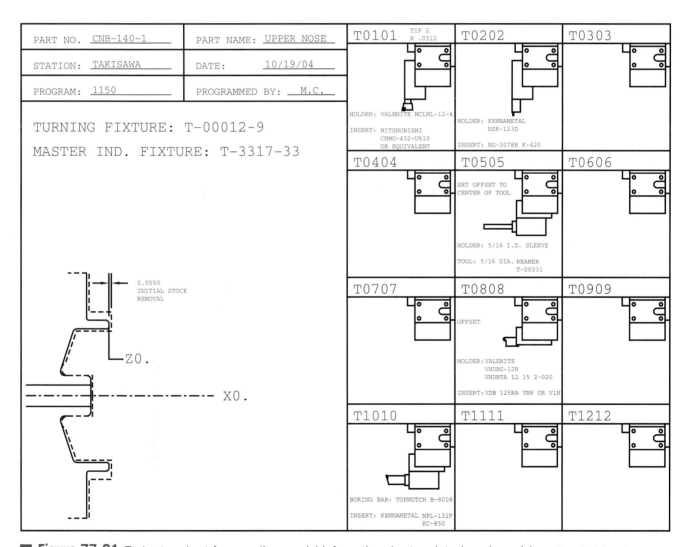

Figure 77-31 Tool setup sheet for recording special information about each tool used on a job. *(Duo-Fast Corp.)*

DIAMETER VERSUS RADIUS PROGRAMMING

The *X* axis (cross-slide) of the turning center can be programmed for either *diameter* or *radius programming*. The method used is determined by preset parameters within the machine control unit or by the use of the correct G-code. For diameter programming, the part print (Fig. 77-32a) is drawn complete, both sides of the centerline with full diameter dimensions. For radius programming, the part print (Fig. 77-32b) is drawn on just one side of the centerline dimensions, the radius of the part.

Most machine control units are automatically preset (default) for diameter programming. If radius programming is necessary or desirable, a correct code must be inserted at the start of the program to inform the MCU that radius programming will be used. Since programming codes may vary slightly with each manufacturer, be sure to refer to the machine's programming manual for the correct code.

ESTABLISH PART ZERO

It is the programmer's choice to place the part zero at the most convenient location. The location of the *X* axis is usually the centerline of the part, while the *Z* axis can be either at the right-hand (*tailstock*) or the left-hand (*chuck*) end of the part (Fig. 77-33). If the part zero is located at the right-hand end (the method most often used), all *Z* movements into the part will require a negative ($-Z$) number. When using the left-hand end of the part, all *Z* movements into the part will be a positive (*Z*) number. Since positive numbers do not require the use of the plus ($+$) sign, there is less chance of error when typing a program.

CODES

On a turning center, the function of some *G*-codes and *M*-codes may differ from the function of those on a machining center, as shown in Unit 75. Table 77.1 shows many of the common turning center *G*-codes that conform to EIA standards, and Table 77.2 shows the common *M*-codes.

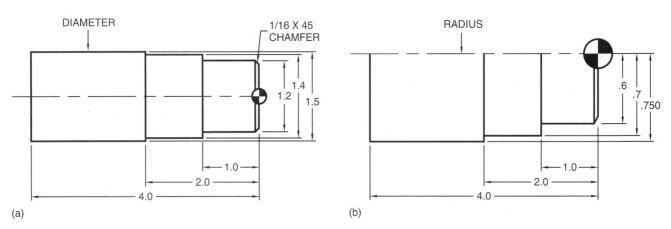

(a)   (b)

■ **Figure 77-32** (a) A sample part dimensioned for diameter programming; (b) a sample part dimensioned for radius programming.   *(Kelmar Associates)*

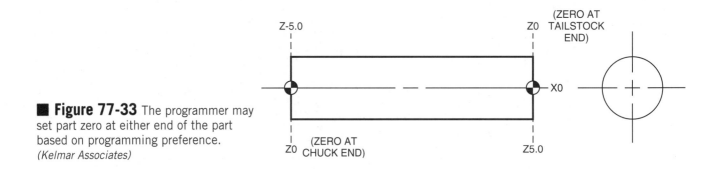

■ **Figure 77-33** The programmer may set part zero at either end of the part based on programming preference. *(Kelmar Associates)*

table 77.1 — Common preparatory codes for turning centers

Code	Function
G00	Rapid positioning
G01	Linear interpolation
G02	Circular interpolation clockwise (CW)
G03	Circular interpolation counterclockwise (CCW)
G04	Dwell
G20	Inch data input
G21	Metric data input
G27	Zero return check
G28	Zero return
G29	Return from zero
G32	Thread cutting
G40	Tool tip R compensation cancel
G41	Tool tip R compensation left
G42	Tool tip R compensation right
G50	Absolute coordinate preset maximum r/min in G96 mode
G52	Local coordinates preset
G54–G59	Work coordinates system selection
G70	Finish cycle
G71	OD, ID rough-cutting cycle
G72	End surface rough-cutting cycle
G73	Closed loop cutting cycle
G74	End surface cutting off cycle
G75	OD, ID cutting off cycle
G76	Thread-cutting cycle
G84	Canned turning cycle
G92	Incremental positioning
G96	Constant surface speed
G97	Constant surface speed cancel
G98	Feed per time
G99	Feed per spindle rotation
X—Address	absolute programming (diameter input)
U—Address	incremental programming (diameter input)
Z—Address	absolute programming parallel to X axis
W—Address	incremental programming parallel to X axis

▶▶ Programming Procedures

There are many manufacturers of CNC turning machine tools throughout the world, and the CNC control units could vary from manufacturer to manufacturer. Therefore, it would be impossible to give an example of a programming procedure that would suit all CNC turning machines.

table 77.2 — Common miscellaneous functions for turning centers

Code	Function
M00	Program stop
M01	Optional stop
M02	End of program
M03	Spindle rotation—forward
M04	Spindle rotation—reverse
M05	Spindle stop
M08	Coolant on
M09	Coolant off
M10	Chuck—clamping
M11	Chuck—unclamping
M12	Tailstock spindle out
M13	Tailstock spindle in
M17	Toolpost rotation normal
M18	Toolpost rotation reverse
M21	Tailstock forward
M22	Tailstock backward
M23	Chamfering on
M24	Chamfering off
M30	Program end and rewind
M31	Chuck bypass on
M32	Chuck bypass off
M41	Spindle speed—low range
M42	Spindle speed—high range
M73	Parts catcher out
M74	Parts catcher in
M98	Call subprogram
M99	End subprogram

Since programming can vary slightly from machine to machine, it is important to follow the programming manual supplied for each machine. In this book, the authors have concentrated on two classes of CNC machines: the bench-top teaching model and the standard turning centers most commonly used in schools and machine shops. This will cover the standard basic programming procedures that, with only slight modifications, suit almost any machine control unit.

BENCH-TOP TEACHING MACHINES

Bench-top teaching machines are well suited for teaching purposes because neither the instructor nor the student should be intimidated by their size or complexity (Fig. 77-34a). They are very easy to program and perform turning operations similar to the larger machines, except with smaller workpieces and lighter cuts. They are relatively inexpensive and ideal for teaching basic programming procedures.

Most of the *G*- and *M*- codes apply to both the teaching bench-top CNC turning lathes and the standard-size turning centers, with a few variations. The cutting tool on bench-top CNC teaching machines is usually at the front of the workpiece, while on standard-size machines, the cutting tool is at the back of the workpiece. As shown in Fig. 77-34b, to cut a radius counterclockwise on a standard machine requires a G03 command, which is the EIA standard. Since the cutting tool on bench-top machines is at the front of the workpiece, to cut the same radius requires a G02 program command.

Simple Programming

The part shown in Fig. 77-35a (roughing) and Fig. 77-35b (finishing) will be used to introduce simple programming in easy-to-understand steps. Each programming step will be explained in detail so that you will have a clear understanding of each code, axis movement, etc. and what happens as a result of each programming step.

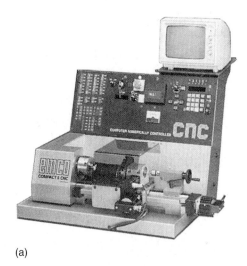

(a)

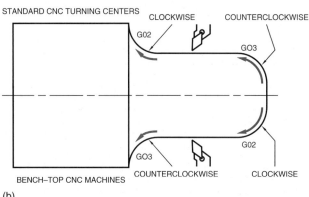

(b)

■ **Figure 77-34** (a) A bench-top CNC teaching lathe *(Emco Maier Corp.)* (b) a comparison of circular interpolation codes for standard CNC turning centers and bench-top CNC lathes *(Kelmar Associates)*

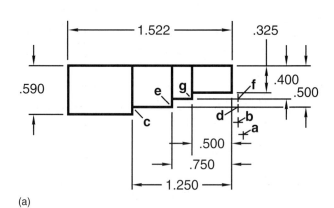

(a)

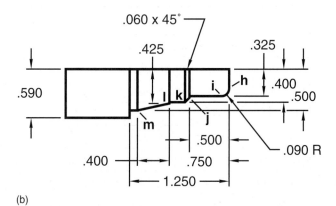

(b)

■ **Figure 77-35** A sample part (radius programming) to be machined on a teaching-size CNC lathe: (a) rough-cut dimensions; (b) finish-cut dimensions. *(Kelmar Associates)*

Program Notes

1. Program in the absolute mode (G90).

2. All programming begins at the *zero*, or *reference*, *point* (*X-Z* zero) at the centerline and the right-hand face of the part.

3. A carbide tool will be used for all operations.

4. Use radius programming, with the toolpost on the operator's side of the part centerline.

5. A position will be established to the right front corner of the machine to provide a safe location for part and tool changes.

6. The material is aluminum. Cutting speed is at 600 sf/min, with feed rate at .010 in.

Programming Sequence

```
%       Rewind stop code/parity check.
N10 G24
    G24             Radius programming.
N20 G92 X.690 Z.1
    G92             Reference point offset.
    X.690           Tool located .100 in. off the outside finish diameter/.690 in.
                    from part centerline (X0)(point a).
    Z.1             Tool located .100 in. to right of part face (Z0).
N30 M03
    M03             Spindle ON clockwise.
```

Note: It may be necessary to turn on and set the speed manually for bench-top machines.

```
N40 G00 X.590 Z.050
    G00             Rapid traverse rate.
    X.590           Tool located .590 in. from part centerline (point b).
    Z.050           Tool located .050 in. from part face.
N50 G84 X.500 Z-1.250 F.010 H.050
    G84             Fixed turning cycle.
    X.500           Tool moves a total of .090 in. toward X0 in two passes
                    (point c).
    Z-1.250         Tool moves 1.250 in. to left of part face.
    F.010           Feed rate .010 in.
    H.050           Incremental inward feed (along X axis) per pass for fixed
                    cycle.
N60 G00 X.500 Z.050
    X.500           Tool located .500 in. from part centerline (point d).
    Z.050           Tool located .050 in. to right of part face.
N70 G84 X.400 Z-.750 F.010 H.050
    X.400           Tool moves a total of .100 in. toward X0 in two passes
                    (point e).
    Z-.750          Tool moves .750 in. to left of part face.
N80 G00 X.400 Z.050
    X.400           Tool located .400 in. from part centerline (point f)
    Z-.050          Tool located .050 in. to right of part face.
N90 G84 X.325 Z-.500 F.010
    X.325           Tool moves .075 in. toward X0 in two passes.
    Z-.500          Tool moves .500 in. to left of part face (point g).
```

Finish Turning

```
N100 G00 Z.050      Tool located .050 in. to right of part face (point f).
N110 G00 X.235 Z0   Tool located at starting position of radius (point h).
N120 G02 X.325 Z-.090 F.010
                    Tool moves clockwise (CW) to end of the radius (point i),
                    positions .325 in. from centerline and .090 in. to the left of
                    part face.
N130 G00 X.340 Z-.500
                    Tool located at start of 45° chamfer (point j).
N140 G01 X.400 Z-.560 F.010
     G01            Linear interpolation (straight-line movement).
     X.400          Tool moves .400 in. from centerline (point k).
     Z-.560         Tool moves .560 in. left of the part face.
N150 G00 X.425 Z-.750
                    Tool located at start of tapered section (point l).
N160 G01 X.500 Z-1.150 F.010
                    Tool moves to end of tapered section (point m).
N170 G00 X.690 Z.100
                    Tool located at the original start position (point a).
N180 M30
     M30            End of program code.
%                   Rewind/stop code.
```

STANDARD-SIZE TURNING CENTER

The part illustrated in Fig. 77-36 is used to introduce additional machining and the use of diameter programming in easy-to-understand steps. An explanation, identical to the one used in the first example, will be provided for clarity.

Program Notes

1. Program in the absolute mode (G90).
2. All programming begins at the *zero*, or *reference*, *point* (*XZ*) at the centerline and the right-hand face of the part.
3. Two carbide tools, a CNMG-434 (roughing) and a DNGA-432 (finishing), will be used.

4. Use diameter programming.
5. Tool change positions are at the top right corner.
6. Material AISI 1018 machine steel:
   a. Roughing 400 ft/min—.012 in./rev feedrate
   b. Finishing 600 ft/min—.008 in./rev for front chamfer, .012 in./rev for rest of finish cut

Programming Sequence

 CAUTION *Be sure* the programming format suits your equipment before machining parts.

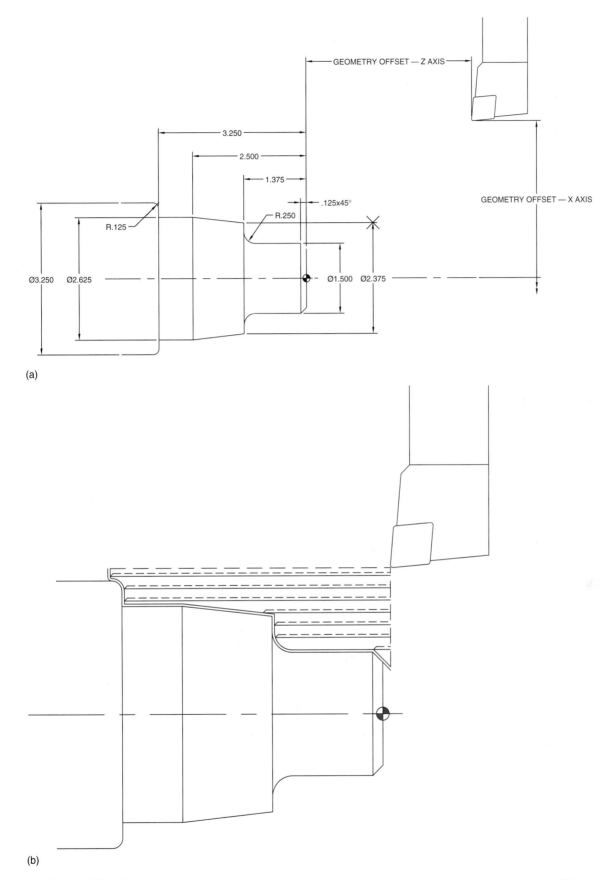

GEOMETRY OFFSET — Z AXIS

GEOMETRY OFFSET — X AXIS

3.250

2.500

1.375

.125x45°

R.250

R.125

Ø3.250 Ø2.625 Ø1.500 Ø2.375

(a)

(b)

■ **Figure 77–36** (a) A sample part (diameter programming) to be machined on a standard-size CNC chucking center; (b) rough-cut the part to shape.

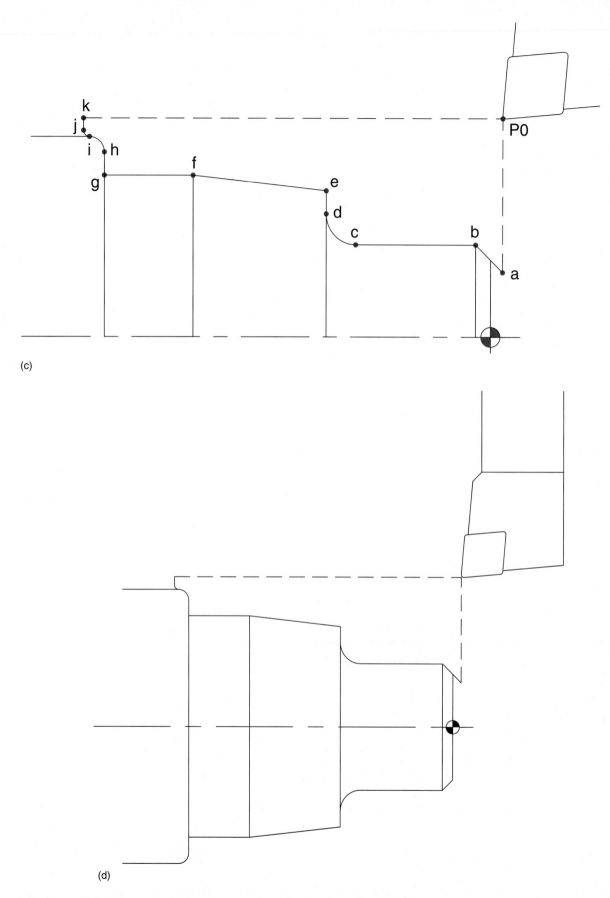

(c)

(d)

■ **Figure 77–36** (c) the finishing cut on the entire part; (d) finished part.

N1 G20;   Inch input data

N2 T0100 M42

     T0100   Call tool station 1—no wear offset

     M42   High spindle gear range

N3 G96 S400 M03

     G96   Constant surface speed selection

     S400   Surface speed 400 ft/min for roughing

     M03   Spindle rotation clockwise (CW)

N4 G42 G00 X3.5 Z0.1 T0101 M08

     G42   Tool nose radius compensation to the right

     G00   Rapid move to start position *XZ* for the G71 cycle (P0)

     T0101   Apply wear offset 1 to the current tool

     M08   Coolant **ON**

N5 G71 U0.1 R0.05

     G71   First block of two—selects roughing cycle

     U0.1   The actual depth of the roughing cut

     R0.05   Amount of retract clearance from each cut

N6 G71 P7 Q17 U0.04 W0.005 F0.012

     G71   Second block of two—selects roughing cycle

     P7   The first block of the finishing contour

     Q17   The last block of the finishing contour

     U0.04   Amount of stock in *X* axis left on diameter

     W0.005   Amount of stock left in *Z* axis on a face

     F0.012   Roughing feed rate in./rev—overrides any feedrate between **p** and **q**

N7 G00 X1.05   Block N7 represent the first point (rapid from **P0** to **a**) on the finished contour

N8 G01 X1.5 Z-0.125 F0.008

     Linear motion **(a)** to **(b)**—feedrate for finish = 0.008 in./rev

N9 Z-1.125 F0.01

     Linear motion **(b)** to **(c)**—feedrate for finish = 0.01 in./rev

N10 G02 X2.0 Z-1.375 R0.25

     Circular motion CW **(c)** to **(d)**—radius 0.250

N11 G01 X2.375   Linear motion **(d)** to **(e)**

N12 X2.625 Z-2.5   Linear motion **(e)** to **(f)**

N13 Z-3.25   Linear motion **(f)** to **(g)**

N14 X3.0   Linear motion **(g)** to **(h)**

N15 G03 X3.25 Z-3.375 R0.125

     Circular motion CCW **(h)** to **(i)**—radius 0.125

N16 G02 X3.35 Z-3.425 R0.05

     Circular motion CW **(i)** to **(j)**—lead-out radius 0.05

N17 G01 U0.2

     Block N17 represent the last point **(j** to **k)** on the finished contour—clear

     Tool return to start point **P0** automatically

N18 G40 G00 X5.0 Z2.0 T0100 M09

     G40   Cancel tool nose radius compensation for the current tool

     G00   Rapid to a tool change position with sufficient clearance

     T0100   Cancel wear offset for the current tool

     M09   Coolant **OFF**

N19 M01	Optional stop
N20 T0200 M42	
T0200	Call tool station 2—no wear offset
M42	High spindle gear range
N21 G96 S600 M3	
G96	Constant surface speed selection
S600	Surface speed 600 ft/min for finishing
M03	Spindle rotation clockwise (CW)
N22 G42 G00 X3.5 Z0.1 T0202 M08	
G42	Tool nose radius compensation to the right
G00	Rapid move to start position XZ
T0202	Apply wear offset 2 to the current tool
M08	Coolant **ON**
N23 G70 P7 Q17	
G70	Selects finishing cycle
P7	First block of finishing contour is N7
Q17	Last block of finishing contour is N17
N24 G40 G00 X5.0 Z2.0 T0200 M09	
G40	Cancel tool nose radius compensation with current tool
G00	Rapid to a tool change position with sufficient clearance
T0200	Cancel wear offset for the current tool
M09	Coolant **OFF**
N25 M30	End of program with return to the top
%	End of transmission during communication

▸▸ Turning Center Setup

Before setting up the turning center, it is necessary for the operator to become familiar with the control panel and the operational procedures. This includes the use of the different modes and how to use the menus to travel safely from one mode to the next. The operator must understand how to establish the work location in relation to machine zero, set tool length offsets, and test run the program.

When the power is turned on to the machine, it is necessary to turn on the servos and zero out/align all the axes so that the control knows the location of the machine home position. This procedure can be performed on each axis individually, or if equipped, some machines use an automatic reference command when starting up. If the program is not stored in memory, it must be loaded by the input media available. Based on the manuscript or separate tool sheet, the operator must prepare the tools and use the part-holding method listed by the programmer. This could include a chuck, collet device, tailstock center, or special holding fixture.

▸▸ Program Test Run

A part should never be machined without test running the program first. Some controls are equipped with a graphics display that allows the operator to go through the program steps on the control screen without cutting the part. Since no slide movement, turret indexing, or tool use is involved, this is a safe way to check the accuracy of a program.

If the machine is not equipped with graphics, another method is to dry run the program without the part in the machine. Use the step/single block mode and feedrate override to slow the programmed rate. Always have a finger on the hold button and know the location of the emergency stop in case there is an error in the program. Always double-check for interference when the tool turret is indexed and make sure that all offsets entered in the control are correctly matched with their tool number. After you are sure that the program and setup are correct, the part can be machined.

1. Why were chucking and turning centers developed?

Types of Turning Centers

2. What type of part is normally machined in a chucking center?

3. What type of part is normally machined in a turning center?

4. What unique machining capability does the turning/milling center have and how is this possible?

Main Operative Parts

5. Name the two categories of main operative parts of the turning center and the parts of each category.

6. Of what types of material can the bed of a turning center be constructed?

7. What is the purpose of the slanted bed on a chucking or turning center?

8. State the purpose of the following CNC parts:
   a. Microcomputer
   b. CRT
   c. Keyboard

Turning Center Capabilities

9. Why can extremely heavy cuts be taken on a turning center with dual turrets?

10. Name three other advantages of a turning center equipped with dual turrets.

Turning Center Accessories

11. Name two methods of measuring tools prior to use in a turning center.

12. Name two accessories used to advance bar stock in a turning center.

Programming Considerations

13. What different cutting-tool materials can be used on the turning center?

14. What is the difference between diameter and radius programming?

15. Where is the Z0 normally established on a turning center part?

Programming Procedure

16. What are some of the items to consider before programming a part for the simple program for the bench-top machine?

17. Name the *G*-codes listed for the simple program and state the function each performs.

18. Name the *M*-codes listed for the simple program and state the function each performs.

Turning Center Setup

19. Why must the turning center operator become familiar with the control panel and operational procedures?

20. What types of devices can be used to hold a part in a turning center?

21. What is the safest method used to find out if a program is correct so that an accident does not occur?

22. Describe an alternate method used to check the accuracy of a program.

UNIT 78

CNC Machining Centers

OBJECTIVES

After completing this unit, you will be able to:

1 Describe the development of the machining center

2 Identify the types and construction of machining centers

3 Explain the operation of the machining center

4 Understand a basic CNC program for a machining center

In the 1960s, industrial surveys showed that smaller machine components requiring several operations took a long time to complete. The reason was that the part had to be sent to several machines before it was finished and often had to wait a week or more at each machine before being processed. In some cases, parts requiring many different operations spent as many as 20 weeks in the shop before completion. Another startling fact was that during all the time in the shop, the part was only in a machine for 5% of the time, and while it was in the machine,

only 30% of the time was spent in machining. Therefore, a part was machined for only 30% of 5%, or 1.5%, of its time in the shop. In contrast, larger parts, such as machine castings, took much less time because the machining was usually done on one machine.

Also in the 1960s, there was much "operator intervention" during the machining process. The operator had to watch the performance of the cutting tool and change spindle speeds and feeds to suit the job and the machine. The operator was frequently changing cutters as they became dull and always had to set depths of roughing and finishing cuts. All of these problems were recognized by machine tool manufacturers, and in the late 1960s and early 1970s, they began to design machines that would perform several operations and probably do about 90% of the machining on one machine. One of the results of this research was the *machining center,* and later the more elaborate version called the *processing center.* The machines can very efficiently perform the operations of drilling, milling, boring, tapping, and precision profiling.

▶▶ Types of Machining Centers

There are three main types of machining centers: the *horizontal, vertical,* and *universal* machines. They are available in many types and sizes, which may be determined by the following factors:

> The size and weight of the largest piece that can be machined

> The maximum travel of the three primary axes (*X, Y, Z*)

> The maximum speeds and feeds available

> The horsepower of the spindle

> The number of tools that the automatic tool changer (ATC) can hold

HORIZONTAL MACHINING CENTER

There are two main types of horizontal spindle machining centers:

1. The *traveling-column* type (Fig. 78-1) is equipped with one or usually two tables on which the work is mounted. With this type, the column and the cutter move toward the work, and while it is machining the work on one table, the operator is changing the workpiece on the other table.

2. The *fixed-column* type (Fig. 78-2) is equipped with a pallet shuttle. The pallet is a removable table on which the workpiece is mounted. With this type, after one workpiece has been machined (Fig. 78-3a), the pallet and workpiece are moved off the receiver onto the shuttle. The shuttle is then rotated, bringing a new pallet into position for the shuttle and the finished work pallet into position for unloading.

As in Fig. 78-3b, the new pallet is shuttled and clamped onto the receiver, while in Fig. 78-3c, the pallet and workpiece are positioned into the

■ **Figure 78-1** A traveling-column machining center.
(Cincinnati Milacron)

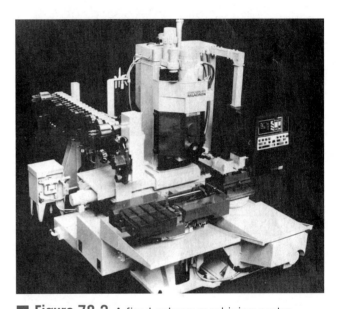

■ **Figure 78-2** A fixed-column machining center.
(Cincinnati Milacron)

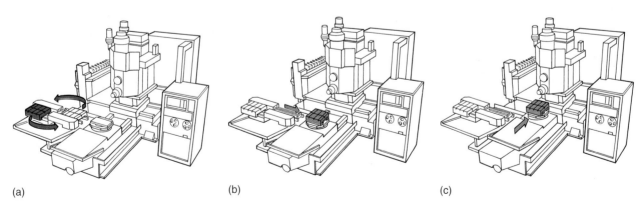

(a) (b) (c)

■ **Figure 78-3** (a) A new pallet ready for loading while the other pallet is ready for unloading; (b) pallet being shuttled onto the receiver; (c) the pallet being moved into machining position. *(Cincinnati Milacron)*

■ **Figure 78-4** A vertical CNC machining center with a tool change storage drum. *(Bridgeport Machines, Inc.)*

■ **Figure 78-5** Typical vertical machining center construction. *(Haas Automation, Inc.)*

programmed machining position. The cycle for changing and shuttling the pallets off and onto the receiver takes only 20 s. If the pallet is equipped with rotary capability, the part can be indexed in 90° increments. This permits machining on multiple surfaces of the part and eliminates the need for separate setups for each surface of cube-shaped parts.

■ **Figure 78-6** A typical rotary accessory. *(Haas Automation, Inc.)*

VERTICAL MACHINING CENTER

The vertical machining center (Fig. 78-4) is a saddle-type construction with sliding bedways that use a sliding vertical head instead of a quill movement. Typical construction is shown in Fig. 78-5. The vertical machining center is generally used to machine flat parts that are held in a vise or simple fixture, and its versatility can be increased by the addition of rotary accessories (Fig. 78-6). The machine control unit must be equipped with a fourth-axis capability to use rotary accessories.

UNIVERSAL MACHINING CENTER

The universal machining center (Fig. 78-7a) combines the features of the vertical and horizontal machining centers. Its spindle can be programmed in both the vertical and horizontal positions (Fig. 78-7b and c). This allows for the machining of all sides of a part in one setup, where normally two machines would be required to finish the part.

Universal machining centers are especially useful for small and medium batch parts such as molds and intricate components. With the addition of accessories such as indexible pallets and rotary-tilt tables, it is possible to machine five or more sides of a part in a single setup. Some examples of the types of parts machined are the mold (Fig. 78-7d) and an aerospace component (Fig. 78-7e).

The advantages of universal machining centers are:

> Eliminate handling and waiting time between machines

> Reduce the number of fixtures and setups required

> Reduced programming time

> Improved product quality

(a)

(b)

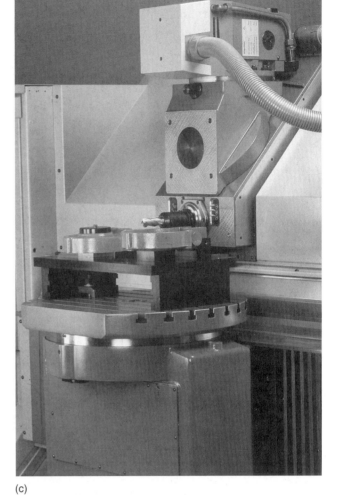

(c)

(d)

■ **Figure 78-7** (a) A universal machining center combines the features of both horizontal and vertical machining centers. Universal machining center spindle positions: (b) vertical; (c) horizontal; (d) Examples of work performed on universal machining centers: (d) mold. *(Deckel Maho, Inc.)*

CNC Machining Centers **645**

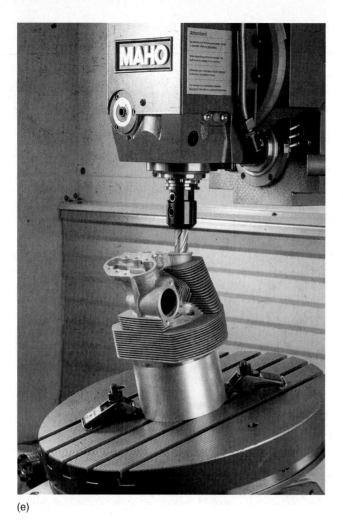

(e)

■ Figure 78-7 (cont.) (e) aerospace component.
(Deckel Maho, Inc.)

> Less work-in-process (WIP) inventory
> Faster product delivery to customers
> Lower manufacturing costs

MAIN OPERATIVE PARTS

The main operative parts of both the vertical and horizontal machining centers (Fig. 78-8a) are basically the same. The position of the machining spindle determines whether it is classified as vertical or horizontal. The main parts of the machine (Fig. 78-8b) are the framework components and those that give the framework its CNC capabilities. The framework components are very similar to the parts of a conventional machine tool and consist of the bed, saddle, column, table, and spindle. The main parts that make the machining center a CNC machine tool are the machine control unit (MCU), servo system, and automatic tool changer (ATC).

▶▶ Machining Center Accessories

There are a number of accessories available for the basic machine that can increase its efficiency and result in an improvement in manufacturing productivity. These accessories may be of two types—those that improve the efficiency or operation of the machine tool and those that involve the holding or machining of the workpiece.

ADAPTIVE CONTROL

When writing a CNC program, it is necessary to include the speeds and feeds for each tool used to manufacture a part. During actual cutting, the operator has the option to override the programmed speeds and feeds, if necessary, when the cutter becomes dull to avoid tool breakage, when excessive material on castings changes the depth of cut, or when the part hardness varies. However, overriding the program should be done only by an experienced operator.

A feature that is becoming very popular is *torque control machining,* which calculates the torque in machining from measurements at the spindle drive motor (Fig. 78-9). This feature increases productivity by preventing or sensing damage to the cutting tool. The torque is measured when the machine is turning but not cutting, and this value is stored in the computer's memory.

As the machining operation begins, the stored value is subtracted from the torque reading at the motor. This gives the net cutting torque, which is compared to the programmed torque or limits stored in the computer. If the net cutting torque goes higher than the programmed torque limits, the computer will reduce the feedrate, turn on the coolant, or even stop the cycle. The feedrate will be lowered whenever the horsepower exceeds the rated motor capacity of the programmed code value.

The system display of three yellow lights advises the operator of the operational conditions in the machine. A left-hand yellow light indicates that the torque control unit is in operation. The middle yellow light indicates that the horsepower limits are being exceeded. The right-hand light comes on when the feedrate drops below 60% of the programmed rate. The meter (Fig. 78-9) indicates the cutting torque (or operational feedrates) as a percentage of the programmed feedrate.

As the tool gets dull, the torque increases and the machine controls will reduce the feedrate and determine what the problem may be. It could be excessive material on the workpiece, or that the tool has become dull or broken. If the tool is dull, the machine will finish the operation, and a new backup tool of the same size will be selected from the storage chain when that operation is to be performed again. If the torque is too great, the machine

■ **Figure 78-8** (a) The main parts of a CNC machining center; (b) the primary components of a machining center. *(Cincinnati Milacron)*

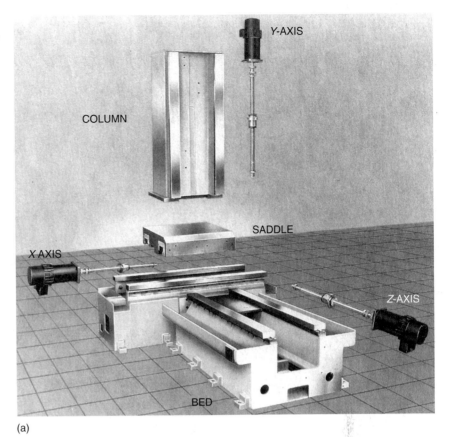

COLUMN

Y-AXIS

SADDLE

X AXIS

Z-AXIS

BED

(a)

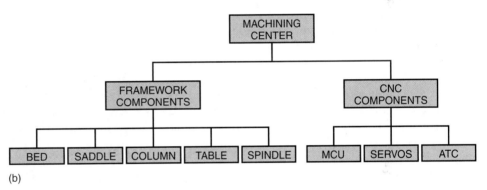

MACHINING CENTER

FRAMEWORK COMPONENTS

CNC COMPONENTS

BED | SADDLE | COLUMN | TABLE | SPINDLE

MCU | SERVOS | ATC

(b)

■ **Figure 78-9** Torque control raises or lowers the feed rate, depending on the depth of cut at any time during the cutting cycle. *(Cincinnati Milacron)*

will stop the operation on the workpiece and program the next part into position for machining.

AUTOMATIC TOOL CHANGERS

Many machining centers used in manufacturing are equipped for automatic numerically controlled tool changing, which is much faster and more reliable than manual tool changing. The larger-capacity horizontal-type tool changers hold up to 200 tools, depending on the type of machine. Tools are held in a storage chain (matrix). Each tool is identified by either the tool number or storage pocket number, and this information is stored in the computer's memory. While one operation is being performed on a workpiece, the tool required for the next

  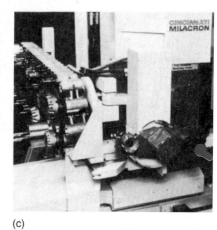

(a)   (b)   (c)

■ **Figure 78-10** (a) The required cutting tool being removed from the storage chain; (b) the tool-change unit swings 90° to change tools in the spindle; (c) the old cutting tool has been removed and replaced with the tool required for the next operation. *(Cincinnati Milacron)*

operation is moved to the pick-up position (Fig. 78-10a), where the tool-change arm removes and holds it.

Immediately after completing the machining cycle, the tool change unit swings 90° to the tool change position (Fig. 78-10b). The tool-change arm then rotates 90° and removes the cutting tool from the spindle. It then rotates 90° and inserts the new cutting tool in the spindle, after which it returns the old cutting tool to its position in the tool carrier (Fig. 78-10c). This whole operation is completed in about 11 seconds or less.

The smaller-capacity, vertical, disk-type tool changer (Fig. 78-11) can hold from 12 to 24 tools. Generally, the next tool is not selected until the current tool is placed back into its assigned pocket on completion of its machining operation. The tool carriage is mounted on a shuttle that slides the carriage next to the tool spindle. The tool pocket for the tool held in the spindle is aligned, and the spindle orients the toolholder (positions the holder key and flange) and the tool lock releases.

The tool changer then rotates to the tool number called. It will rotate in the shortest direction if the tool changer is bidirectional (random selection). The tool lock is energized and the carriage slides out of the way so that machining can continue. Despite the fact that the next tool is not readied until the previous tool has completed its operation, tool-change times range from 2.5 to 6 s.

■ **Figure 78-11** A fully enclosed, completely electric automatic tool changer. *(Haas Automation, Inc.)*

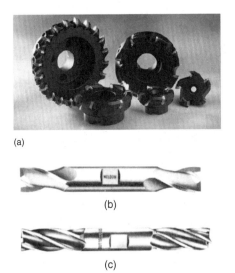

(a)

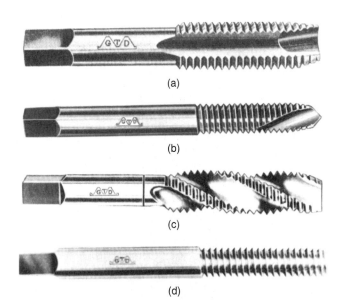

(b)

(c)

■ **Figure 78-12** Common cutting tools used on machining centers. (a) face milling cutters (Carboloy); (b) two-flute end mill (Weldon); (c) four-flute end mill *(Union Butterfield Co.)*

(a)

(b)

(c)

(d)

■ **Figure 78-14** Taps are used to produce a variety of internal threads: (a) gun; (b) stub flute; (c) spiral flute; (d) fluteless. *(Greenfield Industries)*

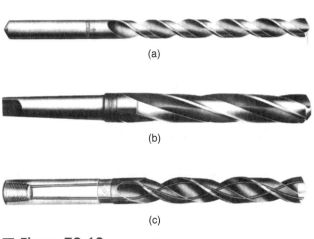

(a)

(b)

(c)

■ **Figure 78-13** Stub drills are widely used on machining centers: (a) a high-helix drill; (b) a core drill; (c) an oil hole drill. *(Cleveland Twist Drill Co.)*

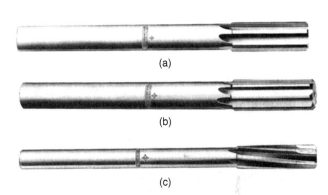

(a)

(b)

(c)

■ **Figure 78-15** Reamers are used to accurately size a hole and produce a good surface finish: (a) rose reamer; (b) fluted reamer; (c) carbide-tipped reamer. *(Cleveland Twist Drill Co.)*

TOOLS AND TOOLHOLDERS

Machining centers use a wide variety of cutting tools to perform various machining operations. These may be conventional high-speed steel tools or insert tools such as cemented carbide, coated carbide, ceramic, cermet, cubic boron nitride (CBN), or polycrystalline diamond. Some of the common tools used on machining centers are end mills (Fig. 78-12), drills (Fig. 78-13), taps (Fig. 78-14), reamers (Fig. 78-15), and boring tools (Fig. 78-16 on p. 650).

Studies show that machining center time consists of 20% milling, 10% boring, and 70% hole-making in an average machine cycle. On conventional milling machines, the cutting tool cuts approximately 20% of the time, while on machining centers, the cutting time can be as high as 75%. The end result is that there is a larger consumption of disposable tools because of decreased tool life, caused by increased tool use.

■ Figure 78-16 Single-point boring tools are used to enlarge a hole and bring it to location. *(Criterion Machine Works)*

■ Figure 78-17 A combination tool is a drilling, chamfering, and threading mill all in one tool. *(Thriller, Inc.)*

COMBINATION TOOLS

In order to improve productivity by applying creative engineering, combination tools have been developed and put into use. If a machining center has helical interpolation capability, one tool can perform drilling, chamfering, and threading operations in a one operational cycle. Fig. 78-17 shows a solid-carbide combination drill/thread tool with a drill tip on the end, a chamfer located at the correct length for the selected application, and a thread mill slightly smaller in diameter than the drill portion. Fig. 78-18 illustrates the sequence of operations for this combination tool, the Thriller®.*

1. The drill point can produce a through hole or a blind hole no deeper than two times the tool diameter [Fig. 78-18(1)].

2. The chamfer is cut, [Fig. 78-18(2)], and the tool is retracted approximately 2½ thread pitches from the bottom of the hole.

3. The tool is fed radially into the wall of the hole to full thread depth during ½ of a turn (180°), while moving ½ of a thread pitch in the −Z axis.

4. Next, the thread is formed by a helical interpolation cycle during one full turn (360°), while moving one thread pitch in −Z axis, [Fig. 78-18(3)].

5. The tool is brought out radially from the wall, to the center of the hole during ½ of a turn (180°), while moving ½ of a thread pitch in the −Z axis.

6. On completion of the cycle, the tool is retracted out of the hole [Fig. 78-18(4)].

This combination drill/thread tool can be used in aluminum, cast iron, and materials that produce easily broken chips. Using a combination tool frees up tool-changer space for other tools and eliminates several tool changes per operation.

TOOLHOLDERS

In order for the wide variety of cutting tools to be inserted into the machine spindle quickly and accurately, all tools must have the same taper shank toolholders to suit the machine spindle. The most common toolholder, with a V-flange and self-releasing taper shank, is used in CNC machining centers (Fig. 78-19a). Only one of the available sizes (which range from No. 30 to 60) can be used on a machine. The size used is determined by the machine capacity and the designed horsepower. The internal receptacle

*®Thriller, Inc.

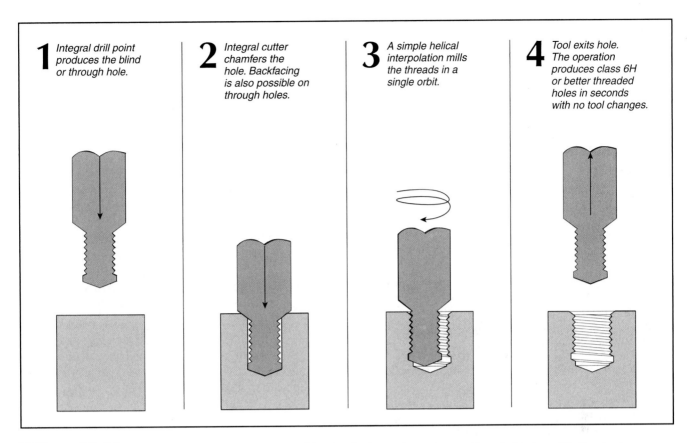

1 Integral drill point produces the blind or through hole.

2 Integral cutter chamfers the hole. Backfacing is also possible on through holes.

3 A simple helical interpolation mills the threads in a single orbit.

4 Tool exits hole. The operation produces class 6H or better threaded holes in seconds with no tool changes.

■ **Figure 78-18** The sequence of operations a combination tool uses to produce a threaded hole.   *(Thriller, Inc.)*

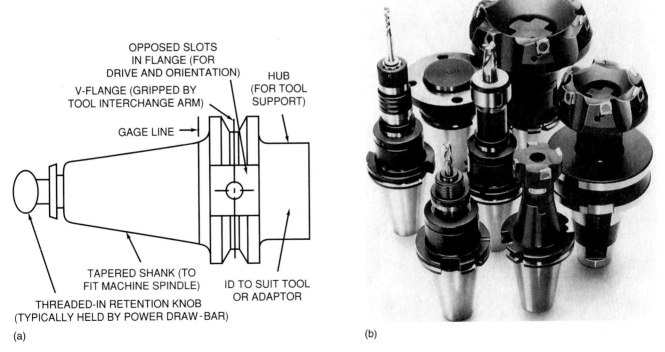

OPPOSED SLOTS
IN FLANGE (FOR
DRIVE AND ORIENTATION)

HUB
(FOR TOOL
SUPPORT)

V-FLANGE (GRIPPED BY
TOOL INTERCHANGE ARM)

GAGE LINE →

TAPERED SHANK (TO
FIT MACHINE SPINDLE)

ID TO SUIT TOOL
OR ADAPTOR

THREADED-IN RETENTION KNOB
(TYPICALLY HELD BY POWER DRAW-BAR)

(a)

(b)

■ **Figure 78-19** CNC toolholders are precision tools designed to be located accurately in a machine spindle by an automatic tool-changing system.   *(Hertel Carbide, Ltd.)*

portion is installed in the quill by the machine tool manufacturer. The toolholder has a V-flange for the tool-change arm to grab, a stud or tapped hole, or some other device for a power draw-bar or other holding mechanism to draw the holder into the spindle (Fig. 78-19b).

When preparing for a machining sequence, the tool assembly drawing is used to select all the cutting tools required to machine the part. Each cutting tool is then assembled off-line in a suitable toolholder and preset to the correct length. Once the cutting tools are assembled and preset, they are loaded into specific pocket locations in the machine's tool-storage magazine, where they are automatically selected as required by the part program.

Work-Holding Devices

Many types of work-holding devices used with conventional machine tools can also be used with machining centers. However, due to the need for quick changeover between parts, certain devices are better choices than others for some applications.

> A *standard step clamp* is used to hold down flat, large parts, but properly locating and tightening are time consuming. A quick-release clamp (Fig. 78-20) may be preferable, especially when clamps have to be temporarily moved during a program stop command (M00) to machine an edge.

> *Plain-style precision vises* (Fig. 78-21a), keyed directly to the table slots, make positioning and clamping accurate and simple. When multiple identical parts must be machined and held in separate vises, a matched set of qualified vises (machined identically to position the solid jaw on the same *Y* axis coordinate) can be used. Qualified vises are also used when a long part requires support on both ends to maintain parallelism.

These vises can be air or hydraulic powered for consistent holding pressure and easy lever or foot pedal control during part changing. When more than one vise is required, especially for holding small parts, as many as 10 vises can be clustered together (Fig. 78-21b). When using double-station cluster vises, a total of up to 20 parts may be held for a machining operation.

> *Vise jaw systems* (Fig. 78-22a) can add versatility and increase the flexibility of a precision vise for holding workpieces. They can be used in both single-station (Fig. 78-22b) and double-station (Fig. 78-22c) vises. Valuable setup and production time is saved because of quick and accurate changeover. A set of master jaws is placed in the vise and the following items are snapped into position as needed: parallels, modular workstops, angle plates, V-jaws, and machinable soft jaws.

> A *table plate* is a flat aluminum plate bolted directly to the machine table. Dowel pin and tapped holes are strategically machined into the plate to permit fastening vises or clamps in many flexible positions other than the limits of the table T-slots.

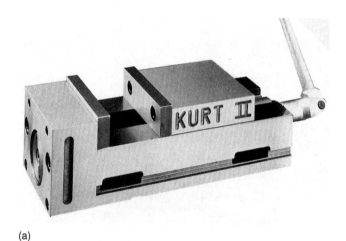

(a)

(b)

■ **Figure 78-20** A quick-release clamping system. *(Royal Products)*

■ **Figure 78-21** (a) A precision machine vise; (b) a double-station cluster vise. *(Kurt Manufacturing)*

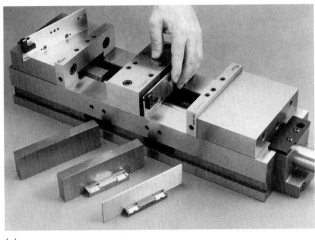

(a)

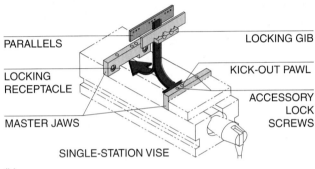

PARALLELS

LOCKING GIB

LOCKING
RECEPTACLE

KICK-OUT PAWL

ACCESSORY
LOCK
SCREWS

MASTER JAWS

SINGLE-STATION VISE

(b)

■ Figure 78-22 A vise jaw fixturing system allows the vise to be quickly changed to suit a variety of different-shaped workpieces: (a) snapping a jaw into place takes only a few seconds; (b) the construction of the vise jaw system; (c) interchangeable vise jaw components. *(Toolex Systems, Inc.)*

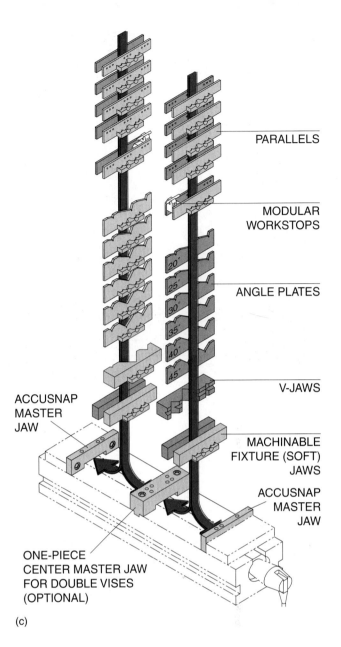

PARALLELS

MODULAR
WORKSTOPS

ANGLE PLATES

V-JAWS

MACHINABLE
FIXTURE (SOFT)
JAWS

ACCUSNAP
MASTER
JAW

ACCUSNAP
MASTER
JAW

ONE-PIECE
CENTER MASTER JAW
FOR DOUBLE VISES
(OPTIONAL)

(c)

> *CNC fixtures* (Fig. 78-23 on p. 654) are used to accurately locate many similar parts and hold them securely for a machining operation. Fixture designs should be kept as simple as possible to make part changes as fast and accurate as possible.

▸▸ Programming Procedures

There are many manufacturers of CNC machine tools throughout the world, and their CNC machine control units could vary from manufacturer to manufacturer. Therefore, it would be impossible to give an example of a programming procedure that would suit all CNC ma-

chines. Since programming can vary slightly from machine to machine, it is important to follow the programming manual supplied for each machine. In this book, the authors have concentrated on two classes of CNC machines: the bench-top teaching model and the standard machine tools most commonly used in schools and machine shops. This will cover the standard basic programming procedures that suit almost any machine control unit, with slight modifications.

 CAUTION *Be sure* the programming format suits your equipment before machining parts.

■ Figure 78-23 CNC fixtures are used to accurately locate a part and hold it securely for machining operations. *(Cincinnati Milacron)*

BENCH-TOP TEACHING MACHINES

Bench-top teaching machines (Fig. 78-24) are ideal for teaching purposes because the instructor and the students find them easy to operate. They are very easy to program and perform similar machining operations as the larger

■ Figure 78-24 Bench-top CNC teaching machines are well suited for teaching and learning purposes. *(Emco Maier Corp.)*

machines but with smaller workpieces and lighter cuts. They are relatively inexpensive and ideal for teaching basic programming procedures.

Simple Programming

The part shown in Fig. 78-25 will be used to introduce *simple programming* in easy-to-understand steps. Each programming step will be explained in detail so that the reader will have a clear understanding of each code, axis movement, etc. and what happens as a result of each programmed step.

Program Notes

1. Program in the absolute mode.
2. All programming begins at the PO *start point*, off the lower edge of the part.

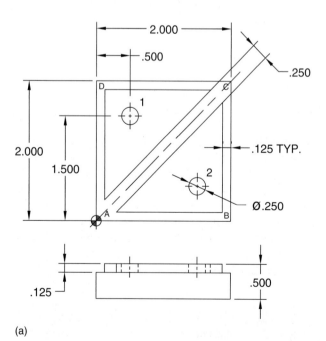

(a)

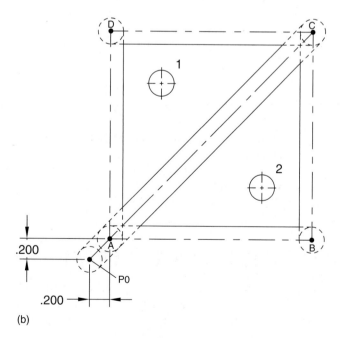

(b)

■ Figure 78-25 A simple CNC programming part; (b) cutting tool path.

3. Use a 2-flute end mill (¼-in. diameter) for all operations, including the two holes.

4. Program counterclockwise (points A-B-C-D-A-C).

5. Terminate program at the last *xy* location and machine zero in the *Z* axis.

6. Material aluminum (CS 500 sf/min).

Programming Sequence

N1 G20	Inch data input	
N2 T01 M06		
	T01	Tool number 1—0.25-dia.—2-flute end mill
	M06	Tool change command
N3 G90 G54 G00 X-0.2 Y-0.2 S2500 M03		
	G90	Absolute programming mode
	G54	Work coordinate system selection
	G00	Rapid motion *XY* to starting point **P0**
	S2500	Spindle speed set to 2500 r/min
	M03	Spindle rotation clockwise (CW)
N4 G43 Z0.1 H01 M08		
	G43	Tool length offset
	Z0.1	Initial level—clearance above top
	H01	Tool length offset 1
	M08	Coolant **ON**
N5 G01 Z-0.125 F10.0		
	G01	Linear interpolation—*Z* axis motion
	Z-0.125	Depth of cut—0.125 in. below the part surface
	F10.0	Cutting federate of 10 in./min
N6 X0 Y0	Tool motion from **P0** to point **A**	
N7 X2.0	Tool motion from point **A** to point **B**	
N8 Y2.0	Tool motion from point **B** to point **C**	
N9 X0	Tool motion from point **C** to point **D**	
N10 Y0	Tool motion from point **D** to point **A**	
N11 X2.0 Y2.0	Tool motion from point **A** to point **C**	
N12 G00 Z0.1	Rapid tool motion 0.1 in. above the part surface	
N13 X0.5 Y1.5	Rapid tool motion to the hole location #1	
N14 G99 G89 R0.1 Z-0.125 P200 F6.0		
	G99	When the bore is completed, retract to level designated by letter R (R0.1)
	G89	Fixed cycle for boring
	R0.1	Start position of cutting motion along *Z* axis
	Z-0.125	Final depth of the bore
	P200	Dwell time at bottom of hole—200 milliseconds
	F6.0	Cutting feedrate of 6 in./min
N15 X1.5 Y0.5	Rapid tool motion to hole #2 + boring cycle	
N16 G80 G00 Z1.0 M09		
	G80	Cancel active fixed cycle (G89)
	G00	Rapid motion to 1.0 in. above the part surface
	M09	Coolant **OFF**

```
N17 G28 Z1.0 M05
    G28          Command to move selected axis to home position
    Z1.0         Z axis selected to move to home position
    M05          Spindle stop
N18 M30          End of program with return to the top
%                End of transmission during communication
```

▸▸ Machining Center Setup

Before setting up the machining center, it is necessary for the operator to become familiar with the control panel and the operational procedures. This includes the use of the different modes and how to use the menus to travel safely from one mode to the next. The operator must understand how to establish the work location to machine zero, set tool length offsets, and test run the program.

When the operator turns on power to the machine, it is necessary to zero out all the axes so that the control knows the location of the machine home position. This can be performed on each axis individually, or machines use an automatic power-up if they are so equipped. If the program is not stored in memory, it must be loaded by the input media available. Based on the manuscript or separate tool sheet, the operator must prepare the tools and use the part-holding method listed by the programmer.

SETTING PART ZERO

Each part has an established part zero, which is not the same as the machine zero. Using the jog mode and an edge finder or dial indicator, locate the part zero position in the *X* and *Y* axes. The distance traveled from machine home is the *work offset distance (position shift offset)* and must be entered on the control's work coordinate page reserved for this information. The distances traveled for the *X* and *Y* axes are entered, while the *Z* axis distance is left at zero (0) if tools are not preset. For preset tools, use the method suggested in the machine's or programmer's manual. Generally, when a machine has multiple work coordinate systems, they are identified by the codes G54 through G59. Such a system defaults automatically to the G54 code when the machine is turned on. If a different work coordinate system is required, this must be indicated at the beginning of the program and the offsets entered under the correct code.

SETTING TOOL LENGTH OFFSET

Starting with an empty automatic tool changer, the operator loads tool #1 by indexing to the proper location of the tool carriage. The tool is placed directly into the spindle and locked in position. Use the jog mode to touch off the tool to Z0 (top surface) of the part. The distance traveled is the Z tool offset and is listed on the control offset page under the offset for tool #1. During actual use, a command of G43 H01 is necessary in the program after the T1 M06 to compensate for this stored information. From there, the tool is loaded into the tool carriage and the process is repeated with each additional tool.

PROGRAM TEST RUN

A part should never be machined without test running the program first. Some controls are equipped with a graphics display, which allows the operator to go through the program steps on the control screen without cutting the part. Since no table movement or tool use is involved, this is a safe way to check a program.

If the machine is not equipped with graphics, another method is to dry run the program without the part in the machine. Use the step/single block mode and feedrate override to slow down the programmed rate. The operator should have a finger on the hold button and know the location of the emergency stop in case there is an error in the program. After the operator is sure that the program is correct, the part can be machined.

STANDARD-SIZE MACHINING CENTER

The part shown in Fig. 78-26 is used to introduce additional machining cycles, including circular and fixed drilling cycles, on a standard-size machining center. An explanation in easy-to-understand steps similar to the one used in the first programming example is provided for clarity. Refer to the *G*- and *M*-code charts in Unit 75.

Program Notes

1. Program in the absolute mode (G90).
2. All programming coordinates should be taken from the zero, or reference, point (*X-Y* zero) at the center of the 1.375-in. radius.
3. Use a 2-flute, 3/8-in. diameter end mill for milling the groove in a clockwise direction (points A-B-C-D-E-F-G-A).

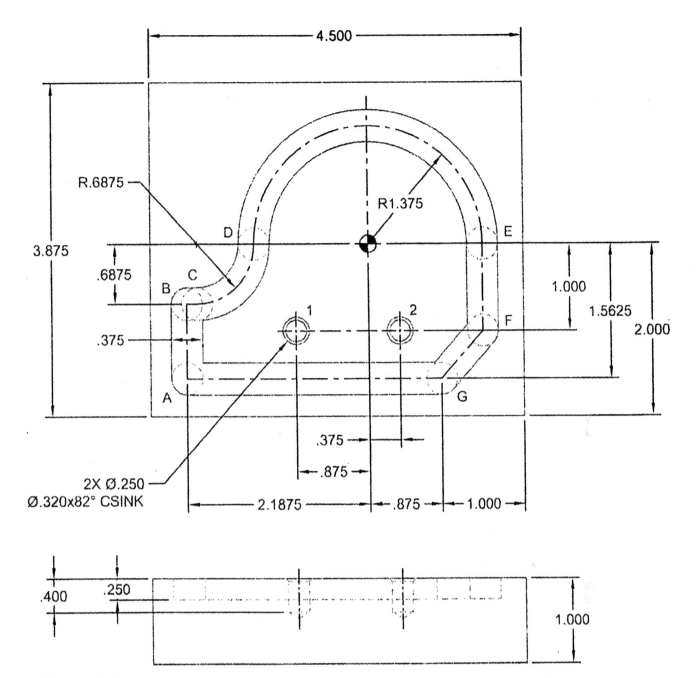

Figure 78-26 A sample machining center part to be used for part programmming. *(Kelmar Associates)*

4. Spot drill, drill, and countersink the holes #1 and #2.

5. Return to the machine *X, Y,* and *Z* zero for part changing at the end of the program (G28 code on some machines).

6. All tool changes should be made at their current location.

7. Material is machine steel (CS 90 sf/min).

Machine Sample Part Program #2

(T01 - 0.375 END MILL)	Comment in the program
N1 G20	Inch data input
N2 G17 G40 G80 G49	Safe block—restores default modes
N3 T01 M06	
T01	Tool number 1—0.375-dia.—2-flute end mill
M06	Tool change command
N4 G90 G54 G00 X-2.1875 Y-1.5625 S1200 M03 T02	
G90	Absolute programming mode
G54	Work coordinate system selection
G00	Rapid motion *XY* to starting point **A**
S1200	Spindle speed set to 1200 r/min
M03	Spindle rotation clockwise (CW)
T02	Prepare next tool for faster tool change
N5 G43 Z0.1 H01 M08	
G43	Tool length offset
Z0.1	Initial level—clearance above top
H01	Tool length offset 1
M08	Coolant **ON**
N6 G01 Z-0.25 F4.0	
N7 Y-0.6875 F10.0	Tool motion from **A** to point **B**
N8 X-2.0625	Tool motion from **B** to point **C**
N9 G03 X-1.375 Y0 I0 J0.6875	Tool motion from **C** to point **D**
N10 G02 X1.375 I1.375 J0	Tool motion from **D** to point **E**
N11 G01 Y-1.0	Tool motion from **E** to point **F**
N12 X0.875 Y-1.5625	Tool motion from **F** to point **G**
N13 X-2.1875	Tool motion from **G** to point **A**
N14 G00 Z0.1 M09	Rapid tool motion 0.1 in. above the part surface—coolant **OFF**
N15 G28 Z0.1 M05	
G28	Command to move selected axis to home position
Z1.0	*Z* axis selected to move to home position
M05	Spindle stop
M16 M01	Optional program stop
(T02 - 0.5 DIA SPOT DRILL)	Comment in the program
N17 T02 M06	
T02	Tool number 2—0.5-dia. spot drill—90°
M06	Tool change command
N18 G90 G54 G00 X-0.875 Y-1.0 S1600 M03 T03	
G90	Absolute programming mode
G54	Work coordinate system selection
G00	Rapid motion *XY* to hole number 1
S1600	Spindle speed set to 1600 r/min

M03	Spindle rotation clockwise (CW)
T03	Prepare next tool for faster tool change

N19 G43 Z1.0 H02 M08

G43	Tool length offset
Z1.0	Initial level—clearance above top
H02	Tool length offset 2
M08	Coolant **ON**

N20 G99 G82 R0.1 Z-0.1 P200 F6.0

G99	When hole is completed, retract to level designated by letter R (R0.1)
G82	Fixed cycle for spot drilling, counterboring, and countersinking
R0.1	Start position of cutting motion along Z axis
Z-0.1	Final depth of the spot drill
P200	Dwell time at bottom of hole—200 milliseconds
F6.0	Cutting feedrate of 6 in./min

N21 X0.375   Repeat fixed cycle at hole location number 2

N22 G80 G00 Z1.0 M09

G80	Cancel active fixed cycle (G82)
G00	Rapid motion to 1.0 in. above the part surface
M09	Coolant **OFF**

N23 G28 Z1.0 M05

G28	Command to move selected axis to home position
Z1.0	Z axis selected as axis to move to home
M05	Spindle stop

N24 M01

(T03 - 0.25 DIA DRILL)   Comment in the program

N25 T03 M06

T03	Tool number 3—0.25-dia. drill
M06	Tool change command

N26 G90 G54 G00 X0.375 Y-1.0 S1500 M03 T04

G90	Absolute programming mode
G54	Work coordinate system selection
G00	Rapid motion XY to hole number 2
S1500	Spindle speed set to 1500 r/min
M03	Spindle rotation clockwise (CW)
T03	Prepare next tool for faster tool change

N27 G43 Z1.0 H03 M08

G43	Tool length offset
Z1.0	Initial level—clearance above top
H03	Tool length offset 3
M08	Coolant **ON**

N28 G99 G81 R0.1 Z-0.475 F8.0

G99	When the bore is completed, retract to level designated by letter R (R0.1)
G81	Fixed cycle for drilling
R0.1	Start position of cutting motion along Z axis
Z-0.475	Final drill depth (0.4 depth + 0.3 * 0.25 dia. = 0.475)
F8.0	Cutting feedrate of 8 in./min

N29 X-0.875	Repeat fixed cycle at hole location number 1
N30 G80 G00 Z1.0 M09	
G80	Cancel active fixed cycle (G81)
G00	Rapid motion to 1.0 in. above the part surface
M09	Coolant **OFF**
N31 G28 Z1.0 M05	
G28	Command to move selected axis to home position
Z1.0	*Z* axis selected to move to home position
M05	Spindle stop
N32 M01	
(T04 - 82-DEG COUNTERSINK)	Comment in the program
N33 T04 M06	
T04	Tool 4 (82° countersink to 0.320 chamfer diameter)
M06	Tool change command
N34 G90 G54 G00 X-0.875 Y-1.0 S1200 M03 T01	
G90	Absolute programming mode
G54	Work coordinate system selection
G00	Rapid motion *XY* to hole number 1
S1200	Spindle speed set to 1200 r/min
M03	Spindle rotation clockwise (CW)
T03	Prepare next tool for faster tool change
N35 G43 Z1.0 H04 M08	
G43	Tool length offset
Z1.0	Initial level—clearance above top
H04	Tool length offset 4
M08	Coolant **ON**
N36 G99 G82 R0.1 Z-0.1841 P200 F5.0	
G99	When the bore is completed, retract to level designated by letter R (R0.1)
G82	Fixed cycle for spot drilling, counterboring, and countersinking
R0.1	Start position of cutting motion along *Z* axis
Z-0.184	Final depth of countersink (0.320 * 0.575 = 0.184)
P200	Dwell time at bottom of hole—200 milliseconds
F5.0	Cutting feedrate of 5 in./min
N37 X0.375	Repeat fixed cycle at hole location number 2
N38 G80 G00 Z1.0 M09	
G80	Cancel active fixed cycle (G82)
G00	Rapid motion to 1.0 in. above the part surface
M09	Coolant **OFF**
N39 G28 Z1.0 M05	
G28	Command to move selected axis to home position
Z1.0	*Z* axis selected to move to home position
M05	Spindle stop
N40 G28 X-0.875 Y-1.0	Return to machine zero in *XY* axes through the current position
N41 M30	End of program with return to the top
%	End of transmission during communication

1. What factors led to the development of the machining center?

Types of Machining Centers

2. Name two types of horizontal spindle machining centers and briefly state the principle of each.

3. How does a vertical machining center provide for Z axis movement?

Main Operative Parts

4. Name two categories of machining center parts.

Machining Center Accessories

5. Briefly describe how a torque-control device operates.

6. For what purpose are automatic tool changers used?

7. List five tools commonly used on machining centers.

8. What is the most common toolholder used on a machining center?

9. How is a tool mounted in the spindle of the machine?

Work-Holding Devices

10. What is the advantage of a quick-release clamp over a standard step clamp?

11. For what application would a set of qualified vises be used?

12. Describe a table plate and state its purpose.

Program Procedures

13. State two main advantages of bench-top teaching-size CNC machines.

14. Name and list the function of each *G*-code found in the sample teaching-size machine program.

15. Name and list the function of each *M*-code in the question #14 program.

Machining Center Setup

16. Name three important factors an operator must understand in order to set up a machining center.

17. What is meant by *work offset distance?*

18. What is the safest method to find out if a program is correct so that an accident does not occur?

UNIT 79

CAD/CAM

OBJECTIVES

After completing this unit, you will be able to:

1 Define CAD and CAM

2 Explain the purposes of solid modeling

3 Know the differences between Cadworthy and Camworthy

4 Describe process chaining and list the steps in the process

5 State the purposes of pocket, contour, drill, and surface toolpaths

CAD/CAM is the *marriage* of the computerized forms of drafting/design and manufacturing. Computer graphics is the CAD part and NC is the CAM (or chip-making part) of CAD/CAM. A personal computer can control banks of CNC-controlled machines, and other tools, moving the product from machine to machine.

CAM is a rapidly spreading technology, linking the programming abilities of the modern computer to material processing CNC machines, lathes, mills, grinders, punches, etc. CAD/CAM is the entire process of the *art to part* concept. It covers the creation of a part from the initial design to the finish-machined part produced on a numerically controlled machine tool.

During the past 10 years, probably no other phase of manufacturing has progressed as quickly as PC-based CAD/CAM technology. The CAD/CAM systems currently in industry have evolved into the technology that excels in handling complex shapes and multi-tasking machines.

CAD/CAM systems allow for detail parts and assemblies to be designed in an interactive environment with design geometry. The amount of mathematical computational power contained in CAD/CAM systems is increasingly exponentially.

▶▶ The CAD Side of CAD/CAM

Computer-Aided Design (CAD) is intended to apply computers to modeling and communication of designs. The combination is to use computers to automate production drawings or diagrams and lists or bills of material in a design—at a more advanced level, to provide new techniques and functionalities that give the designer capabilities to smooth the design process.

▶▶ CAD Development

Commercial CAD system development has primarily centered around modeling techniques to represent conventional drawings and in the production for solid models. Computer-aided design should therefore provide the means for the designer to do the job faster and more accurately. This implies that computer-based techniques for the analysis and simulation of the design, and for the generation of manufacturing code, should be closely integrated with the techniques for modeling the shape and structure of the design.

▶▶ Systems Presently in Use

The CAD/CAM systems presently used no longer consist of only 2D or 3D wireframe entities (units) and simple toolpaths. Software now enables the manufacturers to produce parts faster, more economically, and of more consistent quality. Manufacturers continually revise modeling technology, improved machining capabilities, and automated various machining processes.

▶▶ Solids

Though not everyone is knowledgeable about solid modeling, it is rapidly becoming the industry-modeling standard because solids (Fig. 79-1) provide a more accurate picture of where the material lies, as opposed to wireframe or surface data (Fig. 79-2). Imagine a surface data as a sheet of paper, as opposed to solids, and it is easy to understand that a surface data does not have any thickness. Try wireframe data; it is a geometric model that describes 3D geometry by outlining its edges, similar to a stick figure. Lines that are located in the back of an object show through from the front. All lines are constantly visible in a wireframe (Fig. 79-3).

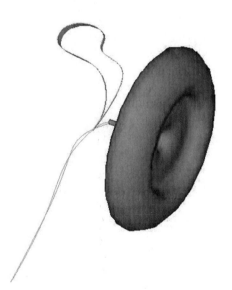

■ **Figure 79-1** Solid model of a flower (helps with visualization of part).

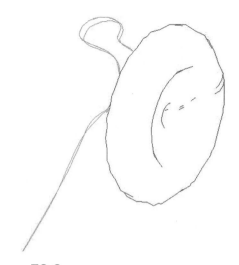

■ **Figure 79-2** Wireframe of a flower with hidden lines removed.

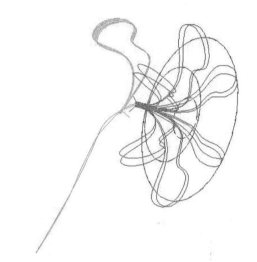

■ **Figure 79-3** Wireframe of a flower.

▶▶ Solid Modeling and CAD/CAM

Solid modeling is no longer a tool available only to large manufacturers with the constant need to design in 3D. PC-based solid modeling and machining has given the CNC machining industry a tool that has helped level the playing field for many shops. As a stand-alone system or as an integrated part of a versatile CAM package, solid modeling capability increases the adaptability of almost any machine shop.

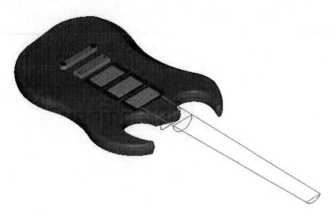

■ Figure 79-4 Solid and wireframe geometry combination.

■ Figure 79-5 Solid model of a cavity for a plastic mold ready for machining.

Many shops are already skilled in machining wireframe and surface models (Fig. 79-4). Adding solids to the mix gives a shop the capability to use the data it wants, a wireframe taken from the solid, surfaces taken from the solid, or the solid itself (Fig. 79-5). This gives the shop more choices for programming a part.

▶▶ How CAD/CAM Works

The following is a generic description of the steps in the CAD/CAM process. This is referred to as the *art-to-part* process. It begins with the art (drawing/design) and ends with a machined part.

GEOMETRY INPUT

The first step of the CAD/CAM processing is the inputting of the geometry.

> This can be accomplished by creating the geometry in the CAD side of the CAD/CAM system.

> The geometry can also be imported via translators (IGES, DXF, ASCII, Parasol, STL, VDA, or SAT) from another CAD system.

> The geometry imported can be in the form of 2D, 3D wireframe, 3D surfaces, or solids.

> Some geometries are imported in the form of *XYZ* coordinates of points. The points are then converted to splines, from which surfaces are formed.

▶▶ Cadworthy Versus Camworthy Geometry

When dealing with geometry in a CAD/CAM system, it is important to understand the difference in *cadworthy* and *camworthy* geometries. Camworthy geometry is the traditional type of geometry, and its purpose is to communicate design intent. A camworthy drawing is created from a machining or production point of view. Cadworthy geometry is created from a CAD, or drafting, point of view. All details and edges of the part are represented (Fig. 79-6).

The purpose of a camworthy drawing is strictly for the CNC manufacturing machine use (Fig. 79-7).

> It contains fewer lines than those needed for a cadworthy drawing. Only the lines needed to create a toolpath are included.

> This makes the toolpath creation easier and sometimes less confusing. Sometimes the geometry is modified to control a toolpath.

> This mainly applies to geometry that consists of 2D, wireframe, and surfaces. Solids usually do not require such modification (Fig. 79-8).

> To make a camworthy drawing from a cadworthy drawing, the geometry is edited to be sure that the end points are connected and that no gap remains.

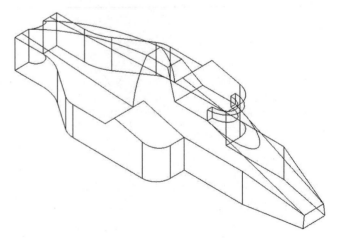

■ Figure 79-6 Cadworthy wireframe of a car body (all lines visible).

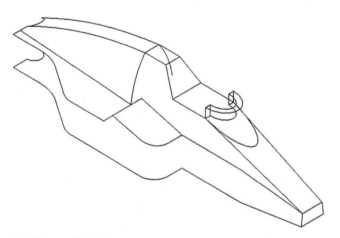

Figure 79-7 Camworthy wireframe of a car body (only the geometry needed for machining is included).

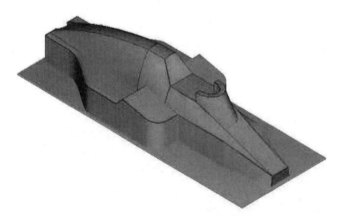

Figure 79-8 Solid model of a car body.

⏩ Defining Where the Cut Takes Place

When the geometry has been entered and prepared for machining, the next step is selecting the order in which the geometry is to be machined. Many software programs call this *process chaining,* or where the cuts are to take place (Fig. 79-9).

After the geometry is prepared for machining (made camworthy), the geometry is evaluated as to the appropriate toolpath type to assign to the segments of geometry. For instance, if a cavity needs to be cut into the stock, it is assigned a pocket toolpath; if the stock needs a hole, a drill toolpath is assigned. There are many different toolpath types in the various CAD/CAM softwares.

HOW THE CUT TAKES PLACE

After the toolpath has been assigned to the geometry, the next step is to tell how the cuts are to take place. Parameters are used to identify things such as what tool to use, the

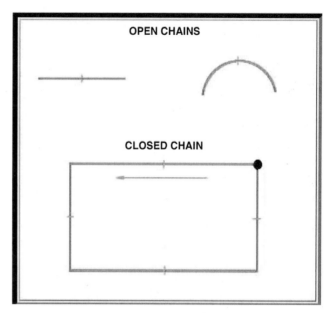

Figure 79-9 Process chaining determines the geometric boundary and cutting direction.

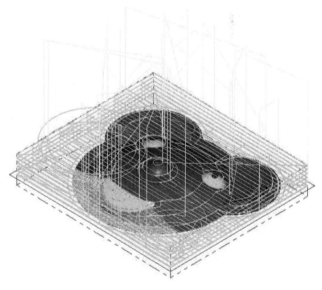

Figure 79-10 Graphic of pocketing parameter screen.

size of the tool, speeds, feeds, the depth of cuts (up and down), the width of cuts (side to side), and cutter compensation.

VIEWING AND VERIFYING THE TOOLPATH

When the parameters have been set, the toolpath is then displayed on the screen in a process called *backplotting* (Fig. 79-10). The toolpath can be displayed showing the

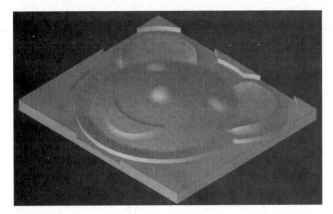

■ Figure 79-11 Graphic of general parameter screen (this is common for all toolpaths).

centerlines or as a solid model with the material removed or being removed. This onscreen visualization helps to catch errors before the actual cutting of a part (Fig. 79-11).

CONVERTING THE TOOLPATH TO *G*- AND *M*-CODES

When all the toolpaths have been assigned and visually verified, the next step is to turn the toolpath files into *G*- and *M*-codes by a process called post processing. The post processor converts a toolpath file to an NC file that is customized for a particular machine control. It is possible to process one toolpath through several different post processors to be run on several different brands of controls.

DOWNLOADING THE PROGRAM TO THE MACHINE

Once the toolpath is post processed, the next step is to download the *G*- and *M*-codes (the program) into the NC controller. This can be done via a direct cable connection if networked. The files can also be loaded via a floppy disk. The drawback of the floppy disk is that many of the CAM-generated programs are much too big to fit on a floppy disk.

SETTING UP THE MACHINE

Once the program is loaded into the controller, the part must be securely held in a holding fixture, and then the machine must be set up (Fig. 79-12).

> This requires setting the *home position* on the machine at the same point as the program home position or origin.

■ Figure 79-12 Special fixture for holding a part for machining.

■ Figure 79-13 Using an edge finder to set the home position or part zero.

> The standard for the location of this point is with the *X* and *Y* zero positions at the part's lower left corner and *Z* zero on the top surface (Fig. 79-13).

> If it is a circular part, the *X* and *Y* zero is on the center and the *Z* zero is on the top surface.

> The tip of the tools or tool zero positions must also be set. The tool zero must also be set for each tool used.

> Next, it is necessary to set the size and tool number of all the tools to be used, as well as any length effects if used.

When the complete machine setup is made, the machine is ready to run the part. Some of the CNC controls also allow the user to graphically verify the program on the control screen, if desired.

▶▶ Running the Part

With the machine setup, it is time to press the cycle start button and begin the machining process. Once the machining is finished, the final steps in the art-to-part, or CAD/CAM, process have been completed.

▶▶ Toolpaths and Functions

TOOLPATHS

Pocket Toolpath

A pocket toolpath's purpose is to remove material from a cavity in the stock (Fig. 79-14). Raised material defined by an enclosed contour inside a pocket, a boss, may be left standing in the cavity, if desired. This martial left standing is commonly referred to as *islands* (Fig. 79-15).

Contour Toolpath

A contour toolpath's purpose is to cut along a line/surface to form a shape. The cutter can follow to the right or left of the line/surface or on the line/surface. A contour toolpath can be used for lettering or engraving if the cutter follows on the line.

■ **Figure 79-14** Graphic representation of pocket toolpath.

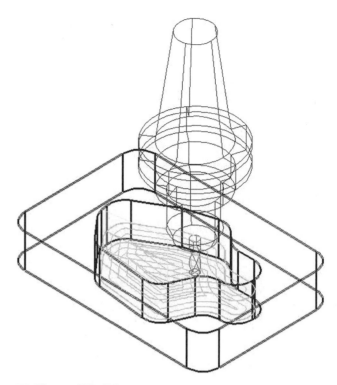

■ **Figure 79-15** Solid model verification of a pocket toolpath with islands.

Drill Toolpath

A drill toolpath's purpose is to create holes and/or enlarge them. With a drill toolpath, the user selects the type of cycle desired, such as a center drill, drill/dwell, peck drill, deep hole drill, tap, or bore.

Surface Toolpath

The purpose of a surface toolpath is to remove the material from a part that is designed as a surface or solid. It allows the user to cut single or multiple surfaces by using the natural surface shape to define the cutter path, resulting in a smoother finish. Flexible surface machining provides finishing suites of tools to allow the user to choose the best method for a specific part. Special cycles are provided for surface cleanup machining, since leftover material requires extra-hard handwork and time. These cycles automate leftover removal, leaving a finer finish and thereby reducing costs.

FUNCTIONS

Constant Volume Removal

One of the functions that affect the cutting process is *constant volume removal*. This function cannot be effectively

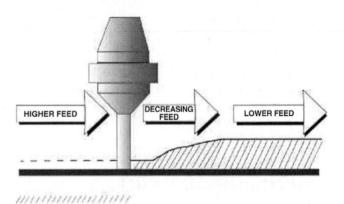

■ **Figure 79-17** High-feed machining.
(Mastercam/CNC software.)

The feedrate optimization feature can greatly reduce cycle time by varying the feedrate based on the volume of material being removed. The parameters can easily set upper and lower limits to take advantage of each machine tool's characteristics (Fig. 79-16).

Optimize the Entire Shop

Each machine tool in the shop has its own maximum efficiency zone, in which it runs as fast as possible without damaging itself or producing off-tolerance parts. Feedrate optimization helps find that zone and allows the user to save all the data for later use.

Smart Cornering

Sharp direction changes at high feedrates can cause excessive servo lag, tool deflection, poor finish, and unnecessary machine tool wear. Feedrate optimization helps solve these problems. Based on the part and machine tool characteristics, it adjusts the feedrate around corners and small radii. It also gradually slows the feedrate as the tool approaches a corner and gradually speeds back up as the cut leaves for a smoother entry and exit (Fig. 79-17).

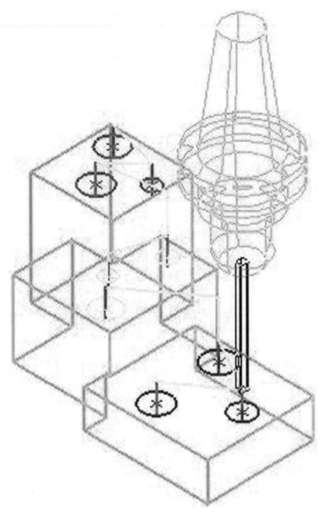

■ **Figure 79-16** Graphic representation of a drill toolpath.

achieved with traditional machining methods. This feature alone can save time and money and can optimize each of the CAD/CAM-controlled machines in the shop. *Machining strategies* refers to toolpath movement and *remachining* refers to material removal and finishing.

1. What is CAD?

2. What is CAM?

3. Describe some of the features of the CAD/CAM systems presently in use.

4. What is *solid modeling* and what is its effect on machine shops?

5. List three ways geometry can be inputted in the CAD/CAM process.

6. What are the differences between *Cadworthy* and *Camworthy* geometries in a CAD/CAM system?

7. Define process chaining and summarize the steps in the process.

8. What is the main purpose of the following?

   a. pocket toolpath
   b. contour toolpath
   c. drill toolpath
   d. surface toolpath

Section 15

Cincinnati Milacron; Brownie Harris/Tony Stone Images; Tom Tracy/FPG International

GRINDING

Early humans used abrasive stones to sharpen tools and produce a smooth surface. Over the years, the manufacture of abrasives and grinding tools has developed gradually to produce stronger and better abrasives. The use of abrasives in modern industry has contributed much to our mass-production methods. Because of abrasives and modern grinding machines, it is possible to produce parts and products to the close tolerances and high surface finishes required by industry and the machine tool trade.

In order to function properly, an abrasive must have certain characteristics:

1. It must be harder than the material being ground.
2. It must be strong enough to withstand grinding pressures.
3. It must be heat-resistant so that it does not become dull at grinding temperatures.
4. It must be friable (capable of fracturing) so that, when the cutting edges become dull, they will break off and present new sharp surfaces to the material being ground.

UNIT 80

Types of Abrasives

OBJECTIVES

After completing this unit, you will be able to:

1 Describe the manufacture of aluminum oxide and silicon carbide abrasives

2 Select the proper grinding wheel for each type of work material

3 Discuss the applications of grinding wheels and abrasive products

Abrasives may be divided into two classes: natural and manufactured.

Natural abrasives, such as sandstone, garnet, flint, emery, quartz, and corundum, were used extensively prior to the early part of the 20th century. However, they have been almost totally replaced by manufactured abrasives with their inherent advantages. One of the best natural abrasives is diamond but, because of the high cost of industrial diamonds (bort), its use in the past was limited mainly to grinding cemented carbides and glass and sawing concrete, marble, limestone, and granite. However, due to the introduction of synthetic or manufactured diamonds, industrial natural diamonds will become cheaper in cost and will be used on many more grinding applications.

Manufactured abrasives are used extensively because their grain size, shape, and purity can be closely controlled. This uniformity of grain size and shape, which ensures that each grain does its share of work, is not possible with natural abrasives. There are several of manufactured abrasives: aluminum oxide, silicon, carbide, boron carbide, cubic boron nitride, and manufactured diamond.

▶▶ Aluminum Oxide

Aluminum oxide is probably the most important abrasive, since about 75% of the grinding wheels manufactured are made of this material. It is generally used for high-tensile-strength materials, including all ferrous metals except cast iron.

Aluminum oxide is manufactured with various degrees of purity for different applications; the hardness and brittleness increase as the purity increases. Regular aluminum oxide (Al_2O_3) is about 94.5% pure and is a tough abrasive capable of withstanding abuse. It is a grayish color and is used for grinding strong, tough materials, such as steel, malleable and wrought iron, and tough bronzes.

Aluminum oxide of about 97.5% purity is more brittle and not as tough as the regular aluminum oxide. This gray abrasive is used in the manufacture of grinding wheels for centerless, cylindrical, and internal grinding of steel and cast iron.

The purest form of aluminum oxide for grinding wheels is a white material that produces a sharp cutting edge when fractured. It is used for grinding the hardest steels and stellite and for die and gage grinding.

Figure 80-1 Aluminum oxide is produced in an arc-type furnace. *(The Carborundum Company)*

MANUFACTURE OF ALUMINUM OXIDE

Bauxite ore, from which aluminum oxide is made, is mined by the open-pit method in Arkansas and in Guyana, Suriname, and French Guiana (South America). The bauxite ore is usually calcined (reduced to powder form) in a large furnace, where most of the water is removed. The calcined bauxite is then loaded into a cylindrical, unlined steel shell about 5 ft (1.5 m) in diameter by 5 ft (1.5 m) deep (Fig. 80-1). This furnace is open at the top and lined with carbon bricks at the bottom. Two or three electrodes of carbon or graphite project into the open top of the furnace. During the operation, the outside of the furnace is cooled by a circumferential spray of water. The furnace is half-filled with a mixture of bauxite, coke screenings, and iron borings in the proper proportions. The coke is used to reduce the impurities in the ore to their metals, which combine with the iron and sink to the bottom of the furnace. The electrodes are then lowered onto the top of the charge, a starting batch of coke screenings is placed between the electrodes, and the current is applied. The coke is rapidly heated to incandescence and the fusion of the bauxite starts. After a small amount of bauxite is fused, it becomes the conductor and carries the current. After the fusion of the bauxite has begun, more bauxite and coke screenings are added, and the process continues until the shell is full. During the entire operation, the height of the electrodes is automatically adjusted to maintain a constant rate of power input. When the furnace is full, the power is shut off, and the furnace is allowed to cool for about 12 h. The fused ingot is removed from the shell and allowed to cool for about a week. The ingot is broken up and fed into crushers; the material then is washed, screened, and graded to size. Stronger aluminum oxide is now available through sintering or seeded gel (SG) abrasives.

►► Silicon Carbide

Silicon carbide is suited for grinding materials that have a low tensile strength (aluminum, brass, and bronze) and high density, such as cemented carbides, stone, and ceramics. It is also used for cast iron and most nonferrous and nonmetal materials. It is harder and tougher than aluminum oxide. Silicon carbide may vary in color from green to black. Green silicon carbide is used mainly for grinding cemented carbides and other hard materials. Black silicon carbide is used for grinding cast iron and soft nonferrous metals such as aluminum, brass, and copper. It is also suitable for grinding ceramics.

MANUFACTURE OF SILICON CARBIDE

A mixture of silica sand and high-purity coke is heated in an electric resistance furnace (Fig. 80-2). Sawdust is added to produce porosity in the finished product and to permit the escape of the large volume of gas formed during the operation. Sodium chloride (salt) is added to assist in removing some of the impurities.

Figure 80-2 Silicon carbide is produced in a resistance-type furnace. *(The Carborundum Company)*

The furnace is a brick-lined rectangle about 50 ft (15 m) long, 8 ft (2.5 m) wide, and 8 ft (2.5 m) high. It is open at the top. One or more electrodes protrude from each end of the furnace. The mixture of sand, coke, and sawdust is loaded into the furnace to the height of the electrodes. A granular core of coke is placed around the electrodes for the full length of the furnace. The core is covered with more sand, coke, and sawdust mixture and heaped to the top of the furnace. The current is then applied to the furnace and the voltage closely regulated to maintain the desired rate of power input. The time required for this operation is about 36 h.

After the furnace has cooled for about 12 h, the brick sidewalls are removed and the unfused mixture falls to the floor; the silicon carbide ingot can then cool more rapidly. After cooling for several days, the ingot is broken up and the silicon carbide removed. Care must be taken to remove the pure silicon carbide, since the outer layer has not been properly fused and is not usable. The inner core surrounding the electrodes is also unusable, since it is only graphitized coke.

The resultant silicon carbide is then crushed, treated with acid and alkalis to remove any remaining impurities, screened, and graded to size.

▶▶ Zirconia-Aluminum Oxide

One of the more recent additions to the abrasive family is *zirconia-aluminum oxide*. This material, containing about 40% zirconia, is made by fusing zirconium oxide and aluminum oxide at extremely high temperatures [3450°F (1900°C)]. This is the first *alloy abrasive* ever produced. It is used for heavy-duty rough and finish grinding in steel mills, for snagging in foundries, and for the rapid rough and finish grinding of welds in metal fabrication plants. The performance of zirconia-aluminum abrasive is superior to that of standard aluminum oxide for rough grinding and snagging operations because the actions of the two grain types are quite different.

During any grinding process, much frictional heat is generated, which breaks down the organic bond (rubber, shellac, or phenalic) on the wheel or the coated disk and lets the wheel "self-dress." Standard aluminum oxide abrasive wheels will, at first, penetrate the surface of the workpiece and remove the metal effectively. Soon the abrasive grains will become dull; the wheel will lose its penetrating power and ride on the surface of the work. Frictional heat builds up, which softens the organic bond, and the entire grain is expelled after about only 25% to 30% of its life.

With the newer zirconia-alumina abrasive, the grain action is quite different. The cutting action starts in the same manner, but when the abrasive grain starts to dull, microfracturing takes place just below the abrasive grain surface. As the small particles break off, sharp new cutting points are generated while the grain remains secured in the bond. As the small broken-off particles of abrasive are ejected, they take with them much of the heat generated by the grinding action. Due to the cooler cutting action of this abrasive, sharp new cutting edges are produced many times before the grain is finally expelled from the bond and only after about 75% to 80% of the abrasive grain has been used. Wheels and disks made with zirconia-alumina base will last from two to five times longer than standard aluminum oxide wheels, depending on the application.

Zirconia-alumina offers several advantages over standard abrasives for heavy-duty grinding:

> Higher grain strength

> Higher impact strength

> Longer grain life

> Maintains its shape and cutting ability under high pressure and temperature

> Higher production per wheel or disk

> Less operator time spent changing wheels or disks

▶▶ Boron Carbide

Another of the new abrasives is *boron carbide*. It is harder than silicon carbide and, next to diamond, it is the hardest material manufactured. Boron carbide is not suitable for use in grinding wheels and is used only as a loose abrasive and a relatively cheap substitute for diamond dust. Because of its extreme hardness, it is used in the manufacture of precision gages and sand blast nozzles. Crush dressing rolls made of boron carbide have proved superior to tungsten carbide rolls for the dressing of grinding wheels on multiform grinders. Boron carbide is also widely accepted as an abrasive used in ultrasonic machining applications.

MANUFACTURE OF BORON CARBIDE

Boron carbide is produced by dehydrated boric acid being mixed with high-quality coke. The mixture is heated in a horizontal steel cylinder, which is completely enclosed except for a hole in each end to accommodate a graphite electrode and vent holes to prevent the escape of the gases formed. The outside of the furnace is sprayed with water to prevent the shell from melting. During the heating process, air must be excluded from the furnace. This is done by dampening the mixture with kerosene, which will volatize and expel the air from the furnace as it is heated. A high current at low voltage is applied for about 24 h, after which the furnace is cooled. The resulting product, boron carbide, is a hard, black, lustrous material.

CUBIC BORON NITRIDE

One of the more recent developments in the abrasive field has been the introduction of cubic boron nitride. This synthetic abrasive has hardness properties between silicon carbide and diamond. The crystal known as Borazon® CBN (cubic boron nitride) was developed by the General Electric Company in 1969. This material is capable of grinding high-speed steel with ease and accuracy and is superior to diamond in many applications.

Cubic boron nitride is about twice as hard as aluminum oxide and is capable of withstanding grinding temperatures of up to 2500°F (1371°C) before breaking down. CBN is cool-cutting and chemically resistant to all inorganic salts and organic compounds. Because of the extreme hardness of this material, grinding wheels made of CBN are capable of maintaining very close tolerances. These wheels require very little dressing and are capable of removing a constant amount of material across the face of a large work surface without having to compensate for wheel wear. Because of the cool-cutting action of CBN wheels, there is little or no surface damage to the work surface.

Manufacture

Cubic boron nitride is synthesized in crystal form from hexagonal boron nitride ("white graphite") with the aid of a catalyst, heat, and pressure (Fig. 80-3a). The combination of extreme heat [2725°F (1496°C)] and tremendous pressure (47,540,000 psi) on the hexagonal boron nitride and the catalyst produces a strong, hard, blocky crystalline structure with sharp corners known as CBN (Fig. 80-3b).

There are two types of CBN:

> *Borazon CBN* is an uncoated abrasive that can be used on plated mandrels and in metal-bonded grinding wheels. This type of wheel is used for general-purpose grinding and for internal grinding on hardened steel.

> *Borazon Type II CBN* is nickel-plated grains of cubic boron nitride used in resin bonds for general-purpose dry and wet grinding of hardened steel. Uses of these wheels range from resurfacing blanking dies to the sharpening of high-speed steel end mills.

►► Manufactured Diamonds

Diamond, the hardest substance known, was primarily used in machine shop work for truing and dressing grinding wheels. Because of the high cost of natural diamonds, industry began to look for cheaper, more reliable sources.

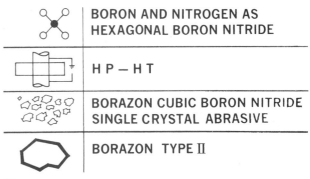

BORON AND NITROGEN AS HEXAGONAL BORON NITRIDE	
H P – H T	
BORAZON CUBIC BORON NITRIDE SINGLE CRYSTAL ABRASIVE	
BORAZON TYPE II	

(a)

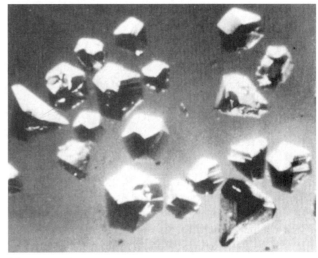

(b)

(c)

■ **Figure 80-3** (a) Making Borazon (cubic boron nitride); (b) crystals of cubic boron nitride *(Diamond Innovations, at www.abrasivesnet.com)*; (c) the SG abrasive grain particles (left) are more uniform in shape and size than the aluminum oxide grain (right). *(Norton Company)*

In 1954, the General Electric Company, after four years of research, produced Man-Made™ diamonds in its laboratory. In 1957, the General Electric Company, after more researching and testing, began the commercial production of these diamonds.

Many forms of carbon were used in experiments to manufacture diamonds. After much experimentation with various materials, the first success came when carbon and iron sulfide in a granite tube closed with tantalum disks were subjected to a pressure of 66,536,750 psi and temperatures between 2550°F and 4260°F (1400°C and 2350°C). Various diamond configurations are produced by using other metal catalysts such as chromium, manganese, tantalum, cobalt, nickel, or platinum in place of iron. The temperatures used must be high enough to melt the metal saturated with carbon and start the diamond growth.

DIAMOND TYPES

Because the temperature, pressure, and catalyst-solvent can be varied, it is possible to produce diamonds of various sizes, shapes, and crystal structure best suited to a particular need.

Type RVG Diamond

The RVG type of manufactured diamond is an elongated, friable crystal with rough edges (Fig. 80-4a). The letters RVG indicate that this type may be used with a resinoid or vitrified bond and is used for grinding ultrahard materials such as tungsten carbide, silicon carbide, and space-age alloys. The Type RVG diamond may be used for wet and dry grinding.

Type MBG-II Diamond

A tough and blocky-shaped crystal, the MBG-II type is not as friable as the RVG type and is used in metal-bonded grinding (MBG) wheels (Fig. 80-4b). It is used for grinding cemented carbides, sapphires, and ceramics as well as in electrolytic grinding.

Type MBS Diamond

The MBS type is a blocky, extremely tough crystal with a smooth, regular surface that is not very friable (Fig. 80-4c). It is used in metal-bonded saws (MBS) to cut concrete, marble, tile, granite, stone, and masonry materials.

Diamonds may be coated with nickel or copper to provide a better holding surface in the bond and to prolong the life of the wheel.

▶▶ Ceramic Aluminum Oxide

In 1988, a new abrasive product, ceramic aluminum oxide, known as SG abrasive, was introduced by the Norton Company. This material greatly outperforms conventional aluminum oxide wheels in the grinding of tough alloys and other hard ferrous and nonferrous metals.

Aluminum oxide grains are made from fused aluminum oxide, which is then crushed to the desired particle size. This produces a grain having very few crystal particles. With this type of grain, as much as one-fifth of the grinding surface of the grain can be lost when a crystal particle breaks away after it becomes dull.

On the other hand, Norton SG abrasive is made by a nonfused process. Thousands of submicron-sized particles are sintered to provide a single abrasive grain of more uniform shape and size, having vastly more cutting edges that remain sharp as they fracture (Fig. 80-3c). This self-sharpening feature combats friction heat caused by the dulling of the standard aluminum oxide grains.

Norton SG abrasive is harder than aluminum oxide and zirconia alumina but not as hard or as long-lasting as

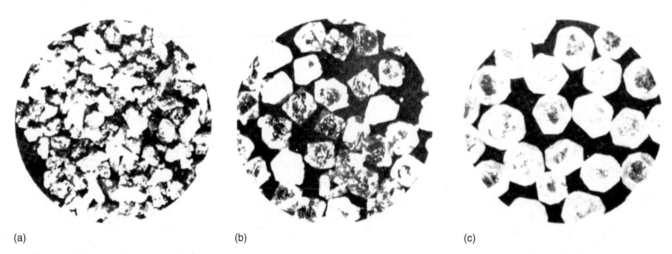

(a)   (b)   (c)

■ **Figure 80-4** (a) Type RVG diamond is used to grind ultrahard materials; (b) Type MBG-II is a tough diamond crystal used in metal-bonded grinding wheels; (c) Type MBS is a very tough, large crystal used in metal-bonded saws. *(Diamond Innovations, at www.abbrasivesnet.com)*

cubic boron nitride or other superabrasive products, which often require special machines for their use. Wheels made of SG abrasive are well suited to CNC grinding because of their cool-grinding feature and their resistance to loading and wear. This combination results in fewer wheel changes, less wheel dressing, higher productivity, and therefore lower labor costs.

When selecting the proper SG wheel for the job, remember that a higher surface finish is created by a dull grain. Since SG grains remain sharp far longer, a finer grit size than that of the aluminum oxide wheels should be selected.

SG AND CBN ABRASIVES

The combination of the technologies of CBN and SG abrasives is used to produce the vitrified CVSG abrasive grinding wheel. This wheel provides most of the high material-removal rates and low wheel wear of CBN yet can be trued with single point diamond tools. CVSG wheels can be used for hard-to-grind materials from 20 to 70 Rc and exotic nickel, and titanium hard-spray flame materials. They are free-cutting, allow increased depths of cuts and feedrates, reduce burning, and lower grinding costs.

Advantages of SG Abrasives over Conventional Abrasives

> They last 5 to 10 times longer than conventional wheels.
> Metal-removal rates are doubled.
> Heat damage to the surface of very thin workpieces is reduced.
> Grinding cycle time is reduced.
> Dressing time is reduced as much as 80%.

▶▶ Abrasive Products

After the abrasive has been produced, it is formed into products such as grinding wheels, coated abrasives, polishing and lapping powders, and abrasive sticks, all of which are used extensively in machine shops.

▶▶ Grinding Wheels

Grinding wheels, the most important products made from abrasives, are composed of abrasive material held together with a suitable bond. The basic functions of grinding wheels in a machine shop are:

1. Generation of cylindrical, flat, and curved surfaces
2. Removal of stock
3. Production of highly finished surfaces
4. Cutting-off operations
5. Production of sharp edges and points

For grinding wheels to function properly, they must be hard and tough, and the wheel surface must be capable of gradually breaking down to expose new sharp cutting edges to the material being ground.

The material components of a grinding wheel are the *abrasive grain* and the *bond;* however, there are other physical characteristics, such as *grade* and *structure,* that must be considered in grinding wheel manufacture and selection.

ABRASIVE GRAIN

The abrasive used in most grinding wheels is either *aluminum oxide* or *silicon carbide.* The function of the abrasive is to remove material from the surface of the work being ground. Each abrasive grain on the working surface of a grinding wheel acts as a separate cutting tool and removes a small metal chip as it passes over the surface of the work. As the grain becomes dull, it fractures and presents a new sharp cutting edge to the material. The fracturing action reduces the heat of friction that would be caused if the grain became dull, producing a relatively cool cutting action. As a result of hundreds of thousands of individual grains all working on the surface of a grinding wheel, a smooth surface can be produced on the workpiece.

One important factor to consider in grinding wheel manufacture and selection is the *grain size.* After the abrasive ingot, or pig, is removed from the electric furnace, it is crushed, and the abrasive grains are cleaned and sized by passing them through screens that contain a certain number of meshes, or openings, per inch. A #8 grain size would pass through a screen having eight meshes per linear inch and would be approximately 1/8 in. across. The sizing of the abrasive grain is rather an important operation, since undersize grains in a wheel will fail to do their share of the work, while oversize grains will scratch the surface of the work.

Commercial grain sizes are classified as follows:

Very Coarse	Coarse	Medium
6	14	30
8	16	36
10	20	46
12	24	54
		60

Fine	Very Fine	Flour Size
70	150	280
80	180	320
90	220	400
100	240	500
120		600

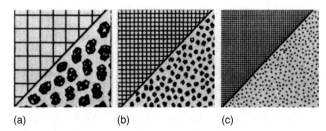

(a) (b) (c)

Figure 80-5 Relative abrasive grain sizes: (a) 8-grain; (b) 24-grain; (c) 60-grain.

Relative grain sizes are shown in Fig. 80-5. General applications for various grain sizes are:

> 8 to 54 for rough grinding operations

> 54 to 400 for precision grinding processes

> 320 to 2000 for ultra precision processes to produce 2 to 4 μ (micron) finish or finer

The factors affecting the selection of grain sizes are:

1. *The type of finish desired.* Coarse grains are best suited for rapid removal of metal. Fine grains are used for producing smooth and accurate finishes.

2. *The type of material being ground.* Generally, coarse grains are used on soft material, while fine grains are used for hard materials.

3. *The amount of material to be removed.* Where a large amount of material is to be removed and surface finish is not important, a coarse-grain wheel should be used. For finish grinding, a fine-grain wheel is recommended.

4. *The area of contact between the wheel and the workpiece.* If the area of contact is wide, a coarse-grain wheel is generally used. Fine-grain wheels are used when the area of contact between the wheel and the work is small.

BOND TYPES

The function of the bond is to hold the abrasive grains together in the form of a wheel. There are six common bond types used in grinding wheel manufacture: vitrified, resinoid, rubber, shellac, silicate, and metal.

Vitrified Bond

Vitrified bond is used on most grinding wheels. It is made of clay or feldspar, which fuses at a high temperature and, when cooled, forms a glassy bond around each grain. Vitrified bonds are strong but break down readily on the wheel surface to expose new grains during the grinding operation. This bond is particularly suited to wheels used for the rapid removal of metal. Vitrified wheels are not affected by water, oil, or acid and may be used in all types of grinding operations. Vitrified wheels should be operated between 6300 and 6500 sf/min (1920 and 1980 m/min).

Resinoid Bond

Synthetic resins are used as bonding agents in resinoid wheels. The majority of resinoid wheels generally operate at 9500 sf/min (2895 m/min); however, the modern trend is toward greater power and faster speeds for faster stock removal. Special resinoid wheels are manufactured to operate at speeds of 12,500 to 22,500 sf/min (3810 to 6858 m/min) for certain applications. These wheels are cool-cutting and remove stock rapidly. They are used for cutting-off operations, snagging, and rough grinding, as well as for roll grinding.

Rubber Bond

Rubber-bonded wheels produce high finishes such as those required on ball bearing races. Because of the strength and flexibility of this type of wheel, it is used for thin cutoff wheels. Rubber-bonded wheels are also used as regulating wheels on centerless grinders.

Shellac Bond

Shellac-bonded wheels are used for producing high finishes on parts such as cutlery, cam shafts, and paper-mill rolls. They are not suitable for rough or heavy grinding.

Silicate Bond

Silicate-bonded wheels are not used to any extent in industry. Silicate bond is used principally for large wheels and for small wheels where it is necessary to keep heat generation to a minimum. The bond (silicate of soda) releases the abrasive grains more rapidly than does the vitrified bond.

Metal Bond

Metal bonds (generally nonferrous) are used on diamond wheels and for electrolytic grinding operations where the current must pass through the wheel.

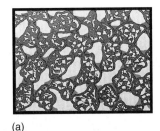
(a)

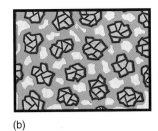
(b)

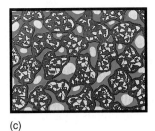
(c)

Figure 80-6 Grinding wheel grades: (a) weak bond posts; (b) medium bond posts; (c) strong bond posts. *(The Carborundum Co.)*

GRADE

The grade of a grinding wheel may be defined as the degree of strength with which the bond holds the abrasive particles in the bond setting. If the bond posts are very strong (Fig. 80-6c), that is, if they retain the abrasive grains in the wheel during the grinding operation, the wheel is said to be of a hard grade. If the grains are released rapidly during the grinding operation, the wheel is classified as a soft grade (Fig. 80-6a).

The selection of the proper grade of wheel is important. Wheels which are too hard do not release the grains readily; consequently, the grains become dull and do not cut effectively. This is known as glazing. Wheels which are too soft release the grain too quickly, and the wheel will wear rapidly.

One of the most difficult features to determine in selecting a grinding wheel for each job is the grade. Trial and error are generally used to decide what grade works best. The characteristics of wheels that are too hard, or too soft, are listed in Table 80.1 as a guide to help select the wheel grade.

It is good to remember that all abrasive grains are hard and that the hardness of a wheel refers to grade (the strength of the bond) and not to the hardness of the grain. Wheel grade symbols are indicated alphabetically, ranging from A (softest) to Z (hardest). The grade selected for a particular job depends on the following factors:

1. *Hardness of the material.* A hard wheel is generally used on soft material and soft grades on hard materials.

table 80.1   Wheel grade faults

Hard-Wheel Characteristics	Soft-Wheel Characteristics
Glaze: Grain wears flat and cannot be discharged from bond post.	*Breaks down too fast:* Wheel is too soft for the grinding operation.
Loading: Material being ground deposits on the wheel face.	*Surface finish gets worse:* Grain breaks from the wheel too rapidly—before the surface can get smooth.
Burn: Flat abrasive grains rub the work, causing burn marks that could result in grinding cracks.	*Cuts freely:* The wheel does not glaze and, therefore, the cutting action is good.
Squeal: Hard wheels emit a high-pitched sound called *squealing.*	*Sparks out quickly:* The free-cutting wheel sparks out because of the low cutting pressure and low machine distortion.
Doesn't cut freely: Cutting action is slow, wheel-to-work pressures are high, and the wheel will not spark out.	*Chatter:* The chatter pattern is usually wide and the surface finish is poor.
Inaccurate work: High cutting pressures and heat distort the workpiece.	*Sizing difficult:* Because of the rapid grain loss, maintaining workpiece size is difficult.
Chatter: Finely spaced chatter marks are often produced in the work.	*Scratches, "fishtails":* Grains breaking out from the wheel roll between the wheel and work, causing surface scratches.

2. *Area of contact.* Soft wheels are used where the area of contact between the wheel and the workpiece is large. Small areas of contact require harder wheels.

3. *Condition of the machine.* If the machine is rigid, a softer grade of wheel is recommended. Light-duty machines or machines with loose spindle bearings require harder wheels.

4. *The speed of the grinding wheel and the workpiece.* The higher the wheel speed in relation to the workpiece, the softer the wheel should be. Wheels that revolve slowly wear faster; therefore, a harder wheel should be used at slow speeds.

5. *Rate of feed.* Higher rates of feed require the use of harder wheels, since the pressure on the grinding wheel is greater than the slower feeds.

6. *Operator characteristics.* An operator who removes the material quickly requires a harder wheel than one who removes the material more slowly. This is particularly evident in offhand grinding and where piece-work programs are involved.

STRUCTURE

The structure of a grinding wheel is the space relationship of the grain and bonding material to the voids that separate them. In brief, it is the *density* of the wheel.

If the spacing of the grains is close, the structure is dense (Fig. 80-7a). If the spacing of the grains is relatively wide, the structure is open (Fig. 80-7c).

Selection of the wheel structure depends on the type of work required. Wheels with open structures (Fig. 80-8) provide greater chip clearance than those with dense structures and remove material faster than dense wheels.

The structure of grinding wheels is indicated by numbers ranging from 1 (dense) to 15 (open). Selection of the proper wheel structure is affected by the following factors:

1. *Type of material being ground.* Soft materials will require greater chip clearance; therefore, an open wheel should be used.

■ **Figure 80-8** An open-structure wheel provides greater chip clearance. *(The Carborundum Company)*

2. *Area of contact.* The greater the area of contact, the more open should be the structure to provide better chip clearance.

3. *Finish required.* Dense wheels will give a better, more accurate finish.

4. *Method of cooling.* Open-structure wheels provide a better supply of coolant for machines using "through the wheel" coolant systems.

In summary, Table 80.2 will serve as a guide to the factors that must be considered in selecting a grinding wheel.

GRINDING WHEEL MANUFACTURE

Most grinding wheels used for machine shop operations are manufactured with vitrified bonds; therefore, the manufacture of only this type of wheel will be discussed. The main operations in the manufacture of vitrified grinding wheels are as follows.

Mixing

The correct proportions of abrasive grain and bond are carefully weighed and thoroughly mixed in a rotary power mixing machine (Fig. 80-9). A certain percentage of water is added to moisten the mix.

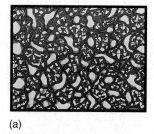
(a)

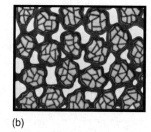
(b)

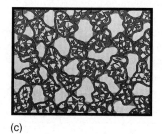
(c)

■ **Figure 80-7** Grinding wheel structure: (a) dense; (b) medium; (c) open. *(The Carborundum Co.)*

table 80.2　Factors to be considered in the selection of a grinding wheel

Grinding Factors	Wheel Considerations				
	Abrasive Type	Grain Size	Bond	Grade	Structure
Material to be ground (high or low tensile strength; hard or soft)	X	X		X	X
Type of operation (cylindrical, centerless, surface, cutoff, snagging, etc.)			X		
Machine characteristics (rugged or light; loose or tight bearings)				X	
Wheel speed (slow or fast)			X	X	
Rate of feed (slow or rapid)				X	
Area of contact (large or small)		X		X	X
Operator characteristics				X	
Amount of stock to be removed (light cut or heavy cut)			X	X	X
Finish required		X	X		X
Use of coolant (wet or dry grinding)			X	X	X

■ **Figure 80-9** Mixing abrasive grain and bond.
(The Carborundum Company)

■ Figure 80-10 Molding grinding wheels using a hydraulic press. *(Milacron, Inc.)*

Molding

The proper amount of this mixture is placed in a steel mold of the desired wheel shape and compressed in a hydraulic press (Fig. 80-10) to form a wheel slightly larger than the finished size. The amount of pressure used varies with the size of the wheel and the structure required.

Shaving

Although the majority of wheels are molded to shape and size, some machines require special wheel shapes and recesses. These are shaped or shaved to size in the green, or unburned, state on a shaving machine, which resembles a potter's wheel.

Firing (Burning)

The green wheels are carefully stacked on cars and are moved slowly through a long kiln 250 to 300 ft (76 to 90 m) long. The temperature of the kiln is held at approximately 2300°F (1260°C). This operation, which takes about five days, causes the bond to melt and form a glassy case around each grain; the product is a hard wheel.

Truing

The cured wheels are mounted in a special lathe and turned to the required size and shape by hardened-steel conical cutters, diamond tools, or special grinding wheels.

Bushing

The arbor hole in a grinding wheel is fitted with a lead or plastic-type bushing to fit a specific spindle size. The edges of the bushing are then trimmed to the thickness of the wheel.

Balancing

To remove vibration that may occur while a wheel is revolving, each wheel is balanced. Generally, small, shallow holes are drilled in the "light" side of the wheel and filled with lead to ensure proper balance.

Speed Testing

Wheels are rotated in a heavy, enclosed case and revolved at speeds at least 50% above normal operating speeds. This ensures that the wheel will not break under normal operating speeds and conditions.

STANDARD GRINDING WHEEL SHAPES

Nine standard grinding wheel shapes have been established by the United States Department of Commerce, the Grinding Wheel Manufacturers, and the Grinding Machine Manufacturers. Dimensional sizes for each of the shapes have also been standardized. Each of these nine shapes is identified by a number, as shown in Table 80.3.

MOUNTED GRINDING WHEELS

Mounted grinding wheels (Fig. 80-11) are driven by a steel shank mounted in the wheel. They are produced in a variety of shapes for use with jig grinders, internal grinders, portable grinders, toolpost grinders, and flexible shafts. They are manufactured in both aluminum oxide and silicon carbide types.

GRINDING WHEEL MARKINGS

The standard marking system chart (Fig. 80-12) is used by the manufacturers to identify grinding wheels. This information is found on the blotter of all small and medium-size grinding wheels. It is stenciled on the side of larger wheels.

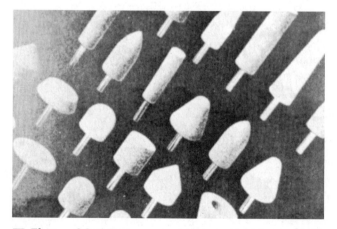

■ Figure 80-11 A variety of mounted grinding wheels. *(The Carborundum Company)*

table 80.3 Common grinding wheel shapes and applications

Shape	Name	Applications
	Straight (Type 1)	Cylindrical, centerless, internal, cutter, surface, and offhand grinding operations
	Cylinder (Type 2)	Surface grinding on horizontal and vertical spindle grinders
	Tapered (both sides) (Type 4)	Snagging operations; the tapered sides lessen the chance of the wheel breaking
	Recessed (one side) (Type 5)	Cylindrical, centerless, internal, and surface grinders; the recess provides clearance for the mounting flange
	Straight cup (Type 6)	Cutter and tool grinder and surface grinding on vertical and horizontal spindle machines
	Recessed (both sides) (Type 7)	Cylindrical, centerless, and surface grinders; the recesses provide clearance for mounting flanges
	Flaring cup (Type 11)	Cutter and tool grinder; used mainly for sharpening milling cutters and reamers
	Dish (Type 12)	Cutter and tool grinder; its thin edge permits it to be used in narrow slots
	Saucer (Type 13)	Saw gumming, gashing, milling cutter teeth

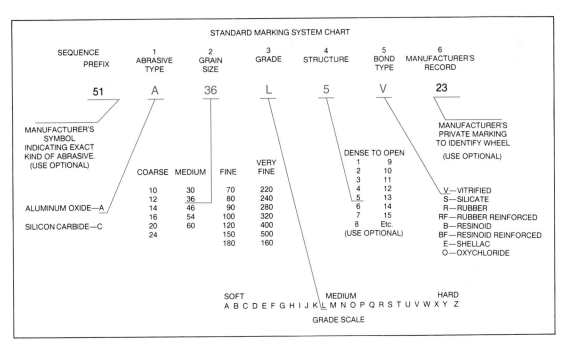

■ **Figure 80-12** Straight grinding wheel marking system chart. *(Grinding Wheel Institute)*

The six positions shown in the standard sequence are followed by all manufacturers of grinding wheels. The prefix shown is a manufacturer's symbol and is not always used by all grinding wheel producers.

Note: This marking system is used only for aluminum oxide and silicon carbide wheels, not for diamond wheels.

▶▶ Selecting a Grinding Wheel for a Specific Job

From the foregoing information, the machinist should be able to select the proper wheel for the job required.

___

EXAMPLE

It is required to rough surface grind a piece of SAE 1045 steel using a straight wheel. Coolant is to be used.

> *Type of abrasive.* Because steel is to be ground, *aluminum oxide* should be used.

> *Size of grain.* Since the surface is not precision-finished, a medium grain can be used: about *46 grit*.

> *Grade.* A *medium-grade* wheel that will break down reasonably well should be selected. Use grade *J*.

> *Structure.* Since this steel is of medium hardness, the wheel should be of medium density: about *7*.

> *Bond type.* Since the operation is standard surface grinding and since coolant is to be used, a *vitrified* bond should be selected.

After all factors have been considered, an A46-J7-V wheel should be selected to rough-grind SAE 1045 steel.

Note: These specifications do not include the manufacturer's prefix or the manufacturer's records.

___

EXAMPLE

It is required to finish-grind a high-speed steel milling cutter on the cutter and tool grinder.

> *Type of abrasive.* Since the cutter is steel, an *aluminum oxide* wheel should be used.

> *Size of grain.* Since the milling cutter must have a smooth finish, a medium to fine grain should be used. About a *60 grit* is recommended for this type of operation.

> *Grade.* It is important that a cool-cutting wheel be used to prevent burning the cutting edge of the

cutter. A wheel that breaks down reasonably well will permit cool grinding. Use a medium-soft grade such as *J*.

> *Structure.* In order to produce a smooth cut, a medium-dense wheel should be used. For this application, use a *#6*.

> *Bond type.* Because most cutter and tool grinders are designed for standard speeds, a *vitrified* bond should be used. When the speed is excessive for the wheel size, a *resinoid* bond should be used.

The wheel selected for this job (disregarding the manufacturer's prefix and records) should be A60-J6-V.

Note: If the cutter is chipped or if considerable metal must be removed to resharpen, it is advisable to first rough-grind using a 46-grit wheel.

___

▶▶ Handling and Storage of Grinding Wheels

Since all grinding wheels are breakable, proper handling and storage are important to protect them from chipping, breaking, or developing hair-line cracks. This care will result in the most productive life for each grinding wheel. Proper handling and storage are necessary for grinding wheels and the following precautions should be observed:

1. *Handling.* Never handle wheels carelessly; they are brittle and break easily.
   > Wheels that are bumped or dropped should be ring tested to make sure they are not damaged.
   > Never lay any tools or objects on wheels; treat them as precision instruments.
2. *Dry at a reasonable temperature.* Some bond-type wheels may be seriously affected by dampness and extreme temperature changes.
3. *Store wheels properly.* Most *straight* or *tapered* wheels are best stored on the edge in individual racks.
   > Thin, organic bonded wheels should be laid on a flat horizontal surface to prevent warping.
   > Large cup and cylindrical wheels should be stored on the flat sides, with packing between wheels.
   > Small cup and internal wheels should be put separately into boxes, bins, or drawers.
   > Wheels stored on their edge should be prevented from rolling.

▶▶ Wheel Grade Faults

The grade of a grinding wheel is probably the most important and difficult specification to select to suit the workpiece material and the grinding operation. Until the grade of a wheel has been proven under grinding conditions, it is wise to start with a medium-grade wheel such as *J*. After noting its performance, adjust the grade until the best grinding conditions are reached. The following characteristics of wheels that are too soft, or too hard, should help in selecting the best grade of wheel for each grinding condition.

SOFT WHEEL CHARACTERISTICS

A wheel grade that is too soft (weak bond posts) will break down quickly, produce a poor surface finish, and have difficulty maintaining work size (Fig. 80-13). The following characteristics may indicate that a wheel is too soft for the material being ground:

1. *Breaks down too fast.* Wheels that are too soft for the material being ground will release abrasive grains too quickly and will break down rapidly.

2. *Poor surface finish.* Because the abrasive grains release so rapidly, there is no chance for the wheel to smooth the work surface and produce a good finish.

3. *Cuts freely.* Because the soft wheel does not have much opportunity to glaze, there is little to slow the cutting action.

4. *Sparks out quickly.* The rapid release of abrasive grains causes the wheel to spark out because of the low cut pressure and low machine distortion.

5. *Difficult to maintain size.* A wheel that loses its grains too quickly makes it very difficult to

maintain workpiece accuracy. The coarser the grain size, the more rapid the loss in wheel diameter.

6. *Scratches (fishtails).* Abrasive grains breaking from the wheel will roll between the wheel and work surface to produce scratches that are deeper than the ground surface. This condition may also be caused by using dirty coolant for the grinding operation.

HARD WHEEL CHARACTERISTICS

A wheel grade that is too hard (strong bond posts) will hold the abrasive grains too securely and will not break down when the grains become dull. A hard-grade wheel tends to glaze and load, doesn't cut freely, and causes excessive grinding heat (Fig. 80-14). The following characteristics may indicate that a wheel is too hard for the material being ground:

1. *Wheel glazes quickly.* The abrasive grains wear flat because they are not released by the strong bond posts. This causes a glazing condition that increases the grinding pressure and creates excessive heat.

2. *Loading.* Because the dull abrasive grains do not release from the wheel, the material ground tends to fill the voids (gaps) between the abrasive grains.

3. *Burned work surface.* Because of the dull abrasive grains, glazing, and loading, the wheel tends to rub instead of cut, and the increased grinding temperature may cause thermal damage to the work.

4. *Squealing noise.* Hard-grade wheels sometimes create a high-pitched sound called "squealing" that is different than the normal sound of a free-cutting wheel.

5. *Doesn't cut freely.* The dull abrasive grains cannot cut as freely as sharp grains, which increases the grinding pressure and reduces the material removal rate.

SOFT WHEEL
CHARACTERISTICS

■ **Figure 80-13** A wheel that is too soft will break down too quickly and cause poor surface finishes. *(Milacron, Inc.)*

HARD WHEEL
CHARACTERISTICS

■ **Figure 80-14** A wheel that is too hard will not cut freely and may produce enough heat to thermally damage the work. *(Milacron, Inc.)*

6. *Inaccurate work dimensions.* Hard wheels create high grinding pressures and excessive heat, which distort the workpiece.

7. *Surface finish gets progressively better.* As the abrasive grain gets smoother (glazes), the surface finish will tend to get better because of the burnishing (rubbing) action of the wheel surface. This burnishing effect may create excessive heat, which can cause thermal damage to the work.

8. *Won't spark out.* A hard wheel will continue to cut, even after the feed of the grinder has been stopped, because of the machine distortion and high cutting pressures.

9. *Heat checks.* These can be caused by hard-grade wheels containing fine abrasive grains. "Grinding cracks" caused by the excessive heat generated may damage the microstructure of the workpiece to a depth of .002 in. (0.05 mm).

▶▶ Inspection of Wheels

After wheels have been received, they should be inspected to see that they have not been damaged in transit. For further assurance that wheels have not been damaged, they should be suspended and tapped lightly with a screwdriver handle for small wheels (Fig. 80-15) or with a wooden mallet for larger wheels. If vitrified or silicate wheels are sound, they give a clear, metallic ring. Organic-bonded wheels give a duller ring, and cracked

■ **Figure 80-15** Testing a grinding wheel for cracks. *(Kelmar Associates)*

wheels do not produce a ring. Wheels must be dry and free from sawdust before testing; otherwise, the sound will be deadened.

▶▶ Diamond Wheels

Diamond wheels are used for grinding cemented carbides and hard vitreous materials, such as glass and ceramics. Diamond wheels are manufactured in a variety of shapes, such as straight, cup, dish, and thin cutoff wheels.

Wheels of 1/2-in. (13-mm) diameter or less have diamond particles throughout the wheel. Wheels larger than 1/2 in. (13 mm) are made with a diamond surface on the grinding face only. The diamonds for this purpose are made in grain sizes ranging from 100 to 400. The proportions of the diamond and bond mixture vary with the application. This diamond concentration is identified by the letter A, B, or C. The C concentration contains four times the number of diamonds in an A concentration. This mixture is coated on the grinding face of the wheel in thicknesses ranging from 1/32 to 1/4 in. (0.8 to 6 mm).

BONDS

There are three types of bonds available for diamond wheels: resinoid, metal, and vitrified.

Resinoid-bonded wheels give a maximum cutting rate and require very little dressing. These wheels remain sharp for a long time and are well suited to grinding carbides.

A recent development in the manufacture of resinoid-bonded diamond wheels has been the coating of the diamond particles with nickel plating by means of an electroplating process. This process is carried out before the diamonds are mixed with the resin. It reduces the tendency of the diamonds to chip and results in cooler-grinding, longer-lasting wheels.

Metal bonds, generally nonferrous, are particularly suited to offhand grinding and cutting-off operations. This type of wheel holds its form extremely well and does not wear on radius work or on small areas of contact.

Vitrified-bonded wheels remove stock rapidly but require frequent cleaning with a boron carbide abrasive stick to prevent the wheel from loading. These wheels are particularly suited for offhand and surface grinding of cemented carbides.

DIAMOND WHEEL IDENTIFICATION

The method used to identify diamond wheels differs from that used for other grinding wheels (Fig. 80-16).

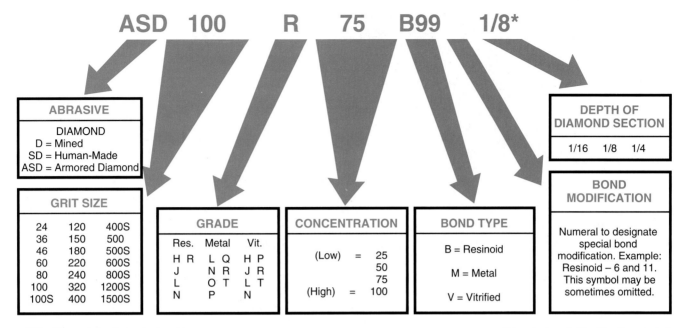

Figure 80-16 Diamond grinding wheel marking system chart. *(Norton Company)*

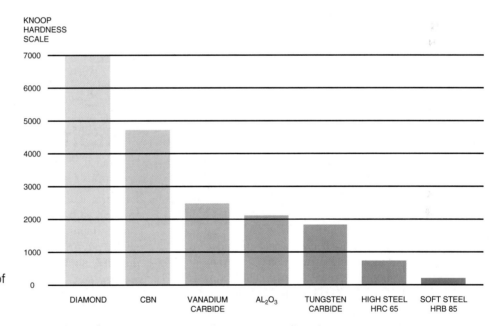

Figure 80-17 Hardness of various metals and abrasives. *(Norton Company)*

▶▶ **Cubic Boron Nitride Wheels**

Cubic boron nitride (CBN) grinding wheels are now recognized as superior cutting tools for grinding difficult-to-machine metals. From their initial use in toolrooms and cutter grinding applications, CBN wheels are really making their presence felt in production grinding operations, where the alternative had been to use less costly conventional abrasives that wear out at much faster rates. The applications for CBN wheels range from the elementary regrinding of high-speed steel cutting tools to the ultra-high-speed grinding of hardened steel components in the automotive industries.

CBN grinding wheels have more than twice the hardness of conventional abrasives for grinding difficult-to-grind ferrous metals (Fig. 80-17). Hardness in an abrasive is

ALUMINUM OXIDE   BORAZON CBN

(a)   (b)

■ **Figure 80-18** (a) Ground with an aluminum oxide wheel shows metallurgical damage, while (b) ground with a cubic boron nitride (CBN) wheel shows no damage. *(Diamond Innovations, at www.abrasivesnet.com)*

meaningless if the abrasive is too brittle to withstand the machining pressures and the heat of production grinding. The CBN abrasive crystal has the toughness to match its hardness so that its cutting edges stay sharp longer with much slower wear rates than those of conventional abrasives.

On difficult-to-grind materials, conventional grinding wheels dull quickly and, as a result, generate high frictional heat. As the abrasive grains dull, the material-removal rate falls and it is difficult to maintain part accuracy and geometry. The CBN wheel's prolonged cutting capacity and high thermal conductivity help prevent uncontrolled heat buildup and therefore reduce the chances of wheel glazing and workpiece metallurgical damage (Fig. 80-18). The CBN abrasive is also thermally and chemically stable at temperatures above 1832°F (1000°C), that is, well above the temperatures generally reached in grinding. This means reduced grinding wheel wear, with easier production of precision workpiece geometry and accuracy.

PROPERTIES OF CUBIC BORON NITRIDE (CBN) WHEELS

Cubic boron nitride is a material that is not found in nature. CBN grinding wheels contain the four main properties necessary to grind extremely hard or abrasive materials at high metal-removal rates: hardness, abrasion resistance, compressive strength, and thermal conductivity (Fig. 80-19).

1. *Hardness.* Cubic boron nitride, next to diamond in hardness, is about twice as hard as silicon carbide and aluminum oxide.

2. *Abrasion resistance.* CBN wheels maintain their sharpness much longer, thereby increasing produc-

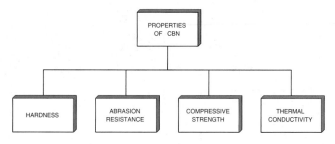

■ **Figure 80-19** The properties of CBN grinding wheels. *(Kelmar Associates)*

tivity while at the same time producing parts that are dimensionally accurate.

3. *Compressive strength.* The high compressive strength of CBN crystals gives them excellent qualities to withstand the forces created during high metal-removal rates.

4. *Thermal conductivity.* CBN wheels have excellent thermal conductivity, which allows greater heat dissipation (transfer), especially when hard, abrasive, tough materials are ground at high material-removal rates.

WHEEL SELECTION

Any successful grinding operation depends to a large extent on choosing the right wheel for the job. The type of wheel selected and how it is used will affect the metal-removal rate (MRR) and the life of a grinding wheel. The selection of a CBN grinding wheel can be a very complex task, and it is always wise to follow the manufacturer's suggestions for each type of wheel.

CBN wheel selection is generally affected by:

> Type of grinding operation

> Grinding conditions

> Surface finish requirements

> Shape and size of the workpiece

> Type of workpiece material

All four types of CBN wheels (resin, vitrified, metal, and electroplated) are highly effective; however, they are designed for specific applications and must be selected accordingly. There is no one type of CBN wheel that is suitable for all grinding operations. Therefore, for the best grinding performance, the characteristics of the wheel abrasive must be matched to the requirements of the specific grinding job. Abrasive characteristics such as concentration, size, and toughness must be considered in the selection of a wheel because they affect metal-removal rates, wheel life, and surface finish of any grinding operation. However, the wheel manufacturer understands the type of CBN abrasive available and will provide the most suitable type for the job.

Wheel Selection Guidelines

Cubic boron nitride (CBN) grinding wheels are available in a complete range of shapes and sizes, including wheels with straight or formed faces, ring wheels, disk wheels, flaring cup wheels, mounted wheels, mandrels, and hones. Individually engineered wheels are also available to suit specific superabrasive machining systems or specific grinding operations. As with all wheels that use high-cost abrasives, the CBN wheel is constructed with a precision, preformed core, with the abrasive portion on the grinding face of the wheel. The abrasive portion is usually between 1/16 to 1/4 in. (1.5 to 6 mm) in depth, and this information is shown on the identification label of the wheel. There generally is a CBN wheel readily available to suit any grinding operation.

The wheel manufacturer will provide the right CBN abrasive for the bond system selected. The manner in which the wheel is actually used, however, will determine whether ideal wear conditions are achieved. Always be sure that the actual operating conditions are within the range of the wheel's capabilities. In order to successfully use CBN grinding wheels, follow these general guidelines:

1. *Select the bond.* Refer to Fig. 80-20 and Table 80.4 for a good first choice in selecting bonds for all major applications.

2. *Specify normal wheel diameters and widths.* When an aluminum oxide wheel is replaced with a CBN wheel, the CBN wheel should be the same diameter and width as the wheel it replaces.

3. *Choose the largest abrasive mesh size that produces the desired finish.* For any given set of grinding operations, a wheel containing coarse (large mesh size) CBN abrasive will have a longer life than a wheel containing fine (small mesh size) abrasive. However, the wheel with fine abrasive will produce better surface finishes at lower downfeed rates.

4. *Choose wheels with the optimum abrasive concentration.* Although low-concentration wheels will generally do the job, they may not always be the most cost-effective. Always select the highest concentration that the grinding machine has the power to drive effectively.

▶▶ Coated Abrasives

Coated abrasives (Fig. 80-21 on p. 692) consist of a flexible backing (cloth or paper) to which abrasive grains have been bonded. Garnet, flint, and emery (natural abrasives) are being replaced by aluminum oxide and silicon carbide

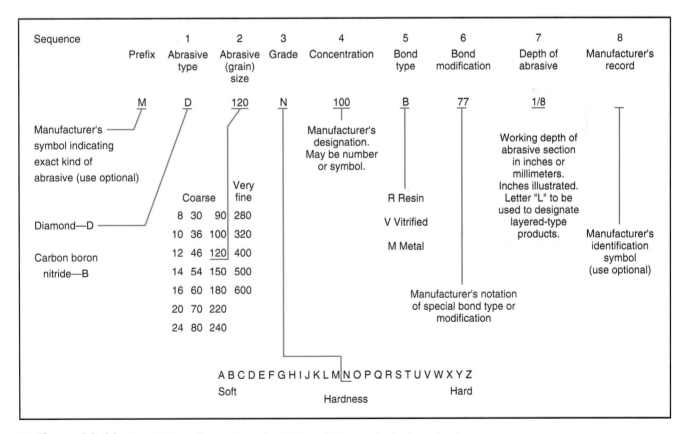

Figure 80-20 The ANSI coding system for CBN and diamond grinding wheels.

Item	Wheel Bond Type			
	Resin	Metal	Vitreous	Electroplated
Rim depth	.079 in. (2 mm) or more	.079 in. (2 mm) or more	.079 in. (2 mm) or more	One abrasive layer
Wheel life	Limited	Long	Long	Limited
Grinding ratio	Medium	High	Medium/high	(Not applicable)
Cutting action	Excellent	Good	Good to excellent	Good to excellent
Material-removal rate	High	Limited	Medium to high	High
Form holding	Good	Excellent	Excellent	Excellent
Precautions for use	None	Rigid machines recommended	Least resistance to damage	None
Grinding	Wet or dry	Wet	Wet	Wet preferred; can be used dry

■ Figure 80-21 A variety of coated abrasives.
(Norton Company)

in the manufacture of coated abrasives. This is due to the greater toughness and the more uniform grain size and shape of manufactured abrasives.

Coated abrasives serve two purposes in the machine shop: metal grinding and polishing. Metal grinding may be done on a belt or disk grinder and, up until the last few years, was a rapid, nonprecision method of removing metal. Coarse-grit coated abrasives are used for rapid removal of metal, whereas fine grits are used for polishing.

Emery, a natural abrasive that is black in appearance, is used to manufacture coated abrasives, such as emery cloth and emery paper. Since the grains are not as sharp as artificial abrasives, emery is generally used for polishing metal by hand.

SELECTION OF COATED ABRASIVES

Aluminum Oxide

Aluminum oxide, gray in appearance, is used for high-tensile-strength materials, such as steels, alloy steels, high-carbon steels, and tough bronzes. Aluminum oxide is characterized by the long life of its cutting edges.

For *hand operations,* 60 to 80 grit is used for fast cutting (roughing), while 120 to 180 grit is recommended for finishing operations.

For *machine operations,* such as on belt and disk grinders, 36 to 60 grit is used for roughing, while 80 to 120 grit is recommended for finishing operations.

Silicon Carbide

Silicon carbide, bluish-black in appearance, is used for low-tensile-strength materials, such as cast iron, aluminum, brass, copper, glass, and plastics. The selection of grit size for hand and machine operations is the same as for aluminum oxide–coated abrasives.

COATED ABRASIVE MACHINING

Over the past few years, coated abrasive machining has become widely used in industry. With improved abrasives and bonding materials, better grain structure, more uniform belt splicing, and new polyester belt backing, abrasive belt machining is being used to a much greater extent. Some types of work that had formerly been performed by milling, turning, and abrasive wheel grinding in machine shops, steel mills, steel fabrication plants, and foundries are now being done more efficiently by belt or disk grinding operations. Coated abrasive operations are now capable of grinding to less than .001 in. (0.03 mm) tolerance and with a surface finish of 10 to 20 μin. (0.3 to 0.5 μm).

As a result of greatly improved coated abrasive products, heavier and better machines requiring much higher horsepower are being produced in this field. The basic machines have become more automated and now include automatic loaders, feeders, and unloaders. Belt machining has been applied very successfully to the centerless grinding concept, where it is used extensively for bar and tube grinding in steel and metalworking plants. In many lumber mills, belt abrasive machining is now used to finish the lumber to size, an operation that previously was done by planing.

unit 80 review questions

1. What characteristics must an abrasive have in order for it to function properly?

Types of Abrasives

2. Name four natural abrasives.

3. What is "bort" and for what purpose is it used?

4. Name three manufactured abrasives and state why they are used extensively today.

5. Why are aluminum oxide and silicon carbide abrasives used?

6. Describe the manufacture of aluminum oxide.

7. Describe the manufacture of silicon carbide.

8. List three uses of zirconia-alumina abrasive.

9. Describe the cutting action of zirconia-alumina abrasive.

10. List four advantages of zirconia-alumina abrasive wheels.

11. How does the manufacture of boron carbide differ from that of the other manufactured abrasives?

12. List the uses for boron carbide.

13. List five advantages of cubic boron nitride grinding wheels.

14. Name two types of Borazon™ (cubic boron nitride) and state where each is used.

15. List three types of manufactured diamonds and state where each is used.

16. a. Name two materials used to coat diamonds.

    b. What is the purpose of coating diamonds?

Grinding Wheels

17. What are the basic functions of a grinding wheel?

18. Describe the function of each abrasive grain.

19. How is grain size determined and why is it important?

20. What factors affect the selection of the proper grain size?

21. What is the function of a bond and how does it affect the grade of a grinding wheel?

22. Name six types of bonds and state the purpose of each type in the manufacture of grinding wheels.

23. Why is the selection of the grade of wheel important to the grinding operation?

24. What factors should be considered when the grade of wheel is being selected for a particular job?

25. Define the structure of a grinding wheel and state how this is indicated.

26. What factors affect the selection of the proper wheel structure?

Grinding Wheel Manufacture

27. Describe briefly the manufacture of a vitrified grinding wheel.

Standard Grinding Wheel Shapes

28. Describe the following grinding wheels and state their purposes: Types 1, 5, 6, 11.

29. For what purpose are mounted grinding wheels used?

Grinding Wheel Markings

30. Explain the meaning of the following grinding wheel marking: A80-G-8-V.

31. What wheel should be selected for grinding:

    a. Machine steel?
    b. Cemented carbide?
    c. Cast iron?

Inspection of Wheels

32. Explain why it is important to inspect a grinding wheel before it is mounted on a grinder.

33. How should grinding wheels be inspected?

Diamond Wheels

34. For what purpose are diamond wheels used?

35. Explain diamond concentration in a grinding wheel.

36. Name three types of bonds used in diamond wheels and state the purpose of each.

37. Define the following diamond wheel markings: D 120–N 100–B 1/8.

Coated Abrasives

38. Name the three common abrasives used in the manufacture of coated abrasives and state their purposes.

39. What grit size is recommended for:

    a. Hand operations?
    b. Machine operations?

UNIT 81

Surface Grinders and Accessories

OBJECTIVES

After completing this unit, you should be able to:

1 Name four methods of surface grinding and state the advantage of each

2 True and dress a grinding wheel

3 Select the proper grinding wheel to be used for each type of work material

Grinding is an important part of the machine tool trade. Improved grinding machine construction has permitted the production of parts to extremely fine tolerances with improved surface finishes and accuracy. Because of the dimensional accuracy obtained by grinding, interchangeable manufacture has become commonplace in most industries.

Grinding has also, in many cases, eliminated the need for conventional machining. With the development of new abrasives and better machines, the rough part is often finished in one grinding operation, thus eliminating the need for other machining processes. The role of grinding machines has changed over the years; initially they were used on hardened work and for truing hardened parts distorted by heat treating. Today, grinding is applied extensively to the production of unhardened parts where high accuracy and surface finish are required. In many cases, modern grinding machines permit the manufacture of intricate parts faster and more accurately than do other machining methods.

▸▸ The Grinding Process

In the grinding process, the workpiece is brought into contact with a revolving grinding wheel. Each small abrasive grain on the periphery of the wheel acts as an individual cutting tool and removes a chip of metal (Fig. 81-1). As the abrasive grains become dull, the pressure and heat created between the wheel and the workpiece cause the dull face to break away, leaving new, sharp cutting edges.

Regardless of the grinding method used, whether it be cylindrical, centerless, or surface grinding, the grinding process is the same, and certain general rules will apply in all cases:

1. Use a silicon carbide wheel for low-tensile-strength material and an aluminum oxide wheel for high-tensile-strength materials.

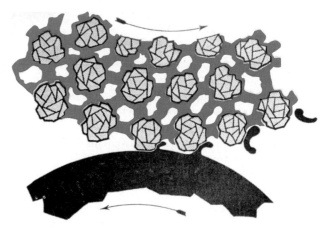

■ **Figure 81-1** Cutting action of abrasive grains. *(The Carborundum Company)*

2. Use a hard wheel on soft materials and a soft wheel on hard materials.

3. If the wheel is too hard, increase the speed of the work or decrease the speed of the wheel to make it act as a softer wheel.

4. If the wheel appears too soft or wears rapidly, decrease the speed of the work or increase the speed of the wheel, but not above its recommended speed.

5. A glazed wheel will affect the finish, accuracy, and metal-removal rate. The main causes of wheel glazing are:

   a. The wheel speed is too fast.
   b. The work speed is too slow.
   c. The wheel is too hard.
   d. The grain is too small.
   e. The structure is too dense, which causes the wheel to load.

6. If a wheel wears too quickly, the cause may be any of the following:

   a. The wheel is too soft.
   b. The wheel speed is too slow.
   c. The work speed is too fast.
   d. The feedrate is too great.
   e. The face of the wheel is too narrow.
   f. The surface of the work is interrupted by holes or grooves.

▶▶ Surface Grinding

Surface grinding is a technical term referring to the production of flat, contoured, and irregular surfaces on a piece of work, which is passed against a revolving grinding wheel.

TYPES OF SURFACE GRINDERS

There are four distinct types of surface grinding machines (Fig. 81-2), all of which provide a means of holding the metal and bringing it into contact with the grinding wheel.

The *horizontal spindle grinder with a reciprocating table* (Fig. 81-2a) is probably the most common type of surface grinder used in the toolroom. The work is reciprocated (moved back and forth) under the grinding wheel, which is fed down to provide the desired depth of cut. Feed is obtained by a transverse movement of the table at the end of each stroke.

The *horizontal spindle grinder with a rotary table* (Fig. 81-2b) is often found in toolrooms for the grinding of flat, circular parts. The surface pattern it produces makes it particularly suitable for grinding parts that must rotate in contact with each other. The work is held on the magnetic chuck of a rotating table and passed under a

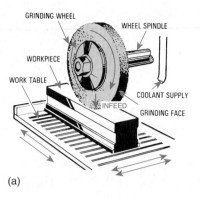

(a)

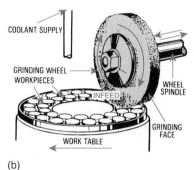

(b)

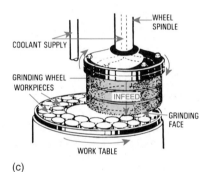

(c)

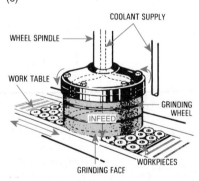

■ **Figure 81-2** (a) Horizontal spindle grinder with a reciprocating table; (b) horizontal spindle grinder with a rotary table; (c) vertical spindle grinder with a rotary table; (d) vertical spindle grinder with a reciprocating table. *(The Carborundum Company)*

grinding wheel. Feed is obtained by the transverse movement of the wheelhead. This type of machine permits faster grinding of circular parts, since the wheel is always in contact with the workpiece.

The *vertical spindle grinder with a rotary table* (Fig. 81-2c) produces a finished surface by grinding with the face of the wheel rather than the periphery, as in horizontal spindle machines. The surface pattern appears as a series of intersecting arcs. Vertical spindle grinders have a higher metal-removal rate than the horizontal-type spindle machines. It is probably the most efficient and accurate form of grinder for the production of flat surfaces.

The *vertical spindle grinder with a reciprocating table* (Fig. 81-2d) grinds on the face of the wheel while the work is moved back and forth under the wheel. Because of its vertical spindle and greater area of contact between the wheel and the work, this machine is capable of heavy cuts. Material up to 1/2 in. (13 mm) thick may be removed in one pass on larger machines of this type. Provision is made on most of these grinders to tilt the wheelhead a few degrees from the vertical. This permits greater pressure where the rim of the wheel contacts the workpiece and results in faster metal removal. When the wheelhead is vertical and grinding is done on the face of the wheel, the surface pattern produced is a series of uniform intersecting arcs. If the wheelhead is tilted, it produces a semicircular pattern.

Horizontal Spindle Reciprocating Table Surface Grinder

The horizontal spindle reciprocating table surface grinder is the most commonly used and will be discussed in detail. Machines of this type may be either hand or hydraulically operated (Fig. 81-3a).

The EZ-SURF® grinder, Fig. 81-3b, can be switched easily from a manual, semi-automatic, or fully automatic surface grinder to suit the grinding application. In the "TEACH" mode, the operator can teach and program up to 100 points for the X and Z coordinates for automatic repetitive grinding operations. The "Intelligent DRO" helps to increase productivity, improve accuracy, and simplify grinding operations. The wheel library allows the operator to create wheel shapes such as tapers and radii with a single or double-diamond dresser.

Parts of a Hydraulic Surface Grinder

The *base* is generally of heavy cast-iron construction. It usually contains the hydraulic reservoir and pump used to operate the table and power feeds. The top of the base has accurately machined ways to receive the saddle.

The *saddle* may be moved in or out across the ways, manually or by automatic feed.

The *table* is mounted on the top of the saddle. The ways for the table are at right angles to those on the base. Thus, the table reciprocates across the upper ways on the saddle, while the saddle (and table) moves in or out on the ways of the base.

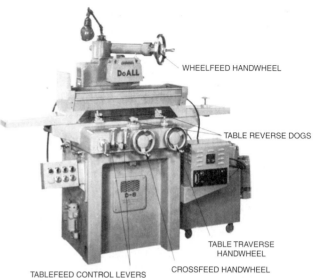

WHEELFEED HANDWHEEL

TABLE REVERSE DOGS

TABLE TRAVERSE HANDWHEEL

TABLEFEED CONTROL LEVERS

CROSSFEED HANDWHEEL

(a)

TWO-AXIS "INTELLIGENT DRO"   TAPER & RADIUS CAPABILITIES   DOUBLE DIAMOND DRESSING

(b)

■ **Figure 81-3** (a) Hydraulic surface grinder *(The DoAll Co.)*; (b) The EZ-Surf® grinder can be used as a conventional and fully automatic surface grinder. *(Hardinge, Inc.)*

The *column*, mounted on the back of the frame, contains the ways for the *spindle housing* and *wheelhead*. The wheelfeed handwheel provides a means of moving the wheelhead vertically to set the depth of cut.

The reciprocating action of the table may be controlled manually by the *table traverse handwheel* or by the *hydraulic control valve lever*.

The direction of the table is reversed when one of the *stop dogs* mounted on the side of the table strikes the *table traverse reverse lever*.

The table may be fed toward or away from the column manually by means of the *crossfeed handwheel* or automatically by the *power crossfeed control*. This operation moves the work laterally under the wheel.

Grinding Wheel Care

In order to ensure the best results in any surface grinding operation, proper care of the grinding wheel must be taken. Follow these guidelines:

1. When not in use, all grinding wheels should be properly stored.

2. Wheels should be tested for cracks prior to use.

3. Select the proper type of wheel for the job.

4. Grinding wheels should be properly mounted and operated at the recommended speed.

MOUNTING A GRINDING WHEEL

After the correct wheel has been selected for the job, proper mounting of the grinding wheel ensures the best grinding performance.

procedure

1. Test the wheel to see that it is not cracked by ring testing with the handle of a screwdriver or hammer.

2. Clean the grinding wheel adapter.

3. Mount the adapter through the wheel and tighten the threaded flange (Fig. 81-4).

   a. Be sure that the blotter is on each side of the wheel prior to mounting. A perforated blotter should be used for through-the-wheel coolant. A rubber washer is sometimes used in place of the blotter on some grinders.

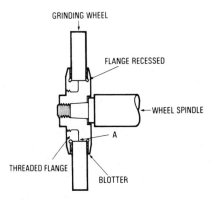

GRINDING WHEEL

FLANGE RECESSED

WHEEL SPINDLE

A

THREADED FLANGE

BLOTTER

■ **Figure 81-4** A grinding wheel properly mounted on the grinder spindle.

b. The wheel should be a good fit on the adapter or spindle. If it is too tight or too loose, the wheel should not be mounted.

c. To comply with the Wheel Manufacturer's Safety Code, the diameter of the flanges should not be less than one-third the diameter of the wheel.

4. Tighten the wheel adapter flanges only enough to hold the wheel firmly. If it is tightened too much, it may damage the flanges or break the wheel.

BALANCING A GRINDING WHEEL

Proper balance of a mounted grinding wheel is very important, since improper balance will greatly affect the surface finish and accuracy of the work. Excessive imbalance creates vibration, which will damage the spindle bearings.

There are two methods of balancing a wheel:

1. *Static balancing.* On some grinders, the wheel is balanced off the machine with the use of a balancing stand and arbor. Counterweights in the wheel flange must be correctly positioned in order to balance the grinding wheel.

2. *Dynamic balancing.* Most new grinding machines are equipped with ball-bearing balancing devices, which automatically balance a wheel in a matter of seconds while it is revolving on the grinder.

After the wheel has been mounted on the adapter, it should be balanced, if provision is made in the adapter for balancing.

To Balance a Grinding Wheel

1. Mount the wheel and adapter on the surface grinder and true the wheel with a diamond dresser.

2. Remove the wheel assembly and mount a special tapered balancing arbor in the hole of the adapter.

3. Place the wheel and arbor on a balancing stand (Fig. 81-5) that has been leveled.

4. Allow the wheel to rotate until it stops. This will indicate that the heavy side is at the bottom. Mark this point with chalk.

5. Rotate the wheel and stop it at three positions, one-quarter, one-half, and three-quarters of a turn, to check the balance. If the wheel moves from any of these positions, it is not balanced.

6. Loosen the setscrews in the wheel counterbalances, in the grooved recess of the flange, and move the counterbalances opposite the chalk mark (Fig. 81-6).

■ Figure 81-5 A grinding wheel balancing stand. *(The DoAll Co.)*

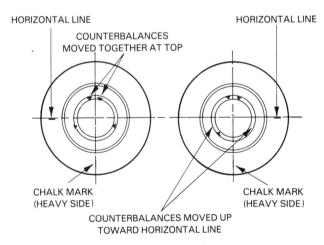

■ Figure 81-6 Adjusting counterbalances to balance a grinding wheel. *(Milacron, Inc.)*

7. Check the wheel in the four positions mentioned in steps 4 and 5.

8. Move the counterbalances around the groove an equal amount on each side of the centerline and check for balance again.

9. Continue to move the balances away from the heavy side until the wheel remains stationary at any position.

10. Tighten the counterbalances in place.

GRINDER SAFETY

All machines and cutting tools are safe when used properly and carefully; however, the grinding wheel has a very dangerous potential, especially since most wheels are operating at a rim speed of 6500 sf/min (73.9 mph). Using a wheel that has been previously damaged, or using it improperly, can be the cause of a serious accident.

> **STOP** The following general precautions should be observed when using any type of grinding wheel.

1. *Use the right wheel.* If a wheel is designed for grinding on its periphery, *do not grind on its side.* If a wheel is designed for grinding on its side, *do not grind on its periphery.*

2. *Test the wheel.* Ring test the wheel before mounting by tapping it with the end of a screwdriver or wooden hammer (Fig. 81-7). Vitrified wheels will give off a clear ring, while resinoid wheels must be inspected by eye.

3. *Blotters.* Always use the mounting blotters supplied with the wheel to ensure even pressure of the flanges on the side of the wheel and around the arbor hole.

4. *Tightening flange bolts.* Clamping nuts on arbor-mounted wheels should be tightened only enough to prevent the wheel from slipping. On large wheels, tighten the flange bolts to 15 ft-lb with a torque wrench (Fig. 81-8). Excessive tightening pressure may spring the flange and collar and possibly crack the wheel.

■ Figure 81-7 Before mounting and using any grinding wheel, be sure to ring test it for soundness. *(Milacron, Inc.)*

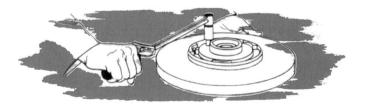

■ Figure 81-8 Tighten the flanges for large wheels in alternate sequence to 15 ft-lb. *(Milacron, Inc.)*

■ Figure 81-9 Always wear safety glasses in a machine shop, especially when grinding. *(Milacron, Inc.)*

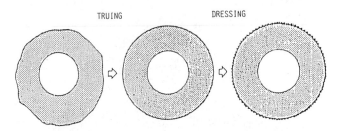

TRUING   DRESSING

■ Figure 81-10 Truing makes the wheel round and true with its axis; dressing sharpens the wheel. *(Kelmar Associates)*

5. *Flanges.* Be sure that flanges are flat and free from burrs and gouges.

6. *Arbor holes.* The wheel should slip freely, but not loosely, onto the spindle arbor. If it is too tight or too loose, do not alter the size of the hole; replace the wheel with a new one.

7. *Maximum speed.* Do not exceed the maximum speed or safe speed marked on the wheel or blotter. Use of speeds higher or lower may result in a serious accident.

8. *Wheel guards.* Always use the wheel guard supplied with each grinder. A proper wheel guard should cover about one-half of the grinding wheel.

9. *Starting the wheel.* Stand to one side whenever a wheel is started; a defective or cracked wheel will break when it reaches full operating speed (approximately 1 min).

10. *Safety glasses.* Always wear safety glasses when grinding to protect your eyes from fine abrasive particles and chips (Fig. 81-9).

TRUING AND DRESSING A GRINDING WHEEL

After mounting a grinding wheel, it is necessary to true the wheel to ensure that it will be concentric with the spindle.

Truing is the process of making a grinding wheel round and concentric with its spindle axis and producing the required form or shape on the wheel (Fig. 81-10). This procedure involves the grinding, or wearing away, of a portion of the abrasive section of a grinding wheel in order to produce the desired form or shape.

Dressing a wheel is the operation of removing the dull grains and metal particles. This operation exposes sharp cutting edges of the abrasive grains to make the wheel cut better (Fig. 81-10). A dull, glazed, or loaded wheel should be dressed for the following reasons:

1. To reduce the heat generated between the work surfaces and the grinding wheel

2. To reduce the strain on the grinding wheel and the machine

■ Figure 81-11 A diamond dresser used to true and dress a grinding wheel. *(Kelmar Associates)*

3. To improve the surface finish and accuracy of the work

4. To increase the rate of metal removal

An industrial diamond, mounted in a suitable holder on the magnetic chuck, is generally used to true and dress a grinding wheel (Fig. 81-11).

To True and Dress a Grinding Wheel

1. Check the diamond for wear and, if necessary, turn it in the holder to expose a sharp cutting edge to the wheel.

Note: Most diamonds are mounted in the holder at an angle of 5° to 15° from the vertical to prevent the possibility of chatter and of the diamond digging into the wheel. It also permits the diamond to wear on an angle so that a sharp cutting edge may be obtained by merely turning the diamond in the holder.

2. Clean the magnetic chuck thoroughly with a cloth and wipe it with the palm of the hand to remove all grit and dirt.

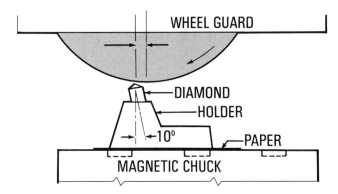

Figure 81-12 Proper positioning of the diamond dresser in relation to the grinding wheel. *(Kelmar Associates)*

3. Place a piece of paper, slightly larger than the base of the diamond holder, on the left-hand end of the magnetic chuck. This prevents scratching of the chuck when the diamond holder is being removed.

4. Place the diamond holder on the paper, covering as many magnetic inserts as possible, and energize the chuck. The diamond should be pointing in the same direction as the grinding wheel rotation (Fig. 81-12).

Note: The diamond dresser should be mounted on the left-hand end of the chuck to prevent flying grit, created during the dressing operation, from pitting and damaging the surface of the magnetic chuck.

5. Raise the wheel above the height of the diamond.

6. Move the table longitudinally so that the diamond is offset approximately 1/2 in. (13 mm) to the left of the centerline of the wheel.

7. Adjust the table laterally so that the diamond is positioned under the high point on the face of the wheel (Fig. 81-13). This is important, since grinding wheels will wear more quickly on the edges of the wheel, leaving the center of the face higher than the edges.

8. Start the wheel revolving and carefully lower the wheel until the high point touches the diamond.

9. Move the table laterally, using the crossfeed handwheel to feed the diamond across the face of the wheel.

10. Lower the grinding wheel about .001 to .002 in. (0.02 to 0.05 mm) per pass, and rough-dress the face of the wheel until it is flat and has been dressed all around the circumference.

11. Lower the wheel .0005 in. (0.01 mm) and take several passes across the face of the wheel. The rate of crossfeed will vary with the structure of the

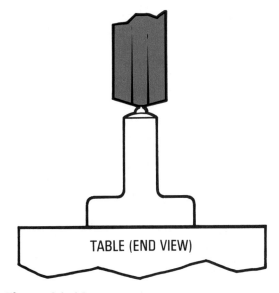

Figure 81-13 Positioning the diamond under the high point of the wheel. *(Kelmar Associates)*

wheel. A rule of thumb is to use a fast crossfeed with coarse wheels and a slow but regular crossfeed for fine, closely spaced grains.

The following additional points may be helpful when truing or dressing a grinding wheel:

1. To minimize wear on the diamond, rough-dress the grinding wheel with an abrasive stick.

2. If coolant is to be used during the grinding operation, use coolant when dressing the wheel. This will protect the diamond and the wheel from excessive heat.

3. A loaded wheel is indicated by a discoloration on the periphery or grinding wheel face. When dressing the wheel, take off sufficient material to remove completely any discoloration on the wheel face.

4. If rapid removal of metal is more important than the surface finish, do not finish-dress the wheel. After the wheel has been rough-dressed, some operators will take a final pass of .001 to .002 in. (0.02 to 0.05 mm) at a high rate of feed. The rough surface produced by this operation will remove the metal more rapidly than a finish-dressed wheel.

▶▶ Work-Holding Devices

MAGNETIC CHUCK

In some surface-grinding operations, the work may be held in a vise, held on V-blocks, or bolted directly to the table. However, most of the ferrous work ground on a surface grinder is held on a *magnetic chuck,* which is clamped to the table of the grinder.

Magnetic chucks may be of two types: the electromagnetic chuck and the permanent magnetic chuck.

The *electromagnetic chuck* uses electromagnets to provide the holding power. It has the following advantages.

> The holding power of the chuck may be varied to suit the area of contact and the thickness of the work.

> A special switch neutralizes the residual magnetism in the chuck, permitting the work to be removed easily.

The *permanent magnetic chuck* provides a convenient means of holding most workpieces to be ground. The holding power is provided by means of permanent magnets. The principle of operation is the same for both electromagnetic and permanent magnetic chucks.

PERMANENT MAGNETIC CHUCK CONSTRUCTION

Refer to Fig. 81-14. The *base plate* provides a base for the chuck and a means of clamping it to the table of the grinder.

The *grid,* or *magnetic pack,* houses the *magnets* and the *grid conductor bars.* It is moved longitudinally by a handle when the chuck is placed in the ON or OFF position.

The *case* houses the grid assembly and permits longitudinal movement of the grid. It also provides an oil reservoir for the lubrication of the moving parts.

The *top plate* contains *inserts* or *pole pieces,* which are separated magnetically from the surrounding plate by means of white metal. This separation provides the poles necessary to conduct the magnetic lines of flux.

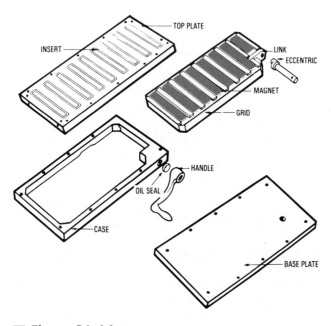

■ **Figure 81-14** Construction of a permanent magnetic chuck. *(James Neill & Co.)*

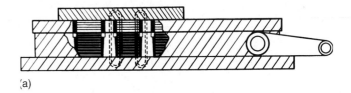

(a)

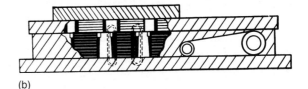

(b)

■ **Figure 81-15** (a) Magnetic chuck in the ON position; (b) magnetic chuck in the OFF position. *(James Neill & Co.)*

When the work is placed on the face of the chuck (top plate) and the handle moved to the ON position (Fig. 81-15a), the grid conductor bars and the inserts in the top plate are in line. This permits the magnetic flux to pass through the work, holding it onto the top plate.

When the handle is rotated 180° to the off position (Fig. 81-15b), the grid assembly moves the grid conductor bars and the inserts out of line. In this position, the magnetic lines of flux enter the top plate and inserts, but not the work.

MAGNETIC CHUCK ACCESSORIES

Often it is not possible to hold all work on the chuck. The size, shape, and type of work will dictate how the work should be held for surface grinding. The holding power of a magnetic chuck is dependent on the size of the workpiece, the area of contact, and the thickness of the workpiece. A highly finished piece will be held better than a poorly machined workpiece.

Very thin workpieces will not be held too securely on the face of a magnetic chuck because there are too few magnetic lines of force entering the workpiece (Fig. 81-16a).

An *adapter plate* (Fig. 81-16b) is used to securely hold thin work [less than 1/4 in. (6 mm)]. The plate's alternate layers of steel and brass convert the wider pole spacing of the chuck to finer spacing with more but *weaker* flux paths. This method is particularly suited to small, thin pieces and reduces the possibility of distortion when thin work is ground.

Magnetic chuck blocks (Fig. 81-17) provide a means of extending the flux paths to hold workpieces that cannot be held securely on the chuck face. V-blocks may be used to hold round or square stock for *light* grinding.

Note: Set the chuck blocks so that the maximum number of magnetic loops pass through the workpiece. They must also be placed so that the laminations are in line with the inner poles or inserts.

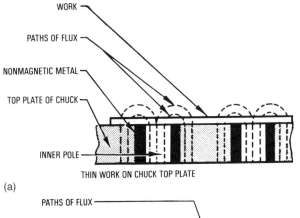

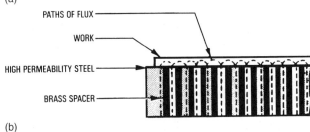

Figure 81-16 (a) Thin work held on a magnetic chuck top plate attracts fewer magnetic lines of force; (b) thin work held on an adapter plate. *(James Neill & Co.)*

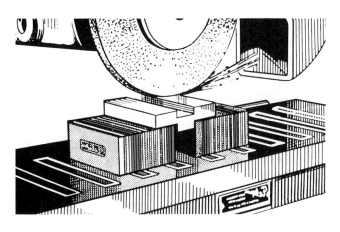

Figure 81-17 Applications of magnetic chuck blocks. *(James Neill & Co.)*

If the following points are observed, magnetic chuck blocks will last longer:

1. Clean thoroughly before and after use.
2. Store in a covered wooden box.
3. Check frequently for accuracy and burrs.
4. If regrinding is necessary to restore accuracy, take light cuts with a dressed wheel. Use coolant when grinding to prevent the magnetic chuck blocks from heating (even slightly).

When it is required to grind an angle on a workpiece, the work may be set up with a sine bar and clamped to an angle plate. Often a *sine chuck* (Fig. 81-18), which is a form of a magnetic sine plate, is used to hold the work. The buildup for the angles is the same as for the sine bar. Compound sine chucks, which have two plates hinged at right angles to each other, are available for grinding compound angles.

Magna-vise clamps (Fig. 81-19 on p. 702) may be used when the workpiece does not have a large bearing area on the chuck or when the work is nonmagnetic. These magnetically actuated clamps consist of comblike bars attached to a solid bar by a piece of spring steel. When work is held with these clamps, the solid bar of one clamp is placed against the backing plate of the magnetic chuck. The work is placed on the chuck surface between the toothed edges of two clamps, as shown in Fig. 81-19. The toothed edges on the bars in contact with the work should be above the magnetic chuck face. When the chuck is energized, the jaws of the clamps are brought down toward the face of the chuck, locking the work in place.

Double-faced tape is often used for holding thin, nonmagnetic pieces on the chuck for grinding. The tape, having two adhesive sides, is placed between the chuck

Figure 81-18 A compound sine chuck. *(The Taft-Peirce Manufacturing Company)*

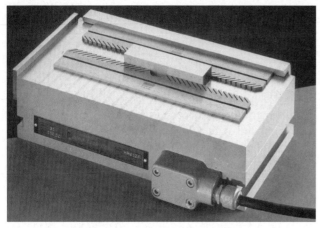

(a)

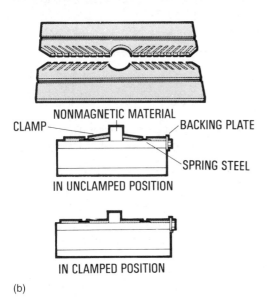

NONMAGNETIC MATERIAL

CLAMP — BACKING PLATE

— SPRING STEEL

IN UNCLAMPED POSITION

IN CLAMPED POSITION

(b)

■ **Figure 81-19** (a) Nonmagnetic workpiece being held for grinding with magna-vise clamps *(Magna Lock Corporation)*; (b) magna-vise clamps used to hold workpiece on a magnetic chuck. *(Brown & Sharpe)*

and the work, causing the work to be held securely enough for light grinding.

Special fixtures are often used to hold nonmagnetic materials and odd-shaped workpieces, particularly when a large number of workpieces must be ground.

▸▸ Grinding Fluids

Although work is ground dry in many cases, most machines have provision for applying grinding fluids or coolants. Grinding fluids serve four purposes:

1. *Reduction of grinding heat,* which affects work accuracy, surface finish, and wheel wear

2. *Lubrication* of the surface between the workpiece and the grinding wheel, which results in a better surface finish

3. *Removal of swarf* (small metal chips and abrasive grains) from the cutting area

4. *Control of grinding dust,* which may present a health hazard

TYPES OF GRINDING FLUIDS

1. *Soluble oil and water,* when mixed, form a milky solution that provides excellent cooling, lubricating, and rust-resistant qualities. This solution is generally applied by flooding the surface of the work.

2. *Soluble chemical grinding fluids and water,* when mixed, form a grinding fluid that may be used with flood cooling or "through-the-wheel" cooling systems. The chemical grinding fluid contains rust inhibitors and bactericides to minimize odors and skin irritation.

3. *Straight oil grinding fluids,* generally applied by the flood system, are used where high finish, accuracy, and long wheel life are required. These fluids have better lubricating qualities than the water-soluble fluids but do not have as high a heat-dissipating capacity.

METHODS OF APPLYING COOLANTS

The *flood system* (Fig. 81-20) is probably the most common form of coolant application. By this method, the coolant is directed onto the workpiece by a nozzle and is recirculated through a system containing a reservoir, a pump, a filter, and a control valve.

■ **Figure 81-20** Many grinding operations use the flood system to keep the workpiece cool. *(DoAll Company)*

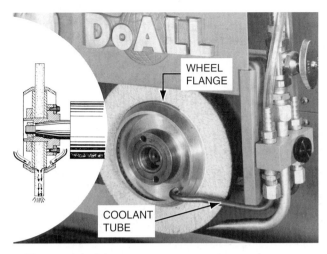

Figure 81-21 Through-the-wheel system applies coolant to the point of grinding contact. *(DoAll Company)*

Figure 81-22 Poor surface finish may be caused by a clogged wheel or by using the wrong grinding wheel. *(Kelmar Associates)*

Through-the-wheel cooling provides a convenient and efficient method of applying coolant to the area being ground. The fluid is pumped through a tube and discharged into a dovetailed groove in the wheel flange (Fig. 81-21). Holes through the flange and corresponding holes in the rubber washer permit the fluid to be discharged into the porous grinding wheel. The centrifugal force, created by the high-speed rotation of the wheel, forces the fluid through the wheel onto the area of contact between the wheel and the work. Some machines have a coolant reservoir above the grinding wheel guard that feeds the coolant into the wheel flange groove by gravity.

The *mist cooling system,* which supplies coolant in the form of a mist, uses the atomizer principle. Air passes through a line containing a T connection that leads to the coolant reservoir. The velocity of the air as it passes through the T connection draws a small amount of coolant from the reservoir and discharges it through a small nozzle in the form of vapor. The nozzle is directed to the point of contact between the work and the wheel. The air and the vapor, as it evaporates, cause the cooling action. The force of the air also blows away the swarf.

▶▶ Surface Finish

The finish produced by a surface grinder is important, and the factors affecting it should be considered. Some parts that are ground do not require a fine surface finish, and time should not be spent producing fine finishes if they are not required.

The following factors affect the surface finish:

> *Material being ground.* Soft material, such as brass and aluminum, will not permit as high a finish as harder ferrous materials. A much finer finish can be produced on hardened-steel workpieces than can be produced on soft steel or cast iron.

> *Amount of material being removed.* If a large amount of material is to be removed, a coarse-grit, open-structure wheel should be used. This will not produce as fine a finish as a fine-grit, dense wheel.

> *Grinding wheel selection.* A wheel containing abrasive grains that are friable (fracture easily) will produce a better finish than a wheel made up of tough grains. A fine-grit, dense-structure wheel produces a smoother surface than a coarse-grit, open wheel. A grinding wheel that is too soft releases the abrasive grains too easily, causing them to roll between the wheel and the work, creating deep scratches in the work (Fig. 81-22).

> *Grinding wheel dressing.* An improperly dressed wheel will leave a pattern of scratches on the work. Care should be taken when finish-dressing the wheel to move the diamond slowly across the wheel face. Always dress the wheel sufficiently to expose new abrasive grains and ensure that all glazing or foreign particles have been removed from the periphery of the wheel. New grinding wheels *that have not been properly balanced and trued* will produce a chatter pattern on the surface of the work.

> *Condition of the machine.* A light machine or one with loose spindle bearings will not produce the accuracy and fine surface finish possible in a rigid machine with properly adjusted spindle bearings. Also, to ensure optimum accuracy and surface finish, the machine should be kept clean.

> *Feed.* Coarse feeds tend to produce a rough finish. If "feed lines" persist when a fine feed is used, the wheel edges should be rounded slightly with an abrasive stick.

Grinding Machines and Process

1. Discuss how grinding has contributed to interchangeable manufacture.

2. How has the role of grinding changed over the years?

3. Outline the action that takes place during a grinding operation.

4. List five important rules that apply to any grinding operation.

Surface Grinding

5. Define surface grinding.

6. Name four different types of surface grinders and briefly outline the principle of each.

Parts of the Hydraulic Surface Grinder

7. Name five *main* parts of the hydraulic surface grinder.

8. Name and state the purposes of five controls found on the hydraulic surface grinder.

Grinding Wheel Care

9. List four points to be observed in grinding wheel care.

To Mount a Grinding Wheel

10. List the steps required to mount a grinding wheel.

To Balance a Grinding Wheel

11. Why is the proper balance of a grinding wheel essential?

12. Describe briefly the procedure for balancing a grinding wheel.

To True and Dress a Grinding Wheel

13. Define truing and dressing.

14. Why are most diamond dressers mounted at an angle of 10° to 15° to the base?

15. Explain how a grinding wheel should be finish-dressed for:
    a. Rough grinding
    b. Finish grinding

Work-Holding Devices

16. List the advantages and disadvantages of an electromagnetic chuck.

17. Describe the construction and operation of a permanent magnetic chuck.

Magnetic Chuck Accessories

18. How are thin workpieces held for surface grinding? Why is this necessary?

19. What precautions must be observed when using magnetic chuck blocks?

20. Describe and state the purpose of magna-vise clamps.

Grinding Fluids

21. State four purposes of grinding fluids.

22. Name and describe three methods of applying coolant.

Surface Finish

23. List any five factors that affect the surface finish on the part being ground. How do these factors affect the surface finish?

Surface Grinding Operations

OBJECTIVES

After completing this unit, you should be able to:

1 Set up various workpieces for grinding

2 Observe the safety rules to operate the grinder

3 Grind flat, vertical, and angular surfaces

The surface grinder is used primarily for grinding flat surfaces on hardened or unhardened workpieces.

Since the workpiece can be held by various methods and the wheel face can be shaped by dressing, it is possible to perform operations such as form, angular, and vertical grinding. Surface grinding brings the work to close tolerances and produces a high surface finish.

Good surface grinding results depend on several factors, such as the proper mounting of the work and the proper wheel selection for the job. Knowledge of the controls and grinding safety practices is of the utmost importance before anyone attempts to use a surface grinder.

▶▶ Mounting the Workpiece for Grinding

The size, shape, and type of work will determine the method by which the work should be held for surface grinding.

FLAT WORK OR PLATES

1. Remove all burrs from the surface of the work.
2. Clean the chuck surface with a clean cloth and then wipe the palm of the hand over the surface to remove dirt.
3. Place a piece of paper slightly larger than the workpiece in the center of the magnetic chuck face.
4. Place the work on top of the paper, and be sure to straddle as many magnetic inserts as possible.

5. If the workpiece is warped and rocks on the chuck face, shim the work where necessary to prevent rocking. This will avoid distortion when the work is removed from the magnetic chuck.
6. Turn the handle to the ON position.
7. Check the work to see that it is held securely by trying to remove the workpiece.

THIN WORKPIECES

Thin workpieces tend to warp because of the heat created during the grinding operation. To minimize the amount of heat generated, it is advisable to mount the workpiece at an angle of approximately 15° to 30° from the side of the chuck (Fig. 82-1 on p. 706). This reduces the length of time the wheel is in contact with the work, which in turn reduces the amount of heat generated per pass. If an adapter is available, it should be used, and the work mounted at an angle.

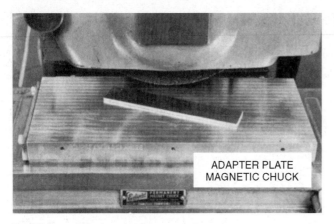

Figure 82-1 A small, thin workpiece should be set at an angle on the adapter plate to minimize warpage caused by grinding heat. *(Kelmar Associates)*

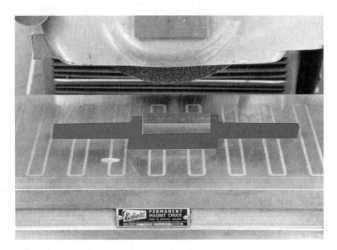

Figure 82-2 Steel blocks or parallels are placed around a short workpiece to prevent it from moving during grinding. *(Kelmar Associates)*

SHORT WORKPIECES

Work that does not straddle three magnetic poles will generally not be held firmly enough for grinding. It is advisable to straddle as many poles as possible and to set parallels or steel pieces around the work to prevent it from moving during the grinding operation (Fig. 82-2). The parallels or steel pieces should be slightly thinner than the workpiece to provide maximum support.

▶▶ Grinder Safety

When operating any type of grinder, it is important that certain basic, time-tested safety precautions be observed. Generally, the safest grinding practice is also the most efficient.

1. Before mounting a grinding wheel, ring test the wheel to check for defects.

2. Be sure that the grinding wheel is properly mounted on the spindle.

3. See that the wheel guard covers at least one-half the wheel.

4. Make sure that the magnetic chuck has been turned on by trying to remove the work.

5. See that the grinding wheel clears the work before starting a grinder.

6. Be sure that the grinder is operating at the correct speed for the wheel being used.

7. When starting a grinder, always stand to one side of the wheel and make sure no one is in line with the grinding wheel in case it breaks on startup.

8. Never attempt to clean the magnetic chuck or mount and remove work until the wheel has stopped completely.

9. *Always* wear safety glasses when grinding.

▶▶ Grinding Operations

The most common operation performed on a surface grinder is the grinding of flat (horizontal) surfaces. Regardless of the type of grinding operation, it is important that the correct wheel be mounted and the work held securely.

TO GRIND A FLAT (HORIZONTAL) SURFACE

1. Remove all burrs and dirt from the workpiece and the face of the magnetic chuck.

2. Mount the work on the chuck, placing a piece of paper between the chuck and the workpiece.

Note: Paper is used so that the work may be easily lifted off the magnetic chuck rather than sliding it. Sliding traps the abrasive grains between the work and the chuck face, which will scratch and damage the chuck.

3. Check to see that the work is held firmly.

4. Set the table reverse dogs so that the center of the grinding wheel clears each end of the work by approximately 1 in. (25 mm).

5. Set the crossfeed for the type of grinding operation—roughing cuts, .030 to .050 in. (0.76 to 1.27 mm); finishing cuts, .005 to .020 in. (0.12 to 0.5 mm).

6. Bring the work under the grinding wheel by hand, *having about ⅛ in. (3 mm) of the wheel edge over the work* (Fig. 82-3).

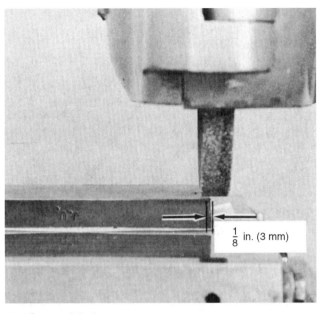

Figure 82-3 The edge of the wheel should overlap the workpiece by about ⅛ in. (3 mm). *(Kelmar Associates)*

$\frac{1}{8}$ in. (3 mm)

7. Start the grinder and lower the wheelhead until the wheel just sparks the work.

8. The wheel may have been set on a low spot of the work. It is good practice, therefore, to always raise the wheel about .005 in. (0.12 mm).

Note: Cutting fluid should be used whenever possible to aid the grinding action and keep the work cool.

9. Start the table traveling automatically and feed the entire width of the work under the wheel to check for high spots.

10. Lower the wheel for every cut until the surface is completed—roughing cuts, .001 to .003 in. (0.02 to 0.07 mm); finishing cuts, .0005 to .001 in. (0.01 to 0.02 mm).

Note: If the surface finish of the work is not satisfactory, refer to Table 82.1 for the possible cause and suggested remedies.

table 82.1   Surface grinding problems, causes, and remedies

Grinding Problem	Cause	Possible Remedy
Burning or discoloration	Wheel is too hard.	Use a softer, free-cutting wheel. Decrease wheel speed. Increase work speed. Coarse-dress the wheel. Take lighter cuts and dress the wheel frequently. Use coolant directed at the point of contact between the wheel and the work.
Burnished work surface (work is highly polished in irregular patches)	Wheel is glazed.	Dress the wheel. Use a coarser-grit wheel. Use a softer wheel. Use a more open-structure wheel.
Chatter or wavy pattern	Wheel is out of balance. Wheel is out of round. Spindle bearing is too loose. Wheel is too hard. Glazing of wheel.	Rebalance. True and dress. Adjust or replace bearings. Use a softer wheel, coarser grit, or more open structure. Increase table speed. Redress the wheel.
Scratches on the work surface	Grinding wheel is too soft. (Abrasive grains break off too readily and catch between the wheel and work surface.) Wheel is too coarse. Loose particles of swarf fall onto the work from the wheel guard. Dirty coolant carries dirt particles onto the work surface. Feed lines	Use a harder wheel. Use a finer-grit wheel. Clean the grinding wheel guard when changing a wheel. Clean the coolant tank and replace the coolant. Slightly round the edges of the wheel.

11. Release the magnet and remove the workpiece, by raising one edge, to break the magnetic attraction. This will prevent scratching the chuck surface.

TO GRIND THE EDGES OF A WORKPIECE

Much work which is machined on a surface grinder must have the edges ground square and parallel so that these edges may be used for further layout or machining operations.

Work which is to be ground all over should be machined to about .010 in. (0.25 mm) over the finished size for each surface. The large, flat surfaces are usually ground first, which then permits them to be used as reference surfaces for further setups.

When the four edges of a workpiece must be ground, clamp the work to an angle plate so that two adjacent sides may be ground square without moving the workpiece.

SETTING UP THE WORKPIECE

1. Clean and remove all burrs from the workpiece, the angle plate, and the magnetic chuck.

2. Place a piece of paper that is slightly larger than the angle plate on the magnetic chuck.

3. Place one end of the angle plate on the paper (Fig. 82-4).

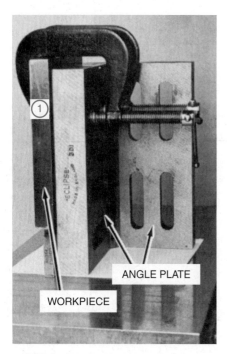

■ **Figure 82-4** The workpiece may be clamped to an angle plate for grinding the edges square. *(Kelmar Associates)*

4. Place a flat-ground surface of the workpiece against the angle plate so that the top and one edge of the workpiece project about ½ in. (13 mm) beyond the edges of the angle plate (Fig. 82-4).

Note: Be sure that the one edge of the work does not project beyond the base of the angle plate. If the work is smaller than the angle plate, a suitable parallel must be used to bring the top surface beyond the end of the angle plate.

5. Hold the work firmly against the angle plate and turn on the magnetic chuck.

6. Clamp the work to the angle plate and set the clamps so that they will not interfere with the grinding operation.

Note: Place a piece of soft metal between the clamp and the work to prevent marring the finished surface.

7. Turn off the magnetic chuck and carefully place the base of the angle plate on the magnetic chuck (Fig. 82-5).

8. Carefully fasten two more clamps on the end of the workpiece to hold the work securely.

Grinding the Edges of a Workpiece Square and Parallel

After the work has been properly set up on the magnetic chuck, the following procedure should be used for grinding the four edges of the workpiece:

1. Raise the wheelhead so that it is about ½ in. (13 mm) above the top of the work.

2. Set the table reverse dogs so that each end of the work clears the grinding wheel by about 1 in. (25 mm).

■ **Figure 82-5** A workpiece set up for grinding the first edge. *(Kelmar Associates)*

3. With the work under the center of the wheel, turn the crossfeed handle until *about ⅛ in. (3 mm) of the wheel edge overlaps the edge of the work* (Fig. 82-3).

4. Start the grinding wheel and lower the wheelhead until the wheel just sparks the work.

5. Move the work clear of the wheel with the crossfeed handle.

6. Raise the wheel about .005 to .010 in. (0.12 to 0.25 mm) in case the wheel has been set to a low spot on the work.

7. Check for high spots by feeding the table by hand so that the entire length and width of the work passes under the wheel. Raise the wheel if necessary.

8. Engage the table reverse lever and grind the surface until all marks are removed. The depth of cut should be .001 to .003 in. (0.02 to 0.07 mm) for roughing cuts and .0005 to .001 in. (0.01 to 0.02 mm) for finishing cuts.

9. Stop the machine and remove the clamps from the right-hand end of the work.

10. Turn off the magnetic chuck and remove the angle plate and workpiece as one unit. Be careful not to jar the work setup.

11. Clean the chuck and the angle plate.

12. Place the angle plate (with the attached workpiece) on its end, with the surface to be ground at the top (Fig. 82-6).

13. Fasten two clamps to the right-hand side of the workpiece and the angle plate.

14. Remove the original clamps from the top of the setup.

15. Repeat steps 1 to 8 and grind the second edge.

16. Remove the assembly from the chuck and remove the workpiece from the angle plate.

Grinding the Third and Fourth Edges

When two adjacent sides have been ground, they are then used as reference surfaces to grind the other two sides square and parallel.

1. Clean the workpiece, the angle plate, and the magnetic chuck thoroughly and remove any burrs.

2. Place a clean piece of paper on the magnetic chuck.

3. Place a ground edge of the workpiece on the paper.

   a. If the workpiece is at least 1 in. (25 mm) thick and long enough to span three magnetic poles on the chuck, and no more than 2 in. (50 mm) high, no angle plate is required (Fig. 82-7).

   b. If the work is less than 1 in. (25 mm) thick and does not span three magnetic poles, it should be fastened to an angle plate (Fig. 82-8 on p. 710).

      1. Place a ground edge on the paper and place an angle plate no higher than the workpiece against the workpiece.

Note: A suitable parallel may be required to raise the edge of the work above the edge of the angle plate.

      2. Turn on the chuck and carefully clamp the work to the angle plate.

■ Figure 82-6 Angle plate and work set for grinding the second edge of the workpiece at 90° to the first edge. *(Kelmar Associates)*

■ Figure 82-7 A workpiece having a sufficient bearing surface may be set on the chuck for finishing the remaining edges. *(Kelmar Associates)*

Figure 82-8 An angle plate may be required to finish the third and fourth edges. *(Kelmar Associates)*

4. Grind the third edge to the required size.

5. Repeat operations 1 to 3 and grind the fourth edge.

▶▶ Grinding a Flat Surface with a CBN Wheel

Cubic boron nitride (CBN) grinding wheels can grind hardened tool and die steels more efficiently than aluminum oxide wheels. CBN wheels increase productivity, improve work quality, and reduce grinding costs. CBN wheel wear is slow and uniform, and wheel life is long. The low wheel wear makes it easier to control part size, and there is no need to adjust the grinder to compensate for wheel wear. Also, it is not necessary to dress the wheel frequently to maintain straightness or form.

CBN grinding wheels grind cool and there is almost no danger of burning the workpiece. Because of the absence of thermal damage, tools ground with CBN wheels stay sharp much longer.

CONDITIONING CBN WHEELS

To get the best performance out of CBN grinding wheels, it is very important that they are conditioned (trued and dressed) properly. *If this is not done correctly, CBN wheels will not cut.*

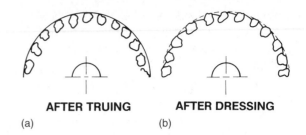

AFTER TRUING   **AFTER DRESSING**
(a)   (b)

Figure 82-9 (a) After truing, the wheel face is smooth; (b) after dressing, the abrasive grits are exposed. *(Norton Co.)*

Truing is the operation of making a grinding wheel round and concentric with the spindle axis. Truing usually leaves the grinding surface of a wheel smooth, with little or no abrasive crystals above the wheel surface for chip removal (Fig. 82-9a). *A wheel in this condition cannot cut* and will burn the workpiece.

Dressing is the operation of removing some of the bond material from the surface of a trued wheel to expose the abrasive crystals and allow the wheel to cut (Fig. 82-9b).

A properly trued and dressed wheel will:

> Produce accurate workpieces and good surface finish

> Use a minimum of grinding power

> Produce work without burn, surface damage, or chatter marks

> Increase material-removal rates and lower grinding costs

TRUING A CBN WHEEL

The most common truing devices used in small industries and school shops are the impregnated diamond nibs and the brake-controlled truing devices. These devices should not be used on CBN grinding wheels over 8 or 10 in. (200 or 250 mm) in diameter. Resin-bond CBN wheels with 100 concentration are the most common wheels used for the general grinding of hardened ferrous materials and most cutting tools. The following example will use an impregnated diamond wheel to true a CBN resin-bond wheel.

step	procedure
1	Mount the CBN wheel on the grinder, snug up the flange nut, and indicate the wheel circumference to within .001 in. (0.02 mm) or less runout.
2	Tighten the flange nut securely.
3	Clean the magnetic chuck and place the diamond holder on the left-hand side of the

Figure 82-10 The diamond nib is set about ½ in. (13 mm) to the left of the wheel center for the truing operation.

chuck, covering as many magnetic inserts as possible; energize the chuck.

4   Adjust the grinder table to locate the diamond nib about ½ in. (13 mm) to the left of the wheel centerline (Fig. 82-10); lock the table in position.

5   Lower the wheelhead by hand until it just touches a piece of paper held between the wheel and the diamond.

6   Turn the grinder crossfeed handle so that the diamond clears the edge of the wheel.

7   Lightly coat the wheel surface to be trued with a wax marking crayon.

8   Start the grinder spindle and feed the wheelhead down in .0004-in. (0.01-mm) increments until contact is made with the diamond.

9   Direct the grinder coolant to the wheel-diamond interface (Fig. 82-10).

10   Feed the diamond across the wheel face at 3 to 12 in./min (75 to 300 mm/min). Be sure that the diamond clears the wheel in every pass.

11   Continue truing passes of .0004 in. (0.01 mm) until the crayon is removed from the entire wheel circumference. *Do not overtrue—it wastes expensive abrasive.*

DRESSING A CBN WHEEL

After the truing operation, the wheel surface is smooth, with no abrasive grains sticking out; this wheel cannot cut. Some of the wheel bond must be removed (lowered) to expose the abrasive crystals so that they can cut the work material. The simplest, least expensive, and most popular method of dressing CBN grinding wheels is with an aluminum oxide dressing stick or block.

step	procedure
1	Select a 200-grit, C-grade aluminum oxide dressing stick.
2	Hold the dressing stick in a vise so that about half of its thickness is above the vise jaws.
3	Mount the vise on the magnetic chuck so that the dressing block is parallel to the table travel (Fig. 82-11).
4	Lower the stationary wheel until it just touches the top of the dressing stick.
5	Use the crossfeed handle to bring the edge of the dressing stick even with the edge of the wheel.
6	Move the table so that the wheel just clears the right-hand end of the dressing stick.
7	Start the grinder spindle and coolant; feed the wheel down about .020 in. (.50 mm).
8	Using a slow but steady feedrate (5 to 10 ft/min or 1.5 to 3 m/min), take one pass lengthwise along the dressing stick.

Figure 82-11 An aluminum oxide dressing stick held in a vise for dressing a CBN wheel.

9   At the end of each pass, index the table slightly less than the width of the grinding wheel.

10   Repeat this procedure until the width of the dressing stick is ground.

11   Downfeed the wheel after the full surface of the dressing stick has been ground.

12   When the grinding wheel appears to be cutting the dressing stick freely, stop and inspect the wheel face. If the surface feels rough, the dressing operation is complete. If it still feels smooth, continue the dressing.

Note: This operation can also be done with the dressing stick held in the hand but should not be attempted by students or inexperienced persons.

GRINDING A FLAT SURFACE

The most common operation on a surface grinder is that of grinding flat surfaces. Fig. 82-12 shows a piece of die steel (AISI M-4) Rc-62 that must be reground. This job will be used to show the procedures which should be followed to set up the machine wheel and grind flat surfaces. The most cost-efficient grinding occurs on a machine in good condition.

step   procedure

1   Select a 100-concentration CBN resin-bond wheel to grind the M-4 steel.

2   Mount the CBN wheel securely on the grinder spindle; *do not use wheel blotters.*

■ **Figure 82-12** A piece of hardened die steel ground with a CBN wheel. *(Diamond Innovations, www.abrasivesnet.com)*

3   True and dress the wheel to ensure that the CBN wheel will perform properly. *Unless this operation is performed correctly, the CBN wheel will not cut.*

Mounting the Workpiece

4   Remove burrs from the magnetic chuck and workpiece, and thoroughly clean the chuck surface.

5   Place a piece of smooth paper between the work and the magnetic chuck; energize the magnetic chuck.

Setting Speeds and Feeds

6   Adjust the table reverse dogs so that the center of the grinding wheel clears each end of the work to be ground by approximately 1 in. (25 mm).

7   Set the table crossfeed:

      a. Rough grinding—one-quarter to one-half the wheel width
      b. Finish grinding—smaller crossfeed increments

8   Set the table speed rate from 50 to 100 ft/min (15 to 30 m/min).

9   Set the spindle speed for the size and type of CBN wheel used.

Setting the Wheel to the Work Surface

10   Set the wheel to the top of the work surface in the usual manner.

11   Traverse the workpiece under the revolving wheel to locate the high spot of the work surface.

12   Move the table so that the wheel clears the edge of the work surface to be ground.

Coolant

13   Use the proper grinding fluids to suit the wheel and workpiece.

14   Stop the grinder spindle and adjust the coolant nozzle so that it is about ¼ in.

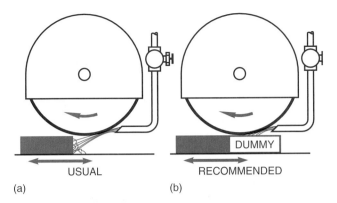

Figure 82-13 A dummy block makes sure that the coolant is at the start of each table pass.

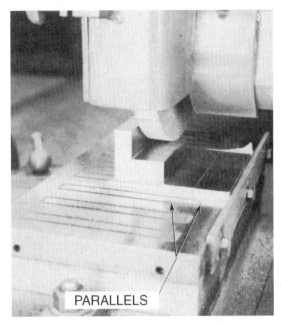

Figure 82-14 The setup for vertical grinding. *(Kelmar Associates)*

(6 mm) above the work surface and as close to the wheel face as possible.

15   Place a dummy block, slightly lower than the work surface, at the right-hand end so that the entire surface receives coolant at all times (Fig. 82-13).

Grinding the Surface

16   Start the grinder spindle and lower the wheelhead .001 in. (0.02 mm) for the first cut.

17   Start the coolant flow, ensuring that a good supply is directed to the point of wheel-workpiece contact.

18   Start the table reciprocating and engage the crossfeed to take a roughing pass across the work surface.

19   Be sure that the edge of the grinding wheel clears the side of the work after each pass.

20   Take as many passes at .001-in. (0.02-mm) depth of cut in order to grind the surface.

21   Set the wheelhead for .0005-in. (0.01-mm) depth of cut for the final pass to improve the surface finish.

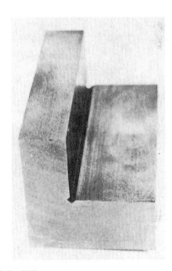

Figure 82-15 The corner of the workpiece is relieved to provide clearance and prevent the corner of the wheel from breaking down. *(Kelmar Associates)*

ing the vertical surface. Before grinding a vertical surface, it is necessary to relieve the corner of the work (Fig. 82-15) to ensure clearance for the edge of the wheel.

TO GRIND A VERTICAL SURFACE

Although most grinding performed on a surface grinder is the grinding of flat, horizontal surfaces, it is often necessary to grind a vertical surface (Fig. 82-14). Extreme care must be taken in the setup of the workpiece when grind-

step	procedure
1	Mount the proper grinding wheel; true, dress, and balance as required.
2	Dress the side of the wheel to give it a slight clearance (Fig. 82-16 on p. 714).

(a)

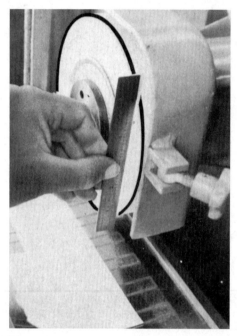

(b)

■ **Figure 82-16** (a) Dressing the side of a wheel for grinding a vertical surface; (b) the side of the wheel should be slightly concave for grinding a vertical surface. *(Kelmar Associates)*

3   Clean the surface of the magnetic chuck and mount the work.

4   With an indicator, align the edge of the work parallel to the table travel.

OR

Place the work against the stop bar of the magnetic chuck that has been aligned. If the work cannot be set against the stop bar, parallels may be used to position the work on the magnetic chuck (Fig. 82-14).

5   Turn on the magnetic chuck and test to see that the work is held securely.

6   Set the reversing dogs, allowing sufficient table travel to permit clearance for the wheel at each end of the stroke.

7   Bring the side of the wheel close to the vertical surface to be ground.

8   Lower the wheel to within .002 to .005 in. (0.05 to 0.12 mm) of the flat or horizontal surface that has been finish ground.

9   Start the table traveling *slowly* and feed the wheel across until it just sparks the vertical surface.

10   Rough-grind the vertical surface to within .002 in. (0.05 mm) of size by feeding the table in approximately .001 in. (0.02 mm) maximum per pass.

11   Redress the side of the wheel if necessary.

12   Finish-grind the vertical surface by feeding the table approximately .0005 in. (0.01 mm) maximum per pass.

TO GRIND AN ANGULAR SURFACE

When it is necessary to grind an angular surface, the work may be held at an angle by a sine bar and an angle plate (Fig. 82-17), a sine chuck (Fig. 82-18), or an adjustable angle vise (Fig. 82-19). When work is held by any of these methods, it is ground with a flat-dressed wheel.

Angular surfaces may also be ground by holding the work flat and dressing the grinding wheel to the required angle with a sine dresser (Fig. 82-20).

When a sine dresser is not available, a parallel set to the desired angle by means of a sine bar may be clamped to an angle plate. This setup is then placed on a magnetic chuck beneath the grinding wheel (Fig. 82-21).

Figure 82-17 The workpiece may be set to a sine bar for grinding an accurate angle. *(Kelmar Associates)*

Figure 82-18 Grinding an accurate angle using a sine chuck. *(DoAll Company)*

▶▶ Form Grinding

Form grinding refers to the production of curved and angular surfaces produced by means of a specially dressed wheel. The reverse form or contour required on the workpiece is dressed on the grinding wheel (Fig. 82-22 on p. 716). Contours and radii may be produced on the grinding wheel by means of a radius wheel dresser (Fig. 82-23a on p. 716).

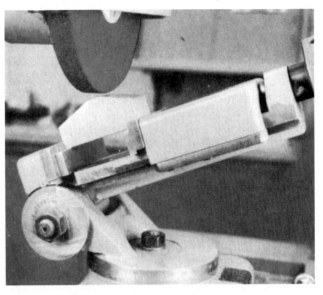

Figure 82-19 Workpiece held for grinding in an adjustable angle vise. *(Kelmar Associates)*

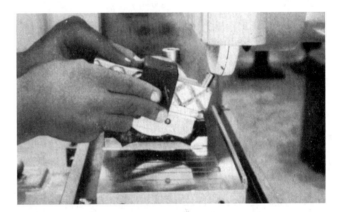

Figure 82-20 Dressing a wheel to an angle using a sine dresser. *(Kelmar Associates)*

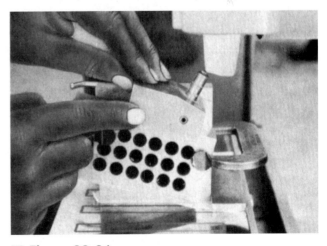

Figure 82-21 Dressing a wheel using a parallel set to an angle. *(Kelmar Associates)*

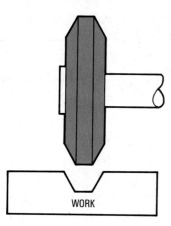

Figure 82-22 For form grinding, the wheel is dressed to the reverse profile of the form required. *(Kelmar Associates)*

TO DRESS A CONVEX RADIUS ON A GRINDING WHEEL

1. Mount the radius dresser (Fig. 82-23a) squarely on a clean magnetic chuck.

2. Set both stops on the radius dresser so that it can be rotated only one-quarter of a turn. The two stops should be 90° apart.

3. Fasten the diamond height-setting bar in the radius dresser. The bottom surface of the height-setting bar is the center of the radius dresser.

4. Place a gage block buildup using wear blocks on each side, equal to the radius required on the grinding wheel, between the height-setting bar and the diamond point.

5. Raise the diamond until it just touches the gage blocks (Fig. 82-23b) and then lock it in this position.

Note: When dressing a *concave radius,* the diamond point must be set *above* the center of the radius dresser a distance equal to the radius desired.

6. Move the table longitudinally until the diamond is under the center of the grinding wheel (Fig. 82-23a).

7. Lock the table to prevent longitudinal movement.

8. Rotate the arm of the radius dresser one-quarter of a turn so that the diamond is in a horizontal position.

9. Start the machine spindle and, using the crossfeed handle, bring the diamond in until it just touches the side of the grinding wheel.

(a)

(b)

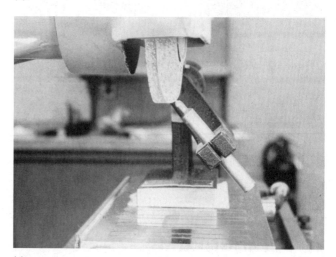

(c)

Figure 82-23 (a) A radius wheel dresser mounted on a magnetic chuck; (b) gage blocks being used to set a diamond to the correct height to dress the radius on a grinding wheel; (c) a convex radius being dressed on a grinding wheel. *(Kelmar Associates)*

10. Lock the table cross-slide in this position.

11. Stop the grinder and raise the wheel until it clears the diamond.

12. Start the grinder and, while slowly rotating the diamond back and forth through the 90° arc, lower the wheel until it just touches the diamond.

13. Feed the wheel down approximately .002 to .003 in. (0.05 to 0.07 mm) for every rotation of the dresser.

14. Continue to dress the radius until the periphery of the wheel just touches the diamond when it is in a vertical position. This indicates that the radius is completely formed (Fig. 82-23c).

15. Stop the grinder, raise the wheel, and remove the radius dresser.

SPECIAL FORMS

When complex profiles must be ground on long production runs, the wheel may often be crush-formed. A tool-steel or carbide roll, having the desired form or contour of the finished workpiece, is forced into the slowly revolving grinding wheel [200 to 300 sf/min (60 to 90 m/min)] (Fig. 82-24). The grinding wheel assumes the reverse form of the crushing roll. The wheel is then used to grind the form or contour on the workpiece (Fig. 82-25). Grinding wheels may be crush-formed to tolerances as close as ±.002 in.(0.05 mm), and to radii as small as .005 in. (0.12 mm), depending on the grit size and structure of the wheel. As the wheel is used, it will gradually wear out of tolerance and the form must be redressed, using the crush roll. When the crush roll wears, as a result of many redressings, it must be reground to the original tolerance.

Some surface grinders are not designed for crush-form dressing of the wheel. It is not advisable to perform

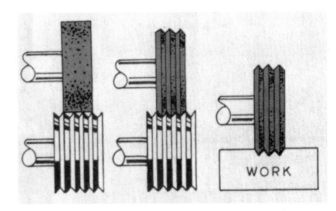

Figure 82-25 Principle of crush-form grinding.

crush-form dressing on any machine equipped with a ball-bearing spindle, since the bearings are subjected to a considerable load in crush dressing and may be damaged. Machines equipped with roller bearings have proved quite satisfactory for crush-dressing operations.

▶▶ Cutting-Off Operations

The surface grinder may be used for cutting off hardened materials by using thin cutoff wheels. The work may be clamped in a fixture or vise and positioned below the wheel (Fig. 82-26). For thin, short pieces, the wheelhead may be fed straight down to cut off the work. If longer pieces are to be cut, the work is properly mounted and the table is reciprocated as in normal grinding, while the wheel is fed down. Diamond wheels may also be mounted for cutting-off operations on carbides.

Figure 82-24 Crush dressing a wheel to grind serrations, with a roll mounted on the machine. *(DoAll Co.)*

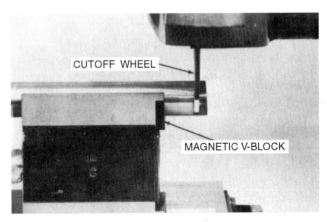

Figure 82-26 Cutting off a workpiece held in a magnetic V-block on a surface grinder. *(Kelmar Associates)*

Mounting the Workpiece for Grinding

1. Describe briefly the procedure for mounting the following for grinding:

   a. Flat work   b. Short work

Grinder Safety

2. How should a wheel be checked for defects?

3. Why is it necessary to see that no one is in line with the grinding wheel before starting the grinder?

4. List five grinder safety rules that you consider to be most important and explain the reason for the selection of each.

To Grind a Flat Surface

5. Explain how the grinding wheel should be set to the surface of the work.

6. How should work be removed from a magnetic chuck?

To Grind the Edges of a Workpiece

7. Why are the edges of workpieces often ground square and parallel?

8. How much allowance should be left on a surface for grinding?

9. What precautions should be observed when mounting a workpiece on an angle plate for grinding two adjacent sides?

10. After the first side of a workpiece has been ground, how is the work set to grind the second side square to the first?

To Grind a Vertical Surface

11. Outline briefly the procedure for grinding a vertical surface.

12. When grinding a vertical surface, why is it necessary first to relieve the corner between the two adjacent surfaces?

13. How is a grinding wheel dressed when grinding a vertical surface? Why is this necessary?

To Grind an Angular Surface

14. By what methods may work be held for grinding an angle when a flat-dressed wheel is being used?

15. Name two methods of dressing a wheel to an angle.

Form Grinding

16. Name two methods of dressing contours and radii on grinding wheels.

17. Describe the principle of crush-form dressing.

18. On what types of surface grinders should crush-form dressing be performed?

Surface Grinding Problems, Causes, and Remedies

19. List three causes of and three remedies for each of the following grinding problems:

   a. Chatter or wavy pattern
   b. Scratches on the work surface

UNIT 83

Cylindrical Grinders

OBJECTIVES

After completing this unit, you should be able to:

1 Set up and grind work on a cylindrical grinder

2 Internal-grind on a universal cylindrical grinder

3 Identify and state the principles of three methods of centerless grinding

The diameter of a workpiece can be ground accurately to size and to a high surface finish on a *cylindrical* grinder. There are two types of machines suitable for cylindrical grinding—the *center type* and the *centerless type*—each with its special applications.

▶▶ Center-Type Cylindrical Grinders

Work which is to be finished on a cylindrical grinder is generally held between centers but may also be held in a chuck. There are two types of cylindrical grinders (center type), the *plain* and the *universal*. The plain-type grinder is generally a manufacturing type of machine. The universal cylindrical grinder (Fig. 83-1 on p. 720) is more versatile, since both the wheelhead and headstock may be swiveled.

PARTS OF THE UNIVERSAL CYLINDRICAL GRINDER

The *base* is of heavy cast-iron construction to provide rigidity. The top of the base is machined to form the *ways* for the table.

The *wheelhead* is mounted on a cross-slide at the back of the machine. The ways on which it is mounted are at right angles to the table ways, permitting the wheelhead to be fed toward the table and the work, either automati-

cally or by hand. On universal machines, the wheelhead may be swiveled to permit the grinding of steep tapers by plunge grinding.

The *table,* mounted on the ways, is driven back and forth by hydraulic or mechanical means. The reversal of the table is controlled by *trip dogs.* The table is composed of the *lower table,* which rests on the ways, and the *upper table,* which may be swiveled for grinding tapers and alignment purposes. The *headstock* and *footstock,* used to support work held between centers, are mounted on the table.

The *headstock* unit is mounted on the left end of the table and contains a motor for rotating the work. A dead center is mounted in the headstock spindle. When work is mounted between centers, it is rotated on *two dead centers* by means of a dog and a driveplate, which is attached to, and revolves with, the headstock spindle. The purpose of a dead center in the headstock is to overcome any spindle inaccuracies (looseness, burrs, etc.) that may be transferred to the workpiece. Grinding work on two dead centers results in truer diameters that are concentric with the centerline of the work. Work may also be held in a chuck that is mounted on the spindle nose of the headstock.

The *footstock* supports the right end of the work and is adjustable along the length of the table. The dead center,

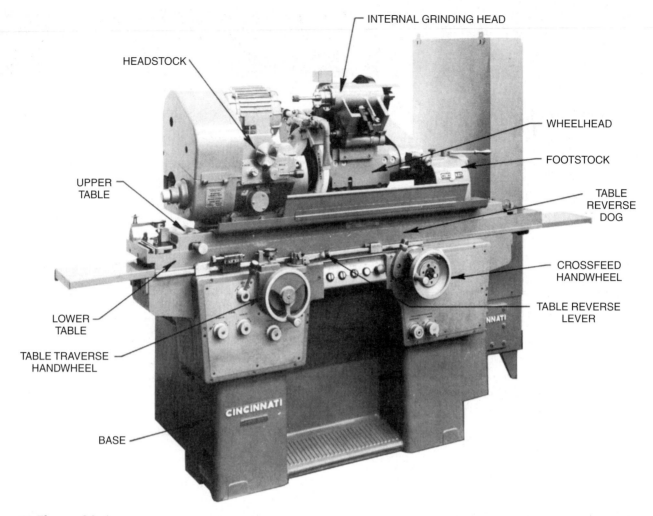

INTERNAL GRINDING HEAD

HEADSTOCK

WHEELHEAD

FOOTSTOCK

UPPER TABLE

TABLE REVERSE DOG

CROSSFEED HANDWHEEL

LOWER TABLE

TABLE REVERSE LEVER

TABLE TRAVERSE HANDWHEEL

BASE

CINCINNATI

■ Figure 83-1 The parts of a universal cylindrical grinder. *(Cincinnati Machine, Inc.)*

on which the work is mounted, is spring-loaded to provide the proper center tension on the workpiece.

The *backrest*, or *steadyrest*, provides support for long, slender work and prevents it from springing. Outward and downward movement of the workpiece is prevented by means of adjustable supports in front of and below the workpiece. It may be positioned anywhere along the length of the table.

The *center rest*, which resembles a lathe steadyrest, may be mounted at any point on the table. It is used to support the right end of the work when external grinding is confined to the end of the workpiece. Long workpieces on which internal grinding is to be performed are also supported on the end by the center rest.

An *internal grinding attachment* may be mounted on the wheelhead on most machines for internal grinding. It is usually driven by a separate motor.

A *diamond wheel dresser* may be clamped to the table to dress the grinding wheel as required. On some types of grinders, diamond dressers may be permanently mounted on the footstock.

The *coolant system* is built into all cylindrical grinders to provide dust control, temperature control, and a better surface finish on the workpiece.

▸▸ Machine Preparation for Grinding

MOUNTING THE WHEEL

All precautions, such as wheel balancing and mounting procedures, used on surface grinders should be observed for cylindrical grinding.

To True and Dress the Wheel

1. Start the grinding wheel to allow the spindle bearings to warm up.

2. Mount the proper diamond in the holder and clamp it to the table. The diamond should be mounted at

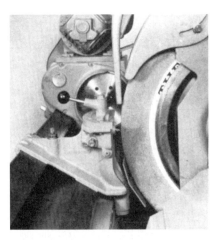

Figure 83-2 Dressing a grinding wheel on a cylindrical grinder. *(Cincinnati Machine, a UNOVA Co.)*

an angle of 10° to 15° to the wheelface and should be held on or slightly below the centerline of the wheel (Fig. 83-2).

3. Adjust the wheel until the diamond almost touches the high point of the wheel, which is usually in the center of the wheelface.

4. Turn on the coolant if it is to be used for the grinding operation.

5. Feed the wheel into the diamond about .001 in. (0.02 mm) per pass, and move it back and forth across the wheelface at a medium rate until the wheelface has been completely dressed.

6. Finish-dress the wheel by using an infeed of .0005 in. (0.01 mm) and a slow traverse feed. When a very fine finish is required, the diamond should be traversed slowly across the wheelface for a few passes without any additional infeed.

To Parallel-Grind an Outside Diameter

Grinding a parallel diameter is the most common operation performed on a cylindrical grinder. If any undesirable work characteristics occur during grinding, refer to Table 83.1 (on pp. 722 and 723) for the possible cause and suggested remedies.

1. Lubricate the machine as required.

2. Start the grinding wheel to warm up the spindle bearings. This will ensure the utmost accuracy when grinding.

3. True and dress the grinding wheel if required and then shut off the spindle.

4. Clean the machine centers and the center holes of the work. If the grinder centers are damaged, they must be reground. On hardened-steel workpieces, the center holes should be honed or lapped to ensure the utmost accuracy.

5. Align the headstock and footstock centers with a test bar and indicator.

6. Lubricate the center holes with a suitable lubricant.

7. Set the headstock and footstock for the proper length of work so that the center of the work will be over the center of the table.

8. Mount the work between centers with the dog mounted loosely on the left end of the work.

9. Tighten the dog on the end of the work and engage the driveplate pin in the fork of the dog.

10. Adjust the table dogs so that the wheel will overrun each end of the work by about one-third of the width of the wheelface. If grinding must be done up to a shoulder, the table traverse must be carefully set so that it reverses just before the wheel touches the shoulder.

11. Set the grinder to the proper speed for the wheel being used. Some machines are provided with a means of increasing the wheel speed as the wheel becomes smaller. If this is not done, the wheel will act softer and wear quickly.

12. Set the work speed for the diameter and type of material being ground. Proper work speed is very important. A slow speed causes heating and distortion of the work. High work speeds will cause the wheel to act softer and break down quickly.

13. Set the headstock spindle to revolve the work in an opposite direction to that of the grinding wheel. When grinding, the sparks should be directed down toward the table.

14. If the machine is so equipped, set the automatic infeed for each table reversal. Also set the dwell, or "tarry," time, which permits the wheel to clear itself at each end of the stroke.

15. Select and set the desired table traverse or speed. This should be such that the table will move one-half to two-thirds of the wheel width per revolution of the work. Finish grinding is done at a slower rate of table traverse.

16. Start the table and move the wheel up to the workpiece until it just sparks.

17. Engage the wheelfeed clutch lever and grind until the work is cleaned up.

18. Check the work for taper and adjust if necessary.

19. Determine the amount of material to be removed and set the feed index for this amount. The infeed of the wheel will stop automatically when the work is at the proper diameter.

20. Stop the machine with the wheel clear of the work and measure the size of the workpiece. If necessary, make a correction on the index setting and grind the piece to size.

table 83.1   Cylindrical grinding faults, causes, and remedies

Fault	Cause	Possible Remedy
"Barber pole" finish	Work loose on centers.	Adjust center tension.
	Centers are a poor fit in the spindle.	Use properly fitting centers.
Burnt work, cracked surfaces	Grinding wheel not trued.	True and dress the wheel.
	Grinding wheel too hard.	Use a softer wheel.
	Grinding wheel structure too dense.	Use a more open wheel.
	Incorrect bond.	Consult manufacturer's handbook.
	Grinding wheel too fast.	Adjust the speed.
	Work revolving too slowly.	Increase the speed.
	Too heavy a cut.	Reduce the depth of cut.
	Insufficient coolant.	Increase coolant supply.
	Wrong type of coolant.	Try another type.
	Dull diamond dresser.	Turn diamond in holder.
Chatter marks	Too heavy a cut.	Try a lighter cut.
	Grinding wheel too hard.	Use a softer wheel.
		Increase work speed.
		Reduce wheel speed.
	Work too slender.	Use a steadyrest.
	Vibrations in machine.	Locate the vibrations and correct.
	External vibrations transferred to machine.	Isolate the machine to prevent vibrations.
Diamond truing lines	Fast dressing feed.	Slow dressing feed.
	Diamond too sharp.	Use a cluster-type nib.
Feed lines	Wrong wheel structure.	Change to suit.
	Improper traverse speed when finish grinding.	Change work speed and wheel speed.
	Coolant not directed properly.	Adjust nozzle.
	Improperly adjusted steadyrest.	Check and adjust.

table 83.1   Cylindrical grinding faults, causes, and remedies (continued)

Fault	Cause	Possible Remedy
Intermittent cutting action	Work too tight on centers.	Adjust center tension.
	Work too hot.	Use coolant.
	Work out of balance.	Counterbalance as required.
Out-of-round work	Work center holes damaged or dirty.	Hone and lap centers.
	Work loose on centers.	Adjust the center tension.
	Loose machine centers.	Clean and reset.
	Machine centers worn.	Regrind centers.
	Loose gibs on the table.	Adjust the gibs.
Rough finish	Wheel too coarse.	Use a finer wheel.
	Wheel too hard.	Use a softer wheel.
	Diamond too sharp.	Use a cluster-type nib.
	Wheel rough-dressed.	Finish-dress wheel.
	Table traverse too fast.	Slow to suitable feed.
	Work speed too fast.	Slow to suitable speed.
	Work springing.	Use a steadyrest.
Wavy marks (long)	Wheel out of balance.	Redress and balance wheel.
	Coolant has been directed against a stationary wheel.	Always shut off coolant well before stopping the wheel.
Wavy marks (short)	Vibrations caused by unmatched belts.	Replace belts with a matched set.
		Check motor and pulleys for balance.

TO GRIND A TAPERED WORKPIECE

Tapered work, held between centers, is ground in the same manner as parallel work, except that the table is swiveled to half the included angle of the taper. The taper should be checked for accuracy after the workpiece is cleaned up and the table adjusted if necessary.

Short, steep tapers may be ground on work held in a chuck or between centers by swiveling the wheelhead to the desired angle and plunge-grinding the tapered surface with the face of the wheel.

PLUNGE GRINDING

When a short tapered or parallel surface is to be ground on a workpiece, it may be plunge-ground by feeding the wheel into the revolving work, with the table remaining stationary. The grinding wheel may be fed in automatically to the setting on the feed index. It then dwells for a suitable time to permit "spark out" and retracts automatically. In the case of tapered work, the wheelhead must be swung to half the included angle. The length of the surface to be ground must be no longer than the width of the grinding wheel face.

▶▶ Internal Grinders

Internal grinding may be defined as the accurate finishing of holes in a workpiece by a grinding wheel. Although internal grinders were originally designed for hardened workpieces, they are now used extensively for finishing holes to size and accuracy in soft material.

Production internal grinding is done on internal grinders designed exclusively for this type of work. The wheel is fed into the work automatically until the hole reaches the required diameter. When the hole is finished to size, the wheel is withdrawn from the hole and automatically dressed before the next hole is ground.

Internal grinding may also be performed on the universal cylindrical grinder, the cutter and tool grinder, and the lathe. Since these machines are not designed primarily for internal grinding, they are not as efficient as the standard internal grinder. For most internal grinding operations, the work is rotated in a chuck mounted on the workhead spindle. Work may also be mounted on a faceplate, collet chucks, or special fixtures. When work is too large to be rotated, internal diameters may be ground by using a planetary grinder. In this operation, the grinding wheel is guided in a circular motion about the axis of the hole and is fed out to the required diameter. The work or the grinding head is fed parallel to the wheel spindle to provide a smooth, uniform surface.

INTERNAL GRINDING ON A UNIVERSAL CYLINDRICAL GRINDER

Although the universal cylindrical grinder is not designed primarily as an internal grinder, it is used extensively in toolrooms for this purpose. On most universal cylindrical grinders, the internal grinding attachment is mounted on the wheelhead column and is easily swung into place when required. One advantage of this machine is that the outside and inside diameters of a workpiece may often be finished in one setup. Although the grade of the wheel used for internal grinding will depend on the type of work and the rigidity of the machine, the wheels used for internal grinding are generally softer than those for external grinding, for the following reasons:

1. There is a larger area of contact between the wheel and the workpiece during the internal grinding operation.

2. A soft wheel requires less pressure to cut than a hard wheel; thus, the spindle pressure and spring are reduced.

To Grind a Parallel Internal Diameter on a Universal Cylindrical Grinder

If any problems occur during internal grinding, refer to Table 83.2 for the possible cause and suggested remedies.

1. Mount the workpiece in a universal chuck, in a collet chuck, or on a faceplate. Care must be taken not to distort thin workpieces.

2. Swing the internal grinding attachment into place and mount the proper spindle in the quill. For maximum rigidity, the spindle should be as large as possible, with the shortest overhang.

3. Mount the proper grinding wheel, as large as possible, for the job.

4. Adjust the spindle height until its center is in line with the center axis of the hole in the workpiece.

5. True and dress the grinding wheel.

6. Set the wheel speed to 5000 to 6500 sf/min (1520 to 1980 m/min).

7. Set the work speed to 150 to 200 sf/min (45 to 60 m/min).

8. Adjust the table dogs so that *only* one-third of the wheel width overlaps the ends of the work at each end of the stroke. On blind holes, the dog should be set to reverse the table just as the wheel clears the undercut at the bottom of the hole.

Note: In order to prevent bell-mouthing, the wheel must *never* overlap the end of the work by more than one-half the width of the wheel.

9. Start the work and the grinding wheel.

10. Touch the grinding wheel to the diameter of the hole.

11. Turn on the coolant.

table 83.2   Internal grinding problems, causes, and remedies

Problem	Cause	Remedy
Bell-mouthed hole	Stroke is too long and wheel overlaps hole too much.	Reduce overlap of wheel at each end of hole.
	Centerlines of workpiece and wheel spindle are at different heights.	Align the centers before setting up work.
Burning or discoloration of work	Wheel too hard.	Use a softer wheel.
		Increase work speed.
		Decrease diameter of wheel.
		Use a narrower wheel.
		Coarse-dress the wheel.
	Insufficient coolant.	Increase coolant supply and direct it at the point of grinding contact.
Chatter marks	Worn spindle bearings.	Adjust if possible or replace.
	Belt slipping.	Adjust tension.
	Defective belts.	Replace the complete set of belts.
	Wheel out of balance.	Balance the wheel.
	Wheel not true.	True and dress.
	Wheel too hard.	Use a softer wheel.
	Incorrect work speed.	Adjust.
Feed lines or spirals	Improper dressing.	Dress the wheel carefully, using a sharp diamond.
	Wheel too hard.	Use a softer wheel.
	Edges of wheel too sharp.	Round edges slightly with an abrasive stick.
	Feed too coarse.	Reduce feed on final passes.
	Wheelhead is tipped or swung.	Align wheelhead and spindle.
Out-of-round hole	Work is distorted during mounting in the chuck or holding device.	Use extreme care when mounting work.
	Work overheated during rough grinding.	Reduce depth of cut and feed. If work is mounted on a faceplate, loosen each clamp slightly and retighten evenly.

table 83.2   Internal grinding problems, causes, and remedies (continued)

Problem	Cause	Remedy
Scratches on ground surface	Wheel too soft and abrasive grains are caught between work surface and wheel.	Use a harder wheel.
	Improperly dressed wheel.	Carefully dress the wheel.
	Dirty coolant deposits particles between the wheel and the work.	Clean coolant tank and replace coolant.
	Wheel too coarse.	Use a finer wheel.
Tapered hole	Workhead set at a slight angle.	Align workhead.
	Wheel too soft to hold size.	Use a harder wheel.
	Feed too fast.	Reduce feed.
Wheel glazing	Wheel too hard.	Use a softer wheel.
		Increase work speed.
	Wheel too dense.	Use a more open wheel.
	Improper dressing.	Use a sharp diamond and rough-dress.
Wheel loading	Wheel too hard	Use a softer wheel.
		Increase work speed.
		Increase traverse feed.
	Wheel too fine.	Use a coarser grain.
	Dirty coolant.	Clean coolant system and replace coolant.
	Truing diamond is dull.	Use a sharp diamond and coarse-dress the wheel.

12. Grind until the hole just cleans up, feeding the wheel in no more than .002 in. (or 0.05 mm) per table reversal.

13. Check the hole size and set the automatic feed (if the machine is so equipped) to disengage when the work is roughed to within .001 in. (0.02 mm) of size.

14. Reset the automatic infeed to .0002 in. (0.005 mm) per table reversal.

15. Finish-grind the work and let the wheel spark out.

16. Move the table longitudinally and withdraw the wheel and spindle from the workpiece.

17. Check the hole diameter and finish-grind if necessary.

TO GRIND A TAPERED HOLE

The same procedures and precautions should be followed for grinding tapered holes as for grinding parallel holes. However, the workhead must be set to one-half the in-

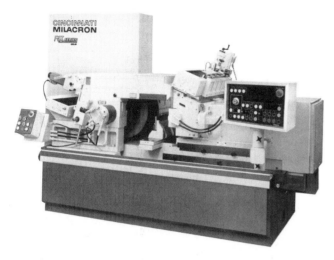

Figure 83-3 RK series 350-20 twin grip centerless grinder. *(Cincinnati Machine, a UNOVA Co.)*

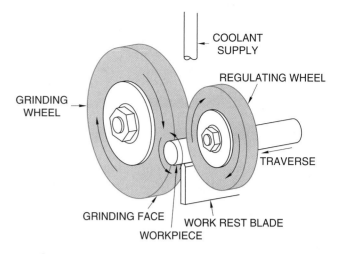

Figure 83-4 Principle of the centerless grinder. *(The Carborundum Co.)*

cluded angle of the taper. It is *very* important that the centerlines of the grinding wheel and the hole be set at the same height in order to produce the correct taper.

▸▸ Centerless Grinders

The production of cylindrical, tapered, and multidiameter workpieces may be achieved on a centerless grinder (Fig. 83-3). As the name suggests, the work is not supported on centers but rather by a work rest blade, a regulating wheel, and a grinding wheel (Fig. 83-4).

On a centerless grinder, the work is supported on the work rest blade, which is equipped with suitable guides for the type of workpiece. The rotation of the grinding wheel forces the workpiece onto the work rest blade and against the regulating wheel, while the regulating wheel controls the speed of the work and the longitudinal feed movement. To provide longitudinal feed to the work, the regulating wheel is set at a slight angle. The rate of feed may be varied by changing the angle and the speed of the regulating wheel. The regulating and grinding wheels rotate in the same direction, and the center heights of these wheels are fixed. Because the centers are fixed, the diameter of the workpiece is controlled by the distance between the wheels and the height of the work rest blade.

The higher the workpiece is placed above the centerlines of the wheels, the faster it will be ground cylindrical. However, there is a limit to the height at which it may be placed, since the work will eventually be lifted periodically from the work rest blade. There is one exception to placing the work above center: when removing slight bends in long, small-diameter work. In this case, the center of the piece is placed below the centerline of the wheels and the rate of traverse is high. This operation eliminates whipping and chattering that might result from bent work and is used primarily for straightening the workpiece. After the work has been straightened, it is ground in the normal manner above centers.

▸▸ Methods of Centerless Grinding

There are three methods of centerless grinding: thru-feed, infeed, and endfeed.

THRU-FEED CENTERLESS GRINDING

Thru-feed centerless grinding (Fig. 83-5) consists of feeding the work between the grinding and regulating wheels. The cylindrical surface is ground as the work is fed by the regulating wheel past the grinding wheel. The speed at which the work is fed across the grinding wheel is controlled by the speed and angle of the regulating wheel.

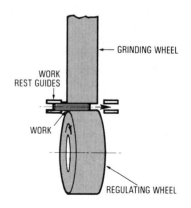

Figure 83-5 Principle of thru-feed centerless grinding. *(Milacron, Inc.)*

Cylindrical Grinders   **727**

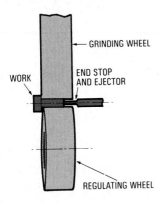

Figure 83-6 Principle of infeed centerless grinding. *(Milacron, Inc.)*

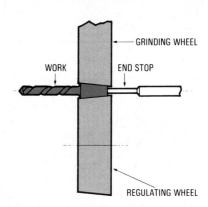

Figure 83-7 Principle of endfeed centerless grinding. *(Milacron, Inc.)*

INFEED CENTERLESS GRINDING

Infeed centerless grinding, a form of plunge grinding, is used when the work being ground has a shoulder or head (Fig. 83-6). Several diameters of a workpiece may be finished simultaneously by infeed grinding. Tapered, spherical, and other irregular profiles are ground efficiently by this method.

With infeed grinding, the work rest blade and the regulating wheel are clamped in a fixed relation to each other. The work is placed on the rest, against the regulating wheel, and is fed into the grinding wheel by moving the infeed lever through a 90° arc. When the lever is at the full end of the travel, the predetermined size has been reached and the part is the desired size. When the lever is reversed, the regulating wheel and work rest move back and the part is ejected either manually or automatically.

If the part to be ground is longer than the wheels, one end is supported on the work rest and the other on rollers mounted on the machine.

ENDFEED CENTERLESS GRINDING

The *endfeed method* (Fig. 83-7) is used mainly for grinding tapered work. The grinding wheel, the regulating

wheel, and the work rest all remain in a fixed position. The work is then fed in from the front, manually or mechanically, up to a fixed stop. When the machine is prepared for endfeed grinding, the grinding wheel and the regulating wheel are often dressed to the required taper. In some cases where only a few parts are required, only the regulating wheel may be dressed.

ADVANTAGES OF CENTERLESS GRINDING

> There is no limit to the length of work being ground.

> There is no axial thrust on the workpiece, which permits the grinding of long workpieces that would be distorted by other methods.

> For truing purposes, less stock is required on the workpiece than if the work is held between centers. This is due to the fact that the work "floats" in the centerless grinder. Work held between centers may run eccentrically and require more stock for truing up.

> Because there is less stock to be removed, there is less wheel wear and less grinding time is required.

Cylindrical Grinders

1. Name two types of cylindrical grinders.

2. List the main parts of a cylindrical grinder and state the purpose of each.

3. What precautions should be observed when truing and dressing the grinding wheel of a cylindrical grinder?

To Parallel-Grind the Outside Diameter

4. List six precautions that must be taken to ensure the utmost accuracy during parallel-grinding the outside diameter of a workpiece.

To Grind a Tapered Workpiece

5. List three ways in which tapers may be ground on a universal cylindrical grinder.

Cylindrical Grinding Faults, Causes, and Remedies

6. List three causes and three remedies for the following cylindrical grinding faults:

   a. Burnt work
   b. Chatter marks
   c. Feed lines
   d. Work out-of-round
   e. Rough finish
   f. Wavy marks

Internal Grinding

7. Describe internal grinding and name four machines on which it may be performed.

Internal Grinding on a Universal Cylindrical Grinder

8. Why are grinding wheels used for internal grinding generally softer than those used for external grinding?

9. What precautions must be taken when setting up and grinding a parallel hole in a workpiece?

To Grind a Tapered Hole

10. At what height should the grinding wheel be set for grinding a taper? Why is this necessary?

Internal Grinding Problems, Causes, and Remedies

11. List four important problems encountered in internal grinding. State at least two causes and two remedies for each of the problems listed.

Centerless Grinding

12. Describe the principle of centerless grinding. Illustrate by means of a suitable sketch.

13. How is the work held during centerless grinding to make it cylindrical as quickly as possible?

Methods of Centerless Grinding

14. Name three methods of centerless grinding and illustrate each of these methods by a suitably labeled sketch.

15. List four advantages of centerless grinding.

UNIT 84

Universal Cutter and Tool Grinder

OBJECTIVES

After completing this unit, you should be able to:

1 Identify and state the purposes of the main parts of a cutter and tool grinder

2 Grind clearance angles on helical and staggered tooth cutters

3 Grind a form-relieved cutter

4 Set up the grinder for cylindrical and internal grinding

The universal cutter and tool grinder is designed primarily for the grinding of cutting tools such as milling cutters, reamers, and taps. Its universal feature and various attachments permit a variety of other grinding operations to be performed. Other operations that may be performed are internal, cylindrical, taper, and surface grinding; single-point tool grinding; and cutting-off operations. Most of the other operations require additional attachments or accessories.

▶▶ Parts of the Universal Cutter and Tool Grinder

Refer to the universal cutter and tool grinder in Fig. 84-1. The *base* is of a heavy, cast-iron, boxlike construction that provides rigidity. The top of the base is machined to provide the *ways* (which are generally hardened) for the saddle.

The *wheelhead* is mounted on a column at the back of the base. It may be raised or lowered by the wheelhead handwheels located on either side of the base. The wheelhead may be swiveled through 360°. The wheelhead spindle is mounted in antifriction bearings and is tapered and threaded at both ends to receive grinding wheel collets. The spindle speed may be varied by stepped pulleys to suit the size of the wheel being used.

The *saddle* is mounted on the ways of the base and is moved in and out by the crossfeed handwheels located at the front and back of the machine. The upper part of the saddle has machined and hardened ways at right angles to the ways on top of the base.

The *table* is composed of two units, the *upper* and *lower* table. The lower table, mounted on the upper ways of the saddle, rests and moves on antifriction bearings. The upper table is fastened to the lower table and may be swiveled for grinding tapers. The table unit (upper and lower tables) may be moved longitudinally by three *table traverse knobs,* one located at the front of the machine and two at the back. The table may also be traversed slowly by means of the *slow table traverse crank.* The table may be locked in place laterally and longitudinally with locking screws.

Stop dogs, mounted in a T-slot on the front of the table, control the length of the table traverse. Each dog has

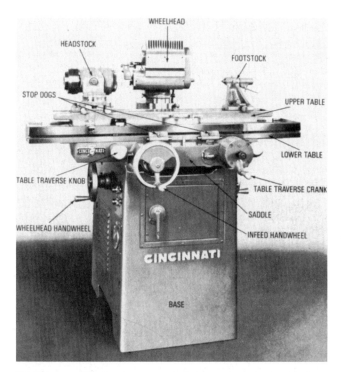

Figure 84-1 A universal cutter and tool grinder. *(Cincinnati Machine, a UNOVA Co.)*

a positive stop pin on one side and a spring-loaded plunger on the other. They are reversible to provide for a positive or a cushioned stop for the table, as desired.

▶▶ Accessories and Attachments

The *right-* and *left-hand tailstocks* are mounted in the T-slot of the upper table and support the work for certain grinding operations. They may be placed at any point along the table.

The *universal workhead* (Fig. 84-1) or headstock is mounted on the left side of the table and used for supporting end mills and face mills for grinding. It may also be equipped with a pulley and motor (motorized headstock; see Fig. 84-19 on p. 741) and used for cylindrical grinding. A chuck may be mounted in the workhead to hold work for internal and cylindrical grinding, as well as cutting-off operations (see Fig. 84-20 on p. 742).

The *centering gage* (see Fig. 84-13 on p. 736) is used to align quickly the tailstock center with the center of the wheelhead spindle. It is also used to align the cutter tooth on center in some grinding setups.

The *adjustable tooth rest* supports the cutter tooth and may be fastened to the wheelhead or table, depending

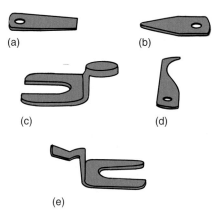

Figure 84-2 Various shapes of tooth rest blades. (a) Plain tooth rest blades; (b) rounded tooth rest blades; (c) offset tooth rest blades; (d) hook, or L-shaped, tooth rest blades; (e) inverted V-tooth rest blades. *(Kelmar Associates)*

on the type of cutter being ground. Another form of tooth rest is the *universal micrometer flicker type,* which has a micrometer adjustment for small vertical movements of the tooth rest.

Plain tooth rest blades (Fig. 84-2a) are used for grinding straight-tooth milling cutters.

Rounded tooth rest blades (Fig. 84-2b) are used for sharpening shell end mills, small end mills, taps, and reamers.

Offset tooth rest blades (Fig. 84-2c) are a universal type suitable for most applications, such as coarse-pitch helical milling cutters and large face mills with inserted blades.

Hook, or *L-shaped, tooth rest blades* (Fig. 84-2d) are used for sharpening slitting saws, straight-tooth plain milling cutters with closely spaced teeth, and end mills.

Inverted V-tooth rest blades (Fig. 84-2e) are used for grinding the periphery of staggered-tooth cutters.

Cutter grinding mandrels and arbors (Fig. 84-3a and b) are used when grinding milling cutters so that they are held in the same manner as they are held for milling. For example, shell end mills should be sharpened on the same arbor as that used for milling.

Plain milling and *side facing cutters,* which are held on the standard milling machine arbor, should be held on a grinding mandrel (Fig. 84-3a) or a cutter grinding arbor (Fig. 84-3b).

A *grinding mandrel* rather than a lathe mandrel should be used to hold the cutter. This is necessary, since a lathe mandrel will hold the cutter only at one end. The straight length of a grinding mandrel is a sliding fit into the cutter, and the slightly tapered end will hold the cutter securely for grinding.

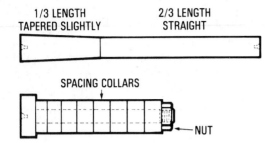

Figure 84-3 (a) A cutter grinding mandrel; (b) a cutter grinding arbor. *(Kelmar Associates)*

Where considerable cutter grinding is done, a cutter grinder arbor will be useful.

▶▶ Milling Cutter Nomenclature

To grind cutters correctly, the cutter parts and their functions should be understood. Milling cutter parts are shown in Fig. 84-4. A brief description of the various parts follows:

> *Primary clearance* is the clearance ground on the land adjacent to the tooth face. It is the angle formed between the slope of the land and a line tangent to the periphery. Primary clearance prevents the land behind the cutting edge from rubbing on the work. The amount of primary clearance on a cutter will vary with the type of material being cut.

> *Secondary clearance* is ground behind the primary clearance and gives additional clearance to the cutter behind the tooth face. When grinding the clearance on a milling cutter, always grind the primary clearance first. The secondary clearance is then used to control the width of the land.

> The *cutting edge* is formed by the intersection of the face of the tooth with the land. This angle formed by the face of the tooth and the primary clearance is called the *angle of keenness*.

On side milling cutters, the cutting edges may be on one or both sides as well as on the periphery. When the teeth are straight, the cutting edge engages along the full width of the tooth at the same moment. This creates a gradual buildup of pressure as the tooth cuts into the work and a sudden release of this pressure as the tooth breaks through, causing a vibration or chatter. This type of cutter produces a poor finish and does not retain its sharp cutting edge as long as a helical cutter does.

When the teeth are helical, the length of the cutting edge contacting the work varies with the helix angle. The number of teeth in contact with the work will vary with the

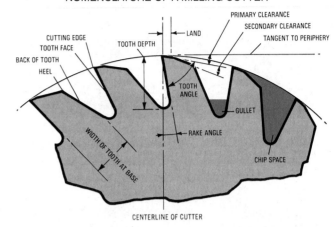

NOMENCLATURE OF A MILLING CUTTER

Figure 84-4 Milling cutter nomenclature.

size of the surface machined, the number of teeth in the cutter, the cutter diameter, and the helix angle. Helical cutters produce a shearing action on the material being cut, which reduces vibration and chatter.

The *helix angle,* sometimes called the *shear angle,* is the angle formed by the angle of the teeth and the centerline of the cutter. It may be measured with a protractor or by bluing the edge of the cutter teeth and rolling the cutter against a straightedge over a sheet of paper (Fig. 84-5). The marks left by the teeth can easily be measured in relation to the axis of the cutter to determine the helix angle.

The *land* is the narrow surface behind the cutting edge on the primary clearance produced when the secondary clearance is ground on the cutter. The width of the

Figure 84-5 One method of measuring the helix angle of a milling cutter. *(Kelmar Associates)*

land varies from about 1/64 in. (0.4 mm) on small cutters to about 1/16 in. (1.5 mm) on large cutters. On face mills, the land is more correctly called the *face edge*.

The *tooth angle* is the included angle between the face of the tooth and the land caused by grinding the primary clearance. This angle should be as large as possible to provide maximum strength at the cutting edge and better dissipation of heat generated during the cutting process.

The *tooth face* is the surface on which the metal being cut forms a chip. This face may be flat, as in straight-tooth plain milling cutters and inserted face-tooth mills, or curved, as in helical milling cutters.

▶▶ Cutter Clearance Angles

To perform efficiently, a milling cutter must be ground to the correct clearance angle. The proper clearance angle on a milling cutter may be determined only by the "cut-and-try" method. The clearance angle will be influenced by factors such as finish, the number of pieces per sharpening, the type of material, and the condition of the machine. Excessive primary clearance produces chatter, causing the cutter to dull quickly.

A general rule followed by a large machine tool manufacturer for grinding cutter clearance angles on high-speed steel cutters is 5° primary clearance plus an additional 5° for the secondary clearance. Thus, the primary clearance is 5° and the secondary clearance is 10° for cutting machine steel. Carbide cutters used on machine steel are ground to 4° primary clearance plus an additional 4°, or a total of 8°, for the secondary clearance.

Table 84.1 provides a rule of thumb for grinding cutter clearance angles on high-speed steel milling cutters. Table 84.2 gives the angles for carbide cutters. It should be remembered that this is only a guide. If the cutter does not perform satisfactorily with these angles, adjustments will have to be made to suit the job.

▶▶ Methods of Grinding Clearance on Cutters

Clearance may be ground on cutters by clearance, hollow, and circle grinding. The type of cutter being ground will determine the method used.

table 84.1 Clearance angles for high-speed steel cutters

Material to Be Machined	Primary Clearance Angle	Secondary Clearance Angle
High-carbon and alloy steels	3° to 5°	6° to 10°
Machine steel	3° to 5°	6° to 10°
Cast iron	4° to 7°	7° to 12°
Medium and hard bronze	4° to 7°	7° to 12°
Brass and soft bronze	10° to 12°	13° to 17°
Aluminum, magnesium, and plastics	10° to 12°	13° to 17°

table 84.2 Primary clearance angles for cemented-carbide cutters

Type of Cutter	Periphery			Chamfer			Face		
	Steel	Cast Iron	Aluminum	Steel	Cast Iron	Aluminum	Steel	Cast Iron	Aluminum
Face or side	4° to 5°	7°	10°	4° to 5°	7°	10°	3° to 4°	5°	10°
Slotting	5° to 6°	7°	10°	5° to 6°	7°	10°	3°	5°	10°
Sawing	5° to 6°	7°	10°	5° to 6°	7°	10°	3°	5°	10°

■ **Figure 84-6** Setup for clearance grinding using a flaring-cup wheel. *(Kelmar Associates)*

CLEARANCE GRINDING

Clearance grinding (Fig. 84-6) produces a flat surface on the land. A 4-in. (100-mm) flaring-cup wheel is used for this method and is offset slightly to permit long cutters to clear the opposite side of the wheel. When clearance grinding, the tooth rest may be set between the center and the top of the wheel, but never below center. The higher the tooth rest is placed, the less will be the clearance between the cutter and the opposite edge of the wheel. When clearance grinding, the tooth rest may be attached to the table or the wheelhead, depending on the type of cutter being ground. For straight-tooth cutters, it may be mounted on the table, but for helical teeth it must be mounted on the wheelhead.

HOLLOW GRINDING

The land produced by hollow grinding (Fig. 84-7) is concave. A 6-in. (150-mm) diameter dish wheel or a 6-in. diameter cutoff wheel is desirable. The cutoff wheel generally produces a better finish and breaks down more slowly because of the resinoid bond. Since all grinding wheels break down in use, it is better to grind diagonally opposite teeth in rotation and to take light cuts. In hollow grinding, the wheel and cutter centers must be aligned. Then, clearance is obtained by raising or lowering the wheel, depending on the method used to set up the cutter.

■ **Figure 84-7** The periphery of the wheel is used for hollow grinding. *(Milacron, Inc.)*

■ **Figure 84-8** The workpiece is revolved against the grinding wheel for circle grinding. *(Milacron, Inc.)*

CIRCLE GRINDING

Circle grinding (Fig. 84-8) provides only a minute amount of clearance and is used mainly for reamers. The reamer is mounted between centers and is rotated *backward* so that the heel of the tooth contacts the wheel first. As the tooth rotates against the grinding wheel, the pressure of the wheel causes the cutter to spring back slightly as each cut progresses. Thus, a very small amount of clearance is produced between the cutting edge and the heel of the tooth. The grinding wheel should be set on center for circle grinding. Secondary clearance must be obtained by clearance or hollow grinding. Circle grinding is also used to obtain concentricity of milling cutters prior to clearance or hollow grinding.

▶▶ Methods of Checking Cutter Clearance Angles

There are three methods of determining tooth clearance on a milling cutter:

1. Dial indicator
2. Brown & Sharpe cutter clearance gage
3. Starrett cutter clearance gage

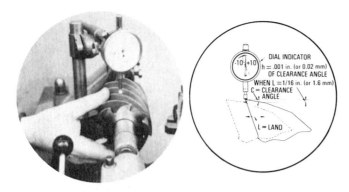

Figure 84-9 Measuring the clearance angle with a dial indicator. *(Kelmar Associates)*

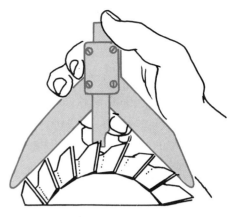

Figure 84-10 Checking the clearance angle with a Brown & Sharpe cutter clearance gage. *(Kelmar Associates)*

TO CHECK CUTTER CLEARANCE WITH A DIAL INDICATOR

When a dial indicator is used, clearance is determined by the movement of the indicator needle from the front to the back of the cutter land (Fig. 84-9). The basic rule used to determine the clearance by this method is as follows.

For a land of ⅟₁₆-in. (or 1.5-mm) width, 1° of clearance is equivalent to .001 in. (or 0.025 mm) on the dial indicator. Thus, 4° of clearance on a ⅟₁₆-in. (or 1.5-mm) land would register .004 in. (or 0.1 mm) on the dial indicator. The cutter diameter does not affect the measurement.

TO CHECK CUTTER CLEARANCE WITH A BROWN & SHARPE CLEARANCE GAGE

When the Brown & Sharpe clearance gage is used (Fig. 84-10), the inside surfaces of the hardened arms (which are at 90°) are placed on top of two teeth of the cutter. The cutter is revolved sufficiently to bring the face of the tooth into contact with the angle ground on the end of the hardened center blade. The clearance angle of the tooth should correspond with the angle marked on the end of the blade. Two gage blades are furnished with each gage and are stamped at each end with the diameters of the cutters for which they are intended. This cutter clearance gage measures all cutters from ½ to 8 in. (13 to 200 mm) in diameter, except those with fewer than eight teeth.

CHECKING THE CUTTER CLEARANCE WITH A STARRETT CUTTER CLEARANCE GAGE

The Starrett gage (Fig. 84-11) may be used to check the clearance on all types of inch cutters from 2 to 30 in. (50 to 750 mm) in diameter, and on small cutters and end mills

Figure 84-11 Checking the clearance angle with a Starrett cutter clearance gage. *(Kelmar Associates)*

from ½ to 2 in. (13 to 50 mm), providing the teeth are evenly spaced. This gage consists of a frame graduated from 0° to 30°, a fixed foot, and a beam. An adjustable foot slides along the beam extension. A blade, which may be adjusted angularly and vertically, is used to check the angle of the land on a tooth. When in use, the feet are positioned on two alternate teeth of the cutter, with the gage at right angles to the tooth face. The adjustable blade is then lowered onto the top of the middle tooth and adjusted until the angle corresponds to the angle of the land being checked. The land angle is indicated on the protractor on the top of the frame.

▶▶ Cutter Grinding Operations and Setups

It is most important that milling cutters be ground properly and to the correct clearance angles. Otherwise, the cutter will not cut efficiently and its life will be shortened considerably.

In order to maintain efficient cutting action, preserve the life of the cutter, and reduce the cost of resharpening, it is important to sharpen cutters when they show signs of wear.

The most common signs of cutter wear are:

> The surface finish on the workpiece gets poorer.

> There is unusual noise and smoking while cutting.

> A large burr appears on the edge of the workpiece.

> The accuracy of the cut changes.

> The chips being produced turn blue.

> The wear land on the teeth is visible.

Cutters should be sharpened when the wear land is .006 in. (0.15 mm) for cutters up to ½-in. (12.7-mm) diameter, and .010 in. (0.25 mm) for cutters over ½-in. (12.7-mm) diameter. Using a cutter beyond this point will damage the cutter, produce inaccurate work, and require more power to remove metal.

TO GRIND A PLAIN HELICAL MILLING CUTTER

Primary Clearance

1. Mount a parallel-ground test bar between the tailstock centers and check the alignment with an indicator. This will ensure that the table travel is parallel to the grinding edge of the wheel.

2. Remove the test bar.

3. Mount a 4-in. (100-mm) flaring-cup wheel (A 60-L 5-V BE) on the grinding head spindle so that the wheel rotates in a counterclockwise direction.

4. Adjust the machine so that the wheel revolves at the proper speed.

5. True the face of the wheel and dress the cutting edge so that it is no more than ¹⁄₁₆ in. (1.5 mm) wide.

6. Swivel the wheelhead to 89° so that the wheel will touch the cutter on the left side of the wheel only.

7. Using a centering gage, adjust the wheelhead spindle to the height of the tailstock centers (Fig. 84-12). Lock the wheelhead spindle.

8. Mount the cutter on a mandrel and place it temporarily between the footstock centers on the machine table.

9. Set up the tooth rest, on which an offset tooth rest blade (Fig. 84-2c) has been mounted, on the

■ Figure 84-12 Setting the wheelhead spindle to center height with a centering gage. *(Milacron, Inc.)*

wheelhead housing. Adjust the top of the tooth rest to approximately center height.

10. Move the table until the cutter is near the tooth rest.

11. Adjust the tooth rest between two teeth at the approximate helix angle of the cutter teeth.

12. Chalk or blue the top of the tooth rest blade.

13. Move the cutter over the tooth rest blade and rotate it until the tooth rests on top of the blade.

14. While holding the tooth face against the rest, traverse the table back and forth to mark the point where the tooth bears on the tooth rest blade.

15. Remove the mandrel and cutter from between the centers.

16. Using the centering gage, adjust the tooth rest so that the *center* of the marked bearing point on the tooth rest is at center height and in the center of the grinding surface of the wheel (Fig. 84-13). The tooth rest blade must be as close to the wheel as possible without touching it.

■ Figure 84-13 Centering the bearing point of the wheel on the tooth rest. *(Kelmar Associates)*

Note: At this point, the center of the grinding spindle, the footstock centers, and the bearing point on the tooth rest blade are in line.

17. Place a dog on the end of the grinding mandrel and mount the work between the tailstock centers.

18. Adjust a cutter tooth onto the top of the tooth rest blade.

19. Set the cutter clearance setting dial (Fig. 84-14) to zero (0) and lock. Adjust the dog into the pin of the cutter clearance gage.

20. Set the wheelhead graduated collar to zero (0).

21. Loosen the wheelhead lock and the cutter clearance setting dial lock.

22. Holding the cutter tooth on the tooth rest blade, carefully *lower* the wheelhead until the required clearance is shown on the cutter clearance dial. See Table 84.3 on p. 738 for the wheelhead adjustment for the required angle. When using a flaring-cup wheel, the distance to lower the wheelhead may also be calculated by either of the following methods:

   a. Distance = .0087 × clearance angle × cutter dia.

   b. Distance = sine of clearance angle × cutter dia. ÷ 2

   If the cutter is being hollow-ground, the distance to lower the wheelhead is

   Distance = .0087 × clearance angle × wheel dia.

23. Remove the dog from the end of the mandrel and unlock the table.

24. Adjust the table stops so that the wheel clears the cutter sufficiently at each end to permit indexing for the next tooth.

25. Start the grinding wheel.

26. Carefully feed the cutter in until it just touches the wheel.

27. Standing at the rear of the machine, turn the table traverse knob with the left hand. At the same time, with the right hand, hold the arbor firmly enough to keep the cutter tooth on the tooth rest (Fig. 84-6).

28. Grind one tooth for the full length and return to the starting position, being careful *at all times* to keep the tooth tight against the tooth rest.

29. Traverse the table until the cutter is clear of the tooth rest and rotate the cutter until the diagonally opposite tooth comes in line with the tooth rest blade.

■ **Figure 84-14** The cutter clearance dial indicates the amount of clearance ground on the cutter. *(Kelmar Associates)*

30. Grind this tooth without changing the infeed setting.

31. Check for taper by measuring both ends of the cutter with a micrometer.

32. Remove any taper, if necessary, by loosening the holding nuts on the upper table and adjusting the table.

33. Grind the remaining teeth.

34. Finish-grind all teeth by using a .0005-in. (0.01-mm) depth of cut.

35. If the land is over ⅟₁₆ in. (1.5 mm) for larger cutters, grind the secondary clearance.

To Grind the Secondary Clearance of a Plain Helical Milling Cutter

1. Reset the dog on the mandrel as in step 17 for grinding the primary clearance.

2. Loosen the clearance dial setscrew.

3. Hold the cutter tooth against the tooth rest and lower the wheelhead until the required secondary clearance is shown on the clearance setting dial.

4. Lock the dial, remove the dog, and proceed to grind the secondary clearance in the same manner as for primary clearance.

5. Grind the secondary clearance until the land is the required width.

table 84.3 — Vertical wheelhead adjustment for cutter clearance angles*

Cutter Diameter (in.)	4° in.	4° mm	5° in.	5° mm	6° in.	6° mm	7° in.	7° mm
1/2	.017	0.45	.022	0.55	.026	0.65	.031	0.8
3/4	.026	0.65	.033	0.85	.040	1	.046	1.15
1	.035	0.9	.044	1.1	.053	1.35	.061	1.55
1 1/4	.044	1.1	.055	1.4	.066	1.65	.077	1.95
1 1/2	.053	1.35	.066	1.65	.079	2	.092	2.35
1 3/4	.061	1.55	.076	1.95	.092	2.35	.108	2.75
2	.070	1.75	.087	2.2	.105	2.65	.123	3.1
2 1/2	.087	2.2	.109	2.75	.131	3.3	.153	3.9
2 3/4	.097	2.45	.120	3.05	.144	3.65	.168	4.25
3	.105	2.65	.131	3.3	.158	4	.184	4.65
3 1/2	.122	3.1	.153	3.9	.184	4.65	.215	5.45
4	.140	3.55	.195	4.95	.210	5.35	.245	6.2
4 1/2	.157	4	.197	5	.237	6	.276	7
5	.175	4.45	.219	5.55	.263	6.7	.307	7.8
5 1/2	.192	4.85	.241	6.1	.289	7.35	.338	8.6
6	.210	5.35	.262	6.65	.315	8	.368	9.35
6 1/2	.226	5.75	.283	7.2	.339	8.6	.396	10.05
7	.244	6.2	.305	7.45	.365	9.25	.426	10.8
7 1/2	.261	6.6	.326	8.3	.392	9.95	.457	11.6
8	.278	7.05	.348	8.85	.418	10.6	.487	12.35
8 1/2	.296	7.5	.370	9.4	.444	11.25	.519	13.2
9	.313	7.95	.392	9.95	.470	11.95	.548	13.9
9 1/2	.331	8.4	.413	10.5	.496	12.6	.579	14.7
10	.348	8.85	.435	11.05	.522	13.25	.609	15.45
11	.383	9.7	.479	12.15	.574	14.55	.670	17
12	.418	10.6	.522	13.25	.626	15.9	.731	18.55
13	.452	11.5	.566	14.35	.679	17.25	.792	20.1
14	.487	12.35	.609	15.45	.731	18.55	.853	21.65
15	.522	13.25	.653	16.6	.783	19.9	.914	23.2
16	.557	14.15	.696	17.65	.835	21.2	.974	24.75

*Inch and metric wheelhead adjustments are approximate equivalents.

Figure 84-15 Setup for clearance grinding a staggered-tooth cutter. *(Kelmar Associates)*

TO GRIND A STAGGERED TOOTH CUTTER

To grind the primary clearance on the periphery of a staggered-tooth cutter (Fig. 84-15), follow this procedure:

1. Carry out steps 1 to 7 for grinding primary clearance on a plain helical milling center.

2. Mount a staggered-tooth cutter tooth rest blade (Fig. 84-2e) in the holder and mount the unit on the wheelhead.

3. Place the high point of the inverted V *exactly* in the center of the width of the grinding wheel cutting face and at center height.

4. Place the centering gage on the table and adjust the wheelhead height until the highest point of the tooth rest blade is at center height.

5. Mount the cutter between centers with the dog loosely on the mandrel and adjust the table until one cutter tooth rests on the blade. Lock the table in position.

6. Set the cutter clearance dial (Fig. 84-14) to zero (0) and tighten the dog on the mandrel.

7. Loosen the cutter clearance dial lock and the wheelhead lock.

8. Lightly holding the cutter tooth onto the tooth rest blade, lower the wheelhead until the required clearance shows on the clearance setting dial.

9. Remove the clearance setting dog and unlock the table.

10. Set the stop dogs so that the wheel clears both sides of the cutter enough to allow indexing for the next tooth.

11. Start the grinding wheel.

12. Adjust the saddle until the cutter just touches the grinding wheel.

13. Grind one tooth and move the cutter clear of the tooth rest.

14. Rotate the next tooth, which is offset in the opposite direction, onto the tooth rest and grind it on the return stroke.

15. After grinding two teeth, check them with a dial indicator to see whether they are the same height. If not, adjust the blade slightly toward the high side and grind the next two teeth. Repeat the process until the teeth are within .0003 in. (0.007 mm).

Secondary Clearance

Because it is necessary to provide adequate chip clearance when milling deep slots, a secondary clearance of 20° to 25° on staggered-tooth cutters is recommended. It is also suggested that enough secondary clearance be ground to reduce the width of the land to approximately $\frac{1}{32}$ in. (0.8 mm). This will permit regrinding of the primary clearance at least once without the need for grinding the secondary clearance.

To Grind the Secondary Clearance on a Staggered-Tooth Cutter

1. Remove the tooth rest from the wheelhead and mount it on the table between the tailstocks. A universal micrometer flicker-type tooth rest and a straight blade (Fig. 84-16 on p. 740) should be used to permit the cutter to be rotated.

2. Place the centering gage on the table and bring the *center* of one tooth to center height. Mark this tooth with layout dye or chalk.

3. Locate the dog on the clearance setting dial pin and tighten it on the mandrel.

4. Rotate the cutter to the desired amount of clearance using the clearance setting dial.

5. Adjust the tooth rest under, or on the side of, the marked tooth.

6. Swivel the table sufficiently to the right or the left (depending on the helix angle of the tooth being ground) to grind a straight land.

7. Grind the secondary clearance on this tooth until the land is $\frac{1}{32}$ in. (0.8 mm) wide.

8. Grind all remaining teeth having the same slope or helix.

Figure 84-16 A flicker-type tooth rest is used for grinding the clearance on the sides of the teeth. *(Kelmar Associates)*

9. Swivel the table in the opposite direction and follow steps 6, 7, and 8 to set up and grind the remaining teeth.

Side Clearance

The side of the teeth of any milling cutter should not be ground unless absolutely necessary, since this reduces the width of the cutter. If the teeth must be ground, follow this procedure:

1. Mount the cutter on a stub arbor in the workhead (Fig. 84-16).
2. Mount a flaring-cup wheel.
3. Tilt the workhead to the desired primary clearance angle. This is generally 2° to 4°. The secondary clearance is about 12°.
4. Place the centering gage on the wheelhead and adjust one tooth of the cutter until it is on center and level. Clamp the workhead spindle.
5. Mount the tooth rest on the workhead using a flicker-type rest and a plain blade.
6. Raise or lower the wheelhead so that the grinding wheel contacts only the tooth resting on the blade.
7. Grind the primary clearance on all teeth.
8. Tilt the workhead to the required angle for the secondary clearance and grind all teeth.

TO GRIND A FORM-RELIEVED CUTTER

Unlike other types of milling cutters, form-relieved cutters are ground on the face of the teeth rather than on the periphery; otherwise, the form of the cutter will be changed when it is sharpened.

When grinding formed cutters for the first time, grind the backs of the teeth before grinding the cutting face to ensure that all teeth are the same thickness. This is necessary because the locating pawl on the grinding fixture bears against *the back of the tooth* when the cutter is being ground.

step	procedure
1	Swing the wheelhead so that the spindle is at 90° to the table travel.
2	Mount a dish wheel and the proper wheel guard.
3	Mount a gear cutter sharpening attachment on the table to the left of the grinding wheel (Fig. 84-17).
4	Place the gear cutter on the stud of the attachment so that the back of each tooth may be ground.

Note: This operation is necessary only when the cutter is being sharpened for the first time.

5	Place the centering gage on the wheelhead and adjust the wheelhead until the center of the tooth face is on center.

Figure 84-17 Grinder setup to sharpen a form-relieved cutter. *(Milacron, Inc.)*

■ Figure 84-18 The back of the tooth is set parallel to the wheel and held in place by the pawl. *(Milacron, Inc.)*

6   Move the table in until the back edge of a tooth is near the grinding wheel. At the same time, rotate the cutter until the back of the tooth is parallel with the face of the wheel (Fig. 84-18).

7   Engage the edge of the pawl on the face of the tooth and clamp the pawl in place (Fig. 84-18).

8   Grind the back of this tooth.

9   Move the table to the left so that the cutter is clear of the grinding wheel.

10   Index the cutter so that the pawl will bear against the next tooth. Hold the tooth face against the pawl when grinding.

11   Grind the backs of all teeth.

12   Reverse the cutter on the stud and adjust the pawl against the back of the tooth, after the face of the tooth has been brought to bear against the centering gage fastened to the attachment. Swing the centering gage out of the way.

13   Adjust the saddle to bring the face of one tooth in line with the grinding wheel. Thereafter, adjust the saddle only to compensate for wheel wear.

14   Loosen one setscrew and tighten the other to rotate the cutter against the grinding wheel.

15   Grind one tooth, traverse the table, and index for the next tooth.

16   Grind all tooth faces.

CYLINDRICAL GRINDING

With the aid of a motorized workhead, the cutter and tool grinder may be used for cylindrical and plunge grinding. Work may be ground between centers or held in a chuck, depending on the type of work.

To Grind Work Parallel Between Centers

1. Mount the motorized workhead on the left end of the table (Fig. 84-19).

2. Examine the centers of the machine and the work to see that they are in good condition.

3. Using the centering gage on the wheelhead, adjust the wheelhead to the tailstock center height.

4. Mount a 6-in. (150-mm) straight grinding wheel on the wheelhead spindle so that the wheel rotates downward at the front of the wheel. This will deflect the sparks downward.

5. Mount a parallel hardened and ground test bar between centers.

6. Using a dial indicator, align the centers for height and then align the side of the bar parallel with the table travel. Remove the bar and indicator.

7. Mount the work between centers.

8. Set the stop dogs so that the wheel overlaps the work by one-third the width of the wheelface at each end.

9. Start the grinding wheel and workhead. The workpiece should revolve in the opposite direction to that of the grinding wheel.

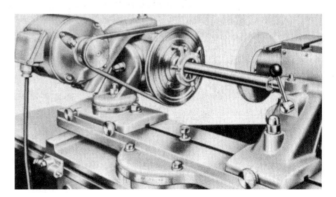

■ Figure 84-19 Setup for cylindrical grinding on a cutter and tool grinder.

10. Bring the revolving work up until it just touches the grinding wheel.

11. Traverse the table slowly and clean up the workpiece. The traverse speed should be such that the work travels approximately one-quarter the width of the wheel for each revolution of the work.

12. Measure each end of the workpiece for size and taper. If a taper exists, adjust as required.

13. After the work is parallel, set the crossfeed graduated collar to zero (0).

14. Feed the work into the grinding wheel approximately .001 in. (0.02 mm) per pass until the work is within .001 in. (0.02 mm) of finished size. Use .0002-in. (0.005-mm) cuts for finishing.

15. Feed in the work until the graduated collar indicates it is the proper size.

Note: Since work expands during grinding, it should never be measured for accurate size when warm.

16. Traverse the table several times to permit the wheel to spark out.

The same procedure is followed for taper grinding, except that the table must be swung to half the angle of the taper. After the work is cleaned up, the taper should be carefully checked for size and accuracy and adjustments made as required. When taper grinding, it is most important that the center height of the wheel and the workpiece be in line.

INTERNAL GRINDING

Light internal grinding may be performed on the tool and cutter grinder by mounting the internal grinding attachment on the wheelhead. The workpiece is held in a chuck mounted on the motorized workhead (Fig. 84-20).

1. Mount a test bar in the workhead spindle and align it both vertically and horizontally. When grinding a tapered hole, the workhead spindle must be aligned vertically and then swung to half the included angle of the taper.

2. Mount the internal grinding attachment on the workhead.

■ **Figure 84-20** Setup for internal grinding on a cutter and tool grinder. *(Milaron, Inc.)*

3. Center the grinding wheel spindle using the centering gage.

4. Mount the proper grinding wheel on the spindle.

5. Mount a chuck on the motorized workhead.

6. Mount the work in the chuck. Care must be taken not to distort the workpiece by gripping it too tightly.

7. Set the rotation of the workhead in the opposite direction to that of the grinding spindle.

8. Start the grinding wheel and the workpiece.

9. Carefully bring the wheel into the hole of the workpiece.

10. Set the table travel so that only one-third of the wheel overlaps the hole at each end.

11. Clean up the inside of the hole and check for size, parallelism, and bell-mouthing. Correct as required.

12. Set the crossfeed graduated collar to zero (0) and determine the amount of material to be removed.

13. Feed the grinding wheel in about .0005 in. (0.01 mm) per pass.

14. When work is close to the finished size, let the wheel spark out to improve the finish and remove the spring from the spindle.

15. Finish-grind the hole to size.

Parts of the Universal Cutter and Tool Grinder

1. Name four main parts of the universal cutter and tool grinder.

2. How many controls are there for each of the following parts of the universal cutter and tool grinder?

    a. Wheelhead   b. Saddle   c. Table

3. Name five accessories used with the universal cutter and tool grinder and state the purpose of each.

4. Name five types of tooth rest blades and indicate the purpose of each.

Cutter Grinding Mandrels and Arbors

5. Make a suitable sketch of a cutter grinding mandrel and a cutter grinding arbor.

6. How does a cutter grinding mandrel differ from a lathe mandrel?

Milling Cutter Nomenclature

7. Make a suitable sketch of at least two teeth on a milling cutter and indicate the following parts:

    a. Primary clearance angle   d. Land
    b. Secondary clearance angle   e. Tooth angle
    c. Cutting angle   f. Tooth face

8. What factors influence the cutter clearance angle?

Methods of Grinding Clearance on Milling Cutters

9. Name and describe briefly three methods of grinding clearance on cutting tools.

Checking Cutter Clearance with a Dial Indicator

10. Describe the principle by which cutter clearance is checked using a dial indicator.

11. Name two other methods of checking cutter clearance angles.

Cutter Grinding Operations and Setups

12. Describe how the tooth edge of a milling cutter is placed on center.

13. Describe how the correct cutter clearance is set on the machine using the clearance setting dial.

14. Explain two methods of determining how far to lower the wheelhead for the proper cutter clearance angle.

15. How far would the wheelhead be lowered to grind the proper clearance on the following cutters?

    a. 5° clearance angle on a 3-in. diameter cutter using a 6-in. straight wheel
    b. 5° clearance angle on a 75-mm diameter cutter using a flaring-cup wheel

16. List four precautions to be taken for grinding a helical milling cutter.

17. Where is the tooth rest mounted when grinding cutters having:

    a. Helical teeth?   b. Straight teeth?

To Grind a Staggered-Tooth Cutter

18. How does the grinding of a staggered-tooth milling cutter differ from the grinding of a plain helical milling cutter?

To Grind a Form-Relieved Cutter

19. On what surface are form-relieved cutters ground when they are being sharpened? Explain why.

20. Why is it advisable to grind the backs of the teeth on a new gear cutter?

21. What type of grinding wheel is used for sharpening gear cutters?

Cylindrical Grinding

22. List the main steps required to *set up* the cutter and tool grinder for cylindrical grinding.

23. How should the stop dogs be set when cylindrical grinding?

24. How fast should the table be traversed when cylindrical grinding on a cutter and tool grinder?

25. What precautions must be taken when setting up the machine for taper grinding?

Internal Grinding

26. What precautions should be taken when setting up the work for internal grinding?

27. List the steps to grind the inside diameter of a bushing using a cutter and tool grinder.

Section 16

Art du Potier d'Étain. *Pl. VIII.*

Carpenter Technology Corporation; Culver service, New York.
(This old picture shows a master workman and his apprentices
making pewter cups and teapots. One apprentice in the background is straining melted pewter to remove impurities. Another is
turning the machine that whirls the cup upon which the master
workman is working. The two appentices at the left are soldering
spouts on teapots; Carpenter Technology Corporation)

METALLURGY

Understanding the properties and heat treatment of metals has become increasingly important to machinists during the past two decades. Study of metal properties and development of new alloys have facilitated reduction in mass and increase in strength of machines, automobiles, aircraft, and many present-day commodities.

The most commonly used metals today are ferrous metals, or those that contain iron. The composition and properties of ferrous materials may be changed by the addition of various alloying elements during manufacture to impart the desired qualities to the material. Cast iron, machine steel, carbon steel, alloy steel, and high-speed steel are all ferrous metals, having different properties.

UNIT 85

Manufacture and Properties of Steel

OBJECTIVES

After completing this unit, you should be able to:

1 Identify six properties of metals

2 Explain the processes by which iron and steel are made

3 Understand how steel is produced in minimills

4 Describe the effect of alloying elements on steel

Metals such as iron, aluminum, and copper are among the most common elements found in nature. Iron, which is found in most parts of the world, was considered a rare and precious metal in ancient times. Throughout the ages, iron was transformed into steel, and today it is one of our more versatile metals. Almost every product made today contains some steel or was manufactured by tools made of steel. Steel can be made hard enough to cut glass, pliable as the steel in a paper clip, flexible as steel in springs, or strong enough to withstand enormous stress. Metals may be shaped in several ways: casting, forging, rolling and bending, drawing and forming, cutting, and joining. To better understand this versatile metal, a knowledge of its properties and manufacture is desirable.

▸▸ Physical Properties of Metals

Physical metallurgy is the science concerned with the physical and mechanical properties of metals. The properties of metals and alloys are affected by three variables:

1. *Chemical properties*—those that a metal attains through the addition of various chemical elements

2. *Physical properties*—those that are not affected by outside forces, such as color, density, conductivity, or melting temperature

3. *Mechanical properties*—those that are affected by outside forces such as rolling, forming, drawing, bending, welding, and machining

To better understand the use of the various metals, one should become familiar with the following terms:

> *Brittleness* (Fig. 85-1a) is the property of a metal that permits no permanent distortion before breaking. Cast iron is a brittle metal; it will break rather than bend under shock or impact.

> *Ductility* (Fig. 85-1b) is the ability of the metal to be permanently deformed without breaking. Metals such as copper and machine steel, which may be drawn into wire, are ductile materials.

> *Elasticity* (Fig. 85-1c) is the ability of a metal to return to its original shape after any force acting on it has been removed. Properly heat-treated springs are good examples of elastic materials.

> *Hardness* (Fig. 85-1d) may be defined as the resistance to forcible penetration or plastic deformation.

Figure 85-1 (a) Brittle metals will not bend but break easily; (b) ductile metals are easily deformed; (c) elastic metals return to their original shape after the load is removed; (d) hard metals resist penetration; (e) malleable metals may be easily formed or shaped; (f) tensile strength is the amount that a metal will resist a direct pull. *(Praxair, Inc.)*

> *Malleability* (Fig. 85-1e) is the property of a metal that permits it to be hammered or rolled into other sizes and shapes.

> *Tensile strength* (Fig. 85-1f) is the maximum amount of pull that a material will withstand before breaking. It is expressed as the number of pounds per square inch (on inch testers) or in kilograms per square centimeter (on metric testers) of pull required to break a bar having a one-square-inch or one-square-centimeter cross section.

> *Toughness* is the property of a metal to withstand shock or impact. Toughness is the opposite condition of brittleness.

▶▶ Manufacture of Ferrous Metals

PIG IRON

Production of pig iron in the blast furnace is the first step in the manufacture of cast iron or steel.

Raw Materials

Iron ore is the chief raw material used to make iron and steel. The main sources of iron ore in North America are at Steep Rock, the Ungava district near the Quebec-Labrador border, and the Mesabi range, situated at the western end of Lake Superior. The most important iron ores are:

> *Hematite* contains about 70% iron and varies in color from black to brick red.

> *Limonite,* a brownish ore similar to hematite, contains water. When the water has been removed by roasting, the ore resembles hematite.

> *Magnetite,* a rich, black ore, contains a higher percentage of iron than any other ore but is not found in large quantities.

> *Taconite,* a low-grade ore containing about 25% to 30% iron, must be specially treated before it is suitable for reduction into iron.

Pelletizing Process

Low-grade iron ores are uneconomical to use in the blast furnace and, as a result, go through a pelletizing process,

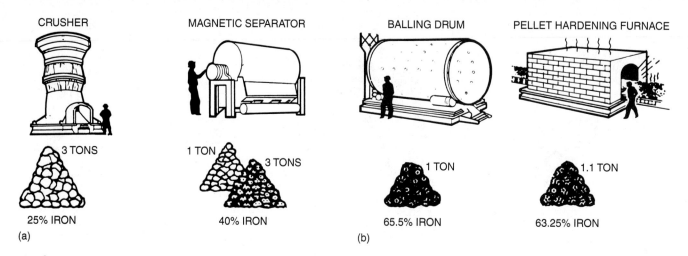

CRUSHER   MAGNETIC SEPARATOR   BALLING DRUM   PELLET HARDENING FURNACE

3 TONS

1 TON   3 TONS

1 TON   1.1 TON

25% IRON   40% IRON   65.5% IRON   63.25% IRON

(a)   (b)

■ **Figure 85-2** (a) The first step in pelletizing separates the iron ore from the rock; (b) iron ore pellets are hardened in the furnace. *(American Iron and Steel Institute)*

where most of the rock is removed and the ore is brought to a higher iron concentration. Some steelmaking firms are now pelletizing most ores at the mine to reduce transportation costs and the problems of pollution and slag disposal at the steel mills.

The crude ore is crushed and ground into a powder and passed through magnetic separators, where the iron content is increased to about 65% (Fig. 85-2a). This high-grade material is mixed with clay and formed into pellets about ½ to ¾ in. (13 to 19 mm) in diameter in a pelletizer. The pellets at this stage are covered with coal dust and sintered (baked) at 2354°F (1290°C) (Fig. 85-2b). The resultant hard, highly concentrated pellets will remain intact during transportation and loading into the blast furnace.

Coal, after being converted to coke, is used to supply the heat to reduce the iron ore. The burning coke produces carbon monoxide, which removes the oxygen from the iron ore and reduces it to a spongy mass of iron.

Limestone is used as a flux in the production of pig iron to remove the impurities from the iron ore.

MANUFACTURE OF PIG IRON

In the blast furnace (Fig. 85-3), iron ore, coke, and limestone are fed into the top of the furnace by means of a skip car. Hot air at 1000°F (537°C) is fed into the bottom of the furnace through the bustle pipe and tuyeres. After the coke is ignited, the hot air makes it burn vigorously. Carbon monoxide, produced by the burning coke, combines with the oxygen in the iron ore, reducing it to a spongy mass of iron. The iron gradually seeps down through the charge and collects in the bottom of the furnace. During this process, the decomposed limestone acts as a flux and unites with the impurities (silica and sulfur) in the iron ore to form a slag, which also seeps to the bottom of the furnace. Since the slag is lighter, it floats on the top of the molten iron. Every 6 h, the furnace is tapped. The slag is drawn off first and then the molten iron is poured into ladles. The iron may be further processed into steel or cast into *pigs,* which are used by foundries in the manufacture of castings.

The manufacture of pig iron is a continuous process, and the blast furnaces are shut down only for repair or rebricking.

▶▶ Direct Ironmaking

In 1989, the American Iron and Steel Institute, which represents most of the iron and steel manufacturers, initiated a five-year developmental program, with Department of Energy funding, to improve the efficiency of ironmaking. The *direct ironmaking* process is based on the smelting of iron ore with coal and oxygen in a liquid bath. It uses a variety of ores, coals, and fluxes in the manufacturing process and is intended to eliminate coke and the pollution problems associated with coke production.

From the reports available, this process, called *bath smelting,* has the potential of leading to a continuous, environmentally sound process with lower energy requirements and lower costs. The bath smelting process will use 27% less energy and have operating costs that are $10.00 per ton lower than the coke-oven and blast furnaces it may replace. The direct ironmaking process recovers the high-temperature heat from the postcombustion of the process offgas and reuses it in the smelting process. This completely enclosed smelting process is intended to make the steel industry more competitive by reducing its capital and operating costs.

MANUFACTURE OF CAST IRON

Most of the pig iron manufactured in a blast furnace is used to make steel. However, a considerable amount is used to manufacture cast-iron products. Cast iron is man-

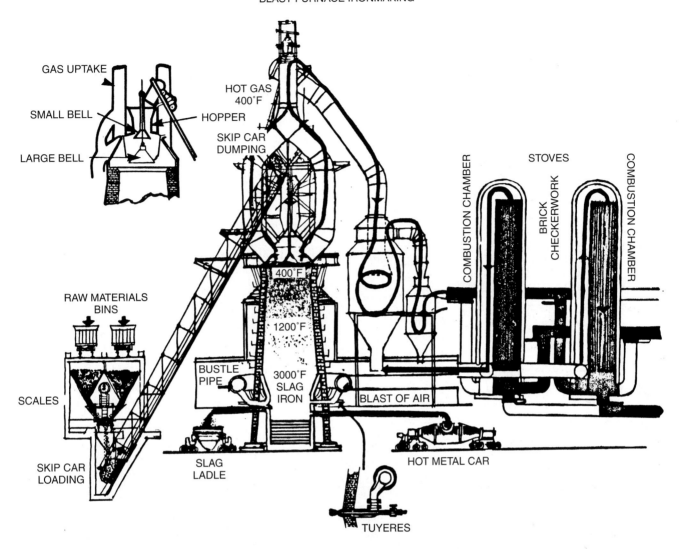

■ Figure 85-3 A blast furnace produces pig iron. *(American Iron and Steel Institute.)*

ufactured in a cupola furnace, which resembles a huge stovepipe (Fig. 85-4 on p. 750).

Layers of coke, solid pig iron, and scrap iron are fed into the top of the furnace. After the furnace is charged, the fuel is ignited and air is forced in near the bottom to aid combustion. As the iron melts, it settles to the bottom of the furnace, where it is tapped into ladles. The molten iron is poured into sand molds of the required shape and the metal assumes the shape of the mold. After the metal has cooled, the castings are removed from the molds.

The principal types of cast-iron castings are:

> *Gray-iron castings,* made from a mixture of pig iron and steel scrap, are the most widely used. They are made into a wide variety of products, including bathtubs, sinks, and parts for automobiles, locomotives, and machinery.

> *Chilled-iron castings* are made by pouring molten metal into metal molds so that the surface cools rapidly. The surface of such castings becomes very hard, and the castings are used for crusher rolls or other products requiring a hard, wear-resistant surface.

> *Alloyed castings* contain certain amounts of alloys such as chromium, molybdenum, and nickel. Castings of this type are used extensively by the automobile industry.

> *Malleable castings* are made from a special grade of pig iron and foundry scrap. After these castings have solidified, they are annealed in special furnaces. This makes the iron malleable and resistant to shock.

Manufacture and Properties of Steel   **749**

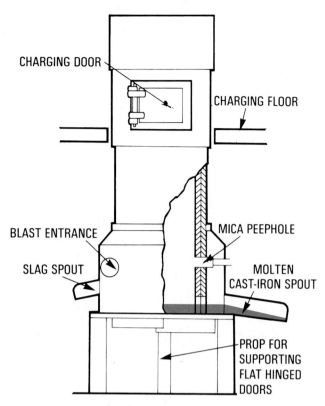

Figure 85-4 A cupola furnace is used to make cast iron. *(Kelmar Associates)*

Labels in Figure 85-4:
- CHARGING DOOR
- CHARGING FLOOR
- BLAST ENTRANCE
- SLAG SPOUT
- MICA PEEPHOLE
- MOLTEN CAST-IRON SPOUT
- PROP FOR SUPPORTING FLAT HINGED DOORS

MANUFACTURE OF STEEL

Since the late 1960s, the methods of manufacturing steel have undergone tremendous changes. The open hearth furnace and Bessemer converter have been phased out and replaced by the more efficient direct current electric arc and the basic oxygen furnaces. New steelmaking plants have been downsized into smaller, more efficient operations, called *minimills,* that produce steel faster and at lower cost. Many new iron- and steelmaking processes have been developed or are in the developmental stages, such as the coal-based direct ironmaking program, direct steelmaking, and the use of iron carbide as a source of iron.

Manufacturing Process

Before molten pig iron from the blast furnace can be converted into steel, most of the impurities must be burned out. This may be done in two types of furnaces: the basic oxygen process furnace and the DC or AC electric arc furnace.

Basic Oxygen Process

The basic oxygen furnace (Fig. 85-5) resembles a Bessemer converter but does not have the air chamber and tuyeres at the bottom to admit air through the charge. Instead

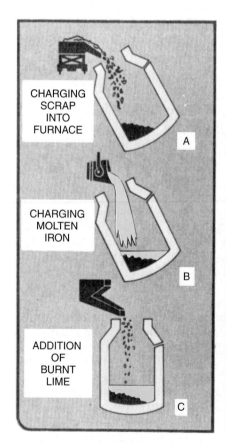

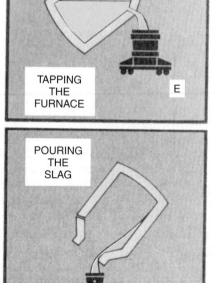

Labels in Figure 85-5:
- CHARGING SCRAP INTO FURNACE — A
- CHARGING MOLTEN IRON — B
- ADDITION OF BURNT LIME — C
- BLOWING WITH OXYGEN — D
- TAPPING THE FURNACE — E
- POURING THE SLAG — F

Figure 85-5 The basic oxygen process. *(Inland Steel Corporation)*

of air being forced through the molten metal, as in the Bessemer process, a high-pressure stream of pure oxygen is directed onto the top of the molten metal.

The furnace is tilted and is first charged with scrap metal (about 30% of the total charge). Molten pig iron is poured into the furnace, after which the required fluxes are added. An oxygen lance with a water-cooled hood is then lowered into the furnace until the tip is within 60 to 100 in. (152 to 254 cm) above the surface of the molten metal, depending on the blowing qualities of the iron and the type of scrap used. The oxygen is turned on and flows at the rate of 5000 to 6000 ft³/min (141 to 169 m³/min) at a pressure of 140 to 160 psi (965 to 1103 kPa).

The introduction of the oxygen causes the temperature of the molten steel (batch) to rise, at which time lime and fluospar may be added to help carry off the impurities in the form of slag. A recent development in this process is the *Lance Bubbling Equilibrium (LBE)*. With this process, inert gases such as argon and nitrogen are introduced by lances through the bottom of the furnace. These gases bubble up through the molten metal, increasing the contact between the metal and the slag, and speed mixing. This results in an increase in yield and a reduction in alloying elements, such as aluminum or silicon, that are used to reduce the oxygen and carbon content in the steel.

The force of the oxygen starts a high temperature churning action and burns out the impurities. After all the impurities have been burned out, there will be a noticeable drop in the flame and a definite change in sound. The oxygen is then shut off and the lance removed.

The furnace is now tilted (Fig. 85-5) and the molten steel flows into a ladle or is taken directly to the strand casting machine. The required alloys are added, after which the molten metal is teemed into ingots or formed into slabs. The refining process takes only about 50 min, and about 300 tons (272 t) of steel can be made per hour.

Electric Furnace

The electric furnace (Fig. 85-6) is used primarily to make fine alloy and tool steels. The heat, the amount of oxygen, and atmospheric conditions can be regulated at will in the electric furnace; this furnace is therefore used to make steels that cannot be readily produced in any other way.

Carefully selected steel scrap, containing smaller amounts of the alloying elements than are required in the finished steel, is loaded into the furnace. The three carbon electrodes are lowered until an arc jumps from them to the scrap. The heat generated by the electric arcs gradually melts all the steel scrap. Alloying materials such as chromium, nickel, and tungsten are then added to make the type of alloy steel required. Depending on the size of the furnace, it takes from 4 to 12 h to make a heat of steel. When the metal is ready to be tapped, the furnace is tilted forward and the steel flows into a large ladle. From the ladle, the steel is teemed into ingots.

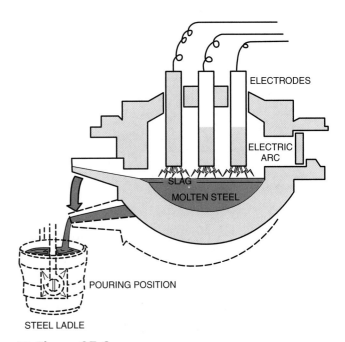

■ **Figure 85-6** Diagram of the electric furnace.

▶▶ Direct Steelmaking

In 1989, the American Iron and Steel Institute (AISI) initiated a developmental program, with Department of Energy funding, to improve the efficiency of steelmaking and reduce manufacturing costs. The process was designed to eliminate the coke production process and make steel directly from iron ore, and it may eventually replace the blast furnace. Figure 85-7 on p. 752 shows the *direct ironmaking* process, which involves four major components, or steps:

1. *Smelting. In-bath smelting* is the heart of the manufacturing process. Oxygen, prereduced iron ore, coal, and flux are gravity-fed into a molten iron-slag bath. Oxygen partially burns the coal to produce some of the heat required to drive the process. The iron oxide melts in the slag and is reduced to molten iron by the carbon in the coal. The rest of the energy required comes from the partial combustion of the carbon monoxide that is produced during the reduction of the iron oxide.

2. *Prereduction.* This process is used to preheat and prereduce the iron ore fed to the smelter, using the offgases from the smelter. Removal of 30% of the oxygen in the iron ore pellets reduces them to *wustite*. The reducing capabilities of the offgas holds the reduction of the iron ore to wustite over a wide range of offgas compositions and flow rates, producing a stable feedstock to the smelter.

3. *Offgas cleaning and handling.* The hot, dust-laden gas produced in the smelter is cooled with a recirculated gas stream and then passed through a

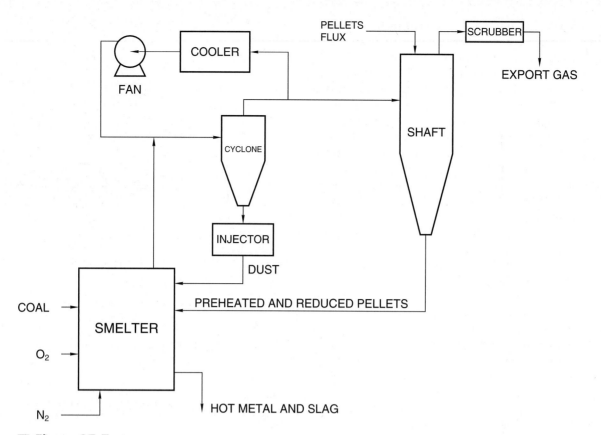

■ Figure 85-7 AISI direct ironmaking flowsheet. *(American Iron and Steel Institute)*

cyclone separator to remove most of the dust. The dust, which consists mostly of carbon and iron oxide, is recycled into the smelter.

The cleaned, cooled gas exiting from the cyclone is split into two streams. One passes through a water scrubber, which cools it for mixing with the hot smelter gasses. The other stream goes to a vertical shaft furnace to heat and partially reduce the iron ore pellets before they are charged into the smelter.

4. *Refining.* The refining process (desulfurization and decarburization) will produce liquid steel, which is suitable for ladle metallurgy treatment (addition of desirable chemical elements). From this point, the liquid steel is processed into steel products by continuous casting, rolling, etc.

The direct steelmaking system is entirely enclosed for the refining stage so that it is almost pollution-free. The only release into the atmosphere is from the combustion of the offgas from the shaft furnace to capture its residual fuel value. This fuel gas has been reduced to less than 40 parts per million (ppm) to hydrogen sulfide. This process is very efficient because the only energy lost is the difference in heat energy from the smelter offgas and the cooler shaft furnace process gas.

STEEL PROCESSING

After the steel has been properly refined in any of the furnaces, it is tapped into ladles where alloying elements and deoxidizers may be added. The molten steel may then be teemed into *ingots* weighing as much as 20 tons (18 t), or it may be formed directly into slabs by the *continuous-casting process* (Fig. 85-8).

The steel is teemed into ingot molds and allowed to solidify. The ingot molds are then removed or *stripped* and the hot ingots are placed in soaking pits at 2200°F (1204°C) for up to 1.5 h to make them a uniform temperature throughout. The ingots are then sent to the rolling mills, where they are rolled and reduced in cross section to form blooms, billets, or slabs.

> *Blooms* are generally rectangular or square and are larger than 36 in.^{2} in cross section. They are used to manufacture structural steel and rails.

> *Billets* may be rectangular or square but are less than 36 in.^{2} in cross-sectional area. They are used to manufacture steel rods, bars, and pipes.

> *Slabs* are usually thinner and wider than billets. They are used to manufacture plate, sheet, and strip steel.

CONTINUOUS CASTING

1. **Molten steel** pours from a ladle into a reservoir called a tundish.

2. **The metal** flows out the bottom of the tundish at a carefully regulated rate into the mold, which is moving up and down to prevent the hot metal from sticking. The interior of the mold is hollow—just the size, in width and thickness, of the slab to be formed. Lining the walls are pipes through which water flows, chilling the metal. A thin shell of steel begins to solidify around the molten metal.

3. **The gradually solidifying slab** moves down through the secondary cooling zone. A series of rollers support the slab and gradually turn it into a horizontal position. Sprays of water under high pressure cool and harden the metal still further.

4. **The ribbon of steel** moves on to a level table.

5. **A flame-cutting torch** slices down through the metal. When the slab is cut off, it is carried on rollers to a cooling bed. The entire trip from the ladle has taken less than one-half hour.

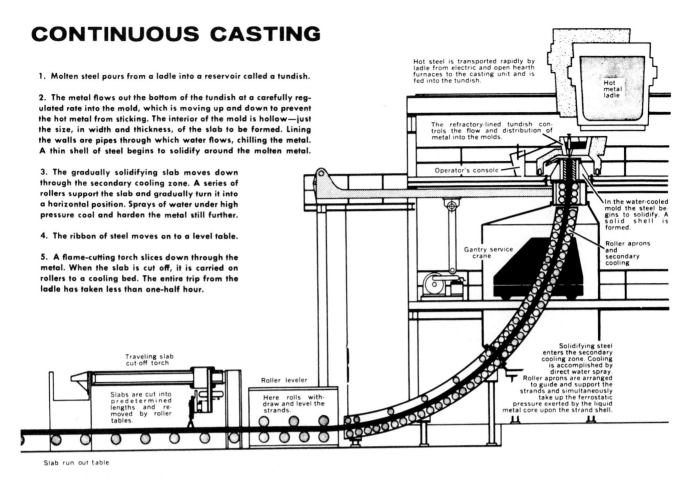

■ **Figure 85-8** The continuous-casting process for making steel blooms or slabs. *(American Iron and Steel Industry)*

STRAND, OR CONTINUOUS, CASTING

Strand, or continuous, casting (Fig. 85-8) is the most modern and efficient method of converting molten steel into semifinished shapes such as slabs, blooms, and billets. This process eliminates ingot teeming, stripping, soaking, and rolling. It also produces higher quality steel, reduces energy consumption, and has increased overall productivity by over 13%.

Continuous casting is fast becoming the major method of producing steel slabs and billets. This, combined with the much faster production of molten steel by the basic oxygen process, has greatly improved the efficiency of steel production. About 95% of the steel produced in the United States, Europe, and Japan is now by the continuous-casting method.

Molten steel from the furnace is taken in a ladle to the top of the strand or continuous caster and poured into the tundish. The *tundish* provides an even pool of molten metal to be fed into the casting machine. It also acts as a reservoir permitting the empty ladle to be removed and the full ladle to be poured without interrupting the flow of molten metal to the caster. The steel is stirred continuously by a nitrogen lance or by electromagnetic devices.

The molten steel drops in a controlled flow from the tundish into the mold section. Cooling water quickly chills the outside of the metal to form a solid skin, which becomes thicker as the steel strand descends through the cooling system. As the strand reaches the bottom of the machine, it becomes solid throughout. The solidified steel is moved in a gentle curve by bending rolls until it reaches a horizontal position. The strand is then cut by a travelling cutting torch into required lengths. In some strand casting machines, the solidified steel is cut when it is in the vertical position. The slab or billet then topples to the horizontal position and is taken away.

Continuous casters are capable of producing strands at up to 15 ft/min (4 m/min). This process is designed for high tonnage operations or small batches, as required. When small batches are required, the molten steel is generally produced by methods other than the basic oxygen or open-hearth furnaces.

After the metal has been made into blooms or slabs by either of the aforementioned processes, it is further rolled into billets and then, while still hot, into the desired shape, such as round, flat, square, or hexagonal (Fig. 85-9 on p. 754). These rolled products are known as *hot-rolled steel* and are easily identified by the bluish-black scale on the outside.

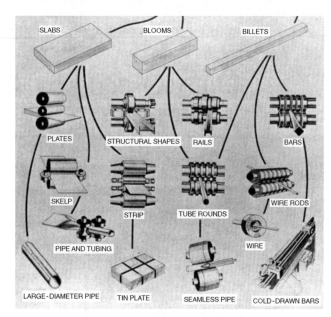

■ Figure 85-9 Various steel shapes produced by rolling. *(American Iron and Steel Institute)*

Hot-rolled steel may be further processed into *cold-rolled* or *cold-drawn steel* by removing the scale in an acid bath and passing the metal through rolls or dies of the desired shape and size.

One of the newest processes in the production of sheet steel is the *Continuous Annealing Line* (CAL), which produces high-strength and ultra-high-strength cold-rolled steel without increased weight, mainly for the automotive industry. This process is capable of altering the steel strength without changing its chemistry. The sheet steel produced by this method is virtually free from defects such as edge damage or sticker breaks. Continuous annealing reduces the time from the 5 to 6 days required for batch annealing to a mere 10 min. It also permits superior steel to be made with lesser amounts of costly alloying elements. Advanced computer controls permit the heating-cooling sequence to "lock in" the grain structure at the desired strength level. It can produce sheet steel with a yield strength of up to 220,000 psi.

VACUUM PROCESSING OF MOLTEN STEEL

Steel used in space and nuclear projects is often processed and solidified in a vacuum to remove oxygen, nitrogen, and hydrogen, which produces a high quality steel.

▶▶ Minimills

Since the late 1960s, there has been a continuous development process in the steel industry to develop newer and better methods of producing steel. The integrated mills (big steelmakers), which have been facing fierce competition from other integrated mills around the world, have been looking for the best technology to enable them to stay profitable and serve their customers. The major questions facing them are:

1. Should they continue to use the coke ovens, blast furnaces, and oxygen furnaces of today?

OR

2. Should they move to the less expensive, more flexible ironmaking, electric-furnace steelmaking, and thin-slab casting that have made the minimills so successful?

THE MINIMILL PROCESS

Minimills, the result of new technologies, provided a flexible and less expensive method of producing steel. The following provides a brief outline of the steps involved in making steel in minimills.

> *Raw materials.* Most of the steel manufactured in minimills is made of selected scrap steel. However, because of the uncertainty of the supply and the variation in cost, many steelmakers are using pig iron, hot-briquetted iron, direct-reduced iron, and iron carbide to supplement the scrap. This has ensured steelmakers of a steady supply of raw materials at fixed prices.

> *Furnaces.* Most furnaces used in minimills are the direct-current electric-arc furnaces. They are 22-ft diameter bottom-tap furnaces that use 24 electrodes and have water-cooled roofs and sidewalls. They provide more flexibility and create less pollution. It is felt by the major steelmakers that electric-arc furnaces will continue to produce more steel, while the blast furnace–basic oxygen furnaces will produce less and less.

> *The melting process.* The charge, consisting of scrap iron and iron supplements, is placed into the DC electric-arc furnace (Fig. 85-10). It is brought up to approximately 2800°F to 2900°F (1538°C to 1593°C) temperature and held there for about one hour.

The molten metal is moved from the furnace to one (or two) *ladle metallurgical station(s)* with a lid-type vacuum degassing unit (Fig. 85-11). The vacuum degassing unit is used only when steel with very low carbon and/or nitrogen are required. The ladle metallurgical stations are used for stirring, removing impurities, alloy additions, and temperature control.

> *Casting tower.* From the ladle station, the metal is taken to the casting tower, where it is fed into the

Figure 85-10 Most minimills use direct-current electric-arc furnaces for steelmaking. *(Nucor Corporation)*

Figure 85-11 The vacuum degassing unit of the ladle metallurgical station is used when steel with low carbon and/or nitrogen is required. *(Nucor Corporation)*

Figure 85-12 Metal from the ladle station is taken to the casting tower, where it is fed into the tundish of the continuous caster. *(Nucor Corporation)*

Figure 85-13 The soaking furnace is used to keep the steel slabs at the proper temperature for the rolling operation. *(Nucor Corporation)*

Figure 85-14 In the finishing mill, the slab goes through various rollers, where it is reduced to about 5% of its original thickness. *(Nucor Corporation)*

tundish of the continuous caster (Fig. 85-12). The steel flows from the tundish into the nozzle by an automatically controlled stopper rod. As the slab comes out of the containment section, its temperature is about 1800°F (980°C) and is traveling at approximately 13 ft/min (4 m/min).

> *Soaking furnace.* As the slabs enter the soaking furnace, they are sheared to lengths of 138 to 150 ft (42 to 46 m) (Fig. 85-13). The soaking furnace can hold up to three slabs at a time, and, when they leave, they are accelerated to the mill entry speed of 66 ft/min (20 m/min). After leaving the soaking furnace, the slab passes a water spray to remove the scale.

> *Finishing mill.* The four-stand finishing mill reduces the slab thickness from 2 to .100 in. (50 to 2.5 mm) (Fig. 85-14). After leaving the fourth

Manufacture and Properties of Steel **755**

stand, the strip passes through a cooling section, where it is cooled from the top and bottom by water sprays. The strip, which is at a temperature of 986°F to 1290°F (530°C to 700°C), is rolled into 76-in. (1930-mm) diameter coils.

Chemical Composition of Steel

Although iron and carbon are the main elements in steel, certain other elements may be present in varying quantities. Some are present because they are difficult to remove, and others are added to impart certain qualities to the steel. The elements found in plain carbon steel are carbon, manganese, phosphorus, silicon, and sulfur.

Carbon is the element that has the greatest influence on the property of the steel, since it is the hardening agent. The hardness, hardenability, tensile strength, and wear resistance will be increased as the percentage of carbon is increased up to about 0.83%. After this point has been reached, additional carbon does not noticeably affect the hardness of the steel but increases wear resistance and hardenability.

Manganese, when added in small quantities (0.30% to 0.60%) during the manufacture of steel, acts as a deoxidizer or purifier. Manganese helps to remove the oxygen, which, if it remained, would make the steel weak and brittle. Manganese also combines with sulfur, which in most cases is considered an undesirable element in steel. The addition of manganese increases the strength, toughness, hardenability, and shock resistance of steel. It will also slightly lower the critical temperature and increase ductility.

When manganese is added in quantities above 0.60%, it is considered an alloying element and will impart certain properties to the steel. When 1.5% to 2% manganese is added to high-carbon steel, it will produce deep-hardening, nondeforming steel that must be quenched in oil. Hard, wear-resistant steels, suitable for use in power shovel scoops, rock crushers, and grinding mills, are produced when up to 15% manganese is added to high-carbon steel.

Phosphorus is generally considered an undesirable element in carbon steel when present in amounts over 0.6%, since it will cause the steel to fail under vibration or shock. This condition is termed *cold-shortness.* Small amounts of phosphorus (about 0.3%) tend to eliminate blow holes and decrease shrinkage in the steel. Phosphorus and sulfur may be added to low-carbon steel (machine steel) to improve the machinability.

Silicon, present in most steels in amounts from 0.10% to 0.30%, acts as a deoxidizer and makes steel sound when it is cast or hot-worked. Silicon, when added in larger amounts (0.60% to 2%) is considered an alloying element. It is never used alone or simply with carbon; some deep-hardening element, such as manganese, molybdenum, or chromium, is usually added with silicon. When added as an alloying element, silicon increases the tensile strength, toughness, and hardness penetration of steel.

Sulfur, generally considered an impurity in steel, causes the steel to crack during working (rolling or forging) at high temperatures. This condition is known as *hot-shortness.* Sulfur may be added purposely to low-carbon steel in quantities ranging from 0.07% to 0.30% to increase its machinability. Sulfurized, free-cutting steel is known as *screw stock* and is used in automatic screw machines.

Classification of Steel

Steel may be classified into two groups: plain carbon steels and alloy steels.

PLAIN CARBON STEELS

Plain carbon steels may be classified as those that contain only carbon and no other major alloying element. They are divided into three categories: low-carbon steel, medium-carbon steel, and high-carbon steel.

Low-carbon steel contains from 0.02% to 0.30% carbon by mass. Because of the low carbon content, this type of steel cannot be hardened but can be case-hardened. *Machine steel* and *cold-rolled steel,* which contain from 0.08% to 0.30% carbon, are the most common low-carbon steels. These steels are commonly used in machine shops for the manufacture of parts that do not have to be hardened. Items such as bolts, nuts, washers, sheet steel, and shafts are made of low-carbon steel.

Medium-carbon steel contains from 0.30% to 0.60% carbon and is used where greater tensile strength is required. Because of the higher carbon content, this steel can be hardened, which makes it ideal for steel forgings. Tools such as wrenches, hammers, and screwdrivers are drop-forged from medium-carbon steel and later heat-treated.

High-carbon steel, also known as *tool steel,* contains over 0.60% carbon and may range as high as 1.7%. This type of steel is used for cutting tools, punches, taps, dies, drills, and reamers. It is available in hot-rolled stock or in finish-ground flat stock and drill rod.

ALLOY STEELS

Often certain steels are needed which have special characteristics that a plain carbon steel would not possess. It is then necessary to choose an *alloy steel.*

Alloy steel may be defined as steel containing other elements, in addition to carbon, that produce the desired qualities in the steel. The addition of alloying elements may impart one or more of the following properties to the steel:

1. Increase in tensile strength
2. Increase in hardness
3. Increase in toughness
4. Alteration of the critical temperature of the steel
5. Increase in wear abrasion
6. Red hardness
7. Corrosion resistance

High-Strength, Low-Alloy Steels

A recent development in the steelmaking industry is that of the high-strength, low-alloy (HSLA) steels. These steels, containing a maximum carbon content of 0.28% and small amounts of vanadium, columbium, copper, and other alloying elements, offer many advantages over the regular low-carbon construction steels. Some of these advantages are:

> Higher strength than medium-carbon steels

> Less expensive than other alloy steels

> Strength properties "built into" the steel and no further heat treating needed

> Bars of smaller cross sections able to do the work of larger, regular-carbon steel bars

> Higher hardness, toughness (impact strength), and fatigue failure limits than carbon steel bars

> May be used unpainted because they develop a protective oxide coating on exposure to the atmosphere

Effects of the Alloying Elements

Elements such as chromium, cobalt, manganese, molybdenum, nickel, phosphorus, silicon, sulphur, tungsten, and vanadium may be added to steel to give it a wide range of desired properties. The properties imparted to the steel by these elements are given in Table 85.1 (on p. 758).

table 85.1 The effect of alloying elements on steel

Effect	Carbon	Chromium	Cobalt	Lead	Manganese	Molybdenum	Nickel	Phosphorus	Silicon	Sulfur	Tungsten	Vanadium
Increases tensile strength	X	X			X	X	X					
Increases hardness	X	X										
Increases wear resistance	X	X			X		X				X	
Increases hardenability	X	X			X	X	X					X
Increases ductility					X							
Increases elastic limit		X				X						
Increases rust resistance		X					X					
Increases abrasion resistance		X			X							
Increases toughness		X				X	X					X
Increases shock resistance		X					X					X
Increases fatigue resistance												X
Decreases ductility	X	X										
Decreases toughness			X									
Raises critical temperature		X	X								X	
Lowers critical temperature					X		X					
Causes hot-shortness										X		
Causes cold-shortness								X				
Imparts red hardness			X			X					X	
Imparts fine grain structure					X							X
Reduces deformation					X	X						
Acts as deoxidizer					X				X			
Acts as desulphurizer					X							
Imparts oil-hardening properties		X			X	X	X					
Imparts air-hardening properties					X	X						
Eliminates blow holes								X				
Creates soundness in casting									X			
Facilitates rolling and forging					X				X			
Improves machinability				X						X		

unit 85 review questions

1. What effect has the study of metals had on modern living?

Physical Properties of Metals

2. Compare hardness, brittleness, and toughness.

Manufacturing of Pig Iron

3. Briefly describe the manufacture of pig iron.

4. What are the advantages of the direct-ironmaking process?

Manufacturing of Cast Iron

5. Explain how cast iron is manufactured.

6. Name four types of cast iron and state one purpose for each.

7. What are the chief differences between the Bessemer and the basic oxygen processes?

8. Briefly describe the basic oxygen process.

Direct Steelmaking

9. Name four advantages of the direct-steelmaking process.

10. List the four steps in the direct-steelmaking process.

11. What is strand or continuous casting, and what are its advantages?

Minimills

12. Name and briefly describe the six steps in the manufacture of steel in minimills.

Chemical Composition of Steel

13. What effect does carbon in excess of 0.83% have on steel?

14. What is the purpose of adding small quantities of manganese to steel?

15. What effect will the addition of larger quantities of manganese (1.5% to 2.0%) have on steel?

16. Why is phosphorus considered an undesirable element in steel?

17. How will steel be affected by the addition of silicon:

   a. In small amounts?　　　b. In large amounts?

18. Why is sulfur considered an undesirable element in steel?

Classification of Steel

19. State the carbon content of and two uses for:

   a. Low-carbon steel
   b. High-carbon steel

20. List six properties that alloying elements may impart to steel.

UNIT 86

Heat Treatment of Steel

OBJECTIVES

After completing this unit, you should be able to:

1 Select the proper grade of tool steel for a workpiece

2 Harden and temper a carbon-steel workpiece

3 Case-harden a piece of machine steel

In order for a steel component to function properly, it is often necessary to heat treat it. Heat treating is the process of heating and cooling a metal in its solid state in order to obtain the desired changes in its physical properties. One of the most important mechanical properties of steel is its ability to be hardened to resist wear and abrasion or be softened to improve ductility and machinability. Steel may also be heat treated to remove internal stresses, reduce grain size, or increase its toughness. During manufacture, certain elements are added to steel to produce special results when the metal is properly heat treated.

▶▶ Heat-Treating Equipment

The heat treating of metal is carried out in specially controlled furnaces, which may use gas, oil, or electricity to provide heat. These furnaces must also be equipped with certain safety devices, as well as control and indicating devices to maintain the temperature required for the job. All furnace installations should be equipped with a fume hood and exhaust fan to take away any fumes resulting from the heat-treating operation or, in the case of a gas installation, to exhaust the gas fumes.

In most gas installations, the exhaust fan, when running, will actuate the air switch in the exhaust duct. The air switch in turn operates a solenoid valve, which permits the main gas valve to be opened. Should the exhaust fan fail for any reason, the air switch will also fail and the main gas supply will close down.

The furnace temperature is controlled by a thermocouple and an indicating pyrometer (Fig. 86-1). After the furnace has been started, the desired temperature is set on

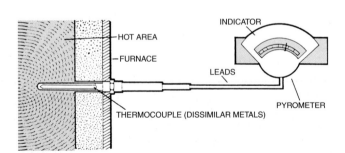

Figure 86-1 A thermocouple and pyrometer are used to indicate and control the temperature of a heat-treating furnace. *(Kelmar Associates)*

the indicating pyrometer. This pyrometer is connected on one side to a thermocouple and on the other side to a solenoid valve, which controls the flow of gas to the furnace.

The thermocouple is made up of two dissimilar metal wires twisted together and welded at the end. The thermocouple is generally mounted in the back of the furnace in a refectory tube to prevent damage and oxidation to the thermocouple wires.

As the temperature in the furnace rises, the thermocouple becomes hot, and due to the dissimilarity of the wires, a small electrical current is produced. This current is conducted to the pyrometer on the wall and causes the pyrometer needle to indicate the temperature of the furnace. When the temperature in the furnace reaches the amount set on the pyrometer, a solenoid valve connected to the gas supply is actuated and the flow of gas to the furnace is restricted. When the furnace temperature drops below the temperature indicated on the pyrometer, the solenoid valve opens, permitting a full flow of gas.

TYPES OF FURNACES

For most heat-treating operations, it is advisable to have a *low-temperature furnace* capable of temperatures up to 1300°F (704°C), a *high-temperature furnace* capable of temperatures up to 2500°F (1371°C), and a *pot-type furnace* (Fig. 86-2). The pot-type furnace may be used for hardening and tempering by immersing the part to be heat treated in the molten heat-treating medium, which may be salt or lead. Carbon-rich mixtures are used for case-hardening operations. One advantage of this type of furnace is that the parts, when being heated, do not come into contact with the air. This eliminates the possibility of oxidation or scaling. The temperature of the pot furnace is controlled by the same method as the high temperature and low temperature furnaces.

▶▶ Heat-Treatment Terms

Plain carbon steel is composed of alternate layers of iron and iron carbide. In this unhardened state, it is known as

A. LOW TEMPERATURE   B. HIGH TEMPERATURE   C. POT TYPE

■ **Figure 86-2** Types of heat-treating furnaces. *(Kelmar Associates)*

pearlite (Fig. 86-3a). Before getting involved in the theory of heat treating, it is good to study and understand a few terms connected with this topic.

> *Heat treatment*—the heating and subsequent cooling of metals to produce the desired mechanical properties

> *Decalescence point*—the temperature at which carbon steel, when being heated, transforms from pearlite to austenite; this is generally at about 1330°F (721°C) for 0.83% carbon steel

> *Recalescence point*—the temperature at which carbon steel, when being slowly cooled, transforms from austenite to pearlite

> *Lower critical temperature point*—the lowest temperature at which steel may be quenched in

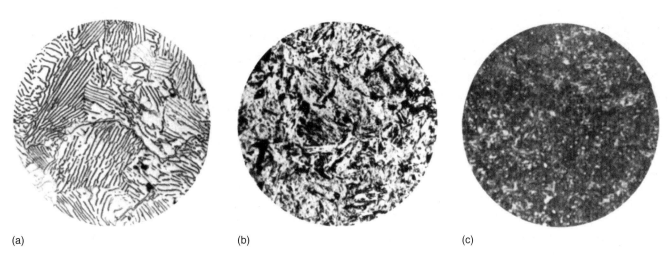

(a)   (b)   (c)

■ **Figure 86-3** (a) Pearlite is usually the condition of carbon steel before heat treating; (b) martensite is the condition of the hardened carbon steel; (c) martensite structure of steel can be altered by tempering.

order to harden it. This temperature coincides with the decalescence point

> *Upper critical temperature point*—the highest temperature at which steel may be quenched in order to attain maximum hardness and the finest grain structure

> *Critical range*—the temperature range bounded by the upper and the lower critical temperatures

> *Hardening*—the heating of steel above its lower critical temperature and quenching in the proper medium (water, oil, or air) to produce martensite (Fig. 86-3b).

> *Tempering* (drawing)—reheating hardened steel to a desired temperature below its lower critical temperature, followed by any desired rate of cooling. Tempering removes the brittleness and toughens the steel. Steel in this condition is called tempered martensite (Fig. 86-3c).

> *Annealing* (full)—heating metal to just above its upper critical point for the required period of time, followed by slow cooling in the furnace, lime, or sand. Annealing will soften the metal, relieve the internal stresses and strains, and improve its machinability.

> *Process annealing*—heating the steel to just below the lower critical temperature, followed by any suitable cooling method. This process is often used on metals that have been work hardened. Process annealing will soften it sufficiently for further cold working.

> *Normalizing*—heating the steel to just above its upper critical temperature and cooling it in still air. Normalizing is done to improve the grain structure and remove the stresses and strains. In general, it brings the metal back to its normal state.

> *Spheroidizing*—the heating of steel to just below the lower critical temperature for a prolonged period of time followed by cooling in still air. This process produces a grain structure with globular-shaped particles (spheroids) of cementite rather than the normal needlelike structure, which improves the machinability of the metal.

> *Alpha iron*—the state in which iron exists below the lower critical temperature. In this state, the atoms form a body-centered cube.

> *Gamma iron*—the state in which iron exists in the critical range. In this state the molecules form face-centered cubes. Gamma iron is nonmagnetic.

> *Pearlite*—a laminated structure of ferrite [iron and cementite (iron carbide)], usually the condition of steel before heat treatment (Fig. 86-3a).

> *Cementite*—a carbide of iron (Fe_3C), which is the hardener in steel

> *Austenite*—a solid solution of carbon in iron, which exists between the lower and upper critical temperatures

> *Martensite*—the structure of fully hardened steel obtained when austenite is quenched. Martensite is characterized by its needlelike pattern (Fig. 86-3b).

> *Tempered martensite*—the structure obtained after martensite has been tempered (Fig. 86-3c). Tempered martensite was formerly known as *troosite* and *sorbite*.

> *Eutectoid steel*—steel containing just enough carbon to dissolve completely in the iron when the steel is heated to its critical range. Eutectoid steel contains from 0.80% to 0.85% carbon. This may be likened to a saturated solution of salt in water.

> *Hypereutectoid steel*—steel containing more carbon than will completely dissolve in the iron when the steel is heated to the critical range. This is similar to a supersaturated solution.

> *Hypoeutectoid steel*—steel containing less carbon than can be dissolved by the iron when the steel is heated to the critical range. Here, there is an excess of iron. This is similar to an unsaturated solution.

▶▶ Selection of Tool Steel

The proper selection and proper heat treatment of a tool steel are both essential if the part being made is to perform efficiently. Problems that may arise in the selection and heat treatment of tool steel include:

1. It may not be tough or strong enough for the job.
2. It may not offer sufficient abrasion resistance.
3. It may not have sufficient hardening penetration.
4. It may warp during heat treatment.

Because of these problems, the steel producers have been forced to manufacture many types of alloy steels to cover the range of most jobs.

To select the proper tool steel to suit the specifications and applications of a part, see Table 86.1. Since tool steels may vary somewhat with each manufacturer, it may be wise to consult the handbook provided by the steel manufacturer, which outlines the heat-treating procedures for each of its steels. For a more detailed description of the qualities and specifications of all types of tool steels, see Table 18 in the appendix.

Tool steels are generally classified as water-hardening, oil-hardening, air-hardening, or high-speed steels. They are usually identified by each manufacturer by a trade name,

table 86.1   **Tool steel selection guide***

Group	Type	Quench Medium	Wear Resistance	Toughness	Warpage Resistance	Hardening Depth	Red Hardness	Machinability (MRR)
High-speed	M	O, A, S	Very high	Low	A, S: low O: medium	Deep	Highest	45–60
	T	O, A, S	Very High	Low	A, S: low O: medium	Deep	Highest	40–55
Hot-work Cr base	H	A, O	Fair	Good	O: fair A: good	Deep	Good	75
W base	W	A, O	Fair to good	Good	O: fair	Deep	Very good	50–60
Mo base	M	O, A, S	High	Medium	A, S: low O: medium	Deep	Good	50–60
Cold-work	D	A, O	Best	Poor	A: best O: lowest	Deep	Good	40–50
	A	A	Good	Fair	Best	Deep	Fair	85
	O	O	Good	Fair	Very good	Medium	Poor	90
Shock resisting	S	O, W	Fair	Best	O: fair W: poor	Medium	Fair	85
Mold steel	P	A, O, W	Low to high	High	Very low	Shallow	Low	75–100
Special-purpose carbon-tungsten	F	B, W	Low to very high	Low to high	High	Shallow	Low	75
Water-hardening	W	B, W	Fair to good	Good	Poor	Shallow	Poor	100

A=air; B=brine; O=oil; S=molten salt; W=water; MRR=material-removal rate.

such as Alpha 8, Keewatin, Nutherm, or Nipigon; which are some of the trade names of the Atlas Steel Company's products.

WATER-HARDENING TOOL STEELS

Water-hardening tool steels generally contain from 0.50% to 1.3% carbon, along with small amounts (about 0.20%) of silicon and manganese. The addition of silicon facilitates the forging and rolling of the material, while manganese helps to make the steel more sound when it is first cast into the ingot. Further addition of silicon (above 0.20%) will reduce the grain size and increase the toughness of water-hardening steel.

Most water-hardening steels achieve the maximum hardness for a depth of about ⅛ in. (3 mm); the inner core remains softer but still tough. Chromium or molybdenum

is sometimes added to increase the hardenability (hardness penetration), toughness, and wear resistance of water-hardening steels. Water-hardening steels are heated to around 1450°F to 1500°F (787°C to 815°C) during the hardening process. These steels are used where a dense, fine-grained outer casing with a tough inner core is required. Typical applications are drills, taps, reamers, punches, jig bushings, and dowel pins.

The problems connected with water-hardening steels are those of distortion and cracking when the material is quenched. Should these problems occur, it would then be wise to select an oil-hardening steel.

OIL-HARDENING STEELS

A typical *oil-hardening steel* contains about 0.90% carbon, 1.6% manganese, and 0.25% silicon. The addition of

Heat Treatment of Steel   763

manganese in quantities of 1.5% or more increases the hardenability (hardness penetration) of the steel up to about 1 in. (25 mm) from each surface. During the quenching of steels with a higher manganese content, the hardening is so rapid that a less severe quenching medium (oil) must be used. The use of oil as a quenching medium retards the cooling rate and reduces the stresses and strains in the steel, which cause warping and cracking. Chromium and nickel, in varying quantities, may be added to oil-hardening steel to increase its hardness and wear resistance. Higher hardening temperatures, from 1500°F to 1550°F (815°C to 843°C), are required for these latter alloy steels.

Often, due to the intricate shape of the part, it may not be possible to eliminate warping or cracking during the quench, and it will be necessary to select air-hardening steel for the particular part. Typical applications of oil-hardening steels are blanking, forming, and punching dies; precision tools; broaches; and gages.

AIR-HARDENING STEELS

Due to the slower cooling rate of *air-hardening steels,* the stresses and strains that cause cracking and distortion are kept to a minimum. Air-hardening steels are also used on parts having large cross sections, where full hardness throughout could not be obtained by using water- or oil-hardening steels.

A typical air-hardening steel will contain about 1.00% carbon, 0.70% manganese, 0.20% silicon, 5.00% chromium, 1.00% molybdenum, and 0.20% vanadium. Air-hardening steels require higher hardening temperatures—from 1600°F to 1775°F (871°C to 968°C).

Typical applications of this steel are large blanking, forming, trimming, and coining dies; rolls; long punches; precision tools; and gages.

HIGH-SPEED STEELS

High-speed steels are used in the manufacture of cutting tools such as drills, reamers, taps, milling cutters, and lathe cutting tools. The analysis of a typical high-speed steel could be as follows: 0.72% carbon, 0.25% manganese, 0.20% silicon, 4% chromium, 18% tungsten, and 1% vanadium. Tools made of high-speed steel will retain their hardness and cutting edges even when operating at red heat.

During heat treatment, high-speed steels must be preheated slowly to 1500°F to 1600°F (815°C to 871°C) in a neutral atmosphere and then transferred to another furnace and quickly brought up to 2300°F to 2400°F (1260°C to 1315°C). They are generally quenched in oil, but small, intricate sections may be air cooled.

▶▶ Classification of Steel

In order to ensure that the composition of various types of steel remains constant and that a certain type of steel will meet the required specifications, the Society of Automotive Engineers (SAE) and the American Iron and Steel Institute (AISI) have devised similar methods of identifying different types of steel, and both are widely used.

SAE-AISI CLASSIFICATION SYSTEMS

The systems designed by the SAE and the AISI are similar in most respects. They both use a series of four or five numbers to designate the type of steel.

The first digit in these series indicates the predominant alloying element. The last two digits (or sometimes three in certain corrosion- or heat-resisting alloys) indicate the average carbon content in points (hundredths of 1%, or 0.01%) (Fig. 86-4).

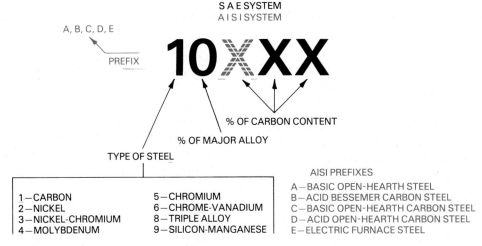

■ **Figure 86-4** The SAE and AISI classification systems.

The main difference in the two systems is that the AISI system indicates the steelmaking process used by the following prefixes:

A—basic open-hearth alloy steel

B—acid-Bessemer carbon steel

C—basic open-hearth carbon steel

D—acid-open-hearth carbon steel

E—electric furnace steel

In the classification charts, the various types of steels are indicated by the first number in the series as follows:

1. Carbon
2. Nickel
3. Nickel-chrome
4. Molybdenum
5. Chromium
6. Chromium-vanadium
7. Triple alloy
8. Manganese-silicon

Table 86.2 indicates the SAE classification of the various steels and alloys. The number 7 does not appear on the chart. It formerly represented tungsten steel, which is no longer listed in this chart, since it is now considered a special steel.

EXAMPLES OF STEEL IDENTIFICATION

Determine the types of steel indicated by the following numbers: 1015, A2360, 4170.

1015—1 indicates plain carbon steel.

—0 indicates there are no major alloying elements.

—15 indicates there is between 0.10% and 0.20% carbon content.

Note: This steel would naturally contain small quantities of manganese, phosphorus, and sulfur.

A2360—A indicates an alloy steel made by the basic open-hearth process.

—23 indicates the steel contains 3.5% nickel (see Table 86.2).

—60 indicates 0.60% carbon content.

4170—41 indicates a chromium-molybdenum steel.

—70 indicates 0.70% carbon content.

table 86.2	SAE classification of steels
Carbon steels	1xxx
Plain carbon	10xx
Free-cutting (resulfurized screw stock)	11xx
Free-cutting manganese	X13xx
High-manganese	T13XX
Nickel steels	2xxx
0.50% nickel	20xx
1.50% nickel	21xx
3.50% nickel	23xx
5.00% nickel	25xx
Nickel-chromium Steels	3xxx
1.25% nickel, 0.60% chromium	31xx
1.75% nickel, 1.00% chromium	32xx
3.50% nickel, 1.50% chromium	33xx
3.00% nickel, 0.80% chromium	34xx
Corrosion- and heat-resisting steels	30xxx
Molybdenum steels	4xxx
Chromium-molybdenum	41xx
Chromium-nickel-molybdenum	43xx
Nickel-molybdenum	46xx and 48xx
Chromium steels	5xxx
Low-chromium	51xx
Medium-chromium	52xx
Chromium-vanadium steels	6xxx
Triple-alloy steels (nickel, chromium, molybdenum)	8xxx
Manganese-silicon steels	9xxx

▶▶ Heat Treatment of Carbon Steel

The proper performance of a steel part depends not only on the correct selection of steel but also on the correct heat-treating procedure and an understanding of the theory behind it. When steel is heated from room temperature to the upper critical temperature and then quenched, several changes take place in the steel. These may be more easily understood if the changes that take place in water, from the frozen state until it is transformed into steam, are considered.

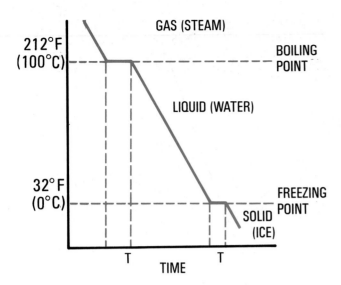

Figure 86-5 The critical points of water.

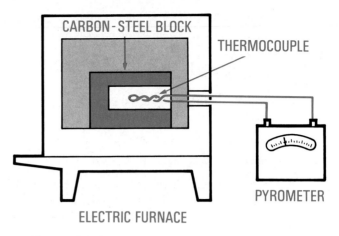

Figure 86-6 Setup to determine the critical points of steel. *(Kelmar Associates)*

By referring to Fig. 86-5, it is noted that water exists as a solid at or below 32°F (0°C). If the ice is heated, the temperature will remain 32°F (0°C) until the ice completely melts. If the water is heated further, it will turn into steam at 212°F (100l°C). Again, the water remains at this temperature for a short time before turning into steam. It should also be noticed that, if the process were reversed and the steam cooled, it would form water at 212°F (100°C) and ice at 32°F (0°C). The points where water transforms to another state are known as the *critical points* of water.

Steel, like water, has critical points that, when determined, will lead to successful heat treatment of the metal.

TO DETERMINE THE CRITICAL POINTS OF 0.83% CARBON STEEL

A simple experiment may be performed to illustrate the critical points and the changes that take place in a piece of carbon steel when heated and slowly cooled.

step	procedure
1	Select a piece of 0.83% (eutectoid) carbon steel about 1½ in. × 1½ in. × 2 in. (38 mm × 38 mm × 50 mm) long and drill a small hole in one end for most of the length.
2	Insert a thermocouple in the hole and seal the end of the hole with fireclay.
3	Place the block in a furnace and run the thermocouple wire to a voltmeter (Fig. 86-6).

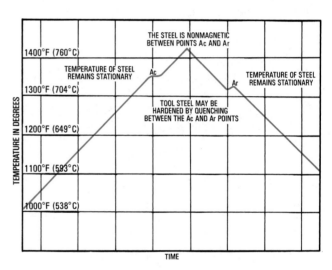

Figure 86-7 A graph illustrating the critical points of steel.

4	Light the furnace and set the temperature for about 1425°F (773°C) on the pyrometer.
5	Plot the readings of the voltmeter needle at regular time intervals (Fig. 86-7).
6	When the furnace reaches 1425°F (773°C), shut it down and let it cool.
7	Continue to plot the readings until the temperature in the furnace drops to approximately 1000°F (538°C).

766   Metallurgy

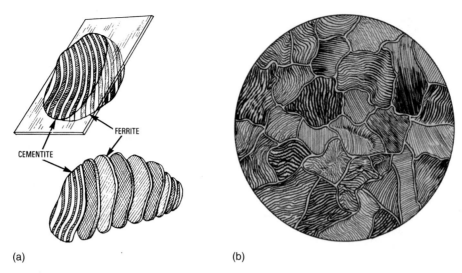

(a)   (b)

■ **Figure 86-8** (a) A pearlite grain is composed of alternate layers of iron (ferrite) and iron carbide (cementite); (b) a photomicrograph of high-carbon steel, showing the pearlite grains surrounded by iron carbide (white lines).   *(Praxair, Inc.)*

OBSERVATIONS AND CONCLUSIONS

Steel at room temperature consists of laminated layers of ferrite (iron) and cementite (iron carbide). This structure is called pearlite (Fig. 86-8). As the steel is heated from room temperature, the time/temperature curve climbs uniformly until a temperature of about 1333°F (722°C) is reached. At this point, Ac (Fig. 86-7), the temperature of the steel drops slightly, although the temperature of the furnace is rising.

The point Ac indicates the *decalescence point.* It is here that several changes take place in the steel:

1. If the steel were observed in the furnace at this time, it would be noticed that the dark shadows in the steel disappear.

2. The steel would be nonmagnetic when tested with a magnet.

3. These changes were caused by a change in the atomic structure of the steel. The atoms rearrange themselves from body-centered cubes (Fig. 86-9a on p. 768) to face-centered cubes (Fig. 86-9b). When the atoms are rearranged, the energy (heat) required for this change is drawn from the metal; thus, a slight drop in the temperature of the workpiece is recorded at the decalescence point. The layers of iron carbide completely dissolve in the iron to form a solid solution known as *austenite.* Thus, the decalescence point marked the transformation point from *pearlite* to *austenite,* or from body-centered cubes to face-centered cubes.

4. It is at this point that the steel, if quenched in water, would also show the first signs of hardening.

5. If the steel could be examined under a microscope, it would be noticed that the grain structure starts to get smaller. As the curve progresses upward past Ac, the grain size would become progressively smaller until the upper critical temperature [1425°F (773°C)] is reached.

As the steel cools, the curve would continue uniformly down and the grain size would gradually get larger until the point Ar, at about 1300°F (704°C), is reached. This is the *recalescence point* and here the needle on the voltmeter would show a slight rise in temperature, although the furnace is cooling. This process is obviously the reverse of the phenomenon that occurs at the decalescence point; the austenite reverts to pearlite, the atoms rearrange themselves into body-centered cubes, and the steel again becomes magnetic.

Another experiment, which demonstrates the decalescence and recalescence points, follows.

Decalescence Point

1. Place a magnet on a firebrick.

2. Select a ½- to ⅝-in. (13- to 15-mm) round piece of 0.90 to 1.00 carbon steel and place it on the magnet (Fig. 86-10 on p. 768).

3. Place a can of cold water under the magnet ends.

4. Heat the piece held to the magnet using a small flame.

Note: Do not allow the flame to come into contact with the magnet.

Heat Treatment of Steel   **767**

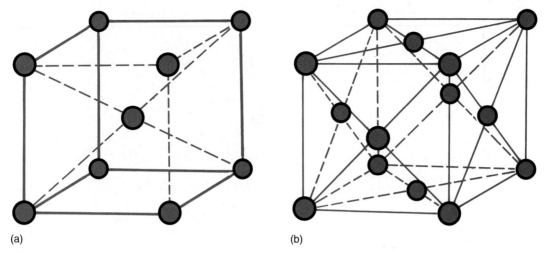

(a)

(b)

■ Figure 86-9 (a) Arrangement of the atoms in a body-centered cube; (b) arrangement of the atoms in a face-centered cube.

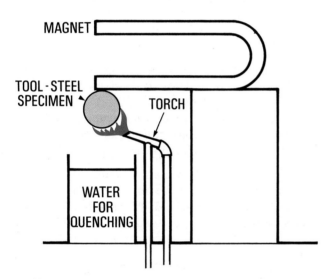

■ Figure 86-10 Decalescence-point experiment. *(Kelmar Associates)*

5. When the temperature reaches its critical point, the steel will drop into the water and become hardened.

Recalescence Point

1. Remove the can of water from under the magnet.
2. Place a flat plate under the work held on the magnet.
3. Heat the steel until it drops from the magnet onto the plate.
4. When the steel cools, it will become attracted by the magnet.

Summary

When the steel loses its magnetic value (decalescence point), it drops into the water and the change in the steel is trapped or stopped. The steel then hardens because it does not have time to revert to another state.

Conversely, when the steel is not quenched but is allowed to cool gradually from the decalescence point, it regains its magnetic value when it has cooled slightly (recalescence point). The steel does not change from magnetic to nonmagnetic; it merely acquires temporary characteristics of being attracted or *not* attracted to the magnet.

HARDENING OF 0.83% CARBON STEEL

Once the critical temperature of a steel has been determined, the proper quenching temperature, which is about 50°F (27°C) over the upper critical temperature, can be determined. Not all steels have the same critical temperature; Fig. 86-11 indicates that the critical temperature of a steel drops as the carbon content increases up to 0.83%, after which it does not change. As a result, steels containing over 0.83% carbon (hypereutectoid) need be heated only to just above the lower critical temperature Ac_1 (Fig. 86-11) to obtain maximum hardness. This makes it possible to use a lower hardening temperature for hypereutectoid steels, thus decreasing the possibility of warping. The increase in the carbon content beyond 0.83% will not increase the hardness of the steel; however, it does increase the wear resistance of the steel considerably.

In order for the steel to be hardened properly, it must be heated uniformly to about 50°F (27°C) above the upper critical temperature and held at this temperature long enough to allow sufficient carbon to dissolve and form a

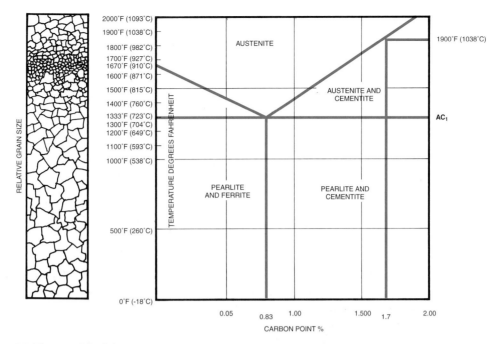

Figure 86-11 Iron carbide diagram illustrating the relationship between the carbon content of steel and critical temperatures.

solid solution, which permits maximum hardness. At this point, the steel will have the smallest grain size and, when quenched, will produce the maximum hardness.

The critical temperature of a steel is also affected by alloying elements such as manganese, nickel, chromium, cobalt, and tungsten.

QUENCHING

When steel has been properly heated throughout, it is quenched in brine, water, or oil (depending on the type of steel) to cool it rapidly. During this operation, the austenite is transferred into *martensite,* a hard, brittle metal. Because the steel is cooled rapidly, the austenite is prevented from passing through the recalescence point (Ar), as in the case of slow cooling, and the small grain size of the austenite is retained in the martensite (Fig. 86-12).

The rate of cooling affects the hardness of steel. If a water-hardening steel is quenched in oil, it cools more slowly and does not attain the maximum hardness. On the other hand, if an oil-hardening steel is cooled too quickly by quenching it in water, it may crack. Cracking may also occur when the quenching medium is too cold.

The method of quenching greatly affects the stresses and strains set up in the metal, which cause warping and cracking. For this reason, long, flat pieces should be held vertically above the quenching medium and plunged straight into the liquid. After immersion, the part should be moved around in a figure 8 motion. This keeps the liquid at a uniform temperature and prevents air pockets from forming on the steel, which would affect the uniformity of hardness.

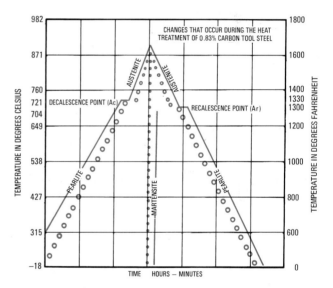

Figure 86-12 Carbon steel, when quenched at the upper critical temperature, retains the smallest grain size. *(Kelmar Associates)*

METCALF'S EXPERIMENT

This simple experiment demonstrates the effect that various degrees of heat have on the grain structure, hardness, and strength of tool steel.

1. Select a piece of SAE 1090 (tool steel) about ½ in. (13 mm) in diameter and about 4 in. (100 mm) long.

2. With a sharp, pointed tool, cut shallow grooves approximately ½ in. (13 mm) apart.

3. Number each section (Fig. 86-13).

4. Heat the bar with an oxyacetylene torch, bringing section 1 to a white heat.

5. Keep section 1 at white heat, and heat sections 4 and 5 to a cherry red. Do *not* apply heat to sections 6 to 8.

6. Quench in cold water or brine.

7. Test each section with the edge of a file for hardness.

8. Break off the sections and examine the grain structure under a microscope.

Results

Sections 1 and 2 have been overheated. They break easily and the grain structure is very coarse (Fig. 86-14a).

Section 3 requires more force to break and the grain structure is somewhat finer. Sections 4 and 5 have greater strength and resistance to shock. These sections have the finest grain structure (Fig. 86-14b).

Sections 6 to 8, where the metal was underheated, require the greatest force to break, and bending occurs. Note that the grain structure becomes coarser toward section 8 (Fig. 86-14c). This section is the original structure of unheated steel (pearlite).

TEMPERING

Tempering is the process of heating a hardened carbon or alloy steel below its lower critical temperature and cool-

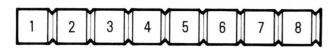

Figure 86-13 A test piece for Metcalf's experiment.
(Kelmar Associates)

ing it by quenching in a liquid or in air. This operation removes many of the stresses and strains set up when the metal was hardened. Tempering imparts toughness to the metal but decreases the hardness and tensile strength. The tempering process modifies the structure of the martensite, changing it to *tempered martensite,* which is somewhat softer and tougher than martensite.

The tempering and drawing temperature is not the same for each type of steel and is affected by several factors:

1. The toughness required for the part
2. The hardness required for the part
3. The carbon content of the steel
4. The alloying elements present in the steel

The hardness obtained after tempering depends on the temperature used and the length of time the workpiece is held at this temperature. Generally, hardness decreases and toughness increases as the temperature is increased (Fig. 86-15).

As the length of the tempering time is increased for a specific part, the hardness of the metal decreases. On the other hand, if the tempering time is too short, the stresses and strains set up by hardening are not totally removed and the metal will be brittle. The cross-sectional size of the workpiece affects its tempering time. The tempering time and temperatures for various steels are always supplied in the steel manufacturer's handbook; these recommended times and temperatures should be followed to obtain the best results.

Tempering Colors

When a piece of steel is heated from room temperature to a red heat, it passes through several color changes, caused by the oxidation of the metal. These color changes indicate the approximate temperature of the metal and are often used as a guide when tempering (Table 86.3).

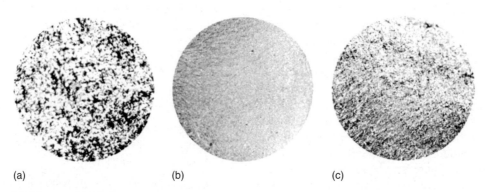

(a)   (b)   (c)

Figure 86-14 (a) Tool steel overheated; (b) tool steel heated to the proper temperature; (c) tool steel underheated. *(Kelmar Associates)*

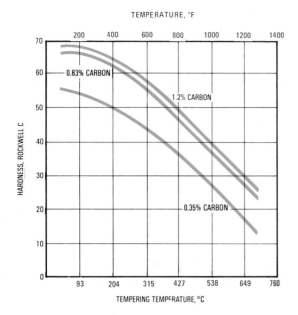

Figure 86-15 As tempering heat increases, hardness decreases.

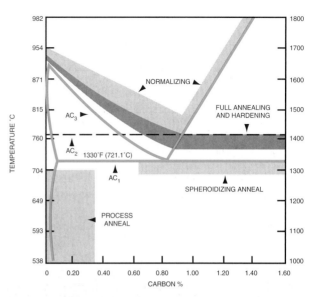

Figure 86-16 Temperature ranges for various heat-treating operations.

table 86.3	Tempering colors and approximate temperatures for carbon steel		

| | Temperature | | |
Color	°F	°C	Use
Pale yellow	430	220	Lathe tools, etc.
Light straw	445	230	Milling cutters, drills, reamers
Dark straw	475	245	Taps and dies
Brown	490	255	Scissors, shear blades
Brownish-purple	510	265	Axes and wood chisels
Purple	525	275	Cold chisels, center punches
Bright blue	565	295	Screwdrivers, wrenches
Dark blue	600	315	Woodsaws

ANNEALING

Annealing is a heat-treating operation used to soften metal and to improve its machinability. Annealing also relieves the internal stresses and strains caused by previous operations, such as forging or rolling.

step	procedure
1	Set the pyrometer approximately 30°F (16°C) above the upper critical temperature and start the furnace (Fig. 86-16).
2	Place the part in the furnace. After the required temperature has been reached, allow it to soak for 1 h per inch (25 mm) of workpiece thickness.
3	Shut off the furnace and allow the part to cool slowly in the furnace, or remove the part from the furnace and pack it immediately in lime or ashes and leave it covered for several hours, depending on the size, until it is cool.

NORMALIZING

Normalizing is performed on metal to remove internal stresses and strains and to improve its machinability.

step	procedure
1	Set the pyrometer approximately 30°F (16°C) above the upper critical temperature of the metal and start the furnace (Fig. 86-16).
2	Place the part in the furnace. After the required temperature has been reached, allow the workpiece to soak for 1 h per inch (25 mm) of thickness.

Heat Treatment of Steel

3   Remove the part from the furnace and allow it to cool slowly in still air. Thin workpieces may cool too rapidly and may harden if normalized in air. It may be necessary to pack them in lime to retard the cooling rate.

SPHEROIDIZING

Spheroidizing is a process of heating metal for an extended period to just below the lower critical temperature. This process produces a special kind of grain structure whereby the cementite particles become spherical in shape. Spheroidizing is generally done on high-carbon steel to improve the machinability.

step	procedure

1   Set the pyrometer approximately 30°F (16°C) below the lower critical temperature of the metal and start the furnace (Fig. 86-16).

2   Place the part in the furnace and allow it to soak for several hours at this temperature.

3   Shut down the furnace and let the part cool slowly to about 1000°F (538°C).

4   Remove the part from the furnace and cool it in still air.

▶▶ Case-Hardening Methods

When hardened parts are required, they may be made from carbon steel and heat treated to the specifications. Often these parts can be made more cheaply from machine steel and then case-hardened. This process produces a hard outer case with a soft inner core, which is often preferable to through-hard parts made from carbon steel. Case hardening may be performed by several methods, such as carburizing, carbonitriding, and nitriding processes.

CARBURIZING

Carburizing is a process whereby low-carbon steel, when heated with some carbonaceous material, absorbs carbon into its outer surface. The depth of penetration depends on the time, the temperature, and the carburizing material used. Carburizing may be performed by three methods: pack carburizing, liquid carburizing, and gas carburizing.

Pack carburizing is generally used when a hardness depth of penetration of .060 in. (1.5 mm) or more is re-

quired. The parts to be carburized are packed with a carbonaceous material such as activated charcoal in a sealed steel box.

step	procedure

1   Place a 1- to 1½-in. (25- to 38-mm) layer of carbonaceous material in the bottom of a steel box that will fit into the furnace.

2   Place the parts to be carburized in the box, leaving about 1½ in. (38 mm) between parts.

3   Pack the carburizer around the parts and cover the parts with about 1½ in. (38 mm) of material.

4   Tap the sides of the box to settle the material and to pack it around the workpieces. This will exclude most of the air.

5   Place a metal cover over the box and seal around the joint with fireclay.

6   Place the box in the furnace and bring the temperature up to about 1700°F (926°C).

7   Leave the box in the furnace long enough to give the required penetration. The rate of penetration is generally about .007 to .008 in./h (0.17 to 0.2 mm/h); however, this decreases as the depth of penetration increases. The proper time (and temperature) for any depth of penetration is usually given in the literature supplied by the manufacturer of the carburizing material (Fig. 86-17).

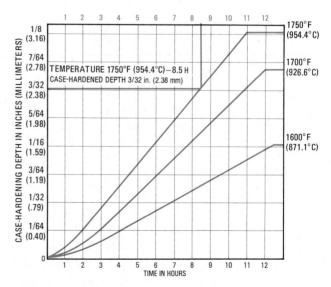

■ **Figure 86-17** Relationship among temperature, time, and depth of case.

HARD OUTER SURFACE

SOFT CORE

■ **Figure 86-18** A broken cross section of a piece of case-hardened machine steel, showing the hardened outer case. *(Kelmar Associates)*

8   Shut down the furnace and leave the box in the furnace until it cools. This may take 12 to 16 h.

9   Remove the box from the furnace and take out the parts and clean them.

Note: The surfaces of the workpieces have now been transformed into a thin layer of carbon steel, which is soft due to the slow cooling of the parts in the carburizing material.

10   Heat the parts to the proper critical temperature in a furnace and quench in oil or water, depending on the shape of the parts and the hardness required. The parts are now surrounded with a hard layer of carbon steel and have a soft inner core (Fig. 86-18).

Liquid carburizing is generally used to produce a thin layer of carbon steel on the outside of low-carbon steel parts. The parts are not usually machined after liquid carburization.

step   procedure

1   Place the carburizing material into a pot furnace. Heat until it is molten and reaches the proper temperature.

2   Preheat the parts to be carburized to approximately 800°F (426°C) in the low temperature drawing furnace. This eliminates the possibility of an explosion due to water or oil on the work when the parts are immersed in the molten liquid.

3   Suspend the parts in the liquid carburizer and leave them for the time required to give the desired penetration. The depth of penetration (depending on the temperature of the liquid) may be from .015 to .020 in. (0.38 to 0.5 mm) for the first hour and about .010 in. (0.25 mm) for each succeeding hour.

4   Use dry tongs to remove the parts; quench the parts immediately in water.

 CAUTION Since some liquid carburizers contain cyanide, extreme care must be taken when these materials are used.

> Avoid letting *any moisture* come into contact with the liquid carburizer. Such contact will cause an explosion.

> Heat the jaws of the tongs before using to remove any moisture or oil.

> Avoid inhaling the fumes; they are toxic.

> Wear protective clothing (gloves, face and arm shields) when removing and quenching parts.

Gas carburizing, like pack carburizing, is used on parts where over .060-in. (1.5-mm) depth of case hardening is required and where it is necessary to grind the parts after carburizing. This method is generally done by specialized heat-treating firms, since it requires special types of furnaces.

The parts are placed in a sealed drum, into which natural gas or propane is introduced and circulated. The workpieces are heated to a carburizing temperature in the gas atmosphere. The gas exhausts at one end of the drum and is burned to prevent air from entering the chamber. The exterior of the drum is heated by a source such as gas or oil. In this process, the carbon from the gas is absorbed by the workpiece.

The parts remain in the drum for the time required to give the desired penetration. Depths of .020 to .030 in. (0.5 to 0.76 mm) are obtained in about 4 h at a temperature of 1700°F (926°C). The parts may then be removed and quenched or allowed to cool, after which they are reheated to the critical temperature and quenched.

CARBONITRIDING PROCESSES

Carbonitriding is a process whereby both carbon and nitrogen are absorbed by the surface of a steel workpiece when it is heated to the critical temperature to produce a hard, shallow outer case. Carbonitriding may be done by liquid or gas methods.

Cyaniding (liquid carbonitriding) is a process that uses a salt bath composed of cyanide-carbonate-chloride salts with varying amounts of cyanide, depending on the application. Liquid cyaniding is generally carried out in a pot-type furnace, and since cyanide fumes are poisonous, extreme care must be taken when using this method.

The parts are suspended in a liquid cyanide bath, which must be at a temperature above the lower critical point of the steel being used. The depth of penetration is about .005 to .010 in. (0.12 to 0.25 mm) in 1 h at 1550°F (843°C). A depth of about .015 in. (0.38 mm) may be obtained in 2 h at the same temperature. The parts may then be quenched in water or oil, depending on the steel being used. After the parts have been hardened, they should be thoroughly washed to remove all traces of the cyanide salt.

Carbonitriding (gas cyaniding) is carried out in a special furnace similar to the gas carburizing furnace. The workpieces are put into the inner drum of the furnace. A mixture of ammonia and a carburizing gas is introduced into and circulated through this chamber, which is heated externally to a temperature of 1350°F to 1700°F (732°C to 926°C). During this process, the workpiece absorbs carbon from the carburizing gas and nitrogen from the ammonia.

The parts are removed from the furnace and quenched in oil, which gives the parts maximum hardness and minimum distortion. The depth of case hardening produced by this method is relatively shallow. Depths of about .030 in. (0.76 mm) are obtained in 4 to 5 h at a temperature of 1700°F (926°C).

NITRIDING PROCESSES

Nitriding is used on certain alloy steels to provide maximum hardness. Most carbon alloy steels can be hardened to only about 62 Rockwell C by conventional means, whereas readings of 70 Rockwell C may be obtained on certain vanadium and chromium alloy steels using a nitriding process. Nitriding may be done in a protected atmosphere furnace or in a salt bath.

In *gas nitriding,* the parts to be nitrided are placed in an airtight drum, which is heated externally to a temperature of 900°F to 1150°F (482°C to 621°C). Ammonia gas is circulated through the chamber. The ammonia, at this temperature, decomposes into nitrogen and hydrogen. The nitrogen penetrates the outer surface of the workpiece and combines with the alloying elements to form hard nitrides. Gas nitriding is a slow process requiring approximately 48 h to obtain a case-hardened depth of .020 in. (0.5 mm). Because of the low operating temperatures used in this process, and since no quenching of the part is required, there is little or no distortion. This method of increasing hardness is used on parts that have been hardened and ground. No further finishing is required on such parts.

Salt bath nitriding is carried out in a salt bath containing nitriding salts. The hardened part is suspended in the molten nitriding salt, which is held at a temperature

from 900°F to 1100°F (482°C to 593°C), depending on the application. Parts such as high-speed taps, drills, and reamers are nitrided to increase surface hardness, which improves durability.

▶▶ Surface Hardening of Medium-Carbon Steels

When selected areas of a part are to be surface-hardened to increase wear resistance and retain a soft inner core, the part must have a medium- or high-carbon content. It may be surface hardened by flame or induction hardening, depending on the size of the part and its application. In both processes, the steel must contain carbon, since no external carbon is added as in other case-hardening methods.

INDUCTION HARDENING

In *induction hardening,* the part is surrounded by a coil through which a high-frequency electrical current is passed. The current heats the surface of the steel to above the critical temperature in a few seconds. An automatic spray of water, oil, or compressed air is used to quench and harden the part, which is held in the same position as for heating. Since only the surface of the metal is heated, the hardness is localized at the surface. The depth of hardness is governed by the current frequency and heating-cycle duration.

The current frequencies vary from 1 kHz to 2 MHz. Higher frequencies produce shallow hardening depths. Lower frequencies produce hardening depths up to ¼ in. (6 mm).

Induction hardening may be used for the selective hardening of gear teeth, splines, crank shafts, camshafts, and connecting rods.

FLAME HARDENING

Flame hardening is used extensively to harden ways on lathes and other machine tools, as well as gear teeth, splines, crank shafts, etc.

The surface of the metal is heated very rapidly to above the critical temperature and is hardened quickly by a quenching spray. Large surfaces, such as lathe ways, are heated by a special oxyacetylene torch, which is moved automatically along the surface, followed by a quenching spray. Smaller parts are placed under the flame and spray-quenched automatically.

Flame-hardened parts should be tempered immediately to remove strains created by the hardening process. Large surfaces are tempered by a special low-tempering torch, which follows the quenching nozzle as it moves along the work. The depth of flame hardening varies from ¹⁄₁₆ to ¼ in. (1.5 to 6 mm), depending on the speed at which the surface is brought up to the critical temperature.

unit 86 review questions

Heat-Treating Equipment

1. Describe a thermocouple and explain how it functions.

2. Sketch a furnace installation showing the furnace, thermocouple, and pyrometer.

3. What is the purpose of:
   a. The pyrometer?   c. The air switch?
   b. The solenoid valve?

Heat-Treatment Terms

4. What is the difference between the decalescence point and the recalescence point of a piece of steel?

5. At what point in relation to the upper and lower critical temperatures are the following heat-treating operations performed?
   a. Hardening   d. Normalizing
   b. Tempering   e. Spheroidizing
   c. Annealing

6. What is the difference between hypereutectoid and hypoeutectoid steel?

Water-Hardening Tool Steel

7. State two problems that often occur with water-hardening steel.

Oil-Hardening Tool Steel

8. Why is oil used to advantage as a quenching medium?

Air-Hardening Steels

9. What are the advantages of air-hardening steels?

10. What elements will give red-hardness quality to a high-speed steel toolbit?

11. Explain the difference in the hardening procedures between plain carbon steel and high-speed steel.

SAE-AISI Classification Systems

12. How does the AISI system differ from the SAE system of identifying steel?

13. What is the composition of the following steels?
   a. 2340   b. 1020   c. E4340

Heat Treatment of Carbon Steel

14. Explain how the critical points of water are determined.

15. List five changes that take place in steel when it is heated to (and above) the decalescence point from room temperature.

16. Describe the changes that take place in a piece of steel when it is cooled to the recalescence point from the upper critical temperature.

Hardening of 0.83% Carbon Steel

17. What is the advantage of using hypereutectoid steel?

18. To what temperature should steel be heated prior to quenching in order to produce the best results?

Quenching

19. Name two quenching media and state the purpose of each.

20. Describe the proper method of quenching long, thin workpieces.

Tempering

21. What is the purpose of tempering a piece of steel?

22. What factors will affect the temperature at which a piece of steel is tempered?

23. What will happen to the steel if the tempering time is:
   a. Too long?   b. Too short?

Annealing, Normalizing, and Spheroidizing

24. What is the difference between annealing, normalizing, and spheroidizing?

Carburizing

25. Describe briefly how pack carburizing is performed.

26. What precautions should be taken when the liquid carburizing process is used?

27. Describe the gas carburizing process.

Carbonitriding Processes

28. Explain the difference between the carburizing and the carbonitriding processes.

Surface Hardening of Carbon Steels

29. What type of steel must be used for the flame-hardening and induction-hardening processes?

Induction Hardening

30. Describe briefly the induction-hardening principle.

Flame Hardening

31. Why is it desirable to temper parts immediately after flame hardening?

32. How may this be done on large surfaces?

UNIT 87

Testing of Metals and Nonferrous Metals

OBJECTIVES

After completing this unit, you should be able to:

1 Explain three methods of hardness testing

2 Perform a Rockwell C hardness test on a workpiece

3 Perform tensile strength and impact tests on a workpiece

4 Describe several nonferrous metals used in industry

After heat treating, certain tests may be performed to determine the properties of the metal. These tests fall into two categories:

1. *Nondestructive testing,* whereby the test may be performed without damaging the sample

2. *Destructive testing,* whereby a sample of the material is broken to determine the qualities of the metal

Nonferrous metals find a variety of uses in the machine trade. Since nonferrous metals contain little or no iron, they are used as bearings to prevent two like metal parts from being in contact with each other, where rust and corrosion are a factor, and where the weight of the product is important.

▶▶ Hardness Testing

Hardness testing is the most common form of nondestructive testing and is used to determine the hardness of a metal. Hardness in steel may be defined as its capacity to resist wear and deformation. The term *hardness* applied to metal is relative and indicates some of its properties. For example, if a piece of steel is hardened, tensile strength increases but ductility is reduced. If the hardness of a metal is known, the properties and performance of the metal can be predicted accurately.

Two types of testing machines are used to measure the hardness of a metal:

1. Those that measure the depth of penetration made by a penetrator under a known load. Rockwell, Brinell, and Vickers hardness testers are examples of this type.

2. Those that measure the height of rebound of a small mass dropped from a known height. The scleroscope is based on this principle.

ROCKWELL HARDNESS TESTER

The Rockwell hardness tester (Fig. 87-1) indicates the hardness value by the depth that a penetrator advances into the metal under a known pressure. A 120° conical diamond penetrator (brale) is used for testing hard materials. A $\frac{1}{16}$- or $\frac{1}{8}$-in. diameter (1.5- or 3-mm) steel ball is used as a penetrator for soft materials (Fig. 87-2 on p. 778).

Rockwell hardness is designated by various letters and numbers. The scales are indicated by the letters A, B, C, and

Figure 87-1 Rockwell hardness tester, showing various anvils. *(Wilson Instrument Division, American Chain and Cable Company)*

D. The C scale, which is the outside scale on the dial, is used in conjunction with the 120° diamond penetrator and a 330-lb (150-kg) major load for testing hardened metals. The B scale, or the red inner scale, is read when the 1/16-in. (1.5-mm) ball penetrator is used along with the 220-lb (100-kg) load for testing soft metals. The other letters, A and D, are special scales and are not used as often as the B and C scales. Rockwell superficial hardness scales are used for testing the hardness of thin materials and case-hardened parts.

To Perform a Rockwell C Hardness Test

Although the various Rockwell-type testers may differ slightly in construction, they all operate on the same principle.

<table>
<tr><td>step</td><td>procedure</td></tr>
<tr><td>1</td><td>Select the proper penetrator for the material to be tested. Use a 120° diamond for hardened materials. Use a 1/16-in. (1.5-mm) steel ball for soft steel, cast iron, and nonferrous metals.</td></tr>
<tr><td>2</td><td>Mount the proper anvil for the shape of the part being tested.</td></tr>
<tr><td>3</td><td>Remove the scale or oxidation from the surface on which the test is to be made. Usually an area of about 1/2 in. (13 mm) in diameter is sufficient.</td></tr>
</table>

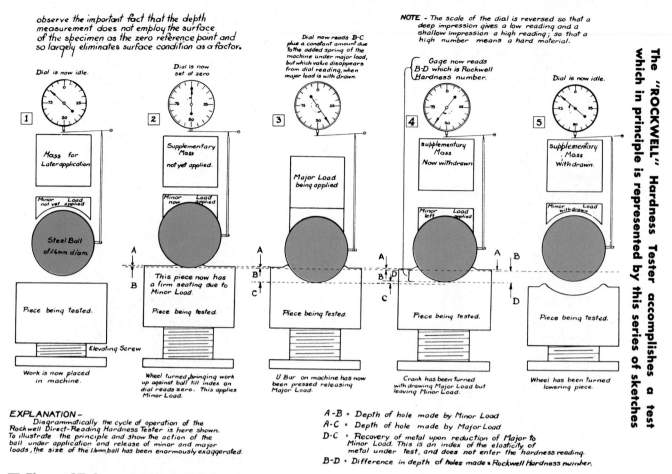

observe the important fact that the depth measurement does not employ the surface of the specimen as the zero reference point and so largely eliminates surface condition as a factor.

NOTE - The scale of the dial is reversed so that a deep impression gives a low reading and a shallow impression a high reading; so that a high number means a hard material.

The "ROCKWELL" Hardness Tester accomplishes a test which in principle is represented by this series of sketches

1 Dial is now idle.
Mass for Later application
Minor Load not yet applied
Steel Ball of 1.6 mm diam.
Piece being tested.
Elevating Screw
Work is now placed in machine.

2 Dial is now set at zero
Supplementary Mass not yet applied.
Minor Load now applied
This piece now has a firm seating due to Minor Load.
Piece being tested.
Wheel turned, bringing work up against ball till index on dial reads zero. This applies Minor Load.

3 Dial now reads B-C plus a constant amount due to the added spring of the machine under major load, but which value disappears from dial reading, when major load is withdrawn.
Major Load being applied
Piece being tested.
U Bar on machine has now been pressed releasing Major Load.

4 Gage now reads B-D which is Rockwell Hardness number.
supplementary Mass Now withdrawn
Minor Load left applied
Piece being tested.
Crank has been turned with drawing Major Load but leaving Minor Load.

5 Dial is now idle.
supplementary Mass With drawn.
Minor Load with drawn
Piece being tested.
Wheel has been turned lowering piece.

EXPLANATION –
Diagrammatically the cycle of operation of the Rockwell Direct-Reading Hardness Tester is here shown. To illustrate the principle and show the action of the ball under application and release of minor and major loads, the size of the 1.6mm ball has been enormously exaggerated.

A -B = Depth of hole made by Minor Load
A -C = Depth of hole made by Major Load
D-C = Recovery of metal upon reduction of Major to Minor Load. This is an index of the elasticity of metal under test, and does not enter the hardness reading.
B-D = Difference in depth of holes made = Rockwell Hardness number.

■ **Figure 87-2** Operating principles of a Rockwell hardness tester—steel-ball type.

4 Place the workpiece on the anvil and apply the minor load (10 kg) by turning the handwheel until the small needle is in line with the red dot on the dial.

5 Adjust the bezel (outer dial) to zero (0).

6 Apply the major load (150 kg).

7 After the large hand stops, remove the major load.

8 When the hand ceases to move backward, note the hardness reading on the C scale (black). This reading indicates the difference in penetration of the brale between the minor and major loads (Fig. 87-3) and indicates the Rockwell C (Rc) hardness of the material.

9 Release the minor load and remove the specimen.

Note: For accurate results, two or three readings should be taken and averaged. If either the brale or

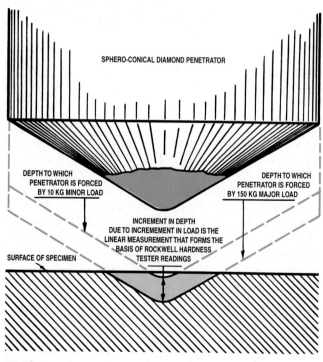

SPHERO-CONICAL DIAMOND PENETRATOR
DEPTH TO WHICH PENETRATOR IS FORCED BY 10 KG MINOR LOAD
DEPTH TO WHICH PENETRATOR IS FORCED BY 150 KG MAJOR LOAD
INCREMENT IN DEPTH DUE TO INCREMENT IN LOAD IS THE LINEAR MEASUREMENT THAT FORMS THE BASIS OF ROCKWELL HARDNESS TESTER READINGS
SURFACE OF SPECIMEN

■ **Figure 87-3** Operating principle of a Rockwell hardness tester—diamond-cone type.

the anvil has been removed and replaced, a "dummy run" should be made to properly seat these parts before performing a test. A test piece of known hardness should be tested occasionally to check the accuracy of the instrument.

BRINELL HARDNESS TESTER

The Brinell hardness tester is operated by pressing a 10-mm hardened steel ball under a load of 3000 kg into the surface of the specimen and measuring the diameter of the impression with a microscope. When the diameter of the impression and the applied load is known, the Brinell hardness number (BHN) can be obtained from Brinell hardness tables. The Brinell hardness value is determined by dividing the load in kilograms that was applied to the penetrator by the area of the impression (in square millimeters).

A standard load of 500 kg is used for testing nonferrous metals. The impression made in softer metals is larger and the Brinell hardness number is lower.

SCLEROSCOPE HARDNESS TESTER

The scleroscope hardness tester is operated on the principle that a small, diamond-tipped hammer, when dropped from a fixed height, will rebound higher from a hard surface than from a softer one. The height of the rebound is converted to a hardness reading.

Scleroscopes are available in several models. On some models, the hardness reading is taken directly from the vertical barrel or tube (Fig. 87-4). On others, the hardness numbers are marked on the dial (Fig. 87-5). These models may also show the corresponding Brinell and Rockwell numbers. See Table 16 in the appendix for a hardness conversion chart.

▶▶ Destructive Testing

There is a close relationship between the properties of a metal; for example, the tensile strength of a metal increases as the hardness increases, and the ductility decreases as the hardness increases. Thus, the tensile strength of a metal may be determined with reasonable accuracy if the hardness and the composition of the metal are known. A more accurate method of determining the tensile strength of a material is that of tensile testing. The tensile strength, or the maximum amount of pull (force) that a material can withstand before breaking, is determined on a tensile testing machine (Fig. 87-6 on p. 780).

A sample of the metal is pulled or stretched in a machine until the metal fractures. This test indicates not only the tensile strength but also the elastic limit, the yield

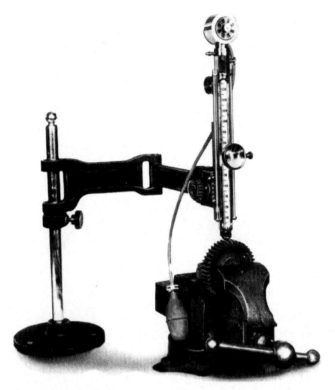

■ **Figure 87-4** A vertical scale scleroscope. *(The Shore Instrument and Manufacturing Company)*

■ **Figure 87-5** A dial recording scleroscope. *(The Shore Instrument and Manufacturing Company)*

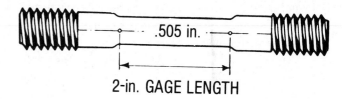

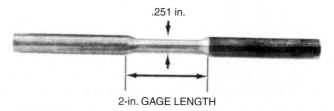

2-in. GAGE LENGTH

■ **Figure 87-7** A test sample .505 in. in diameter.

.251 in.

2-in. GAGE LENGTH

■ **Figure 87-8** A test sample .251 in. in diameter.
(Kelmar Associates)

a definite cross-sectional area that can be used conveniently in calculating the tensile strength. For example, most samples are machined to .505 in. in diameter (Fig. 87-7), which is 0.2 in^2. Smaller tensile testing machines use a sample that has been machined to .251 in. in diameter (Fig. 87-8), which is an area of $\frac{1}{20}$ in^2.

For a load of 10,000 lb using .505-in. diameter sample (0.2 in.^{2}), the tensile strength is:

$$\text{Tensile strength} = \frac{10,000}{0.2}$$

$$= 50,000 \text{ psi}$$

For a load of 3000 lb using a .251-in. diameter sample $\frac{1}{20}$ in.^{2}), the tensile strength is:

$$\text{Tensile strength} = \frac{3000}{\frac{1}{20}}$$

$$= 60,000 \text{ psi}$$

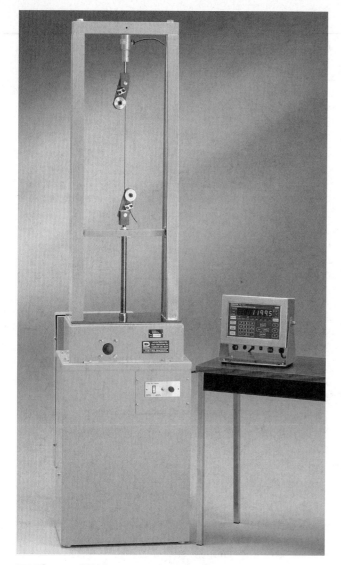

■ **Figure 87-6** A tensile testing machine. *(Dillon, Inc., A Division of Avery Weigh–Tronix)*

point, the percentage of area reduction, and the percentage of elongation of the material.

TENSILE TESTING (INCH EQUIPMENT)

The tensile strength of a material is expressed in terms of pounds per square inch and is calculated as follows:

$$\text{Tensile strength} = \frac{\text{load, lb}}{\text{area, in.}^2}$$

Machines capable of extremely high tension loads are required to test a steel sample having a cross-sectional area of 1 in^2. For this reason, most samples are reduced to

TO DETERMINE THE TENSILE STRENGTH OF STEEL

1. Turn a sample of the steel to be tested to the dimensions shown in Fig. 87-8, and place on it two center-punch marks exactly 2 in. apart.

2. Mount the specimen in the machine (Fig. 87-9a), and make sure that the jaws grip the sample properly by jogging the motor switch until the black needle just starts to move. If necessary, remove any tension by reversing the motor.

3. Turn the red hand back until it bears against the black hand on the dial.

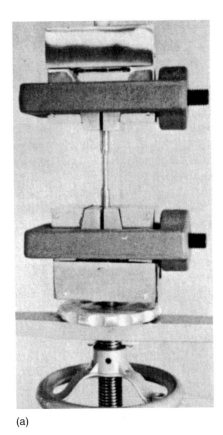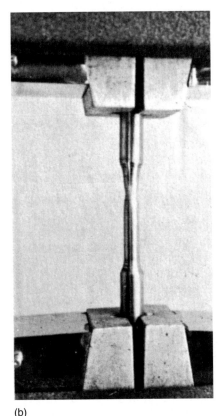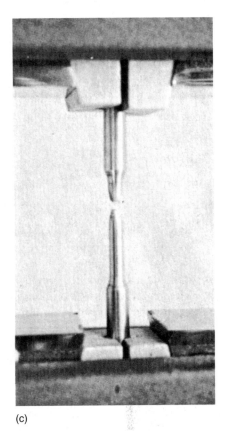

(a)   (b)   (c)

■ **Figure 87-9** (a) A .251-in. diameter specimen mounted and ready for testing; (b) specimen "necking down"; (c) fractured specimen. *(Kelmar Associates)*

4. Set a pair of dividers to the center-punch marks on the sample (2 in.).

5. Start the machine and apply the load to the specimen.

6. Observe and record the readings at which there are any changes in the uniform movement of the needle.

Note: At this time, it is possible to determine the elastic limit of the metal, or the maximum stress that can be developed without causing permanent deformation. This is done by checking the distance between the two center-punch marks with the preset dividers. Increased loads must be applied to the specimen and removed several times. After each load has been removed, check that the distance between the center-punch marks remains at 2 in. When this distance increases even slightly, the elastic limit of the metal has been reached. An *extensometer* may also be used to indicate the elastic limit.

7. Continue to exert the pull on the sample until it "necks down" (Fig. 87-9b) and finally breaks (Fig. 87-9c). From this procedure, several properties of the metal may be determined (Fig. 87-10).

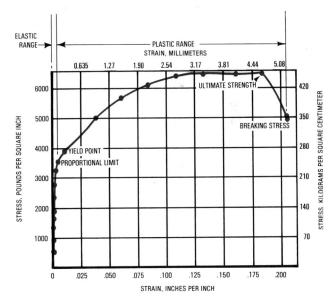

■ **Figure 87-10** Stress-strain graph.

8. Remove the sample pieces, place the broken ends together, and clamp them in this position.

9. Measure the distance between the center-punch marks to determine the amount of elongation.

10. Measure the diameter of the specimen at the break to determine the reduction in diameter.

Observations

Refer to Fig. 87-10.

1. The needle continued to move uniformly until about 3600 lb showed on the scale, after which it slowed down slightly. The point at which it began to slow down indicated the *proportional limit*. It is here that the metal reached its *elastic limit* and no longer returned to its original size or shape. At this point, the stress and strain were no longer proportional and the curve changed.

   The *yield point* was reached just beyond the proportional limit and here the metal started to stretch or yield; the strain increased without a corresponding stress increase.

2. The needle continued to move slowly up to about 6300 lb and then it remained stationary.

3. After a short time, the metal began to show a reduction in diameter (necking down), at which time the needle began to show a rapid drop. The highest travel of the needle indicated the *ultimate strength*, or the *tensile strength*, of the metal. This was the maximum pull to which the metal may be subjected before breaking.

4. The needle continued to move backward (leaving the red hand stationary) and suddenly the metal broke at the point where it had necked down. The point at which it broke is known as the *breaking stress*.

5. The position of the red hand was at 6300 lb. This indicated the load required to break a cross-sectional area of $\frac{1}{20}$ in.^{2}. The ultimate strength of the metal, or the tensile strength, was $6300 \times 20 = 126,000$ lb.

6. When the pieces were placed back together, clamped, and measured, the distance between the center-punch marks was about 2.185 in. This was an elongation of about 9%.

7. When the diameter of the metal at the fracture was measured, it was found to be .170 in. This indicated a reduction in area of .080 in., or about 32%.

TENSILE TESTING (METRIC EQUIPMENT)

Metric tensile testers are graduated in kilograms per square centimeter, and the cross-sectional area of the sample is calculated in square centimeters or square millimeters. Metric extensometers are graduated in millimeters.

The calculations involved in determining the tensile strength, percentage elongation, and area of reduction are the same as for inch calculations.

Although not the accepted SI unit of pressure, most metric tensile testing machines available and in use at the time of publication were graduated in kilograms per square centimeter (kg/cm^2). For conversion to pascals, use the formula $1 \text{ kg/cm}^2 = 980.6$ Pa.

EXAMPLE

A sample piece is 1.3 cm in diameter. The ultimate pull exerted on it during a tensile test was 4650 kg. What was the tensile strength of this metal?

$$\text{Tensile strength} = \frac{\text{load, kg}}{\text{area, cm}_2}$$
$$= \frac{4650}{1.3}$$
$$= 3577 \text{ kg/cm}^2$$

Percentage of Elongation

If the punch marks were 50 mm apart at the start of the pull and 54 mm after the sample was broken, what was the percentage of elongation?

$$\% \text{ of elongation} = \frac{\text{amount of elongation}}{\text{original length}} \times 100$$
$$= \frac{4}{50} \times 100$$
$$= 8\%$$

Percentage Reduction of Area

The original diameter was 1.3 cm and after breaking, the diameter was 0.95 cm. What was the percentage of reduction of area?

$$\text{Amount of reduction} = 1.3 - 0.95$$
$$= 0.35 \text{ cm}$$
$$= \frac{0.35}{1.3} \times 100$$
$$= 27\%$$

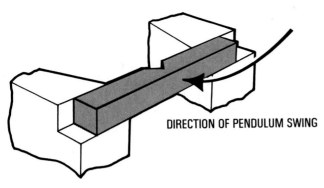

Figure 87-11 Principle of the Charpy impact test.
(Kelmar Associates)

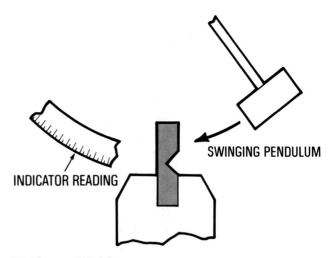

Figure 87-12 Principle of the Izod impact test.
(Kelmar Associates)

IMPACT TESTS

The toughness of metal, or its ability to withstand a sudden shock or impact, may be measured by the *Charpy impact test* or the *Izod test*. In both tests, a 10-mm-square specimen is used; it may be notched or grooved, depending on the test used.

Both tests use a swinging pendulum of a fixed mass that is raised to a standard height, depending on the type of specimen. The pendulum is released, and as it swings through an arc, it strikes the specimen in the pendulum's path.

In the Charpy test (Fig. 87-11), the specimen is mounted in a fixture and supported at both ends. The V or notch is placed on the side opposite the direction of the pendulum's swing. When the pendulum is released, the knife edge strikes the sample in the center, reducing the travel of the pendulum. The difference in height of the pendulum at the beginning and end of the stroke is shown on the gage, and this indicates the amount of energy used to fracture the specimen.

The Izod test (Fig. 87-12) is similar in principle to the Charpy test. One end of the work is gripped in a clamp with the notched side toward the direction of the pendulum's swing. The amount of energy required to break the specimen is recorded on the scale (Fig. 87-13 on p. 784).

▸▸ Nonferrous Metals and Alloys

Nonferrous metals, as the name implies, contain little or no iron and are generally nonmagnetic. Since all pure nonferrous metals do not offer the qualities required for industrial applications, they are often combined to produce alloys having the desired qualities for a particular job. The most widely used nonferrous metals for industrial use are aluminum, copper, lead, magnesium, nickel, tin, and zinc.

ALUMINUM

Aluminum is a light, soft, white metal produced from bauxite ore. It is resistant to atmospheric corrosion and is a good conductor of electricity and heat. It is malleable and ductile and can easily be machined, forged, rolled, and extruded. It has a low melting point of 1220°F (660°C) and can be cast easily. It is used extensively in transportation vehicles of all types, the construction industry, transmission lines, cooking utensils, and hardware.

Aluminum is not generally used in a pure state, since it is too soft and weak. It is alloyed with other metals to form strong alloys that are used extensively in industry.

Aluminum-Base Alloys

Duralumin, an alloy of 95% aluminum, 4% copper, 0.05% manganese, and 0.05% magnesium, is widely used in the aircraft and transportation industries. This is a naturally aging alloy; that is, it hardens as it ages. Due to this peculiarity, duralumin must be kept at a subzero temperature until ready for use. When it is brought to room temperature, the hardening process begins.

Other alloys contain varying amounts of copper, magnesium, and manganese.

Aluminum-silicon-magnesium alloys have excellent corrosion resistance, are heat treatable, and are easily worked and cast. They are used for small boats, furniture, bridge railings, and architectural applications.

Aluminum-zinc alloys contain zinc, magnesium, and copper, with smaller amounts of manganese and chromium. They have high tensile strength, have good corrosion resistance, and may be heat treated. They are used for aircraft structural parts when great strength is required.

COPPER

Copper is a heavy, soft, reddish-colored metal refined from copper ore (copper sulfide). It has high electrical and thermal conductivity, good corrosion resistance and strength, and is easily welded, brazed, or soldered. It is very ductile and is easily drawn into wire and tubing. Since copper work hardens readily, it must be heated at about 1200°F (648°C) and quenched in water to anneal.

Because of its softness, copper does not machine well. The long chips produced in drilling and tapping tend to clog the flutes of the cutting tool and must be cleared frequently. Sawing and milling operations require cutters with good chip clearance. Coolant should be used to minimize heat and aid the cutting action.

Copper-Base Alloys

Brass, an alloy of copper and zinc, has good corrosion resistance and is easily formed, machined, and cast. There are several forms of brass. Alpha brasses containing up to 36% zinc are suitable for cold working. Alpha + beta brasses containing 54% to 62% copper are used in hot working of this alloy. Small amounts of tin or antimony are added to alpha brasses to minimize the pitting effect of salt water on this alloy. Brass alloys are used for water and gasoline line fittings, tubing, tanks, radiator cores, and rivets.

Bronze originally referred to an alloy of copper and tin; now it has been extended to include all copper-base alloys, except copper-zinc alloys, which contain up to 12% of the principal alloying element.

Phospor-bronze contains 90% copper, 10% tin, and a very small amount of phosphorus, which acts as a hardener. This metal has high strength, toughness, and corrosion resistance and is used for lock washers, cotter pins, springs, and clutch discs.

Silicon-bronze (a copper-silicon alloy) contains less than 5% silicon and is the strongest of the work-hardenable copper alloys. It has the mechanical properties of machine steel and the corrosion resistance of copper. It is used for tanks, pressure vessels, and hydraulic pressure lines.

■ **Figure 87-13** Impact testing machine. *(Ametek Testing Equipment)*

Aluminum-manganese alloys have good formability, good resistance to corrosion, and good weldability. These alloys are used for utensils, gasoline and oil tanks, pressure vessels, and piping.

Aluminum-silicon alloys are easily forged and cast. They are used for forged automotive pistons, intricate castings, and marine fittings.

Aluminum-magnesium alloys have good corrosion resistance and moderate strength. They are used for architectural extrusions and automotive gas and oil lines.

Aluminum-bronze (a copper-aluminum alloy) contains between 4% and 11% aluminum. Other elements, such as iron, nickel, manganese, and silicon, are added to aluminum bronzes. Iron (up to 5%) increases the strength and refines the grain. Nickel, when added (up to 5%), has effects similar to those of iron. Silicon (up to 2%) improves machinability. Manganese promotes soundness in casting.

Aluminum-bronzes have good corrosion resistance and strength and are used for condenser tubes, pressure vessels, nuts, and bolts.

Beryllium-bronze (copper and beryllium), containing up to about 2% beryllium, is easily formed in the annealed condition. It has high tensile strength and fatigue strength in the hardened condition. Beryllium-bronze is used for surgical instruments, bolts, nuts, and screws.

LEAD

Lead is a soft, heavy metal that has a bright, silvery color when freshly cut but turns gray quickly when exposed to air. It has a low melting point, low strength, low electrical conductivity, and high corrosion resistance. It is used extensively in the chemical and plumbing industries. Lead is also added to bronzes, brasses, and machine steel to improve their machinability.

Lead Alloys

Antimony and *tin* are the most common alloying elements of lead. Antimony, when added to lead (up to 14%), increases its strength and hardness. This alloy is used for battery plates and cable-sheathing.

The most common lead-tin alloy is solder, which may be composed of 40% tin and 60% lead, or 50% of each. Antimony is sometimes added as a hardener.

Lead-tin-antimony alloys have been used as type metals in the printing industry.

MAGNESIUM

Magnesium is a lightweight element that, when alloyed, produces a light, strong metal used extensively in the aircraft and missile industries. Magnesium plates are used to prevent corrosion by salt water in underwater fittings on ship hulls. Magnesium rods, when inserted in galvanized domestic water tanks, will prolong the life of the tank. Other uses for this metal are in photographic flashbulbs and thermite bombs.

NICKEL

Nickel, a whitish metal, is noted for its resistance to corrosion and oxidation. It is used extensively for electroplating, but its most important application is in the manufacture of stainless and alloy steels.

Nickel Alloys

Nickel-chromium-iron-base alloys (containing about 60% nickel, 16% chromium, and 24% iron) are widely used for electric heating elements in toasters, percolators, and water heaters.

Monel metal, containing about 60% nickel, 38% copper, and small amounts of manganese or aluminum, is a tough, ductile metal with good machining qualities. It is corrosion-resistant, is nonmagnetic, and is used in valve seats, chemical marine pumps, and nonmagnetic aircraft parts.

Hasteloy, containing about 87% nickel, 10% silicon, and 3% copper, is widely used in the chemical industry because of its noncorrosive qualities.

Inconel, a strong, tough alloy containing about 76% nickel, 16% chromium, and 8% iron is used in food-processing equipment, milk pasteurizers, exhaust manifolds for aircraft, and heat-treating furnaces and equipment.

TIN

Pure *tin* has a silvery-white appearance, has good corrosion resistance, and melts at about 450°F (232°C). It is used as a coating on other metals, such as iron, to form tin plate.

Tin Alloys

As mentioned, tin forms solder when it is alloyed with lead.

Babbitt is an alloy containing tin, lead, and copper.

Pewter, another tin-base alloy, is composed of 92% tin and 8% antimony and copper.

ZINC

Zinc is a coarse, crystalline, brittle metal used mainly for die-casting alloys and as a coating for sheet steel, chain, wire, screws, and piping. Zinc alloys, containing approximately 90% to 95% zinc, 5% aluminum, and small amounts of copper or magnesium, are widely used in the die-casting field to produce automotive parts, building hardware, padlocks, and toys.

BEARING METALS

Bearing metals may be divided into two groups: leaded bronzes and babbitt.

Leaded Bronzes

The composition of bronze bearings varies according to their use. Bearings used to support heavy loads contain about 80% copper, 10% tin, and 10% lead. For lighter loads and faster speeds, the lead content is increased. A typical bearing of this type might contain 70% copper, 5% tin, and 25% lead.

Babbitt

Babbitt bearing materials may be divided into two groups: lead-base and tin-base.

Lead-base babbitt may contain 75% lead, 10% tin, and 15% antimony, depending on the application. A small amount of arsenic is often added to permit the bearing to carry heavier loads. Applications of lead-base babbitt bearings are in automotive connecting rods, main and crankshaft bearings, and diesel engine bearings.

Tin-base babbitt may contain up to 90% tin with copper and antimony added, or 65% tin, 15% antimony, 2% copper, and 18% lead. Since tin has become less plentiful, tin-base babbitts are used in high-grade bearing applications, such as steam turbines.

unit 87 review questions

Testing of Metals

1. Explain the difference between nondestructive and destructive testing.

2. What information can be determined by hardness testing?

Rockwell Hardness Tester

3. Name two scales found on a Rockwell-type hardness tester and state the penetrator used in each case.

4. Briefly describe how a Rockwell C hardness test is performed.

Brinell Hardness Tester

5. Compare the principles of the Brinell hardness tester and the Rockwell hardness tester.

Scleroscope

6. Describe the principle of the scleroscope.

Destructive Testing

7. What effect does hardening have on the tensile strength and ductility of a metal?

8. Explain the principle of tensile testing.

9. What other properties may be determined by a tensile test?

10. How is the tensile strength of a metal calculated?

11. Explain the procedure for performing a tensile test.

12. How may the elastic limit of a metal be determined?

13. Define proportional limit, yield point, ultimate strength, and breaking stress.

14. Describe the principle of impact testing.

Nonferrous Metals and Alloys

15. Define *nonferrous metal*.

16. Name four nonferrous metals commonly used as base metals in alloys.

17. Name four aluminum-base alloys and state one application of each.

18. Name two types of brasses and state the composition of each.

19. What is the difference between brass and bronze alloys?

20. Why is lead added to steel?

21. Where are magnesium alloys used extensively?

22. Name three nickel-base alloys and state two applications of each.

23. Name two tin-base alloys.

24. For what purpose are zinc alloys used extensively?

25. What are the three metals used for bearing alloys?

26. What is the basic difference between bronze and babbitt?

27. Explain the use of the following:

    a. Lead-base babbitt

    b. Tin-base babbitt

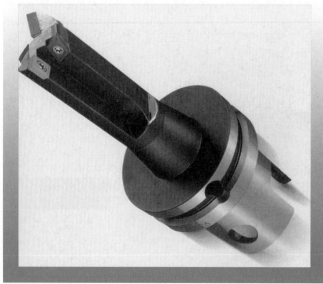

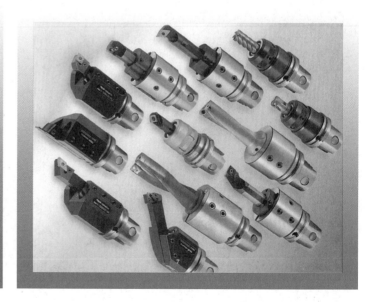

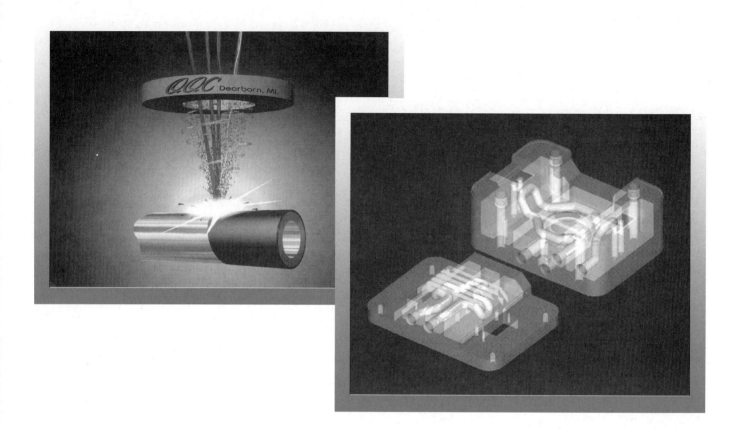

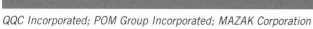

QQC Incorporated; POM Group Incorporated; MAZAK Corporation

INSTANT MANUFACTURING TECHNOLOGY

For today's manufacturers, it matters greatly if they cannot compete with the price and quality of the goods. Only by streamlining the operation and using the Internet can a company be better, faster, and still remain competitive. The Internet and computers have simply accelerated the effort; gone are the days when a plant operated autonomously.

The plant floor is the point where companies must start if they are going to compete globally. It must be based on Information Technology (IT) among the consumer, manufacturing operations, and suppliers. IT has arrived on the factory floor, and those who accept it will produce high quality goods and be around to share in tomorrow's wealth.

E-manufacturing can assist in lowering the cost, reducing errors and inefficiencies, increasing productivity, and producing quality products. It is not something that requires a complete redesign of the plant floor; it is a way of getting operational excellence out of an organization.

Tiered supply chains and contract manufacturers are making companies manage themselves to produce quality products at prices consumers are willing to pay. It is important for a company to resolve unscheduled work stoppages to maintain nonstop operations. The only way to achieve these goals is to use the technology available to any manufacturer in the world. To be successful, it is important to provide a link between the factory floor and the enterprise business system.

UNIT 88

Instant Digital Manufacturing

OBJECTIVES

After completing this unit, you will be able to:

1 Define Instant Manufacturing and give some of its features

2 Explain indirect manufacturing and provide some examples of its use

3 Describe stereolithography (SLA®*) apparatus and list its benefits

4 Describe Selective Laser Sintering (SLS®*) and list its benefits

5 Describe Multi-Jet-Modeling (MJM) and list its benefits

In the *produce or perish* world of manufacturing, companies compete to build products better and cheaper. If one manufacturer does not meet that high standard of performance, it is safe to say that somebody else will. Time to market is critical for any manufacturer, and everything that reduces product-development time has a definite effect on how soon a product gets to the market and how much of the marketplace a company can capture.

▶▶ A Progression of Technology

In the past, technical part prints were made and the part manufactured using conventional machine tools and machining processes. This traditional method was used for many years until the age of computers and the introduction of Numerical Control (NC) (Fig. 88-1).

It was not long after NC was introduced to machine tools that Computer-Aided Design (CAD) started to replace drafting as a means of producing technical part prints. Eventually, this evolved into CAD/CAM where the information data on CAD generated prints was used in CAM (Computer-Assisted Manufacturing) to manufacture a part.

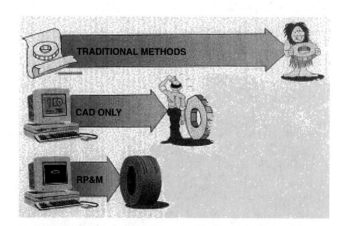

■ **Figure 88-1** ADM solid-imaging systems provide a wide variety of benefits to manufacturers. *(3D Systems)*

---

*SLA® and SLS® are registered trademarks of 3D Systems.

The next logical step in the manufacturing process was the introduction of Rapid Prototyping and Manufacturing (RP&M) for the design and production of prototype models, to reduce or eliminate manufacturing errors and bring products to the market faster and at lower cost.

►► Solid Imaging Solutions

The current line of solid imaging systems, software, equipment, and materials produced by Rapid Prototyping manufacturers is now used as instant manufacturing tools with automotive, dental, biomedical, motor sports, consumer electronics, and military aerospace applications. They are used to speed the production of customized/specialized end-use parts. The ability to manufacture a product using additive fabrication techniques will radically alter designs and manufacturing methods over the next decade and beyond. Using solid imaging systems such as selective laser sintering and other solid imaging solutions, like those offered by 3D Systems, existing designs can be manufactured without the costs and lead-time associated with hard tooling, and more complex designs will become easier to manufacture.

►► Instant Digital Manufacturing

Selective laser sintering with solid imaging systems are reliable and cost-effective methods of making end-use parts for preproduction or production applications. Solid imaging systems are expected to become a key enabling method for the customization of design and manufacturing, also called mass customization.

> *Solid imaging is a comprehensive suite of customer-driven solutions that covers the production of a part starting at the design stage, progressing through the prototype development, and ending with the manufacturing stage. It uses solid-imaging technology to directly or indirectly produce end-use components or product.*

Direct manufacturing is the method for creating end-use products directly on a solid imaging system. Recent advancements in Selective Laser Sintering (SLS) and Stereolithography (SL) technology have made solid imaging systems an alternative to some conventional manufacturing methods.

The aerospace industry has used LS technology to manufacture nonstructional aircraft components. Manufacturers of hearing aids have recognized the value of the LS and SL technology in the production of custom-fitted in-the-ear (ITE) devices. The fundamental benefits of direct solid imaging solutions are no tooling required; the ability to design for function, not for a conventional and limiting manufacturing process; significant cost savings for low production runs; and the ability to make design changes quickly at a very low cost.

Indirect manufacturing is the method for creating end-use parts from a mold, pattern, or tool that is generated on a solid-imaging system. One of the best examples of an innovative utilization of indirect manufacturing is the way Align Technology manufactures invisible orthodontic treatment devices called Aligners®. The company thermoforms a thin sheet of polycarbonate over accurate individual molds created on the SLA® 7000 system. Another example of indirect manufacturing systems is using solid imaging systems to produce a pattern for investment casting and then create a metal part. Another example of indirect manufacturing is generating a tool on a solid imaging system that can be used on an injection mold machine to produce plastic parts.

Some of the main features of Instant Manufacturing are:

> Designers and engineers are able to add custom features and complexity to designs not currently feasible with present manufacturing techniques.

> Solid imaging systems opens up new product design possibilities that were not possible to manufacture using traditional tooling, molding, and casting methods.

> The number of steps in the engineering and manufacturing phases is reduced, saving time and money from the elimination of tooling, thereby reducing part and product cost and reducing the time it takes to get a product to the market.

> The use of 3D Systems' solid imaging systems to speed the production of customized/specialized parts

> The ability to manufacture a product using additive fabrication techniques will greatly change the present design and manufacturing methods.

> The costs and lead time associated with hard tooling is eliminated and more complex designs will be easier to manufacture.

►► How Instant Manufacturing Got Started

Founded in 1986, 3D Systems, the solid imaging company[SM], provides solid imaging products and systems solutions that reduce the time and cost of designing products and facilitate direct and indirect manufacturing. Its systems utilize patented proprietary technologies to create physical objects from digital input that can be used in design communication, prototyping, and as functional end-use parts.

Figure 88-2 shows the most common solid-modeling imaging systems and materials used for ADM systems.

> **Concept modeling, three-dimensional printing:**
Solid imaging solutions are used for concept-modeling

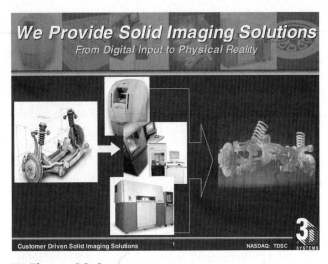

Figure 88-2 Common solid-imaging systems and materials. *(3D Systems)*

and three-dimensional printing applications, to produce three-dimensional shapes, primarily for visualizing and communicating mechanical design applications, as well as for other applications including supply-chain management, architecture, art, surgical medicine, marketing and entertainment.

> **Rapid prototyping:** Solid imaging solutions are used for rapid prototyping applications, including the generation of product concept models, functional prototypes and master-casting and tooling patterns that are often used as an efficient, cost-effective means of evaluating product designs.

> **Instant manufacturing:** Solid imaging solutions are used for instant manufacturing applications to manufacture end-use parts directly from a digital image without the need for expensive tooling or molds and without lengthy set-ups, resulting in significant flexibility and mass customization capabilities.

A typical solid imaging center may contain multiple solid-imaging technologies such as SLA® (Stereolithography Apparatus) systems, SLS® (Selective Laser Systems), and an MJM (Multi-Jet Modeling) printer supporting solid imaging applications. Solid imaging solutions is the shifting from mass produced off-the-shelf goods (products) to goods specifically customized to individual customers' tastes, offering a greater range of product choices.

Instant Manufacturing consists of a range of integrated technologies capable of providing solid-imaging solutions, speeding the production of smaller-volume customized parts.

To accomplish this change requires the following:

> A strong focus on developing and acquiring superior materials capable of wider applications for manufacturing

> Solid imaging systems offering greater choices in material-delivery systems

> Software required to grasp fully the opportunities developing for instant manufacturing applications

▶▶ Solid Imaging Hardware Components

The major components of solid imaging systems are as follows:

1. **Stereolithography or (SLA®) systems** (Fig. 88-3) use a laser to convert photosensitive resins into solid cross-sections, layer by layer, until the desired objects are complete. SLA® systems are capable of making multiple parts at the same time and are designed to produce prototype or end use parts that have a wide range of sizes and shapes.

   > Parts created from digital data using computer-aided design and manufacturing or CAD/CAM software utilities and related computer applications.
   > A laser beam exposes and solidifies successive layers of liquid photosensitive polymers into solid cross sections, layer by layer, until the desired part is formed.
   > SL-produced parts are used for concept models, engineering prototypes, patterns and masters for molds, consumable tooling, and short-run manufacturing of a final product.

   The benefits of the SL process are:

   > A reduction in product-development and design time
   > Improved part quality
   > Durable parts that can be used for rapid manufacturing
   > An ability to make multiple objects at the same time

2. **Selective Laser Sintering (SLS®) systems** (Fig. 88-4) use heat from a laser to melt and fuse, or sinter, powdered materials into solid cross-sections, layer-by-layer, until the desired parts are complete. SLS® systems can create parts from a variety of plastic and metal powders and are capable of processing multiple parts within the same build cycle.

   > Parts are created from digital data using computer-aided design and manufacturing or CAD/CAM software utilities and related computer applications.
   > Laser energy is used to melt and fuse, or sinter

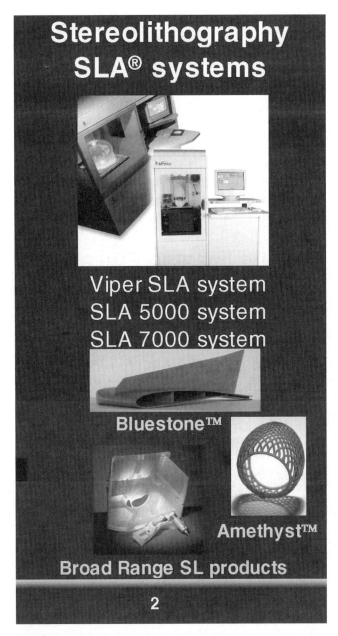

The benefits of SLS systems are:

> Reduces the product-development time from months or weeks to days or even hours.
> Produces functional models from plastic or metal powdered sintering material.
> Multiple objects can be made at the same time.
> Produces durable parts that can be used for rapid manufacturing.
> Parts can be used in final product assemblies.

3. **Multi-Jet Modeling technology ("MJM"), or 3D printing,** uses hot-melt jetting technology to print three-dimensional physical parts by accumulating proprietary solid imaging materials ("SIMs")

Figure 88-4 The major components of a Selective Laser Sintering system. *(3D Systems)*

in successive layers. MJM technology is the basis for affordable three-dimensional solutions for printing any three-dimensional part from digital data.

> The ThermoJet® printer and Invision™ 3D printer produce models used to verify CAD model geometry, to communicate design intent, and to obtain design feedback from others.
> It is used as marketing models for design review meetings, customer and prospect presentation, and preliminary assembly analysis.
> The ThermoJet® printer produces wax-based patterns for investment casting applications.

The main steps in the MJM technology process are:

> Parts are created from digital data using computer-aided design and manufacturing or CAD/CAM software utilities and related computer applications.
> Models are created by depositing material onto a build platform, layer-by-layer, using an ink-jet-style print head.

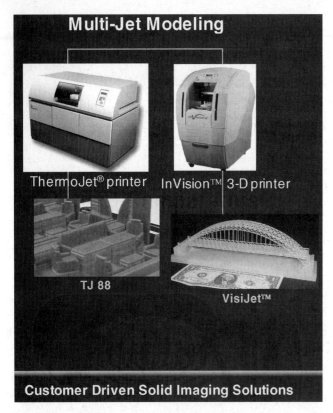

Multi-Jet Modeling

ThermoJet® printer   InVision™ 3-D printer

TJ 88

VisiJet™

Customer Driven Solid Imaging Solutions

■ **Figure 88-5** The solid-imaging system and the material used for Multi-Jet Modeling printers. *(3D Systems)*

> The print head scans back and forth, following the information from the CAD design until the model is completed.

The benefits of the MJM process are:

> A convenient networked 3D printer that can be used easily by almost everyone in the organization
> It requires very little training and is easily understood by technical or nontechnical people.
> Models are built unattended.
> ThermoJet printers use standard office power and are about the same size as office copiers.

Solid Imaging Software Components

The part-preparation software for personal computers and engineering workstations provides an interface between digital data and the solid-imaging equipment. Digital data, such as CAM/CAM, is converted within the software package and, depending on the software package, the object can be viewed, rotated, and scaled, and model structures can be added. The software then generates the information that will be used by the SLA, SLS, and MJM systems to produce the desired object.

Solid Imaging Processing Materials

Various types of materials are used in solid-imaging systems. They fall into two general classes:

> Photosensitive liquid resins for stereolithography
> Sintering powdered materials, including functional plastics, nylon, and metal powders, to suit the SLS system

3D Systems blend, market and distribute material products under a variety of brand names that they sell for use in all of their solid imaging systems. The families of engineered products are designed for use with their systems and processes to produce high-quality models, prototypes and parts. They market the stereolithography products under the Accura® brand, selective laser sintering products under the DuraForm™, LaserForm™ and CastForm™ brands, and multi-jet modeling products under the ThermoJet® and VisiJet® brands.

A golf club manufacturer, a mold and die company, and the Armed Forces are using an SLS (Selective Laser Sintering) system to produce a prototype. The process time was reduced by 50% to 75%, or from one to four weeks.

unit 88 review questions

1. Define Instant Manufacturing/or solid imaging.
2. List five of the main features of Instant Manufacturing or solid imaging.
3. Define stereolithography (SL) apparatus and give three of its benefits.
4. Define Selective Laser Sintering (SLS) and give three of its benefits.
5. Define Multi-Jet-Modeling (MJM) and give three of its benefits.
6. What are the two general classes of materials used in solid imaging systems?

UNIT 89

Cryogenic Treatment/Tempering

OBJECTIVES

After completing this unit, you will be able to:

1 Explain the cryogenic treatment/tempering process

2 Describe the typical cycle of the cryogenic process

3 Discuss the changes in metallurgy in a typical cryogenic treatment

4 List the advantages of cryogenic tempering

Deep cryogenic treatment/tempering is a one-time, permanent process that improves the physical and mechanical properties of various materials, such as ferrous and nonferrous metals, aluminum, and their alloys. It uses subzero temperatures to dimensionally stabilize, refine, and close grain structures; to release internal stresses for the life of the material; and to produce longer wear life to parts subject to wear and abrasion. The cryogenic treatment reduces downtime, improves performance, and increases the life of metal tools.

While not a magic wand, cryogenic treatment can extend the life of products such as drills, taps, reamers, broaches, end mills, dies, gears, slicers, and cutting knives (Table 89.1 on p. 796). It can create a premium, more profitable tool and can reduce the cost of tooling for manufacturers. Unlike a coating, it is a through treatment, which treats the entire tool and keeps its benefits even after repeated resharpening until the tool is completely worn out.

▶▶ Technical Data

In austenite, a crystalline form of steel, each unit is a face-centered cubic structure with iron atoms at the corners and center of each face of the cube. These face-centered atoms (Fig. 89-1a on p. 796) form an octahedron where carbon atoms can occupy any of the spaces indicated between iron atoms. After quenching from high temperature, austenite becomes martensite, a different crystalline form of steel. Cryogenic treatment can cause most of the austenite retained after heat treatment to transform to martensite.

A unit cell of the martensite crystal is not cubic but is slightly elongated. Iron atoms still occupy the corners, but those that were face-centered in the austenite cell move as shown in Fig. 89-1b, leaving the no. 2 atom in the center of the cell, rather than on one of the faces. The no. 1 iron atom occupies the center of the next cell (indicated by dotted lines). Austenite crystals are ductile but soft; martensite crystals are hard but must be tempered to overcome brittleness.

▶▶ Cold Treatment Versus Cryogenics

COLD TREATMENT

It has been recognized for years that the properties of many materials could be enhanced by cooling them to below room temperature. Before liquid nitrogen was readily

table 89.1 — Cryogenic treatment improves the life of metal tools

Tool Type	Company	Tool Material	Results
Drills	Aircraft manufacturer	M42,M7,C2	300%
Milling	Aircraft manufacturer	M7	250%
Gear cutter	Major manufacturer	Ti-N coated	350%
Broach	Metal milling co.	Carbide	300%
Punching	Major manufacturer	M7	600%
End mill	Aerospace mfg.	M42	450%
Hob	Turbine manufacturer	M2-M7	300%
Face mill	Aerospace	C2 carbide	400%
Tap	Toolmaker	C2 carbide	600%
Die	Casting co.	Hi-ni alloy	300%
Broach	Auto manufacturer	Hi nickel	250%
Milling	Machine shop	347 stainless	375%
Stamping die	Steel furniture manufacturer	D2	1000%

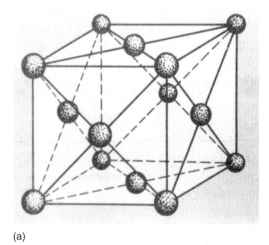
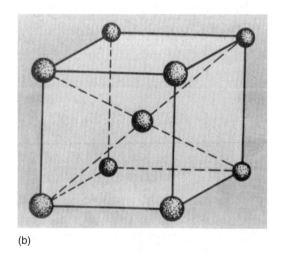

(a)　　　　(b)

■ **Figure 89-1A** (a) A face-centered cubic austenite cell with iron atoms at the corners and center of each face of the cube; (b) a slightly elongated unit cell of martensite crystal. *(300° Below, Inc.)*

available, parts were cooled to about −110°F in vats of alcohol cooled by dry ice. The results showed that there was a good improvement in the quality of the parts. This process is called *Cold Treatment*. Common practice for cold treatment is generally in the range of −120°F (−84°C) that refrigeration units can reach.

CRYOGENICS

In the early 1990s, methods were developed to reduce temperatures even further to –300°F, making it possible to improve the properties even more; this process is called *Cryogenic Treatment*. In cryogenic treating, parts are

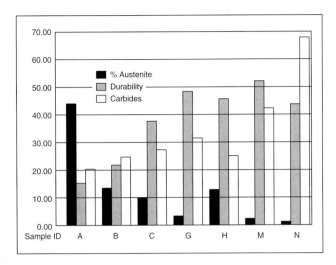

Figure 89-2 Cryogenic treatment changes retained austenite to the denser martensite.

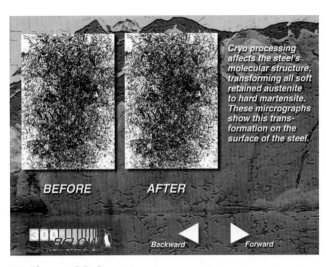

Cryo processing affects the steel's molecular structure, transforming all soft retained austenite to hard martensite. These mircrographs show this transformation on the surface of the steel.

BEFORE   AFTER

Figure 89-3 The microstructure of steel before and after cryogenic treatment. *(300° Below, Inc.)*

chilled to approximately −300°F (−185°C). The benefits of cold treatment range from enhancing the transformation from austenite to martensite to improving the stress relief of casting and machined parts. In each case, even greater benefits are realized as a result of cryogenic treatment.

Heat treating gives steel its hardness, toughness, wear resistance, and ductility. *Heat treating* is really a misnomer; it should be called *cold treating*. The changes that heat treating imparts to metal do not actually take place from the heating but, rather, from the cooling or *quenching* from a high temperature. The changes do not stop at room temperature but continue well below cryogenic practice. Deep cryogenic processing can be thought of as *an extension of heat treating.*

Cryogenic Process

The cryogenic temperatures of −300°F (−185°C) is required to create a complete molecular change in most alloy steels, making most retained austenite turn to martensite, a denser, refined mix, smaller and more uniform than austenite (Fig. 89-2). Dry cryogenic processing physically transforms the microstructure into a new, more refined, uniform substructure, which may be stronger and more wear-resistant. The dry cryogenic process does not expose the material to liquid nitrogen, thus eliminating the risk of thermal shock. During this process, the following occurs:

> Over a period of several hours, nitrogen is circulated into a vacuum-sealed chamber.

> Once the desired temperature of −300°F (−185°C) (the bottom temperature) is reached, the workpiece will be kept at this temperature for up to 36 hours.

> This depends upon the shape of the metal treated and the total weight of the product in the chamber.

> Cross-sectional area, material, and bottom line are the factors that determine the rate and uniformity of the temperature penetration of the product in the chamber.

> Changes in the material structure take place at the molecular level when a part is subjected to long periods of deep cryogenic temperatures.

> The molecules are tightly packed together at −300°F (−185°C).

> As the material is brought back to room temperature, the molecules change back to their normal separation but with a difference.

> The molecules as well as the complex carbides are evenly spaced, eliminating pockets of high density (Fig. 89-3).

> When the cycle is complete, the temperature in the chamber is slowly brought back to room temperature over a 12-hour period.

> A typical cycle of the cryogenic process runs about three days.

▸▸ Stress Relief

Stress in steel comes from the cooling of uneven sections and machining that create complex, invisible (to the naked eye), random patterns. The cryogenic process will thermally stress relieve a part. As parts expand from the heat generated during operation, the retained stress causes uneven expansion, increased dimensional instability, less fatigue life, and increased wear along with decreased performance. Stress boundary areas are susceptible to microcracking, that lead to fatigue and eventual failure. Residual stresses, those that remain in a part, exist in engine parts from the original steel forming, casting, or forging operations and the machining operations used to finish the part.

An engine part expands from the heat generated by running and stress impedes expansion. The steel part will then warp as it is heated from running. Residual stress will therefore cause a part to progressively warp, such as a cylinder head from overheating. Residual stresses are uneven and located throughout the structure. Deep cryogenic processing is an effective method for decreasing residual stress, in addition to increasing the durability, or *wear life,* of steels. Normally, when parts are assembled, they *move* when heated, causing problems with fit and creating undue eccentric wear as a result of warpage. Parts *finish machined* after deep cryogenic processing do not move; as a result, there is less wear from abnormal tensions.

DEPTH OF CRYOGENIC STRESS RELIEVING

The motor sports racing applications for cryo-processed steel products are numerous and can be of great benefit to racers. Aluminum after market engine blocks used by racers can be treated cryogenically. Machinists report significant gains in machinability and in the finish of the aluminum after cryo-processing. Compressive tensile stresses in steels are also created by the mechanical methods of machining, boring, and forming. Thermal stresses are created in steels after heat treating through the quench hardening process. An ice cube, when dropped into a cup of hot coffee, illustrates this effect. The hot coffee creates expansive stress on the exterior of the ice cube while the core is still frozen. The result is that the ice cube cracks from stress shear imparted due to the differing rates of thermal *growth.* This is called the *differential of coefficients of expansion.*

Dropping a motor part into liquid nitrogen has the same effect of actually creating stress. Stress relief (the opposite effect) takes place when the *entire mass* is at an *equal* temperature (core and surface) and cycled slowly through a wide temperature range. Taking a mass to extremely low temperatures also creates a very dense molecular state. If the rate of temperature change is slow enough, thermal compression and expansion take place equally from the core to the surface, releasing internal stresses. The result is a homogenously stabilized material. This process takes a long period of time to keep the entire mass in equilibrium through the temperature cycling.

CARBIDE INCREASE

Studies found that, in some material, the number of countable small carbides increased throughout a heat treatable

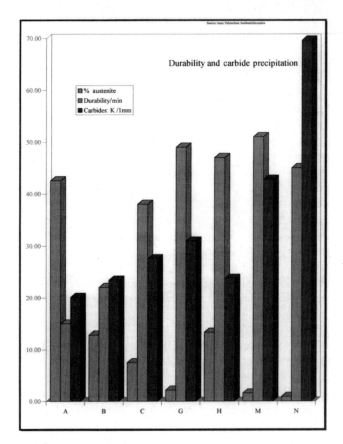

■ Figure 89-4 Durability and carbide perception. *(300° Below, Inc.)*

steel from 33,000 per square millimeter to over 80,000 per square millimeter as a result of cryogenic processing (Fig. 89-4). This increase in carbides adds greatly to the wear resistance of a part. The carbides make a refined, flat, *superhard* surface on the metal. A refined surface structure not only is more wear-resistant but also reduces friction and heat, allowing more rapid movement and greater horsepower.

Scientists found that, for various metal samples processed at −300°F (−185°C), the wear resistance was approximately two to five times greater than that for samples processed at −120°F (−84°C).

▶▶ Wear Resistance

Deep cryogenic strengthening is a permanent, one-time process that creates stronger, more durable tools that resist abrasive wear and show dramatic results in tool performance. Cryogenic treatment actually changes the microstructure into a more refined, uniform grain structure. The treatment improves dimensional stability, minimizes retained austenite levels, increases surface hardness, and improves wear properties. As studies show (Tables 89.2 and 89.3 on p. 800), the wear resistance of

| AISI# | Description | Materials That Showed Significant Improvement | |
		At −110°F	At −310°F
D-2	High carbon/chromium die steel	316%	817%
S-7	Silicon tool steel	241%	503%
52100	Standard steel	195%	420%
O-1	Oil hardening cold work die steel	221%	418%
A-10	Graphite tool steel	230%	264%
M-1	Molybdenum high-speed steel	145%	225%
H-13	Chromium/moly hot die steel	164%	209%
M-2	Tungsten/moly high-speed steel	117%	203%
T-1	Tungsten high-speed tool steel	141%	176%
CPM-10V	Alloy steel	94%	131%

some metals can increase from 100% to over 800% using the cryogenic treatment vs. cold treatment.

▶▶ Metallurgy Changes

A typical cryogenic treatment consists of a slow-down rate from ambient (normal) temperature to near the temperature of the boiling point of liquid nitrogen. By using gaseous nitrogen, any desired cool-down cycle can be programmed to avoid thermal shock and obtain the desired properties.

KINETICS OF CRYOGENIC TREATMENT

According to one theory, with cryogenic treatment, the transformation of retained austenite is nearly complete—a conclusion that has been verified by X-ray diffraction measurements. Another theory is based on strengthening a material via the precipitation of submicroscopic carbides. An added benefit is said to be a reduction in internal stresses in the martensite developed during carbide precipitation. Lower residual stresses may also reduce tendencies to microcrack.

Cryogenic temperatures, −300°F (−185°C), are required to effect a complete molecular change in most ferrous alloys, nonferrous alloys, and polymers.

Dimensional stabilization and rigidity are maximized by the atom and the molecules new crystalline state of arrangement from cryogenic treatment. The E modulus, or modulus of elasticity, of ferrous metals has increased without changing the maco-hardness of conversion hardness testing. All ferrous metals have the same E modulus; cryogenics increases the range of the E modulus.

E modulus is a measure of the rigidity of metal—the ratio of stress, within proportional limit, to corresponding strain.

> The modulus obtained in tension or compression is Young's modulus, stretch modulus, or modulus of extensibility.

> The modulus obtained in torsion or shear is modulus of rigidity, shear modulus, or modulus of torsion.

> The modulus covering the ratio of mean normal stress to the change in volume per unit volume is the bulk modulus.

> The tangent modulus is the slope of the stress-strain curve at a specified point.

table 89.3 — Documented gains for all types of metals: comparison between −120°F shallow quenching vs. −310°F deep cryogenic tempering

AISI #	Description of steel	(−110°F)	(−310°F)
D-2	High carbon/chromium die steel	316%	817%
A-2	Chromium cold work die steel	204%	560%
S-7	Silicon tool steel	241%	503%
52100	Bearing steel	195%	420%
O-1	Oil hardening cold work die steel	221%	418%
A-10	Graphite tool steel	230%	264%
M-1	Molybdenum high speed steel	145%	225%
H-13	Chromium/moly hot die steel	164%	209%
M-2	Tungsten/moly high speed steel	117%	203%
T-1	Tungsten high steel steel	141%	176%
CPM-10V	Alloy steel	94%	131%
P-20	Mold steel	123%	130%
440	Martensitic stainless	128%	121%
430	Ferritic stainless	116%	119%
303	Austenitic stainless	105%	110%
8620	Nickel-chromium-moly steel	112%	104%
C1020	Carbon steel	97%	98%
AQS	Graphitic cast iron	96%	97%
A-6	Manganese air work cold die steel	73%	97%
T-2	Tungsten high speed steel	72%	92%

> The secant modulus is the slope of a line from the origin to a specified point on the stress-strain curve.

The mechanical and physical properties of a material reveal its elastic and inelastic behavior where force is applied, thereby indicating its suitability for:

> *Mechanical applications,* such as modulus of elasticity, tensile strength, elongation, hardness, and fatigue limit

> *Physical properties* that relate to the physics of a material, such as density, electrical conductivity, heat conductivity, and thermal expansion

> *Elastic limits,* the maximum stress to which a material may be subjected without any permanent strain remaining upon complete release of stress

> Nearly complete stress release of ferrous, non-ferrous, and polymers by cryogenic treatment, can be achieved.

table 89.4   TEST RESULTS: Percent of increase in wear resistance after cryogenic treatment

Materials That Showed Significant Improvement

AISI#	Description	At −110°F (−79°C)	At −310°F (−190°C)
D-2	High carbon/chromium die steel	316%	817%
S-7	Silicon tool steel	241%	503%
52100	Standard steel	195%	420%
O-1	Oil hardening cold work die steel	221%	418%
A-10	Graphite tool steel	230%	264%
M-1	Molybdenum high-speed steel	145%	225%
H-13	Chromium/moly hot die steel	164%	209%
M-2	Tungsten/moly-speed steel	117%	203%
T-1	Tungsten high-speed tool steel	141%	176%
CPM-10V	Alloy steel	94%	131%
P-20	Mold steel	123%	130%
440	Martensitic stainless	128%	121%

▶▶ Advantages of Cryogenic Tempering

Deep cryogenic strengthening is a permanent, one-time process that creates stronger, more durable metals. The cryogenic process improves performance and durability on high-speed steel; high-strength, high-alloy steels; aluminum; martensitic stainless steel; titanium; composites; and polymers. The following are some of the main benefits of cryogenic treatment:

> Increases the wear resistance of the material, resulting in longer tool life and increased production (Table 89.4).

> The cost of the cryogenic process is very small, compared with that of replacement tools.

> One permanent through treatment changes the entire structure, not just the surface of the material.

> Refinishing or regrinding does not affect the permanent improvements.

> Closes and refines grain structures to create a denser molecular structure, that results in a larger contact surface area, that reduces friction and wear (Fig. 89-5).

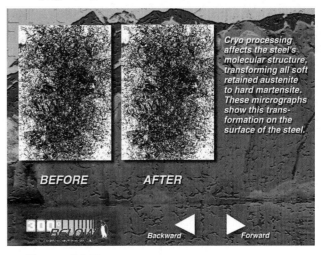

Figure 89-5 The grain structure of steel before and after cryogenic treatment. *(300° Below, Inc.)*

> Transforms almost all the soft-retained austenite to martensite

> Forms microfine complex carbide to strengthen larger carbide structures and add wear resistance

> Increases the performance and durability or wear life of the treated material up to 500% or more

> Decreases the residual stresses in tool steels

> May decrease the brittleness of the material

> May increase tensile strength, toughness, and stability, combined with the release of internal stresses

Deep cryogenic treatment is used in the aircraft industry, die castings, cutting tools, motor sports, and wherever else it is desirable to stabilize the geometry and microstructure of a material.

unit 89 review questions

1. Explain why *heat treating* should really be called *cold treating*.

2. Compare the cryogenic process with the cold treatment process.

3. Summarize the typical cycle of the cryogenic process.

4. The modulus that is the slope of a line from the origin to a specified point on the stress-strain curve is the _____ modulus.

5. The modulus covering the ratio of mean normal stress to the change in volume per unit volume is called the _____ modulus.

6. Provide seven of the main benefits of cryogenic treatment.

UNIT 90

QQC Diamond Coating

OBJECTIVES

After completing this unit, you will be able to:

1 Define the role of diamonds in industrial applications

2 Compare the HP/HT and CVD processes

3 Describe the QQC process and summarize the steps in the process

4 Provide QQC applications and describe the benefits of diamond coating of tooling and machinery components

5 Discuss diamondlike carbon (DLC) and its main application

QQC, a revolutionary process, can deposit a uniform layer of diamond on almost any type of material ranging from glass and plastic to metals. It is done using the carbon dioxide from the air as the carbon source and subjecting it to a combination of lasers. This process can do in seconds what takes conventional chemical vapor deposition (CVD) processes hours. This laser process creates pure diamond and bonds it to a surface of a material with the ease of paint on a brush.

Imagine having a pair of eyeglasses and windshields where the lens or surface never scratches or a kitchen knife that never dulls. It is possible to coat the cutting edges of all types of tools that will last much longer and dull only after prolonged use. Consider valves and casings, as well as blades on rotating machinery that would be considered wear-resistant in comparison with today's already high standards. Longer-lasting tools, instruments, windshields, and everyday goods are only a few of the available applications for diamond coating (Fig. 90-1 on p. 804).

Diamond is an incredible material with properties that make it suitable for many industrial applications and consumer goods (Fig. 90-2 on p. 804). Nature has concentrated in one material all of the following qualities: highest hardness, highest transparency near the visible region, highest thermal conductivity, highest electronic mobility, and highest sound velocity. Most people think of diamonds as naturally occurring stones that must be mined, but Man-Made (manufactured) diamond has been used in manufacturing for machining and grinding hard, abrasive, and difficult-to-cut materials since it was invented in the 1950s.

▸▸ Diamond Tool Development

Diamond-tipped precision lathe tools were used by J. Ramsden in 1771, and in 1819 an English patent was granted to William Brockendon for a diamond wire-drawing die. Around 1900, large, circular saw blades were set with diamonds to cut architectural stone. About the same time, grinding wheels were being developed by impregnating metal bodies with diamond particles.

■ Figure 90-1 Diamond coating jet-fighter canopies provide resistance to pitting.

■ Figure 90-2 Diamonds are used in industry to cut hard, abrasive, and difficult-to-cut nonferrous materials.

MANUFACTURED INDUSTRIAL DIAMOND

The development of diamond tools within the past 50 or so years has been dramatic, especially with the invention of manufactured diamond in the 1950s by the General Electric Co.'s high temperature, high-pressure process (Fig. 90-3).

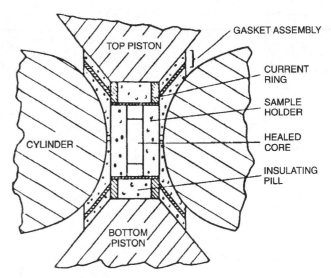

■ Figure 90-3 Diamond crystals are manufactured in a high temperature, high-pressure belt-type apparatus.

DIAMOND DEPOSITION

Over the years many experiments were conducted on the concepts of diamond deposition by chemical vapor deposition. There are many advantages of chemical vapor deposition (CVD) of diamond over the high-pressure, high temperature (HP/HT) synthesis process discovered and commercialized by the General Electric Co.

> The HP/HT process requires pressures of 1 million pounds per square inch and temperatures of about 2000°F in the presence of metal catalysts.

> The diamond produced by the HP/HT method requires further steps to be made into useful tools and products.

> The coating must be bonded to saws and grinding wheels by resin, impregnation, or plating.

> For cutting tools, diamond powder is sintered to a tungsten carbide substrate (base), then cut into shapes that are brazed to tool bodies or inserts and then ground to the finished size and shape.

> Applications to geometrical shapes are severely limited.

In spite of this labor-intensive and costly fabrication, polycrystalline diamond (PCD) is a superior cutting tool because of its resistance to abrasion and the strength of its carbide backing.

Chemical Vapor Deposition (CVD) Process

CVD can be directly synthesized (coated) to a cutting tool substrate, eliminating many steps in the PCD fabrication process. It also allows diamond to be used on intricate shapes such as cutting-tool inserts with molded-in chip breaker geometry, twist drills, and taps. The process and most of its variants consist of the following steps:

1. The first step is to produce atomic hydrogen from the diatomic (two-atom) hydrogen molecule. This can be done by any method that adds enough energy to the gas for dissociation (separation). The typical methods are thermally assisted and plasma-assisted processes. The hydrogen passes over a hot filament or through a plasma and separates into atomic hydrogen.

2. Then the gaseous hydrocarbon source, such as methane or propane, is mixed with the hydrogen and passed over a substrate.

3. Diamond condenses and falls onto the substrate as crystals, which fuse into a polycrystalline layer.

▶▶ Diamond Characteristics

Although cutting tools have been the primary market focus for developers of the CVD process, the superior properties of diamond are desirable for a wide variety of applications. Its properties are mainly due to the short, strong bonds between its carbon atoms. Its high-molar density makes it the hardest known material and the best conductor of heat, with extraordinarily high electrical resistivity.

Diamond's hardness makes it an ideal material for coating parts that are subject to sliding friction.

PREVIOUS LIMITATIONS

Diamond coatings are still relatively expensive, and the CVD process is still a batch processing method. Deposition rates range from less than 1 micron to 5 microns per hour. The substrate to be coated must be thoroughly cleaned and preheated in a vacuum chamber that limits the size of articles to be coated. It is restricted to limited substrate chemistry. A batch of articles in a reactor may take 24 hours or more to be coated.

The major difficulty with cutting-tool applications is the adhesion of the diamond coating to the tungsten carbide substrate. This has caused producers to use costly methods of substrate preparation to achieve good bonding.

▶▶ A Diamond Coating Breakthrough

A major breakthrough in diamond deposition technology occurred when Pravin Mistry, a metallurgist, was doing research. He was trying to fabricate hard materials, using lasers to synthesize ceramics and metal-matrix compos-

ites (MMC) on aluminum extrusion dies to improve their performance and longevity. During the laser synthesis of titanium diboride, Mistry switched carbon dioxide for nitrogen and produced a coating speckled with some black particulate inclusions.

Analysis of the coating's surface indicated the presence of polycrystalline diamond and found a radical method for synthesizing polycrystalline diamond films. The QQC diamond coating process uses the carbon dioxide from the atmosphere as the carbon source and subjects it to multiplexed lasers to produce a diamond film that can be deposited onto almost any material.

THE QQC PROCESS

The QQC process creates diamond in an ordinary atmosphere, not the high temperature vacuum used in standard diamond manufacture. Multiple laser beams are directed through a cloud of carbon dioxide at a tungsten carbide surface. The lasers break the carbon dioxide into oxygen and carbon. Diamond is formed from the bonding of this carbon with carbon atoms that the laser energy has put into motion from the rotating object surface.

The QQC Process

1. Laser energy is directed at a substrate (part) to mobilize, vaporize, and react a primary element (e.g., carbon) contained within the substrate.

   > This changes the composition (e.g., crystalline structure) of the basic element and diffuses the modified constituent back into the part, as an addition to fabricating a coating (e.g., diamond or diamondlike carbon) on the part surface.

   > This creates a conversion zone immediately beneath the substrate and results in diffusion bonding of the coating to the substrate.

2. Additional (secondary) similar (e.g., carbon) or different elements may be introduced in a reaction zone on and above the part surface to expand the fabrication and the composition of the coating.

3. The laser energy is provided by a combination of lasers: excimer, Nd:YAG, and CO_2 (Fig. 90-4 on p. 806).

   > The output beams are directed through a nozzle delivering the secondary element to the reaction zone.

   > The reaction zone is shielded by an inert (nonreactive) shielding gas (e.g., N2) delivered through the nozzle.

   > Flat plasma is created by the lasers, constituent element and secondary element on the surface of the substrate, and the flat plasma optionally extends around the edges of the substrate to fabricate the coating.

wear and tear, and in some cases to replace chromium and cadmium plating. Some tests found QQC coated tools to be the best in terms of performance, wear, and adherence on carbide tool inserts.

ADVANTAGES

The key advantages of the QQC process over existing technology include:

> Superior adhesion and reduced interfacial stress result from a graded metallurgical bond between the diamond and the part.

> The process is carried out without the restrictions of a vacuum chamber. Controlling movements of the lasers or workpiece can coat almost any size or shape. Operations such as coating continuous wire, fiber, or coiled stock are possible.

> Pretreatment and/or preheating of the substrate is not required, permitting coating of the substrate of as-manufactured components and elimination of wet chemistry pretreatment.

> Only carbon dioxide is used as a primary/ secondary source for carbon with nitrogen acting as a shield and possible stoichiometric (stockpiling) process ingredient. This replaces the use of dangerous gases such as hydrogen and methane, critical ingredients in the CVD process.

> Deposition rates are dramatically increased, with linear growth rates exceeding 1 micron per second, as opposed to 1 to 5 microns per hour by CVD. This is a key economic factor in commercialization of the process.

> The process can be applied to almost any substrate such as stainless steel, high-speed steel, iron, plastic, glass, copper, aluminum, titanium, and silicon.

> Cobalt content limitations for CVD require special substrates for tungsten carbide cutting tools that can compromise the inserts' toughness. The multiplexed laser process can accommodate any percentage of cobalt without affecting the diamond synthesis.

> Unlike CVD, the process can be changed automatically to control crystal size, orientation, and morphology. The system can produce tetrahedrally

Figure 90-4 The QQC process uses a combination of lasers: excimer, Nd:YAG, and CO_2 to produce a diamond coating. *(QQC, Inc.)*

Certain advantageous metallurgical changes are created in the substrate due to the pretreatment. The processes (pretreatment and coating fabrication) are suitably performed in ambient, without preheating the substrate and without a vacuum.

DIAMOND COATING

The object to be coated can be moved around by a robotic arm under the laser to control the deposition of the diamond. Adjustment of the lasers can control crystal size and structure. Most synthetic diamond is made by CVD (chemical vapor deposition). In spite of years of effort, the CVD process can still coat only a few, coin-sized shapes and requires a vacuum chamber that must be heated to 800°C.

DIAMOND THICKNESS

The thickest layer of diamond made so far by the QQC process has been 1000 microns, compared with the 7 to 22 micron layers usually created by CVD. Most amazing is how fast the diamond forms, at a rate of about 1 micron per second, while it bonds metallurgically to the surface below. This compared with a few microns per hour for CVD.

FIELD TESTS

The initial claims of field tests stated that tools coated both in diamond and in TNC (tetrahedrally bonded noncrystalline carbon) are being used for some production automotive power train and chassis components (Fig 90-5). These include gears, shock-rods and struts, and brake rotors to provide corrosion-proof properties, to improve bonded noncrystalline carbon, hydrogenated diamondlike carbon, superlattice hard coatings, and other coatings to achieve desired properties.

> Most important, the system is production engineered to permit the economical coating of production components with 24-hour unmanned operation.

▸▸ Diamondlike Carbon

A near relative of diamond is commonly referred to as *diamondlike carbon (DLC)*. It can be described as tetrahedrally bonded noncrystalline diamond (TNC). This coating material shows many of the desirable properties of diamond but may contain some SP^2 bonds (weaker than the SP^3 bonds that give diamond its superior properties). It is amorphous, glassy, and very smooth and is an ideal material for supplementing ever-increasingly cost-prohibitive, environmentally maligning chromium plating. QQC provides a beneficial metallurgical bond to this TNC coating, making it an economical alternative to chrome.

▸▸ Looking Ahead

Cubic boron nitride (CBN), another superlattice material with crystal morphology similar to diamond's, is the preferred cutting-tool coating material for machining cast iron (diamond is best for nonferrous materials but is soluble in iron). In the cutting-tool arena, CBN could enjoy an even greater market than diamond. CVD processes have not been able to effectively coat cutting tools with CBN, but scientists have already managed to synthesize pure and composite CBN on tungsten carbide inserts using their multiplexed laser system.

▸▸ Cutting-Tool Technology

Diamond cutting tools have provided major benefits for the machining of aluminum alloys. QQC's diamond deposition process provides the machine tool industry with a cost-effective, high-performance dry machining capability.

The ability of diamond coatings deposition on a variety of shapes has also allowed for revolutionary round and flat cutting-tool inserts.

▸▸ Diamond Coating Technology

QQC has recently developed a revolutionary and proprietary process for high-speed diamond deposition on a variety of material substrates. The diamond coating is metallurgically bonded to the substrates. The diamond coatings are proving to be the final key for advanced dry machining of aluminum alloys and possibly applications in the automotive and aerospace industries. The benefits of diamond coating of tooling and machinery components include increased production speeds, high product quality, and the elimination of coolant requirements.

unit 90 review questions

1. What is the main advantage of the QQC process over the CVD process?

2. Describe three of the qualities of diamonds.

3. Give three of the disadvantages of the HP/HT process when compared with the CVD process.

4. What are the three steps in the CVD process?

5. List five of the advantages of the QQC process over the present technology.

6. What is diamondlike carbon (DLC) and what is its main use?

UNIT 91

Direct Metal Deposition

OBJECTIVES

After completing this unit, you will be able to:

1 Describe DMD and be able to compare it with conventional rapid prototyping processes

2 Explain what is meant by "The Big Three of Manufacturing"

3 Discuss how the DMD process is used in various industrial applications

Over the past 10 years, designers and engineers have recognized the value of seeing models of the products they are developing. A rapid prototyping process, Direct Metal Deposition (DMD) is a form of rapid tooling that makes parts and molds from metal powder that is melted by a laser to a computer-aided design (CAD) of the part and is then solidified in place (Fig. 91-1). The rate of solidification is dependent on the small HAZ (heat-affected zone) of the laser and the metallurgical properties of the powder. This process closely resembles conventional rapid prototyping processes (material processed by laser under computer control) but differs in that metal powder, and even tool steel, can be melted, rather than plastic polymers. DMD allows the production or reconfiguration of parts, molds, and dies that are made out of the actual end material, such as tool steel or aluminum. It always produces a new part or part reconfiguration directly from a CAD drawing.

▶▶ DMD Technology

The DMD direct fabrication technology represents the first major advance in metalworking in decades. DMD produces improved material properties in less time and at a lower cost than is possible with traditional fabrication technologies.

DMD is the blending of five common technologies: lasers, computer-aided design, computer-aided manufacturing (CAM), sensors, and powder metallurgy. The resulting process creates parts by focusing an industrial CO_2 laser beam onto a flat tool-steel workpiece or part shape to create a molten pool of metal.

A small stream of powdered tool steel is then injected into the melt pool to increase the size of the molten pool. By moving the laser beam back and forth, under CNC control, and tracing out a pattern controlled by a CAD design, the solid metal part is built, line-by-line, one layer at a time. With this process and its focused laser beam, the molten pool cools and solidifies, rapidly producing metal parts of superior quality and strength, with no material waste as in conventional machining operations.

The parts have consistent, fine microstructures that yield superior quality and tool strength. More important, with DMD, the metallic composition can be altered on-the-fly by injecting different types of metal powders into the melt pool. This creates a mixture of graded metallic compositions that have never before been available.

This laser-based technology provides the manufacture of 3D metal components or tooling with closer tolerances and ideal properties directly from a CAD drawing. The benefits are shorter time to market, lower tooling costs, and improved productivity.

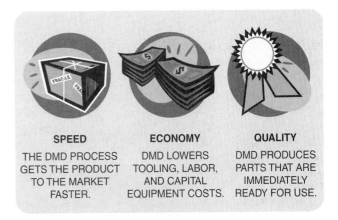

Figure 91-2 DMD provides the Big Three of Manufacturing. *(POM Company, Inc.)*

Figure 91-1 Direct Metal Deposition offers manufacturers new alternatives to conventional rapid prototyping processes. *(POM Company, Inc.)*

▶▶ From CAD to Steel

A typical job starts on Precision Optical Manufacturing's (POM's) secure FTP website (*www.pom.net*), where customers post CAD files. Engineers download and edit the CAD file, adding machining stock or the hard-face surface geometry. The updated solid CAD model is then sliced, and toolpaths, identical to that used for CNC machining, are created.

The data is post-processed, inverting the file and embedding laser and powder commands. CNC machining processes from the top down, while DMD deposits material from the bottom up. Information is downloaded to the three-axis machine similar to a CNC setup with travel in the *X, Y,* and *Z* axes.

DMD uses a variety of metal powders and metal matrix composite materials, including conventional tool-steel alloys, stainless steel, copper, Stellite alloys, Inconel, tungsten carbide, and titanium diboride. The fast solidification rate benefit result in a very-fine-grain part microstructure.

Hard faces are applied within an inert-atmosphere box, filled with pure argon that surrounds the workpiece and the machine nozzle. Without the inert atmosphere, the metal would probably be exposed to excessive oxidation and porosity.

▶▶ The Big Three of Manufacturing

The Direct Metal Deposition process for processing molds, dies, and prototype parts provides what can be thought of as the Big Three of Manufacturing (Fig. 91-2):

> *Speed:* faster product to market. A study shows that die production time can be reduced by 40% with DMD.

> *Economy:* lower tooling costs due to factors including the reduction of labor (this is an automated process) and capital equipment costs (there is one machine that does the lion's share of the work)

> *Quality:* the parts produced are generally .001 in. oversized, so, after a quick clean-up to obtain required tolerances and surface finish, they are ready for use

▶▶ The DMD Process

A CNC-controlled overhead gantry is used to control a nozzle and focusing optics associated with a CO_2 laser according to CAM toolpath data associated with a solid model CAD geometry. Metallic powder, typically a tool alloy (H13, P20, S7, or SS) or OFHC pure copper, is moved from on-board powder feeders by an inert gas at a predefined rate to the nozzle. The metallic powder is added to the dynamic melt pool established by a traversing-beam energy source (Fig. 91-3 on p. 810).

The result is a 3D deposition of tool steel, identical in shape and geometry to that of the CAD model. As a result, the cooling rates associated with the process (in excess of 103°F per second), the tool-steel deposition is martinsitic in structure, typical to that of an as-quenched condition. The end result is a near-net-shape tool-steel

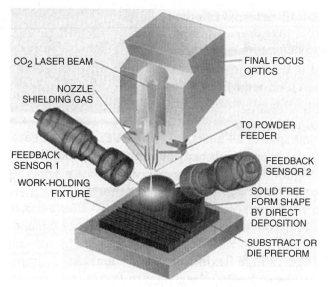

Figure 91-3 DMD is the blending of five technologies: lasers, computer-aided design, computer-aided manufacturing, sensors, and powder metallurgy. *(POM Company, Inc.)*

Labels in figure:
- CO₂ LASER BEAM
- NOZZLE SHIELDING GAS
- FEEDBACK SENSOR 1
- WORK-HOLDING FIXTURE
- FINAL FOCUS OPTICS
- TO POWDER FEEDER
- FEEDBACK SENSOR 2
- SOLID FREE FORM SHAPE BY DIRECT DEPOSITION
- SUBSTRACT OR DIE PREFORM

deposition, which is later tempered and finished using conventional CNC machining and Electrical Discharge Machining (EDM) processes.

▶▶ DMD Applications

The DMD process can be used for prototype or production tooling in a variety of industrial applications:

> *Die repair and refurbishment.* Downtime costs can mount quickly when a mold or die cracks or becomes worn. Conventional repair processes are time-consuming and can warp or damage the mold or die. The DMD repair produces a durable, high quality repair with a small heat-affected zone (HAZ) that does not weaken or damage the part. The result is a mold or die with tool life, strength, and heat resistance comparable to tooling produced by regular machining methods.

  The DMD process is a method that can repair, reconfigure, or resurface existing parts, molds, or dies by adding metal that matches the original tool. It can also make a totally new part directly from a CAD drawing.

> *Thermal management.* The DMD process provides the ability to produce cooling channels or CoolMold technology, for injection molding and aluminum die cast cavities with inserts that follow the part shape (conformal cooling) (Fig. 91-4). The use of conformal cooling and imbedded heat sinks maintain uniform die temperatures that promote dimensional stability and reduce cycle time by 50% or more.

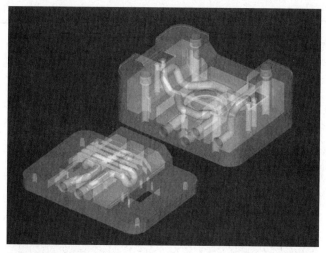

Figure 91-4 Thermal channels conforming to the part shape can be fabricated with DMD. *(POM Company, Inc.)*

> *It's cool.* The molding process consists of five steps: (1) close the mold, (2) fill the mold, (3) pack and hold, (4) cool the part, and (5) eject the part. The biggest time consumer in the process is part cooling, which can account for 44% of the cycle time. Software programs allow the creation of thermal models. These models permit designers and engineers to locate the hot spots, that are then addressed through the generation of cooling channels.

> *Direct metal prototypes.* Manufacturing companies can produce rapid metal prototypes instead of plastic SLA (Stereolithography) models. Using DMD, it is possible to modify or add material to existing aluminum, steel, or other metallic parts or to make a fully functional prototype directly from the CAD design.

> *Surface modification and coatings.* DMD can improve wear resistance, corrosion resistance, and heat checking of part surfaces through the deposition of a wear-resistant hard-facing layer (Fig. 91-5). The DMD process achieves this by providing the capability to deposit surface coatings of greater thickness than can be achieved with other deposition processes. The thickness of each hard-facing layer ranges between .005 and .015 in. (0.13 mm and 0.38 mm). The process can deposit up to 1 cubic inch of metal per hour at a travel speed of 23.6 in./min (600 mm/min).

> *Aerospace and aircraft component repair.* The DMD process is ideally suited for repair work in the aerospace industry, due to the strong metallurgical bond and fine, uniform microstructure it produces. The process uses a CO₂ laser, allowing much shallower heat penetration of the substrate, and provides the ability to deposit aluminum and copper powders.

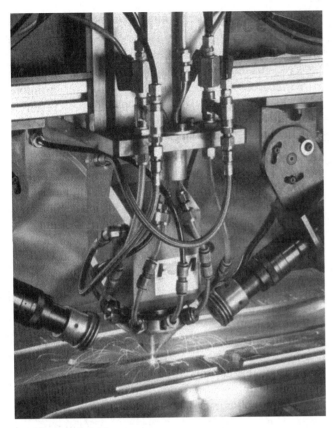

Figure 91-5 Hard facing a pump to provide a wear-resistant surface. *(POM Company, Inc.)*

▶▶ From Prototype to Production

The DirecTool process offers short lead times, compared with other additive processes. The time saved is because deposition can begin as soon as the CAD file is ready. A CAD file an be turned into production tooling in 24 hours, regardless of the complexity of the design:

> *First-stage tooling.* The production tooling comes first, and it can be used as a prototype.

> *Design change flexibility.* Material can be seamlessly added to the tooling by DMD without leaving an interface boundary, unlike welding extra stock onto a machined mold.

> *Suitability for complex designs.* This process works the same without regard for size or complexity.

▶▶ Overview

The DMD process is proving to be among the most promising metalworking advances to come into the marketplace in decades. The benefits can directly impact manufacturers' bottom line by reducing time to market, improving productivity, minimizing tooling costs, and reducing environmental waste:

> *Die repair and refurbishment.* Downtime costs can mount quickly when a die cracks or breaks, and die replacement using traditional methods may take three to four weeks.

> *Direct metal prototyping.* Using DMD, manufacturers can produce metal prototypes with the same functions as a finished component instead of conventional plastic stereolithography models.

> *Thermal management.* The DMD process has the ability to produce conformal cooling channels and imbed high conductivity heat sinks essential to die cavities to promote uniform die temperatures and reduce the possibility of warpage.

The rapid solidification rate of DMD also results in finer microstructures within parts for strength improvements over other processes.

DMD can save energy through shorter part-to-part molding cycle times and reduce scrap-metal pollution and waterborne contamination.

unit 91 review questions

1. What is the difference between DMD and conventional rapid prototyping processes?

2. DMD is the blending of what five common technologies?

3. List five of the metal powders and metal matrix composite materials used in DMD.

4. Regarding DMD, what is meant by "The Big Three of Manufacturing"?

5. The DMD process is used in a variety of industrial applications. List five of these applications.

6. Explain how the DMD process benefits manufacturers.

UNIT 92

e-Manufacturing

OBJECTIVES

After completing this unit, you will be able to:

1 Discuss the goal of e-Manufacturing

2 Define InterNetworking and describe its effects on machine tools

3 Identify the benefits of extending Ethernet connectivity to the factory floor

4 Define e-Service Portal and list the benefits of Platinum Maintenance Service

e-Manufacturing is about the increasing need for communications to and from the factory floor and the customer. In today's extremely competitive world, manufacturers that use machine tools need to improve their productivity by taking advance of any new automation technology available. Every metal-working manufacturer must look for ways to reduce machining time, optimize labor efficiency, and reach higher levels of quality. Until now, automation technologies have been the key to minimizing costs and maintaining consistent quality. Now the question is, How can the automation be automated?

▶▶ Internetworking Standard

Even today, modern machine tools remain largely closed *islands of automation*. Plant machinery needs to be networked into the enterprise-wide information system. Presently, this lack of connectivity represents a huge constraint as far as productivity is concerned. e-Manufacturing's goal is to create an InterNetworking standard that makes every machine tool a piece on the corporate network in order to identify, monitor, and optimize production throughput on the factory floor (Fig. 92-1). Then, every machine tool becomes a vital part of management's information system and is integrated into the enterprise-wide profit process. The point of production is then linked directly to the supply chain, on the one hand, and the demand chain of fulfillment, on the other. This integration among management planning, purchasing, production, operations, sales, and service is currently missing the direct link to the machine tool.

▶▶ Connecting Machine Tools

Connecting machine tools on the shop floor into an overall *plant nervous system* will release (unleash) the information from each machine and allow management to increase profitability. This dynamic infrastructure extends information related to production beyond the factory floor. Then, machine tools become servers of information in real time, feeding their information to other functions within the corporation anywhere in the world.

Until now, the missing element has been the universal bridge between Industrial Automation and the Information Technology sector. Manufacturers need a network that integrates all points of production into a secure, browser-based, queriable information system. Better information such as event monitoring, tool diagnostics, and productivity information from the bottom up can only now be enabled by a nearby open architecture. Unfortunately, largely because of proprietary applications, Computerized

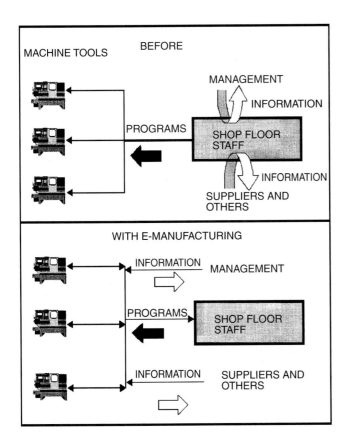

Figure 92-1 An analysis of Root Problems in manufacturing. *(e-Manufacturing Networks, Inc.)*

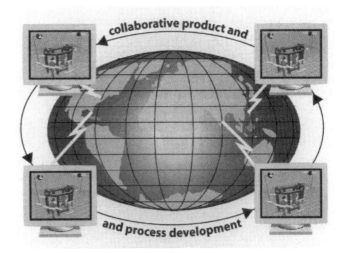

Figure 92-2 e-Manufacturing opens up the world. *(Delmia Robotics, Inc.)*

Numerical Controls (CNCs) do not communicate with each other or with management adequately. Rather, they are primarily receivers of part program data. An open architecture platform is being introduced by e-Manufacturing Networks Inc. to address this general problem. Different machine tool systems can then be configured (arranged) into a contiguous, enterprise-wide management information system.

⏩ Machine Tools as Web-Enabled Appliances

Very few manufacturers have ever considered the result of machine tools as web-enabled appliances on their factory floor. In many cases, CNC machine tools are not used to their fullest potential and they act as stand-alone islands of technology. They are:

> Dumb islands of technology—not networked

> Made to run only part programs well

> Cut off from the management information system

> Mostly closed embedded systems.

Giving every machine tool a hardware and software upgrade to permit it to host Internet Protocol (IP) ad-

dresses shatters the glass wall between the factory floor and the world that depends upon it (Fig. 92-2).

The technology of networking machine tools over the Internet has been around since the mid-1990s. And many specific applications that control and monitor equipment remotely via this method have grabbed everyone's attention and inspired them since then.

What e-Manufacturing is doing expands those possibilities through *InterNetworking*. The hardware and software technology (either direct Original Equipment Manufacturers or aftermarket) turns machine tool controls into Web servers, each with its own unique Web address and complete communicative functionality.

Once so equipped, machine tools are integrated into the supply chain in ways limited only by the manufacturing imagination. Any process that contributes to a machine tool's effectiveness and productivity—monitoring, diagnostics, repair, planned maintenance training, customer service, inventory control, warranty issues—can see dramatic efficiency improvements within the *InterNetwork*.

InterNetworking creates a secure, open architecture platform that turns every machine tool into a node on the corporate network. A machine tool as a node becomes a *Web appliance* that connects the point of production to management's information system—the supply and demand chains—in real time (Fig. 92-3 on p. 814). Just as *e-mail* has dramatically changed the way we communicate, and *e-commerce* has fundamentally changed the way we do business, *e-Manufacturing* will prove to maintain the rapid rate of change while leveraging the investment industry has made in automation and people.

To quote from a 1998 ARC Advisory Group survey report, "The largest reservoir of untapped operational information is locked in the machine tools on the manufacturing floor. Using open architecture CNCs in a plant is

Figure 92-3 CNC machine tools become Web appliances on an InterNetworking network. *(e-Manufacturing, Inc.)*

fundamental in gaining a competitive advantage. Open architecture CNCs tied into the information technology mix is equally critical in optimizing production in both job shops and high production lines."

▶▶ Standard Operating System

The machine tool industry is moving to a standard operating system that will integrate with other levels in the corporation. The OMAC (Open Modular Architecture Controls) Users Group, with its counterparts in Europe and Japan, has been working on an open standard since December 1991. The focus is to extend upon architecture to CNCs.

Every item in a plant with a microprocessor can, will, and should be connected. The value of extending Ethernet connectivity to the factory floor will result in the following benefits:

> Cost savings in terms of reduced inventory and operational expense

> Faster production as bandwidth and speed are increased tenfold

> Improved service as a result of remote diagnostics, which means less downtime

This results in better, more informed decisions by management as they link CNCs to information systems such as Enterprise Resource Planning and standard accounting packages.

▶▶ Theory of Constraints

Why is a Web-enabled factory floor so important? It allows for the Theory of Constraints (TOC) to be applied in an automated way. TOC talks about identifying, monitoring, and optimizing bottlenecks in the value chain process. Manufacturers should not simply improve the operations in an isolated way; rather, they should strive to find the weakest link in their production system and ex-

ploit it. Once found, bottlenecks must be elevated so that they become the best-managed links in the process.

The operations on the factory floor are linear, dependent, and statistically variable in time. Machining is currently a start-stop-wait-repeat process. It is important to reduce wait periods by making those key machine tools productive during those times. Real-time data is necessary to accomplish this type of integrated systems optimization. Automation in the future will eliminate or significantly reduce the wait-periods that are also in the manufacturing information process. Instead of having to wait for program delivery, tool offsets by hand, setup, management approval, and the front office accounting department to realize that inventory is ready to ship, a seamless integration will speed the *Just-in-Time* process into *Just-in-Seconds*.

▶▶ Machine Tools as Smart Partners

Making the machine tool a smart partner instead of a dumb box introduces a natural feature to the process. By introducing the concept of an ever-optimizing *neural network* onto the shop floor, e-Manufacturing concepts will bring a new class of service to the entire corporation (Fig. 94-4). Remote monitoring can be done from anywhere at any time, through a standard TCP/IP connection—which means diagnostics and parts program recovery can be done from any location. Even better, these services can be provided automatically, if and when the machine tool control itself sends out an alert to the corporate network. This concept of *self-healing* allows the operator to focus on more important details.

Let machines do what they do best—repetitive, high-volume, and even dangerous tasks. This frees hu-

Figure 92-4 The results of combining CNC machine tool technology and e-Manufacturing. *(e-Manufacturing, Inc.)*

Figure 92-5 The importance of machines in manufacturing.

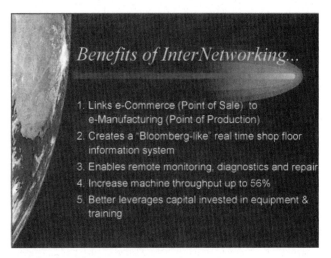

Figure 92-6 The benefits of InterNetworking factory floors. *(e-Manufacturing Networks, Inc.)*

mans to do what they do best—create, design, build, and dream. *e-Manufacturing* will prove to continue the rapid rate of change while leveraging the investment made in automation and people. A whole new industry is emerging as e-Manufacturing *InterNetwork* factory floors to the world.

▶▶ Ethernet Connectivity

The goal is to make every machine tool a node on the corporate network, an IP address available on the Internet, in order to identify, monitor, and optimize throughput on the InterNetwork factory floor (Fig. 92-5). This will create a cost-effective global Web-centered communications standard that joins together the factory floor and the rest of the Net-based world.

Giving every machine tool a hardware and software upgrade, to enable it to host an IP address, shatters the glass wall between the factory floor and the world that depends upon it. Universal open architecture allows applications such as real-time monitoring, remote diagnostics, and 24x7 maintenance via the Internet.

▶▶ Direct CNC Internetworks for the Shop Floor

The future of shop floor information systems, which includes Direct Numerical Control (DNC), Manufacturing Execution Systems (MES), Supervisory Control and Data Acquisition (SCADA), Statistical Process Control (SPC), and even Cellular Manufacturing, requires high-speed Ethernet technology extended to the CNC.

A visionary company has extended the high-speed, robust, and universal TCP/IP network protocol to every machine tool in the shop. This leading-edge technology makes any CNC into a node on the corporate network—an open factory automation platform that could even extend the access globally via the Internet. The network optimization product line features:

> Ethernet DNC systems (with file "Push" and "Pull")
> Ethernet-based Fanuc® memory upgrades
> Ethernet to serial convertors for CNCs
> Next-generation Ethernet Behind Tape Readers (BTRs) with memory
> Ethernet-based FMS cell controller solutions
> NT-based remote CNC monitoring software
> Fanuc® Open Systems Interconnection (OSI) and FTP Ethernet connectivity solutions

The following are in various stages of development:

> Remote Internet maintenance capabilities
> The new Enhanced Technology Architecture Fanuc® Compatible 2, three and five axes open CNC

The benefits of InterNetworking are shown in Figure 92-6.

▶▶ Internetworking Factory Floors to the World

e-Manufacturing networks will make the machine tool a node on the corporate network. The unique CNC knowledge and capabilities provide Ethernet-based machine tool connectivity. NetDNC transforms each machine tool into an on-line server of information, giving managers the ability to monitor and remotely control the manufacturing process via the Internet (see Fig. 92-3).

e-SERVICE PORTAL

The remote controls and monitors provide diagnostic maintenance for the factory floor in real-time. Platinum Maintenance Service provides end users with:

> Backup of CNC parameters
> Fast reinstallation
> Notification of downtime
> Service requirements
> Remote diagnostics
> Automatic code repair
> Enterprise Resource Planning (ERP)/supply chain linkage

e-COMMERCE PORTAL

CNCpartslocator.com (CPL) is the fastest, most useful on-line source of new and used CNC parts. The goal is to have a comprehensive list of CNC parts that allows a person to find a critical part quickly. CPL brings together a global network of dealers, suppliers, and OEMs. New products and information services are offered, putting anyone in touch with the best in the CNC world. If there are spare CNC parts in the inventory, CPL is able to market them to the world.

▶▶ Summary

> The goal of e-Manufacturing is to create an Inter-Networking standard that makes every machine tool a node on the corporate network.

> Connecting machine tools on the shop floor into an overall *plant nervous system* will release (unleash) the information from each machine and result in increased profitability.

> Better information such as event monitoring, tool diagnostics, and productivity information from the bottom up can only now be enabled by an adjoining open architecture.

> In many cases, CNC machine tools are not used to their fullest potential and they act as stand-alone islands of technology.

> If information is "it"—then manufacturers should get it once, get it right, get it digitally, get it to everybody who needs to know, and get it in real time.

unit 92 review questions

1. What is e-Manufacturing and what is its goal?
2. Describe some of the disadvantages of not using CNC machine tools as Web-enabled appliances.
3. What are three benefits of extending Ethernet connectivity to the factory floor?
4. List five of the features of the network optimization line.
5. List five of the benefits of Platinum Maintenance Service.

UNIT 93

STEP-NC and Internet Manufacturing

OBJECTIVES

After completing this unit, you will be able to:

1 Define STEP and STEP NC

2 Discuss the advantages to manufacturers of STEP NC

3 Compare STEP with IGES

4 Summarize the results reported by STEP users

5 Describe the Super Model Project

STEP-NC, a process under development since 1984, has the potential for dramatically changing the way products are manufactured in the world (Fig. 93-1 on p. 818). STEP-NC is a worldwide standard developed by the International Standards Organization (ISO) to extend STEP (*ST*andard for the *E*xchange of *P*roduct) model data so that it can be used to define data for NC (numerical control) machine tools. STEP became a full ISO standard in 1994, and since then all leading CAD software vendors have implemented STEP data translation into their products. It has drawn the interest and support of the Department of Energy, Boeing, Lockheed Martin, General Electric, General Motors, Daimler-Chrylser, and others.

▶▶ What Is STEP?

STEP is an extensible, comprehensive, international data standard for product data created by an international team of data experts. It is designed to give a complete representation of product data throughout its entire life cycle. There has always been a mismatch between the lifetime of software systems and the information they produce. To see this clearly, look at the number of outdated word processing files in any computer. This problem is greatly magnified when comparing technical, design, and engineering software files. CAD and analysis systems change every 2 to 3 years, while technical prints and manufacturing plans for things such as aircraft, ships, and building must be kept for 30 to 50 years.

▶▶ Working STEPS

With the development of STEP-NC, what is happening is not simply the reshaping of CNC; it is the reshaping of manufacturing. In the vision that is emerging, the CNC machine tool will be more important than ever. STEP-NC changes the way manufacturing is done by defining data as *working steps:* a library of specific operations that might be performed on a CNC machine tool (Fig. 93-2 on p. 818). STEP-NC will make *G*- and *M*-codes obsolete.

Numerically controlled machine tools are programmed using a language called RS274. This language is almost 40 years old and severely restricts the range of information that can be communicated to the machine tool controller. STEP-NC (ISO 14649) replaces RS274 with a

Figure 93-1 STEP-NC can change the way products are manufactured. *(Modern Machine Shop)*

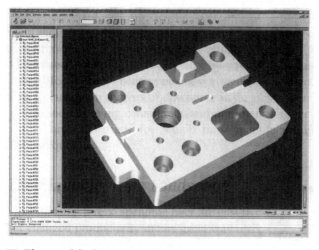

Figure 93-2 STEP defines data as working steps for operations performed on a CNC machine tool. *(STEP Tools, Inc.)*

rich data format that integrates the manufacturing process plan with the manufacturing features and geometry that were used to derive that process plan.

▶▶ Advantages of STEP

With the concept of working steps in place, the manufacturing process becomes streamlined. All manufacturers will be able to share information reliably and instantaneously. A STEP-NC–converted CAD file that is com-

table 93.1   **A few sample STEP application protocols for engineering and manufacturing**

STEP Application Protocols	
• Part 204	Mechanical Design Using Boundary Representation
• Part 207	Sheet Metal Dies and Blocks
• Part 208	Life Cycle Product Change Process
• Part 213	Numerical Control Process Plans for Machined Parts
• Part 214	Core Data for Automotive Mechanical Design Process
• Part 219	Dimensional Inspection Process Planning for CMMs
• Part 224	Mechanical Product Definition for Process Planning
• Part 235	Materials Information for Products

STEP Tools, Inc.

pleted on the East Coast can be sent over the Internet to a machine shop on the West Coast, and the shop can immediately start machining the part.

The design of STEP concentrates the standardization effort on information content, rather than on implementation technology. This ensures that the standard will not have to be discarded whenever computing technology changes. STEP can and will be continually expanded and refined so that it will never be out of date.

The common catalog covers geometry, topology, tolerances, attributes, assemblies, configuration, manufacturing processes and more (Table 93.1). It is very likely that, over a period of time, many industries will develop their own protocols.

▶▶ STEP versus Initial Graphics Exchange Specifications (IGES)

To understand STEP-NC and where it is headed, it is important to look at STEP and its relationship to IGES. IGES is about exchanging data and only the data contained in graphics files. STEP is about sharing data, allowing parties to work together by communicating information interactively.

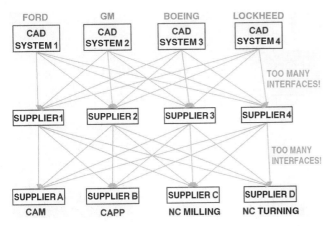

Figure 93-3 3-D model data is difficult to use in manufacturing because a supply chain may have to rely on too many interfaces among unlike computer systems. (*STEP Tools, Inc.*)

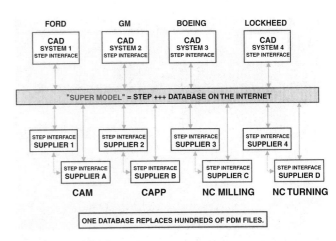

Figure 93-4 The STEP Super Model will replace hundreds of product-management files with one database, shared via the Internet. (*STEP Tools, Inc.*)

IGES first appeared about 25 years ago, when designers and engineers started using computers to create product designs. Although producing the original design file could take longer than preparing the engineering drawing on paper, the design file could be quickly copied, modified, printed and otherwise manipulated.

The biggest problem was that the computer-aided design (CAD) systems used to create these digital design files were not compatible with each other (Fig. 93-3). A design created on a Computervision system was meaningless to an Applicon system, for example. Companies with unlike CAD systems could not exchange CAD data until the IGES was agreed upon.

IGES became a workable but not perfect approach to exchanging CAD files. It allowed one system to communicate the lines and symbols of a computerized engineering drawing, but IGES failed to communicate the meaning of the information the drawing was intended to convey. It did not provide a reliable means by which product features could be transmitted with the geometry so that computer-based applications could *understand* the engineering drawing.

While IGES was being developed and became more functional as it moved through the standards formation process, efforts to develop a true *product data exchange specification* were launched. The goal of this effort was to capture and present *logical* information about product features and provide *physical* mechanisms for data exchange.

By 1984, this international effort to develop a Product Data Exchange Specification had been established under the auspices of ISO, the international standards-making body. The goal was to define the methods for creating product data models that could be understood by computers.

The international standards covering these product data models became known as STEP. For the past 15 years, various groups and committees have been meeting regularly to develop standards for product data models. They have made considerable progress, and the STEP standards are now sufficiently developed to cover all of the original purposes of IGES. STEP has officially taken its place (Fig. 93-4).

By July 2000, almost every major and minor CAD system vendor had STEP translators in the latest releases of their CAD products. Moreover, these translators were tested for conformance and interoperability. One of the innovative features of the STEP formation process was the early commitment to include testing procedures for assuring that STEP-compliant systems would truly function as intended. STEP is working and, according to industry analysts, more than 1 million STEP-enabled CAD stations are in place around the world.

►► Developments

Imagine calling up a Web browser on the PC-based CNC at the machine tool by going to a certain Web site and, from a menu on the home page, selecting one of the databases it contains. A 3D imagine of a part such as the one shown in Fig. 93-5 on p. 820 comes up. Clicking on the proper icon in the task bar, checking a few parameters and default settings on a pop-up window, and clicking on the CYCLE START button will start the CNC machine to produce the required part.

The CNC does not use *G*-codes; everything it must know about how to move the cutting tool is in the product model's database. It is not necessary to create a new, separate file of toolpath data. Toolpaths are figured out in the CNC itself, based on the product model; this can eliminate the need for post-processors. Data is formatted for execution by the machine within the CNC. Because the product

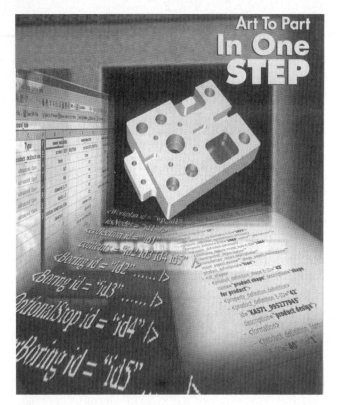

Figure 93-5 The information to manufacture the 3D part can be downloaded by selecting the correct database over the Internet. *(Modern Machine Shop)*

THERE ARE NOW MORE THAN 1 MILLION STEP-ENABLED CAD STATIONS IN THE WORLD.

Figure 93-6 Many STEP-enabled CAD systems are being used throughout the world. *(STEP Tools, Inc.)*

model does not change, it is available for machining *hard copies* when and wherever required.

▶▶ The First STEP

STEP replaces IGES as the means by which graphical information is shared among unlike computer systems around the world. The big difference is that STEP is designed so that virtually all essential information about a product, not just CAD files, can be passed back and forth among users.

The STEP standard is designed for the long term. It is being maintained and extended by product data experts, who meet three times a year to design extensions and fold in new technologies. Users, not vendors, develop STEP because user-driven standards are results-oriented, while vendor-driven standards are technology-oriented.

STEP, a standard that can grow, uses its own language, called *EXPRESS,* and can be extended to any industry shown in Table 93.2. EXPRESS is the only widely available data language that describes the complexities of solid geometry. A standard that grows will not be outdated as soon as it is published. Because STEP is a neutral standard, not owned by any CAD, CAM, or NC Control vendor, it will be the first to span the whole *art-to-part* design and manufacturing process.

▶▶ STEP in U.S. Manufacturers

STEP standards are being used daily in many of the largest companies in the world. Parts lists, assembly information, and other types of Product Data Management (PDM) information are routinely shared using STEP. Many large engineering companies are using STEP to standardize the product information supplied by vendors.

Major industries throughout the world, such as the aerospace, automotive, electronics, and shipbuilding industries, are using STEP to overcome problems with data transfer and manufacturing (Fig. 93-6). Some of the reports from STEP users indicate the following positive results:

> Ten percent improvement in data exchange reliability

> Ten percent process saving for noncomposite parts

> Fifty percent process saving for composites

> Twenty-seven percent tool design saving on CAD/CAM systems

> Thirty-nine percent saving for NC CAM systems

> Seventy-five percent reduction for manufacturing process plan visuals

> Fifty hours saved in the exchange of 12 data files by STEP

> Improved product quality and reduced cycle time

> Elimination of many compatibility problems and data archival issues

table 93.2   STEP application protocols for design and manufacturing

The following Application Protocols are available or in development within ISO.

- Part 201   Explicit Drafting
- Part 202   Associative Drafting
- Part 203   Configuration Controlled Design
- Part 204   Mechanical Design Using Boundary Representation
- Part 205   Mechanical Design Using Surface Representation
- Part 206   Mechanical Design Using Wireframe Representation
- Part 207   Sheet Metal Dies and Blocks
- Part 208   Life Cycle Product Change Process
- Part 209   Design Through Analysis of Composite and Metallic Structures
- Part 210   Electronic Printed Circuit Assembly, Design and Manufacturing
- Part 211   Electronics Test Diagnostics and Remanufacture
- Part 212   Electromechanical Plants
- Part 213   Numerical Control Process Plans for Machined Parts
- Part 214   Core Data for Automotive Mechanical Design Processes
- Part 215   Ship Arrangement
- Part 216   Ship Molded Forms
- Part 217   Ship Piping
- Part 218   Ship Structures
- Part 219   Dimensional Inspection Process Planning for CMMs
- Part 220   Printed Circuit Assembly Manufacturing Planning
- Part 221   Functional Data and Schematic Representation for Process Plans
- Part 222   Design Engineering to Manufacturing for Composite Structures
- Part 223   Exchange of Design and Manufacturing DPD for Composites
- Part 224   Mechanical Product Definition for Process Planning
- Part 225   Structural Building Elements Using Explicit Shape Rep
- Part 226   Shipbuilding Mechanical Systems
- Part 227   Plant Spatial Configuration
- Part 228   Building Services
- Part 229   Design and Manufacturing Information for Forged Parts
- Part 230   Building Structure frame steelwork
- Part 231   Process Engineering Data
- Part 232   Technical Data Packaging
- Part 233   Systems Engineering Data Representation
- Part 234   Ship Operational logs, records and messages
- Part 235   Materials Information for products
- Part 236   Furniture product and project data

(STEP Tools, Inc.)

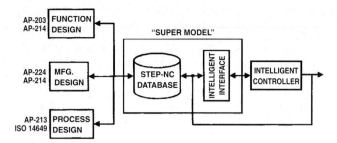

■ Figure 93-7 Model-driven Intelligent Control of manufacturing—architecture of the Super Model Project. *(STEP Tools, Inc.)*

■ Figure 93-8 A number of parts produced by rapid prototyping. *(300° Below, Inc.)*

▶▶ Super Model

Fig. 93-7 shows the NC-STEP Super Model Project supporting a three-stage process-functional design and delivering data produced by the process to an Intelligent Controller. The output can be described using AP-203 of STEP for the aerospace industry and AP-214 of STEP for the automotive industry with other *Application Protocols (APs)* to be available for other industries.

Manufacturing design describes the model with features suitable for the manufacturing processes on a shop floor. Process design defines a manufacturing process for a specific type of machine. In STEP language, this model is to be described using a range of Application Protocols that have not yet been assigned numbers within the STEP framework. However, in the STEP-NC framework (developed by a *sister* committee of STEP in the International Standards Organization) the models are ISO 14649-11 for milling machine processing, ISO 14649-12 for turning machine processing, and ISO 14649-13 for EDM machine processing. Also, AP-213 of STEP captures the *macro* process plan showing the production order between the machines.

Some manufacturing outlines that can be supported by these three systems include:

1. *Rapid prototyping.* The functional and feature design is produced using an integrated CAD/CAM system at an Original Equipment Manufacturer (OEM) (Fig. 93-8).

   > The result is output as an ISO 14649 (STEP-NC) model and is read into an Intelligent machine tool Controller containing a shop floor path planning system.
   > The tool Controller is located at the in-house shop of the OEM or belongs to a supplier.
   > The path planning system in the Controller dynamically defines a toolpath for a tool selected by the operator, and the part is cut.

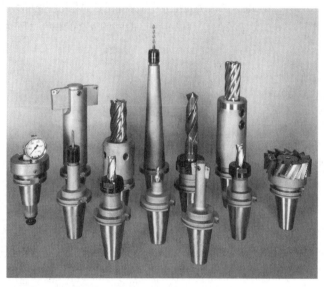

■ Figure 93-9 Types of tooling used to manufacture products. *(Giddings and Lewis, Inc.)*

2. *Tooling.* Tooling for a production line such as a mold, die, or fixture is designed by an OEM (Fig. 93-9).

   > The result is output as an AP-203 or AP-214 file and sent to a job shop.
   > The job shop supervisor reads the file into a CAM system and uses that system to define manufacturing features that can be produced using the machines available in the shop.
   > The result is output as a set of AP-224 files. Skilled operators read the AP-224 files into their own CAM systems and produce ISO 14649 files to make the selected features on their machine tools.
   > Each ISO 14649 file is read into a machine tool Controller and used to cut a part.

3. *Production.* The complete manufacture process for a production line is defined as an integrated AP-213/ISO 14649 database containing the manufacturing sequence and the control files for each machine in the line (Fig. 93-10).

> The database is used to configure the production run at setup time and as an archive after the end of the production run.
> The database is developed using a combination of CAD, CAM, and process planning tools.

For more information on STEP-NC, visit the Web site www.steptools.com.

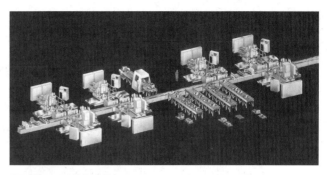

■ **Figure 93-10** A CNC number of machine tools and machining sequences included in a production line. *(Cincinnati Machine, a UNOVA Co.)*

unit 93 review questions

1. What is STEP NC and when was it developed?
2. Discuss the advantages of STEP.
3. Summarize the differences between STEP and IGES.
4. List six of the positive results reported by STEP users.
5. Model ISO 14649-11 is used for _____ machine processing.
6. Model ISO 14649-12 is used for _____ machine processing.
7. Model ISO 14649-13 is used for _____ machine processing.

UNIT 94

Optical/Laser/Vision Measurement

OBJECTIVES

After completing this unit, you will be able to:

1 Define both contact measurement and non-contact measurement

2 Describe laser measurement and identify the equipment used for measurement

3 Explain scanning probe technology

4 List the advantages of video-measuring systems

Since the early 1990s, manufacturers have been using a variety of optical measuring instruments and inspection techniques to get enhanced views of their parts, goods, and products to assist in inspection, quality control, and research and development

(R&D). The collection and analysis of dimensional data is the most effective technique for maintaining control of manufacturing processes and evaluating the accuracy of the finished part.

Analysis of this information can indicate many conditions, as well as ways of modifying a process to improve workpiece quality. Accurately collecting enough dimensional information for effective process control has always been a challenge, particularly for contoured or irregular shapes.

There are video microscopes, laser measuring and tool-settling instruments, optical comparators, and coordinate measuring machines (CMMs) (Fig. 94-1) that can automatically measure 3D forms and use that information along with computer-aided design (CAD) systems to provide insights into the processes involved in the workpiece manufacturing operation.

▶▶ Measurement: Contact or Noncontact

Contact measurement occurs when the measuring probe touches the workpiece at specified points or remains in constant contact with the workpiece while recording data points.

Noncontact measuring, as the name implies, measures with no contact or probing of the workpiece. Scanning enables high-density data inspection of irregularities or identifies out-of-tolerance characteristics, which could

pass using the touch-trigger probe (TTP) method, the most commonly used probe (Fig. 94-2). It has been used on coordinate measuring machines (CMMs) for many years.

Two-dimensional (2D) probing head systems have two modes of operation for the measurement and inspection of parts:

> The *Axial mode* (Fig. 94-3a) is used for profiling. In axial scanning only the *Z* axis is free to move while the *X* and *Y* axes are locked, for scanning profiles lying in orthogonal planes containing the probe stylus.

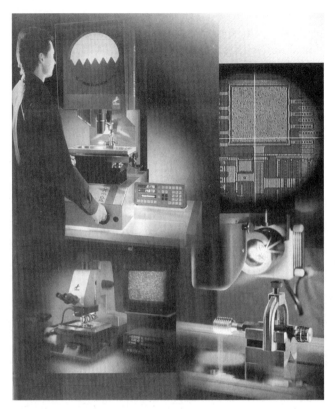

Figure 94-1 Laser, video, and optical equipment is used for measuring. *(Brown & Sharpe)*

Figure 94-2 The touch-trigger probe is the most widely used for contact measuring. *(Brown & Sharpe)*

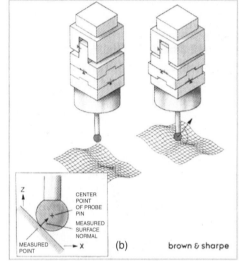

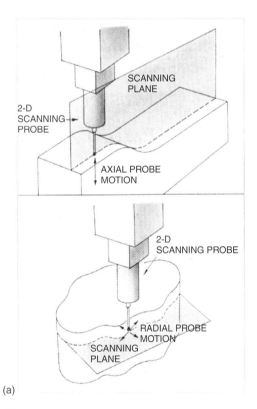

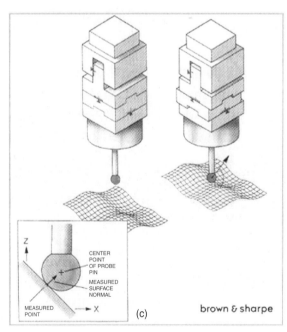

Figure 94-3 (a) During the axial-scanning mode, the Z axis is free to move while the X and Y axes are locked; (b) during the radial-scanning mode, the Z axis is locked and the X and Y axes are free to move; (c) the 3D probe head is capable of simultaneously measuring in three axes—X, Y, and Z. *(Brown & Sharpe)*

Optical/Laser/Vision Measurement   **825**

> In the *Radial mode* (Fig. 94-3b), the axial probe axis *Z* is locked and the *X* and *Y* axes are free to move. Radial scanning is used with contour surfaces for scanning perimeters in orthogonal planes.

The *three-dimensional (3D)* probing head (Fig. 94-3c) is used for measuring and inspecting convoluted (complex) part surfaces such as gears, cams, and rotor blades for which measurements must be made with 3D movements and not in machine axes movements.

▶▶ Optical Comparators

The optical comparator, or shadowgraph (Fig. 94-4), is a fast, accurate means of measuring or comparing the shape or accuracy of a workpiece with a master. Light from a lamp passes through a condenser lens and is projected against the workpiece. The shadow caused by the workpiece is transmitted through a projecting-lens system, which magnifies the image and projects it onto a mirror. The image is further magnified and reflected to the view screen. The enlarged image may be compared with a master form that indicates the limits of size and the dimensions or the contour of the part being checked. Charts are available for special jobs, but the most commonly used are linear measuring, radius, and angular charts. A vernier protractor screen is also available for checking and measuring angles.

The surface of the workpiece can also be checked by a surface illuminator, which lights up the face of the workpiece, allowing the image to be projected onto the screen.

Image or surface finish magnification depends on the lens used. Interchangeable lenses for optical comparators are available in magnifications of 5X, 10X, 31.25X, 50X, 62.5X, 90X, 100X, and 125X.

Optical comparators are used when the workpiece is small, odd-shaped, or difficult to check by other methods.

▶▶ Laser Measurement

Industry needs the ability and flexibility of noncontact measuring and checking of products and workpieces. In highly competitive industries such as aerospace, automotive, and advanced manufacturing, lasers are proving to be a valuable asset. Laser systems are generally used in manufacturing for welding, machining, heat treatment, measuring, inspection, laser marking, and bar coding.

LASER-TOOL MEASUREMENT

Tool measurement is as critical to the metal-cutting process as inspection is to the final product. Knowing the length, diameter, and profile of a cutting tool and the ability to measure and monitor these dimensions can reduce variability and help to optimize the machining process.

Noncontact laser-tool measuring systems (Fig. 94-5) allow verification of tool integrity, tool wear compensa-

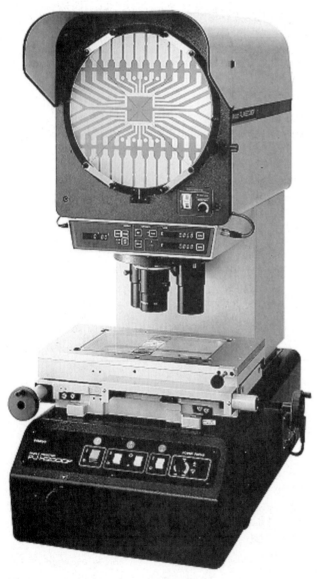

■ **Figure 94-4** Optical comparators are used for measuring and comparing a part with a master form. *(Mitutoyo Mfg. Co.)*

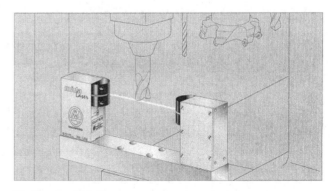

■ **Figure 94-5** Laser noncontact measuring provides static and dynamic measuring of the tool diameter. *(Marposs Corp.)*

tion, tool length measurement, static and dynamic measurement of the tool diameter, tool identification, thermal drift compensation of the machine axes, and profile checking of the rotating cutting tools on milling machines and machining centers.

The system uses a high-precision laser beam and a photodiode to carry out the measuring operations. The CNC machine uses the signals generated by the system following the interruption of the laser beam.

The cutting tool represents the end effect of a long chain of process parameters. Knowing the exact location and the physical properties of the cutting tool as well as monitoring wear characteristics enhance the manufacturer's ability to produce quality products at a competitive price.

LASER SCAN MICROMETER

The laser scan micrometer is a highly accurate method of measurement, with an accuracy of ± 2 μm (± 0.08 Mil) and a repeatability of ± 0.15 μm (± 0.005 Mil). The workpiece is located in the center of the laser beam, creating a shadow in the path of the scanning beam, which is detected, enabling the unit to determine the edges of the part.

The optical micrometer is a very precise and reliable tool for performing high-speed measurements on stationary or moving targets on production lines. The optical micrometer (Fig. 94-6) uses a uniform high-intensity LED (*Light-Emitting Diode*) light source and an HL-CCD (*High-speed Linear Charged-Coupled Device*). The signal is digitally processed to provide the high-speed and accuracy in measuring. This system is capable of performing high-speed sampling at a rate of 2400 samples per second with a repeatability of ± 0.15 μm. With no moving parts the system is maintenance free, problems caused by vibration and heat generated by electric motors are eliminated.

LASER SCAN MICROSCOPE

In its most basic form, a microscope is two magnifying lenses, located a specific distance apart and at a set distance from the eye of the observer. A light source is used to illuminate the sample part. Laser scan microscopes (Fig. 94-7) have an accuracy of 1/2 millionth of an inch (0.01 micron).

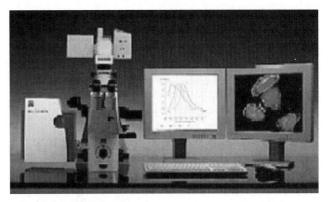

■ **Figure 94-7** Laser scan microscopes have an accuracy of 1/2 millionth of an inch (0.01 micron). (*Keyence Corp.*)

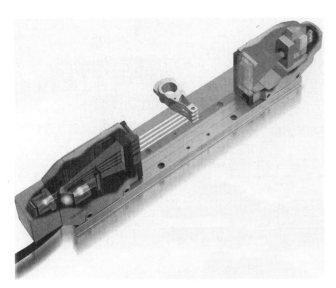

■ **Figure 94-6** The optical micrometer uses a high-intensity LED light source and the camera provides the image on the viewer. (*Keyence Corp.*)

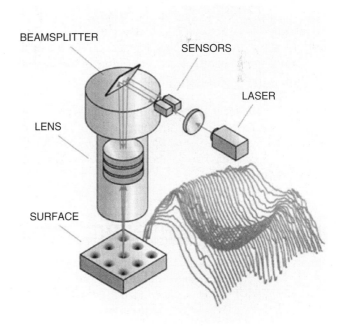

■ **Figure 94-8** Laser surface-finish measurement permits precise, high quality measurement and verification of contours and surface roughness. (*Optical Gaging Products, Inc.*)

LASER SURFACE-FINISH MEASUREMENT

Laser surface-finish measurement (Fig. 94-8) permits precise, high quality measurement and verification of contours and surface roughness from a profile. The hardware is based on the laser interferometric probe system with a resolution of 5 nm over the entire probing range. Up to 150,000 measuring points can be captured at speeds between 0.02 mm/s and 2 mm/s. State-of-the-art technologies and materials eliminate environmental influences such as temperature fluctuations. The software compensates for any deviations of the probe arm geometry and the probe tip radius.

▶▶ Scanning Probes

Many industrial applications require precise and accurate measurement of free-form objects in a wide range of settings. These measurements can be used for design, assembly of prototypes, reverse engineering, quality control, production assembly, measurement on assembly lines, and quality assurance.

3D scanning measuring systems collect large amounts of data in a very short time, then analyze and virtually display the data. Scanning probes read hundreds of points in the time that it takes a touch-trigger probe to probe a few points. When measuring, high-speed scanning can produce hundreds of measuring points in seconds. A 30-second scan of a diameter milled on a CNC machining center provides data on about 3,000 measured points. Technology is now available to allow for 3D noncontact measurement on the shop floor that is unaffected by vibration and industrial lighting conditions.

Scanning probes (Fig. 94-9) are becoming more popular because this technology allows the scanning machine to be used outside of a clean, environment-controlled room. This allows it to be used on the shop floor, where measurements and corrections can be made while the workpiece is still in the machine.

▶▶ Reverse Engineering

Reverse engineering is the ability to reproduce parts direction from samples. Instead of having to digitize the workpiece, parts can be scanned using scanning CMMs to probe the surface or the contour of the part to be copied. High-speed scanning can provide hundreds of data points, allowing for a better quality part to be reproduced.

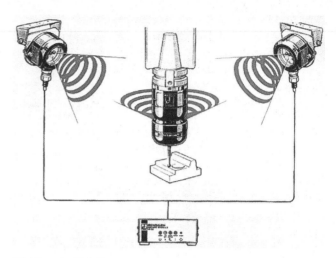

■ **Figure 94-9** Optical probes measure using a 360° signal, allowing part inspection while still in the machine.

▶▶ Video Measuring

The video-based imaging system (Fig. 94-10) is a powerful measuring tool providing the user with on-line documentation and archiving, repeatability, image processing, and manipulation.

Video systems have other advantages as well:

> A fully automated video-measuring system with a computerized vision system will locate randomly oriented parts.

> The system will adapt to the varying orientations, take the required measurements, and compare them with preset tolerances for evaluation.

■ **Figure 94-10** Video measuring systems provide fast, accurate optical inspection. *(Optical Gaging Products, Inc.)*

- The system is used for measuring, inspection, R&D, teaching, and documentation.
- The user is looking at a monitor, making these systems extremely comfortable for use over long periods of time.
- Graphical part display is provided for measuring sequences.

VIDEO MICROSCOPE

As video microscopes (Fig. 94-11) become more advanced, they are starting to replace traditional optical microscopes. Some video microscopes allow the image to be magnified onto large monitors or projected, making them an effective teaching tool for large groups. A microscope, equipped with a video output device, can easily record and permanently store the digital images for future use. Engineers working at two separate locations can both view the workpiece image at the same time. Although video microscopes make viewing easier for the user, they are limited by the magnification of current video systems. Video systems cannot produce an image that is as clear as that of binocular optics.

As microscope technology continues to increase, video-based microscope measuring systems are quickly becoming the standard for industry, teaching, research, and many other applications.

VIDEO MAGNIFIERS

Video magnifiers are unique in the industry for their ease of use and versatility when viewing magnified images of parts or components. Video magnifiers (Fig. 94-12) provide a means of electronically enhanced optical-image magnification for small parts with three-dimensional characteristics, such as components mounted on a circuit

■ **Figure 94-11** Video microscopes are rapidly becoming the industry standard. *(Keyence Corp.)*

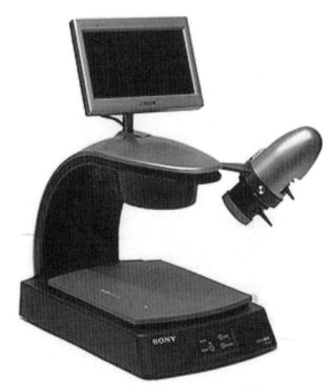

■ **Figure 94-12** Video magnifiers provide quick, easy magnification for inspection, training, and documentation. *(Sony Corp.)*

board. Magnification range is between 4X and 50X (or higher) in black and white or in color.

Video magnifiers can be used for a wide range of applications:

- In-process inspection
- Quality control
- Examination of engineering prototypes
- ISO training programs

Video magnifiers have several advantages:

- Monitor screens provide ergonomic benefits.
- More than one person can view the component.
- They reduce inspection time.

▸▸ Scanning CMMs

Scanning is a simple way of automatically collecting a large number of data location points to accurately measure or define the shape of an object or a workpiece. The coordinate measuring machine is as vital to inspection and measurement as CNC machine tools are to the manufacturing process. CMMs equipped with scanning heads (Fig. 94-13 on p. 830)

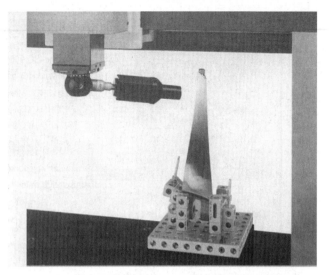

■ Figure 94-13 Scanning technology being used to check the dimensional accuracy of a part.

can automatically scan parts for inspection, for digitizing, or for creating a graphical representation of a part.

▶▶ Inspection for the Future

Process control simplification, inspection, and quality control are important in any competitive business. The trend toward nano-technology products that are smaller, lighter, and simpler is forcing more manufacturers to inspect parts that are too fragile to measure with conventional mechanical contact equipment. In such cases, measurement and inspection require a system that can accommodate the workpieces, while providing a quick and accurate cycle time. Noncontact scanning will help provide the ability to perform these operations. When purchasing a measuring system, it is important that not only present but also future requirements be considered.

unit 94 review questions

1. In two-dimensional (2D) probing head systems, what are the two modes of operation for the measurement and inspection of parts?

2. Identify three types of laser measurement.

3. Why are scanning probes becoming more popular?

4. Give three advantages of video-measuring systems.

5. List three applications for video magnifiers.

6. State three advantages of video magnifiers.

Electrical Discharge Machining

OBJECTIVES

After completing this unit, you should be able to:

1 Define electrical discharge machining and state its principle

2 Summarize the EDM process

3 Identify the advantages and the limitations of electrical discharge machining

4 Name the main operating systems of wire-cut electrical discharge machines

Electrical discharge machining, commonly known as EDM, is a process that is used to remove metal through the action of an electrical discharge of short duration and high current density between the tool or wire and the workpiece (Fig. 95-1a on p. 832). The EDM process can be compared to a miniature version of a lightning bolt striking a surface, creating a localized intense heat and melting away the work surface.

▶▶ Use of EDM

Electrical discharge machining die sinking or wire has proved valuable in the machining of supertough, electrically conductive materials, such as the new space-age alloys. These metals would have been difficult to machine by conventional methods. EDM has made it relatively simple to machine intricate shapes that would be impossible to produce with conventional cutting tools. It is used extensively in the plastics industry to produce cavities of almost any shape in the steel molds.

▶▶ Principle of EDM

Electrical discharge machining is a controlled metal-removal technique whereby an electric spark is used to cut (erode) the workpiece, which takes a shape opposite to that of the cutting tool or electrode (Fig. 95-1b). The *cutting tool (electrode)* is made from electrically conductive

material, usually carbon. The die sinking electrode, made to the shape of the cavity required, and the workpiece are both submerged in a *dielectric fluid* (a light lubricating oil). This dielectric fluid should be a nonconductor (or poor conductor) of electricity. A *servo mechanism* maintains a gap of about .0005 to .001 in. (0.01 to 0.02 mm) between the electrode and the work, preventing them from coming into contact with each other. A *direct current* of low voltage and high amperage is delivered to the electrode at the rate of approximately 20,000 hertz (Hz). These electrical energy impulses vaporize the oil at this point. This permits the spark to jump the gap between the electrode and the workpiece through the dielectric fluid (Fig. 95-2 on p. 832). Intense heat is created in the localized area of the spark impact; the metal melts and a small particle of molten metal is expelled from the surface of the workpiece. The dielectric fluid, which is constantly being circulated, carries away the eroded particles of metal during the off-cycle of the pulse and assists in dissipating the heat caused by the spark.

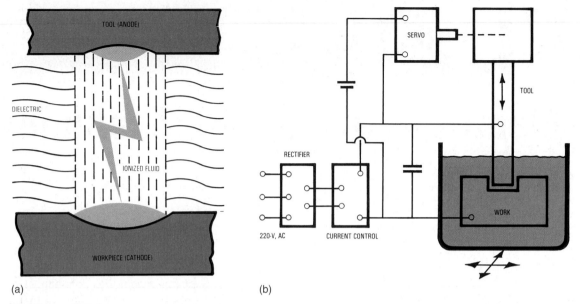

(a)   (b)

Figure 95-1 (a) A controlled spark removes metal during electrical discharge machining (EDM); (b) basic elements of an electrical discharge system.

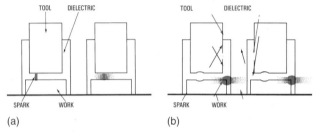

(a)   (b)

Figure 95-2 Stages of single spark.

▶▶ Types of EDM Circuits

Several types of electrical discharge power supply have been used for EDM. Although there are many differences among them, each type is used for the same basic purpose, that is, the precise, economical removal of metal by electrical spark erosion.

The two most common types of power supplies are the:

1. Resistance-capacitance power supply
2. Pulse-type power supply

Resistance-capacitance power supply, also known as the *relaxation-type power supply,* was widely used on the first EDM machines. It is still the power supply used on many of the machines of foreign manufacture.

The capacitor is charged through a resistance from a direct-current voltage source that is generally fixed. As soon as the voltage across the capacitor reaches the breakdown value of the dielectric fluid in the gap, a spark occurs. A relatively high voltage (125 V), high capacitance of over 100 μF (microfarads) for roughing cuts, low spark frequency, and high amperage are characteristics of the resistance-capacitance power supply.

In resistance-capacitance circuits, an increase in metal-removal rates depends more on larger amperage and capacitance than on increasing the number of discharges per second. The combination of low frequency, high voltage, high capacitance, and high amperage results in:

1. A rather coarse surface finish
2. Large overcut around the electrode (tool)
3. Larger metal particles being removed and more space being required to flush out particles

The advantages of the resistance-capacitance power supply are:

> The circuit is simple and reliable.

> It works well at low amperages, especially with milliampere currents required for holes under .005 in. (0.12 mm) in diameter.

Pulse-type power supply is similar to the resistance-capacitance type; however, vacuum tubes or solid-state devices are used to achieve an extremely fast pulsing switch effect. The pulse width and intervals may also be accurately controlled by switching devices. The switching is extremely fast and the discharges per second are 10 or more times greater than with the resistance-capacitance power supply at low frequencies. The results of more discharges per second are illustrated in Fig. 95-3. With more discharges per second, and using the same current (10 A), it is clear that smaller craters are created, producing a finer surface finish while still maintaining the same metal-removal rate.

Pulse-type power supply circuits are usually operated on low voltages (70 to 80 V), high frequency (sparks at the rate of 260,000 Hz), low capacitance (50 μF or less), and low-energy spark levels.

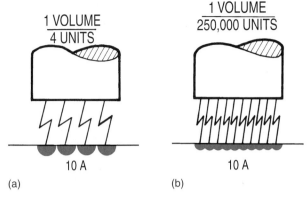

Figure 95-3 Effects on cratering and surface finish using various frequencies of spark discharge.

The main advantages of the pulse-type circuit are:

> It is extremely versatile and can be accurately controlled for roughing and finishing cuts.

> Better surface finish is produced as less metal is removed per spark, since there are many sparks per unit of time.

> There is less overcut around the electrode (tool).

▶▶ The Electrode

The *electrode* in die sinking EDM is formed to the shape of the cavity desired. As in conventional machining, some materials have better cutting and wearing qualities than others. Electrode materials must, therefore, have the following characteristics:

1. Be good conductors of electricity and heat
2. Be easily machined to shape at a reasonable cost
3. Produce efficient metal removal from the workpiece
4. Resist deformation during the erosion process
5. Exhibit low electrode (tool) wear rates

The more common electrode materials are graphite, copper, copper graphite, copper tungsten, brass, and steel. None of these electrode materials has general-purpose application; each machining operation dictates the selection of the electrode material. *Yellow brass* has been used primarily as electrode material for pulse-type circuits because of its good machinability, electrical conductivity, and relatively low cost. *Copper* produces better results in the resistance-capacitance circuits where higher voltages are employed.

High-density and high-purity carbon, or *graphite,* is an electrode material that is gaining wide acceptance. It is commercially available in various shapes and sizes, is relatively inexpensive, can be machined easily, and makes an excellent electrode. Its tool wear rate is much less and its high metal-removal rate is almost double that of any other electrode material.

▶▶ The EDM Process

The use of EDM is increasing as more and more applications are found for this process. As important technological advances in equipment and application techniques become available, more industries are adopting the process.

SERVO MECHANISM

It is important that there be no physical contact between the electrode (tool) and the workpiece; otherwise, arcing will occur, causing damage to both the electrode and the workpiece. Electrical discharge machines are equipped with a servo control mechanism that automatically maintains a constant gap of approximately .0005 to .001 in. (0.01 to 0.02 mm) between the electrode and the workpiece. The mechanism also advances the tool into the workpiece as the operation progresses, and it senses and corrects any shorted condition by rapidly retracting and returning the tool. If the gap is too large, ionization of the dielectric fluid does not occur and machining cannot take place. If the gap is too small, the tool and workpiece may weld together.

When chips in the spark gap reduce the voltage below a critical level, the servo mechanism causes the tool to withdraw until the chips are flushed out by the dielectric fluid. The servo system should not be too sensitive to "short-lived" voltages caused by chips being flushed out; otherwise, the tool would be constantly retracting, thereby seriously affecting machining rates.

Servo feed control mechanisms can be used to control the vertical movement of the electrode (tool) for sinking cavities. It can also be applied to the table of the machine for work requiring horizontal movement of the electrode (tool).

CUTTING CURRENT (AMPERAGE)

The EDM power supply provides the direct current electrical energy for the electrical discharges that occur between the tool and the workpiece. As the pulse-type power supply is the more commonly used in North America, only the characteristics of this type will be listed.

Characteristics of Pulse-Type Circuits

1. Low voltages (normally about 70 V, which drops to about 20 V after the spark is initiated)
2. Low capacitance (about 50 μF or less)
3. High frequencies (usually 20,000 to 30,000 Hz but may be as high as 260,000 Hz)
4. Low-energy spark levels

THE DISCHARGE PROCESS

Upon application of sufficient electrical energy between the electrode (cathode) and the workpiece (anode), the dielectric fluid changes into a gas, allowing a heavy discharge of current to flow through the ionized path and strike the workpiece. The energy of this discharge vaporizes and decomposes the dielectric fluid surrounding the column of electric conduction. As the conduction continues, the diameter of the discharge column expands and the current increases. The heat between the electrode and the work surface causes a small pool of molten metal to be formed on the work surface. When the current is stopped, usually only for microseconds, the molten metal particles solidify and are washed away by the dielectric fluid.

These electrical discharges occur at the rate of 20,000 to 30,000 Hz between the electrode and the workpiece. Each discharge removes a minute amount of metal. Since the voltage during discharge is constant, the amount of metal removed will be proportional to the amount of charge between the electrode and the work. For fast metal removal, high amounts of current should be delivered as quickly as possible to melt the maximum amount of metal. This, however, produces large craters in the workpiece, resulting in a rough surface finish. To obtain smaller craters and therefore finer surface finishes, smaller charges of energy can be used. This results in slower stock removal rates. If the current is maintained but the frequency (number of hertz) is increased, this also results in smaller craters and better surface finish.

DIELECTRIC FLUID

The dielectric fluid used in the electrical discharge process has several main functions:

1. Serves as an insulator between the tool and the workpiece until the required voltage is reached

2. Vaporizes (ionizes) to initiate the spark between the electrode and the workpiece

3. Confines the spark path to a narrow channel

4. Flushes away the metal particles to prevent shorting

5. Acts as a coolant for both the electrode and the workpiece

Types of Dielectrics

The types of fluids used as dielectrics must be able to ionize (vaporize) and deionize rapidly and have a low viscosity that will allow them to be pumped through the narrow machining gap. The most commonly used EDM fluids, proven to be satisfactory dielectrics, have been various petroleum products, such as light lubricating oils, transformer oils, silicon-base oils, and kerosene. These all perform reasonably well, especially with graphite electrodes, and are reasonable in cost.

The selection of the dielectric is important to the EDM process, since it affects the metal-removal rate and electrode wear. A fluid consisting of triethylene glycol, water, and monoethyl ether of ethylene glycol has been used in research with superior results, especially with metallic electrodes. It is quite likely that many new dielectric fluids will be developed to improve the EDM process.

Methods of Circulating Dielectrics

The dielectric fluid must be circulated under constant pressure if it is to flush away efficiently the metal particles and assist in the machining process. The pressure used generally begins with 5 psi (34 kPa) and is increased until optimum cutting is attained. Too much dielectric fluid will remove the chips before they can assist in the cutting action and thereby cause slower machining rates. Too little pressure will not remove the chips quickly enough and thereby cause short circuits.

Four methods are generally used to circulate the dielectric fluid. All must use fine filters in the system to remove the metal particles so that they are not recirculated.

Down Through the Electrode

A hole (or holes) is drilled through the electrode, and the dielectric fluid is forced through the electrode and between it and the workpiece (Fig. 95-4a). This rapidly flushes away the metal particles from the machining area. On cavities, a small standing slug or core remains and must be ground away after the machining operation has been completed.

Up Through the Workpiece

Another common method is to cause the fluid to be circulated up through the workpiece (Fig. 95-4b). This type of flushing is limited to through-hole cutting applications and to cavities having holes for core or ejector pins.

Vacuum Flow

A negative pressure (vacuum) is created in the gap, which causes the dielectric to flow through the normal .001-in. (0.02-mm) clearance between the electrode and the workpiece. (Fig. 95-4c). The flow can be either up through a hole in the electrode or down through a hole in the workpiece. Vacuum flow has several advantages over other methods; it improves machining efficiency, reduces smoke and fumes, and helps to reduce or eliminate taper in the workpiece.

Vibration

A pumping and sucking action is used to cause the dielectric to disperse the chips from the spark gap. (Fig. 95-

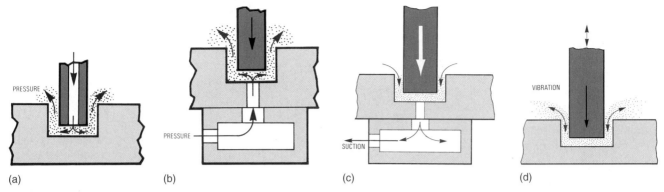

(a)   **(b)**   **(c)**   **(d)**

■ **Figure 95-4** (a) Circulating dielectric fluid down through the electrode; (b) circulating dielectric fluid up through the workpiece; (c) circulating dielectric fluid down through the workpiece by vacuum; (d) circulating dielectric fluid by vibration.

4d). The vibration method is especially valuable for very small holes, deep holes, or blind cavities where it would be impractical to use other methods.

METAL-REMOVAL RATES

Metal-removal rates for EDM are somewhat slower than with conventional machining methods.

The rate of metal removal is dependent on the following factors:

1. Amount of current in each discharge
2. Frequency of the discharge
3. Electrode material
4. Workpiece material
5. Dielectric flushing conditions

The normal metal-removal rate is approximately 1 in.^{3} (16 cm^3) of work material per hour for every 20 A of machining current. However, metal-removal rates of up to 15 in.^{3}/h (245 cm^3/h) are possible for roughing cuts with special power supplies.

ELECTRODE (TOOL) WEAR

During the discharge process, the electrode (tool), as well as the workpiece, is subject to wear or erosion. As a result, it is difficult to hold close tolerances as the tool gradually loses its shape during the machining operation. At times it is necessary to use many electrodes to produce a cavity of the required shape and tolerance. For through-hole operations, stepped electrodes are used to produce roughing and finishing cuts in one pass.

Fortunately, the rate at which the tool wears is considerably less than that of the workpiece. An *average wear ratio* of the workpiece to the electrode is 3:1 for metallic tools, such as copper, brass, and zinc alloys. With graphite electrodes, this wear ratio can be greatly improved to 10:1.

Reverse-polarity machining, a relatively new development, promises to be a major breakthrough in reducing electrode wear. With this method, molten metal from the workpiece is deposited on a graphite electrode about as fast as the electrode is worn away. Thus, minute electrode wear is continually being replaced by a deposit of the work material. Reverse-polarity machining operates best on low spark-discharge frequencies and high amperage. It improves the metal-removal rates and greatly reduces electrode wear.

OVERCUT

Overcut is the amount the cavity in the workpiece is cut larger than the size of the electrode used in the machining process. The distance between the surface of the work and the surface of the electrode (overcut) is equal to the length of the sparks discharged, which is constant over all areas of the electrode.

The amount of overcut in EDM ranges from .0002 to .007 in. (0.005 to 0.1 mm) and is dependent on the amount of gap voltage. As illustrated in Fig. 95-5, the overcut distance increases with the increased gap voltage. The amount of overcut is generally varied to suit the metal-removal rate and the surface finish required, which in turn determines the size of the chip removal.

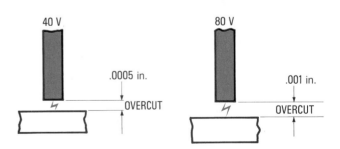

■ **Figure 95-5** Example of overcut produced by different voltages in EDM.

Most manufacturers of EDM machines provide overcut charts to show the amount of clearance produced with various currents. The charts make it possible to determine accurately the electrode size required to machine an opening to within .0001 in. (0.002 mm).

The size of the chips removed is an important factor in setting the amount of overcut because:

1. Chips in the space between the electrode and the work serve as conductors for the electrical discharges.

2. Large chips produced with higher amperages require a larger gap to enable them to be flushed out effectively.

SURFACE FINISH

In recent years, major advances have been made with regard to the surface finishes that can be produced. With low metal-removal rates, surface finishes of 2 to 4 μin. (0.05 to 0.10 μm) are possible. With high metal-removal rates [as much as 15 in.^{3}/h (245 cm^3/h)], finishes of 1000 μin. (24 μm) are produced.

The type of finish required determines the number of amperes that can be used and the capacitance, frequency, and voltage setting. For fast metal removal (roughing cuts), high amperage, low frequency, high capacitance, and minimum gap voltage are required. For slow metal removal (finish cut) and good surface finish, low amperage, high frequency, low capacitance, and the highest gap voltage are required.

ADVANTAGES OF EDM

Electrical discharge machining has many advantages over conventional machining processes, including the following:

> Any material that is electrically conductive can be cut, regardless of its hardness. EDM is especially valuable for cemented carbides and the new super-tough space-age alloys that are extremely difficult to cut by conventional means.

> Work can be machined in a hardened state, thereby overcoming the deformation caused by the hardening process.

> Broken taps or drills can readily be removed from workpieces.

> It does not create stresses in the work material, since the tool (electrode) never comes into contact with the work.

> The process is burr-free.

> Thin, fragile sections can be easily machined without deforming.

> Secondary finishing operations are generally eliminated for many types of work.

> The process is automatic in that the servo mechanism advances the electrode into the work as the metal is removed.

> One person can operate several EDM machines at one time.

> Intricate shapes, impossible to produce by conventional means, are cut out of a solid with relative ease (Fig. 95-6a and b).

> Better dies and molds can be produced at lower cost.

> A die punch can be used as the electrode to reproduce its shape in the matching die plate, complete with the necessary clearance.

(a)

(b)

■ **Figure 95-6** (a) Examples of work produced by EDM; (b) tool movement required to produce cavities of various shapes.

LIMITATIONS OF EDM

Electrical discharge machining has found many applications in the machine tool trade. However, it does have some limitations:

> Metal-removal rates are low.

> The material to be machined must be electrically conductive.

> Cavities produced are slightly tapered but can be controlled for most applications to as little as .0001 in. (0.002 mm) in every .250 in. (6 mm).

> Rapid electrode wear can become costly in some types of EDM equipment.

> Electrodes smaller than .003 in. (0.07 mm) in diameter are impractical.

> The work surface is damaged to a depth of .0002 in. (0.005 mm) but it is easily removed.

> A slight case hardening occurs. This, however, may be classed as an advantage in some instances.

▸▸ Wire-Cut EDM Machine

Electrical discharge machining has advanced quickly with the addition of computer numerical control (CNC). Today EDM is used for a wide variety of precision metalworking applications which would have been almost impossible just a few years ago. Cutting tolerances, cutting speeds, and surface finish quality have been greatly improved.

Another application of electrical discharge machining is the wire-cut EDM machine (Fig. 95-7). Unlike other EDM applications that use an electrode in the shape and size of the cavity or hole required, this machine generally uses a thin brass or copper wire as the electrode, making it possible to cut most shapes and contours from flat plate materials.

Wire-cut EDM can do things older technologies cannot do as well, as quickly, as inexpensively, and as accurately. Most parts can now be programmed and produced as a solid, rather than in sections and then assembled as a unit, as was necessary previously. Wire-cut EDM is capable of producing complex shapes such as tapers, involutes, parabolas, and ellipses (Fig. 95-8). This process is now commonly used to machine tungsten carbide, difficult-to-machine material, polycrystalline diamond, polycrystalline cubic boron nitride, and pure molybdenum.

THE PROCESS

The wire-cut EDM machine uses CNC to move the workpiece along the X and Y axes (backward, forward, and sideways) in a horizontal plane toward a vertically moving wire

■ **Figure 95-7** The wire-cut electrical discharge machine is used for machining complex forms. *(The Makino Co.)*

■ **Figure 95-8** Tapers, involutes, parabolas, ellipses, and many other shapes can be cut easily on wire-cut electrical discharge machines. *(The Makino Co.)*

■ Figure 95-9 The workpiece is moved along the *X* and *Y* axes by numerical control to cut the desired shape. *(The Makino Co.)*

electrode (Fig. 95-9). The wire electrode does not contact the workpiece but operates in a stream of dielectric fluid (usually deionized water), which is directed to the spark area between the work and the electrode. When the machine is in operation, the dielectric fluid in the spark area breaks down, forming a gas that permits the spark to jump between the workpiece and the electrode. The eroded material caused by the spark is then washed away by the dielectric fluid.

Movement of the wire is controlled continuously in minimum increments of .00001 in. (0.0002 mm) in two- to five-axis positions. Along with the *XY* axis, the head can be tilted up to 30° (*UV* axis) for cutting tapered sections and can be raised or lowered (*Z* axis) to suit various workpiece thicknesses. Any contour, within the size of the machine, can be cut very accurately. During the cutting action, an arc gap of .001 to .002 in. (0.02 to 0.05 mm) is maintained between the workpiece and the wire electrode.

OPERATING SYSTEMS

The four main operating systems, or components, of the wire-cut electrical discharge machines are the servo mechanism, the dielectric fluid, the electrode, and the machine control unit.

Servo Mechanism

The EDM servo mechanism controls the cutting current levels, the feed rate of the drive motors, and the travelling speed of the wire. The servo mechanism automatically maintains a constant gap of approximately .001 to .002 in.

(0.02 to 0.05 mm) between the wire and the workpiece. It is important that there be no physical contact between the wire (electrode) and the workpiece; otherwise, arcing will occur, which could damage the workpiece and break the wire. The servo mechanism also advances the workpiece into the wire as the operation progresses, senses the work-wire spacing, and slows or speeds up the drive motors, as required, to maintain the proper arc gap. If the gap is too large, the dielectric fluid between the wire (electrode) and the workpiece will not break down into a gas, a spark cannot be conducted between the wire and the workpiece, and therefore machining cannot take place. If the gap is too small, the wire will touch the workpiece, causing the wire to short and break.

Dielectric Fluid

One of the most important factors in a successful EDM operation is the removal of the particles (chips) from the working gap. Flushing these particles out of the gap with the dielectric fluid will produce good cutting conditions, while poor flushing will cause erratic cutting and poor machining conditions.

The dielectric fluid in the wire-cut EDM process is usually deionized water. This is tap water that is circulated through an ion-exchange resin. The deionized water makes a good insulator, while untreated water is a conductor and is not suitable for the electrical discharge machining process. The amount of deionization of the water determines its resistance. For most operations, the lower the resistance, the faster will be the cutting speed. However, the resistance of the dielectric fluid should be much higher when carbides and high-density graphites are being cut.

The dielectric fluid used in the wire-cut EDM process serves several functions:

1. It helps to initiate the spark between the wire (electrode) and the workpiece.

2. It serves as an insulator between the wire and the workpiece.

3. It flushes away the particles of disintegrated wire and workpiece to prevent shorting.

4. It acts as a coolant for both the wire and the workpiece.

Electrode

The electrode in wire-cut EDM may be a spool of brass, copper, tungsten, molybdenum, or zinc wire ranging from .002 to .012 in. (0.05 to 0.3 mm) in diameter and from 2 to 100 lb (0.90 to 45.36 kg) in weight. The electrode continuously travels from a supply spool to a takeup spool, so that new wire is always in the spark area. With this type of electrode, the wear on the wire does not affect the accuracy of the cut because new wire is being constantly fed past the workpiece at rates from a fraction of an inch to several inches per minute. Both the electrode wear and the material-removal rate from the workpiece depend on factors such as the material's electrical and thermal conductivity, its melting point, and the duration and intensity of the electrical pulses. Electrode materials must have the following characteristics:

1. Be a good conductor of electricity
2. Have a high melting point
3. Have a high tensile strength
4. Have a good thermal conductivity
5. Produce efficient metal removal from the workpiece

Machine Control Unit

The control unit for the wire-cut EDM can be separated into three individual operator panels:

> The control panel for setting cutting conditions (servo mechanism)

> The control panel for machine setup and the data required to produce the part (numerical control)

> The control panel for manual data input (MDI) and a cathode ray tube display

Although some of the newer wire-cut electrical discharge machines eliminate some of these controls and incorporate them as part of the machine's automatic cutting cycle, a knowledge of what is being controlled during the cutting cycle should give the operator a better overall understanding of the wire-cut machine and cutting process.

unit 95 review questions

1. Define electrical discharge machining (EDM).
2. Give three of the characteristics required of electrode materials used in die-sinking EDM.
3. List three of the main functions of the dielectric fluid used in the electrical discharge process.
4. Identify five of the advantages of electrical discharge machining over conventional machining processes.

5. What are four limitations of electrical discharge machining?
6. What are the four main operating systems of wire-cut electrical discharge machines?
7. Name three of the characteristics required of electrode materials used in wire-cut electrical discharge machines.

UNIT 96

Robotics

OBJECTIVES

After completing this unit, you will be able to:

1 Describe industrial robots and provide applications of their use

2 Identify the different types of robots, describe their basic principle, and list their main parts

3 Understand vision-based robotics

4 State the four modes of programming robots

5 List the advantages of robots

6 Recognize the safety precautions required when working with robots

Robots are a source of fascination for humans because of the ease with which they carry out work that is too difficult, monotonous, or hazardous for humans. From the first American industrial robot application in the automotive industry in the early 1960s, robots have improved tremendously and found thousands of other applications. They are used in all types of manufacturing, assembly, medicine, and so on. Robots provide thousands of hours of service without tiring, complaining, or breaking down in dangerous or boring environments that are not suitable for humans. Robots are good pickers and placers on automated assembly tasks, excellent welders and painters, and useful for performing many repetitive tasks. Robots were one of the prime factors in the growth of industrial production in the 1980s and 1990s.

Advances in robotic design have simplified the automation of complex manufacturing processes while finding many more applications outside the automotive field. Humanlike senses of sight, touch, and hearing have been added to allow robots to perform more complicated and sensitive tasks. Early robots were *taught* by leading them by hand through the operation to be performed and saving the program on a cassette tape. Today, off-line computers using 3D simulation software (Fig. 96-1) create robot programs.

▶▶ Robot Requirements

An industrial robot is a programmable, multi-functional manipulator designed to move material parts, tools, and devices through various programmed motions for the performance of a variety of tasks.

To be effective in a variety of applications, robots must meet the following criteria: they can be adaptable for many applications, reliable, easy to program, safe to operate, and capable of working in hazardous places. The rate of production of many of today's products would not have been possible without the use of robot technology.

Figure 96-1 Reducing up to 70% of programming time may be possible using off-line programming. *(Technomatix Technologies Inc.)*

▶▶ Application Flexibility

The most useful industrial robot would be capable of performing many different operations and tasks. The main manufacturing areas where robots find wide use are the following:

1. *Handling.* Robots for handling materials must be quipped with gripping fingers or vacuum cups to pick up or move the material.

   > They are used to load or unload machine tools, injection-molding machines, forging, and stamping presses and also for inspection or gaging equipment.
   > Robots can handle material (moving it from one place to another), retrieving parts from storage areas or conveyor systems, packaging, and palletizing.

2. *Processing.* Robots used for this purpose carry out operations such as metalizing, seam welding, spot welding, cleaning, and other operations where the robot can use tools to carry out a manufacturing operation.

3. *Assembly.* Robots find wide application in the assembly of a great number of components such as printed circuit boards. They perform operations such as riveting, drilling, reaming, and inserting and tightening fastening devices.

4. *Sealing and painting.* Operations such as spray painting, sealing automotive parts, and applying grease and adhesives are some of the areas where robots are widely used.

▶▶ Robot Types

There are many types of industrial robots used to perform a wide variety of tasks, and they all work on the same basic principal. The simplest type of robot has a single arm equipped with a set of *grippers (end-effectors)* that are used to perform an operation or to pick and move something.

> They must be able to reach all locations in the work area with speed and fluid motion.

> The robot system must be very accurate in following a programmed path and positioning, and it must be able to direct the robot through a series of directions in sequence.

> Versatility, high repeatability, ease of programming, compactness, and quick change-over capabilities are key requirements for robots.

▶▶ Parts of a Robot

The three main parts of a robot are the *arm* for using tools and moving parts, the *power supply* that provides electricity for the motors, and the *computer* that controls the actions of the robot.

Industrial robots are equipped with an arm (manipulator) that can move in 6 or more axes, including rotary motion to imitate the human arm (Fig. 96-2). Spray-painting robots may have up to 12-axis movements to allow the arm to be positioned to suit almost any curved surface. Most robots used in industry are electrically powered because they are fast, accurate, and easy to maintain. Electric robots find wide use because of low power consumption, smooth motion, and reliability.

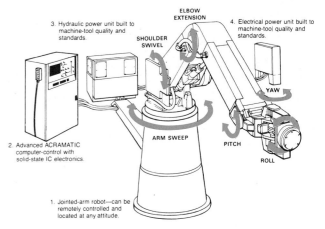

Figure 96-2 The robot arm may have six or more axes of movement. *(Cincinnati Machine, a UNOVA Co.)*

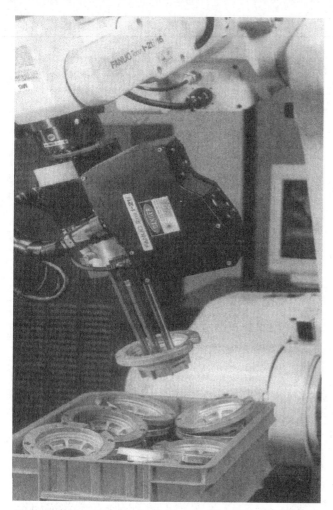

■ Figure 96-3 Tactile sensors mounted on a robot gripper are used during the assembly of parts. *(Manufacturing Engineering Magazine)*

▶▶ Vision-Based Robotics

Since the earliest days of robots, great things have been expected of them; some of those things have materialized but others have not. Vision-based assembly systems have progressed to the point where robots can be used to automate high-speed assembly work cells. Improvements in robotic photoelectric diodes play an important role in the robot's controllers, safety standards, and work-cell simulation systems, enabling robot manufacturers to assist their customers in getting products to the market faster. Solid-state, closed-circuit black and white video cameras are used in robot vision and, by a system of template matching, the images are processed by computer analysis. This analysis can also evaluate grasp orientation and automatically find variations in the part by comparing them with stored images in the computer memory.

INTELLIGENT VISION FOR ROBOTS

Robotics vision systems are a building block for the next generation of manufacturing systems that incorporate communications, flexible intelligent behavior, and increased productivity while reducing manufacturing costs. More than one-half of the human brain is involved with visual intelligence and the gathering of information. This visual information is essential to the flexible, creative, and productive processes of humans.

Adding intelligent vision systems to robots improves their present senses (mechanical, thermal, magnetic, acoustic, etc.), allowing them to have greater flexibility and to perform more complex tasks more quickly. Researchers are trying to duplicate human eye movements and are paying particular attention to how the human vision system processes information and makes predictions.

▶▶ Types of Sensors

Most robots use some type of vision or sensory system to collect data, to enable the control system to predict or make decisions, to locate and position components, and to control the operation of the robot.

For a robot to feel or fit parts properly, tactile (touch) sensors are located in the gripper and/or in combination with other sensors or a vision system in the wrist (Fig. 96-3). These are used to control the amount and direction of gripping pressure. Vision sensory systems enable the robot to identify the type and rotational position of the object. Force or torque sensors may also be located in the gripper or the robot's wrist to indicate the resistance force (weight) of the object being held and to adjust gripping pressure to avoid crushing the object.

VISION SYSTEM DEVELOPMENT

Human-style binocular vision systems will result in better robotic positioning and awareness of surroundings. These systems will provide sensory input and will reduce programming requirements because the robot will be able to make its own decisions based on clues (sensor feedback) about the surrounding environment. Vision systems that use smart cameras to mimic human vision are becoming reality.

Software under development will allow engineers to design motion control; vision processing; analog and digital input/output, including 3-D simulation; graphical programming; and Web-based access and control—all from one control center. This will result in low-cost programming and design platforms, widening the use of robotic systems in all phases of manufacturing.

▸▸ Artificial Intelligence

The speed at which industrial robots move is tied to the command and control languages and the software that works with the robot's sensory systems. All this activity is stored in the computer program for the robot.

There are four ways of programming robots: manual mode, teach mold, automatic mode, and off-line programming:

1. *Manual mode.* In the manual mode, a robot can be taught by leading the arm through the necessary movements by an operator. The robot learns from this experience, and the movements are stored in its memory to be used again in the future. Manual mode programming is suited for performing operations that are difficult to do or are difficult to explain or program.

2. *Teach mode.* The axes of the robot arm can be moved in a coordinated manner by the use of a portable teach pendant, (Fig. 96-4). Movement within the coordinate system is controlled by the pendant's positional pushbuttons for the operation and programming of the robot. The pushbuttons on the teach pendant make it very simple for the human operator to *teach* the robot to perform specific tasks within the coordinate system.

3. *Automatic mode.* Robotic movement is controlled by the computer program. Various types of control languages are used for programming robots, the most common being *VAL* (*versatile assembly language*) and *KAREL* (named after the person who introduced the term *robot*). VAL and KAREL generally cover range of motion, grippers, input, output, and so on.

4. *Off-line programming.* Newer controls allow the programming of the robot and the simulation of its movements on the computer screen (Fig. 96-5). The simulation and 3D visualization provide a check on the accuracy of the programming and, if necessary, an opportunity to change it to avoid costly crashes.

Although teaching by example may still be used, it will soon be replaced by Artificial Intelligence and vision systems. With the tremendous technological changes taking place, it will be difficult to know exactly to what extent robots will be used in the future.

▸▸ Next-Generation Manufacturing

Robotics vision systems are a building block for the so-called next-generation manufacturing systems, incorporating open

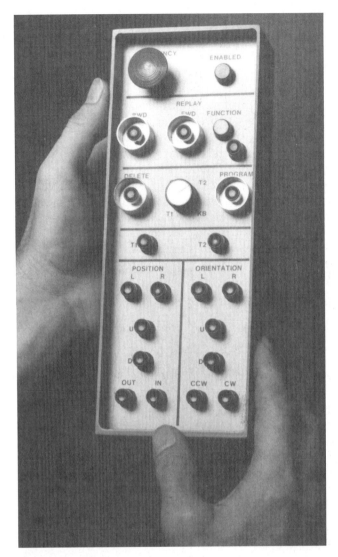

■ **Figure 96-4** The portable teach pendant makes it easy to teach a robot various tasks. *(Cincinnati Machine, a UNOVA Co.)*

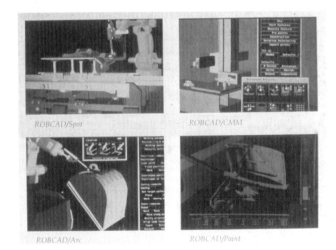

■ **Figure 96-5** Some controls allow the programming of a robot and the simulation of its movements on the computer screen. *(Technomatix Technology, Inc.)*

communications, flexible adaptive intelligent behavior, and benefits spanning productivity, cost, and human safety.

INTELLIGENT VISION

More than 50% of the human brain is used for visual intelligence. This visual information supports one of the most flexible, creative and productive species in the known universe, human beings.

In a similar way, adding vision systems to robotics solutions enhances the senses (to the other mechanical, thermal, electrical, magnetic and acoustic sensors), allowing more complex tasks to be accomplished more quickly, with more flexibility in the usually dirty and dangerous application environments.

ROBOTICS AND VISION WORKING TOGETHER

The most common applications uses for industrial robotics since the 1960s have been described by the three *D*s— (dull, dirty, and dangerous) or three *H*s (hot, heavy, and hazardous)—applications that improve productivity, increase quality, and reduce costs while protecting workers and improving safety. Most robotic systems have a multi-linked manipulator and an end-effector to hold tools for welding, painting, gripping, and measuring, along with a control system with feedback devices. Applications include materials handling, spot welding, machining operations, application of adhesives and sealants, spray painting, automated assembly, and inspection.

BENEFITS OF NEXT-GENERATION SYSTEMS

Robotics and improved vision systems will be the core of the next-generation manufacturing systems that will have a direct and positive impact on manufacturing, but manufacturers still need to justify the cost of a robotics vision system on the basis of a number of factors:

> Whether or not 100% inspection Statistical Process Control (SPC) is required

> Number of parts produced per month

> Number of production lines/shifts/operators

> Current manual inspection time per piece and labor rate per inspector/supervisor

> Number of rejects per month, including the value of a reject/rework and rate/materials

> Monthly warranty/liability costs, scrap/waste, inventory costs, and monthly units sales and price (also profit after cost of money)

When training savings, increased quality, customer satisfaction, uptime, safety, and monotony are considered, many robotics vision systems have a payback of less than one year.

▸▸ Advantages of Robots

The computer-controlled industrial robot has many advantages that make it a very important industrial manufacturing tool:

1. The computer controls the position, orientation, velocity, and acceleration of the path of the tool center point (TCP) at all times.

2. The robot can easily be taught because it can be moved to different coordinate locations with the teach pendant. The console display provides communication between the teacher and the control.

3. The computer can store system software and taught data in its memory. This allows editing of the taught program at any time, especially when changes are necessary in the program to suit a different application.

4. The program in the computer's memory can be easily expanded to add more taught points or extra functions.

5. The robot can be interfaced with other equipment or machine tools, and it can receive and interpret signals from various sensor devices. It can also provide valuable data to a supervisory or executive computer on manufacturing conditions, quantities, and problems.

6. The system can be easily changed to suit production changes. The robot computer can be linked to a supervisory computer that can direct the robot to convert to other programmed operations as required.

7. The computer is able to monitor the robot and the equipment in service, and it displays diagnostic messages on the console screen to indicate various error conditions. If necessary, the computer can direct the robot to stop an operation.

▸▸ Safety Precautions

Robots, like other pieces of machinery, must be handled carefully in order to prevent accidents. Always observe the basic safety rules used with other machines, as well as the following that apply specifically to working with robots:

1. The robot working area should be enclosed by a barrier to prevent people from entering while the

robot is working. Entrance gates to the enclosure should have controls that automatically stop the robot if anyone enters accidentally.

2. Emergency stop buttons, which stop the robot and cut off the power, should be available outside the working range of the robot.

3. Robots should be programmed, serviced, and operated only by fully trained workers who understand the robot's action.

4. Extreme care should be used during the programming cycle, especially if it is necessary to have humans within the robot's working area.

5. The work done by a robot should not be a health or safety hazard to human workers in nearby areas.

6. Hydraulic and electrical cables should be place so that they cannot be damaged by the operation of the robot or any related equipment.

unit 96 review questions

1. What is an industrial robot and what criteria does it need to meet to be effective?

2. State the four main manufacturing areas where robots find wide use.

3. What are the three main parts of a robot?

4. What are the four modes of programming robots?

5. List four advantages of robots.

6. Give three safety precautions that should be followed when working with robots.

Manufacturing Intelligence:
Can a Company Survive Without Real-Time Knowledge?

OBJECTIVES

After completing this unit, you will be able to:

1 Identify the main factors involved in manufacturing intelligence

2 Provide the main components of the Cyber Management System

3 Summarize the operation of the Cyber Management System

4 Explain Visual Plant

5 Describe PolyCAPP and list its advantages

Manufacturing is undergoing a revolution that is fueled by a global economy that requires greater speed, better accuracy, and lower manufacturing costs. All companies are facing the demands of trying to meet this challenge and exceed the pace of these demands. There is an almost frantic search for methods of improving productivity and product quality and reducing manufacturing costs. Progressive companies are embracing the Web; open, network-based intelligence tools; and new manufacturing technologies to convert their companies into the next generation of manufacturing (e-Manufacturing).

The perfect factory is one that uses real-time* manufacturing intelligence™ to achieve the best performance with a minimum of unscheduled downtime (Fig. 97-1). As market leaders, these model plants are highly efficient, productive, and profitable. By implementing new technology strategies that yield real-time manufacturing intelligence, manufacturers can achieve high levels of optimization and overall equipment effectiveness.

▸▸ Manufacturing Intelligence Factors

Manufacturing intelligence enables managers to improve the effectiveness and efficiency of the operations of their enterprise by providing them with easy access to key enterprise-wide information. Manufacturing intelligence provides measures to analyze, control, and manage operations performance and efficiency (Fig. 97-2).

INFORMATION SOURCES

Each machine tool should become an information node and be a valuable resource in the corporate information network. A direct connection to machine controls is used to continuously monitor, track, compare, and analyze production parameters. This enables manufacturers to

---

*Real-time refers to having access to production events, over a network, the moment they occur.

Advanced Manufacturing Technology

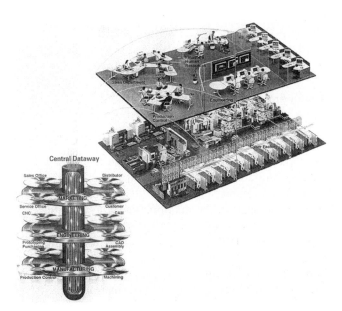

■ Figure 97-1 Real-time production parameters are important to survive in today's manufacturing world. *(Mazak Corp.)*

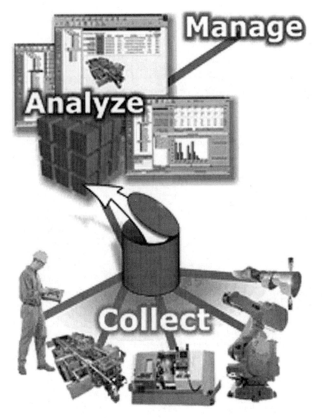

■ Figure 97-2 Manufacturing intelligence enables a company to improve the effectiveness and efficiency of its operations. *(Executive Manufacturing Technologies Inc.)*

uncover opportunities for improvement and to ensure that machines are operating within the defined expectations. Being able to view and analyze monitored production data from anywhere, at any time, is what provides real-time manufacturing intelligence.

Continuous control-system monitoring can save manufacturers millions of dollars. The ability to monitor dozens of process-line parameters, such as temperature and line speed, and alert plant staff to changing conditions can greatly reduce downtime. Continuous monitoring gives remote engineers a direct line of sight into process line activity, enabling them to solve production problems quickly and to perform predictive and preventive maintenance. Real-time manufacturing intelligence also enables manufacturers to focus on making process improvements during scheduled downtime.

TURNING DATA INTO KNOWLEDGE

Web-enabling process-line functions using continuous monitoring and data analysis are highly effective in increasing productivity and Overall Efficiency Effectiveness (OEE). These services allow plant personnel, supervisors, and off-site technical specialists to view and manage data directly from any Web-enabled PC. This provides them with a comprehensive view of process-line activity that is unprecedented in scope, detail, and accessibility. It also provides the manufacturer with the knowledge and means to gain a competitive advantage.

DECREASING DOWNTIME

Because downtime is one of the most costly conditions a manufacturer can experience, a proactive technical support program can generate significant cost savings. To diagnose problems effectively and make process improvements, remote engineers must have a good understanding of the manufacturing industries in which their customers specialize.

PROTECTING MANUFACTURING INVESTMENTS

Continuous-monitoring services can also generate cost savings by protecting existing investments. Regardless of their size, manufacturers face demands for performance, cost, and delivery:

> Shop Floor Manufacturing (SFM) is a Web-based solution that bridges the gap between traditional Manufacturing Execution Systems (MES) and the systems that manage business and supply activities across multiple facilities.

Manufacturing Intelligence: Can a Company Survive Without Real-Time Knowledge?   **847**

> Manufacturing intelligence delivers better information for decision making and a framework for continuous process improvements for manufacturers.

▶▶ Management Intelligence Systems

Many companies offer intelligence improvement systems that are designed to suit certain aspects of a manufacturing operation. The following will briefly cover three systems that seem to offer good management intelligence solutions for an entire factory or a part of the manufacturing process. They are the *Cyber Management System* by the Mazak Corporation (www.mazak.com); the *VisualPlant* by Executive Manufacturing Technologies, Inc. (www.visualplant.com); and *PolyCAPP* software from POLYPLAN Technologies (www.polycapp.com).

MAZAK CYBER MANAGEMENT SYSTEM

The Cyber Management System (Fig. 97-3) provides a comprehensive information-gathering system that starts with the design of a product, continues through every stage of the manufacturing process, and includes customer service. The information assembled through the *Central Dataway* is available to everyone involved in the manufacturing process, such as marketing, engineering, and manufacturing. It also includes everyone who plays a part in the manufacturing process, such as sales, service, suppliers, and distributors. With this type of information available, it is possible to manage a factory in real time by providing accessibility to machine, manufacturing, and tool data; production schedules; and many other types of information required to run a first-class manufacturing operation.

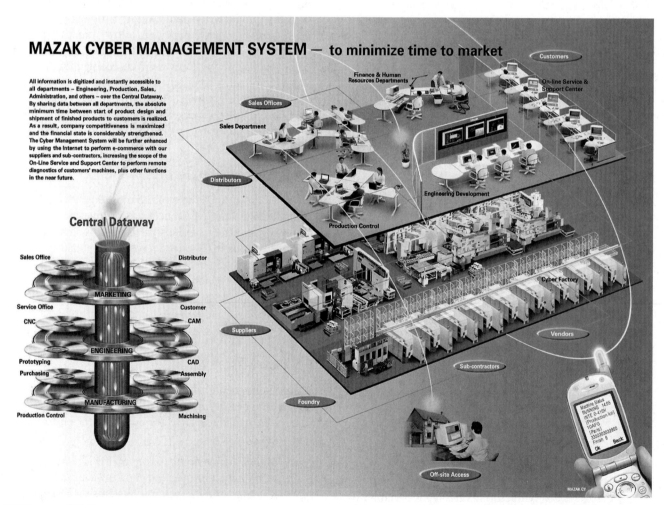

■ Figure 97-3 The Cyber Management System provides industry with a comprehensive information-gathering system. *(Mazak Corporation)*

Some of the main components of the Cyber Management System include:

1. *CAMWARE* software can easily create machining programs on a Windows personal computer. CAD files can be imported and their data used to reduce the time required to prepare machining programs. Other modules of the Cyber Production Center share the data contained in programs generated by CAMWARE, such as tooling list and cycle times.

2. The *Cyber Scheduler* is used for the on-line collection of shop floor data and production schedule monitoring. It displays current workloads on all machines and the estimated completion time when new jobs are added.

   > The Scheduler has the ability to respond quickly to an unplanned schedule change to maximize all the shop equipment.
   > This advanced DNC system and its shop floor data collection provides a comprehensive system that is easy to operate.

3. *Cyber Tool Management* provides for on-line tool status monitoring, tool life management, and the required tooling list. This system eliminates possible mistakes in tool loading and greatly reduces the time for inputting tool data. It also includes a tool identification system and tool life monitoring.

   > The Tool ID chip is mounted in the retention stud of each tool.
   > The required tool data is tied to each tool and registered in the factory network, allowing tool data management even when a tool is unloaded from a machine.
   > The tools, their machining time for each workpiece, plus the number of workpieces in each production lot is available over the Cyber Production Center network.
   > The Cyber Tool Management monitors the tool data status for each machine's tool magazine and then determines the required tools for each upcoming production lot.
   > This data is then accessed in the toolroom so that all the required tools are prepared in order to meet the production schedule.

4. The *Cyber Monitor* shows the real-time operation status of the networked machines, whether they are in the operation, alarm, or idle mode. It also provides status information such as current work number and part counts. The Cyber Monitor also provides information to the production scheduler and maintenance personnel, so the highest machine time efficiency is possible.

5. The Cyber Production Center network allows management to operate a factory in real time, providing accessibility to machine data, machining programs, fixture data, tool data, production schedules, and other data. The same data is available over a company network to sales, finance, engineering, production control, suppliers, and distributors, greatly increasing the efficiency and productivity of a factory.

VISUALPLANT

An increasingly competitive environment, tolerating nothing short of manufacturing excellence, is driving the requirements for plant and enterprise-wide decision-making tools. Information has evolved into the most potent business force on the planet. However gathering information is not enough, especially in manufacturing facilities where daily operations are generally remote from management decision makers. These people need accessible real-time information about what is happening on the plant floor. When manufacturing intelligence is communicated instantly to all points along the supply chain, it stimulates fast response to operational challenges. It becomes a powerful business tool—one that can bring a razor-sharp edge to competitive advantage.

VisualPlant, a mission-critical manufacturing software platform, is designed to collect real-time data from all plant floor machines and to provide the tools to visualize, analyze, and report on it in a Web browser (Fig. 97-4). It is the analytical layer on top of the plant floor, providing true operational visibility and enabling a fact-based decision-making process.

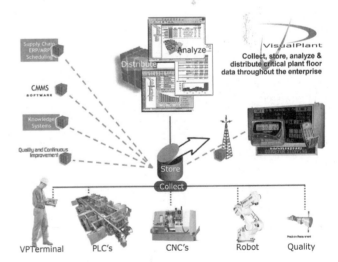

■ **Figure 97-4** VisualPlant, a mission-critical manufacturing software platform, is designed to collect real-time data from all plant operations. *(Executive Manufacturing Technologies, Inc.)*

Improving Productivity

VisualPlant contains a complete suite of advanced analytical tools that deliver complex data from across the plant floor to decision makers to help them make timely, accurate decisions. It can greatly improve production and reduce manufacturing costs at many levels throughout a company.

Manufacturing Intelligence Requirements

A manufacturing software solution must act as a manufacturing intelligence platform and a bi-directional pipeline between the plant floor and various business systems. It must:

> Collect, visualize, analyze, and report on any plant floor data in real time.

> Be mission-critical, guaranteeing data integrity along every step of the way, and designed to be running 24/7/365.

> Interface simply and seamlessly with the devices on the plant floor and with the business systems that benefit from the information it provides.

Achieving Operational Visibility

Even though much money has been spent over the past decade to improve quality, increase efficiencies, and reduce costs, technology has virtually by-passed the plant floor. Effectively connecting business systems to the plant floor, focusing on sources of downtime in the plant, identifying and removing constraints, and improving product quality are challenges for which most manufacturing plants require a solution.

The Manufacturing Landscape

Most manufacturers' spending over the past five years has not changed productivity in any great way.

> Many have been disappointed by Enterprise Resource Planning (ERP) implementations that have not delivered the promised manufacturing analysis tools.

> Supply Chain Management (SCM) implementations have not delivered the value that was anticipated.

> While Six Sigma, Lean Manufacturing, Continuous Improvement and others were designed to drive out inefficiencies and improve quality, their value seems to have come in the early stages of their activity.

The Challenge

Most attempts to improve productivity and plant floor efficiency have suffered from the availability, timelessness, and accuracy of the plant floor data. Data plays the following roles:

1. Most ERP systems depend upon manually entered data for their connection to the plant floor. The time delay between the occurrence of an event causes most manufacturers to limit the amount and type of data to strategic or problematic areas.

2. Among the aspects of Shop Floor Management (SFM) related to the plant floor, there is a critical dependence on manually entered data for event monitoring and management. It becomes a difficult, and often impossible, task to build a forward-looking model of an operation with only a limited and delayed view of the past.

3. Current schedules are created based on large numbers of assumptions and calculations that include the actual availability of all required assets, resources, and machines.

   > The actual (real) time required to complete a defined volume of production, scrap percentages, change-over, or setup times is of tremendous value to show exactly what is being produced.

4. Most predictive and preventive maintenance is done based on time and the manual total of downtime sheets once a shift or once a day. By the time this data is aggregated and reported, it is often too late to react. Critical to effective maintenance is clear and dynamic information about how every asset is performing.

5. Knowing how much is being produced and scrapped, where the plant problems are, how much inventory is tied up on the plant floor, and the top sources of downtime are all critical to a plant's decision-making process.

6. The same issues and problems that exist at the plant level exist at the enterprise level. A manufacturer may be producing the same or similar products in multiple plants, and it is necessary to compare production statistics and efficiencies between plants.

POLYCAPP

PolyCAPP® is a Manufacturing Process Management (MPM) desktop solution that captures and promotes the reuse of manufacturing to everyone involved in the development and execution of the manufacturing process. MPM allows manufacturers and their suppliers to develop and optimize manufacturing strategies that define the manufacturing steps and processes for each product. This is based on its design, its product specifications, and the availability of physical resources and human experience.

The Challenges

To be successful in a changing environment, dynamic and competitive manufacturers must be able to quickly adapt their manufacturing processes to:

> Promote innovation in the design and manufacturing of a product

> Make the product-development cycle more agile and flexible for shorter time to market

> Integrate product-development processes with other corporate sectors, as well as with external partners

Shortening the Product-Development Cycle

To get new products to the market quickly requires not only designing products but also defining and improving manufacturing processes or methods in parallel. A total product-development cycle specifies what, how, and when to produce.

The PolyCAPP MPM software solution allows design, manufacturing, and process engineering users to collaborate in *real-time* in a Design-For-Manufacturability environment, through the life cycle, the company, and the supply chain (Fig. 97-5). It uses 3D/2D information from the CAD/Product Data Management (PDM) environment to enable multi-user/multi-site/multi-company interactive-secured collaboration for the creation and sharing of geometry, specifications, manufacturing processes, resources, and documentation, resulting in reduced time to market.

This software can also integrate with any CAD/PDM and Enterprise Resource Planning (ERP)/ Manufacturing Execution Systems (MES) system, to en-

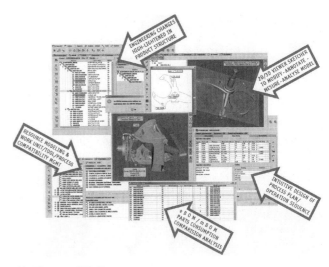

■ **Figure 97-5** The PolyCAPP software allows design, manufacturing, and process engineering users to collaborate in real time. *(POLYPLAN Technologies)*

sure that the information created at the beginning of the design cycle flows through all processes down to the shop floor with no re-entry of information. Even though some manufacturers feel they spend only 5% to 10% of their costs in the product design phase, once the final design has been approved, they have more likely committed 60% to 80% of their profitability. Therefore, it is important that there is close collaboration between design and process engineers before the final design approval.

PolyCAPP Advantages

For manufacturers and component suppliers, the results of implementing PolyCAPP are the following:

> Reduced costs and time to market for the product

> Shorter product development by improved communications

> Reduced ramp-up and production costs

> Improved product quality

> Increased engineering productivity and creativity

> Better management of manufacturing know how.

1. What are five main factors involved in manufacturing intelligence?

2. List three Management Intelligence Systems.

3. _____ is a Web-based solution that bridges the gap between traditional manufacturing execution systems (MES) and the systems that manage business and supply activities across multiple facilities.

4. What are three of the five main components of the Cyber Management System?

5. What is Visual Plant?

6. Give five of the six advantages to manufacturers of implementing PolyCAPP.

UNIT 98

Multi-Tasking Machines

OBJECTIVES

After completing this unit, you will be able to:

1 State the purposes of the Cyber Management System

2 Define Multi-Tasking Machines

3 List the advantages and describe the construction of Multi-Tasking Machines

4 Identify the accessories and optional equipment available for Multi-Tasking Machines

Manufacturing is undergoing a revolution fueled by a global economy that requires shorter lead time, better accuracy, and lower manufacturing costs. Thousands of companies are searching for methods of improving productivity, product quality and lower manufacturing costs. They are embracing the Web; open, network-based intelligence tools; multi-tasking machines, and new manufacturing technologies to convert their companies into the next generation of manufacturing (e-Manufacturing).

In today's rapid-paced changing technology, it is important to understand that the future is not based upon the past but, rather, on the ability to adapt to the continuously changing present. As Woodrow Wilson, the 28th U.S. president once said, "If you ever want to make enemies, just try not changing, as not changing will result in failure."

It is important to look at and update the entire process of part production through the use of the new technological tools that are available to any company in the world. Two important technologies worth considering are the Cyber Management System and multi-tasking machines.

▶▶ Cyber Management System

The Cyber Management System (Fig. 98-1 on p. 854) provides a comprehensive information-gathering system that starts with the design of a product, continues through every stage of the manufacturing process, and includes customer service. The information assembled through the Central Dataway is available to everyone involved in the manufacturing process such as marketing, engineering, and manufacturing. It also includes everyone who plays a part in the manufacturing process such as sales, service, suppliers, and distributors. With this type of information available, it is possible to manage a factory in real time by providing accessibility to machine, manufacturing, and tool data; production schedules; and many other types of information required to run a first-class manufacturing operation.

The Cyber Production Center network allows management to operate a factory in real time, providing accessibility to machine data, machining programs, fixture data, tool data, production schedules, and other data. The same data is available over a company network to sales and marketing, engineering, production and control, suppliers, and distributors, increasing the efficiency and productivity of a factory.

▶▶ Multi-Tasking Machining Defined

Multi-tasking is defined as the ability to perform various manufacturing operations without manual intervention. It also applies to the newest multi-tasking machine tools that can complete all the operations required to completely manufacture a part in a single machine setup. These include turning, milling, drilling, gear hobbing, and handling operations, such as automatic part transfer, and loading and unloading operations that may be required to complete a part (Fig. 98-2). Multi-tasking machines combine all the operations normally performed on two or three machines: the vertical and/or horizontal machining center and the turning center. Multi-tasking machines allow industries to change manufacturing processes in order to reduce multiple setups, throughput time, work-in-process and to improve productivity.

■ **Figure 98-1** The Cyber Production Center provides factory management support by providing real-time production data. *(Mazak Corporation)*

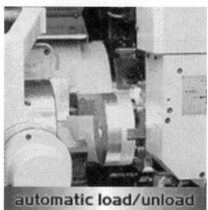

■ **Figure 98-2** Multi-tasking machines can perform many operations without human intervention. *(Mazak Corporation)*

Machine Virtually any Workpiece by the INTEGREX

Angle drilling
The milling spindle can be rotated to any required angle by positioning the B-axis. In addition to drilling holes at any angle, inclined face machining can be carried out in the same workpiece.

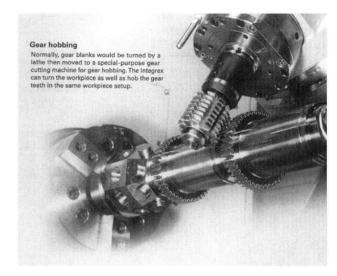

Gear hobbing
Normally, gear blanks would be turned by a lathe then moved to a special-purpose gear cutting machine for gear hobbing. The Integrex can turn the workpiece as well as hob the gear teeth in the same workpiece setup.

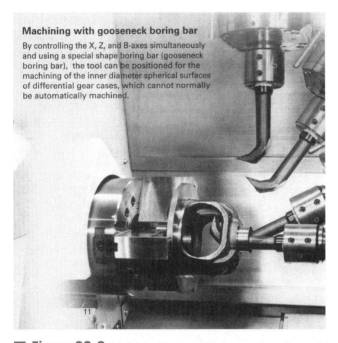

Machining with gooseneck boring bar

By controlling the X, Z, and B-axes simultaneously and using a special shape boring bar (gooseneck boring bar), the tool can be positioned for the machining of the inner diameter spherical surfaces of differential gear cases, which cannot normally be automatically machined.

Cylindrical grinding
By using the optional dressing unit, the grinding wheel can be dressed automatically.

■ **Figure 98-3** Multi-tasking machines can handle a wide variety of part designs, shapes, and sizes. *(Mazak Corporation)*

PURPOSE OF MULTI-TASKING MACHINES

The main purpose of multi-tasking machines is to reduce the number of machine tools required to manufacture a part to one. Their main advantage comes from being able to machine complex or special workpieces that normally require two or more machines to produce a finished part (Fig. 98-3). This can result in the following advantages to a manufacturer:

> Fewer machines are required, resulting in fewer tools, fewer setups, fewer operators, and less floor space required.

> Time is saved because parts do not have to be moved from machine to machine, reducing in-process inventory.

> Parts are more accurate, because there are fewer chances of setup errors because of the single setup fixtures.

> The overall cycle time, lead-time, and nonvalue-added time are greatly reduced.

> There is increased flexibility to handle many different part designs, shapes, and operations.

> The manufacturing process improves cash flows.

■ Figure 98-4 The Integrex Model 200-III is one of the most popular multi-tasking machines available. *(Mazak Corporation)*

TYPES OF MULTI-TASKING MACHINES

A wide variety of multi-tasking machines are available to suit the needs of any manufacturer. These can range from the basic single-spindle, single vertical turret machine to the most comprehensive model that contains twin turning spindles and twin machining turrets.

Since the concept of multi-tasking machines has common features with conventional machines, the Mazak Integrex Model 200-III will be used to discuss why these machines are bringing a production revolution to the factory floor (Fig. 98-4).

Construction

> First and second turning spindles are used for front and back operations.

> The lower turret, equipped with as many as nine tools, can be used for turning operations on work held in a chuck or bar and shaft workpieces.

> The upper turret, with a spindle motor, is equipped with 20 tools or as many as 80 and is used for milling-type operations. Its *B* axis can be positioned up to a range of 225° for flat or angular surface machining.

> The Fusion 640 MT Pro CNC Control 64-bit RISC processor provides complete shop floor communication and efficiency in real time.

Machining a Complex Part

The Integrex 200-III is very versatile and ideal for machining complex workpieces. It is a turning center that can perform secondary machining operations, and it is a powerful multi-tasking machine tool with great flexibility. Even workpeices that do not require turning operations can be machined efficiently:

1. The *first process* consists of gripping the part in the pre-bored chuck soft jaws of the left spindle (Fig. 98-5a).

   > All turning, milling, and drilling operations are performed on the part extending beyond the chuck jaws.

2. The next step in the sequence is automatically transferring the part from the first spindle to the second spindle (Fig. 98-5b).

   > The finished right hand of the part is gripped in the pre-bored jaws of the second spindle.

3. The *second process* consists of performing all turning, milling, and drilling operations on the unfinished left-hand end of the part (Fig. 98-5c).

   > All operations on a part can be completed in a single machine setup from material supply to the completed workpiece. This results in higher

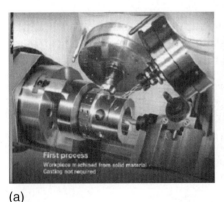

(a)

(b)

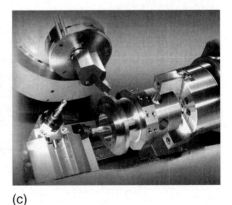

(c)

■ Figure 98-5 The stages a complex part goes through from the first operation to the final part: (a)   *(Mazak Corporation)*

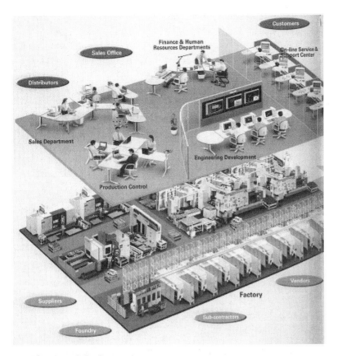

■ **Figure 98-6** The Cyber Production Module combines a wide range of Information Technology to manage an entire factory. *(Mazak Corporation)*

productivity due to the combining of all machining processes and the continuous operation from the main to the second spindle.

▶▶ Accessories

Many accessories and optional equipment are available to make multi-tasking machines a true combination of Information Technology (IT) and Manufacturing Technology (MT).

Cyber Production Module

The e-series multi-tasking machines are not just advanced machine tools but feature communication gears to be used for a machine operator as well as production management. These communication gears include programming, ordering of the required fixturing and tooling, notification of alarms, maintenance and scheduling, production control, generation of production status reports, and numerous other functions (Fig. 98-6).

Tool Data Management

The management of cutting tools is made possible by using Information Technology along with the power of the Mazatrol Fusion CNC. Tool data is tied to a *tool identification chip,* that is mounted in each tool's retention stud (Fig. 98-7). Once all tool data is registered, tool data is

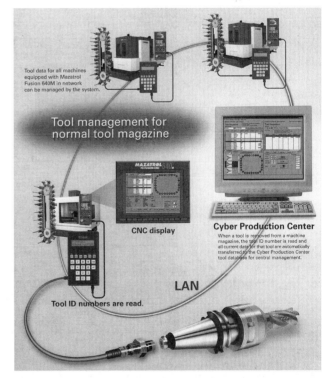

■ **Figure 98-7** The Tool Data Management System provides a comprehensive record of all cutting-tool data. *(Mazak Corporation)*

automatically downloaded over the network when the tool ID chip is read.

When a tool is removed from a machine magazine, the tool ID number is read and all current data for that tool is automatically transferred to the Cyber Production Center tool database for central management.

Flash Tool Holder

The Flash Tool Holder is a cutting tool that can perform 12 machining operations without a tool change. It is designed to complement the multi-tasking Integrex machine tools. This is a turning-center-based machine that uses a swivel head to perform fixed tool operations while rotating the part or rotary tool operations with the workpiece stationary.

Machine tools such as turning centers and machining centers normally use many tool changes in their operation. Reducing the number of tool changes is an efficient way of reducing cycle times and increasing productivity. The Flash Tool Holder uses a four-flute shank, with each flute containing a different insert (Fig. 98-8a). The Flash Tool Holder is designed to work with the indexing and angling capabilities of the turret spindle in the Integrex multi-tasking machine. Orienting the various inserts allows this single tool to perform up to 12 machining operations instead of using 12 different cutting tools (Fig. 98-8b).

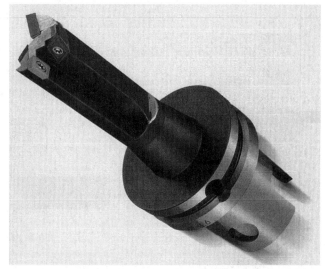
(a)

(b)

■ **Figure 98-8** The programmable Flash Tool Holder can perform the machining operations that would normally require 12 cutting tools. *(Mazak Corporation)*

unit 98 review questions

1. In a Cyber Management System, the _____ allows management to operate a factory in real time providing accessibility to machine data, machining programs, fixture data, tool data, production schedules, and other data.

2. What is the first process in machining a complex part?

3. What are four of the six advantages of using Multi-Tasking Machines?

4. List three accessories available for Multi-Tasking Machines.

5. In Tool Data Management, tool data is tied to a(n) _____ that is mounted in each tool's retention stud.

6. The _____ is a cutting tool that can perform 12 machining operations without a tool change.

Section 18

GLOSSARY
APPENDIX OF TABLES
INDEX

glossary

A

abrasion resistance The ability of a material to resist wear.

abrasive The material used in making grinding wheels or abrasive cloth; it may be either natural or artificial. The natural abrasives are emery and corundum. The artificial abrasives are silicon carbide and aluminum oxide.

abrasive cutoff saw A sawing machine that uses a thin, rubber-bonded grinding wheel to cut material to length.

absolute programming A CNC programming system where all locational dimensions are given from one common datum or zero point.

active cutting oils These oils contain sulfur that is not firmly attached to the oil. This sulfur is released during the machining operation and reacts with the work surface.

adapter A tool that provides a means of fitting the shank of a cutting tool or arbor into the spindle of a machine.

adaptive control Also known as torque control which measures the cutting forces. When the cutting forces exceed the stored value for each operation, the control may reduce the feedrate, turn on coolant, or stop the operation.

alignment Linear accuracy, uniformity, or coincidence of the centers of a lathe; a straight line of adjustment through two or more points. Setting the lathe in alignment means adjusting the tailstock in line with the headstock spindle to produce parallel work.

allowance As applied to the fitting of machine parts, a difference in dimensions prescribed in order to secure classes of fits. It is the minimum or maximum interference intentionally permitted between mating parts.

alloy A mixture of two or more metals melted together. As a rule, when two or more metals are melted together to form an alloy, the substance formed is a new metal.

aluminum A very light, silvery-white metal used independently or in alloys with copper and other metals.

aluminum oxide An abrasive material made from bauxite, coke screenings, and iron borings; used for grinding ferrous materials.

annealing Heating metal to just above its upper critical point for a certain period of time followed by slow cooling. This process softens metal, relieves internal stresses, and improves its machinability.

apprentice A person who is bound by an agreement to learn a trade or business under the guidance of a skilled craftsperson.

arbor A short shaft or spindle on which an object may be mounted.

Spindles or supports for milling machine cutters and saws are called arbors.

assembly drawing This type of drawing is used to show how various components (parts) fit into a product in its completed stage.

austenite A solid solution of carbon in iron that exists between the lower and upper critical temperatures.

B

backlash eliminator A device used on machines that eliminates backlash (play) between the nut and the table leadscrew; it permits climb-milling operations.

basic size The size of a dimension from which the limits for that dimension are derived.

bauxite A white to red earthy aluminum hydroxide. It is used primarily in the preparation of aluminum and alumina and for the lining of furnaces that are exposed to intense heat.

blind hole A hole that does not go through a workpiece.

body clearance The undercut portion of the drill circumference, extending from the back of the margin to the start of the flute.

bond The media or glue that holds abrasive grains together in the form of a wheel.

bore The internal diameter of a pipe, cylinder, or hole.

boring The operation of making or finishing circular holes in metal; usually done with a boring tool.

brittleness The property of a metal that allows no permanent distortion before breaking.

broach A hardened, multitooth form tool used to produce a similar form on the inside or outside of a surface.

built-up edge Small metal particles that form on a cutting tool's edge as a result of high temperature, high pressure, and high frictional resistance during machining.

burr A thin edge on a machined or ground surface left by the cutting tool.

bushing A sleeve or liner for a bearing. Some bushings can be adjusted to compensate for wear.

butt welder A unit found on contour bandsaws that joins two ends of a blade to make a flexible continuous saw band.

C

cam A device that converts rotary motion to straight-line or reciprocating motion; it transfers this motion through a follower to other parts of a machine or assembly.

carbon The element in steel that allows it to be hardened.

carburizing The process of increasing the carbon content of low-carbon steel by heating the metal below the melting point while it is in contact with carbonaceous material.

cartesian coordinate system A method used to locate any point or position in relation to a set of axes at right angles to each other.

case hardening A process by which a thin, hard film is formed on the surface of low-carbon steel.

CBN (Cubic Boron Nitride) A superhard, super wear-resistant abrasive used to machine or grind hard, abrasive, difficult-to-cut ferrous metals.

cemented carbide A very hard metal carbide cemented together with a little cobalt as a binder, to form a cutting edge nearly as hard as a diamond.

center drill A short drill, used for centering work so that it may be supported by lathe centers. Center drills are usually made in combination with a countersink, which permits a double operation with one tool.

center gage A gage used to align a threading toolbit for thread cutting in a lathe.

center head A tool used for finding the center of a circle or an arc of a circle. It is frequently used to find the center of a cylindrical piece of metal.

cermet A cutting-tool material consisting of ceramic particles bonded with a metal. Cermets, more shock-resistant than ceramics, are used for high-speed machining.

chamfer A bevelled edge or a cutoff corner.

chatter Caused, while machining work, by lack of rigidity in the cutting tools or in machine bearings or parts.

chip per tooth The amount removed by each tooth of a cutter in one revolution as it advances along the work. This amount is governed by the feedrate set on the machine.

chip pressure The force exerted on a cutting tool when removing material during machining.

chip-producing machines Machines that produce parts to shape by turning, milling, grinding, or any operations that remove metal in the form of chips.

chip-tool interface The area where the cutting tool contacts the workpiece and a chip is produced.

chisel edge The end of a drill, formed at the web where the two cutting edges meet, that does not have a very efficient cutting action.

circle grinding A method of grinding reamers or similar tools where the tool is rotated backwards so its heel touches the grinding wheel first to produce a small amount of clearance.

circular interpolation A mode of contouring control that allows arcs or circles up to 360° to be produced using only one block of programmed information.

clearance grinding A method of producing a flat surface on the land of a cutter.

climb milling The milling operation when the table feed and the rotation of the cutter are going in opposite directions.

clogging The condition where the spaces between the file teeth (gullets) are full of filings and the file cannot cut.

closed-loop system A CNC system where the program output, or the distance a machine slide moves, is measured and compared to the program input. The system automatically adjusts the output to be the same as the input.

CNC programmer A person capable of using coded instructions of letters, numbers, and symbols to activate servomotors, spindle drives, etc. to automatically control various machining functions.

coarseness The pitch or spacing of the saw teeth. A coarse-pitch blade will cut faster than a fine-pitch blade.

coated carbide tools Cemented carbide tools with a thin coating of titanium nitride, titanium carbide, or ceramic oxide to improve wear resistance and increase their life.

cold circular cutoff saw This saw uses a thin, circular blade, similar to a wood-cutting table saw, to cut material to length.

compressive strength The maximum stress in compression that a material will take before it ruptures or breaks.

computer-aided design (CAD) The use of computers and special software in the various stages in the design of a product or component.

computer-aided manufacturing (CAM) The use of computers to control all phases of machining and manufacturing.

computer-integrated manufacturing (CIM) A computer system controls all phases of production from management, design, quality control, and manufacturing to sales and order processing.

computer numerical control (CNC) A machine tool microprocessor that permits the creation or modification of parts. The programmed numerical control activates the machine's servos and spindle drives and controls the machining operation.

computer program A series of instructions, in language a machine control unit can understand, that outlines each step a machine must make to produce a part.

concentricity The condition where all diameters of a shaft have a common center or axis.

continuous blade These blades, which provide a continuous cut in one direction, are used because they are flexible enough to bend around the contour bandsaw's pulleys.

continuous chip A continuous ribbon produced when the metal flow is not greatly restricted by a built-up edge or friction at the chip-tool interface.

continuous chip with built-up edge A rough continuous chip produced by the friction of the metal particles that have welded themselves to the tool's cutting edge.

continuous path positioning Also called contouring; a situation in which two or more axes are controlled simultaneously to keep a constant cutter-work relationship.

contour bandsaw A vertical sawing machine, using an endless saw band, that can cut intricate shapes and contours on a workpiece.

conventional milling The milling operation where the table feed and the rotation of the cutter are going in the same direction.

coordinate measuring machine (CMM) An inspection machine capable of making many very accurate measurements of three-dimensional objects in a short period of time.

corundum A natural abrasive material used in place of emery.

counterbore A tool used to enlarge a hole through part of its length.

countersink A tool used to recess a hole conically for the head of a screw or rivet.

critical temperature The temperature at which certain changes take place in the chemical composition of steel during heating and cooling.

crossfeed A transverse feed; in a lathe, it usually operates at right angles to the axis of the work.

crush grinding A method of producing intricate forms in a part by using a grinding wheel dressed to the shape of the form required.

cutting fluids Various fluids used in machining and grinding operations to assist the cutting action by cooling and lubricating the cutting tool and workpiece.

cutting speed The speed in feet or meters per minute at which the cutting tool passes the work, or vice versa.

D

detail drawing This type of drawing is used to provide all the information required to manufacture a component (part).

dial calipers These instruments generally have a dial indicator mounted on the movable jaw and provide direct readings in inches or millimeters.

dial indicator A comparison-measuring instrument that is used to compare sizes and measurements to a known standard; commonly used to check the alignment of machine tools, fixtures, and workpieces.

diameter programming A programming format used on CNC lathes where the diameter of the round part is used for programming purposes.

diamond (manufactured) Developed by the General Electric Co. in 1954. A form of carbon and a catalyst are put under high pressure and high temperature to form diamond.

dielectric fluid The fluid used in EDM to control the arc discharge in the erosion gap.

digital readout (DRO) A measuring device, consisting of linear spars and optical reading heads, that accurately measures the machine-slide movements and displays the dimensions on a screen.

digital tools A class of measuring tools that can download to computers.

direct indexing A dividing head attachment that uses a slotted or drilled index plate to divide the circumference of a workpiece into equally spaced divisions.

direct ironmaking This process consists of the smelting of iron ore with coal, fluxes, and oxygen, in a liquid bath to eliminate the coke process and its pollution.

direct numerical control (DNC) A number of CNC machines in a manufacturing process controlled by one master computer.

direct-reading tools Mechanical measuring tools that give a direct measurement readout by digital numbers.

direct steelmaking A procedure that eliminates the coke production process and makes steel directly from iron ore in one step.

discontinuous chip A segmented chip produced when brittle metals such as cast iron and bronze are cut; can be some ductile materials under poor cutting conditions.

dividing head A milling machine attachment used to divide the circumference of a workpiece into equally spaced divisions for milling gears, splines, etc.

dressing The operation of making a grinding wheel cut better by removing dull grains and metal particles in the wheel surface.

drill chuck A holding device, generally with three jaws, used to hold and drive straight-shank cutting tools.

drill press (sensitive) A bench-type drill press used for light drilling operations on small parts.

dry run A term that applies to testing the accuracy of a CNC program by going through each block of information without actually cutting the part.

ductility The ability of a metal to be permanently deformed without breaking.

dust/mist respirator A protective screen, often made of fine gauze, worn over the mouth and nose to prevent a person from inhaling dust or vapors.

E

eccentric A shaft that may have two or more turned diameters parallel to each other but not concentric with the normal axis of the part.

electrical discharge machining (EDM) A process that removes metal by controlled electric spark erosion.

electro-machining processes A process that vaporizes conductive materials by controlled applications of pulsed electrical current that flows between a workpiece and the electrode (tool) in a dielectric fluid.

electronic-digital tools Microelectronic measuring tools with digital readouts that are easy to use and calibrate and are very reliable.

electro-optical tools These tools use a combination of electronic or optical magnification to make very precise measurements. Electronic tools are commonly used to measure parts after the machining is completed, while optical tools can measure a part while it is being machined.

emulsifiers Any liquid that mixes with water to form a cutting fluid to improve machinability, prolong tool life, and cool the cutting tool and workpiece.

end mill A milling cutter, usually smaller than 1 in. (25 mm) in diameter, with straight or tapered shanks. The cutting portion is cylindrical in shape, so that it can cut both on the sides and on the end.

engineer A person who has graduated from a college or university with a degree in a certain phase of engineering such as mechanical, electrical, electronics, metallurgy, aerospace, etc.

extreme-pressure oils Cutting fluid additives (chlorine, sulfur, or phosphorous compounds) that react with the work material to reduce built-up edges and are good for high-speed machining.

F

facing The operation of producing a flat surface, 90° to the lathe spindle axis, on the side of a shoulder or face.

feed The longitudinal movement of a tool in inches per minute or thousandths of an inch per revolution. Metric feeds are stated in millimeters per minute or hundredths of a millimeter per revolution.

feedback system Used in CNC closed-loop systems to compare the CNC program signal with the actual movement of the machine slides. If necessary, the system makes adjustments to make the table position the same as the programmed dimension.

ferrous metals Metals containing ferrite or iron.

fillet A concave or radius surface joining two adjacent faces of a part to strengthen the joint, as between two diameters on a shaft.

finishing Machining a surface to size with a fine feed produced in a lathe, milling machine, or grinder.

fit The variation between two mating parts with respect to the amount of clearance or interference that is present.

fixture A device designed and built for holding a particular piece of work for machining operations.

flaw Irregularities, such as scratches, holes, cracks, ridges, or hollows, that do not follow a regular pattern such as waviness and roughness.

flexible manufacturing system (FMS) An automated manufacturing system consisting of a number of CNC machine tools, serviced by a material handling system, under the control of one or more dedicated computers. It is designed to produce parts with a minimum of production changeover time.

fly cutter A unit, holding two or more replaceable cutting tools, that is used for milling large, flat surfaces.

friction sawing A contour bandsaw operation where metal is removed by the heat built up by the friction of a high-velocity saw blade.

friction thimble A control mechanism built into the micrometer thimble that ensures uniform contact pressure for every measurement.

G

gage A tool used for checking dimensions of a job. A surface gage can also be used to lay out for machining.

gage blocks Gage blocks, made of seasoned hardened and ground steel, carbide, or ceramic, are the world acceptable physical standard of measurement. An 83-piece gage block set can make up to 120,000 different measurements in increments of .0001 in.

general shop A general shop, usually equipped with conventional machine tools, makes parts for all types of tools and equipment.

grade The degree of strength with which the bond holds the abrasive particles in the bond setting.

graduated collar (diameter-reduction) A graduated micrometer collar where the amount of material removed from the diameter is the same as the collar setting.

graduated collar (radius-reduction) A graduated micrometer collar where the amount of material removed from a diameter is twice the collar setting.

graduated micrometer collar Collars found on the crossfeed and compound rest screws that may be graduated in inches or millimeters for making accurate cutting-tool settings.

grain size The nominal size of the abrasive particles contained in a grinding wheel.

grit The characteristic of a grinding wheel that is used to designate the nominal size of the abrasive grain in a wheel.

grooving The operation of producing square, round, or V-shaped forms at the end of a thread or the side of a shoulder.

guide-bar setover [OFFSET] The amount the guide bar of a taper attachment is set off parallel alignment to cut a taper on a round shaft.

H

half center A lathe center with a flat ground on one side to allow a facing tool to machine the full end of a part mounted between centers.

hardening The process of heating and cooling metals to produce the desired mechanical properties.

hardness The resistance of a metal to forcible penetration or plastic deformation.

heat treating A process that combines the controlled heating and cooling of a metal or an alloy to change its microstructure and produce desired results.

height gage A precision instrument used in toolrooms and inspection rooms to lay out and measure vertical distances to an accuracy of .001 in. or 0.02 mm.

high-speed steel A hard steel, made of carbon, manganese, silicon, chromium, tungsten, and vanadium, used for the manufacture of cutting tools.

high-tensile strength Refers to materials that can withstand a heavy load before they break or rupture.

high-velocity machining A system developed to assist manufacturers to respond quickly to fast market changes. It uses linear axis drives, fluid-spindle bearings, laser feedback, high-precision tooling, and other technologies.

horizontal bandsaw A horizontal sawing machine that uses an endless saw band to cut workpieces to length.

horizontal-spindle grinder A surface grinder with its spindle in a horizontal position and a reciprocating table.

hydraulics The use of water, oil, or other type of liquid, which is under pressure, to convert a small force into a larger force to operate mechanical devices.

I

inactive cutting oils These oils contain sulfur that is firmly attached to the oil. Therefore, a very small amount of sulfur is released during machining to react with the work surface.

inch system The standard of measurement used in the United States and Canada; uses the inch as its base linear unit. An inch can be divided into fractions such as halves, quarters, or eighths or decimals of tenths, thousandths, or ten-thousandths.

inch vernier scale This scale is used primarily on vernier calipers where 24 or 49 bar divisions occupy the same space as 25 or 50 vernier divisions; only one line of the vernier scale will match a line on the bar at any one setting.

incremental programming A CNC programming system in which each dimension is taken from the immediate past position and not from a common data point.

indexable-insert tools Cutting-tool inserts, made of cemented carbide, ceramic, or superabrasives, that can be mounted in a holder and quickly replaced or indexed as they are required.

index plate A circular plate, with a series of equally spaced hole circles, that is used on dividing heads to move complete turns or fractions of a turn.

inspection room A climate- and humidity-controlled room, set at 68°F (20°C), used to check the accuracy of manufactured parts or measuring tools.

interchangeable manufacture A system based on fixed measurement standards where parts made anywhere in the world can be assembled and will operate properly as a unit.

interpolation A function of a control where data points between two coordinate locations are generated in the order they will be used in the program.

isometric projection These drawings revolve an object 45° to the horizontal to bring the front corner toward the viewer and show the top, front, and side views at a 30° angle.

J

jig A device that holds and locates a piece of work and guides the tools that operate upon it.

job shop This shop is generally equipped with standard and CNC machine tools to produce jigs, fixtures, dies, molds, or special parts to customer specifications.

jog mode A method used by the operator to run through a CNC program one block at a time to check the accuracy of the program.

Just-In-Time A system where materials are available at the time they are needed for production. Its purpose is to improve productivity, reduce costs, reduce scrap and rework, overcome the shortage of machines, reduce inventory and work-in-process, and use manufacturing space efficiently.

K

keyway A rectangular groove cut along a shaft or hub into which a metal part fits to provide a positive drive between a shaft and a hub.

knurling The operation of impressing diamond- or straight-shaped indentations on a work surface to improve its appearance and provide a better grip.

L

lapping An abrading process used to remove minute amounts of material from a surface using fine abrasive powders.

laser An acronym for *L*ight *A*mplification by *S*timulated *E*mission of *R*adiation. Some common uses for lasers are cutting, measuring, and guidance systems.

lathe A machine tool used for turning cylindrical forms on workpieces. Modern lathes are often equipped with digital readouts and numerical control.

lay The direction of the predominant surface pattern caused by the machining process.

laying out The operation of marking straight or curved lines on metal surfaces to indicate the shape of the part, or the amount of material to be removed.

layout (precision) A layout performed with tools such as vernier and dial height gages where the layout accuracy must be less than ¼ in.

layout (semiprecision) A layout performed with tools such as a rule, divider, combination set, and surface gage, where the accuracy of the layout is acceptable.

lead The distance a thread advances along its axis in one complete revolution.

leadscrew A threaded shaft that runs longitudinally in front of the lathe bed.

limits or tolerance The limits of accuracy, oversize or undersize, within which a part being made must be kept to be acceptable.

linear interpolation A control function where data points are generated between coordinate positions to allow two or more axes of motion in a straight (linear) path.

lip clearance The amount, in degrees, that the heel of the drill is

lower than the lip or cutting edge. It allows the lip of the drill to cut into metal.

low-tensile strength Refers to materials that cannot withstand a heavy load before they fracture or break.

M

machinability The ease with which a metal is machined or ground.

machine control unit A system of hardware and software that controls the operation of a CNC machine.

machine home position Most machine tools have a default coordinate system (home or zero position). This is a fixed position built into the machine by the manufacturer.

machining center A computer numerically controlled milling machine.

machinist A skilled worker who can read technical drawings, use precision measuring instruments, and operate all machine tools in a shop.

magnetic chuck A device used to hold ferrous (magnetic) materials by permanent magnet or electro-magnetic force for machining.

major diameter The largest diameter of an external or internal thread. The minor or root diameter is the smallest diameter of an external or internal thread.

malleability The property of a metal that allows it to be hammered or rolled into other sizes and shapes.

mandrel A shaft or spindle on which an object may be fixed for rotation, such as that used when a piece is to be machined in a lathe between centers.

manufactured diamond Diamond manufactured by subjecting carbon along with other materials such as iron sulfide, chromium, and nickel to high temperatures and high pressures.

manufacturing cell A group of machines combined to perform all the manufacturing operations on a part before it leaves the cell.

manuscript Generally a copy of a CNC program, written in symbolic form, that contains all the data required for a machine to produce a part.

martensite The needlelike structure of fully hardened steel obtained when austenite is quenched.

MCU (master control unit) The CNC master control unit that transfers data from the program to the CNC machine tool.

mesh The engagement of teeth or gears of a sprocket and chain.

metallurgy The science of the composition, structure, manufacturing, and properties of metals.

metal-removal rate The rate at which metal is removed from an unfinished part, measured in cubic inches or cubic centimeters per minute.

metric system This system, sometimes called the Système International (SI), uses the meter as its base linear unit. All multiples or divisions of the meter are directly related to the meter by a factor of 10.

metric vernier scale Used primarily on vernier calipers where 49 bar divisions and 50 vernier scale divisions occupy the same space so that only one vernier line will match a bar line at any one setting.

microprocessor The building block of all computers that contains the arithmetic logic, the instruction register and decoding logic, and the data registers.

microstructure The structural characteristics such as size, shape, and arrangement of the crystals present in a metal or alloy.

minimill A small, more efficient steel mill that produces steel and thin-slab casting faster, and at less cost, than the large integrated steel mills.

minor diameter The smallest diameter of an internal or external thread; formerly known as the root diameter.

modal code A code that stays in effect in a program until it is replaced or changed by another code of the same group number.

modular tooling A complete tooling system that combines the flexibility and versatility to build a series of tools necessary to produce a part. Modular systems combine accuracy and quick-change capabilities to increase productivity.

moiré fringe pattern A high-resolution measurement system formed by light passing through dark lines on a series of equally spaced lines on two pieces of plastic set at angles to each other.

multiple thread A situation where there are more than one thread starts around the periphery of a workpiece. These threads are used to increase the lead of a thread when it is not possible to cut a deep, coarse thread.

N

new-generation machines This generally refers to machines that shape a part to form and size by electrical, chemical, or laser energy.

non-chip-producing machines Machines that form metal to size and shape by a pressing, drawing, bending, extruding, or shearing action. Examples of these machines are the punch press, forming press, and hobbing press.

nonferrous metal Metal that contains little or no iron, such as brass, copper, and aluminium.

normalizing The process of removing internal stresses and strains from a metal by heating and slow cooling to improve its machinability.

numerical control (NC) Any controlled equipment that allows an operator to program its movements through a series of coded instructions consisting of numbers, letters, symbols, etc.

O

open-loop system A CNC system where there is no method of checking that the servomotors have accurately moved the machine slides to the programmed distance.

orthographic view The representation of a three-dimensional view of an object (width, height, depth) in a single plane on a sheet of paper.

overcut A common term used in EDM to indicate the amount a cavity or cut is larger than the electrode used for machining.

P

parallel A straight, rectangular bar of uniform thickness or width, used for setting up work in the same plane as a fixed surface.

parallel turning The operation where a diameter is machined to size with the same diameter along its entire length.

part zero A reference point chosen by the CNC programmer that best suits the part, operation, or tool-change position. It is generally one corner of the part or any other place that is most convenient.

PCBN An acronym for *Polycrystalline Cubic Boron Nitride*; it is a crystalline body of many small crystals randomly or partially randomly oriented to form a material for cutting hard ferrous metals.

PCD An acronym for *PolyCrystalline Diamond*; a cutting tool material, consisting of a layer of diamond fused to a cemented-carbide substrate; used for cutting hard, abrasive nonferrous materials.

pearlite The structure of a metal that has a combination of ferrite (iron) and iron carbide.

pelletizing A process where rock is removed from iron ore to bring the ore to a higher iron concentration.

periphery The line bounding a round surface, such as the circumference of a wheel.

pilot hole A small hole drilled to guide and allow free passage for the thickness of the web of a twist drill.

pinion The smaller of two gears in mesh; generally the driving gear.

pinning The condition where metal particles stick or almost weld themselves in the spaces between the file teeth. This prevents the file from cutting properly and causes scratches on the work surface.

pitch The distance from the center of one thread or gear tooth to the corresponding point on the next thread or tooth. For threads, it is measured parallel to the axis. For gear teeth, it is measured on the pitch circle.

plastic deformation A permanent change in dimensions or set that occurs in a metal body as a result of excessive tensile stress, shear, or compressive stresses.

plunge cutting The operation where a cutting tool is fed vertically into the workpiece.

point-to-point positioning A numerical control system used to go quickly from one programmed point to another with no control over the path taken.

polar coordinates A system used to dimension angles and radii to reduce the need of converting polar dimensions to Cartesian coordinates.

polishing The operation of improving the surface finish of a part using fine abrasive cloth or other abrasive materials.

powder metallurgy A process where metal powders along with binders are blended together, compacted, formed to the desired shape in a die, and then sintered.

production shop This type of shop, which is usually associated with a large plant or factory, makes many identical parts for a product or machinery.

productivity The output realized when human and material resources are used most efficiently.

program format Generally refers to absolute and incremental programming modes. Most CNC systems are capable of handling both at any time, provided the proper codes are used.

Q

quality control A formal program of monitoring product quality by applying statistical process control methods.

quenching The process of rapidly cooling heated steel to trap the microstructure in the state that gives it the desired qualities.

quick-change gearbox A unit found on most engine lathes that can vary the speed of the feed rod and leadscrew. This can be done by moving levers to change the gear ratio and speed.

quick-change toolholder A style of toolholder capable of quickly and accurately clamping holders with preset tools on a dovetailed toolpost. It is used when multiple machining operations are required and to increase productivity.

R

rack A straight strip of metal having teeth to engage with those of a gear wheel, as in a rack and pinion.

radius programming A programming format used on CNC lathes where the radius of a round part is used for programming purposes.

rake angle The angle that provides keenness to the tool's cutting edge.

rapid traverse A method of positioning the workpiece and the cutter close to the approximate location rapidly (150 to 400 in./min) before a machining operation begins.

reamer A tool used to enlarge, smooth, and size a hole that has been drilled or bored.

reference line A machined edge from which all layouts begin. For most layouts, two machined edges at 90° to each other are used as reference lines.

revolutions per minute (r/min) The number of complete turns a workpiece or cutter makes in 1 minute.

ring testing A method of checking a grinding wheel for cracks or damage by striking its side with a wooden or plastic tool handle. A sound wheel will produce a sharp ring.

robot A single-arm device that can be programmed to automatically move tools, parts, and materials or to perform a variety of tasks.

rotary table A circular milling attachment that makes it possible to machine radii and circular forms.

roughness Finely spaced patterns caused by the cutting tool or machine feed.

rough turning An operation done before finishing to remove surplus stock rapidly when fine surface finish is not important.

run-out The amount a saw blade will run out of square as it cuts through a material. Dull blades will have more run-out than sharp blades.

S

SAE (Society of Automotive Engineers) These letters are used to indicate that the article or measurement is approved by the Society of Automotive Engineers.

scale A thin surface on castings or rolled metal caused by burning, oxidizing, or cooling.

scale size A code used on technical or engineering drawings to indicate whether the part is being shown in full size, half-size, one-quarter size, one-eighth size, etc.

scroll plate A flat plate, usually found inside a three-jaw or universal chuck, with a gear-type form cut in its face that causes the chuck jaws to move simultaneously.

section view A view that shows the interior detail of a part that is too complicated to show by outside views or hidden lines.

sequence of operations The order of steps to be taken to successfully machine a part or perform an operation.

serrations A series of grooves produced in metal to provide a grip or locking action. Vise jaws are often serrated.

servomechanism A control mechanism used in EDM machines to feed the electrode into the part, maintain the work/electrode gap, and slow or speed up the drive motors as required.

shear angle The angle or plane of the area of material where plastic deformation occurs as a chip is being produced.

shoulder turning The operation of producing square, filleted, or chamfered forms on a shoulder or step of a part.

SI (Système International) The standardized metric system that defines 1 meter as 1,650,763.73 wave lengths in a vacuum of orange-red light spectrum of the krypton-86 atom.

silicon carbide An abrasive material made from silica sand, high-purity coke, sawdust, and salt; used for grinding nonferrous materials.

simple indexing Uses a 40-tooth worm wheel on the dividing head spindle that is connected to an index crank single-threaded worm to divide workpieces into any number of turns or fractions of a turn.

sine bar A precision bar, with two cylinders exactly 5 or 10 in. apart, used with gage blocks for setting work to within a few seconds of a degree.

soaking Heating metal at a uniform heat for a period of time for complete penetration.

spindle nose (cam lock) Three cam locks match three cam-lock studs on the spindle accessory that holds the accessory securely on the spindle.

spindle nose (tapered) A taper and aligning key on the spindle match a corresponding taper and keyway in the accessory. A lock ring tightens the accessory to the spindle.

spindle nose (threaded) A right-hand thread on the end of the spindle allows threaded-type accessories to be fastened on the headstock.

spindle speed The number of r/min that is made by the spindle (cutting tool) of a machine.

statistical process control (SPC) A quality assurance method of using performance data to identify product and process errors that lead to the production of faulty goods. Correct analysis of this data should lead to correcting the errors and producing only acceptable goods.

straddle milling The use of two side milling cutters to machine the opposite sides of a workpiece parallel in one cut.

structure The space relationship of the abrasive grain and the bonding material to the voids (spaces) that separate them.

superabrasives Manufactured diamond and cubic boron nitride abrasives that are used to make superhard, super wear-resistant cutting tools and grinding wheels.

surface deviations Any departure from the nominal surface in the form of waviness, roughness, flaws, lay, and profile.

surface finish The degree of variation of surface roughness or waviness from a reference or nominal plane, usually measured in microinches or micrometers.

surface gage A machinist's gage consisting of a heavy base and a scriber for marking in layout for machining.

surface plate A cast-iron scraped plate used for layout work. Granite plates are also used.

T

tailstock center (revolving) A lathe center that contains antifriction bearings and revolves with the workpiece to reduce friction and eliminate the need to adjust the center tension due to heat expansion.

tap An accurate, hardened cutting tool used to produce internal threads. Taps generally come in sets of three, consisting of taper, plug, and bottoming.

tap drill size The size of the drill that should be used to leave the proper amount of material in a hole for a tap to produce a 75% full thread.

taper A uniform increase or decrease in the diameter of a piece of work measured along its length. Self-holding tapers are held in position by the wedging action of the taper; self-releasing tapers are held by a draw-bar.

taper-ring gage An internal master taper gage used to check the accuracy of external tapers.

taper-shank tools Tools which generally have a taper shank that fits into the spindle of a drill press or milling machine.

taper turning The operation of producing a tapered form by offsetting the tailstock, the taper attachment, or the compound rest.

technician A person who works at a level between an engineer and a machinist and is specialized in one area of technology but has a working knowledge of a number of technologies.

technologist A person who works at a level between an engineer and a technician to perform design studies, production planning, laboratory experiments, and quality and cost control studies.

tempering Reheating hardened steel to a temperature below its critical point, followed by quenching. This process removes the brittleness and toughens the steel.

tensile strength The resistance of steel or iron to a lengthwise pull.

test bar A hardened and ground parallel-diameter bar used to check the alignment of centers.

test run The procedure of running a computer program one block at a time without cutting metal, to check the program's accuracy.

thermal conductivity The characteristic of a material that defines its ability to transfer heat.

thread A helical ridge of uniform section formed on the inside or outside of a cylinder or cone. The American National thread form is common in America; the ISO thread is the standard metric thread form.

thread-chasing dial A revolving dial that is connected to a gear that meshes with the leadscrew. This dial indicates when to engage the split-nut lever for taking successive cuts during thread cutting.

threading The operation of producing internal or external threads on a workpiece.

thread-ring gage A master gage used to check the accuracy of an external thread.

through hole A hole that goes through a workpiece.

throw The distance that a set of centers of an eccentric has been offset from the normal work axis.

titanium nitride A thin coating, applied to high-speed and carbide tools, which improves the tools' wear resistance and increases productivity.

tolerance The amount of interference required for two or more parts that are in contact. The amount of variation, over or under the required size, permitted on a piece of machined work.

tool and diemaker A skilled craftsperson who has above-average mechanical ability, can operate all machine tools, and has a broad knowledge of shop mathematics, print reading, metallurgy, and process planning.

tool-change position Usually the machine home or zero position. However, the programmer may choose a convenient position that clears the part and allows tools to be changed.

tool deflection The amount a cutting tool springs under the pressure of a cut.

tool-length offset A value input manually to compensate for the different lengths of each cutting tool used. This allows the operator to program all tools as though they were the same length.

tool-nose radius compensation Used to compensate for the differences in the tool-nose radius to prevent deviations from the programmed work surface on circular and angular cuts.

toolpost (quick-change) This unit consists of a dovetail toolpost onto which preset tools and their holders can be quickly, accurately, and securely clamped.

toolpost (turret-type) Generally holds four different cutting tools that can be indexed quickly and accurately to perform various machining operations.

toolpost (vertical-index) A rotary-indexing turret that may contain up to eight different types of cutting tools. It is mounted on the lathe compound rest and can be indexed to the next tool in less than a second.

tooth form The shape, or form, of the saw teeth; the most common are precision, claw, and buttress.

tooth sets The offset of the saw teeth from the band that provides free-cutting action; straight, wave, and raker sets are the most common.

toughness The property of a metal to withstand shock or impact.

trial cut A short cut made on a workpiece in a lathe or milling machine to check the accuracy of the depth of cut setting.

truing The operation of making the periphery of a job or tool concentric with the axis of rotation.

turning center A computer numerically controlled lathe.

turret toolholder Most models hold four cutting tools, held in their holders, that can be quickly and accurately indexed into cutting position.

UV

variable speed drive A spindle-speed mechanism that allows speeds to be increased or decreased while the machine spindle is revolving.

vernier caliper Precision measuring instrument with a vernier scale mounted on its movable jaw. Used to make accurate internal and external measurements in thousandths of an inch or hundredths of a millimeter.

vision systems AI (Artificial Intelligence) technology, combined with computers, software, television cameras, and optical sensors, that allows machines to perform jobs normally done by humans.

W

wave lengths of light A means of accurately defining linear measurement.

waviness Widely spaced surface irregularities in the form of waves caused by vibrations in the machine or work.

wear resistance The ability of a metal to resist abrasion and wear.

word-address format The alphabetical letter used in each block of programmed information that identifies the meaning of the word.

work hardening The result of the action of cutting edges deforming or compressing the surface of a material, causing a change in its structure that increases hardness.

XYZ

zero reference point The point, on or off the part, chosen by the programmer as being most suitable for programming and machining purposes.

table 1 — Decimal inch, fractional inch, and millimeter equivalents

Decimal Inch	Fractional Inch	Millimeter	Decimal Inch	Fractional Inch	Millimeter
.015625	1/64	0.397	.515625	33/64	13.097
.03125	1/32	0.794	.53125	17/32	13.494
.046875	3/64	1.191	.546875	35/64	13.891
.0625	1/16	1.588	.5625	9/16	14.288
.078125	5/64	1.984	.578125	37/64	14.684
.09375	3/32	2.381	.59375	19/32	15.081
.109375	7/64	2.778	.609375	39/64	15.478
.125	1/8	3.175	.625	5/8	15.875
.140625	9/64	3.572	.640625	41/64	16.272
.15625	5/32	3.969	.65625	21/32	16.669
.171875	11/64	4.366	.671875	43/64	17.066
.1875	3/16	4.762	.6875	11/16	17.462
.203125	13/64	5.159	.703125	45/64	17.859
.21875	7/32	5.556	.71875	23/32	18.256
.234375	15/64	5.953	.734375	47/64	18.653
.25	1/4	6.35	.75	3/4	19.05
.265625	17/64	6.747	.765625	49/64	19.447
.28125	9/32	7.144	.78125	25/32	19.844
.296875	19/64	7.541	.796875	51/64	20.241
.3125	5/16	7.938	.8125	13/16	20.638
.328125	21/64	8.334	.828125	53/64	21.034
.34375	11/32	8.731	.84375	27/32	21.431
.359375	23/64	9.128	.859375	55/64	21.828
.375	3/8	9.525	.875	7/8	22.225
.390625	25/64	9.922	.890625	57/64	22.622
.40625	13/32	10.319	.90625	29/32	23.019
.421875	27/64	10.716	.921875	59/64	23.416
.4375	7/16	11.112	.9375	15/16	23.812
.453125	29/64	11.509	.953125	61/64	24.209
.46875	15/32	11.906	.96875	31/32	24.606
.484375	31/64	12.303	.984375	63/64	25.003
.5	1/2	12.7	1	1	25.4

table 2 — Conversion of inches to millimeters, conversion of millimeters to inches

Conversion of Inches to Millimeters						Conversion of Millimeters to Inches					
Inches	Milli-meters	Inches	Milli-meters	Inches	Milli-meters	Milli-meters	Inches	Milli-meters	Inches	Milli-meters	Inches
.001	0.025	.290	7.37	.660	16.76	0.01	.0004	0.35	.0138	0.68	.0268
.002	0.051	.300	7.62	.670	17.02	0.02	.0008	0.36	.0142	0.69	.0272
.003	0.076	.310	7.87	.680	17.27	0.03	.0012	0.37	.0146	0.7	.0276
.004	0.102	.320	8.13	.690	17.53	0.04	.0016	0.38	.0150	0.71	.0280
.005	0.127	.330	8.38	.700	17.78	0.05	.0020	0.39	.0154	0.72	.0283
.006	0.152	.340	8.64	.710	18.03	0.06	.0024	0.4	.0157	0.73	.0287
.007	0.178	.350	8.89	.720	18.29	0.07	.0028	0.41	.0161	0.74	.0291
.008	0.203	.360	9.14	.730	18.54	0.08	.0031	0.42	.0165	0.75	.0295
.009	0.229	.370	9.4	.740	18.8	0.09	.0035	0.43	.0169	0.76	.0299
.010	0.254	.380	9.65	.750	19.05	0.1	.0039	0.44	.0173	0.77	.0303
.020	0.508	.390	9.91	.760	19.3	0.11	.0043	0.45	.0177	0.78	.0307
.030	0.762	.400	10.16	.770	19.56	0.12	.0047	0.46	.0181	0.79	.0311
.040	1.016	.410	10.41	.780	19.81	0.13	.0051	0.47	.0185	0.8	.0315
.050	1.27	.420	10.67	.790	20.07	0.14	.0055	0.48	.0189	0.81	.0319
.060	1.524	.430	10.92	.800	20.32	0.15	.0059	0.49	.0193	0.82	.0323
.070	1.778	.440	11.18	.810	20.57	0.16	.0063	0.5	.0197	0.83	.0327
.080	2.032	.450	11.43	.820	20.83	0.17	.0067	0.51	.0201	0.84	.0331
.090	2.286	.460	11.68	.830	21.08	0.18	.0071	0.52	.0205	0.85	.0335
.100	2.54	.470	11.94	.840	21.34	0.19	.0075	0.53	.0209	0.86	.0339
.110	2.794	.480	12.19	.850	21.59	0.2	.0079	0.54	.0213	0.87	.0343
.120	3.048	.490	12.45	.860	21.84	0.21	.0083	0.55	.0217	0.88	.0346
.130	3.302	.500	12.7	.870	22.1	0.22	.0087	0.56	.0220	0.89	.0350
.140	3.56	.510	12.95	.880	22.35	0.23	.0091	0.57	.0224	0.9	.0354
.150	3.81	.520	13.21	.890	22.61	0.24	.0094	0.58	.0228	0.91	.0358
.160	4.06	.530	13.46	.900	22.86	0.25	.0098	0.59	.0232	0.92	.0362
.170	4.32	.540	13.72	.910	23.11	0.26	.0102	0.6	.0236	0.93	.0366
.180	4.57	.550	13.97	.920	23.37	0.27	.0106	0.61	.0240	0.94	.0370
.190	4.83	.560	14.22	.930	23.62	0.28	.0110	0.62	.0244	0.95	.0374
.200	5.08	.570	14.48	.940	23.88	0.29	.0114	0.63	.0248	0.96	.0378
.210	5.33	.580	14.73	.950	24.13	0.3	.0118	0.64	.0252	0.97	.0382
.220	5.59	.590	14.99	.960	24.38	0.31	.0122	0.65	.0256	0.98	.0386
.230	5.84	.600	15.24	.970	24.64	0.32	.0126	0.66	.0260	0.99	.0390
.240	6.1	.610	15.49	.980	24.89	0.33	.0130	0.67	.0264	1	.0394
.250	6.35	.620	15.75	.990	25.15	0.34	.0134				
.260	6.6	.630	16.	1	25.4						
.270	6.86	.640	16.26								
.280	7.11	.650	16.51								

(Automatic Electric Company.)

table 3 — Letter drill sizes

Letter	in.	mm	Letter	in.	mm
A	.234	5.9	N	.302	7.7
B	.238	6	O	.316	8
C	.242	6.1	P	.323	8.2
D	.246	6.2	Q	.332	8.4
E	.250	6.4	R	.339	8.6
F	.257	6.5	S	.348	8.8
G	.261	6.6	T	.358	9.1
H	.266	6.7	U	.368	9.3
I	.272	6.9	V	.377	9.5
J	.277	7	W	.386	9.8
K	.281	7.1	X	.397	10.1
L	.290	7.4	Y	.404	10.3
M	.295	7.5	Z	.413	10.5

table 4 — Drill gage sizes

No.	in.	mm	No.	in.	mm	No.	in.	mm
1	.2280	5.8	34	.1110	2.8	66	.0330	0.84
2	.2210	5.6	35	.1100	2.8	67	.0320	0.81
3	.2130	5.4	36	.1065	2.7	68	.0310	0.79
4	.2090	5.3	37	.1040	2.65	69	.0292	0.74
5	.2055	5.2	38	.1015	2.6	70	.0280	0.71
6	.2040	5.2	39	.0995	2.55	71	.0260	0.66
7	.2010	5.1	40	.0980	2.5	72	.0250	0.64
8	.1990	5.1	41	.0960	2.45	73	.0240	0.61
9	.1960	5	42	.0935	2.4	74	.0225	0.57
10	.1935	4.9	43	.0890	2.25	75	.0210	0.53
11	.1910	4.9	44	.0860	2.2	76	.0200	0.51
12	.1890	4.8	45	.0820	2.1	77	.0180	0.46
13	.1850	4.7	46	.0810	2.05	78	.0160	0.41
14	.1820	4.6	47	.0785	2	79	.0145	0.37
15	.1800	4.6	48	.0760	1.95	80	.0135	0.34
16	.1770	4.5	49	.0730	1.85	81	.0130	0.33
17	.1730	4.4	50	.0700	1.8	82	.0125	0.32
18	.1695	4.3	51	.0670	1.7	83	.0120	0.31
19	.1660	4.2	52	.0635	1.6	84	.0115	0.29
20	.1610	4.1	53	.0595	1.5	85	.0110	0.28
21	.1590	4	54	.0550	1.4	86	.0105	0.27
22	.1570	4	55	.0520	1.3	87	.0100	0.25
23	.1540	3.9	56	.0465	1.2	88	.0095	0.24
24	.1520	3.9	57	.0430	1.1	89	.0091	0.23
25	.1495	3.8	58	.0420	1.05	90	.0087	0.22
26	.1470	3.7	59	.0410	1.05	91	.0083	0.21
27	.1440	3.7	60	.0400	1	92	.0079	0.2
28	.1405	3.6	61	.0390	0.99	93	.0075	0.19
29	.1360	3.5	62	.0380	0.97	94	.0071	0.18
30	.1285	3.3	63	.0370	0.94	95	.0067	0.17
31	.1200	3	64	.0360	0.92	96	.0063	0.16
32	.1160	2.95	65	.0350	0.89	97	.0059	0.15
33	.1130	2.85						

table 5 — Tap drill sizes

Nominal Diameter (mm)	Thread Pitch (mm)	Tap Drill Size (mm)	Nominal Diameter (mm)	Thread Pitch (mm)	Tap Drill Size (mm)
1.6	0.35	1.2	20	2.5	17.5
2	0.4	1.6	24	3	21
2.5	0.45	2.05	30	3.5	26.5
3	0.5	2.5	36	4	32
3.5	0.6	2.9	42	4.5	37.5
4	0.7	3.3	48	5	43
5	0.8	4.2	56	5.5	50.5
6.3	1	5.3	64	6	58
8	1.25	6.8	72	6	66
10	1.5	8.5	80	6	74
12	1.75	10.2	90	6	84
14	2	12	100	6	94
16	2	14			

table 6 — ISO metric pitch and diameter combinations

Nominal Dia. (mm)	Thread Pitch (mm)	Nominal Dia. (mm)	Thread Pitch (mm)
1.6	0.35	20	2.5
2	0.4	24	3
2.5	0.45	30	3.5
3	0.5	36	4
3.5	0.6	42	4.5
4	0.7	48	5
5	0.8	56	5.5
6.3	1	64	6
8	1.25	72	6
10	1.5	80	6
12	1.75	90	6
14	2	100	6
16	2		

table 7 — Tap drill sizes, american national form thread

NC National Coarse			NF National Fine		
Tap Size	Threads per Inch	Tap Drill Size	Tap Size	Threads per Inch	Tap Drill Size
# 5	40	#38	# 5	44	#37
# 6	32	#36	# 6	40	#33
# 8	32	#29	# 8	36	#29
#10	24	#25	#10	32	#21
#12	24	#16	#12	28	#14
$\frac{1}{4}$	20	# 7	$\frac{1}{4}$	28	# 3
$\frac{5}{16}$	18	F	$\frac{5}{16}$	24	I
$\frac{3}{8}$	16	$\frac{5}{16}$	$\frac{3}{8}$	24	Q
$\frac{7}{16}$	14	U	$\frac{7}{16}$	20	$\frac{25}{64}$
$\frac{1}{2}$	13	$\frac{27}{64}$	$\frac{1}{2}$	20	$\frac{29}{64}$
$\frac{9}{16}$	12	$\frac{31}{64}$	$\frac{9}{16}$	18	$\frac{33}{64}$
$\frac{5}{8}$	11	$\frac{17}{32}$	$\frac{5}{8}$	18	$\frac{37}{64}$
$\frac{3}{4}$	10	$\frac{21}{32}$	$\frac{3}{4}$	16	$\frac{11}{16}$
$\frac{7}{8}$	9	$\frac{49}{64}$	$\frac{7}{8}$	14	$\frac{13}{16}$
1	8	$\frac{7}{8}$	1	14	$\frac{15}{16}$
$1\frac{1}{8}$	7	$\frac{63}{64}$	$1\frac{1}{8}$	12	$1\frac{3}{64}$
$1\frac{1}{4}$	7	$1\frac{7}{64}$	$1\frac{1}{4}$	12	$1\frac{11}{64}$
$1\frac{3}{8}$	6	$1\frac{7}{32}$	$1\frac{3}{8}$	12	$1\frac{19}{64}$
$1\frac{1}{2}$	6	$1\frac{11}{32}$	$1\frac{1}{2}$	12	$1\frac{27}{64}$
$1\frac{3}{4}$	5	$1\frac{9}{16}$			
2	$4\frac{1}{2}$	$1\frac{25}{32}$			
NPT National Pipe Thread					
$\frac{1}{8}$	27	$\frac{11}{32}$	1	$11\frac{1}{2}$	$1\frac{5}{32}$
$\frac{1}{4}$	18	$\frac{7}{16}$	$1\frac{1}{4}$	$11\frac{1}{2}$	$1\frac{1}{2}$
$\frac{3}{8}$	18	$\frac{19}{32}$	$1\frac{1}{2}$	$11\frac{1}{2}$	$1\frac{23}{32}$
$\frac{1}{2}$	14	$\frac{23}{32}$	2	$11\frac{1}{2}$	$2\frac{3}{16}$
$\frac{3}{4}$	14	$\frac{15}{16}$	$2\frac{1}{2}$	8	$2\frac{5}{8}$

The major diameter of an NC or NF number size tap or screw = (N × .013) + .060.
EXAMPLE: The major diameter of a #5 tap equals (5 × .013) + .060 = .125 diameter.

table 8 — Three wire thread measurement (60° metric thread)

$M = PD + C \qquad PD = M - C$

M = Measurement over wires
PD = Pitch diameter
C = Constant

Pitch		Best Wire Size		Constant	
Inches	mm	Inches	mm	Inches	mm
.00787	0.2	.00455	0.1155	.00682	0.1732
.00886	0.225	.00511	0.1299	.00767	0.1949
.00934	0.25	.00568	0.1443	.00852	0.2165
.01181	0.3	.00682	0.1732	.01023	0.2598
.01378	0.35	.00796	0.2021	.01193	0.3031
.01575	0.4	.00909	0.2309	.01364	0.3464
.01772	0.45	.01023	0.2598	.01534	0.3897
.01969	0.5	.01137	0.2887	.01705	0.433
.02362	0.6	.01364	0.3464	.02046	0.5196
.02756	0.7	.01591	0.4041	.02387	0.6062
.02953	0.75	.01705	0.433	.02557	0.6495
.03150	0.8	.01818	0.4619	.02728	0.6928
.03543	0.9	.02046	0.5196	.03069	0.7794
.03937	1	.02273	0.5774	.03410	0.866
.04921	1.25	.02841	0.7217	.04262	1.0825
.05906	1.5	.03410	0.866	.05114	1.299
.06890	1.75	.03978	1.0104	.05967	1.5155
.07874	2	.04546	1.1547	.06819	1.7321
.09843	2.5	.05683	1.4434	.08524	2.1651
.11811	3	.06819	1.7321	.10229	2.5981
.13780	3.5	.07956	2.0207	.11933	3.0311
.15748	4	.09092	2.3094	.13638	3.4641
.17717	4.5	.10229	2.5981	.15343	3.8971
.19685	5	.11365	2.8868	.17048	4.3301
.21654	5.5	.12502	3.1754	.18753	4.7631
.23622	6	.13638	3.4641	.20457	5.1962
.27559	7	.15911	4.0415	.23867	6.0622
.31496	8	.18184	4.6188	.27276	6.9282
.35433	9	.20457	5.1962	.30686	7.7942
.39370	10	.22730	5.7735	.34095	8.6603

table 9 — Commonly used formulas

Code

c.p.t = Chip per tooth
CS = Cutting speed
D = Large diameter
d = Small diameter
G.L. = Guide bar length

N = Number of threads per inch
n = Number of teeth in cutter
O.L. = Overall length of work
P = Pitch
r/min = Revolutions per minute

T.D.S. = Tap drill size
T.L. = Taper length
t/ft = Taper per foot
t/mm = Taper per millimeter
T.O. = Tailstock offset

Inch	Metric
$T.D.S. = D - \left(\dfrac{1}{N}\right)$	$T.D.S. = D - P$
$r/min = \dfrac{CS\,(ft) \times 4}{D\,(in.)}$	$r/min = \dfrac{CS\,(m) \times 320}{D\,(mm)}$
$t/ft = \dfrac{(D - d) \times 12}{T.L.}$	$t/mm = \dfrac{(D - d)}{T.L.}$
$T.O. = \dfrac{t/ft \times O.L.}{24}$	$T.O. = \dfrac{t/mm \times O.L.}{2}$
$\text{Guide bar setover} = \dfrac{(D - d) \times 12}{T.L.}$	$\text{Guide bar setover} = \dfrac{(D - d)}{2} \times \dfrac{G.L.}{T.L.}$
$\text{Milling feed (in./min)} = N \times c.p.t \times r/min$	$\text{Milling feed (mm/min)} = N \times c.p.t \times r/min$

table 10   Formula shortcuts

For the correct formula, block out (cover) the unknown; the remainder is the formula. In each diagram the horizontal line is the division line; the vertical line(s) is the multiplication line.

Code: A = Area L = Length b = Base
      C = Circumference R = Radius h = Height
      CS = Cutting speed   r/min = Revolutions/minute   m = Meters
      D = Diameter S = Strokes/minute   mm = Millimeters
      W = Width

Circle

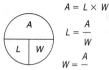

$$C = \pi \times D$$
$$D = \frac{C}{\pi}$$

Four-Element Formulas

1. Block out unknown.
2. Cross-multiply diagonally opposite elements.
3. Divide by remaining element.

Triangles

$$A = \frac{b \times h}{2}$$
$$b = \frac{A \times 2}{h}$$
$$h = \frac{A \times 2}{b}$$

Area

Squares and Rectangles

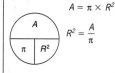

$$A = L \times W$$
$$L = \frac{A}{W}$$
$$W = \frac{A}{L}$$

Circles

$$A = \pi \times R^2$$
$$R^2 = \frac{A}{\pi}$$

Revolutions per Minute (r/min)
(Lathe, Drill, Mill, Grinder)

Inch

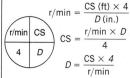

$$\text{r/min} = \frac{CS\ (\text{ft}) \times 4}{D\ (\text{in.})}$$
$$CS = \frac{\text{r/min} \times D}{4}$$
$$D = \frac{CS \times 4}{\text{r/min}}$$

Metric

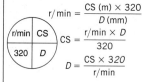

$$\text{r/min} = \frac{CS\ (\text{m}) \times 320}{D\ (\text{mm})}$$
$$CS = \frac{\text{r/min} \times D}{320}$$
$$D = \frac{CS \times 320}{\text{r/min}}$$

table 11   Morse tapers*

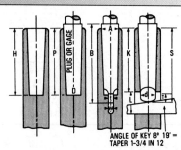

ANGLE OF KEY 8° 19′ =
TAPER 1-3/4 IN 12

Number of Taper	Diameter of Plug at Small End	Diameter at End of Socket	Whole Length of Shank	Shank Depth	Depth of Hole	Standard Plug Depth	Thickness of Tongue	Length of Tongue	Diameter of Tongue	Width of Keyway	Length of Keyway	End of Socket to Keyway	Taper per Foot
	D	A	B	S	H	P	t	T	d	w	L	K	
0	.252	.356	2-11/32	2-7/32	2-1/32	2	5/32	1/4	.235	.160	9/16	1-15/16	.6246
1	.369	.475	2-9/16	2-7/16	2-3/16	2-1/8	13/64	3/8	.343	.213	3/4	2-1/16	.5986
2	.572	.700	3-1/8	2-15/16	2-5/8	2-9/16	1/4	7/16	17/32	.260	7/8	2-1/2	.5994
3	.778	.938	3-7/8	3-11/16	3-1/4	3-3/16	5/16	9/16	23/32	.322	1-3/16	3-1/16	.6023
4	1.020	1.231	4-7/8	4-5/8	4-1/8	4-1/16	15/32	5/8	31/32	.478	1-1/4	3-7/8	.6232
5	1.475	1.748	6-1/8	5-7/8	5-1/4	5-3/16	5/8	3/4	1-13/32	.635	1-1/2	4-15/16	.6315
6	2.116	2.494	8-9/16	8-1/4	7-3/8	7-1/4	3/4	1-1/8	2	.760	1-3/4	7	.6256
7	2.750	3.270	11-1/4	11-5/8	10-1/8	10	1-1/8	1-3/8	2-5/8	1.135	2-5/8	9-1/2	.6240

All measurements are in inches.

table 12   Standard milling machine taper*

Milling Machine Spindles	Milling Machine Arbors

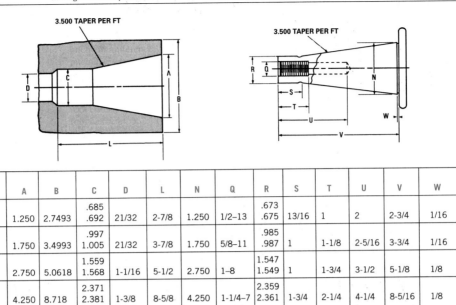

Taper No.	A	B	C	D	L	N	Q	R	S	T	U	V	W
30	1.250	2.7493	.685 .692	21/32	2-7/8	1.250	1/2–13	.673 .675	13/16	1	2	2-3/4	1/16
40	1.750	3.4993	.997 1.005	21/32	3-7/8	1.750	5/8–11	.985 .987	1	1-1/8	2-5/16	3-3/4	1/16
50	2.750	5.0618	1.559 1.568	1-1/16	5-1/2	2.750	1–8	1.547 1.549	1	1-3/4	3-1/2	5-1/8	1/8
60	4.250	8.718	2.371 2.381	1-3/8	8-5/8	4.250	1-1/4–7	2.359 2.361	1-3/4	2-1/4	4-1/4	8-5/16	1/8

*All measurements are in inches.

table 13   Tapers and angles

Taper per Foot	Included Angle		With Center Line		Taper per Inch	Taper per Inch from Centerline
	Degree	Minute	Degree	Minute		
1/8	0	36	0	18	.010416	.005208
3/16	0	54	0	27	.015625	.007812
1/4	1	12	0	36	.020833	.010416
5/16	1	30	0	45	.026042	.013021
3/8	1	47	0	53	.031250	.015625
7/16	2	05	1	02	.036458	.018229
1/2	2	23	1	11	.041667	.020833
9/16	2	42	1	21	.046875	.023438
5/8	3	00	1	30	.052084	.026042
11/16	3	18	1	39	.057292	.028646
3/4	3	35	1	48	.062500	.031250
13/16	3	52	1	56	.067708	.033854
7/8	4	12	2	06	.072917	.036458
15/16	4	28	2	14	.078125	.039063
1	4	45	2	23	.083330	.041667
1 1/4	5	58	2	59	.104166	.052083
1 1/2	7	08	3	34	.125000	.062500
1 3/4	8	20	4	10	.145833	.072917
2	9	32	4	46	.166666	.083333
2 1/2	11	54	5	57	.208333	.104166
3	14	16	7	08	.250000	.125000
3 1/2	16	36	8	18	.291666	.145833
4	18	56	9	28	.333333	.166666
4 1/2	21	14	10	37	.375000	.187500
5	23	32	11	46	.416666	.208333
6	28	04	14	02	.500000	.250000

(Morse Twist Drill & Machine Co.)

table 14 Allowances for fits*

Running Fits

Shaft Diameter	For Shafts with Speeds Under 600 r/min— Ordinary Working Conditions	For Shafts with Speeds over 600 r/min Heavy Pressure— Severe Working Conditions
Up to ½	−.0005 to −.001	−.0005 to −.001
½ to 1	−.00075 to −.0015	−.001 to −.002
1 to 2	−.0015 to −.0025	−.002 to −.003
2 to 3½	−.002 to −.003	−.003 to −.004
3½ to 6	−.0025 to −.004	−.004 to −.005

Sliding Fits

Shaft Diameter	For Shafts with Gears, Clutches, or Similar Parts That Must Be Free to Slide
Up to ½	−.0005 to −.001
½ to 1	−.00075 to −.0015
1 to 2	−.0015 to −.0025
2 to 3½	−.002 to −.003
3½ to 6	−.0025 to −.004

Push Fits

Shaft Diameter	For Light Service Where Part Is Keyed to Shaft and Clamped Endwise—No Fitting	With Play Eliminated—Part Should Assemble Readily—Some Fitting and Selecting May Be Required
Up to ½	Standard to −.00025	Standard to +.00025
½ to 3½	Standard to −.0005	Standard to +.0005
3½ to 6	Standard to −.00075	Standard to +.00075

Driving Fits

Shaft Diameter	For Permanent Assembly of Parts So Located That Driving Cannot Be Done Readily	For Permanent Assembly and Severe Duty and Where There Is Ample Room for Driving
Up to ½	Standard to +.00025	+.0005 to +.001
½ to 1	+.00025 to +.0005	+.0005 to +.001
1 to 2	+.0005 to +.00075	+.0005 to +.001
2 to 3½	+.0005 to +.001	+.00075 to +.00125
3½ to 6	+.0005 to +.001	+.001 to +.0015

Forced Fits

Shaft Diameter	For Permanent Assembly and Very Severe Service— Hydraulic Press Used for Larger Parts
Up to ½	+.00075 to +.001
½ to 1	+.001 to +.002
1 to 2	+.002 to +.003
2 to 3½	+.003 to +.004

*All measurements are in inches.

table 15 Rules for finding dimensions of circles, squares, etc.

D is diameter of stock necessary to turn shape desired.
E is distance "across flats," or diameter of inscribed circle.
C is depth of cut into stock turned to correct diameter.

Triangle
E = side × .57735
D = side × 1.1547 = 2E
Side = D × .866
C = E × .5 = D × .25

Square
E = side = D × .7071
D = side × 1.4142 = diagonal
Side = D × .7071
C = D × .14645

Pentagon
E = side × 1.3764 = D × .809
D = side × 1.7013 = E × 1.2361
Side = D × .5878
C = D × .0955

Hexagon
E = side × 1.7321 = D × .866
D = side × 2 = E × 1.1547
Side = D × .5
C = D × .067

Octagon
E = side × 2.4142 = D × .9239
D = side × 2.6131 = E × 1.0824
Side = D × .3827
C = D × .038

(Morse Twist Drill & Machine Co.)

table 16 Hardness conversion chart

10 mm Ball 3000 kg	120° Cone 150 kg	1/16 in. Ball 100 kg	Model C	Mpa	10 mm Ball 3000 kg	120° Cone 150 kg	1/16 in. Ball 100 kg	Model C	Mpa
Brinell	Rockwell C	Rockwell B	Shore Scleroscope	Tensile Strength	Brinell	Rockwell C	Rockwell B	Shore Scleroscope	Tensile Strength
800	72		100		276	30	105	42	938
780	71		99		269	29	104	41	910
760	70		98		261	28	103	40	889
745	68		97	2530	258	27	102	39	876
725	67		96	2460	255	26	102	39	862
712	66		95	2413	249	25	101	38	848
682	65		93	2324	245	24	100	37	820
668	64		91	2248	240	23	99	36	807
652	63		89	2193	237	23	99	35	793
626	62		87	2110	229	22	98	34	779
614	61		85	2062	224	21	97	33	758
601	60		83	2013	217	20	96	33	738
590	59		81	2000	211	19	95	32	717
576	57		79	1937	206	18	94	32	703
552	56		76	1862	203	17	94	31	689
545	55		75	1848	200	16	93	31	676
529	54		74	1786	196	15	92	30	662
514	53	120	72	1751	191	14	92	30	648
502	52	119	70	1703	187	13	91	29	634
495	51	119	69	1682	185	12	91	29	627
477	49	118	67	1606	183	11	90	28	621
461	48	117	66	1565	180	10	89	28	614
451	47	117	65	1538	175	9	88	27	593
444	46	116	64	1510	170	7	87	27	579
427	45	115	62	1441	167	6	87	27	565
415	44	115	60	1407	165	5	86	26	558
401	43	114	58	1351	163	4	85	26	552
388	42	114	57	1317	160	3	84	25	538
375	41	113	55	1269	156	2	83	25	524
370	40	112	54	1255	154	1	82	25	517
362	39	111	53	1234	152		82	24	510
351	38	111	51	1193	150		81	24	510
346	37	110	50	1172	147		80	24	496
341	37	110	49	1158	145		79	23	490
331	36	109	47	1124	143		79	23	483
323	35	109	46	1089	141		78	23	476
311	34	108	46	1055	140		77	22	476
301	33	107	45	1020	135		75	22	462
293	32	106	44	993	130		72	22	448
285	31	105	43	965					

table 17 Solutions for right-angled triangles

	Sine∠ $= \dfrac{\text{Side opposite}}{\text{Hypotenuse}}$	Cosecant∠ $= \dfrac{\text{Hypotenuse}}{\text{Side opposite}}$
	Cosine∠ $= \dfrac{\text{Side adjacent}}{\text{Hypotenuse}}$	Secant∠ $= \dfrac{\text{Hypotenuse}}{\text{Side adjacent}}$
	Tangent∠ $= \dfrac{\text{Side opposite}}{\text{Side adjacent}}$	Cotangent∠ $= \dfrac{\text{Side adjacent}}{\text{Side opposite}}$

	Knowing	Formulas to Find	
	Sides a and b	$c = \sqrt{a^2 - b^2}$	$\sin B = \dfrac{b}{a}$
	Side a and angle B	$b = a \times \sin B$	$c = a \times \cos B$
	Sides a and c	$b = \sqrt{a^2 - c^2}$	$\sin C = \dfrac{c}{a}$
	Side a and angle C	$b = a \times \cos C$	$c = a \times \sin C$
	Sides b and c	$a = \sqrt{b^2 + c^2}$	$\tan B = \dfrac{b}{c}$
	Side b and angle B	$a = \dfrac{b}{\sin B}$	$c = b \times \cot B$
	Side b and angle C	$a = \dfrac{b}{\cos C}$	$c = b \times \tan C$
	Side c and angle B	$a = \dfrac{c}{\cos B}$	$b = c \times \tan B$
	Side c and angle C	$a = \dfrac{c}{\sin C}$	$b = c \times \cot C$

table 18 Tool steel types

High-Speed Tool Steels (Molybdenum)

AISI	Description
M1	Used primarily for twist drills, reamers, threading tools.
M2	Most widely used; (general-purpose, high-speed uses).
M3-Class 1	Good wear resistance and grindability.
M3-Class 2	Good red hardness, edge toughness, and wear resistance.
M4	Exceptional abrasion resistance (machining abrasive alloys, castings, heat-treated forgings).
M6	Cobalt-bearing type with good red hardness (cutting hard materials, heat-treated forgings).
M7	Good abrasion resistance and grindability.
M10	Good toughness and abrasion resistance (small cutting tools).
M30	Cobalt-bearing type; good balance of red hardness, wear resistance, and edge strength.
M33	Excellent red hardness (machining hard, heat-treated steel parts).
M34	Excellent red hardness, wear resistance, and edge strength.
M41, M42 M43, M44 M46, M47	High-carbon cobalt-bearing types; can be heat treated to RC 68–70; (machining high-strength steels, high temperature alloys, titanium).

(Continued)

table 18   Tool steel types (Continued)

High-Speed Tool Steels (Tungsten)

AISI	Description
T1	General-purpose, high-speed steel.
T2	Finishing cuts at high speed to produce fine surface finishes.
T4	Roughing cuts on hard materials at higher speeds and feeds.
T5	Single-point tools taking heavy cuts on hard, abrasive materials.
T6	High red hardness, heavy-duty lathe tools.
T8	High red hardness, abrasion resistance, and toughness, machining stainless steels.
T15	Maximum wear resistance (single-point tools, blanking, forming, trimming dies).

Hot Work Tool Steels (Chromium)

AISI	Description
H10	Air hardening, excellent resistance to softening at high temperatures, resists heat cracking.
H11	Air hardening, resists heat checking and cracking when water cooled (hot forging tools used at high temperatures).
H12	Air hardening, resists thermal shock, (hot forging and piercing dies, extrusion tools).
H13	Good wear resistance, toughness, and excellent hot hardness (die-casting dies, extrusion tools).
H14	Good high temperature and heat resistance (long-run die-casting dies, brass forging dies, extrusion tools).
H19	Good red hardness, resists shock and abrasion at high temperatures (forging dies, brass extrusion tools).

Hot Work Tool Steels (Tungsten)

AISI	Description
H21	High red hardness, good wear resistance and toughness at high temperatures.
H22	High compressive strength, red hardness, wear resistance (hot forming dies, hot punches, brass forging dies).
H23	High resistance to heat checking, good wear resistance (molds for brass and bronze, copper, and brass extrusion tools).
H24	Very high red hardness, moderate toughness (hot forging punches and dies, hot forming rolls).
H25	General-purpose hot work steel with excellent toughness, good resistance to softening at high temperatures.
H26	Very high red hardness and wear resistance (springs operating at high temperatures).

Hot Work Tool Steels (Molybdenum)

AISI	Description
H41	Low-carbon, high-speed steel (hot-work applications that require increased toughness and wear-resistance).
H42	Similar to H41 with higher heat resistance.
H43	High red hardness, excellent wear resistance, gives long wear under severe conditions.

Cold Work Tool Steels (High-Carbon, High-Chromium)

AISI	Description
D2	Air hardening, high hardness, toughness, dimensional accuracy (long-run tools and dies).
D3	Oil hardening, outstanding wear resistance, deep hardenability and size stability (long production-run tools and dies).
D4	Excellent wear resistance, dimensional stability (dies and wear parts).
D5	With cobalt for red hardness and resistance to galling (semi-hot work applications).
D7	Excellent abrasion wear resistance (brick molds, liner and die plates, briquetting dies, wear-resisting parts).

Cold Work Tool Steels (Oil-Hardening)

AISI	Description
O1	Good wear resistance and toughness (general-purpose tool steel).
O2	Good machinability and wear resistance (tool and die applications).
O6	Excellent machinability, good wear resistance (drawing, forming, shaping applications).
O7	High hardness, good toughness, retains keen cutting edge (taps, threading tools, reamers).

(Continued)

table 18 Tool steel types (Continued)

Cold Work Tool Steels (Medium Alloy, Air Hardening)

AISI	Description
A2	Excellent dimensional accuracy (tooling with intricate sections).
A3	Wear resistance, toughness, dimensional stability (dies, machine ways, gages, punches).
A4	Minimum distortion during heat treating (dies, slitting cutters, gages, arbors, bushings).
A6	Low hardening temperatures, minimum distortion (gages and tools with close tolerances).
A7	Exceptional abrasion resistance (ceramic, foundry, refractory applications).
A8	High toughness, good wear resistance (heavy shearing, blanking, forming).
A9	High toughness, size stability, resistant to heat checking (heavy blanking, punching, forming operations).
A10	Excellent machinability, minimum distortion (nonuniform dies, parts with varying crosssections).

Shock-Resisting Tool Steels

AISI	Description
S1	Good toughness and hardness (chisels, pneumatic tools, shear blades).
S2	Water hardening, extreme toughness (battering tools, heavy-duty dies and punches).
S5	High toughness, high hardenability (heavy-duty shear blades).
S7	Air hardening, high shock resistance and strength, good distortion resistance (general-purpose applications).

Mold Steels

AISI	Description
P2	Outstanding cold-hobbing steel (injection or compression molding).
P4	Cold-hobbing steel, higher wear strength than PC, can be machined.
P5	Case-hardening steel, easy hobability, good machinability, high strength.
P6	Cavity mold steel, large cavities and high pressures.
P20	Designed for zinc die-casting dies, plastic molds.
P21	High-strength steel (injection molding of thermoplastics, holding blocks, die-casting dies).

Special-Purpose Tool Steels (Low Alloy)

AISI	Description
L2	Water hardening chromium–vanadium steel, medium hardenability.
L3	Chromium-vanadium steel, high hardenability.
L6	Oil-hardening steel, good hardness, toughness, wear resistance (tools, dies, machine parts).

Special-Purpose Tool Steels (Carbon-Tungsten)

AISI	Description
F1	Water hardening, with tungsten to improve wear resistance (threading dies, heading applications).
F2	Water hardening, very hard case, tough core (drawing dies).

Water-Hardening Tool Steels

AISI	Description
W1	Wear-resistant surface, tough core, easily machined.
W2	Carbon-vanadium steel similar to W1 with higher toughness.
W5	Water hardening with slight alloy addition for deeper hardness penetration.

table 19 Sine bar constants (5-in. bar)
(multiply constants by 2 for a 10-in. sine bar)

Min.	0°	1°	2°	3°	4°	5°	6°	7°	8°	9°	10°	11°	12°	13°	14°	15°	16°	17°	18°	19°	Min.
0	.00000	.08725	.17450	.26170	.34880	.43580	.52265	.60935	.69585	.78215	.86825	.95405	1.0395	1.1247	1.2096	1.2941	1.3782	1.4618	1.5451	1.6278	0
2	.00290	.09015	.17740	.26460	.35170	.43870	.52555	.61225	.69875	.78505	.87110	.95690	.0424	.1276	.2124	.2969	.3810	.4646	.5478	.6306	2
4	.00580	.09310	.18030	.26750	.35460	.44155	.52845	.61510	.70165	.78790	.87395	.95975	.0452	.1304	.2152	.2997	.3838	.4674	.5506	.6333	4
6	.00875	.09600	.18320	.27040	.35750	.44445	.53130	.61800	.70450	.79080	.87685	.96260	.0481	.1332	.2181	.3025	.3865	.4702	.5534	.6361	6
8	.01165	.09890	.18615	.27330	.36040	.44735	.53420	.62090	.70740	.79365	.87970	.96545	.0509	.1361	.2209	.3053	.3893	.4730	.5561	.6388	8
10	.01455	.10180	.18905	.27620	.36330	.45025	.53710	.62380	.71025	.79655	.88255	.96830	1.0538	1.1389	1.2237	1.3081	1.3921	1.4757	1.5589	1.6416	10
12	.01745	.10470	.19195	.27910	.36620	.45315	.54000	.62665	.71315	.79940	.88540	.97115	.0566	.1417	.2265	.3109	.3949	.4785	.5616	.6443	12
14	.02035	.10760	.19485	.28200	.36910	.45605	.54290	.62955	.71600	.80230	.88830	.97405	.0594	.1446	.2293	.3137	.3977	.4813	.5644	.6471	14
16	.02325	.11055	.19775	.28490	.37200	.45895	.54580	.63245	.71890	.80515	.89115	.97690	.0623	.1474	.2322	.3165	.4005	.4841	.5672	.6498	16
18	.02620	.11345	.20065	.28780	.37490	.46185	.54865	.63530	.72180	.80800	.89400	.97975	.0651	.1502	.2350	.3193	.4033	.4868	.5699	.6525	18
20	.02910	.11635	.20355	.29070	.37780	.46475	.55155	.63820	.72465	.81090	.89685	.98260	1.0680	1.1531	1.2378	1.3221	1.4061	1.4896	1.5727	1.6553	20
22	.03200	.11925	.20645	.29365	.38070	.46765	.55445	.64110	.72755	.81375	.89975	.98545	.0708	.1559	.2406	.3250	.4089	.4924	.5755	.6580	22
24	.03490	.12215	.20940	.29655	.38360	.47055	.55735	.64400	.73040	.81665	.90260	.98830	.0737	.1587	.2434	.3278	.4117	.4952	.5782	.6608	24
26	.03780	.12505	.21230	.29945	.38650	.47345	.56025	.64685	.73330	.81950	.90545	.99115	.0765	.1615	.2462	.3306	.4145	.4980	.5810	.6635	26
28	.04070	.12800	.21520	.30235	.38940	.47635	.56315	.64975	.73615	.82235	.90830	.99400	.0793	.1644	.2491	.3334	.4173	.5007	.5837	.6663	28
30	.04365	.13090	.21810	.30525	.39230	.47925	.56600	.65265	.73905	.82525	.91120	.99685	1.0822	1.1672	1.2519	1.3362	1.4201	1.5035	1.5865	1.6690	30
32	.04655	.13380	.22100	.30815	.39520	.48210	.56890	.65550	.74190	.82810	.91405	.99970	.0850	.1700	.2547	.3390	.4228	.5063	.5893	.6718	32
34	.04945	.13670	.22390	.31105	.39810	.48500	.57180	.65840	.74480	.83100	.91690	1.0016	.0879	.1729	.2575	.3418	.4256	.5091	.5920	.6745	34
36	.05235	.13960	.22680	.31395	.40100	.48790	.57470	.66130	.74770	.83385	.91975	.0054	.0907	.1757	.2603	.3446	.4284	.5118	.5948	.6772	36
38	.05525	.14250	.22970	.31685	.40390	.49080	.57760	.66415	.75055	.83670	.92260	.0082	.0935	.1785	.2631	.3474	.4312	.5146	.5975	.6800	38
40	.05820	.14540	.23265	.31975	.40680	.49370	.58045	.66705	.75345	.83960	.92545	1.0110	1.0964	1.1813	1.2660	1.3502	1.4340	1.5174	1.6003	1.6827	40
42	.06110	.14835	.23555	.32265	.40970	.49660	.58335	.66995	.75630	.84245	.92835	.0139	.0992	.1842	.2688	.3530	.4368	.5201	.6030	.6855	42
44	.06400	.15125	.23845	.32555	.41260	.49950	.58625	.67280	.75920	.84530	.93120	.0168	.1020	.1870	.2716	.3558	.4396	.5229	.6058	.6882	44
46	.06690	.15415	.24135	.32845	.41550	.50240	.58915	.67570	.76205	.84820	.93405	.0196	.1049	.1898	.2744	.3586	.4423	.5257	.6085	.6909	46
48	.06980	.15705	.24425	.33135	.41840	.50530	.59200	.67860	.76495	.85105	.93690	.0225	.1077	.1926	.2772	.3614	.4451	.5285	.6113	.6937	48
50	.07270	.15995	.24715	.33425	.42130	.50820	.59490	.68145	.76780	.85390	.93975	1.0253	1.1106	1.1955	1.2800	1.3642	1.4479	1.5312	1.6141	1.6964	50
52	.07565	.16285	.25005	.33715	.42420	.51105	.59780	.68435	.77070	.85680	.94260	.0281	.1134	.1983	.2828	.3670	.4507	.5340	.6168	.6991	52
54	.07855	.16580	.25295	.34010	.42710	.51395	.60070	.68720	.77355	.85965	.94550	.0310	.1162	.2011	.2856	.3698	.4535	.5368	.6196	.7019	54
56	.08145	.16870	.25585	.34300	.43000	.51685	.60355	.69010	.77645	.86250	.94835	.0338	.1191	.2039	.2884	.3726	.4563	.5395	.6223	.7046	56
58	.08435	.17160	.25875	.34590	.43290	.51975	.60645	.69300	.77930	.86540	.95120	.0367	.1219	.2068	.2913	.3754	.4591	.5423	.6251	.7073	58
60	.08725	.17450	.26170	.34880	.43580	.52265	.60935	.69585	.78215	.86825	.95405	1.0395	1.1247	1.2096	1.2941	1.3782	1.4618	1.5451	1.6278	1.7101	60

(Brown & Sharpe)

table 19

Sine bar constants (5-in. bar) (Continued)
(multiply constants by 2 for a 10-in. sine bar)

Min.	20°	21°	22°	23°	24°	25°	26°	27°	28°	29°	30°	31°	32°	33°	34°	35°	36°	37°	38°	39°	Min.
0	1.7101	1.7918	1.8730	1.9536	2.0337	2.1131	2.1918	2.2699	2.3473	2.4240	2.5000	2.5752	2.6496	2.7232	2.7959	2.8679	2.9389	3.0091	3.0783	3.1466	0
2	.7128	.7945	.8757	.9563	.0363	.1157	.1944	.2725	.3499	.4266	.5025	.5777	.6520	.7256	.7984	.8702	.9413	.0114	.0806	.1488	2
4	.7155	.7972	.8784	.9590	.0390	.1183	.1971	.2751	.3525	.4291	.5050	.5802	.6545	.7280	.8008	.8726	.9436	.0137	.0829	.1511	4
6	.7183	.8000	.8811	.9617	.0416	.1210	.1997	.2777	.3550	.4317	.5075	.5826	.6570	.7305	.8032	.8750	.9460	.0160	.0852	.1534	6
8	.7210	.8027	.8838	.9643	.0443	.1236	.2023	.2803	.3576	.4342	.5100	.5851	.6594	.7329	.8056	.8774	.9483	.0183	.0874	.1556	8
10	1.7237	1.8054	1.8865	1.9670	2.0469	2.1262	2.2049	2.2829	2.3602	2.4367	2.5126	2.5876	2.6619	2.7354	2.8080	2.8798	2.9507	3.0207	3.0897	3.1579	10
12	.7265	.8081	.8892	.9697	.0496	.1289	.2075	.2855	.3627	.4393	.5151	.5901	.6644	.7378	.8104	.8821	.9530	.0230	.0920	.1601	12
14	.7292	.8108	.8919	.9724	.0522	.1315	.2101	.2881	.3653	.4418	.5176	.5926	.6668	.7402	.8128	.8845	.9554	.0253	.0943	.1624	14
16	.7319	.8135	.8946	.9750	.0549	.1341	.2127	.2906	.3679	.4444	.5201	.5951	.6693	.7427	.8152	.8869	.9577	.0276	.0966	.1646	16
18	.7347	.8162	.8973	.9777	.0575	.1368	.2153	.2932	.3704	.4469	.5226	.5976	.6717	.7451	.8176	.8893	.9600	.0299	.0989	.1669	18
20	1.7374	1.8189	1.8999	1.9804	2.0602	2.1394	2.2179	2.2958	2.3730	2.4494	2.5251	2.6001	2.6742	2.7475	2.8200	2.8916	2.9624	3.0322	3.1012	3.1691	20
22	.7401	.8217	.9026	.9830	.0628	.1420	.2205	.2984	.3755	.4520	.5276	.6025	.6767	.7499	.8224	.8940	.9647	.0345	.1034	.1714	22
24	.7428	.8244	.9053	.9857	.0655	.1447	.2232	.3010	.3781	.4545	.5301	.6050	.6791	.7524	.8248	.8964	.9671	.0369	.1057	.1736	24
26	.7456	.8271	.9080	.9884	.0681	.1473	.2258	.3036	.3807	.4570	.5327	.6075	.6816	.7548	.8272	.8988	.9694	.0392	.1080	.1759	26
28	.7483	.8298	.9107	.9911	.0708	.1499	.2284	.3061	.3832	.4596	.5352	.6100	.6840	.7572	.8296	.9011	.9718	.0415	.1103	.1781	28
30	1.7510	1.8325	1.9134	1.9937	2.0734	2.1525	2.2310	2.3087	2.3858	2.4621	2.5377	2.6125	2.6865	2.7597	2.8320	2.9035	2.9741	3.0438	3.1125	3.1804	30
32	.7537	.8352	.9161	.9964	.0761	.1552	.2336	.3113	.3883	.4646	.5402	.6149	.6889	.7621	.8344	.9059	.9764	.0461	.1148	.1826	32
34	.7565	.8379	.9188	.9991	.0787	.1578	.2362	.3139	.3909	.4672	.5427	.6174	.6914	.7645	.8368	.9082	.9788	.0484	.1171	.1849	34
36	.7592	.8406	.9215	2.0017	.0814	.1604	.2388	.3165	.3934	.4697	.5452	.6199	.6938	.7669	.8392	.9106	.9811	.0507	.1194	.1871	36
38	.7619	.8433	.9241	.0044	.0840	.1630	.2414	.3190	.3960	.4722	.5477	.6224	.6963	.7694	.8416	.9130	.9834	.0530	.1216	.1893	38
40	1.7646	1.8460	1.9268	2.0070	2.0867	2.1656	2.2440	2.3216	2.3985	2.4747	2.5502	2.6249	2.6987	2.7718	2.8440	2.9153	2.9858	3.0553	3.1239	3.1916	40
42	.7673	.8487	.9295	.0097	.0893	.1683	.2466	.3242	.4011	.4773	.5527	.6273	.7012	.7742	.8464	.9177	.9881	.0576	.1262	.1938	42
44	.7701	.8514	.9322	.0124	.0920	.1709	.2492	.3268	.4036	.4798	.5552	.6298	.7036	.7766	.8488	.9200	.9904	.0599	.1285	.1961	44
46	.7728	.8541	.9349	.0150	.0946	.1735	.2518	.3293	.4062	.4823	.5577	.6323	.7061	.7790	.8512	.9224	.9928	.0622	.1307	.1983	46
48	.7755	.8568	.9376	.0177	.0972	.1761	.2544	.3319	.4087	.4848	.5602	.6348	.7085	.7815	.8535	.9248	.9951	.0645	.1330	.2005	48
50	1.7782	1.8595	1.9402	2.0204	2.0999	2.1787	2.2570	2.3345	2.4113	2.4874	2.5627	2.6372	2.7110	2.7839	2.8559	2.9271	2.9974	3.0668	3.1353	3.2028	50
52	.7809	.8622	.9429	.0230	.1025	.1814	.2596	.3371	.4138	.4899	.5652	.6397	.7134	.7863	.8583	.9295	.9997	.0691	.1375	.2050	52
54	.7837	.8649	.9456	.0257	.1052	.1840	.2621	.3396	.4164	.4924	.5677	.6422	.7158	.7887	.8607	.9318	3.0021	.0714	.1398	.2072	54
56	.7864	.8676	.9483	.0283	.1078	.1866	.2647	.3422	.4189	.4949	.5702	.6446	.7183	.7911	.8631	.9342	.0044	.0737	.1421	.2095	56
58	.7891	.8703	.9510	.0310	.1104	.1892	.2673	.3448	.4215	.4975	.5727	.6471	.7207	.7935	.8655	.9365	.0067	.0760	.1443	.2117	58
60	1.7918	1.8730	1.9536	2.0337	2.1131	2.1918	2.2699	2.3473	2.4240	2.5000	2.5752	2.6496	2.7232	2.7959	2.8679	2.9389	3.0091	3.0783	3.1466	3.2139	60

table 19 Sine bar constants (5-in. bar) (Continued)
(multiply constants by 2 for a 10-in. sine bar)

Min.	40°	41°	42°	43°	44°	45°	46°	47°	48°	49°	50°	51°	52°	53°	54°	55°	56°	57°	58°	59°	Min.
0	3.2139	3.2803	3.3456	3.4100	3.4733	3.5355	3.5967	3.6567	3.7157	3.7735	3.8302	3.8857	3.9400	3.9932	4.0451	4.0957	4.1452	4.1933	4.2402	4.2858	0
2	.2161	.2825	.3478	.4121	.4754	.5376	.5987	.6587	.7176	.7754	.8321	.8875	.9418	.9949	.0468	.0974	.1468	.1949	.2418	.2873	2
4	.2184	.2847	.3499	.4142	.4774	.5396	.6007	.6607	.7196	.7773	.8339	.8894	.9436	.9967	.0485	.0991	.1484	.1965	.2433	.2888	4
6	.2206	.2869	.3521	.4163	.4795	.5417	.6027	.6627	.7215	.7792	.8358	.8912	.9454	.9984	.0502	.1007	.1500	.1981	.2448	.2903	6
8	.2228	.2890	.3543	.4185	.4816	.5437	.6047	.6647	.7235	.7811	.8377	.8930	.9472	4.0001	.0519	.1024	.1517	.1997	.2464	.2918	8
10	3.2250	3.2912	3.3564	3.4206	3.4837	3.5458	3.6068	3.6666	3.7254	3.7830	3.8395	3.8948	3.9490	4.0019	4.0536	4.1041	4.1533	4.2012	4.2479	4.2933	10
12	.2273	.2934	.3586	.4227	.4858	.5478	.6088	.6686	.7274	.7850	.8414	.8967	.9508	.0036	.0553	.1057	.1549	.2028	.2494	.2948	12
14	.2295	.2956	.3607	.4248	.4879	.5499	.6108	.6706	.7293	.7869	.8433	.8985	.9525	.0054	.0570	.1074	.1565	.2044	.2510	.2963	14
16	.2317	.2978	.3629	.4269	.4900	.5519	.6128	.6726	.7312	.7887	.8451	.9003	.9543	.0071	.0587	.1090	.1581	.2060	.2525	.2978	16
18	.2339	.3000	.3650	.4291	.4921	.5540	.6148	.6745	.7332	.7906	.8470	.9021	.9561	.0089	.0604	.1107	.1597	.2075	.2540	.2992	18
20	3.2361	3.3022	3.3672	3.4312	3.4941	3.5560	3.6168	3.6765	3.7351	3.7925	3.8488	3.9039	3.9579	4.0106	4.0621	4.1124	4.1614	4.2091	4.2556	4.3007	20
22	.2384	.3044	.3693	.4333	.4962	.5581	.6188	.6785	.7370	.7944	.8507	.9058	.9596	.0123	.0638	.1140	.1630	.2107	.2571	.3022	22
24	.2406	.3065	.3715	.4354	.4983	.5601	.6208	.6805	.7390	.7963	.8525	.9076	.9614	.0141	.0655	.1157	.1646	.2122	.2586	.3037	24
26	.2428	.3087	.3736	.4375	.5004	.5621	.6228	.6824	.7409	.7982	.8544	.9094	.9632	.0158	.0672	.1173	.1662	.2138	.2601	.3052	26
28	.2450	.3109	.3758	.4396	.5024	.5642	.6248	.6844	.7428	.8001	.8562	.9112	.9650	.0175	.0689	.1190	.1678	.2154	.2617	.3066	28
30	3.2472	3.3131	3.3779	3.4417	3.5045	3.5662	3.6268	3.6864	3.7448	3.8020	3.8581	3.9130	3.9667	4.0193	4.0706	4.1206	4.1694	4.2169	4.2632	4.3081	30
32	.2494	.3153	.3801	.4439	.5066	.5683	.6288	.6883	.7467	.8039	.8599	.9148	.9685	.0210	.0722	.1223	.1710	.2185	.2647	.3096	32
34	.2516	.3174	.3822	.4460	.5087	.5703	.6308	.6903	.7486	.8058	.8618	.9166	.9703	.0227	.0739	.1239	.1726	.2201	.2662	.3111	34
36	.2538	.3196	.3844	.4481	.5107	.5723	.6328	.6923	.7505	.8077	.8636	.9184	.9720	.0244	.0756	.1255	.1742	.2216	.2677	.3125	36
38	.2561	.3218	.3865	.4502	.5128	.5744	.6348	.6942	.7525	.8096	.8655	.9202	.9738	.0262	.0773	.1272	.1758	.2232	.2692	.3140	38
40	3.2583	3.3240	3.3886	3.4523	3.5149	3.5764	3.6368	3.6962	3.7544	3.8114	3.8673	3.9221	3.9756	4.0279	4.0790	4.1288	4.1774	4.2247	4.2708	4.3155	40
42	.2605	.3261	.3908	.4544	.5169	.5784	.6388	.6981	.7563	.8133	.8692	.9239	.9773	.0296	.0807	.1305	.1790	.2263	.2723	.3170	42
44	.2627	.3283	.3929	.4565	.5190	.5805	.6408	.7001	.7582	.8152	.8710	.9257	.9791	.0313	.0823	.1321	.1806	.2278	.2738	.3184	44
46	.2649	.3305	.3950	.4586	.5211	.5825	.6428	.7020	.7601	.8171	.8729	.9275	.9809	.0331	.0840	.1337	.1822	.2294	.2753	.3199	46
48	.2671	.3326	.3972	.4607	.5231	.5845	.6448	.7040	.7620	.8190	.8747	.9293	.9826	.0348	.0857	.1354	.1838	.2309	.2768	.3213	48
50	3.2693	3.3348	3.3993	3.4628	3.5252	3.5866	3.6468	3.7060	3.7640	3.8208	3.8765	3.9311	3.9844	4.0365	4.0874	4.1370	4.1854	4.2325	4.2783	4.3228	50
52	.2715	.3370	.4014	.4649	.5273	.5886	.6488	.7079	.7659	.8227	.8784	.9329	.9861	.0382	.0891	.1386	.1870	.2340	.2798	.3243	52
54	.2737	.3391	.4036	.4670	.5293	.5906	.6508	.7099	.7678	.8246	.8802	.9347	.9879	.0399	.0907	.1403	.1886	.2356	.2813	.3257	54
56	.2759	.3413	.4057	.4691	.5314	.5926	.6528	.7118	.7697	.8265	.8820	.9364	.9896	.0416	.0924	.1419	.1902	.2371	.2828	.3272	56
58	.2781	.3435	.4078	.4712	.5335	.5947	.6548	.7138	.7716	.8283	.8839	.9382	.9914	.0433	.0941	.1435	.1917	.2387	.2843	.3286	58
60	3.2803	3.3456	3.4100	3.4733	3.5355	3.5967	3.6567	3.7157	3.7735	3.8302	3.8857	3.9400	3.9932	4.0451	4.0957	4.1452	4.1933	4.2402	4.2858	4.3301	60

(Brown & Sharpe)

table 20a   Coordinate factors and angles, 3-hole division

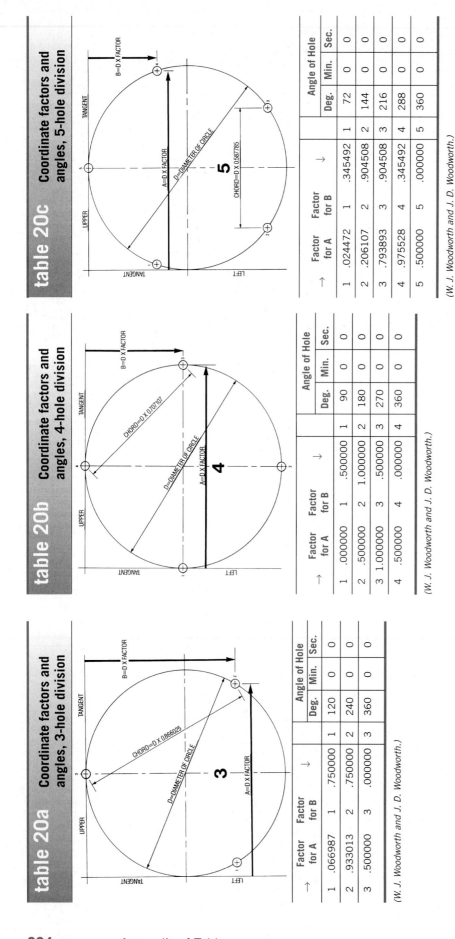

	Factor for A		Factor for B	→		Angle of Hole		
						Deg.	Min.	Sec.
1	.066987	1	.750000	→	1	120	0	0
2	.933013	2	.750000		2	240	0	0
3	.500000	3	.000000		3	360	0	0

(W. J. Woodworth and J. D. Woodworth.)

table 20b   Coordinate factors and angles, 4-hole division

→	Factor for A		Factor for B	→		Angle of Hole		
						Deg.	Min.	Sec.
1	.000000	1	.500000	→	1	90	0	0
2	.500000	2	1.000000		2	180	0	0
3	1.000000	3	.500000		3	270	0	0
4	.500000	4	.000000		4	360	0	0

(W. J. Woodworth and J. D. Woodworth.)

table 20c   Coordinate factors and angles, 5-hole division

→	Factor for A	Factor for B			→		Angle of Hole		
							Deg.	Min.	Sec.
1	.024472	1	.345492		→	1	72	0	0
2	.206107	2	.904508			2	144	0	0
3	.793893	3	.904508			3	216	0	0
4	.975528	4	.345492			4	288	0	0
5	.500000	5	.000000			5	360	0	0

(W. J. Woodworth and J. D. Woodworth.)

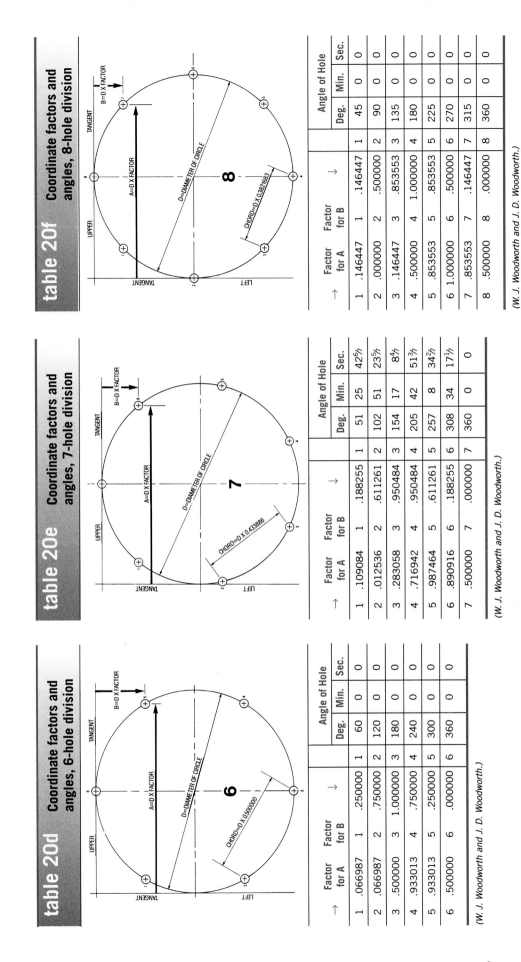

table 20d — Coordinate factors and angles, 6-hole division

→	Factor for A		Factor for B	→	Angle of Hole		
					Deg.	Min.	Sec.
1	.066987	1	.250000	1	60	0	0
2	.066987	2	.750000	2	120	0	0
3	.500000	3	1.000000	3	180	0	0
4	.933013	4	.750000	4	240	0	0
5	.933013	5	.250000	5	300	0	0
6	.500000	6	.000000	6	360	0	0

(W. J. Woodworth and J. D. Woodworth.)

table 20e — Coordinate factors and angles, 7-hole division

→	Factor for A		Factor for B	→	Angle of Hole		
					Deg.	Min.	Sec.
1	.109084	1	.188255	1	51	25	42⁶⁄₇
2	.012536	2	.611261	2	102	51	23⁵⁄₇
3	.283058	3	.950484	3	154	17	8⁴⁄₇
4	.716942	4	.950484	4	205	42	51³⁄₇
5	.987464	5	.611261	5	257	8	34²⁄₇
6	.890916	6	.188255	6	308	34	17¹⁄₇
7	.500000	7	.000000	7	360	0	0

(W. J. Woodworth and J. D. Woodworth.)

table 20f — Coordinate factors and angles, 8-hole division

→	Factor for A		Factor for B	→	Angle of Hole		
					Deg.	Min.	Sec.
1	.146447	1	.146447	1	45	0	0
2	.000000	2	.500000	2	90	0	0
3	.146447	3	.853553	3	135	0	0
4	.500000	4	1.000000	4	180	0	0
5	.853553	5	.853553	5	225	0	0
6	1.000000	6	.500000	6	270	0	0
7	.853553	7	.146447	7	315	0	0
8	.500000	8	.000000	8	360	0	0

(W. J. Woodworth and J. D. Woodworth.)

Appendix of Tables   **885**

table 20g — Coordinate factors and angles, 9-hole division

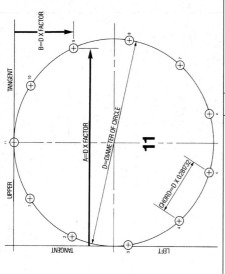

	Factor for A		Factor for B	Angle of Hole Deg.	Min.	Sec.
1	.178606	1	.116978	40	0	0
2	.007596	2	.413176	80	0	0
3	.066987	3	.750000	120	0	0
4	.328990	4	.969846	160	0	0
5	.671010	5	.969846	200	0	0
6	.933013	6	.750000	240	0	0
7	.992404	7	.413176	280	0	0
8	.821394	8	.116978	320	0	0
9	.500000	9	.000000	360	0	0

(W. J. Woodworth and J. D. Woodworth.)

table 20h — Coordinate factors and angles, 10-hole division

	Factor for A		Factor for B	Angle of Hole Deg.	Min.	Sec.
1	.206107	1	.095492	36	0	0
2	.024472	2	.345492	72	0	0
3	.024472	3	.654508	108	0	0
4	.206107	4	.904508	144	0	0
5	.500000	5	1.000000	180	0	0
6	.793893	6	.904508	216	0	0
7	.975528	7	.654508	252	0	0
8	.975528	8	.345492	288	0	0
9	.793893	9	.095492	324	0	0
10	.500000	10	.000000	360	0	0

(W. J. Woodworth and J. D. Woodworth.)

table 20i — Coordinate factors and angles, 11-hole division

	Factor for A		Factor for B	Angle of Hole Deg.	Min.	Sec.
1	.229680	1	.079373	32	43	$38\frac{2}{11}$
2	.045184	2	.292293	65	27	$16\frac{4}{11}$
3	.005089	3	.571157	98	10	$54\frac{6}{11}$
4	.122125	4	.827430	130	54	$32\frac{8}{11}$
5	.359134	5	.979746	163	38	$10\frac{10}{11}$
6	.640866	6	.979746	196	21	$49\frac{1}{11}$
7	.877875	7	.827430	229	5	$27\frac{3}{11}$
8	.994911	8	.571157	261	49	$5\frac{5}{11}$
9	.954816	9	.292293	294	32	$43\frac{7}{11}$
10	.770320	10	.079373	327	16	$21\frac{9}{11}$
11	.500000	11	.000000	360	0	0

(W. J. Woodworth and J. D. Woodworth.)

table 21 Natural trigonometric functions

′	0° sin	0° cos	0° tan	0° cot	0° sec	0° cosec
0	.00000	1.0000	.00000	Infinite	1.0000	Infinite
1	.00029	.0000	.00029	3437.7	.0000	3437.7
2	.00058	.0000	.00058	1718.9	.0000	1718.9
3	.00087	.0000	.00087	1145.9	.0000	1145.9
4	.00116	.0000	.00116	859.44	.0000	859.44
5	.00145	1.0000	.00145	687.55	.0000	687.55
6	.00174	.0000	.00174	572.96	.0000	572.96
7	.00204	.0000	.00204	491.11	.0000	491.11
8	.00233	.0000	.00233	429.72	.0000	429.72
9	.00262	.0000	.00262	381.97	.0000	381.97
10	.00291	0.99999	.00291	343.77	.0000	343.77
11	.00320	.99999	.00320	312.52	.0000	312.52
12	.00349	.99999	.00349	286.48	.0000	286.48
13	.00378	.99999	.00378	264.44	.0000	264.44
14	.00407	.99999	.00407	245.55	.0000	245.55
15	.00436	.99999	.00436	229.18	.0000	229.18
16	.00465	.99999	.00465	214.86	.0000	214.86
17	.00494	.99999	.00494	202.22	.0000	202.22
18	.00524	.99999	.00524	190.98	.0000	190.99
19	.00553	.99998	.00553	180.93	.0000	180.93
20	.00582	.99998	.00582	171.88	.0000	171.89
21	.00611	.99998	.00611	163.70	.0000	163.70
22	.00640	.99998	.00640	156.26	.0000	156.26
23	.00669	.99998	.00669	149.46	.0000	149.47
24	.00698	.99997	.00698	143.24	.0000	143.24
25	.00727	.99997	.00727	137.51	.0000	137.51
26	.00756	.99997	.00756	132.22	.0000	132.22
27	.00785	.99997	.00785	127.32	.0000	127.32
28	.00814	.99997	.00814	122.77	.0000	122.78
29	.00843	.99996	.00844	118.54	.0000	118.54
30	.00873	.99996	.00873	114.59	.0000	114.59
31	.00902	.99996	.00902	110.89	.0000	110.90
32	.00931	.99996	.00931	107.43	.0000	107.43
33	.00960	.99995	.00960	104.17	.0000	104.17
34	.00989	.99995	.00989	101.11	.0000	101.11
35	.01018	.99995	.01018	98.218	.0000	98.223
36	.01047	.99994	.01047	95.489	.0000	95.495
37	.01076	.99994	.01076	92.908	.0000	92.914
38	.01105	.99994	.01105	90.463	.0001	90.469
39	.01134	.99993	.01134	88.143	.0001	88.149
40	.01163	.99993	.01164	85.940	.0001	85.946
41	.01193	.99993	.01193	83.843	.0001	83.849
42	.01222	.99992	.01222	81.847	.0001	81.853
43	.01251	.99992	.01251	79.943	.0001	79.950
44	.01280	.99992	.01280	78.126	.0001	78.133
45	.01309	.99991	.01309	76.390	.0001	76.396
46	.01338	.99991	.01338	74.729	.0001	74.736
47	.01367	.99991	.01367	73.139	.0001	73.146
48	.01396	.99990	.01396	71.615	.0001	71.622
49	.01425	.99990	.01425	70.153	.0001	70.160
50	.01454	.99989	.01454	68.750	.0001	68.757
51	.01483	.99989	.01484	67.402	.0001	67.409
52	.01512	.99988	.01513	66.105	.0001	66.113
53	.01542	.99988	.01542	64.858	.0001	64.866
54	.01571	.99988	.01571	63.657	.0001	63.664
55	.01600	.99987	.01600	62.499	.0001	62.507
56	.01629	.99987	.01629	61.383	.0001	61.391
57	.01658	.99987	.01658	60.306	.0001	60.314
58	.01687	.99986	.01687	59.266	.0001	59.274
59	.01716	.99985	.01716	58.261	.0001	58.270
60	.01745	.99985	.01745	57.290	.0001	57.299
′	cos	sin	cot	tan	cosec	sec
	89°					

′	1° sin	1° cos	1° tan	1° cot	1° sec	1° cosec
0	.01745	.99985	.01745	57.290	1.0001	57.299
1	.01774	.99984	.01775	56.350	.0001	56.359
2	.01803	.99984	.01804	55.441	.0001	55.450
3	.01832	.99983	.01833	54.561	.0002	54.570
4	.01861	.99983	.01862	53.708	.0002	53.718
5	.01891	.99982	.01891	52.882	.0002	52.891
6	.01920	.99981	.01920	52.081	.0002	52.090
7	.01949	.99981	.01949	51.303	.0002	51.313
8	.01978	.99980	.01978	50.548	.0002	50.558
9	.02007	.99980	.02007	49.816	.0002	49.826
10	.02036	.99979	.02036	49.104	.0002	49.114
11	.02065	.99979	.02066	48.412	.0002	48.422
12	.02094	.99978	.02095	47.739	.0002	47.750
13	.02123	.99977	.02124	47.085	.0002	47.096
14	.02152	.99977	.02153	46.449	.0002	46.460
15	.02181	.99976	.02182	45.829	.0002	45.840
16	.02210	.99975	.02211	45.226	.0002	45.237
17	.02240	.99975	.02240	44.638	.0002	44.650
18	.02269	.99974	.02269	44.066	.0002	44.077
19	.02298	.99974	.02298	43.508	.0003	43.520
20	.02326	.99973	.02327	42.964	.0003	42.976
21	.02356	.99972	.02357	42.433	.0003	42.445
22	.02385	.99971	.02386	41.916	.0003	41.928
23	.02414	.99971	.02415	41.410	.0003	41.423
24	.02443	.99970	.02444	40.917	.0003	40.930
25	.02472	.99969	.02473	40.436	.0003	40.448
26	.02501	.99969	.02502	39.965	.0003	39.978
27	.02530	.99968	.02531	39.506	.0003	39.518
28	.02559	.99967	.02560	39.057	.0003	39.069
29	.02589	.99966	.02589	38.618	.0003	38.631
30	.02618	.99966	.02618	38.188	.0003	38.201
31	.02647	.99965	.02648	37.769	.0003	37.782
32	.02676	.99964	.02677	37.358	.0003	37.371
33	.02705	.99963	.02706	36.956	.0004	36.969
34	.02734	.99963	.02735	36.563	.0004	36.576
35	.02763	.99962	.02764	36.177	.0004	36.191
36	.02792	.99961	.02793	35.800	.0004	35.814
37	.02821	.99960	.02822	35.431	.0004	35.445
38	.02850	.99959	.02851	35.069	.0004	35.084
39	.02879	.99958	.02880	34.715	.0004	34.729
40	.02908	.99958	.02910	34.368	.0004	34.382
41	.02937	.99957	.02939	34.027	.0004	34.042
42	.02967	.99956	.02968	33.693	.0004	33.708
43	.02996	.99955	.02997	33.366	.0004	33.381
44	.03025	.99954	.03026	33.045	.0004	33.060
45	.03054	.99953	.03055	32.730	.0005	32.745
46	.03083	.99952	.03084	32.421	.0005	32.437
47	.03112	.99951	.03113	32.118	.0005	32.134
48	.03141	.99951	.03143	31.820	.0005	31.836
49	.03170	.99950	.03172	31.528	.0005	31.544
50	.03199	.99949	.03201	31.241	.0005	31.257
51	.03228	.99948	.03230	30.960	.0005	30.976
52	.03257	.99947	.03259	30.683	.0005	30.699
53	.03286	.99946	.03288	30.411	.0005	30.428
54	.03315	.99945	.03317	30.145	.0005	30.161
55	.03344	.99944	.03346	29.882	.0005	29.899
56	.03374	.99943	.03375	29.624	.0006	29.641
57	.03403	.99942	.03405	29.371	.0006	29.388
58	.03432	.99941	.03434	29.122	.0006	29.139
59	.03461	.99940	.03463	28.877	.0006	28.894
60	.03490	.99939	.03492	28.636	.0006	28.654
′	cos	sin	cot	tan	cosec	sec
	88°					

′	2° sin	2° cos	2° tan	2° cot	2° sec	2° cosec
0	.03490	.99939	.03492	28.636	1.0006	28.654
1	.03519	.99938	.03521	28.399	.0006	28.417
2	.03548	.99937	.03550	28.166	.0006	28.184
3	.03577	.99936	.03579	27.937	.0006	27.955
4	.03606	.99935	.03608	27.712	.0006	27.730
5	.03635	.99934	.03638	27.490	.0007	27.508
6	.03664	.99933	.03667	27.271	.0007	27.290
7	.03693	.99932	.03696	27.056	.0007	27.075
8	.03722	.99931	.03725	26.845	.0007	26.864
9	.03751	.99930	.03754	26.637	.0007	26.655
10	.03781	.99928	.03783	26.432	.0007	26.450
11	.03810	.99927	.03812	26.230	.0007	26.249
12	.03839	.99926	.03842	26.031	.0007	26.050
13	.03868	.99925	.03871	25.835	.0007	25.854
14	.03897	.99924	.03900	25.642	.0008	25.661
15	.03926	.99923	.03929	25.452	.0008	25.471
16	.03955	.99922	.03958	25.264	.0008	25.284
17	.03984	.99921	.03987	25.080	.0008	25.100
18	.04013	.99919	.04016	24.898	.0008	24.918
19	.04042	.99918	.04045	24.718	.0008	24.739
20	.04071	.99917	.04075	24.542	.0008	24.562
21	.04100	.99916	.04104	24.367	.0008	24.388
22	.04129	.99915	.04133	24.196	.0008	24.216
23	.04158	.99913	.04162	24.026	.0009	24.047
24	.04187	.99912	.04191	23.859	.0009	23.880
25	.04217	.99911	.04220	23.694	.0009	23.716
26	.04246	.99910	.04249	23.532	.0009	23.553
27	.04275	.99908	.04279	23.372	.0009	23.393
28	.04304	.99907	.04308	23.214	.0009	23.235
29	.04333	.99906	.04337	23.058	.0009	23.079
30	.04362	.99905	.04366	22.904	.0009	22.925
31	.04391	.99903	.04395	22.752	.0010	22.774
32	.04420	.99902	.04424	22.602	.0010	22.624
33	.04449	.99901	.04453	22.454	.0010	22.476
34	.04478	.99900	.04483	22.308	.0010	22.330
35	.04507	.99898	.04512	22.164	.0010	22.186
36	.04536	.99897	.04541	22.022	.0010	22.044
37	.04565	.99896	.04570	21.881	.0010	21.904
38	.04594	.99894	.04599	21.742	.0010	21.765
39	.04623	.99893	.04628	21.606	.0011	21.629
40	.04652	.99892	.04657	21.470	.0011	21.494
41	.04681	.99890	.04687	21.337	.0011	21.360
42	.04711	.99889	.04716	21.205	.0011	21.228
43	.04740	.99888	.04745	21.075	.0011	21.098
44	.04769	.99886	.04774	20.946	.0011	20.970
45	.04798	.99885	.04803	20.819	.0011	20.843
46	.04827	.99883	.04832	20.693	.0012	20.717
47	.04856	.99882	.04862	20.569	.0012	20.593
48	.04885	.99881	.04891	20.446	.0012	20.471
49	.04914	.99879	.04920	20.325	.0012	20.350
50	.04943	.99878	.04949	20.205	.0012	20.230
51	.04972	.99876	.04978	20.087	.0012	20.112
52	.05001	.99875	.05007	19.970	.0012	19.995
53	.05030	.99873	.05037	19.854	.0013	19.880
54	.05059	.99872	.05066	19.740	.0013	19.766
55	.05088	.99870	.05095	19.627	.0013	19.653
56	.05117	.99869	.05124	19.515	.0013	19.541
57	.05146	.99867	.05153	19.405	.0013	19.431
58	.05175	.99866	.05182	19.296	.0013	19.322
59	.05204	.99864	.05212	19.188	.0013	19.214
60	.05234	.99863	.05241	19.081	.0014	19.107
′	cos	sin	cot	tan	cosec	sec
	87°					

′	3° sin	3° cos	3° tan	3° cot	3° sec	3° cosec	′
0	.05234	.99863	.05241	19.081	1.0014	19.107	60
1	.05263	.99861	.05270	18.975	.0014	19.002	59
2	.05292	.99860	.05299	18.871	.0014	18.897	58
3	.05321	.99858	.05328	18.768	.0014	18.794	57
4	.05350	.99857	.05357	18.665	.0014	18.692	56
5	.05379	.99855	.05387	18.564	.0014	18.591	55
6	.05408	.99854	.05416	18.464	.0015	18.491	54
7	.05437	.99852	.05445	18.365	.0015	18.393	53
8	.05466	.99850	.05474	18.268	.0015	18.295	52
9	.05495	.99849	.05503	18.171	.0015	18.198	51
10	.05524	.99847	.05532	18.075	.0015	18.103	50
11	.05553	.99846	.05562	17.980	.0015	18.008	49
12	.05582	.99844	.05591	17.886	.0016	17.914	48
13	.05611	.99842	.05620	17.793	.0016	17.821	47
14	.05640	.99841	.05649	17.701	.0016	17.730	46
15	.05669	.99839	.05678	17.610	.0016	17.639	45
16	.05698	.99837	.05707	17.520	.0016	17.549	44
17	.05727	.99836	.05737	17.431	.0016	17.460	43
18	.05756	.99834	.05766	17.343	.0017	17.372	42
19	.05785	.99832	.05795	17.256	.0017	17.285	41
20	.05814	.99831	.05824	17.169	.0017	17.198	40
21	.05843	.99829	.05853	17.084	.0017	17.113	39
22	.05872	.99827	.05883	16.999	.0017	17.028	38
23	.05902	.99826	.05912	16.915	.0017	16.944	37
24	.05931	.99824	.05941	16.832	.0018	16.861	36
25	.05960	.99822	.05970	16.750	.0018	16.779	35
26	.05989	.99820	.05999	16.668	.0018	16.698	34
27	.06018	.99819	.06029	16.587	.0018	16.617	33
28	.06047	.99817	.06058	16.507	.0018	16.538	32
29	.06076	.99815	.06087	16.428	.0018	16.459	31
30	.06105	.99813	.06116	16.350	.0019	16.380	30
31	.06134	.99812	.06145	16.272	.0019	16.303	29
32	.06163	.99810	.06175	16.195	.0019	16.226	28
33	.06192	.99808	.06204	16.119	.0019	16.150	27
34	.06221	.99806	.06233	16.043	.0019	16.075	26
35	.06250	.99804	.06262	15.969	.0019	16.000	25
36	.06279	.99803	.06291	15.894	.0020	15.926	24
37	.06308	.99801	.06321	15.821	.0020	15.853	23
38	.06337	.99799	.06350	15.748	.0020	15.780	22
39	.06366	.99797	.06379	15.676	.0020	15.708	21
40	.06395	.99795	.06408	15.605	.0020	15.637	20
41	.06424	.99793	.06437	15.534	.0021	15.566	19
42	.06453	.99791	.06467	15.464	.0021	15.496	18
43	.06482	.99790	.06496	15.394	.0021	15.427	17
44	.06511	.99788	.06525	15.325	.0021	15.358	16
45	.06540	.99786	.06554	15.257	.0021	15.290	15
46	.06569	.99784	.06583	15.189	.0022	15.222	14
47	.06598	.99782	.06613	15.122	.0022	15.155	13
48	.06627	.99780	.06642	15.056	.0022	15.089	12
49	.06656	.99778	.06671	14.990	.0022	15.023	11
50	.06685	.99776	.06700	14.924	.0022	14.958	10
51	.06714	.99774	.06730	14.860	.0023	14.893	9
52	.06743	.99772	.06759	14.795	.0023	14.829	8
53	.06772	.99770	.06788	14.732	.0023	14.765	7
54	.06801	.99768	.06817	14.668	.0023	14.702	6
55	.06830	.99766	.06846	14.606	.0023	14.640	5
56	.06859	.99764	.06876	14.544	.0024	14.578	4
57	.06888	.99762	.06905	14.482	.0024	14.517	3
58	.06918	.99760	.06934	14.421	.0024	14.456	2
59	.06947	.99758	.06963	14.361	.0024	14.395	1
60	.06976	.99756	06993	14.301	.0024	14.335	0
′	cos	sin	cot	tan	cosec	sec	′
	86°						

(Bethlehem Steel Corporation)

(Continued)

table 21 Natural trigonometric functions (Continued)

4°

′	sin	cos	tan	cot	sec	cosec
0	.06976	.99756	.06993	14.301	1.0024	14.335
1	.07005	.99754	.07022	14.241	.0025	14.276
2	.07034	.99752	.07051	14.182	.0025	14.217
3	.07063	.99750	.07080	14.123	.0025	14.159
4	.07092	.99748	.07110	14.065	.0025	14.101
5	.07121	.99746	.07139	14.008	.0025	14.043
6	.07150	.99744	.07168	13.951	.0026	13.986
7	.07179	.99742	.07197	13.894	.0026	13.930
8	.07208	.99740	.07226	13.838	.0026	13.874
9	.07237	.99738	.07256	13.782	.0026	13.818
10	.07266	.99736	.07285	13.727	.0026	13.763
11	.07295	.99733	.07314	13.672	.0027	13.708
12	.07324	.99731	.07343	13.617	.0027	13.654
13	.07353	.99729	.07373	13.563	.0027	13.600
14	.07382	.99727	.07402	13.510	.0027	13.547
15	.07411	.99725	.07431	13.457	.0027	13.494
16	.07440	.99723	.07460	13.404	.0028	13.441
17	.07469	.99721	.07490	13.351	.0028	13.389
18	.07498	.99718	.07519	13.299	.0028	13.337
19	.07527	.99716	.07548	13.248	.0028	13.286
20	.07556	.99714	.07577	13.917	.0029	13.235
21	.07585	.99712	.07607	13.146	.0029	13.184
22	.07614	.99710	.07636	13.096	.0029	13.134
23	.07643	.99707	.07665	13.046	.0029	13.084
24	.07672	.99705	.07694	12.996	.0029	13.034
25	.07701	.99703	.07724	12.947	.0030	12.985
26	.07730	.99701	.07753	12.898	.0030	12.937
27	.07759	.99698	.07782	12.849	.0030	12.888
28	.07788	.99696	.07812	12.801	.0030	12.840
29	.07817	.99694	.07841	12.754	.0031	12.793
30	.07846	.99692	.07870	12.706	.0031	12.745
31	.07875	.99689	.07899	12.659	.0031	12.698
32	.07904	.99687	.07929	12.612	.0031	12.652
33	.07933	.99685	.07958	12.566	.0032	12.606
34	.07962	.99682	.07987	12.520	.0032	12.560
35	.07991	.99680	.08016	12.474	.0032	12.514
36	.08020	.99678	.08046	12.429	.0032	12.469
37	.08049	.99675	.08075	12.384	.0032	12.424
38	.08078	.99673	.08104	12.339	.0033	12.379
39	.08107	.99671	.08134	12.295	.0033	12.335
40	.08136	.99668	.08163	12.250	.0033	12.291
41	.08165	.99666	.08192	12.207	.0033	12.248
42	.08194	.99664	.08221	12.163	.0034	12.204
43	.08223	.99661	.08251	12.120	.0034	12.161
44	.08252	.99659	.08280	12.077	.0034	12.118
45	.08281	.99656	.08309	12.035	.0034	12.076
46	.08310	.99654	.08339	11.992	.0035	12.034
47	.08339	.99652	.08368	11.950	.0035	11.992
48	.08368	.99649	.08397	11.909	.0035	11.950
49	.08397	.99647	.08426	11.867	.0035	11.909
50	.08426	.99644	.08456	11.826	.0036	11.868
51	.08455	.99642	.08485	11.785	.0036	11.828
52	.08484	.99639	.08514	11.745	.0036	11.787
53	.08513	.99637	.08544	11.704	.0036	11.747
54	.08542	.99634	.08573	11.664	.0037	11.707
55	.08571	.99632	.08602	11.625	.0037	11.668
56	.08600	.99629	.08632	11.585	.0037	11.628
57	.08629	.99627	.08661	11.546	.0037	11.589
58	.08658	.99624	.08690	11.507	.0038	11.550
59	.08687	.99622	.08719	11.468	.0038	11.512
60	.08715	.99619	.08749	11.430	.0038	11.474
′	cos	sin	cot	tan	cosec	sec

85°

5°

′	sin	cos	tan	cot	sec	cosec
0	.08715	.99619	.08749	11.430	1.0038	11.474
1	.08744	.99617	.08778	11.392	.0038	11.436
2	.08773	.99614	.08807	11.354	.0039	11.398
3	.08802	.99612	.08837	11.316	.0039	11.360
4	.08831	.99609	.08866	11.279	.0039	11.323
5	.08860	.99607	.08895	11.242	.0039	11.286
6	.08889	.99604	.08925	11.205	.0040	11.249
7	.08918	.99601	.08954	11.168	.0040	11.213
8	.08947	.99599	.08983	11.132	.0040	11.176
9	.08976	.99596	.09013	11.095	.0040	11.140
10	.09005	.99594	.09042	11.059	.0041	11.104
11	.09034	.99591	.09071	11.024	.0041	11.069
12	.09063	.99588	.09101	10.988	.0041	11.033
13	.09092	.99586	.09130	10.953	.0041	10.998
14	.09121	.99583	.09159	10.918	.0042	10.963
15	.09150	.99580	.09189	10.883	.0042	10.929
16	.09179	.99578	.09218	10.848	.0042	10.894
17	.09208	.99575	.09247	10.814	.0043	10.860
18	.09237	.99572	.09277	10.780	.0043	10.826
19	.09266	.99570	.09306	10.746	.0043	10.792
20	.09295	.99567	.09335	10.712	.0043	10.758
21	.09324	.99564	.09365	10.678	.0044	10.725
22	.09353	.99562	.09394	10.645	.0044	10.692
23	.09382	.99559	.09423	10.612	.0044	10.659
24	.09411	.99556	.09453	10.579	.0044	10.626
25	.09440	.99553	.09482	10.546	.0045	10.593
26	.09469	.99551	.09511	10.514	.0045	10.561
27	.09498	.99548	.09541	10.481	.0045	10.529
28	.09527	.99545	.09570	10.449	.0046	10.497
29	.09556	.99542	.09599	10.417	.0046	10.465
30	.09584	.99540	.09629	10.385	.0046	10.433
31	.09613	.99537	.09658	10.354	.0046	10.402
32	.09642	.99534	.09688	10.322	.0047	10.371
33	.09671	.99531	.09717	10.291	.0047	10.340
34	.09700	.99528	.09746	10.260	.0047	10.309
35	.09729	.99525	.09776	10.229	.0048	10.278
36	.09758	.99523	.09805	10.199	.0048	10.248
37	.09787	.99520	.09834	10.168	.0048	10.217
38	.09816	.99517	.09864	10.138	.0048	10.187
39	.09845	.99514	.09893	10.108	.0049	10.157
40	.09874	.99511	.09922	10.078	.0049	10.127
41	.09903	.99508	.09952	10.048	.0049	10.098
42	.09932	.99505	.09981	10.019	.0050	10.068
43	.09961	.99503	.10011	9.9893	.0050	10.039
44	.09990	.99500	.10040	9.9601	.0050	10.010
45	.10019	.99497	.10069	9.9310	.0050	9.9812
46	.10048	.99494	.10099	9.9021	.0051	9.9525
47	.10077	.99491	.10128	9.8734	.0051	9.9239
48	.10106	.99488	.10158	9.8448	.0051	9.8955
49	.10134	.99485	.10187	9.8164	.0052	9.8672
50	.10163	.99482	.10216	9.7882	.0052	9.8391
51	.10192	.99479	.10246	9.7601	.0052	9.8112
52	.10221	.99476	.10275	9.7322	.0053	9.7834
53	.10250	.99473	.10305	9.7044	.0053	9.7558
54	.10279	.99470	.10334	9.6768	.0053	9.7283
55	.10308	.99467	.10363	9.6493	.0053	9.7010
56	.10337	.99464	.10393	9.6220	.0054	9.6739
57	.10366	.99461	.10422	9.5949	.0054	9.6469
58	.10395	.99458	.10452	9.5679	.0054	9.6200
59	.10424	.99455	.10481	9.5411	.0055	9.5933
60	.10453	.99452	.10510	9.5144	.0055	9.5668
′	cos	sin	cot	tan	cosec	sec

84°

6°

′	sin	cos	tan	cot	sec	cosec
0	.10453	.99452	.10510	9.5144	1.0055	9.5668
1	.10482	.99449	.10540	.4878	.0055	.5404
2	.10511	.99446	.10569	.4614	.0056	.5141
3	.10540	.99443	.10599	.4351	.0056	.4880
4	.10568	.99440	.10628	.4090	.0056	.4620
5	.10597	.99437	.10657	9.3831	.0057	9.4362
6	.10626	.99434	.10687	.3572	.0057	.4105
7	.10655	.99431	.10716	.3315	.0057	.3850
8	.10684	.99428	.10746	.3060	.0057	.3596
9	.10713	.99424	.10775	.2806	.0058	.3343
10	.10742	.99421	.10805	9.2553	.0058	9.3092
11	.10771	.99418	.10834	.2302	.0058	.2842
12	.10800	.99415	.10863	.2051	.0059	.2593
13	.10829	.99412	.10893	.1803	.0059	.2346
14	.10858	.99409	.10922	.1555	.0059	.2100
15	.10887	.99406	.10952	9.1309	.0060	9.1855
16	.10916	.99402	.10981	.1064	.0060	.1612
17	.10944	.99399	.11011	.0821	.0060	.1370
18	.10973	.99396	.11040	.0579	.0061	.1129
19	.11002	.99393	.11069	.0338	.0061	.0890
20	.11031	.99390	.11099	9.0098	.0061	9.0651
21	.11060	.99386	.11128	8.9860	.0062	.0414
22	.11089	.99383	.11158	.9623	.0062	.0179
23	.11118	.99380	.11187	.9387	.0062	8.9944
24	.11147	.99377	.11217	.9152	.0063	.9711
25	.11176	.99373	.11246	8.8918	.0063	8.9479
26	.11205	.99370	.11276	.8686	.0063	.9248
27	.11234	.99367	.11305	.8455	.0064	.9018
28	.11262	.99364	.11335	.8225	.0064	.8790
29	.11291	.99360	.11364	.7996	.0064	.8563
30	.11320	.99357	.11393	8.7769	.0065	8.8337
31	.11349	.99354	.11423	.7542	.0065	.8112
32	.11378	.99350	.11452	.7317	.0065	.7888
33	.11407	.99347	.11482	.7093	.0066	.7665
34	.11436	.99344	.11511	.6870	.0066	.7444
35	.11465	.99341	.11541	8.6648	.0066	8.7223
36	.11494	.99337	.11570	.6427	.0067	.7004
37	.11523	.99334	.11600	.6208	.0067	.6786
38	.11551	.99330	.11629	.5989	.0067	.6569
39	.11580	.99327	.11659	.5772	.0068	.6353
40	.11609	.99324	.11688	8.5555	.0068	8.6138
41	.11638	.99320	.11718	.5340	.0068	.5924
42	.11667	.99317	.11747	.5126	.0069	.5711
43	.11696	.99314	.11777	.4913	.0069	.5499
44	.11725	.99310	.11806	.4701	.0069	.5289
45	.11754	.99307	.11836	8.4489	.0070	8.5079
46	.11783	.99303	.11865	.4279	.0070	.4871
47	.11811	.99300	.11895	.4070	.0070	.4663
48	.11840	.99296	.11924	.3862	.0071	.4457
49	.11869	.99293	.11954	.3655	.0071	.4251
50	.11898	.99290	.11983	8.3449	.0071	8.4046
51	.11927	.99286	.12013	.3244	.0072	.3843
52	.11956	.99283	.12042	.3040	.0072	.3640
53	.11985	.99279	.12072	.2837	.0073	.3439
54	.12014	.99276	.12101	.2635	.0073	.3238
55	.12042	.99272	.12131	8.2434	.0073	8.3039
56	.12071	.99269	.12160	.2234	.0074	.2840
57	.12100	.99265	.12190	.2035	.0074	.2642
58	.12129	.99262	.12219	.1837	.0074	.2446
59	.12158	.99258	.12249	.1640	.0075	.2250
60	.12187	.99255	.12278	8.1443	.0075	8.2055
′	cos	sin	cot	tan	cosec	sec

83°

7°

′	sin	cos	tan	cot	sec	cosec	′
0	.12187	.99255	.12278	8.1443	1.0075	8.2055	60
1	.12216	.99251	.12308	.1248	.0075	.1861	59
2	.12245	.99247	.12337	.1053	.0076	.1668	58
3	.12273	.99244	.12367	.0860	.0076	.1476	5
4	.12302	.99240	.12396	.0667	.0076	.1285	56
5	.12331	.99237	.12426	8.0476	.0077	8.1094	55
6	.12360	.99233	.12456	.0285	.0077	.0905	54
7	.12389	.99229	.12485	.0095	.0078	.0717	53
8	.12418	.99226	.12515	7.9906	.0078	.0529	52
9	.12447	.99222	.12544	.9717	.0078	.0342	51
10	.12476	.99219	.12574	7.9530	.0079	8.0156	50
11	.12504	.99215	.12603	.9344	.0079	7.9971	49
12	.12533	.99211	.12633	.9158	.0079	.9787	48
13	.12562	.99208	.12662	.8973	.0080	.9604	47
14	.12591	.99204	.12692	.8789	.0080	.9421	46
15	.12620	.99200	.12722	7.8606	.0080	7.9240	45
16	.12649	.99197	.12751	.8424	.0081	.9059	44
17	.12678	.99193	.12781	.8243	.0081	.8879	43
18	.12706	.99189	.12810	.8062	.0082	.8700	42
19	.12735	.99186	.12840	.7882	.0082	.8522	41
20	.12764	.99182	.12869	7.7703	.0082	7.8344	40
21	.12793	.99178	.12899	.7525	.0083	.8168	39
22	.12822	.99174	.12928	.7348	.0083	.7992	38
23	.12851	.99171	.12958	.7171	.0084	.7817	37
24	.12879	.99167	.12988	.6996	.0084	.7642	36
25	.12908	.99163	.13017	7.6821	.0084	7.7469	35
26	.12937	.99160	.13047	.6646	.0085	.7296	34
27	.12966	.99156	.13076	.6473	.0085	.7124	33
28	.12995	.99152	.13106	.6300	.0085	.6953	32
29	.13024	.99148	.13136	.6129	.0086	.6783	31
30	.13053	.99144	.13165	7.5957	.0086	7.6613	30
31	.13081	.99141	.13195	.5787	.0087	.6444	29
32	.13110	.99137	.13224	.5617	.0087	.6276	28
33	.13139	.99133	.13254	.5449	.0087	.6108	27
34	.13168	.99129	.13284	.5280	.0088	.5942	26
35	.13197	.99125	.13313	7.5113	.0088	7.5776	25
36	.13226	.99121	.13343	.4946	.0089	.5611	24
37	.13254	.99118	.13372	.4780	.0089	.5446	23
38	.13283	.99114	.13402	.4615	.0089	.5282	22
39	.13312	.99110	.13432	.4451	.0090	.5119	21
40	.13341	.99106	.13461	7.4287	.0090	7.4957	20
41	.13370	.99102	.13491	.4124	.0090	.4795	19
42	.13399	.99098	.13520	.3961	.0091	.4634	18
43	.13427	.99094	.13550	.3800	.0091	.4474	17
44	.13456	.99090	.13580	.3639	.0092	.4315	16
45	.13485	.99086	.13609	7.3479	.0092	7.4156	15
46	.13514	.99083	.13639	.3319	.0092	.3998	14
47	.13543	.99079	.13669	.3160	.0093	.3840	13
48	.13571	.99075	.13698	.3002	.0093	.3683	12
49	.13600	.99071	.13728	.2844	.0094	.3527	11
50	.13629	.99067	.13757	7.2687	.0094	7.3372	10
51	.13658	.99063	.13787	.2531	.0094	.3217	9
52	.13687	.99059	.13817	.2375	.0095	.3063	8
53	.13716	.99055	.13846	.2220	.0095	.2909	7
54	.13744	.99051	.13876	.2066	.0096	.2757	6
55	.13773	.99047	.13906	7.1912	.0096	7.2604	5
56	.13802	.99043	.13935	.1759	.0097	.2453	4
57	.13831	.99039	.13965	.1607	.0097	.2302	3
58	.13860	.99035	.13995	.1455	.0097	.2152	2
59	.13888	.99031	.14024	.1304	.0098	.2002	1
60	.13917	.99027	.14054	7.1154	.0098	7.1853	0
′	cos	sin	cot	tan	cosec	sec	′

82°

(Continued)

table 21 Natural trigonometric functions (Continued)

′	8°						9°						′
′	sin	cos	tan	cot	sec	cosec	sin	cos	tan	cot	sec	cosec	′
0	.13917	.99027	.14054	7.1154	1.0098	7.1853	.15643	.98769	.15838	6.3137	1.0215	6.3924	60
1	.13946	.99023	.14084	.1004	.0099	.1704	.15672	.98764	.15868	.3019	.0125	.3807	59
2	.13975	.99019	.14113	.0854	.0099	.1557	.15701	.98760	.15898	.2901	.0125	.3690	58
3	.14004	.99015	.14143	.0706	.0099	.1409	.15730	.98755	.15928	.2783	.0126	.3574	57
4	.14032	.99010	.14173	.0558	.0100	.1263	.15758	.98750	.15958	.2665	.0126	.3458	56
5	.14061	.99006	.14202	7.0410	.0100	7.1117	.15787	.98746	.15987	6.2548	.0127	6.3343	55
6	.14090	.99002	.14232	.0264	.0101	.0972	.15816	.98741	.16017	.2432	.0217	.3228	54
7	.14119	.98998	.14262	.0117	.0101	.0827	.15844	.98737	.16047	.2316	.0128	.3113	53
8	.14148	.98994	.14291	6.9972	.0102	.0683	.15873	.98732	.16077	.2200	.0128	.2999	52
9	.14176	.98990	.14321	.9827	.0102	.0539	.15902	.98727	.16107	.2085	.0129	.2885	51
10	.14205	.98986	.14351	6.9682	.0102	7.0396	.15931	.98723	.16137	6.1970	.0129	6.2772	50
11	.14234	.98982	.14380	.9538	.0103	.0254	.15959	.98718	.16167	.1856	.0130	.2659	49
12	.14263	.98978	.14410	.9395	.0103	.0112	.15988	.98714	.16196	.1742	.0130	.2546	48
13	.14292	.98973	.14440	.9252	.0104	6.9971	.16017	.98709	.16226	.1628	.0131	.2434	47
14	.14320	.98969	.14470	.9110	.0104	.9830	.16045	.98704	.16256	.1515	.0131	.2322	46
15	.14349	.98965	.14499	6.8969	.0104	6.9690	.16074	.98700	.16286	6.1402	.0132	6.2211	45
16	.14378	.98961	.14529	.8828	.0105	.9550	.16103	.98695	.16316	.1290	.0132	.2100	44
17	.14407	.98957	.14559	.8687	.0105	.9411	.16132	.98690	.16346	.1178	.0133	.1990	43
18	.14436	.98952	.14588	.8547	.0106	.9273	.16160	.98685	.16376	.1066	.0133	.1880	42
19	.14464	.98948	.14618	.8408	.0106	.9135	.16189	.98681	.16405	.0955	.0134	.1770	41
20	.14493	.98944	.14648	6.8269	.0107	6.8998	.16218	.98676	.16435	6.0844	.0134	6.1661	40
21	.14522	.98940	.14677	.8131	.0107	.8861	.16246	.98671	.16465	.0734	.0135	.1552	39
22	.14551	.98936	.14707	.7993	.0107	.8725	.16275	.98667	.16495	.0624	.0135	.1443	38
23	.14579	.98931	.14737	.7856	.0108	.8589	.16304	.98662	.16525	.0514	.0136	.1335	37
24	.14608	.98927	.14767	.7720	.0108	.8454	.16333	.98657	.16555	.0405	.0136	.1227	36
25	.14637	.98923	.14796	6.7584	.0109	6.8320	.16361	.98652	.16585	6.0296	.0136	6.1120	35
26	.14666	.98919	.14826	.7448	.0109	.8185	.16390	.98648	.16615	.0188	.0137	.1013	34
27	.14695	.98914	.14856	.7313	.0110	.8052	.16419	.98643	.16644	.0080	.0137	.0906	33
28	.14723	.98910	.14886	.7179	.0110	.7919	.16447	.98638	.16674	5.9972	.0138	.0800	32
29	.14752	.98906	.14915	.7045	.0111	.7787	.16476	.98633	.16704	.9865	.0138	.0694	31
30	.14781	.98901	.14945	6.6911	.0111	6.7655	.16505	.98628	.16734	5.9758	.0139	6.0588	30
31	.14810	.98897	.14975	.6779	.0111	.7523	.16533	.98624	.16764	.9651	.9139	.0483	29
32	.14838	.98893	.15004	.6646	.0112	.7392	.16562	.98619	.16794	.9545	.9140	.0379	28
33	.14867	.98889	.15034	.6514	.0112	.7262	.16591	.98614	.16824	.9439	.0140	.0274	27
34	.14896	.98884	.15064	.6383	.0113	.7132	.16619	.98609	.16854	.9333	.0141	.0170	26
35	.14925	.98880	.15094	6.6252	.0113	6.7003	.16648	.98604	.16884	5.9228	.0141	6.0066	25
36	.14953	.98876	.15123	.6122	.0114	.6874	.16677	.98600	.16914	.9123	.0142	5.9963	24
37	.14982	.98871	.15153	.5992	.0114	.6745	.16705	.98595	.16944	.9019	.0142	.9860	23
38	.15011	.98867	.15183	.5863	.0115	.6617	.16734	.98590	.16973	.8915	.0143	.9758	22
39	.15050	.98862	.15213	.5734	.0115	.6490	.16763	.98585	.17003	.8811	.0143	.9655	21
40	.15068	.98858	.15243	6.5605	.0115	6.6363	.16791	.98580	.17033	5.8708	.0144	5.9554	20
41	.15097	.98854	.15272	.5478	.0116	.6237	.16820	.98575	.17063	.8605	.0144	.9452	19
42	.15126	.98849	.15302	.5350	.0116	.6111	.16849	.98570	.17093	.8502	.0145	.9351	18
43	.15155	.98845	.15332	.5223	.0117	.5985	.16878	.98565	.17123	.8400	.0145	.9250	17
44	.15183	.98840	.15362	.5097	.0117	.5860	.16906	.98560	.17153	.8298	.0146	.9150	16
45	.15212	.98836	.15391	6.4971	.0118	6.5736	.16935	.98556	.17183	5.8196	.0146	5.9049	15
46	.15241	.98832	.15421	.4845	.0118	.5612	.16964	.98551	.17213	.8095	.0147	.8950	14
47	.15270	.98827	.15451	.4720	.0119	.5488	.16992	.98546	.17243	.7994	.0147	.8850	13
48	.15298	.98823	.15481	.4596	.0119	.5365	.17021	.98541	.17273	.7894	.0148	.8751	12
49	.15328	.98818	.15511	.4472	.0119	.5243	.17050	.98536	.17303	.7794	.0148	.8652	11
50	.15356	.98814	.15540	6.4348	.0120	6.5121	.17078	.98531	.17333	5.7694	.0149	5.8554	10
51	.15385	.98809	.15570	.4225	.0120	.4999	.17107	.98526	.17363	.7594	.0150	.8456	9
52	.15413	.98805	.15600	.4103	.0121	.4878	.17136	.98521	.17393	.7495	.0150	.8358	8
53	.15442	.98800	.15630	.3980	.0121	.4757	.17164	.98516	.17423	.7396	.0151	.8261	7
54	.15471	.98796	.15659	.3859	.0122	.4637	.17193	.98511	.17453	.7297	.0151	.8163	6
55	.15500	.98791	.15689	6.3737	.0122	6.4517	.17221	.98506	.17483	5.7199	.0152	5.8067	5
56	.15528	.98787	.15719	.3616	.0123	.4398	.17250	.98501	.17513	.7101	.0152	.7970	4
57	.15557	.98782	.15749	.3496	.0123	.4279	.17279	.98496	.17543	.7004	.0153	.7874	3
58	.15586	.98778	.15779	.3376	.0124	.4160	.17307	.98491	.17573	.6906	.0153	.7778	2
59	.15615	.98773	.15809	.3257	.0124	.4042	.17336	.98486	.17603	.6809	.0154	.7683	1
60	.15643	.98769	.15838	6.3137	.0125	6.3924	.17365	.98481	.17633	5.6713	.0154	5.7588	0
′	cos	sin	cot	tan	cosec	sec	cos	sin	cot	tan	cosec	sec	′
	81°						80°						

′	10°						11°						′
′	sin	cos	tan	cot	sec	cosec	sin	cos	tan	cot	sec	cosec	′
0	.17365	.98481	.17633	5.6713	1.0154	5.7588	.19081	.98163	.19438	5.1445	1.0187	5.2408	60
1	.17393	.98476	.17663	.6616	.0155	.7493	.19109	.98157	.19468	.1366	.0188	.2330	59
2	.17422	.98471	.17693	.6520	.0155	.7398	.19138	.98152	.19498	.1286	.0188	.2252	58
3	.17451	.98465	.17723	.6425	.0156	.7304	.19166	.98146	.19529	.1207	.0189	.2174	57
4	.17479	.98460	.17753	.6329	.0156	.7210	.19195	.98140	.19559	.1128	.0189	.2097	56
5	.17508	.98455	.17783	5.6234	.0157	5.7117	.19224	.98135	.19589	5.1049	.0190	5.2019	55
6	.17537	.98450	.17813	.6140	.0157	.7023	.19252	.98129	.19619	.0970	.0191	.1942	54
7	.17565	.98445	.17843	.6045	.0158	.6930	.19281	.98124	.19649	.0892	.0191	.1865	53
8	.17594	.98440	.17873	.5951	.0158	.6838	.19309	.98118	.19680	.0814	.0192	.1788	52
9	.17622	.98435	.17903	.5857	.0159	.6745	.19338	.98112	.19710	.0736	.0192	.1712	51
10	.17651	.98430	.17933	5.5764	.0159	5.6653	.19366	.98107	.19740	5.0658	.0193	5.1636	50
11	.17680	.98425	.17963	.5670	.0160	.6561	.19395	.98101	.19770	.0581	.0193	.1560	49
12	.17708	.98419	.17993	.5578	.0160	.6470	.19423	.98095	.19800	.0504	.0194	.1484	48
13	.17737	.98414	.18023	.5485	.0161	.6379	.19452	.98090	.19831	.0427	.0195	.1409	47
14	.17766	.98409	.18053	.5393	.0162	.6288	.19480	.98084	.19861	.0350	.0195	.1333	46
15	.17794	.98404	.18083	5.5301	.0162	5.6197	.19509	.98078	.19891	5.0273	.0196	5.1258	45
16	.17823	.98399	.18113	.5209	.0163	.6107	.19537	.98073	.19921	.0197	.0196	.1183	44
17	.17852	.98394	.18143	.5117	.0163	.6017	.19566	.98067	.19952	.0121	.0197	.1109	43
18	.17880	.98388	.18173	.5026	.0164	.5928	.19595	.98061	.19982	.0045	.0198	.1034	42
19	.17909	.98383	.18203	.4936	.0164	.5838	.19623	.98056	.20012	4.9969	.0198	.0960	41
20	.17937	.98378	.18233	5.4845	.0165	5.5749	.19652	.98050	.20042	4.9894	.0199	5.0886	40
21	.17966	.98373	.18263	.4755	.0165	.5660	.19680	.98044	.20073	.9819	.0199	.0812	39
22	.17995	.98368	.18293	.4665	.0166	.5572	.19709	.98039	.20103	.9744	.0200	.0739	38
23	.18023	.98362	.18323	.4575	.0166	.5484	.19737	.98033	.20133	.9669	.0201	.0666	37
24	.18052	.98357	.18353	.4486	.0167	.5396	.19766	.98027	.20163	.9594	.0201	.0593	36
25	.18080	.98352	.18383	5.4396	.0167	5.5308	.19794	.98021	.20194	4.9520	.0202	5.0520	35
26	.18109	.98347	.18413	.4308	.0168	.5221	.19823	.98016	.20224	.9446	.0202	.0447	34
27	.18138	.98341	.18444	.4219	.0169	.5134	.19851	.98010	.20254	.9372	.0203	.0375	33
28	.18166	.98336	.18474	.4131	.0169	.5047	.19880	.98004	.20285	.9298	.0204	.0302	32
29	.18195	.98331	.18504	.4043	.0170	.4960	.19908	.97998	.20315	.9225	.0204	.0230	31
30	.18223	.98325	.18534	5.3955	.0170	5.4874	.19937	.97992	.20345	4.9151	.0205	5.0158	30
31	.18252	.98320	.18564	.3868	.0171	.4788	.19965	.97987	.20375	.9078	.0205	.0087	29
32	.18281	.98315	.18594	.3780	.0171	.4702	.19994	.97981	.20406	.9006	.0206	.0015	28
33	.18309	.98309	.18624	.3694	.0172	.4617	.20022	.97975	.20436	.8933	.0207	4.9944	27
34	.18338	.98304	.18654	.3607	.0172	.4532	.20051	.97969	.20466	.8860	.0207	.9873	26
35	.18366	.98299	.18684	5.3521	.0173	5.4447	.20079	.97963	.20497	4.8788	.0208	4.9802	25
36	.18395	.98293	.18714	.3434	.0174	.4362	.20108	.97957	.20527	.8716	.0208	.9732	24
37	.18424	.98288	.18745	.3349	.0174	.4278	.20136	.97952	.20557	.8644	.0209	.9661	23
38	.18452	.98283	.18775	.3263	.0175	.4194	.20165	.97946	.20588	.8573	.0210	.9591	22
39	.18481	.98277	.18805	.3178	.0175	.4110	.20193	.97940	.20618	.8501	.0210	.9521	21
40	.18509	.98272	.18835	5.3093	.0176	5.4026	.20222	.97934	.20648	4.8430	.0211	4.9452	20
41	.18538	.98267	.18865	.3008	.0176	.3943	.20250	.97928	.20679	.8359	.0211	.9382	19
42	.18567	.98261	.18895	.2923	.0177	.3860	.20279	.97922	.20709	.8288	.0212	.9313	18
43	.18595	.98256	.18925	.2839	.0177	.3777	.20307	.97916	.20739	.8217	.0213	.9243	17
44	.18624	.98250	.18955	.2755	.0178	.3695	.20336	.97910	.20770	.8147	.0213	.9175	16
45	.18652	.98245	.18985	5.2671	.0179	5.3612	.20364	.97904	.20800	4.8077	.0214	4.9106	15
46	.18681	.98240	.19016	.2588	.0179	.3530	.20393	.97899	.20830	.8007	.0215	.9037	14
47	.18709	.98234	.19046	.2505	.0180	.3449	.20421	.97893	.20861	.7937	.0215	.8969	13
48	.18738	.98229	.19076	.2422	.0180	.3367	.20450	.97887	.20891	.7867	.0216	.8901	12
49	.18767	.98223	.19106	.2339	.0181	.3286	.20478	.97881	.20921	.7798	.0216	.8833	11
50	.18795	.98218	.19136	5.2257	.0181	5.3205	.20506	.97875	.20952	4.7728	.0217	4.8765	10
51	.18824	.98212	.19166	.2174	.0182	.3124	.20535	.97869	.20982	.7659	.0218	.8967	9
52	.18852	.98207	.19197	.2092	.0182	.3044	.20563	.97863	.21012	.7591	.0218	.8630	8
53	.18881	.98201	.19227	.2011	.0183	.2963	.20592	.97857	.21043	.7522	.0219	.8563	7
54	.18909	.98196	.19257	.1929	.0184	.2883	.20620	.97851	.21073	.7453	.0220	.8496	6
55	.18938	.98190	.19287	5.1848	.0184	5.2803	.20649	.97845	.21104	4.7385	.0220	4.8429	5
56	.18967	.98185	.19817	.1767	.0185	.2724	.20677	.97839	.21134	.7317	.0221	.8362	4
57	.18995	.98179	.19347	.1686	.0185	.2645	.20706	.97833	.21164	.7249	.0221	.8296	3
58	.19024	.98174	.19378	.1606	.0186	.2566	.20734	.97827	.21195	.7181	.0222	.8229	2
59	.19052	.98168	.19408	.1525	.0186	.2487	.20763	.97821	.21225	.7114	.0223	.8163	1
60	.19081	.98163	.19438	5.1445	.0187	5.2408	.20791	.97815	.21256	4.7046	.0223	4.8097	0
′	cos	sin	cot	tan	cosec	sec	cos	sin	cot	tan	cosec	sec	′
	79°						78°						

(Continued)

table 21

Natural trigonometric functions (Continued)

	12°							13°							14°							15°						
′	sin	cos	tan	cot	sec	cosec	′	sin	cos	tan	cot	sec	cosec	′	sin	cos	tan	cot	sec	cosec	′	sin	cos	tan	cot	sec	cosec	′

(Continued)

table 21 Natural trigonometric functions (Continued)

′	16° sin	16° cos	16° tan	16° cot	16° sec	16° cosec	17° sin	17° cos	17° tan	17° cot	17° sec	17° cosec
0	.27564	.96126	.28674	3.4874	1.0403	3.6279	.29237	.95630	.30573	3.2708	1.0457	3.4203
1	.27592	.96118	.28706	.4836	.0404	.6243	.29265	.95622	.30605	.2674	.0458	.4170
2	.27620	.96110	.28737	.4798	.0405	.6206	.29293	.95613	.30637	.2640	.0459	.4138
3	.27648	.96102	.28769	.4760	.0406	.6169	.29321	.95605	.30668	.2607	.0460	.4106
4	.27675	.96094	.28800	.4722	.0406	.6133	.29348	.95596	.30700	.2573	.0461	.4073
5	.27703	.96086	.28832	3.4684	.0407	3.6096	.29376	.95588	.30732	3.2539	.0461	3.4041
6	.27731	.96078	.28863	.4646	.0408	.6060	.29404	.95579	.30764	.2505	.0462	.4009
7	.27759	.96070	.28895	.4608	.0409	.6024	.29432	.95571	.30796	.2472	.0463	.3977
8	.27787	.96062	.28926	.4570	.0410	.5987	.29460	.95562	.30828	.2438	.0464	.3945
9	.27815	.96054	.28958	.4533	.0411	.5951	.29487	.95554	.30859	.2405	.0465	.3913
10	.27843	.96045	.28990	3.4495	.0412	3.5915	.29515	.95545	.30891	3.2371	.0466	3.3881
11	.27871	.96037	.29021	.4458	.0413	.5879	.29543	.95536	.30923	.2338	.0467	.3849
12	.27899	.96029	.29053	.4420	.0413	.5843	.29571	.95528	.30955	.2305	.0468	.3817
13	.27927	.96021	.29084	.4383	.0414	.5807	.29598	.95519	.30987	.2271	.0469	.3785
14	.27955	.96013	.29116	.4346	.0415	.5772	.29626	.95511	.31019	.2238	.0470	.3754
15	.27983	.96005	.29147	3.4308	.0416	3.5736	.29654	.95502	.31051	3.2205	.0471	3.3722
16	.28011	.95997	.29179	.4271	.0417	.5700	.29682	.95493	.31083	.2172	.0472	.3690
17	.28039	.95989	.29210	.4234	.0418	.5665	.29710	.95485	.31115	.2139	.0473	.3659
18	.28067	.95980	.29242	.4197	.0419	.5629	.29737	.95476	.31146	.2106	.0474	.3627
19	.28094	.95972	.29274	.4160	.0420	.5594	.29765	.95467	.31178	.2073	.0475	.3596
20	.28122	.95964	.29305	3.4124	.0420	3.5559	.29793	.95459	.31210	3.2041	.0476	3.3565
21	.28150	.95956	.29337	.4087	.0421	.5523	.29821	.95450	.31242	.2008	.0477	.3534
22	.28178	.95948	.29368	.4050	.0422	.5488	.29848	.95441	.31274	.1975	.0478	.3502
23	.28206	.95940	.29400	.4014	.0423	.5453	.29876	.95433	.31306	.1942	.0478	.3471
24	.28234	.95931	.29432	.3977	.0424	.5418	.29904	.95424	.31338	.1910	.0479	.3440
25	.28262	.95923	.29463	3.3941	.0425	3.5383	.29932	.95415	.31370	3.1877	.0480	3.3409
26	.28290	.95915	.29495	.3904	.0426	.5348	.29959	.95407	.31402	.1845	.0481	.3378
27	.28318	.95907	.29526	.3868	.0427	.5313	.29987	.95398	.31434	.1813	.0482	.3347
28	.28346	.95898	.29558	.3832	.0428	.5279	.30015	.95389	.31466	.1780	.0483	.3316
29	.28374	.95890	.29590	.3795	.0428	.5244	.30043	.95380	.31498	.1748	.0484	.3286
30	.28401	.95882	.29621	3.3759	.0429	3.5209	.30070	.95372	.31530	3.1716	.0485	3.3255
31	.28429	.95874	.29653	.3723	.0430	.5175	.30098	.95363	.31562	.1684	.0486	.3224
32	.28457	.95865	.29685	.3687	.0431	.5140	.30126	.95354	.31594	.1652	.0487	.3194
33	.28485	.95857	.29716	.3651	.0432	.5106	.30154	.95345	.31626	.1620	.0488	.3163
34	.28513	.95849	.29748	.3616	.0433	.5072	.30181	.95337	.31658	.1588	.0489	.3133
35	.28541	.95840	.29780	3.3580	.0434	3.5037	.30209	.95328	.31690	3.1556	.0490	3.3102
36	.28569	.95832	.29811	.3544	.0435	.5003	.30237	.95319	.31722	.1524	.0491	.3072
37	.28597	.95824	.29843	.3509	.0436	.4969	.30265	.95310	.31754	.1492	.0492	.3042
38	.28624	.95816	.29875	.3473	.0437	.4935	.30292	.95301	.31786	.1460	.0493	.3011
39	.28652	.95807	.29906	.3438	.0438	.4901	.30320	.95293	.31818	.1429	.0494	.2981
40	.28680	.95799	.29938	3.3402	.0438	3.4867	.30348	.95284	.31850	3.1397	.0495	3.2951
41	.28708	.95791	.29970	.3367	.0439	.4833	.30375	.95275	.31882	.1366	.0496	.2921
42	.28736	.95782	.30001	.3332	.0440	.4799	.30403	.95266	.31914	.1334	.0497	.2891
43	.28764	.95774	.30033	.3296	.0441	.4766	.30431	.95257	.31946	.1303	.0498	.2861
44	.28792	.95765	.30065	.3261	.0442	.4732	.30459	.95248	.31978	.1271	.0499	.2831
45	.28820	.95757	.30096	3.3226	.0443	3.4698	.30486	.95239	.32010	3.1240	.0500	3.2801
46	.28847	.95749	.30128	.3191	.0444	.4665	.30514	.95231	.32042	.1209	.0501	.2772
47	.28875	.95740	.30160	.3156	.0445	.4632	.30542	.95222	.32074	.1177	.0502	.2742
48	.28903	.95732	.30192	.3121	.0446	.4598	.30569	.95213	.32106	.1146	.0503	.2712
49	.28931	.95723	.30223	.3087	.0447	.4565	.30597	.95204	.32138	.1115	.0504	.2683
50	.28959	.95715	.30255	3.3052	.0448	3.4532	.30625	.95195	.32171	3.1084	.0505	3.2653
51	.28987	.95707	.30287	.3017	.0448	.4498	.30653	.95186	.32203	.1053	.0506	.2624
52	.29014	.95698	.30319	.2983	.0449	.4465	.30680	.95177	.32235	.1022	.0507	.2594
53	.29042	.95690	.30350	.2948	.0450	.4432	.30708	.95168	.32267	.0991	.0508	.2565
54	.29070	.95681	.30382	.2914	.0451	.4399	.30736	.95159	.32299	.0960	.0509	.2535
55	.29098	.95673	.30414	3.2879	.0452	3.4366	.30763	.95150	.32331	3.0930	.0510	3.2506
56	.29126	.95664	.30446	.2845	.0453	.4334	.30791	.95141	.32363	.0899	.0511	.2477
57	.29154	.95656	.30478	.2811	.0454	.4301	.30819	.95132	.32395	.0868	.0512	.2448
58	.29181	.95647	.30509	.2777	.0455	.4268	.30846	.95124	.32428	.0838	.0513	.2419
59	.29209	.95639	.30541	.2742	.0456	.4236	.30874	.95115	.32460	.0807	.0514	.2390
60	.29237	.95630	.30573	3.2708	.0457	3.4203	.30902	.95106	.32492	3.0777	.0515	3.2361
′	cos	sin	cot	tan	cosec	sec	cos	sin	cot	tan	cosec	sec
	73°						72°					

18° sin	18° cos	18° tan	18° cot	18° sec	18° cosec	19° sin	19° cos	19° tan	19° cot	19° sec	19° cosec	′
.30902	.95106	.32492	3.0777	1.0515	3.2361	.32557	.94552	.34433	2.9042	1.0576	3.0715	60
.30929	.95097	.32524	.0746	.0516	.2332	.32584	.94542	.34465	.9015	.0577	.0690	59
.30957	.95088	.32556	.0716	.0517	.2303	.32612	.94533	.34498	.8987	.0578	.0664	58
.30985	.95079	.32588	.0686	.0518	.2274	.32639	.94523	.34530	.8960	.0579	.0638	57
.31012	.95070	.32621	.0655	.0519	.2245	.32667	.94514	.34563	.8933	.0580	.0612	56
.31040	.95061	.32653	3.0625	.0520	3.2216	.32694	.94504	.34595	2.8905	.0581	3.0586	55
.31068	.95051	.32685	.0595	.0521	.2188	.32722	.94495	.34628	.8878	.0582	.0561	54
.31095	.95042	.32717	.0565	.0522	.2159	.32749	.94485	.34661	.8851	.0584	.0535	53
.31123	.95033	.32749	.0535	.0523	.2131	.32777	.94476	.34693	.8824	.0585	.0509	52
.31150	.95024	.32782	.0505	.0524	.2102	.32804	.94466	.34726	.8797	.0586	.0484	51
.31178	.95015	.32814	3.0475	.0525	3.2074	.32832	.94457	.34758	2.8770	.0587	3.0458	50
.31206	.95006	.32846	.0445	.0526	.2045	.32859	.94447	.34791	.8743	.0588	.0433	49
.31233	.94997	.32878	.0415	.0527	.2017	.32887	.94438	.34824	.8716	.0589	.0407	48
.31261	.94988	.32910	.0385	.0528	.1989	.32914	.94428	.34856	.8689	.0590	.0382	47
.31289	.94979	.32943	.0356	.0529	.1960	.32942	.94418	.34889	.8662	.0591	.0357	46
.31316	.94970	.32975	3.0326	.0530	3.1932	.32969	.94409	.34921	2.8636	.0592	3.0331	45
.31344	.94961	.33007	.0296	.0531	.1904	.32996	.94399	.34954	.8609	.0593	.0306	44
.31372	.94952	.33039	.0267	.0532	.1876	.33024	.94390	.34987	.8582	.0594	.0281	43
.31399	.94942	.33072	.0237	.0533	.1848	.33051	.94380	.35019	.8555	.0595	.0256	42
.31427	.94933	.33104	.0208	.0534	.1820	.33079	.94370	.35052	.8529	.0596	.0231	41
.31454	.94924	.33136	3.0178	.0535	3.1792	.33106	.94361	.35085	2.8502	.0598	3.0206	40
.31482	.94915	.33169	.0149	.0536	.1764	.33134	.94351	.35117	.8476	.0599	.0181	39
.31510	.94906	.33201	.0120	.0537	.1736	.33161	.94341	.35150	.8449	.0600	.0156	38
.31537	.94897	.33233	.0090	.0528	.1708	.33189	.94332	.35183	.8423	.0601	.0131	37
.31565	.94888	.33265	.0061	.0539	.1681	.33216	.94322	.35215	.8396	.0602	.0106	36
.31592	.94878	.33298	3.0032	.0540	3.1653	.33243	.94313	.35248	2.8370	.0603	3.0081	35
.31620	.94869	.33330	.0003	.0541	.1625	.33271	.94303	.35281	.8344	.0604	.0056	34
.31648	.94860	.33362	2.9974	.0542	.1598	.33298	.94293	.35314	.8318	.0605	.0031	33
.31675	.94851	.33395	.9945	.0543	.1570	.33326	.94283	.35346	.8291	.0606	.0007	32
.31703	.94841	.33427	.9916	.0544	.1543	.33353	.94274	.35379	.8265	.0607	2.9982	31
.31730	.94832	.33459	2.9887	.0545	3.1515	.33381	.94264	.35412	2.8239	.0608	2.9957	30
.31758	.94823	.33492	.9858	.0546	.1488	.33408	.94254	.35445	.8213	.0609	.9933	29
.31786	.94814	.33524	.9829	.0547	.1461	.33435	.94245	.35477	.8187	.0611	.9908	28
.31813	.94805	.33557	.9800	.0548	.1433	.33463	.94235	.35510	.8161	.0612	.9884	27
.31841	.94795	.33589	.9772	.0549	.1406	.33490	.94225	.35543	.8135	.0613	.9859	26
.31868	.94786	.33621	2.9743	.0550	3.1379	.33518	.94215	.35576	2.8109	.0614	2.9835	25
.31896	.94777	.33654	.9714	.0551	.1352	.33545	.94206	.35608	.8083	.0615	.9810	24
.31923	.94767	.33686	.9686	.0552	.1325	.33572	.94196	.35641	.8057	.0616	.9786	23
.31951	.94758	.33718	.9657	.0553	.1298	.33600	.94186	.35674	.8032	.0617	.9762	22
.31978	.94749	.33751	.9629	.0554	.1271	.33627	.94176	.35707	.8006	.0618	.9738	21
.32006	.94740	.33783	2.9600	.0555	3.1244	.33655	.94167	.35739	2.7980	.0619	2.9713	20
.32034	.94730	.33816	.9572	.0556	.1217	.33682	.94157	.35772	.7954	.0620	.9689	19
.32061	.94721	.33848	.9544	.0557	.1190	.33709	.94147	.35805	.7929	.0622	.9665	18
.32089	.94712	.33880	.9515	.0558	.1163	.33737	.94137	.35838	.7903	.0623	.9641	17
.32116	.94702	.33913	.9487	.0559	.1137	.33764	.94127	.35871	.7878	.0624	.9617	16
.32144	.94693	.33945	2.9459	.0560	3.1110	.33792	.94118	.35904	2.7852	.0625	2.9593	15
.32171	.94684	.33978	.9431	.0561	.1083	.33819	.94108	.35936	.7827	.0626	.9569	14
.32199	.94674	.34010	.9403	.0562	.1057	.33846	.94098	.35969	.7801	.0627	.9545	13
.32226	.94665	.34043	.9375	.0563	.1030	.33874	.94088	.36002	.7776	.0628	.9521	12
.32254	.94655	.34075	.9347	.0565	.1004	.33901	.94078	.36035	.7751	.0629	.9497	11
.32282	.94646	.34108	2.9319	.0566	3.0977	.33928	.94068	.36068	2.7725	.0630	2.9474	10
.32309	.94637	.34140	.9291	.0567	.0951	.33956	.94058	.36101	.7700	.0632	.9450	9
.32337	.94627	.34173	.9263	.0568	.0925	.33983	.94049	.36134	.7675	.0633	.9426	8
.32364	.94618	.34205	.9235	.0569	.0898	.34011	.94039	.36167	.7650	.0634	.9402	7
.32392	.94608	.34238	.9208	.0570	.0872	.34038	.94029	.36199	.7625	.0635	.9379	6
.32419	.94599	.34270	2.9180	.0571	3.0846	.34065	.94019	.36232	2.7600	.0636	2.9355	5
.32447	.94590	.34303	.9152	.0572	.0820	.34093	.94009	.36265	.7575	.0637	.9332	4
.32474	.94580	.34335	.9125	.0573	.0793	.34120	.93999	.36298	.7550	.0638	.9308	3
.32502	.94571	.34368	.9097	.0574	.0767	.34147	.93989	.36331	.7525	.0639	.9285	2
.32529	.94561	.34400	.9069	.0575	.0741	.34175	.93979	.36364	.7500	.0641	.9261	1
.32557	.94552	.34433	2.9042	.0576	3.0715	.34202	.93969	.36397	2.7475	.0642	2.9238	0
cos	sin	cot	tan	cosec	sec	cos	sin	cot	tan	cosec	sec	′
71°						70°						

(Continued)

table 21 Natural trigonometric functions (Continued)

	20°						21°						22°						23°						
′	sin	cos	tan	cot	sec	cosec	sin	cos	tan	cot	sec	cosec	sin	cos	tan	cot	sec	cosec	sin	cos	tan	cot	sec	cosec	′
0	.34202	.93969	.36397	2.7475	1.0642	2.9238	.35837	.93358	.38386	2.6051	1.0711	2.7904	.37461	.92718	.40403	2.4751	1.0785	2.6695	.39073	.92050	.42447	2.3558	1.0864	2.5593	60
1	.34229	.93959	.36430	.7450	.0643	.9215	.35864	.93348	.38420	.6028	.0713	.7883	.37488	.92707	.40436	.4730	.0787	.6675	.39100	.92039	.42482	.3539	.0865	.5575	59
2	.34257	.93949	.36463	.7425	.0644	.9191	.35891	.93337	.38453	.6006	.0714	.7862	.37514	.92696	.40470	.4709	.0788	.6656	.39126	.92028	.42516	.3520	.0866	.5558	58
3	.34284	.93939	.36496	.7400	.0645	.9168	.35918	.93327	.38486	.5983	.0715	.7841	.37541	.92686	.40504	.4689	.0789	.6637	.39153	.92016	.42550	.3501	.0868	.5540	57
4	.34311	.93929	.36529	.7376	.0646	.9145	.35945	.93316	.38520	.5960	.0716	.7820	.37568	.92675	.40538	.4668	.0790	.6618	.39180	.92005	.42585	.3482	.0869	.5523	56
5	.34339	.93919	.36562	2.7351	.0647	2.9122	.35972	.93306	.38553	2.5938	.0717	2.7799	.37595	.92664	.40572	2.4647	.0792	2.6599	.39207	.91993	.42619	2.3463	.0870	2.5506	55
6	.34366	.93909	.36595	.7326	.0648	.9098	.36000	.93295	.38587	.5916	.0719	.7778	.37622	.92653	.40606	.4627	.0793	.6580	.39234	.91982	.42654	.3445	.0872	.5488	54
7	.34393	.93899	.36628	.7302	.0650	.9075	.36027	.93285	.38620	.5893	.0720	.7757	.37649	.92642	.40640	.4606	.0794	.6561	.39260	.91971	.42688	.3426	.0873	.5471	53
8	.34421	.93889	.36661	.7277	.0651	.9052	.36054	.93274	.38654	.5871	.0721	.7736	.37676	.92631	.40673	.4586	.0795	.6542	.39287	.91959	.42722	.3407	.0874	.5453	52
9	.34448	.93879	.36694	.7252	.0652	.9029	.36081	.93264	.38687	.5848	.0722	.7715	.37703	.92620	.40707	.4565	.0797	.6523	.39314	.91948	.42757	.3388	.0876	.5436	51
10	.34475	.93869	.36727	2.7228	.0653	2.9006	.36108	.93253	.38720	2.5826	.0723	2.7694	.37730	.92609	.40741	2.4545	.0798	2.6504	.39341	.91936	.42791	2.3369	.0877	2.5419	50
11	.34502	.93859	36760	.7204	.0654	.8983	.36135	.93243	.38754	.5804	.0725	.7674	.37757	.92598	.40775	.4525	.0799	.6485	.39367	.91925	.42826	.3350	.0878	.5402	49
12	.34530	.93849	.36793	.7179	.0655	.8960	.36162	.93232	.38787	.5781	.0726	.7653	.37784	.92587	.40809	.4504	.0801	.6466	.39394	.91913	.42860	.3332	.0880	.5384	48
13	.34557	.93839	.36826	.7155	.0656	.8937	.36189	.93222	.38821	.5759	.0727	.7632	.37811	.92576	.40843	.4484	.0802	.6447	.39421	.91902	.42894	.3313	.0881	.5367	47
14	.34584	.93829	.36859	.7130	.0658	.8915	.36217	.93211	.38854	.5737	.0728	.7611	.37838	.92565	.40877	.4463	.0803	.6428	.39448	.91891	.42929	.3294	.0882	.5350	46
15	.34612	.93819	.36892	2.7106	.0659	2.8892	.36244	.93201	.38888	2.5715	.0729	2.7591	.37865	.92554	.40911	2.4443	.0804	2.6410	.39474	.91879	.42963	2.3276	.0884	2.5333	45
16	.34639	.93809	.36925	.7082	.0660	.8869	.36271	.93190	.38921	.5693	.0731	.7570	.37892	.92543	.40945	.4423	.0806	.6391	.39501	.91868	.42998	.3257	.0885	.5316	44
17	.34666	.93799	.36958	.7058	.0661	.8846	.36298	.93180	.38955	.5671	.0732	.7550	.37919	.92532	.40979	.4403	.0807	.6372	.39528	.91856	.43032	.3238	.0886	.5299	43
18	.34693	.93789	.36991	.7033	.0662	.8824	.36325	.93169	.38988	.5649	.0733	.7529	.37946	.92521	.41013	.4382	.0808	.6353	.39554	.91845	.43067	.3220	.0888	.5281	42
19	.34721	.93779	.37024	.7009	.0663	.8801	.36352	.93158	.39022	.5627	.0734	.7509	.37972	.92510	.41047	.4362	.0810	.6335	.39581	.91833	.43101	.3201	.0889	.5264	41
20	.34748	.93769	.37057	2.6985	.0664	2.8778	.36379	.93148	.39055	2.5605	.0736	2.7488	.37999	.92499	.41081	2.4342	.0811	2.6316	.39608	.91822	.43136	2.3183	.0891	2.5247	40
21	.34775	.93758	.37090	.6961	.0666	.8756	.36406	.93137	.39089	.5583	.0737	.7468	.38026	.92488	.41115	.4322	.0812	.6297	.39635	.91810	.43170	.3164	.0892	.5230	39
22	.34803	.93748	.37123	.6937	.0667	.8733	.36433	.93127	.39122	.5561	.0738	.7447	.38053	.92477	.41149	.4302	.0813	.6279	.39661	.91798	.43205	.3145	.0893	.5213	38
23	.34830	.93738	.37156	.6913	.0668	.8711	.36460	.93116	.39156	.5539	.0739	.7427	.38080	.92466	.41183	.4282	.0815	.6260	.39688	.91787	.43239	.3127	.0895	.5196	37
24	.34857	.93728	.37190	.6889	.0669	.8688	.36488	.93105	.39189	.5517	.0740	.7406	.38107	.92455	.41217	.4262	.0816	.6242	.39715	.91775	.43274	.3109	.0896	.5179	36
25	.34884	.93718	.37223	2.6865	.0670	2.8666	.36515	.93095	.39223	2.5495	.0742	2.7386	.38134	.92443	.41251	2.4242	.0817	2.6223	.39741	.91764	.43308	2.3090	.0897	2.5163	35
26	.34912	.93708	.37256	.6841	.0671	.8644	.36542	.93084	.39257	.5473	.0743	.7366	.38161	.92432	.41285	.4222	.0819	.6205	.39768	.91752	.43343	.3072	.0899	.5146	34
27	.34939	.93698	.37289	.6817	.0673	.8621	.36569	.93074	.39290	.5451	.0744	.7346	.38188	.92421	.41319	.4202	.0820	.6186	.39795	.91741	.43377	.3053	.0900	.5129	33
28	.34966	.93687	.37322	.6794	.0674	.8599	.36596	.93063	.39324	.5430	.0745	.7325	.38214	.92410	.41353	.4182	.0821	.6168	.39821	.91729	.43412	.3035	.0902	.5112	32
29	.34993	.93677	.37355	.6770	.0675	.8577	.36623	.93052	.39357	.5408	.0747	.7305	.38241	.92399	.41387	.4162	.0823	.6150	.39848	.91718	.43447	.3017	.0903	.5095	31
30	.35021	.93667	.37388	2.6746	.0676	2.8554	.36650	.93042	.39391	2.5386	.0748	2.7285	.38268	.92388	.41421	2.4142	.0824	2.6131	.39875	.91706	.43481	2.2998	.0904	2.5078	30
31	.35048	.93657	.37422	.6722	.0677	.8532	.36677	.93031	.39425	.5365	.0749	.7265	.38295	.92377	.41455	.4122	.0825	.6113	.39901	.91694	.43516	.2980	.0906	.5062	29
32	.35075	.93647	.37455	.6699	.0678	.8510	.36704	.93020	39458	.5343	.0750	.7245	.38322	.92366	.41489	.4102	.0826	.6095	.39928	.91683	.43550	.2962	.0907	.5045	28
33	.35102	.93637	.37488	.6675	.0679	.8488	.36731	.93010	.39492	.5322	.0751	.7225	.38349	.92354	.41524	.4083	.0828	.6076	.39955	.91671	.43585	.2944	.0908	.5028	27
34	.35130	.93626	.37521	.6652	.0681	.8466	.36758	.92999	.39525	.5300	.0753	.7205	.38376	.92343	.41558	.4063	.0829	.6058	.39981	.91659	.43620	.2925	.0910	.5011	26
35	.35157	.93616	.37554	2.6628	.0682	2.8444	.36785	.92988	.39559	2.5278	.0754	2.7185	.38403	.92332	.41592	2.4043	.0830	2.6040	.40008	.91648	.43654	2.2907	.0911	2.4995	25
36	.35184	.93606	.37587	.6604	.0683	.8422	.36812	.92978	.39593	.5257	.0755	.7165	.38429	.92321	.41626	.4023	.0832	.6022	.40035	.91636	.43689	.2889	.0913	.4978	24
37	.35211	.93596	.37621	.6581	.0684	.8400	.36839	.92967	.39626	.5236	.0756	.7145	.38456	.92310	.41660	.4004	.0833	.6003	.40061	.91625	.43723	.2871	.0914	.4961	23
38	.35239	.93585	.37654	.6558	.0685	.8378	.36866	.92956	.39660	.5214	.0758	.7125	.38483	.92299	.41694	.3984	.0834	.5985	.40088	.91613	.43758	.2853	.0915	.4945	22
39	.35266	.93575	.37687	.6534	.0686	.8356	.36893	.92945	.39694	.5193	.0759	.7105	.38510	.92287	.41728	.3964	.0836	.5967	.40115	.91601	.43793	.2835	.0917	.4928	21
40	.35293	.93565	.37720	2.6511	.0688	2.8334	.36921	.92935	.39727	2.5171	.0760	2.7085	.38537	.92276	.41762	2.3945	.0837	2.5949	.40141	.91590	.43827	2.2817	.0918	2.4912	20
41	.35320	.93555	.37754	.6487	.0689	.8312	.36948	.92924	.39761	.5150	.0761	.7065	.38564	.92265	.41797	.3925	.0838	.5931	.40168	.91578	.43862	.2799	.0920	.4895	19
42	.35347	.93544	.37787	.6464	.0690	.8290	.36975	.92913	.39795	.5129	.0763	.7045	.38591	.92254	.41831	.3906	.0840	.5913	.40195	.91566	.43897	.2781	.0921	.4879	18
43	.35375	.93534	.37820	.6441	.0691	.8269	.37002	.92902	.39828	.5108	.0764	.7026	.38617	.92242	.41865	.3886	.0841	.5895	.40221	.91554	.43932	.2763	.0922	.4862	17
44	.35402	.93524	.37853	.6418	.0692	.8247	.37029	.92892	.39862	.5086	.0765	.7006	.38644	.92231	.41899	.3867	.0842	.5877	.40248	.91543	.43966	.2745	.0924	.4846	16
45	.35429	.93513	.37887	2.6394	.0694	2.8225	.37056	.92881	.39896	2.5065	.0766	2.6986	.38671	.92220	.41933	2.3847	.0844	2.5958	.40205	.91531	.44001	2.2727	.0925	2.4829	15
46	.35456	.93503	.37920	.6371	.0695	.8204	.37083	.92870	.39930	.5044	.0768	.6967	.38698	.92209	.41968	.3828	.0845	.5841	.40301	.91519	.44036	.2709	.0927	.4813	14
47	.35483	.93493	.37953	.6348	.0696	.8182	.37110	.92859	.39963	.5023	.0769	.6947	.38725	.92197	.42002	.3808	.0846	.5823	.40328	.91508	.44070	.2691	.0928	.4797	13
48	.35511	.93482	.37986	.6325	.0697	.8160	.37137	.92848	.39997	.5002	.0770	.6927	.38751	.92186	.42036	.3789	.0847	.5805	.40354	.91496	.44105	.2673	.0929	.4780	12
49	.35538	.93472	.38020	.6302	.0698	.8139	.37164	.92838	.40031	.4981	.0771	.6908	.38778	.92105	.42070	.3770	.0849	.5787	.40381	.91484	.44140	.2655	.0931	.4764	11
50	.35565	.93462	.38053	2.6279	.0699	2.8117	.37191	.92827	.40065	2.4960	.0773	2.6888	.38805	.92164	.42105	2.3750	.0850	2.5770	.40408	.91472	.44175	2.2637	.0932	2.4748	10
51	.35592	.93451	.38086	.6256	.0701	.8096	.37218	.92816	.40098	.4939	.0774	.6869	.38832	.92152	.42139	.3731	.0851	.5752	.40434	.91461	.44209	.2619	.0934	.4731	9
52	.35619	.93441	.38120	.6233	.0702	.8074	.37245	.92805	.40132	.4918	.0775	.6849	.38859	.92141	.42173	.3712	.0853	.5734	.40461	.91449	.44244	.2602	.0935	.4715	8
53	.35647	.93431	.38153	.6210	.0703	.8053	.37272	.92794	.40166	.4897	.0776	.6830	.38886	.92130	.42207	.3692	.0854	.5716	.40487	.91437	.44279	.2584	.0936	.4699	7
54	.35674	.93420	.38186	.6187	.0704	.8032	.37299	.92784	.40200	.4876	.0778	.6810	.38912	.92118	.42242	.3673	.0855	.5699	.40514	.91425	.44314	.2566	.0938	.4683	6
55	.35701	.93410	.38220	2.6164	.0705	2.8010	.37326	.92773	.40233	2.4855	.0779	2.6791	.38939	.92107	.42276	2.3654	.0857	2.5681	.40541	.91414	.44349	2.2548	.0939	2.4666	5
56	.35728	.93400	.38253	.6142	.0707	.7989	.37353	.92762	.40267	.4834	.0780	.6772	.38966	.92096	.42310	.3635	.0858	.5663	.40567	.91402	.44383	.2531	.0941	.4650	4
57	.35755	.93389	.38286	.6119	.0708	.7968	.37380	.92751	.40301	.4813	.0781	.6752	.38993	.92084	.42344	.3616	.0859	.5646	.40594	.91390	.44418	.2513	.0942	.4634	3
58	.35782	.93379	.38320	.6096	.0709	.7947	.37407	.92740	.40335	.4792	.0783	.6733	.39019	.92073	.42379	.3597	.0861	.5628	.40620	.91378	.44453	.2495	.0943	.4618	2
59	.35810	.93368	.38353	.6073	.0710	.7925	.37434	.92729	.40369	.4772	.0784	.6714	.39046	.92062	.42413	.3577	.0862	.5610	.40647	.91366	.44488	.2478	.0945	.4602	1
60	.35837	.93358	.38386	2.6051	.0711	2.7904	.37461	.92718	.40403	2.4751	.0785	2.6695	.39073	.92050	.42447	2.3558	.0864	2.5593	.40674	.91354	.44523	2.2460	.0946	2.4586	0
′	cos	sin	cot	tan	cosec	sec	cos	sin	cot	tan	cosec	sec	cos	sin	cot	tan	cosec	sec	cos	sin	cot	tan	cosec	sec	′
	69°						68°						67°						66°						

(Continued)

table 21 Natural trigonometric functions (Continued)

	24°						25°						26°						27°						
′	sin	cos	tan	cot	sec	cosec	sin	cos	tan	cot	sec	cosec	sin	cos	tan	cot	sec	cosec	sin	cos	tan	cot	sec	cosec	′
0	.40674	.91354	.44523	2.2460	1.0946	2.4586	.42262	.90631	.46631	2.1445	1.1034	2.3662	.43837	.89879	.48773	2.0503	1.1126	2.2812	.45399	.89101	.50952	1.9626	1.1223	2.2027	60
1	.40700	.91343	.44558	.2443	.0948	.4570	.42288	.90618	.46666	.1429	.1035	.3647	.43863	.89867	.48809	.0488	.1127	.2798	.45425	.89087	.50989	.9612	.1225	.2014	59
2	.40727	.91331	.44593	.2425	.0949	.4554	.42314	.90606	.46702	.1412	.1037	.3632	.43889	.89854	.48845	.0473	.1129	.2784	.45451	.89074	.51026	.9598	.1226	.2002	58
3	.40753	.91319	.44627	.2408	.0951	.4538	.42341	.90594	.46737	.1396	.1038	.3618	.43915	.89841	.48881	.0458	.1131	.2771	.45477	.89061	.51062	.9584	.1228	.1989	57
4	.40780	.91307	.44662	.2390	.0952	.4522	.42367	.90581	.46772	.1380	.1040	.3603	.43942	.89828	.48917	.0443	.1132	.2757	.45503	.89048	.51099	.9570	.1230	.1977	56
5	.40806	.91295	.44697	2.2373	.0953	2.4506	.42394	.90569	.46808	2.1364	.1041	2.3588	.43968	.89815	.48953	2.0427	.1134	2.2744	.45528	.89034	.51136	1.9556	.1231	2.1964	55
6	.40833	.91283	.44732	.2355	.0955	.4490	.42420	.90557	.46843	.1348	.1043	.3574	.43994	.89803	.48989	.0412	.1135	.2730	.45554	.89021	.51172	.9542	.1233	.1952	54
7	.40860	.91271	.44767	.2338	.0956	.4474	.42446	.90544	.46879	.1331	.1044	.3559	.44020	.89790	.49025	.0397	.1137	.2717	.45580	.89008	.51209	.9528	.1235	.1939	53
8	.40886	.91260	.44802	.2320	.0958	.4458	.42473	.90532	.46914	.1315	.1046	.3544	.44046	.89777	.49062	.0382	.1139	.2703	.45606	.88995	.51246	.9514	.1237	.1927	52
9	.40913	.91248	.44837	.2303	.0959	.4442	.42499	.90520	.46950	.1299	.1047	.3530	.44072	.89764	.49098	.0367	.1140	.2690	.45632	.88981	.51283	.9500	.1238	.1914	51
10	.40939	.91236	.44872	2.2286	.0961	2.4426	.42525	.90507	.46985	2.1283	.1049	2.3515	.44098	.89751	.49134	2.0352	.1142	2.2676	.45658	.88968	.51319	1.9486	.1240	2.1902	50
11	.40966	.91224	.44907	.2268	.0962	.4418	.42552	.90495	.47021	.1267	.1050	.3501	.44124	.89739	.49170	.0338	.1143	.2663	.45684	.88955	.51356	.9472	.1242	.1889	49
12	.40992	.91212	.44942	.2251	.0963	.4395	.42578	.90483	.47056	.1251	.1052	.3486	.44150	.89726	.49206	.0323	.1145	.2650	.45710	.88942	.51393	.9458	.1243	.1877	48
13	.41019	.91200	.44977	.2234	.0965	.4379	.42604	.90470	.47092	.1235	.1053	.3472	.44177	.89713	.49242	.0308	.1147	.2636	.45736	.88928	.51430	.9444	.1245	.1865	47
14	.41045	.91188	.45012	.2216	.0966	.4363	.42630	.90458	.47127	.1219	.1055	.3457	.44203	.89700	.49278	.0293	.1148	.2623	.45761	.88915	.51466	.9430	.1247	.1852	46
15	.41072	.91176	.45047	2.2199	.0968	2.4347	.42657	.90445	.47163	2.1203	.1056	2.3443	.44229	.89687	.49314	2.0278	.1150	2.2610	.45787	.88902	.51503	1.9416	.1248	2.1840	45
16	.41098	.91164	.45082	.2182	.0969	.4332	.42683	.90433	.47199	.1187	.1058	.3428	.44255	.89674	.49351	.0263	.1151	.2596	.45813	.88888	.51540	.9402	.1250	.1828	44
17	.41125	.91152	.45117	.2165	.0971	.4316	.42709	.90421	.47234	.1171	.1059	.3414	.44281	.89661	.49387	.0248	.1153	.2583	.45839	.88875	.51577	.9388	.1252	.1815	43
18	.41151	.91140	.45152	.2147	.0972	.4300	.42736	.90408	.47270	.1155	.1061	.3399	.44307	.89649	.49423	.0233	.1155	.2570	.45865	.88862	.51614	.9375	.1253	.1803	42
19	.41178	.91128	.45187	.2130	.0973	.4285	.42762	.90396	.47305	.1139	.1062	.3385	.44333	.89636	.49459	.0219	.1156	.2556	.45891	.88848	.51651	.9361	.1255	.1791	41
20	.41204	.91116	.45222	2.2113	.0975	2.4269	.42788	.90383	.47341	2.1123	.1064	2.3371	.44359	.89623	.49495	2.0204	.1158	2.2543	.45917	.88835	.51687	1.9347	.1257	2.1778	40
21	.41231	.91104	.45257	.2096	.0976	.4254	.42815	.90371	.47376	.1107	.1065	.3356	.44385	.89610	.49532	.0189	.1159	.2530	.45942	.88822	.51724	.9333	.1258	.1766	39
22	.41257	.91092	.45292	.2079	.0978	.4238	.42841	.90358	.47412	.1092	.1067	.3342	.44411	.89597	.49568	.0174	.1161	.2517	.45968	.88808	.51761	.9319	.1260	.1754	38
23	.41284	.91080	.45327	.2062	.0979	.4222	.42867	.90346	.47448	.1076	.1068	.3328	.44437	.89584	.49604	.0159	.1163	.2503	.45994	.88795	.51798	.9306	.1262	.1742	37
24	.41310	.91068	.45362	.2045	.0981	.4207	.42893	.90333	.47483	.1060	.1070	.3313	.44463	.89571	.49640	.0145	.1164	.2490	.46020	.88781	.51835	.9292	.1264	.1730	36
25	.41337	.91056	.45397	2.2028	.0982	2.4191	42920	.90321	.47519	2.1044	.1072	2.3299	.44489	.89558	.49677	2.0130	.1166	2.2477	.46046	.88768	.51872	1.9278	.1265	2.1717	35
26	.41363	.91044	.45432	.2011	.0984	.4176	.42946	.90308	.47555	.1028	.1073	.3285	.44516	.89545	.49713	.0115	.1167	.2464	.46072	.88755	.51909	.9264	.1267	.1705	34
27	.41390	.91032	.45467	.1994	.0985	.4160	.42972	.90296	.47590	.1013	.1075	.3271	.44542	.89532	.49749	.0101	.1169	.2451	.46097	.88741	.51946	.9251	.1269	.1693	33
28	.41416	.91020	.45502	.1977	.0986	.4145	.42998	.90283	.47626	.0997	.1076	.3256	.44568	.89519	.49785	.0086	.1171	.2438	.46123	.88728	.51983	.9237	.1270	.1681	32
29	.41443	.91008	.45537	.1960	.0988	.4130	.43025	.90271	.47662	.0981	.1078	.3242	.44594	.89506	.49822	.0071	.1172	.2425	.46149	.88714	.52020	.9223	.1272	.1669	31
30	.41469	.90996	.45573	2.1943	.0989	2.4114	.43051	.90258	.47697	2.0965	.1079	2.3228	.44620	.89493	.49858	2.0057	.1174	2.2411	.46175	.88701	.52057	1.9210	.1274	2.1657	30
31	.41496	.90984	.45608	.1926	.0991	.4099	.43077	.90246	.47733	.0950	.1081	.3214	.44646	.89480	.49894	.0042	.1176	.2398	.46201	.88688	.52094	.9196	.1275	.1645	29
32	.41522	.90972	.45643	.1909	.0992	.4083	.43104	.90233	.47769	.0934	.1082	.3200	.44672	.89467	.49931	.0028	.1177	.2385	.46226	.88674	.52131	.9182	.1277	.1633	28
33	.41549	.90960	.45678	.1892	.0994	.4068	.43130	.90221	.47805	.0918	.1084	.3186	.44698	.89454	.49967	.0013	.1179	.2372	.46252	.88661	.52168	.9169	.1279	.1620	27
34	.41575	.90948	.45713	.1875	.0995	.4053	.43156	.90208	.47840	.0903	.1085	.3172	.44724	.89441	.50003	1.9998	.1180	.2359	.46278	.88647	.52205	.9155	.1281	.1608	26
35	.41602	.90936	.45748	2.1859	.0997	2.4037	.43182	.90196	.47876	2.0887	.1087	2.3158	.44750	.89428	.50040	1.9984	.1182	2.2346	.46304	.88634	.52242	1.9142	.1282	2.1596	25
36	.41628	.90924	.45783	.1842	.0998	.4022	.43208	.90183	.47912	.0872	.1088	.3143	.44776	.89415	.50076	.9969	.1184	.2333	.46330	.88620	.52279	.9128	.1284	.1584	24
37	.41654	.90911	.45819	.1825	.1000	.4007	.43235	.90171	.47948	.0856	.1090	.3129	.44802	.89402	.50113	.9955	.1185	.2320	.46355	.88607	.52316	.9115	.1286	.1572	23
38	.41681	.90899	.45854	.1808	.1001	.3992	.43261	.90158	.47983	.0840	.1092	.3115	.44828	.89389	.50149	.9940	.1187	.2307	.46381	.88593	.52353	.9101	.1287	.1560	22
39	.41707	.90887	.45889	.1792	.1003	.3976	.43287	.90145	.48019	.0825	.1093	.3101	.44854	.89376	.50185	.9926	.1189	.2294	.46407	.88580	.52390	.9088	.1289	.1548	21
40	.41734	.90875	.45924	2.1775	.1004	2.3961	.43313	.90133	.48055	2.0809	.1095	2.3087	.44880	.89363	.50222	1.9912	.1190	2.2282	.46433	.88566	.52427	1.9074	.1291	2.1536	20
41	.41760	.90863	.45960	.1758	.1005	.3946	.43340	.90120	.48091	.0794	.1096	.3073	.44906	.89350	.50258	.9897	.1192	.2269	.46458	.88553	.52464	.9061	.1293	.1525	19
42	.41787	.90851	.45995	.1741	.1007	.3931	.43366	.90108	.48127	.0778	.1098	.3059	.44932	.89337	.50295	.9883	.1193	.2256	.46484	.88539	.52501	.9047	.1294	.1513	18
43	.41813	.90839	.46030	.1725	.1008	.3916	.43392	.90095	.48162	.0763	.1099	.3046	.44958	.89324	.50331	.9868	.1195	.2243	.46510	.88526	.52538	.9034	.1296	.1501	17
44	.41839	.90826	.46065	.1708	.1010	.3901	.43418	.90082	.48198	.0747	.1101	.3032	.44984	.89311	.50368	.9854	.1197	.2230	.46536	.88512	.52575	.9020	.1298	.1489	16
45	.41866	.90814	.46101	2.1692	.1011	2.3886	.43444	.90070	.48234	2.0732	.1102	2.3018	.45010	.89298	.50404	1.9840	.1198	2.2217	.46561	.88499	.52612	1.9007	.1299	2.1477	15
46	.41892	.90812	.46136	.1675	.1013	.3871	.43471	.90057	.48270	.0717	.1104	.3004	.45036	.89285	.50441	.9825	.1200	.2204	.46587	.88485	.52650	.8993	.1301	.1465	14
47	.41919	.90790	.46171	.1658	.1014	.3856	.43497	.90044	.48306	.0701	.1106	.2990	.45062	.89272	.50477	.9811	.1202	.2192	.46613	.88472	.52687	.8980	.1303	.1453	13
48	.41945	.90778	.46206	.1642	.1016	.3841	.43523	.90032	.48342	.0686	.1107	.2976	.45088	.89258	.50514	.9797	.1203	.2179	.46639	.88458	.52724	.8967	.1305	.1441	12
49	.41972	.90765	.46242	.1625	.1017	.3826	.43549	.90019	.48378	.0671	.1109	.2962	.45114	.89245	.50550	.9782	.1205	.2166	.46664	.88444	.52761	.8953	.1306	.1430	11
50	.41998	.90753	.46277	2.1609	.1019	2.3811	.43575	.90006	.48414	2.0655	.1110	2.2949	.45140	.89232	.50587	1.9768	.1207	2.2153	.46690	.88431	.52798	1.8940	.1308	2.1418	10
51	.42024	.90741	.46312	.1592	.1020	.3796	.43602	.89994	.48449	.0640	.1112	.2935	.45166	.89219	.50623	.9754	.1208	.2141	.46716	.88417	.52836	.8927	.1310	.1406	9
52	.42051	.90729	.46348	.1576	.1022	.3781	.43628	.89981	.48485	.0625	.1113	.2921	.45191	.89206	.50660	.9739	.1210	.2128	.46741	.88404	.52873	.8913	.1312	.1394	8
53	.42077	.90717	.46383	.1559	.1023	.3766	.43654	.89968	.48521	.0609	.1115	.2907	.45217	.89193	.50696	.9725	.1212	.2115	.46767	.88390	.52910	.8900	.1313	.1382	7
54	.42103	.90704	.46418	.1543	.1025	.3751	.43680	.89956	.48557	.0594	.1116	.2894	.45243	.89180	.50733	.9711	.1213	.2103	.46793	.88376	.52947	.8887	.1315	.1371	6
55	.42130	.90692	.46454	2.1527	.1026	2.3736	.43706	.89943	.48593	2.0579	.1118	2.2880	.45269	.89166	.50769	1.9697	.1215	2.2090	.46819	.88363	.52984	1.8873	.1317	2.1359	5
56	.42156	.90680	.46489	.1510	.1028	.3721	.43732	.89930	.48629	.0564	.1120	.2866	.45295	.89153	.50806	.9683	.1217	.2077	.46844	.88349	.53022	.8860	.1319	.1347	4
57	.42183	.90668	.46524	.1494	.1029	.3706	.43759	.89918	.48665	.0548	.1121	.2853	.45321	.89140	.50843	.9668	.1218	.2065	.46870	.88336	.53059	.8847	.1320	.1335	3
58	.42209	.90655	.46560	.1478	.1031	.3691	.43785	.89905	.48701	.0533	.1123	.2839	.45347	.89127	.50879	.9654	.1220	.2052	.46896	.88322	.53096	.8834	.1322	.1324	2
59	.42235	.90643	.46595	.1461	.1032	.3677	.43811	.89892	.48737	.0518	.1124	.2825	.45373	.89114	.50916	.9640	.1222	.2039	.46921	.88308	.53134	.8820	.1324	.1312	1
60	.42262	.90631	.46631	2.1445	.1034	2.3662	.43837	.89879	.48773	2.0503	.1126	2.2812	.45399	.89101	.50952	1.9626	.1223	2.2027	.46947	.88295	.53171	1.8807	.1326	2.1300	0
′	cos	sin	cot	tan	cosec	sec	cos	sin	cot	tan	cosec	sec	cos	sin	cot	tan	cosec	sec	cos	sin	cot	tan	cosec	sec	′
	65°						64°						63°						62°						

(Continued)

table 21 Natural trigonometric functions (Continued)

′	28°						29°						30°						31°						′
′	sin	cos	tan	cot	sec	cosec	sin	cos	tan	cot	sec	cosec	sin	cos	tan	cot	sec	cosec	sin	cos	tan	cot	sec	cosec	′
0	.46947	.88295	.53171	1.8807	1.1326	2.1300	.48481	.87462	.55431	1.8040	1.1433	2.0627	.50000	.86603	.57735	1.7320	1.1547	2.0000	.51504	.85717	.60086	1.6643	1.1666	1.9416	60
1	.46973	.88281	.53208	.8794	.1327	.1289	.48506	.87448	.55469	.8028	.1435	.0616	.50025	.86588	.57774	.7309	.1549	1.9990	.51529	.85702	.60126	.6632	.1668	.9407	59
2	.46998	.88267	.53245	.8781	.1329	.1277	.48532	.87434	.55507	.8016	.1437	.0605	.50050	.86573	.57813	.7297	.1551	.9980	.51554	.85687	.60165	.6621	.1670	.9397	58
3	.47024	.88254	.53283	.8768	.1331	.1266	.48557	.87420	.55545	.8003	.1439	.0594	.50075	.86559	.57851	.7286	.1553	.9970	.51578	.85672	.60205	.6610	.1672	.9388	57
4	.47050	.88240	.53320	.8754	.1333	.1254	.48583	.87405	.55583	.7991	.1441	.0583	.50101	.86544	.57890	.7274	.1555	.9960	.51603	.85657	.60244	.6599	.1674	.9378	56
5	.47075	.88226	.53358	1.8741	.1334	2.1242	.48608	.87391	.55621	1.7979	.1443	2.0573	.50126	.86530	.57929	1.7262	.1557	1.9950	.51628	.85642	.60284	1.6588	.1676	1.9369	55
6	.47101	.88213	.53395	.8728	.1336	.1231	.48633	.87377	.55659	.7966	.1445	.0562	.50151	.86515	.57968	.7251	.1559	.9940	.51653	.85627	.60324	.6577	.1678	.9360	54
7	.47127	.88199	.53432	.8715	.1338	.1219	.48659	.87363	.55697	.7954	.1446	.0551	.50176	.86500	.58007	.7239	.1561	.9930	.51678	.85612	.60363	.6566	.1681	.9350	53
8	.47152	.88185	.53470	.8702	.1340	.1208	.48684	.87349	.55735	.7942	.1448	.0540	.50201	.86486	.58046	.7228	.1562	.9920	.51703	.85597	.60403	.6555	.1683	.9341	52
9	.47178	.88171	.53507	.8689	.1341	.1196	.48710	.87335	.55774	.7930	.1450	.0530	.50226	.86471	.58085	.7216	.1564	.9910	.51728	.85582	.60443	.6544	.1685	.9332	51
10	.47204	.88158	.53545	1.8676	.1343	2.1185	.48735	.87320	.55812	1.7917	.1452	2.0519	.50252	.86457	.58123	1.7205	.1566	1.9900	.51753	.85566	.60483	1.6534	.1687	1.9322	50
11	.47229	.88144	.53582	.8663	.1345	.1173	.48760	.87306	.55850	.7905	.1454	.0508	.50277	.86442	.58162	.7193	.1568	.9890	.51778	.85551	.60522	.6523	.1689	.9313	49
12	.47255	.88130	.53619	.8650	.1347	.1162	.48786	.87292	.55888	.7893	.1456	.0498	.50302	.86427	.58201	.7182	.1570	.9880	.51803	.85536	.60562	.6512	.1691	.9304	48
13	.47281	.88117	.53657	.8637	.1349	.1150	.48811	.87278	.55926	.7881	.1458	.0487	.50327	.86413	.58240	.7170	.1572	.9870	.51827	.85521	.60602	.6501	.1693	.9295	47
14	.47306	.88103	.53694	.8624	.1350	.1139	.48837	.87264	.55964	.7868	.1459	.0476	.50352	.86398	.58279	.7159	.1574	.9860	.51852	.85506	.60642	.6490	.1695	.9285	46
15	.47332	.88089	.53732	1.8611	.1352	2.1127	.48862	.87250	.56003	1.7856	.1461	2.0466	.50377	.86383	.58318	1.7147	.1576	1.9850	.51877	.85491	.60681	1.6479	.1697	1.9276	45
16	.47357	.88075	.53769	.8598	.1354	.1116	.48887	.87235	.56041	.7844	.1463	.0455	.50402	.86369	.58357	.7136	.1578	.9840	.51902	.85476	.60721	.6469	.1699	.9267	44
17	.47383	.88061	.53807	.8585	.1356	.1104	.48913	.87221	.56079	.7832	.1465	.0444	.50428	.86354	.58396	.7124	.1580	.9830	.51927	.85461	.60761	.6458	.1701	.9258	43
18	.47409	.88048	.53844	.8572	.1357	.1093	.48938	.87207	.56117	.7820	.1467	.0434	.50453	.86339	.58435	.7113	.1582	.9820	.51952	.85446	.60801	.6447	.1703	.9248	42
19	.47434	.88034	.53882	.8559	.1359	.1082	.48964	.87193	.56156	.7808	.1469	.0423	.50478	.86325	.58474	.7101	.1584	.9811	.51977	.85431	.60841	.6436	.1705	.9239	41
20	.47460	.88020	.53919	1.8546	.1361	2.1070	.48989	.87178	.56194	1.7795	.1471	2.0413	.50503	.86310	.58513	1.7090	.1586	1.9801	.52002	.85416	.60881	1.6425	.1707	1.9230	40
21	.47486	.88006	.53957	.8533	.1363	.1059	.49014	.87164	.56232	.7783	.1473	.0402	.50528	.86295	.58552	.7079	.1588	.9791	.52026	.85400	.60920	.6415	.1709	.9221	39
22	.47511	.87992	.53995	.8520	.1365	.1048	.49040	.87150	.56270	.7771	.1474	.0392	.50553	.86281	.58591	.7067	.1590	.9781	.52051	.85385	.60960	.6404	.1712	.9212	38
23	.47537	.87979	.54032	.8507	.1366	.1036	.49065	.87136	.56309	.7759	.1476	.0381	.50578	.86266	.58630	.7056	.1592	.9771	.52076	.85370	.61000	.6393	.1714	.9203	37
24	.47562	.87965	.54070	.8495	.1368	.1025	.49090	.87121	.56347	.7747	.1478	.0370	.50603	.86251	.58670	.7044	.1594	.9761	.52101	.85355	.61040	.6383	.1716	.9193	36
25	.47588	.87951	.54107	1.8482	.1370	2.1014	.49116	.87107	.56385	1.7735	.1480	2.0360	.50628	.86237	.58709	1.7033	.1596	1.9752	.52126	.85340	.61080	1.6372	.1718	1.9184	35
26	.47613	.87937	.54145	.8469	.1372	.1002	.49141	.87093	.56424	.7723	.1482	.0349	.50653	.86222	.58748	.7022	.1598	.9742	.52151	.85325	.61120	.6361	.1720	.9175	34
27	.47639	.87923	.54183	.8456	.1373	.0991	.49166	.87078	.56462	.7711	.1484	.0339	.50679	.86207	.58787	.7010	.1600	.9732	.52175	.85309	.61160	.6350	.1722	.9166	33
28	.47665	.87909	.54220	.8443	.1375	.0980	.49192	.87064	.56500	.7699	.1486	.0329	.50704	.86192	.58826	.6999	.1602	.9722	.52200	.85294	.61200	.6340	.1724	.9157	32
29	.47690	.87895	.54258	.8430	.1377	.0969	.49217	.87050	.56539	.7687	.1488	.0318	.50729	.86178	.58865	.6988	.1604	.9713	.52225	.85279	.61240	.6329	.1726	.9148	31
30	.47716	.87882	.54295	1.8418	.1379	2.0957	.49242	.87035	.56577	1.7675	.1489	2.0308	.50754	.86163	.58904	1.6977	.1606	1.9703	.52250	.85264	.61280	1.6318	.1728	1.9139	30
31	.47741	.87868	.54333	.8405	.1381	.0946	.49268	.87021	.56616	.7663	.1491	.0297	.50779	.86148	.58944	.6965	.1608	.9693	.52275	.85249	.61320	.6308	.1730	.9130	29
32	.47767	.87854	.54371	.8392	.1382	.0935	.49293	.87007	.56654	.7651	.1493	.0287	.50804	.86133	.58983	.6954	.1610	.9683	.52299	.85234	.61360	.6297	.1732	.9121	28
33	.47792	.87840	.54409	.8379	.1384	.0924	.49318	.86992	.56692	.7639	.1495	.0276	.50829	.86118	.59022	.6943	.1612	.9674	.52324	.85218	.61400	.6286	.1734	.9112	27
34	.47818	.87826	.54446	.8367	.1386	.0912	.49343	.86978	.56731	.7627	.1497	.0266	.50854	.86104	.59061	.6931	.1614	.9664	.52349	.85203	.61440	.6276	.1737	.9102	26
35	.47844	.87812	.54484	1.8354	.1388	2.0901	.49369	.86964	.56769	1.7615	.1499	2.0256	.50879	.86089	.59100	1.6920	.1616	1.9654	.52374	.85188	.61480	1.6265	.1739	1.9093	25
36	.47869	.87798	.54522	.8341	.1390	.0890	.49394	.86949	.56808	.7603	.1501	.0245	.50904	.86074	.59140	.6909	.1618	.9645	.52398	.85173	.61520	.6255	.1741	.9084	24
37	.47895	.87784	.54559	.8329	.1391	.0879	.49419	.86935	.56846	.7591	.1503	.0235	.50929	.86059	.59179	.6898	.1620	.9635	.52423	.85157	.61560	.6244	.1743	.9075	23
38	.47920	.87770	.54597	.8316	.1393	.0868	.49445	.86921	.56885	.7579	.1505	.0224	.50954	.86044	.59218	.6887	.1622	.9625	.52448	.85142	.61601	.6233	.1745	.9066	22
39	.47946	.87756	.54635	.8303	.1395	.0857	.49470	.86906	.56923	.7567	.1507	.0214	.50979	.86030	.59258	.6875	.1624	.9616	.52473	.85127	.61641	.6223	.1747	.9057	21
40	.47971	.87742	.54673	1.8291	.1397	2.0846	.49495	.86892	.56962	1.7555	.1508	2.0204	.51004	.86015	.59297	1.6864	.1626	1.9606	.52498	.85112	.61681	1.6212	.1749	1.9048	20
41	.47997	.87728	.54711	.8278	.1399	.0835	.49521	.86877	.57000	.7544	.1510	.0194	.51029	.86000	.59336	.6853	.1628	.9596	.52522	.85096	.61721	.6202	.1751	.9039	19
42	.48022	.87715	.54748	.8265	.1401	.0824	.49546	.86863	.57039	.7532	.1512	.0183	.51054	.85985	.59376	.6842	.1630	.9587	.52547	.85081	.61761	.6191	.1753	.9030	18
43	.48048	.87701	.54786	.8253	.1402	.0812	.49571	.86849	.57077	.7520	.1514	.0173	.51079	.85970	.59415	.6831	.1632	.9577	.52572	.85066	.61801	.6181	.1756	.9021	17
44	.48073	.87687	.54824	.8240	.1404	.0801	.49596	.86834	.57116	.7508	.1516	.0163	.51104	.85955	.59454	.6820	.1634	.9568	.52597	.85050	.61842	.6170	.1758	.9013	16
45	.48099	.87673	.54862	1.8227	.1406	2.0790	.49622	.86820	.57155	1.7496	.1518	2.0152	.51129	.85941	.59494	1.6808	.1636	1.9558	.52621	.85735	.61882	1.6160	.1760	1.9004	15
46	.48124	.87659	.54900	.8215	.1408	.0779	.49647	.86805	.57193	.7484	.1520	.0142	.51154	.85926	.59533	.6797	.1638	.9549	.52646	.85020	.61922	.6149	.1762	.8995	14
47	.48150	.87645	.54937	.8202	.1410	.0768	.49672	.86791	.57232	.7473	.1522	.0132	.51179	.85911	.59572	.6786	.1640	.9539	.52671	.85004	.61962	.6139	.1764	.8986	13
48	.48175	.87631	.54975	.8190	.1411	.0757	.49697	.86776	.57270	.7461	.1524	.0122	.51204	.85896	.59612	.6775	.1642	.9530	.52695	.84989	.62003	.6128	.1766	.8977	12
49	.48201	.87617	.55013	.8177	.1413	.0746	.49723	.86762	.57309	.7449	.1526	.0111	.51229	.85881	.59651	.6764	.1644	.9520	.52720	.84974	.62043	.6118	.1768	.8968	11
50	.48226	.87603	.55051	1.8165	.1415	2.0735	.49748	.86748	.57348	1.7437	.1528	2.0101	.51254	.85866	.59691	1.6753	.1646	1.9510	.52745	.84959	.62083	1.6107	.1770	1.8959	10
51	.48252	.87588	.55089	.8152	.1417	.0725	.49773	.86733	.57386	.7426	.1530	.0091	.51279	.85851	.59730	.6742	.1648	.9501	.52770	.84943	.62123	.6097	.1772	.8950	9
52	.48277	.87574	.55127	.8140	.1419	.0714	.49798	.86719	.57425	.7414	.1531	.0081	.51304	.85836	.59770	.6731	.1650	.9491	.52794	.84928	.62164	.6086	.1775	.8941	8
53	.48303	.87560	.55165	.8127	.1421	.0703	.49823	.86704	.57464	.7402	.1533	.0071	.51329	.85821	.59809	.6720	.1652	.9482	.52819	.84912	.62204	.6076	.1777	.8932	7
54	.48328	.87546	.55203	.8115	.1422	.0692	.49849	.86690	.57502	.7390	.1535	.0061	.51354	.85806	.59849	.6709	.1654	.9473	.52844	.84897	.62244	.6066	.1779	.8924	6
55	.48354	.87532	.55241	1.8102	.1424	2.0681	.49874	.86675	.57541	1.7379	.1537	2.0050	.51379	.85791	.59888	1.6698	.1656	1.9463	.52868	.84882	.62285	1.6055	.1781	1.8915	5
56	.48379	.87518	.55279	.8090	.1426	.0670	.49899	.86661	.57580	.7367	.1539	.0040	.51404	.85777	.59928	.6687	.1658	.9454	.52893	.84866	.62325	.6045	.1783	.8906	4
57	.48405	.87504	.55317	.8078	.1428	.0659	.49924	.86646	.57619	.7355	.1541	.0030	.51429	.85762	.59967	.6676	.1660	.9444	.52918	.84851	.62366	.6034	.1785	.8897	3
58	.48430	.87490	.55355	.8065	.1430	.0648	.49950	.86632	.57657	.7344	.1543	.0020	.51454	.85747	.60007	.6665	.1662	.9435	.52942	.84836	.62406	.6024	.1787	.8888	2
59	.48455	.87476	.55393	.8053	.1432	.0637	.49975	.86617	.57696	.7332	.1545	.0010	.51479	.85732	.60046	.6654	.1664	.9425	.52967	.84820	.62446	.6014	.1790	.8879	1
60	.48481	.87462	.55431	1.8040	.1433	2.0627	.50000	.86603	.57735	1.7320	.1547	2.0000	.51504	.85717	.60086	1.6643	.1666	1.9416	.52992	.84805	.62487	1.6003	.1792	1.8871	0
′	cos	sin	cot	tan	cosec	sec	cos	sin	cot	tan	cosec	sec	cos	sin	cot	tan	cosec	sec	cos	sin	cot	tan	cosec	sec	′
	61°						60°						59°						58°						

(Continued)

table 21 Natural trigonometric functions (Continued)

	32°						33°						34°						35°						
′	sin	cos	tan	cot	sec	cosec	sin	cos	tan	cot	sec	cosec	sin	cos	tan	cot	sec	cosec	sin	cos	tan	cot	sec	cosec	′
0	.52992	.84805	.62487	1.6003	1.1792	1.8871	.54464	.83867	.64941	1.5399	1.1924	1.8361	.55919	.82904	.67451	1.4826	1.2062	1.7883	.57358	.81915	.70021	1.4281	1.2208	1.7434	60
1	.53016	.84789	.62527	.5993	.1794	.8862	.54488	.83851	.64982	.5389	.1926	.8352	.55943	.82887	.67493	.4816	.2064	.7875	.57381	.81898	.70064	.4273	.2210	.7427	59
2	.53041	.84774	.62568	.5983	.1796	.8853	.54513	.83835	.65023	.5379	.1928	.8344	.55967	.82871	.67535	.4807	.2067	.7867	.57405	.81882	.70107	.4264	.2213	.7420	58
3	.53066	.84758	.62608	.5972	.1798	.8844	.54537	.83819	.65065	.5369	.1930	.8336	.55992	.82855	.67578	.4798	.2069	.7860	.57429	.81865	.70151	.4255	.2215	.7413	57
4	.53090	.84743	.62649	.5962	.1800	.8836	.54561	.83804	.65106	.5359	.1933	.8328	.56016	.82839	.67620	.4788	.2072	.7852	.57453	.81848	.70194	.4246	.2218	.7405	56
5	.53115	.84728	.62689	1.5952	.1802	1.8827	.54586	.83788	.65148	1.5350	.1935	1.8320	.56040	.82822	.67663	1.4779	.2074	1.7844	.57477	.81832	.70238	1.4237	.2220	1.7398	55
6	.53140	.84712	.62730	.5941	.1805	.8818	.54610	.83772	.65189	.5340	.1937	.8311	.56064	.82806	.67705	.4770	.2076	.7837	.57500	.81815	.70281	.4228	.2223	.7391	54
7	.53164	.84697	.62770	.5931	.1807	.8809	.54634	.83756	.65231	.5330	.1939	.8303	.56088	.82790	.67747	.4761	.2079	.7829	.57524	.81798	.70325	.4220	.2225	.7384	53
8	.53189	.84681	.62811	.5921	.1809	.8801	.54659	.83740	.65272	.5320	.1942	.8295	.56112	.82773	.67790	.4751	.2081	.7821	.57548	.81781	.70368	.4211	.2228	.7377	52
9	.53214	.84666	.62851	.5910	.1811	.8792	.54683	.83724	.65314	.5311	.1944	.8287	.56136	.82757	.67832	.4742	.2083	.7814	.57572	.81765	.70412	.4202	.2230	.7369	51
10	.53238	.84650	.62892	1.5900	.1813	1.8783	.54708	.83708	.65355	1.5301	.1946	1.8279	.56160	.82741	.67875	1.4733	.2086	1.7806	.57596	.81748	.70455	1.4193	.2233	1.7362	50
11	.53263	.84635	.62933	.5890	.1815	.8775	.54732	.83692	.65397	.5291	.1948	.8271	.56184	.82724	.67917	.4724	.2088	.7798	.57619	.81731	.70499	.4185	.2235	.7355	49
12	.53288	.84619	.62973	.5880	.1818	.8766	.54756	.83676	.65438	.5282	.1951	.8263	.56208	.82708	.67960	.4714	.2091	.7791	.57643	.81714	.70542	.4176	.2238	.7348	48
13	.53312	.84604	.63014	.5869	.1820	.8757	.54781	.83660	.65480	.5272	.1953	.8255	.56232	.82692	.68002	.4705	.2093	.7783	.57667	.81698	.70586	.4167	.2240	.7341	47
14	.53337	.84588	.63055	.5859	.1822	.8749	.54805	.83644	.65521	.5262	.1955	.8246	.56256	.82675	.68045	.4696	.2095	.7776	.57691	.81681	.70629	.4158	.2243	.7334	46
15	.53361	.84573	.63095	1.5849	.1824	1.8740	.54829	.83629	.65563	1.5252	.1958	1.8238	.56280	.82659	.68087	1.4687	.2098	1.7768	.57714	.81664	.70673	1.4150	.2245	1.7327	45
16	.53386	.84557	.63136	.5839	.1826	.8731	.54854	.83613	.65604	.5234	.1960	.8230	.56304	.82643	.68130	.4678	.2100	.7760	.57738	.81647	.70717	.4141	.2248	.7319	44
17	.53411	.84542	.63177	.5829	.1828	.8723	.54878	.83597	.65646	.5233	.1962	.8222	.56328	.82626	.68173	.4669	.2103	.7753	.57762	.81630	.70760	.4132	.2250	.7312	43
18	.53435	.84526	.63217	.5818	.1831	.8714	.54902	.83581	.65688	.5223	.1964	.8214	.56353	.82610	.68215	.4659	.2105	.7745	.57786	.81614	.70804	.4123	.2253	.7305	42
19	.53460	.84511	.63258	.5808	.1833	.8706	.54926	.83565	.65729	.5214	.1967	.8206	.56377	.82593	.68258	.4650	.2107	.7738	.57809	.81597	.70848	.4115	.2255	.7298	41
20	.53484	.84495	.63299	1.5798	.1835	1.8697	.54951	.83549	.65771	1.5204	.1969	1.8198	.56401	.82577	.68301	1.4641	.2110	1.7730	.57833	.81580	.70891	1.4106	.2258	1.7291	40
21	.53509	.84479	.63339	.5788	.1837	.8688	.54975	.83533	.65813	.5195	.1971	.8190	.56425	.82561	.68343	.4632	.2112	.7723	.57857	.81563	.70935	.4097	.2260	.7284	39
22	.53533	.84464	.63380	.5778	.1839	.8680	.54999	.83517	.65854	.5185	.1974	.8182	.56449	.82544	.68386	.4623	.2115	.7715	.57881	.81546	.70979	.4089	.2263	.7277	38
23	.53558	.84448	.63421	.5768	.1841	.8671	.55024	.83501	.65896	.5175	.1976	.8174	.56473	.82528	.68429	.4614	.2117	.7708	.57904	.81530	.71022	.4080	.2265	.7270	37
24	.53583	.84433	.63462	.5757	.1844	.8663	.55048	.83485	.65938	.5166	.1978	.8166	.56497	.82511	.68471	.4605	.2119	.7700	.57928	.81513	.71066	.4071	.2268	.7263	36
25	.53607	.84417	.63503	1.5747	.1846	1.8654	.55002	.83469	.65980	1.5156	.1980	1.8158	.56521	.82495	.68514	1.4595	.2122	1.7693	.57952	.81496	.71110	1.4063	.2270	1.7256	35
26	.53632	.84402	.63543	.5737	.1848	.8646	.55090	.83453	.66021	.5147	.1983	.8150	.56545	.82478	.68557	.4586	.2124	.7685	.57975	.81479	.71154	.4054	.2273	.7249	34
27	.53656	.84386	.63584	.5727	.1850	.8637	.55121	.83437	.66063	.5137	.1985	.8142	.56569	.82462	.68600	.4577	.2127	.7678	.57999	.81462	.71198	.4045	.2276	.7242	33
28	.53681	.84370	.63625	.5717	.1852	.8629	.55145	.83421	.66105	.5127	.1987	.8134	.56593	.82445	.68642	.4568	.2129	.7670	.58023	.81445	.71241	.4037	.2278	.7234	32
29	.53705	.84355	.63666	.5707	.1855	.8620	.55169	.83405	.66147	.5118	.1990	.8126	.56617	.82429	.68685	.4559	.2132	.7663	.58047	.81428	.71285	.4028	.2281	.7227	31
30	.53730	.84339	.63707	1.5697	.1857	1.8611	.55194	.83388	.56188	1.5108	.1992	1.8118	.56641	.82413	.68728	1.4550	.2134	1.7655	.58070	.81411	.71329	1.4019	.2283	1.7220	30
31	.53754	.84323	.63748	.5687	.1859	.8603	.55218	.83372	.66230	.5099	.1994	.8110	.56664	.82396	.68771	.4541	.2136	.7648	.58094	.81395	.71373	.4011	.2286	.7213	29
32	.53779	.84308	.63789	.5677	.1861	.8595	.55242	.83356	.66272	.5089	.1997	.8102	.56688	.82380	.68814	.4532	.2139	.7640	.58118	.81378	.71417	.4002	.2288	.7206	28
33	.53803	.84292	.63830	.5667	.1863	.8586	.55266	.83340	.66314	.5080	.1999	.8094	.56712	.82363	.68857	.4523	.2141	.7633	.58141	.81361	.71461	.3994	.2291	.7199	27
34	.53828	.84276	.63871	.5657	.1866	.8578	.55291	.83324	.66356	.5070	.2001	.8086	.56736	.82347	.68899	.4514	.2144	.7625	.58165	.81344	.71505	.3985	.2293	.7192	26
35	.53852	.84261	.63912	1.5646	.1868	1.8569	.55315	.83308	.66398	1.5061	.2004	1.8078	.56760	.82330	.68942	1.4505	.2146	1.7618	.58189	.81327	.71549	1.3976	.2296	1.7185	25
36	.53877	.84245	.63953	.5636	.1870	.8561	.55339	.83292	.66440	.5051	.2006	.8070	.56784	.82314	.68985	.4496	.2149	.7610	.58212	.81310	.71593	.3968	.2298	.7178	24
37	.53901	.84229	.63994	.5626	.1872	.8552	.55363	.83276	.66482	.5042	.2008	.8062	.56808	.82297	.69028	.4487	.2151	.7603	.58236	.81293	.71637	.3959	.2301	.7171	23
38	.53926	.84214	.64035	.5616	.1874	.8544	.55388	.83260	.66524	.5032	.2010	.8054	.56832	.82280	.69071	.4478	.2153	.7596	.58259	.81276	.71681	.3951	.2304	.7164	22
39	.53950	.84198	.64076	.5606	.1877	.8535	.55412	.83244	.66566	.5023	.2013	.8047	.56856	.82264	.69114	.4469	.2156	.7588	.58283	.81259	.71725	.3942	.2306	.7157	21
40	.53975	.84182	.64117	1.5596	.1879	1.8527	.55436	.83228	.66608	1.5013	.2015	1.8039	.56880	.82247	.69157	1.4460	.2158	1.7581	.58307	.81242	.71769	1.3933	.2309	1.7151	20
41	.53999	.84167	.64158	.5586	.1881	.8519	.55460	.83211	.66650	.5004	.2017	.8031	.56904	.82231	.69200	.4451	.2161	.7573	.58330	.81225	.71813	.3925	.2311	.7144	19
42	.54024	.84151	.64199	.5577	.1883	.8510	.55484	.83195	.66692	.4994	.2020	.8023	.56928	.82214	.69243	.4442	.2163	.7566	.58354	.81208	.71857	.3916	.2314	.7137	18
43	.54048	.84135	.64240	.5567	.1886	.8502	.55509	.83179	.66734	.4985	.2022	.8015	.56952	.82198	.69286	.4433	.2166	.7559	.58378	.81191	.71901	.3908	.2316	.7130	17
44	.54073	.84120	.64281	.5557	.1888	.8493	.55533	.83163	.66776	.4975	.2024	.8007	.56976	.82181	.69329	.4424	.2168	.7551	.58401	.81174	.71945	.3899	.2319	.7123	16
45	.54097	.84104	.64322	1.5547	.1890	1.8485	.55557	.83147	.66818	1.4966	.2027	1.7999	.57000	.82165	.69372	1.4415	.2171	1.7544	.58425	.81157	.71990	1.3891	.2322	1.7116	15
46	.54122	.84088	.64363	.5537	.1892	.8477	.55581	.83131	.66860	.4957	.2029	.7992	.57023	.82148	.69415	.4406	.2173	.7537	.58448	.81140	.72034	.3882	.2324	.7109	14
47	.54146	.84072	.64404	.5527	.1894	.8468	.55605	.83115	.66902	.4947	.2031	.7984	.57047	.82131	.69459	.4397	.2175	.7529	.58472	.81123	.72078	.3874	.2327	.7102	13
48	.54171	.84057	.64446	.5517	.1897	.8460	.55629	.83098	.66944	.4938	.2034	.7976	.57071	.82115	.69502	.4388	.2178	.7522	.58496	.81106	.72122	.3865	.2329	.7095	12
49	.54195	.84041	.64487	.5507	.1899	.8452	.55654	.83082	.66986	.4928	.2036	.7968	.57095	.82098	.69545	.4379	.2180	.7514	.58519	.81089	.72166	.3857	.2332	.7088	11
50	.54220	.84025	.64528	1.5497	.1901	1.8443	.55678	.83066	.67028	1.4919	.2039	1.7960	.57119	.82082	.69588	1.4370	.2183	1.7507	.58543	.81072	.72211	1.3848	.2335	1.7081	10
51	.54244	.84009	.64569	.5487	.1903	.8435	.55702	.83050	.67071	.4910	.2041	.7953	.57143	.82065	.69631	.4361	.2185	.7500	.58566	.81055	.72255	.3840	.2337	.7075	9
52	.54268	.83993	.64610	.5477	.1906	.8427	.55726	.83034	.67113	.4900	.2043	.7945	.57167	.82048	.69674	.4352	.2188	.7493	.58590	.81038	.72299	.3831	.2340	.7068	8
53	.54293	.83978	.64652	.5467	.1908	.8418	.55750	.83017	.67155	.4891	.2046	.7937	.57191	.82032	.69718	.4343	.2190	.7485	.58614	.81021	.72344	.3823	.2342	.7061	7
54	.54317	.83962	.64693	.5458	.1910	.8410	.55774	.83001	.67197	.4881	.2048	.7929	.57214	.82015	.69761	.4335	.2193	.7478	.58637	.81004	.72388	.3814	.2345	.7054	6
55	.54342	.83946	.64734	1.5448	.1912	1.8402	.55799	.82985	.67239	1.4872	.2050	1.7921	.57238	.81998	.69804	1.4326	.2195	1.7471	.58661	.80987	.72432	1.3806	.2348	1.7047	5
56	.54366	.83930	.64775	.5438	.1915	.8394	.55823	.82969	.67282	.4863	.2053	.7914	.57262	.81982	.69847	.4317	.2198	.7463	.58684	.80970	.72477	.3797	.2350	.7040	4
57	.54391	.83914	.64817	.5428	.1917	.8385	.55847	.82952	.67324	.4853	.2055	.7906	.57286	.81965	.69891	.4308	.2200	.7456	.58708	.80953	.72521	.3789	.2353	.7033	3
58	.54415	.83899	.64858	.5418	.1919	.8377	.55871	.82936	.67366	.4844	.2057	.7898	.57310	.81948	.69934	.4299	.2203	.7449	.58731	.80936	.72565	.3781	.2355	.7027	2
59	.54439	.83883	.64899	.5408	.1921	.8369	.55895	.82920	.67408	.4835	.2060	.7891	.57334	.81932	.69977	.4290	.2205	.7442	.58755	.80919	.72610	.3772	.2358	.7020	1
60	.54464	.83867	.64941	1.5399	.1922	1.8361	.55919	.82904	.67451	1.4826	.2062	1.7883	.57358	.81915	.70021	1.4281	.2208	1.7434	.58778	.80902	.72654	1.3764	.2361	1.7013	0
′	cos	sin	cot	tan	cosec	sec	cos	sin	cot	tan	cosec	sec	cos	sin	cot	tan	cosec	sec	cos	sin	cot	tan	cosec	sec	′
	57°						56°						55°						54°						

(Continued)

table 21 Natural trigonometric functions (Continued)

′	36°						37°						38°						39°						′
′	sin	cos	tan	cot	sec	cosec	sin	cos	tan	cot	sec	cosec	sin	cos	tan	cot	sec	cosec	sin	cos	tan	cot	sec	cosec	′
0	.58778	.80902	.72654	1.3764	1.2361	1.7013	.60181	.79863	.75355	1.3270	1.2521	1.6616	.61566	.78801	.78128	1.2799	1.2690	1.6243	.62932	.77715	.80978	1.2349	1.2867	1.5890	60
1	.58802	.80885	.72699	.3755	.2363	.7006	.60205	.79846	.75401	.3262	.2524	.6610	.61589	.78783	.78175	.2792	.2693	.6237	.62955	.77696	.81026	.2342	.2871	.5884	59
2	.58825	.80867	.72743	.3747	.2366	.6999	.60228	.79828	.75447	.3254	.2527	.6603	.61612	.78765	.78222	.2784	.2696	.6231	.62977	.77678	.81075	.2334	.2874	.5879	58
3	.58849	.80850	.72788	.3738	.2368	.6993	.60251	.79811	.75492	.3246	.2530	.6597	.61635	.78747	.78269	.2776	.2699	.6224	.63000	.77660	.81123	.2327	.2877	.5873	57
4	.58873	.80833	.72832	.3730	.2371	.6986	.60274	.79793	.75538	.3238	.2532	.6591	.61658	.78729	.78316	.2769	.2702	.6218	.63022	.77641	.81171	.2320	.2880	.5867	56
5	.58896	.80816	.72877	1.3722	.2374	1.6979	.60298	.79776	.75584	1.3230	.2535	1.6584	.61681	.78711	.78363	1.2761	.2705	1.6212	.63045	.77623	.81219	1.2312	.2883	1.5862	55
6	.58920	.80799	.72921	.3713	.2376	.6972	.60320	.79758	.75629	.3222	.2538	.6578	.61703	.78693	.78410	.2753	.2707	.6206	.63067	.77605	.81268	.2305	.2886	.5856	54
7	.58943	.80782	.72966	.3705	.2379	.6965	.60344	.79741	.75675	.3214	.2541	.6572	.61726	.78675	.78457	.2746	.2710	.6200	.63090	.77586	.81316	.2297	.2889	.5850	53
8	.58967	.80765	.73010	.3697	.2382	.6959	.60367	.79723	.75721	.3206	.2543	.6565	.61749	.78657	.78504	.2738	.2713	.6194	.63113	.77568	.81364	.2290	.2892	.5854	52
9	.58990	.80747	.73055	.3688	.2384	.6952	.60390	.79706	.75767	.3198	.2546	.6559	.61772	.78640	.78551	.2730	.2716	.6188	.63135	.77549	.81413	.2283	.2895	.5839	51
10	.59014	.80730	.73100	1.3680	.2387	1.6945	.60413	.79688	.75812	1.3190	.2549	1.6552	.61795	.78622	.78598	1.2723	.2719	1.6182	.63158	.77531	.81461	1.2276	.2898	1.5833	50
11	.59037	.80713	.73144	.3672	.2389	.6938	.60437	.79670	.75858	.3182	.2552	.6546	.61818	.78604	.78645	.2715	.2722	.6176	.63180	.77513	.81509	.2268	.2901	.5828	49
12	.59060	.80696	.73189	.3663	.2392	.6932	.60460	.79653	.75904	.3174	.2554	.6540	.61841	.78586	.78692	.2708	.2725	.6170	.63203	.77494	.81558	.2261	.2904	.5822	48
13	.59084	.80679	.73234	.3655	.2395	.6925	.60483	.79635	.75950	.3166	.2557	.6533	.61864	.78568	.78739	.2700	.2728	.6164	.63225	.77476	.81606	.2254	.2907	.5816	47
14	.59107	.80662	.73278	.3647	.2397	.6918	.60506	.79618	.75996	.3159	.2560	.6527	.61886	.78550	.78786	.2692	.2731	.6159	.63248	.77458	.81655	.2247	.2910	.5811	46
15	.59131	.80644	.73323	1.3638	.2400	1.6912	.60529	.79600	.76042	1.3151	.2563	1.6521	.61909	.78532	.78834	1.2685	.2734	1.6153	.63270	.77439	.81703	1.2239	.2913	1.5805	45
16	.59154	.80627	.73368	.3630	.2403	.6905	.60552	.79582	.76088	.3143	.2565	.6514	.61932	.78514	.78881	.2677	.2737	.6147	.63293	.77421	.81752	.2232	.2916	.5799	44
17	.59178	.80610	.73412	.3622	.2405	.6898	.60576	.79565	.76134	.3135	.2568	.6508	.61955	.78496	.78928	.2670	.2739	.6141	.63315	.77402	.81800	.2225	.2919	.5794	43
18	.59201	.80593	.73457	.3613	.2408	.6891	.60599	.79547	.76179	.3127	.2571	.6502	.61978	.78478	.78975	.2662	.2742	.6135	.63338	.77384	.81849	.2218	.2922	.5788	42
19	.59225	.80576	.73502	.3605	.2411	.6885	.60622	.79530	.76225	.3119	.2574	.6496	.62001	.78460	.79022	.2655	.2745	.6129	.63360	.77365	.81898	.2210	.2926	.5783	41
20	.59248	.80558	.73547	1.3597	.2413	1.6878	.60645	.79512	.76271	1.3111	.2577	1.6489	.62023	.78441	.79070	1.2647	.2748	1.6123	.63383	.77347	.81946	1.2203	.2929	1.5777	40
21	.59272	.80541	.73592	.3588	.2416	.6871	.60668	.79494	.76317	.3103	.2579	.6483	.62046	.78423	.79117	.2639	.2751	.6117	.63405	.77329	.81995	.2196	.2932	.5771	39
22	.59295	.80524	.73637	.3580	.2419	.6865	.60691	.79477	.76364	.3095	.2582	.6477	.62069	.78405	.79164	.2632	.2754	.6111	.63428	.77310	.82043	.2189	.2935	.5766	38
23	.59318	.80507	.73681	.3572	.2421	.6858	.60714	.79459	.76410	.3087	.2585	.6470	.62092	.78387	.79212	.2624	.2757	.6105	.63450	.77292	.82092	.2181	.2938	.5760	37
24	.59342	.80489	.73726	.3564	.2424	.6851	.60737	.79441	.76456	.3079	.2588	.6464	.62115	.78369	.79259	.2617	.2760	.6099	.63473	.77273	.82141	.2174	.2941	.5755	36
25	.59365	.80472	.73771	1.3555	.2427	1.6845	.60761	.79424	.76502	1.3071	.2591	1.6458	.62137	.78351	.79306	1.2609	.2763	1.6093	.63495	.77255	.82190	1.2167	.2944	1.5749	35
26	.59389	.80455	.73816	.3547	.2429	.6838	.60784	.79406	.76548	.3064	.2593	.6452	.62160	.78333	.79354	.2602	.2766	.6087	.63518	.77236	.82238	.2160	.2947	.5743	34
27	.59412	.80437	.73861	.3539	.2432	.6831	.60807	.79388	.76594	.3056	.2596	.6445	.62183	.78315	.79401	.2594	.2769	.6081	.63540	.77218	.82287	.2152	.2950	.5738	33
28	.59435	.80420	.73906	.3531	.2435	.6825	.60830	.79371	.76640	.3048	.2599	.6439	.62206	.78297	.79449	.2587	.2772	.6077	.63563	.77199	.82336	.2145	.2953	.5732	32
29	.59459	.80403	.73951	.3522	.2437	.6818	.60853	.79353	.76686	.3040	.2602	.6433	.62229	.78279	.79496	.2579	.2775	.6070	.63585	.77181	.82385	.2138	.2956	.5727	31
30	.59482	.80386	.73996	1.3514	.2440	1.6812	.60876	.79335	.76733	1.3032	.2605	1.6427	.62251	.78261	.79543	1.2572	.2778	1.6064	.63608	.77162	.82434	1.2131	.2960	1.5721	30
31	.59506	.80368	.74041	.3506	.2443	.6805	.60899	.79318	.76779	.3024	.2607	.6420	.62274	.78243	.79591	.2564	.2781	.6058	.63630	.77144	.82482	.2124	.2963	.5716	29
32	.59529	.80351	.74086	.3498	.2445	.6798	.60922	.79300	.76825	.3016	.2610	.6414	.62297	.78224	.79639	.2557	.2784	.6052	.63653	.77125	.82531	.2117	.2966	.5710	28
33	.59552	.80334	.74131	.3489	.2448	.6792	.60945	.79282	.76871	.3009	.2613	.6408	.62320	.78206	.79686	.2549	.2787	.6046	.63675	.77107	.82580	.2109	.2969	.5705	27
34	.59576	.80316	.74176	.3481	.2451	.6785	.60968	.79264	.76918	.3001	.2616	.6402	.62342	.78188	.79734	.2542	.2790	.6040	.63697	.77088	.82629	.2102	.2972	.5699	26
35	.59599	.80299	.74221	1.3473	.2453	1.6779	.60991	.79247	.76964	1.2993	.2619	1.6396	.62365	.78170	.79781	1.2534	.2793	1.6034	.63720	.77070	.82678	1.2095	.2975	1.5694	25
36	.59622	.80282	.74266	.3465	.2456	.6772	.61014	.79229	.77010	.2985	.2622	.6389	.62388	.78152	.79829	.2527	.2795	.6029	.63742	.77051	.82727	.2088	.2978	.5688	24
37	.59646	.80264	.74312	.3457	.2459	.6766	.61037	.79211	.77057	.2977	.2624	.6383	.62411	.78134	.79876	.2519	.2798	.6023	.63765	.77033	.82776	.2081	.2981	.5683	23
38	.59669	.80247	.74357	.3449	.2461	.6759	.61061	.79193	.77103	.2970	.2627	.6377	.62433	.78116	.79924	.2512	.2801	.6017	.63787	.77014	.82825	.2074	.2985	.5677	22
39	.59692	.80230	.74402	.3440	.2464	.6752	.61084	.79176	.77149	.2962	.2630	.6371	.62456	.78097	.79972	.2504	.2804	.6011	.63810	.76996	.82874	.2066	.2988	.5672	21
40	.59716	.80212	.74447	1.3432	.2467	1.6746	.61107	.79158	.77196	1.2954	.2633	1.6365	.62479	.78079	.80020	1.2497	.2807	1.6005	.63832	.76977	.82923	1.2059	.2991	1.5666	20
41	.59739	.80195	.74492	.3424	.2470	.6739	.61130	.79140	.77242	.2946	.2636	.6359	.62501	.78061	.80067	.2489	.2810	.6000	.63854	.76958	.82972	.2052	.2994	.5661	19
42	.59762	.80177	.74538	.3416	.2472	.6733	.61153	.79122	.77289	.2938	.2639	.6352	.62524	.78043	.80115	.2482	.2813	.5994	.63877	.76940	.83022	.2045	.2997	.5655	18
43	.59786	.80160	.74583	.3408	.2475	.6726	.61176	.79104	.77335	.2931	.2641	.6346	.62547	.78025	.80163	.2475	.2816	.5988	.63899	.76921	.83071	.2038	.3000	.5650	17
44	.59809	.80143	.74628	.3400	.2478	.6720	.61199	.79087	.77382	.2923	.2644	.6340	.62570	.78007	.80211	.2467	.2819	.5982	.63921	.76903	.83120	.2031	.3003	.5644	16
45	.59832	.80125	.74673	1.3392	.2480	1.6713	.61222	.79069	.77428	1.2915	.2647	1.6334	.62592	.77988	.80258	1.2460	.2822	1.5976	.63944	.76884	.83169	1.2024	.3006	1.5639	15
46	.59856	.80108	.74019	.3383	.2483	.6707	.61245	.79051	.77475	.2907	.2650	.6328	.62615	.77970	.80306	.2452	.2825	.5971	.63966	.76865	.83218	.2016	.3010	.5633	14
47	.59879	.80090	.74064	.3375	.2486	.6700	.61268	.79033	.77521	.2900	.2653	.6322	.62638	.77952	.80354	.2445	.2828	.5965	.63989	.76847	.83267	.2009	.3013	.5628	13
48	.59902	.80073	.74809	.3367	.2488	.6694	.61290	.79015	.77568	.2892	.2656	.6316	.62660	.77934	.80402	.2437	.2831	.5959	.64011	.76828	.83317	.2002	.3016	.5622	12
49	.59926	.80056	.74855	.3359	.2491	.6687	.61314	.78998	.77614	.2884	.2659	.6309	.62683	.77915	.80450	.2430	.2834	.5953	.64033	.76810	.83366	.1995	.3019	.5617	11
50	.59949	.80038	.74900	1.3351	.2494	1.6681	.61337	.78980	.77661	1.2876	.2661	1.6303	.62706	.77897	.80498	1.2423	.2837	1.5947	.64056	.76791	.83415	1.1988	.3022	1.5611	10
51	.59972	.80021	.74946	.3343	.2497	.6674	.61360	.78962	.77708	.2869	.2664	.6297	.62728	.77879	.80546	.2415	.2840	.5942	.64078	.76772	.83465	.1981	.3025	.5606	9
52	.59995	.80003	.74991	.3335	.2499	.6668	.61383	.78944	.77754	.2861	.2667	.6291	.62751	.77861	.80594	.2408	.2843	.5936	.64100	.76754	.83514	.1974	.3029	.5600	8
53	.60019	.79986	.75037	.3327	.2502	.6661	.61405	.78926	.77801	.2853	.2670	.6285	.62774	.77842	.80642	.2400	.2846	.5930	.64123	.76735	.83563	.1967	.3032	.5595	7
54	.60042	.79968	.75082	.3319	.2505	.6655	.61428	.78908	.77848	.2845	.2673	.6279	.62796	.77824	.80690	.2393	.2849	.5924	.64145	.76716	.83613	.1960	.3035	.5590	6
55	.60065	.79951	.75128	1.3311	.2508	1.6648	.61451	.78890	.77895	1.2838	.2676	1.6273	.62819	.77806	.80738	1.2386	.2852	1.5919	.64160	.76698	.83662	1.1953	.3038	1.5584	5
56	.60088	.79933	.75173	.3303	.2510	.6642	.61474	.78873	.77941	.2830	.2679	.6267	.62841	.77788	.80786	[illegible]	.2855	.5913	.64189	.76679	.83712	.1946	.3041	.5579	4
57	.60112	.79916	.75219	.3294	.2513	.6636	.61497	.78855	.77988	.2822	.2681	.6261	.62864	.77769	.80834	.2371	.2858	.5907	.64212	.76660	.83761	.1939	.3044	.5573	3
58	.60135	.79898	.75264	.3286	.2516	.6629	.61520	.78837	.78035	.2815	.2684	.6255	.62887	.77751	.80882	.2364	.2861	.5901	.64234	.76642	.83811	.1932	.3048	.5568	2
59	.60158	.79881	.75310	.3278	.2519	.6623	.61543	.78819	.78082	.2807	.2687	.6249	.62909	.77733	.80930	.2356	.2864	.5896	.64256	.76623	.83860	.1924	.3051	.5563	1
60	.60181	.79863	.75355	1.3270	.2521	1.6616	.61566	.78801	.78128	1.2799	.2690	1.6243	62932	.77715	.80978	1.2349	.2867	1.5890	.64279	.76604	.83910	1.1917	.3054	1.5557	0
′	cos	sin	cot	tan	cosec	sec	cos	sin	cot	tan	cosec	sec	cos	sin	cot	tan	cosec	sec	cos	sin	cot	tan	cosec	sec	′
	53°						52°						51°						50°						

(Continued)

table 21 Natural trigonometric functions (Continued)

	40°						41°						42°						43°						
′	sin	cos	tan	cot	sec	cosec	sin	cos	tan	cot	sec	cosec	sin	cos	tan	cot	sec	cosec	sin	cos	tan	cot	sec	cosec	′
0	.64279	.76604	.83910	1.1917	1.3054	1.5557	.65606	.75471	.86929	1.1504	1.3250	1.5242	.66913	.74314	.90040	1.1106	1.3456	1.4945	.68200	.73135	.93251	1.0724	1.3673	1.4663	60
1	.64301	.76586	.83959	.1910	.3057	.5552	.65628	.75452	.86980	.1497	.3253	.5237	.66935	.74295	.90093	.1100	.3460	.4940	.68221	.73115	.93306	.0717	.3677	.4658	59
2	.64323	.76567	.84009	.1903	.3060	.5546	.65650	.75433	.87031	.1490	.3257	.5232	.66956	.74276	.90146	.1093	.3463	.4935	.68242	.73096	.93360	.0711	.3681	.4654	58
3	.64345	.76548	.84059	.1896	.3064	.5541	.65672	.75414	.87082	.1483	.3260	.5227	.66978	.74256	.90198	.1086	.3467	.4930	.68264	.73076	.93415	.0705	.3684	.4649	57
4	.64368	.76530	.84108	.1889	.3067	.5536	.65694	.75394	.87133	.1477	.3263	.5222	.66999	.74236	.90251	.1080	.3470	.4925	.68285	.73056	.93469	.0699	.3688	.4644	56
5	.64390	.76511	.84158	1.1882	.3070	1.5530	.65716	.75375	.87184	1.1470	.3267	1.5217	.67021	.74217	.90304	1.1074	.3474	1.4921	.68306	.73036	.93524	1.0692	.3692	1.4640	55
6	.64412	.76492	.84208	.1875	.3073	.5525	.65037	.75356	.87235	.1463	.3270	.5212	.67043	.74197	.90357	.1067	.3477	.4916	.68327	.73016	.93578	.0686	.3695	.4635	54
7	.64435	.76473	.84257	.1868	.3076	.5520	.65759	.75337	.87287	.1456	.3274	.5207	.67064	.74178	.90410	.1061	.3481	.4911	.68349	.72996	.93633	.0680	.3699	.4631	53
8	.64457	.76455	.84307	.1861	.3080	.5514	.65781	.75318	.87338	.1450	.3277	.5202	.67086	.74158	.90463	.1054	.3485	.4906	.68370	.72976	.93687	.0674	.3703	.4626	52
9	.64479	.76436	.84357	.1854	.3083	.5509	.65803	.75299	.87389	.1443	.3280	.5197	.67107	.74139	.90515	.1048	.3488	.4901	.68391	.72956	.93742	.0667	.3707	.4622	51
10	.64501	.76417	.84407	1.1847	.3086	1.5503	.65285	.75280	.87441	1.1436	.3284	1.5192	.67129	.74119	.90568	1.1041	.3492	1.4897	.68412	.72937	.93797	1.0661	.3710	1.4617	50
11	.64523	.76398	.84457	.1840	.3089	.5498	.65847	.75261	.87492	.1430	.3287	.5187	.67150	.74100	.90621	.1035	.3495	.4892	.68433	.72917	.93851	.0655	.3714	.4613	49
12	.64546	.76380	.84506	.1833	.3092	.5493	.65869	.75241	.87543	.1423	.3290	.5182	.67172	.74080	.90674	.1028	.3499	.4887	.68455	.72897	.93906	.0649	.3718	.4608	48
13	.64568	.76361	.84556	.1826	.3096	.5487	.65891	.75222	.87595	.1416	.3294	.5177	.67194	.74061	.90727	.1022	.3502	.4882	.68476	.72877	.93961	.0643	.3722	.4604	47
14	.64590	.76342	.84606	.1819	.3099	.5482	.65913	.75203	.87646	.1409	.3297	.5171	.67215	.74041	.90780	.1015	.3506	.4877	.68497	.72857	.94016	.0636	.3725	.4599	46
15	.64612	.76323	.84656	1.1812	.3102	1.5477	.65934	.75184	.87698	1.1403	.3301	1.5166	.67237	.74022	.90834	1.1009	.3509	1.4873	.68518	.72837	.94071	1.0630	.3729	1.4595	45
16	.64635	.76304	.84706	.1805	.3105	.5471	.65956	.75165	.87749	.1396	.3304	.5161	.67258	.74002	.90887	.1003	.3513	.4868	.68539	.72817	.94125	.0624	.3733	.4590	44
17	.64657	.76286	.84056	.1798	.3109	.5466	.65978	.75146	.87801	.1389	.3307	.5156	.67280	.73983	.90940	.0996	.3517	.4863	.68561	.72797	.94180	.0618	.3737	.4586	43
18	.64679	.76267	.84806	.1791	.3112	.5461	.66000	.75126	.87852	.1383	.3311	.5151	.67301	.73963	.90993	.0990	.3520	.4858	.68582	.72777	.94235	.0612	.3740	.4581	42
19	.64701	.76248	.84856	.1785	.3115	.5456	.66022	.75107	.87904	.1376	.3314	.5146	.67323	.73943	.91046	.0983	.3524	.4854	.68603	.72757	.94290	.0605	.3744	.4577	41
20	.64723	.76229	.84906	1.1778	.3118	1.5450	.66044	.75088	.87955	1.1369	.3318	1.5141	.67344	.73924	.91099	1.0977	.3527	1.4849	.68624	.72737	.94345	1.0599	.3748	1.4572	40
21	.64745	.76210	.84956	.1771	.3121	.5445	.66066	.75069	.88007	.1363	.3321	.5136	.67366	.73904	.91153	.0971	.3531	.4844	.68645	.72717	.94400	.0593	.3752	.4568	39
22	.64768	.76191	.85006	.1764	.3125	.5440	.66087	.75049	.88058	.1356	.3324	.5131	.67387	.73885	.91206	.0964	.3534	.4839	.68666	.72697	.94455	.0587	.3756	.4563	38
23	.64790	.76173	.85056	.1757	.3128	.5434	.66109	.75030	.88110	.1349	.3328	.5126	.67409	.73865	.91259	.0958	.3538	.4835	.68688	.72677	.94510	.0581	.3759	.4559	37
24	.64812	.76154	.85107	.1750	.3131	.5429	.66131	.75011	.88162	.1343	.3331	.5121	.67430	.73845	.91312	.0951	.3542	.4830	.68709	.72657	.94565	.0575	.3763	.4554	36
25	.64834	.76135	.85157	1.1743	.3134	1.5424	.66153	.74992	.88213	1.1336	.3335	1.5116	.67452	.73826	.91366	1.0945	.3545	1.4825	.68730	.72637	.94620	1.0568	.3767	1.4550	35
26	.64856	.76116	.85207	.1736	.3138	.5419	.66175	.74973	.88265	.1329	.3338	.5111	.67473	.73806	.91419	.0939	.3549	.4821	.68751	.72617	.94675	.0562	.3771	.4545	34
27	.64878	.76097	.85257	.1729	.3141	.5413	.66197	.74953	.88317	.1323	.3342	.5106	.67495	.73787	.91473	.0932	.3552	.4816	.68772	.72597	.94731	.0556	.3774	.4541	33
28	.64900	.76078	.85307	.1722	.3144	.5408	.66218	.74934	.88369	.1316	.3345	.5101	.67516	.73767	.91526	.0926	.3556	.4811	.68793	.72577	.94786	.0550	.3778	.4536	32
29	.64923	.76059	.85358	.1715	.3148	.5403	.66240	.74915	.88421	.1309	.3348	.5096	.67537	.73747	.91580	.0919	.3560	.4806	.68814	.72557	.94841	.0544	.3782	.4532	31
30	.64945	.76041	.85408	1.1708	.3151	1.5398	.66262	.74895	.88472	1.1303	.3352	1.5092	.67559	.73728	.91633	1.0913	.3563	1.4802	.68835	.72537	.94896	1.0538	.3786	1.4527	30
31	.64967	.76022	.85458	.1702	.3154	.5392	.66284	.74876	.88524	.1296	.3355	.5087	.67580	.73708	.91687	.0907	.3567	.4797	.68856	.72517	.94952	.0532	.3790	.4523	29
32	.64989	.76003	.85509	.1695	.3157	.5387	.66305	.74857	.88576	.1290	.3359	.5082	.67602	.73688	.91740	.0900	.3571	.4792	.68878	.72497	.95007	.0525	.3794	.4518	28
33	.65011	.75984	.85559	.1688	.3161	.5382	.66327	.74838	.88628	.1283	.3362	.5077	.67623	.73669	.91794	.0894	.3574	.4788	.68899	.72477	.95062	.0519	.3797	.4514	27
34	.65033	.75965	.85609	.1681	.3164	.5377	.66349	.74818	.88680	.1276	.3366	.5072	.67645	.73649	.91847	.0888	.3578	.4783	.68920	.72457	.95118	.0513	.3801	.4510	26
35	.65055	.75946	.85660	1.1674	.3167	1.5371	.66371	.74799	.88732	1.1270	.3369	1.5067	.67666	.73629	.91901	1.0881	.3581	1.4778	.68941	.72437	.95173	1.0507	.3805	1.4505	25
36	.65077	.75927	.85710	.1667	.3170	.5366	.66393	.74780	.88784	.1263	.3372	.5062	.67688	.73610	.91955	.0875	.3585	.4774	.68962	.72417	.95229	.0501	.3809	.4501	24
37	.65100	.75908	.85761	.1660	.3174	.5361	.66414	.74760	.88836	.1257	.3376	.5057	.67709	.73590	.92008	.0868	.3589	.4769	.68983	.72397	.95284	.0495	.3813	.4496	23
38	.65121	.75889	.85811	.1653	.3177	.5356	.66436	.74741	.88888	.1250	.3379	.5052	.67730	.73570	.92062	.0862	.3592	.4764	.69004	.72377	.95340	.0489	.3816	.4492	22
39	.65144	.75870	.85862	.1647	.3180	.5351	.66458	.74722	.88940	.1243	.3383	.5047	.67752	.73551	.92116	.0856	.3596	.4760	.69025	.72357	.95395	.0483	.3820	.4487	21
40	.65166	.75851	.85912	1.1640	.3184	1.5345	.66479	.74702	.88992	1.1237	.3386	1.5042	.67773	.73531	.92170	1.0849	.3600	1.4755	.69046	.72337	.95451	1.0476	.3824	1.4483	20
41	.65188	.75832	.85963	.1633	.3187	.5340	.66501	.74683	.89044	.1230	.3390	.5037	.67794	.73511	.92223	.0843	.3603	.4750	.69067	.72317	.95506	.0470	.3828	.4479	19
42	.65210	.75813	.86013	.1626	.3190	.5335	.66523	.74664	.89097	.1224	.3393	.5032	.67816	.73491	.92277	.0837	.3607	.4746	.69088	.72297	.95562	.0464	.3832	.4474	18
43	.65232	.75794	.86064	.1619	.3193	.5330	.66545	.74644	.89149	.1217	.3397	.5027	.67837	.73472	.92331	.0830	.3611	.4741	.69109	.72277	.95618	.0458	.3836	.4470	17
44	.65254	.75775	.86115	.1612	.3197	.5325	.66566	.74625	.89201	.1211	.3400	.5022	.67859	.73452	.92385	.0824	.3614	.4736	.69130	.72256	.95673	.0452	.3839	.4465	16
45	.65276	.75756	.86165	1.1605	.3200	1.5319	.66588	.74606	.89253	1.1204	1.3404	1.5018	.67880	.73432	.92439	1.0818	.3618	1.4732	.69151	.72236	.95729	1.0446	.3843	1.4461	15
46	.65298	.75737	.86216	.1599	.3203	.5314	.66610	.74586	.89306	.1197	.3407	.5013	.67901	.73412	.92493	.0812	.3622	.4727	.69172	.72216	.95785	.0440	.3847	.4457	14
47	.65320	.75718	.86267	.1592	.3207	.5309	.66631	.74567	.89358	.1191	.3411	.5008	.67923	.73393	.92547	.0805	.3625	.4723	.69193	.72196	.95841	.0434	.3851	.4452	13
48	.65342	.75700	.86318	.1585	.3210	.5304	.66653	.74548	.89410	.1184	.3414	.5003	.67944	.73373	.92601	.0799	.3629	.4718	.69214	.72176	.95896	.0428	.3855	.4448	12
49	.65364	.75680	.86368	.1578	.3213	.5299	.66675	.74528	.89463	.1178	.3418	.4998	.67965	.73353	.92655	.0793	.3633	.4713	.69235	.72156	.95952	.0422	.3859	.4443	11
50	.65386	.75661	.86419	1.1571	.3217	1.5294	.66697	.74509	.89515	1.1171	.3421	1.4993	.67987	.73333	.92709	1.0786	.3636	1.4709	.69256	.72136	.96008	1.0416	.3863	1.4439	10
51	.65408	.75642	.86470	.1565	.3220	.5289	.66718	.74489	.89567	.1165	.3425	.4988	.68008	.73314	.92763	.0780	.3640	.4704	.69277	.72115	.96064	.0410	.3867	.4435	9
52	.65430	.75623	.86521	.1558	.3223	.5283	.66740	.74470	.89620	.1158	.3428	.4983	.68029	.73294	.92817	.0774	.3644	.4699	.69298	.72095	.96120	.0404	.3870	.4430	8
53	.65452	.75604	.86572	.1551	.3227	.5278	.66762	.74450	.89672	.1152	.3432	.4979	.68051	.73274	.92871	.0767	.3647	.4695	.69319	.72075	.96176	.0397	.3874	.4426	7
54	.65474	.75585	.86623	.1544	.3230	.5273	.66783	.74431	.89725	.1145	.3435	.4974	.68072	.73254	.92926	.0761	.3651	.4690	.69340	.72055	.96232	.0391	.3878	.4422	6
55	.65496	.75566	.86674	1.1537	.3233	1.5268	.66805	.74412	.89777	1.1139	.3439	1.4969	.68093	.73234	.92980	1.0755	.3655	1.4686	.69361	.72035	.96288	1.0385	.3882	1.4417	5
56	.65518	.75547	.86725	.1531	.3237	.5263	.66826	.74392	.89830	.1132	.3442	.4964	.68115	.73215	.93034	.0749	.3658	.4681	.69382	.72015	.96344	.0379	.3886	.4413	4
57	.65540	.75528	.86775	.1524	.3240	.5258	.66848	.74373	.89882	.1126	.3446	.4959	.68136	.73195	.93088	.0742	.3662	.4676	.69403	.71994	.96400	.0373	.3890	.4408	3
58	.65562	.75509	.86826	.1517	.3243	.5253	.66870	.74353	.89935	.1119	.3449	.4954	.68157	.73175	.93143	.0736	.3666	.4672	.69424	.71974	.96456	.0367	.3894	.4404	2
59	.65584	.75490	.86878	.1510	.3247	.5248	.66891	.74334	.89988	.1113	.3453	.4949	.68178	.73155	.93197	.0730	.3669	.4667	.69445	.71954	.96513	.0361	.3898	.4400	1
60	.65606	.75401	.86929	1.1504	.3250	1.5242	.66913	.74314	.90040	1.1106	.3456	1.4945	.68200	.73135	.93251	1.0724	.3673	1.4663	.69466	.71934	.96569	1.0355	.3902	1.4395	0
′	cos	sin	cot	tan	cosec	sec	cos	sin	cot	tan	cosec	sec	cos	sin	cot	tan	cosec	sec	cos	sin	cot	tan	cosec	sec	′
	49°						48°						47°						46°						

(Continued)

table 21 **Natural trigonometric functions (Continued)**

44°

′	sin	cos	tan	cot	sec	cosec	′
0	.69466	.71934	.96569	1.0355	1.3902	1.4395	60
1	.69487	.71914	.96625	.0349	.3905	.4391	59
2	.69508	.71893	.96681	.0443	.3909	.4387	58
3	.69528	.71873	.96738	.0337	.3913	.4382	57
4	.69549	.71853	.96794	.0331	.3917	.4378	56
5	.69570	.71833	.96850	1.0325	.3921	1.4374	55
6	.69591	.71813	.96907	.0319	.3925	.4370	54
7	.69612	.71792	.96963	.0313	.3929	.4365	53
8	.69633	.71772	.97020	.0307	.3933	.4361	52
9	.69654	.71752	.97076	.0301	.3937	.4357	51
10	.69675	.71732	.97133	1.0295	.3941	1.4352	50
11	.69696	.71711	.97189	.0289	.3945	.4348	49
12	.69716	.71691	.97246	.0283	.3949	.4344	48
13	.69737	.71671	.97302	.0277	.3953	.4339	47
14	.69758	.71650	.97359	.0271	.3957	.4335	46
15	.69779	.71630	.97416	1.0265	.3960	1.4331	45
16	.69800	.71610	.97472	.0259	.3964	.4327	44
17	.69821	.71589	.97529	.0253	.3968	.4322	43
18	.69841	.71569	.97586	.0247	.3972	.4318	42
19	.69862	.71549	.97643	.0241	.3976	.4314	41
20	.69883	.71529	.97700	1.0235	.3980	1.4310	40
21	.69904	.71508	.97756	.0229	.3984	.4305	39
22	.69925	.71488	.97813	.0223	.3988	.4301	38
23	.69945	.71468	.97870	.0218	.3992	.4297	37
24	.69966	.71447	.97927	.0212	.3996	.4292	36
25	.69987	.71427	.97984	1.0206	.4000	1.4288	35
26	.70008	.71406	.98041	.0200	.4004	.4284	34
27	.70029	.71386	.98098	.0194	.4008	.4280	33
28	.70049	.71366	.98155	.0188	.4012	.4276	32
29	.70070	.71345	.98212	.0182	.4016	.4271	31
30	.70091	.71325	.98270	1.0176	.4020	1.4267	30
31	.70112	.71305	.98327	.0170	.4024	.4263	29
32	.70132	.71284	.98384	.0164	.4028	.4259	28
33	.70153	.71264	.98441	.0158	.4032	.4254	27
34	.70174	.71243	.98499	.0152	.4036	.4250	26
35	.70194	.71223	.98556	1.0146	.4040	1.4246	25
36	.70215	.71203	.98613	.0141	.4044	.4242	24
37	.70236	.71182	.98671	.0135	.4048	.4238	23
38	.70257	.71162	.98728	.0129	.4052	.4233	22
39	.70277	.71141	.98786	.0123	.4056	.4229	21
40	.70298	.71121	.98843	1.0117	.4060	1.4225	20
41	.70319	.71100	.98901	.0111	.4065	.4221	19
42	.70339	.71080	.98958	.0105	.4069	.4217	18
43	.70360	.71059	.99016	.0099	.4073	.4212	17
44	.70381	.71039	.99073	.0093	.4077	.4208	16
45	.70401	.71018	.99131	1.0088	.4081	1.4204	15
46	.70422	.70998	.99189	.0082	.4085	.4200	14
47	.70443	.70977	.99246	.0076	.4089	.4196	13
48	.70463	.70957	.99304	.0070	.4093	.4192	12
49	.70484	.70936	.99362	.0064	.4097	.4188	11
50	.70505	.70916	.99420	1.0058	.4101	1.4183	10
51	.70525	.70895	.99478	.0052	.4105	.4179	9
52	.70546	.70875	.99536	.0047	.4109	.4175	8
53	.70566	.70854	.99593	.0041	.4113	.4171	7
54	.70587	.70834	.99651	.0035	.4117	.4167	6
55	.70608	.70813	.99709	1.0029	.4122	1.4163	5
56	.70628	.70793	.99767	.0023	.4126	.4159	4
57	.70649	.70772	.99826	.0017	.4130	.4154	3
58	.70669	.70752	.99884	.0012	.4134	.4150	2
59	.70690	.70731	.99942	.0006	.4138	.4146	1
60	.70711	.70711	1.00000	1.0000	.4142	1.4142	0
′	cos	sin	cot	tan	cosec	sec	′

45°

Index

A

Abilities, self-assessed, 27
Abrasion
  cutting tool factors, 214–215
  milling cutter failure and, 478–479
Abrasion resistance
  CBN grinding wheels, 688
  PCBN, 253
Abrasive(s), 672–692
  aluminum oxide, 672–673, 677
  boron carbide, 674–675
  CBN, 592, 675, 677, 688, 807
  for CBN grinding wheels, 592
  ceramic aluminum oxide, 676–677
  cloths, 172
  coated, 689–691
  cutoff saw, 275
  cutting, 274
  grains, 677–678
  grinding wheels. *See* Grinding wheels
  for lapping, 184
  manufactured diamonds. *See* Manufactured diamonds
  metals, 255, 490
  products, 677
  SG, 677
  silicon carbide, 673–674, 677
  vitrified CVSG, 677
  zirconia-aluminum oxide, 674
Absolute position, 615
Absolute programming system, 608–609
Absolute squareness, 68
Absolute system, engineering drawings, 44
Accessories
  for comparators, 120
  for drilling machines. *See* Drilling machines
  for horizontal milling machines. *See* Horizontal milling machines
  for jig borers, 581–583
  for lathes. *See* Lathes
  for layout, 147–148
  for multi-tasking machines, 857–858
  rack indexing, 512

rack milling, 512
slotting attachment, 512
for universal cutter/tool grinder, 731–732
universal milling attachment, 512
vertical milling attachment, 511–512
Accidents, 32
Accuracy, 6–7
  CNC and, 603, 605
  contour bandsaws, 288
  depth of cut, lathes, 390–392
  drilling to layout, 328–329
  facing workpieces to length, 387
  of gage block sets, 97
  of machine tools, 5
  micrometers and, 75–76
Acid method, broken hand taps, 178
Acme thread, cutting, 434
Active cutting oils, 262–263
Actual size, threads, 422
Acute-angle attachment, protractors, 102
Adapters
  collet, 512
  horizontal milling machines, 512–513
  magnetic chucks, 700
  vertical mill cutters, 483
Adaptive control, 14, 646–647
ADAPT programming language, 13
Addendum gears, 550, 551, 557
Addresses, machines axes, 606
Adhesion, of gage blocks, 98
Adjustable center rest, 537
Adjustable hand reamers, 182
Adjustable lap, 185
Adjustable micrometer square, 69
Adjustable positive stops, 587
Adjustable reamers, 182, 334
Adjustable roll snap gage, 113
Adjustable screw plate die, 178–179
Adjustable snap gage, 113
Adjustable split die, 178
Adjustable squares, 66, 68–69, 145
Adjustable stop for hole depths, 580
Adjustable thread ring gage, 112
Adjustable tooth rest, 731
Adjustable wrenches, 165
Advanced Digital Manufacturing, 791
Aerospace component repair, 810
Aerospace engineering, 24
Agile manufacturing, 14

Aircraft component repair, 810
Air gages, 12, 123–125
Air-hardening steels, 764
Air nozzle, contour bandsaws, 281
Alignment
  of lathe centers, 378–380
  of milling machine vises, 528–529
  of tables, horizontal milling machines, 527–528
  of vertical heads, on vertical mills, 482–483
  of vises, on vertical mills, 483
Allen setscrew wrenches, 165
Allowances, 42
  dimensional, 108
  grinding, 589
  ISO metric threads, 422–423
  reaming, 334–335
  threads, 421–423
Alloys. *See also* Alloy steels
  alloyed castings, 749
  alloying elements, effects of, 757–758
  aluminum-based, 783–784
  carbon-alloy bandsaw blades, 282
  cast alloy toolbits, 209
  copper-based, 205, 784–785
  heat-resistant, 255
  lead-tin-antimony, 785
  nickel, 785
  nonferrous metals and, 783–786
  superalloys, 255
  tin, 785
  zinc, 785–786
Alloy steels, 756–758
  cutting fluids for, 269t
  high-strength, low-alloy steels (HSLA), 757
  machinability of, 204
  milling, 490
  tungsten-alloy steel, 168
Alpha alumina, 244
Alpha brasses, 784
Alpha iron, 762
Alphanumeric keyboard, 611
Aluminum/Aluminum-based materials, 783–784
  aluminum-bronze, 205, 785
  aluminum-magnesium, 784
  aluminum-manganese, 784
  aluminum oxide, ceramic, 676–677

aluminum oxide abrasives, 672–673, 677, 690
aluminum-silicon alloys, 784
aluminum-silicon-magnesium alloys, 784
aluminum-zinc alloys, 784
cutting fluids for, 269t
files, 172
machinability of, 204–205
preparing for layout, 142
American National Acme thread, 420
American National Standard thread, 419–420
American Society of Mechanical Engineers (ASME), 45
American Standard Code for Information Interchange (ASCII), 611
Ampere, 59
Amplifier, 118, 135
AMT (Association for Manufacturing Technology), 25
Analog computers, 602
Angle plates, 147, 308
Angle(s)
  cemented-carbide cutting tools, 231–232
  of chamfer, reamers, 332
  drill point, 313
  of the helix. *See* Helix angle
  of keenness, 213–215, 217, 732
  symbols, 59
  of thread, 418
Angle vise, 307
Angular (beveled) shoulders, 395–396
Angular bevel gears, 547
Angular cutting, contour bandsaws, 281
Angular holes, drilling, 491–492
Angular indexing, 541, 543–544
Angularity, 43, 45
Angular measurement, 102–107
  compound sine plate, 106
  compound sine table, 106
  metric tapers, 105
  sine bars, 103–106
  tapers, 105
  universal bevel protractor, 102–103
Angular milling cutters, 519–520